W0055309

Acoustical Imaging
Volume 22

Acoustical Imaging

Recent Volumes in This Series:

A Continuation Order Plan is available for this series. A continuation order will bring delivery of each
new volume immediately upon publication. Volumes are billed only upon actual shipment. For further
information please contact the publisher.

Acoustical Imaging
Volume 22

Edited by

Piero Tortoli
and
Leonardo Masotti

University of Florence
Florence, Italy

Volume I

SPRINGER SCIENCE+BUSINESS MEDIA, LLC

The Library of Congress cataloged the first volume of this series as follows:

International Symposium on Acoustical Holography.

Acoustical holography; proceedings. v. 1–
New York, Plenum Press, 1967–

v. illus. (part col.), ports. 24 cm.

Editors: 1967– . A. F. Metherell and L. Larmore (1967 with H. M. A. el-Sum)
Symposium for 1967– held at the Douglas Advanced Research Laboratories, Huntington
Beach, Calif.

1. Acoustic holography—Congresses—Collected works. I. Metherell. Alexander A., ed.
II. Larmore, Lewis, ed. III. el-Sum, Hussein Mohammed Amin, ed. IV. Douglas Advanced
Research Laboratories, v. Title.
QC244.5.I.5   69-12533

Proceedings of the 22nd International Symposium on Acoustical Imaging,
held September 3 – 7, 1995, in Florence, Italy

ISBN 978-1-4613-4687-6   ISBN 978-1-4419-8772-3 (eBook)
DOI 10.1007/978-1-4419-8772-3

© 1996 Springer Science+Business Media New York
Originally published by Plenum Press, New York in 1996
Softcover reprint of the hardcover 1st edition 1996

All rights reserved

No part of this book may be reproduced, stored in a retrieval system, or transmitted in any form or by any
means, electronic, mechanical, photocopying, microfilming, recording, or otherwise, without written
permission from the Publisher

International Advisory Board

Chairman
Piero Tortoli, ITALY

President of the Scientific Commitee
Leonardo Masotti, ITALY

Scientific Commitee
Pierre M. Alais, FRANCE
Adriano Alippi, ITALY
Yoshinao Aoki, JAPAN
Walter Arnold, GERMANY
Carlo Atzeni, ITALY
Jeffrey C. Bamber, UNITED KINGDOM
Elena Biagi, ITALY
Valentin A. Burov, RUSSIA
Noriyoshi Chubachi, JAPAN
Helmut Ermert, GERMANY
Katherine W. Ferrara, USA
Leonard Ferrari, USA
Mathias Fink, FRANCE
Woon Siong Gan, SINGAPORE
Emilio Gatti, ITALY
James F. Greenleaf, USA
Benli Gu, CHINA
Christopher R. Hill, UNITED KINGDOM
Joie P. Jones, USA
Hugh A.C. Jones, CANADA
Larry W. Kessler, USA
Hua Lee, USA
Sidney Leeman, UNITED KINGDOM
Frederic L. Lizzi, USA
Vernon L. Newhouse, USA
Andrzej Nowicki, POLAND
William O'Brien jr., USA
Song Bai Park, KOREA
Leandre Pourcelot, FRANCE
John P. Powers, USA
John M. Reid, USA
Santina Rocchi, ITALY
Johan M. Thijssen, THE NETHERLANDS
Olaf T. Von Ramm, USA
Robert C. Waag, USA
Glen Wade, USA
Peter N.T. Wells, UNITED KINGDOM

PREFACE

This volume contains 131 of the papers presented at the 22nd International Symposium on Acoustical Imaging. This meeting, which was held for the first time in Florence, Italy, on September 3-6, 1995, allowed an intense and friendly exchange of ideas between over 150 researchers from 26 different countries of Europe (70%), America (20%), Asia and Australia (10%).

The Symposium started on Sunday, September 3, with the opening Session held in the magnificent 'Salone dei 500' in Palazzo Vecchio; this included invited talks by Peter Wells and Hua Lee, who reviewed the State of the Art in Acoustical Imaging research. One hundred and forty papers, selected from the nearly 200 submitted Abstracts, were presented in 11 non-parallel oral Sessions and one Poster Session.

This year a 'Best Poster' award was introduced, which was won by V.Miette, M.Fink and F.Wu. Also, a special session on Acoustical Microscopy was organized by Walter Arnold, in which invited speakers Joie Jones, Oleg Kolosov, Andrew Briggs and Ute Rabe reviewed the capabilities of this emerging topic.

The success of the Symposium was of course due to many contributions. First, all we would like to thank the authors for their qualified and enthusiastic participation. For three and a half days they created an atmosphere combining science, culture and fellowship. In addition, an important role was played by the members of the International Advisory Board, who not only reviewed the Abstracts, but also carefully assisted each step of the conference organization. Special thanks are due to Joie Jones for his continuous help and encouragement.

Finally, the Editors wish to acknowledge support received from the University of Florence, Area di Ricerca-CNR, CESVIT, Regione Toscana and EsaOte Biomedica. The enthusiastic help of Gabriele Guidi and colleagues of the Dipartimento di Ingegneria Elettronica of the University of Florence, is also gratefully acknowledged.

The 23rd International Symposium on Acoustical Imaging will be held in Boston, USA, in April 1997. We would like to express our best wishes to Sidney Lees as chairman of this upcoming symposium.

Piero Tortoli
Leonardo Masotti

University of Florence
Firenze, Italy

PREFACE

CONTENTS

MATHEMATICS AND PHYSICS OF ACOUSTICAL IMAGING

NOVEL APPROACHES IN BIOMEDICAL IMAGING

TISSUE CHARACTERIZATION

FLOW IMAGING

TRANSDUCERS AND ARRAYS

IMAGING SYSTEMS AND TECHNIQUES

UNDERWATER AND INDUSTRIAL APPLICATIONS

ACOUSTICAL MICROSCOPY

NONDESTRUCTIVE TESTING

MATCHED FILTER IMAGING THROUGH INHOMOGENEOUS MEDIA

Christian Dorme and Mathias Fink

Laboratoire Ondes et Acoustique. Université Paris VII / URA CNRS 1503
E.S.P.C.I. 10, rue Vauquelin. 75231 Paris Cedex 05. France

INTRODUCTION

In echographic mode, as a reflective point-like target backscatters a pulsed wave, the presence of an aberrating layer in the medium usually generates diffraction effects. If this layer is close to the recording transducer array, diffraction does not occur and the medium can even be considered as homogeneous by using an adaptive technique based on time shift compensation. On the other hand, through media including a distant aberrating layer, the diffraction provided between the layer and the array results in a wavefield distorted in phase and amplitude. Time reversal has been presented as an original way to focus on a reflective target in such media [1,2,3,4].

In a signal processing point of view, previous papers have shown that time reversal realizes the spatio-temporally matched filter to the target position [5,6]. Focusing on this position is then closely related to the transmission of the signals constituting its proper Green's function. Therefore, these signals are stored, before their transmission, in a data bank which makes the time reversal process locked on the target position.

However, imaging a zone of interest in a medium requires to focus not only on the brightest point (for example, a dominant reflector), but also on each elementary position of the zone. The knowledge of each Green's function corresponding to each position is then fundamental but is not experimentally available.

This paper extends the time reversal technique to focus not only on the reflector, but also on areas surrounding the reflector. For this purpose, conventional steering consists to incline the wavefront focused on the reflector in order to focus nearby. If the aberrator is very close to the array, conventional steering of the data bank matched to the reflector provides good focusing around the reflector. However, this focusing degrades as the array-aberrator distance increases. We first present the degradation of the focusing due to a conventional steering process in presence of a distant aberrator. We present then a new method that allows to solve the steering focusing problem for aberrating layers located at any distance from the array. It consists to numerically synthesize the Green's functions of neighboring positions from the initial Green's function of the reflector position.

This original steering process can be coupled with an extension of the matched filter concept in the receive mode. In the second part of this paper, we create thus a real transmit-receive focusing on each point of the zone of interest. We remind this extension and finally present B-scans images of a simple reflector through an inhomogeneous medium.

I. STEERING A FOCUSED BEAM THROUGH A THIN ABERRATING LAYER

In a preliminary step, a single transducer element is located at a distance F=90 mm from the transducer array. This element acts as an active point-like source, transmitting an impulse wave through an aberrating medium made of a silicone rubber layer. This thin layer is located at 15 mm from the array. This distance has been choosen as generating clearly diffraction effects. Of course, any other distance could be processed and the generated diffraction effects compensated.

The transducer array is constituted by 64 elements for a total aperture of 48 mm. Its central frequency is 3 MHz. The distorted pressure field recorded by the array is time-reversed, transferred to the buffer memories and then fired from the transducer array. The initial source element is then used to measure the directivity pattern in the focal plane with a 0.5 mm scanning step. The observed pattern is presented in Figure 1 where it is compared to a time reversal focusing through pure water and then, to a cylindrical focusing through the layer.

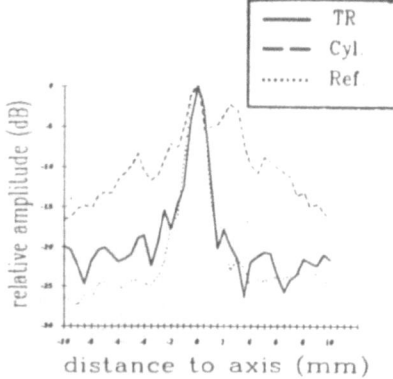

Figure 1. *Directivity patterns obtained by time reversal focusing and by cylindrical focusing through the aberrator. The reference curve is a time reversal focusing through pure water.*

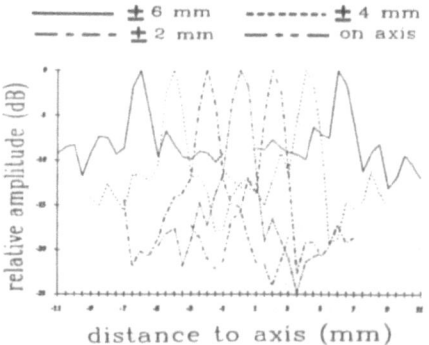

Figure 2. *Directivity patterns obtained by inclining the initial data bank matched to the reflector position. Focusing degrades with scan angle. Conventional steering permits only very small angles.*

We notice a perfect compensation of the diffraction effects with respect to the time reversal achieved through water. The transmitted data bank is therefore exactly matched to the initial focal point.

To investigate the medium around this initial focal point, imaging systems have to focus the ultrasound beam in many new focal points. Indeed, they require as many proper Green's functions as new focal points needed in the zone of interest. But, focusing at r_1 when the reflector is located at r_0 requires to know how to synthesize the Green's function associated with r_1 from the one associated with r_0. If r_1 is very near r_0, their Green's functions can be strongly correlated, and the new one can even be made by inclination of the initial one. However, the time reversal focusing on r_1, by using an inclined version of the initial data bank, is quickly degraded as $\overline{r_0 r_1}$ increases and is valid only for very small angles (< 1 degree for J.W. Goodman [7]).

I.1 Focusing Degradation by Conventional Steering

From the initial data bank matched to the reflector position, a conventional inclination, with a linear set of complementary time delays, allows us to create six new data banks in order to focus in the focal plane at ± 2 mm, ± 4 mm, and ± 6 mm from the reflector position. The directivity patterns observed in the reflector plane as transmitting these inclined versions of the initial data bank are presented in Figure 2.

If the focusing is good until ± 2 mm from the reflector, the side lobes raise highly for other positions (more than 10 dB). Moreover, the inclination of the initial data bank unbalances the directivity pattern. As a conclusion of this experiment, we notice that it is useless to try to focus elsewhere by adding time delays if the original Green's function is not separable in space and time (if the signals constituting this Green's function are not similar).

An ideal calculation of the Green's function associated with the new focal point would provide, for each transducer of the array, the amplitude and the phase for each frequency component contained in the new Green's function Fourier spectrum. Our aim is simpler: it consists of applying conventional beamsteering using time delays to the case of distant layers. For this purpose, we have to reduce the spatiotemporally complicated wave coming from the reflector (the initial data bank) to the wavefront originating from the output plane of the aberrating layer. This is possible by numerically backpropagating the initial data bank in a homogeneous medium until a distance equal to the array-aberrator distance. This procedure is equivalent to a time reversal and removes the differences in shape between the individual signals. Thus, we obtain the echo that would have been received by the transducer array if it was close to the layer. The following algorithm describes how to obtain the optimized backpropagation distance, and how to synthesize a new Green's function matched to a new focal point.

I.2 Green's Function Synthesis

The distance d between the array and the layer could be determined with the arrival time of the layer echo. However, this echo is extended in time and thus could not provide an accurate distance d. The optimized backpropagation distance d_{opt} is defined as the range at which the backpropagated waveforms are most alike in the sense that the waveforms are only time shifted with a possible amplitude factor. Thus, d_{opt} corresponds to the location of the output plane of the layer where diffraction has not occured.

The backpropagation process consists to calculate the pressure signals coming from a plane P, parallel to the array and located at a distance d from the array. The plane P is spatially sampled as the transducer array. Each backpropagated signal is determined by summation of the signals $p_{r_0}(r_i, T-t)$ constituting the initial data bank matched to the reflector. Because the medium is considered homogeneous between the transducer array and plane P, this summation takes into account the convolution with the Green's function in free space:

$$p_{r_0}(r_{i,d}, t) = \sum_{j=1}^{N \text{ transducers}} p_{r_0}(r_j, T-t) \times \frac{1}{\sqrt{r_j r_{i,d}}} \otimes \delta\left(t - \frac{\overline{r_j r_{i,d}}}{c}\right) \tag{1}$$

where the second term in the right-hand side is linked to cylindrical loss, and the last one is the delay due to the propagation. Thus, we calculate a set of pressure signals $p_{r_0}(r_{i,d}, t)$ and compare them to each other for similarity. A waveform similarity is computed, based on cross-correlations between these backpropagated waveforms. In this way, couples of two signals, $p_{r_0}(r_{i,d}, t)$ and $p_{r_0}(r_{j,d}, t)$, are imposed a zero mean value and are then correlated by a normalized cross-correlation written:

$$R_{i,j}^d(t') = \frac{\int p_{r_0}(r_{i,d}, t') \cdot p_{r_0}(r_{j,d}, t - t') \, dt'}{\sqrt{\int p_{r_0}^2(r_{i,d}, t) \, dt \times \int p_{r_0}^2(r_{j,d}, t) \, dt}} \tag{2}$$

The maximum of this cross-correlation function is noted

$$M_{ij}(d) = \max\left[R_{i,j}^d(t)\right] \tag{3}$$

Thus, at the distance d, we obtain a set of maxima $M_{ij}(d)$ whose $\overline{M}(d)$ is the average. $\overline{M}(d_{opt})$ is attempted to be ideally equal to 1 due to the similarity of the waveforms in the output plane of the layer. Because neighbor signals (spaced by about 1 wavelength) always look alike, it is most useful to use a larger spacing such as j = i + 6 or 7 or more. This speeds the process, improves the sensitivity of selection of the maximum and increases the precision on d_{opt}.

With the set of pressure signals calculated for $d = d_{opt}$, we have thus a configuration equivalent to the one where the transducer array would be close to the aberrating layer. Then, focusing at a point other than the position of the reflector requires only complementary geometric time delays, just as in conventional beamsteering. The wavefront found by numerical backpropagation is then inclined to converge at the new focal point. Now, we have to determine the Green's function of this point, that's to say, acquire in the array plane the wavefront corresponding to this new focal point. For this purpose, we just propagate the inclined wavefront back to the transducer array. The time reversal of this new wavefront focuses at the new focal point through the aberrating layer. This process is illustrated in Figure 3. Notice that the new wavefront, numerically synthesized, constitutes the new data bank matched in both transmit and receive modes to the new focal point.

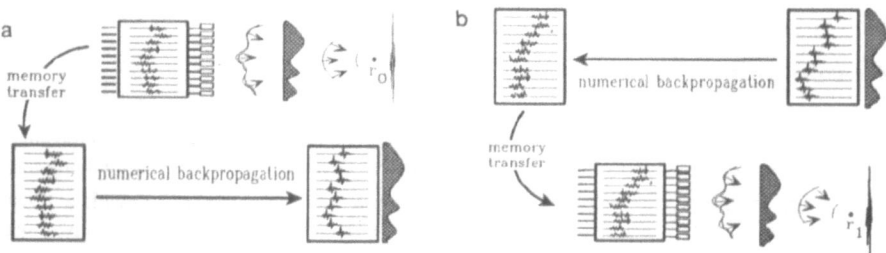

Figure 3. *Illustration of the complete algorithm determining the Green's function (the data bank of signals) associated with a new desired focal point. In a), the recorded wave volume is backpropagated until it becomes a wave surface in the aberrator plane. In b), this wave surface is inclined by conventional beamsteering and then propagated back to the array plane. All this process is numerically achieved by computer. The new wave volume is matched to the new desired focal point.*

I.3 Discussion

In fact, all this process is absolutely not real-time. If the propagation through a homogeneous medium is very quickly achieved, the cross-correlations operations take a lot of time and have to be calculated at each distance determined by the algorithm. For example, calculating 15 similarity factors for 64 backpropagated signals (one signal is defined on about 500 points due to the sampling rate in reception) takes about 25 minutes on a UNIX workstation. We obtain certainly an optimized backpropagation distance, but we could try to know it a priori by transmitting a pulsed wave by the central transducer and by deducing the array-aberrator distance from the received echo.

A second limitation is that the process we present assumes that the layer is thin and parallel to the array. The choice of a model with a parallel layer is related to the configuration of the experiment and to the simplicity of implementation. Future studies will tend to demonstrate the validity of this model for more severe layers up to thick biological tissue samples. Finally, whatever the layer, the algorithm allows to define the distance of the equivalent assumed phase screen. In our model, we have just to determine a low threshold for $\overline{M}(d)$, for example 0.8, above which the medium can be considered having an equivalent parallel phase screen. Moreover, the delays induced by the layer have not to be defined in our process. This is an advantage with respect to a numerical focusing such as the one described by Liu and Waag [8]. These authors have already numerically backpropagated waveforms until they are identical except to a time delay. They can thus estimate the delays induced by the layer and numerically focus on the source. In this paper, we propose an experimental imaging procedure with an accurate focusing on each point thanks to the synthesis of each Green's function. This procedure is self-adaptive and does not require the estimation of the delays induced by the layer. Its efficiency is presented in the following part.

II. EXPERIMENTAL RESULTS

II.1 Transmit Focusing by Synthesized Data Banks

In this part, the experimental configuration is the same as the one described in part I. To study the backpropagation of the initial data bank, our algorithm works in a range of 5 mm to 30 mm from the array. The algorithm reduces more and more this range to find the maximum similarity factor with a required precision of 0.3 mm. d_{opt} is found equal to 14.7 mm. The waveforms obtained are shown in Figure 4. The similarity factor $\overline{M}(14.7 \text{ mm})$ is equal to 0.892. This wavefront is then successively inclined to focus, in the focal plane, at new focal points located at ± 2 mm, ± 4 mm, and ± 6 mm from the initial reflector position. To incline the wavefront, we consider of course a reduced focal length equal to $F - d_{opt} = 75.3 \text{ mm}$.

Backpropagation to the array gives six new wavefronts that, time-reversed and retransmitted, achieve the directivity patterns presented Figure 5. The improvement is obvious with respect to time delay inclination of the initial data bank (compare with Figure 2). We can observe well-balanced patterns without rise in the sidelobes. Thus, it is possible to generate, from the backpropagated wavefront, a complete set of data banks matched to each point of the zone of interest. However, we can synthesize accurate Green's functions only in the neighborhood of the reflector. Imaging the neighborhood of the reflector consists then to use each data bank in both transmit and receive modes.

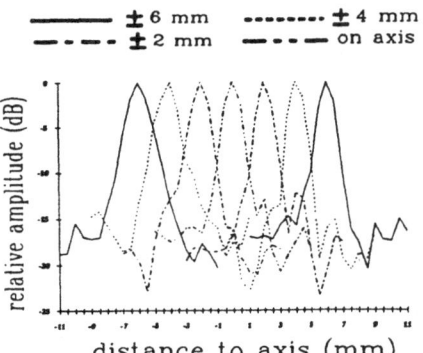

Figure 5. *Directivity patterns obtained by transmitting the synthesized wave volumes matched to the new desired focal points. From the data bank of signals matched to each desired focal point, a matched filter transmit focusing is achieved at each of these positions. The matched filter process is achieved without requiring a reflector at the desired focal position.*

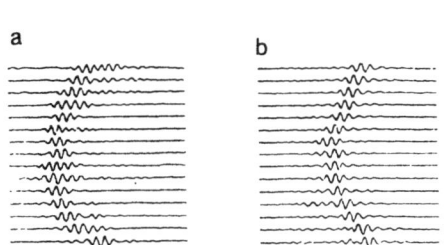

Figure 4. *Backscattered echo from a reflective point-like target through an inhomogeneous layer located at 15 mm from the transducer array (a). This wave volume is numerically backpropagated to the optimal distance found by our algorithm. Waveforms obtained are displayed in (b). This wave surface presents identical signals (time and space are dissociated).*

II.2 Imaging Procedure

We have seen that the transmit focusing on a desired focal point is achieved by transmitting the appropriate data bank. The propagation of this set of matched signals through the medium insures a matched filter process on the desired focal point in transmit mode. However, if the matched filter is automatically achieved in the transmit mode, it has to be processed in the receive mode [6,9]. Thus, after transmission of a matched data bank, the received echo from the focal point has to be convolved line to line with this same matched data bank. For others focal points than the initial reflector position, the efficiency of this transmit-receive focusing depends on the accuracy of the synthesized Green's functions. For this purpose, experiments were done in order to analyze the limits of this matched transmit-receive focusing mode.

We consider only 4 synthesized data banks corresponding to the following positions: the initial position of the reflector, and positions located at +2 mm, +4 mm and +6 mm from the initial position. Then, we measure the directivity patterns in transmit-receive mode for each one of these positions. In this experiment, the transducer array remains always matched to the focal position by transmitting the signals stored in the data bank. A needle hydrophone, used as a reflective point-like target, scans the focal plane through the desired focal point with a 0.5 mm step. In this way, the transducer array acquires in receive mode a set of backscattered wavefronts corresponding to each position of the hydrophone. When the hydrophone is located at the focal point associated to the synthesized data bank, the convolution of the echo with the transmitted data bank achieves, as previously seen, a matched filter process and yields a maximum signal for the focal position.

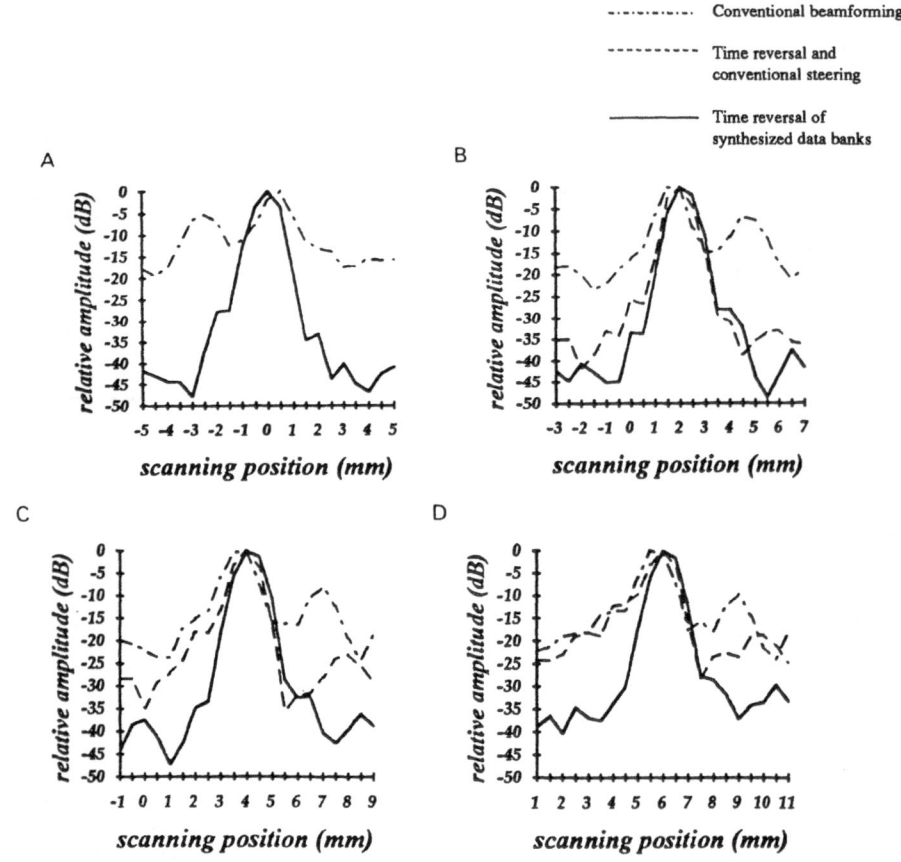

Figure 6. *Directivity patterns obtained in transmit-receive mode. (a), (b), (c) and (d) present respectively for three techniques the patterns observed as the system desires to focus on axis (at the reflector position), at +2 mm, at +4 mm and at +6 mm from this initial focal point. Focusing by inclination of the initial matched data bank degrades rapidly. On the other hand, each synthesized data bank is well matched to each desired focal point.*

Figure 6 displays the directivity patterns obtained in transmit-receive mode for each synthesized data bank and compares the results with the conventional procedure (cylindrical focusing) and with the inclination of the initial data bank. At +6 mm from the reflector position (see Figure 6d), we notice that the transmit-receive focusing obtained with the matched synthesized data bank is quite as good as the one obtained for the reflector position (see Figure 6a). The focusing is only altered by a weak rise of the sidelobes. Building an image of a little zone around the reflector is then possible, with the same focusing quality for all the resolution cells of the zone. However, farther desired focal points would certainly provide a more important rise in the side lobes. This limitation in the synthesis of Green's functions is related to the coherence length of the aberrating layer at the work frequency. Thus, a Green's function synthesized for a focal point far from the initial reflector should not be identical to the Green's function that this point would have generated by transmitting a pulsed wave.

Figure 7 displays B-scans achieved for the head of the hydrophone (1 mm wide) located at +6 mm from the initial reflector position. In this experiment, Green's functions are synthesized to focus on this new position and also on positions located until ± 10 mm from the new reflector position with a 0.5 mm step. A transmit-receive focusing is then achieved for each one of these positions. The representation in gray level of the amplitude of the signals shows the improvement due to our technique with respect to an inclination of the initial data bank and to the conventional imaging achieved in actual echographs. Now, we know how to focus laterally with synthesized data banks. Focusing at different depths would modify only in curvature the wavefront backpropagated in the aberrator plane. Thus, we hope to present soon C-scans images, through inhomogeneous layers, of a reflector embedded in a scattering medium or of an extended target.

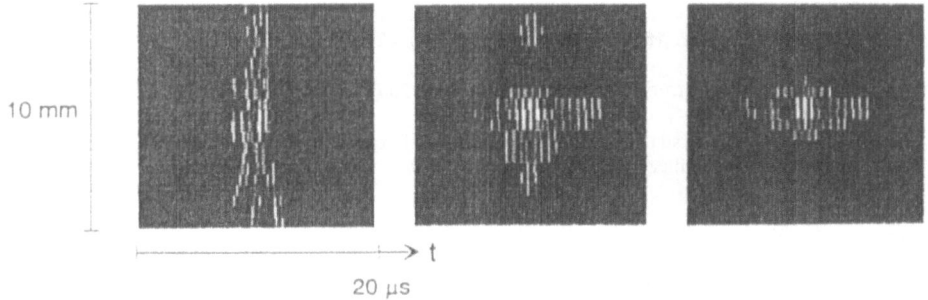

10 mm

20 µs

Figure 7. *B-scans achieved in transmit-receive mode with data banks focusing at +6 mm from the initial reflector position and also on surrounding positions. In (a), an initial cylindrical data bank is inclined to scan the zone of interest. In (b), the inclined data bank is the one acquired by time reversal and matched to the initial reflector position. In (c), this data bank is processed to synthesize new data banks matched to each point of the zone. Focusing with these appropriate data banks provides a clear B-scan of the hydrophone head.*

CONCLUSION

From a theoretical point of view, the matched filter approach optimizes the ultrasonic focusing through inhomogeneous media. In cases where the diffraction impulse response of a reflector is not dissociated in space and time, the conventional time delay law has to be replaced by the Green's function of the reflector. This achieves a matched filter process in both transmit and receive modes. Thus, focusing in any point of an inhomogeneous zone requires the knowledge of the Green's function matched to this point. If time reversal is self-adaptive to only a reflective point, we are however able to determine the Green's functions matched to points surrounding this reflector. Now, biological

7

tissues or material samples usually present either many diffusers or, what is the worse for our system, a pure scattering inhomogeneous medium without any particular reflector. In a multi-diffuser medium, the data bank of signals matched to the dominant reflector can be found by iterating the time reversal process. The extension of our self-adaptive technique to scattering media is still in study in our laboratory.

However, in the presence of a reflector in the inhomogeneous medium, perfect images are achieved of the reflector zone thanks to a matched filter process for each resolution cell of the zone. This steering method, valid for media that can be modeled with one phase screen located at a given distance from the transducer array, is being studied for media equivalent to N layers (phase screens) distributed over the whole medium and also for thick slices of biological tissue. The applications of this technique are obvious in medical imaging or in the treatment of a tissue area. The interest is great also in Non Destructive Testing to visualize defects in material samples. We look forward to a quick evolution of practical systems for real-time imaging based on this new technique.

REFERENCES

1. M. Fink, C. Prada, F. Wu, D. Cassereau. " Self Focusing in Inhomogeneous Media with Time Reversal Acoustic Mirrors". *1989 IEEE Ultras. Symposium Proc.* Vol.2, pp. 681-686.

2. M. Fink. " Time Reversal of Ultrasonic Fields. Part I : Basic Principles". *IEEE trans. Ultras. Ferroelec. Freq. Con.* vol. 39, pp.555-567 (1992).

3. F. Wu, J.L. Thomas, and M. Fink, " Time Reversal of Ultrasonic Fields - Part II: Experimental Results" *IEEE trans. Ultras. Ferroelec. Freq. Con.* vol. 39, pp.567-578 (1992).

4. D. Cassereau, M. Fink. "Time Reversal of Ultrasonic Fields. Part III: Theory of the Closed Time-Reversal Cavity". *IEEE trans. Ultras. Ferroelec. Freq. Con.* vol. 39, pp. 579-593 (1992).

5. M. Fink "Time Reversal Mirrors." *J. Phys. D: Appl. Phys.* 26 (1993), pp. 1333-1350.

6. C. Dorme, M. Fink and C Prada."Focusing in the Transmit-Receive Mode through Inhomogeneous Media: the Matched Filter Approach". *1992 IEEE Ultras. Symposium Proc.* vol.1, pp. 629-634.

7. J.W. Goodman, W.H. Huntley, D.W. Jackson, M. Lehmann. "Wavefront-Reconstruction Imaging through Random Media." *Applied Physics Letters* vol.8, pp. 311-313, 1966.

8. D. L. Liu and R. C. Waag. "Correction of Ultrasonic Wavefront Distortion Using Backpropagation and a Reference Waveform Method for Time-Shift Compensation." *Journ. Acoust. Soc. Am.* vol.96, pp. 649-660 (August 1994).

9. C. Dorme and M. Fink. "Focusing in Transmit-Receive Mode through Inhomogeneous Media: the Time Reversal Matched Filter Approach." *Journ. Acoust. Soc. Am.*, vol. 98, No. 2, Pt. 1, pp. 1155-1162 (August 1995).

DIRECTIONAL FIELD OF A POINT SOURCE FOR CALCULATION OF THREE-DIMENSIONAL HARMONIC WAVES IN LAYERED MEDIA

Elfgard Kühnicke

Dresden University of Technology
Institute of Technical Acoustics
D-01062 Dresden, Germany

INTRODUCTION

The description of the propagation of elastic waves emitted by a transducer in a solid is of fundamental importance in many applications of ultrasonics. Since there exist many problems with two kinds of geometric dimensions - dimensions much larger than the wavelength and dimensions in the range of the wavelength - the application of the geometric ray theory and the finite element method, resp., is difficult. That's why semianalytical methods are a good alternative for such problems with a mixed geometry. For a 2-dimensional geometry, such as a half-space, a plate with parallel surfaces or a horizontally layered medium solutions exist in form of Green's functions[1,2]. In [3] the displacement in the interior of a test object for a disk-shaped single-element transducer is obtained by spatial convolution of the solution for a point force with the amplitude distribution of the point sources. This approach, however, is restricted to axisymmetrical sources.

The paper presents an algorithm for the calculation of the monofrequent sound field of ultrasonic transducers applied to layered media with non-parallel interfaces. The medium is decomposed into different layers and the wave propagation is calculated separately in each layer by means of a point source synthesis. Using the generalized ray theory the fundamentals for a separate sound field calculation in each layer are derived. The directional fields for a concentrated body force acting at the interface between two half-spaces of different materials and for normal stress caused by an incident wave at the same interface are given.

SOURCE FUNCTION FOR NORMAL FORCE UNDER LOAD

For the calculation a linear elastic isotropic medium is supposed. The wave equation for the displacements is solved introducing the scalar potentials ϕ, ψ and χ. A vertical point

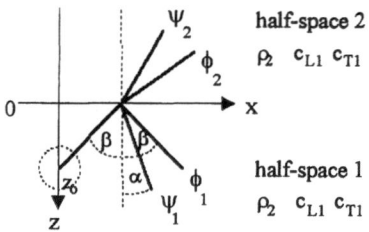

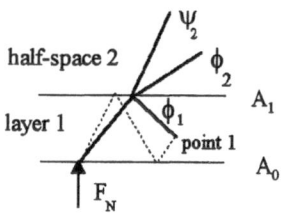

Fig.1. Potentials in 2 adjoining half-spaces   Fig.2. Introduction of an additional interface

source with the time function f(t) acting at point $r=0$, $z=z_0$ in the interior of a half-space is discussed (fig.1). According to the formalism of the generalized ray theory the potentials of two adjoining half-spaces in the twice transformed area - with respect to time and space - result in:

$$\hat{\phi}_1 (w,z,s) = a[S^P e^{sp_1|z-z_0|} + S^P R^{PP} e^{-sp_1|z_0+z|} + S^S R^{SP} e^{-s[p_1z+q_1z_0]}] \tag{1a}$$

$$\hat{\psi}_1 (w,z,s) = -\frac{a}{sw}(S^S e^{sq_1|z-z_0|} + S^P R^{PS} e^{-s[p_1z_0+q_1z]} + S^S R^{SS} e^{-sq_1[z+z_0]}) \tag{1b}$$

$$\hat{\phi}_2 (w,z,s) = a[S^P T^{PP} e^{s[p_1z_0+p_2z]} + S^S T^{SP} e^{s[p_2z+q_1z_0]}] \tag{1c}$$

$$\hat{\psi}_2 (w,z,s) = -\frac{a}{sw}(S^P T^{PS} e^{s[p_1z_0+q_2z]} + S^S T^{SS} e^{s[q_1z_0+q_2z]}) \tag{1d}$$

ϕ_1 and ψ_1 are the potentials of the longitudinal wave (L-wave) and the transversal wave (T-wave) propagating in half-space 1, respectively; ϕ_2 and ψ_2 are the potentials of the waves in half-space 2. $S^P = -1$ and $S^S = w/q_1$ are the source functions of an axisymmetrical force in the interior for the L-wave or for the T-wave. The parameter of transform is w, p and q are given by $p = \sqrt{w^2 + c_L^{-2}}$ and $q = \sqrt{w^2 + c_T^{-2}}$ with the wave velocities c_L and c_T. R are the generalized reflection coefficients, T are the generalized transmission coefficients. These coefficients were determined by means of the boundary conditions at the interface between the two half-spaces. The index P indicates the pressure wave, S the shear wave. The first index explains the shape of the incident wave and the second one the shape of the reflected and refracted wave, respectively.

The potential ϕ_1 for the L-wave in the half-space 1 is composed of the following ray terms (equ.1a): 1 - the L-wave component propagated from the source which is neither reflected nor refracted, 2 - the component of the emitted L-wave, again reflected as an L-wave, and 3- the component of the emitted T-wave and reflected as L-wave. In the case of a normal force acting at the interface between the two half spaces, these three components coalesce forming the new source term. We obtain the source terms S^P for the L-wave and analogously S^S from equ.(1b) for the T-wave for a normal force acting at the interface between half-space 1 and half-space 2 result in:

$$S^P(w) = -1 + R^{PP} + \frac{w}{q_1} R^{SP} \qquad\qquad S^S(w) = \frac{w}{q_1} + \frac{w}{q_1} R^{SS} + R^{PS} \tag{2}$$

Applying the reflection coefficients of a solid-solid interface given in[1], or of a solid-solid interface with liquid coupling given in[4] the source function for a force acting at the surface covered with a solid (transducer material coupled to specimen) is obtained.

The potentials for the half-space with a normal force at the interface are then the first terms in equ.(1a) and (1b) with the source functions (2). Now an additional interface A_0 is introduced (fig.2) and we get a layer with an overlaying half-space. The point source acts at the interface A_0. Now we use the potentials for the half-space together with the source terms (2) for a force at the interface. The 2nd term and the 3rd term in equ.(1a) and (1b) describe the single reflected waves in layer 1. The application of the potentials of the half-space   for a layer leads to correct results only before the wave parts which are reflected at the source area A_0 again reach the point 1 of observation (dashed line).

The directional field of the L-wave for the point source is obtained using the generalized reflection coefficients for two solids coupled by a liquid layer given in[4] and replacing w by $w = i \sin(\beta_1)\, c_{L1}^{-1}$ in equ.(2).

$$S^P(\beta_1) = \frac{1}{N_1 + N_2}\, 2\cos\beta_1 \left(1 - 2\left(\frac{c_{T1}}{c_{L1}}\right)^2 \sin^2\beta_1\right) \tag{3a}$$

$$N_1(\beta_1) = \left\{1 - 2\left(\frac{c_{T0}}{c_{L1}}\right)^2 \sin^2\beta_1\right\}^2 \frac{\rho_0}{\rho_1}\cos\beta_1 \sqrt{\left(\frac{c_{L1}}{c_{L0}}\right)^2 - \sin^2\beta_1}^{\;-1} \tag{3b}$$

$$+ 4\left(\frac{c_{T0}}{c_{L1}}\right)^4 \frac{\rho_0}{\rho_1}\sin^2\beta_1 \sqrt{\left(\frac{c_{L1}}{c_{T0}}\right)^2 - \sin^2\beta_1}\;\cos\beta_1$$

$$N_2(\beta_1) = 4\left(\frac{c_{T1}}{c_{L1}}\right)^2 \sin^2\beta_1 \sqrt{\left(\frac{c_{L1}}{c_{T1}}\right)^2 - \sin^2\beta_1}\;\cos\beta_1 + \left\{1 - 2\left(\frac{c_{T1}}{c_{L1}}\right)^2 \sin^2\beta_1\right\}^2 \tag{3c}$$

and for the T-wave w by $w = i \sin(\alpha_1)\, c_{T1}^{-1}$

$$S^S(\alpha_1) = \frac{1}{N}\, 4 \sin\alpha_1\, \cos\alpha_1 \sqrt{\left(\frac{c_{T1}}{c_{L1}}\right)^2 - \sin^2\alpha_1} \tag{4a}$$

$$N_1(\alpha_1) = \left\{1 - 2\left(\frac{c_{T0}}{c_{T1}}\right)^2 \sin^2\alpha_1\right\}^2 \frac{\rho_0}{\rho_1} \sqrt{\left(\frac{c_{T1}}{c_{L1}}\right)^2 - \sin^2\alpha_1} \sqrt{\left(\frac{c_{T1}}{c_{L0}}\right)^2 - \sin^2\alpha_1}^{\;-1} \tag{4b}$$

$$+ 4\left(\frac{c_{T0}}{c_{T1}}\right)^4 \frac{\rho_0}{\rho_1}\sin^2\alpha_1 \sqrt{\left(\frac{c_{T1}}{c_{T0}}\right)^2 - \sin^2\alpha_1} \sqrt{\left(\frac{c_{T1}}{c_{L1}}\right)^2 - \sin^2\alpha_1}$$

$$N_2(\alpha_1) = 4 \sin^2\alpha_1\, \cos\alpha_1 \sqrt{\left(\frac{c_{T1}}{c_{L1}}\right)^2 - \sin^2\alpha_1} + \left\{1 - 2\sin^2\alpha\right\}^2 \tag{4c}$$

β_1 and α_1 are the angles between the vector from the source point to the point of observation and the normal vector in the source point. The directivity patterns of a point source at a free surface become simpler since $\rho_0 = 0$ and thus $N_1 = 0$ in equ. (3b) and (4b).

SOURCE FUNCTION FOR NORMAL STRESS AT THE INTERFACE

To find out the source function of normal stress at an interface the following method is applied[4]. Assuming that between two solid media a thin water layer exists, so that only normal stress and no shear stress is transferred at the interfaces, the reflection and transmission coefficients given in[4] may be used. At first the normal stress at the interface $1 \rightarrow 2$ generated by an incident wave in layer 1 is calculated by means of the potentials (1) for the case of a plane interface.

$$\sigma_{zz1 \rightarrow 2}(w, 0, s) = a\rho_2 c_{T2}^2 \left\{\left[\left(2w^2 + \frac{1}{c_{T2}^2}\right) T^{PP} - 2wq_2 T^{PS}\right] S^P\, e^{s\, p_1\, z_0} + \left[\left(2w^2 + \frac{1}{c_{T2}^2}\right) T^{SP} - 2wq_2 T^{SS}\right] S^S\, e^{s\, q_1\, z_0}\right\} \tag{5}$$

Then the potentials for the half-space 2 are written according to equ.(6).

$$\hat{\phi}_2 = \sigma_{zz1 \rightarrow 2}(w, 0, s)\, S^{P*}(w) e^{s p_2 z} \qquad \hat{\psi}_2 = \sigma_{zz1 \rightarrow 2}(w, 0, s)\, S^{S*}(w) e^{s q_2 z} \tag{6}$$

S^* are the source functions for normal stress at the interface between 2 solids with liquid coupling and σ_{zz} is the stress calculated before according to equ.(5). Replacing ϕ_2 and ψ_2 by the equ.(1c) and (1d) the source functions for normal stress are obtained:

$$S^{P*}(w) = \frac{c_{T2}^2\left(2w^2 + \frac{1}{c_{T2}^2}\right)}{\rho_2\left[-4c_{T2}^4 q_2 p_2 w^2 + c_{T2}^4\left(2w^2 + \frac{1}{c_{T2}^2}\right)^2\right]} \qquad S^{S*}(w) = \frac{2c_{T2}^2 p_2 w}{\rho_2\left[-4c_{T2}^4 q_2 p_2 w^2 + c_{T2}^4\left(2w^2 + \frac{1}{c_{T2}^2}\right)^2\right]} \qquad (7)$$

The directional fields for normal stress are obtained replacing w.

$$S^{P*}(\beta_2) = \frac{\cos(\beta_2)\left(1 - 2\left(\frac{c_{T2}}{c_{L2}}\right)^2 \sin^2(\beta_2)\right)}{\rho_2\left[\left(1 - 2\left(\frac{c_{T2}}{c_{L2}}\right)^2 \sin^2(\beta_2)\right)^2 + 4\left(\frac{c_{T2}}{c_{L2}}\right)^4 \sin^2(\beta_2)\cos(\beta_2)\sqrt{\left(\frac{c_{L2}}{c_{T2}}\right)^2 - \sin^2(\beta_2)}\right]}$$

$$(8)$$

$$S^{S*}(\alpha_2) = \frac{2i\cos(\alpha_2)\sin(\alpha_2)\sqrt{\left(\frac{c_{T2}}{c_{L2}}\right)^2 - \sin^2(\alpha_2)}}{\rho_2\left[\left(1 - 2\sin^2(\alpha_2)\right)^2 + 4\sin^2(\alpha_2)\cos(\alpha_2)\sqrt{\left(\frac{c_{T2}}{c_{L2}}\right)^2 - \sin^2(\alpha_2)}\right]}$$

The source functions S* are independent on the wave propagation in medium 1. The source functions for normal stress S* are different from those of an normal force under load and are equal to those for a normal force at a free-surface. The formulae of the free surface are obtained substituting ρ_0 and c_{T0} by 0 in the equ.(3) and (4).

The normal stress at the interface contains sufficient information to calculate the transmitted waves in the adjacent layer. This is the supposition for a stepwise calculation in each layer. Using the potentials of the half-space (1) with the source functions (2) or (7) the forward motion of the wave may be simulated in the partial layers, and hence the stress generated by the wave at the interface can be calculated. The point source functions are derived for plane interfaces.

The formulae also may be applied for curved and non-parallel interfaces if the interfaces are discretized. For this purpose the transducer element and the interfaces are uniformly covered with point sources. The sound field results from the superposition of all elementary waves of the point sources. Using the source function for a normal force under load (2) for the active element, the forward motion of the wave may be simulated in layer 1. For the point sources at the following interfaces the source functions for normal stress (7) have to be applied. The formulae for a point source synthesis with harmonic waves are given in [5] and calculation examples of the sound field with this method are given in [6].

INFLUENCE OF SOURCE FUNCTION ON THE SOUND FIELD

In this section the directional fields of the two point sources are compared. Fig.3 shows the calculated directivity pattern of the L-wave (solid line) and the T-wave (dashed line)

- ◆ a and b - for a normal force under load at the interface (equ.(3), (4))
- ◆ c and d - for a normal stress at the interface (equ.(8))

in steel and water. The directional field of a force acting at a surface covered with a solid (transducer element coupled to specimen) depends on the parameters of the specimen and also of the element. In example 3a and 3b the active element (medium 0) is PZT[*] with the parameters $\rho = 7{,}6$ g/cm^3, $c_L = 3875$m/s, $c_T = 2027$m/s.

In steel the difference between the directional fields of the L-wave for a point force at a free surface and for a point force under load at an interface is very small (compare fig.3a, c). That's why the point source function for a force at a free surface is a good approximation. In contrary there is a great difference between the two directional fields in water (compare fig.3b, d), i.e. using the point source function for the free surface instead of

---

[*] The given velocities are Hills averages calculated by Bergner (Dresden University of Technology) from the tabulated elastic constants.

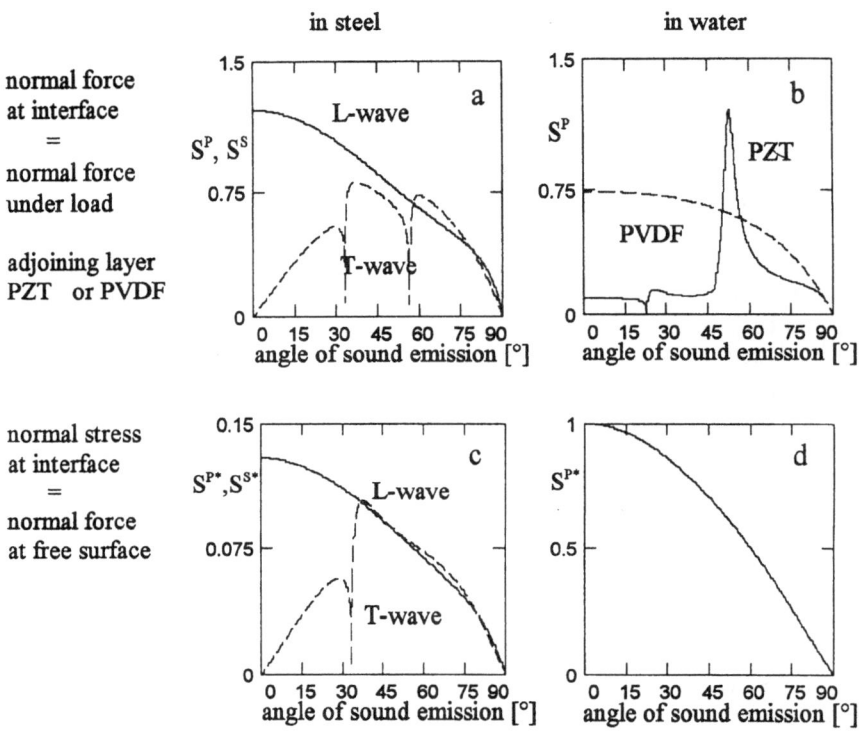

Fig.3. Directivity pattern for L- and T-wave in water and steel for the two different point sources

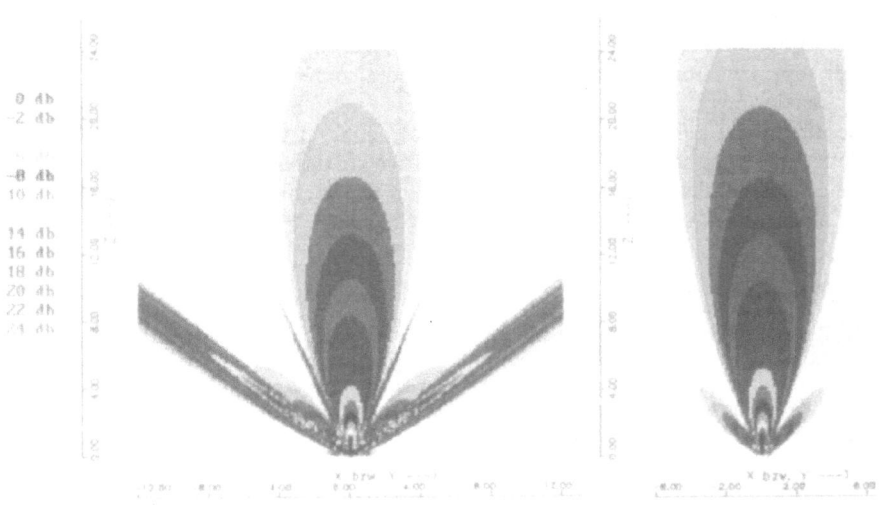

normal force under load (element of PZT)   normal force at free surface

Fig.4. Sound field in water for extended source (element size : 2λ, f=10MHz) for different point sources

the source function of a force under load with boundary conditions of PZT-water gives incorrect calculation results.

The zeros in the directivity pattern graphs are caused by roots with the value zero. This corresponds to the critical angle for interface waves, i.e. the refracted or transmitted waves propagate along the interface. In steel the zeros in the directivity pattern of the T-wave occur when the reflected LT-wave propagates along the interface ($\sin\alpha_1 = c_{TSt}/c_{LSt}$).

The comparison of the L-wave propagating in water for the two cases - force under load and force at the free surface - shows different characteristics. While the directional field for the free surface force results in a cos-function ($c_{T2} = 0$ in equ. (8), fig.3d) the field for normal force under load becomes more sophisticated due to the dependence of the source function on both media. For the latter case the component N_2 in equ. (3c) is equal 1. The directivity pattern for the boundary conditions PZT-water has a zero for an angle of $23°$, ($\sin\beta_1 = c_{LW}/c_{LPZT}$ and the 1st component in N_1 in (3b) becomes infinite). The LL-wave transmitted from water to PZT propagates along the interface. For the angle $48°$ the 2.term in N_1 is equal 0. The radiation has a maximum at an angle of $52.7°$ which results from the superposition of both terms of N_1. The measured sound amplitudes in water for a small strip of PZT, which is a good realisation of a line force, show a similar behaviour[7].

The directivity pattern shows a great variety in dependence on the transducer material. A point source acting at the interface PVDF-water, for instance, with the parameters of PVDF $\rho = 1.8$ g/cm³, $c_L = 1400$ m/s, $c_T = 870$ m/s generates a more cos-like function with no zeros (fig.3b-dashed line).

As outlined before, the sound field of an extended source results from the superposition of the elementary waves of all point sources. The influence of the directivity pattern of a single point source (fig.3b, 3d) on the resulting sound field is examined for the cases: normal force at the interface PZT-water and normal force at free surface (fig.4). The elements are rectangular ones with a frequency of $f = 10$MHz and the dimensions of 2λ (0.3×0.3 mm²). It can be seen that for such a small element the directional fields of the point sources highly affect the transducer generated sound field.

CONCLUSIONS

The generalized ray theory is applied to obtain the displacements generated by a concentrated body force acting at the interface between two half-spaces of different materials and by normal stress at the same interface caused by an incident wave. Using a Fraunhofer approximation the directional fields for both sources are derived. There are two important conclusions:

 * The directional fields of a concentrated force acting at the interface and of normal stress at the interface caused by an incident wave are different.
 * The directional field for normal stress acting at the interface and the directional field of a normal force acting at a free surface are equivalent.

REFERENCES

1. Y.-H.Pao, R.R.Gajewski: The generalized ray theory and transient response of layered elastic solids, in: Physical Acoustics, W.P.Mason, G.H.Thurston, ed., Academic New York, New York, (1977)
2. J. Kutzner: Grundlagen der Ultraschallphysik, B.G. Teubner, Stuttgart, (1983)
3. L.F. Bresse, D.A. Hutchins: Transient generation of elastic waves by a disk-shaped normal force source, J.Acoust.Soc.Am., 86(2), 810, (1989)
4. E. Kühnicke: Semianalytical method to calculate acoustic waves in layered elastic media, Ultrasonic International, Edinburgh,.July 4-7, (1995)
5. E. Kühnicke: Berechnung der Ausbreitung harmonischer Wellen in geschichteten Medien, Fortschritte der Akustik - DAGA 95, 21. Deutsche Jahrestagung für Akustik Saarbrücken, March 13-17, (1995)
6. E. Kühnicke: Simulation calculations for monofrequent sound fields in layered media, 21st International Symposium on Acoustical Imaging, Laguna Beach, March 28-30, (1994)
7. B. Köhler: Sound fields of ultrasonic circumferential arrays, 21st International Symposium on Acoustical Imaging, Laguna Beach, March 28-30, (1994)

PULSE-ECHO ULTRASOUND SIMULATION IN 3D ABERRATING MEDIA

A.P. Berkhoff*, L.A.F. Ledoux[+], P.M. van den Berg[#], and J.M. Thijssen*

* Biophysics Laboratory, Department of Ophthalmology
University Hospital Nijmegen
P.O.Box 9101, 6500 HB Nijmegen
The Netherlands

[+] Department of Biophysics
University of Limburg
P.O.Box 616, 6200 MD Maastricht
The Netherlands

[#] Center of Technical Geoscience
Laboratory of Electromagnetic Research
Department of Electrical Engineering
Delft University of Technology
P.O.Box 5031, 2600 GA Delft
The Netherlands

INTRODUCTION

Intervening tissue layers (Fig. 1) can be a major obstacle for obtaining clear ultrasonic images in echographic systems with large array transducers. Various correction methods have been proposed to obtain better images for these cases[1]-.[4] In order to test the applicability and limits of the correction methods, it is important to have an accurate and efficient computational model describing the propagation through these distorting tissue layers. The sound propagation through tissue is often modeled as the propagation through a layer consisting of a cascade of thin sublayers, where each sublayer acts upon the incident wave as a random time-shift operator.[5] Various papers[5]-,[7] however, report that the wave modification can not be described sufficiently with these models and that refraction effects also should be taken into account. In particular, ultrasound propagation through female breast[5,7] suffers from severe refractive errors. In abdominal animal tissue[6] strong multi-path components were noticed, with refraction as a possible cause. Recently, therefore, we have tried to come up with an efficient numerical method[8] to model the process of refraction, which is attacked by solving the numerical problem of wave propagation through an irregular interface between two uniform media. Of course, the present model should be extended with distributed wave aberrations to get a more complete description of wave propagation through human tissue.

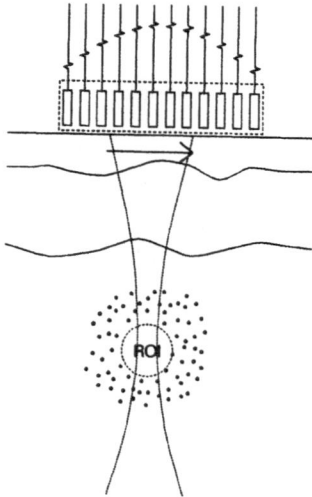

Figure 1. Configuration showing an array transducer making a linear scan of a medium consisting of point scatterers in a layered and attenuating medium, with the exception of the anechoic lesion (no point scatterers) in the Region Of Interest, where the compressibility and density in each of the layers are assumed to be different.

Simulation approach

The simulation of the sound propagation in absorbing media can be described with continuous Fourier-type descriptions representing a summation of homogeneous and inhomogeneous monochromatic plane waves. More efficient numerical implementations result if the continuous transforms are replaced by spatial and temporal FFT's. In order to arrive at increased robustness of the spectral propagation operators, we developed weak forms of the k-space Green's functions.[9] Especially for waves reflected from small inhomogeneities, with a large content at high spatial frequencies, this leads to considerably improved numerical results. Waveforms (in the time-domain) are obtained by analyzing the problem at several discrete frequencies in a temporal frequency range of interest and by subsequent evaluation of the temporal inverse Fourier transforms. Rigorous descriptions of absorption in media with continuous relaxation losses were also developed.[9]

Rough media interfaces: Conjugate Gradient Rayleigh method

A method for the computation of reflection and transmission of acoustic waves at a rough interface between uniform media[10,11] has been presented.[8] The method is based on a wave-function expansion technique in which the chosen elementary wave functions satisfy the wave equation and the radiation condition analytically, while the boundary conditions at the rough interface are satisfied numerically. An integrated squared error criterion can be defined for the deviations in the boundary conditions in the pressure and the surface normal component of particle velocity. The integrated squared error is minimized with a continuous form of a conjugate gradient technique suitable for the non-selfadjoint nature of the numerical problem. It was shown[8] that for typical media contrasts found in human tissue,[12] the method gives accurate results and good convergence properties if the surface roughness, expressed in the maximum surface slope, is not too large. The excellent convergence properties of the iterative scheme are for a large part based on the observation that evanescent waves can safely be neglected in most cases of interest. A comparison of the method with iterative and direct integral equation methods has also been presented.[13,14]

CGR-extensions

If the contrast between the media becomes larger, or if the surface roughness is very large, with surface slopes much larger than unity, evanescent components can not be neglected anymore.[14] However, the evanescent components cause severe ill-conditioning of the underlying system of linear equations,[16] with a resulting convergence degradation of the iterative process. For these cases, a preconditioned conjugate gradient scheme was developed as well as a suitable preconditioner[14] (see also texts on discrete systems of linear equations[15]). The preconditioned scheme shows superior convergence to the non-preconditioned scheme if evanescent components are included. Furthermore, the preconditioned scheme shows no degradation in performance as compared to the normal scheme if evanescent components are absent. In addition, an efficient scheme was developed for 3D sound fields with essentially 2D rough media interfaces, the 2.5 CGR method.[17] A formal derivation of the 2.5D method can be given.[18] If the rough interface is assumed to be described by piecewise polynomials and if the strong form of the integrated squared error criterion is replaced by a weak form, then the resulting sub-integrals can be evaluated analytically. A combination of the latter techniques: the preconditioned scheme, the 2.5D method, the weak forms of the numerically evaluated spatial sub-integrals, with a reverberating multi-layered configuration have been shown to be effective for describing propagation through acoustic lens structures occuring in 1D linear arrays where a lens is used for increased beam narrowing in the elevation direction.[19]

It was shown by Ledoux et al.[17] that measured sound fields show excellent agreement with theoretical predictions with the 2D, 2.5D and 3D implementations. In the 2.5D method,[18] the total error in 3D for the pertinent cylindrical interfaces can be written as an integral over k_y of the errors of simple 2D calculations. Therefore, the minimization in 3D is equivalent to repeated 2D minimizations for all values of k_y that occur in our problem. As a result, the analysis of cylindrical interfaces in 3D fields can be computed without the integration in the y-direction, thereby reducing the computational dimensionality of the problem by one.

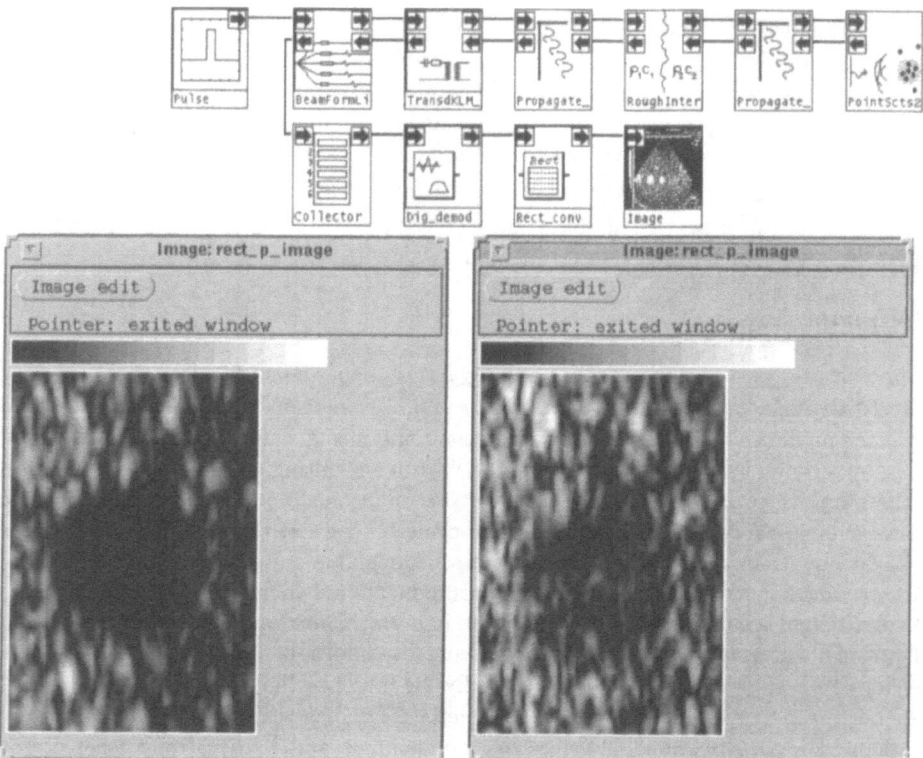

Figure 2. Upper: Block diagram created in simulation software[20] showing flexible environment for modeling of scanner electronics, transducers, and inhomogeneous acoustic media. Lower left image: Simulation examples for imaging with a 3.5 MHz linear array transducer having a pitch of 0.4 mm and an aperture size of 64 elements. The medium consists of an anechoic lesion with a diameter of 6 mm embedded in a region consisting of randomly positioned point scatterers. The displayed tissue has dimensions 14 mm x 10 mm. Lower right image: As left image except for the medium where an aberrating rough interface between two media is included with an equivalent rms time shift of 70 ns and a correlation length of 2 mm.

Simulation of anechoic lesions in aberrating media

A flexible software package[20] was developed in which the techniques described above were incorporated. In addition, various other aberrating phase-screen models,[7] reverberating medium models and transducer models[21] are available. The software uses a graphical shell around a large collection of separate simulation and calculation programs receiving data from each other. The simulation programs can be run independently on a server in Internet-based client-server systems just by clicking a menu-button. Most programs reflect the two-way character of continuous media and transducer systems by integrating transmitting and receiving behaviour in a single module (Fig. 2). However, simulation of one-way operation is also possible, such as calculation of beam plots or pressure pulses in the acoustic media. In Fig. 2 an example is given of the images obtained with a linear array transducer in aberrating and non-aberrating media. The software is designed to incorporate, and will be equiped with, adaptive wave-field inversion schemes such as Time-Reversal mirrors,[22] correlation-based aberration correction[23] and other,[24] corrective imaging techniques, with received and processed signals possibly re-emitted into the aberrated medium for improved detectability of tissue geometry and characteristics.

REFERENCES

1. L. Nock, G.E. Trahey, and S.W. Smith, Phase aberration correction in medical ultrasound using speckle brightness as a quality factor, *J. Acoust. Soc. Am.* 85:1819-1833 (1989).

2. C. Gambetti and F.S. Foster, Correction of phase aberrations for sectored annular array ultrasound transducers, *Ultrasound Med. Biol.* 19:763-776 (1993).

3. M. Karaman, A. Atalar, H. Kömen, and M. O'Donnell, A phase aberration correction method for ultrasound imaging, *IEEE Trans. Ultrason. Ferroel. Freq. Contr.* 40:275-282 (1993).

4. D.L. Liu and R.C. Waag, Time-shift compensation of ultrasonic pulse focus degradation using least-mean square error estimates of arrival time, *J. Acoust. Soc. Am.* 95:542-555 (1994).

5. Q. Zhu and B.D. Steinberg, Modeling, measurement and correction of wavefront distortion produced by breast specimens, *in* proc. Ultrasonics Symposium, M. Levy, ed., IEEE, New York, 1613-1617 (1994).

6. B.S. Robinson, A. Shmulewitz and T.M. Burke, Waveform aberrations in an animal model, *in* proc. Ultrasonics Symposium, M. Levy, ed., IEEE, New York, 1619-1624 (1994).

7. Q. Zhu and B.D. Steinberg, Wavefront amplitude distribution in the female breast, *J. Acoust. Soc. Am.* 96:1-9 (1994).

8. A.P. Berkhoff, P.M. van den Berg, and J.M. Thijssen, Iterative calculation of reflected and transmitted acoustic waves at a rough interface, *IEEE Trans. Ultrason. Ferroel. Freq. Control* 42:663-671 (1995).

9. A.P. Berkhoff, J.M. Thijssen and R.J.F. Homan, Simulation of ultrasonic imaging with linear arrays in causal absorptive media" *Ultrasound Med. Biol.*, in press.

10. J.T. Fokkema and P.M. van den Berg, "Seismic Applications of Acoustic Reciprocity," Elsevier, Amsterdam (1993).

11. J.A. Ogilvy, "Theory of wave scattering from random rough surfaces," IOP Publishing Ltd, London (1991).

12. P. Altmeyer, S. el-Gammal, and K. Hoffmann, " Ultrasound in Dermatology," Springer Verlag, Berlin (1992).

13. A.P. Berkhoff, P.M. van den Berg, J.M. Thijssen, Simulation of wave propagation through aberrating layers of biological media, *in* proc. Ultrasonics Symposium, M. Levy, ed., IEEE, New York, 1797-1800 (1994).

14. A.P. Berkhoff, P.M. van den Berg, and J.M. Thijssen, Ultrasound wave propagation through rough interfaces: iterative methods, submitted to *J. Acoust. Soc. Am.* (1995).

15. G.H. Golub and C.F. van Loan, "Matrix Computations, 2nd edition," John Hopkins University Press, Baltimore, 527-529 (1989).

16. J-M. Chesneaux and A. Wirgin, Reflection from a corrugated surface revisited, *J. Acoust. Soc. Am.* 96:1116-1129 (1994).

17. L.A.F. Ledoux, A.P. Berkhoff, and J.M. Thijssen, Ultrasound wave propagation through aberrating layers: experimental verification of the Conjugate Gradient Rayleigh method, submitted to *IEEE Trans. Ultrason. Ferroel. Freq. Control* (1995).

18. A.P. Berkhoff, P.M. van den Berg, J.M. Thijssen, "Numerical implementation of the Conjugate Gradient Rayleigh method," to be presented at the IEEE Int. Ultrasonics Symposium, Seattle, 7-10 Nov. (1995).

19. A.P. Berkhoff, L.A.F. Ledoux, P.M. van den Berg, J.M. Thijssen, Rigorous analysis of propagation through acoustic lens structures, *Ultrasonic Imaging* 17:59 (1995).

20. Simplan user's guide. University of Nijmegen (1995).

21. R. Krimholtz, D.A. Leedom, G.L. Matthaei, New equivalent circuits for elementary piezoelectric transducers, *Electronics Letters* 6:398-399 (1970).

22. M. Fink, Time reversal of ultrasonic fields - part I: basic principles, *IEEE Trans. Ultrason. Ferroel. Freq. Control* 39:555-566 (1992).

23. S.W. Flax, M. O'Donnell, Phase-aberration correction using signals from point reflectors and diffuse scatterers: basic principles, *IEEE Trans. Ultrason. Ferroel. Freq. Control* 35:758-767 (1988).

24. G.C. Ng, S.S. Worrell, P.D. Freiburger, G.E. Trahey, A comparative evaluation of several algorithms for phase aberration correction, *IEEE Trans. Ultrason. Ferroel. Freq. Control* 41:631-643 (1994).

(*) Supported by Philips Medical Systems, the Netherlands, and the Technological Science Branch of the Netherlands' Organization for Scientific Research, project no. NGN 11.2427.

ANALYSIS OF SCATTER FIELDS IN DIFFRACTION
TOMOGRAPHY EXPERIMENTS

Helmar S. Janée[1,2], Joie P. Jones[3], and Michael P. André[1,2]

[1]Department of Radiology, University of California, San Diego, CA
92093-9114, Radiology Service, [2]Veterans Affairs Medical Center, San
Diego, CA, and [3]Department of Radiological Sciences, University of
California, Irvine, CA

INTRODUCTION

The propagation of acoustic waves through complex, inhomogeneous materials such as tissue is characterized by a number of processes that include scattering, diffraction, refraction and reflection. Reconstructing images of such objects from measured transmit or scatter fields provides a considerable challenge and the reconstruction approach that is most suitable may depend on which of these processes dominate. In order to gain a better understanding of these factors, we have initiated a detailed investigation of the nature of scatter fields observed *in vivo* in human breast tissue for a number of diffraction tomography experiments.

A recently developed ultrasound diffraction tomography system for breast imaging[1,2] was used to investigate the nature of scatter fields from both simple and complex objects. Statistical parameters of the amplitude distributions of the scatter fields were analyzed following procedures used by several researchers[3-6] to assess and potentially correct distortion introduced in the wave-front by a strong scatterer. These workers report measurements using a single transmitter and a linear array of receivers over relatively narrow angles (on the order of 20-30°) at frequencies above 3 MHz. We report measurements of the total scatter field (360°) obtained with 1024 transceivers at frequencies ranging from 0.3 to 1.3 MHz. We assessed the scatter fields and their amplitude distributions for simple, calculable scatterers such as small diameter nylon cylinders ($a<\lambda$) and solution-filled latex balloons ($a>\lambda$) in order to verify our experimental and analytical procedures. We report scatter fields for breast tissue from four patients with different breast classifications. Since these analyses characterize the nature and degree of scattering in the object to be imaged, it was natural to try to interpret these data in terms of the Born and Rytov approximations which are generally employed in the solution of inverse scattering problems. In particular, we briefly review the utility and range of applicability of the Born and Rytov approximations and present a new operator-based formalism that illustrates their physical and mathematical origins.

BREAST IMAGING SYSTEM

The ultrasound CT diffraction tomography system incorporates both 512 and 1024-element transducer rings operated at center frequencies of 0.5 and 1.0 MHz, respectively.[2,7] Both rings are 20 cm in diameter and the transducer elements are spaced less than $\lambda/2$ with dimensions of 0.62 x 12 mm and 0.3 x 12 mm, respectively. Data are recorded at 10 discrete frequencies ranging from 0.7-1.3 MHz for the 1 MHz ring and at 20 frequencies ranging from 0.3-0.7 MHz for the 0.5 MHz ring. Amplitude and phase are recorded in quadrature after a suitable delay time to allow for system stabilization.

A photograph of the 1 MHz transducer ring assembly is shown in Figure 1a and a schematic diagram of the data acquisition sequence in Figure 1b. The raw data of waveform peak voltage and phase for each transducer are parsed and gain-corrected and can be displayed in a number of ways. One display is in the form of a sinogram which was used to determine the range of receivers subtended by the test object. Our results in all cases are the mean values of n (n=512 or 1024) observations, that is, each measured scatter field of n receivers on a complete circle surrounding the test object is the average of n transmitters.

SCATTER AMPLITUDE DISTRIBUTION ANALYSIS

Wave propagation in the breast, according to a number of researchers,[3-6] is characterized principally by scatter and refraction and these processes are the primary causes of amplitude wavefront distortion. In the case of weak scattering, phase distortion can be adequately addressed by various phase-aberration correction schemes. Amplitude distortion, however, is observed when scattering becomes strong and its appearance can be thought to define the limit of applicability of weak-scattering theories. The evidence for this transition from weak to strong scattering is the shape of the observed probability density function (pdf) of the complex wavefront the modulus of which is approximated by a Rayleigh distribution. If scattering is weak, the observed distribution should be Rician and ultimately Gaussian.

Let the scatter field with an object in the path and the incident field, at a particular location x be given by:

$$E(x) = A(x)\exp(j\phi(x)); \qquad E_0(x) = A_0(x)\exp(j\phi_0(x)) \tag{1}$$

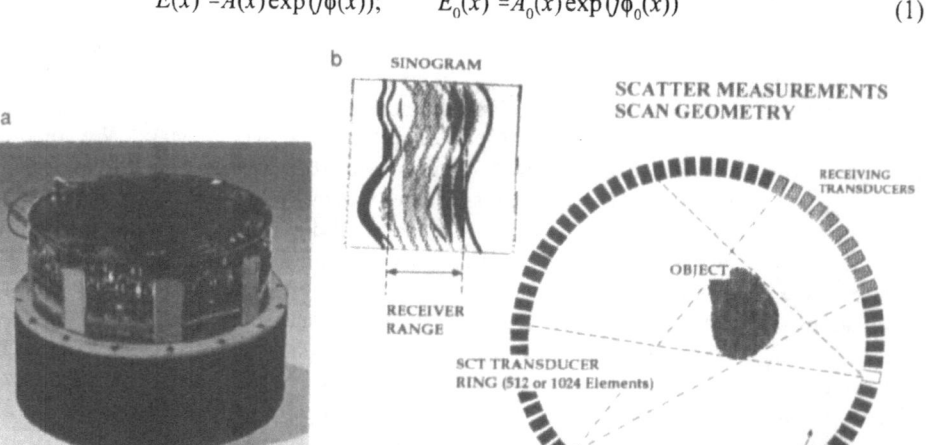

Figure 1. a) Transducer array assembly. b) Schematic of data acquisition procedure.

The wavefront amplitude distortion term is defined in general and for our experimental geometry in particular as:

$$a(x) = A(x)/A_0(x) \; ; \quad a_j = \overline{A_j^x}/\overline{A_j^0} \quad where \quad \overline{A_j} = \frac{1}{N}\sum_{k=0}^{N} A_{j,k} \; ; \quad j \neq k \tag{2}$$

By forming a histogram of the a(x) values, plots of pdf vs. amplitude are obtained and compared to Rayleigh distributions of identical standard deviation. If there is close agreement between the two, that is, if the means of the computed Rayleigh and the experimentally determined distributions are close, then the scatter field is considered to be fully developed.

The general form of the distribution to which our data were fitted is given by the expression:[8]

$$p(A) = \frac{A}{\sigma^2} \exp\left[\frac{-A^2 + A_0^2}{2\sigma^2}\right] I_0 \left(\frac{AA_0}{\sigma^2}\right) \tag{3}$$

where p(A) is the probability density for amplitude A, the random term. A_0 is a constant phasor, has integral values and when $A_0 = 0$, equation 3 tends toward a Rayleigh distribution. For large values of A_0, p(A) becomes Gaussian. I_0 is the modified Bessel function. The ratio of the mean value of the distribution p(A) to its standard deviation is found to increase with A_0 and thus provides a useful index to indicate whether a scatterer is mostly Rayleighian or Rician. We define the Tissue Index, $<a>/\sigma_a$, by:[9]

$$\frac{\overline{a}}{\sigma_a} = \sqrt{\frac{\pi}{2}} \; \exp\left(-\frac{A_0^2}{4}\right)\left[(1 + \frac{A_0^2}{2}) I_0 \left(\frac{A_0^2}{2}\right) + \frac{A_0^2}{2} I_1 \left(\frac{A_0^2}{2}\right)\right] \tag{4}$$

This ratio is unity for an exponential distribution, 1.24 for a Rayleigh distribution and becomes arbitrarily large with A_0 for a Rician distribution.

BREAST TISSUE SCATTERING

We have extended this analysis to frequencies lower than previously reported by studying the scatter amplitude distributions from four patients who had been imaged as part of our clinical trials series at the UCSD Center for Women's Health. Two patients had been imaged with the 1 MHz SCT ring and two with the new low-frequency ring. The former were classified as dense/glandular (Patient 5) and fatty with a prior history of radiation and chemotherapy treatment (Patient 8). The latter breast types were classified as normal (Patient 14) and fatty (Patient 15). One slice from each patient was analyzed according to the procedures outlined above. The initial analyses used only those receivers of the transducer ring that were in the shadow of the object, using the sinograms of the data to select the minimum range of receivers subtended by the object. This is essentially comparable to the technique used by [3,4] who used a single transmitter and a linear receiving array to record only pulses transmitted through the object. We compare this subset of the scatter data with the amplitude distributions for the whole scatter field, which we obtain in a single measurement.

Figure 2 shows the scatter data over all angles for both the scan tank alone and for the case with a breast object. Data shown are for Patient 5 at several lines about the 1 MHz center frequency. The amplitude data for the empty water tank can be fairly well fitted with a $\cos^2(\theta)$ distribution which one would expect if the transducer radiation pattern is indeed a dipole term. The high spatial frequency fluctuations are indeed acoustical signals of unknown origin but may be due to diffraction effects and multiple reflections. The same data obtained with a breast object show significant attenuation through the breast and significant energy scattered at angles larger than those subtended by the object. This suggests that such large-angle scattering is due to compressibility fluctuations which gives rise to monopole terms.

The scatter amplitude distributions computed according to equation 2 for the data of Figures 3b for the case of using only receivers subtended by the object and for the total scatter field are compared in Figure 3. It is seen that the restricted angular distribution is clearly non-Rayleighian, being most closely fitted by a Rician distribution with a value for A_0 of 2. The tissue index ($<a>/\sigma$) for this case is 2.6 while it is 1.9 for the second case

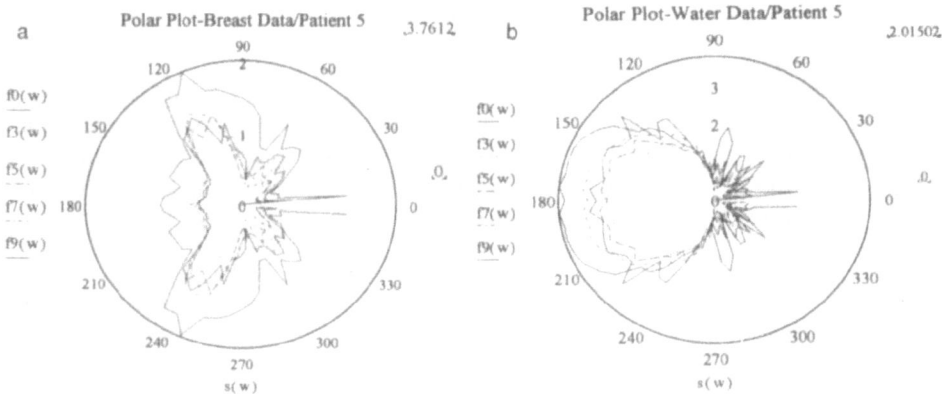

Figure 2. Scatter amplitude for 1023 receivers averaged over 1024 transmitters. a) Water medium. b) Patient 5 breast data.

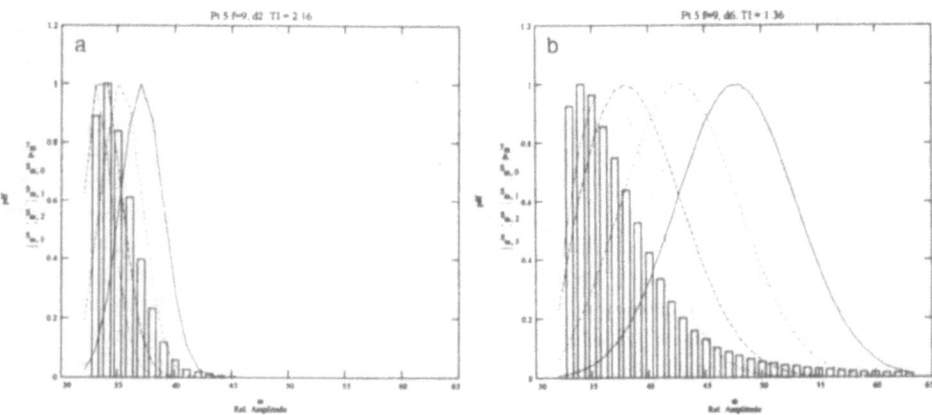

Figure 3. Measured scatter amplitude pdf for Patient 5 compared to computed Rician distributions. a) Limited angular distribution of the scatter field. b) Total scatter field.

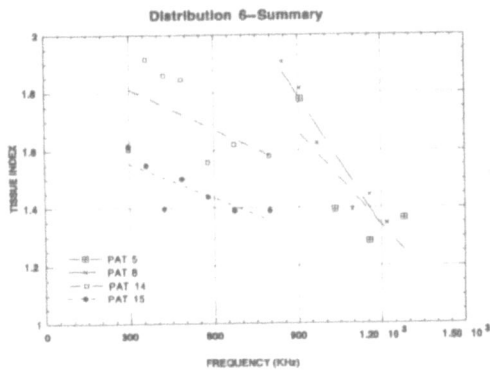

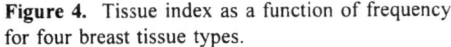

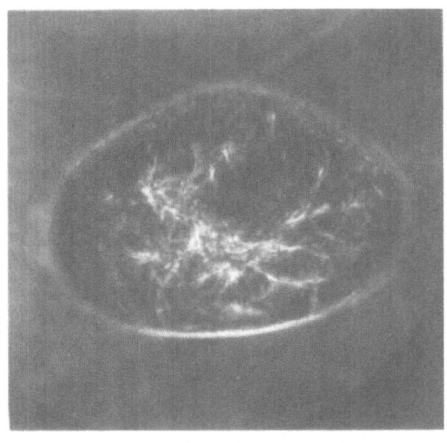

Figure 4. Tissue index as a function of frequency for four breast tissue types.

Figure 5. Tomographic slice image of a dense, glandular breast (Patient 5).

where the total scatter field is recorded. This distribution is much closer to being Rayleighian which follows expectations since the contribution of transmitted and back-reflected energy is much smaller.

An analysis of the data such as shown in Figure 3 for several frequency bands in terms of the Tissue Index defined earlier showed that a clear differentiation between some breast tissue types could be made. Thus, as seen in Figure 4, a clear difference between normal and fatty breast type can be discerned, with the latter being the much stronger scatterer since the index is lower, suggesting a more nearly Rayleighian scatter amplitude distribution. It is noteworthy that an analysis of the limited-angle scatter data yielded breast tissue attenuation values as a function of frequency that ranged roughly between 0.5 and 1 db/cm/MHz. Shown in Figure 5 is the image of a tomographic slice of the breast for Patient 5, reconstructed with techniques based on the first-order Born approximation.

BORN AND RYTOV APPROXIMATION OPERATOR FORMALISM

We briefly summarize a new operator formalism for the Born and Rytov approximations that permits the expression for one or the other to be derived from a single expression through appropriate setting of an operator variable. This formalism shows both approximations to be not only functionally similar but to be part of perhaps a continuum of possible solutions to the Helmholtz wave equation for inhomogeneous media.

Let the inhomogeneous Helmholtz wave equation with constant density for a medium characterized by $k_0 = \omega / c_0$ with ω the frequency and c_0 the speed of sound of the medium be given by:

$$\nabla^2 \psi(\bar{r}) + k_0^2 \psi(\bar{r}) = -k_0^2 f(\bar{r}) \psi(\bar{r}) \tag{5}$$

where $\psi(r)$ is the total field and $f(r)$ is an object with speed of sound fluctuations given by $f(\mathbf{r}) = [(c(\mathbf{r})/c_0)^2 - 1]$. With $\psi_0 = A_0 e^{j\omega t}$ as the incident field and generalized transformations given by: $\chi = \psi_0 \, g(\alpha)$; $\alpha = \psi / \psi_0$; and finally, $g(\alpha) = m(\alpha^{1/m} - 1)$, it can be shown that equation 5 can be expressed in the general form:

$$\nabla^2 \chi + k_0^2 \chi = -k_0^2 f \psi_0 \alpha^{1/m} - \psi_0 (1 - \frac{1}{m}) \alpha^{1/m} (\nabla \ln \alpha)^2 \qquad (6)$$

When m=1, equation 7 defaults to the Born approximation and when m=∞ it becomes the Rytov approximation. This suggests that the Born and Rytov approximations represent opposite limits to a more general approximation and that the variable m, if properly chosen, might yield an optimal solution to the inverse scattering problem. We are not prepared at this stage to suggest the form of that compromise but find it an intriguing subject for further exploration.

CONCLUSIONS

Analysis of the amplitude distributions of scatter fields from breast tissue in diffraction tomography experiments suggests that different tissue types can be described by a simple index based on the distribution mean and standard deviation. The nature of the distribution is an indication of the strength of the scattering and in general it is found that scattering in breast tissue is well established even at frequencies below 1 MHz. A new operator-based formalism for the Born and Rytov approximations is developed and it is concluded that the general ranges of validity for the Born and Rytov approximations are consistent with the scatter amplitude distributions ranges. That is, a Rayleigh distribution implies strong scattering and thus suggests that the Born approximation would be inappropriate and that the Rytov approximation might be more useful. On the other hand, dominant reflection and weak scattering which result in a Rician distribution suggest that the Rytov approximation might be inappropriate due to the presence of sharp index of refraction gradients or discontinuities. In most cases, however, neither extreme will be optimal and an intermediate approximation might be called for. The form such an approximation would take is the subject of ongoing research by the authors and would lead to an inversion formalism that is object-specific.

REFERENCES

1. P.J. Martin, M.P.André, B.A. Spivey and D.A. Palmer, Sonic computed tomography for breast imaging, *in:* "Breast Imaging Update," Madjar H, Teubner J, Hackeloer BJ, eds., Karger, Basel (1993).
2. M.P. André, P.J. Martin, G.P. Otto, L.K. Olson, T.K. Barrett and B.A. Spivey, A new consideration of diffraction computed tomography for breast imaging: Studies in phantoms and patients, *Acoustical Imaging 21(1994)*.
3. Q. Zhu and B.D. Steinberg, Wavefront amplitude distribution in the female breast, *J. Acoust. Soc. Amer. 96:1 (1994)*.
4. Q. Zhu and B.D. Steinberg, Wavefront amplitude distortion and image sidelobe levels: Part I--theory and computer simulations, *IEEE Trans. Ultrason. Ferroelect. Frequ. Control 40:754 (1993)*.
5. G.E. Trahey, P.D. Freiburger, L.F. Nock and D.C. Sullivan, In vivo measurements of ultrasonic beam distortion in the breast, *Ultrason. Imag. 13:71, (1991)*.
6. B.D. Steinberg, A discussion of two wavefront aberration correction procedures, *Ultrason. Imag. 14:398 (1992)*.
7. B.A. Spivey, P.J. Martin, D.A. Palmer, "Acoustic Imaging Device", *U.S. Patent Number 5,305,752* (April 26, 1994).
8. A. Ishimaru, "Wave Propagation and Scattering in Random Media, Vol.1", Academic Press, N.Y. (1978).
9. J.W. Goodman, "Statistical Optics", J. Wiley & Sons, N.Y. (1983).

DIFFERENCES OF ULTRASOUND PROPAGATION IN TISSUE AND TISSUE MIMICKING MATERIALS

Jaroslav D. Satrapa[1] and Ivan Zuna[2]

[1]TCC, A-4850 Timelkam, Lenaustr. 10
[2]DKFZ, D-69120 Heidelberg, Im Neuenheimer Feld 280

INTRODUCTION

The ultrasound images taken on homogeneous and isotropic tissue mimicking phantoms cannot be directly compared to the images taken from inhomogeneous and anisotropic tissues. For this purpose special comparison and measuring methods for the recognition and quantitative embracement of the propagation phenomena in inhomogeneous media were developed. Propagation phenomena appear in many cases as special attenuation issues or as characteristic frequency dependent back scattering. By severe speed of sound inhomogeneities appears serious reduction of image contrast accompanied by reverberations and multiple scattering.

It is known that causality between tissue structure and propagation parameters, respectively image statistical parameters, are especially important. The study of these causalities is associated with difficulties, if the parameters of the reference tissue mimicking objects, are not precisely known.

Besides the methods for measuring attenuation, speed of sound and back scattering, the new methods are developed for estimating the contrast reduction, anisotropy and inhomogeneities for tissue and tissue mimicking materials. The new materials are proposed as propagation reference objects. The methods are supported with mathematic models and simulations. Some applications of novel methods in vivo will be illustrated.

The ultrasound propagation is characterized by a very complex interaction of ultrasound waves and tissue structure. The complex interaction may be described by physical parameters of the scanner, as well as by propagation parameters of the tissue. Both parameter groups influence the generation of the B-image. The determination of scanner parameters is the basic condition for the estimation of propagation parameters and tissue characterization. Mostly the interrogating tissue is not in immediate connection with the scanner. The tissue, skin, fat or muscles, between scanner and object, influences the image appearance in some cases essentially. Generally, there is a certain *tissue "stand off"* with different propagation properties. In many cases this "stand off" is homogeneous and can be compared with many tissue mimicking materials. However, some abdominal walls and breasts show full complexity of nonhomogeneous propagation coming from irregular interlacing of fat and collagen.

The nonhomogeneities are in the speed of sound and attenuation and cause the phase aberrations and additional generation of side lobes. In a strong nonhomogeneous environment the reverberation and multiple scattering is observed. It blurs the image and reduces the image contrast. In many cases, only reduced contrast is visible in B-image.

In the effort to improve the B-scanners diagnostic efficiency, the special phantoms or reference objects are developed as a tool for estimation of unknown scanner parameters and quantification of attenuation, backscattering, reverberation and contrast reduction.

The reference object image, delivers the parameters for gray level histogram normalization and gray level calibration. Normalization is equivalent to the restoration of statistical parameters of the first order. The image correction is still object of research, in the sense of essentially improving of contrast and elimination of reverberation.

In order to estimate the system parameters accurately, the phantom materials parameters such as speed of sound, attenuation and back scattering with their characteristic temperature and frequency dependencies must be known. The parameters of homogeneity and anisotropy help also for the estimation of propagation parameters.

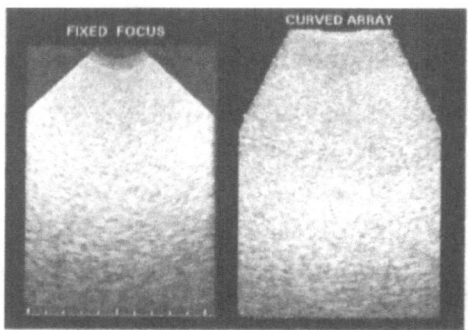 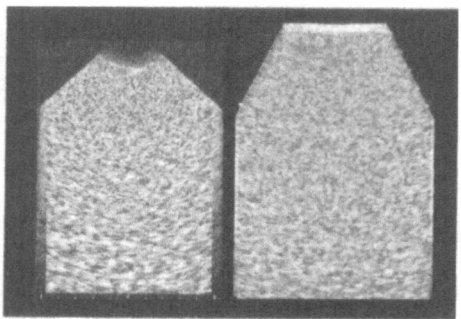

Fig.1 - There is the noticeable difference in speckle pattern for the same object and different scanners.

Fig.2 - Normalization of images, shown in Fig.1, using the equalization of *mean* and *standard deviation* for all depths.

The scanner is characterized with parameters as axial, lateral and elevation resolution. The better alternative to this parameters is the detectability of small cysts. The cysts detectability correspond substantially to the contrast resolution as a detectability for focal lesions. All three aspects of resolution, respectively, axial, lateral and elevations resolution, are involved in the mentioned cysts detectability. The B-imaging is accompanied with large depth variations of cysts detectability. Fortunately, the cysts detectability is directly measurable under specific conditions using phantoms with spherical or cylindrical voids acoustically equivalent to small cysts.

BASIC OBSERVATIONS OF PROPAGATION PHENOMENA

The propagation phenomena may be directly observed by the B-scanner and by suitable objects immersed in water. A convenient way to show the propagations phenomena is a sequence of examples.

Fig.1 shows the B-images of artificial foam immersed in water with a fixed focus scanner and a curved array scanner.

Although both images represent the same object, there is a fundamental difference in the speckle pattern. If the nonlinear distortion of the local dynamic range is eliminated, the histogram analysis shows, that the gray levels of both images follow the Rayleigh distribution.

The gray level distribution is independent from the position in the image and is the same for the near field, far field or focus. The shape of the speckle corn characterizes the system resolution, but the system resolution cannot be unambiguously estimated from the shape of the speckle corn or from the autocorrelation function.

The aprior knowledge about the Rayleigh distribution is used for gray level calibration. It means that the histogram of the homogeneous foam image may be normalized using PC and frame grabber and equalized in mean and standard deviation for all depth, Fig.2.

What supplementary information can be extracted from the B-image of the foam? The optical microscope image of the foam shows the spatial lattice as short threads in space, mutually connected in knots. The position of lattice cells in space is random and thickness of the short fibers (fibrils) are almost equal. The number of fibers per volume unit and their orientation, changes across the space. It may be expected that interference of ultrasound waves and described lattice structure will have the consequence in some resonance and aliasing effects normally not noticeable in the B-image.

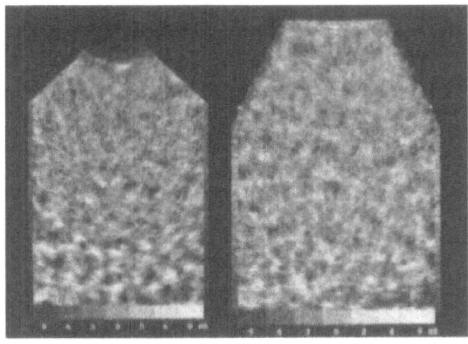

Fig.3 - Enhancement of normalized images, shown in Fig.2, by weighting and summing (WS) processing.

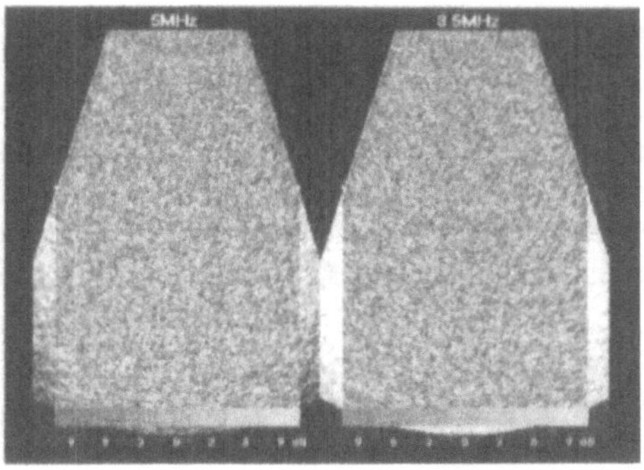

Fig.4 - The normalized images, taken at the same B-plane, with frequency switchable scanner, do not show obvious differences.

In the homogeneous artificial foam B-image, the strong changes of scatterer density, which causes large deviation from Rayleigh distributions cannot be observed. Only the weak differences in gray levels, like darker and brighter areas may be observed. The B-image decomposed in several spatial frequency bands and composed the frequency bands back by *Weighting* and *Summing* (WS) processing, let us see these weak differences better, Fig.3.

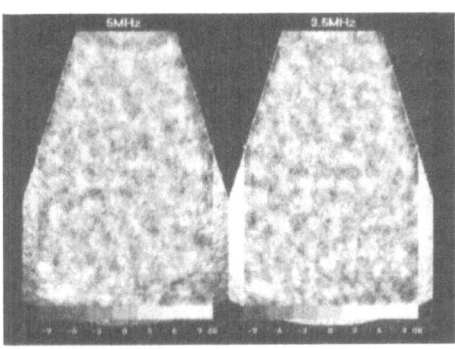

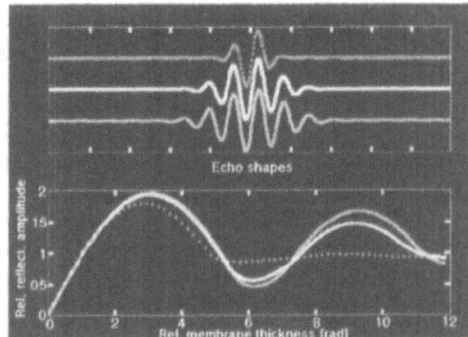

Fig.5 - The same images, shown in Fig.4, processed by WS present clearly the frequency dependent areas of backscattering.

Fig.6 - Resonance of the thin membrane dependent from the echo shape and membrane thickness.

The processed image shows indeed patterns of different gray level areas corresponding to different scatterer densities or aliasing effects due to the anisotropy in the lattice structure. Which in both effects is responsible for areas, looking like lesions or specific texture, can be found by insonification of the same B-plane by different scanner frequency bands. With frequency switchable scanners, the normalized images do not show obviously the differences Fig.4. The same images processed by WS shows clearly the frequency dependent areas of backscattering, Fig.5. Dark areas in both images at the same places indicate the lower scatterer density.

Due to the scatterer fiber shape of he presented foam structure it has the typical frequency dependent backscattering, increasing with 6 db/octave or 20dB/decade, corresponding to the frequency dependency of single fiber or thin membrane. The membrane shows the typical frequency dependency behavior with resonance effect by frequency, corresponding to $\lambda/4$ of membrane thickness, Fig.6.

PROPAGATION IN LIVING TISSUE

In comparison with foam immersed in water, the living tissue is characterized by inhomogeneities given by anatomical structure or caused by a pathological process. In the case of a diffused disease, the tissue structure will be changed, but the homogeneous character is maintained. The tissue frequency dependent backscattering is strongly structure dependent and may differ from values found in foam and liver.

The inhomogeneities and changes of tissue structure may be observed in B-images but not measured with available diagnostic instruments. Deficient conception of propagation in tissue may be improved by:

- computer modeling of propagation in tissue and
- experimental measurement of propagation phenomena

The most popular method of modeling by computer is the multidimensional convolution model successfully applicable for non attenuating and non reverberating media. The convolution model which much more simplifies the propagation and many essentially important propagation phenomena cannot be studied on such model.

More suitable is the RAY-tracing model, primarily used for sliced structures and plane waves. The RAY model applied, on general structures, can be very intensive for computing and time consuming. Due to such shortcomings it finds the application only in special cases.

Experimental proof is unavoidable in any case for models and simulations. The comparison between tissue and reference object is not limited in any sense, but needs the careful choice of experimental set up, taking in to account that a prior knowledge and possible large set of quasi artifactual appearances and propagation explicitness.

Let us see the basic observations and the explicitness of propagation.

Fig.7. shows the typical "sound amplification" and "shadowing" effected by gap (opening) and bar (obstruct). The propagation in non homogeneous tissue in the sense of attenuation have randomly distributed "gap and bars".

For experimental purposes the regular array of steel bars with diameters of 0.5 mm and 1mm distance put at the front of the transducer, Fig.8, generates the grating lobes and reduces the contrast of image. Images show the phantom voids with 5mm diameter and the image of the same object influenced by an array of steel bars .

The slice of animal fat put on the front of a transducer evokes also contrast reduction, Fig.9. The "random array" of fat and collagen with interlaced muscles produces the side lobes caused from inhomogeneities in attenuation and in speed of sound.

It is coming primarily from the acoustical difference between fat and collagen. Strong reverberations and multiple scattering can be observed as diminishing brightness behind the slice of fat immersed in water, Fig.10.

In order to correct exactly the image, the correction of *amplitude and phase* for any RAY must be accomplished and for all ranges of images separately. It is not known how to correct the complex reverberations.

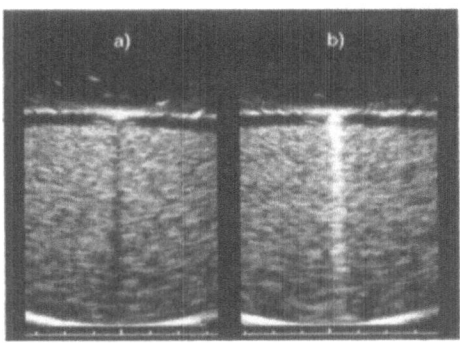

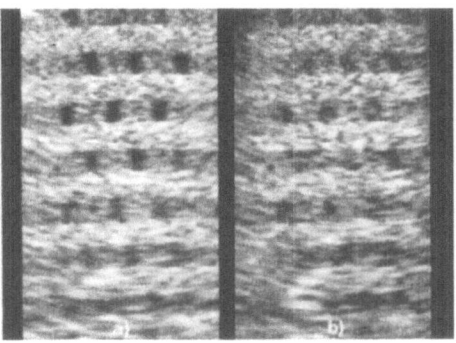

Fig.7 - The typical "shadowing" (a) and "sound amplification" (b), effected by gap (opening) and bar (obstruct), are basic appearances of propagation in attenuating and nonhomogeneous media.

Fig.8 - Regular array of steel bars put at the front of the transducer, reduce the contrast of image; a) Image of small cylindrical voids included in phantom; b) Image taken with the same phantom, with the steel bars array, put at the front of the transducer.

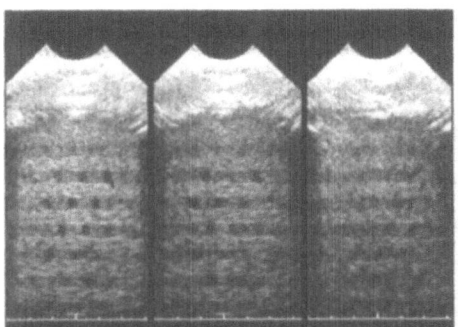

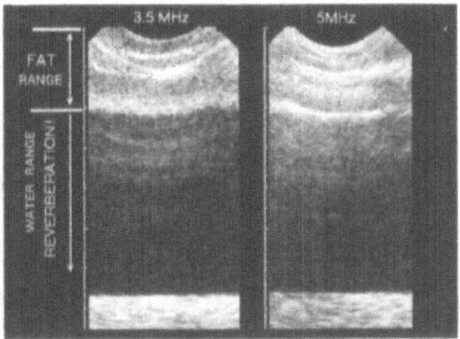

Fig.9 - The slice of animal fat put on the front of the transducer evokes contrast reduction. The "random array" of fat and collagen with interlaced muscles produces the side lobes caused from inhomogeneities in attenuation and in speed of sound. Images show the phantom with cylindrical voids, with three different positions of animal fat slice used as stand off.

Fig.10 - Strong reverberations and multiple scattering can be observed as diminishing brightness, behind the slice of animal fat immersed in water.

PROPAGATION IN OTHER ARTIFICIAL MEDIA

The graphite powder in gelatin shows a similar attenuation characteristic as foam. Only the back scattering follows other frequency dependence other then it is found in foam or liver. The reason for such behavior is relative high acoustic impedance of graphite (4 MRayl) compared to collagen (1.6 MRayl) and polyurethane (1.7 MRayl), and small dimensions of graphite corn (40μ). In order to mimic the liver frequency dependency of backscattering, the glass beads are usually added to the graphite corn. Glass beads, as spherical targets do not show the typical anisotropy effect as some tissues or foam.

With rubber based phantoms it was possible to avoid the water desiccation. The main disadvantage of polyurethane rubber is the low speed of sound from 1450 m/s. The attenuation presents itself mainly as an absorption. It is not know how to produce the typical tissue frequency dependent back scattering in such absorbing media.

PRACTICAL WAYS OF B-IMAGE PROCESSING

The correction of the B-image may be calculated by using the data from a reference object image. The condition for image pixels correction is the small difference between the attenuation of the reference object and the tissue. Taking this condition in to account, the diffraction error will be minimized. The new image pixel value will be given by the shown expression:

$$p_n(j,i) = \mu_n + \sigma_n *[\ p_o(j,i) - \mu_o(j,i)\ ] / \sigma_o(j,i) \qquad (1)$$

where

p_n = new pixel value

μ_n = new mean

σ_n = new standard deviation

p_o = pixel of original object image

μ_o = mean of reference object image

σ_o = standard deviation of reference object image

The $\mu_o(j,i)$ and $\sigma_o(j,i)$ may be prepared previously as moving averages in 2D look up tables. The result in general approach is 2D look up table $LUT_{j,i}$

$$p_n(j,i) = LUT_{j,i} [p_o(j,i)] \qquad (2)$$

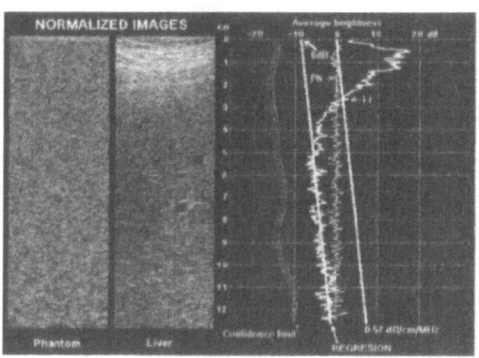

Fig.11 - Normalized images of reference object (phantom) and liver. The reverberation of abdominal wall is very strong and progresses deep in the liver parenchyma diminishing at the depth of 4cm. The averages of brightness, presented as curves from reference object and liver, show clearly the reverberations range of abdominal wall. The brightness regression line for liver deliver the estimation for backscattering and attenuation of liver.

Fig.11 shows already the normalized reference object and liver image. From the normalized images follow the estimation of liver attenuation and backscattering. The backscattering and reverberation of abdominal wall is very strong and progresses deep in the liver parenchyma. By TGC adjustment, the mostly strong abdominal wall backscattering, will be adjusted unconsciously and wrongly away. The image normalization recovers this wrong adjustment and gives the information and impression about abdominal backscattering and quasi hidden reverberation. The possible reduction of contrast by the abdominal wall is already shown.

Typical pathologic liver appearance is accumulation of fat and collagen as fibrosis. Eight cases of liver interrogation show the spreading of attenuation and backscattering,

Fig.12. The normalization of B-images and WS processing cannot improve the detectability of small cysts. For this reason it is of serious significance to find out, for any scanner, the cyst detectability for different cyst sizes and usable scanning depths. The best illustration of such measurements presents the phantom with cylindrical voids taken with 3D scanner in sequence of C-images, Fig.13.

A special problem is the estimation of contrast reduction due to the tissue inhomogeneities.

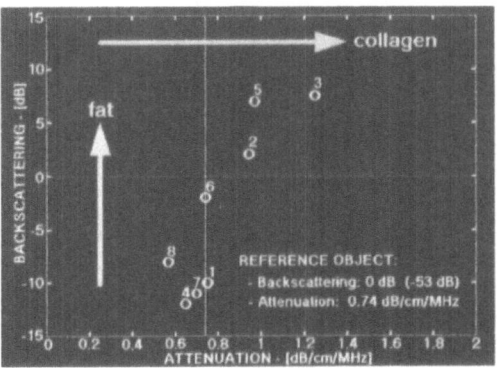

Fig.12 - Typical pathologic liver appearance is accumulation of fat and collagen as fibrosis. Eight cases of liver interrogation show the spreading of attenuation and backscattering.

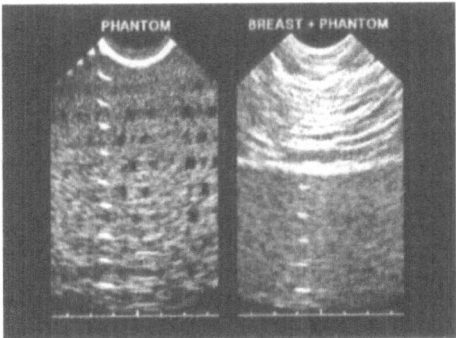

Fig.13 - The normalization of B-images and WS processing cannot improve the detectability of small cysts. For this reason it is of serious significance to find out, for any scanner, the cyst detectability for different cyst sizes and usable scanning depths. A 3D scan shows, two B-images and seven C-images of phantom with cylindrical voids from 5, 4 and 3 mm diameter in depth from 1cm to 7cm. The 3mm voids are clearly visible only at 3cm depth.

A unique possibility for the direct measurement of contrast reduction reveals the breast. The breast when pressed, as usual in mammography, creates the shape like "stand off" and the contrast reduction may be estimated in the same way as shown with animal fat Fig.14 and 15. The phantom is provided with nylon threads and cylindrical voids. Only the nylon threads are visible and the voids disappeared. The nylon threads are specular reflectors and behave obviously differently than voids.

The contrast and contrast-reduction can be treated quantitatively using normalized images. Haralick introduced the co-ocurrence matrix (CM) for calculation of texture contrast. It is applicable for ultrasound normalized images:

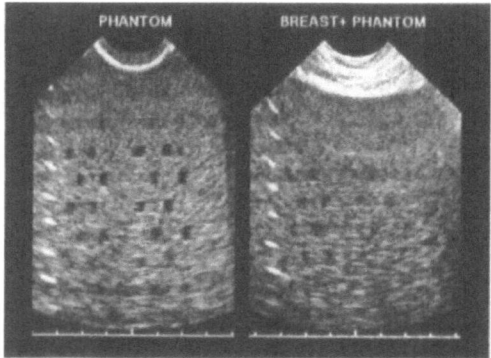

Fig.14 - The breast when pressed, as usual in mammography, creates the shape like "stand off" and the contrast reduction may be estimated in the same way as shown with animal fat. Only the nylon threads are visible and the voids disappeared. The nylon threads are specular reflectors and behave obviously differently then voids.

$$C = \sum_{i,j} | j - i |^k * p(j,i)^m \qquad (3)$$
$$d,\theta$$

where j and i are the pixel values on distance d and angle θ. The $p(j,i)$ is the probability of coocurrence. For k=1 and m=1 texture contrast definition have the best correspondence to other contrast definitions. For spherical targets and spherical reference the CR may be introduced

$$CR = C/C_r \qquad (4)$$

where C_r is contrast of a reference. For practical test object applications, spherical or cylindrical voids are the best choice.

The inhomogeneities of a reference object could be understood also as random distributed targets in volume. However, these targets are random in shapes and dimensions and the interpretation of contrast need some changes.

STATISTICAL IMAGE PROCESSING

The normalized images with a test of scanner contrast ability, and a test of tissue reduction contrast, is prepared for any other speckle reduction, wavelet and fractal analysis, or any other statistical image processing. It is obvious that any other or statistical processing of the image, or part of the image, may be useless if the contrast in the region of interest (ROI) is lost or it is to much reduced. As shown, the special 3D phantoms have to be used for measurement of contrast and contrast reduction to find out the spites of propagation in inhomogeneous media.

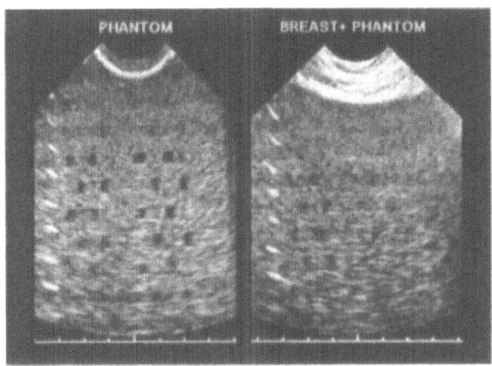

Fig.15 - Thinner slice of breast tissue causes less contrast distortion and phantom voids are still visible.

CONCLUSION

The differences of ultrasound propagation in tissue and tissue mimicking materials are characterized mostly with fat interlacing of normal or pathological origin. As shown, the fat slice structures reduce the contrast essentially by inhomogenities in speed of sound and attenuation. The contrast will be additionally reduced by reverberations and multiple scattering. In order to avoid the inefficient post processing of images, the contrast and contrast-reduction estimation will be suggested. The test on tissue mimicking materials is not sufficient as proof if the differences in propagation are not taken into account.

REFERENCES

Garra, Madsen, Parks, Skelly,Zagzebski: AIUM Quality Assurance for Gray- Scale Ultrasound Scanners, Handbook (1994)

J.A. Zagzebski, Z.F. Lu, L.X. Yao: Quantitative Ultrasound Imaging: In Vivo Results in Normal Liver. Ultrasonic Imaging 15, 335-351 (1993)

J.D.Satrapa: Qualitätskontrolle von ultraschalldiagnostischen Geräten mit Testphantomen, Ultraschal-Diagnostik 1985, Drei-Ländertreffen, Zürich 1985

J.D. Satrapa,, I. Zuna: Deterioration of B-Scan Liver Images by Abdominal Wall, The 6th WFUMB Congress, Sapporo, Japan 1994

J.C. Bamber, R.J. Dickinson: Ultrasonic B-scanning: a computer simulation, *Phys. Med. Biol.*, 1980, Vol. 25. No.3, 463-479.

D.R. Foster, M. Arditi, F.S. Foster, M.S. Patterson, J.W.Hunt: Computer Simulations of Speckle in B-scan Images, *Ultrasonic Imaging* 5, 308-330 (1983)

J.A. Zagzebski, E.L. Madsen M.M. Goodsitt: Quantitative Tests of a Three-dimensional Gray scale Texture Model, *Ultrasonic Imaging* 7, 252-263 (1985)

S. Finette: Synthetic B-scan Images by Numerical Solution of wave equation, *Ultrasound Imaging* 10, 220-228 (1988)

D.E. Robinson, L.S. Wilson, T. Bianchi: Beam pattern (Diffraction) Correction for Ultrasonic Attenuation Measurement, *Ultrasonic Imaging* 6, 293-303 (1984)

K.J. Parker, M.S.Asztely, L.M. Lerner, E.A. Schenk, R.C. Waag: In Vivo Measurements of Group and Benign Breast Diseases, *Ultrasonic Imaging* 12, 47-57, (1990)

M. Fein, I. Zuna, W.J. Lorenz: New Method of Estimating the Ultrasonic Beam Ratio Parameter for Characterizing Sound Propagation in Tissue, *Ultrasound in Med.&Biol.* Vol. 18, No.10, pp. 881-889, 1992

O. Basset, Z. Sun, J.L. Mestas, G. Gimenez: Texture Analysis of Ultrasonic Images of the Prostate by Means of Co-occurrence Matrices, *Ultrasonic Imaging* 15, 218-237 (1993)

Y.W. Yuan, K.K. Shung: The Effect of Focusing on Ultrasonic Backscatterer Measurement, *Ultrasonic Imaging* 8, 121-130 (1986)

W.E. Howell, S.K. Numrich, H. Überall: Selective observation of elastic-body resonances via ringing in transient acoustic scattering, *J. Acoust. Soc. Am* 78(3), Sept. 1985

M.S. Patterson, F.S.Foster: The Improvement and Quantitative Assessment of B-Mode Images Produced by an Annular Array/Cone Hybrid, *Ultrasonic Imaging* 5, 195-213 (1983)

Standards Methods for measuring Performance of pulse-Echo ultrasound Imaging Equipment, *AIUM Standard* 1993

M.Ueda, M.Tabei, K.Nagata: Image Quality of ultrasonic B-mode images, Tokio Institute of Technology 1993.

M.Ueda: Evaluation and Improvement of Image Quality by means of Contrast Reduction Rate for Spherical test Object in Tissue-Mimicking Phantom and 3D Echo Simulations, Tokio Inst. of Tech. 1994.

R.M. Haralick: Statistical and Structural Approaches to Texture, *Proceedings of the IEEE*, Vol.67, No. 5, May 1979.

G.E.Trachey, J.W Allison, S.W. Smith, O.T. von Ramm: A Quantitative Approach to Speckle Reduction via Frequency Compounding, *Ultrasonic Imaging* 8, 151-164 (1986)

J.M. Thijsen, B.J. Ostervald, R.F. Wagner: Gray Level Transforms and Lesion Detectability in Echographic Images, *Ultrasonic images*, 10, 171-195 (1988)

HIGH-RESOLUTION IMAGE RECONSTRUCTION TECHNIQUES FOR CIRCULAR-APERTURE ARRAY IMAGING SYSTEMS

Wayne R. Lewis and Hua Lee

Department of Electrical and Computer Engineering
University of California
Santa Barbara, California 93106

ABSTRACT

The data acquisition of most conventional acoustic imaging systems is conducted by planar aperture arrays, in either stationary or synthetic-aperture modes. This allows us to utilize the property of shift invariance, which provides significant improvement and simplification for system analysis and computational complexity, especially for backward propagation or Fourier-transform based image reconstruction algorithms.

In practice, circular aperture array is also one of the common data acquisition configurations, especially in medical imaging or NDE applications. Due to the fundamental difference, many well documented properties such as resolution criterion, shift invariance, wavefield orthogonality, system transfer functions, and sampling rate, for planar-array systems are not valid for circular configurations. Therefore it is of great importance to establish the fundamental structure of system analysis and image formation for acoustic imaging systems with circular apertures.

In terms of algorithm structures, there are two effective approaches to the image formation problem for circular-array apertures. One is to decompose the wavefield pattern over the circular aperture in terms of circular harmonics in order to directly construct the spatial spectral content for image reconstruction. The other approach is to formulate the backward propagation method along concentric circular paths. The image formation is then modeled as a circular convolution and the process is angular-shift invariant, which greatly enhances the computational efficiency.

In this paper, we present the theoretical analysis and experimental results of these approaches. In addition, a study of resolution limit and sampling requirements will be included in the presentation.

INTRODUCTION

This paper investigates methods to reconstruct an object function, $a(r, \phi)$, via knowledge of the wavefield impinging on a circular aperture of radius R_a. The object function describes the relative scattering potential of an active source which emits scalar waves satisfying Helmholtz's equation and the free-space radiation criterion. A circle of radius

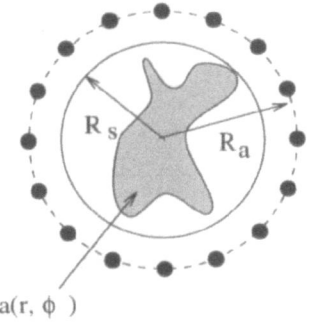

a(r, φ)

Figure 1. System Geometry

R_s, $R_s < R_a$, restricts the support of $a(r, \phi)$ to its interior. In Figure 1 the geometry is shown with transducer elements designated by solid circles.

The active source model allows the development of the reconstruction methods to be discussed without reference to a particular type of insonification. The section on demodulation techniques describes the necessary adaptations for realization of a passive target.

Chosing $e^{-j\omega t}$ time harmonics and a linearized scattering model allows the wavefield on the aperture to be expressed as a 2-dimensional convolution of the object function and the zero order Hankel function of the first kind. The Hankel function serves as the Green's function in the 2-dimensional scattering integral expression.

$$U(k, \theta) = \int_0^{2\pi} \int_0^{R_s} a(r, \phi) H_0(kR) r \, dr \, d\phi \qquad (1)$$

where $R = \sqrt{R_a^2 + r^2 - 2R_a r \cos(\theta - \phi)}$ and $k = \frac{\omega}{c}$.

This paper describes two methods for solving the stated problem. First, an inverse filter method is derived specifically for the circular array. It yields $a(r, \phi)$ without error for the hypothetical perfect noiseless system with continuous aperture and an infinite bandwidth. Second, the generalized backward-propagation method [2] is applied to this problem. Generalized backward-propagation is applicable to arbitrary aperture geometries, but produces an approximation to $a(r, \phi)$, even with perfect data.

During the development, the two methods assume a continuous aperture and an infinite system bandwidth. The sections on sampling requirements and resolution analysis discuss the applicability to realistic systems where these assumptions are obviously invalid. Experimental results conclude with an illustration using real data.

INVERSE FILTER SOLUTION

Due to the linear model used, a modal decomposition will prove useful in solving the inverse problem. The Hankel function generates modes when expressed by its addition formula, [1]

$$H_0\left(\sqrt{X^2 + x^2 - 2Xx\cos(\theta)}\right) = H_0(X)J_0(x) + 2\sum_{n=1}^{\infty} H_n(X)J_n(x)\cos(n\theta), \qquad (2)$$

valid for $|X| > |x|$.

Rewrite this expression by applying Euler's formula, combining the two resulting summations, substituting into the integral Eq. (1) and reordering terms and operators.

$$U(k, \theta) = \sum_{n=-\infty}^{\infty} H_{|n|}(kR_a)e^{jn\theta} \int_0^{2\pi} \int_0^{R_s} a(r, \phi) J_n(kr)e^{-jn\phi} r \, dr \, d\phi. \tag{3}$$

Introducing the notation $F_n(\rho)$ for the Fourier-Hankel transform of $f(r, \theta)$ and performing a Fourier series decomposition of Eq. (3) gives

$$U_n(k) = H_{|n|}(kR_a)A_n(k). \tag{4}$$

The term $H_{|n|}(kR_a)$ plays the role of transfer function for the scattering filter. The expression $\sqrt{\frac{2}{\pi kR_a}}$ [1] lower bounds the magnitude of this filter, making the inverse filter, $\frac{1}{H_{|n|}(kR_a)}$, stable for all n and typically bounded kR_a. Division and application of the inverse Fourier-Hankel transform completes the solution.

$$A_n(k) = \frac{U_n(k)}{H_{|n|}(kR_a)} \tag{5}$$

BACKWARD-PROPAGATION SOLUTION

The generalized backward-propagation method applies to any size space and aperture geometry. It consists of two steps. For each wavelength, match-filter the observed wavefield to form a holographic approximation of the object function. Then superimpose each of these images into a multi-frequency tomograhic image.

The equation

$$a^{(bp)}(r, \theta) = \int_0^{\infty} \int_0^{2\pi} U(k, \phi)H_0^*(kR) \, d\phi \, k \, dk \tag{6}$$

describes the backward-propagation reconstuction. The superscript (bp) distinguishes the inexact reconstruction of the backward-propagation method from the actual values. The inner integral performs the single-frequency matched filtering while the exterior integral superimposes the holographic images into a complete tomographic reconstruction.

The inverse filter method's derivation motivates application of the addition formula and the Fourier-Hankel transform which yields

$$A_n^{(bp)} = H_{|n|}^*(kR_a)U_n(k). \tag{7}$$

Eq. (5) and (7) follow the familiar forms of inverse and matched filtering, respectively. The backward-propagation method fulfills the expectation of preserving phase, but inverting magnitude.

The behavior of $|H_n(kR_a)|$ dictates the backward-propagation method's deviation from ideal. The section on resolution analysis shows that the object function's energy is negligible for n near and greater than kR_s, reducing the pertinent part of $|H_n(kR_a)|^2$ to $n < kR_s$. This function increases monotonically in n slowly for n near zero and with great acceleration as $n \to x$. A useful approximation developed by Meissel and described in [1], allows the creation of a bound on the deviation in the filter magnitude in the pertinent area.

$$\frac{|H_{kR_s}(kR_a)|^2}{|H_0(kR_a)|^2} \approx \frac{1}{\sqrt{\frac{R_a^2}{R_s} - 1}} \tag{8}$$

Clearly as the aperture moves into the far field, the magnitude of the transfer function in the pertinent area flattens and the difference between the backward-propagation and inverse filter methods diminish.

SAMPLING REQUIREMENTS

The Nyquist criterion does not hold for sampling on a circular aperture. Eq. (4) makes this apparent since there exists no N such that $H_{|n|}(kR_a) = 0$ for $|n| > N$. An N exists, however, that shows a significant drop in aliasing power.

For a numerical experiment, a single point source at radius R_s illuminates the aperture. Figure 2 illustrates the aliased power for an aperture of N elements for various values of R_s. Although k and R_a are fixed at 40 and 1, respectively, similarly shaped curves appear in plots with other values.

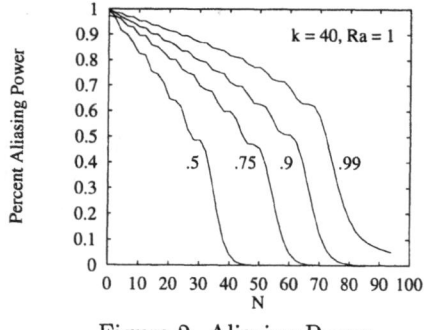

Figure 2. Aliasing Power
for $R_s = 0.5, 0.75, 0.9, 0.99$

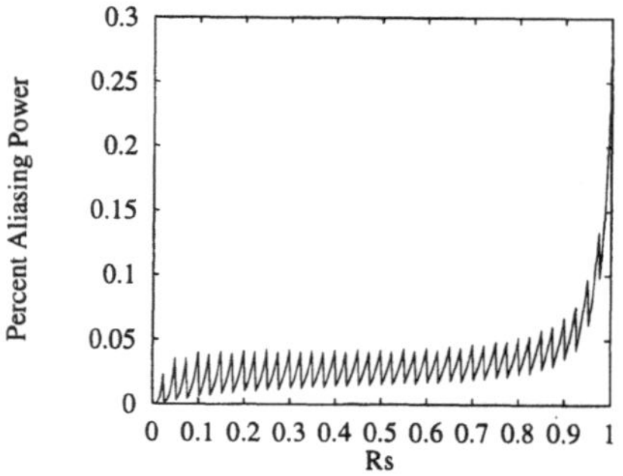

Figure 3. Aliasing Power for $N = 2kR_s$

The sharp knee in each plot near $2kR_s$ motivates its choice for the value of N. Figure 3 shows how this choice performs as a function of R_s. Note the steep rise in aliasing power as R_s approaches R_a. A spacing of $\frac{\lambda}{2}$ between elements on a radius of R_s coincides with this choice of N.

RESOLUTION ANALYSIS

Both bandwidth limitations and a discretized aperture reduce the resolution capabilities of the system. Here the resolution analysis is performed for a system with wavenumbers, $k_{\min} < k < k_{\max}$ and aperture element count, $N = \lceil 2k_{\max}R_s \rceil$. The results above motivate the choice of N.

The inner product of the kernels of the 2-dimensional Fourier and Fourier-Hankel transforms gives the conversion between the two domains. A circle of radius ω_ρ in the former converts to the line $\rho = \omega_\rho$ in the latter. Eq. (4) makes obvious the fact that the aperture is sensitive only to components of the object function $A_n(k)$. Thus the range of k translates to an annulus of inner radius $k_{\min}$ and outer radius $k_{\max}$ in the 2-dimensional Fourier domain. This annulus is termed "maximum coverage" and corresponds to the spectral coverage of a far-field planar aperture system with the same range of wavenumbers. The coverge obtainable lies in a rectangle in the Fourier-Hankel domain, bound along the ρ axis by $k_{\min}$ and $k_{\max}$ and along the n axis by $\pm\lceil k_{\max}R_s \rceil$. Figure 4 illustrates the maximum coverage and coverage obtainable.

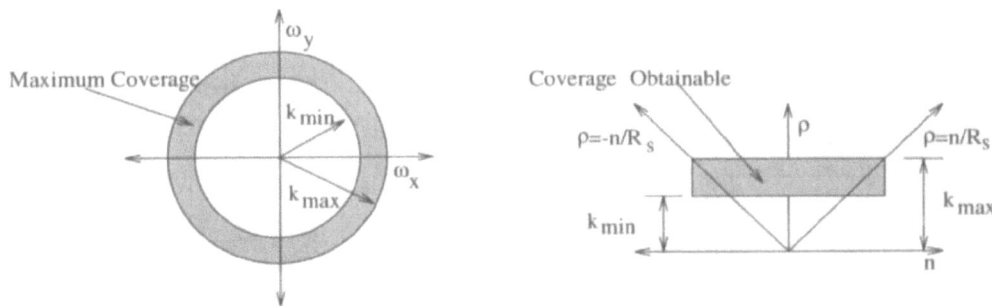

Figure 4. Maximum Coverage and Coverage Obtainable

A far-field planar aperture with the same range of wavenumbers serves as the benchmark for resolution capability. The geometry of the circular aperture fails to provide the same spatial invariant reconstruction as the benchmark, however the full array does provide invariance with respect to θ. For a point source at a given distance from the origin, the ratio of the power in the coverage obtainable to the power in the maximum coverage region quantifies the relative resolution performance of the circular aperture. Furthermore, since both apertures are similarly dense with respect to ρ, the measure can be refined to the ratio of the respective powers for a given ρ. As this ratio approaches unity for all $k_{min} < \rho < k_{max}$, the system reaches maximal resolution capability for the given the range of wavenumbers.

The Fourier-Hankel transform of a point source at (r, θ) is simply $J_n(\rho r)e^{-jn\theta}$, which yields the following as the appropriate power ratio.

$$\epsilon_\rho(r) = \frac{\sum_{|n| < \lceil k_{max} R_s \rceil} J_n^2(\rho r)}{\sum_n J_n^2(\rho r)} \tag{9}$$

This expression is simplified by noting that the denominator is identically one leaving an easily tabulated function. Figure 5 plots $\psi(x, M) = \sum_{|n| < M} J_n^2(x)$ with respect to x for various values of M.

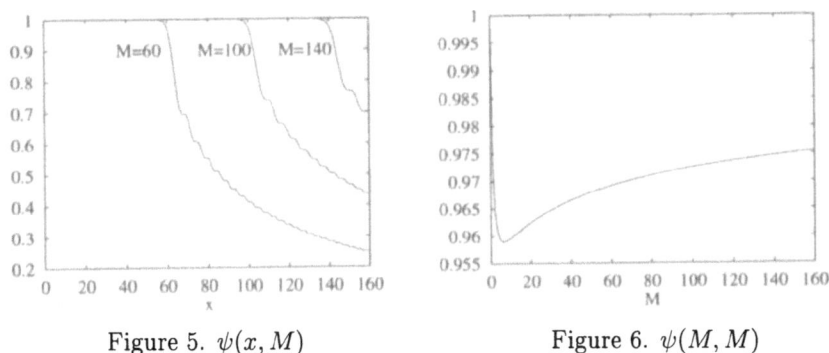

Figure 5. $\psi(x, M)$ Figure 6. $\psi(M, M)$

Since $\rho r \leq \lceil k_{max} R_s \rceil R_s$, $\epsilon_\rho(r)$ takes values of $\psi(x, M)$ that are nearly one. Figure 6 emphasizes this observation by plotting $\psi(M, M)$, the worst case power ratio for a given $\lceil k_{max} R_s \rceil R_s$.

This analysis shows that the choice of N made in the section on sampling requirements provides nearly complete coverage of the maximum coverage region. Thus adherence to $N = \lceil 2k_{\max}R_s \rceil$ allows the resolution of the circular aperture to approach that of the far-field planar aperture.

DEMODULATION TECHNIQUES

The inverse filter and backward-propagation techniques have been sucessfully applied to both monostatic and bistatic systems. For sake of brevity, only the monostatic technique is discussed here.

For a monostatic system, the transmitter and receiver move together around the target, synthesizing a circular aperture. Insonfication with a wide-band pulse and recording the echo provides a time delay profile at each point on the aperture. Scaling this profile by the propagation velocity yields a distance delay profile denoted by $d(\theta, x)$. Convolution with the appropriate Green's function simulates the wavefield that would appear on the aperture if the targets were, in fact, active sources.

$$U(k, \theta) = \int_0^\infty d(\theta, \frac{x}{2})H_0(kx)dx. \tag{10}$$

EXPERIMENTAL RESULTS

Data taken from a monostatic FMCW radar system gave the reconstruction in Figure 7 via the inverse filter method. The target consisted of two metallic cylinders with equal scattering potential. The image generated by application of the backward-propagation method shows negligible difference and is not shown.

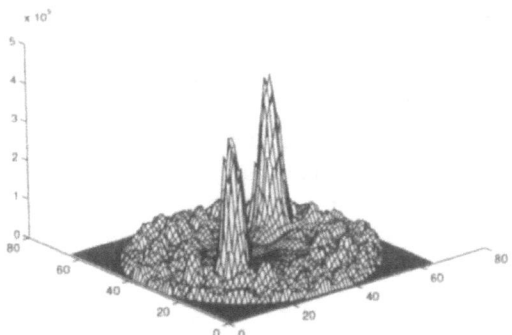

Figure 7. Reconstruction of Two Metallic
Cylinders Via the Inverse Filter Method

CONCLUSION

Two methods of solving the linearized scattering problem on a circular aperture were discussed. An inverse filter method, derived specifically for the problem, and backward-propagation, a general algorithm, both were sucessful. The sampling requirements were considered and the choice for the number of receiver elements, $N = \lceil 2k_{\max}R_s \rceil$, proved useful. The resolution capabilities were determined to be dictated by the range of wavelengths available and not degraded significantly by the discretization of the aperture as long as the above choice for N is used. Finally, a reconstruction based on real monostatic data was performed using the inverse filter method.

ACKNOWLEDGMENTS

This research is supported by the National Science Foundation and National Research Council.

REFERENCES

1. G.N. Watson, "A Treatise on the Theory of Bessel Functions," Cambridge University Press, Cambridge (1944).

2. Hua Lee, "Formulation of the Generalized Backward Projection Method for Acoustical Imaging," *IEEE Transactions on Sonics and Ultrasonics,* vol. SU-31, no.3, pp. 157-167, May 1984.

A NOVEL TECHNIQUE FOR IMAGING PULSED ULTRASOUND FIELDS

Andy Healey[1] and Sidney Leeman[2]

[1]Present Address Dept. of Electrical Engineering, Imperial College
London U.K.
[2]Dep. of Medical Engineering and Physics
King's College School of Medicine and Dentistry, London U.K.

INTRODUCTION

Conventionally, a knowledge of the acoustic field is gained via point-to-point measurements performed with a point hydrophone, usually in a water bath. A complete knowledge of the field is then comprised of a four dimensional function, $p(\mathbf{r}, t)$. Due to the large number of measurements required, it is convenient to restrict data acquisition to a subset of points, usually acquired in a plane. This measurement set, combined with some appropriate theory (conventionally based on the angular spectrum formalism) allows subsequent prediction of the field at other spatial locations. A number of problems arise in obtaining a general measurement knowledge of the field. The hydrophone may change the field it is measuring (particularly in the very near field). The field varies rapidly at some spatial locations. Point hydrophones are imperfect devices. No easy predictive quality exists, either at other spatial locations, or within other media.

This paper presents an alternative, general and computationally efficient method of predicting the ultrasound field, based on a set of measurements performed with a Large Aperture Hydrophone (LAH), as opposed to the conventional point devices. The technique is general in that is may be used to predict 'ideal' point measurements at any location in the space-time field, in homogenous lossy and non-linear media. The approach is valid for continuous wave, pulsed, unfocused and focused fields, is inherently three dimensional in nature, and does not require recourse to problematic evanescent waves.

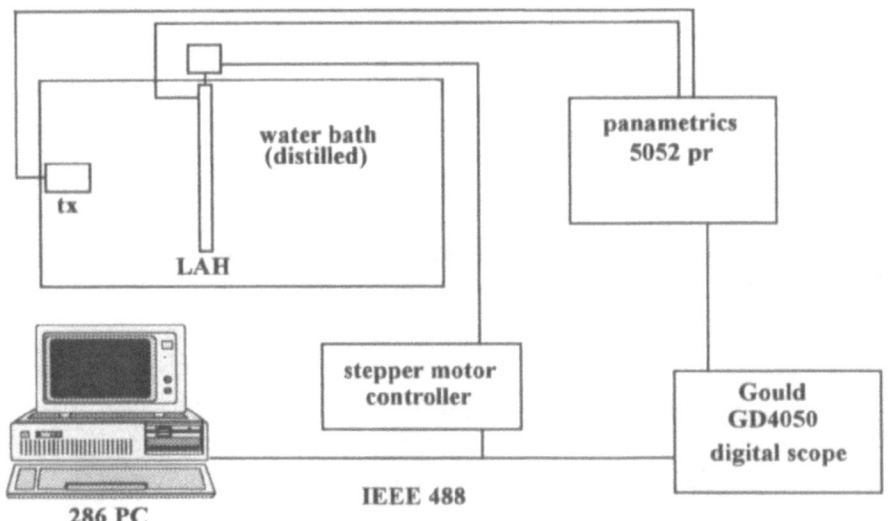

Figure 1. Data acquisition schematic.

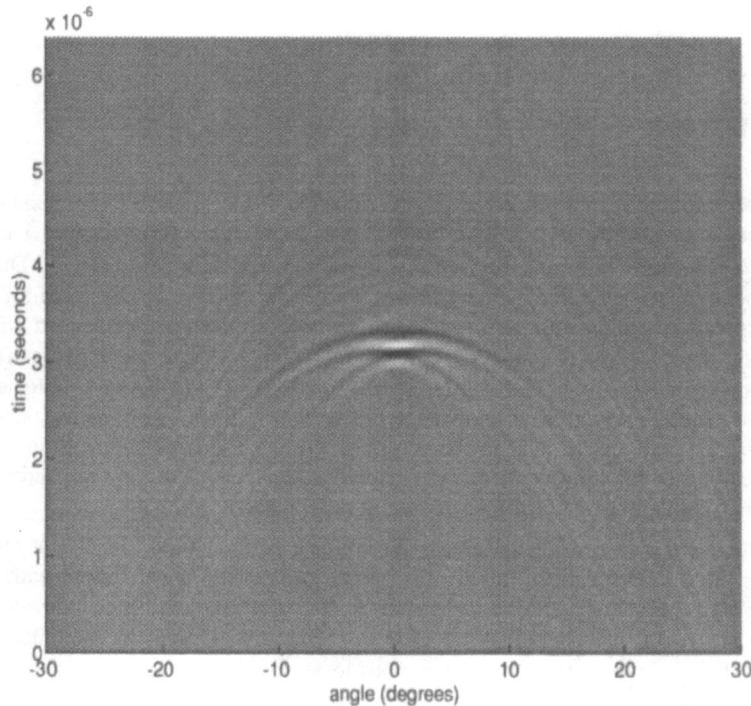

Figure 2. Temporal signals recorded from the LAH as a function of angle (relative to the acoustic axis).

THEORY

An alternative approach to the point hydrophone measurement scheme is possible based on large, (effectively 'infinite') aperture hydrophones for characterising and extrapolating fields (including pulsed fields). The technique is grounded on a formalism we have called the 'directivity spectrum' approach. Consider a forward travelling acoustic (pressure) field $p(\mathbf{r}, t)$ propagating in an ideal lossless medium and satisfying the canonical lossless wave equation. The pressure field may be expressed as

$$p(\mathbf{r}, t) = \frac{1}{(2\pi)^3} \int d^3 k F(\mathbf{k}) \exp\left(i\{\mathbf{k} \bullet \mathbf{r} - \omega t\}\right) \tag{1}$$

where $F(\mathbf{k})$ is the spatial Fourier transform of the field at some convenient time, say $t = 0$

$$F(\mathbf{k}) = \int_{-\infty}^{\infty} d^3 \mathbf{r}\, p(\mathbf{r}, 0) \exp\left(-i\{\mathbf{k} \bullet \mathbf{r}\}\right), \tag{2}$$

and is comprised of a set of travelling, continuous plane waves. Note that $F(\mathbf{k})$ does not depend on $\mathbf{r}$ or t (for the canonical lossless wave equation). A complete knowledge of the pulse may be equated to either an entire (measurement) knowledge of $p(\mathbf{r}, t)$ or alternatively $F(\mathbf{k})$. Hence a direct measurement knowledge of $F(\mathbf{k})$ encodes all the information regarding the field. The description is also applicable to lossy media with the field pressure $p_\alpha(\mathbf{r}, t)$, being given by the equation

$$p_\alpha(\mathbf{r}, t) = \frac{1}{(2\pi)^3} \int_{-\infty}^{\infty} d^3 k F_\alpha(\mathbf{k}) \exp\left(i\{\mathbf{k}' \bullet \mathbf{r} - \omega' t\}\right) \tag{3}$$

with complex k formalism $\mathbf{k}' = \mathbf{k} + i\alpha(k)\hat{\mathbf{k}}$, and $\omega' = \omega(\mathbf{k})$. The directivity spectrum approach may be regarded as an extension of the angular spectrum formalism.

The LAH has a desirable property. If the planar face of the LAH membrane intercepts the entire acoustic field, then the signal produced is related directly only to those Fourier components of the interrogated acoustic field which are orthogonal to the plane of the LAH membrane. The LAH thus produces a one dimensional (1D) signal, the (1D) Fourier transform of which relates directly to the directivity signature, $F(\mathbf{k})$, evaluated along a straight line in k space which intercepts the origin and has a direction which is orthogonal to the LAH face [Leeman et al, 1985] [Costa et al, 1986]. Hence a measurement knowledge of $F(\mathbf{k})$ may be directly obtained by a set of measurements performed with the LAH about a single spatial point, over all angulations of the face (which replace the point hydrophone measurements in a plane incorporated into the angular spectrum approach). Note that the measurements may be performed around *any* convenient spatial location, the values of $|F(\mathbf{k})|$ being constant and the phase related by a simple linear factor (a slightly more complicated relation holding for $F_\alpha(\mathbf{k})$. Hence Equation (3) provides a means of predicting the field at *any* point in the space-time acoustic field. Note that no additional information is required such as transducer dimensions, location, and centre frequency, and that problematic evanescent waves are not incorporated in the formalism (as with the angular spectrum approach). Also Equation (3) lends itself amicably to standard efficient numerical computational techniques.

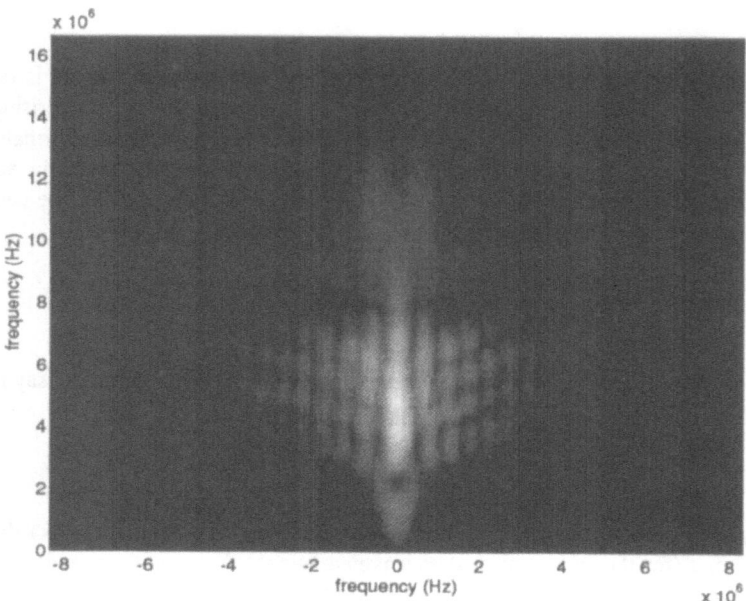

Figure 3. Amplitude of the directivity spectrum for K&B Aerotech transducer 6mm dia. aperture 5MHz centre frequency 1-4cm focus commercial transducer.

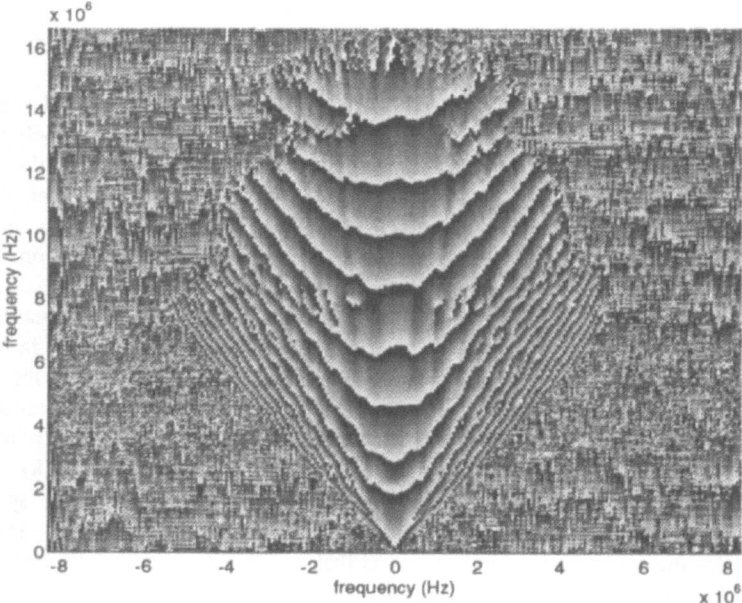

Figure 4. Phase of directivity spectrum associated with figure 3.

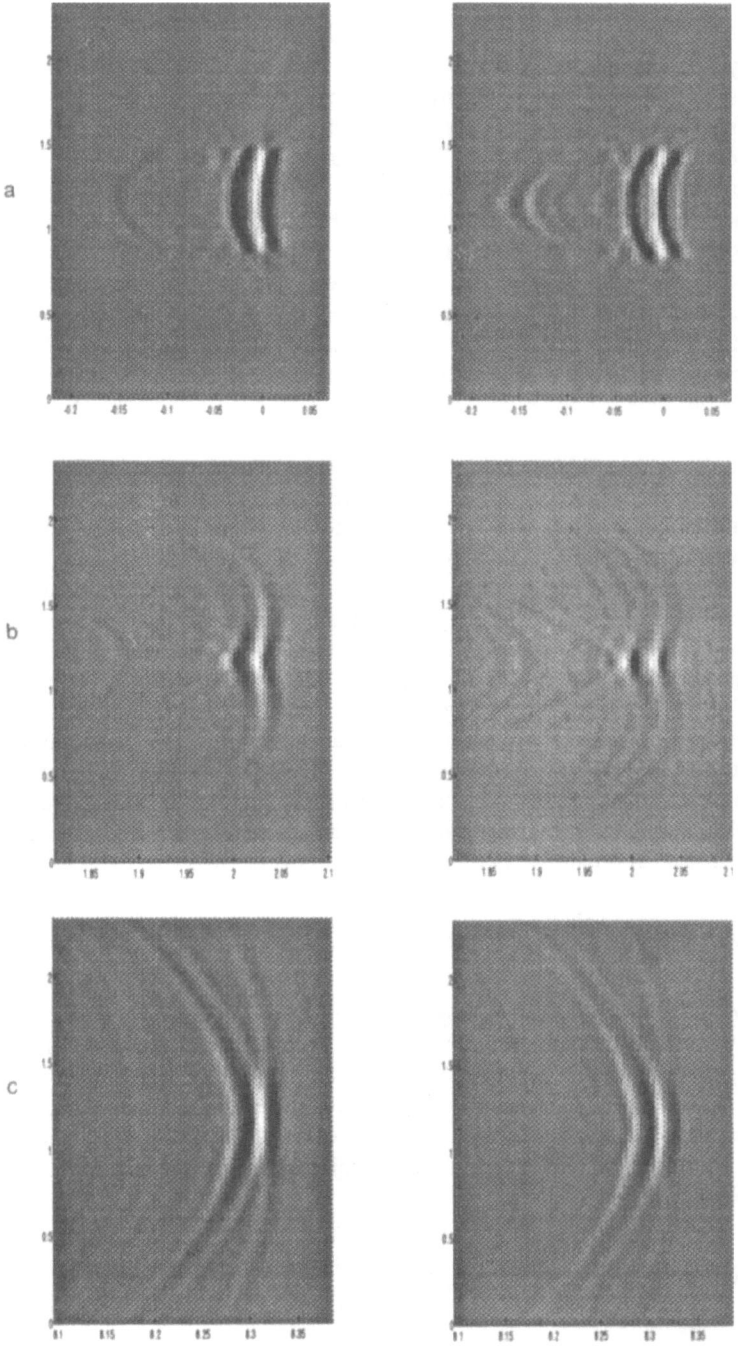

Figure 5. left - projection and right - pressure field at: (a) transducer face, (b) approximately at focus and (c) in the far field (image dimensions in cm).

MEASUREMENTS

A typical example of (a two dimensional plane through) the (three dimensional) directivity spectrum is shown in Figures (2) (amplitude) and (3) (phase), which were calculated from a set of time domain experimental data from a K&B Aerotech 5MHz centre frequency, 6mm dia. circular aperture, 1-4cm focused transducer recorded to a nominal 8 bits at 100 MHz sampling frequency, measured in a water bath approximately 4.2 cm from the transducer face. Circular symmetry of the transducer face was assumed (implying cylindrical symmetry of the ultrasound beam). The experimental set-up is shown in Figure (1). Working with the assumption of cylindrical symmetry of the acoustic field (a reasonable assumption as may be seen from Figure (4), measurements were restricted to a single set of recordings (301 rf A-line signals) produced via an angulation of a LAH (75mm dia. 27 micron thickness PVDF), in steps of 0.2 degrees, -30 to 30 degrees from the acoustic axis, in a plane intersecting the entire acoustic axis. Alignment of the LAH face normal to the acoustic axis was performed via an angulation of the LAH in order to obtain maximum signal output. The time domain measurements are shown in Figure (4). In order to generate the directivity spectrum the (one dimensional) Fourier transforms of (each of) the (analytic) temporal signals were calculated, and placed into a Cartesian grid (from the polar measurement topology) with appropriate interpolation (nearest neighbour) where necessary. The resulting matrix represents a two dimensional slice (the plane of which intersects the entire acoustic axis) through the three dimensional directivity spectrum, $F(\mathbf{k})$, (with assumed cylindrical symmetry). The directivity signature encodes *all* the information regarding the acoustic pulse (in a loss-less medium) and it is interesting to note that various features of the directivity signature relate directly to the generating transducer. For example the obvious vertical modulations of the directivity signature relate directly to the finite spatial dimensions (and apodisation) of the transducer, visible in Figure (5a). The inverse two dimensional Fourier transform of this slice, then represents a projection of the pressure field. Figure (5) shows the projections and actual reconstructed pressure fields at the transducer face, approximately at the focus and in the far field.

DISCUSSION AND CONCLUSIONS

A computationally efficient and general method for predicting acoustic fields based on Large Aperture Hydrophone measurements has been outlined. The LAH has a number of advantages over the conventional point hydrophone, and measurements may be performed at any convenient location. The algorithm has efficient space and time complexities which implementation on a standard IBM PC compatible (66MHz 486) feasible (in the order of one second for the projections, and one minute for the actual pressure fields shown in Figure 5).

REFERENCES

Leeman S., Seggie D. A., Ferrari L. A., Sankar P. V., and Doherty M., "Diffraction-free attenuation estimation". In : *Ultrasonics International 85*, Ed.: Novak Z. [Butterworth, Guilford], pp 128-132, 1985.

[2] Costa E. T., Leeman S., Richardson P. A. C., and Seggie D. A., Measurement and calibration of transducer fields. In: *Proc. Inst. Acoustics*, **8**, Ed.: Lawrence R. [Inst. of Acoustics Edinburgh (UK)], pp. 113-118, 1986.

EXPERIMENTAL COMPARISON OF MEASURED ULTRASOUND PRESSURE FIELDS WITH THEORETICAL PREDICTION

John P. Powers; LT Benito Baylosis, USN; LTjg Peter Gatchell, USN; and LT William Reid, USN

Department of Electrical and Computer Engineering (Code EC/Po)
Naval Postgraduate School
833 Dyer Road, Room 437
Monterey CA 93943-5121 USA

INTRODUCTION

We have presented a technique in the past [1–3] to calculate the acoustic potential and pressure from knowledge of the excitation spatial and temporal velocity function. In this paper we wish to present a comparison of a predicted field from using the technique with experimentally measured data.

REVIEW OF PROPAGATION SIMULATION TECHNIQUE

Our propagation simulation uses fast Fourier spatial transforms to rapidly calculate the spatial impulse response wave, $h(x, y, z, t)$, at a location z in front of the source. The complete temporal response can be found by convolving the impulse response $h(x, y, z, t)$ with the time excitation waveform $T(t)$.

Figure 1 shows the geometry of the problem. The source is assumed to be located in a planar rigidly baffled region shown at the bottom of the figure. The normal velocity of the source is assumed to be known and separable ($v(x_0, y_0, 0, t) = s(x, y)T(t)$); we want to find the excited wave at the observation point (x, y, z) (or in the entire parallel plane located a distance z away from the source plane) as a function of time. The medium is assumed to be linear, lossless, and homogeneous and has a velocity of 1500 m/s (i.e., water). Based on these assumptions, the propagation simulation technique [1,2,3] follows linear systems theory.

The acoustic potential $\phi(x, y, z, t)$ is found [1,2,3] from

$$\phi(x, y, z, t) = T(t) {\overset{*}{\underset{t}{}}} \mathcal{F}^{-1} \left\{ \tilde{s}(f_x, f_y) \tilde{g}(f_x, f_y, z, t) \right\}, \tag{1}$$

where $T(t)$ is the known time dependence of the source (see Fig. 2), $\mathcal{F}_{xy}^{-1}$ is the two-dimensional inverse spatial Fourier transform operator, $\tilde{s}(f_x, f_y)$ is the two-dimensional spatial Fourier transform of the spatial portion of the separable excitation function, and $\tilde{g}(f_x, f_y, z, t)$ is the two-dimensional spatial Fourier transform of the Green's function. This transform of the Green's function is usually solvable for simple aperture

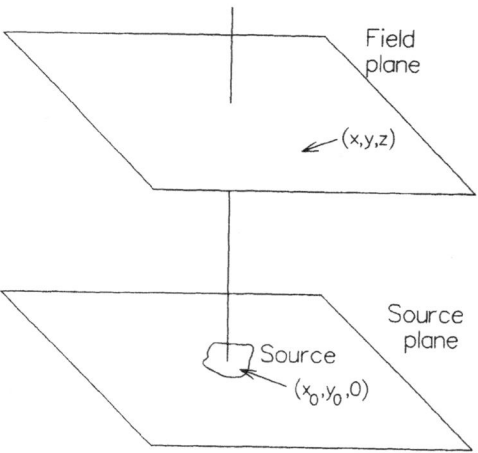

Figure 1. Geometry of source and receiving plane.

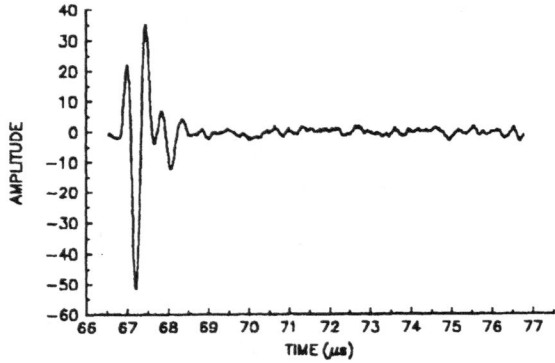

Figure 2. Measured temporal response of source, using wideband PDVF transducer.

geometries. For the rigid baffle and propagation into the half-space of lossless media, this transfer function is known [2] to be

$$\tilde{g}(f_x, f_y, z, t) = J_0\left(\rho\sqrt{c^2t^2 - z^2}\right) H(ct - z) \tag{2}$$

where $\rho = \sqrt{f_x^2 + f_y^2}$ and $H(\cdot)$ is the Heaviside step function.

The method for simulating propagation, then, is

1. Find the two-dimensional Fourier transform of the assumed $s(x, y)$.

2. For each desired value of z and t, multiply the result with $\tilde{g}(f_x, f_y, z, t)$ (Eq. 2).

3. Take the inverse two-dimensional inverse transform of the product to find the spatial impulse response, $h(x, y, z, t)$.

4. If desired, find the output for various $T(t)$ by convolving $T(t)$ with $h(x, y, z, t)$.

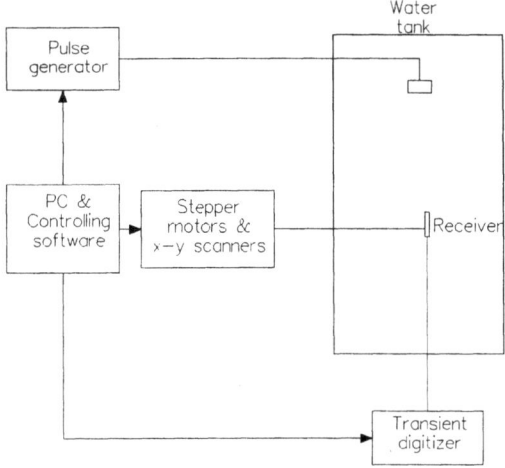

Figure 3. Block diagram of experimental measurement system.

5. If desired, find the wave pressure $p(x, y, x, t)$ from

$$p(x, y, z, t) \;\; = \;\; \rho_0 \frac{\partial \phi}{\partial t}, \qquad (3)$$

where ρ_0 is the density of the medium.

This simulation technique has been implemented in Fortran [4] and in MATLAB [5,6]. The following studies were produced with the MATLAB models.

EXPERIMENTAL MEASUREMENTS

Figure 3 shows a block diagram of the experimental system [7] that was used to measure the time-varying acoustic field. The measurement was under the control of a computer running a LabView program that controlled all aspects of the measurement. This setup scanned a piezoelectric point receiver (or, alternatively, a PDVF receiver) in an x-y raster pattern. The pattern was generated by stepper motors under the control of the LabView software. At each measurement position, the software triggered the pulsed source and recorded the digitized received signal.

The source was a circular piston transducer that was 1.27 cm in diameter. The sample area was 5.12 cm x 5.12 cm (31 samples). There were 128 x 128 samples taken (at a sample spacing of 0.4 mm). (In the figures, only a limited portion of the sample space is shown, samples 33 through 96.) There were 436 samples in the time domain for each spatial sample location.

RESULTS

Figure 4 show perspective views of these spatial excitation functions for a circular source with a diameter of 31 samples at a distance of 0.2 m from the source [9]. Due to the four-dimensional nature of the recorded waveform (amplitude, two lateral spatial dimensions, and time), only a slice of the recorded waveform $h(r, 0, t)$ is shown. There are 128 samples in the lateral dimension and 106 samples along the time axis. The

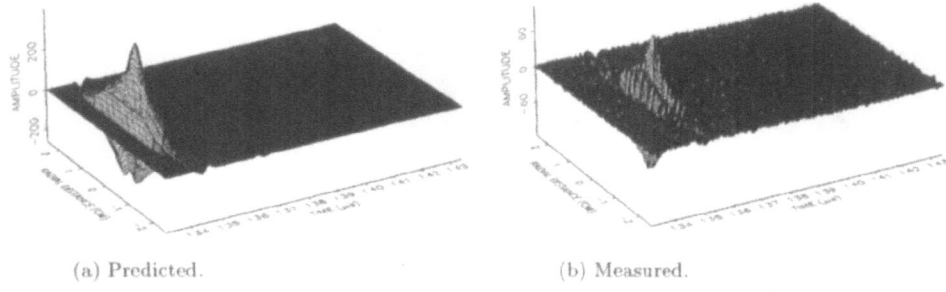

(a) Predicted. (b) Measured.

Figure 4. Comparison of (a) predicted and (b) measured amplitude values for a propagation distance of 0.2 m.

plotted time interval begins just before $t = z/c$, i.e., just before the first arrival of the first part of the wave at the observation plane.

Figures 5 and 6 show the predicted and measured time response curves at $r = 0$ and $r = 0.65$ cm lines, respectively, for the same 0.2 m distance from the source [9]. Again, the plotted time interval begins just before $t = z/c$.

We observe from Figure 4 that the waves are qualitatively very similar. The pressure wave shows a positive rise followed by a negative portion, returning gain to a positive value. Figure 5 shows a more quantitative view. (This view is a time history taken through the $r = 0$ sample point.) It shows general agreement in the shape of the waves. The predicted first downward spike is not as deep as the measured downward spike. Similarly, the measured positive spike is not as high as predicted. Similar behavior is seen in Fig. 6 which shows the predicted and measured time-varying behavior taken at the $x = 0.635cm, y = 0cm$ sample location. The general shape of the predicted waveform appears correct, but the values of the peaks of the waves do not match exactly.

The source of the errors is felt to lie predominantly in the alignment of the sample plane with the source plane. As shown in Fig. 1, the predictive model requires that the source and sample plan be parallel; this is difficult to achieve in practice. One possible solution would be to more carefully align the planes using a cw source and the resulting phase measurement to perform an interferometric alignment.

SUMMARY

This paper reported a comparison of experimentally measured ultrasound pressure fields from a test transducer with predictions from an acoustic diffraction model. The diffraction model uses a computationally efficient technique combining two-dimensional spatial Fourier transforms with temporal convolution to predict the acoustic potential and pressure fields from a pulsed source with an arbitrary spatial excitation function. The propagation model has been implemented in the MATLAB programming environment and run on PCs and UNIX workstations. A pulsed ultrasound data collection facility that scanned a point receiver in a plane located an arbitrary distance from a circular source collected the experimental data. The collection process is automated and controlled by a LabView program. The source was excited as a piston with a single cycle tone burst and the measured data was digitized and recorded under the control

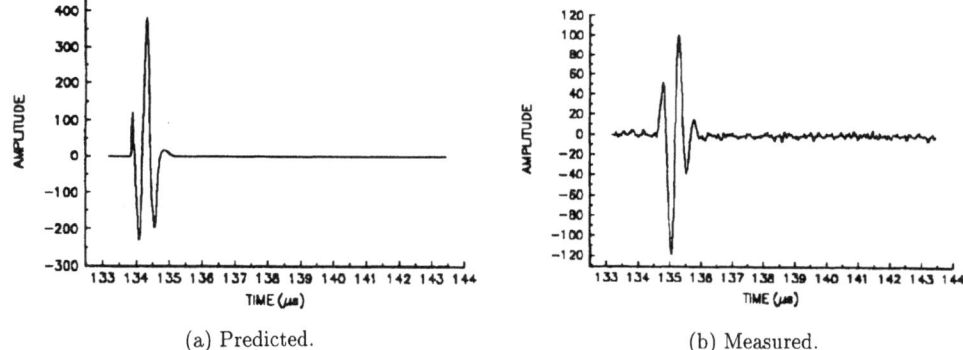

(a) Predicted.   (b) Measured.

Figure 5. Comparison of (a) predicted and (b) measured values of time response along the $r = 0$ line.

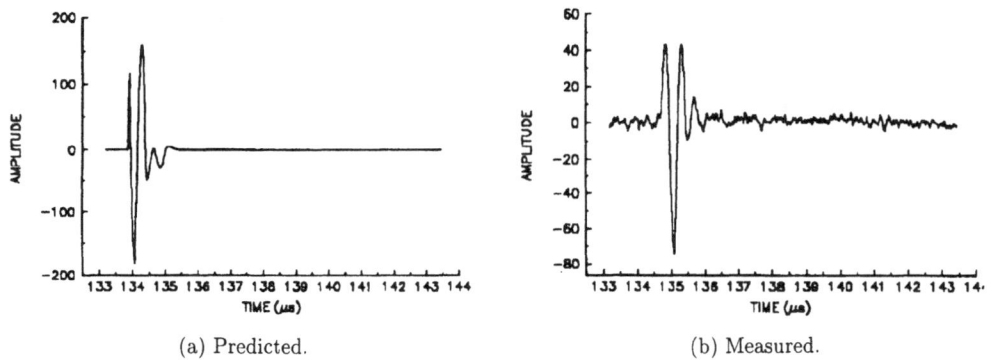

(a) Predicted.   (b) Measured.

Figure 6. Comparison of (a) predicted and (b) measured values of time response along the $r = 0.635$ cm line.

program. The predicted pressure fields and the measured pressure fields were displayed in a three-dimensional graphical format for comparison. Qualitative comparison of the measured and predicted fields appear to be in good agreement.

ACKNOWLEDGEMENTS

This work was sponsored by the Direct-Funded Research Program of the Naval Postgraduate School with the cooperation of the Office of Naval Research. The Space and Naval Warfare Command also provided partial support.

REFERENCES

[1] D. Guyomar and J. Powers, "Propagation of transient acoustic waves in lossy and lossless media," *Acoustical Imaging, Vol. 14,* A.J. Berkhout, J. Ridder, and L.F. van der Wal, Eds., Plenum Press: New York, pp. 521–531, 1985.

[2] D. Guyomar and J. Powers, "A Fourier approach to diffraction of pulsed ultrasonic waves in lossless media," *J. Acoustical Society of America,* vol. 82, no. 1, pp. 354–359, 1987.

[3] D. Guyomar and J. P. Powers, "Boundary effects on transient radiation fields from vibrating surfaces," *J. Acoustical Society of America,* vol. 77, no. 3, pp. 907–915, 1985.

[4] T. Merrill, *A transfer function approach to scalar wave propagation in lossy and lossless media,* Master's thesis, Naval Postgraduate School, Monterey, California, March 1987.

[5] W. R. Reid, *Microcomputer simulation of a Fourier approach to ultrasonic wave propagation,* Master's thesis, Naval Postgraduate School, Monterey, California, December 1992.

[6] J. P. Powers, "Acoustic propagation modeling using MATLAB," Tech. Rep. NPS EC–93–104, Naval Postgraduate School, Monterey, California, September 1993.

[7] P. A. Gatchell, *Experimental hardware development for a pulsed ultrasonic data collection facility,* Master's thesis, Naval Postgraduate School, Monterey, California, June 1994.

[8] B. E. Baylosis, *Acoustic imaging of ultrasonic wave propagation,* Master's thesis, Naval Postgraduate School, Monterey, California, December 1994.

COMPARISON OF PLANE WAVE DECOMPOSITION
AND RAY-TRACING IN SIMULATION OF B-MODE
IMAGING IN LAYERED MEDIA[*]

Martin Krueger and Helmut Ermert

Institut fuer Hochfrequenztechnik
Ruhr-Universitaet
D-44780 Bochum
Germany

INTRODUCTION

To obtain optimal design criteria for parameters of conventional and enhanced B-mode imaging systems simulations based on precise modeling are of great interest. Using a combination of the finite element method and Huygens' principle[1] and considering attenuation with a perturbation approach[2] the propagation in a homogeneous medium can be computed in a satisfactory way. However, the right way to predict pulse propagation through inhomogeneities is still a discussed topic.

The sonified biological object is a complex compound of different tissue types. Due to the small wavelength of ultrasonic pulses the propagation cannot be computed with a discretized model using finite element or finite difference methods. A first approach to analyze wave propagation through biological tissue is the modeling of the structure by planar homogeneous layers and adaption of the Huygens-integral to this type of medium. Depending on contrast propagation from layer to layer demands consideration of refraction, reflection, and diffraction.

In frequency domain two approaches have been established: (1) ray-tracing and (2) plane wave decomposition. These models can be also used for time domain simulations, which has been performed in medical ultrasound for the ray-tracing approach[3] and in geophysical applications for the plane wave decomposition[4]. Ray-tracing is simple, quick and flexible but it is derived for harmonic waves with infinitely small wavelengths. Therefore, even in analytical solutions a certain error level will remain. Decomposition into plane waves would perform an exact solution if all integrations were done continuously. Depending on the chosen degree of precision it takes more calculation effort and the method is constrained to planar boundaries.

The purpose of this paper is to establish the plane wave decomposition method

---

[*]This work was partially supported by Siemens Medical Systems Inc. We are grateful to R. Lerch, University of Linz, for providing finite element analysis data.

for medical ultrasound and to find criteria when ray-tracing can be used instead.

PLANE WAVE DECOMPOSITION

Simulation of wave propagation through layered media is done in temporal frequency domain using an expansion into plane waves[5,6,7]:

$$P(\omega|\,x,y,z) = \frac{1}{4\pi^2}\iint\limits_{-\infty}^{+\infty}\tilde{P}(\omega|\,k_x,k_y,z)e^{-j(k_x x+k_y y)}\,dk_x dk_y, \tag{1}$$

$$\tilde{P}(\omega|\,k_x,k_y,z) = \iint\limits_{-\infty}^{+\infty}P(\omega|\,x,y,z)e^{j(k_x x+k_y y)}\,dxdy, \tag{2}$$

where $P(\omega|\,x,y,z)$ is the temporal spectrum of the space and time dependent acoustic pressure $p(t|\,x,y,z)$. The propagation of each plane wave can be calculated with its simple laws of reflection and transmission at planar interfaces. In the plane of observation the pressure field is given by the composition of all plane waves.

The frequency domain method is used for a wide range of applications in the calculation of the pressure field of the transducer. However, to calculate the point spread function, no underrepresentation of the phase is allowed. To compute the propagation of a single 7MHz sine cycle through 60mm biological tissue, taking approximately $40\mu s$, a spectral sampling rate of 25kHz has to be considered. In a frequency range of at least 5MHz bandwidth this leads to 200 complex spectral sampling points per pixel. In time domain the same pulse is represented by 20 samples even with the much higher sampling interval of 10ns. Hence, the introduction of time domain methods can significantly reduce memory and CPU efforts in comparison to frequency domain methods.

Transient plane wave decomposition

To derive the transient plane wave decomposition the wave numbers k_x and k_y are expressed in terms of the angular frequency ω and slowness parameters s_x and s_y:

$$k_x = \omega s_x, \quad k_y = \omega s_y. \tag{3}$$

The spatial Fourier transform $\tilde{P}$ is substituted by

$$\hat{P}(\omega|\,s_x,s_y,z) = \omega^2\tilde{P}(\omega|\,\omega s_x,\omega s_y,z). \tag{4}$$

The inverse temporal Fourier transform of the modified Eq. 1 leads to the definition of the transient plane wave decomposition:

$$p(t|\,x,y,z) = \frac{1}{4\pi^2}\iint\limits_{-\infty}^{+\infty}\hat{p}(t-s_x x-s_y y|\,s_x,s_y,z)ds_x ds_y. \tag{5}$$

In the same way Eq. 2 can be transformed to time domain:

$$\hat{p}(t|\,s_x,s_y,z) = -\frac{\partial^2}{\partial t^2}\iint\limits_{-\infty}^{+\infty}p(t+s_x x+s_y y|x,y,z)dxdy. \tag{6}$$

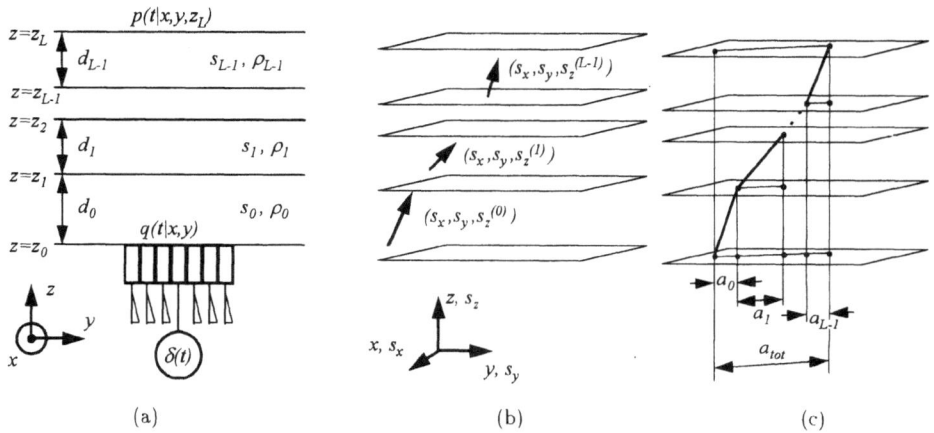

Figure 1 (a) Ultrasonic array (below $z = z_0$) sonifying a multi-layered medium (L layers), (b) plane wave symbolized by slowness vectors $(s_x, s_y, s_z^{(l)})$ passing layered medium, (c) ray passing layered medium, lateral distances of intersection points are defined with a_l.

The resulting transient plane wave function $\hat{p}$ is a two-dimensional Radon transform[8]. To exploit the plane wave decomposition for radiation problems the Huygens-integral is Radon transformed[4]:

$$p(t\,|\,x, y, z) = \iint\limits_{-\infty}^{+\infty} \frac{q(t - sR\,|\,x', y')}{2\pi R} dx' dy' \quad (R = [(x - x')^2 + (y - y')^2 + (z - z_0)^2]^{\frac{1}{2}}) \quad (7)$$

$$\Rightarrow \quad \hat{p}(t\,|\,s_x, s_y, z) = \frac{\partial}{\partial t} \frac{\hat{q}(t - s_z z\,|\,s_x, s_y)}{2\pi s_z} \quad (s_z = [s^2 - s_x^2 - s_y^2]^{\frac{1}{2}}). \quad (8)$$

$q(t\,|\,x, y)$ denotes the force density directly in front of the aperture plane $z = z_0$ and s is the propagation slowness which is reciprocal to the phase velocity c. $\hat{p}$ becomes singular in case of $s_x^2 + s_y^2 = s^2$ and $s_x^2 + s_y^2 > s^2$ leads to evanescent waves. However, in our simulations the homogeneous plane waves far beyond the critical angle represented the fields sufficiently.

Propagation through layered media

Plane wave propagation through layered media is very simple since they are eigenfunctions of planar interfaces. The laws of reflection and transmission are presented in common literature, e. g. [9]. A plane wave

$$p_1^+ = f(t - s_x x - s_y y - s_z^{(1)} z) \quad (s_z^{(1)} = [s_1^2 - s_x^2 - s_y^2]^{\frac{1}{2}}) \quad (9)$$

which is propagating in a halfspace with mass density ρ_1 and slowness s_1 and which is incident to the interface causes a reflected wave p_1^- and a transmitted wave

$$p_2^+ = \frac{2\rho_2 s_z^{(1)}}{\rho_2 s_z^{(1)} + \rho_1 s_z^{(2)}} f(t - s_x x - s_y y - s_z^{(2)} z) \quad (s_z^{(2)} = [s_2^2 - s_x^2 - s_y^2]^{\frac{1}{2}}) \quad (10)$$

in the half-space possessing the material parameters ρ_2 and s_2. Negligence of multiple reflections leads to a simplified algorithm to simulate the propagation of a plane wave component $\hat{p}(t|s_x, s_y, z_0)$ through a multi-layered medium (Fig. 1a-b):

$$\hat{p}(t|s_x, s_y, z_L) = \prod_{l=0}^{L-1} \frac{2\rho_{l+1}s_z^{(l)}}{\rho_{l+1}s_z^{(l)} + \rho_l s_z^{(l+1)}} \, \hat{p}\left(t - \sum_{l=0}^{L-1} s_z^{(l)}d_l, s_x, s_y, z_0\right). \tag{11}$$

Numerical implementation

To enable a numerical implementation of the plane wave decomposition a finite and discrete representation of $p(t|x, y, z)$ and $\hat{p}(t|s_x, s_y, z)$ has to be found. The temporal dependence is represented by an equidistant sampling with an interval of 7.2ns which satisfies Shannon's theorem and allows to calculate focused and steered beams. In space domain the mesh given by the results of the FEM simulation is not equidistant and varies from approximately $6 \ldots 60\mu m$. Investigations of simple examples revealed that an equidistant sampling of s_y from values $|s_y| \le 250\mu sm^{-1}$ leads to an adequate representation for simulations of 2D-B-mode imaging.

RAY-TRACING

B-scan simulation using ray-tracing is performed mainly in the way as proposed in [3]. Due to the given locations of the $q(t|y)$-samples the processing order is different: The path starting at each point of integration and ending at each point of observation is derived with Fermat's principle (see Fig. 1c):

$$\tau = \min\left\{ \sum_{l=0}^{L-1} s_l [d_l^2 + a_l^2]^{\frac{1}{2}} \middle| \sum_{l=0}^{L-1} a_l = a_{tot} \right\}. \tag{12}$$

The Lagrange method solution of this optimization problem is

$$a_l = d_l \sigma [s_l^2 - \sigma^2]^{-\frac{1}{2}}, \tag{13}$$

whereas the Lagrange parameter σ is given by the constraint to a_l. The constraint equation is nonlinear and can be solved using a third order Taylor-approximation.

B-MODE IMAGING SIMULATIONS

The point spread functions of either model were calculated in the following order: The finite element data describe the $q(t|y)$-impulse response of a voltage (transmit mode) and current (receive mode) driven single element (Fig. 1a). The radiation field of the single mode array is calculated in the described way. Then the transmit and reciproc receive multi-channel modes are derived using superposition. The resulting pressure fields describe "one half" of the point spread function if a Born type scatterer is assumed. If $h_T(t|y, z_L)$ describes the pressure field resulting from transmit mode and $h_R(t|y, z_L)$ from receive mode the point spread function (PSF) equals

$$h_{PSF}(t|y, z_L) = \frac{h_T(t|y, z_L) * h_R(t|y, z_L)}{\rho_L c_L}. \tag{14}$$

To compare the different parameters an absolute integration with respect to time was performed:

$$\overline{|h_{T|R|PSF}(y)|} = \int_{-\infty}^{+\infty} |h_{T|R|PSF}(t|y, z_L)| dt \tag{15}$$

and the temporal centroid was calculated.

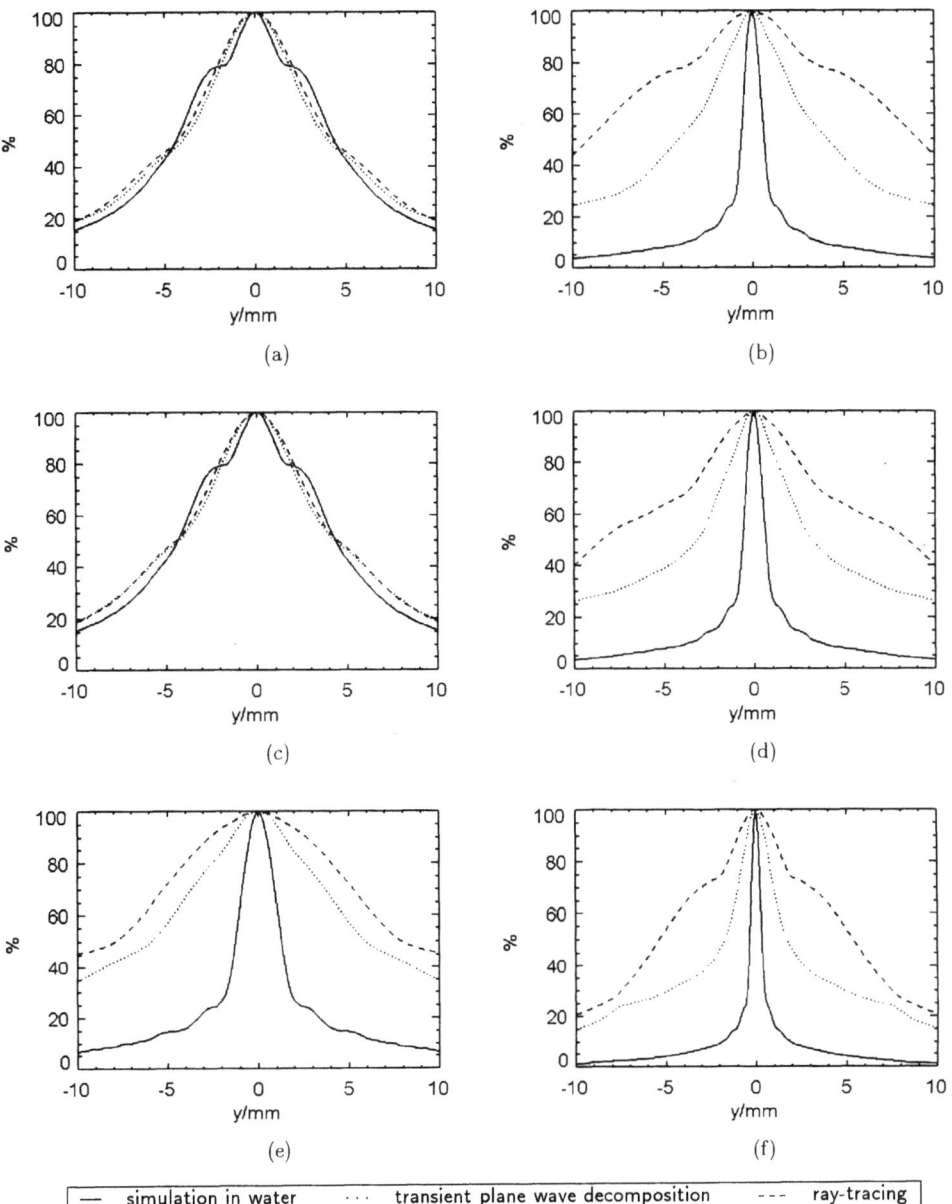

Figure 2 Simulation results of transmit and reciprocal receive mode pressure field, absolutely integrated with respect to time and normalized to 100% versus lateral displacement y/mm. (a) and (b): Medium 1, (a) transmit mode, 64 channels focusing 50mm, (b) receive mode, 128 channels, focusing 80mm. (c) and (d) same with Medium 2. (e) and (f): medium 2, receive mode, 64 and 128 channels, respectively, focusing 80mm.

The array was excited by a single cylcle sine pulse. In transmit mode a depth of $z = 50$mm was focused by 64 channels. In receive mode dynamic focusing with $64 \ldots 128$ channels was simulated. The plane of observation was located in $z = 80$mm. Two compounds of layers were considered: (1) 20mm of fat, 30mm of muscle, and 30mm of liver and (2) 5mm of plexiglas (shearwaves neglected), 20mm of fat, 30mm of muscle, and 25mm of liver. All parameters, i. e. mass density, phase velocity and attenuation, were taken from [10,11]. The frequency dependence was modeled linearly and a minimum phase term considering the Kramers-Kronig relation[2] was implemented. For comparison purposes the PSF in pure water was calculated for the same operating modes.

Due to the different phase velocities the temporal centroids of the PSF for water and for the media (1) and (2) were different. However, concerning this subject there was no significant dependence on the simulation method. Representative simulation results of $\overline{|h_T|}$ and $\overline{|h_R|}$ are shown in Fig. 2. Remarkable is the weak dependence on the type of medium and the strong dependence on the number of active channels and the location of the focus.

DISCUSSION

In the simulation of planarly stratified media the main criterion for the choice of simulation method is the sonifying aperture. If the aperture is not focused to the plane of observation or the number of active elements is not too large (with respect to this simulation up to 64 channels) ray-tracing can be used as well as the plane wave decomposition. This is valid for all typical compounds of materials. Further investigations will have to find out criteria for rough interfaces, 3D-B-mode imaging, beam steering, and inhomogeneous layers.

REFERENCES

1. R. Lerch, H. Landes, and H. T. Kaarmann. Finite element modeling of the pulse-echo behavior of ultrasound transducers. In *Proc. 1994 Ultrason. Symp.*, pages 1021–1025, 1994.
2. J. A. Jensen, D. Gandhi, and William D. O'Brien Jr. Ultrasound fields in an attenuating medium. In *Proc. 1993 Ultrason. Symp.*, pages 943–946, 1993.
3. L. Odegaard, S. Holm, F. Teigen, and T. Kleveland. Acoustic field simulation for arbitrary shaped transducers in a stratified medium. In *Proc. 1994 Ultrason. Symp.*, pages 1535–1538, 1994.
4. Peter Hybral and Martin Tygel. Transient response from a planar acoustic interface by a new point source decomposition into plane waves. *Geophysics*, 50(5):766–774, May 1985.
5. A. P. Berkhoff, P. M. van den Berg, and J. M. Thijssen. Iterative calculation of reflected and transmitted acoustic waves at a rough interface. *IEEE Trans. Ultrason. Ferroel. Freq. Cont.*, 42(4):663–671, July 1995.
6. W. C. Chew. *Waves and Fields in Inhomogeneous Media.* IEEE Press, New York, 1995.
7. C. J. Vecchio, M. E. Schafer, and P. A. Lewin. Prediction of ultrasonic field propagation through layered media using the angular spectrum method. *Ultrasound in Med. & Biol.*, 20(8):611–622, 1994.
8. T. S. Durrani and D. Bisset. The Radon transform and its properties. *Geophysics*, 49(8):1180–1187, August 1984.
9. M. Tygel and P. Hubral. *Transient waves in layered media.* Elsevier, Amsterdam, NL, 1987.
10. P. N. T. Wells. *Biomedical Ultrasonics.* Academic Press, London, 1988.
11. J. and H. Krautkraemer. *Ultrasonic testing of materials.* Springer, Berlin, 1986.

ULTRASONIC VELOCITY MAPPING OF MULTILAYERED MEDIA

Wagner C. A. Pereira, Ana Valéria D. Greco and João Carlos Machado

Biomedical Engineering Program
Federal University of Rio de Janeiro
RJ -Brazil
P.O. Box 68510 - zip code 21945-970
electronic-mail: wagner@serv.peb.ufrj.br

INTRODUCTION

The geometrical acoustics theory is still an approach widely used when it comes to applications of ultrasonic methods to non destructive tests and characterization of acoustic properties of materials. In biomedical applications, however, results have demonstrated limited success, particularly in attempts to construct images of ultrasonic biological tissue parameters (wave velocity, amplitude attenuation, scattering, etc.). The main reason for this may be the great complexity of ultrasonic wave interactions with the living matter, not adequately covered by the assumptions of geometrical acoustics. Nevertheless this theory must still be explored in the biomedical field, starting from simple situations where a small number of variables can be considered (Shung, 1990, Liu, 1991).

Following this principle we have already developed geometrical acoustics models to simultaneously estimate the thickness and the wave velocity of multilayered media (Pereira, 1992). The first model assumed a focused beam and was based on the measurement of the time-of-flight (TOF) collected at two different locations. Experiments with a 1.9 MHz/19 mm diameter transducer and a 0.6 mm diameter PVDF hydrophone pointed to the potentiality of the method but also for the need of an accurate alignment of the experimental apparatus (Pereira, Leeman and Machado, 1992; Pereira and Machado, 1992).

In this work we present a new geometrical acoustics model that estimates the thickness and the wave velocity of multilayered media, simultaneously. This model simplifies the experimental conditions to measure the TOF's that are collected at different locations.

THEORY

The model considers the situation of figure 1 where a transmitting transducer (T_x) emits an ultrasonic (US) wave towards a target that is away from its face by a distance Z. After reflection the wave front travels back to T_x and also to another transducer, called receiver (R_x). The beam is assumed to be focused with focus at F and a least mean square method to estimate the focal point is used (Pereira, Simpson and Machado, 1992).

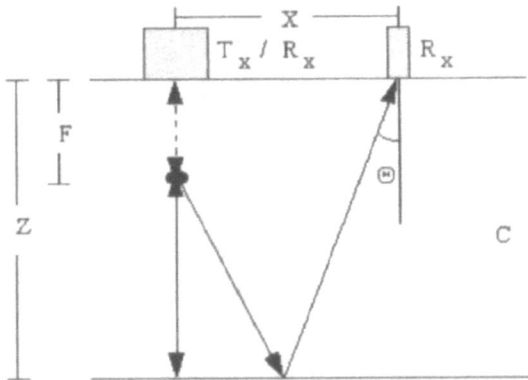

Figure 1. Ray path assumed by the model.

By geometrical considerations one can see that T_0 (TOF of the US wave to leave and come back to T_x) can be written as:

$$T_0 = \frac{2 \cdot Z}{c} \tag{1}$$

where c is the US wave velocity in the medium. In a similar way T_1 (TOF for the wave to reach the receiver R_x) can be written as:

$$T_1 = \frac{F}{c} + \frac{2 \cdot Z - F}{c \cdot \cos\theta} \tag{2}$$

After dividing T_1 by T_0, this results in equation (3) where c has been eliminated and $\cos\theta$ has also been taken from the figure.

$$\frac{T_1}{T_0} = \frac{F + \sqrt{X^2 + (2 \cdot Z - F)^2}}{2 \cdot Z} \tag{3}$$

Parameter X is the distance T_x-R_x. So equation (3) can be used to estimate the thickness Z, once the parameters X and F are known and T_0 and T_1 are measured. After obtaining Z, the velocity c can be taken from equations (1) or (2). The equations developed here can be extended to a multilayered medium, bearing recursive equations.

EXPERIMENTAL METHODS

The experiments use an 8.0 MHz/10 mm diameter ultrasonic circular transducer with an epoxy acoustic lens (spherical - 20 mm radius and conical -17 degrees between lens and transducer's face), and a PVDF hydrophone of 0.5 mm face diameter immersed in a water tank. Figure 2 presents a block diagram of the experimental set up.

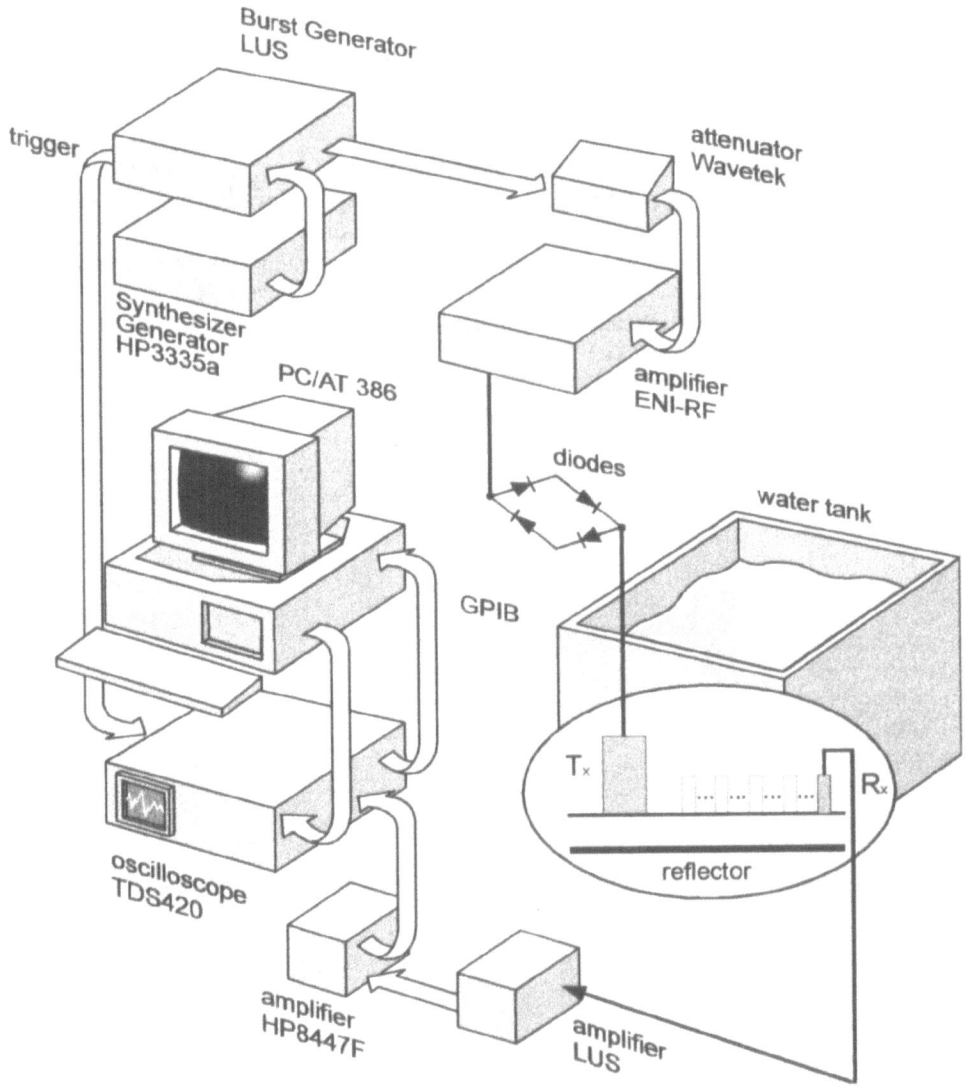

Figure 2. Experimental configuration. Notice R_x placed in different positions.

The Synthesizer Generator (HP335A) inputs a 8.0 MHz sinusoidal wave in a burst generator. The sinusoidal bursts are 0.8 μseconds long and are amplified before exciting the

65

transducer. The echoes are amplified in 20 dB by the HP8447F and in 25 dB by the LUS/PEB (made in our lab.) and then digitized in an oscilloscope (TDS-420) with a sampling frequency of 2.5 GHz. After that a IBM/486 compatible microcomputer receives the wave form via a GPIB interface. Figure 3 shows two examples of the wave forms.

The TOF is estimated via a time cross correlation process. The TOF for T_0 is obtained by detecting the beginning of the wave form. T_0 is taken as a reference from which all the other TOF's are estimated. A time gate of 1 μs is opened to embrace the reference wave. Then the cross correlation is proceeded sliding this gate along each of the echoes that reached R_x. From the peak of the correlation is taken the delay estimate of T_1.

RESULTS

Ten experiments for the spherical lens were made with a layer 67.5 mm thick. The point focus of the beam was 27.53 mm from the face of T_x. For each experiment the receiver was placed at ten different positions resulting in ten different values of T_1. So each experiment estimates 10 values for Z and for c. Another ten similar experiments used the conical lens but with 50 mm water layer. The focal point of the beam was 22.7 mm from the T_x face. It is important to stress that before each experiment the set up was rebuilt to give an idea of alignment sensibility.

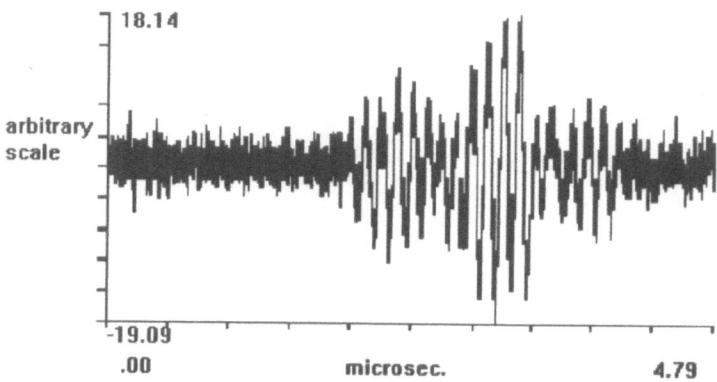

Figure 3(a). Examples of the 12,000 point wave forms, taken in the tenth position. Sampling frequency of 2.5 GHz, with spherical lens

In figure 3 one can see examples of typical wave forms used in the processing. Notice the signal-to-noise ratio and the change in the shape of the two waves (see figures 3(a) and 3(b)).

For each experiment a mean value for the thickness Z and a Coefficient of Variation were obtained. Table I shows this result for both lenses. For the velocity estimates c, we obtained similar results. Once by equation (1) it can be seen that Z and c are related by a constant. The c estimates are not shown here.

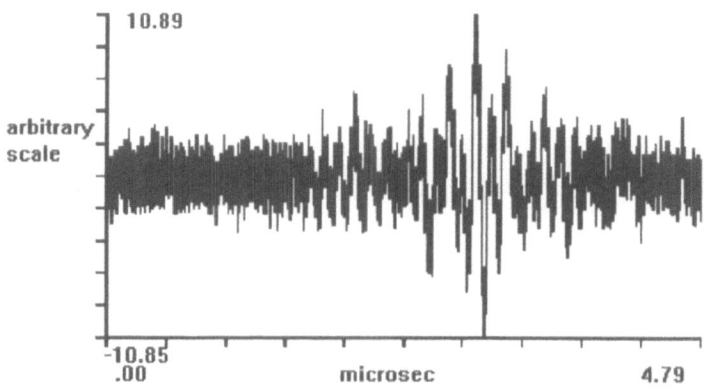

Figure 3(b). Examples of the 12,000 point wave forms, taken in the tenth position. Sampling frequency of 2.5 GHz, with conical lens.

Table I. Thickness estimates for the 10 experiments with each lens.

Thickness	Spherical	Conical
Real Z (mm)	67.50	50
Estimated Z (mm)	66.25	47.79
Coef. of Variation (%)	15.61	6.82
% Error	1.88	4.62

DISCUSSION AND CONCLUSION

For the conical lens we obtained for the thickness estimates a mean value of 47.79 mm (4.62 % error, compared to the real thickness of 50 mm) and a Coefficient of Variation (CV) of 6.82 %, showing acceptable accuracy and repeatability.

For the spherical lens the mean value was 66.25 mm (1.88 % error compared to the real thickness of 67.5 mm) showing good accuracy and CV was 15.61 % meaning that precision was poor. This may be due to the poor signal-to-noise ratio of the echoes during the experiments with the spherical one. This fact compromised the quality of the cross correlation process.

These results are more significant than the first ones (Pereira and Machado, 1992) and it is probably due to the simpler experimental conditions. In the first ones the model demanded echoes from two positions of R_x and presently only one position is needed, so alignment is less influent. The differences in the equations of the two models cannot explain the improvement of the results by themselves. The decision of rebuilding the experimental set up before each experiment could show that an important dependency of the estimates on the alignment of the experimental devices inside the tank is still present.

Presently we are proceeding experiments with phantoms of 2 and 3 layers and studying the influence of amplitude attenuation and multiple reflection on the model.

As a conclusion we can say that the model shows potential for the mapping of ultrasonic wave velocity in multilayered media and the first biomedical application devised is the characterization of the human skin.

ACKNOWLEDGEMENTS

We would like to thank the electronic engineer, Mr. Gustavo S. Chelles, for his drawings. This work was supported by the Brazilian agencies CNPq and CAPES

REFERENCES

LIU, D.L. (1991), "Sound Velocity Inversion in Layered Media with Band-Limited and Noise-Corrupted Data", IEEE - Transactions on Biomedical Engineering, vol. 38, no. 10, pp. 1042 - 1047.

PEREIRA, W.C.A. (1992), *Método Ultra-sônico de Pulso - Eco para a Determinação Simultânea da Velocidade de Propagação da Onda e Espessuras em Meios Multicamadas*, D.Sc. Thesis Biomedical Engineering Program/COPPE/UFRJ, Rio de Janeiro.

PEREIRA, W.C.A, LEEMAN, S. and MACHADO, J.C. (1992), "Ultrasound Velocity Imaging by a Pulse-Echo Technique", Acoustical Imaging, vol. 19, pp 475 - 480.

PEREIRA, W.C.A & MACHADO, J.C. (1992), "Ultrasonic Focused Pulse-Echo Method for Determination of Propagation Velocity and Layer Thickness of Multilayered Media", Proceedings of 14th Annual International Conference of the IEEE Engineering in Medicine and Biological Society, part 5 of 7, pp 2110 - 2111, Paris, France.

PEREIRA, W.C.A.; SIMPSON, D.M. e MACHADO, J.C. (1992), "An Estimator of Focal Position based on Geometric Acoustics", *IEEE - Ultrasonics Symposium*, pages 323 -325.

SHUNG, K.K. (1990), Basic Principles of Ultrasound Tissue Characterization. In: Noninvasive Techniques in Biology and Medicine, S.E. Freeman, E. Fukushima and E.R. Greene, Eds., San Francisco Press, pp.205 - 217.

AN ANALYTICAL FORMULATION FOR WIDE-BAND
BEAM PATTERNS USING SPARSE ARRAYS

Vittorio Murino and Andrea Trucco

Department of Biophysical and Electronic Engineering (DIBE)
University of Genoa
Via Opera Pia 11A
I - 16145 Genova, Italy

INTRODUCTION

In this paper the evaluation of beam patterns in active beamforming systems using wide-band signals and sparse arrays is addressed. Beamforming is a linear technique aimed at processing the signals sampled by an array of sensors[1]. The goal of beamforming is to enhance the signals incoming from a selected steering direction and to attenuate signals incoming from any other direction. In active sonar systems, a scene is insonified first by an adequate acoustic pulse transmitted toward it. After the acquisition of the backscattered echoes by an array, the image of the scene[2,3,4,5,6] can be created by computing and displaying several beam signals, steered in the directions of interest.

After fixing a steering direction, the far-field beam pattern shows the attenuation with which the system transfers a plane wave to the beam signal, for each possible incidence angle[1]. The knowledge of the beam pattern is very important to analyze the performance of a system in terms of resolution. In active systems, when the spatial aperture of the array is larger than the equivalent spatial length of the pulse used for the scene insonification (i.e., $c \cdot T$, where T is the time duration of the pulse and c is the propagation speed), the beam pattern profile depends on the pulse envelope shape[5,7]. This condition is usually referred to as the wide-band condition, since it requires pulses of short time durations.

Usually, far-field beam patterns are computed under narrow-band conditions, that is, the equivalent spatial lengths of the signals are supposed to be wider than the array aperture. The narrow-band hypothesis is useful, as it allows one to formulate the beam pattern in an analytical way that does not depend on the transmitted pulse and, when unit weight coefficients are applied, to compute the beam pattern by a simple closed-form equation[1]. However, this assumption is not correct in several practical applications. Signals of very short durations are often adopted[2,6,7,8] as they permit one to overcome a classic problem related to narrow-band conditions: when high spatial resolution and a small number of array elements are required, the spacing between elements becomes wider than $\lambda/2$ (λ being the wavelength related to the carrier frequency) and this causes the presence of grating lobes in the beam pattern. The use of wide-band signals is a very suitable method to reduce the elevation of grating lobes[2,4,5,6] by transforming them into normal side lobes. Moreover, wide-band signals are useful in improving the performances of imaging systems, as they make it possible to limit speckle effects and improve the range resolution. In general, images obtained by using wide-band signals are of better quality and potentially contain more information than those obtained by means of narrow-band signals[5,6].

Previous papers dealt with two possible criteria to define wide-band beam patterns[7], so allowing to estimate the resulting reduction in the grating-lobe level[2,5,6]. Beam patterns were computed for different pulse shapes and different weighting coefficients by means of numerical simulations, assuming the array elements to be spaced less[3,4,6,7,9] or more,[4,5,6,10] than $\lambda/2$. Simulations were useful, as, although complicated analytical formulations of the beam pattern were achieved, depending on time and the pulse shape, a closed-form equation was developed exclusively for an inter-element spacing less than $\lambda/2$ and for a rectangular pulse shape[7]. Unfortunately, this closed form was derived by adopting a criterion (for defining the wide-band beam pattern) that is not well suited for imaging applications.

The goal of this paper is to propose a quasi-closed form able to compute the far-field beam pattern, also for an array in which the spacing between elements is more than $\lambda/2$. The importance and the novelty of this contribution lie in the opportunity to face the case of sparse arrays[4,10] (i.e., arrays in which the spacing is sharply wider than $\lambda/2$) by means of a quasi-closed form well suited for imaging applications. In this sense, the proposed work bridges a literature gap regarding the application of wide-band signals to sparse arrays, which represents an interesting topic in current acoustic imaging.

WIDE-BAND BEAM PATTERN DEFINITION

Since, under wide-band conditions, the conventional formulation of the beam pattern is not valid any more, a redefinition of it is needed. In the literature, two methods were proposed[7] that assume the knowledge of the signals received by the array elements when a plane wave reaches the array at a given incidence angle. In the simplest case, such signals can be modelled as a replica of the transmitted pulse, adequately delayed in accordance with the incidence angle and the positions of the array elements[3,6,7,10]. After fixing a steering direction, the knowledge of these signals allows the computation of the beam signal. The value of the beam pattern for a given angle of incidence can be considered as the maximum of the beam signal over time[3,4,6,7,10], or as the total energy of the beam signal[7,8,9].

Under a far-field hypothesis and with reference to the linear equispaced array shown in Fig. 1, consisting of M punctiform and omnidirectional elements, the signal $x_i(t)$ received in quadrature[1] by the i-th element of the array and produced by a plane wave at an incidence angle α can be expressed as:

$$x_i(t) = A\left(t + \frac{idsen\alpha}{c}\right) \cdot \exp\left(j\omega \frac{idsen\alpha}{c}\right)$$

(1)

where $A(t)$ is the envelope of the adopted pulse (possibly, after pulse compression), d is the distance between the array elements, and $\omega = 2\pi f$ is the angular frequency of the carrier wave. According to the definition of quadrature beamforming[1], the beam signal $b_\theta(t)$ steered in the direction θ (see Fig. 1) is given by:

$$b_\theta(t) = \sum_{i=0}^{M-1} x_i\left(t - \frac{idsen\theta}{c}\right) \cdot \exp\left(-j\omega\frac{idsen\theta}{c}\right)$$

(2)

where unit weight coefficients are assumed. The beam pattern is often described in terms of $u = sen\alpha - sen\theta$, over the range $-2 \le u \le 2$, which contains all possible combinations of α and θ. By including Eq. (1) in Eq. (2), the beam signal can be rewritten in terms of t and u on the basis of the adopted pulse:

$$b(t,u) = \sum_{i=0}^{M-1} A\left(t + \frac{idu}{c}\right) \cdot \exp\left(j\omega\frac{idu}{c}\right)$$

(3)

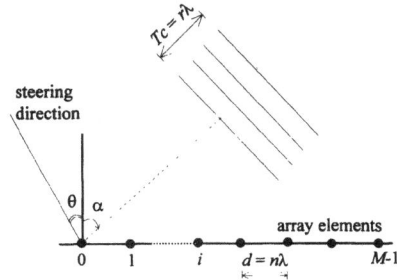

Figure 1. Geometry and notation used for a linear array and a propagating pulse.

As previously stated, under wide-band conditions, the beam pattern $BP(u)$ can be defined as the maximum of the beam signal over time or on the basis of the total energy of the beam signal. One can prove that, in both cases, when the wide-band hypothesis is relaxed until to a narrow-band condition is achieved, the resulting beam pattern is equal to the conventional narrow-band beam pattern[3,9].

Depending on the application considered, one can choose the more suitable definition of the beam pattern. In some applications, the feature extracted from the computed beam signals is the energy contained in each of them[1,7,8,9], whereas, in other applications, the behaviour of the beam signal envelope is used directly or indirectly to obtain the expected results[3,4,5,6,10]. For instance, in acoustic imaging the envelopes of the beam signals are displayed in order to visualize the internal structure of a scene[2]. In this case, the more suitable definition of the beam pattern is the maximum of the beam signal over time, as suggested in some papers[3,4,6,10]. Accordingly, we shall adopt the following relation:

$$BP(u) = \max_t\left\{ \left|b(t,u)\right| \right\}$$

(4)

to define the wide-band beam pattern. As a consequence, we cannot exploit the closed-form equation proposed by Lurton[7] as it was developed for the total-energy case[8,9].

THE QUASI-CLOSED FORM FOR THE BEAM PATTERN

To achieve an analytical formulation of the beam pattern, a rectangular shape of unit amplitude and duration T is assumed for the pulse envelope $A(t)$. If the inter-element spacing d is written in terms of carrier wavelength (i.e., $d = n\cdot\lambda$) and the pulse $A(t)$ contains r cycles of the carrier frequency f (i.e., $T = r / f = r\lambda / c$), then Eq. (3) can be rewritten as:

$$b(t,u) = \sum_{i=0}^{M-1} A\left(t + \frac{iT}{(r/nu)}\right) \cdot \exp\left(j\omega \frac{iT}{(r/nu)}\right)$$

$$(5)$$

One can notice that the arguments of the summation are M vectors different from zero within the time interval $[-iT/(r/nu),\ T-iT/(r/nu)]$, characterized by unit modulus, and having phase contributions linearly function of i. For instance, for $M = 6$, $r = 5$ (i.e., the pulse $A(t)$ contains 5 carrier cycles), $n = 2$ (i.e., $d = 2\lambda$), and $u = 0.84$, the six vectors that are added by the summation in Eq. (5) to compute the beam signal $b(t, u)$ are shown in Fig. 2(a), versus time. The time duration of $A(t)$ is $T = 50$ time units and, for each vector, the value of the phase is constant and is given on the right side of the modulus profile.

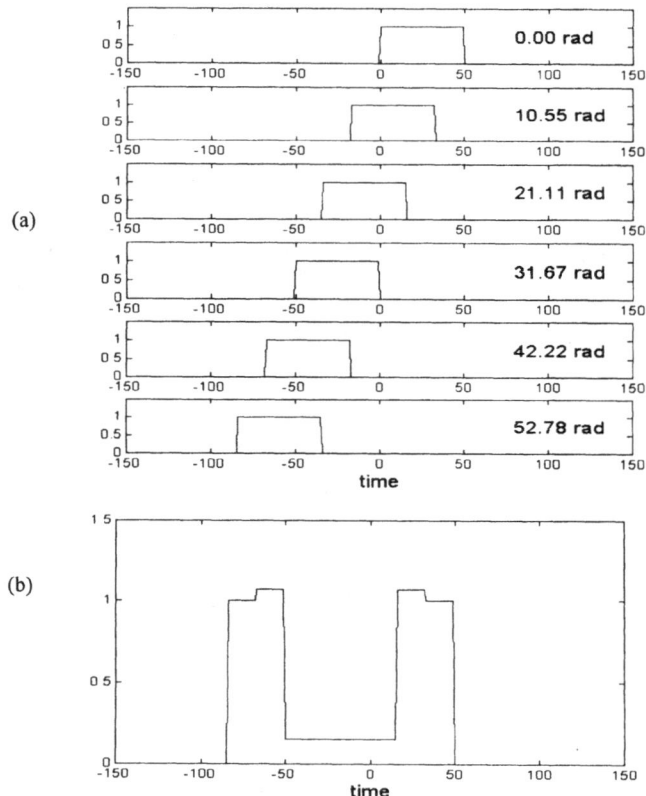

Figure 2. Amplitude of the modulus versus time for a collection of complex signals. (a) Signals that summed form a beam signal for $M = 6$, $r = 4.5$, $n = 2$, $u = 0.84$. (b) Modulus of the beam signal given as the sum of the signals in (a).

As depicted in Fig. 2, one can find a time instant t^* at which the beam signal $b(t^*, u)$ consists of the sum of l vectors, where l is an integer included in the range $[0,\ \text{ceil}(r/nu)]$ and the function ceil(q) returns the smallest integer larger than q (e.g., in the above example, ceil$(r/nu) = 3$). If ceil(r/nu) is wider than M, then the admissible range of l becomes $[0,\ M]$. Due to phase effects in complex additions, there is no guarantee that max$_t\{\ |b(t, u)|\ \}$ will occur at a time instant at which the maximum number of vectors (i.e., ceil(r/nu) or M) are added. As depicted in Fig. 2(b), which shows the modulus of the addition of the complex signals shown in Fig. 2(a), the max$_t\{\ |b(t, u)|\ \}$ for the above example occurs when the beam signal results from the addition of 2 vectors. Moreover, as the phase of the vectors is a linear function of i, one can notice that the modulus of the sum of a given number of subsequent vectors is equal for all possible choices. Therefore, as we are only interested in the envelope of the beam signal, we have to take into account the number of subsequent vectors summed at each time instant (to find the maximum envelope value) but not the time instant at which this fact occurs. For the above example, Fig. 2(b) shows two time intervals over which the complex addition of two subsequent vectors result in the same maximum of the envelope amplitude (i.e., 1.07).

From these considerations, one can conclude that the maximum of $|b(t, u)|$ over time is equal to the maximum modulus obtained by summing l subsequent vectors, where l is an integer included in the range $[1,\ l_{max}]$ and l_{max} is the minimum between ceil(r/nu) and M. Hence:

$$BP(u) = \max_{t}\left\{ \left| b(t,u) \right| \right\} = \max_{l \in [1,\ l_{max}]}\left\{ \left| \sum_{i=0}^{l-1} \exp\left(j\omega \frac{iT}{(r/nu)} \right) \right| \right\}$$

(6)

By using a conventional mathematical relation[1], one can write the above summation in closed form, obtaining the wide-band beam pattern for a rectangular pulse in a quasi-closed form:

$$BP(u) = \max_{l \in [1,\ l_{max}]}\left\{ \frac{\left| \sin(l\pi un) \right|}{\left| \sin(\pi un) \right|} \right\}$$

(7)

where l is an integer and l_{max} is given by $l_{max} = \min\{\ \text{ceil}(r/nu), M\ \}$.

One can notice that the wide-band beam pattern depends on the number of carrier cycles contained in the pulse, on the inter-element spacing, and on the number of array elements. Moreover, it is very similar to the closed-form equation for the narrow-band beam pattern[1]: when the use of continuous waves is assumed, l is always equal to M, thus Eq. (7) becomes perfectly equal to the narrow-band case.

RESULTS

Figure 3 shows two beam power patterns computed by the proposed quasi-closed form for an array with 25 equispaced elements and a rectangular pulse having a time duration equal to 4.5 carrier cycles. The results are visualized on a logarithmic scale normalized to 0 dB. One can notice that the function $BP(u)$ is equal to $BP(-u)$, hence it is sufficient to study the beam pattern over the range $0 \le u \le 2$. Figure 3(a) presents the beam power pattern for a 2λ inter-element spacing, whereas in Fig. 3(b) the spacing is increased to 5λ.

The first consideration concerns the grating lobes: in the narrow-band case, they are present for $u = k\cdot(1/n)$, where k is an integer[1,5]. In the wide-band case, the grating lobes are replaced by side lobes whose levels increase while approaching the main lobe (i.e., $u = 0$). By simple substitutions one can compute the exact value of the highest side lobe (HSL) and notice that such a value depends only on the number of cycles contained in the pulse and on the number of elements (in particular, HSL is equal to -13.98 dB for the case in Fig. 3).

Another consideration concerns the lowest side lobe level (LSL), for each value of u. Note that $\max_t\{\ |b(t, u)|\ \}$ cannot be smaller than 1, as a time instant always exists at which the beam signal is formed by only one unit vector, which is not added to the other vectors (see Fig. 2). As a consequence the LSL depends only on the total number of array elements (in particular, LSL is equal to -27.96 dB for the case in Fig. 3).

Finally, regarding the width of the main lobe, measured at -3 dB, each wide-band beam pattern shown has the same width as the corresponding narrow-band beam pattern[3,5], provided by classical relations[1].

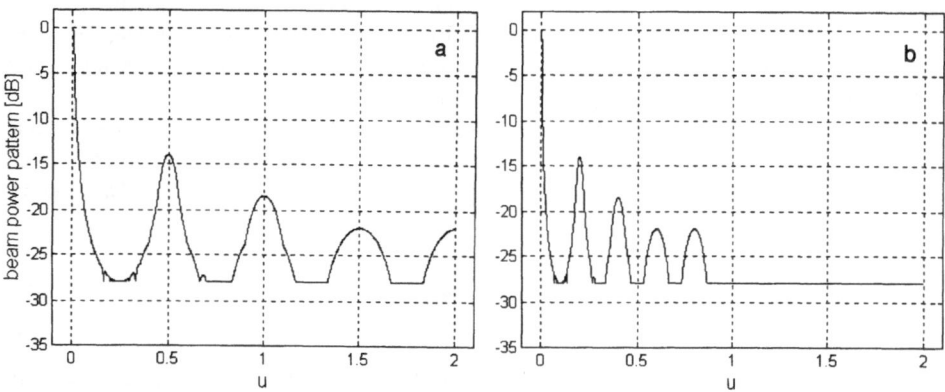

Figure 3. Wide-band beam power patterns obtained by the quasi-closed form for an array made up of 25 elements and for $r = 4.5$. (a) 2λ spacing ($n = 2$). (b) 5λ spacing ($n = 5$).

CONCLUSIONS

A wide-band beam pattern definition for imaging applications has been presented, and a quasi-closed form equation has been obtained for a rectangular pulse. Such an equation is valid for whatever inter-element spacing and signal bandwidth. The main characteristics of the obtained beam patterns have been analyzed, placing special emphasis on the invariance of the grating-lobe level to the inter-element spacing.

The use of sparse arrays and wide-band signals seems an interesting way of obtaining a good lateral resolution and avoiding grating lobes, while maintaining a small number of array elements. Future work will be devoted to extend the quasi-closed form to two-dimensional arrays and pulse envelope with a shape different from the rectangular one.

REFERENCES

1. R. O Nielsen. "Sonar Signal Processing," Artech House, Boston (1991).
2. M.H. Bae, I.H. Sohn, and S.B. Park, Grating lobes reduction in ultrasonic synthetic focusing,*Electronics Letters* 27:1225 (1991).
3. P. Challande and P. Cervenka, Theoretical influence of various parameters on the behaviour of a linear antenna, *Journal d'Acoustique* 3:17 (1990).
4. J. Lu and J.F. Greenleaf, A study of two-dimensional array transducers for limited diffraction beams,*IEEE Trans. Ultras., Ferroel., Freq. Control* 41:724 (1994).
5. A. Macovski, Ultrasonic imaging using arrays, *IEEE Proceedings*, 67:484 (1979).
6. D.H. Turnbull and F.S. Foster, Beam steering with pulsed two-dimensional transducer arrays,*IEEE Trans. Ultras., Ferroel., Freq. Control* 38:320 (1991).
7. X. Lurton and S. Coatelan, Array directivity for wide band signals: a time-domain approach,*Proc. Europ. Conf. on Underwater Acoustics* 721 (1992).
8. R.F. Follett and J.P. Donohoe, A wideband, high-resolution, low-probability-of-detection FFT beamformer, *IEEE Jour. Oceanic Engin.* 19:175 (1994).
9. A.P. Goffer, M. Kam, and P.R. Herczfeld, Design of phased arrays in terms of random subarrays,*IEEE Trans. Antennas and Propag.* 42:820 (1994).
10. G.R. Lockwood and F.S. Foster, Optimizing sparse two-dimensional transducer arrays using an effective aperture approach, *Proc. IEEE Ultrasonics Symp.* 3:1497 (1994).

MULTIPLE SCATTERING AND
DIFFRACTION TOMOGRAPHY

Sidney Leeman[1], Mark Betts[1], and Leonard Ferrari[2]

[1]Department of Medical Engineering and Physics
King's College School of Medicine and Dentistry
London SE22 8PT, U.K.
[2]The Bradly Department of Electrical Engineering
Virginia Polytechnic Institute and State University
Blacksburg, VA 24061

INTRODUCTION

Multiple scattering processes are often considered to be the limiting factor in the application of diffraction tomography techniques in practice. However, it is relatively easy to conceive idealised examples where such a statement does not strictly apply. For example, consider a typical (one dimensioanl) impediography setup, where a varying impedance is to be recovered from a backscattering experiment. In this case, the contributions of the various orders of multiple scattering processes to the backscattering may be explicitly calculated, and even summed. The standard impediography inversion technique may be applied, and, provided the appropriate boundary conditions are known, and the impedance profile may be recovered from the backscattering. However, the extension of the technique to higher dimensions is fraught with ambiguities, and it is widely felt that a convincing case cannot be built on such an idealised example.

One intuitive feature of the impediography example carries over to higher dimensions, viz. that multiple scattering effects are identical to pulse-, or wave-, 'bounce' processes, in which the actual wave 'sings' around in the scattering medium. It is easy to dispel this notion, by considering a simple Helmholtz wave equation, for scattering of an incident plane wave by a *uniform* sphere of constant wave velocity, embedded in an ambient medium of a different constant wave velocity. In fact, it is not difficult to show multiple scattering effects may be significant in this case, even though there can be no serious possibility of considering real wave-bounce diagrams to illustrate their various contributions.

In order to understand multiple scattering effects, a clear definition for the term will be needed, and it is shown below that the concept is tied to a particular formalism for describing the scattered field. Armed with that insight, it will also be suggested that diffraction tomography (or the application of inverse scattering techniques) may indeed be implemented, even when multiple scattering effects are significant -- at least within the context of a certain class of scattering potentials.

THE SCATTERING EQUATION

It is convenient to write the Lippmann-Schwinger equation[1], which describes the general relationship between the scattering 'potential' and the scattered wave, in a concise symbolic operator notation:

$$T = V + VGT \tag{1}$$

Here, G denotes the free space Green's function, V the scattering potential, and T the so-called 'transition operator', whose matrix elements, $\langle \mathbf{k'}|T|\mathbf{k} \rangle$, relate to the scattering amplitude between the (generally continuum) *plane wave* states denoted by the wave vectors $\mathbf{k'}$ (scattered) and $\mathbf{k}$ (incident). Note that Eq. (1) is, in fact, an integral equation, and that its apparently simple form hides some unexpected complexity, since its kernel, K=VG, is not necessarily square-integrable, as many approaches towards its solution demand. T is generally introduced as the operator which transforms the the incident field, ψ_0, into the final total field, $\psi = \psi_0 + \psi_{sc}$ (where the subscript 'sc' denotes the scattered field), via the equation

$$T\psi_0 = V\psi \tag{2}$$

There are two contexts in which the solution of Eq. (1) is usually attempted. In the *forward problem*, V is known and T is desired (with G calculable). On the other hand, in the *inverse problem*, T is known, and V is desired (with G calculable). Diffraction tomography is clearly the attempt to image V by solving a particular inverse problem, even though, in practice, the form of G in human tissues can only be approximated.

THE FORWARD PROBLEM

Definition of Multiple Scattering

The most direct approach, arguably, towards solving the forward problem is to apply an iterative scheme (Born-Neumann expansion), to give:

$$T = V + VGV + VGVGV + \ldots\ldots \tag{3}$$

Note that this provides an exact solution, provided that the series converges. The commonly used first Born approximation ('1BA') consists of dropping all but the first term in the expansion: in that case, the scattering amplitude and the scattering potential are merely the Fourier transforms of one another. The other terms in the series are generally referred to as higher order Born terms (with the Nth in the series denoted as 'NBA').

The individual terms in the series in Eq. (3) allow a very interesting interpretation. The presence of a 'V' in any term is is interpreted as a local (point) interaction/scattering site, and (the two-point function) 'G' as denoting a 'propagator' of a (virtual) wave between two interaction/scattering sites. The integration of any term over all appropriate interaction sites represents the contribution of that Born term to the scattering amplitude. Terms involving multiple V's are called 'multiple scattering' terms, and an important point to note is that this defines precisely what is meant by multiple scattering. Moreover, it is clear that the whole notion of multiple scattering is bound up with the expansion expressed in Eq. (3). *Any expansion will not provide multiple scattering terms -- only terms with the structure of the higher Born terms are, by definition, multiple scattering terms.*

Other Approaches

A rather different approach towards solving the direct problem is to apply the method of Rytov[2], which is based on writing the field in the form:

$$\psi(\mathbf{r}) = \psi_0(\mathbf{r}) \exp[\phi(\mathbf{r})] \tag{4}$$

The essence of the Rytov method is to find an expression for ϕ. This is not done exactly, but in the context of an approximation, the so-called 'Rytov approximation', which results in

$$\phi(\mathbf{r}) = [1/\psi_0(\mathbf{r})] \, GV\psi_0(\mathbf{r}) \tag{5}$$

Note that the Rytov approach gives an approximation to the field, and, as such, does not represent an exact solution. Moreover, as Eq. (4) emphasises, the field ψ vanishes wherever ψ_0 is zero, so that the Rytov approximation is particularly suited to calculations of the forward scattered field, and, under certain circumstances, may not be at all appropriate for the determination of scattering into large angles.

We have been able to show that the Rytov approximation leads to the following expression for the scattering amplitude:

$$T\psi_0(\mathbf{r}) = V\psi_0(\mathbf{r}) \times \left\{ \sum_{N=0}^{\infty} \psi_0^{-N}(\mathbf{r})[GV\psi_0(\mathbf{r})]^N \right\} \tag{6}$$

Eq. (6) represents what has been called the first Rytov approximation. Given the precise meaning that has been ascribed to multiple scattering terms, it appears from this formalism that even the first Rytov approximation ('1RA') contains elements of higher order multiple scattering contributions, but not in an exact way. Indeed, a little algebra shows that the 1RA contains the 1BA and 2BA exactly, but that analogues of the higher order Born terms occur in a somewhat corrupted way.

Yet another approach towards solving the forward problem relies on Fredholm's solution[1] to equations with the structure of Eq. (1). Although the classical Fredholm technique cannot be applied directly, because the kernel is not square-integrable, a solution with the following structure (reminiscent of the original treatment) can be obtained:

$$T = V + \{N/D\}V \tag{7}$$

where the $\{N/D\}$ operator is explicitly given as an expansion involving both V and G. However, each term in the expansion does not correspond to a specific order of multiple scattering. The Fredholm solution is an exact one, with a much wider range of convergence than the Born expansion, and termination of the Fredholm series at successasive terms does provide a sequence of higher order approximations to T. These approximations, to any order, will incorporate multiple scattering (as defined above) terms, but will not necessarily encapsulate the total contribution of any particular multiple scattering process to the scattering amplitude.

THE INVERSE PROBLEM

As has been pointed before, diffraction tomography is essentially an attempt to solve the inverse problem, given a particular set of experimental data, within the context of a specific physical model. In principle, the formulation of a solution to the inverse problem presents the same structure as for the forward case. Indeed, the Lippmann-Schwinger equation may be formally rewritten as

$$V = T - VGT = T - TGV \tag{8}$$

and the parallel with Eq. (1) is immediately apparent. However, there are some severe difficulties in attempting to solve this equation using the same methods as for the forward problem. The main obstacle is that only measured values of the T-matrix are available, and these represent only a subset of all the values that are required for a complete solution to Eq. (8). In practice, many inverse scattering solutions depend on some form of analytic continuation of the measured data from the 'physical region' (i.e., accessible from real measurements) to the T-matrix values in an 'unphysical region'. Even small amounts of noise in the measurements may consequently have disastrous consequences for the accuracy of the solution. This problem is not significant when the assumption is made that the scattering is accurately approximated by the 1BA: in that case, it is relatively straightforward to devise experiments which access all the required elements of the T-matrix, and the inversion procedure is no more complicated than taking the Fourier transform. It is shown below that this relative simplicity may be exploited, together with the established techniques for solving the forward problem, in order to devise an iterative scheme that, in principle, allows the inverse problem to be solved.

An Iterative 1BA Inversion Scheme

Let the desired potential be expressed as an infinite series, such that

$$V = V_0 + V_1 + V_2 + \ldots \tag{9}$$

where

$$V_0 \Leftrightarrow T_E \tag{10}$$

and where T_E relates to the experimentally measured values of the scattering amplitude. The relationship expressed in Eq. (10) implies that a 1BA-type inversion ($\equiv$ Fourier transform) holds between T_E and some suitably chosen V_0. Let it be further assumed that

$$T = T_1 + T_2 + T_3 + \ldots \tag{11}$$

where

$$T_N = V_{N-1} + V_{N-1}GT_N \tag{12}$$

with

$$V_N = V_{N-1} - \sum_{K=1}^{N} T_K \tag{13}$$

Once V_0 is known, all the higher V_N may be successively found from known entities, since the T_N are all successively calculable via a solution of the forward problem. Providing the structure of the potential is suitable, and provided that all the series implied in the above equations converge, then it can be shown that successively higher resolution estimates of V are obtained with each iteration.

Iterative 1BA Inversion: Explicit Example

The above assertions may be proven directly[3] in the context of a scattering potential that can be expressed in the form:

$$V(r) = \int_{\mu}^{\infty} ds\, \sigma(s)\, e^{-sr}/s \quad \text{with} \quad \mu > 0 \tag{14}$$

For simplicity, and with no loss of generality, the scattering potential has been chosen to be spherically symmetric. In this case, the properties of the scattering amplitude in the complex scattering angle, θ, space may be exploited to provide an exact reconstruction of $V(r)$, from (angle) scattering measurements, for a fixed direction of incident wave, as would be conducted in a typical diffraction tomography setup. This comes about because of the particular nature of the analyticity properties of the various Born terms in κ-space, where $\kappa = -2k^2(1 - \cos\theta)$, and k is the magnitude of the wave vector of an incident plane wave. Each higher Born term has a branch-cut, on the positive real axis, starting at a successively higher κ-value: the NBA term has a branch cut starting at $(N\mu)^2$. Note that the physical region of κ -values, that can be accessed by real measurement, lies on the negative real axis, between $-4k^2$ and 0. The discontinuity across each successive branch cut, as calculated from the analytic continuation of data from the measured physical region, allows a successively better ($\equiv$ higher resolution) approximation to the scattering potential to be made, if the inversion scheme outlined above is followed. V_0 is obtained from the 1BA inversion of the experimentally determined (via analytic continuation) discontinuity across the branch cut between μ^2 and $(2\mu)^2$.

Since the specific technique for this potential relies on analytic continuation, it may be quite delicate to apply in practice. On the other hand, this analytically exact example demonstrates that the inverse problem may be solved even when multiple scattering is not negligible, by repeated application of the forward problem and 1BA inversion, and lends credence to the hope that the general scheme outlined above could apply to a more general class of potentials -- without the need to have recourse to continuation in a complex domain.

It should be borne in mind that, in general, even when no analytic continuation is necessary, the 1BA inversion scheme is quite computationally intensive, since it involves multiple applications of the forward scattering problem. However, in principle, the latter is always feasible, even if such complex techniques as the Fredholm approach have to be resorted to.

CONCLUSIONS

It has been emphasised that the concept of multiple scattering is essentially based on a 'pictorial' view of the Born series solution to the forward scattering problem. Multiple scattering must not be interpreted as a real wave 'singing' around in the scattering medium, although there may be some situations in which such a physical picture is, indeed, correct. Other approaches towards the solution of the forward scattering problem do not necessarily lead naturally to the notion that the scattered field may be regarded as an ordered hierarchy of contributions from successively higher multiple scattering processes. In particular, the first Rytov approximation to the field does not any terms higher than the 2BA (first order multiple scattering), although there is some approximation to them.

It has proved possible to devise an iterative inversion scheme that utilises only 1BA inversion, albeit at the computational cost of repeated application of the forward problem. The approach has been carried out in a theoretically exact example (although the calculational details are only hinted at in this communucation), but, in practice, great accuracy in both measured data and (forward) computation will be called for, if good resolution is to be obtained. Nonetheless, it does demonstrate that the presence of significant multiple scattering processes need not be seen as a limiting factor in the application of diffraction tomography.

ACKNOWLEDGEMENTS

The EPSRC and The Wellcome Trust are thanked for their support.

REFERENCES

1. S. Leeman, Nonlocal potentials and two-body scattering, Proc. Roy. Soc. Lond., A315:497 (1970)
2. A. Ishimaru, "Wave Propagation and Scattering in Random Media", Academic Press, New York (1978)
3. S. Leeman, P.R. Chandler, and L.A. Ferrari, Diffraction tomography with multiple scattering, in: "Acoustical Imaging", 15:29, Plenum Press, New York (1987)

SIMULATION OF ACOUSTIC WAVE PROPAGATION WITHIN FLOWING MEDIA USING SIMPLIFIED BOUNDARY ELEMENT TECHNIQUES

Peter-Christian Eccardt,[1] Hermann Landes,[2] Reinhard Lerch,[2] and Valentin Mágori[1]

[1]Siemens AG, ZFE T KM 1, D-81730 München, email: eccardt@zfe.siemens.de
[2]Dep. of Meßtechnik, University of Linz, A-4040 Linz

INTRODUCTION

The development of a novel ultrasonic gas meter[1] for household applications demonstrated that ultrasound technique can be a very precise and cost effective method for flow meters. Fig. 1 shows the basic principle of an ultrasound meter based on the delay time measurement principle. The time of flight between the two transducers positioned along the flow channel is influenced by the mean flow along the measurement path. The mean flow rate can be calculated as

$$\bar{q} = \bar{v} \cdot S = \frac{L}{2\cos(\alpha)} \frac{t_{up} - t_{down}}{t_{up} \cdot t_{down}} \cdot S. \tag{1}$$

with: L: Length of sound path with vector components in flow direction
      α: angle between sound path and mean flow direction
      S: sectional area,   v: mean flow velocity
t_{down}, t_{up}:   time of flight in and against flow direction

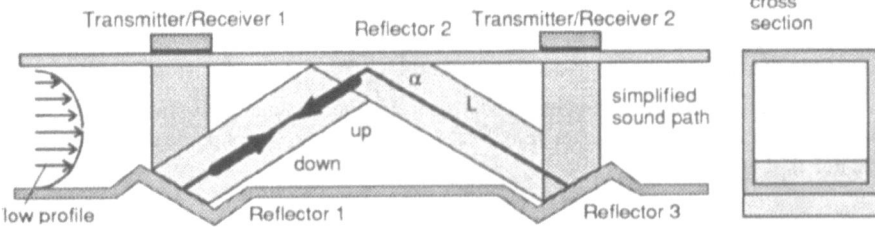

Fig 1. Measuring principle of a flow meter for fluids

This formula describes only the integral effect in a first order approach. The developments in the recent years revealed that precise meters can be developed only by utilizing appropriate models that are considering the influence of flow profile, transducer design and sound path. These models become essential, if the properties of the measured medium, especially it's viscosity are unknown and cannot be compensated by calibration. Today's transducer development is based on finite element programs used for the optimization of the dynamic behavior and the radiation characteristics. Finite element (FE) programs are well established in the design of flow channels with respect to the optimization of parameters like pressure loss. For the calculation of acoustic wave propagation the existing programs are based on the finite element, boundary element or ray tracing method, but no commercial program was found which takes the influence of flow into account. So special algorithms had to be developed. The approach presented in this paper is based on Helmholtz integration and boundary element techniques. The other approach is based on raytracing techniques and is also published in this book.[2] To minimize development efforts existing FE programs for transducer and flow simulation are integrated in our calculation scheme. The article gives a short overview over the possible simulation methods and presents theory and results of the implemented solution.

SIMULATION TECHNIQUES FOR WAVE PROPAGATION PROBLEMS

In a homogeneous infinite medium the wave equation can be described as a function of the velocity potential Φ:

$$\nabla^2\Phi - \frac{1}{c^2}\frac{\partial^2\Phi}{\partial t^2} = 0 \tag{2}$$

With the wave number $k=2\pi f/c$, the particle velocity vector $v = -\nabla\Phi$, the speed of sound c, the sound pressure $p = jkZ_0\Phi$ and the acoustic impedance Z_0 the wave equation for the steady state (Helmholtz equation) can be written as[3,4]

$$\nabla^2\Phi + k^2\Phi = 0. \tag{3}$$

In order to consider the influence of an existing flow profile in the area of sound propagation different solutions exist. One approach for small Mach numbers and incompressible media adds a velocity potential $\overline{\Phi}$ of the flow vector w (where $w = -\nabla\overline{\Phi}$) to the wave equation (2).[5,6]

$$\nabla^2\Phi - \frac{1}{c^2}\left(\frac{\partial}{\partial t} + \frac{\partial\overline{\Phi}}{\partial y}\frac{\partial}{\partial y}\right)^2\Phi = 0 \tag{4}$$

Finite element or finite difference methods

Finite element or finite difference programs allow the calculation of acoustic wave propagation. Some implementations are using the velocity potential Φ as a scalar degree of freedom per node, but no modified wave equation for the influence of flow profiles like Eqn.(4) were implemented in commercial programs so far. To get proper results in FEM programs the element size has to be small against the wavelength (for elements with linear shape functions the typical element size is about 1/10th of the wavelength).[7] In a flow meter for water with a speed of sound of 1500 m/s and an operating frequency of 1.5 MHz the wavelength is 1 mm. The tube has a volume of approx. 100 000 mm^3 and a surface of approx. 15 000 mm^2. This means that a simulation of the tube in Fig.1 would ask for $100*10^6$ 3D elements. This number of elements is not computable with today's hardware within a reasonable time. Finite element method is therefore only used for the calculation of the velocity potential at the transducer's surface including it's surrounding fluid area.

Helmholtz integral and boundary element methods

For a homogeneous medium without sources inside the medium the Helmholtz integral (5) describes the velocity potential Φ_P on a point P as a function of the vector potential and it's derivative on all points at the surface

$$C_o\cdot\Phi_P = \oiint_s\left(G\cdot\frac{\partial\Phi}{\partial n} - \Phi\cdot\frac{\partial G}{\partial n}\right)\cdot ds, \qquad v_n = -\frac{\partial\Phi}{\partial n} \tag{5}$$

with $C_o=1$ for points inside the medium, $C_o=1/2$ for points on the surface and $C_o=0$ elsewhere and $G(R)$ as the Green's function for the three-dimensional space given by

$$G(R) = \frac{e^{-jkR}}{4\pi R} \qquad \text{R: distance between source point and target point} \tag{6}$$

The pressure $p=jkZ_0\Phi$ and the normal velocity v_n in points on the surface can be calculated by integrating over the velocity potential of all other grid points. In boundary element technique discretisation of the surrounding surface leads to a matrix equation based on Eqn. (5). The resulting matrices are full and unsymmetric having a dimension equal to the number of discretisation points. The boundary element method in it's classical form requests a piecewise homogeneous medium with sources only at the boundaries.[8] For the flow meter a reasonable discretisation leads to $1.5*10^6$ linear 2D elements or a matrix of $18*10^{12}$ Byte. Therefore, this approach is also not feasible.

Succesive calculation of the Helmholtz integrals

If diffraction phenomena may be split into different areas, the velocity potential Φ is calculable separately for each area. At first, the transducer, it's surrounding housing and fluid can be calculated with finite element programs. For a short time pulse the influence of the reflectors is neglectable because of their large distance to the transducer. Now the velocity potential at the first reflector is calculated from the results on the transducer area by calculating the Helmholtz integral.[9] Assuming total reflection the velocity potential at the second reflector can be calculated from the known velocity potential at the first reflector, etc..

If the wavelength is small compared to the dimensions of a plane reflecting surface, the reflections can be described by virtual mirror sources. The results of a two-dimensional FE simulation for a wave transmitted from a single piston and reflected by two perpendicular plane surfaces are shown in Fig.2. A comparison of the reflected wave and the corresponding one transmitted from a virtual mirror source shows only minor differences. Accordingly, most of the flow meter's reflecting areas may be treated by this method and in this way it is possible to calculate the flow channel by the Helmholtz integral method within a reasonable amount of computation time.

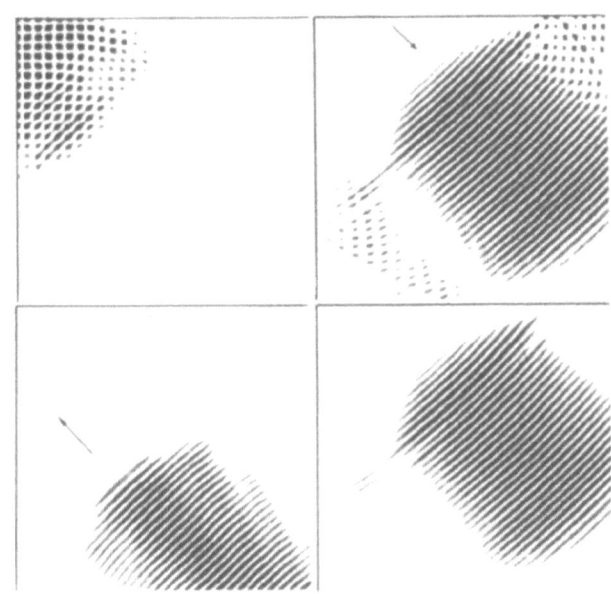

Fig.2. Plane wave, reflected on two plane surfaces positioned perpendicular to each other
top left: (2) wave after first reflection   top right: (3) wave after second reflection
bottom left: (1) wave before reflection   bottom right: (3a) direct wave from mirrored source

APPROACHES FOR BOUNDARY ELEMENT METHODS IN FLOWING MEDIA

Due to the dimensions of the flow meter finite element methods as well as straight forward applications of boundary element methods are not practicable. Ray tracing methods on the other hand would neglect existing diffraction effects on the transducers and the small reflectors. Since the change of the flow profile over one wavelength is small in comparison to the wavelength (except for the laminar boundary flow in turbulent profiles) we have developed a computational scheme combining raytracing methods, known from the wave theory and calculating the propagation of sphere waves in flowing media, with a step by step Helmholtz integral method for the interaction between the boundary points. The method interprets the Green's function from it's physical meaning and does not claim to be an exact solution of the governing wave equation. The results, however, showed that the main flow effects we expect like changes in delay time, focusing and diffraction are covered.

In a channel of a constant flow vector v each point of the wave's surface is shifted within the time dt by the vector $v*dt$. Furthermore, for each point at the surface the normal vector remains constant during the propagation of the wave. With the presence of gradients in the flow profile the direction of the propagation vector changes according to fermat's principle, thus minimizing the delay time between two points on this path. Equivalently the propagation vector is always perpendicular to the wavefront and so allows a fast calculation of the new propagation vector. We define two points on the surface of the wave P_a and P_b such that

$$\delta \cdot a = P_a - P, \quad |a| = |b| = 1,$$
$$\delta \cdot b = P_b - P, \quad a \times b = n, \quad \delta \to 0. \tag{7}$$

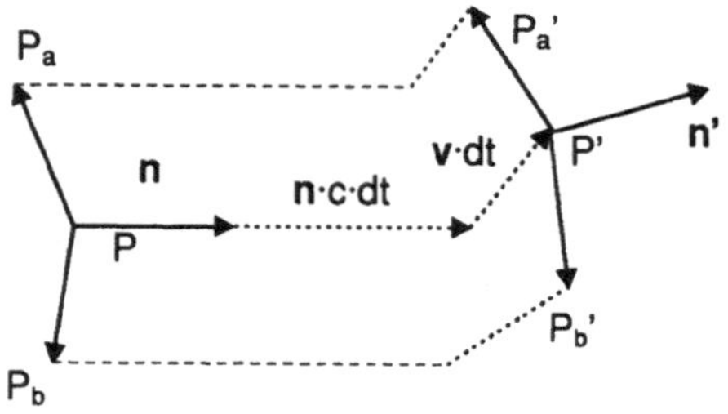

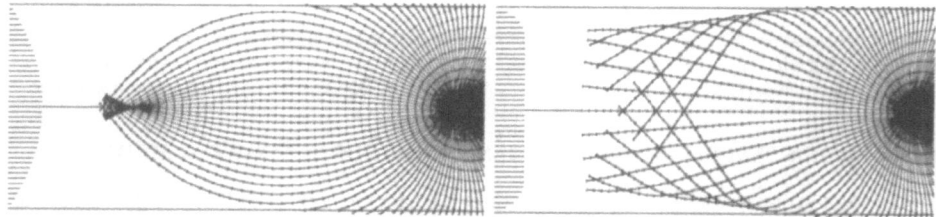

Fig. 3. Change in ray vector induced by flow gradients

The flow vector at P_a and P_b is equal to the sum of the flow vector v at P and the vector gradient of v in respect to the vectors a and b:

$$v_a = v + (a\nabla)v$$
$$v_b = v + (b\nabla)v \tag{8}$$

The new normal vector can be calculated from the vector product of a' and b' by

$$\delta \cdot a' = P_a' - P' = \delta(a + (a\nabla)v\,dt)$$
$$\delta \cdot b' = P_b' - P' = \delta(b + (b\nabla)v\,dt) \tag{9}$$

$$n' = (a' \times b') = \left(a + \frac{dv}{da}dt\right) \times \left(b + \frac{dv}{db}dt\right)$$

Fig. 4 shows the propagation of a sphere wave influenced by a laminar and a turbulent flow profile, respectively. The algorithm also covers focusing effects. To illustrate the focusing effect v_{max} was set to 30% of the speed of sound. The curvature of the ray is directly proportional to the gradient of the flow. Turbulent profiles exhibit large gradients in the area of the laminar boundary layers. The limits of the method are reached, when rays are crossing. In the crossing points the wavelength cannot be neglected any longer.[10]

Fig. 4. Focusing effect of a laminar (left) and a turbulent (right) profile to a sphere wave (v_{max}=0.3*c)

The above wave equation (2) and thus the Helmholtz integral (5) is only valid for homogeneous media. Accordingly, a wave propagation method had to be developed taking the influence of the flow profile into account. It was obvious to modify directly Green's function (6) for the 3D space, which contains a term G_{delay} describing the delay time $\tau=R/c$ between two points and a term G_{div} describing the divergence a sphere wave. The term $1/R$ of G_{div} can be interpreted from energy conservation and means, that the power as an integral over the sphere has to be constant.

$$G_{delay} = e^{-jkR} = e^{-j\omega\frac{R}{c}} = e^{-j\omega\tau}, \qquad G_{div} = \frac{1}{4\pi R} = \frac{1}{4\pi c\tau} \tag{10}$$

The actual delay and divergence are used for the Green's function instead of the actual distance between source and target point. Considering the curvature of the ray, this can be done by calculating the intersection points of the sphere wave (starting from each source point and influenced by the flow profile) with the bounding surface. Dividing the surface of a sphere wave into partial areas dS each area can be represented by a distinct ray. Energy conservation for the propagating wave means for each partial ray

$$p \cdot \sqrt{dS} = const. \tag{11}$$

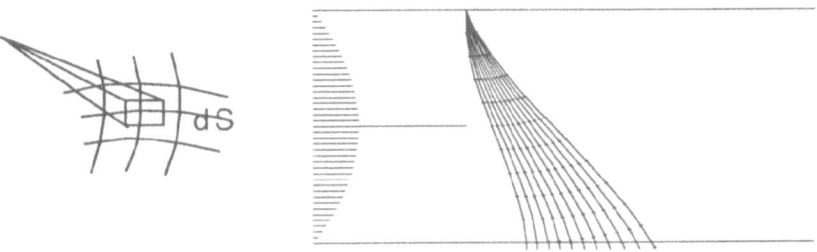

Fig. 5. Areas dS representing a partial ray, distorted sphere wave crossing the bounding surface

From Eqn. 11 the divergence term can be calculated from the area dS' (derived from the distance to the neighboring rays) and the area dS (derived from the area of a sphere wave after the delay time τ). With the corrected distance $R^* = \tau \cdot c$ this leads to the modified Green's function

$$G^*(R^*) = \frac{e^{-jkR^*}}{4\pi \frac{\sqrt{dS'}}{\sqrt{dS}} R^*}. \tag{12}$$

For small distortions of the wave the divergence term can be set to dS'≈dS. Thus, the influence of the flow can be taken into account by a correcting factor α_R to the Green's function

$$\alpha_R = \frac{R^*}{R} = \frac{\tau \cdot c}{R} \quad \text{and} \quad G(R^*) = G(\alpha_R R) = \frac{e^{-jk\alpha_R R}}{4\pi \alpha_R R}. \tag{13}$$

A further approach for the calculation of α_R is possible for small flow velocities. With the average flow vector **v** along the line **d** the delay time τ can be found by determining the point P which is shifted by the flow within τ to the target point E. The delay time τ can be calculated using vector geometry as:

$$c \cdot \tau = |d - v \cdot \tau| \quad \Rightarrow \quad \tau = \frac{-d \cdot v + \sqrt{(d \cdot v)^2 + d^2(c^2 - v^2)}}{c^2 - v^2} \tag{14}$$

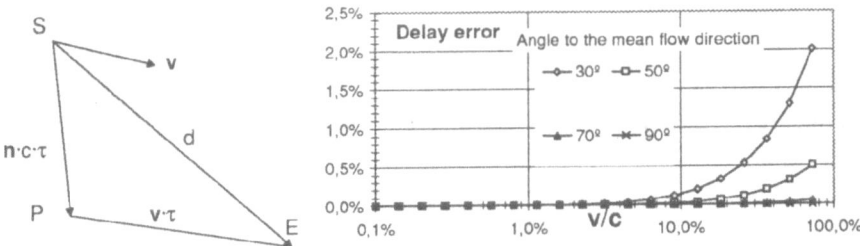

Fig.6 Linear delay time approach: principle and error (angle to the flow direction as a parameter)

Fig.6 shows, that the linear approach is sufficient for flow velocities up to 1-10% of the speed of sound. Main advantage of the linear approach is, that the correcting factor is derived directly from the mean flow between two boundary points.

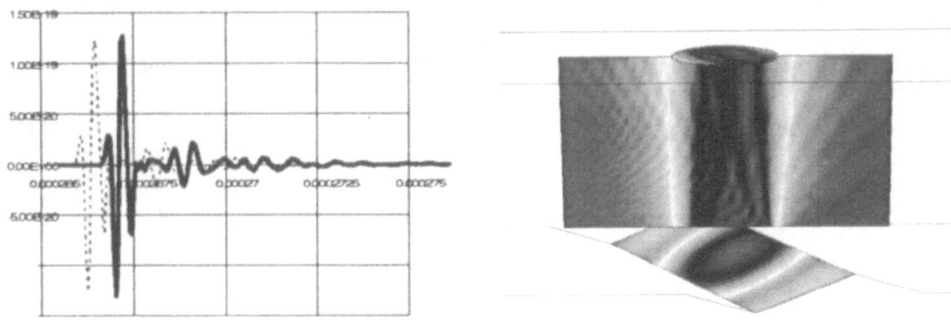

Fig.7. Near field of an ultrasonic transducer distorted by a laminar flow profile (v_{max}=0.3*c) defocussing effect in flow direction (top),   focusing effect against flow direction (bottom)

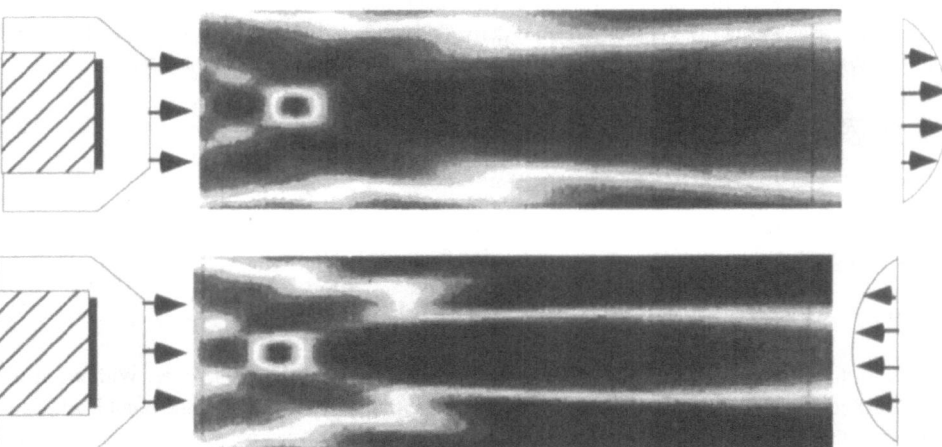

Fig.8. Calculated time signals for the upward and downward sound path (left)
sound pressure on and between right reflector and right (receiving) transducer (right)

RESULTS

Fig.7 illustrates the focusing and defocusing effect of a laminar flow profile in the near field of an ultrasonic transducer (v_{max}=0.3*c for Illustration). Fig.8 shows the time signals for the upward and downward sound path, as well as the drifted and diffracted pattern of sound pressure between the right reflector and the right (receiving) transducer. The mean flow velocity is 15 m/s.

CONCLUSION

A simulation program was developed which calculates step by step the Helmholtz integrals with the „refraction" or the „linear" approach. The velocity potential of the transmitter with it's surrounding fluid is calculated with finite element programs or is taken from mathematical assumptions. The diffraction effects of the small reflectors are considered by calculating the Helmholtz integral from one reflector to the next. The time signal on the receiving transducer is calculated with finite element programs using the velocity potential as a boundary condition. If necessary, the displacements and velocities at the reflectors can be calculated with FEM as well.

REFERENCES

[1] Jena, A.v.; Mágori, V. Rußwurm, W.: Ultrasound Gas Flow Meter for Household Application. Sensors ans Actuators A Vol. 37-38 (1993), pp 135-140

[2] Garssen, H.G.v.; Mágori, V.: Modelling of ultrasonic flow meters by acoustical raytracing, *Acoustical Imaging*, Florence (1995)

[3] Kino, G.: „Acoustic Waves", Springer (1987)

[4] Goodmann, J: „Introduction to Fourier Optics", McGraw-Hill 1968

[5] Dowling, A.P.: Effects of motion on acoustic sources, *Modern Methods in analytical acoustics*, 406-427, Springer lecture notes (1992)

[6] Howe, M.S.: The generation of sound by aerodynamic sources in an inhomogeneous steady flow, *J.Fluid Mech.*, 67:597-610 (1975)

[7] Zienkewicz, O.C.: „Finite Element Method in engineering science", Mc Graw Hill, London (1971)

[8] Brebbia, C.A.; Telles, J.C.F; Wrobel, L.C.: „Boundary Element Techniques, Theory and Applications In Engineering", Springer 1984

[9] Lerch, R.; Landes, H.; Kaarmann, H.: Finite Element Modelling of the pulse-echo behavior of ultrasound transducers; *Ultrasonic Symposium*, Cannes (1994)

[10] Bloom, C.; Kazarinoff, N.: „Short Wave Radiation Problems in Inhomogeneous Media", Lecture Notes in Mathematics No. 522, Springer (1976)

THREE-DIMENSIONAL ULTRASONIC IMAGING
OF TEMPERATURE DISTRIBUTIONS

V.I. Mirgorodsky, V.V. Gerasimov, and S.V. Peshin[1]

Institute of Radioengineering and Electronics of RAS
[1] International Scientific Research Center of Acoustic
Technologies
141120 Fryazino of the Moscow region, Russia
E-mail: vim288@ire216.msk.su.

A new principle of the investigation of space distributions of incoherent radiation sources based on the correlation processing of signals accepted by the receivers disposed at different positions is offered. The possibility to obtain the tomographic type information by this principle was experimentally shown.

It is known that the stationary interference picture of the incoherent radiation is observed when the path difference of the interfering beams is less then the coherence length of radiation. This peculiarity is used in particular at measurements of the linear sizes of standards of measure[1] by Fizeau or Kosters types of the interferometers.

The offered paper is devoted to the new principle based on this phenomenon destined for getting the information about a space distribution of incoherent radiation sources of the acoustic electromagnetic or any other nature.

The principle offered permits to receive the information about 3-dimensional distributions in the space of incoherent radiation intensity while one can measure the size of objects along only one dimension by the interferometers of Fizeau or Kosters types. The main requirement to the parameters of radiation for such sounding realization is the coherence length L_k less then the required spatial resolution Δr.

We shall consider the region of space, in which the radiation is spreading with the velocity "v" and attenuation "α". For simplicity of the analysis we shall concentrate all sources of a incoherent radiation inside the area "V" (Fig. 1). For the sake of simplicity we shall present instant amplitudes of radiation sources by a scalar function of sources $N(r,t)$, the properties of which will be discussed below. Let the receivers of radiation be located in various points of space r_i (i=1,2.... N) not belonging to the region "V". Then the amplitude of radiation of sources near the sensitive elements of the receivers will be determined by the following expression:

$$S_i(t) = \int_V N(r, t - \frac{|r - r_i|}{v}) \frac{\exp(-\alpha|r - r_i|)}{|r - r_i|} d^3r \tag{1}$$

The action of the receivers of signals is to make the linear transformation of an amplitude of the radiation $S_i(t)$ to an electrical signal. Usually such a transformation results in the restriction of the frequency and spatial spectra of the received signal. However without an essential decrease of a consideration generality let us consider ideal receivers. The effects of the restriction of frequency and spatial spectra can be taken into account by the appropriate function of sources $N(r,t)$ choice as well as the assumption that the radiation receivers have sensitivity diagrams close to the isotropic type. In this case it is possible to consider that the electrical signals on the outputs of the receivers are the same as on their inputs - $S_i(t)$.

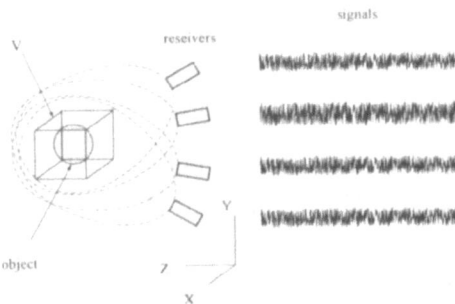

Figure 1. An arrangement of the region under investigation and the receivers (on the left) and forms of signals on the outputs of the receivers (on the right).

For the analysis of signals $S_i(t)$ we shall use the expression for the correlation function of the second order:

$$C_{ij}(\tau_{ij}) = \int_{-\infty}^{\infty} S_i(t)S_j(t-\tau_{ij})dt. \tag{2}$$

Where τ_{ij} - is the delay between i-th and j-th signals. Substituting the expression $S_i(t)$ (1) into (2) and changing the integration order results in the following expression:

$$\int_V d^3r \int_V d^3r' \frac{\exp\left[-\alpha\left(|r-r_i|+|r'-r_j|\right)\right]}{|r-r_i||r'-r_j|} \int_{-\infty}^{\infty} N(r,t-\frac{|r-r_i|}{v})N(r',t-\frac{|r'-r_j|}{v}-\tau_{ij})dt \tag{3}$$

Where integral:

$$I = \int_{-\infty}^{\infty} N(r,t-\frac{|r-r_i|}{v})N(r',t-\frac{|r'-r_j|}{v}-\tau_{ij})dt \tag{4}$$

is calculated on the base of frequency and spatial coherency parameters of the radiation, which are determined by the function of the sources $N(r,t)$.

Generally the correlation parameters of the radiation can be rather diverse therefore for the simplification of the analysis we shall consider the case of thermal radiation, which is rather general and is of practical interest.

The correlation length of the thermodynamic fluctuation of the temperature of any substance is determined by the known expression $L_t = \sqrt{2\chi/\omega}$ where -χ is thermal diffusivity of a substance and "ω"- is the circular frequency of fluctuations[2].

Even for relatively low frequencies - order 1 MHz and at high for a condensed matter thermal diffusivity $\chi = 1$ cm^2/s one can obtain the value L_t 6 10^{-3} cm. This value is less than the required spatial resolution Δr for many practically useful cases of sounding, which gives the base for the analysis simplification in the case of the required spatial resolution exceeding L_t.

In this case the radiation emitted by various points of the substance "r" and "r'" (spaced from each other more than on L_t) can be considered as not correlated and the "I" dependence on "r" and "r'" will be proportional to $\delta(r - r')$.. Similarly it is possible to solve the problem of the influence on the "I" temporal coherence of the radiation. If one is limited as before to the case of $\Delta r > L_k$ where L_k - is the coherence length determined by the expression - - $L_k = v/\Delta f$ ("v" - is the of radiation velocity and Δf- is its frequency band) the "I" dependence on τ_{ij}, which is the delay can be expressed as $\delta(\dfrac{|r - r_i|}{v} - \dfrac{|r' - r_j|}{v} - \tau_{ij})$. Thus as mentioned before the integral (4) can be evaluated as follows:

$$I \sim \langle N^2(r) \rangle \delta(r - r') \delta(\frac{|r - r_i|}{v} - \frac{|r' - r_j|}{v} - \tau_{ij}) \qquad (5)$$

Where the notation <> means the time averaging[*] . Substituting (5) in the initial expression (3) and integrating over d r' we shall obtain:

$$\int\limits_{V_{ij}} d^3 r \frac{\exp\{-\alpha(|r - r_i| + |r - r_j|)\}}{|r - r_i||r - r_j|} \langle N^2(r) \rangle = k_i k_j C_{ij}(\tau_{ij}) \ i, j = 1, 2 \ldots n; \ i \neq j, \qquad (6)$$

Where k_i -is the factor of sensitivity i -th channel and V_{ij} is the region of the space (hyperboloid of revolution) satisfying to the equation:

$$|r - r_i| - |r - r_j| = v\tau_{ij} \qquad (7)$$

The sense of this is clear - the value of the correlation function is defined by sources, which are arranged on the surface V_{ij}. A path difference from points of V_{ij} up to receivers located at the points "r" and "r'" is equal to - $v\tau_{ij}$. The availability of the "n" receivers located at the various points of space permits to obtain different signals combinations of 2 and hence $\dfrac{n!}{2(n-2)!}$ various correlation functions $C_{ij}(\tau_{ij})$ forming the system of equations (6) where the required function is $\langle N^2(r) \rangle$. The obtained system belongs to the type of Fredholm equations of the 1- st genus, which solving is usually difficult.

This difficulty appeared to be overcome by changing a method of the processing of accepted signals namely with the help of the replacement of the correlation integral of 2 -d of the order (2) for the correlation integral of 4 -th order:

---

[*] It should note, that the mentioned procedure allows generalization on cases of partial coherence of radiation, when integral "I" is not already expressed through - function, and has a more complex form[3].

$$C_{ijkl}(\tau_{ij}, \tau_{ik}, \tau_{il}) = \int_{-\infty}^{\infty} S_i(t) S_j(t + \tau_{ij}) S_k(t + \tau_{ik}) S_l(t + \tau_{il}) dt \tag{8}$$

Where $\tau_{ij}, \tau_{ik}, \tau_{il}$ - are the times of delay j-th, k-th and l-th of channels relatively i-th. In this case the use of the derivation similar to the mentioned above permits to transform the expression (8) in to a form:

$$\int_{V_{ijkl}} \langle N^4(r) \rangle \frac{\exp\{-\alpha(|r - r_i| + |r - r_j| + |r - r_k| + |r - r_l|)\}}{|r - r_i||r - r_j||r - r_k||r - r_l|} d^3r \tag{9}$$

Where "V_{ijkl}" - is the area of the space "V" satisfying to the system of equations:

$$\begin{aligned}
|r - r_i| - |r - r_j| &= v\tau_{ij} \\
|r - r_i| - |r - r_k| &= v\tau_{ik} \\
|r - r_i| - |r - r_l| &= v\tau_{il}
\end{aligned} \tag{10}$$

The equations of the system (10) as has already been noted describe hyperbolic surfaces. Therefore the area "V_{ijkl}" at a non-coincident location of receivers r_i, r_j, r_k, r_l is determined by crossing of hyperbolic surfaces, the parameters of which are defined by delays $\tau_{ij}, \tau_{ik}, \tau_{il}$.

For the case of the receivers location in one plane say XY; the picture of crossings will evidently be symmetric concerning this plane. As far as sounding region is located at Z >0 the solution of a system (10) should be taken into account only at Z >0. Because the area "V" presents only one point it follows that:

$$\begin{aligned}
\langle N^4(r) \rangle &= k_i k_j k_k k_l |r - r_i||r - r_j||r - r_k||r - r_l| * \\
&* C_{ijkl}(\tau_{ij}, \tau_{ik}, \tau_{il}) \exp\{\alpha(|r - r_i| + |r - r_j| + |r - r_k| + |r - r_l|)\}
\end{aligned} \tag{11}$$

Hence the value $\langle N^4(r) \rangle$ in any point of the area "V" is determined by the correlation function $C_{ijkl}(\tau_{ij}, \tau_{ik}, \tau_{il})$. The factors k_i, k_j, k_k, k_l as well as before are determined by the sensitivities of i-th, j-th, k-th and l-th channels. Setting various values of delay times $\tau_{ij}, \tau_{ik}, \tau_{il}$ one can consistently look through points of the area under investigation and thus to obtain 3-D space distributions of the intensities of incoherent radiation sources. In the case of the availability more than 4 receivers the formation of the $\frac{n!}{4!(n-4)!}$ images from different points of view is possible by using various combinations of the signals $S_i(t)$ of 4.

For checking the possibilities of the offered principle the mathematical modeling of process of the signals reception and recover the space distribution of intensity of radiation sources was conducted. The modeling was to sum signals coming to receivers from space according to expression (1). The space is filled with statistically independent sources. As a result the restored distributions appeared to be close to the form of initial distribution.

However some smoothing of spatial gradients and random (not repeated from experiment to experiment) spatial noise are seen on restored distributions compared with initial ones. The analysis of the reasons of smoothing of the picture has shown that it is connected with the limitation of a used signal spectrum. As additional calculations have shown the reduction of a used signal digitization period results in the reduction of the effect of smoothing. It also results in an appropriate increase of the volume of the information, which needs to be processed.

The spatial noise as the calculations have shown arises at an insufficient averaging during correlation processing - they decrease with an increase of the number of readouts N under the law close to $/\sqrt{N}$.

For the check of the realization of the offered principle experimental investigations of the process of spatial distribution reception of the source intensity of the acoustic radiation were conducted. The experiment was to receive acoustic signals with the help four microphones irradiated by dynamic loudspeaker excited by a noise electrical signal. The volume level of excitation was established sufficient for maintenance in receiving channels of the signal to noise ratio exceeding 10.

The loudspeaker had the characteristic sizes about 50 cm. The processing of signals was to transformation them in the digital form with the help of 8 digit ADC sequential interrogation working in mode of 4 - of channels with clock frequency about 24 KHz. Signals with width of a frequency spectrum about 500 Hz were registered, which was basically determined by a frequency band of the loudspeaker. During the measurement the signals were recorded in the RAM the computer whence after the ending of the measurements were transferred for the storage on a hard disk.

The process of construction of one section consisting of $\sim 10^3$ spatial points on the basis of realizations of accepted signals of 10^5 temporary points takes about 10 minutes on a computer of type "AT486".

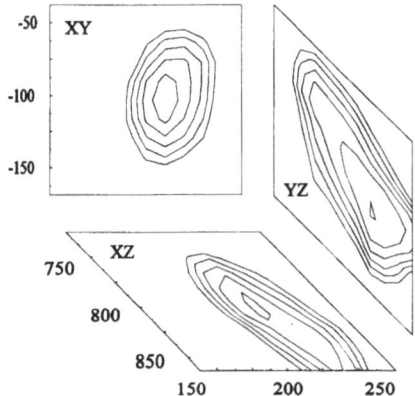

Figure 2. Experimentally obtained spatial distribution of the acoustical incoherent radiation source. The sections of the image by the planes, which are parallel to the planes "XY", "XZ" and "YZ" are shown. The extreme contours of the sections correspond to a level of half-height of a maximum.

An arrangement of the region under investigation with the receivers are presented on the left part of the Fig. 1. The forms of signals on outputs of receivers are presented on the right part of the Fig. 1. The distance from a plane, in which the receivers were located up to the source was about 8m. 3 receivers were placed on the triangle tops and 4-th in the triangle center. The triangle was close to the equilateral triangle with the legs, which size was about 6 m. On Fig. 2 typical image is presented as the sections by planes parallel to "XY", "XZ" and "YZ" (for reflection tomographic character of the picture). The extreme contours of the sections correspond to a level of half-height of a maximum. As it is seen the received distribution has a character of a spatial maximum having the form resembling an ellipsoid. The size along a direction "X" was about 50 cm along a direction " Y " - 70 cm and along a direction " Z " - about 100 cm. The maximum size of sections on a level 0.5 from a maximum was about 120 cm and was observed in a plane "YZ".

The analysis of the reasons of broadening of observable distribution has shown that the deterioration of the spatial resolution is connected with the limitation of a signal frequency spectrum that results in reasonably large lengths L_k of accepted signals. The same picture was observed in the computer experiments. The valuations have shown that under the conditions of the experiment the coherence length of signals was about 1 m that is close to the size of the received spatial resolution.

In the conclusion it should be pointed out one essential difference of the given principle of getting spatial information from the known ones. The difference is that 4 of receivers of radiation for obtaining 3-D space of distributions of radiation's sources with large (>> 4 resolute elements) are enough. While known principles of passive mapping of the spatial information, on one of which sight is based require the quantities of receivers equal or exceeding the necessary quantity of elements or resolution. The physical basis for such a difference is to our opinion that the principle given can be realized only with incoherent signals not possessing the period of recurrence, the autocorrelation function of which has one maximum. The known principles will be also realized with periodic signals, autocorrelation functions of which have periodic character.

Also it should be noted the most probable areas of use of the given principle as we see it. First of all they are acousto- and radio-thermometry as well as applications requiring determinations the site of aperiodic perturbations such as earthquakes signals of the acoustic emission previous to the destruction of designs stormy lightning, etc.

REFERENCES

1. M. Born, E. Wolf, Principles of optics, Pergamon press, 1964.
2. H.S. Carslaw, J.C. Jaeger, Conduction of heat in solids, Oxford at the Clarendon Press.
3. B.F. Corn, B.C. Hassell, and F.J. Keltonie, J. Acoust. Soc. Am., V 37, 523 (1965).

ACOUSTICAL IMAGING AND POINT PROCESSES II : EXACT SOLUTION

J. M. Richardson, G. Flesher and S. Isakson

Department of Electrical and Computer Engineering
University of California, Santa Barbara, CA 93106 U.S.A.

and

G. Wade and R. Duarte

Centro de Investigaciones en Optica, A.C.
Apdo. Postal 1 - 948, 37000 Leon, GTO MEXICO

ABSTRACT

There are many situations in acoustical imaging (active or passive) where the possible sources or scatterers to be detected (or discriminated against) are points, or, more generally, are represented by models in which a point process is embedded. A point process is a random set of points (random both in number and positions) in an appropriate state space. Examples of acoustical imaging problems involving point processes (in the static case) may be found in non destructive evaluation (including acoustical emission), underwater surveillance (e.g., bearing estimation), medical imaging (e.g., detection of echogenic nodules in breast tissues), etc.

At a much earlier symposium in the present series, a paper [1] with the same general title was presented, where several problems were considered but with approximate solutions. In particular, the point-process aspect of the problems was treated by a complex, but approximate, solution methodology. In this paper, the same problems are considered, but with exact solutions. Here the point-process aspect of the problems is treated by a relatively simple, but exact, solution methodology. A brief discussion of this methodology is given and numerical test examples based upon synthetic data are presented.

INTRODUCTION

As is well known, imaging with no a priori information yields images whose quality (i.e., resolution) is degraded by diffraction, whose severity is inversely related to the size of the aperture. A priori information used in the framework of Bayesian decision theory can significantly improve the image quality. In the absurd case in which the a priori information is complete one can obtain a perfect image without an imaging system. In the intermediate cases one can obtain significant improvement, assuming, of course, that the a priori information is limiting but not inconsistent with the likely images to be encountered, e.g., positivity.

The case of point process models is more restrictive than positivity and at the same time is relevant to a large number of categories of types of imagery in the real world. In an earlier volume of this series, a paper with the same general title as the present paper, gives examples of what we have in mind here [1].

In this earlier paper we applied some approximation techniques that provided significant enhancements but not probabilistically exact solutions. In the present paper we present such an exact solution for a time independent imaging problem with a point process model representing the a priori statistics. It turns out that the mathematical structure of this methodology is essentially identical to that of another inverse scattering problem involving the determination of the unknown boundary of an inclusion of known material [2].

The treatment is divided into four sections: Point Processes, Formulation of Inverse Scattering Problem, Solution, and Computational Example.

POINT PROCESSES

A point process is a random set of points in a given state space. The set contains a random number of points and each point is randomly positioned. The state space can be continuous with a variety of dimensions. For example, a random set of spherical voids in a solid would be represented by a point process in four-dimensional space corresponding to three position variables and the void radius. This space is, in general, called the single-object state space. The state of a set of many objects is represented by a set of points in the single-object state space, each one of which corresponds to the state of each object. There is a very large number of examples in the physical world represented by point processes and thus point processes are among the most pervasive random processes with the exception of Gaussian.

There is a difficulty associated with the representation of point processes in single-object state space. A conventional approach to the representation is shown in Fig. 1. The single-object state space is assumed here to be two-dimensional and an arbitrary position is given by the two-dimensional vector z with the components z_1 and z_2. The physical meaning of z_1 and z_2 is unspecified here. If the number of objects is not random and equal to 6, for example, then their random positions could be given by $z^{(1)}$, $z^{(2)}$, $z^{(3)}$, $z^{(4)}$, $z^{(5)}$, and $z^{(6)}$ as shown in Fig. 1. This choice of numbers is completely arbitrary.

The case in which the total number of objects is random is awkward to handle by the present conventional approach. A far more convenient and elegant approach is the occupation number representation shown in Fig. 2. The z-space is divided into cells and the occupation number of each cell is 0 or 1 in a manner that is clearly equivalent to the conventional representation in Fig. 1. In the case of a random number of objects in three-dimensional physical space we consider a regular array of cells whose cell centers form a cubic lattice spanning the localization volume.

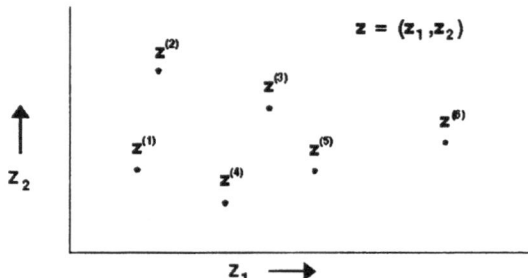

Fig. 1 Representative points in two dimensional single object state space.

Fig. 2 Occupation number representation. The representative points are unlabelled. The cells are labelled (addressed) by the value of z at the cell center. $\Gamma(z) = 1$ implies occupancy of the z-cell. $\Gamma(z) = 0$ implies vacancy of the z-cell.

FORMULATION OF INVERSE SCATTERING PROBLEM

We consider a random number of identical spherical objects randomly positioned in three-dimensional physical space in which **r** denotes possible cell-center positions in a cubic lattice spanning the localization volume. The volume of the cell is denoted by δr. The appropriate measurement model for longitudinal-wave pulse-echo scattering from the above point process is given by the expression

$$f(t, e) = \sum_{r} \delta r \, p(t - 2c^{-1} e \cdot r) \, \Gamma(r) + v(t, e) \tag{1}$$

where $f(t,e)$ is the waveform received at time t through the transducer whose incident wave-propagation direction is **e** and where $v(t,e)$ is the associated error. $\Gamma(r)$ is the occupation number of the cell centered at **r** ; c is the velocity of the longitudinal waves in the physical space; and the function $p(t)$ is the noiseless waveform obtained from a preliminary experiment in which a spherical scatterer is placed at the center of the localization volume. The term $-2c^{-1}$ **e•r** gives the time shift of the waveform when a spherical scatterer is in a non-central position within the localization volume. In the context of the far-field assumption the corresponding correction of the inverse square of the distance is negligible. In all cases we assume that the transducers are equidistant from the center of the localization domain.

To complete the description of the measurement model the a priori statistical properties of v and Γ must be defined. We assume that v and Γ are statistically independent of each other. The error v is assumed to be a set of Gaussian random variables with the properties

$$E v(t, e) = 0 \tag{2a}$$

$$E v(t, e) \, v(t', e') = \delta_{ee'} \delta_{tt'} \sigma^2 \, . \tag{2b}$$

The values of the characteristic function at two different positions are assumed to be statistically independent with $\Gamma(r)$ taking the values 0 and 1 with probabilities $1 - P$ and P, respectively. In the present treatment we will assume that P is independent of **r** and thus $\Gamma(r)$ is an example of a stationary non-Gaussian random process.

SOLUTION

We first calculate the joint probability function $P(\Gamma,v)$ for the characteristic function $\Gamma(r)$ for all points r on the lattice and the measurement error $v(t, e)$ for all t and e values. The statistical assumptions, discussed in the previous section, imply the relation

$$\log P(\Gamma,v) = -\frac{1}{2\sigma^2} \sum_{t,e} v^2(t,e) + \sum_r [\Gamma(r) \log P + [1 - \Gamma(r)] \log(1-P)] \tag{3}$$

in which an ignorable additive constant has been neglected. Our procedure is to maximize the above expression with respect to Γ and v while regarding the measurement model as a set of constraints. Using a somewhat nonstandard form of the Lagrange multiplier method for handling constraints we obtain the following variational function

$$\phi \equiv \phi(\Gamma, v, w, f)$$

$$= \log P(\Gamma, v) - \sum_{t,e} w(t,e) \left[f(t,e) - \sum_r \delta r p(t - 2c^{-1} e \cdot r) \Gamma(r) - v(t,e) \right] \tag{4}$$

where $w = w(t, e)$ is the Lagrange multiplier vector. It is to be noted that setting the variation of ϕ with respect to w equal to zero implies the measurement model (1). Thus, the vanishing of the variations with respect to Γ, v and w implies a maximum with respect to Γ and v constrained by (1).

Our procedure is first to maximize ϕ with respect to Γ and v keeping w fixed. It is possible to perform this maximization analytically with the result

$$\phi(\hat{\Gamma}, \hat{v}, w, f) \equiv \phi(w, f)$$

$$= \sum_{t,e} \left[\frac{1}{2} \sigma^2 w^2(t,e) - f(t,e) w(t,e) \right] + \sum_r g[\lambda(r)] \tag{5}$$

where the function $g(\lambda)$ is given by

$$g(\lambda) = \frac{1}{2} [\lambda + \log P + \log(1-P)] + \frac{1}{2} |\lambda + \log P - \log(1-P)| \tag{6}$$

and λ is defined by

$$\lambda(r) = \sum_{t,e} \delta r p (t - 2c^{-1} e \cdot r) w(t,e). \tag{7}$$

It can be readily shown that ϕ is a convex function of w, i.e.

$$\phi(\beta_1 w_1 + \beta_2 w_2, f) \leq \beta_1 \phi(w_1, f) + \beta_2 \phi(w_2, f) \tag{8}$$

where β_1 and β_2 are positive real constants subject to the condition $\beta_1 + \beta_2 = 1$ and where w_1 and w_2 are any two values of the vector w. Thus, a minimum must exist and relative minima in other locations cannot exist. However, this minimum may not be unique, but then the non-uniqueness must be of a special kind. Informally speaking, if non-unique, the minimum must be like a flat region at the "bottom of the valley" and this region must have a convex boundary.

In any case, if a unique minimum exists, then the minimization ψ (f, w) on $\hat{w}$ yields a best estimate w from which the corresponding estimate of the characteristic function is given by the relation

$$\hat{\Gamma}(r) = 1 \left[\hat{\lambda}(r) + \log P - \log(1-P) \right] \tag{9}$$

where $1[\cdot]$ is the unit step function. In the last expression λ is giving by substituting $\hat{w}$ into (7). This minimization must be carried out by computational means.

COMPUTATIONAL EXAMPLE

Several test runs were made for the two-dimensional case using synthetic test waveforms $f(t, e)$ derived on the assumption that the spherical scatterers are replaced by solid circular cylinders with their axes in the z-direction and with their cross-sections in the xy-plane. Various levels of noise were added. Four simulated transducers were assumed to lie in the xy-plane aimed at the center of the localization volume (now replaced by a suitable two-dimensional localization area). The incident directions, taking the two-dimensional form $e = e_x \cos \theta + e_y \sin \theta$, were assumed to have the directions given by $\theta = 0°$, $45°$, $90°$, and $135°$. The function p(t) was assumed to be a sinusoid with a Gaussian envelope. The grid in the localization area was assumed to be 11 x 11. In each scattered waveform 80 data point were taken. The a priori occupancy probability was assumed to be 0.05.

Assuming two adjacent occupied cells in the synthetic data we obtained the deduced images for SNR = 10 dB in Fig. 4 and for SNR = 4 dB in Fig. 5. Both results are to be compared with the true image presented in Fig. 3. The perfect deduced image in Fig. 4 is indeed gratifying and the imperfect deduced image in Fig. 5 is not unexpected.

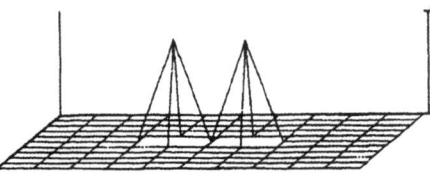

Fig. 3 Image that would be reconstructed from perfect data and no noise. The grid intersections represent the positions of the cell centers of Fig. 2. The synthetic data assumed that only two cells, two lattice lengths appart, were occupied.

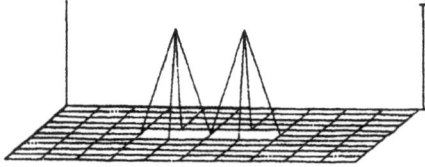

Fig. 4 Image reconstructed with simulated data for an SNR of 10 dB.

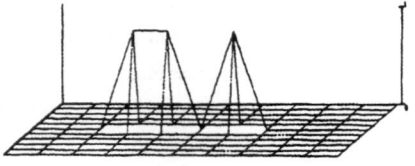

Fig. 5 Image reconstructed with simulated data for an SNR
of 4 dB.

REFERENCES

1. J. M. Richardson and K. A. Marsh, "Acoustical Imaging and Point Process," Acoustical Imaging, Vol. 15, pp. 615-633, (1987).

2. J. M. Richardson, "The Inverse Problem for the Scattering of Elastic Waves from Inclusions with Unknown Boundaries," 1982 Ultrasonics Symposium, pp. 985 - 987.

LOCATION AND SHAPE RECONSTRUCTION VIA DIFFRACTED WAVES AND CANONICAL SOLUTIONS

T. Scotti and A. Wirgin

Laboratoire de Mécanique et d'Acoustique
31 chemin Joseph Aiguier 13402 Marseille cedex 20, France

INTRODUCTION

This work deals with the location and shape reconstruction of a penetrable body from measurements of the scattered field when the body is exposed to a plane acoustic wave. The basic assumptions are : 1) the measured (not necessarily far) field is available as concerns both phase and amplitude, on a part or on the totality of a surface completely enclosing the body, 2) the scattering surface is penetrable (transmission boundary conditions), 3) the spaces outside and within the body are filled with linear, homogeneous, isotropic fluids, 4) the incident monochromatic scalar field is that of a plane longitudinal wave such that the incident wavevector lies in a plane perpendicular to the generators of the boundary, the latter being cylindrical and of infinite extent in one direction.

The inverse problem will be solved using a new approach to the direct problem, the local approach, which has the following characteristics : for each scattering direction, and if the shape of the body is not too different from a circular cylinder with center at the origin of the laboratory system, we assume that we can use the exact solution of the canonical problem of diffraction of a plane wave by a particular circular cylinder to compute an approximation of the diffracted field by the real body.

Thus, the inverse problem simply reduces to the search, in each scattering direction, for the radius of a circular cylinder having the same known composition as that of the real body which gives the same scattered field as the measured scattered field, and to the identification of this radius with the local radius of the unknown body.

The same method has already been employed for shape reconstruction of acoustically soft and hard bodies (Scotti and Wirgin, 1995 a,b).

PROBLEM INGREDIENTS

The exterior fluid medium, noted M_0, is homogeneous and unbounded. The unknown cylinder (occcuped by medium M_1) is also homogeneous, fluid filled, and bounded by a surface whose trace in the xy plane is Γ. Its axis is the z axis of the $Oxyz$ cartesian coordinate system where O is assumed, for convenience, to be located within Γ. The incident wave vector lies in the Oxy plane so that the wavefields do not depend on z ($2D$ problem). The $e^{-i\omega t}$ time dependence is omitted. Ψ^i will represent the incident field, Ψ^0 and Ψ^1 the total fields respectively in M_0 and M_1. These fields do not depend on z. Ψ^0 and Ψ^1 are square integrable in M_0 and M_1, governed by the Helmholtz equations ($(\Delta + k_j^2)\Psi^j = 0$ in M_j with k_j the wave number in M_j $j = 0,1$), satisfy the outgoing wave condition at infinity (as concerns Ψ^0) and obey the transmission boundary conditions on Γ:

$$\begin{cases} \alpha_0 \Psi_0 = \alpha_1 \Psi_1 \\ \beta_0 \partial_n \Psi_0 = \beta_1 \partial_n \Psi_1 \end{cases} \tag{1}$$

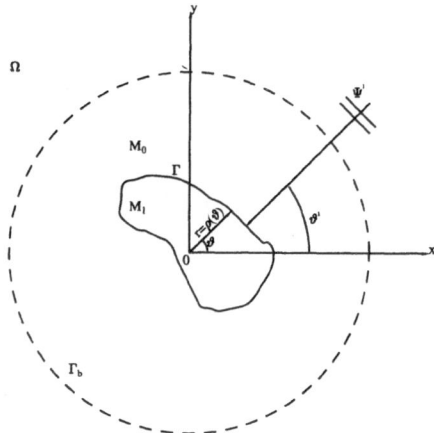

Figure 1. Cross section of scattering configuration

with $\alpha_1 \, \alpha_0 \, \beta_0 \, \beta_1$ known constants. For the inverse problem, the measured field Ψ^0 is supposed to be known at given points on a circular surface Γ_b (whose radius is $r_b \geq max\{\rho(\vartheta)\}$, with $r = \rho(\vartheta)$ the parametric equation description of Γ_b) enclosing the body. This data is actually computed by the rigorous Waterman theory (Waterman, 1969; Bolomey and Wirgin, 1974; Wirgin, 1978).

Thus, given Ψ^0 on Γ_b, ϑ^i, k_0 and k_1, the objective is to fully or partially, determine the bounding curve Γ of the body.

FIELD REPRESENTATION

It can be shown (Bolomey and Wirgin, 1974) from Green's theorem, irregardless of the shape of the scattering cylinder, that the total wave field can be represented in $\Omega^+ = \{r \geq \bar{r} = Max[\,\rho(\vartheta)\,; 0 \leq \vartheta < 2\pi]\,\}$ by :

$$\Psi^0(r, \vartheta) = \Psi^i(r, \vartheta) + \sum_{n=-\infty}^{+\infty} A_n \, H_n(k_0 r) \, e^{in\vartheta}, \tag{2}$$

with

$$\Psi^i(r, \vartheta) = \sum_{n=-\infty}^{+\infty} J_n(k_0 r) \, e^{-in(\vartheta' + \frac{\pi}{2})} = e^{-ikr\cos(\vartheta - \vartheta')}. \tag{3}$$

Similarly, in $\Omega^- = \{r \leq \underline{r} = Min[\,\rho(\vartheta)\,; 0 \leq \vartheta < 2\pi]\,\}$,

$$\Psi^1(r, \vartheta) = \sum_{n=-\infty}^{+\infty} C_n \, J_n(k_1 r) \, e^{in\vartheta}, \tag{4}$$

wherein J_n and H_n are the n-th order Bessel and Hankel functions of the first kind respectively.

DETERMINATION OF THE RADIUS OF A CIRCULAR CYLINDER WITH CENTER AT THE ORIGIN

Forward problem

The problem is to determine the field on Γ_b for known $\Gamma, \vartheta^i, k_1, k_0$. Let Γ be the circular cylinder of radius a. The starting points are Eqs.(1)-(4) wherein $\rho(\vartheta) = a$. Introducing Eqs.(2)-(4) into Eq.(1) gives :

$$A_n(a, \vartheta^i) = \frac{\alpha_0\,\beta_1\,k_1\,J_n(k_0a)\,\dot{J}_n(k_0a) - \beta_0\,\alpha_1\,k_0\,\dot{J}_n(k_0a)\,J_n(k_1a)}{-\alpha_0\,\beta_1\,k_1\,H_n(k_0a)\,\dot{J}_n(k_1a) + \beta_0\,\alpha_1\,k_0\,\dot{H}_n(k_0a)\,J_n(k_1a)}\; e^{-in(\vartheta' + \frac{\pi}{2})}, \quad (5)$$

$$\Psi^0(r = r_b, \vartheta) = \Psi^i(r_b, \vartheta) + \sum_{n=-N}^{+N} A_n(a, \vartheta^i)\,H_n(kr_b)\,e^{in\vartheta}. \quad (6)$$

wherein $\dot{H}_n(\xi)$ and $\dot{J}_n(\xi)$ are the derivatives of J_n and H_n with respect to ξ and we have reduced the infinite series in Eq.(2) to a finite series for computational purposes.

Inverse problem

The problem for the same circular cylinder body is to determine a, given k_1, k_0 and Ψ^0 on Γ_b. For a particular scattering direction ϑ^i, we match the expression of Ψ^0, given by Eq.(6), with the given data Ψ^0:

$$\underbrace{\Psi^0(r_b, \vartheta)}_{\text{given data}} - \underbrace{\left[\Psi^i(r_b, \vartheta^i) + \sum_{n=-N}^{+N} A_n(\eta, \vartheta^i)\,H_n(kr_b)\,e^{in\vartheta}\right]}_{\text{theoretical values}} = 0\,;\; l = 1, 2 ... L. \quad (7)$$

Since the A_n are known, analytically speaking (see Eq.(5)) to within the single parameter η, which is the radius of the circular cylinder that gives, for this particular scattering direction, the same diffracted field as the data, Eq.(7) enables one to determine η.

Remarks

1) From a theoretical point of view the inverse problem is trivial in that Eq.(7) should give rise to the expected solution $\eta = a$. But, from a numerical point of view, the problem is much less trivial in that there are (measurement or calculational) errors in the first member of the equation and the series in the second member is necessarily limited to a finite number of terms;

2) if we take $2L + 1$ measured samples of the diffracted field at angles ϑ^l, we have to solve a system of $2L + 1$ non-linear equations (one for each sample) in $2L + 1$ unknowns (one "radius" η^l for each scattered direction ϑ^l);

3) these equations are not coupled in terms of η^l, so that the system can be solved equation-by-equation instead of globally;

4) η^l should be real, but, because errors always exist in numerical computations, the solution η of Eq.(7) is, in fact, complex;

5) we keep only the real part of η^l to test our results;

6) for each equation, the solution is not unique and we have developed an algorithm to eliminate spurious roots.

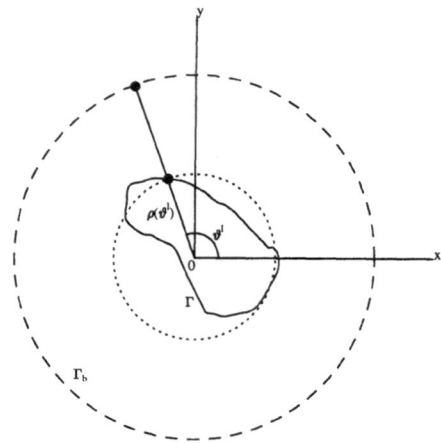

Figure 2. Construction of the local canonical body (circular cylinder)

DETERMINATION OF THE SHAPE OF A NON CIRCULAR CYLINDER

Direct problem

The process is the same as previously, except that now we do not dispose of the solutions (i.e., the A_n) of the forward problem. Therefore we make the following approximation of the forward scattering problem : if the body is not much different from a circular cylinder, we assume that, for each particular scattering direction, the field is given by Eq.(6) wherein the A_n are those of an appropriate circular cylinder with center at the origin. More specifically, for a given point on Γ_b, the field diffracted by the real body is assumed to be approximately the same as that diffracted by a circular cylinder of radius η^l where :

$$\underbrace{\eta^l}_{\text{circular cylinder}} = \underbrace{\rho(\vartheta^l)}_{\text{local radius of body}} \quad ,$$

The previous expression (6) is now replaced by the approximation :

$$\Psi^0(r = r_b, \vartheta^l) \simeq \Psi^i(r_b, \vartheta^l) + \sum_{n=-N}^{+N} A_n^l(\eta^l, \vartheta^i) \, H_n(kr_b) \, e^{in\vartheta^l} \ , \ l = 1, 2, ...L, \qquad (8)$$

wherein the $A_n^l(\eta^l, \vartheta^l)$ are given by Eq.5 with a replaced by η^l.

Inverse (shape reconstruction) problem

The inverse problem reduces to
1. searching, for each scattering direction ϑ^l, the radius η^l of a circular cylinder which gives the same diffracted field as that given by the measurements and
2. associating η^l with the local radius of the body :

$$\Psi^0(r_b, \vartheta^l) - \left[\Psi^i(r_b, \vartheta^l) + \sum_{n=-N}^{+N} A_n(\eta^l, \vartheta^l) \, H_n(kr_b) \, e^{in\vartheta^l} \right] \simeq 0 \ ; \ l = 1, 2...L. \qquad (9)$$

If this is done for a set of measurements in an angular sector (or all around the body) then the shape function $\rho(\vartheta)$ is thereby partially (or totally) reconstructed.

COMPUTATIONAL PROCEDURES

The Bessel and Hankel functions with (real) arguments $k_0 b$ were computed by means of the IMSL (IMSL, 1991) subroutines DBSJS and DBSYS. The Bessel and Hankel functions having (complex) arguments k_1 were computed by means of the IMSL (IMSL, 1991) subroutines DCBJS and DCBYS .

The nonlinear systems Eq.(7) or (9) were solved, equation by equation, by means of the (IMSL, 1991) subroutine DZANLY. The latter computes the complex zeros of a complex function by the so-called Müller method which, presumably, is a variant of the Newton-Raphson scheme. It is important to point out that each equation possesses an infinite number of roots. Practically, the number is finite and user-specified for each call to DZANLY. This number must be chosen large enough to make sure that one has not left out the sought-for root.

We chose (unless specified otherwise) : $L = 40$, $N = 6$, $k_0 = 1.5$ and $k_1 = 1 + 1.1i$, $r_b = 2$ and $\alpha_0 = 1$, $\alpha_1 = 1.5$ $\beta_0 = 1$, $\beta_1 = 1.5$, and tested the algorithm for various decentered circles and centered ellipses.

POST PROCESSING TO CHOOSE THE 'RIGHT' PROFILE

We first eliminated profiles for which the real part of $\rho(\vartheta^l)$ is negative, unreasonably large (or larger than Γ_b in the near field measurements), then we proceeded as follows. For measurements in the near field, we chose the profile corresponding to the smallest imaginary part of $\int \rho(\vartheta)d\vartheta / \int d\vartheta$ whereas for measurements in the the far field, we used the scattering diagram to get an idea of the larger dimension of the body (by measurements of the maximal amplitude of the main lobe), thus eliminating larger bodies, and then chose profiles corresponding to the smallest imaginary part of $\int \rho(\vartheta)d\vartheta / \int d\vartheta$.

RESULTS

We employed "measurements" in either the near ($r_b \ll \infty$) or far field ($r_b \to \infty$) and chose $\vartheta^l = \vartheta^i$ (backscattering), the incident angles being equally spaced all around Γ_b. Figs.3 and 4 show that the reconstructions using near or far field measurements are of comparable accuracy and quite acceptable, even for an ellipse of rather high excentricity. Fig.5 shows that the algorithm not only succeeds in reconstructing the shape but also in locating the body with respect to the laboratory reference system.

CONCLUSION

The local canonical body approximation for the forward problem enables to locate and reconstruct the shape of a penetrable cylinder in a simple manner by minimizing a cost functional involving only the measured and trial fields on the measurement surface, with no need of penalisation terms as, for instance, in (Angell, et al., 1986).

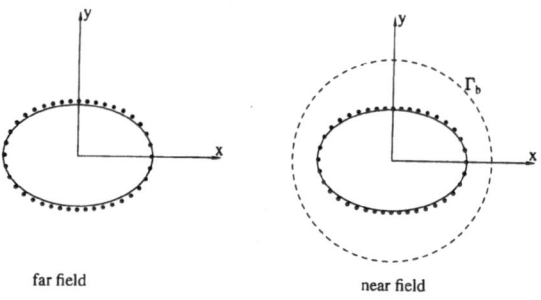

far field   near field

Figure 3. Reconstruction using backscattering measurements for an ellipse with horizontal and vertical semi-axis $a_h = 1.5$, $a_v = 1$. The actual boundary is the full line curve, the reconstructed boundary is the set of points.

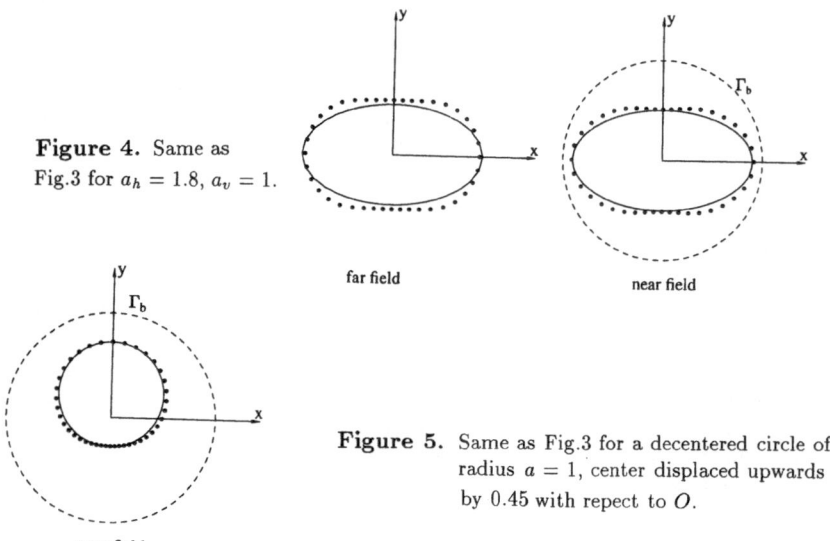

Figure 4. Same as
Fig.3 for $a_h = 1.8$, $a_v = 1$.

far field   near field

near field

Figure 5. Same as Fig.3 for a decentered circle of radius $a = 1$, center displaced upwards by 0.45 with repect to O.

References

Angell, T.S., Kleinman, R.E., and Roach, G.F., 1986, An inverse transmission problem for the Helmholtz equation, *Inverse Prob.*, 3:1987.

Bolomey J.C. and Wirgin A., 1974, Numerical comparaison of the Green's function and the Waterman and Rayleigh theories of scattering from a cylinder with arbitrary cross section, *Proc.IEE*, 121:794.

IMSL, 1991 User's manual Fortran subroutines for mathematical applications: MATH/LIBRARY, special functions, Version 2.0 *Ref.SFLB and MALB-USM-PERFECT-EN9104-2.0, IMSL, Houston.*

Scotti T. and Wirgin A., 1995a, Shape reconstruction using diffracted waves and canonical solutions, *Inverse Prob.*, to appear.

Scotti T. and Wirgin A., 1995b, Location and shape reconstruction of a soft body by means of canonical solutions and measured scattered sound fields, *Comptes Rend.Acad.Sci. II*, 320, to appear.

Waterman P.C, 1969, Scattering by dielectric obstacles, *Alta Freq.*, 38:348.

Wirgin A., 1978, New theoretical approach to scattering from a periodic interface, *Opt.Commun.*, 27:189.

INFLUENCE OF THE SCATTERING DATA REDUNDANCY ON UNIQUENESS AND STABILITY IN RECONSTRUCTION OF STRONG AND COMPLICATED SCATTERERS

V.A. Burov and O.D. Rumiantseva

Moscow State University, Faculty of Physics, Department of Acoustics
Moscow, 119899, Russia

Our goal is to draw attention to connection between the uniqueness and stability of solution of the acoustical inverse scattering problem and experimental data redundancy for arbitrary strong and complicated scatterers without any mathematical strictness and completeness.

NONUNIQUENESS AND INSTABILITY OF STRONG SCATTERER RECONSTRUCTION IN POINTS-LIKE DESCRIPTION

The simplest way to describe a scatterer $\varepsilon(\mathbf{r})$ is to represent it as a combination of S point scatterers localized at points $\mathbf{r}_n$, $n = \overline{1,S}$: $\varepsilon(\mathbf{r}) = \sum_{n=1}^{S} \varepsilon(\mathbf{r}_n)\delta(\mathbf{r} - \mathbf{r}_n)$. Proportion between the number S of the scatterer parameters that has to be reconstructed and the total number D of experimental scattering data is important for analysis of the uniqueness and stability of the $\varepsilon(\mathbf{r})$ reconstruction for such problem. Let U_0^α be an incident field with parameters (probing direction and the frequency) that are described by the index $\alpha = \overline{1,A}$, $A = X\Omega$, where X and Ω are the total number of the probing directions and frequencies correspondingly. Scattered field u_{sc}^α are observed for B directions or for B observation points at each fixed U_0^α, i.e. D=AB.

Reconstruction Nonuniqueness for Nonredundant Set of the Scattering Data

Let's suppose that experimental scheme is organized in such a way that weak scatterer is uniquely reconstructed from the nonredundant scattering data (D=S). In the same time the *strong scatterer* $\varepsilon(\mathbf{r})$ is reconstructed nonuniquely from such data. [1] It is simple to illustrate this fact by an example of the "invisible" scatterer existence. This "invisible" scatterer existence can be shown by using next expression:

$$\hat{Q}_{out}^+ u_{sc}^\alpha = \hat{Q}_{out}^+ \hat{Q}_{out} [\hat{E} - \hat{\varepsilon}\hat{Q}_{in}]^{-1}\hat{\varepsilon}U_0^\alpha.$$

Here $\hat{E}$ is a unite operator. Operators $\hat{Q}_{out}$ and $\hat{Q}_{in}$ have kernels in the Green function form and act from the scattering domain to the observation domain and within the scattering domain correspondingly. $\hat{Q}_{out}^+$ is the Hermitian conjugated operator and $\hat{\varepsilon}$ is a diagonal operator combined from the point scatterers amplitudes $\varepsilon(\mathbf{r}_n)$. In this scheme operator $\hat{Q}_{out}^+\hat{Q}_{out}$ is degenerate, the order of this operator is equal to S and the rang is equal to B. The scatterer $\varepsilon(\mathbf{r}) = \varepsilon_{inv}(\mathbf{r})$ is "invisible" if

$u_{sc}^{\alpha} = 0$ for all D measurements i.e. following relation for the secondary sources complex amplitude should take place:

$$J(r) = [\hat{E} - \hat{\varepsilon}_{inv}\hat{Q}_{in}]^{-1}\hat{\varepsilon}_{inv}U_0^{\alpha} = \sum_{i=1}^{S-B} a_i^{\alpha}\psi_i(r), \qquad \alpha = \overline{1,A}.$$

Here $\{\psi_i(r_n)\}_{n=1,S}^{i=\overline{1,S-B}}$ is S-B linearly-independent eigenvectors of the operator $\hat{Q}_{out}^{+}\hat{Q}_{out}$ corresponding to the zero eigenvalues; $\{a_i^{\alpha}\}$ is a set of expansion coefficients. Therefore for each fixed point r_n one can get:

$$\varepsilon_{inv}(r_n) = \left[\sum_{i=1}^{S-B} a_i^{\alpha}\psi_i(r_n)\right] // \left[U_0^{\alpha}(r_n) + \sum_{i=1}^{S-B} a_i^{\alpha}\varphi_i(r_n)\right], \qquad \varphi_i \equiv \hat{Q}_{in}\psi_i. \tag{1}$$

Here notation // means the division of vector component by vector component one by one. The values $\{a_i^{\alpha}\}_{i=1,S-B}^{\alpha=\overline{1,A}}$ can be found from the equation system obtained by equating the fraction (1) for different pairs ($\alpha = \alpha_1$, $\alpha = \alpha_2$):

$$\sum_{i=1}^{S-B} Z_i^{\alpha_2}a_i^{\alpha_1} - \sum_{j=1}^{S-B} Z_j^{\alpha_1}a_j^{\alpha_2} + \sum_{i=1}^{S-B}\sum_{j=1}^{S-B} N_{ij}a_i^{\alpha_1}a_j^{\alpha_2} = 0, \ \alpha_2 = \overline{1,A}, \ \alpha_2 \neq \alpha_1 \text{ at one fixed } \alpha_1. \tag{2}$$

The values $Z_i^{\alpha} = U_0^{\alpha}(r_n)\psi_i(r_n)$; $N_{ij} = \psi_i(r_n)\varphi_j(r_n) - \psi_j(r_n)\varphi_i(r_n)$ are known. For the nonredundant data set AB=S and the number (A-1)S of independent equations (2) is equal to the number (S-B)A of independent unknown coefficients $\{a_i^{\alpha}\}$. Simultaneously Eq.1 allows to find the "invisible" scatterer value corresponding to the given defined set $\{a_i^{\alpha}\}$.

Thus, the number of the "invisible" scatterers is defined by and coincides with the number of solutions $\{a_i^{\alpha}\}$ of the bilinear equation system (2) in the given discrete framework. The number of solutions can be evaluated from two following requirements for physically realizable passive scatterer. First, the imaginary part of the refraction coefficient must have the certain sign corresponding to the wave absorption into scatterer. Secondly, for any marked out subdomain R_{sub} of the scattering domain R the total energy of the field $\hat{Q}_{in}\hat{\varepsilon}U_0$ scattered within R as the result of *single* scattering can not exceed the total energy of the field U_0 stored in R before this single act. It means that

$$\|U_0\|_{L_2 \ (\forall \ R_{sub} \in R)} \geq \|\hat{Q}_{in}\hat{\varepsilon}U_0\|_{L_2 \ (\forall \ R_{sub} \in R)}, \quad \text{or} \quad \|\hat{Q}_{in}\hat{\varepsilon}\|_{L_2 \ (\forall \ R_{sub} \in R)} \leq 1$$

at $\varepsilon(r) \equiv 0$, $r \notin R_{sub}$;

$\|.\|$ - the L_2-space norm. It should be note that this requirement does not inhibit both the accumulation of the scattered field energy within R_{sub} as a result of *multiple* scattering (as in semi-opened resonators) and the *local* energy focusing under single scattering (as in concave mirrors).

Let's introduce some space of parameters that can completely describe scatterers. The certain domain in this space (with boundary schematically shown on Fig.1 as a circle) corresponds to the class of physically realizable scatterers. Each "invisible" scatterer is depicted as a point in this space. Around each of these points there exists some subdomain that corresponds to the scatterers of intermediate force with respect to the given "invisible" scatterer as a ground. The boundaries of such subdomains are shown on Fig.1 as closed lines. These subdomains have no intersections. Really, for the intermediate force scatterers the nonredundant inverse problem has a unique solution. For example, the iterative process that begins with the parameter values of the fixed "invisible" scatterer leads to the unique solution in the class of the intermediate force scatterers with respect to the given "invisible" one[1] (oriented lines on Fig.1).

Suppose that the "volumes" of such subdomains are approximately equal to each other then the estimation of the number of scatterers ("candidates" for the nonredundant problem solution without the additional requirement on the scatterer force) is limited by the ratio of the volume of domain for the physically realizable scatterers to the volume of one subdomain for intermediate force scatterers. For the large dimension ($10^4 - 10^8$) parameter space this ratio can be very high.

It should be underlined once more that in the class of the intermediate force scatterers with respect to the homogeneous "0" medium the system (2) has the unique solution $\{a_i^\alpha\} \equiv 0$ $\forall i, \forall \alpha$, that corresponds to $\varepsilon_{inv}(r_n) \equiv 0$ $\forall r_n$. Because of this fact the other "invisibles" are the strong scatterers with respect to the homogeneous medium and to each other. They cardinally distort the incident field within the scattering domain creating the additional wave phase displacement comparable with or even grater than 2π.

Role of Redundancy: Anomalous Errors

The situation with the solution uniqueness can be clearly described by conception of discrepancy Dis. For example, $\text{Dis} = \sum_{d,d'=1}^{D} (u_d^{sc} - \overline{u}_d^{sc})^+ K_{dd'}^{-1} (u_{d'}^{sc} - \overline{u}_{d'}^{sc})$, where u_d^{sc} and $\overline{u}_d^{sc}$ are the scattering data observed experimentally and obtained for the certain reference scatterer with variable parameters correspondingly, $K_{dd'}$ is the matrix of dispersion of estimation. Let the horizontal axis schematically represents all multidimensional space of parameters describing the physically realizable scatterers; the state "0" corresponds to the homogeneous nonscattering medium. The value $W=1/(1+\text{Dis})$ is placed along the vertical axis. $W=1$ at $\text{Dis}=0$.

Reconstruction of the multipoint scatterer on the base of the nonredundant scattering data is represented on Fig.2. For all possible solutions (the true scatterer and false ones) strictly obeying to the experimental scattering data $W=1$, $\text{Dis}=0$. The role of redundancy of experimental data is that it makes the solution of inverse problem unique without restriction on the scatterer force. However, the way of removing of nonuniqueness with redundancy increasing is very important for practice. Thus addition of one supplementary measurement to the nonredundant data already makes the system (2) overfilled and compatible with guaranty only for $\varepsilon_{inv} \equiv 0$. Then $\text{Dis}=0$ only for the true scatterer. False scatterers that had $\text{Dis}=0$ for the nonredundant data now get small nonzero values $\text{Dis} \neq 0$ (the continuous line on Fig.3). Consequently, the solution uniqueness is formally provided. Nevertheless, if character of the scattering data is very close to nonredundant (for example it can happen, when the additional receiver is placed near one of already available receivers) then the false "peaks" of the value W become only slightly lower than the true "peak". Therefore even small errors of measurement can easily lead to "jumping" from the true solution to the false one. Because picture of false "peaks" in the scatterer parameter space can be very complicated, these "jumps" can lead to very different false "peaks". Such "jumping" gives the extremely unstable solution and leads to so-called anomalous errors of the scatterer reconstruction and to scatterer reconstruction nonuniqueness as a practical consequence. In the case of anomalous errors the characteristics of a false reconstructed scatterer can be anyhow strong different from the true scatterer parameters.

The natural way to fight with the nonuniqueness and instability phenomena is increasing the redundancy coefficient of the scattering data. In this case the methods to reveal the true scatterer may be different. For example, from $D=S+S_{add}$ scattering data (S is the nonredundant number) C_D^S subsets of nonredundant scattering data can be composed (C_D^S is the binomial coefficient). For every subset of nonredundant data there exists picture of Fig.2 type where false "peaks" correspond to the different scatterer parameters in each subset and the true "peak" always corresponds to the same parameters for any subset (continuous and dotted lines on Fig.3). Compatibility of all nonredundant subsets for the true scatterer is provided by physics of the process. In principle one could find the true solution by comparing these pictures. However, on practice it is more convenient to use whole set of redundant data simultaneously. As the redundancy coefficient increases, false "peaks" of the value W are distributed more evenly and the ratio of their amplitudes to the true peak decreases.

INVERSE SCATTERING PROBLEM IN FUNCTIONAL DESCRIPTION

Comparison of Functional and Points-Like Descriptions

In the previous case of points-like scatterer description the secondary sources arise exclusively at fixed points. Because of this fact the number of the point degrees is the same as for scatterer as for secondary sources independently from the scatterer force. In the functional description the scatterer function $\varepsilon(r)$ is reconstructed for all continual manifold of arguments $r \in R$. Taking into consideration m-fold scattering the space spectrum of the secondary sources is m times wider than that of the scatterer [2,3]. Therefore to make the physical essence of the scattering process and its mathematical description consistent with each other two ways of sampling are possible. First way is to sample both the scatterer and the secondary sources in accordance to space spectrum width of the secondary sources. Second way is to introduce two sampling grids with different steps; one grid describes the scatterer and another (more frequent) does the secondary sources or the inner fields [4].

Let us suppose that the scatterer can be described with desirable accuracy by the finite sum $\varepsilon(r) = \sum_{s=1}^{S} \varepsilon_s \xi_s(r)$, where functional basis $\{\xi_s(r)\}$ is agreed with space spectrum of the scatterer (it could be for example Sinc-basis or Karhunen-Loève basis). To describe the secondary sources structure it is necessary to have more complete functional basis $\{\chi_n(r)\}$ corresponding to wider space spectrum of the sources

$$J(r) = \sum_{n=1}^{N} J_n \chi_n(r) .$$

For N basis functions there are N-B combinations that create invisible radiation *for our receiving system*. Redundancy of type (2) system is a condition of absence of "invisible" solutions: (N-B)A < S(A-1). Taking into account that for multiple number of rescattering the relation N>>S is valid and the numbers of probing and observation directions are generally approximately equal A≈B>>1 we have:

$$N < S+B \le AB+B \approx A^2 .$$

If at the same time the angle between nearest directions is not less than the Rayleigh's limit $2\pi / (k_0 L)$ (k_0 is the wave number) of angle resolution for aperture of scattering domain size L then the relation $N < S+B_{Rel}$ is a condition of uniqueness and stability of solution and the relation $N < (A_{Rel})^2$ is a condition of solution stability only. These simple conditions give restrictions on the permitted scatterer force and completeness that are essentially different for inverse problems in 2 and 3 dimensions. The latter relations remain valid if type (2) equations are rewritten for N-space points and restrictions on the width of scatterer space spectrum are taken into account. In that case the redundancy of whole system is restored.

So, the functional description allows not only to analyze the physical essence of the nonuniqueness and instability phenomena but also to establish the connection of these phenomena with the whole set of factors: the dimension of space , scatterer force, complication degree of scatterer space structure i.e. space spectrum width and space size of scatterer. It is known that in the functional description the conception of the scattering data redundancy is closely connected with the dimension of coordinate space in which the reconstructed scatterer is localized.

Two-Dimensional Inverse Scattering Problem

A number of theorems proved in the functional approach let state that two-dimensional **monochromatic** inverse scattering **problem** allows the unique solution for weak and intermediate force scatterers [5]. There is no unique solution for strong scatterers. Consequently, the situation with discrepancy is analogous to the situation with nonredundant scattering data in the points-like description (Fig.2). Such analogy is connected with the fact that the two-dimensional inverse scattering problem is functionally-nonredundant, i.e. the dimensions of the coordinate space and the independent scattering data space coincide. Nevertheless, the number of independent scattering data samples must be at least in m^2 times more than the number of scatterer samples [2]. This circumstance shows the difference between the functional scatterer description and points-like one. The redundancy of data samples is required to compensate effects of the reconstruction accuracy loss which takes place if the sampling- and functional models are not adequate.

At the same time the two-dimensional monochromatic problem solution becomes unstable already for the intermediate force scatterers with complicated space structure. The space spectrum of the secondary sources for these scatterers spreads over boundary of $2k_0$-radius circle and therefore back scattering arises. The approximate (for the scatterers with sufficiently simple space structure) and strict (for scatterers with complicated one) Grinevich-Novikov algorithms[6,2] in the complex k-space formalism can be taken for clear illustration of these conclusions. So, for weak scatterers (m=1) the exact reconstruction stability takes place only under the condition

$$s_0 \le 2\sqrt{k_0^2 + L^{-2}} , \tag{3}$$

where L is the scatterer's linear size, s_0 is the domain radius of the scatterer space spectrum's concentration [6]. For tomography experiment the correction $L^{-2} << k_0^2$, so that the restriction (3) allows to extend the reconstructed scatterer space spectrum only slightly compared with the case of the back scattering absence ($s_0 < 2k_0$). Here there is an evident analogy with the analytical extension operation of the space spectrum of functions with finite support in the coordinate space. This operation has a very strong instability. For this reason the extension is, in practice, possible only for one or two samples in the frequency space. Similar situation occurs also for nonweak scatterers, though the estimation like (3) is, of course, different and depends on the number m defined by the scatterer force.

The more the secondary sources spectrum spreads over the $2k_0$-radius circle the more these instability effects. Indeed, every component of scatterer space spectrum can generate wave field spectrum components on the other combinative space frequencies both higher and lower [3]. This leads to appearance of fine scale structure of the wave field inside the scattering domain. These processes of space frequency migration amplify instability of the reconstruction of complicated inner field structure. The inner field reconstruction in explicit or implicit form (as an intermediate step of the scatterer reconstruction) is necessary independently of the method of inverse scattering problem solution and is omitted only for weak scatterers.

Similar situation can also be observed in the iterative solution method [3]. For each new iteration the sampling dimension of the solved problem (in which the scatterer- and secondary sources space spectra are evaluated) increases, and its conditionality becomes worse. The stability loss takes place when the secondary source space spectrum extends over the $2k_0$-radius circle.

Thus, in the two-dimensional monochromatic problem two classes of scatterers are reconstructed with sufficient accuracy. The first class is represented by weak scatterers which may have sufficiently wide space spectrum; the second class - by the intermediate force scatterers, space spectrum of which is mainly localized within the narrow frequency diapason. In other cases the scatterer reconstruction attempts lead to the strong problem instability.

The stability increasing in the two-dimensional problem can be obtained only by using the **pulse** (multifrequencies) **regime**. It is the unique source to obtain the functionally-redundant scattering data. The pulse regime role may be illustrated by the two-dimensional variant of the Marchenko-Newton-Rose equation [7] generalized for the probing pulses with a finite duration in the approximation of the working frequency band with finite width [8].

Three-Dimensional Inverse Scattering Problem

Three-dimensional problem has the unique solution in both pulse and monochromatic regimes for any force scatterers if manifold of parameters of the functionally-redundant scattering data has nonzero measure [9,10,11]. For the monochromatic problem there is a reserve of functional redundancy conditioned by the space dimension (dimension of the full scattering data space is four-fold against three-fold coordinate space of scatterer). In the case of the pulse problem the redundancy is intensified by the frequency degree of freedom. The investigation of the appearance of the solution instability with increasing of scatterer size, contrast and degree of space structure complication could be done by using more perspective models. The Novikov-Nachman algorithm [12] in the monochromatic regime and the three-dimensional Marchenko-Newton-Rose equation generalized to the finite duration pulses[8] could be chosen for this.

Situation with discrepancy is analogous to that of redundant scattering data in the points-like description. The scatterer reconstruction in the three-dimension monochromatic problem is already possible by using the functionally-nonredundant scattering data for the two-dimensional manifold of the incident wave directions and the one dimensional manifold of the scattered field on an arc of circle. This functionally-nonredundant case is analog of the two-dimensional problem (Fig.2). If now the data arc is converted into a narrow band around it or the monochromatic regime is transited to the narrow-band pulse regime, then the discrepancy picture will correspond to Fig.3. In spite of the formal availability of the solution uniqueness the anomalous error possibility is very large. As the redundancy measure of the scattering data increases, the revealing of the true solution becomes easier. In each physical case the numerical values of the parameters for this stabilization process are different and in need of special analysis. In this connection it is important but sufficiently difficult to investigate solution stability of the additional equations of the characterization [12,13]. These equations give strong restrictions on the manifold of the redundant data because all these data were obtained for the same scatterer.

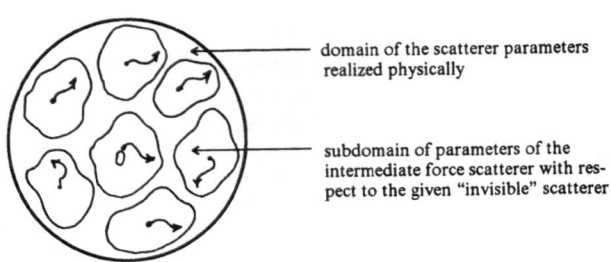

Figure 1. To the question on the number of "invisible" scatterers in nonredundant problem.

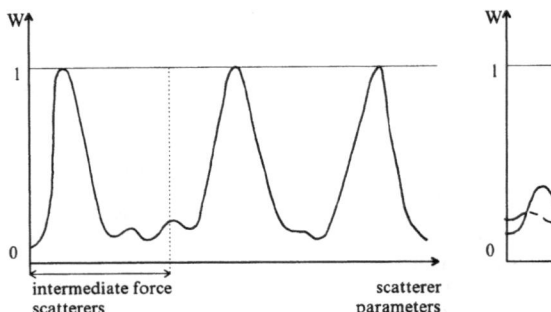

Figure 2. The discrepancy conduction for nonredundant scattering data.

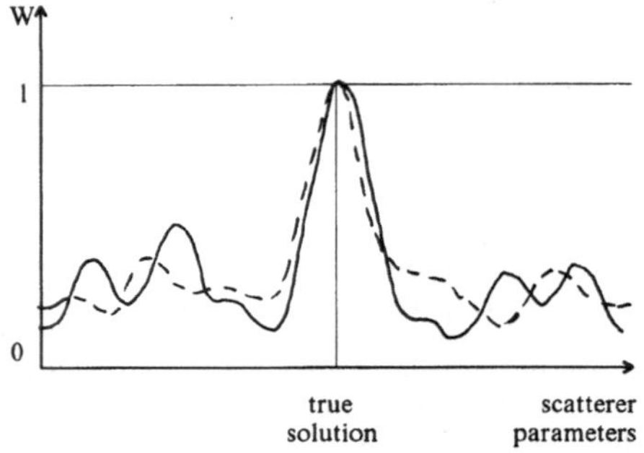

Figure 3. The discrepancy conduction for different sets of redundant scattering data

REFERENCES

1. V.A.Burov, M.N.Rychagov, and A.V.Saskovets, Account of multiple scattering in acoustic inverse problems of tomographic type, *in*: "Acoustical Imaging-19", H.Ermert and H.-P.Harjes, ed., Plenum Press, New York (1992).
2. V.A.Burov, and O.D.Rumiantseva, Solution of two-dimensional inverse problem of acoustic scattering based on functional-analytical methods II. Efficient application domain, *Acoust.Phys.* 39:419 (1993).
3. V.A.Burov, and O.D.Rumiantseva, The functional-analytical methods for the scalar inverse scattering problems, *in*: "Analit. Methods for Opt. Tomography", G.G.Levin, ed.,Proceed.SPIE, Washington (1992).
4. Benli Gu, Dongyun Deng, Inversion of acoustic velocity with moment multi-grid algorithm, *in*: "Acoustical Imaging-20", Y.Wei and B.Gu, ed., Plenum Press, New York (1993).
5. R.G.Novikov, Multidimensional inverse spectral problem for the equation $-\Delta\psi+(v(\mathbf{x})-E\,u(\mathbf{x}))\psi=0$, *Func.Anal. Appl.*22:263 (1988).
6. V.A.Burov, and O.D.Rumiantseva, The solution stability and restrictions on the space scatterer spectrum in the two-dimensional monochromatic inverse scattering problem, *in*: "Ill-Posed Problems in Natural Sciences", A.N.Tikhonov, ed., VSP/TVP, Moscow (1992).
7. D.E.Budreck, and J.H.Rose, Three-dimensional inverse scattering in anisotropic elastic media, *Inverse Probl.* 6:331 (1990).
8. V.A.Burov, E.E.Kasatkina, and O.D.Rumiantseva, Statistical estimations in inverse scattering problems, *in*: "Acoustical Imaging-22", P.Tortoli, ed., Plenum Press, New York (1996).
9. Yu. M.Berezansky, On uniqueness theorem in inverse problem of spectral analysis for Schrödinger equation, *Mosc. Math. Soc. Proc.* 7:3 (1958) [in Russian].
10. A.G.Ramm, Completeness of the products of solutions to PDE and uniqueness theorems in inverse scattering, *Inverse Probl.* 3:L77 (1987).
11. R. Weder, Global uniqueness at fixed energy in multidimensional inverse scattering theory, *Inverse Probl.* 7:927 (1991).
12. A.I.Nachman, Reconstruction from boundary measurements, *Annals Math.*128:531 (1988).
13. R.G.Newton, The Marchenko and Gel'fand methods in the inverse scattering problem in one and three dimensions, *in*: "Conf. on Inverse Scattering: Theory and Applications", J.B.Bednar et al., ed., SIAM, Philadelphia (1983).

STATISTICAL ESTIMATIONS IN INVERSE SCATTERING PROBLEMS

V.A.Burov, E.E.Kasatkina, and O.D.Rumiantseva

Moscow State University, Faculty of Physics, Department of Acoustics
Moscow, 119899, Russia

In this paper we discuss some approaches to obtain the statistical estimations of scatterer characteristics with interference arising during the measurements.

1. FIRST APPROACH. USING OF THE LIPPMANN-SCHWINGER (LS) EQUATION FOR INNER FIELD

First approach is based on the system of the LS integral equations for scattering data in coordinate (or wave vector K-space) representation:

$$U (y | x) = U_0 (y | x) + \omega^2 \int_R G_0 (y - r' | r') \varepsilon_c (r') U (r' | x) d r' + n (y | x) \qquad (1)$$

$$U (r | x) = U_0 (r | x) + \omega^2 \int_R G_0 (r - r') \varepsilon_c (r') U (r' | x) d r' \qquad (2)$$
$$y \in Y; \ x \in X; \ r \in R;$$

where X and R are the regions of sources and scatterers localization, correspondingly; Y is a region of receiving; U_0 is the primary field; G_0 is Green function; $\varepsilon_c (r) = (c_0^{-2} - c^{-2} (r))$ is a the function characterizing the inhomogeneity of sound phase velocity $c (r)$ with respect to homogeneous medium c_0; n – additional interference.

Maximum a posteriori probability method for the statistical estimation $\varepsilon_c (r)$ is based on minimization of corresponding nonquadratic functional. This functional includes quadratic form F_0 from discrepancy between estimations and experimental data, with inverse interference correlation matrix as a kernel. It should also depend on a priori scatterer statistics F_a and restrictions that follow from the LS equations for the inner fields F_r. These restrictions are regarded as equations of constrain with Lagrangian functional like multipliers. (Constraints based on direct using of wave equation for fields was regarded in [1] .) If interference and a priori distribution of inhomogeneity are considered as a normal processes characterized by the matrix N and ε, correspondingly, than the maximum a posteriori estimation method problem leads to minimization of the following functional

$$F \equiv F_0 + F_a + F_r , \qquad (3)$$
where

$$F_0 = \iint_{X R} [U^+ (y_1' | x_1') - U_0^+ (y_1' | x_1') - \omega^2 \int_R G_0^+ (y_1' - r_1') \varepsilon_c^+ (r_1') U^+ (r_1' | x_1') d r_1'] \times$$

$$\times N^{-1} (y_1' , y_2' , x_1' , x_2') [U (y_2' | x_2') - U_0 (y_2' | x_2') - \omega^2 \int_R G_0 (y_2' - r_2') \varepsilon_c (r_2') \times$$

$$\times U (r_2' | x_2') d r_2'] d y_1' d y_2' d x_1' d x_2' ;$$

$$F_a = \int_R \varepsilon_c^+ (r_1') \varepsilon^{-1} (r_1' , r_2') \varepsilon_c (r_2') d r_1' d r_2' ;$$

$$F_r = \iint_{XR} \lambda(\mathbf{r}', \mathbf{x}')[\,U(\mathbf{r}'\,|\,\mathbf{x}') - U_0(\mathbf{r}'\,|\,\mathbf{x}') - \omega^2 \int_R G_0(\mathbf{r}' - \mathbf{r}'')\varepsilon_c(\mathbf{r}'') \times$$

$$\times U(\mathbf{r}''\,|\,\mathbf{x}')\,d\mathbf{r}''\,]\,d\mathbf{r}'\,d\mathbf{x}' + \iint_{XR}\lambda^+(\mathbf{r}', \mathbf{x}')[\,U^+(\mathbf{r}'\,|\,\mathbf{x}') - U_0^+(\mathbf{r}'\,|\,\mathbf{x}') - \omega^2 \int_R G_0^+(\mathbf{r}' - \mathbf{r}'') \times$$

$$\times \varepsilon_c^+(\mathbf{r}'')\,U^+(\mathbf{r}''\,|\,\mathbf{x}')\,d\mathbf{r}''\,]\,d\mathbf{r}'\,d\mathbf{x}'\;;$$

where "+" denotes a Hermitian conjugation operation.

The functional variation over variable $\varepsilon_c^+(\mathbf{r})$ leads to equation, that after some integrations, becomes:

$$-\omega^2 \iint_{XY} G_0^+(\mathbf{y}_1' - \mathbf{r})\,U^+(\mathbf{r}\,|\,\mathbf{x}_1')\,N^{-1}(\mathbf{y}_1', \mathbf{y}_2', \mathbf{x}_1', \mathbf{x}_2')\,[\,U(\mathbf{y}_2'\,|\,\mathbf{x}_2') - U_0(\mathbf{y}_2'\,|\,\mathbf{x}_2') -$$

$$-\omega^2 \int_R G_0(\mathbf{y}_2' - \mathbf{r}_2')\,\varepsilon_c(\mathbf{r}_2')\,U(\mathbf{r}_2'\,|\,\mathbf{x}_2')\,d\mathbf{r}_2'\,]\,d\mathbf{y}_1'\,d\mathbf{y}_2'\,d\mathbf{x}_1'\,d\mathbf{x}_2' + \int_R \varepsilon^{-1}(\mathbf{r}, \mathbf{r}_2')\,\varepsilon_c(\mathbf{r}_2')\,d\mathbf{r}_2' -$$

$$-\omega^2 \iint_{XR} \lambda^+(\mathbf{r}', \mathbf{x}')\,G_0^+(\mathbf{r}' - \mathbf{r})\,U^+(\mathbf{r}\,|\,\mathbf{x}')\,d\mathbf{r}'\,d\mathbf{x}' = 0. \tag{4}$$

Within the first Born approximation, the equation (4) is reduced to the relation:

$$\hat{\varepsilon}_{Born}(\mathbf{r}) = [\,\omega^4 \iint_{XY} G_0^+(\mathbf{y}_1' - \mathbf{r})\,U_0^+(\mathbf{r}\,|\,\mathbf{x}_1')\,N^{-1}(\mathbf{y}_1', \mathbf{y}_2', \mathbf{x}_1', \mathbf{x}_2')\,\int_R G_0(\mathbf{y}_2' - \mathbf{r}_2') \times$$

$$\times U_0(\mathbf{r}_2'\,|\,\mathbf{x}_2')\,d\mathbf{r}_2'\,d\mathbf{y}_1'\,d\mathbf{y}_2'\,d\mathbf{x}_1'\,d\mathbf{x}_2' + \int_R \varepsilon^{-1}(\mathbf{r}, \mathbf{r}_2')\,d\mathbf{r}_2']^{-1} \iint_{XY} G_0^+(\mathbf{y}_1' - \mathbf{r})\,U_0^+(\mathbf{r}\,|\,\mathbf{x}_1') \times$$

$$\times N^{-1}(\mathbf{y}_1', \mathbf{y}_2', \mathbf{x}_1', \mathbf{x}_2')\,u^{sc}(\mathbf{y}_2'\,|\,\mathbf{x}_2')\,d\mathbf{y}_1'\,d\mathbf{y}_2'\,d\mathbf{x}_1'\,d\mathbf{x}_2'\;; \tag{5}$$

here $\quad u^{sc}(\mathbf{y}\,|\,\mathbf{x}) = U(\mathbf{y}\,|\,\mathbf{x}) - U_0(\mathbf{y}\,|\,\mathbf{x})$.

The Eq. (5) gives the Wiener's type regularized estimation, in accordance with a priori information about interference and inhomogeneity. To solve the whole problem, considering rescattering processes, it is necessary to add to the Eq.(4) results of the variation over the Lagrangian multipliers $\lambda^+(\mathbf{r}, \mathbf{x})$ and over the additional unknown quantities $U^+(\mathbf{r}, \mathbf{x})$. First gives the Eq. (2) and second leads to equation:

$$-\omega^2 \iint_{XY} G_0^+(\mathbf{y}_1' - \mathbf{r})\,\varepsilon_c^+(\mathbf{r})\,N^{-1}(\mathbf{y}_1', \mathbf{y}_2', \mathbf{x}_1', \mathbf{x}_2')[\,U(\mathbf{y}_2'\,|\,\mathbf{x}_2') - U_0(\mathbf{y}_2'\,|\,\mathbf{x}_2') - \omega^2 \int_R G_0(\mathbf{y}_2' - \mathbf{r}_2') \times$$

$$\times \varepsilon_c(\mathbf{r}_2')\,U(\mathbf{r}_2'\,|\,\mathbf{x}_2')\,d\mathbf{r}_2'\,]\,d\mathbf{y}_1'\,d\mathbf{y}_2'\,d\mathbf{x}_2' + \lambda^+(\mathbf{r}, \mathbf{x}) - \omega^2 \int_R \lambda^+(\mathbf{r}', \mathbf{x})\,\varepsilon_c^+(\mathbf{r})\,G_0^+(\mathbf{r}' - \mathbf{r})\,d\mathbf{r}' = 0. \tag{6}$$

Let us assume, that: $\quad N^{-1}(\mathbf{y}_1, \mathbf{y}_2, \mathbf{x}_1, \mathbf{x}_2) \approx n_0^{-1}\delta(\mathbf{y}_1 - \mathbf{y}_2)\,\delta(\mathbf{x}_1 - \mathbf{x}_2)$; $\tag{7}$
then the equation (4) becomes:

$$-n_0^{-1}\omega^2 \iint_{XY} G_0^+(\mathbf{y}_2' - \mathbf{r})\,U^+(\mathbf{r}\,|\,\mathbf{x}_2')[\,U(\mathbf{y}_2'\,|\,\mathbf{x}_2') - U_0(\mathbf{y}_2'\,|\,\mathbf{x}_2') - \omega^2 \int_R G_0(\mathbf{y}_2' - \mathbf{r}_2')\,\varepsilon_c(\mathbf{r}_2') \times$$

$$\times U(\mathbf{r}_2'\,|\,\mathbf{x}_2')\,d\mathbf{r}_2'\,]\,d\mathbf{y}_2'\,d\mathbf{x}_2' + \int_R \varepsilon^{-1}(\mathbf{r}, \mathbf{r}_2')\,\varepsilon_c(\mathbf{r}_2')\,d\mathbf{r}_2' - \omega^2 \iint_{XR} \lambda^+(\mathbf{r}', \mathbf{x}')\,G_0^+(\mathbf{r}' - \mathbf{r}) \times$$

$$\times U^+(\mathbf{r}\,|\,\mathbf{x}')\,d\mathbf{r}'\,d\mathbf{x}' = 0. \tag{4a}$$

First term of the Eq. (4a) is reduced to zero by the exact deterministic solution. If the interference discrepancy (expression in square brackets) is not equal to zero, then in this term the weight discrepancy averaging is carried out. If probability distribution of interference is large enough than the last two terms of the Eq. (4a) will have the main regularization influence.

With condition (7) the equation (6) can be rewritten in the form:

$$-n_0^{-1}\omega^2 \int_Y G_0^+(\mathbf{y}_1' - \mathbf{r})\,\varepsilon_c^+(\mathbf{r})[\,U(\mathbf{y}_1'\,|\,\mathbf{x}) - U_0(\mathbf{y}_1'\,|\,\mathbf{x}) - \alpha_i\,\omega^2 \int_R G_0(\mathbf{y}_1' - \mathbf{r}_2')\,\varepsilon_c(\mathbf{r}_2') \times$$

$$\times U(\mathbf{r}_2'\,|\,\mathbf{x})\,d\mathbf{r}_2'\,]\,d\mathbf{y}_1' + \lambda^+(\mathbf{r}, \mathbf{x}) - \omega^2 \int_R \lambda^+(\mathbf{r}, \mathbf{x})\,\varepsilon_c^+(\mathbf{r})\,G_0^+(\mathbf{r}' - \mathbf{r})\,d\mathbf{r}' = 0. \tag{6a}$$

Nonlinear equation system (4a),(6a),(2) can be solve by the iterative procedure. First estimation is the filtered Born's one (5). The rescattering effects are gradually introduced in terms describing rescattering effects by additional multiplier α_i, that is gradually changed from 0 to 1 during the iteration process)[2]. The convergence of iterative procedure for reconstruction of high contrast inhomogeneities may be provided by this way. The result of given scheme is the (regularized) statistical scatterer estimation in which the rescattering processes are taken into account.

The advantages of iterative scheme based on equations (1)-(2) are:

– possibility to use fragmentary data and to introduce any a priori information in variation process;

– the method is universal enough in the sense that it can be used for primary illumination by waves with any spectrum or by pulses of any time form (in that case the procedures of T and λ estimations are separately built for each frequency, but the estimation ε is general).

However the large number of additional unknown functions (which require the solution of nonlinear equations for its determination) leads to complexity of iterative process. Moreover, it is necessary to take care of convergence in iterative procedure by regulation of the rate of introducing the rescattering effects in the iterative process.

2. SECOND APPROACH. USING OF THE MARCHENKO-NEWTON-ROSE (MNR) EQUATION

Unlike the first approach based solely on the LS equation, the second approach uses the MNR equation to find inner field. In contrast to the LS equation, the MNR determines direct connection (i.e. without involving the scatterer function) between the external scattering data and the inner field values. This equation requires pulse probing regime, point sources $x \in S$ and point receivers $y \in S$, where S is a surface surrounding the scattering domain R. Though in the initial strict method [3] the MNR equation requires delta-pulse probing signal, it may be generalized to the case of arbitrary finite duration probing field $U_A^0(t, r, x)$ with spectrum $A(\omega)$ [4]. The generalized equation is the second kind Fredholm equation for unknown inner scattered fields $u_A^{sc}(t, r, x)$ with spectrum $A(\omega)$ and $u_{A^2}^{sc}(t, r, x)$ with spectrum $A^2(\omega)$; for three-dimension it is written as

$$\int_S d^2\hat{y}|y|^2 \int dt' \left(u_A^{sc}(t'-t, y, r) \frac{\partial}{\partial n_y} U_A(t', y, x) - \frac{\partial}{\partial n_y} u_A^{sc}(t'-t, y, r) U_A(t', y, x) \right) +$$

$$+ u_{A^2}^{sc}(t, r, x) = f(t, r, x), \qquad t \geq \tau = \min\{t_A, t_{A^2}\}; \tag{8}$$

$$u_A^{sc}(t, r, x) = 0, \quad t < t_A = |r - x| / c_m + t_{init}; \quad u_{A^2}^{sc}(t, r, x) = 0, \quad t < t_{A^2} = |r - x| / c_m + 2 t_{init};$$

i.e. $\quad u_A^{sc}(t, r, x) = u_{A^2}^{sc}(t, r, x) = 0$ at $t < \tau \; \forall x \in S$. $\tag{8a}$

Here c_m is a maximum value of phase velocity $c(r)$ within the volume bounded by the surface S; t_{init} is an initial moment of the probing signal. The field $U_A(t', y, x) = u_A^{sc} + U_A^0$ is the experimental data, and function f is known:

$$f(t, r, x) = -\int_S d^2\hat{y}|y|^2 \int dt' \left(U_A^0(t'-t, y, r) \frac{\partial}{\partial n_y} u_A^{sc}(t', y, x) - \frac{\partial}{\partial n_y} U_A^0(t'-t, y, r) u_A^{sc}(t', y, x) \right);$$

$d^2\hat{y}$ is a solid angle element, $|\hat{y}| = 1$. The total inner field $U_{in} \equiv U_A(t, r, x)$ is uniquely connected with the inhomogeneity functions of the phase velocity $\varepsilon_c(r)$, of the density $\varepsilon_\rho(r) = \sqrt{\rho(r)} \Delta(1/\sqrt{\rho(r)})$ and of the absorption $\varepsilon_\alpha(r) = 2\alpha(r)/c(r)$ (amplitude absorption coefficient $\alpha(r, \omega) = \omega^2 \alpha(r)$ for any fixed frequency) by the wave equation

$$\left[\frac{\partial^2}{\partial r^2} - \left(\frac{1}{c_0^2} - \varepsilon_c(r) \right) \frac{\partial^2}{\partial t^2} - \varepsilon_\rho(r) + \varepsilon_\alpha(r) \frac{\partial^3}{\partial t^3} \right] U_A(t, r, x) = A(t)\delta(r - x), \tag{9}$$

where
$$A(t) = \frac{1}{2\pi}\int A(\omega)e^{-i\omega t}d\omega .$$

The restriction in (8) is posed on the duration $\tau^0_{A^2}$ of the probing signal $U^0_{A^2}(t,\mathbf{r},\mathbf{x}) = U^0_A(t,\mathbf{r},\mathbf{x}) \otimes A(t)$ ($\otimes$ means convolution) corresponding to the spectrum $A^2(\omega)$:

$$\tau^0_{A^2} < 2|\mathbf{r} - \mathbf{x}| / c_m \quad \forall \mathbf{r} \in R, \ \forall \mathbf{x} \in S. \tag{10}$$

The condition (10) leads to $-u^{sc}_{A^2}(-t,\mathbf{r},\mathbf{x}) = 0$ for the Eq.(8) [34]. Moreover, if $\tau \geq 0$ then the Eq. (8) is of the type of the Volterra equation for the temporal variable. This fact facilitates the procedure of the equation solution. Because $\tau = \min\{t_A, t_{A^2}\}$, the condition $\tau \geq 0$ always takes place for $t_{init} \geq 0$; in case of $t_{init} < 0$ it is valid if $|t_{init}| \leq |\mathbf{r} - \mathbf{x}| / 2c_m$.

It is straightforward to find solution of the Eq.(8) for the signal with rectangular spectrum $A^2(\omega) = A(\omega)$, $\omega \in \Omega$, and $A(\omega) = 0$, $\omega \notin \Omega$, though in this case this equation (8) is only approximation [4] and $t_{init} \approx -\tau^0_{A^2} / 2$ up to the certain accuracy.

Errors (interference) in the scattering data measurements turn into the values $u^{sc}_{in} \equiv u^{sc}_A(t,\mathbf{r},\mathbf{x})$ that were reconstructed from the Eq.(8). This Eq.(8) in the operator form is:

$$\hat{B}\, u^{sc}_{in} = f . \tag{11}$$

The errors appear both in the matrix $\hat{B}$ by $\Delta\hat{B}$ and in the right side f by Δf:

$$(\hat{B} + \Delta\hat{B})(u^{sc}_{in} + \Delta u^{sc}_{in}) = f + \Delta f . \tag{12}$$

It should be noted that sampling process of Eqs. (11),(12) requires additional research. Supposing that $\hat{B}^{-1}$ exists and the errors are small, Eqs. (11),(12) lead to

$$\Delta u^{sc}_{in} \approx \hat{B}^{-1}\Delta f - \hat{B}^{-1}\Delta\hat{B}\hat{B}^{-1}f ,$$

and error statistics of the inner scattered field reconstruction described by the correlation matrix $W = \langle (\Delta u^{sc}_{in})^+ (\Delta u^{sc}_{in}) \rangle$ may be obtained.

Thus if some regularized estimation U^{est}_{in} of the total field U_{in} is found by the Eq.(8) and W is known then this estimation may be used by different ways.

2.1. Scatterer Estimation from the Pulse Lippmann-Schwinger Equation for Observed Data

The field U^{est}_{in} is introduced into the pulse LS equation for observed scattering data $u^{sc}_{obs} = u^{sc}_A(t,\mathbf{y},\mathbf{x})$. Obtained system is solved as the first kind Fredholm equation with inexactly known right side and integral kernel:

$$\int_R d\mathbf{r}' \int_{-\infty}^{t} dt'\, g_0(t - t'; \mathbf{y},\mathbf{r}') \left[-\varepsilon_c(\mathbf{r}')\frac{\partial^2}{(\partial t')^2} + \varepsilon_\rho(\mathbf{r}') - \varepsilon_\alpha(\mathbf{r}')\frac{\partial^3}{(\partial t')^3} \right] U^{est}_{in}(t',\mathbf{r}',\mathbf{x}) =$$
$$= u^{sc}_{obs}(t,\mathbf{y},\mathbf{x}), \tag{13}$$

g_0 is pulse Green function of the homogeneous medium. The system (13) in the operator form is

$$\hat{\Gamma}_0 U^{est}_{in}\varepsilon' \equiv \hat{D}\varepsilon' = u^{sc}_{obs}, \tag{14}$$

where $\varepsilon'(\mathbf{r})$ is a column-vector-meaning function formed by $\varepsilon_i(\mathbf{r})$, $i = c,\rho,\alpha$; $\hat{\Gamma}_0$ is a generalized operator formed by Green function operator and by time differential operators of corresponding order. A "primitive" solution of (14) is

$$\varepsilon'^{est} = (\mu\hat{E} + \hat{D}^+\hat{D})^{-1}\hat{D}^+ u_{obs}^{sc}, \tag{15}$$

where $\hat{E}$ is unit operator, μ is regularizing multiplier. More exact estimation for ε' is obtained by generalization of the Eq.(15) on the case of the known statistics for $\hat{D}$, u_{obs}^{sc} and ε'.

The "optimal" scatterer estimation is reached by minimization of the certain functional. It is necessary to take into consideration both the measurement errors statistics and a priori statistical information about the estimated scatterer. The main part of all functionals is similar to F_0+F_a from (3), but for the pulse regime. In the first mode U_{in} is considered as fixed, $U_{in} = U_{in}^{est}$, and the functional is varied by ε' only. In the other two modes the statistics of the estimation U_{in}^{est} is taken into account by the additional term

$$(U_{in}^{est} - U_{in})^+ W^{-1}(U_{in}^{est} - U_{in}) \quad \text{or} \quad (\hat{B}u_{in}^{sc} - f)^+ W^{-1}(\hat{B}u_{in}^{sc} - f) \quad \text{(see (8), (11))}$$

correspondingly, and the functional is varied by ε' and U_{in}. However, because of the presence in the LS Eq.(13) both the inner field itself and its high order temporal derivatives the variational problem becomes extremely cumbersome. The latter mode is close to the first approach, though used in the second approach MNR equations are varied only by U_{in}, in contrast to the LS equations.

2.2. The Scatterer Estimation Directly from the Wave Equation

The Eq.(9) with zero right side is locally valid for any fixed point $r \in R$ for fixed t and x. To find $\varepsilon_i(r)$ for the fixed r from (9) the values $U_{in}(t,r,x) = U_{in}^{est}$ and their derivatives have to be known for three different time moments. If $\varepsilon_\rho(r) = \varepsilon_\alpha(r) = 0$ then the scatterer $\varepsilon_c(r)$ is given by

$$\varepsilon_c(r) = c_0^{-2} - \left(\frac{\partial^2}{\partial r^2} U_{in}(t,r,x)\right)\Big/\left(\frac{\partial^2}{\partial t^2} U_{in}(t,r,x)\right). \tag{16}$$

In case of determinate (i.e. ideally exact) scattering data reconstruction the scatterer from (16) is possible by calculation of nonzero field value U_{in} for any time moment t and any source position x. This phenomenon is known as a "miracle" [5,6]. If the interference $\Delta U_A(t,y,x)$ is present in the experimental data then the profit of using of the values U_{in} for different t and x depends on the correlation properties of ε_c-reconstruction errors. In the linear approximation these correlation properties repeat those of the inner field reconstruction errors $\Delta U_{in}(t,r,x)$. For non-strong scatterers the correlation function K of the field reconstruction errors has been estimated:

$$K(r = 0; t, t_1; x, x_1) \equiv < \Delta U_{in}^+(t, r = 0, x)\Delta U_{in}(t_1, r = 0, x_1) > \sim \tag{17}$$

$$\sim \int_S d^2\hat{y}\int_S d^2\hat{y}_1 \; <\frac{\partial}{\partial y}(\Delta U_A^+(t + \frac{y}{c_0}, \; y, x) \; \frac{\partial}{\partial y_1}(\Delta U_A(t_1 + \frac{y_1}{c_0}, \; y_1, x_1)>, \; y \equiv |y|, \; y_1 \equiv |y_1|.$$

It follows from (17) that K depends on the given model of the interference of experiment. The correlation degree of the field reconstruction errors at the fixed point $r \in R$ repeats the correlation degree of the experimental errors for different positions of sources. Scatterer power is increasing, the inner field errors correlation becomes weaker as a result of multiple rescattering.

Therefore, if the observed fields errors have a finite correlation radius in t- and x-space, then the inner field reconstruction errors have a finite one, within of which "miracle" takes place. Outside of this domain "miracle" is valid only within accuracy of the reconstruction errors, and all reconstructed field values are statistically informative. Then the scatterer ε_c can be estimated as

$$\varepsilon_c^{est}(r) = c_0^{-2} - \frac{1}{M}\sum_{\{t,x\}}\left(\frac{\partial^2}{\partial r^2} U_{in}(t,r,x)\right)\Big/\left(\frac{\partial^2}{\partial t^2} U_{in}(t,r,x)\right), \tag{18}$$

where M is the number of terms.

In the general case when $\varepsilon_i(r)$, $i = c, \rho, \alpha$ are present the scatterer estimation can be obtained from the Eq.(9) considered for all manifold of parameters t and x at each fixed r by the least square method.

The approaches 1 and 2.1 are very close and must give the scatterer estimations with the similar accuracy. In the first place, both of them use all experimental information for the *global* estimation $\varepsilon_c(\mathbf{r})$ taking into consideration the scatterer statistics. In the second place the LS equation and the MNR one are derived from the same wave equation. The approach 2.2 has a larger error dispersion in the estimation ε_i, because it gives the *local* estimation. Further, such an estimation can be improved by the secondary space filtration of the Wiener's type:

$$\varepsilon_i^{sec.\ est} = \Phi^{-1}\left(\widetilde{\varepsilon}_i^{est}\ \widetilde{S}_{\varepsilon_i}\ /\ (\widetilde{S}_{\varepsilon_i} + \widetilde{N}_{\varepsilon_i})\right).$$

Here $\widetilde{\varepsilon}_i^{est}$ is the space spectrum of ε_i^{est}; $\widetilde{S}_{\varepsilon_i}$ and $\widetilde{N}_{\varepsilon_i}$ are the averaged power space spectra of the scatterer and interference correspondingly; Φ^{-1} is the inverse Fourier transformation.

Advantages and Disadvantages of the Second Approach

The common advantage of all modes of the second approach is the application of the MNR equation. First, this equation supposes traditional in echoscopy devices pulse regime. Secondly, it provides the possibility to estimate the inner field independently from the scatterer. The dimension of the inner field estimation problem is the product of the source positions number and the dimension of the temporal samplings.

The method 2.1 with known fixed inner field and the method 2.2 have as an advantage that the solution is obtained in the direct form with the statistics taken into account.

As the advantages of the method 2.2 it should be noted the space locality of the obtained estimation and the possibility to vary the averaging character of the type (18).

The common disadvantage of the second approach is the necessity to have the complete volume of the experimental scattering data. The particular disadvantage of the method 2.1 is the necessity to estimate the scatterer within the whole scattering domain simultaneously.

Remarks on the Possibility to Use the MNR Equation for Incomplete Data

In the case of the incomplete experimental scattering data to direct using of the MNR equation (8) is not possible. First way to overcome this difficulty is to compose the discrepancy functional, in which the MNR-equations for the complete data are involved. The functional should be varied by the scatterer functions , by the inner field function and by the lacking scattering data - the additional variables.

Second way is to iteratively recover the lacking data by the Eq.(8). The additional surface S_1 surrounding the scattering domain R is introduced. This surface S_1 is outside the domain R, but it is inside the volume bounded by S. At the first iterative step the lacking data are supposed to be equal to zero. The field values are reconstructed by (8) only on the surface S_1 (during the solution of the Eq.(8) the co-ordinate $\mathbf{r} \in S_1$ is fixed). Since S_1 lies outside R these reconstructed on S_1 values can be extended as far as the surface S. At the second iterative step to reconstruct the field on S_1 the extended field values are taken instead of the lacking scattering data, but the experimentally known scattering data are taken without change, and so on. Thus, the lacking scattering data will be restored as the result of the described iterative procedure; then the second approach is valid. The question of the minimal known data volume, which is necessary to provide the uniqueness and stability of the iterative solution is in need of future research.

REFERENCES

1. Longji Tang, Iterative method for acoustical wave inversion with sparse data, *in*: " Acoustical Imaging-20", Y.Wei and B.Gu, ed.,Plenum Press, New York (1993).
2. V.A.Burov, A.V.Saskovets, and I.O.Fatkullina, Local convergence of iterative solutions of inverse scattering problems with step-by-step inclusion of multiple scattering, *Acoust.Phys.* 37:14-16 (1991).
3. D.Budreck, and J.H.Rose, Three-dimensional inverse scattering in anisotropic elastic media, *Inverse Probl.* 6:331 (1990).
4. V.A.Burov, and O.D.Rumiantseva, Exact inverse scattering solutions in multi-dimensions (perspective of using in acoustical imaging), *in*: " Acoustical Imaging-21", J.Jones, ed., Plenum Press, New York (1995).
5. J.H.Rose, M.Cheney, and B.DeFacio, The connection between time- and frequency-domain three-dimensional inverse scattering methods, *J.Math.Phys.* 25(10): 2995 (1984).
6. M.Cheney, G.Beylkin, E.Somersalo, and R.Burridge, Three-dimensional inverse scattering for the wave equation with variable speed: near-far formulae using point sources, *Inverse Problems,* 5:1 (1989).

OCEAN TOMOGRAPHY BY VERTICAL ARRAYS

Valentin A. Burov and Sergei N. Sergeev

Faculty of Physics - Department of Acoustics
Moscow State University
119899 Moscow, Russia

THEORETICAL ASPECTS OF OCEAN TOMOGRAPHY

In general case classical tomographic scheme can be reduced to the solution of the parametric manifold of Fredholm integral equations of first kind

$$\int P_\alpha(\mathbf{r}, \mathbf{r}')g(\mathbf{r}')\mathrm{d}\mathbf{r}' = f(\mathbf{r}|\alpha), \tag{1}$$

Here $\mathbf{r}$ and $\mathbf{r}'$ - the spatial coordinates, $P(\mathbf{r}', \mathbf{r})$ - integral transformation kernel, $f(\mathbf{r})$ - measured characteristics, $g(\mathbf{r}')$ -reconstructed image.

For fixed source and receiver location the parameters of equation (1) have the following sense: $P(\mathbf{r}', \mathbf{r})$ determines the exploring ray trajectory, $f(\mathbf{r}) \rightarrow f_k = t_k$ – time delay of signal propagation on k-th ray, $g(\mathbf{r}) = 1/c(\mathbf{r})$ – distribution of reverse sound speed. The ray trajectory itself is a function of $c(\mathbf{r})$, so that the tomographic problem in this case is nonlinear. It is also peculiar to the inverse wave problems of strong inhomogeneity reconstruction. The problem is roughly linearized relatively to the small perturbations $\delta c(\mathbf{r})$. This can achieved by the choice of zero approximation for sound velocity $c_0(\mathbf{r})$ which is closed to the true one:

$$t_k - t_k^0 = \int_{l_k^0} \frac{\delta c(\mathbf{r})}{c_0^2(\mathbf{r})}\mathrm{d}l$$

where l_k^0 is the ray trajectory of k-th ray calculated for unperturbated sound speed profile. In this approximation it is assumed that the single ray trajectory could be resolved (it takes place, for example, in the case with deep sound channel).

An opposite approach is the mode tomography that uses the reconstruction of the propagating mode characteristics (for example, their phase velocities). As a result the information about ocean inhomogeneities can be obtained. The question about relation between rays and modes was studied by Munk and Wunch[1]. When a distance between source and observer's point increases the number of constructive interfering modes in every ray group decreases and at some distance even neighboring modes cease to constructively interfere. In this case the mode description of sound is the only possible way. The using of the mode description is most justified in the case of adiabatic approximation validity.

Let's assume that the region under investigation is surrounded by S radiating and R receiving vertical antennae, their numbers are approximately equal to each other ($S \approx R$) and $S \times R \geq M$ where M is the number of cells contained in the surface of studied region. The number of radiating elements in each antenna can be small (1-3) because for definite depth locations even one element can excite all considered or counted modes in the wavequide.

If vertical receiving antennae make available separation of all counted modes then for reconstruction of phase velocity "map" in the investigated region it will be possible to form the independent inverse problem for every mode number. This solution can be found by using general methods that have been developed for such problems[2-3]. If the inhomogeneity under study is smooth enough so that the maximum lateral size of the ray tube between any pair of emitter-receiver systems is less than spatial cell the further simplification can be done by using of the combined presentation of type "vertical modes, horizon rays".

The influence of scatterer force to the methods of solution and its uniqueness has specific features for this type of problems. For Born approximation the solution of the inverse problem can be done in monochrome regime. Many-frequency measurements result in the improvement of solution accuracy due to increasing of the data redundancy. However it is not necessary to have this regime for uniqueness of solution. The using of the monochrome regime is also possible if phase changes of receiving modes are in the limit $\pm\pi$. This is a case of intermediate force scattering[3]. For this case the ray path distortions are essential and corrections of these paths should be done in the iteration cycle. Now the role of the data redundancy is more important because of its influence on the speed and region of convergence.

Further increasing of scatterer force resulting in phase distortion of receiving modes more then 2π makes the estimated perturbation strong. In this case the solution ambiguity (ununiqueness) gets evident form of phase ambiguity. A manyfrequency and pulse regimes restore the uniqueness. Especially clear it can be seen in pulse tomography case where the group time delay of modes propagation is measured. The delay of n-th mode propagating time on the i-th trajectory equals to

$$\delta t_n = -\int\limits_{l_k^0} \frac{\delta v_n(x,y)}{v_n^2(x,y)} \mathrm{d}l$$

where v_n is the group velocity, δv_n - it's perturbation caused by sound speed perturbation δc.

In a case of strong frequency dispersion (for frequencies close to critical one) a blurring of receiving packet demands more complete data measuring in a form of more detailed frequency or time description estimation.

At the end of such many-frequency measurements obtained for disperse data allow to reconstruct the hydrology profile by using the method similar to the one that was described[4] for the internal waves.

Such sequence of reconstruction operations (measurement data → modes separation → hydrology reconstruction) is not absolutely necessary: the mode separation of the receiving signals can be omitted and the connection "hydrology perturbation → signal perturbation" can be used directly although mode representation remains useful for the corresponding operator construction. The exclusion of modes structure reconstruction operation is necessary in case when the number of receiving hydrophones is less then the number of modes participating in sound field forming.

If the transition to the group of high frequency neighboring modes can be done then this limit case returns us to the initial pulse-rays ocean tomography.

Development of the convenient tomographic scheme in purposes of wave ocean tomography has an important feature, which consists in an influence of hydrology of the whole ocean depth on received signal that makes the problem 3-D in principal i.e. the task can be treated as a hydrology "map" reconstruction. In the other words the ocean tomography is a vector-parametric problem.

Let's divide the investigated region of the ocean on vertical bars-cells and examine the sound field transformation during passing throw one bar. The fields in its "exit" and "entrance" are connected by the propagation operator $\hat{A}^0$: $U(z_k, \mathbf{r} + \Delta\mathbf{r}) \cong U_k = \hat{A}_{ki}^0 U(z_i, \mathbf{r})$. The summing over repeat indexes is implied.

The hydrology perturbation $\delta c(z_j)$ causes the operator's perturbation: $\delta\hat{A}_{ki} = \frac{\partial \hat{A}_{ki}}{\partial c(z_j)} \cdot \delta c(z_j) = \hat{B}_{kij} \cdot \delta c_j$. Let the signal pass throw P bars between radiating and receiving antenna. In this case the main field transformation is $U_k = \left[\prod\limits_{p=1}^{P} \hat{A}^{0^p}\right]_{ki} U_i$. The perturbation presence results in observable profile distortion that in the linearized case is

$$\delta U_k = \sum_i \sum_{p \in P_l} \left[\prod_{t \neq p}^{P} \hat{A}^{0^t} \sum_j \hat{B}_j^p \cdot \delta c_j^p\right]_{ki} U_i = \sum_i \left[\sum_{p \in P_l} \hat{D}_p \sum_j \hat{B}_j^p \cdot \delta c_j^p\right]_{ki} U_i =$$
$$= \sum_{p \in P_l} \sum_j \sum_i \left[\hat{D}_p \hat{B}_j^p\right]_{ki} U_i \cdot \delta c_j^p = \sum_{p \in P_l} \sum_j \hat{F}_{kj}^p \delta c_j^p, \tag{2}$$

where the sense of introduced designations is seen from the transformation sequence. In the adiabatic approximation and in mode representation operator $\hat{D}$ is diagonal and P_l describes the l-th ray trajectory.

In this formula the mentioned "volumness" of ocean tomography is reflected. It means that it is necessary to take into account and to reconstruct the vertical hydrology for description of each ocean cross section. That gives the second sum on the depth coordinate which appears in formula (2).

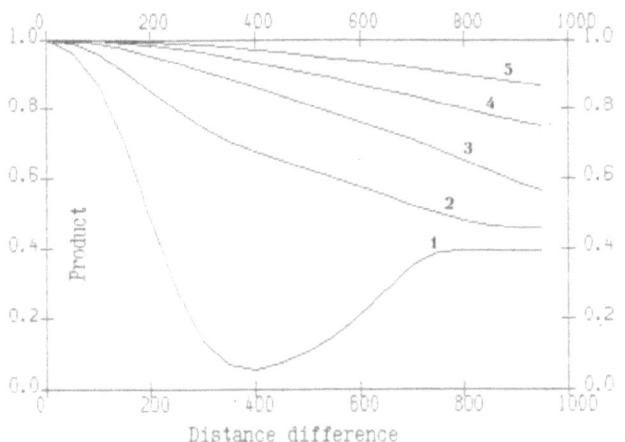

Figure 1. The illustration of using of the 4-frequency algorithm with small distance difference

The situation becomes simpler in the case of the adiabatic approximation. The analogous consideration gives the next relation for the vector of complex amplitudes:

$$\delta\psi_k = \sum_{p\in P_l}\sum_i\sum_j \left[\dot{D}_p \dot{B}_j^p \delta c_j^p\right]_{ki}\psi_i,$$

i.e. every vertical bar in such case gives equal contribution to every ray which passing throw it. This relation generalizes the scalar tomography case (where every cell contributes equally to all passing rays) to the vector-parameter case.

Due to diagonality of the operator $\hat{A}$ in adiabatic approximation the $\hat{B}$ and $\hat{D}$ become diagonal also and it allows us to get separate tomographic problem for phase velocity of each mode by use only one depth of source location such that all modes of investigated waveguide are excited. In general case the nonsingularity of operator $\hat{F}$ is not guaranteed. It is possible to restore it by data redundancy coming from using of different radiator depth.

DECLINATION OF THE VERTICAL ANTENNA AND THE 4-FREQUENCY ALGORITHM

The obtained data can be used only if the space location of each antenna elements is exactly known. However the real declination of antenna elements from vertical line can reach hundred meters. Standard methods of improvement of this disadvantage are the using of heavy cargo to strength antenna or estimation of true profile by additional devices.

As a new approach we have proposed "4-frequency" method which allowed to compensate the unknown antenna distortion[5].

The deflection of k-th antenna element with unknown value Δx_k results in the appearance of additional phase for l-th mode $\kappa_l\delta x_k$, where κ_l - horizontal wave number. If the value of deflection is smaller than the space resolution ability (which is defined by the difference between reverse phase velocities of lowest and highest modes) then the difference for phase velocities of separate modes can be ignored and the phase factor is equal to $\exp(i(\kappa_l\Delta x_k)) \approx \exp(i\omega\Delta x_k/c)$, where c is the some average phase velocity of the accounted modes. In this case the additional phase will be compensated by the combination of received field magnitudes into the product of 4-th order for the 4 frequencies:

$$M_k = U_{\omega_1 k}U_{\omega_2 k}^* U_{\omega_3 k}^* U_{\omega_4 k}, \tag{3}$$

where k is the vertical number of hydrophone, 3 frequencies are chosen arbitrary and the fourth one is defined by the rule $\omega_1 + \omega_4 = \omega_2 + \omega_3$

The fig. 1 illustrates the 4-frequency algorithm for waveguide with depth 1 km and symmetric parabola sound speed profile changing from 1.4550 to 1.5052 km/s which has been chosen as a model. The product of 4-th order that was built by the use of described method was compared with reference product M_4, calculated by the same way as (3) for some set of estimated parameters. The comparison

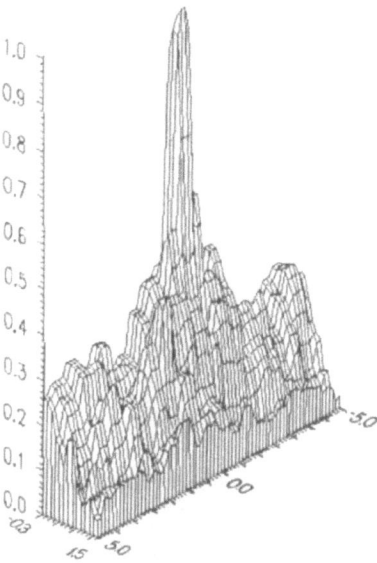

Figure 2. The ambiguity function for subtropics hydrology type

was done by the correlation formula

$$R = \left| \sum_k M_{k_0} M_k^* \right| \bigg/ \left(\sum_k |M_k|^2 \sum_k |M_{k_0}|^2 \right)^{1/2}$$

the closeness of this expression to the maximum value is the criterion of the correspondence of the estimated parameters to the true ones. The curves on Fig. 1 are corresponded to the frequency set 17, 20, 20, 23 Hz (1); 27, 30, 30, 33 Hz (2); 77, 80, 80, 83 Hz (3); 157, 160, 160, 163 Hz (4); 197, 200, 200, 203 Hz (5) and sound source location at the distance of 10 km from antenna. From this picture it is clear from the picture that when the chosen frequencies increase the result becomes less dependent on the declination lower then 0.5 km so that the antenna distortions with shifts up to hundreds meters don't cause the deterioration. But the averaged product is sensitive to the large (≥ 1 km) distance and hydrology mismatches. The peak corresponding to the true source location is well distinguished. This property of 4-th order product allows us to use it for tomography purposes with reasonable precession of space resolution and accuracy of navigational knowledge about antennae coordinates.

In order to study the opportunity of practical use of 4-th order product the ambiguity function was calculated for base types of sound speed stratification in the ocean. Example for subtropics hydrology is represented in Fig. 2. This function is the response to the source distance and depth mismatch. The discrepancy in the distance from -5 to +5 km was plotted along axis X and the one in the depth from -0.25 to +2 km) along axis Y. The source was located in the point (0,0).

RECONSTRUCTION OF THE REFRACTION TYPE INHOMOGENEITY

The next important question is about the choice of work frequencies in the case where reconstruction is done with using mode representation. It is so important because of for single work frequency it is possible to reconstruct the map of mode phase velocities but it is not possible to reconstruct the profile $c(z)$ due to the fact that the perturbation of phase velocity of n-th mode ψ_n: $\delta v \sim \int \delta c(z) \psi_n^2(z)\, dz$ and the set $\{\psi_n^2\}$ doesn't form the complete system for one frequency. So the reconstruction of $\delta c(z)$ is possible only by data received on several frequencies.

In general case for solving the direct problem and constructing the operator of correspondence (the matrix of translation from measured data to the hydrology characteristics) which is the base of inverse problem solution it is possible to employ the mode approach with using of first members of perturbation theory expansion. For this purpose the next equation should be regarded

$$\psi_l''(z) + k^2(z)\psi_l(z) = \kappa_l^2 \psi_l(z), \tag{4}$$

where κ_l^2 is the l-th eigenvalue (square of horizontal wave number), ψ_l - the corresponding eigenfunction (mode) and $k(z) = \omega/c(z)$, $c(z)$ - the sound speed profile. The field in the homogeneous waveguide on depth z and distance r from the source that has been located on depth z_0, has a form

$$U(r, z) = \sum_{l=1}^{\infty} \sqrt{\frac{2}{\pi \kappa_l r}} \psi_l(z_0)\psi_l(z) \exp\left(i(\kappa_l r - \pi/4)\right). \tag{5}$$

The equation (4) has the exact solution only for limited number of occasions. For the fields calculation it is necessary to use the approximate methods. One of them is the perturbation theory method that is used when the square of wavenumber $k(z)$ has a form $k^2(z) = k_0^2(z) + k_1^2(z)$. The profile of nonperturbated wavenumber $k_0(z)$ is chosen so that the nonperturbated wave equation $\psi_l^{0\prime\prime}(z) + k_0^2(z)\psi_l^0(z) = \kappa_l^{0\,2}\psi_l^0(z)$ has known analytical or numerical solution. The consistent calculation of these corrections results in the expansion into series for some formal parameter λ.

The small perturbation of wavenumber profile $k_1(z)$ results in the perturbation of sound field[6], which is approximately proportional to $k_1(z)$:

$$\Delta U(r, z_k) = \sum_{j=1}^{N} k_{1,j}^2 Q_j(r, z_k), \tag{6}$$

where $Q_j(\mathbf{r}, z)$ is the matrix which calculated by perturbation theory formulae, j is the horizon of hydrophone location number.

The formula (6) sets up the linearized relation which corresponds to the simplest tomographic scheme for the volume consisting of one elementary bar and allows to solve inverse refraction problem immediately[7].

The review of modern modification of perturbation theory for hydroacoustics has been representeed[6]. That let us to exclude some disadvantages of standard theory such as necessity to know all spectrum of nonperturbated operator for calculation of corrections to every mode. The exclusion of these disadvantages is made by "delinearization" of equation (4) to the nonlinear first order equation $y_l' - y_l^2 = k^2 - \kappa_l^2$. It was done by substitution $y_l = -\psi_l'/\psi_l = -(\ln \psi_l)'$ into the primary equation. This approach called "nonlinear perturbation theory" or "delinearization method". The expansion upon the formal parameter λ results in the correction formulae which contain only nonperturbated eigenfunctions and eigenvalues of correspondent number. Moreover, the result can be represented in quadratures that makes the method be suitable for computer calculations.

Account of only first linear corrections also results in the lineariezed relation (6) but now the matrix Q is calculated by using of nonlinear perturbation theory. The solution of this system gives the corrections to the local eigenvalues. Such process can be repeated iterationly, so that the gradually more precise approximation to the right solution is built.

For modeling of the described algorithm the numeric calculation was created. At the first step of program we have chosen the sampling theorem basis functions $\theta_j(z)$ as $\sin x/x$ and the discrete profile of initial hydrology has been expanded upon them. At the second step the field that has to be "measured" by the vertical array has been calculated. The simulation of measuring process has been carried out with assignment of "true" hydrology profile. At the next stage of the calculations the matrix Q has been formed. After that system of linear equations (6) was solved, and the corrections κ_1 in the points of hydrophone location were founded so that the refined hydrology was rebuilt and the next iteration was based on this result. The process of the convergence to the true profile was controlled by the discrepancy – sum of the squares of the reconstructed profile values declination from the true one $S = \sum_{i=1}^{N} (c_{iter} - c_{true})^2$. Here N is the number of points where the velocity had been distorted.

The results of these calculations for the parabolic model of ocean with depth 1 km, monochrome (50 rad/s) source located at the depth 250 m are present in the tab where profile values were taken during every 100 m. First line of this table contains the values of "true" profile, second – that was chosen as initial one during calculations. The third line contains the results of third iteration. The discrepancy changed from $1.6271 \cdot 10^{-5}$ to $3.9200 \cdot 10^{-8}$.

Vertical profile of sound speed, km/s

1.5032	1.4850	1.4714	1.4620	1.4566	1.4550	1.4573	1.4634	1.4736	1.4881	1.5072
1.5052	1.4866	1.4725	1.4627	1.4569	1.4550	1.4569	1.4627	1.4725	1.4866	1.5052
1.5033	1.4851	1.4714	1.4620	1.4566	1.4550	1.4572	1.4633	1.4736	1.4881	1.5072

The gradient method used here was based on the nonlinear perturbation theory.

The 4-frequencies algorithm for distorted antenna reduces the inverse refraction problem in the simplest form to the equation system $\Delta M_4(r, z_k) = \sum_{j=1}^{N} k_{1_j}^2 A_j(r, z_k)$ where A is the matrix determined by the matrix Q for chosen frequencies[5].

At the conclusion let's briefly describe the complete scheme of ocean tomography using the set of vertical antenna distorted by the ocean flows.

For this purpose determine the hydrology profile $c_0(i, k, j)$ at every cell of investigated region. The presence of the inhomogeneity $\delta c(i, k, j)$ leads to the field perturbation at the "exit" of every bar which is calculated with help of perturbation theory:

$$\delta U(i, k, j) = \sum_{n=1}^{N} \delta k_n^2(i, k) Q_n(i, k, j) \approx -2 \sum_{n=1}^{N} k_0^2 Q_n(i, k, j) \frac{\delta c_n}{c_0},$$

where $n = 1, \ldots, N$ – index of functional basis, for example, the basis of sampling theorem. If there are P vertical bars with coordinates (i, k) between radiating and receiving antenna then the passed field has the perturbation

$$\delta U_P = \sum_{p=1}^{P} \left(-2 \cdot \sum_{n=1}^{N} \frac{k^{p^2}}{c_0} Q_n^p(j) \delta c^p(n) \right).$$

This formula is the analog of linearized computer tomography formula where field perturbation on P-th ray δU^p plays the role of single point in single cross section and $\delta c^p(j)$ - of reconstructed image.

This perturbation through the P_l-ray gives possibility to regard the complete perturbation of 4-th order product along the ray. Irradiating the investigated region by intersecting rays gives in these approach the algebra equation system. Solution of this system allows us to reconstruct the inhomogeneity. However it is only first step of iteration procedure. The next one begins from a new value $c_0^1(z) \rightarrow c_0^0(z) + \delta c_1(z)$.

REFERENCES

1. W. Munk and C. Wunch, Ocean acoustic tomography: Rays and modes *Rev. of Geoph. and Space Phys.*. 21:777 (1983).
2. V.A. Burov, M.N. Rychagov and A.V. Saskovets, Account of multiple scattering in acoustic inverse problems of tomographic type, *Acoust. Imaging*. 19:35 (1992).
3. V.A. Burov, M.N. Rychagov and A.V. Saskovets, Iterative methods for the reconstruction of characteristics of strong inhomogeneities by the data of acoustic scattering, in Proc. Ultrason. Int. Conf. (1991).
4. S.V. Baykov and V.A. Burov, The estimation of buoyancy frequency profile by dispersive characteristics of the internal waves, *Marine hydrophys. mag.*. 3:35 (1983) [in Russian].
5. V.A. Burov, S.N. Sergeev and N.P. Sergievskaya, Acoustical tomography of the ocean using data from a vertical mode array subjected to arbitrary banding by underwater currents, *Acoustical Physics*. 2:187 (1992).
6. V.A. Burov and S.N. Sergeev, Modern methods of perturbation theory for calculating hydroacoustic fields, *Moscow University Physics Bulletin(Vestnik Moskovskogo Universiteta. Fizika)*. 47(2):45 (1992).
7. V.A. Burov and S.N. Sergeev, Solution of inverse refraction problem by nonlinear perturbation theory, *Acoustical Physics*. 3:221 (1991).

NONLINEAR ACOUSTICAL TOMOGRAPHY IN INHOMOGENEOUS MEDIA

V.A. Burov, I.E. Gurinovich, O.V. Rudenko, and E. Ya. Tagunov

Moscow State University, Faculty of Physics, Acoustics Department
Moscow,119899, Russia

Today most methods of nonlinear acoustic tomography are based on nonlinear interaction of collinear acoustic waves[1,2]. In some works an interaction of low frequency or impulse power pump wave with a test wave of high frequency and lower amplitude are used also. The first of the two methods is close to the traditional tomography ones. The second method is more simple and allows to trace the distribution of nonlinear parameter ε along a signal propagation line, independently on measurements in other directions.

The values of other acoustic parameters (sound velocity and density) are assumed to be constant everywhere in the propagation region, and the theoretical treatment is performed in the ray approximation. The latter condition poses a principal limitation on the spatial resolution which is defined by the transverse size of the probing beam for this method. For the second group of nonlinear acoustic tomography the most studied method is the method of phase modulation of high frequency test wave by a power pump impulse or low frequency wave[3,4].

In posing the problem of reconstruction of the full acoustic parameters picture for an inhomogeneous medium, we should keep in mind that, as a rule, the medium's parameters are interrelated. In the case when all the parameters are changing simultaneously, it is expedient to determine the distributions of acoustic parameters from the data of linear and nonlinear acoustic scattering[5]. This work presents a step that may lead to the solution of this problem for the medium with inhomogeneous sound velocity c, density ρ, and nonlinearity parameter ε.

As a next step, we will consider the $c\varepsilon$-problem, i.e., the problem of step-by-step reconstruction of the inhomogeneities of velocity $c(\mathbf{r})$, and nonlinearity parameter $\varepsilon(\mathbf{r})$ inhomogeneities when they are localized in either completely or partially overlapping regions.

In paper[5], we have demonstrated the possibility of reconstructing the distribution of the nonlinearity parameter $\varepsilon(\mathbf{r})$ from the results of the nonlinear interaction of acoustic waves, which is accompanied by the emission of combination frequencies from the interaction region. In this paper we have also considered the scattering of all acoustic waves from c-inhomogeneities. It is necessary also to take into account the scattering of both interacting waves and combination frequency waves emitted from the region of nonlinear interaction.

Let us assume as a basis the equation of Westerwelt:

$$\Delta p - \frac{1}{c^2}\frac{\partial^2 p}{\partial t^2} = -\frac{\varepsilon}{\rho_0 c^4}\frac{\partial p^2}{\partial t^2} . \tag{1}$$

In accordance with the problem formulation, let us assume that $c = c(\mathbf{r})$ and $\varepsilon = \varepsilon(\mathbf{r})$ are the functions of spatial coordinate $\mathbf{r}$. We will solve equation (1) by the method of successive approximations. Let us rewrite it in the form:

$$\Delta p - \frac{1}{c_0^2}\frac{\partial^2 p}{\partial t^2} = -F(p), \text{where} \tag{2}$$

$$F(p) = (\varepsilon/\rho_0 c^4)(\partial^2 p^2/\partial t^2) + (\chi/c_0^2)(\partial^2 p/\partial t^2), c \equiv c(\mathbf{r}), \quad \chi \equiv \chi(\mathbf{r}) = c_0^2[1/c_0^2 - 1/c^2(\mathbf{r})].$$

The first term in the expression for F describes the generation of the waves with combination frequencies $\Omega = \omega_1 - \omega_2$ and $\nu = \omega_1 + \omega_2$, and the emission of second harmonics at frequencies $2\omega_1$ and $2\omega_2$; the second term represents the scattering of all field components from the inhomogeneities of sound velocity. We will search a solution of this problem in the form of the superposition of the waves with frequencies $\omega_1, \omega_2, \Omega$ and ν, each being written as follows:

$$p_{\omega_j}(\mathbf{r},t)=[p_j(\mathbf{r},t)\exp(-i\omega_jt)/2]+c.c.; \quad j = 1,2,$$

$$p_\Omega(\mathbf{r},t)=[p_-(\mathbf{r},t)\exp(-i\Omega t)/2]+c.c., \qquad p_\nu(\mathbf{r},t)=[p_+(\mathbf{r},t)\exp(-i\nu t)/2]+c.c. \tag{3}$$

We discard the waves at frequencies $2\omega_j$ (which also should be considered, in general, for this approximation) because they have no effect on the amplitude of combination frequency (Ω and ν) waves in question. In this case, equation (2) falls into a set of equations for each of the complex amplitudes:

$$\Delta p_1 + \frac{\omega_1^2}{c_0^2}p_1 = \frac{\omega_1^2}{c_0^2}\chi\,p_1, \qquad\qquad \Delta p_2 + \frac{\omega_2^2}{c_0^2}p_2 = \frac{\omega_2^2}{c_0^2}\chi\,p_2$$

$$\Delta p_- + \frac{\Omega^2}{c_0^2}p_- = \frac{\varepsilon\,\Omega^2}{\rho_0 c^4}p_1p_2^* + \frac{\Omega^2}{c_0^2}\chi\,p_-, \quad \Delta p_+ + \frac{\nu^2}{c_0^2}p_+ = \frac{\varepsilon\,\nu^2}{\rho_0 c^4}p_1p_2 + \frac{\nu^2}{c_0^2}\chi\,p_+ \tag{4}$$

The first two equations of system (4) allow us to find the inhomogeneities of sound velocity from the measuring of the waves scattered amplitudes. The waves, scattered by the sound velocity inhomogeneities, interact and form combination frequency waves, which, in turn, are scattered from the sound speed inhomogeneities. These processes are described by the last two equations of system (4). Thus, we can write an expression for the wave of difference frequency:

$$p_-^0 = \frac{\Omega^2}{\rho_0}\hat{g}_{-(y,r)}[\beta p_1 p_2^*], \text{ where } \beta = \beta(\mathbf{r}) = \varepsilon(\mathbf{r})/c^4(\mathbf{r}); \ \hat{g}\text{-operator with the Green}$$

function as kernel. Finally, the total scattered field of the difference frequency is presented in the form:

$$p_-(\bar{\mathbf{y}}) = p_-^0(\bar{\mathbf{y}}) + \frac{\Omega^2}{c_0^2}\hat{g}_{-(y,r)}[E - \frac{\Omega^2}{c_0^2}\chi\,\hat{g}_{-(r,r')}]^{-1}\chi\,p_-^0(\mathbf{r}') , \tag{5}$$

where $\mathbf{r},\mathbf{r}' \in \mathcal{R}, \mathcal{R}$ is the scattering region; $\mathbf{y} \notin \mathcal{R}, \mathbf{y}$ denotes the observation point.

The expression (5) connects the primary irradiating fields with the measured field of operators, which are nonlinear in χ and based on Green's function of the homogeneous medium. For the wave of sum frequency we get an expression, analogous to (5).

Let us consider the first terms in expressions for the fields of the difference and sum frequencies. In the far field, there hold the approximate equalities:

$$g_-(\mathbf{y},\mathbf{r}) \approx g_-(\mathbf{y})\exp(-i\mathbf{k}_{s-}\mathbf{r}), \qquad \mathbf{k}_{s-} = \frac{\Omega}{c_0}\frac{\mathbf{y}}{|\mathbf{y}|};$$

$$g_+(\mathbf{y},\mathbf{r}) \approx g_+(\mathbf{y})\exp(-i\mathbf{k}_{s+}\mathbf{r}), \qquad \mathbf{k}_{s+} = \frac{\nu}{c_0}\frac{\mathbf{y}}{|\mathbf{y}|} ,$$

where $g_\pm(\mathbf{y}) = (-1/4\pi)\exp(i\mathbf{k}_\pm|\mathbf{y}|)/|\mathbf{y}|$. Then we have, respectively:

$$p_-(\mathbf{y}) = \frac{\Omega^2}{\rho_0}p_{01}p_{02}^*g_-(\mathbf{y})\tilde{\beta}(\mathbf{K}_-), \qquad\qquad \mathbf{K}_- = \mathbf{k}_{s-} - \mathbf{k}_1 + \mathbf{k}_2 \qquad\qquad \text{and}$$

$$p_+(\mathbf{y}) = \frac{\nu^2}{\rho_0}p_{01}p_{02}g_+(\mathbf{y})\tilde{\beta}(\mathbf{K}_+), \qquad\qquad \mathbf{K}_+ = \mathbf{k}_{s+} - \mathbf{k}_1 - \mathbf{k}_2 ,$$

where $\tilde{\beta}(\mathbf{K})$ is the Fourier transform of $\beta(\mathbf{r})$; $p_1^0 = p_{01}\exp(i\mathbf{k}_1\mathbf{r}), p_2^0(\mathbf{r}) = p_{02}\exp(i\mathbf{k}_2\mathbf{r})$; p_1^0 and p_2^0 are the primary irradiating fields.

If $c(\mathbf{r}) = c_0$, then the amplitudes of the waves scattered at difference and sum frequencies are respectively:

$$p_-(\mathbf{y}) = \frac{\Omega^2}{\rho_0 c_0^4} p_{01} \dot{p}_{02} g_-(\mathbf{y}) \tilde{\varepsilon}(\mathbf{K}_-), \qquad p_+(\mathbf{y}) = \frac{v^2}{\rho_0 c_0^4} p_{01} p_{02} g_+(\mathbf{y}) \tilde{\varepsilon}(\mathbf{K}_+) . \qquad (6)$$

Equations (6) provide the basis for analysis of the function $\tilde{\varepsilon}(\mathbf{K})$. Practical realization of different data acquisition schemes, used to solve the inverse problem of nonlinear scattering on the basis of expressions (6), produces a substantial redundancy of data. Indeed, let us first consider the case of fixed frequencies ω_1 and ω_2 and arbitrary oriented vectors $\mathbf{k}_1$ and $\mathbf{k}_2$. The function $\tilde{\varepsilon}(\mathbf{K})$ can be determined in each individual measurement for $\mathbf{K}$ values that lie on the spheres of radii $\mathbf{k}_{s-}$ and $\mathbf{k}_{s+}$ for the difference and sum frequencies, respectively. The centers of these spheres are defined by the vectors $-(\mathbf{k}_1 - \mathbf{k}_2)$ and $-(\mathbf{k}_1 + \mathbf{k}_2)$ (see Figure 1 , where, for simplicity, the two-dimensional sections are shown).

In the case of collinear interaction, we can measure the values of $\mathbf{K}$ inside the spheres of radii $2\mathbf{k}_{s-}$ and $2\mathbf{k}_{s+}$ for difference and sum frequencies, respectively. If $\Omega << \omega_1, \omega_2$, then the measurements at the difference frequency provide the reconstruction of the low-frequency spatial spectrum of the function $\varepsilon(\mathbf{r})$. In the limiting case, when all the possible directions of $\mathbf{k}_1$ and $\mathbf{k}_2$ are used, we can reconstruct $\tilde{\varepsilon}(\mathbf{K})$ for all $\mathbf{K}$ inside the spheres of radii $2\mathbf{k}_1$ and $2\mathbf{k}_{s+}$ while working at the difference and sum frequencies. It is evident that the use of all irradiation directions leads to a redundancy of information. One should choose the matched set of irradiation directions to optimize the determination of the function $\tilde{\varepsilon}(\mathbf{K})$. Figure 2 shows the wave scheme for organization of the minimal informative experiment at the difference frequency.

An interaction of two short acoustic impulses with flat fronts that are parallel to the OZ axis and make an angle of 2θ has a huge practical interest. Let them be written as follows:

$$p_1 = A\, f_1\left(t - \frac{x}{c}\cos\theta - \frac{y}{c}\sin\theta\right), \qquad p_2 = A\, f_2\left(t - \frac{x}{c}\cos\theta + \frac{y}{c}\sin\theta\right). \qquad (7)$$

The interaction region that makes a parallelepiped, which is infinite along the OZ axis, moves along the OX axis with velocity $v = \dfrac{c}{\cos\theta}$. The section by the plane that is orthogonal to the OZ axis makes a rhomb with a side of $\dfrac{c\tau}{\cos\theta}$, where τ is the duration of the impulses. Let us limit the interaction region by the OZ axis, then the interaction region will make a limited parallelepiped, which will be a moving nonlinear acoustic source, and an acoustic field, which satisfies the Westerwelt equation in fixed coordinate system. Let the source lie in point with coordinates (ξ_0, y_0, ζ_0), in the coordinate system, connected with the interaction region. Then in a fixed coordinate system the equation of Westerwelt (1) can be written in the following form:

$$\Delta p - \frac{1}{c^2}\frac{\partial^2 p}{\partial t^2} = -4\pi \frac{\varepsilon(x,y,z)}{4\pi\rho_0 c_0^4} \frac{\partial^2 p_0^2(x,y,z,t)}{\partial t^2} \delta(x - vt - \xi_0)\delta(y - y_0)\delta(z - \zeta_0), \qquad (8)$$

where $p_0 = p_1 + p_2$. The solution of the wave equation (8) is this:

$$p(x,y,z,t) =$$

$$= \iiint \frac{\dfrac{\varepsilon(x',y',z')}{4\pi\rho_0 c_0^4} \dfrac{\partial^2 p_0^2\left(x',y',z',t - \dfrac{r}{c}\right)}{\partial t^2}\delta(x' - vt - \xi_0)\delta(y' - y_0)\delta(z' - \zeta_0)dv'}{r}, \qquad (9)$$

127

where $r = \sqrt{(x - x')^2 + (y - y')^2 + (z - z')^2}$ is the distance between the source of sound (point (x', y', z')) and the point of observation. Let us introduce a new variables to carry out the integral calculus:

$$X = x' - v\left(t - \frac{r}{c}\right) - \xi_0, \qquad Y = y' - y_0, \qquad Z = z' - \zeta_0.$$

Then $dx'dy'dz' = IdXdYdZ$, where $I = \dfrac{1}{\left|1 - \dfrac{v_R}{c}\right|}$, v_R is projection on the r-direction, taken at the

moment of time $\left(t - \dfrac{r}{c}\right)$. Making an integral calculation on X,Y,Z we obtain the following relation:

$$p(x, y, z, t) = \sum \left(\frac{\frac{\varepsilon}{\rho_0 c_0^4} \frac{\partial^2 p_0^2}{\partial t^2}}{r} I \right), \tag{10}$$

where the sum is taken on the points, where X=Y=Z=0. From these conditions we can easily define these points:

$$x' - x = v\left(t - \frac{r}{c}\right) + \xi_0 - x; \qquad y' - y = y_0 - y; \qquad z' - z = \zeta_0 - z. \tag{11}$$

If we square these equations and sum them, we get an equation for finding values of r=R at point X=Y=Z=0:

$$R^2 = \left(x - v\left(t - \frac{r}{c}\right) - \xi_0\right)^2 + (y - y_0)^2 + (z - \zeta_0)^2 \tag{12}$$

Since v>c, we have two positive solutions of this equation : $R_{1,2} = \dfrac{\pm R^* - \gamma \xi^*}{v\gamma^2 - 1}$,

where $R^* = \sqrt{\xi^{*2} - \rho^2}$; $\quad \xi^* = \dfrac{x - vt - \xi_0}{\sqrt{\gamma^2 - 1}}$; $\quad \gamma = \dfrac{v}{c}$; $\quad \rho^2 = (y - y_0)^2 + (z - \zeta_0)^2$

All the answers lie in a cone $\xi^{*2} - \rho^2 \geq 0$, elements of this cone $\xi^{*2} - \rho^2 = 0$ come from point $(vt + \xi_0, y_0, \zeta_0)$, in which the source is, and are inclined to the velocity v with the angle φ, $\sin \varphi = \dfrac{c}{v}$. In this way

$$p(x, y, z, t) = \frac{\varepsilon\left(v\left(t - \frac{R_1}{c}\right) + \xi_0, y_0, \zeta_0\right) \frac{\partial^2 p_0^2}{\partial t^2}\left(v\left(t - \frac{R_1}{c}\right) + \xi_0, y_0, \zeta_0, t - \frac{R_1}{c}\right)}{\rho_0 c_0^4 R^* \sqrt{\gamma^2 - 1}} +$$

$$+ \frac{\varepsilon\left(v\left(t - \frac{R_2}{c}\right) + \xi_0, y_0, \zeta_0\right) \frac{\partial^2 p_0^2}{\partial t^2}\left(v\left(t - \frac{R_2}{c}\right) + \xi_0, y_0, \zeta_0, t - \frac{R_2}{c}\right)}{\rho_0 c_0^4 R^* \sqrt{\gamma^2 - 1}} \tag{13}$$

Since there must not be any disturbance in front of the source, we have to limit the region: $x - vt - \xi_0 < 0$. Physically this means that two disturbances come to each point of this cone. Thus, there are two effective places for the source, from which sound comes to the point of observation at time t. The common field is defined by expression:

128

$$p(x,y,z,t) = \iiint \frac{d\xi_0 dy_0 d\zeta_0}{R^* \rho_0 c_0^4 \sqrt{\gamma^2-1}} [\varepsilon(v(t-\frac{R_1}{c})+\xi_0,y_0,\zeta_0)]*$$

$$* \frac{\partial^2 p_0^2}{\partial t^2}(v(t-\frac{R_1}{c})+\xi_0,y_0,t-\frac{R_1}{c})+\varepsilon(v(t-\frac{R_2}{c})+\xi_0,y_0,\zeta_0)* \qquad (14)$$

$$* \frac{\partial^2 p_0^2}{\partial t^2}(v(t-\frac{R_2}{c})+\xi_0,y_0,\zeta_0,t-\frac{R_2}{c}).$$

With $R^* \gg V_0^{1/3}$, R^* may be taken out of the integral. The equation (14) is an integral equation relatively the unknown function ε (V_0 is the interaction region). Let us consider interaction of two short impulses :

$$p_1 = A\frac{\alpha}{\sqrt{\pi}}\exp(-\alpha^2(t-\frac{x}{c}\cos\theta-\frac{y}{c}\sin\theta)^2), p_2 = A\frac{\alpha}{\sqrt{\pi}}\exp(-\alpha^2(t-\frac{x}{c}\cos\theta+\frac{y}{c}\sin\theta)^2),$$

where $\qquad p_1 \to \delta(\tau_1,\alpha), \qquad p_2 \to \delta(\tau_2,\alpha) \qquad$ with $\qquad \alpha \to \infty, \qquad$ and

$\tau_1 = (t-\frac{x}{c}\cos\theta-\frac{y}{c}\sin\theta), \qquad \tau_2 = (t-\frac{x}{c}\cos\theta+\frac{y}{c}\sin\theta)$. Interaction region reduces to point with a $\to 0$ (a is the limit along the OZ axis) with greater α-s, coordinates of this point-like region in movable system of coordinates are $\xi_0 = y_0 = \zeta_0 = 0$. Then values of the function with greater, but finite, α-s are the following:

$$\frac{\partial^2 p_0^2}{\partial t^2}(v(t-\frac{R_1}{c})+\xi_0,y_0,\zeta_0,t-\frac{R_1}{c}) = -A\frac{12\alpha^4}{\pi},$$

$$\frac{\partial^2 p_0^4}{\partial t^2}(v(t-\frac{R_2}{c})+\xi_0,y_0,\zeta_0,t-\frac{R_2}{c}) = -A\frac{12\alpha^4}{\pi},$$

and the common field is this:

$$p(x,y,z,t) = -\frac{12\alpha^4}{\pi}\frac{A^2}{\rho_0 c_0^4 R^* \sqrt{\gamma^2-1}}[\varepsilon(vt-R_1\gamma,0,0)+\varepsilon(vt-R_2\gamma,0,0)], \qquad (15)$$

where

$$R^* = \frac{\sqrt{(x-vt)^2-(y^2+z^2)(\gamma^2-1)}}{\sqrt{\gamma^2-1}} = \sqrt{\frac{(x-vt)^2}{\gamma^2-1}-y^2z^2},$$

$$R_{1,2} = \frac{\pm\sqrt{(x-vt)^2-(y^2+z^2)(\gamma^2-1)}-\gamma(x-vt)}{\gamma^2-1}.$$

In particular, making measurements at point x=y=z=0, we have

$$p(0,0,0,t) = -A^2\frac{12\alpha^4 A^2}{\pi}\frac{tg\theta}{\rho_0 c_0^4 vt}[\varepsilon(vt\frac{\cos\theta}{1+\cos\theta})+\varepsilon(vt\frac{\cos\theta}{\cos\theta-1})]. \qquad (16)$$

Thus, field being measured at moment of time t is equal to the sum of values of nonlinear parameter ε at points $(1\frac{\cos\theta}{1+\cos\theta})$ and $(1\frac{\cos\theta}{\cos\theta-1})$, where l is a distance in a straight line. One of arguments can be disappeared if the values of ε for this argument is known a priori.

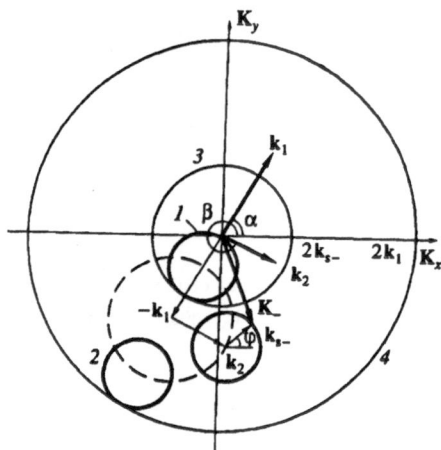

Figure 1. Diagram of the wave vectors for the process of nonlinear scattering at the difference frequency. Thick lines denote the range of vectors $\mathbf{K}_-$ used to measure the Fourier transform $\tilde{\varepsilon}(\mathbf{K}_-)$ for various directions $\mathbf{k}_1$ and $\mathbf{k}_2$:(1)$\beta = \alpha$,(2)$\beta = \alpha + 2\pi$. Thin lines are the boundaries of information regions:(3)$\beta = \alpha$,(4) use of all the possible directions $\mathbf{k}_1$ and $\mathbf{k}_2$.

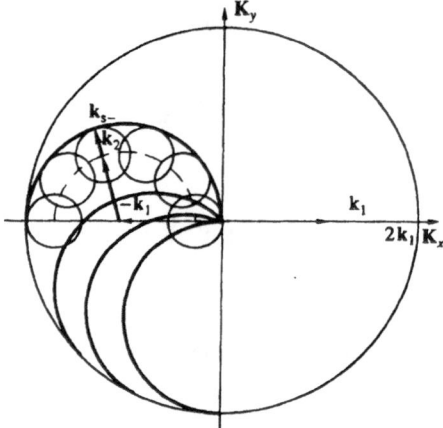

Figure 2. The wave scheme for organization of the minimal informative experiment at the difference frequency. Thick lines present the semicircles where the observation points are located.

REFERENCES

1. Y. Nakagava, Ultrasonic nonlinear parameter CT by nonlinear interaction, *Trans.IECE*,69:1215 (1988).
2. Y. Nakagava, W. Aou, A. Cai, et al., Imaging the acoustic nonlinearity parameter with Sound Waves, *Trans. IECE*, E. 71:799 (1988).
3. T. Sato, A. Fukushima, N. Ichida, Y. Mina, et al., Nonlinear parameter tomography using counterpropagating probe and pump waves, *Ultrasonic Imaging*,7: 49 (1985).
4. T. Sato, K. Yamashita, H. Ninoyu, et al.,Imaging of acoustical nonlinear parameters and its medical and industrial applications, *in:* "Acoustical Imaging-20" , Y.Wey and B.Gu, ed., Plenum Press, Ney York (1993).
5. V.A.Burov, I. E. Gurinovich, O. V. Rudenko, and E. Ya. Tagunov, Reconstruction of the spatial distribution of the nonlinearity parameter and sound velocity in acoustic nonlinear tomography, *Acoustical Physics*, 40:816 (1994).
6. Tjotta, J. Nase and S. Tjotta, Interaction of sound waves, part 1: Basic equations and plane waves, *J. Acoust. Soc. Am.*, 82:1425 (1987).

A HIGHER-ORDER STATISTICAL REFORMULATION OF THE THEORY OF DIFFRACTION TOMOGRAPHY

Woon S. Gan

Acoustical Services Pte Ltd
29 Telok Ayer Street
Singapore 0104
Republic of Singapore

INTRODUCTION

This paper is the first application of higher-order statistics (HOS) to diffraction. Previous works on HOS to acoustical imaging (holography and tomography) are using HOS to represent incoming nonGaussian signals without involving diffraction. That is only conventional (straight-line) tomography is concerned. In this work, HOS is used to describe the scattered wavefield. So far the theory of diffraction tomography is limited to the linear case, using Born and Rytov approximation and only second-order statistics such as convolution and correlation is involved. This is known as first order diffraction tomography. For real world situation, such as sound propagation in inhomogeneous medium, nonlinear effects have to be included.

THE FORWARD PROBLEM

The inhomogeneous Helmholtz wave equation is used:

$$(\nabla^2 + K_0^2)u(\vec{r}) = -o(\vec{r})u(\vec{r}) \tag{1}$$

where $\quad o(\vec{r}) = K_0^2[n^2(\vec{r}) - 1] \tag{2}$

and K_0 = wave number, $u(\vec{r})$ = acoustic pressure field at position $\vec{r}$, $o(\vec{r})$ = object function and $n(\vec{r})$ = refractive index of the medium.

Using Green's function technique and linear summation of scattered fields due to each individual point scatterer, the solution of equation (1) can be given as :

$$u_s(\vec{r}) = \int g(\vec{r} - \vec{r'})o(\vec{r'})d\vec{r'} \qquad (3)$$

Now $u = u_0 + u_s$ where u_0 = incident field and u_s scattered field. Eq (3) is a Fredholm equation of the second kind. This equation cannot be solved directly. So far the Born and Rytov approximations have been used. These approximations ignored the interaction of sound wave within the object and they assume that the field inside the object is equal to the incident field.

The purpose of this paper is to find a better estimate for the field inside the object. A new scheme of multiple scattering of sound wave within the object is proposed as follows. In this model, the object is assumed to consist of several scattering points where the incident field u_0 is first scattered by object point one and then the scattered field propagates to the object point two where it is scattered again. This is illustrated in Fig.1

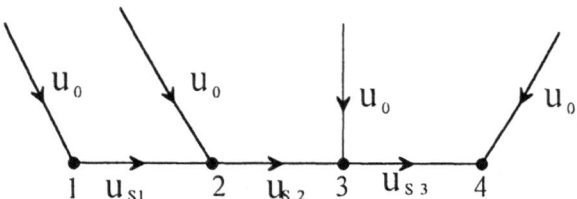

Fig.1 The Multiple Scattering Scheme Within the Object.

The Eq.(1) can be written as

$$u_s(\vec{r}) = \int g(\vec{r} - \vec{r'})o(\vec{r'})u_0(\vec{r'})d\vec{r'} +$$
$$\int g(\vec{r} - \vec{r'})o(\vec{r'})u_s(\vec{r'})d\vec{r'} \qquad (4)$$

if $u_s(\vec{r}) \ll u_0(\vec{r})$

Hence after the first scattering by object point one, the scattered field can be written as

$$u_{s_1}(\vec{r}) = \int g(\vec{r} - \vec{r'})o_1(\vec{r'})u_0(\vec{r'})d\vec{r'} \qquad (5)$$

After the second scattering by object point two, the scattered field can be written as

$$u_{s_2}(\vec{r}) = \int g(\vec{r} - \vec{r'})o_2(\vec{r'})u_{s_1}(\vec{r'})d\vec{r'} \, d\vec{r''}$$
$$= \int\int g^2(\vec{r} - \vec{r'})u_0(\vec{r'})o_1(\vec{r'})o_2(\vec{r''})d\vec{r'} \, d\vec{r''} \qquad (6)$$

After the third scattering by object point three, the scattered field can be written as

$$u_{s_3}(\vec{r}) = \int g(\vec{r} - \vec{r}')o_3(\vec{r}''')u_{s_2}(\vec{r}')d\vec{r}' \, d\vec{r}'' d\vec{r}'''$$
$$= \iiint g^3(\vec{r} - \vec{r}')u_0(\vec{r}')o_1(\vec{r}')o_2(\vec{r}'')o_3(\vec{r}''')d\vec{r}' \, d\vec{r}'' d\vec{r}''' \quad (7)$$

and so on.

One recognizes that eq.(5) denotes single correlation between $o_1(\vec{r}')$ and $u_0(\vec{r}')$
Fig.(6) denotes bicorrelation between $u_0(\vec{r}'), o_1(\vec{r}')$ and $o_2(\vec{r}'')$. Eq(7) denotes tricorrelation between $u_0(\vec{r}')$, $o_1(\vec{r}')$, $o_2(\vec{r}'')$ and $o_3(\vec{r}''')$. One finds that these multiple correlations can be expressed in terms of higher order statistics (HOS). The moments functions will be used here because sound propagation in inhomogeneous medium will give rise to chaotic phenomenon[1] and moments functions are suitable to describe deterministic signals. The advantage of using HOS is that phase information between various wave components are preserved and this provides extra information on the object.

To express in terms of HOS, eq(5) can be written as

$$u_{s_1}(\vec{r}) = m_2^{u_s}(\tau_1) = \text{Second-Order Moments} \quad (8)$$

Eq.(6) can be written as

$$u_{s_2}(\vec{r}) = m_3^{u_s}(\tau_1, \tau_2) = \text{Third-Order Moments} \quad (9)$$

Eq.(7) can be written as

$$u_{s_3}(\vec{r}) = m_4^{u_s}(\tau_1, \tau_2, \tau_3) = \text{Fourth-Order Moments}$$

In order to avoid Born approximation, the first scattered field by the first object point one, i.e. u_{s_1} is obtained from the exact scattered field due to P.M.Morse and K.U. Ingard[2] This is restricted to simple shape such as sphere. The simplest case would be that of a rigid, motionless sphere of radius a, centred at the origin. The expression for the exact wave scattered from a sphere of radius a whose centre is the polar region is

$$u_s = -A \sum_{m=0}^{\infty}(2m + 1)i^{m+1}e^{-i\delta_m}P_m(\cos\theta)\times$$
$$[j_m(kr) + in_m(kr)]e^{-2\pi ift} \quad (11)$$

where $\delta_m \approx \dfrac{-m(ka)^{2m+1}}{1^2 \bullet 3^2 \bullet 5^2 \cdots (2m+1)^2(2m+1)(m+1)}$

P_m = Legendre function of order m

j_m = Spherical Bessel function of order m

n_m = Spherical Neumann function of order m

f = frequency

INVERSE PROBLEM

The method of iteration will be used. The purpose here is to obtain the reconstructed image from the scattered field. The following basic equations are used here[3].

$$c_k^{-2}(x) = c_{k-1}^{-2}(x) + \gamma_k(x) \tag{12}$$

where c = sound velocity, x = position, γ = disturbance.
and

$$u_k(x', t) - u_{k-1}(x', t) = \iint G_{k-1}(x, x', t, t')$$
$$[u_{k-1}(x, t') + \mu u_{sk-1}(x, t')] \gamma_k(x) dx dt' \tag{13}$$

where x' = position of measurement, u_k = total wavefield = incident field + scattered field.
and $u_{sk} = \mu u_{sk-1}$.

The nonlinear iteration procedure could display chaotic behaviours such as containing some phases with complicated phase transitions[4].

The expression for the higher order scattered fields from the forward problem will be used here. One can write

$$\int g(\vec{x} - \vec{x}') o_1(\vec{x}') u_0(\vec{x}') d\vec{x}' = u_{sk1}(\vec{x}') \tag{14}$$

$$\int g(\vec{x} - \vec{x}') u_0(\vec{x}') o_1(\vec{x}') o_2(\vec{x}'') d\vec{x}' d\vec{x}'' = u_{sk2}(\vec{x}', \vec{x}'') \tag{15}$$

$$\int g(\vec{x} - \vec{x}') u_0(\vec{x}') o_1(\vec{x}') o_2(\vec{x}'') o_2(\vec{x}''') d\vec{x}' d\vec{x}'' d\vec{x}'''$$
$$= u_{sk3}(\vec{x}', \vec{x}'', \vec{x}''') \tag{16}$$

etc.
and

$$u_0(\vec{x}') = u_k(\vec{x}) \tag{17}$$

Eq(12) is also the Poincare Map for describing the chaotic sequence. Equations (12) and (13) can be used for successive iteration. Given initial model and data, one can calculate initial G and u, then find γ_1, from eq. (13) by setting $u_{sk} = 0$. The next step would be to calculate the kth values of u, G and u_s and use these values to solve eq.(13) for γ_k. The iteration produces a sequence of velocity estimates $c_k(x), k = 1, 2, \cdots$, The resulting reconstructed image will be the velocity image of the object.

NUMERICAL WORK

One of the key points in the numerical work is the computation of the multiple correlation functions. The FFT algorithm will be used to implement these. The computation will be up to the tricorrelation only. The computation procedure for FFT of finite-length functions is given as follows.

1. Let $x(t)$, $h_1(t)$, $h_2(t)$ and $h_3(t)$ be finite-length functions shifted from the origin by a, b, c and d respectively.

2. Let P be the number of samples defining $x(t)$, Q be the number of samples defining $h_1(t)$, R be the number of samples defining $h_2(t)$, S be the number of samples defining $h_3(t)$.

3. Choose N to satisfy the relationships

$$N \geq P + Q + R + S - 1$$

$$N = 2^\gamma \qquad \gamma \text{ integer valued}$$

4. Define $x(k), h_1(k), h_2(k)$ and $h_3(k)$ as follows:

$$
\begin{aligned}
x(k) &= 0 & k &= 0, 1, \cdots, N - P \\
x(k) &= x(kT + a) & k &= N - P + 1, N - P + 2, \cdots, N - 1
\end{aligned}
$$

$$
\begin{aligned}
h_1(k) &= h_1(kT + b) & k &= 0, 1, \cdots, Q - 1 \\
h_1(k) &= 0 & k &= Q, Q + 1, \cdots, N - 1
\end{aligned}
$$

$$
\begin{aligned}
h_2(k) &= h_1(kT + c) & k &= 0, 1, \cdots, R - 1 \\
h_2(k) &= 0 & k &= R, R + 1, \cdots, N - 1
\end{aligned}
$$

$$
\begin{aligned}
h_3(k) &= h_3(kT + d) & k &= 0, 1, \cdots, S - 1 \\
h_3(k) &= 0 & k &= S, S + 1, \cdots, N - 1
\end{aligned}
$$

5. Compute the FFT $x(k), h_1(k), h_2(k)$ and $h_3(k)$:

$$X(n) = \sum_{k=0}^{N-1} X(k) e^{-j2\pi nk/N}$$

$$H_1(n) = \sum_{k=0}^{N-1} h_1(k) e^{-j2\pi nk/N}$$

$$H_2(n) = \sum_{k=0}^{N-1} h(k) e^{-j2\pi nk/N}$$

$$H_3(n) = \sum_{k=0}^{N-1} h_3(k)e^{-j2\pi nk/N}$$

6. Change the sign of the imaginary part of $H_1(n)$ to obtain $H_1^*(n)$, $H_2(n)$ to obtain $H_2^*(n)$ and $H_3(n)$ to obtain $H_3^*(n)$.

7. Compute the product

$$Z(n) = X(n)H_1^*(n)H_2^*(n)H_3^*(n)$$

8. Compute the inverse FFT using the forward FFT: (note scaling by 1/N):

$$Z(k) = \sum_{n=0}^{N-1} [\tfrac{1}{N}Z^*(n)]e^{-j2\pi nk/N}$$

9. Scale the results by sample interval T

FORMATION OF CHAOTIC IMAGE

The reconstructed velocity image of the object formed will have chaotic feature due to chaotic nature of the nonlinear inversion. We call it chaotic image. For our computation, the multiple correlation will stop at tricorrelation. We call this higher order diffraction tomography because of the higher order scattered field used. The output sequence $c_k(x)$ of eq(12) would become disorder or chaotic when k as well as the inner entropy becomes large. The inner entropy for such a nonlinear system, corresponding to inversion errors, increases with k

CONCLUSIONS

The use of higher order statistics in diffraction tomography has an advantage over the B/A nonlinear parameter diffraction tomography in that phase information between wavefield components can be preserved. This gives extra information on the reconstructed image. The higher order scattered field obtained can give more accurate information for the reconstructed chaotic image because chaotic phenomenon occurs only when nonlinear effect is taken into account.

REFERENCES

1. W.S.Gan, "Application of Chaos to Sound propagation in Random Media," Acoustical Imaging, Vol.19, 99, Plenum Press, New York and London (1991).

2. P.M.Morse and K.U.Ingard, Theoretical Acoustics, 419, McGraw-Hill, New York (1968).

3. W.C.Yang, "Nonlinear Chaotic Inversion of Seismic Traces," Chinese Journal of Geophysics, 36:241 (1993).

4. H.G.Schuster, Deterministic Chaos, Cambridge University Press (1987).

STATE-OF-THE-ART OF ULTRASOUND IMAGING IN MEDICINE AND BIOLOGY

P.N.T. Wells

Department of Medical Physics and Bioengineering
Bristol General Hospital
Bristol BS1 6SY, UK

INTRODUCTION

The history of medical ultrasonic imaging can be traced back over half a century. Amongst the pioneers, Howry and Bliss (1952) in Denver and Wild and Reid (1952) in Minneapolis developed two-dimensional scanners which produced remarkable pictures despite the limited state of knowledge and the primitive nature of the instrumentation that existed at the time. From these beginnings, ultrasonic imaging has evolved so that it is no longer a laboratory curiosity but is in the first rank of radiological science. It has been estimated that about 15 per cent of all radiological work is carried out using ultrasound and that the proportion is increasing.

IMPROVEMENTS IN IMAGE PROCESSING AND DISPLAY

Two-dimensional ultrasonic imaging became a practical reality with the introduction of manually-operated scanners in which the transducer probe was moved in contact with the patient's skin (Wells, 1966). The scanning process required skill and the image was not acquired in real time, since it took about three seconds to move the probe around the patient. This limitation on scanning speed was overcome with the development of real-time imaging systems, in which the direction of the ultrasonic beam within the scan plane was controlled either mechanically (Griffith and Henry, 1974) or by means of a stationary transducer array (Bom, 1972). Although a few modern scanners still employ mechanical probes, the trend towards array systems is likely to continue. The images that are produced by a state-of-the-art commercially-available scanning instrument are extremely informative and have resolutions approaching those which are theoretically possible according to diffraction theory. An example of such a scan is shown in Figure 1. Amongst the ultimate limits to ultrasonic imaging resolution, it is the inhomogeneity of the tissues themselves that pose the most formidable problems at the present time (Harris and Wells, 1993).

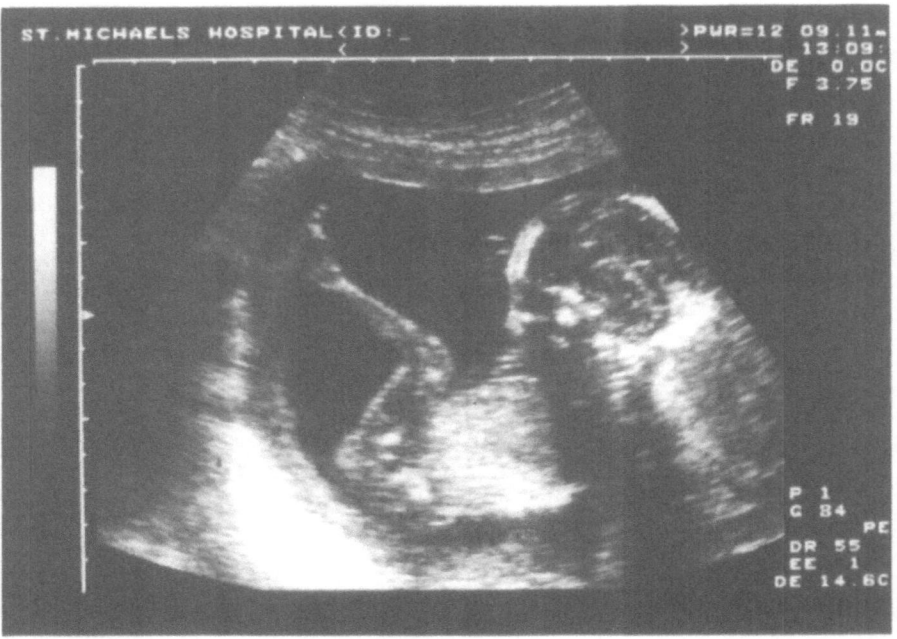

Figure 1. An example of a typical modern ultrasonic scan. The image shows a fetus in longitudinal section within the uterus. The head is on the right; a leg and foot can be seen on the left.

TRANSDUCERS AND PROBES

Modern ultrasonic scanning equipment embodies the most recent developments in materials science and digital electronics. Traditional ferroelectric ceramic transducers (Jaffe et al., 1955) are being replaced by piezoelectric composites (Gururaja et al., 1985) which have consistent and appropriate properties and which permit the construction of arrays with small elements. Small probes are in routine use for intracavitary and intravascular imaging. A typical family of contemporary probes is shown in Figure 2. The development of special-purpose probes, for example, for laparoscopic use, is proceeding rapidly.

Intracavitary and Endoscopic Systems

The most widespread use of intracavitary scanning is probably for transvaginal imaging to obtain high-resolution pictures of the uterus, its contents and neighbouring structures (Fried and Cosgrove, 1993). The subject of transoesophageal scanning of the heart has reached a high level of refinement (Salustri and Roelandt, 1995), with probes capable of transverse, longitudinal and multiplane image acquisition. Currently, the majority of probes for intravascular scanning (Gussenhoven et al., 1991) are mechanical devices producing radial images, but there have been spectacular achievements in the development of miniature cylindrical arrays for this purpose (Nissen and Gurley, 1991).

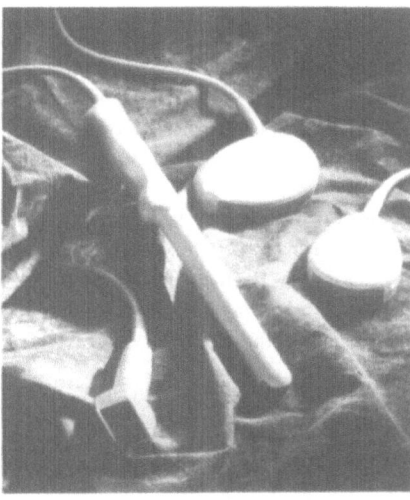

Figure 2. Some typical modern ultrasonic probes, using transducer arrays for two-dimensional scanning. Top-to-bottom: linear array for superficial structures; endovaginal curved array; and two curved arrays for abdominal imaging. (From ATL promotional material.)

IMAGING SYSTEMS

Analogue or Digital Electronics?

Almost without exception, advanced modern ultrasonic scanners operate under digital control. Much of the signal pathway is also digital. As the dynamic range and sampling speed of analogue-to-digital converters have improved, so it has become possible to digitise the received ultrasonic signal at earlier stages in the signal processing chain. Although some high-performance ultrasonic scanners still employ analogue delay lines for steering the ultrasonic beam produced by the transducer array, even this function can now be achieved digitally. Application-specific integrated circuits have been developed which allow compact devices to perform the functions previously only possible with large and complicated circuit boards. The advantages of digitisation are that, once digitised, further signal processing is carried out without distortion and the signal processes themselves can be changed by simple software changes.

High Resolution Imaging

Ultimately, the resolution of an ultrasonic imaging system is limited by the wavelength of ultrasound (Wells, 1977). At a frequency of 3 MHz, for example, the wavelength is about 0.5 mm. Increasing the frequency to, say, 20 MHz, results in a reduction in the wavelength to 75 μm. Of course, the price paid for the shorter wavelength is the corresponding increase in the attenuation of ultrasound in tissue (Wells, 1975), which limits the possible penetration. Nevertheless, imaging systems operating at frequencies of around 20 MHz can produce remarkable pictures of accessible tissues such as skin and the epithelial linings of blood vessels and other cavities.

FLOW AND MOTION

When ultrasound is reflected or backscattered by moving structures or flowing blood, the frequency of the received ultrasound is shifted from that of the transmitter by the Doppler effect (Satomura, 1957; Wells, 1989). The Doppler shift frequency usually lies in the audible range.

As a result of tissue motion or blood flow, consecutive echo wavetrains carry information about the changing range of the targets, in addition to Doppler shift frequencies. Therefore, it is possible to measure velocity by time domain processing (Bonnefous and Pesque, 1986) as well as by the Doppler effect, which depends on frequency domain processing. This is achieved by cross-correlation.

Functional Studies

One of the first clinical applications of the ultrasonic Doppler effect was for the study of blood flow in peripheral arteries in health and disease (Woodcock, 1970). The development of a collateral circulation as a result of stenosis due to atherosclerosis modifies the distal Doppler shift frequency spectrum, reducing its pulsatility and delaying its arrival time. In the case of localised arterial disease, such as is commonly found in the carotid arteries, blood flow is disturbed and the extent of disease can be gauged from the broadening of the Doppler frequency spectrum (Zwiebel, 1995).

Blood flow volume rate can be estimated from the average blood flow velocity measured by Doppler ultrasound and the cross-sectional area of the vessel, determined from pulse-echo imaging. For example, in the superior mesenteric artery, the blood flow volume rate increases in the period immediately following feeding, primarily as the result of decrease in the peripheral vascular impedance which leads to a higher level of blood flow during diastole. There is an approximately linear relationship between the calorific value of a meal and the peak increase in blood flow volume rate which follows (Braatvedt et al., 1991).

Tumour Blood Flow

The growth of malignant tumours is dependent on angiogenesis to maintain metabolic function at the advancing edge. Neovascular blood flow is characterised by low impedance, so that flow continues throughout diastole, and high-velocity components due to jet flow through arterio-venous shunts. These characteristics can be observed by Doppler ultrasound (Burns et al., 1982), as illustrated in Figure 3.

COLOUR FLOW IMAGING

The first ultrasonic Doppler imaging instruments used either continuous waves or were multi-gated, the probe being moved over the patient's skin by hand to form the picture (Reid and Spencer, 1972). Some of these early pictures were colour-coded according to the velocity of blood flow (Curry and White, 1978). The scanning speed, however, was very slow. The possibility of producing colour flow images in real time was demonstrated by Kasai et al. (1985), using autocorrelation Doppler signal detection. The two-dimensional flow images were coded in colour according to the velocity of blood flow and superimposed on real-time grey scale images of the corresponding anatomical structures. Subsequently, similar images were produced using time domain processing.

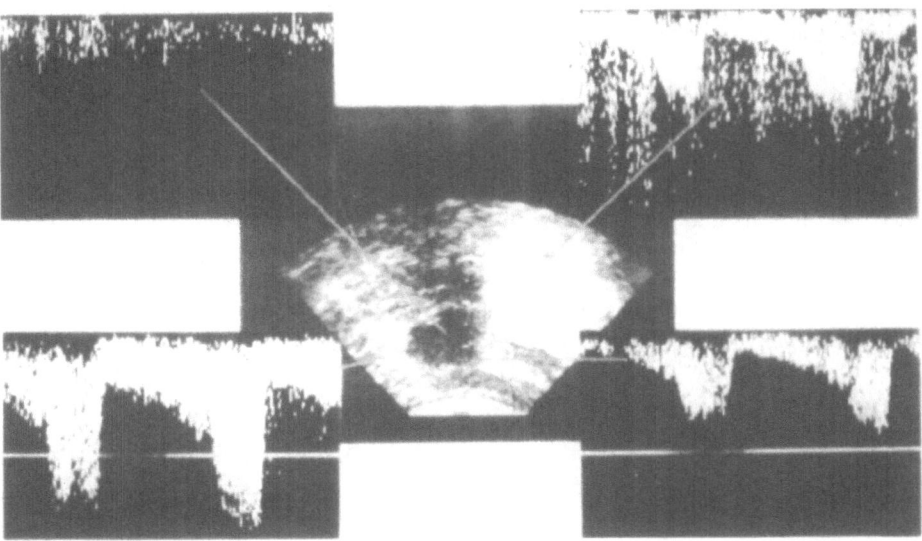

Figure 3. Two-dimensional image of malignant breast tumour, showing origins of corresponding Doppler frequency-shift spectra. Note the high-velocity signals from the advancing edge of the tumour and the relative absence of Doppler signals from the necrotic centre.

Colour flow images are quite easy to interpret, provided that account is taken of the relative directions of the ultrasonic beam and, for example, the vessel within which flow is occurring (Wells, 1994). Even this difficulty can be avoided, however, if the colour coding is made to be dependent on the strength of the flow signal, rather than on its magnitude (Burns, 1995). Such pictures are sometimes called "power Doppler" images.

Although often considered to be real time, colour flow imaging is typically at least five times slower than traditional grey scale scanning. This is because sufficient time is required for the acquisition of signals along each line-of-sight within the image to allow the flow velocity to be estimated or, at least, to allow its power to be measured (Wells, 1995). Commercial imaging instruments employ a number of stratagems to reduce the impact of this problem but its fundamental existence needs to be remembered when colour flow images are being interpreted.

CONTRAST AGENTS

The echogenicity of blood can be substantially enhanced by the administration of a contrast agent consisting of a suspension of small bubbles (Goldberg et al., 1994). Such bubbles are best if they are small enough to cross over capillary beds and if they have sufficient lifetime to be observed in the systemic circulation following intravenous injection. The bubbles are most effective if they resonate at the frequency of the irradiating ultrasound.

In addition to resonating at the fundamental frequency, bubbles also resonate at harmonic frequencies. This means that contrast agents can be detected as the second harmonic of the transmitted frequency (Schrope and Newhouse, 1993), a characteristic not possessed by soft tissues and blood. Consequently, it should be possible to develop a new family of ultrasonic imaging systems using contrast agents to improve the sensitivity of the blood flow and with the rejection of clutter signals by second harmonic detection.

IMAGE MANIPULATION AND DISPLAY

Ultrasonic imaging in contemporary practice is a two-dimensional process, albeit carried out in real time. Thus, it is often easy, at least in principle, to collect two-dimensional images of contiguous scan planes, so that three-dimensional data sets can be acquired (Halliwell et al., 1989; Masotti and Pini, 1993). Such a three-dimensional data set can be displayed in three dimensions or explored, using an image processing workstation, from arbitrary viewpoints. Figure 4 gives a typical example.

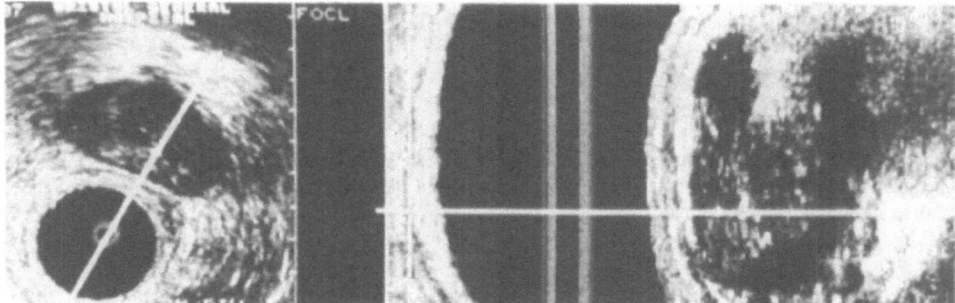

Figure 4. Example of a typical transverse scan of a prostate gland (left panel) with reconstruction of longitudinal scan (right panel) from multiple contiguous transverse scans. The white lines indicate the planes of the corresponding scans.

SAFETY OF ULTRASONIC DIAGNOSIS

The fact that ultrasound can produce biological changes and that such changes may, under some circumstances, be damaging, is well known (Wells, 1987). These effects are what make ultrasonic physiotherapy and surgery possible. Therefore, it is prudent to consider whether or not the exposure levels used in ultrasonic diagnosis are safe (Wells, 1986). If the estimate of risk is set at too high a level, a more dangerous or otherwise inappropriate diagnostic test may be used. If the estimate is too low, there may be inappropriate use of ultrasound or there may be a failure to use a more appropriate test. There is no evidence that the exposure levels used in contemporary ultrasonic diagnosis are dangerous but the application of the ALARA (as low as reasonably achievable) principle is generally accepted to be the correct approach.

It is the implications of misdiagnosis which are likely to be the principal hazard in the use of diagnostic ultrasound. Of the various kinds of misdiagnosis, false-negative results are likely to be the most dangerous, since they may result in failure to provide appropriate treatment.

CONCLUSIONS

State-of-the-art ultrasound scanning provides high-resolution two- and three-dimensional images in real time, complemented by information about blood flow. Developments in transducer technology, microelectronics, digital signal processing and image display will lead to further improvements in diagnostic capability. The use of contrast agents will enhance the clinically useful information and suppress clutter signals. Exposure conditions in contemporary practice appear to be safe. It can confidently be predicted that ultrasonic imaging will become an increasingly important branch of radiological science.

REFERENCES

Bom, N., 1972, "New Concepts in Echocardiography". Stenfert Kroese. Leiden.

Bonnefous, O., and Pesque, P., 1986. Time domain formulation of pulse-Doppler ultrasound and velocity estimation by cross correlation. *Ultrason. Imag.* 8: 75.

Braatvedt, G.D., Halliwell. M., Wells. P.N.T., Read. A., and Corrall. R.J.M., 1991, Postprandial mesenteric blood flow, *Gut* 32: 1428.

Burns, P.N., Halliwell, M., Wells. P.N.T., and Webb. A.J., 1982. Ultrasonic Doppler studies of the breast, *Ultrasound Med. Biol.* 8: 127.

Burns, P.N., 1995. Interpreting and analyzing the Doppler examination, *in* "Clinical Applications of Doppler Ultrasound", 2nd. edn., K.J.W. Taylor. P.N. Burns. and P.N.T. Wells, eds., Raven Press, New York.

Curry, G.R., and White, D.N., 1978. Color coded ultrasonic differential velocity arterial scanner (Echoflow), *Ultrasound Med. Biol.* 4:27.

Fried, A.M., and Cosgrove, D.O., 1993. Uterus and ovaries. *in* "Textbook of Abdominal Ultrasound", B.B. Goldberg, ed., Williams & Wilkins. Baltimore.

Goldberg, B.B., Liu, J.-B. and Forsberg. F., 1994. Ultrasound contrast agents: a review. *Ultrasound Med. Biol.* 20: 319.

Griffith, J.M., and Henry, W.L., 1974. A sector scanner for real time two-dimensional echocardiography, *Circulation* 49: 1147.

Gururaja, T.R., Schulze, W.A., Cross. L.E. and Newnham. R.E., 1985. Piezoelectric composite materials for transducer applications. *I.E.E.E. Trans. Sonics Ultrason.* 32: 499.

Gussenhoven, W.J., Bom, N., and Roelandt. J., eds., 1991. "Intravascular Ultrasound 1991", Kluwer, Dordrecht.

Halliwell, M., Key, H., Jenkins, D., Jackson. P.C., and Wells. P.N.T., 1989. New scans from old: digital reformatting of ultrasonic images. *Br. J. Radiol.* 62: 824.

Harris, R.A., and Wells, P.N.T., 1993. Ultimate limits in ultrasound image resolution. *in* "Advances in Ultrasound Techniques and Instrumentation". P.N.T. Wells. ed., Churchill Livingstone. New York.

Howry, D.H., and Bliss, W.R., 1952. Ultrasonic visualization of soft tissue structures of the body, *J. Lab. Clin. Med.* 40: 579.

Jaffe, B., Roth, R.S., and Marzullo. S., 1955. Properties of piezoelectric ceramics in solid-solution series lead titanate - lead zirconate - lead oxide: tin oxide and lead titanate - lead hafnate, *J. Res. Nat. Bur. Stand.* 55: 239.

Kasai, C., Namekawa, K., Koyano. A., and Omoto. R., 1985. Real-time two-dimensional blood flow imaging using an autocorrelation technique. *I.E.E.E. Trans. Sonics Ultrason.* 32: 458.

Masotti, L., and Pini, R., 1993, Three-dimensional imaging. *in* "Advances in Ultrasound Techniques and Instrumentation". P.N.T. Wells. ed., Churchill Livingstone. New York.

Nissen, S.E., and Gurley, J.C., 1991. Application of intravascular ultrasound for detection and quantification of coronary atherosclerosis *in* "Intravascular Ultrasound 1991", W.J. Gussenhoven, N. Bom, and J Roelandt. eds., Kluwer. Dordrecht.

Reid, J.M., and Spencer, M., 1972. Ultrasonic Doppler technique for imaging blood vessels, *Science* 176: 1235.

Salustri, A., and Roelandt, J.R.T.C., 1995. Ultrasonic three-dimensional reconstruction of the heart, *Ultrasound Med. Biol.* 21: 281.

Satomura, S., 1957, Ultrasonic Doppler method for the inspection of cardiac functions, *J. Acoust, Soc. Am.* 29: 1181.

Schrope, B.A., and Newhouse, V.L., 1993. Second harmonic ultrasonic blood perfusion measurement, *Ultrasound Med. Biol.* 19: 567.

Wells, P.N.T., 1966, Developments in medical ultrasonics. *Wld. Med. Electron.* 4: 272.

Wells, P.N.T., 1975, Absorption and dispersion of ultrasound in biological tissues, *Ultrasound Med. Biol.* 1: 369.

Wells, P.N.T., 1977, "Biomedical Ultrasonics". Academic Press, London.

Wells, P.N.T., 1986, The prudent use of diagnostic ultrasound, *Br. J. Radiol.* 59: 1143.

Wells, P.N.T., ed., 1987, "The Safety of Diagnostic Ultrasound", *Br. J. Radiol.* suppl. 20.

Wells, P.N.T., 1989, Doppler ultrasound in medical diagnosis, *Br. J. Radiol.* 62: 399.

Wells, P.N.T., 1994, Ultrasonic colour flow imaging, *Phys. Med. Biol.* 39: 2113.

Wells, P.N.T., 1995, Today's state-of-the-art: does colour velocity imaging overtake colour Doppler?, *J. Vasc .Invest.* 1: 38.

Wild, J.J., and Reid, J.M., 1952, Further pilot echographic studies of the histologic structure of the living intact human breast, *Am. J. Path.* 28: 839.

Woodcock, J.P., 1970, The significance of changes in velocity/time waveform in occlusive arterial disease in the leg, *in* "Ultrasonics in Biology and Medicine". L. Filipczynski, ed., Polish Scientific, Warsaw.

Zwiebel, W.J., 1995, Cerebrovascular Doppler applications. *in* "Clinical Applications of Doppler Ultrasound". 2nd. edn.. K.J.W. Taylor. P.N. Burns. and P.N.T. Wells. eds.. Raven Press, New York.

REMOVING THE AMBIGUITY FROM SINGLE IMAGE SPECKLE REDUCTION TECHNIQUES

Andy Healey[1], Sidney Leeman[2] and Leonard A. Ferrari[3]

[1]Present Address Dept. of Electrical Engineering, Imperial College
London U.K.

[2]Dep. of Medical Engineering and Physics
King's College School of Medicine and Dentistry, London U.K.

[3]Dept. of Electrical Engineering and Computing, University of California
Irvine, U.S.A.

INTRODUCTION

Almost all Medical Ultrasound pulse-echo signals suffer from interference artefacts such as speckle, which are generally attributed to the coherent nature of the imaging pulse. There is universal agreement that in a number of imaging applications a dramatic improvement in image quality would be achieved if speckle could be effectively removed from the B-mode image. In an attempt to achieve this goal a plethora of reduction techniques have been devised, most of which regard the artefact as a form of (usually multiplicative) noise. Such procedures may be split broadly into the two groups of multi— and single— image techniques. Multi-image techniques compound a number of images (of the same object region) obtained via a variation of some imaging parameter(s) in order to obtain a different realisation of the speckle component. As a result of compounding, the speckle 'noise' is generally averaged to a lower level, whereas the 'true' structure component is assumed to remain stable to the parameter change and therefore reinforced. Single-image techniques operate in the framework of a single speckle noise realisation. The standard approach is to form some (statistical) model of the speckle and / or signal components and then (optimally) estimate the signal preferentially over the noise. Such an approach is essentially statistical in nature and invariably involves some trade-off of other factors of image quality for effective speckle reduction (usually resolution). Due to factors of data acquisition time and system complexity, the single image problem is of more practical importance in real-time B-mode imaging systems.

Single-image techniques may be further classified into two classes as to whether they incorporate an explicit speckle recognition stage (adaptive) or not (non-adaptive). For many techniques the precise values of algorithm parameter are not explicitly predicted by theory, and coupled with the fact that (for many algorithms) relatively small parameter changes have an effect on the processed image, considerable ambiguity exists as to whether the image has been processed optimally. This paper specifically addresses : the nature of the resolution trade-off and ambiguities for frequency diversity techniques; ambiguities associated with the adaptive filtering methodology; and optimal parameter selection for the ZAP (Zero Adjustment Processing) algorithm.

FREQUENCY DIVERSITY PROCESSING

Frequency diversity processing [Gehlbach and Sommer, 1987] is a single image technique which operates by segmenting the received signal into a number of (partially overlapping) spectral bands, each of which are envelope detected in the temporal domain and compounded in the final image. The procedure provides a reduction in speckle at a cost of (axial) resolution. The parameters required for implementation are the shape, number and overlap of the spectral windows. It is interesting to note that a quantitative determination of the algorithm parameters may be achieved via an analysis identical to those of the Welch Periodogram - which is essentially the same procedure applied in the inverse domain.

The frequency diversity filters are applied to each rf A-line signal, and are one dimensional in operation. Consider a simple two dimensional convolutional model of the 2D rf A-line data set $rf(x,y)$, between an imaging pulse with envelope $b(x,y)$ modulating a plane wave, $\exp(i\mathbf{k}_x x)$, and a function representing the structure component $s(x,y)$,

$$rf(x,y) = b(x,y)\exp(i\mathbf{k}_x x) \otimes s(x,y) \tag{1}$$

where $\otimes$ represents the convolution operator. For a structure comprised of two equal amplitude scatterers positioned at (x_1,y_1) and $(x_1 + 1/2\mathbf{k}_x, y_1)$ destructive interference (speckle) results in artefactual fine detail in the image, see Figure 1, which is removed by frequency diversity processing at a cost of axial resolution. For a structure comprised of two equal amplitude scatterers, but of opposing sign, placed at (x_1,y_1) and (x_1,y_2) the resulting signal is given by

$$rf(x,y) = b(x-x_1,y-y_1)\exp(i[\mathbf{k}_x(x-x_1)]) + b(x-x_1,y-y_2)\exp(i[\mathbf{k}_x(x-x_1)+\pi]) \tag{2}$$

Note that for this situation there will always be a phase difference of π radians between pulse components for any centre frequency $\mathbf{k}_x$, and hence all outputs of the frequency diversity processing filters retain the speckle component. In this situation axial resolution is lost with no reduction of speckle corruption (see Figure 2). This problem may be overcome by applying frequency diversity processing on a 2D basis, where 2D filters are applied to the rf data set, the outputs of each being envelope detected and compounded in the final image (see Figures 2), allowing greater flexibility in the resolution trade-off.

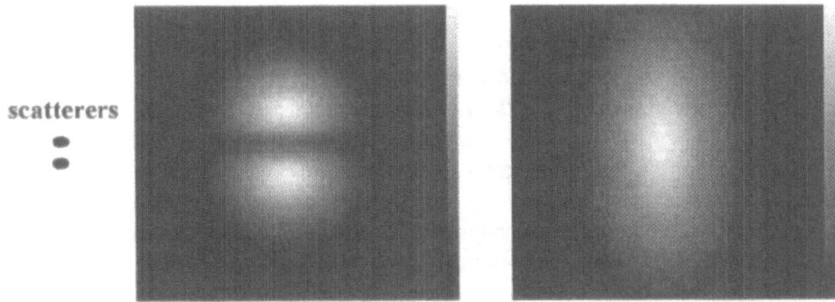

Figure 1. (image left) original showing artefactial fine detail due to speckle interference; (image right) results of frequency diversity processing showing a loss in axial resolution for effective speckle suppression.

Figure 2. (left image) original showing artefactual fine detail due to speckle interference; (middle image) results of 1D frequency diversity processing; (right image) results of 2D frequency diversity processing.

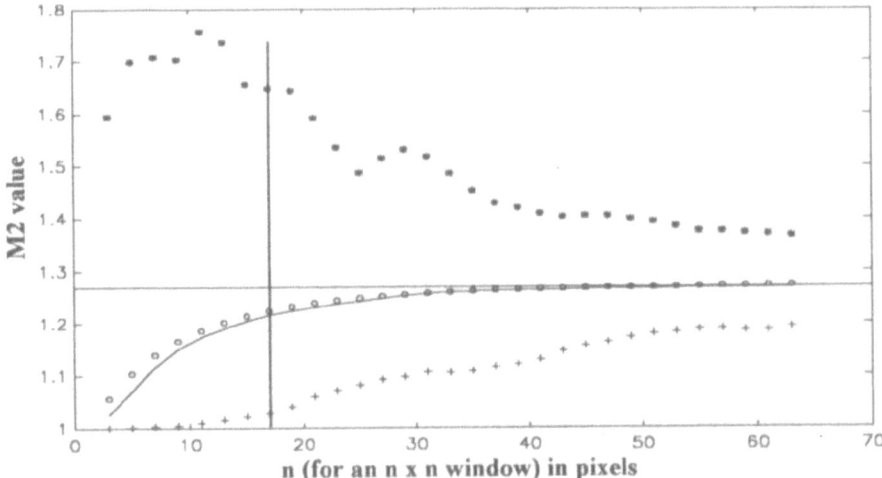

Figure 3. $M2$ value calculated as a function of window size, n, for an n by n window: mean value: 'o', 0 percentile: '+', 100 percentile: '*', and 50 percentile: solid curve. A window area of 5 speckle cells corresponds to a value for n of approximately 17.

ADAPTIVE FILTERING

Many adaptive single image post image formation filtering techniques exist. For this approach the deterministic speckle interference pattern is modelled as a stochastic problem due to ambiguity regarding the underlying structure. In order to determine optimal pixel replacement algorithms precise statistical models are required. Unfortunately it is difficult to define models which are appropriate for the complex structures being imaged and hence a number of techniques take a more heuristic approach. We will concentrate on one such technique, originally proposed by [Bamber and Daft, 1986] for ultrasound images, as it has received a great deal of development and use by a number of groups. As currently implemented, this approach relies on some statistical feature(s), in order to recognise the local presence of speckle. A common speckle recognition parameter, due to its ease of calculation, is the second moment of the amplitude distribution (M2) defined as

$$M2 = \frac{\langle p^2 \rangle}{\langle p \rangle^2} \qquad (3)$$

where p represents the amplitude image value and $\langle p \rangle$ is the mean of p. $M2$ has a theoretically predicted value of 1.27 (3 s.f.) for fully developed speckle. The value for a region of 'full structure' (no speckle present) is chosen depending on the object being imaged (see for example [Crawford *et al*, 1993]) and is around 2.0. A speckle similarity coefficient, k, is then calculated via

$$k = \frac{(M2 - M2_{speckle})}{(M2_{structure} - M2_{speckle})} \qquad (4)$$

where $M2$ is the local segment value. A window is then passed over the image where the similarity coefficient is calculated. The window centre pixel, p_c, at each location is then replaced via,

$$p_c = \langle p \rangle + k(p_c - \langle p \rangle) . \qquad (5)$$

However, as the speckle noise component is correlated, the statistic $M2$ depends upon the dimensions of the local image region for which it is calculated, and a bias is introduced for the window sizes commonly implemented (minimum size approximately 5-6 point spread functions [Crawford *et al*, 1993]). Hence current filter implementations may tend to overestimate the speckle content of an image, and an appropriate correction factor should be introduced. Figure 3 shows a plot of $M2$ values as a function of window size for a simulated fully developed speckle pattern image of 256 by 256 pixels, and an $M2$ for the entire image of 1.27 (3 s.f.).

ZERO ADJUSTMENT PROCESSING

Zero Adjustment Processing [Healey *et al*, 1991] is a single image speckle reduction technique which utilises a signals based definition of the artefact, and claims not to suffer the usual resolution trade-off associated with conventional single image techniques. In this definition speckle is associated with destructive interference effects.

The procedure is two stage in nature. The first stage utilises a deterministic marker for speckle via a consideration of the signals phase (localised large deviations in instantaneous frequency (IF)). The second stage provides an envelope correction to compensate for the local destructive interference, based on a simple few scatterer model. There are a number of algorithm parameters which need to be fixed, for application of the ZAPping technique. In contrast to certain other single image techniques, where freedom is given to the operator to arbitrarily change parameters (based on the *expected* appearance of the speckle free image, rather than any real knowledge of the 'true' result), parameter selection is rather tightly specified in this case. The ZAPping parameter set comprises:

1)   A threshold for IF deviation to be used as the interference marker — This threshold may be set to zero (all IF maxima processed). However, for very small IF deviations the envelope correction is minimal and hence a computational saving may be achieved with little difference to the processed image, by implementing a threshold of an appropriate level. A typical value is one tenth of the signal centre frequency.

2)   The size of data window utilised in order to encapsulate the local interference effects — This is chosen to be one and a half pulse lengths, for applicability of the few (2-3) scatterer model and may be calculated via a consideration of the BT product.

3)   The amount of rotation of the speckle or 'ZAP' zero — One way to determine an optimal value may be calculated with recourse to computer simulation. A two scatterer model was employed with a gaussian pulse of 2MHz centre frequency and 2 MHz bandwidth.   rf signals were generated for a (temporal) separation of pulses of equal magnitude, from 0 to 0.83 μs in steps of 0.0416 μs. At each separation the relative phase of the scatterers was varied uniformly between 0 and 2π (32 steps). The speckle-free signal was defined as the convolution between the envelope of the imaging pulse and the magnitude of the scattering structure. An error function was defined as the sum of the absolute difference between the processed and the speckle-free signals. Figure 2 shows a plot of the error function against the ZAP zero rotation angle normalised to the pulse bandwidth. The optimal value for this simulation is approximately one fifth of the pulse bandwidth.

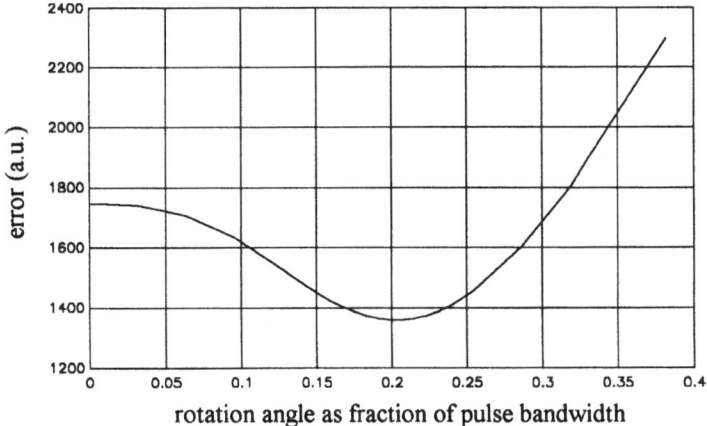

Figure 4. Plot of error function (see text) against zero rotation angle, normalised by the pulse bandwidth. A minimum occurs at approximately one fifth of the pulse bandwidth.

CONCLUSIONS

Ambiguities exist in the operation and implementation of conventional single image speckle reduction techniques. For frequency diversity processing an extension to two dimensions is proposed, which provides greater flexibility in the speckle reduction-resolution trade-off. Statistical parameters utilised in current adaptive filtering techniques, such as $M2$, are biased for typical window sizes, and an appropriate correction factor should be incorporated. The ZAPping technique has effectively only one parameter — the zero rotation angle. A method to quantitatively fix this parameter has been proposed, based on a simple computer simulation.

REFERENCES

Crawford D. C., Bell D. S., and Bamber J. C., Compensation for the signal processing characteristics of ultrasound B-mode scanners in adaptive speckle reduction, Ultrasound in Med. & Biol., **19**, no. 6, pp. 469-485, 1993.

Gehlbach S. M. and Sommer F. G., Frequency diversity speckle processing, Ultrasonic Imaging **9**, 92, 1987.

Healey A. J., Leeman S. and Forsberg F., Turning Off Speckle, Acoustical Imaging **19**, 1991.

REDUCTION OF PHASE ABERRATION IN A DIFFRACTION TOMOGRAPHY SYSTEM FOR BREAST IMAGING

Michael P. André[1,2], Helmar S. Janée[1,2], Gregory P. Otto[3], Peter J. Martin[1,3], and Joie P. Jones[4]

[1]Radiology Service, Veterans Affairs Medical Center, [2]Department of Radiology, University of California, San Diego, CA 92093-9114, [3]ThermoTrex Corporation, San Diego, CA, and [4]University of California, Irvine, CA

INTRODUCTION

In the past three years we have developed a diffraction tomography system for research in medical imaging.[1,2] The system was designed to address some of the aspects of current medical ultrasound that limit its use in many applications. For example, high-resolution ultrasound is restricted to a small field of view which can make it difficult to locate the area of interest and which does not allow ready comparison of normal and suspected abnormal tissues. As a result, the quality of medical sonography greatly depends on the skill of the operator who must recognize the important images and capture them on film for review by physicians. The findings of sonography are therefore not easily reproduced.

One of the most challenging applications of high-resolution sonography is in the breast where it is widely employed, but its use in the United States is restricted to determining whether a breast mass is cystic or solid. Breast ultrasound is not accepted as a screening modality particularly since it is unreliable for detection of microcalcifications. However, several recent reports have suggested that when expertly applied, sonography in conjunction with mammography may be able to increase the accuracy of diagnosis and reduce the need for biopsy.[3,4]

Computed tomography applied to ultrasound promises many advantages over conventional ultrasound but it has proven difficult to implement *in vivo*. Our approach employs advanced instrumentation that acquires a rich data set encompassing the entire scattered field, $S(\vec{x})$, around the breast. Several solutions to invert $S(\vec{x})$ have been described using the Born approximation, but in the breast these image reconstructions produce artifacts due to "phase aberration" error.[5,6] This paper describes methods we developed to address phase aberration which have improved the image quality in our experimental system. The instrument is now employed in a new clinical evaluation to systematically study a wide range of breast types and known masses.

MATERIALS AND METHODS

Breast sonography requires an imaging system capable of accommodating a range of tissue sound speeds of about ±5% relative to water, attenuation of 1 dB/cm/MHz and it is desirable to have resolution below 1 mm. At 1 MHz, the average breast is ~60λ in diameter, presents ~10 dB of attenuation and induces two to three "waves" of aberration. The aberration problem has been a difficult one to solve in previous breast tomography research.[7,8]

Our system uses expanding cylindrical wavefronts from a toroidal array to probe the medium. The reconstruction transforms the measured set of data into a geometrical form analogous to plane wave transmitters and receivers sequentially rotated around the medium. The acoustic wave equation for a spatially varying sound speed and density reduces to the familiar Helmholz relation for the field ψ: $[\nabla^2+S(\vec{x})]\psi(\vec{x})=0$. $S(\vec{x})$ is the complex scattering potential which can be derived as $S(\vec{x})=k^2(\vec{x})-k_0^2-\rho(\vec{x})^{\frac{1}{2}}\nabla^2\rho(\vec{x})^{-\frac{1}{2}}$, where k is the complex wave number and ρ is density. For our experimental design, the complex amplitude of the scattered acoustic wave, m_{jk} (j,k=1,2,...N, N=512 or 1024), measured at transducer j due to insonification of the medium by transducer k is described by

$$m_{jk} = \sum_{p,q=-\infty}^{\infty} \int d\vec{x}\ H_p'(k_0 r_0)\ H_q'(k_0 r_0)\ J_p(k_0 r)\ J_q(k_0 r)\ e^{ip(\theta-\theta_j)+iq(\theta-\theta_k)}\ S(\vec{x}) \qquad (1)$$

In Equation 1, J_n and H_n' are the Bessel function and first derivative of the Hankel function, respectively. H_n' is indicative of a dipole (cos θ) radiation beam pattern and is refined for our toroidal arrays by applying a gain calibration process. This first-order approach to reconstruction utilizes the Born approximation to the wave equation. The scatter field in the empty water bath, or incident wave, is subtracted from the measured object data which contains the incident plus scattered field, $m_{jk} = m_{jk}^{object} - m_{jk}^{incident}$. Equation 1 is inverted to obtain

$$\tilde{S}(k_x,k_y) = \tilde{S}(-2k_0\ \sin\bar{\theta}\ \cos\frac{\Delta\theta}{2},\ 2k_0\ \cos\bar{\theta}\ \cos\frac{\Delta\theta}{2}) = \mathscr{F}_2[\frac{\mathscr{F}_2[(m_{jk})_{pq}]}{H_p'\ H_q'}]_{\alpha\beta} \qquad (2)$$

where $\tilde{S}$ is the discrete two-dimensional Fourier transform of the true scattering potential of the medium, $\mathscr{F}_2[\ ]$ represents the discrete two-dimensional Fourier transform operation, $\bar{\theta} = (\theta_\alpha+\theta_\beta)/2$ and $\Delta\theta = \theta_\alpha-\theta_\beta$. The discrete inverse two-dimensional Fourier transform of equation 2 recovers $S(\vec{x}) = \mathscr{F}_2^{-1}[\tilde{S}(\vec{x})]$. This algorithm produces an image or map of $S(\vec{x})$ with a spatial bandwidth of $2k_0$ which is equivalent to a diffraction limit of $\lambda/4$ or 0.37 mm at 1 MHz and 0.75 mm at 0.5 MHz in water. We represent the solution to Equation 1 as proportional to equation 2 expressed in terms of $k^2(\vec{x})$, where the real component represents sound speed and the imaginary component represents attenuation. This paper reports results to extend the Born approximation by solving Equation 2 in an iterative fashion described in Reference 1. Briefly, this method synthesizes pulses using data from a multiple-frequency acquisition and sums them as a Fourier series. A map of the time-of-flight, $\tau(x,y)$, to each point in the object is extracted for each transmit-receive pair. The phase correction, $e^{-i\omega\tau(x,y)}$, is applied to each single frequency image which are then summed.

The instrumentation in this research employs large-scale (20-cm diameter) toroidal arrays of 1024 and 512 transducer elements which operate with center frequencies of 1.0 and 0.5 MHz, respectively.[9] The array is installed on a modified stereotactic breast biopsy table (Lorad, Danbury, CT). Each element is spaced at less than $\lambda/2$ and acts in turn as a transmitter while the remaining transducers record the signal scattered from the object. The two arrays are operated in either broad-band (340-660 and 687-1250 kHz) or discrete frequency modes. The large 20-cm ring of transducers illuminates the tissue via a heated coupling bath the characteristics of which may be altered. The array is controlled through a high-speed 16-channel multiplexing network (Thermotrex Corporation, San Diego, CA) that acquires 128 MB of data in three seconds for the 1 MHz array and 32 MB in less than one second for the 0.5 MHz array. Low acoustic intensities are employed. In this experimental configuration, the patient lies prone on the table with the breast in the dependent position suspended in the water bath. The transducer elements are long in the elevation direction to ensure cylindrical wavefronts. This produces a slice thickness of about 10 mm in water. The resulting image is either a 1024x1024 or 512x512 matrix of the complex scattering potential, $S(\vec{x})$, of the medium.

RESULTS

Four major approaches were studied to reduce aberration error which results in part from the neglected terms in the approximate wave equation. These methods were: 1) computing the magnitude image, 2) combining multiple frequencies across the bandwidth for the two arrays, 3) applying an iterative phase aberration correction algorithm, and 4) varying the sound speed of the water bath, k_0.

In computer simulations reported previously, the useful range of the Born approximation was increased by calculating the magnitude value of $S(\vec{x})$.[1] For comparison, experimental data was generated on the 1 MHz array for thin-walled latex tubes 1.5 cm in diameter that were filled with saline solutions varying in sound speed over a wide range from -11 to +8%. Attenuation in these solutions was negligible. Figure 1 shows the mean pixel value in the tubes for the first-order Born reconstruction versus measured sound speed. Aberration is evident in that the real value is not linear with sound speed and the imaginary component is not zero (no attenuation). The magnitude of the data in Figure 1 is plotted in Figure 2 as object contrast-to-noise ratio and shows significantly better linearity. Data from the iterative reconstruction employing aberration correction is also plotted in Figure 2 and appears to improve linearity. The sound speed contrast sensitivity for this corrected data is approximately 0.1%.

The performance of the iterative algorithm is demonstrated in Figure 3 for a breast-

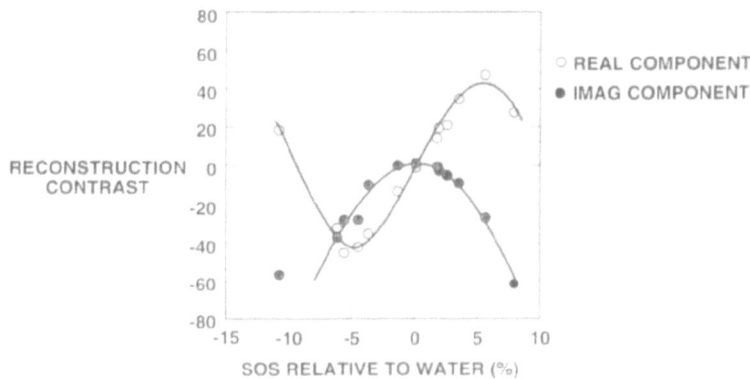

Figure 1. Born reconstructed values for saline solutions from 1 MHz array.

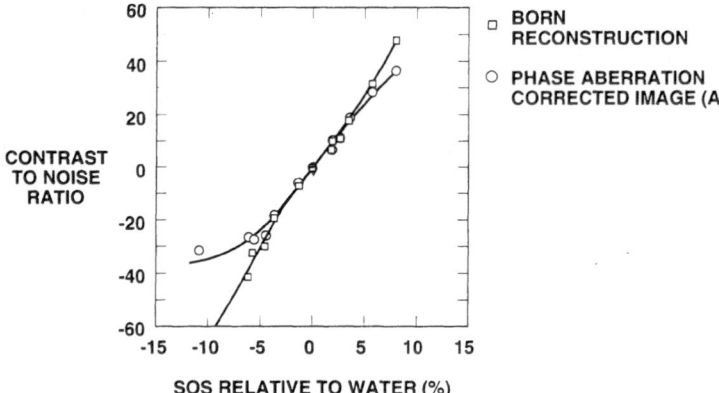

Figure 2. Data for magnitude and aberration-corrected reconstruction.

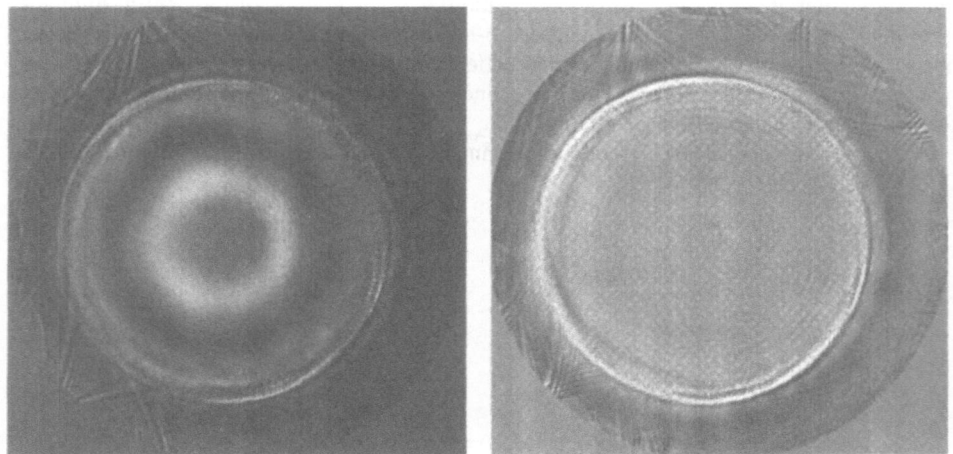

Figure 3. 13 cm diameter balloon containing saline with +2% relative sound speed. Left image is uncorrected, right image is iterative reconstruction of the same data.

sized (13 cm diameter) latex balloon. The balloon was filled with saline solution with sound speed +2% relative to the surrounding water bath. The left image is a single-frequency reconstruction at 500 kHz using the simple Born method and the right image is the same data reconstructed with phase aberration correction. At least two "waves" of aberration are apparent in the left image while the corrected image shows good uniformity.

The effect of averaging multiple frequencies across the transducer bandwidth was examined using a fresh porcine kidney. After surgery, the kidney was kept immersed in physiologic saline to minimize air bubbles, the capsule was removed and data was acquired using the 1.0 MHz array. In Figure 4, the left and center magnitude images were reconstructed with the simple Born algorithm. The left-hand image is from single-frequency data acquired at 987 kHz and the center image is the combination of ten single-frequency images from 687 to 1250 kHz. We speculate that the clear improvement in image quality is in part due to the fact that the spatial characteristics of the artifacts and image speckle apparent in the single-frequency image are frequency dependent. Combining multiple frequencies smooths these phenomena but reinforces the object data which is relatively independent of frequency. In general, we find that there is no significant artifact reduction

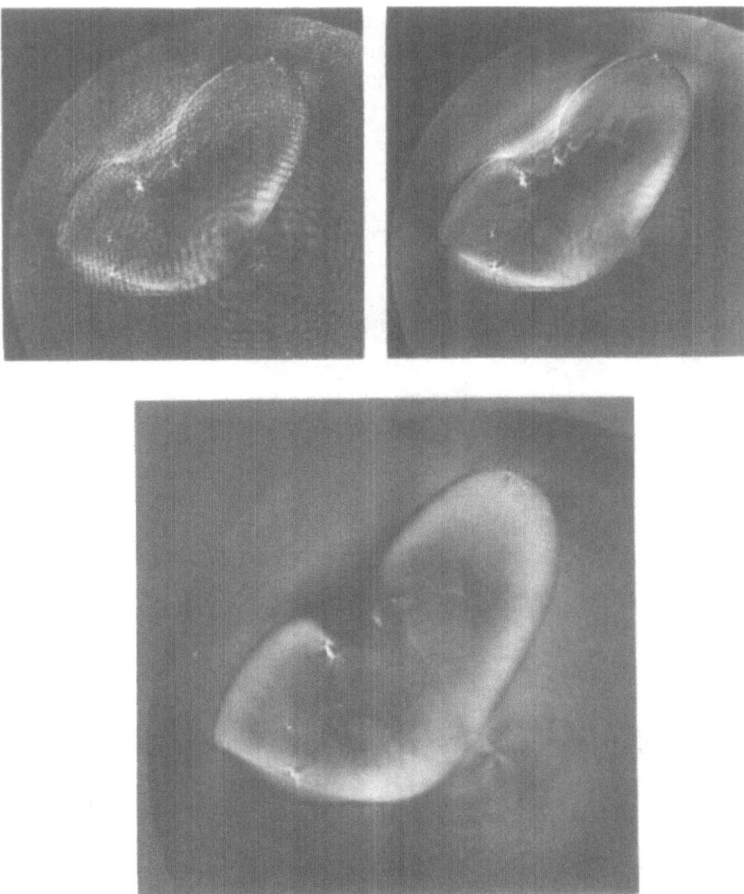

Figure 4. Magnitude images of a fresh porcine kidney: left acquired at 987 kHz, right from ten images acquired at 687-1250 kHz, bottom aberration correction of right.

with more than six frequencies at 1 MHz and 12 at 0.5 MHz. We have also shown[9] that this level of frequency summing does not significantly reduce the point spread function. The right-hand image in Figure 4 was reconstructed employing the iterative algorithm with aberration correction for the same ten frequencies as the center image. Some of the large-scale artifacts, which are possibly due to strong reflections, are substantially reduced. Small air bubbles are visible in the kidney images as bright point scatterers. These points are increasingly better resolved with frequency summing and abberration correction.

It is apparent from the expression for $S(\bar{x})$ that one may be able to reduce the range of perturbations induced in the incident wave by the object if k_0 is adjusted to better match the object. This effect--which is a complex function of frequency, object size and impedance--was examined in a 10-cm diameter tissue-equivalent gel phantom. The material was made following the method of Madsen, et al,[10] and was measured to be 1650 m/sec and 0.5 dB/cm at 500 kHz. The phantom contains two 1-cm tubes and was placed in the water coupling bath. The concentration of saline in the bath and in one of the tubes was increased from 0 to 2.09 molality to achieve sound speeds from 0 to +8% relative to distilled water. The second tube contained distilled water in all cases. The images were reconstructed without iterative aberration correction at a single frequency (500 kHz). Four images are shown in Figure 5. The upper right-hand image, in which k_0 was adjusted to +2%, shows the least aberration effects with better uniformity in the homogenous gel and the correct relationship between the two tubes; the tube with water (0%) is dark while the saline tube is bright (higher sound speed). The other images exhibit "cupping" artifacts in the gel and in the lower two images the relationship between the two tubes appears to be reversed. Adjusting k_0 may be a useful method to improve imaging for some objects but its effects are complicated and require further study.

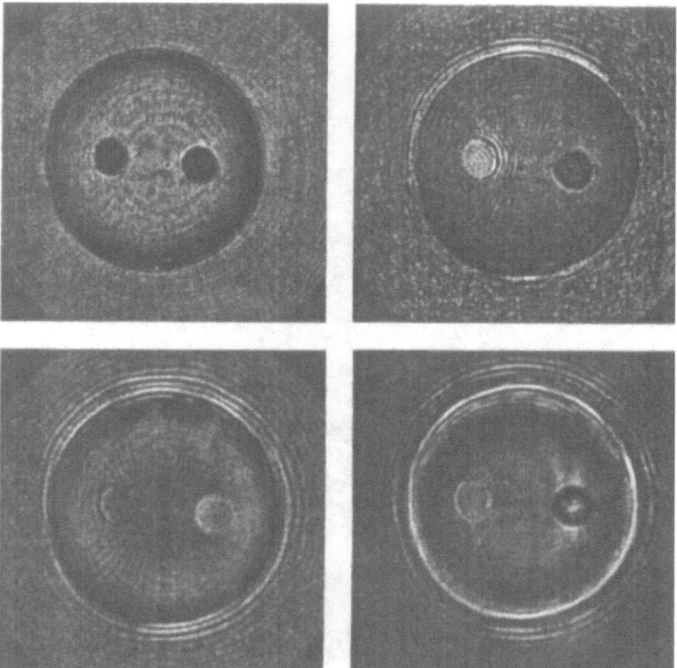

Figure 5. Single-frequency reconstructions of a tissue-equivalent gel. Sound speed of bath increases from 0-6%, left to right, top to bottom.

CONCLUSIONS

Several experiments were performed to examine methods to manage phase aberration in a diffraction tomography system. Image quality was significantly improved and artifacts were significantly reduced by using a multiple-frequency iterative algorithm. The lower operating frequencies (0.5 and 1.0 MHz) of the large 20-cm arrays, in part, make it practical to implement these methods.

The following points were demonstrated: 1) calculating the magnitude of the complex scattering potential extends the range of the Born approximation, 2) frequency averaging reduces spatial artifacts with minimal degradation of resolution, 3) the iterative reconstruction algorithm corrects large distortions, and 4) adjusting the sound speed of the coupling bath (k_0) can reduce aberrations depending on the properties of the object.

Twenty-one women, both asymptomatic and symptomatic, have been successfully imaged by this technique and the results show promise for clinical evaluation. We have begun a systematic study of a wide variety of breast sizes and types and are exploring the utility of this diffraction tomography system for detection and characterization of known breast masses. It will be interesting to further explore optimization of the system for breast imaging.

REFERENCES

1. M.P. André, P.J Martin, G.P. Otto, L.K. Olson, T.K. Barrett, B.A. Spivey, D. A. Palmer. "A new consideration of diffraction computed tomography for breast imaging: Studies in phantoms and patients," *Acoustical Imaging* 21, (1995), in press.
2. P.J. Martin, M.P. André, B.A. Spivey, D.A. Palmer, "Sonic computed tomography for breast imaging." In: *Breast Ultrasound*, B.J. Hackeloer, H. Madjar, J. Taubner, Eds. (S. Karger, Freiburg, 1994), pp 52-58.
3. D. Kopans, J. Meyer, K. Lindfors, "Whole breast US imaging; four-year follow-up," *Radiology* 157, 505-507 (1985).
4. E. Sickles, R. Filly, P. Callen, "Breast cancer detection with sonography and mammography," *Amer J Roentgenol* 140, 843-845 (1983).
5. M. Kaveh, R. Mueller, R. Rylander, T. Coulter, M. Soumekh, "Experimental results in ultrasonic diffraction tomography," *Acoustical Imaging* 9, 433-450 (1979).
6. A. Devaney, "A filtered backpropagation algorithm for diffraction tomography," *Ultrasonic Imaging* 4, 336-350 (1982).
7. J.F. Greenleaf, S.A. Johnson, W.F. Samayoa, F.A. Duck, Algebraic reconstruction of spatial distributions of acoustic velocities in tissue from their time-of-flight profiles, *in:* "Acoustic Holography," N. Booth, ed., Plenum, New York, (1975).
8. P.L. Carson, C.R. Meyer, A.L. Scherzinger, T.V. Oughton, "Breast imaging in coronal planes with simultaneous pulse-echo and transmission ultrasound," *Science* 214:1141-1143 (1981).
9. M.P. André, H.S. Janée, G.P. Otto, P.J. Martin, "Diffraction tomography with large-scale toroidal arrays," *Intl J Imaging Systems & Technol* 8, (1996), in press.
10. E.L. Madsen, J.A. Zagzebski, G.R. Frank, "Oil-in-gelatin dispersions for use as ultrasonically tissue-mimicking materials," *Ultrasound in Med & Biol* 8(3):277-287 (1982).

TWO-DIMENSIONAL FREQUENCY DOMAIN PHASE ABERRATION CORRECTION

Mahmoud E. Allam and James F. Greenleaf

Biodynamics Research Unit
Department of Physiology & Biophysics
Mayo Clinic/Foundation
Rochester, MN 55905

INTRODUCTION

Phase aberration due to tissues with inhomogeneous acoustic speeds is a major source of image degradation in medical ultrasound. In most phased array pulse-echo ultrasound systems, the delays used to focus and steer the beams are calculated assuming constant acoustic speed. However, it is known that the acoustic speed varies for different types of tissue, resulting in defocusing, loss of resolution, and blurring of the image. The issue of phase aberration measurement and correction has been the focus of several research groups in recent years[1,2,3]. Many of the existing techniques correct for aberration by estimating the phase error profile and using the estimate to align the rf signals.

In this paper, we present a new technique for phase deaberration for linear arrays. The deaberration is performed in the two-dimensional spatial-temporal frequency domain without the need to estimate a phase error profile. The correction is therefore applicable only for the receive end of the imaging system and cannot be applied to improve the focusing by retransmission. The 2D spatial-temporal frequency domain representation of signals has been used successfully in radar and sonar array processing for direction of arrival estimation[4,5], and also has been used in Doppler ultrasound to estimate blood flow velocity[6,7]. In the context of phase aberration correction, a modified version of the 2-D domain representation was used to estimate the aberration delays between signals received by pairs of adjacent array elements[8]. The 2D frequency domain representation also allows the detection and correction of multiple images artifact. The proposed method is simple and does not require retransmission. It can potentially be implemented in real time as it is neither adaptive nor iterative and its main computational load consists of Fast Fourier Transforms (FFTs).

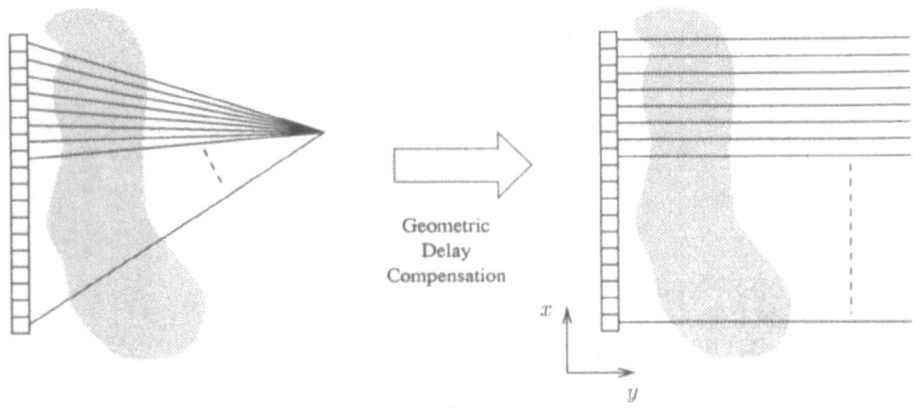

Figure 1. The geometric delay compensation can be seen as a mapping of the circular wavefronts into plane waves parallel to the array.

THEORY

In coherent imaging systems, the first step consists of aligning the radio-frequency (rf) signals received by the array elements. This is realized by compensating for the geometric delays which are calculated assuming a constant speed of sound. In an ideal homogeneous medium of propagation, the geometric delay compensation can be seen as mapping the spherical wavefronts into plane waves parallel to the array (Fig. 1). In this case, there is no difference in the time of arrival of these wavefronts and they will have zero spatial frequency. However, in the presence of inhomogeneous media, the difference in the acoustic speed causes the signals to arrive to array elements at different times and the wavefront will have non-zero spatial frequency components.

To represent the rf echoes received by an array of elements in the 2D spatial-temporal frequency domain, a 2D Discrete Fourier Transform (DFT) is applied after the geometric delay correction. Let f_t be the temporal frequency of the received signal and f_s the spatial frequency along the array axis. The difference in phase of the signal received at element i in the array with respect to the adjacent element, $i-1$ is given by,

$$\Delta\phi = 2\pi f_t \frac{[y(x_i) - y(x_{i-1})]}{v},\tag{1}$$

where $y(x_i)$ is the thickness of the aberrating layer is the path of the signal received by the i th element and v is the speed of propagation inside the aberrating layer. The same phase delay as a function of spatial frequency is written as

$$\Delta\phi = 2\pi f_s D,\tag{2}$$

where D is the distance between adjacent array elements. From Eqns. (1) and (2), we can write

$$f_s = f_t \frac{[y(x_i) - y(x_{i-1})]}{vD}\tag{3}$$

This means that the relationship between the spatial and temporal frequencies is proportional to the differential change in phase along the array axis. For a fixed inter-

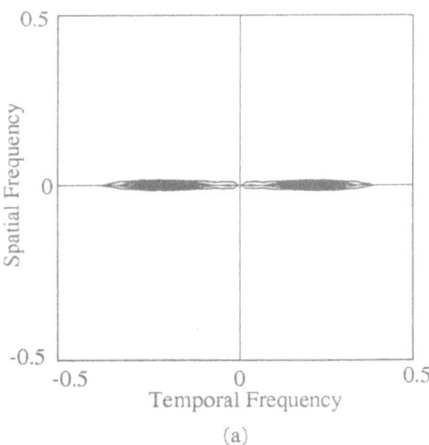
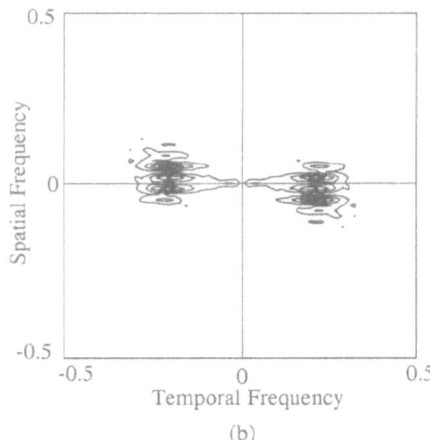

Figure 2. The 2D spatial-temporal Fourier spectrum of the rf data after geometric delay compensation without aberration (a), and with aberration (b). The temporal and spatial frequencies are normalized by multiplying them by the sampling period and inter-element distance respectively.

element distance, the slope of the linear relationship is only a function of the thickness of the aberrating layer and its acoustic speed. Figure 2 shows the 2-D FFT of the echoes received by a 64 element array with and without aberration. It is clear that, for the aberration-free snapshot, the spatial frequency consists of a narrow strip centered at zero. On the other hand, for the phase aberrated snapshot, the spatial frequency is spread out. Therefore, to correct for phase aberration, we need to reduce the variation in spatial frequency.

A possible phase aberration correction method would be to collapse the spatial frequency spectrum on the temporal frequency axis. Analytically, this can be is implemented by calculating the spatial Fourier transform of the received snapshot then summing over spatial frequency to yield a 1D spectrum. In this case, the phase corrected A-line is the inverse Fourier transform of this 1D spectrum. However, if the relationship between the spatial and temporal frequencies is expressed as a function of the direction of arrival, it can be shown that[9],

$$f_s = \frac{f_t \sin \theta}{c}. \tag{4}$$

where θ is the direction of arrival of the wavefront. An echo returning from the focusing region will have a zero angle of arrival and zero spatial frequency. This implies that non-zero spatial frequencies could also correspond to echos from scatterers outside the focusing region. It is now clear that by simply collapsing the spatial frequency spectrum onto the temporal frequency axis, contributions from all scatterers in the vicinity of the focusing region will be included. This obviously reduces the lateral resolution of the imaging system. Therefore, there is a trade-off between aberration correction and lateral resolution. To preserve the lateral resolution of the system, only small variations in spatial frequency components can be included in the construction of the corrected A-lines.

Because the 2D Fourier representation reveals information about the direction of arrival of the echoes, it can be used to detect the occurrence of multiple images arti-

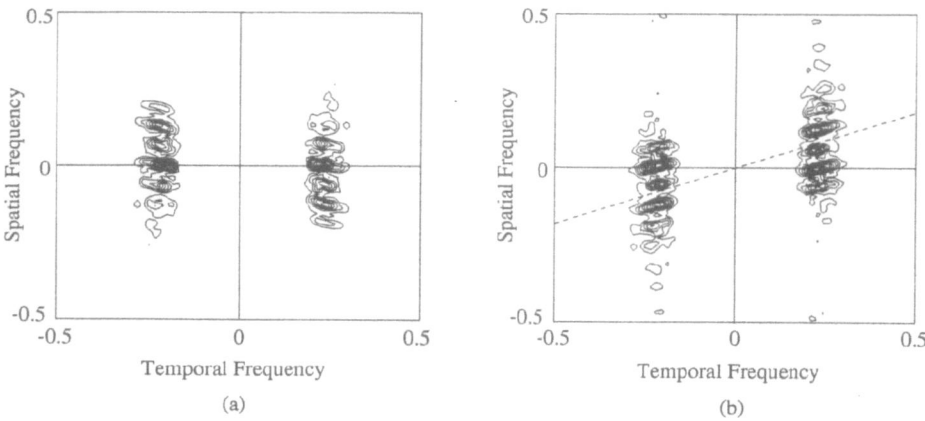

Figure 3. 2D spectrum of rf echoes in the presence of an aberrator causing multiple images artifact, (a) represents a patch corresponding to a true image, and (b) a patch corresponding to a ghost image.

fact. This artifact, also known as the multipath phenomenon, occurs when refractions at the surface between different types of tissue cause a change in the direction of the beam and/or beam splitting. As a result, the echoes received by the array will arrive from directions other than that corresponding of the intended focusing direction. Such signals will have a non-zero angle of arrival, θ, and from Eqn.(4) will appear as radial lines in the 2D frequency spectrum. To make use of this property, a radial projection is performed to accumulate all signal components at all temporal frequencies, while preserving their direction of arrival. Two different methods were proposed to implement the radial projection[4]. After performing the radial projection, the direction of arrival of the echoes is determined by calculating the centroid of the projected spectrum. The A-line is then reconstructed by applying a 1D inverse Fourier transform of the frequency components along the line of projection. This projection operation is the same as that used in wideband radar and sonar signal processing, and recently used by us for blood flow measurements using ultrasound[8].

RESULTS

Experiments were performed to test the proposed technique and its ability to correct multiple images artifacts. A strong aberrator placed between the array and the phantom caused the ultrasound beam to split into three components, thus creating three images of each wire in the phantom; a *true* image and two *ghost* images. An ATL-Ultramark®8 scanner was used. The array consisted of 48 elements with inter-element distance of 0.279 mm. The frequency was 2.75 MHz with a 50 % relative bandwidth. The rf signals were sampled at 12 MHz. Figure 3 shows the 2D spectra from two patches in the image; the first one corresponds to a true image, having a spatial frequency centered at zero, and the second is for a patch corresponding to a ghost image with a non-zero spatial frequency centroid. Fig. 4-(a) shows the aberrated B-scan image. The 2D spectrum of Fig. 3-(b) corresponds to the rf data of the marked white line in Fig. 4-(a), with the projection line indicated by the dashed line in Fig.

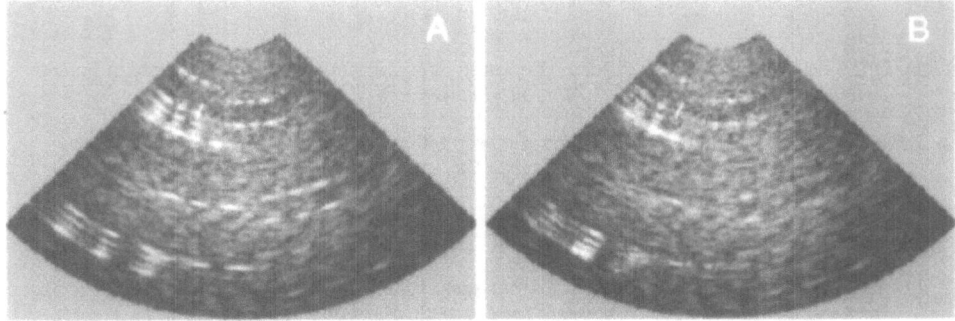

Figure 4. B-scan with multiple images artifact without (A) and with correction(B). The white line represents the patch analyzed in Fig. 3-(b).

3-(b). The corrected A-lines were obtained by calculating the inverse Fourier transform of the 1D spectrum corresponding to the projection line of the centroid for each patch in the image. The corrected B-scan image is shown in Fig. 3-(b).

CONCLUSION

In this paper, a novel approach to the phase aberration problem is proposed. Unlike other correction methods, no delay profiles are estimated and no retransmission is involved. The method consists of an additional step to be applied during beamforming, after compensating for the geometric delays. The compensation for the remaining phase errors and the coherent summation operation are both performed in the spatial-temporal frequency domain without explicit evaluation for such delays. This can be viewed as frequency domain beamforming. The proposed correction method is labeled as 'receive only'. In other words, defocusing caused by the aberrating layers during transmission is not accounted for. In that sense, the correction accomplished by this method can not be considered optimal. However, there are several advantages of such an approach. First, the frequency domain representation of the echoes reveals some characteristics that time domain methods do not explicitly show. For example, the multiple image artifacts are easily detected in the two dimensional spectrum and can be corrected by slightly modifying the frequency domain beamforming process. Another important advantage of the proposed method is that its main computational load consists of calculating the 2D spectra of the received echoes. With the currently available technology and high speed Digital Signal Processing (DSP) and Fast Fourier Transform (FFT) chips, simple and fast hardware implementation of this approach becomes possible, thus making it suitable for real time applications.

ACKNOWLEDGMENTS

This work was supported in part by grant CA 43920 from the National Institutes of Health. The authors would also like to thank ATL for supplying the experimental data.

REFERENCES

1. M. O'Donnell, and S. W. Flax, "Phase aberration measurements in medical ultrasound: human studies," *Ultrasonic Imaging*, 10(1):1-11 (1988).

2. L. Nock, and G. E. Trahey, "Phase aberration correction in medical ultrasound using speckle brightness as a quality factor," *J. Acoust. Soc. Am.*, 85(5):1819-1833 (1989).

3. G. C. Ng, S. S. Worrell, P. D. Freiburger and G. E. Trahey, "A comparative evaluation of several algorithms for phase aberration correction," *IEEE Trans. UFFC* 41(5):631-643 (1994).

4. M. Allam and A. Moghaddamjoo, "Spatial-temporal DFT projection for wideband array processing," *IEEE Signal Processing Letters*, 1(2):35-37 (1994).

5. M. Allam and A. Moghaddamjoo, "Two-dimensional DFT projection for wideband direction-of-arrival estimation," *IEEE Trans. on Signal Processing*, 43(7):1728-1732 (1995).

6. W. T. Mayo and P. M. Embree, "Two dimensional processing of pulsed doppler signals," U.S. Patent 4,930,513.

7. L. S. Wilson, "Description of broad-band pulsed doppler ultrasound processing using the two-dimensional fourier transform," *Ultrasonic Imaging*, 13:301-315 (1991).

8. M. Allam and J. F. Greenleaf, "Phase aberration correction for ultrasound imaging in temporal-spatial frequency domain," *Proc. IEEE Int. Conf. Acoust. Speech and Sig. Proc, ICASSP-95*, (1995).

9. M. Allam and J.F. Greenleaf, "Isomorphism between pulsed-wave doppler ultrasound and direction-of-arrival estimation–part I: basic principles," submitted to *IEEE Trans. UFFC*, (in review).

PHASE ABERRATION MEASUREMENT AND CORRECTION BY COMPLEX CROSS CORRELATION OF ANALYTIC SIGNALS IN ULTRASONIC IMAGING SYSTEMS

Mok Kun Jeong, Song Bai Park, and Jong Beom Ra

Department of Electrical Engineering
Korea Advanced Institute of Science & Technology
373-1 Kusong-dong, Yusong-gu, Taejon, 305-701, Korea

INTRODUCTION

This paper is concerned with phase aberration measurement and aberration correction suitable for ultrasonic linear array systems. We assume that the aberration is caused mainly by a layer of velocity inhomogeneity located near the transducer surface, such as a fatty layer under the skin surface. The phase aberration profile along the array is measured in the following way.

Using a certain length of aperture, transmit focusing is formed at an axial point in the near region consisting of random scatterers, beyond the inhomogeneity layer. The received signals at the central two elements are compared for their delay time difference by applying the complex cross correlation method to the respective analytic signals. In this way we obtain a local phase aberration profile in front of the pair elements. Repeating this procedure by shifting the aperture (electronically, of course) by one element at a time, we obtain the whole phase aberration profile along the array. Since initially we have no a priori knowledge of the aberration profile, transmission focusing is formed solely from the geometrical consideration in the usual way, and hence the focusing may not be sharp enough so that the received signals at the center pair elements may have lost similarity to some extent. This results in an erroneous estimate of the relative delay time. The estimate can, however, be improved iteratively, that is, by repeating the measurement with improved transmit focusing utilizing the aberration profile obtained in the previous step. Once an accurate aberration profile is obtained, this information can be utilized both in the transmit focusing and in the receive dynamic focusing in actual imaging of any ROI assuming homogeneity of the intervening medium.

RELATIVE DELAY TIME ESTIMATION BY COMPLEX CROSS CORRELATION

The cross correlation method for time delay estimation is well developed in radar, speech processing or sonar. The phase difference and hence the relative time delay between two signals of similar waveform can be directly calculated from the cross correlation of their complex analytic signals in the following way. Let us express two RF signals of the same waveform except some time delay τ_d by their analytic signals:

$$
\begin{aligned}
x_o(t) &= w(t)\exp(j\omega_o t)\\
&= r(t)\exp(j\Phi(t))\exp(j\omega_o t)
\end{aligned}
\tag{1}
$$

and the delayed replica

$$
\begin{aligned}
x_d(t) &= w(t+\tau_d)\exp(j\omega_o(t+\tau_d))\\
&= r(t+\tau_d)\exp(j\Phi(t+\tau_d))\exp(j\omega_o(t+\tau_d))
\end{aligned}
\tag{2}
$$

where $r(t)$ and $\Phi(t)$ are respectively the envelope and phase of the baseband signal. Then, the cross correlation of zero lag can be calculated by

$$
\begin{aligned}
R(0) &= \frac{1}{T_w}\int_{T_w}[\ x_o^*(t)x_d(t)]\ dt\\
&= \frac{\exp(j\omega_o\tau_d)}{T_w}\int_{T_w}r(t)r(t+\tau_d)\exp(j\Phi(t+t_d)-j\Phi(t))dt
\end{aligned}
\tag{3}
$$

where * represents the conjugate of the complex signal and T_w is the time window contributing to the integral. If the phase variation of the baseband signal within the time interval t_d is small, the exponential is approximated as

$$
\exp(j\Phi(t+\tau_d)-j\Phi(t)) \approx 1
\tag{4}
$$

in which case the integral takes a real value and the relative time delay τ_d is obtained from the arctangent of the real and imaginary parts of $R(0)$:

$$
\tau_d = \frac{1}{\omega_o}\tan^{-1}[\ \frac{Im(R(0))}{Re(R(0))}\ ]
\tag{5}
$$

For the sampled version of $x_o(t)$ and $x_d(t)$, $R(0)$ can be calculated by

$$
R(0) = \frac{1}{M}\sum_{n=0}^{M}[\ x_o^*(nT)x_d(nT)]
\tag{6}
$$

where M is the number of sampled data and T is the sampling period which can be as large as 1/(bandwidth of the signal).

In order to apply the above method for the delay estimation, several points must be considered. (1) The phase difference must be within $\pm\pi$; otherwise, aliasing will occur. O'Donnell et al [2] reported the r.m.s. phase error in-vivo is of the order of several tens degrees. Therefore, aliasing may not arise in actual ultrasonic medical imaging. (2) The S/N of the RF signal must be high because phase is very sensitive to noise. (3) A proper choice of ω_o is critical, since it is inversely related to the time delay estimate as seen in Eq. (5). For a symmetrical spectrum, the best choice of ω_o is obviously the center frequency. For an asymmetrical spectrum the choice is not obvious. Extensive computer simulation with actual ultrasound echo pulses with different asymmetric spectra has shown that the best delay time estimation can be obtained by the average of the peak frequency and the mean frequency.

ANALOG PSEUDO HILBERT TRANSFORMER

Though the digital Hilbert transformer is well described in many standard text-books on digital signal processing or digital filters [4], the analog counterpart is very difficult to realize. In this section, we propose an analog pseudo-Hilbert transformer (APHT), which consists of two all-pass filters(APF) having a phase difference of approximately 90 degrees over the frequency range of interest $\omega_1 \sim \omega_2$. The outputs of the two APF's with the same input signal applied are then regarded as the in-phase and quadrature components, $I'(t)$ and $Q'(t)$, of some complex analytical signal, to be called pseudo-analytic signal of the input signal. By optimizing the four coefficients of two third-order APF's, it is possible to get an error within only ± 0.5 degrees over a normalized frequency range of 1 to 10. When the outputs, $I'(t)$ and $Q'(t)$, of this APHT are used to estimate time delay difference of two identical wideband pulses(with asymmetric spectra, perhaps) by the complex cross-correlation method described previously, it turns out that the error is practically negligible although it increases linearly with the delay time. Therefore, we used the APHT instead of the digital Hilbert transformer for hardware simplicity

PHASE ABERRATION MEASUREMENT SCHEME

In applying the correlation method to the relative time delay estimation, obviously it is desirable for an obvious reason that the pair signals to be correlated be fully correlated. The best way to achieve this is to use the signals at two adjacent elements and to form a sharp transmit focusing with a large aperture. In the past, the phase difference profile was obtained by comparing the phase at each element with the reference phase at the center element [1-3]. As the interval between the pair elements is increased, the signals are less correlated. To make the situation worse, the estimation error propagates along the array in this method.

In the linear array system, the estimation error can be reduced and its propagation can be avoided by using signals at the central two elements of the transmit aperture and independent measurement of phase aberration at each scanline. Fig. 1 shows the architecture of the data acquisition scheme for this purpose. From the signals at the center pair elements at the kth scanline, phase difference $\Delta p(k)$ is calculated by using the complex cross correlation, which gives the local phase difference, due to the presence of the aberrating layer. For the best estimate of the local phase difference, the transmit focal point must be made close to the aperture, because then the received signals become stronger and more correlated due to the reduced sample volume. This also reduces the phase aberration measurement time. In the experiments, an aberrating medium is interposed between the aperture and the transmit focal depth. The local phase difference is calculated at each scanline independently and the actual phase difference profile $P(n)$ at the nth element position is obtained by summing the local phase differences:

$$P(n) = \sum_{k=0}^{n} \Delta p(k) \tag{7}$$

with DC offset adjustment. The hardwares needed for this measurement are two APHT, four AD converters, high speed memory for storing the sampled data, one complex cross correlator, and one adder (in the experiment the stored data are transferred to the computer and actual delay estimation was done on the computer). The correlator inputs are switched to the center pair elements of the transmit aperture as the scanline is shifted. As the local phase differences are added, the resulting phase aberration profile is successively obtained. All these steps can be executed in real time using high speed digital signal devices. In this scheme, the sampling time quantization error is not severe, because signals at the center pair elements are sampled simultaneously with an uniform sampling interval [5]. Once the whole aberration profile is obtained in this way, we can utilize it in both transmit focusing and receive dynamic digital focusing in actual imaging of ROI which may be deeply located from the array.

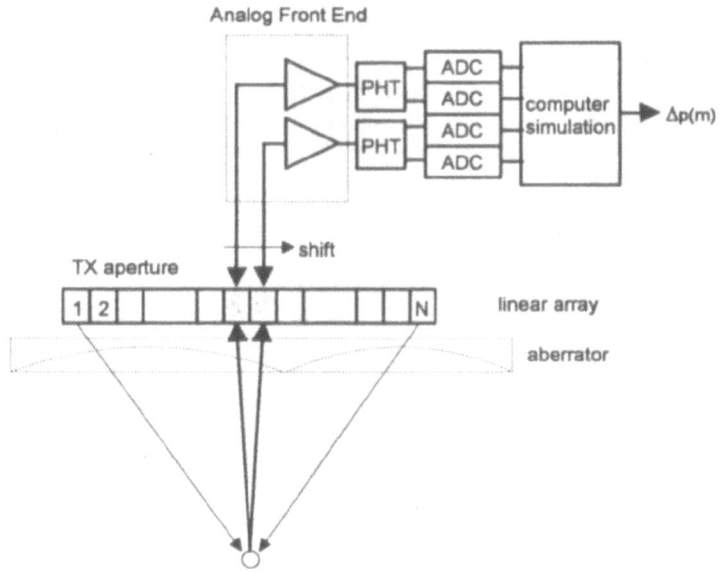

Fig. 1 Data acquisition scheme for phase aberration measurement.

EXPERIMENTAL RESULTS

In order to verify the proposed method of phase aberration measurement and correction, we performed several experiments using a 3.5 MHz linear array and a transmit aperture of 32 elements out of 128 elements. First, in order to examine if any erroneous phase aberration is introduced to the measurement in the absence of an aberrator, the transducer was directly applied to a cyst phantom model 539(I) of ATS laboratory, which contains three cysts of different scatterer densities (diameter = 15mm each) located at 50 mm depth and surrounded by a homogeneous medium of random scatterers with higher density, and focusing was formed at 3 cm depth. The received signals at the central pair elements were sampled at a rate of 12.6 MHz and the usual complex correlation method was applied to calculate the phase difference of the pair signals with a time window of 6 mm. This measurement was repeated for different scan lines. The standard deviation of the individual phase differences was about 2.8 degrees, which is small enough compared to the equivalent of the sampling time resolution (± 10 ns) and actual phase aberration expected. This phase error is inherent to the random nature of scatterer distribution and unavoidable in any aberration correction algorithm. It can be reduced, however, by sharper transmit focusing, higher sampling time resolution, time averaging of the received data within a large time window, and spatial averaging of the data from multiple elements.

In the next experiment, an artificial aberrator was interposed between the transducer and the cyst phantom, and the whole unit was immersed in a water tank. The aberrator used was made of silicon rubber with a rippled surface, the grooves being filled with a different silicon material; the average interval of grooves was 15mm(see Fig. 1). In estimating the local phase difference, the transmit beam was focused at 3 cm depth. Since it was found that the received signal at the center element was too weak, the spatial averaged data from two elements on each side of the aperture center were used to measure the local phase difference (which is justifiable because the signals at two neighboring elements had a strong correlation), discarding low level signals of -15 dB below the maximum envelope within a time window of about 6 mm at the focal depth. Curve (a) of Fig. 2 shows the resulting phase aberration profile with the phase offset adjustment. Since this profile was obtained without a priori knowledge of the aberrator characteristics, transmit focusing may not be sharp enough. A better focusing and hence a more accurate aberration profile will be obtained if we repeat transmit focusing utilizing the initially obtained aberration profile. We repeated the iteration once more but the resulting profile did not change appreciably from the second one, so we simply stopped at this point. Note that both curves of Fig. 2 have about the same shape except amplitude scaling(≈ 1.1 in the present case)

Now that we have a supposedly accurate aberration profile of the aberrator, we are ready to do the final experiment, that is, actual imaging for the cysts located at 50 mm, by fixing the transmit focusing at this point and applying dynamic focusing in the receive mode employing the sampling delay focusing scheme [5]. A master clock of 50.4 MHz was used and hence the transmit pulse clocking and the receive sampling time can be controlled within $\pm 10 ns$ accuracy(the sampling rate for for the center element was 12.6 MHz). The proper transmit clocking for each array element of the aperture was computed for the case of aberration correction intended and not intended and the data were stored in the high speed RAM. The same was done for the receive mode using the same measured aberration profile. Fig. 3 shows the log-compressed cyst images obtained after different aberration corrections made by computer simulation. Fig. 3(a) shows the image obtained without an aberrator interposed between the transducer and the phantom surface. In other images, the aberrator, whose characteristics was previously measured, was interposed. Fig. 3(b) shows the image obtained without any aberration correction. Fig. 3(c) shows the image obtained after aberration correction made in the receive mode only. Fig. 3(d) shows the image obtained after aberration correction made in both of the transmit and receive modes. We note an impressive improvement of the image quality in the last image over Fig. 3(b) and an appreciable improvement over Fig. 3(c) (only "appreciable", because transmit focusing was fixed at a point whereas the receive focusing was dynamic). All this indicates the effectiveness of the proposed technique of aberration profile measurement and aberration correction.

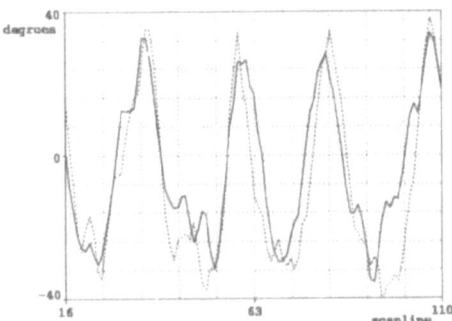

Fig 2. Phase aberration profile of the aberrator obtained in the first stage (solid curve), and in the second stage (dotted curve). Note that dotted curve is increased approximately by a factor 1.1 compared with solid curve.

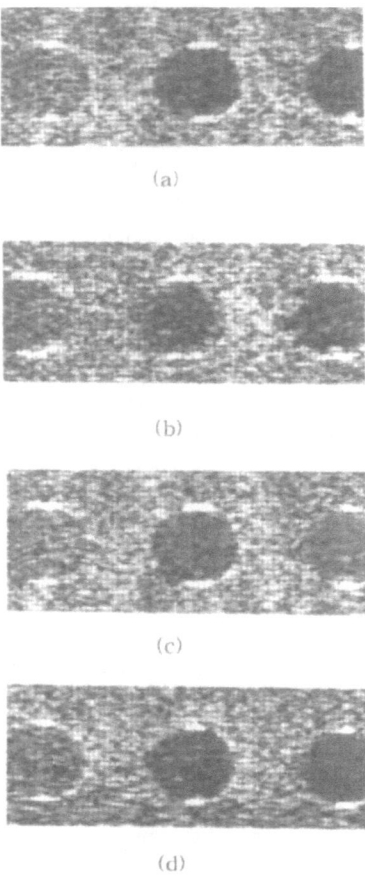

(a)

(b)

(c)

(d)

Fig. 3 Images of the cyst phantom. The aberrator was not interposed in (a) and interposed in all other images. (b) No aberration correction made, (c) correction made in receive mode only, (d) correction made both the transmit and receive mode. Transmit focusing was fixed at the cyst position and the receive focusing was dynamic.

DISCUSSION AND CONCLUSIONS

We reported a preliminary experimental study on a new phase aberration measurement technique as applied to a linear array system assuming the aberration is mainly caused by a velocity inhomogeneity layer close to the transducer surface, and using complex cross correlation method to estimate the relative phase difference of the signals at the central two elements or central two small group of elements of the aperture to obtain waveforms of maximum similarity except delay time difference. Since the accuracy of the phase aberration profile improves with the transmit focusing, the focusing was iteratively made using the previously obtained aberration profile. Actual images of a cyst phantom with an aberrator interposed between the phantom and the array show an impressive image quality improvement when the aberration measured by the proposed technique was corrected both in the transmit and in the receive modes, even if the aberration of the aberrator is small (standard deviation of 5.6 degrees). In the case of more severe aberration, the effect of aberration correction by the proposed technique should manifest itself more vividly.

Regarding the possibility of real time aberration measurement and correction, we would like to point out that the computation of the delay time difference by the complex cross correlation can be executed in real time with the use of APHT and high speed digital devices. If the data acquisition for iterative transmit focusing at 3 cm depth is repeated twice in the aberration profile measurement stage, the overall frame rate may be decreased only by 25% for a maximum imaging depth of 18cm. This percentage decrease can be reduced by using the first aberration profile simply multiplied by some constant (between 1.1 ~ 1.5) instead of trying the second iteration. Obviously, the technique can also be applied for the convex array system with success. In the near future, we hope to present in-vivo images perhaps using a wider aperture, a higher bit ADC with a higher sampling frequency and a higher sampling time resolution, and using strong transmit pulses.

REFERENCES

1. M. O'Donnell, and S.W. Flax, Phase Aberration Measurements in Medical Ultrasound: Human Studies, Ultrasonic Imaging, 10, 1-11, (1988).

2. S.W. Flax, and M. O'Donnell, Phase Aberration Correction Using Signals from Point Reflectors and Diffuse Scatterers:Basic Principles, IEEE Trans. UFFC, vol. 35, no. 6, (1988).

3. M. O'Donnell, and W.E. Engeler, Correlation-Based Aberration Correction in the Presence of Inoperable Elements, IEEE Trans. UFFC, vol. 39, no. 6, (1992).

4. R.R. Lawrence, and G. Rabiner, Theory and Application of Digital Signal Processing, Prentice-Hall, (1975).

5. J. H. Kim, T. K. Song, and S. B. Park, A Pipelined Sampled Delay Focusing in Ultrasound Imaging Systems, Ultrasonic Imaging, 9, pp. 75-91. (1987)

FUNDAMENTAL MECHANICAL LIMITATIONS ON THE VISUALIZATION OF ELASTICITY CONTRAST IN ELASTOGRAPHY*

Hari Ponnekanti, Jonathan Ophir, Yijun Huang and Ignacio Céspedes

Ultrasonics Laboratory, Dept. of Radiology
The University of Texas Medical School
Houston, Texas 77030, USA

ABSTRACT

Elastography is a new ultrasonic imaging technique that produces images (elastograms) of the elastic properties of compliant tissue. The ultrasonically measured quantity is the normal strain component in the direction of the applied load, and the three normal components of stress may be estimated using the modified Love's analytical models while assuming a value close to 0.5 (incompressible) for Poisson's ratio. The distribution of Young's moduli can thus be computed and displayed in the form of two dimensional images called elastograms. The analytical models used for the estimation of the three normal components of stress assume that the target is semi-infinite and homogeneous in composition. The objective of this paper is to determine some of the errors associated with the assumption of homogeneity of the target. Experiments using finite-element simulations were performed to study the efficiency with which elastograms display the contrast in the Young's modulus of a lesion or target, with respect to its background under certain conditions. It was observed (using the definition of contrast-transfer efficiency of elastography as the ratio of the elasticity contrast as measured from an elastogram, to the true contrast), that elastograms were consistently efficient in quantitatively depicting the elasticity contrast of hard lesions; however, they showed suboptimal contrast-transfer efficiency in cases of soft lesions in a hard background. In general, elastograms are efficient in displaying the elasticity contrast of hard or soft lesions which have a low contrast level with respect to the surround, irrespective of their size and location.

INTRODUCTION

In a previous paper Ophir *et al.* (1991) have described a technique known as elastography for imaging the elasticity of biological soft tissues. The technique produces images of the internal strains produced in a soft-tissue target resulting from a small external displacement of the tissue. Elastography is performed by obtaining a set of ultrasonic RF A-lines from a target, subjecting the target to a small deformation and obtaining a second set of RF A-lines. Time shift estimations along the direction of the applied loads are computed by performing piecewise crosscorrelation on congruent pairs of A-line segments. The time-shift estimations are then converted to strain information that is displayed in the form of a two-dimensional longitudinal strain image.

Strain by itself is an incomplete indicator of the actual elasticity distribution in the target. Although the spatial variation in strain may partially reflect the changes in local Young's moduli, the knowledge of the distribution of stresses is essential in order to characterize the true contrast in elasticity (Ophir et al., 1991). Ponnekanti et al. (1994) have shown a technique for the estimation of the stress distribution in an isotropic, homogeneous, finite elastic target that is being deformed by a rectangular punch using a modification of Love's (1929) analytical models. The estimated distribution of stresses is used along with the measured values of strain, while assuming the target tissue to be nearly incompressible (Poisson's ratio $\simeq 0.5$), to estimate the distribution of Young's moduli in the target. The distribution of Young's moduli in the plane of interest is then displayed in the form of a two-dimensional gray-scale image known as an elastogram.

The modified Love's models that are used to determine the stress distribution assume that the target is homogeneous and semi-infinite. Since real targets are seldom homogeneous or semi-infinite, the elastograms suffer from certain errors. The errors associated with the assumption of the semi-infiniteness of real targets have been analyzed by Konofagou et al., (1995). Additional errors incurred due to the assumption of target homogeneity are investigated in this paper. It has been observed from preliminary simulation data that the lesion-to-background contrast of elastographic lesions changes with the true modulus contrast. The principal objective of this paper is to characterize these changes under a restricted set of experimental conditions of target composition and boundary conditions. Simple one-dimensional models have been used previously by Céspedes and Ophir (1993) and Belaid et al. (1994) to simulate elastograms. The use of these models was restricted to simple applications where detailed analysis of the mechanical response of the target to complex boundary conditions was not essential. Therefore, these models were not adequate to accurately describe the complex three-dimensional nature of the mechanical problem in elastography. For example, experimental results in the past have revealed several artifacts associated with the three-dimensional nature of the mechanical response of a target to external boundary conditions, such as the nonuniform distribution of stress in an elastic target that is being acted upon by a finite size punch (Ponnekanti et al., 1993; Ponnekanti et al., 1994). Furthermore, the experiments of Belaid et al. (1994) involved very low contrast lesions. Such lesions are expected to minimally interact with the background, resulting in excellent contrast-transfer efficiency (defined as η = observed modulus contrast / true modulus contrast). In the present work we investigate this efficiency of elastography over a much wider, $\pm$ 40 dB modulus contrast dynamic range. The results reveal a markedly different behavior of the contrast-transfer efficiency of hard and soft lesions, while confirming the excellent contrast-transfer efficiency associated with low contrast lesions. These are expected results that stem from some fundamental mechanical limitations affecting elasticity imaging, which are discussed later.

MATERIALS AND METHODS

Elastograms were generated using several steps. First, mechanical models were generated using a commercial finite element analysis program (MSC-PAL2, version 4,

McNeal-Schwendler Corp., Los Angeles, CA). An acoustic Rayleigh scattering model was added to the displacement information obtained as an output of the FEA program. Pre- and post-loading RF-signals were generated using a convolutional model and a simulated acoustic transducer (Céspedes and Ophir, 1990). The strain distribution was estimated by applying crosscorrelation techniques to congruent pairs of these RF-signals (Ophir *et al.*, 1991). Modified Love's models were applied with appropriate boundary conditions to estimate the distribution of all three normal components of stress at all locations. The strain and stress data were then converted to Young's moduli and displayed as elastograms.

The finite-element (FE) mechanical models were generated with some typical conditions for clinical elastography in mind. The punch used for the application of the load was chosen to have a lateral dimension of one fifth the lateral dimension of the target (0.2 a), where a is the lateral dimension of the target, and the depth of the target was chosen to be equal to 3a. These dimensions are typically encountered in elastography of the breast. The punch at the base of the target was assumed to have an area larger than the area of the base of the target. These boundary conditions were used to simulate a situation where the breast is placed on a large flat table while being examined by a small rectangular linear array transducer (Céspedes *et al.*, 1993).

The entire target was simulated with a total of 1353 nodes. In accordance with common practice (Beaupré and Carter, 1991), the number of nodes per unit area was varied over the target model for optimum utilization of the number of nodes available in the FE program. A higher nodal density was used in areas where the distribution of stresses was expected to be more nonuniform, such as in areas close to the punch and in the area of the lesion. A mesh with quadrilateral elements was generated to connect the nodes. FE techniques require that the choice of the nodal locations and meshing be made carefully in order to have a stable model. Therefore, simple trial models were initially generated with different nodal and mesh arrangements to check for the stability of the model.

The mechanical problem in elastography is scaleable in terms of the punch size. Therefore, the distances were calibrated in terms of an arbitrary constant called a 'unit' such that the punch had a lateral dimension of 80 units. The target contained a rectangular arrangement of quadrilateral elements that was 400 units wide and 240 units deep. It is to be noted that a unit is not necessarily equal to the distance between nodes; it was a constant that was arbitrarily chosen to scale the distances in the models. The background of the target had a Young's modulus of 21 kPa that was arbitrarily chosen to represent normal tissue, and a Poisson's ratio of 0.495. A small vertical deformation of 1 unit (0.4 % compressive strain) was provided by a punch. The nodes corresponding to the lower surface were restricted to move in the lateral direction only while the sides were left free. Circular lesions were simulated inside the target by nodes and elements arranged in concentric circles. Two sizes for the circular lesion were chosen; the diameters of the smallest and the largest lesions were 1/20 (4 units) and 1/2 (40 units) the lateral dimension of the punch, respectively. Given a typical compressing aperture size of 40 mm reported by Céspedes *et al.* (1993), these values correspond to lesion diameters of 2 mm and 20 mm, respectively. Six locations were chosen for the center of the lesion at three different depths, three were along the axis of the punch and the other three were along an axis parallel to and 20 units to the right of the punch axis. The locations for these lesions were chosen in a region under the punch which corresponds to the insonified region of interest. The depths of the lesions were chosen as 40 units, 120 units and 200 units to represent situations where the lesions are within distances that are equal to 0.5a, 1.5a and 2.5a. Symmetry about the axis of the punch was assumed. Six different lesion-to-background elasticity contrasts were chosen: 0.01:1 (-40 dB), 0.1:1 (-20 dB), 0.5:1 (-6 dB), 2:1 (6 dB), 10:1 (20 dB) and 100:1 (40 dB). The theoretical contrast of 1:1 (0 dB) was also considered.

RESULTS

We define the contrast-transfer efficiency of elastography as the ratio of the modulus contrast in elasticity as measured from an elastogram, to the true modulus contrast in the corresponding target, viz.

$$\eta \ = \ C_0 / C_t \ , \tag{1}$$

where C_0 is the observed contrast, $C_t > 0$ is the true contrast, and η is the contrast-transfer efficiency ($\eta = 1$ indicates an ideal 100 % efficiency; $\eta = 0$ indicates 0 % efficiency).

Since we have investigated a wide dynamic range, i.e. $0.01 \leq C_t \leq 100$, it is convenient to represent the results on a logarithmic (dB) scale. By taking the logarithm of both sides of Eq. (7), we get

$$\eta (dB) = C_0 (dB) - C_t (dB) \ . \tag{2}$$

Using this presentation, $\eta = 0$ (dB) means 100% efficiency, and $\eta < 0$ (dB) signifies a reduced efficiency.

Typical results are plotted in figure 1, which shows plots of the observed lesion modulus contrast from the elastograms versus the true contrast on a dB scale. The plot contains two curves, one each for the small (4 units in diameter) and large (40 units in diameter) lesions. A diagonal line was drawn to represent ideal performance($\eta = 0$ (dB), or $C_0 = C_t$). The Figure clearly demonstrates that the behavior of the observed contrast is different for hard and soft lesions in all cases considered. In order to get a clearer picture, we combine the data from all the lesion sizes and locations, and plot in figure 2 the mean contrast-transfer efficiency $\bar{\eta}$ (in dB) vs. true contrast C_t, where the error bars represent one standard deviation. This figure clearly shows several interesting results:

1. for hard or soft lesions with low contrast ($|C_t| \leq 6$ dB), the mean contrast-transfer efficiency $\bar{\eta}$ of elastography is always better than -3 dB (> 0.7);

2. for soft lesions with high contrast ($C_t < -6$ dB), the mean contrast-transfer efficiency gets progressively poorer as the contrast increases, achieving a very low mean transfer efficiency of - 33.4 dB (~ 0.02) at a contrast level $C_t = - 40$ dB;

3. for hard lesions with high contrast ($C_t > 6$ dB), the mean transfer efficiency declines to a moderate fixed level of about -5 dB (~ 0.56) for lesions with a contrast level of 20 to 40 dB; and

4. These results were consistent for all cases of lesion size and location, as demonstrated by the modest values of the standard deviations (shown in figure 2), which were on the order of ± 1 to ± 2 dB.

DISCUSSION AND CONCLUSIONS

The major factor that influences the continuously declining contrast-transfer efficiency of elastography of soft lesions is the phenomenon by which soft lesions that are confined by a harder background appear harder than they really are, when the Poisson's

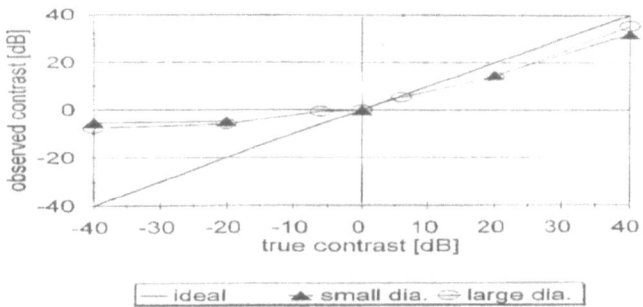

Figure 1. Typical transfer characteristic of observed elastographic contrast vs. true modulus contrast, over an input range of ±40 dB. The diagonal line represents ideal transfer characteristic. Note that for positive true contrast (hard lesions in soft background) the transfer characteristic parallels the ideal one. For soft lesions in a hard background, the characteristic departs from the ideal one. Note also that for low contrast lesions, the characteristic approaches the ideal one.

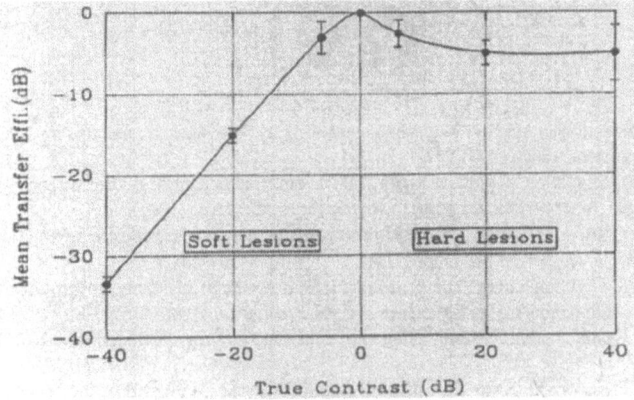

Figure 2. Mean contrast-transfer efficiency $\overline{\eta}$ (±one s.d.). The ideal efficiency is 0 dB everywhere. Note low efficiency for soft lesions, relatively high efficiency for hard lesions, and very high efficiency for low contrast lesions.

ratio is nonzero. This results in an effective contrast dependent hardening of the constrained lesion. This phenomenon represents a fundamental limitation on the elastographic imaging of soft lesions. For soft lesions with low contrast ($C_t > $ -6 dB), the mean contrast-transfer efficiency is quite good ($\geq$ -3 dB).

On the other hand, for all cases that were investigated, the visualization of hard lesions had efficiencies that were consistently better than about -5 dB. It is also interesting to note that for contrast levels $\geq$ 20 dB, the mean transfer efficiency remains at this -5 dB contrast level, which raises the possibility of applying a constant correction factor to the observed contrast level, in order to correct it for this bias. For low contrast levels ($<$ 6 dB), the contrast-transfer efficiency is rather good, ranging between 0 dB and -3 dB. We postulate that the source of this behavior is the perturbation of the background by the lesion, which is not accounted for by Love's modified model that assumes a uniform target. Indeed, the data of figure 2 demonstrate that for hard lesions of low contrast, the contrast-transfer efficiency is high (i.e. $C_0 \cong C_t$); these cases would be expected to cause the least amount of perturbation of the background, and hence be more efficient. For hard lesions with high contrast($C_t \geq$ 20 dB), the lesions would tend to behave as rigid bodies, exerting a fixed perturbation on the softer background material regardless of contrast; this would tend to result in a fixed bias in the efficiency such as that which is demonstrated in figure 2. It is interesting to note that the contrast-transfer efficiency of elastography is generally high for hard lesions, in light of the fact that most cancerous lesions tend to be harder than normal tissue.

ACKNOWLEDGEMENTS

*Supported in part by National Institutes of Health grants R01-CA38515, R01-CA60520, and P01-CA64597, and by a grant from Diasonics Ultrasound, Milpitas, CA.

REFERENCES

N. Belaid, Lesion detection in elastography versus echography: a quantitative analysis, Master's Thesis, University of Houston, (1993).

G.S. Beaupré, and D. R. Carter, "Finite Element Analysis in Biomechanics: structures and Systems", A.A. Biewener, ed., pp 149-174, Oxford University Press, Oxford (1992).

I. Céspedes, and J. Ophir, Diffraction correction methods for pulse-echo acoustic attenuation estimation, *Ultras. Med. Biol.*, 16:707-717 (1990).

I. Céspedes, J.Ophir, H. Ponnekanti, and N. Maklad, Elastography: elasticity imaging using ultrasound with application to muscle and breast in vivo, *Ultrasonic Imaging*, 15:73-88 (1993).

I. Céspedes, and J. Ophir, J. Reduction of image noise in elastography, *Ultrasonic Imaging*, 15: 89-102 (1993a).

A.E.H. Love, The stress produced in a semi-infinite solid by pressure on part of the boundary, *Trans. Royal Soc.*, London, Series A, 228:337 (1929).

J. Ophir, I. Céspedes, H. Ponnekanti, Y. Yazdi, and X. Li, Elastography: A method for imaging the elasticity in biological tissues, *Ultrasonic Imaging*, 13:111-134 (1991).

H. Ponnekanti, J. Ophir, and I. Céspedes, Axial stress distributions between coaxial compressors in elastography: an analytical model, *Ultrasound Med. Biol.*, 18:667-673 (1992).

H. Ponnekanti, J. Ophir, and I. Céspedes, Ultrasonic imaging of the stress distribution in elastic media due to an external compressor, *Ultrasound Med. Biol.*, 21: (4), 533-543 (1995).

RENAL FUNCTION ASSESSED BY MONITORING THE SIZE OF ACOUSTIC SCATTERERS: MAP ESTIMATORS AND THE USE OF PRIOR INFORMATION

Michael F. Insana, Timothy J. Hall, Mehmet Bilgen,
Pawan Chaturvedi, John G. Wood[1,2], and Glendon G. Cox

Departments of Radiology, [1]Surgery, and [2]Physiology
University of Kansas Medical Center
Kansas City, KS 66160-7234

INTRODUCTION

Many kidney disease processes begin with changes in the microscopic tissue structure, particularly the vasculature. Ultrasonic analysis can detect the changes [8, 6] that lead to earlier treatment and improved patient outcome. To characterize the microstructure, we estimate the average size of tissue scatterers D by analyzing the change in the backscatter coefficient $\sigma(u)$ versus frequency u. The expression $\sigma(u) = \sigma_0(u)\Gamma(u, D)$ summarizes the measurement technique, assuming the tissue structure is random [7]. An acoustic form factor is estimated from the echo spectrum using $\hat{\Gamma}(u, D) = \sigma(u)/\sigma_0(u)$. This estimate is compared to model functions $\Gamma(u, D)$ for a range of D values. The model function that produces the smallest mean-square error when compared to the estimate $\hat{\Gamma}$ determines the scatterer size estimate. Form factors provide a link between measurement and theory that may be used to estimate scatterer sizes for random media in the range of 25 μm to 400μm using diagnostic frequencies between 1 and 15 MHz.

Measurement precision as low as 10% of the mean size has been observed when $\hat{\Gamma}$ is determined from large tissue volumes (>1 cm^3). To generate scatterer size images, however, each form factor must be estimated from a more limited data set, and the concomitant reduction in precision must be acceptable if we are to obtain the spatial resolution required to form a scatterer size image. Bayesian analysis is a method for reducing image noise by incorporating valid prior information about the object under investigation into the reconstruction process [3, 4]. In forming an image, we seek a solution that maximizes the posterior probability, i.e., the probability of a particular object given the data used to form the image. Therefore, the data reduction expressions that result are known as maximum a posteriori (MAP) estimators.

BAYESIAN ESTIMATION

Prior information about stochastic objects is introduced into the analysis in the form of the probability density function (pdf) for the feature to be determined. In our measurements to determine the average size of kidney microstructures, we use information about the morphology of the nephron to guide solutions. Only those values that are consistent with renal scattering under the experimental conditions are considered. For example, previous studies [5] have shown that the echo spectrum obtained from renal cortex between 2 and 5 MHz is largely determined by properties of the glomerulus — an important component of the nephron that determines function. Morphological analysis of normal glomeruli, e.g., Fig. 2a in [5], indicates that the diameters are normally distributed, $\mathcal{N}(\bar{D}, \sigma_D^2)$, with mean $\bar{D} \simeq 200\mu m$ and standard deviation $\sigma_D \simeq 25\mu m$. Constraining the search to include only those values consistent with the prior information reduces uncertainty and therefore image noise. If the prior information is accurate, the mean square error is reduced. However, if the prior information is inaccurate, bias is increased to reduce variance. Ultimately the imaging task determines the merit of such tradeoffs.

Standard MAP Formulation

Consider an incident pressure field scattered by inhomogeneities in the propagating medium. Ultrasonic echo signals are formed by sensing this scattered field with a phase-sensitive transducer. Under most imaging conditions, the process of forming echo signals is linear and may be expressed as

$$\mathbf{z} = \mathbf{Hf} + \mathbf{n} \ . \tag{1}$$

Echo-signal waveforms are represented by the column vector $\mathbf{z}$ (an $M \times 1$ matrix); the corresponding object function is represented by $\mathbf{f}$ (an $L \times 1$ matrix); $\mathbf{H}$ is an $M \times L$ transformation matrix; and $\mathbf{n}$ is an $M \times 1$ noise matrix. (Note that Eq. 1 holds for non-additive noise if $\mathbf{n}$ is a function of $\mathbf{f}$.)

Before acquiring any data, our knowledge about the object $\mathbf{f}$ is summarized by the *prior* pdf $p(\mathbf{f})$. Once a meaningful measurement $\mathbf{z}$ is acquired, however, our knowledge of $\mathbf{f}$ is improved. The pdf of $\mathbf{f}$ conditioned on $\mathbf{z}$, viz., $p(\mathbf{f}|\mathbf{z})$, is called the *posterior* pdf and is related to the prior pdf by Bayes' rule:

$$p(\mathbf{f}|\mathbf{z}) = p(\mathbf{z}|\mathbf{f}) \, p(\mathbf{f})/p(\mathbf{z}) \ . \tag{2}$$

The posterior pdf is highly significant because it specifies our state of certainty about an estimate of the object $\mathbf{f}$ obtained after analyzing a particular data set $\mathbf{z}$. The conditional probability $p(\mathbf{z}|\mathbf{f})$ is the *likelihood* of observing $\mathbf{z}$ given that the data were acquired from a particular object $\mathbf{f}$. The likelihood pdf contains the new information provided by the data. The constraint on $p(\mathbf{z})$ is that it does not depend on $\mathbf{f}$.

It is difficult for human decision makers to assimilate all of the information contained in the multidimensional posterior pdf. So we choose one value for each pixel to form a single image. Specifically, we select the estimate $\hat{\mathbf{f}}$ that maximizes the posterior probability:

$$\hat{\mathbf{f}} = \arg \max_f \ p(\mathbf{f}|\mathbf{z}) \ . \tag{3}$$

At this time, the noise is modeled as a zero-mean Gaussian random variable $\mathcal{N}(\mathbf{0}, \mathbf{K_n})$, where $\mathbf{K_n} = \langle \mathbf{nn}^t \rangle_n$ is the covariance matrix of the noise and $\mathbf{n}^t$ is the transpose of $\mathbf{n}$. Assuming the noise is additive and white, then $\mathbf{K_n} = \sigma_n^2 \mathbf{I}$ where $\mathbf{I}$ is the identity matrix. The object is assumed to be an uncorrelated Gaussian process $\mathcal{N}(\bar{\mathbf{f}}, \mathbf{K_f})$, where $\bar{\mathbf{f}} = (\bar{f}(1), \cdots, \bar{f}(L))^t$ is the mean vector and the covariance matrix

$$\mathbf{K_f} = \begin{bmatrix} \sigma^2(1,1) & & 0 \\ & \ddots & \\ 0 & & \sigma^2(L,L) \end{bmatrix} \tag{4}$$

is diagonal. Consequently, the data are also a Gaussian process $\mathcal{N}(\bar{\mathbf{z}}, \mathbf{K_z})$, where from Eq. 1 we find $\bar{\mathbf{z}} = \langle \mathbf{z} \rangle_{n|f} = \mathbf{Hf}$ and $\mathbf{K_z} = \mathbf{K_{n|f}} = \sigma_n^2 \mathbf{I}$. [1]

---

[1] The notation $\langle \mathbf{z} \rangle_{n|f}$ indicates the conditional mean over all n but for a fixed value of f. Alternatively, $\langle \mathbf{z} \rangle_{n,f}$ indicates the mean obtained by averaging over all n and f.

It is more convenient, and yet equivalent, to study the monotonic transformation of Eq. 2, $-\ln p(\mathbf{f}|\mathbf{z})$, which for the Gaussian processes described above is given by

$$-\ln p(\mathbf{f}|\mathbf{z}) = \frac{1}{2}(\mathbf{z} - \bar{\mathbf{z}})^t \mathbf{K}_{\mathbf{z}}^{-1}(\mathbf{z} - \bar{\mathbf{z}}) + \frac{1}{2}(\mathbf{f} - \bar{\mathbf{f}})^t \mathbf{K}_{\mathbf{f}}^{-1}(\mathbf{f} - \bar{\mathbf{f}}) + \ln p(\mathbf{z}) \ . \tag{5}$$

MAP estimates are solutions to the set of linear equations that result by taking the gradient of Eq. 5 with respect to $\mathbf{f}$, i.e., $\nabla_{\mathbf{f}} \equiv (\partial/\partial f_1, \cdots \partial/\partial f_L)^t$, and setting the result equal to zero:

$$\nabla_{\mathbf{f}}(-\ln p(\mathbf{f}|\mathbf{z})) = -\frac{1}{\sigma_n^2}\mathbf{H}^t(\mathbf{z} - \mathbf{H}\mathbf{f}) + \mathbf{K}_{\mathbf{f}}^{-1}(\mathbf{f} - \bar{\mathbf{f}}) = 0 \ . \tag{6}$$

Assuming $\sigma_n^2 \neq 0$ and solving for $\mathbf{f}$, the MAP estimate of the object function is

$$\hat{\mathbf{f}} = (\mathbf{H}^t\mathbf{H} + \sigma_n^2\mathbf{K}_{\mathbf{f}}^{-1})^{-1}(\mathbf{H}^t\mathbf{z} + \sigma_n^2\mathbf{K}_{\mathbf{f}}^{-1}\bar{\mathbf{f}}) \ . \tag{7}$$

Estimating Acoustic Form Factors

The autocorrelation function of the tissue reflectivity profile characterizes the structure of random scattering media such as soft biological tissues [1, 7, 8]. The Fourier transform of the tissue autocorrelation function is the form factor. It is a frequency-space characterization of the tissue structure that includes information about the average scatterer size. We now use Bayesian techniques to estimate form factors from which scatterer size measurements D are then obtained.

The vector $\mathbf{z}$ contains echo signals from a volume of tissue obtained in a C-scan format, i.e., the sound beam is perpendicular to the image plane (Fig. 1). Pixel values in a D image represent estimates obtained by analyzing individual echo segments that are each NI samples in length. There are NJ horizontal and $N\ell$ vertical pixels in the image. The data vector is organized as follows: $\mathbf{z} = (z(1,1,1), \cdots, z(NI,1,1), z(1,2,1), \cdots \cdots, z(NI, NJ, N\ell))^t$, and $M = NI \times NJ \times N\ell$. The noise vector has the same organization. In general, $\mathbf{f}$ may not be the same dimension as $\mathbf{z}$. However, the sampled functions $\mathbf{z}$, $\mathbf{f}$, and $\mathbf{n}$ may be extended by padding with zeroes so that all have dimension M. Consequently, the transformation matrix $\mathbf{H}$ is square and it consists of shifted copies of an Nh-point pulse-echo point spread function, h:

$$\mathbf{H} = \begin{bmatrix} h(1) & h(2) & \cdots & h(Nh) & & \\ & h(1) & h(2) & \cdots & h(Nh) & \\ & & & \ddots & & \\ & & h(Nh) & \cdots & h(2) & h(1) \end{bmatrix} \ . \tag{8}$$

Because $\mathbf{H}$ is a circulant matrix [2], we can write

$$\mathbf{H} = \mathbf{W}\mathbf{\Lambda}\mathbf{W}^{-1} \ , \tag{9}$$

where $\mathbf{W}$ is a matrix containing the eigenvectors of $\mathbf{H}$, $W(n,m) = \exp(i2\pi nm/M)$, and $\mathbf{\Lambda}$ is a diagonal matrix of elements that are the eigenvalues of $\mathbf{H}$. Also, it can shown that $\mathbf{H}^t = \mathbf{W}\mathbf{\Lambda}^*\mathbf{W}^{-1}$, where $\mathbf{\Lambda}^*$ is the conjugate of $\mathbf{\Lambda}$, $\mathbf{W}^* = \mathbf{W}^{-1}$, and $\mathbf{K}_{\mathbf{f}} = \mathbf{W}\mathbf{S}_{\mathbf{f}}\mathbf{W}^{-1}$, where the elements of $\mathbf{S}_{\mathbf{f}}$ are given by the Fourier transform of the elements in $\mathbf{K}_{\mathbf{f}}$ [2]. Substituting the above expressions for $\mathbf{H}$, $\mathbf{H}^t$, and $\mathbf{K}_{\mathbf{f}}$ into Eq. 7 gives

$$\hat{\mathbf{f}} = (\mathbf{W}\mathbf{\Lambda}^*\mathbf{\Lambda}\mathbf{W}^{-1} + \sigma_n^2\mathbf{W}\mathbf{S}_{\mathbf{f}}^{-1}\mathbf{W}^{-1})^{-1}(\mathbf{W}\mathbf{\Lambda}^*\mathbf{W}^{-1}\mathbf{z} + \sigma_n^2\mathbf{W}\mathbf{S}_{\mathbf{f}}^{-1}\mathbf{W}^{-1}\bar{\mathbf{f}}) \ . \tag{10}$$

Multiplying both sides of Eq. 10 by $\mathbf{W}^{-1}$, it is straightforward to find

$$\mathbf{W}^{-1}\hat{\mathbf{f}} = (\mathbf{\Lambda}^*\mathbf{\Lambda} + \sigma_n^2\mathbf{S}_{\mathbf{f}}^{-1})^{-1}(\mathbf{\Lambda}^*\mathbf{W}^{-1}\mathbf{z} + \sigma_n^2\mathbf{S}_{\mathbf{f}}^{-1}\mathbf{W}^{-1}\bar{\mathbf{f}}) \ . \tag{11}$$

181

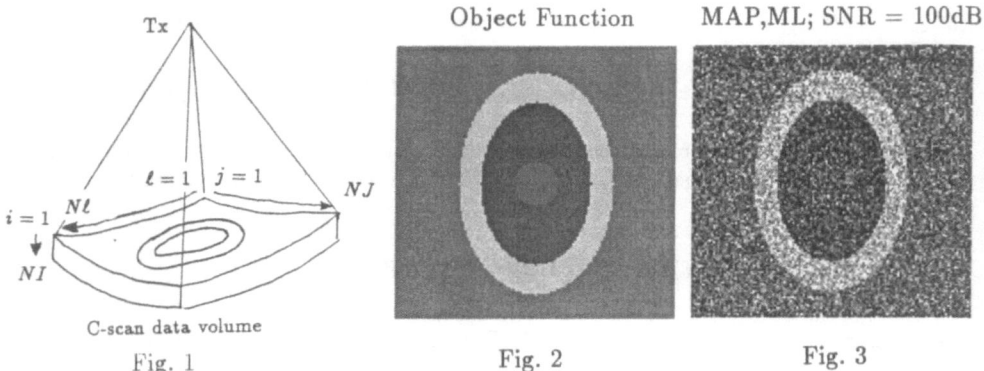

Tx

$\ell = 1 \quad j = 1$

$i = 1$ $N\ell$ NJ

NI

C-scan data volume

Fig. 1

Object Function

MAP,ML; SNR = 100dB

Fig. 2

Fig. 3

Finally, we note that the m-th element of $\mathbf{W}^{-1}\mathbf{f}$, which we denote using $F(m)$, is given by

$$F(m) = \frac{1}{M} \sum_{n=0}^{M-1} f(n) \exp\left(-i2\pi nm/M\right) \quad , \tag{12}$$

so that $\mathbf{W}^{-1}\mathbf{f}$ is the Fourier transform of the object vector: $F = \mathcal{F}\{f\}$ [2]. Applying this relation to the vectors $\hat{\mathbf{f}}$, $\bar{\mathbf{f}}$, and $\hat{\mathbf{z}}$ in Eq. 11 and using the fact that $\Lambda^*\Lambda = |H(u)|^2$, where $H = M \times \mathcal{F}\{h\}$, then Eq. 11 reduces to

$$\hat{F}(u) = \frac{H^*(u)Z(u) + \sigma_n^2 \bar{F}(u)/S_f(u)}{|H(u)|^2 + \sigma_n^2/S_f(u)} \quad , \tag{13}$$

where S_f is the spectral density of the object function $\mathbf{f}$ and $u = m\Delta u$ is frequency.

The MAP estimate for acoustic form factors is $\hat{\Gamma}(u) = \frac{1}{M}|\hat{F}(u)\hat{F}^*(u)|$. This estimate is compared with modeled functions to determine $\hat{D}$. We can demonstrate how Eq. 13 adapts to different noise levels by examining the following limiting conditions. When the noise power is small compared to that of the object, i.e, $\sigma_n^2/S_f(u) \simeq 0$, prior information is ignored and Eq. 13 reduces to the simple inverse filter, $\hat{F}(u) \simeq Z(u)/H(u)$. In this limit the MAP estimator reduces to the maximum likelihood (ML) estimator. The ML solution is obtained directly by setting $p(\mathbf{f})$ in Eq. 2 to a constant. We have used the ML estimator in our previous work since prior information was unavailable. At the other extreme, when the noise power overwhelms the signals power, then Eq. 13 reduces to $\hat{F}(u) \simeq \bar{F}(u)$. As noise power increases, the MAP solution responds by placing greater emphasis on the prior pdf according to the ratio $\sigma_n^2/S_f(u)$, subject to the constraint that the posterior pdf is maximum. Finally, note that when $\bar{F}(u) = 0$, Eq. 13 reduces to the Weiner filter [2].

RESULTS

Echo signals with properties similar to those of kidney tissues were simulated. Using the linear process of Eq. 1 and the C-scan geometry of Fig. 1, we generated a "software phantom" of an object with regions having scattering properties similar to that of several kidney tissues. The object function, whose mean values are displayed in Fig. 2, has four regions that vary in their respective means and standard deviations for scatterer size: in the central circular area $D = 75$ μm and $\sigma_D = 15$ μm; in the dark elliptical area (simulating the renal medulla) $D = 50$ μm and $\sigma_D = 10$ μm; in the bright ring area (simulating the renal cortex) $D = 200$ μm and $\sigma_D = 25$ μm; and in the surrounding background region $D = 100$ μm and $\sigma_D = 15$ μm. Means and standard deviations for object regions simulating the medulla and cortex were based on histological observations [5]. The point spread function of the transformation matrix was that of a 5 MHz, Gaussian-shaped pulse (50% bandwidth).

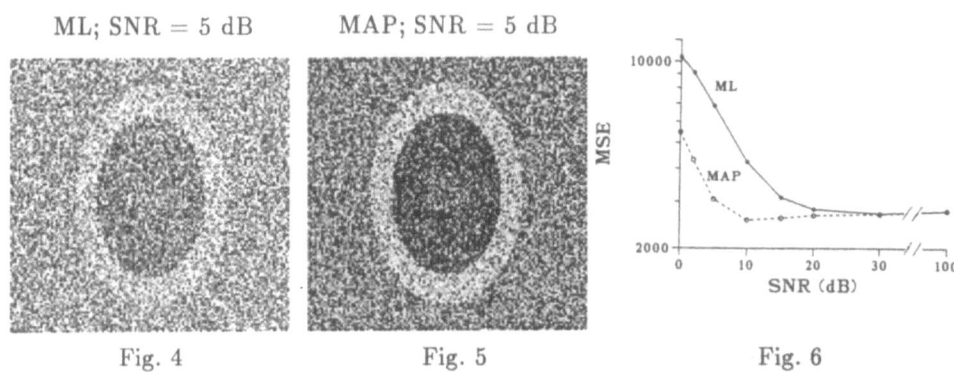

ML; SNR = 5 dB   MAP; SNR = 5 dB

Fig. 4   Fig. 5   Fig. 6

The echo data were then processed using either MAP or ML estimators to produce the scatterer size images of Figs. 3 - 5. As in the mean object function of Fig. 2, image brightness increases linearly with $\hat{D}$. The image in Fig. 3 was reconstructed from low-noise data; the signal-to-noise ratio (SNR) was 100 dB. (SNR is defined as the peak in the spectral density of the signal to that of the noise.) The variability in this image is due primarily to acoustic speckle. To a lesser degree, variability is caused by the random nature of the object function. Because there was negligible additive noise, the ML and MAP estimators produced the same image, Fig. 3. Figs. 4 and 5 are, respectively, the ML and MAP reconstructions of the data at SNR = 5 dB. Under noisy conditions, the four regions of different D are more clearly distinguished in the MAP image than in the ML image. The improvement is shown quantitatively in Fig. 6, where the mean-square error, MSE $= \left\langle (\hat{D} - \bar{D})^2 \right\rangle_{n|f}$, versus SNR is plotted for data in the background regions, including those Figs. 3-5. The MAP estimator provides an advantage over the ML estimator when SNR < 20 dB.

DISCUSSION

The MAP estimator of Eq. 13 pulls the result toward the ML solution when the noise in the data is small compared with the object variability, and toward the solution given by the prior pdf when the noise is relatively large. Obviously completeness and accuracy of the prior information are important, as is the weighting given to it in the analysis, if the reconstruction is to be successful. In Fig. 5, the prior information was the shape of the objects and the mean scatterer sizes in those regions. How might we obtain this information in clinical imaging where the macrostructure of biological media is so diverse? Some prior information about shape of a particular organ is available from the B-mode image. B-mode images might be segmented based on the large differences in backscattered intensity that exists among renal tissues. It may also be possible to develop prior models based on the average 3-D shape of the organ, but include some geometric flexibility to allow the prior model to be warped based on the B-mode data. The feasibility of using flexible prior models in tomographic reconstructions was demonstrated by Hanson (1993b). Other types of prior information, such as structural anisotropy, may also be useful.

This preliminary study is very limited in the sense that only additive white noise, e.g., electronic noise, was considered. Although electronic noise can be significant, particularly in the low-scattering renal medulla, the largest source of noise is acoustic speckle. Future work will include the covariance from speckle noise $\mathbf{K_s}$ [9], such that $\mathbf{K_z} = \mathbf{K_s} + \sigma_n^2 \mathbf{I}$ and $\mathbf{K_f} = \mathbf{W} \mathbf{S_f} \mathbf{W}^{-1} + \sigma_f^2 \mathbf{I}$.

The value of Bayesian techniques cannot be judged from the MSE error alone. Performance must be assessed in the context of the diagnostic task, e.g., detecting changes in D that indicate renal transplant rejection. The prior information used and the performance of the resulting image ultimately depend entirely on the purpose for acquiring the image.

ACKNOWLEDGEMENTS

This work was supported by NIH grant R01 DK43007 and the Clinical Radiology Foundation at KUMC.

References

[1] Campbell, J.A. and Waag, R.C., 1983, Normalization of ultrasonic scattering measurements to obtain average differential scattering cross sections for tissues, *J. Acoust. Soc. A,.* 74:393-399.

[2] Gonzalez, R.C. and Wintz, P. "Digital Image Processing," Addison-Wesley, Reading, MA (1977).

[3] Hanson, K.M., 1993a, Introduction to Bayesian image analysis, *Proc. SPIE* 1898:716-731.

[4] Hanson, K.M., 1993b, Bayesian reconstruction based on flexible prior models, *J. Opt. Soc. Am. A* 10:997-1004.

[5] Insana, M.F., Hall, T.J., and Fishback, J.L., 1991, Identifying acoustic scattering sources in normal renal parenchyma from the anisotropy in acoustic measurements, *Ultrasound Med. Biol.* 17:613-626.

[6] Insana, M.F., Wood, J.G., Hall, T.J., 1992, Identifying acoustic scattering sources in normal renal parenchyma in vivo by varying arterial and ureteral pressures, *Ultrasound Med. Biol.* 18:587-599.

[7] Insana, M.F. and Brown, D.G., 1993, Acoustic scattering theory applied to soft biological tissues, *in:* "Ultrasonic Scattering in Biological Tissues," K.K. Shung and G.A. Thieme, eds., CRC Press, Boca Raton, pp. 75-124.

[8] Lizzi, F.L., Ostromogilsky, M., Feleppa, E.J., Rorke, M.C., and Yaremko, M.M., 1987, Relationship of ultrasonic spectral parameters for features of tissue microstructure, *IEEE Trans. Ultrason. Ferroelect. Freq. Contrl.* 34:319-329.

[9] Wagner, R.F., Insana, M.F., and Brown, D.G., 1987, Statistical properties of radio-frequency and envelope-detected signals with applications to medical ultrasound, *J. Opt. Soc. Am. A* 4:910-922.

ACOUSTIC PROPERTIES OF RENAL CELL CARCINOMA TISSUES

Hidehiko Sasaki , Yoshifumi Saijo, Hiroaki Okawai,
Yoshio Terasawa, Shinichi Nitta and Motonao Tanaka

Department of Medical Engineering and Cardiology
Division of Organ Pathophysiology
Institute of Development, Aging and Cancer
Tohoku University, Japan

INTRODUCTION

The early detection of many renal cell carcinoma is now available by the development of imaging methodology.Ultrasonography is one of the most useful tool for early diagnosis of the cancer and the screenings because of its non-invasiveness and repeatability. However, diagnosis is difficult in some cases because clinical echographic features of the tumor showed various patterns. The comparison between the echographic appearance and the pathohistological pattern of the renal cell carcinoma has been studied in detail[1-5], but the relation has not been concluded at the present time.

Therefore, measurements of the acoustic properties at the microscopic level are very important to understanding the clinical echographic images.

The purpose of the present study is to characterize renal cell carcinoma tissue by the measurement of micro acoustic properties with a Scanning Acoustic Microscope (SAM) system.

METHODS

Surgically excised renal cell carcinoma tissues from 15 patients (8 clear cell subtype and 7 granular cell subtype) were investigated. The tissues were formalin-fixed, paraffin-embedded, sectioned 10μm in thickness and mounted on glass slides for SAM studies. The paraffin was removed by the graded alcohol method, just before the ultrasonic measurement. A neighboring section of the SAM specimen was sectioned and stained with Elastica-Masson staining. The region of interest (ROI) for acoustic microscopy was determined after observation of the optical microscopic image.

Figure. 1 is a block diagram of the SAM system[4-6]. The system is consisted of five parts; (1) ultrasonic transducers, (2) mechanical scanner, (3) analogue signal processor, (4) image signal processor, and (5) display unit.

The ultrasonic frequency is variable over the range 100-200 MHz for SAM

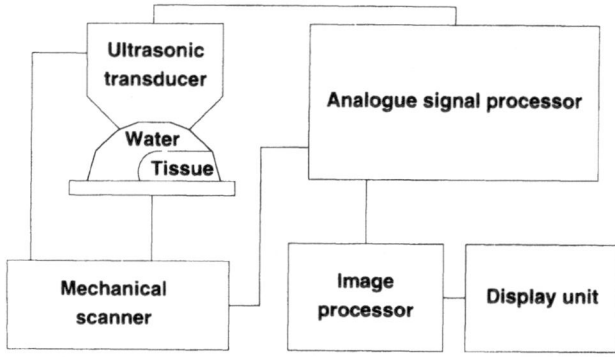

Figure 1. Block diagram of the scanning acoustic microscope system.

measurement. (The Transducer is equipped with a lens for which the dimensions of diameter, radius of curvature and aperture angle are 1.25mm, 1.25mm and 60 degrees, respectively, and having a -3dB, beam width in water at 20°C between 5μm (at 200MHz) and 10μm (at 100MHz)). The transducer is mechanically scanned at 60Hz, in the lateral direction (X) above the specimen, which remains stationary on the sample holder, while the holder is scanned in the other lateral direction (Y). The mechanical scanner is so arranged that the ultrasonic beam is transmitted for every 4μm interval over a 2mm width. The number of sampling points are 480 in one scanning line and 480 ×480 points make one frame in this system.

The SAM system can display two-dimensional distributions of the attenuation constant and the sound speed. These acoustic parameters in this system are related by the color bar scale showing in Table 1.

Table 1. Relationship of color coded scales and values of sound speed and attenuation constant.

Attenuation constant (dB/mm/MHz)	Color	Sound Speed (m/s)
1.9 ~	Red	1690 ~
1.7 ~ 1.9	Magenta	1670 ~ 1690
1.5 ~ 1.7	Orange	1650 ~ 1670
1.3 ~ 1.5	Brown	1630 ~ 1650
1.1 ~ 1.3	Yellow	1610 ~ 1630
0.9 ~ 1.1	Green	1590 ~ 1610
0.7 ~ 0.9	Olive Green	1570 ~ 1590
0.5 ~ 0.7	Cyan	1550 ~ 1570
0.3 ~ 0.5	Royal blue	1530 ~ 1550
0.1 ~ 0.3	Blue	1510 ~ 1530
~ 0.1	Black	~ 1510

RESULTS

(1) Normal renal cortex and medulla:

Fig.2 shows the optical and acoustic images of normal renal kidney.

The values of attenuation constant and sound speed are 1.3dB/mm/MHz and 1636m/s in renal corpuscles and 0.6dB/mm/MHz and 1558m/s in renal tubules, respectively. Both values are 1.6dB/mm/MHz and 1657m/s in the interstitial collagenous fiber. Average values of whole renal cortex are 1.1dB/mm/MHz, 1610m/s, and in whole renal medulla are 1.0dB/mm/MHz and 1588m/s, respectively.

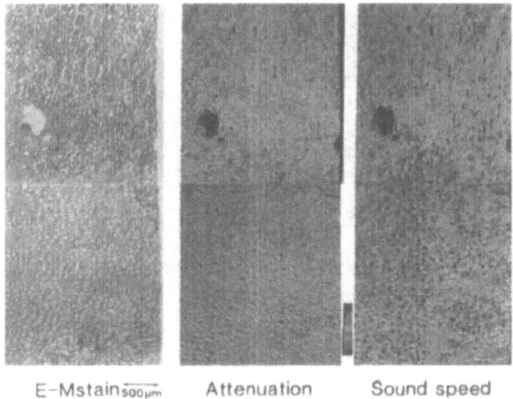

E−Mstain 500μm   Attenuation   Sound speed

Figure 2. Optical and acoustic images of a specimen of normal renal cortex and medulla.E-Mstain: Elastica-Masson staining.

(2) clear cell carcinoma:

Fig.3 shows a case of clear cell subtype of renal cell carcinoma.

Optical microscopy shows that the hemorrhage from capillary vessels are observed around the clear cell. Hemorrhage shows very high values of both acoustic parameters, 1.9dB/mm/MHz and 1666m/s, respectively. Fibrosis also shows high values of both acoustic parameters, while the values of attenuation constant (0.4dB/mm/MHz) and sound speed (1536m/s) are low in the clear cell (Fig.3).

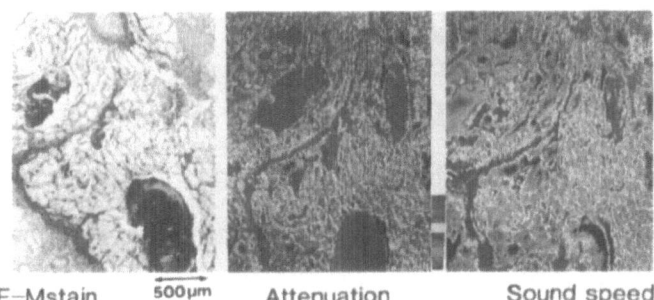

E−Mstain   500μm   Attenuation   Sound speed

Figure 3. Optical and acoustic images of a specimen of clear cell subtype of renal cell carcinoma.E-M stain: Elastica-Masson staining.

(3) Granular cell carcinoma:

Fig.4 demonstrates a case of granular cell subtype of renal cell carcinoma.

Both acoustic parameters are also,viz., 0.5dB/mm/MHz and 1541m/s in the granular cell (arrows). And both values are high in the fibrous lesion.

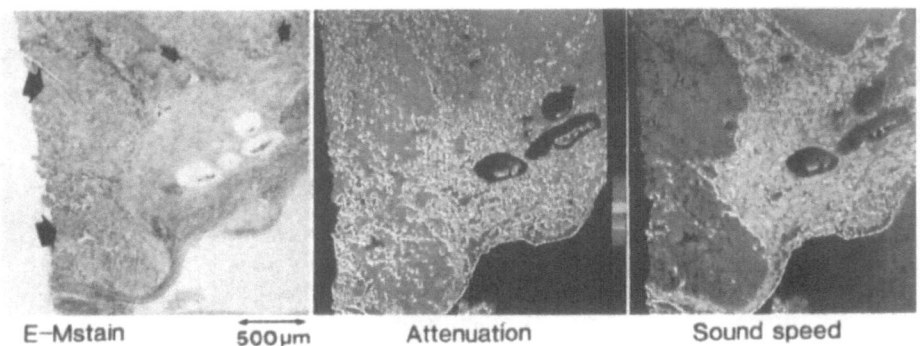

E—Mstain   500μm   Attenuation   Sound speed

Figure 4. Optical and acoustic images of a specimen of granular cell subtype (arrows)of renal cellcarcinoma. E-M stain: Elastica-Masson staining.

Fig.5-6 shows the means and the standard deviation of acoustic parameters of all the renal cell carcinoma tissues.

The values of attenuation constant and sound speed are lower in both kinds of tumor cells than those in normal kidney, although significant difference is not found between the clear cell and the granular cell. And the both acoustic parameters of cancer cells are significantly lower than those in surrounding tissue elements, such as fibrosis or hemorrhage.

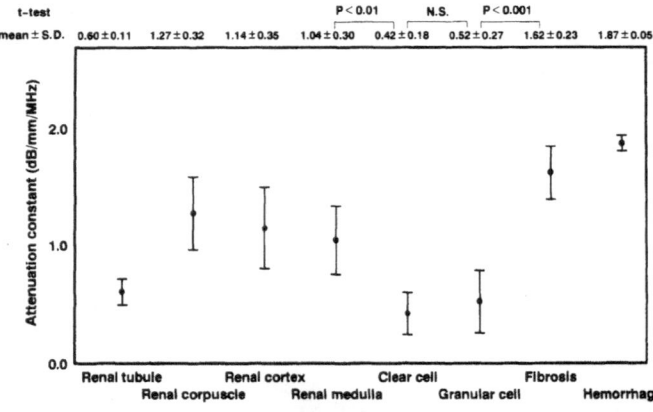

ATTENUATION CONSTANT OF TISSUE ELEMENTS IN RENAL CELL CARCINOMA

Figure 5. Graph showing the means and standard deviations of the attenuation constant data of all the specimens.

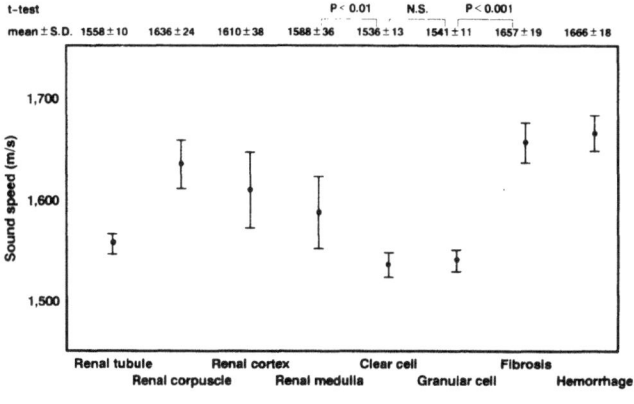

Figure 6. Graph showing the means and standard deviations of the sound speed data of all the specimens.

DISCUSSION

The frequency used for the clinical ultrasonic diagnosis is 2.5-5.0MHz, and its beam width at the focal volume ranges level is about 1mm. Therefore, with higher frequency ultrasonic waves are necessary to measure the acoustic properties at the microscopic level[7]. Ultrasonic frequency used in the SAM system is 100-200MHz[4-6], so the attenuation constant and sound speed of the micro acoustic field can be measured.

It is generally considered that the interface echo separating a pair of media is explained in terms of the magnitude difference of the two specific acoustic impedances.The specific acoustic impedance is demonstrated by the following formula,

$$z = \rho c$$

(z:the acoustic impedance, ρ:the density, c:the sound speed) and depends mainly on the sound speed. Therefore, strong echo may be produced at the interface between the clear cell and hemorrhage or fibrosis because the difference of the acoustic impedance between the two types of tissue element is very large. On the other hand,the attenuation constant is comprised of two factors, viz., Absorption and scattering. Absorption is increased, when the macromolecular content of the medium is large. Scattering may be considered to be related to the structural features of the medium. So, in collagenous fiber such as high density and large molecular weight media, both absorption and scattering are high, leading to high attenuation constant. Clear cell is considered exhibit low attenuation constant because the cancer cell contains less amount of macromolecular contents and its density is low.

As the interface of pairs of tissue elements is not smooth, the ideal reflection does not always occur in biological tissues. And it is not clear that the relationship between the size of the hemorrhage in Figure 3 and the reflection. But our data suggest that the high intensity echo in clinical echography of the tumor may be related to the heterogeneity of the micro acoustic field in the carcinoma tissue.

Furthermore, it is known that the sound speed in a medium modeled as a fluid may be given by

$$c = \sqrt{K} / \sqrt{\rho}$$

(c:the sound speed (m/s), K:the elastic bulk modulus (N/m^2), ρ:the density (kg/m^3)). Thus the measured sound speed reflects elasticity of the material. In the present study, reduced bulk elasticity of the clear cell carcinoma tissue is associated with the looser intercellular junction.

Electron microscopic studies demonstrated that the cells of clear cell carcinoma are almost identical to normal epithelial cells of the proximal convoluted tubule[8]. However, the junctional surface of the clear cell is flat, without showing severe complication which can be detected in the normal proximal convoluted tubule. And the number of the intercellular tight junction or the desmosome decreased remarkably. Similarly with the clear cell, characteristics originate in the proximal convoluted tubule such as microvilli, etc., are also found in the granular cell. Therefore, both the clear and the granular cell which are essentially same. These lesser stiffness may be accounted for reduced sound speed values.

In this study, it was considered that elasticity of renal cell carcinoma tissue may be lower than that of normal renal tissue, because the value of sound speed are significantly lower in cancer cells than that of normal renal tissue.

CONCLUSIONS

Acoustic properties of renal cell carcinoma tissue were measured with a scanning acoustic microscope system.

The characteristics of the cancerous tissue are described as follows.

1) The values of attenuation constant and sound speed were lower in both kinds of cancer cells than those in the normal kidney.

2) No significant difference was found between the clear cell and the granular cell.

3) The values of attenuation constant and sound speed are significantly higher in fibrosis or hemorrhage than those in cancer cells.

4) The elasticity of renal cell carcinoma tissue may be lower than that of normal renal tissue.

5) The high intensity echo in clinical echography may be related to the heterogeneity of the micro acoustic field in the carcinoma tissues.

REFERENCES

1. Jinzaki M, Hisa N, Fujikura Y, Ohkuma K, Tashiro Y, Sugiura H, Comparative study between ultrasonographic and pathohistological findings of small renal cell carcinoma. Jpn. J. Med. Ultrasonics 17:280-287 (1990).

2. Fujii H, Kaneko S, Hashimoto H, Sasaki M, Yachiku S, Ulutrasonotomograms of Renal Cell Carcinoma-Retrospective Study of the Internal Echogram of Tumor-, Jpn.J.Med. Ultrasonics. 16:375-382 (1989).

3. Mihara S, Early Detection of Renal Cell Carcinoma by Ultrasonics Mass Survey, Jpn.J.Med.Ultrasonics 20:482-490 (1993).

4. Okawai H, Tanaka M, Dunn F, Chubachi N, Honda K, Quantitative display of acoustic properties of biological tissue elements. Acoustical Imaging 17: 193-201 (1988).

5. Okawai H, Tanaka M and Dunn F, Non-contact acoustic method for the simultaneous measurement of thickness and acoustic properties of biological tissues. Ultrason 28: 401-410 (1990).

6. Y. Saijo, M. Tanaka, H. Okawai, F. Dunn, The ultrasonic properties of gastric cancer tissues obtained with a scanning acoustic microscope system. Ultrasound Med & Biol. 17: 709 (1991).

7. O'Brien WD, Olerud J and Shung KW, Quantitative acoustic assessment of wound maturation with acoustic microscopy. J.Acoust. Soc. Am. 69: 575-579 (1981).

8. Oberling C, Riviere M, Hagueman F, Ultrastracture of clear cells in renal carcinoma and its importance for the demonstration of their renal origin. Nature 186: 402 (1960).

PULSE-ECHO IMAGING WITH X WAVE

Jian-yu Lu, Mostafa Fatemi, and James F. Greenleaf

Biodynamics Research Unit
Department of Physiology and Biophysics
Mayo Clinic/Foundation
Rochester, MN 55905, U.S.A.

INTRODUCTION

Limited diffraction beams have a large depth of field. They could have applications in medical imaging,[1,2] tissue characterization,[3] nondestructive evaluation (NDE) of materials,[4] Doppler velocity measurement,[5] as well as other areas such as optics[6] and electromagnetics.[7]

In this paper, we perform the first experimental study of pulse-echo imaging with a recently discovered X wave.[8] Results are compared to those of pulse-echo imaging with a Bessel beam[6] and a conventional focused Gaussian beam. It is shown that pulse-echo imaging with limited diffraction beams have a larger depth of field than that with a conventional focused beam.

THEORY

A limited diffraction beam in a three-dimensional space is given by

$$\Phi(x, y, z - c_1 t), \tag{1}$$

where Φ represents a wave and is a solution to the isotropic/homogeneous wave equation, x, y, and z are spatial variables in rectangular coordinates, t is time, c_1 is real and is the phase velocity of the wave, and $z - c_1 t$ is the propagation term. X waves,[8]

$$\Phi_{XBB_0} = \frac{a_0}{\sqrt{(r \sin \zeta)^2 + [a_0 - i \cos \zeta (z - c_1 t)]^2}}, \tag{2}$$

and Bessel beams,[6]

$$\Phi_{J_0} = J_0(\alpha r)e^{i\beta(z-c_1 t)}, \tag{3}$$

have the form of Eq. (1), where $c_1 = c/\cos\zeta$ and $c_1 = \omega/\beta$ for Eq. (2) and Eq. (3), respectively, a_0, c, ζ, and α are constants, $r = \sqrt{x^2 + y^2}$ is a radial distance, J_0 is the zero-order Bessel function of the first kind, $\beta = \sqrt{(\omega/c)^2 - \alpha^2}$, and ω is an angular frequency.

EXPERIMENT

Because a limited diffraction beam has the same lateral beam profile at any axial distance, z, it can be constructed for $z > 0$ if it is produced at $z = 0$. Limited diffraction beams at $z = 0$ are given by

$$\Phi(x, y, z - c_1 t)|_{z=0}, \tag{4}$$

which is a function of t when observed at any transverse position, (x, y). If limited diffraction beams are rotary symmetric around the axis, such as those in Eqs. (2) and (3), they can be approximated with a stepwise function of r and can be approximately produced with an annular array. In this paper, we study only this case.

A block diagram of pulse-echo imaging using an X wave is shown in Fig. 1. A 13–element, 3.5 MHz central frequency, and 22.6 mm diameter ultrasonic annular array transducer was used to produce the X wave. Waveforms to drive the transducer were calculated from Eq. (2) with $z = 0$ at various radial distances that corresponded to the central radii of the transducer elements. Radio frequency (rf) echo signals were received by the same transducer and the outputs of the transducer elements were amplified and digitized. To obtain a B-mode image, the transducer was scanned linearly over an object to produce multiple A-lines. A modified AIUM standard 100 mm test object (Fig. 2) was used for the imaging study.

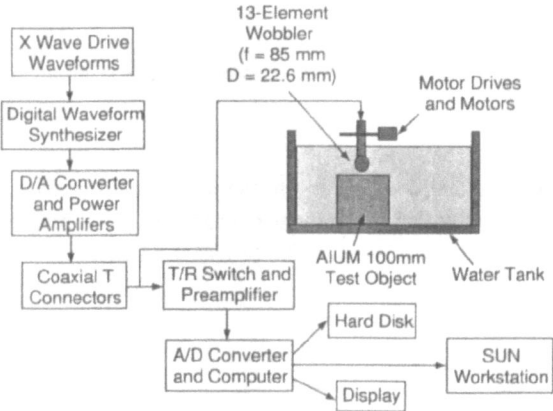

Figure 1. Block diagram of pulse-echo imaging with X wave.

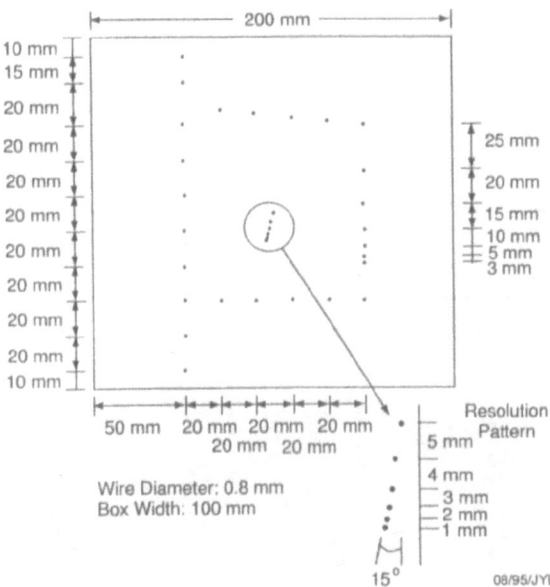

Figure 2. A cross section of a modified AIUM standard 100 mm test object. The object was obtained by adding four additional wires on top (two wires) and bottom (two wires) of the standard AIUM test object. The additional wires are useful for studying the large depth of field of limited diffraction beams.

To produce an X wave response in receive, rf data obtained from the transducer elements were filtered, summed, and then envelope detected. The impulse responses of the filters were calculated from Eq. (2) with $z = 0$ and $r = r_i$, where r_i was the central radius of the ith element (Fig. 3). Similarly, the same filters can also be used in transmit to replace the multichannel waveform synthesizer (Fig. 1) for producing drive waveforms (Fig. 3).

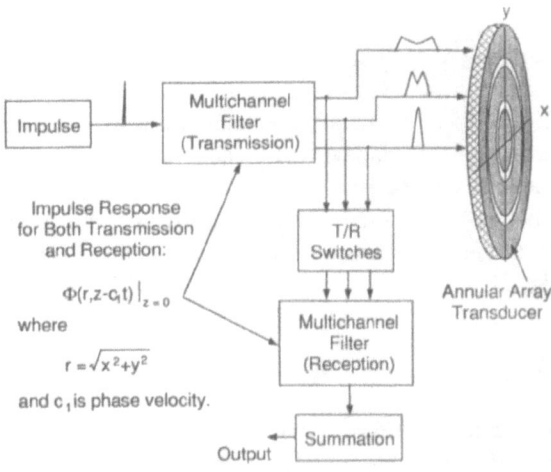

Figure 3. Filters for the X wave pulse-echo imaging system.

RESULTS

Experiment results for X wave pulse-echo imaging of the modified AIUM test object are shown in the left panel of Fig. 4. For comparison, pulse-echo images with a Bessel beam and a focused Gaussian beam are also shown. These images were obtained with the same conditions except that the transmit beams and the filters in receive were different. Parameters of these beams are the same as those in reference,[9] and are shown on the bottom of Fig. 4. For images obtained with the X wave and the Bessel beam, a dynamic aperture was used in receive to reduce sidelobes near the surface of the transducer. The size of aperture was increased from 5.46 mm (three elements) to 22.6 mm (13 elements) linearly as the axial distance where echoes returned was increased from 10.6 mm to 104.7 mm.

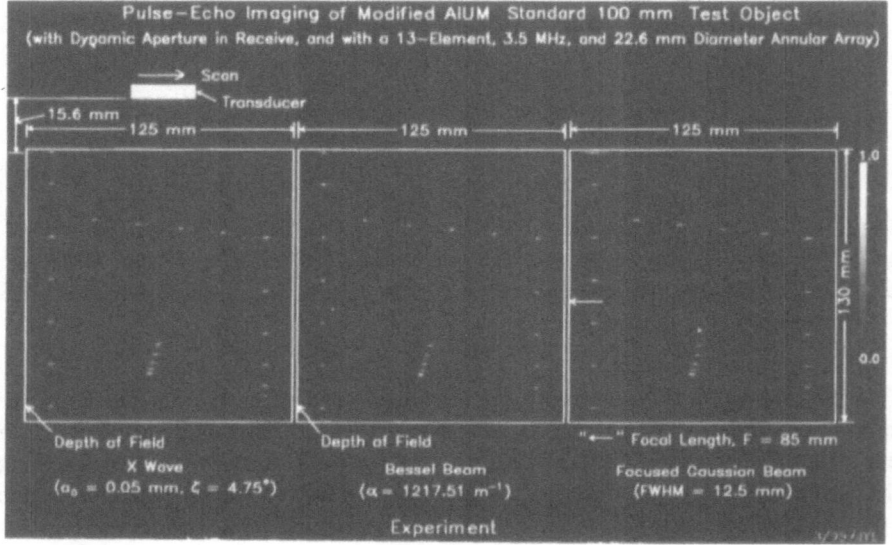

Figure 4. Pulse-echo images of the modified AIUM standard 100 mm test object with an X wave (left panel), Bessel beam (middle panel), and focused Gaussian beam (right panel).

To see clearly the lateral resolution at different axial distances, cross-sectional images of lines (line spread function or LSF) on the left-most column of each image in Fig. 4 were cut with a window of 24 mm (lateral) × 10 mm (vertical). The peak of each vertical line (A-line) of the image in the window was plotted versus the lateral distance (Fig. 5) to show the highest sidelobes of the LSF. The −6dB width of the LSF is shown in Panel (8) of Fig. 5 as a function of the axial distance from 0 to 200 mm. The depth of field for both the X wave and the Bessel beam is about 136 mm that is indicated by a vertical line (Panel (8) of Fig. 5).

DISCUSSION

From the experiment results (Figs. 4 and 5), one sees that the pulse-echo images obtained with limited diffraction beams have a larger depth of field but have higher sidelobes than those obtained with a conventional focused Gaussian beam. It is noted that the sidelobes shown in Fig. 5 are that of the LSFs and are much higher than those of point spread functions (PSFs). The results also show that the −6dB width of the LSF of the X wave is larger than that of the Bessel beam in the depth of field (136 mm). This is because the actual central frequency of the transducer is about 2.9 MHz instead of 3.5 MHz indicated by the manufacturer[1]. For Bessel beams, the change of central frequency will not affect the −6dB beam width that is determined by the width of the central lobe of the Bessel function. For conventional focused Gaussian beams, the −6dB beam width is determined by the apodization of aperture, the f-number, as well as the central frequency. It is seen that the variation of the width of the LSF over distance is the largest for the focused Gaussian beam (Panel (8) of Fig. 5) that has an f-number of about 3.76 (focal length is 85 mm) and a full width at half maximum (FWHM) of about 12.5 mm at the surface of transducer.

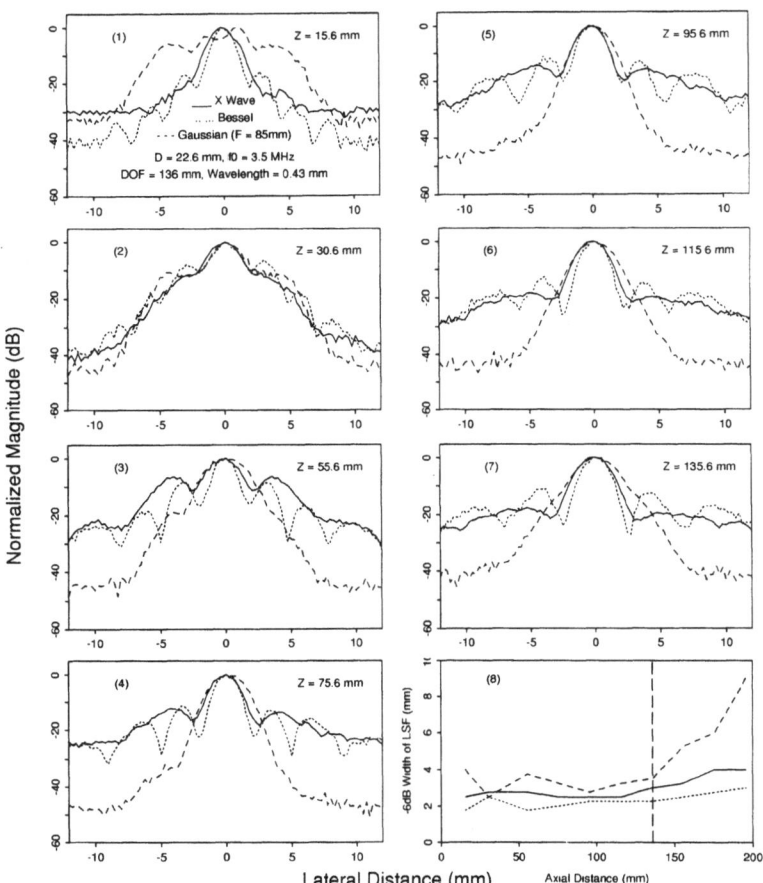

Figure 5. Line spread functions (LSF) of the pulse-echo imaging systems with the X wave (full lines), Bessel beam (dotted lines), and focused Gaussian beam (dashed lines) at several axial distances, z = 15.6 mm (Panel (1)), 30.6 mm (Panel (2)), 55.6 mm (Panel (3)), 75.6 mm (Panel (4)), 95.6 mm (Panel (5)), 115.6 mm (Panel (6)), and 135.6 mm (Panel (7)). The −6dB width of the LSF versus the axial distances is shown in Panel (8) where the vertical line (long dashed line) indicates the depth of field of the X wave and the Bessel beam.

---

[1]   Echo Ultrasound, Reedsville, PA, U.S.A.

Theoretically, beam width of limited diffraction beams should be constant over a large distance (large depth of field). The fluctuation of the width of the LSFs in Panel (8) of Fig. 5 could be caused by several factors in our experiment. For example, the transducer is not rotary symmetric; the curvature (85 mm) of transducer can not be completely compensated because each element has a certain width; the transfer function, bandwidth, and the central frequency of each element is different from each other; the weightings may deviate from theoretical values due to the variation of sensitivity from element to element. In addition, the plastic cap (wobbler cap) that is in front of the transducer may also cause some distortion.

CONCLUSION

We have successfully produced a pulse-echo image with an X wave. Like Bessel beams, X waves have a large depth of field but higher sidelobes as compared to conventional focused beams. However, sidelobes of limited diffraction beams can be reduced by various methods.[10]

ACKNOWLEDGMENTS

The authors appreciate the secretarial assistance of Elaine C. Quarve. This work was supported in part by grants CA 54212 and CA 43920 from the National Institutes of Health.

REFERENCES

1. Jian-yu Lu, H. Zou, and J.F. Greenleaf, Biomedical ultrasound beamforming, *Ultrasound Med. Biol.* 20(5):403–428 (July, 1994).
2. Jian-yu Lu and J.F. Greenleaf, Ultrasonic nondiffracting transducer for medical imaging, *IEEE Trans. Ultrason., Ferroelec., Freq. Contr.* 37(5):438–447 (Sept., 1990).
3. Jian-yu Lu, and J.F. Greenleaf, Evaluation of a nondiffracting transducer for tissue characterization, *IEEE 1990 Ultrason. Symp. Proc.* 90CH2938-9, 2:795–798 (1990).
4. Jian-yu Lu and J.F. Greenleaf, Producing deep depth of field and depth-independent resolution in NDE with limited diffraction beams, *Ultrason. Imag.* 15(2):134–149 (April, 1993).
5. Jian-yu Lu, X-L. Xu, H. Zou, and J.F. Greenleaf, Application of Bessel beam for Doppler velocity estimation, *IEEE Trans. Ultrason., Ferroelec., Freq. Contr.* 42(4):649–662 (July, 1995).
6. J. Durnin, Exact solutions for nondiffracting beams. I. The scalar theory, *J. Opt. Soc. Am.* 4(4):651–654 (1987).
7. J.N. Brittingham, Focus wave modes in homogeneous Maxwell's equations: transverse electric mode, *J. Appl. Phys.* 54(3):1179–1189 (1983).
8. Jian-yu Lu and J.F. Greenleaf, Nondiffracting X waves — exact solutions to free-space scalar wave equation and their finite aperture realizations, *IEEE Trans. Ultrason., Ferroelec., Freq. Contr.* 39(1):19–31 (Jan., 1992).
9. Jian-yu Lu, T.K. Song, R.R. Kinnick, and J.F. Greenleaf, *In vitro* and *in vivo* real-time imaging with ultrasonic limited diffraction beams, *IEEE Trans. Med. Imag.* 12(4):819–829 (Dec., 1993).
10. Jian-yu Lu, Bowtie limited diffraction beams for low-sidelobe and large depth of field imaging, *IEEE Trans. Ultrason., Ferroelec., Freq. Contr.* (In press).

EXTRACTION OF ENDOCARDIAL BOUNDARY FROM ECHOCARDIOGRAPHIC IMAGES BY MEANS OF THE KOHONEN SELF-ORGANIZING MAP

Marek Belohlavek,[1] Armando Manduca,[1] Jean Buithieu,[2] James F. Greenleaf,[1] and James B. Seward[2]

[1]Department of Physiology and Biophysics
[2]Division of Cardiovascular Diseases and Internal Medicine
Mayo Clinic and Foundation
Rochester, MN 55905

INTRODUCTION

There is considerable clinical interest in development of algorithms for reproducible determination of endocardial boundaries in echocardiographic images.[1-9] This is feasible in the era of computer-aided analysis of cardiac morphology and function. However, ultrasound images are notoriously difficult to process because they are typically incomplete (dropouts, noise, etc.). Thus, automatic endocardial detection techniques require image enhancement to deal with discontinuous border definition. Our initial experiences with self-organizing maps (SOM) for the delineation of endocardial echoes is very encouraging and discussed in this manuscript. The objective was to determine whether this form of neural network combined with algorithms for edge detection can perform reproducible automated endocardial boundary delineation in artifact-prone echocardiographic images. The SOM has been preferred because: 1) no external operator is necessary to oversee the learning process of the unsupervised neural net, 2) it can be initialized with certain target-relevant shapes, 3) topological relationships are maintained between the neural net lattice nodes (the nodes define the vertices of surface tiles which may be useful for curved distances, surface area, and volume calculation), and 4) similar SOM algorithms have been successfully applied to other complex images.[10]

BACKGROUND

The Kohonen SOM algorithm was introduced in the early 1980s.[11,12] It is a type of neural network which maps input data, defined by input vectors, from an n-dimensional space onto a lattice of nodes in an m-dimensional space (where m ≤ n) in an ordered fashion. The geometry of the lattice is fixed, and the nodes have neighborhood relationships defined on the lattice. Each node has an associated reference vector, called an internal state vector, which can be considered to be a position (i.e., a coordinate of a node) in the n-dimensional input space. The usefulness of the network is that during training the internal state vectors automatically fit themselves to the probability distribution of the input vectors while preserving local neighborhood relations—i.e., nearby points in the input space tend to map to nearby points on the lattice. The starting value of the internal state vectors may be random or approximate some (a priori known) target shape, such as a sphere or ellipsoid in the case of the left ventricle (LV). A priori knowledge enhances the network training speed and provides appropriate spatial distribution of individual nodes in the input space. The self-organization process can be intentionally influenced by assigning weights to individual input vectors thus controlling attraction of SOM nodes to them. The network functioning is further explained in Figure 1.

METHODS

Image acquisition: A conventional echocardiographic system equipped with a mechanically rotatable phased array transducer (SONOS 1500, Omniplane, Hewlett Packard, Andover, MA) was employed for image acquisition during clinically indicated echocardiographic examinations. Two image sets collected by sequential rotation of the transducer were used for testing the SOM. The sets contained 8 and 12 two-dimensional (2D) slices suitable for three–dimensional (3D) reconstruction. Typical artifacts such as dropouts due to signal attenuation and specular and random noise were present in the images.

Image processing: Algorithms for noise and speckle reduction were combined with techniques for image gradient magnitude and direction calculation.[13] The gradients were weighted according to their inner product with the positional vector directed from a user-approximated center of gravity of the LV cavity. The resulting weighted gradient magnitudes were considered to represent a probability of occurrence of the LV endocardial boundary because: 1) most gradients generated by noise had a direction different from that of the positional vector, such that the inner product would be small, and 2) gradients related to (unwanted) epicardial pixels, although nearly parallel to the positional vectors representing endocardial gradients, had opposite direction so the inner product had opposite sign and could be discarded (i.e., probability set to zero). Figure 2 shows the analysis of gradients in an echo tomogram. The SOM algorithm has been employed to "traverse" across the peaks and approximate the "true" LV endocardial boundary. The appropriate distribution of SOM nodes was influenced by their attraction to each peak, which was proportional to the peak amplitude (i.e., to the probability of occurrence of the endocardium). The self-organizing map algorithm was adapted from the freeware package SOM–PAK.[14]

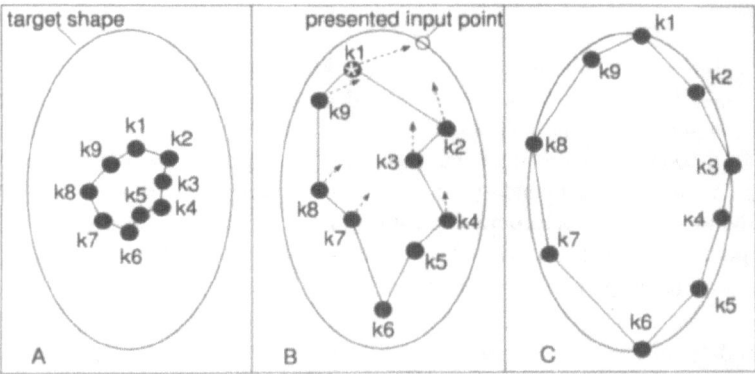

Figure 2 Analysis of gradients in an echo tomogram. A) Original echo tomogram positioned to match the gradient plot. Note image noise and dropouts in the apical and posterior wall regions. B) Planar estimation of the probability of the occurrence of the endocardium by an automated edge detection algorithm. The 3D plot illustrates positive gradients calculated from a computer-enhanced 2D echo scan. Peak locations provide clues about the position of the LV cavity boundary. (APX = apex, IVS = interventricular septum, PW = posterior wall)

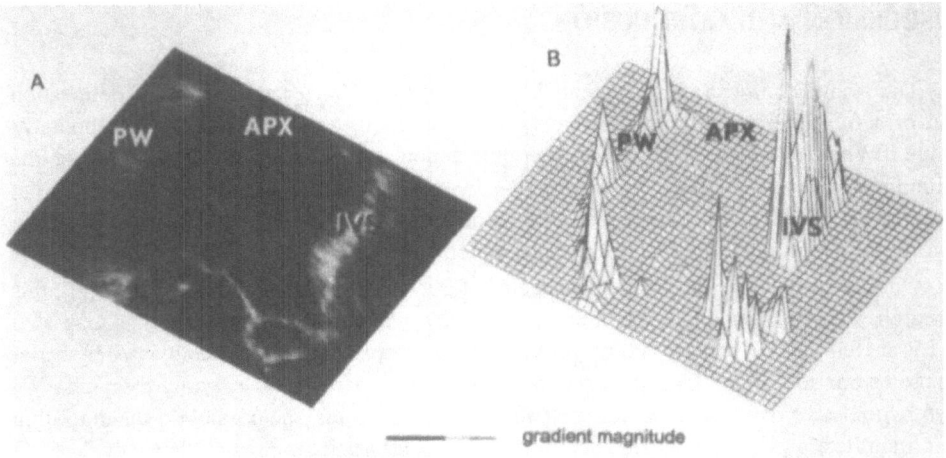

Figure 1 Principles of the self-organizing map (SOM). The goal of SOM training is to distribute the nodes to best reproduce the target shape (represented by an ellipse above). Panel A: A one-dimensional map with nodes k1-k9 organized in approximate circular initial shape. Panel B: The node nearest to a presented input vector is declared the winning node (*) and moved part of the distance toward the input vector. Neighboring nodes are also moved, depending on their distance from the winning node (the most distant neighbor moves least). Panel C: After multiple training iterations, all input points have been presented to the map many times, and the map conforms to the shape of the feature space (i.e., an ellipse).

RESULTS

The algorithm was first tested on sequential 2D echo tomograms. An example is in Figure 3. Each tomogram contained regions of a large apical dropout and an area with multiple edges in the APX and LVOT regions in Figure 3, respectively. The interdependence of neighboring nodes prevented the nodes traversing the region of the dropout from escaping outside the LV endocardial boundary. In the LVOT region, the presence of multiple edges led the SOM to adopt a "compromise" position. These results indicated that the SOM can handle dropouts suitably but a lack of information in regions with little data or confusing information due to multiple edges may result in a somewhat inaccurate outline of the LV cavity.

To provide more spatial information about the LV endocardial boundary, we presented to an initially cylindrical SOM a 3D reconstructed image of the human heart where probability of values corresponding to right ventricle and both atria were suppressed (zeroed) by user interaction (Figure 4). Similar to the previously described experiments with 2D tomograms, this 3D volume image also contained dropouts (see D1 and D2, Figure 4) and a complicated shape with multiple edges (see MV region in Figure 4). However, interdependence of the SOM nodes combined with the full 3D information about LV shape resulted in a smooth, naturally-looking representation of the LV cavity boundary. The SOM underwent 50,000 training steps to form this shape, which required approximately one minute on a SPARCstation 10.

DISCUSSION AND CONCLUSIONS

The results demonstrate the potential benefits of the SOM for automated LV delineation and of a full 3D echocardiographic image reconstruction to provide more clues about LV shape. When developing the SOM, user interaction was required to adjust some of the training parameters, such as the size of the interdependent neighborhood and number of training cycles. This adjustment, however, is necessary only during the development of the algorithm. Once tuned, the SOM operates automatically.

Our 3D images typically contained regions far away from or unrelated to the LV. In this research, we constrained the algorithm to the LV by interactively suppressing data unrelated to LV anatomy. This process is time-consuming, subjective, and should be obviated by better image estimation of the LV boundary location. This is part of our ongoing research. We also hypothesize that improved automated border detection may be possible if the algorithm can assimilate a priori knowledge about texture as well as shape characteristics of the LV.

We conclude that the Kohonen SOM is useful for delineation of the left ventricle boundary in images where data are incomplete. The mutual connection of individual SOM notes is always preserved, and the fitted image provides a continuous delineation even in situations where the ultrasound signal is attenuated. This technique represents an advanced step in the development of an accurate automated boundary tracing algorithm suitable for shape analysis and volume quantitation without geometric assumptions. Such an algorithm should ultimately function as a trainable electronic expert.

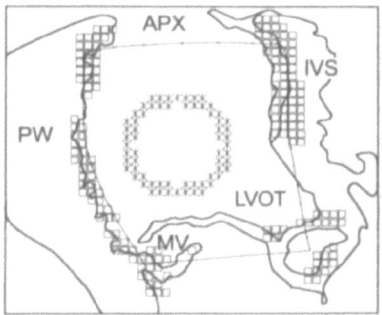 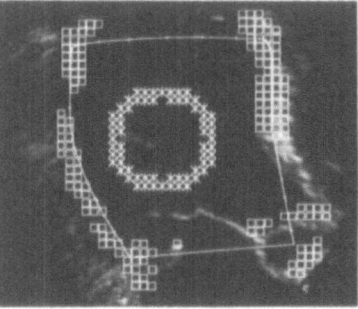

Figure 3 2D SOM delineation of the LV cavity boundary with image dropouts. The self-organizing map (SOM) begins with an approximately circular shape of nodes denoted by asterisks (*). The automatically generated endocardial tracing is shown by the continuous red line with small dots indicating the final distribution of the SOM nodes. Squares illustrate a bird-eye view of input vectors representing the probability of occurrence of the endocardium (Figure 2A). Note the fit even in regions where there is little data. The network does not escape through the large dropouts (such as the one in the apical area) because of the interdependence of the SOM nodes. The SOM solution conforms to regions with high gradients. (APX = apex, IVS = interventricular septum, LVOT = left ventricular outflow, MV = mitral valve, PW = posterior wall)

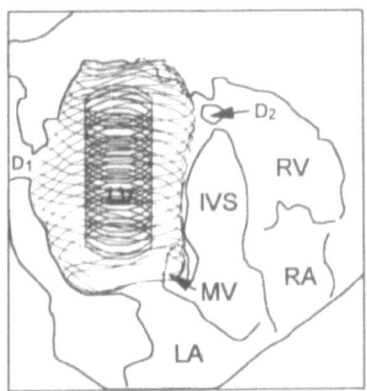 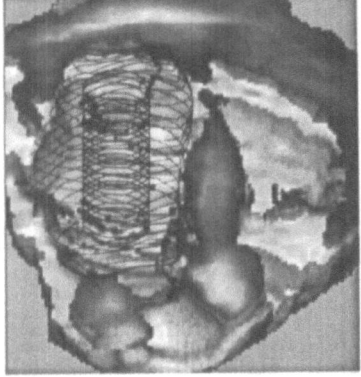

Figure 4 3D SOM delineation of the LV cavity. The initial SOM shape is cylindrical. During training, the SOM expands itself within the cavity thus generating a smooth 3D outline of the endocardium. Spatial interdependence of nodes (defined by the neighborhood relationship on the lattice - see Background) combined with fully 3D information about LV shape prevented escape of the nodes through the dropouts D1 and D2 and resulted in smooth, natural outlines in the regions as well as in the region of the mitral valve (MV) containing multiple edges. For better appreciation, the fit of the map is illustrated using a 3D plot of the SOM superimposed over 3D rendering of the corresponding, electronically dissected heart. An SOM with a rectangular lattice was used; only horizontal lines are shown in this illustration for ease of comprehension. (IVS = interventricular septum, LA = left atrium, LV = left ventricle, RA = right atrium, RV = right ventricle)

REFERENCES

1. C.H. Chu, E.J. Delp, and A.J. Buda. Detecting left ventricular endocardial and epicardial boundaries by digital two-dimensional echocardiography. IEEE Trans. Med Imag. 7:81–90 (1988).

2. N. Friedland and D. Adam. Automatic ventricular cavity boundary detection from sequential ultrasound image using simulated annealing. IEEE Trans. Med. Imag. 8:344–353 (1989).

3. L.H. Staib and J.S. Duncan. Deformable Fourier models for surface finding in 3D images. SPIE Visualization in Biomedical Computing 1808:90–104 (1992).

4. T.L. Faber, E.M. Stokely, R.M. Peshock, and J.R. Corbet. A model-based four-dimensional left ventricular surface detector. IEEE Trans. Med. Imag. 10:321–329 (1991).

6. T. Brotherton, T.G. Pollard, P. Simpson, and A.N. DeMaria. Classifying tissue and structure in echocardiograms. IEEE Eng. Med. Biol. pp. 754–760 (Nov./Dec. 1994).

7. G. Coppini, R. Poli, and G. Valli. Recovery of the 3-D shape of the left ventricle from echocardiographic images. IEEE Trans. Med. Imag. 14:301–317 (1995).

8. T. Gustavsson, P. Pascher, and K. Caidahl. Model based dynamic 3D reconstruction and display of the left ventricle from 2D cross-sectional echocardiograms. Comput. Med. Imag. Graphics 17:273–278 (1993).

9. J.G. Bosch, J.H.C. Reiber, G. vanBurken, J.J. Gerbrands, A. Kostov, A. J. van de Goor, M.E.R.M. van Daele, and J.R.T.C. Roelandt. Developments toward real-time frame-to-frame automatic contour detection on echocardiograms. Proc. Comput. Cardiol. pp. 435–438 (1991).

10. C. Manhaeghe, I. Lemahieu, D. Vogelaers, and F. Colardyn. Automatic initial estimation of the left ventricular myocardial midwall in emission tomograms using Kohonen maps. IEEE Trans. Patt. Anal. Machine Intell. 16:259–266 (1994).

11. T. Kohonen. The self-organizing map. Proc. IEEE 78:1464–1480 (1990).

12. T. Kohonen. "Self-Organization and Associative Memory," Springer-Verlag, New York, NY (1987).

13. A.K. Jain. "Fundamentals of Digital Image Processing," Prentice-Hall, Englewood Cliffs, NJ (1989).

14. University of Helsinki, Laboratory of Computer and Information Science, Rakentajanaukio 2 C, SF-02150 Espoo, Finland. SOM_PAK. The self-organizing map program package, release: 1.2 edition (Nov. 2, 1992).

RENAL PERFUSION EVALUATED WITH A NEW
ULTRASOUND CONTRAST AGENT

F. Forsberg,[1] J.B. Liu,[1] S. Kim,[1] M. Thakur,[1] N.M. Rawool,[1] D. Johnson,[2]
and B. B. Goldberg[1]

[1] Department of Radiology
   Thomas Jefferson University
   Philadelphia, PA 19107
   USA
[2] Nycomed R & D Inc.
   Collegeville, PA 19426
   USA

INTRODUCTION

Extensive efforts are currently underway to develop efficacious ultrasound contrast
agents. The ideal agent would be (1) nontoxic; (2) injectable intravenously; (3) capable of
traversing the pulmonary, cardiac and capillary circulations; and (4) stable for
recirculation. It is only within the last decade that pharmaceutical companies and
researchers have approached this goal. Contrast agents have the potential to improve the
diagnostic capabilities of ultrasound imaging. Enhancement of Doppler signals from small
and/or deep lying vessels have been investigated, as well as increases in the difference in
echogenicity between normal and adjacent abnormal (e.g. tumor) tissue in organs such as
the liver and spleen. Further details on the many different types of contrast agents and
their potential clinical applications can be found in several, recent review articles and their
associated bibliographies[1,2].

This study was performed to assess the ability of one such novel agent to evaluate
normal renal perfusion as well as to improve the diagnosis of renal abnormalities. The
contrast agent employed was NUS (Nycomed Imaging, Oslo, Norway), which consists of
stabilized microbubbles with a mean diameter of 3 to 5 µm and a concentration of 7.5×10^6
bubbles/ml. The animal model used was the rabbit, while the abnormalities studied were
VX-2 tumors and regions of ischemia in the kidneys.

EXPERIMENTAL PROCEDURE

Fourteen New Zealand white rabbits weighing 3.0 kg to 4.5 kg (mean weight: 3.5 kg) were studied. Each animal was sedated with 0.65 mg/kg of a mixture of ketamine hydrochloride (Ketaset; Aveco, Fort Dodge, IA) and xylazine hydrochloride (Gemini; Rugby Laboratory, Rockville Centre, NY) given intramuscularly. Anesthesia was maintained by IV injections of ketamine. Seven of the rabbits had renal tumors grown after injection of 0.5 ml of 3×10^6 VX-2 tumor cells (Biomeasure Inc., Hopkinton, MA). This cell line does not spread by natural means to other rabbits or humans. An anaplastic vascular carcinoma, ranging in size from 7x8 mm to 20x19 mm, developed over 7 to 21 days. An 18 gauge angiocatheter was placed in the left jugular vein of each animal. At this site 2 to 9 (on average 4) injections of contrast media per animal were administered in dosages of 0.02 to 1.0 ml/kg. The catheter was flushed with isotonic saline solution after each injection. Arterial blood pressure and heart rate were monitored throughout the experiments. Each animal was wet-shaved in appropriate areas to facilitate ultrasound scanning. After the initial imaging a kidney was surgically exposed, and a segmental renal artery ligated to produce a region of ischemia.

Pulsed Doppler, Color Doppler imaging (CDI), color velocity imaging (CVI™ i.e., velocity estimation using time domain correlation), and color amplitude imaging (CAI) signals were recorded on videotape during the entire study. A Spectra VST Master's Series Plus (Diasonics, Santa Clara, CA) or a P700 scanner (Philips Ultrasound, Irvine, CA) with either a 5 or a 10 MHz linear array transducer were employed. To limit the color blooming artifact associated with ultrasound contrast agent studies[3] the pulse repetition frequency (PRF) was adjusted to display only limited flow information prior to injection. All system parameters were kept constant during injection.

An *in vivo* dose-response curve was produced from two rabbits. A custom-made silex 10 MHz cuff transducer was placed directly around a surgically exposed vessel (the distal aorta) and stabilized with sutures. This allowed the quadrature audio outputs from a pulsed Doppler instrument (SDD 600; Vingmed Sound, Oslo, Norway) to be recorded with minimal interference from clutter noise and respiratory movements. The receiver is linear over 80 dB and operates with a sample volume size of 1.9 mm at a PRF of 20 kHz. The Doppler signals were digitized using a commercial software package (LabView; National Instruments, Austin, TX) and a PowerPC. Serial fast Fourier transforms were employed to calculate the mean signal power in segments of 256 samples. The signal power (in dB) over time is a measure of the contrast enhancement.

Finally, in three rabbits (four kidneys) the gold standard of perfusion measurement i.e., radioactive isotopes, was compared to the results on ischemia obtained with ultrasound. The rabbits were placed under a designated gamma camera (CX 250; Picker International, Cleveland, OH), and a bolus dose of 200 µCi/kg of Tc-99m-MAG-3 (Tc-99m mercaptoacetyl-glycylglycylglycine) was administered. The protein binding of MAG-3 is approximately 90 % and the extraction efficiency by normal adult humane kidneys is 70 to 80 %. Peak activity in each kidney was reached about 5 minutes post administration. Dynamic images were obtained in posterior position with the gamma camera coupled to a dedicated computer (PCS-TTU using gamma 11; Digital Equipment Corporation, Maynard, MA) for the first 30 minutes, and then followed by delayed static images. By choosing appropriate regions of interest in each kidney renograms were generated and the effective renal plasma flow was calculated.

In all cases gross pathological correlation was obtained after sacrificing the rabbit. The animal studies reported in this paper were carried out in an ethical and humane fashion

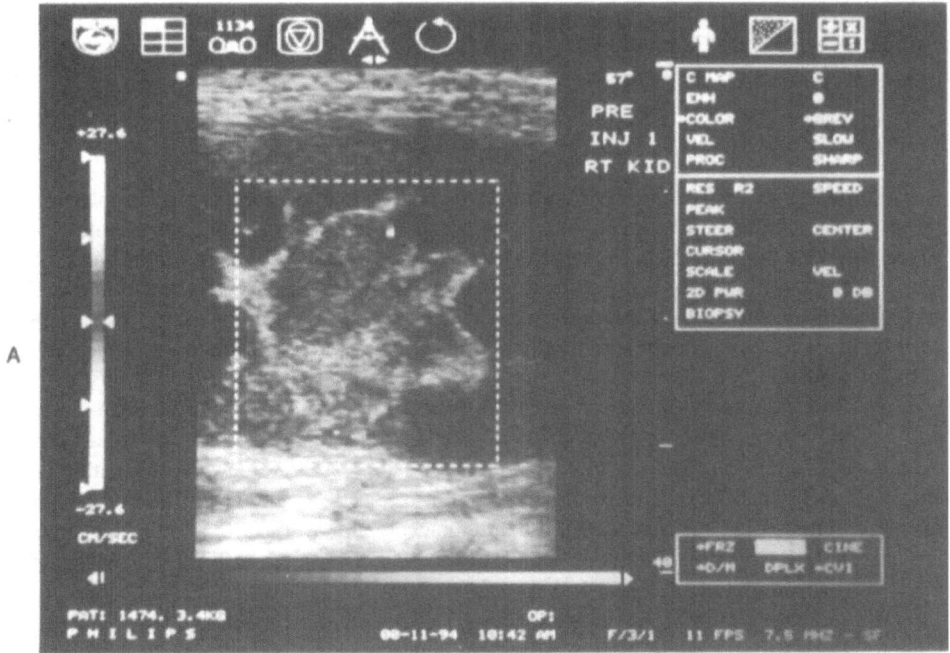

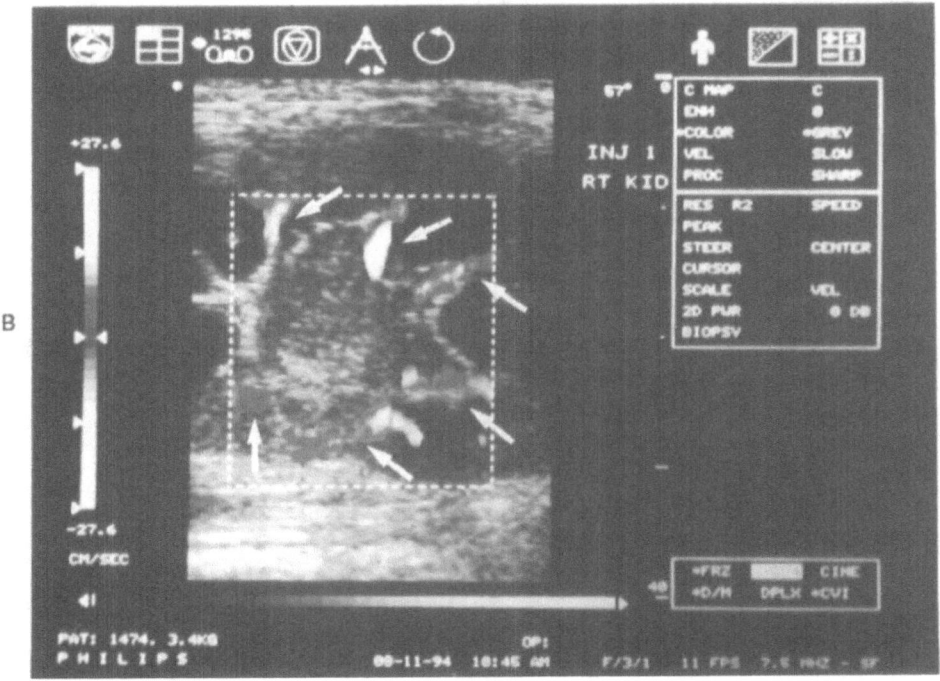

Figure 1. Example of kidney with a VX-2 tumor imaged in CVI mode. Images reproduced in B/W. (A): pre injection. (B): post injection of 0.05 ml/kg of NUS.

under supervision of a veterinarian and fully conformed to the National Institutes of Health guidelines for use of laboratory animals. All protocols were approved by the University's Animal Use and Care Committee.

RESULTS

Following injection of 0.02 to 0.1 ml/kg of NUS a marked increase in Doppler signal intensities (including tumor vessels) was observed over a period of 4 to 6 minutes. The blood flow enhancement allowed better visualization of the VX-2 tumors in all three color Doppler modes. In Figure 1 tumor feeding vessels are clearly seen with CVI after injection of 0.05 ml/kg of NUS. No tumors were detected post contrast administration that were not seen pre contrast. No changes in blood pressure or heart rate occurred at the dosages used in this study.

An *in vivo* dose-response curve was also produced based on flow in the distal aorta of two rabbits. Doses ranging from 0 to 1.0 ml/kg were injected, and the enhancement calculated (Figure 2). The response is nonlinear, and varies somewhat from animal to animal (there is a difference in peak enhancement of 8.6 dB). However, the overall shape of the two curves is similar with significant enhancement (over 10 dB) occurring for very low doses ($\leq$ 0.1 ml/kg). The mean increase was 18.3 dB $\pm$ 6.1 dB for a 0.1 ml/kg dose, while the maximum enhancement was 22.6 dB.

The regions of renal ischemia were clearly delineated after injection of NUS. In color Doppler modes the ischemic region was seen as an area without color flow (Figure 3A). Both nuclear medicine and contrast enhanced ultrasound visualized and localized the regions of ischemia. The two modalities correlated well with a reduction in effective renal blood flow as low as 6 % being detectable (see Table 1). An example from the lower left pole of a kidney, where the renal perfusion was reduced by 8 %, is given in Figure 3.

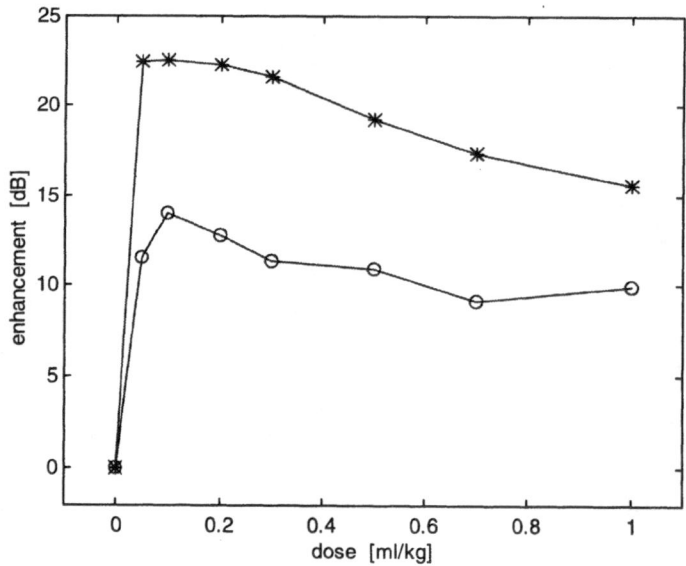

Figure 2. *In vivo* dose response curves from the distal aorta of two rabbits. Maximum enhancement was 22.6 dB; mean enhancement, for a 0.1 ml/kg dose, was 18.3 dB $\pm$ 6.1 dB.

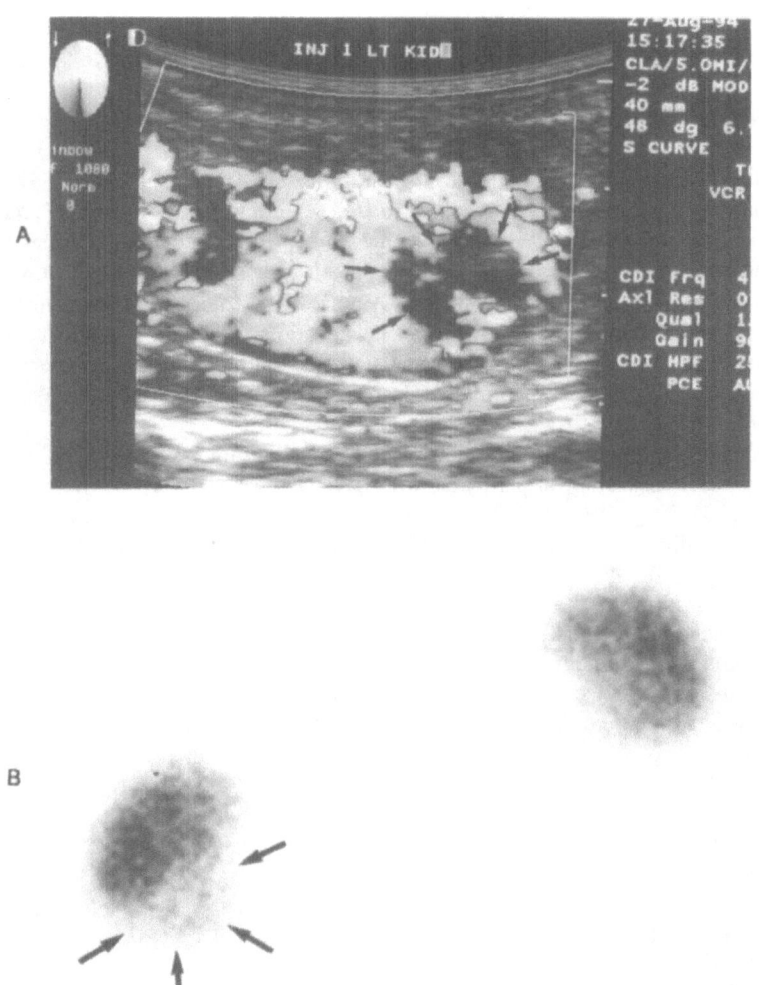

Figure 3. Kidney with ischemia in the lower left pole. (A): CDI post injection of 0.1 ml/kg of NUS (in B/W). (B): post injection of radioisotopes. The 8 % reduction in flow is seen with both modalities (arrows).

Table 1. Effective reduction in renal plasma flow calculated from renograms.

Rabbit No.	Region of ischemia	Flow reduction %
A-5	Left; upper pole	6
A-8	Left; lower pole	8
A-9	Right; lower pole	64
A-9	Left; upper pole	36

DISCUSSION AND CONCLUSION

Vascular contrast agents increase the intensity of Doppler signals. Hence, the expectation that contrast agents will enable better detection of blood flow in small, deep vessels and that this in turn will improve the ability to differentiate between areas of normal and abnormal perfusion; e.g. associated with ischemia, vessel occlusion or tumor vascularity[2]. Some contrast agents, such as Albunex™ (Molecular Biosystems Inc., San Diego, CA), do not recirculate following an IV injection[4]. They are effectively one-pass-only agents. The ability of NUS to traverse both the pulmonary and systemic capillary circulations, in multiple recirculations, while producing significant enhancement makes it potentially more useful.

The location and continuity of small vessels (such as tumor vessels) were seen better post contrast administration; in particular with CAI. However, CAI was more susceptible to flash artifacts and had limited temporal resolution[5]. The quantitative analysis of NUS resulted in a nonlinear dose response curve with maximum enhancement of almost 23 dB. This is in line with the 14 to 25 dB of contrast induced increase in Doppler signal intensities reported in the literature[6,7]. Both renograms and contrast enhanced ultrasound detected the ischemic regions. However, due to the difference between the two modalities (projection images vs. tomographic images) no further comparisons can be made.

In conclusion, visualization of renal tumors and ischemia in rabbits was improved by using a new ultrasound contrast agent. These preliminary results suggest that NUS may improve the accuracy of diagnosing renal abnormalities. Clinical efficacy in humans has yet to be established.

REFERENCES

1. J. Ophir and K.J. Parker, Contrast agents in diagnostic ultrasound, *Ultrasound Med. Biol.* 15:319-333 (1989).
2. B.B. Goldberg, J.B. Liu and F. Forsberg, Ultrasound contrast agents: a review, *Ultrasound Med. Biol.* 20:319-333 (1994).
3. F. Forsberg, J.B. Liu, P.N. Burns, D.A. Merton and B.B. Goldberg, Artifacts in ultrasonic contrast agent studies, *J. Ultrasound Med.* 13:357-365 (1994).
4. B.B. Goldberg, P.L. Hilpert, P.N. Burns, et al. Hepatic tumors: signal enhancement after intravenous injection of a contrast agent, *Radiology* 177:713-717 (1990).
5. J.M. Rubin, R.O. Bude, P.L. Carson, R.L. Bree and R.S. Adler, Power Doppler US: a potentially useful alternative to mean-frequency-based color Doppler sonography, *Radiology* 190:853-856 (1994).
6. B.B. Goldberg, J.B. Liu, P.N. Burns, D.A. Merton, and F. Forsberg, Galactose-based intravenous sonographic contrast agent: experimental studies, *J. Ultrasound Med.* 12:463-470 (1993).
7. P.N. Burns, P. Hilpert and B.B. Goldberg, Intravenous contrast agent for ultrasound Doppler: in vivo measurement of small vessel dose-response, *Proc. IEEE Eng. Med. Biol. Soc.* 322-324 (1990).

ACOUSTICAL TISSUE IMAGES FOR DETECTION OF ATHEROSCLEROTIC CHANGES IN BLOOD VESSELS

Wolfram Schmidt[1], Detlef Behrend[1], Olaf Skerl[2], Heiner Martin[1], Wilhelm Urbaszek[3], and Klaus-Peter Schmitz[1]

[1]Institute of Biomedical Engineering
[2]Institute of Animal Physiology
[3]Clinic for Internal Medicine, Department of Cardiology
University of Rostock
D-18055 Rostock, Germany

INTRODUCTION

Recent developments of intravascular ultrasound (IVUS) offer some new features for the detection and quantification of vascular disease which are very valuable for clinical diagnosis. Making possible visualization of the full circumference of the vessel, IVUS allows the inspection of lumen and plaque shape by one single tomogram. Such views are successfully used for planimetry to overcome special problems such as bifurcations or overlapping vessels. In addition, IVUS is used to obtain plaque thickness and echogenicity for determination of its area and composition. It is established that intravascular ultrasonic imaging provides best results in quantifying atherosclerotic changes in vessels[1,2]. Decisions for optimal treatment are usually made in terms of grey scaled video images, which are processed from the amplitude of ultrasonic echo signals. However, there is much more information in radio-frequency (RF) echo signal which should be used for tissue characterization and identification[3] as well as special experiences in soft tissue classification[4,5].

Our approach to analyze RF-signals was to create well-determined acoustic models for in vitro measurements of ultrasonic backscatter. These models should prove comparable in dimension and composition to normal and pathologic vessels. In a second step we introduced the Wigner-Ville distribution for high resolution time-frequency analysis of sampled data to obtain additional information related to standard tissue images.

MATERIALS AND METHODS

Morphology, structure and components of normal as well as of atherosclerotic arterial vessels are well known from histologic studies. Some in depth reviews were published in the past[6,7]. The earliest lesion in atherosclerotic progression is the fatty streak, which is a lipid-rich region. Advanced atherosclerosis is characterized by fibrous plaques being surrounded by connective tissue and containing intracellular and extracellular lipid. Underneath this cellular structure there may be an area of necrotic tissue, cholesterol and calcification[6].

Ultrasonic phantoms: We developed cylindrical tube phantoms with at least two different layers each. As carrier material, we used a polyurethane elastomer (Pellethane 2363-80A, Young's modulus = 10.2 Mpa) due to its good handling qualities like flexibility, self-adherent layers without any gas bubbles, and acoustic properties which are comparable to soft tissue. In addition, we included materials which were similar to the atherosclerotic changes, as there is collagen and cholesterol for simulation of lipids and fibrotic tissue, and powdered calcium carbonate ($CaCO_3$), hydroxyapatite and granulated silicon dioxide[13,14] to mimic different types of calcified plaques. Collagen and cholesterol were dissolved in water respectively ethanol to get a paste. $CaCO_3$ hydroxyapatite or silicon dioxide were submerged in dental wax. The completed phantoms consist of concentric layers with determined dimensions (see fig.1).

Fig. 1.   Ultrasonic phantoms, consisting of a cylindrical PUR tube and embedded
layers of pure dental wax, calcium carbonate and cholesterol (from left to right)

Ultrasonic equipment: All ultrasonic measurements in vitro were done by applying an intravascular ultrasound device for clinical use (CVIS Insight, CVIS Corporation, Sunnyvale, CA) with cardiovascular catheters (MicroView 2,9F or MicroRail 3,2F), each operating with a 30 MHz rotating crystal. With an additional RF-box (CVIS), it was possible to record the RF echo signal before intitiating signal processing. Recording was supported by an additional frame trigger signal that was related to one radial segment of the whole image. The latter could be marked on the screen to select the region of interest for analysis.

Digitizer: All RF-data were digitized for computer-based analysis by a digital scope (TEK524A, Tektronix, 500 MHz bandwidth, 500 MS/s maximum). Every shot was sampled with 5000 points of 8 bit data. Measurements were triggered by the previously described frame trigger signal in order to avoid phase errors. The gain adjustment of the scope was fixed at a high level obtain a sufficient resolution of the backscatter amplitude. Effects of overmodulation during the transmission pulse had to be accepted.

Signal processing: The stored 8 bit data were used to specify characteristic parameters of ultrasonic backscatter signals. The Wigner-Ville distribution was applied to these signals to overcome the limitations of standard procedures for frequency or time-frequency analysis, e.g. FFT or spectrogram.

The Wigner distribution (WD) for signals is given by:

$$W(t,\omega) = \int_{-\infty}^{\infty} s(t+\tfrac{\tau}{2})s^*(t-\tfrac{\tau}{2})e^{-j\omega\tau}d\tau$$
$$= F_\tau \left\{ s(t+\tfrac{\tau}{2})s^*(t-\tfrac{\tau}{2}) \right\} \tag{1}$$

It is very important for most signal processing that WD can be considered as a time-varying power spectrum of stationary as well as of instationary signals. Special properties and implementations of the WD have been reported and applied in the last years[8,910,11]. Computer implementation leads to discrete time signals. Thus, to avoid aliasing, the Nyquist rate has to be changed to a more strict rule for WD:

$$f_s \geq 4f_{max}, \tag{2}$$

where f_s is the minimum sample frequency and f_{max} denotes the bandwidth of the signal.

The extraction of characteristic parameters of ultrasonic backscatter from vessel or phantom structures requires a special postprocessing of the WD time-frequency distribution. The first order frequency moment of every time slice was calculated by a discrete form of the following analytic definition:

$$\overline{\omega}(t) = \frac{\int_{-\infty}^{\infty}\omega W(t,\omega)d\omega}{\int_{-\infty}^{\infty} W(t,\omega)d\omega}. \tag{3}$$

For this, integral limits had to be changed to finite record length and integration was performed by a discrete approximation. The width of frequency steps depends on the length of the time window, as known from Fourier transformation.

Experimental approach: Signal processing was evaluated by making two different approaches. Ultrasonic phantom samples were investigated to obtain basic relations between signal characteristics and the different materials and structures of the samples. Secondly, although it is reported, that acoustic properties of tissue specimens alter in vitro due to shrinkage[1] and protein denaturation, some specimens taken from autopsy were investigated.

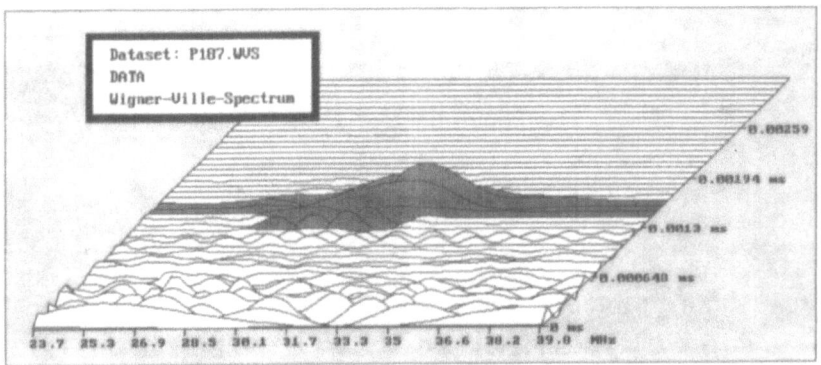

Fig. 2.   Wigner-Ville distribution of echo signal from hydroxyapatite phantom in PUR matrix (Δf=240kHz, Δt=80ns)

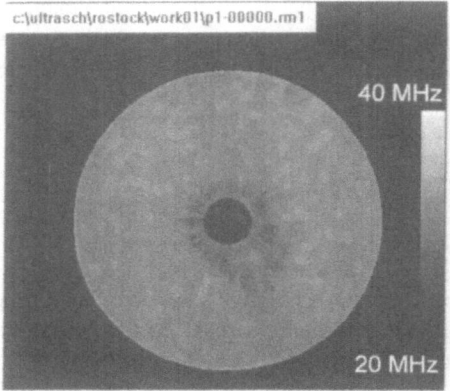

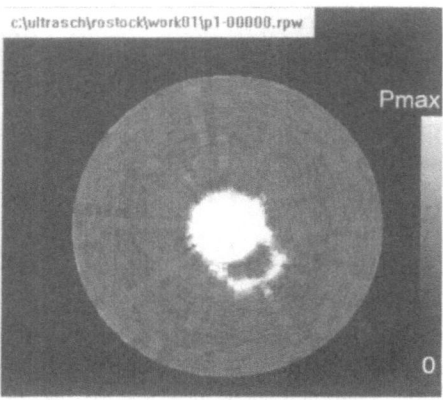

Fig 3a. Distribution of first order moment calculated from WD of RF data from wax phantom

Fig 3b. Power distribution derived from WD data of the phantom from fig. 3a

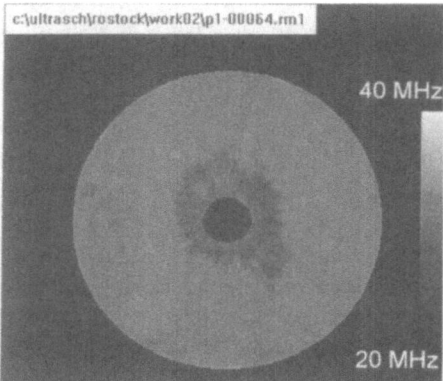

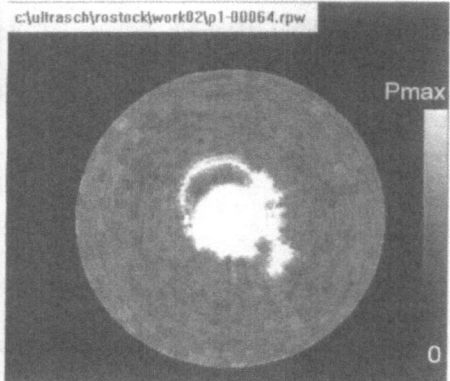

Fig 4a. Distribution of first order moment derived from WD of RF data from phantom with $CaCO_3$ and wax layers

Fig 4b. Power distribution derived from WD data of the phantom from fig. 4a

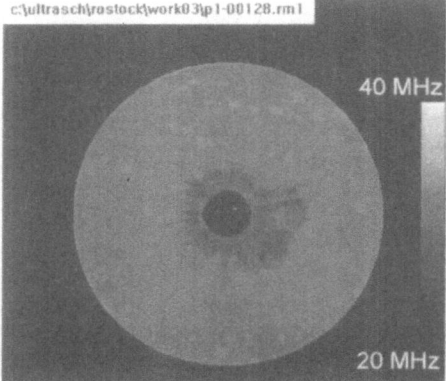

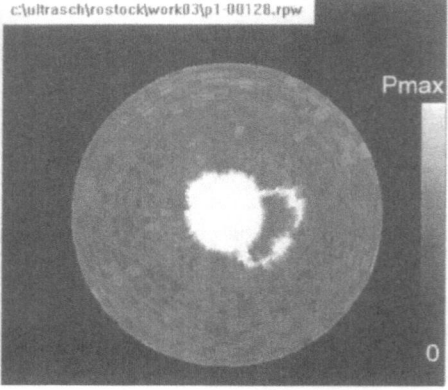

Fig 5a. Distribution of first order moment, calculated from WD of RF data from phantom with hydroxyapatite and wax layers

Fig 5b. Power distribution derived from WD data of the phantom from fig. 4a

212

RESULTS

Wigner-Ville distribution calculated from every single shot was found to provide excellent data for analysis in the frequency domain. Fig. 2 shows a WD calculated from a hydroxyapatite phantom in PUR matrix. The echo region with high echogenicity is marked in this figure as a dark grey area. One can see that - after a short period of relaxation - there is only low energy in backscatter signal except in the high amplitude echo pulse. Changes in frequency content caused by various effects, e.g. frequency dependent attenuation or scattering, can be observed and quantified by numerical methods. Specific parameters were extracted from such distributions and visualized in fig. 3 to fig. 5 for dental wax, $CaCO_3$ and hydroxylapatite phantoms, respectively. Every image is based on 64 single shots and all parameter sets were standardized unsing constant limits and linear scales.

First order moment plots demonstrate typical changes in frequency content of the backscatter signals. Concentric areas of lower mean values were caused by several effects, as there are overmodulation during transmission pulse time, near-field phenomena around the rotating catheter (ring-down artefacts) and the coupling behavior of the ultrasonic device. Artificial layers of the phantoms also show regions with changed mean frequency. The distribution of signal power can be considered to be quite similar to classic echograms. The pictures show clear and sharp contours of highly echogenic layers. Low energy parts in signals could not be discriminated by this way, because there is no signal compression or adaptation carried out at all.

The visualization on its own does not give quantitative parameters. Therefore, a numerical analysis is necessary. For this purpose, regions of interest (ROI) were defined on the basis of power distribution images of IVUS as well as of the known geometry of phantoms. The result is shown in fig. 6, where the mean frequency is drawn as a function of distance from the ultrasonic transducer, assuming a constant sound velocity of 1500 m/s. Fig. 6 shows typical slopes of first order moments which depend on stiffness and structure of the

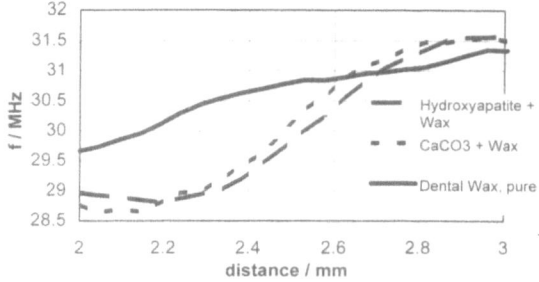

Fig. 6. Distribution of first order moments of the three types of ultrasonic phantoms from fig. 3-5, each averaged over ROI

scattering material. The wax layer, consisting of virtually homogeneous material with low Young's modulus, causes only slow changes in mean frequency. $CaCO_3$ and hydroxyapatite material are powdered substances (grain size 0.5 to 10 μm) with high elasticity and high acoustic impedance giving high echogenicity and much higher slope values than wax. These effects were supported by comparable measurements with other material combinations noted above. Due to the low number of verifications, no statistic calculations, for instance for significance analysis, were performed.

DISCUSSION

The results show that there is additional information in the RF backscatter signal of IVUS catheters. The current problem is to determine additional and independent parameters without any loss in local resolution, because the latter is very important for recognition of special anatomic structures and cannot be neglected at all. Our initial hypothesis of using WD function as a convenient tool for extracting these parameters was found to be successful. Some special problems of WD were not the scope of this work but should be mentioned to be complete. Most of all it is difficult to take WD as a power spectrum because of the existence of cross terms. These terms were described in related literature and cannot be removed from actual spectrum-like parts. Because of their positive and negative signs they do not change the total power of the signal, but the calculation of the signal bandwidth in the usual way is not possible.

The typical differences found in the slope of the first order moment of frequency are based on different physical effects related to scatter and transmission phenomena of ultrasound propagation. In comparison to these results in frequency domain, attenuation slopes are discussed for specification of tissue[12]. Additionally, frequency analysis of the envelope of ultrasonic backscatter is done to specify tissue types[5]. Up to now no studies are reported wether these different approaches are uncorrelated and can be used to describe tissue types by classification operators. Thus, further studies will have do deal with statistical verification of extracted parameters as well as with implementation of suitable classificators for material or tissue recognition.

REFERENCES

1. S.E. Nissen, J.C. Gurley, D.C. Booth, and A.N. DeMaria, Intravascular ultrasound of the coronary arteries: current applications and future directions, Am J Cardiol 69:18H-29H (1992)
2. J.F. Benenati, Intravascular ultrasound: the role in diagnostics and therapeutic procedures, Radiol. Clin. North America 33; 31-50 (1995)
3. D.T. Linker, A. Kleven, A. Gronningsaeter, P.G. Yock, and B.A.J. Angelsen, Tissue characterization with intra-arterial ultrasound: special promise and problems, Int. J. of Cardiac Imag. 6: 225-263 (1991)
4. J.A. Jensen, A model for propagation and scattering of ultrasound in tissue, JASA 89:182-190 (1991)
5. M. Lang, H. Ermert, and L. Heuser, In vivo study of online liver tissue classification based on envelope power spectrum analysis, Ultrasonic Imaging 16:77 - 86 (1994)
6. R. Ross, The pathogenesis of atherosclerosis - an update; The New England J. of Med. 314:488-500 (1986)
7. J. Honye, D.J. Mahon, A. Jain, C.J. White, S.R. Ramee, J.B. Wallis, A. Al-Zarka, and J.M. Tobis, Morphological effects of coronary balloon angioplasty in vivo assessed by intravascular ultrasound imaging, Circulation 85:1012 - 1025 (1992)
8. T.A.C.M. Claasen and W.F.G. Mecklenbräuker, The Wigner distribution - a tool for time-frequency signal analysis, Philips Journ. Res. 35:217-250(I), 276-300(II), 372-389 (III) (1980)
9. B. Boashash, Note on the use of the Wigner distribution for time-frequency signal analysis; IEEE Trans. ASSP 36:1518-1521 (1988)
10. O. Skerl, and I. Hartmann, The Wigner distribution in Doppler sonography, Acoust. Imag. 19:335-339 (1992)
11. O. Skerl, W. Schmidt, and O. Specht, Wigner-Verteilung als Werkzeug zur Zeit-Frequenz-Analyse nichtstationärer Signale; Technisches Messen 61:7-15 (1994)
12. L.S. Wilson, M.L. Neale, H.E. Talhami, and M. Appleberg, Preliminary results from attenuation-slope mapping of plaque using intravascular ultrasound; Ultrasound in Med. & Biol. 20:529 - 542 (1994)
13. D. Werner, D. Behrend, M. Schröder., K.-P. Schmitz, and W. Urbaszek, Interaktionen zwischen PTCA-Ballon und Koronarstenose beim Dilatationsprozeß - eine rasterelektronmikroskopische und röntgenmikroanalytische Untersuchung, Biomedizinische Technik 39:91 - 92 (1994)
14. D. Werner, D. Behrend, K.-P. Schmitz, W. Urbaszek, Impression koronarer Plaquepartikel in die PTCA-Ballonoberfläche durch den Dilatationsprozeß, Z. Kardiol. 84: 377 - 384 (1995)

GREY SCALE IMAGING OF ULTRASONIC SHADOW BEHIND RIGID, ELASTIC AND GASEOUS SPHERES IN A TISSUE-LIKE ATTENUATING MEDIUM

Leszek Filipczyński, Tamara Kujawska, and Janusz Wójcik

Ultrasound Department, Insitute of Fundamental Technological Research
Polish Academy of Sciences
Świętokrzyska 21
00-049 Warsaw
Poland

INTRODUCTION

The purpose of the paper is to answer the question if and what kind of information can be obtained from the grey scale shadow of targets detected by means of short ultrasonic pulses. The theory of wave reflection from spherical obstacles was applied for determination of the shadow caused by plane wave pulses incident on rigid, steel, gaseous spheres as well as on spheres made of kidney stones immersed in water.

Water was chosen in the beginning as a tissue-like medium for the following reason. Densities and wave speeds of water and soft tissues are similar while on the contrary attenuation is entirely different. However, we wanted at first to simplify the complex phenomenon of shadow by neglecting the frequency dependent attenuation of tissues. Water enabled us also to measure pressure distributions in the shadow range point by point to verify the applied theoretical and numerical procedures.

THE GREY SCALE IMAGING OF THE SHADOW

Acoustic pressure distributions behind the spheres were determined for ka values equal to 21, 53 and 73 corresponding at the frequency of $5MHz$ to spheres with the radius of $1mm$, $2mm$ and $3.5mm$, respectively (k - wave length, a - sphere radius). The computed pressure distributions were verified experimentally at the frequency of $5MHz$ for the steel sphere $2.5mm$ in radius immersed in water. The experimental and theoretical pulses had the shape of a Hanning function being composed of about 3 ultrasonic frequency periods. Acoustic pressure distributions in the shadow zone of all the spheres were presented in the grey scale with the dynamics of $40dB$ in $2dB$ steps. Details of the applied theoretical and experimental procedures are described elsewhere[4].

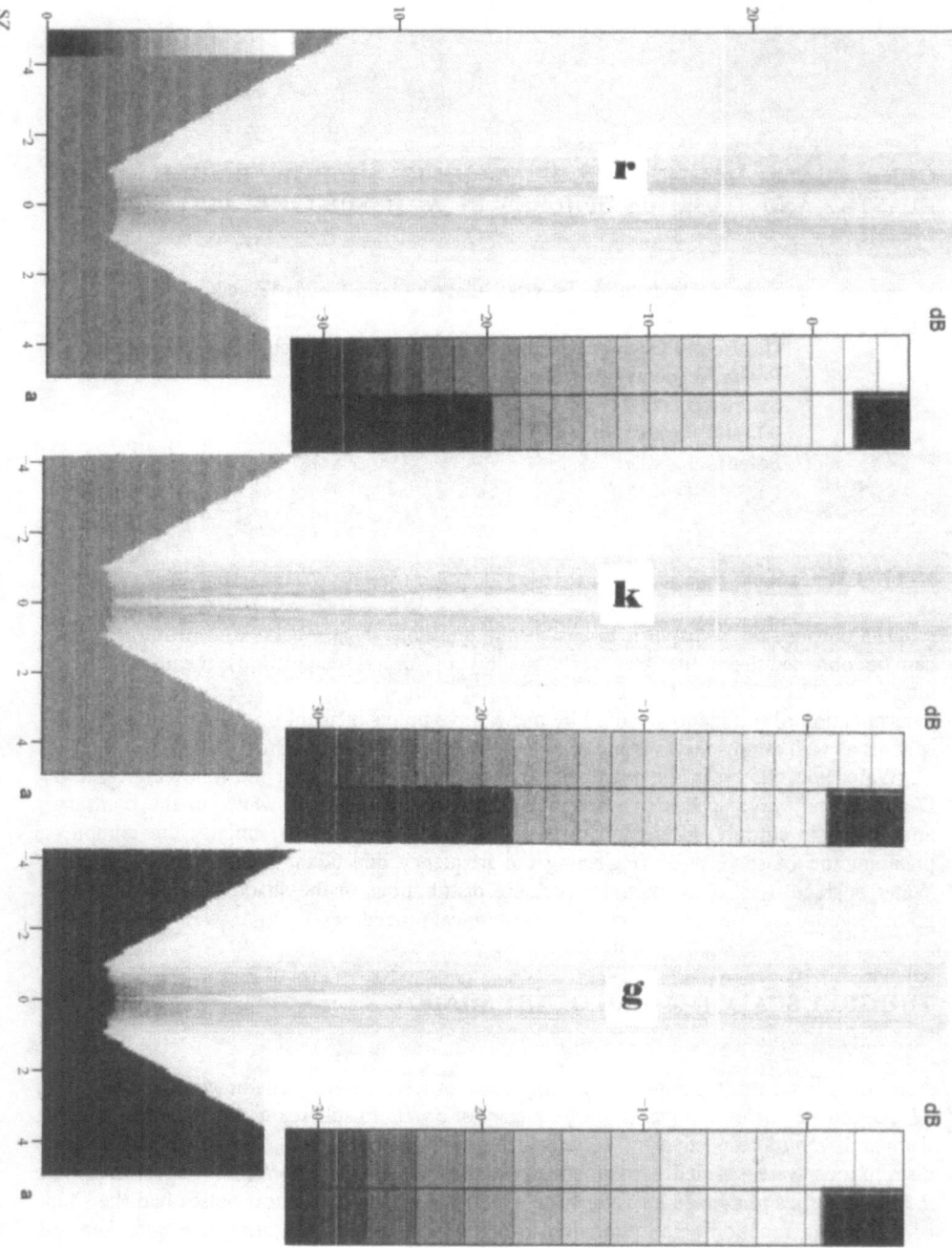

Fig.1 Grey scale images of shadows computed behind rigid (*r*), gaseous (*g*) and kidney stone (*k*) spheres *3.5 mm* in radius (*f= 5MHz, ka = 73*) together with double grey scales (see the text). Sphere center located at *0,0*. Wave propagation from left to right.

Fig. 1 presents as an example grey scale imaging of the shadow computed behind three different spheres; rigid, gaseous and made of kidney stone. The radius of the spheres was equal to $a = 3.5mm$ giving at the frequency of 5 MHz the value of $ka = 73$. The shadow images show some subtle differences in their shape and in their grey distributions. However, the significant difference occurs in their greyness dynamics. To prove it we showed in the lower portion of the every shadow image the full 40dB grey scale and below the adjacent scale of greynesses which are represented in the actual beam image. One should notice that the grey dynamics of the gaseous sphere is equal to 40 dB while only to 21 dB in cases of rigid, kidney stone and also steel spheres (not shown here). A similar effect was observed for spheres 2.5 and 1 mm in radius. In this way it was shown that there exists a theoretical possibility of differentiating between gaseous and solid inclusions when comparing the shadow grey dynamics of detected spherical inclusions.

The next interesting result is connected with the determination of the shadow length l_{-6dB} which is defined as the distance behind the sphere at which the acoustical pressure drop equals 6 dB in relation to the acoustical pressure of the incident wave[2,3]. The shadow length can be found from acoustical isobar patterns which can be drawn from the computed acoustical pressure distributions. Fig. 2 presents such a case for a sphere 3.5mm in radius made of kidney stone at the frequency of 5 MHz. It was found[4] that the length of the shadow l_{-6dB} determined in this way for rigid, steel, kidney stone and gaseous spheres 3.5, 2.5 and 1 mm in radius (immersed in water) equals

$$L_{-6dB} = x \cdot a^2 / \lambda \tag{1}$$

where $x = 3.7 \pm 0.2$

Relation (1) makes it possible to determine the radius a of the spherical sphere in water when knowing the shadow length.

THE EFFECT OF TISSUE ATTENUATION ON THE SHADOW

The obtained results are valid for water, however, real tissues show attenuation which in the first approximation is proportional to frequency. There arises a question if in such a situation the conclusions obtained for water are still valid. The shadow arises as the interference of the wave incident and reflected from the sphere. Since there does not exist any analytical solution of the wave reflection from the spherical obstacle in an attenuating medium we are able only to estimate approximately such a case.

The average value of the atenuation coefficient in soft tissue equals $A_1 = 0.5dB / (cm \cdot MHz) = 0.058 Nep / (cm \cdot MHz)$.

Then the values of attenuation of the incident and reflected waves are, respectively, equal to (see also Fig. 3)

$$e^{A_i} = e^{A_1 f r \cos\theta} = e^{A_1 \frac{cr}{2\pi a} ka \cos\theta} \tag{2}$$

$$e^{A_r} = e^{A_1 f r} = e^{A_1 \frac{cr}{2\pi a} ka} \tag{3}$$

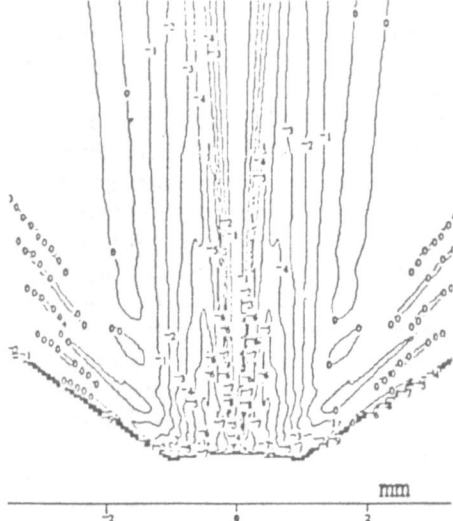

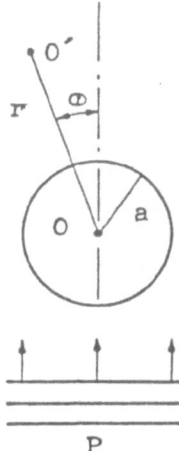

Fig.2 Acoustical isobars. Horizontal dimentions are expanded 2.6 times. Numbers show half the *dB* level in relation to the incident wave pressure. Sphere center at *0*.

Fig.3 Coordinate system. *P* - incident plane wave pulse, *O* - sphere center, *O'* - observation point, *a* - sphere radius, r, θ - coordinates.

The reflection itself on the boundary tissue-sphere is not influenced by the tissue attenuation since the tissue acoustic impedance and therefore the boundary conditions do not depend on attenuation as long as $A_f(kr) = 0.0014 \ll 1$.

One should notice that we neglect the small attenuation difference between the incident and reflected waves [see eqs (2) and (3)] in the vicinity of the sphere surface. The difference is small for small angles of θ in the main direction of the shadow. Therefore one can say that the reflecion itself is here considered as in the case of surrounding water. The error is rather small as long as the sphere radius is small in relation to the shadow length.

Then the incident pulse pressure p_i expressed in the form of the inverse Fourier transform[4] can be expanded by the attenuation term (2), so

$$p_i(\tau') = p_0 \frac{1}{2\pi} \int_{-\infty}^{\infty} S_i(ka) e^{jka\tau'} e^{A_1 \frac{cr}{2\pi a} ka \cos\theta} \, d(ka) \qquad (4)$$

where $\tau' = (ct - r\cos\theta)/a$, p_0 - amplitude of the incident wave, S_i - spectrum of the incident pulse in an attenuationfree medium.

Similarly the attenuated pulse of the reflected wave can be expressed in the form

$$p_r(\tau) = p_0 \frac{1}{2\pi} \int_{-\infty}^{\infty} \frac{a}{2r} f_f(ka) S_i(ka) e^{jka\tau} e^{A_1 \frac{cr}{2\pi a} ka} \, d(ka) \qquad (5)$$

where $\tau = (ct - r)/a$, f_f - is the reflection form function for the case of an attenuationfree medium. It is a complex function of many parameters $f_f = f(r, a, k, \theta, \rho, \rho_s, c, c_1, c_2)$. r, c denote here density and wave speed in the surrounding medium, ρ_s - density of the sphere, c_1, c_2 - longitudinal and transverse wave speeds in the sphere. The expressions of f_f for all spheres under consideration were computed in our former paper [4]. The interference of the reflected (4) and incident (5) pulses forms finally the ultrasonic field around the sphere. Hence we computed the shadow length l_{-6dB} as the function of ka for spheres with extremely different elastic properties namely for rigid and gaseous ones. Fig.4 presents the results for the case of ultrasonic pulses (dashed lines) used in ultrasonography with typical ultrasonic carrier frequencies 3.5, 5 and 7.5 MHz.

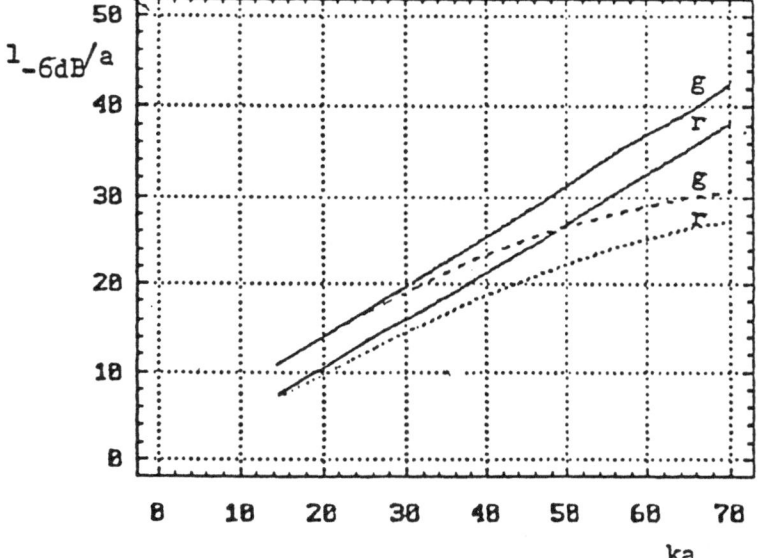

Fig.4 The relative shadow length l_{-6dB}/a behind rigid (r) and gaseous (g) spheres computed in an attenuating tissue for pulses (dotted lines) and for continuous waves (solid lines). g - denotes gaseous spheres, r- rigid spheres.

For small values of ka the dependence is linear, however, at higher values the curves start to bend. So instead of eqn (1) we found in this case the approximated nonlinear dependence

$$l_{-6dB} = (1.2 \pm 2)a + 3.5a^2 / \lambda - 7.9 \cdot 10^{-3} a^4 / \lambda^3 \qquad (6)$$

where the sign "+" should be used for gaseous spheres and "-" for rigid ones. For other elastic materials the first term in eq.(6) will be between 3.4 and -0.8. However, in practical cases the first term will be much smaller than the second one.

The nonlinear relation between the shadow length and ka can be explained by the fact that attenuation terms [see eqs.(4)(5)] change to a great extend the spectra of reflected and incident pulses. Their higher components are more attenuated than the lower ones, moreover, attenuation increases as the carrier frequency increases. To confirm this reasoning computations were carried out for continuous waves (see Fig.4, solid lines). As expected they have shown a linear dependence.

CONCLUSIONS

The theory of wave reflections from spherical obstacles was applied for determination of the shadow caused by plane wave pulses incident on rigid, steel, gaseous spheres and on spheres made of kidney stones. The theory and numerical results obtained previously for water as surrounding medium were expanded for a case of an attenuating tissue-like medium assuming a 3 cycle pressure pulse of frequencies used in ultrasonography.

The analysis of the obtained grey scale shadow images showed a significantly higher dynamics for gaseous spheres than for solid ones. Thus a theoretical possibility of differentiating between gaseous and other inclusions was shown. The grey scale shadow images in the tissue-like attenuating medium are not significantly changed in relation to ones obtained in water.

Also the determination of the size of spherical inclusions is theoretically possible from the shadow length (see eqn.6) in the range of $ka=15 - 70$. In a general case, to omit the limitation of the highest ka value, one should use continuous waves or pulses of much longer duration time.

REFERENCES

1. F.Duck," Physical properties of tissue",Academic Press, London, (1990)
2. L.Filipczyński, T.Kujawska, Acoustic shadow of a sphere immersed in water.Part I and II, Archives of Acoustics, 14, 29 - 43 and 181-190, (1989)
3. L.Filipczyński, T.Kujawska, T.Waszczuk, The shadow behind a sphere immersed in water - measured, estimated and computed, IEEE Trans. UFFC,38, 35-39, (1991)
4. L.Filipczyński, T.Kujawska, R.Tymkiewicz, J.Wójcik, Amplitude, isobar and grey scale imaging of ultrasonic shadows behind rigid, elastic and gaseous spheres, Ultrasound Med.Biol., in print (1995)

ACOUSTICAL IMAGING OF LIVING CELLS DURING VOLUME REGULATION

Samir Yastas and Jürgen Bereiter-Hahn

Cinematic Cell Research, Biozentrum,
Marie-Curie-Str. 9, 60439 Frankfurt am Main

ABSTRACT

Information about the morphomechanical changes during cell swelling could be a basis for understanding sensoring in volume regulatory signal transduction. We therefore observed living human keratinocytes of the line Hacat (Boukamp et al. 1988) during RVD with reflection scanning acoustic microscopy using an ELSAM operated at 1 Ghz. The aquired images clearly showed topographic changes during cell swelling and RVD. A transient, reversible decrease in ultrasound velocity profiles could be determined from extrema of interference fringes using an iterative algorithm method (Litniewski and Bereiter-Hahn, 1990). This indicates a reversible change in physical properties of the cytocortex which could be the mechanical basis for the unknown volume sensor of animal cells.

INTRODUCTION

Cellular architecture and mechanics are involved in a variety of morphomechanical reactions of living cells, including movement, adhesion (Davies et al. 1994), Cytokinesis (Nicklas et al.1995), neuronal process projection (Heidemann and Buxbaum 1994), embryonic development (Klymkowsky and Karnovsky 1994), substrate topography orientation (Chou et al. 1995) and volume regulation. A steady state equilibrium of forces leads to a certain three dimensional cellular geometry at every timepoint which serves as mechano-biochemical transducer connecting environmental or cellular topography dynamics to different biochemical reaction cascades.

In this context of cellular biophysics volume regulation is a typical reaction in connecting environmental mechanical changes associated with volume pertubations to defined cellular reactions. It is an evolutionary early evolved mechanism which enables every cell to counteract osmolarity changes which lead to swelling or shrinking. This is done by the activation of certain ion transporters for K, Cl and Na which in turn lead to passive, volume restoring water movements. The reduction of cellvolume after swelling is named RVD (regulatory volume decrease), the opposite reaction RVI (regulatory volume increase). Both reactions are found in many celltypes, although RVI occurs not in everyone, which lead to the assumption that RVD is the more common reaction serving as a burst protection in early cellular environments.

In higher organisms, even though medium osmolarity is quite well defined, with exceptions e.g. in glands or kidney cells, osmolarity changes occur in many celltypes by metabolism or transport of osmolytes.

One important question concerning volume control is how cells sense deviations from their normal volume. Mechanosensitive channels, embedded in the cortical actingel and connected to the cytoskeleton, were suggested as candidates for this unknown volume sensor.

Our aim was to use acoustic microscopy to monitor the micromechanical behaviour of cells during volume regulation.

MATERIALS AND METHODS

Cells

The human keratinocyte line HaCat (Boukamp et al. 1988) has been used to study the mechanical responses of cells in culture to a sudden decrease of the osmolarity of the culture medium. The cells were grown on glass slides using Hank´s medium with 5%

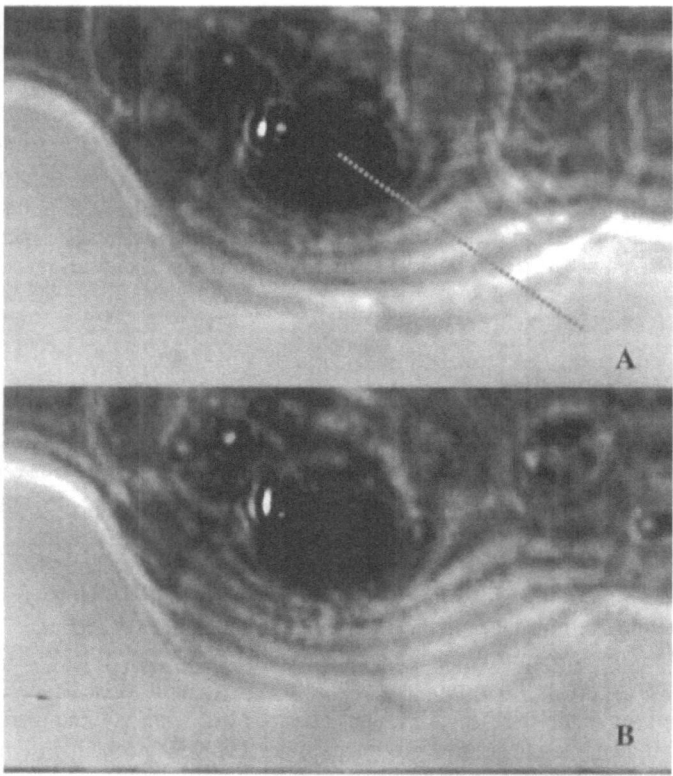

Figure 1. HaCat Cell imaged immediately before (A) and 1 min after (B) hypotonic treatment. In A the course of the linescan used for iterative evaluation of acoustical properties is indicated. Topography changes are clearly visible, the increase of number of interference fringes indicates the higher volume. After RVD the interference pattern returns to a state similar to the initial pattern. Horizontal size of an image corresponds to 97 µm.

newborn calf serum, 1% antibiotic mixture (penicillin, streptomycin, neomycin). For SAM investigations the glass slides were placed on a warm plate with controlled temperature (37°C). The culture medium containing 10mM Hepes buffer (pH 7.2) provided acoustic coupling between the lens and the specimen.

Acoustic Microscopy

The acoustic microscope (ELSAM®, Leica, Wetzlar F.R.G.) was operated at 1 GHz, 256 line images (512 pixels wide, 8 bit each) were taken with a frame grabber AT FS-100 (Imaging Technology Inc., Tucson, Arizona) and stored on a hard disk for further processing. Cell volume and cytoplasmic stiffness were registered under control conditions (immediately before hypotonicity), 1 min after addition of 1ml of water to 3ml of culture medium and after RVD (5 to 10 min). Sound velocity values were calculated by an iterative algorithm from minina and maxima of consecutive interference fringes of different order as previously described (Litniewski and Bereiter-Hahn, 1990).

RESULTS

1. Topography
To identify changes in physical properties of cells during RVD we analyzed topography of the cells and interference pattern dynamics during swelling with an acoustical microscope. Cells were observed immediately before and 1 min after 33% hypotonicity and later after RVD. Cell swelling can be clearly seen from the change in interference pattern indicating topographic changes in the cell (Fig.1 and Fig. 2).

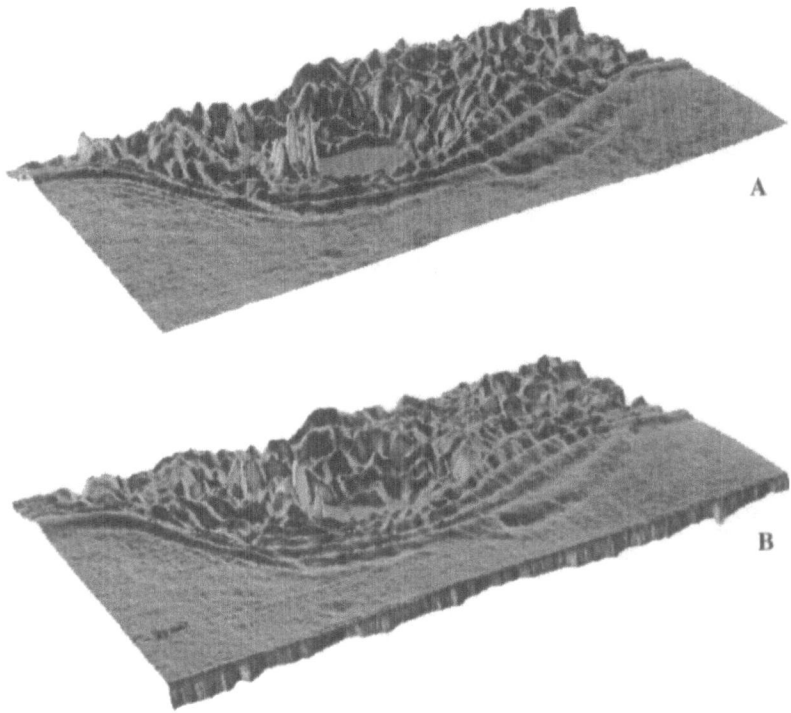

Figure 2. A and B show a three dimensional representation of the data corresponding to Figures 1, A and B.

2. Sound velocity

The quantiative evaluation of grey level profiles utilizes an iterative algorithm which operates based on the intensity values at minima and maxima of different order interferences. Sound velocity profiles are calculated from this extrema as previously described (Litniewski and Bereiter-Hahn, 1990). The sound velocity profile is limited to the resolution of minima an maxima of interference fringes. In Fig. 3 sound velocity profiles are shown for the different stages of swelling.

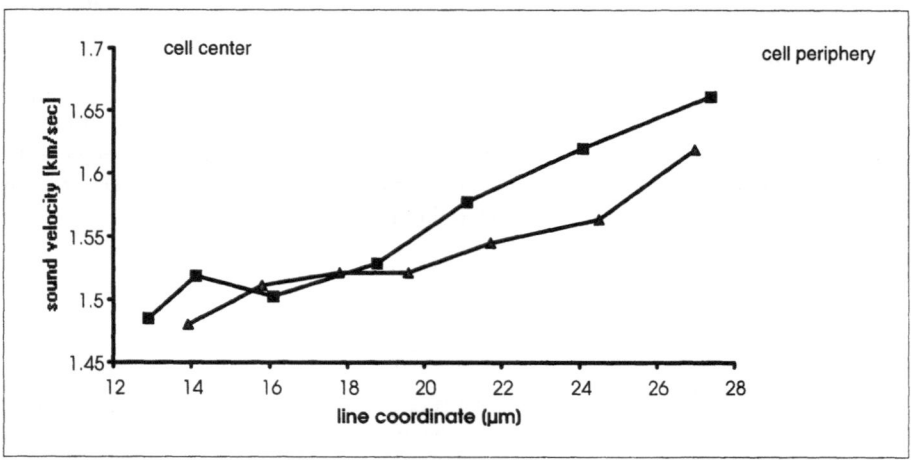

Figure 3. P-wave sound velocity distribution before hypotonicity (■) and at maximal swelling (Δ) plotted against relative cell position.

The profiles show higher values for sound velocity at the cell margins. A decrease during swelling (Δ) can be seen.

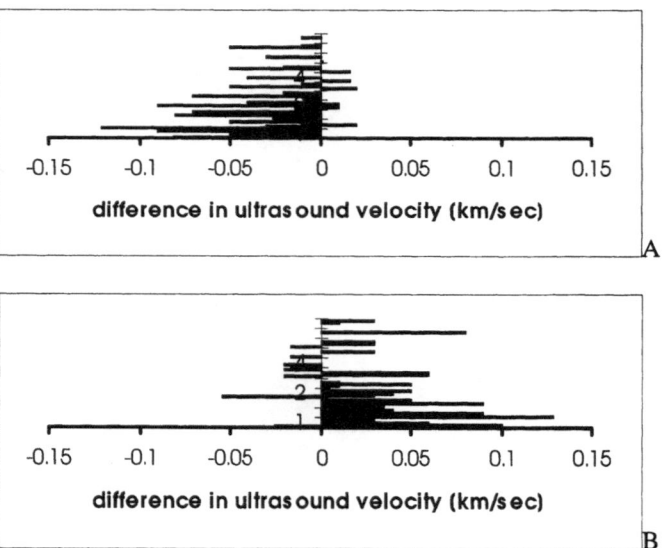

Figure 4. Change in ultrasound velocity during swelling plotted against order of fringes. Sound velocity transiently decreased (A) and was restored during RVD (B). Sound velocity difference values from 5 independent experiments are shown.

Sound velocity changes at different timepoints of volume control are shown in Fig. 4. Sound velocity was measured in the resting cell, during swelling and after RVD. A decrease in sound velocity values is found during swelling, which is restored during RVD.

DISCUSSION

We measured topography and sound velocity in living cells before and during cellular swelling and RVD to investigate the mechanical basis of physicobiochemical coupling which was suggested as a possible explanation for volume control. We found a decrease in sound velocity values during swelling, which was reversible during RVD.

Sound velocity changes in cells indicate changes in the stiffness of the cytoplasm.

A cell is enveloped by a lipid membrane which is supported by an underlying cortical actingel. This gel is built by actin and actin binding proteins and can be regulated by different pathways (Dufort and Lumsden 1993). The actual state of the gel is determined by many parameters as are number of filaments, length distribution of filaments, crosslink number and distance, ionic composition of the cytosol and is mediated by the activity of gel-sol regulator proteins like gelsolin (Ito et al.1992). In addition contraction and relaxation of actomyosin would alter the stress pattern in the gel. Hence, the observed change in physical properties of the cytocortex indicates a volume sensing mechanism involving the change in activity of one or more gel-sol modulators like Calcium and PIP_2 and/or changes in contractile elements during volume control.

Further experiments should aim to reveal the interaction of physiological processes with mechanical properties of cells in more detail to confine the beginning understanding of fundamental mechanobiochemical feedbackcycles like volume control in living systems.

REFERENCES

Boukamp P., Petrussevska R.T., Breitkreutz D., Hornung J., Markham A. 1988. Normal keratinization in a spontaneously immortalized aneuploid human keratinocyte cell line. Journ Cell Biol. 106: 761-771

Chou, L., Firth, J.D., Uitto, V.J. and Brunette, D.M. 1995. Substratum topography alters cell shape and regulates fibronectin mRNA level, mRNA stability, secretion and assembly in human fibroblasts. Journ Cell Sci 108:1536-1573

Davies PF., Robotewskyj A. and Griem ML. 1994. Quantitative studies of endothelial cell adhesion. Directional remodeling of focal adhesion sites in response to flow forces. Journal of Clinical Investigation. 93(5):2031-8

Dufort PA. and Lumsden CJ.1993. Cellular automaton model of the actin cytoskeleton. Cell Motility & the Cytoskeleton. 25(1):87-104

Heidemann SR and Buxbaum RE. 1994. Mechanical tension as a regulator of axonal development. Neurotoxicology 15(1):95-107

Ito T. Suzuki A. and Stossel TP.1992. Regulation of water flow by actin-binding protein-induced actin gelatin. Biophysical Journal. 61(5):1301-5, 1992

Klymkowsky MW. Karnovsky A. 1994. Morphogenesis and the cytoskeleton: studies of the Xenopus embryo. Developmental Biology. 165(2):372-84

Litniewski J., Bereiter-Hahn J. 1990. Measurements of cells in culture by scanning acoustic microscopy. Journ Microsc. 158: 95-107

Nicklas, R.B., Ward, S.C and G.J. Gorbsky. 1995. Kinetochore chemistry is censitive to tension and may link mitotic force to a cell cycle checkpoint Journ Cell Biol 130:929-939

OPTICAL WAVELENGTH EFFECTS OBSERVED IN PHOTOACOUSTIC SIGNALS FROM BIOLOGICAL TISSUE

P.A. Payne, R.J. Dewhurst, A. Kuhn, K.F. Pang and Q. Shan

Department of Instrumentation and Analytical Science, University of Manchester Institute of Science and Technology (UMIST), P O Box 88, Manchester M60 1QD, UK

ABSTRACT

We have previously reported our observations that, at laser wavelengths of 1,064 and 532 nm, following repeated exposure to short pulses at power density levels well below those reported to cause permanent damage to biological tissue, the photoacoustic signal detected using an ultrasound probe diminishes significantly. Examination by electron microscopy of such tissue at the site at which repeated laser pulse exposures have been made reveals little evidence of structural changes to the tissue. The effect that we have observed remains unexplained.

The experiments referred to above were carried out as part of a programme to develop a new, combined form of diagnostic and therapeutic probe in which the laser power is delivered via an optical fibre which is concentric with a miniature forward and sideways looking ultrasound transducer. At the laser wavelengths mentioned previously signal decay on repetitive laser shots is of great concern since it reduces signal-to-noise ratios. We have carried out further experiments at a wavelength of 266 nm. At this wavelength, the diminution of the photo-acoustic signal does not occur, although the energy level that we are currently able to operate at is not sufficient to give ultrasonic echoes from structure within samples of arterial wall.

This paper reports on these recent experiments and on the development of the new form of intra-arterial probe.

INTRODUCTION

Intravascular ultrasound imaging has become well established as a means of examining the arterial system of patients suffering from atherosclerosis[1]. The technique is to introduce a diagnostic probe in conjunction with some form of guided catheter which is moved along the arterial system to the point of interest and cross-sectional images of the artery are obtained using normal pulse echo ultrasound techniques. The transducers employed can be either rotating devices or phased array multi-element transducers have been used employing 64 or more individual elements. The ultrasound frequencies employed are well in excess of 10 MHz and good clarity images are produced by the commercial systems now available.

Having produced an image that indicates the presence of atherosclerotic plaque, in most cases, the imaging probe is withdrawn and a device capable of ablating, heating, cutting, grinding, vacuuming or burning the plaque is then employed. In many cases, the use of balloon angioplasty continues to be preferred. However, the disadvantage with this approach is that between removal of the imaging probe and introduction of the therapeutic device, some movement may occur and treatment, therefore, may be less effective.

For some while, the research group in Manchester have been developing an alternative probe system. This is based on the use of a concentric optical fibre used to conduct laser energy to the point of interest within an artery. We then exploit the photoacoustic effect produced by the laser pulse impinging on biological tissue and receive, via a small piezoelectric polymer transducer ultrasound signals which are characteristic of the tissue being interrogated. Further description of the new probe is given later. The major advantage of this approach is that, with careful choice of optical fibre, we can switch from diagnostic laser energy to therapeutic laser energy and ablate tissue which is diagnosed as atherosclerotic plaque. By moving between diagnostic and therapeutic lasers, the progress of the ablation can be monitored and there is, for example, far less chance of puncturing the arterial wall. The approach may also have advantages in cases where the artery is completely blocked, which frequently produces major difficulties with the conventional approach.

PROBE DESIGN

The general layout of the new probe is shown in Figure 1 which also shows the manner in which the optical fibre is fed through the centre of a polyvinylidene fluoride (PVDF) based receiving transducer. This design is intended to be used for forward looking applications. We are currently working on a sideways looking device which will be added to the present probe in due course. This is also based on a PVDF transducer and at present the approach is to spin-

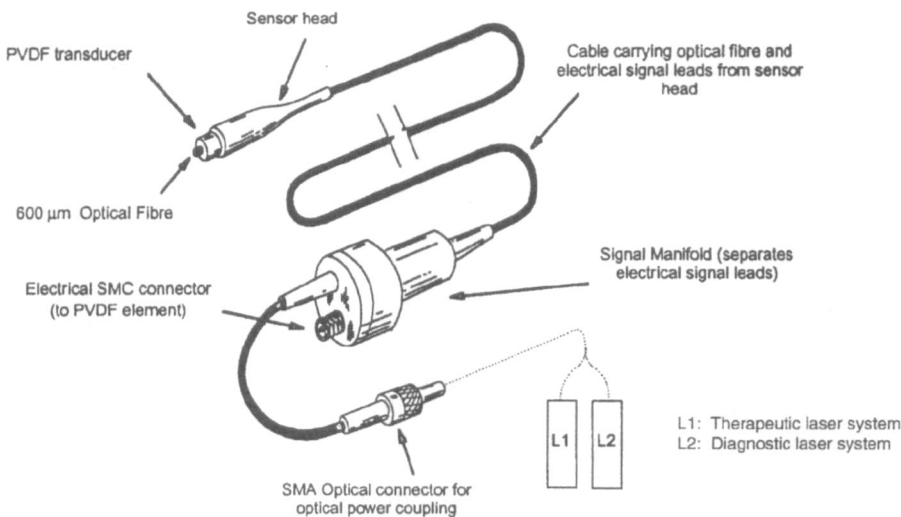

Figure 1. General arrangement of the new probe.

coat a section of the transducer probe with PVDF and to use conventional photolithography to define the 32 transducer elements. Some experiments have been performed to determine whether the multi-element transducer will need to transmit an ultrasound signal in order to obtain cross-sectional images of the artery, or whether the forward going laser pulse produces sufficient ultrasound energy for it to be "bounced" off the artery walls and received by the 32 elements of the sideways looking transducer. The experiments indicate that this may be possible and we hope to report on further work in this area in the near future.

WAVELENGTH EFFECTS

For some time now we have been conducting experiments in parallel with considerable effort in the area of mathematical modelling, the aim of this work being to characterise the probe and the photoacoustic effects that we observe. The tissue experiments have been conducted using an arrangement similar to that shown in Figure 2(a), whilst the mathematical modelling work has been conducted in parallel with experiments based on the arrangement shown in Figure 2(b).

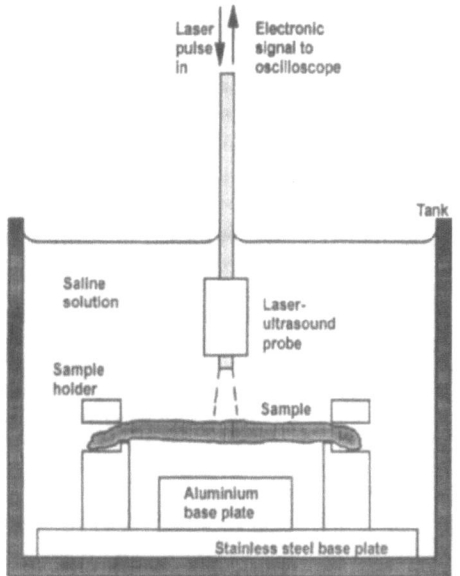

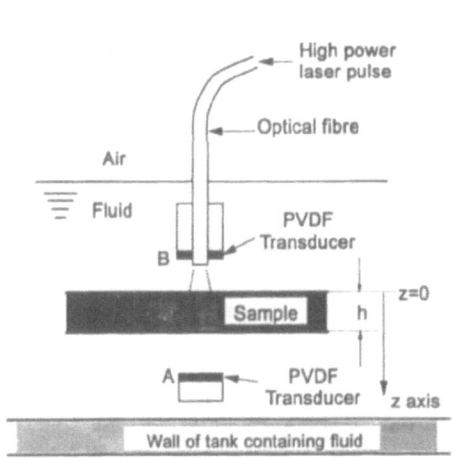

Figure 2(a). Tissue experimental arrangement.

Figure 2(b). Experimental arrangement for mathematical modelling studies.

Figure 3 shows the results of a series of measurements of photoacoustic signals from a sample of porcine arterial wall. The signal labelled A is the received ultrasound signal produced by the photoacoustic effect at the surface of the artery wall. As can be seen, the amplitude of this signal decays rapidly with successive laser pulses, yet the energy being used, 2.9 mJ, is well below the figure at which any damage to tissue might be expected. These were the initial experiments that gave us cause for concern. The data were collected at a laser wavelength of 1.064 nm.

We took two immediate steps. Firstly, we began a further set of experiments at a shorter wavelength, in this case 532 nm, and a pulse energy of 2.5 mJ. As is shown in Figure 4, the effect is still seen. The other step that was taken was to make contact with colleagues in the University of Manchester Medical School in order to undertake a series of electron

229

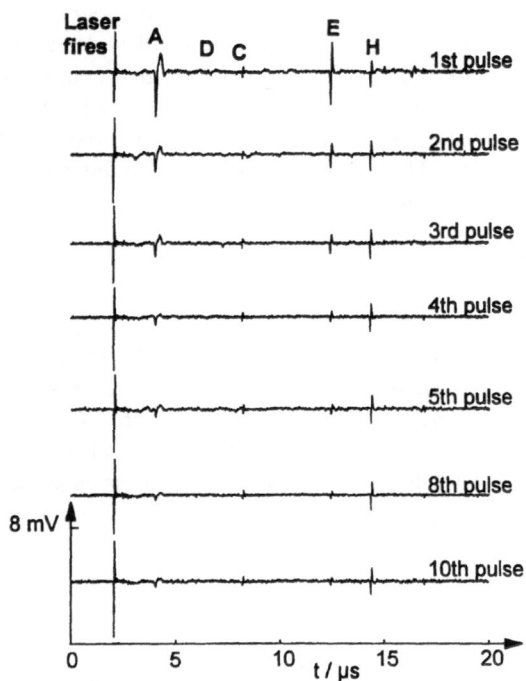

Figure 3. Photoacoustic signals, A, from surface of porcine artery.

Laser wavelength: 1.064 nm
Laser energy: 2.9 mJ

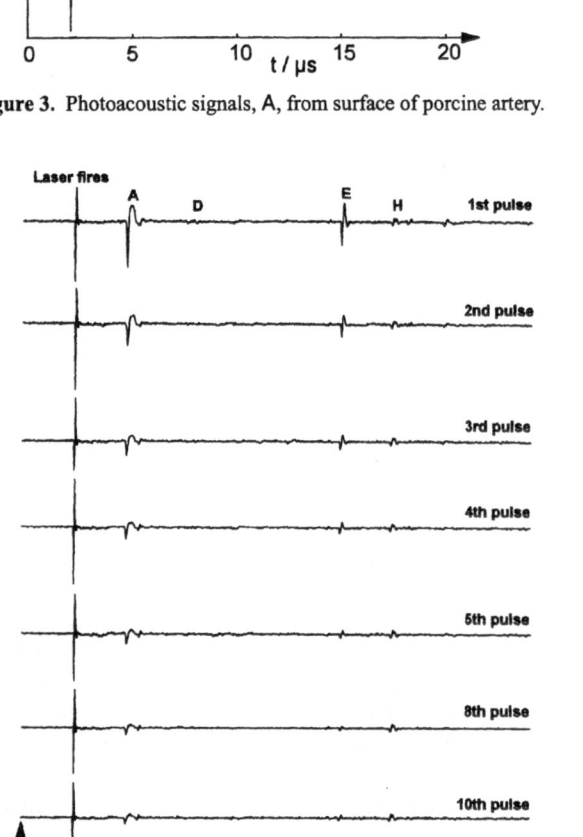

Figure 4. Photoacoustic signals, A, from surface of porcine artery.

Laser wavelength: 532 nm
Laser energy: 2.5 mJ

microscopy studies. This work is ongoing and to date there is very little evidence of any tissue damage that can be found in tissue that has been irradiated in the manner indicated at either 1,064 or 532 nm.

We next chose to conduct experiments using ultraviolet wavelengths (266 nm), but here the limitations on our optical fibre allowed us only to use 0.5 mJ. A set of experiments were conducted and some of these are shown in Figure 5 and, as can be seen, even after ten successive laser pulses, at this wavelength the arterial wall signal remains unchanged. Unfortunately, using only 0.5 mJ means that we have insufficient ultrasonic energy to receive echoes from interfaces within the tissue. We are currently commissioning new optical systems which will enable us to operate at each of the wavelengths mentioned previously and will produce about the same energy level at 266 nm as we as used for the longer wavelength experiments.

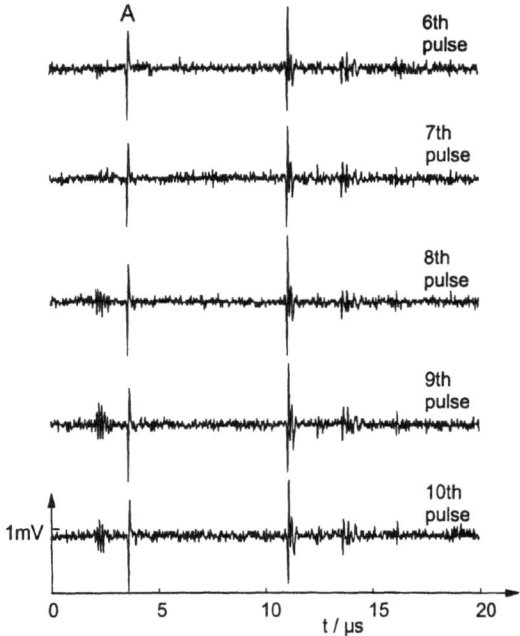

Figure 5. Photoacoustic signals, A, from porcine artery wall.

MODELLING THE NEW PROBE

It is absolutely essential that we fully understand the mechanisms associated with the use of the new probe and, to this effect, a mathematical modelling exercise has been undertaken, details of which have recently been reported[2]. Further work is being undertaken to refine this, but as an indication of the accuracy with which the model fits measured data, see Figures 6(a) and 6(b).

CONCLUSIONS

A new form of combined diagnostic and therapeutic intra-arterial probe has been described. Experiments have been performed at wavelengths of 1,064, 532 and 266 nm. At the two longer wavelengths an unexplained effect causes the photoacoustic signal from the tissue to diminish with successive laser pulses which are at energy levels well below those reported to cause any tissue damage. So far, electron microscopy studies are inconclusive

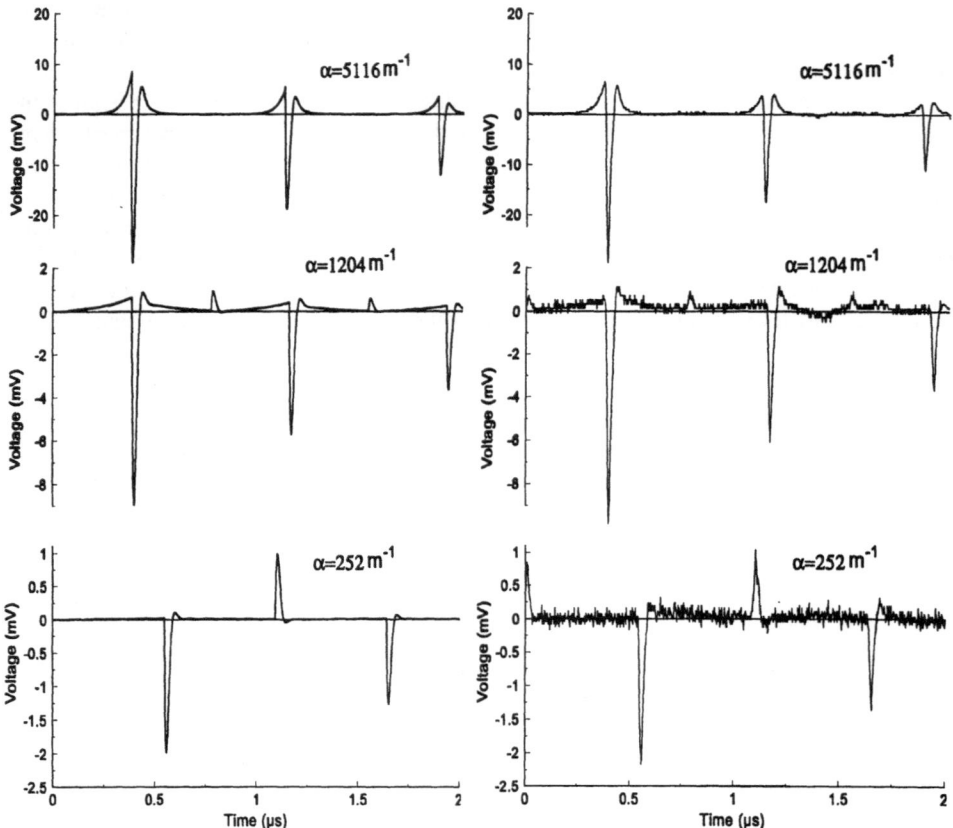

Figure 6(a). Predictions of photoacoustic signals from three glass plates possessing different optical absorption coefficients, α.

Figure 6(b). Experimentally obtained photoacoustic signals in corresponding plates.

regarding any evidence of tissue damage, but this work is on-going. The fact that there is no signal reduction at 266 nm may be a wavelength related issue. It may also be related to the reduction in energy used (0.5 mJ rather than 2.5 to 2.9 mJ).

We have also conducted considerable theoretical work in order to obtain a mathematical model of the new probe and considerable success at predicting probe performance has been obtained.

ACKNOWLEDGEMENTS

Funding for this work is provided by the National Heart Research Fund and the UK Engineering and Physical Sciences Research Council.

REFERENCES

1. J.M. Tobis and P.G. Yock (eds). "Intravascular Ultrasound Imaging," Churchill Livingstone Inc, New York (1992).
2. Q. Shan, R.J. Dewhurst, A. Kuhn, K.F. Pang and P.A. Payne, Modelling of a photoacoustic probe designed for medical applications. Presented at 16th Ultrasonics International Conf., Edinburgh, July 1995.

IMAGING OF SPATIAL DISTRIBUTION AND FLOW OF MICROBUBBLES USING NONLINEAR ACOUSTIC PROPERTIES

Volkmar Uhlendorf and Frank-Detlef Scholle

Research Laboratories of Schering AG
13342 Berlin, Germany

INTRODUCTION

Today, medical ultrasound images can reveal subtle details of anatomy or blood flow with high spatial and temporal resolution. This has been made possible by the rapid progress in acoustical imaging during the last decades. Various ultrasound contrast agents are now becoming additional tools which allow special problems in imaging and functional diagnostics to be solved. They have a number of applications with state-of-the-art imaging techniques since most agents became sufficiently adapted to "conventional" imaging modes during their development. A certain class of contrast agents (first and foremost SHU 508 A, i.e. Levovist®) shows a very marked response at the second harmonic of the transmitted frequency f_0, which is now used for selective detection in Harmonic Imaging - a class of imaging modes specially adapted to contrast media.[1]

Here we present some acoustical properties of a new ultrasound contrast agent of the third generation (SHU 563 A), which consists of micrometer-sized bubbles stabilized by a thin polymeric shell. Several minutes after intravenous injection the coated microbubbles are cleared from the blood and collect in liver and spleen tissue. Besides their unique pharmacokinetic characteristics[2] they have a most remarkable nonlinear acoustical response.[3,4] It will be shown that detection and imaging of this contrast agent is possible with the ultimate sensitivity, i.e. even as single particles. This can be achieved when a characteristic signal termed "acoustically stimulated Acoustic Emission" (as-AE) is used. As the name already suggests, a kind of AE can be excited by short ultrasound pulses like those used in medical diagnostics. This as-AE is a transient response generated by cracking hollow microparticles, and is strong enough to be detected across thick layers of attenuating tissue (bright, minute spots in B-mode, aliasing-like pattern in CD-mode even without flow).[2-4] When used at low concentrations there is practically no additional attenuation caused by coated microbubbles. One can thus avoid acoustical shadowing, which has been addressed as the main limiting factor in previous attempts at tissue contrast imaging by conventional techniques based on ordinary backscattering.[5] Some of the potential applications are illustrated in a provisional way using state-of-the-art equipment, while significantly improved imaging performance can be expected from novel, specially developed "as-AE detection modes".

NON-ACOUSTIC CHARACTERIZATION OF MATERIALS

SHU 563 A (Schering AG, Berlin) was resuspended with physiological saline solution and well mixed immediately before a sample was taken from a vial. The specific density of this concentrated suspension was determined with a DMA-38 (Chempro, Hanau, FRG) before and after mechanical destruction of the polymer-coated microbubbles in an ultrasonic cleaning bath. 100 µl of a freshly resuspended solution were diluted in 100 ml of 0.01 percent (w/w) aqueous detergent solution to prevent particle aggregation. A commercial optical instrument (MasterSizer, Malvern, UK) was employed to measure the total volume fraction of particles, and their volume distribution. For evaluation of the light scattering data the polymer wall was ignored and the Mie theory for spherical air bubbles in water applied as approximation. The resulting volume fraction $4 \cdot 10^{-5}$ was consistent with density measurements and with the assumption of very thin walls. The known absolute volume fraction together with the normalized volume distribution allowed these values to be converted to absolute numbers of hollow spheres in each size class. For diameters smaller than about 0.5 µm experimental accuracy was insufficient because of the small total volume (and thus relatively large fluctuations) in these classes. From the sum of particles in all classes with diameters above 0.5 µm a concentration of $2 \cdot 10^7$ p./ml was calculated for the diluted sample. A conventional counting method using a light microscope gave results within 8 percent of the stated value, on which dilution factors for subsequent acoustical experiments on single particles were based. Typical final concentrations were in the range 500 - 5000 particles/ml.

ACOUSTICAL METHODS

For experiments on single particles there must be just 1 particle per pulse volume on average (but due to statistical fluctuations occasionally two or none will be found). The required concentration therefore depends on pulse length and transducer beam geometry.

Beam geometry and sensitivity of the single-element broadband transducers (V319, Panametrics) were determined by an AIMS-01 with calibrated needle- and membrane-type hydrophones (NP-1000 and TMA001, all from NTR, Seattle) for at least 6 frequencies between 2 MHz and 8 MHz with high spatial resolution.[4] This was done for RF bursts of less than 20 cycles at several driving voltages from an HP3314A function generator, all below the 1dB compression point of a linear power amplifier (A-300, ENI). Two separate transducers of the same type were adjusted to have a common focus point about 8 mm behind a thin coupling window of the sample cell. This set-up allows as-AE "events" to be produced at typical peak pressure values which diagnostic ultrasound scanners may generate in situ above 2 MHz, but at most about 1 MPa. Peak positive pressures exceeding 0.12 MPa could not be used at 2 MHz since they caused distortions of the transmitted ultrasound pulse. The receiver was calibrated between 1 MHz and 20 MHz, extrapolating the flat frequency response function of the membrane-type hydrophone into the range beyond 12 MHz.

From experiments reported elsewhere[3,4] it is known, that the hollow microparticles of SHU 563 A are only stable in pulsed ultrasound of sufficiently low amplitude. They "immediately" rupture when hit by a microsecond pulse at peak pressure above a particular threshold (as do stabilized microbubbles of SHU 508 A).[6] Therefore, pulse timing (120 ms interval in manual trigger mode) and sample flow rate (80 ml per minute) were controlled in addition to frequency, pulse duration, and pulse amplitude to allow a flow of fresh contrast agent into the sample cell before the next pulse arrived.

The acoustic signal received by the second transducer was amplified (AU-4A-0110, Miteq, +61 dB) and 20 MHz low-pass filtered. The split signal was simultaneously recorded with different sensitivities on two channels of a LeCroy 9310L oscilloscope. Up to 1000 segments per channel were acquired, each with 1000 sample points per 10 µs. Excitation voltage was limited when the received signals approached the 1dB compression point of the receiver. Further processing of the whole segment from the respective channel with better signal level was done with ASYST® (Keithley) on an 80486-based PC, which also controlled function generator and receiver via GPIB bus.

A small average background signal from spurious reflections was subtracted in the time domain. Spectra were then calculated by FFT and corrected for the interpolated smooth frequency response function of the receiver to reduce the influence of the experimental equipment. Finally, corrected time domain signals of as-AE "events" were obtained by IFFT. Additional effects from spatial variation of both acoustic peak pressure and receiver sensitivity have not yet been taken into account.

Commercial ultrasound scanners which allow sufficient variation of the transmit amplitude (UM9 HDI and HDI 3000, both from ATL, and Sonos 1000 from HP) were used in a provisional way to demonstrate some aspects of the novel contrast imaging techniques proposed here. This was done by capturing sequences of conventional B-mode and - even more sensitive - conventional Color Doppler mode images[2-4] either with cineloop functions or by recording on video tape while switching between previously determined transmit levels.

RESULTS AND DISCUSSION

A single as-AE event near the focal point in water can be seen in Fig.1 (n = 4 cycle sine pulse, f_0 = 3.00 MHz, peak positive pressure p_+ = 0.6 MPa, almost symmetrical). The RF signal from SHU 563 A and corresponding power spectra are shown both before (Fig.1a) and after frequency domain filtering (Fig.1b, where the time scale has been expanded). The strongly nonlinear response, presumably from a single coated microbubble, consists of short spikes, each of them followed by some ringing. The sequence of spikes has approximately the same period as the driving frequency (within a few percent). Second

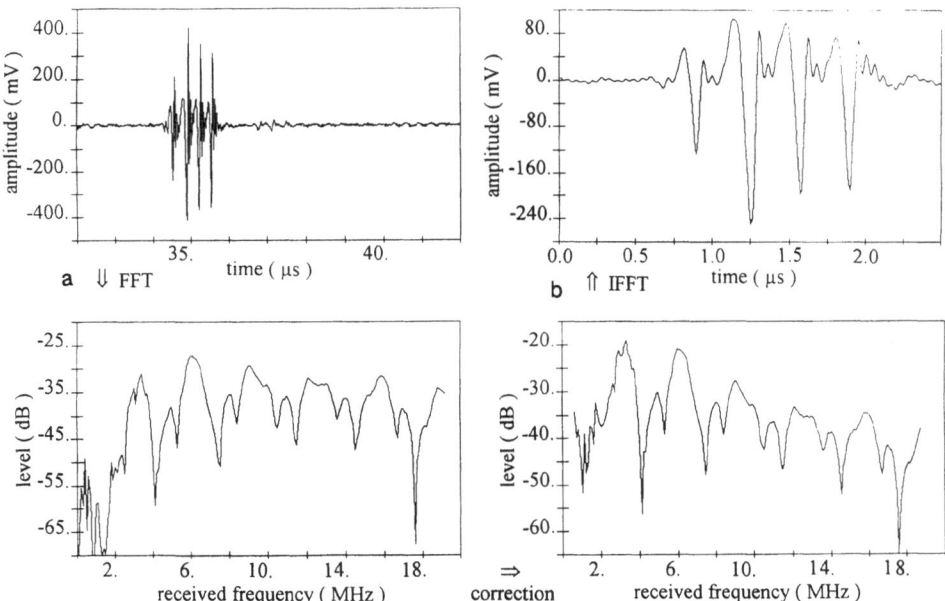

Figure 1. Time domain signal and spectrum of single as-AE event from an SHU 563 A microbubble near the focus (a) before and (b) after correction. Background subtracted from the time domain signal before frequency domain filtering was slightly larger than noise. Note expanded time scale of filtered curve.

harmonics are much stronger than in the spectrum of the transmit pulse (not shown), where they are typically more than 20 dB below the level of the fundamental frequency.

As already mentioned, the acoustically stimulated AE of a microbubble is a transient phenomenon.[3] This means that subsequent transmit pulses sent (as specified below) shortly after a strong one into a quasi-stationary dilute sample usually do not cause a detectable second response, since the coated microbubbles have been irreversibly cracked during the previous as-AE event. The gas core of about 1 μm diameter (a value resulting from optical measurements) must have vanished rapidly and scattering from the remains of the shell is too weak to be detected with our set-up. This is the current explanation and accords with numerous experimental observations reported elsewhere.[4,3,6]

At present it is known that the microbubble lifetime ΔT after an as-AE event typically is shorter than about 40 ms, but may well be as short as the pulse duration Δt or even as short as half a period of the driving frequency f_0 (i.e. about 100-200 ns) in some cases. The lifetime depends on the mechanism(s) by which a microbubble disappears after an as-AE event. A slightly modified experimental set-up will allow determination of the distribution of lifetimes within the upper and lower limits given here. Exact values may turn out to be important for imaging modes with pulse (or even frame) intervals within these limits.

The upper traces in Fig.1 suggest the possibility of localizing as-AE events with higher precision and higher axial resolution than the fundamental frequency of the transmit pulse would allow (of course, signal correction need not be neccessary for imaging, it is merely used to show intrinsic properties of as-AE signals). Currently, it is not quite clear whether this finer resolution - which indeed is *apparent* from B-mode images (Fig.2a) - has any meaning when single events are used for imaging. It could be made obsolete by a large and/or uncertain response delay time δ between the rising edges of the stimulating acoustic pulse and the as-AE response of a coated microbubble of SHU 563 A. Large values of δ would cause an unacceptably large spatial offset δx between the true location of the point source and the apparent position of this event in the image, though a constant δx probably could be compensated. Fortunately, all observations are consistent with $\delta x < 1$ mm, and δ seems to be shorter than the pulse duration $\Delta t = n / f_0$, typically at most a fraction of a microsecond. This guarantees proper localization of as-AE events with the same resolution as provided by conventional diagnostic ultrasound. In some cases fluctuations of δ values of different events may be as large as δ itself and this would defeat improved resolution near the lower end of the useful concentration range, where speckles cannot develop. In practice, an approximately 10-fold concentration might be a good compromise to reduce the effect of uncertainties in δ by increasing the chance of a prompt response and yet avoid

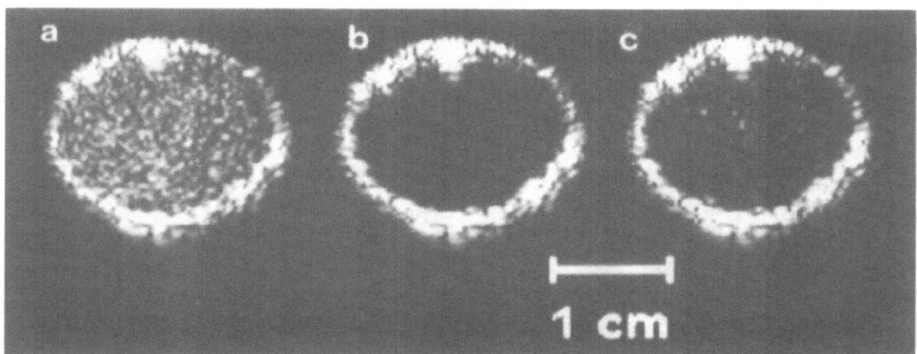

Figure 2. Images of a stationary sample obtained by "acoustically stimulated Acoustic Emission" from very low concentrations of SHU 563 A in water. Fig.2a) and 2b) are two frames in direct sequence showing the transient nature of the response (initial concentration 1200 microbubbles per ml). Fig.2c) corresponds to a) but was taken at about 10 times lower concentration, showing isolated as-AE events.

speckle patterns caused by higher concentrations of the contrast agent (Fig.2a). Regardless of an ultimate solution to this complication, it is possible, and could still be useful, just to detect the *presence* of a few hollow microspheres, even if their true location may be known only within about one millimeter (Fig.2c).

5 MHz B-mode images of a stationary sample shown in Fig.2 have been recorded while suddenly increasing the acoustic pressure amplitude. Prior to acquisition of the first frame the microbubble concentration was just appropriate to give homogeneous enhancement (1200 microbubbles per ml, Fig.2a). All microbubbles seen (by their as-AE) in the first frame after the amplitude step are destroyed when this frame has been completely acquired, thus no response from SHU 563 A can be detected at the same location in subsequent frames (second frame shown in Fig.2b). At much lower initial concentrations individual particles can be distinguished, appearing at random positions in the first frame after the increase in the acoustic peak pressure (Fig.2c). At higher concentrations speckles begin to develop due to interference between the as-AE signals of several events at different positions within the range gate of the receiver (not shown). A frame acquired at low transmit amplitude, where even prolonged ultrasound exposure does not have a measurable effect on coated microbubbles, would look similar to Fig.2b) for concentrations below about $1 \cdot 10^6$ microbubbles per ml, even when the receiver gain is compensating the reduced transmit level.

Closely related to imaging of spatial distributions of contrast agents is a novel way of imaging flow patterns without recourse to Doppler methods. For this purpose the almost instantaneous destruction of (coated) microbubbles by suitable microsecond pulses is used as a means of "tagging" flow within or across the scan plane at some instant. Subsequent images then show cross sections of the marked volume (dark) as its shape evolves and becomes deformed by the specific flow pattern (Fig.3). This was first noticed some years ago during contrast imaging of the left ventricle, but only recently has an explanation for the basic phenomenon been given.[6] Its controlled use by special pulse sequences could be a new tool for diagnostic ultrasound, although analogous methods are well-known from Magnetic Resonance Imaging. A similar principle has been developed independently for contrast imaging and demonstrated on a kidney phantom.[7]

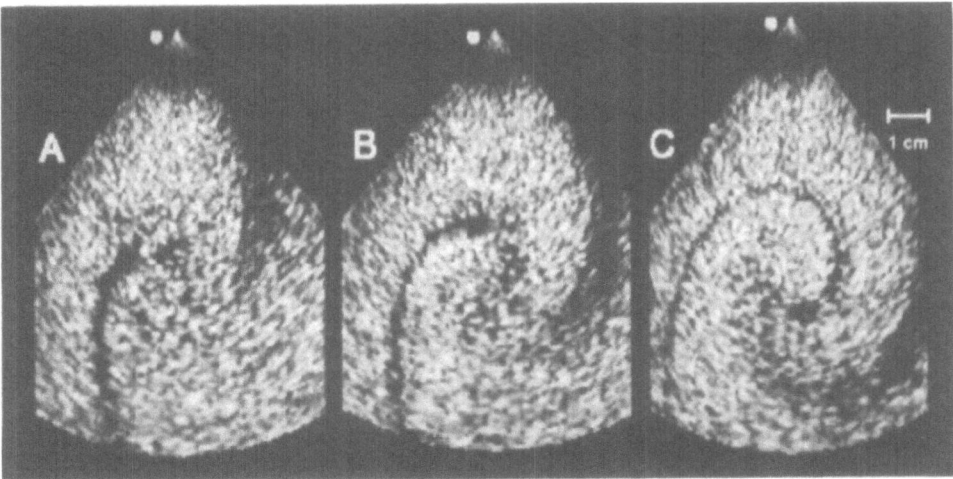

Figure 3. Image sequence of evolving flow pattern created by acoustical tagging of contrast agent in water.

In a further application ultrasound beam patterns can be made visible in a gel containing a sufficiently high amount of homogeneously distributed SHU 563 A.[3] Ultrasound pulses give rise to a kind of self-induced acoustical (but also optical) transparency in regions where the peak pressure exceeds the threshold.

A "clean" background as seen in Figs.1-3 will only be found in weakly scattering media like water or blood, but as-AE events of SHU 563 A in very low concentration are easily detectable in scattering tissue like liver,[2] as is also demonstrated in Fig.4.

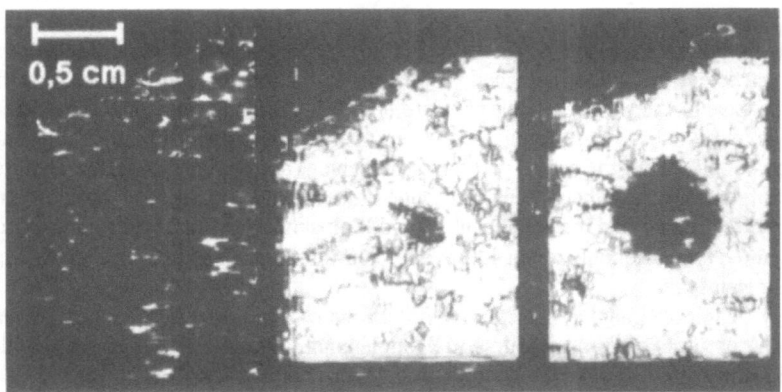

Figure 4. Image at low amplitude (left, standard B-mode) detects neither tumour nor presence of small amounts of SHU 563 A in rabbit liver ex vivo. Images using as-AE clearly show tumour in two adjacent cross sections (middle and right, CD-mode colour signals are caused by as-AE, here reproduced in white).

"Acoustic Emission Imaging" as proposed here is expected to solve the long-standing problem of tissue contrast. Moreover, these techniques will allow i) high spatial resolution not limited by speckle patterns, ii) flow imaging and quantification by methods which resemble event counting techniques used in nuclear medicine (one important difference is that the "decay" can be acoustically controlled by transmit parameters), and iii) a significant dose reduction of the contrast agent (several orders of magnitude).

REFERENCES

1. P.N. Burns, J.E. Powers, D. Hope Simpson , A. Brezina, A. Kolin, C.T. Chin, V. Uhlendorf, T. Fritzsch, Harmonic power mode doppler using microbubble contrast agents: an improved method for small vessel flow imaging, Proceedings 1994 IEEE Ultrasonics Symposium, p. 1547 (1994).
2. B.B. Goldberg, F. Forsberg, T. Fritzsch, J.B. Liu, D.A. Merton, Induced acoustic emision as a contrast mechanism for detection of hepatic abnormalities, J. Ultrasound Med. 14:S7 (1995),
   T. Fritzsch, P. Hauff, D. Heldmann, F. Lüders, M. Reinhardt, V. Uhlendorf, W. Weitschies, A new tumour-specific imaging method based on acoustic emission, (in preparation).
3. V. Uhlendorf, C. Hoffmann, Nonlinear acoustical response of coated microbubbles in diagnostic ultrasound, Proceedings 1994 IEEE Ultrasonics Symposium, p.1559 (1994).
4. C. Hoffmann, Die mechanischen Eigenschaften von kleinen Hohlkugeln in Wasser und deren Verhalten unter Einwirkung diagnostischer Ultraschallwellen, Thesis (in German), Technische Universität Berlin 1994, Shaker, Aachen (1995).
5. V. Uhlendorf, Physics of ultrasound contrast imaging: scattering in the linear range, IEEE Trans. UFFC 41:70 (1994).
6. F.-D. Scholle, V.Uhlendorf, T. Fritzsch, Physical mechanism of inhomogeneous left ventricular echo-contrast, Proceedings 1994 IEEE Ultrasonics Symposium, p.1563 (1994).
7. J.B. Fowlkes, E.A. Gardner, P.L Carson, J.A. Ivey, J.M. Rubin, Use of contrast interruption in the measurement of blood flow, Ultrasonic Imaging 17:67 (1995).

POTENTIAL FOR TISSUE MOVEMENT COMPENSATION IN CONFORMAL CANCER THERAPY

Jeffrey C. Bamber, Jorn A.A. Verwey, Robert J. Eckersley,
Christopher R. Hill, and Gail R. ter Haar

Joint Department of Physics
Institute of Cancer Research and Royal Marsden NHS Trust
Downs Road, Sutton
Surrey, SM2 5PT, UK

INTRODUCTION

In conformal approaches to minimally invasive cancer therapy, such as focused ultrasound surgery[1,2], body movement during the surgical procedure may limit the precision and speed with which a tumour volume may be treated. This is particular relevant to the treatment of tumours such as liver metastases since the liver may move substantially with respiration and cardiovascular pulsation. We report on a preliminary investigation of the potential and limitations of ultrasonic speckle decorrelation and speckle tracking for achieving on-line compensation for body movement associated with the respiratory and cardiac cycles. Figure 1 illustrates the concept that we have in mind for eventually employing data from an ultrasound echo imaging system to allow the focus of the high intensity treatment beam to automatically track a moving tissue target.

EXPERIMENTAL

Two dimensional correlation pattern-matching algorithms[3] were applied to track regions of interest (ROI) in frames from sequences of real-time B-scans of the liver. Correlation coefficients rather than smoothed absolute difference (SAD) values[4,5] were used because speed of calculation is not important at present and because the normalized coefficient provides a measure of reliability of tracking for comparing alternative search algorithms. Both fixed reference pattern (in which the speckle pattern from the first ROI is searched for in all subsequent frames) and incremental tracking (in which the reference pattern is replaced by the best match located in each subsequent frame) algorithms were studied. A graphic overly onto the first frame of the sequence of images in the loop,

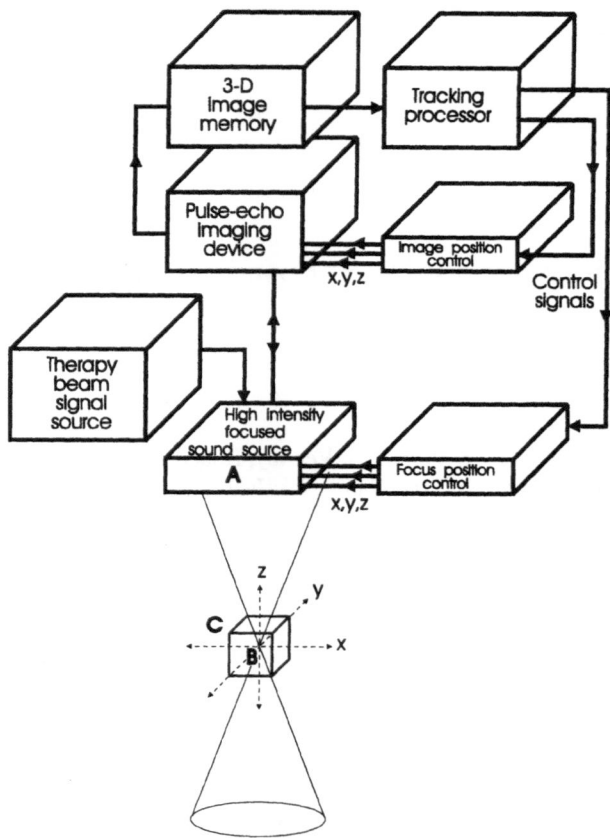

Figure 1. Schematic illustration of the principle underlying automatic ultrasound therapy target tracking. The source of high intensity sound, A, produces a therapy beam focused at position B(x,y,z). A pulse-echo source/receiver, which may be at A or some other position, is used to obtain 3-D echographic data at the volume C, which is spatially linked to the focal position B. The tracking processor analyses the echo data at two times and produces control signals which are used to alter the positions of B and C, to follow the tissue motion.

showing the locus of the tracked path and the variation in the value of the correlation coefficient (in grey scale), was used to aid interpretation of the results.

RESULTS

It was found possible to locate scan planes in which the motion was predominantly within the plane, so that two dimensional tracking was feasible (Figures 2 and 3). This result is encouraging since three dimensional tracking in real time would place extreme demands on technology. Regions could also be located which would cyclically return to positions close to the starting point, both in terms of the locus (Figure 3) and the correlation value (Figure 4). This suggests potential for a method of speckle correlation "gating" of the surgical procedure.

The alternative tracking algorithms were, in this part of the study, found to display complementary advantages and disadvantages. Speckle decorrelation, possibly due to rotational motion or movement of tissue in the direction orthogonal to the scan plane,

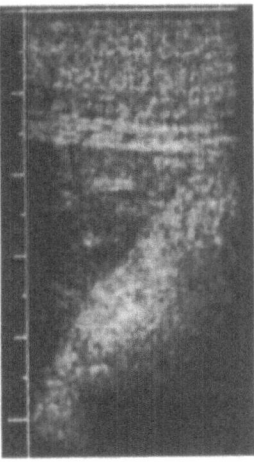

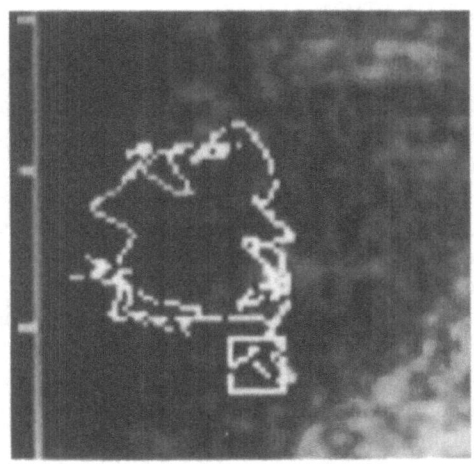

Figure 2. Example of a linear B-scan of the liver from a sequence captured from an Acuson™ scanner by successively displaying and frame grabbing each image in the scanner's cine-loop memory, whilst the volunteer subject was quietly breathing but otherwise motionless.

Figure 3. Example locus of tracked motion demonstrating a scan plane in the liver where the combination of cardiac and respiratory induced motion appeared to be within the scan plan and also cyclical, causing the tissue to return approximately to the same position.

would sometimes lead to failure of the fixed pattern search. On the other hand, the incremental search, although it handles gradual speckle decorrelation, in the process of doing this it loses information about the original region to be tracked and may wander onto false tracks (Figure 5). Some combination of the two approaches may provide an appropriate solution. Also, there is potentially very useful information available from initiating multiple searches from many closely positioned starting points. Continuity constraints might then be applied to compute the best statistical estimate of a whole field of motion loci.

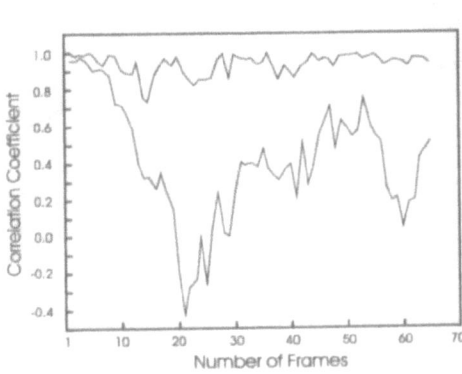

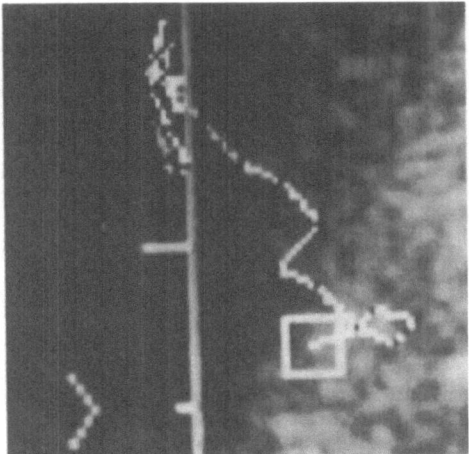

Figure 4. Examples of the correlation coefficients computed along tracked loci for fixed pattern (lower curve) and incremental tracking (upper curve) methods, for the same tissue region.

Figure 5. A rather extreme example of how the incremental track may wander away from the direction of tissue motion, which visually appeared to remain within the image field.

SIMULATION STUDY

Although the clinical study described above was useful for gaining a feeling for the range of behaviour of these simple tracking algorithms, for a more thorough evaluation it is necessary to know the true motion of the structure being tracked. Both experiments with phantoms and computer simulations present possible approaches to conducting such an evaluation. We have made a start using the latter approach. Simulated B-scan images were generated by convolving a two-dimensional model for an imaging radio frequency point response function with a two-dimensional model for the tissue backscattering impulse response, obtained from an inhomogeneous continuum described by a Gaussian autocorrelation function[6]. Backscattering impulse responses were generated for tissue models under varying degrees of rotation, resulting in sequences of B-scan images suitable for tracking the motion, so long as the tracking algorithms did not fail due to the speckle decorrelation associated with rotation. At this preliminary stage of the study we did not attempt to include in the simulation the artefactual speckle motion which has been noted for tissue rotation in association with point response functions that have a strong curvature[7].

SIMULATION RESULTS

In this test the incremental search tended to display the most favourable overall properties, the locus tending to better match the true locus and searches extending over the largest angles of tissue rotation (Figures 6 and 7). Performance of the fixed pattern search improved with size of the reference ROI (Figure 8). Both rotational step angle (Figure 9) and the size of the search region were important for the incremental search (too large a search area generated false loci).

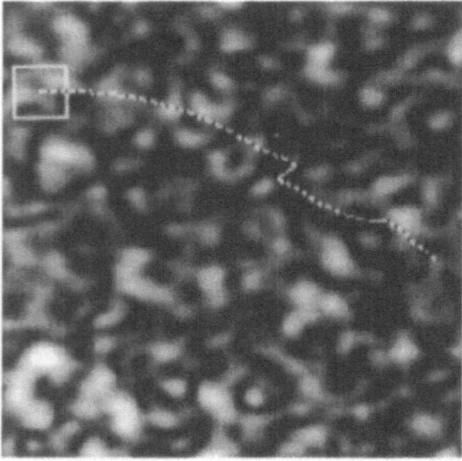

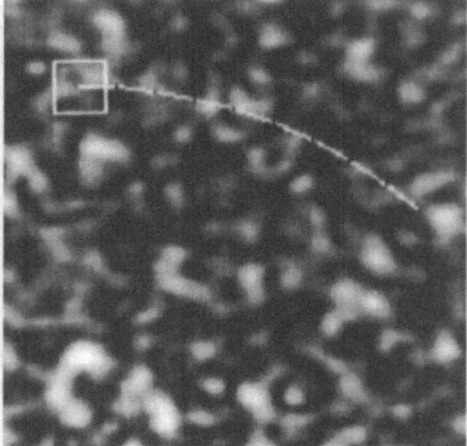

Figure 6. Example showing the locus generated by the fixed pattern search for a 45 degree simulated tissue rotation generated in steps of 1 degree. The size of the window for the reference echo pattern was 20x20 pixels. The true tissue movement was followed until about 26 degrees rotation, after which the speckle decorrelation appeared to influence the result although the tracked locus still moved in broadly the right direction.

Figure 7. Example of the performance of the incremental tracking algorithm for the same simulated tissue region and rotation as shown in Figure 6. The tissue, however, was rotated in steps of 6 degrees. The tracked locus, as shown, followed the true tissue path for the whole of its length and the algorithm performed as well as it did when the tissue was rotated in steps of 1 degree (not shown).

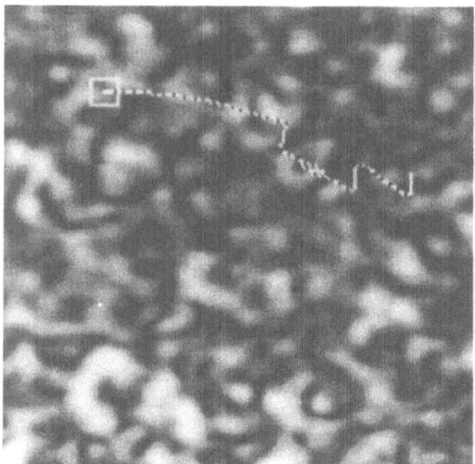 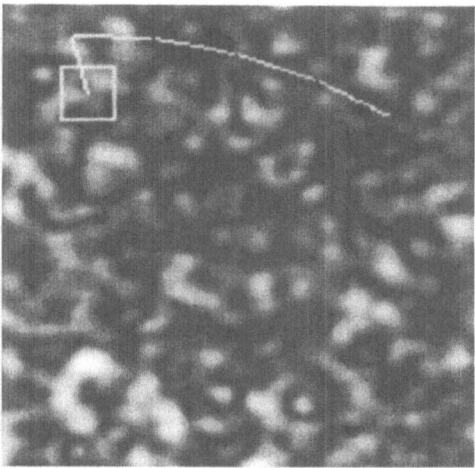

Figure 8. The locus resulting from the fixed reference pattern search and the same simulation as shown in Figure 6 but with a reference window size of 10x10 pixels. The tracked locus followed the true tissue path only until about 20 degrees rotation. However, once again the remaining portion of the tracked path is broadly in the correct direction.

Figure 9. Example for comparison with Figure 7, showing that the incremental tracking method began to fail when the tissue was rotated in steps of 8 degrees.

CONCLUSION

A preliminary study has been conducted of the properties of several approaches to the use of speckle tracking techniques for target tracking in high intensity focused ultrasound therapy of cancer in the liver. Further research is required to evaluate, develop and optimise these procedures. We have been able to show, however, that the methods are feasible, both in principle and in practice.

Eventually, three dimensional real-time ultrasound scanning and echo acquisition will be required for these methods to be fully functional. The success of the incremental tracking algorithm suggests, however, that real-time three dimensional ultrasound imaging may be required only over a relatively small volume which changes its location so that it remains centred around the echo pattern being tracked. In the shorter term it seems feasible that the value of the speckle correlation coefficient for comparing the echo pattern at a fixed point in space with subsequent echoes at the same location may provide a signal for gating the therapy beam so that treatments are given only when the tissue cyclically returns to the same location. Alternatively, in some as yet undetermined number of cases, it appears that target tracking may be possible using current two dimensional real-time scanners by locating a scan plane within which the motion is predominantly confined.

REFERENCES

1. G.R. ter Haar, D. Sinnet, and I. Rivens, High intensity focused ultrasound - a surgical technique for the treatment of discrete liver tumours. *Physics Med Biol.* 34:1743 (1989)
2. N.L. Bush, J.C. Bamber, I. Rivens, and G.R. ter Haar, Acoustic properties of lesions generated with an ultrasound therapy system, *Ultrasound Med Biol.* 19:789 (1993)
3. G.E. Trahey, J.W. Allison, and O.T. von Ramm, Angle independent ultrasonic detection of blood flow. *IEEE Trans Biomed Eng.* 34:965 (1987)

4. L.N. Bohs, B.H. Friemel, B.A. McDermott, and G.E. Trahey, A real time system for quantifying and displaying two-dimensional velocities using ultrasound. *Ultrasound Med Biol.* 19:751 (1993)
5. J.C. Bamber, C. Dance, D.S. Bell, N.L. Bush, Imaging flow and shear by speckle tracking with a neural network. *Brit J Radiol.* 64:651 (1991)
6. J.C. Bamber and R.J. Dickinson, Ultrasonic B-scanning: a computer simulation. *Physics Med Biol.* 25:463 (1980)
7. F. Kallel, M. Bertrand, J. Meunier, Speckle motion artifact under tissue rotation. *IEEE Trans Ultrason Ferroelec and Frequ Cont.* 41:105 (1994)

ON THE POSSIBILITY OF COMPUTER AIDED DIAGNOSIS
OF SILICONE BREAST IMPLANT RUPTURE

F. Forsberg,[1] E. F. Conant,[1] K. M. Russell,[1] and J. H. Moore, Jr.[2]

[1] Department of Radiology
[2] Department of Surgery
   Thomas Jefferson University
   Philadelphia, PA 19107

INTRODUCTION

Silicone breast implants, used for augmentation and reconstruction, were first developed in 1963; an estimated 1 to 2 million implants have been placed in the United States since that time[1]. Recent concern about the safety and complications of breast implants has focused interest on methods to evaluate implant integrity. The incidence of implant failure and silicone leakage is unknown. Rupture may be intracapsular i.e. contained by the surrounding fibrous capsule, or extracapsular meaning free extravasation of silicone into the soft tissues of the breast or axilla. Unfortunately, the ability to accurately diagnose implant rupture by physical examination or mammography is limited[2] Magnetic resonance imaging (MRI) is able to detect both intracapsular and extracapsular silicone leakage[3], but the procedure is time consuming and expensive.

Recent studies have shown that ultrasound is useful in detecting implant rupture[2,3] Both intracapsular and extracapsular silicone leakage can be detected. Ultrasound evaluation of breast implants are rapidly performed and inexpensive relative to MRI. However, ultrasound assessment of implants is operator dependent; there is a learning curve in recognizing the often subtle acoustical changes indicative of silicone leakage. To reduce this operator dependability a system for computer aided diagnosis (CAD) of implant rupture is being developed. The system relies on the clinical observation that the acoustic properties of implants change with the integrity of the devices. Feasibility has been examined by *in vitro* measurement of the speed of sound, the attenuation, and the integrated backscatter of explanted prostheses. These acoustic parameters were then correlated with implant integrity. To assess the potential diagnostic performance of CAD receiver operating characteristic (ROC) analysis was performed.

MATERIALS AND METHODS

In total 45 explanted silicone breast implants were obtained from 28 women; 26 intact prostheses and 19 ruptured. Silicone gel from each was placed in individual acoustic test chambers of known dimensions (3.0 x 3.8 x 3.8 cm) to obtain a standardized test environment. The chambers were sealed with an acoustically transparent membrane. Since silicone gel is highly viscous it is impossible to avoid randomly introducing a few, small airbubbles into the test chamber. Additionally, debris from the surgical procedure will also be present in the case of ruptured implants. Care was taken not to image areas with visible debris, but some extra variation in the ultrasonic properties will exist.

The system used for acquiring RF A-lines consists of an Ultramark 9 scanner (ATL, Bothell, WA) equipped with a digital interface to a data acquisition module (DAM). The scanner utilizes a modified beamformer controller card to transfer RF data to the DAM memory (16 MB). The DAM in turn connects to a PC via an RS232 interface, for control commands, and an RS422 interface for data transfer. The RF A-lines are sampled at 12 MHz (16 bit precision). All measurements were performed with a 5 MHz linear array using a single focal zone and a flat TGC. A standoff pad was used to position the implant sample in the focal region, while a rubber backing layer reduced reverberations from the far end chamber interface. The system was configured to acquire 3 images per implant, each containing 128 A-lines of 1024 samples. In each image 20 equidistant A-lines were selected for post processing. Thus, a total of 60 A-lines per implant were averaged to obtain stable estimates of the ultrasonic properties.

Estimation of Acoustic Parameters

Standard time-of-flight techniques were used to estimate the speed of sound based on the true size of the test chamber. The attenuation parameters of a first order linear model were estimated using the spectral difference method originally developed by Kuc and Schwartz[4]. Segments from the near and far end of the test chamber (at depths z_1 and z_2) were smoothed with a Hamming window, and their power spectra denoted $S_1(f)$ and $S_2(f)$, respectively, calculated. Assuming the implant sample to be a linear system only subject to attenuation, one gets:

$$-\frac{10[\log\{S_2(f_i)\} - \log\{S_1(f_i)\}]}{2(z_2 - z_1)} = \alpha_0 + \alpha_1(f_i - f_c) \tag{1}$$

where f_i consists of equidistant frequencies within the transducer's pass band and the center frequency of the transducer is denoted f_c. The attenuation can be found from Eq. (1) using standard linear regression techniques.

The intensity of a reflected ultrasound wave can be characterized by the integrated backscatter (IB), defined[5] as the frequency average of the backscatter transfer function over the bandwidth of the transducer (BW). This study employed a frequency compensated narrow-bandwidth method[6]. First the power spectrum from echoes originating within the implant ($S_I(f)$) was calculated, over a gate length of approximately 10.7 µs, using an FFT and a Hanning time window. The frequency content of the backscattered signal depends on both the scattering properties of the implant sample and the transducer employed. To compensate for the transducer's electromechanical response a reference power spectrum $S_R(f)$ was obtained from a perfect planar reflector (a steel plate 12.5 mm thick). Hence, the IB is found as:

$$IB(f_c, BW) = \frac{1}{BW} \sum_{i=0}^{N-1} \frac{S_I(f_i)}{S_R(f_i)} \qquad (2)$$

where the customary integration has been substituted for a summation. The BW used in these calculations was 2.2 MHz, which is approximately equivalent to a -7 dB bandwidth.

ROC Analysis

To estimate the potential accuracy of diagnosing breast implant rupture using quantitative ultrasound parameters ROC analysis was performed[7]. Since all implants in this study were surgically removed, their true integrity is known. A commercial software package (LABROC 1; Charles Metz, Chicago, IL) was used to calculate the area under the ROC curve (A_z), which is a measure of the performance of the diagnostic test. The program assumes a binormal model and produces a maximum likelihood estimate of the ROC curve from continuously distributed data.

Combining the results from the different acoustic parameters should increase the overall performance (i.e., the value of A_z). Two methods were employed to determine the optimal combination of acoustic parameters: multiple linear regression and logistic regression. Both techniques establish a quantitative relationship between a group of predictors (the acoustic parameters) and an outcome (i.e., ruptured versus intact). The logistic model predicts the probability of rupture, and the coefficients in this model provide the parameters needed for the ROC curve. Since the outcome is binary the logistic regression is expected to perform the best[8].

RESULTS

All velocities measured were smaller than the average speed of sound in tissue by 200 to 600m/s. Thus direct *in vivo* measurement of the implant size will produce errors up to 40%. In Table 1 the overall mean velocity of intact and ruptured implants is presented. The speed of sound in ruptured implants is larger than in intact ones, which have had much less tissue contact. The difference in mean speed of sound values was assessed with a standard, one-sided t-test assuming equal variances (the variances could not be considered different using an F-test) and was highly statistically significant ($p=0.003$).

Estimating the attenuation coefficients proved difficult. There was no statistically significant difference between intact and ruptured prostheses. The variations encountered were so large that quite often negative attenuation, i.e. enhancement, was measured. The very small signals reflected from the silicone was obscured by large (artifactual) echoes.

The results from the integrated backscatter measurements are presented in Table 2. As with the speed of sound an increase in IB from ruptured implants was found. The difference between the mean IB of ruptured and intact implants was highly statistically significant ($p=0.006$). The range of values measured (-58 to -97 dB) show how weak the silicone signals are; supporting the argument for the failure in measuring attenuation.

To assess the possible performance of diagnosing implant rupture based on the speed of sound and the IB ROC curves were calculated. The curves are presented in Figure 1, and the A_z's were 0.73 for velocity and 0.70 for IB. The combination of speed of sound and IB obtained from a linear regression and a logistic regression, respectively, are also depicted in Figure 1. The A_z's increased to 0.81 and 0.82 as shown in Table 3, where the quantitative relationships between the velocity and the IB can be found as well.

Table 1. Overall mean velocities and the standard deviation of the means.

integrity	E[v] m/s	σ m/s	range m/s	t-test
intact	1060	50.1	966 to 1133	
ruptured	1115	74.3	1015 to 1346	p=0.003

Table 2. Overall mean IB and the standard deviation of the means.

integrity	E[IB] dB	σ dB	range dB	t-test
intact	-83.9	7.94	-96.7 to -68.6	
ruptured	-77.2	9.07	-89.6 to -58.3	p=0.006

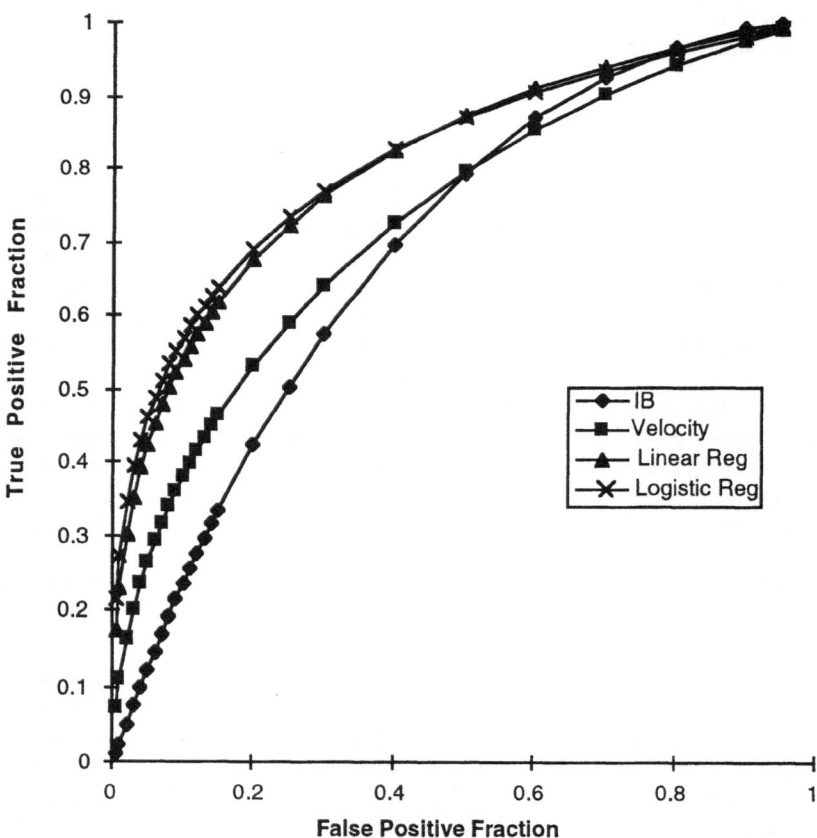

Figure 1. Comparison of ROC curves for diagnosing rupture based on velocity, IB, a combination of the two based on linear regression and a combination of the two based on logistic regression.

Table 3. Parameters used to produce the ROC curves and the resulting A_z's.

Method of combination	ROC curve	A_z
v alone	v	0.73
IB alone	IB	0.70
linear regression	0.0028 v + 0.0188 IB	0.81
logistic regression	0.0194 v + 0.1007 IB	0.82

DISCUSSION AND CONCLUSION

All measurements in this study were performed with a focused transducer. Since the frequency and power of the received signal varies with the focal setting, this may cause problems for the IB and attenuation measurements. To avoid problems all calculations of the IB were performed on data obtained from the transducers focal zone, where the received power is at its maximum. Moreover, the position of the focal zone was kept constant throughout the experiments. Another problem was the random introduction of airbubbles into the test chambers. To compensate for any bias this may have caused in the velocity and IB measurements averaging was performed over 60 A-lines.

Attenuation was clearly not a reliable parameter for *in vitro* implant characterization in this study. Due to the very noisy and weak signals received, it soon became clear that only very limited information could be obtained from estimates of implant attenuation. The problems encountered stem from the implant preparation in general, and the ruptured breast implants in particular. There might still be some value to *in vivo* attenuation estimation in silicone breast implants. To determine this will require further studies.

In the literature only one mentioning of speed of sound measurements in silicone breast implants has been found[2]. Harris and co-workers found the velocity of one intact implant to be 997 m/s. Since no experimental details were given, it is difficult to perform any comparison with the values obtained in this study. However, the velocity 997 m/s does fall within the range measured in these experiments (cf. Table 1).

The reason for the change in acoustic appearance occurring with implant rupture is not currently known[9]. We assume it is caused by the free silicone binding protein to form aggregates[10]. The end result is the same: the effective backscattering cross-section increases due to the increase in scatterer size. This in turn produces the so called "echogenic noise" associated clinically with rupture[2].

The ROC analysis will, obviously, have to be validated on an independent data-set obtained *in vivo* before the results can provide more than an indication of possibilities. Nonetheless, the areas under the ROC curve which were obtained (0.70 to 0.82) are encouraging for the feasibility of quantitative ultrasonic diagnosis of implant rupture. The results are better than the 0.54 value of A_z reported by a group of relatively inexperienced radiologists reading standard ultrasound images[11], and en par with the A_z of 0.77 obtained by a more experienced group of physicians under very controlled circumstances (in an animal model)[12].

The long term objective of this project is to develop a CAD system for *in vivo* detection of breast implant rupture. The statistically significant differences between ultrasonic properties of intact and ruptured implants support the feasibility of using quantitative ultrasound. However, the large overlap of the measured parameters (cf. Tables 1 and 2) will make *in vivo* implant diagnosis on an individual basis more difficult to perform. Most likely, the CAD system will have to include other components, such as patient demographics, to approach the diagnostic performance of MRI. Expansion of the CAD as well as *in vivo* measurements of acoustic parameters are currently being pursued.

In conclusion, changes in speed of sound and integrated backscatter of silicone breast implants, depending on implant integrity, have been demonstrated *in vitro*. The changes were statistically significant; $p=0.003$ and $p=0.006$, respectively. However a good deal of overlap in ultrasonic properties between intact and ruptured implants exists. The ROC analysis suggested that an improvement in diagnostic accuracy (compared to conventional ultrasound) might be achievable; provided the *in vitro* measurements can be repeated *in vivo*. These results indicate the potential for computer aided diagnosis of silicone breast implant rupture performed *in vivo*.

ACKNOWLEDGMENT

Thanks to E. J. Halpern, MD for performing the logistic regression analysis.

REFERENCES

1. D.A. Kessler, The basis for the FDA's decision on breast implants, *N Eng J Med* 326:1713-1715 (1992)
2. K.M. Harris; M.A. Ganolt; K.C. Shestak, H.W. Losken, H Tobon, Silicone implant rupture: detection with US, *Radiology* 187:761-768 (1993).
3. D.P. Gorczyca, S. Sinha; C.Y. Ohn, et al. Silicone breast implants in vivo: MR imaging. *Radiology* 185:407-410 (1992).
4. R. Kuc and M. Schwartz, Parametric estimation of the acoustic attenuation coefficient slope for liver from reflected ultrasound signal,. *IEEE Trans. Sonics. Ultrason.* 26:353-362 (1979).
5. M. O'Donnell; D. Bauwens, J.W. Mimbs, J.G. Miller, Broad band integrated backscatter: an approach to spatially localized tissue characterization in vivo *Proc. IEEE Ultrason. Symp.* 175-178 (1979).
6. H. Rijsterborgh, F. Mastik, C.T. Lancee, P. Verdouw, J. Roelandt, N. Bom, Ultrasound myocardial integrated backscatter signal processing: frequency domain versus time domain, *Ultrasound in Med. & Biol.* 19:211-219 (1993).
7. C.E. Metz, ROC methodology in radiologic imaging, *Invest. Radiol.* 21:720-733 (1986).
8. D.W. Hosmer Jr., and S. Lemeshow. "Applied Logistic Regression," John Wiley & Sons Inc., New York (1989).
9. J.M. Rubin, M.A. Helvie, R.S. Adler, D. Ikeda, US appearance of ruptured silicone breast implants (letter), *Radiology* 190:583-584 (1994).
10. F. Forsberg, E.F. Conant, K.M. Russell and J.H. Moore, Jr., Quantitative ultrasonic diagnosis of silicone breast implant rupture: an in vitro feasibility study, *Ultrasound in Med. & Biol.* (1995). In press.
11. W.A. Chilcote, R. Dowden, D. Paushter, et al. Detection with US of silicone gel breast implant leaks: a prospective analysis (abstract), *Radiology* 189(P):155 (1993).
12. D.P. Gorczyca, N.D. DeBruhl, C.Y. Ahn, A. Hoyt; J.W. Sayre, P. Nudell, M. McCombs, W.W. Shaw, L.W. Basset, Silicone breast implant ruptures in an animal model: comparison of mammography, MR imaging, US, and CT, *Radiology* 190:227-232 (1994).

DETECTION OF IMMUNE REACTIONS IN RENAL TRANSPLANTS BASED ON ULTRASONIC IMAGE ANALYSIS

F.W. Albert,[1] N. Berndt,[1] U. Schmidt[1],
H. Höfer,[2] D. Keuser,[2] and M. Pandit[2]

[1]Klinikum Kaiserslautern, 67653 Kaiserslautern
[2]Universität Kaiserslautern, 67653 Kaiserslautern

1. INTRODUCTION

Transplantation is the preferred therapy for terminal renal insufficiency. In case a transplant is subjected to immunological rejection reactions, timely detection is essential in order to initiate appropriate therapeutic measures. Sonographic investigation, which usually consists of visual examination of the ultrasound scan of a transplant, offers a non-invasive method of detection. To enhance the reliability of the diagnosis, it is expedient to develop methods to quantify the image properties for classifying imaged tissues and tissue changes as a basis for detection of immune reactions. In this paper, a classification method is described in which a vector of parameters characterizing image features of the ultrasound scan of the transplant is processed to obtain a diagnosis. The parametrized image features are chosen first on the basis of the criteria employed by the expert physician for diagnosis and secondly by correlating various parameters with the result of diagnosis by other (invasive) means. The method could be used to classify 83 % of the scans correctly.

2. IMMUNE REACTIONS IN RENAL TRANSPLANTS

The human kidney is bean shaped, 10 - 12 cm. long and 4 - 5 cm broad and weighs about 120 - 200 gm. Fig. 1 shows a section through a kidney. The interior can be divided into two regions: an outer ridge (3) and the inner cortex.

Fig. 2 shows the ultrasound scans of 4 renal transplants two of which correspond to healthy transplants and two to transplants undergoing immune reactions. The various layers in Fig. 1 are also discernible here. Depending on the intensity and the period of the immune reactions it is usual to distinguish three stages [Sandritter, 1983] : a) *hyper-acute rejection* which occurs immediately after transplantation, b) *acute rejection* which develops over a period of several days or weeks subsequent to the transplantation, caused by the infiltration of the globular cells into the renal tissue and c) *chronic immune reaction* which takes place over a period of several weeks. The first and the last types of immune reactions are not recognizable with methods available currently in the ultrasound scans. Consequently, we are concerned only with the acute rejection in the sequel.

Visual Indicators of Acute Immune Reactions in the Ultrasound Image
An acute immune reaction is accompanied by certain changes in the ultrasound image which can be

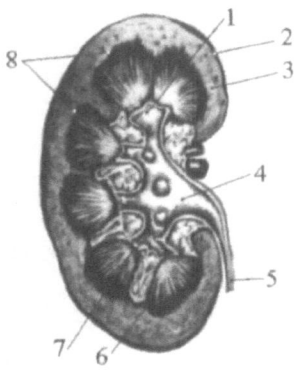

Figure 1   Anatomy of the kidney

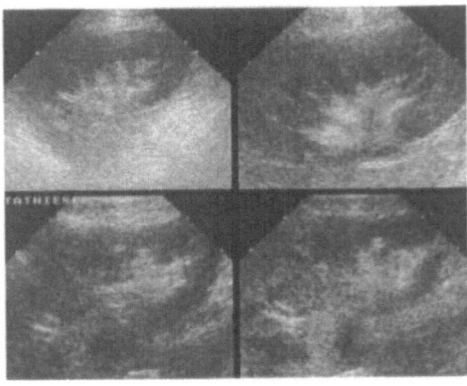

Figure 2   Ultrasound scans of renal transplants

considered under two aspects:

a) variations of certain features of the ultrasound scan of a patient over a period of time

b) deviations of the features with respect to features of scans of healthy patients.

Whereas changes under aspect a) can be identified by comparing the scans of the patient taken at regular time intervals, the deviations of the features of a given scan from the features corresponding to healthy patients is difficult.

The changes can be broadly described as follows

1. An acute reaction is accompanied by an oedema in the transplant which results in a swelling. *This leads to an enlargement of the ultrasound image of the renal transplant.*

2. In a healthy transplant the echogenity of the cortex differs from that of the surrounding tissue. This difference decreases with the onset of immune reactions. *Thus the lowering of the difference in the mean grey level is an indication of an immune reaction.*

3. As the immune reaction affects both the cortex and the surrounding tissue in the same manner, causing a decrease in the difference of the echogenity of these tissues. *This manifests itself as a loss of sharpness of the boundary between the regions.*

4. A further visual criterion is the texture of the region depicting the cortex. *Whereas a fine structure indicates a healthy renal transplant, a coarse structure indicates immune reactions.*

However, these features are not always easy to recognize as can be seen in the scans in Fig. 2. The two scans to the left are those of healthy transplants and those to the right of transplants with acute immune reactions. The features 2. to 4. are considered to develop a method of analysis of the image texture for diagnosing the state of health of a transplant.

3. ULTRASOUND IMAGE ACQUISITION

The success of image processing methods are dependent on the reproducibility of the conditions of imaging. It is necessary to define the image acquisition and standardizing procedures before examining the image features. All ultrasound images of the renal transplants referred to in this paper were obtained with an Acuson 128 XP Ultrasound Scanner and a sector transducer (V 328) with a center frequency of 3.5 mhz. The scans were collected in the clinic with a PC and a frame grabber on magnetic tape and transferred to a Sun workstation equipped with Khoros image processing software. For standardizing the imaging conditions and selecting the region of interest *the setting of the parameters of the scanning device* and *the transducer position* are maintained constant. *Fluctuations due to factors dependent on individual patients* which arise due to voluntary and involuntary movements of tissues and physiological activities of the investigated patient cannot be suppressed. Heart-beat caused cyclic fluctuations which at first sight seem to be avoidable by synchronizing the imaging process with the heart beat. However actual tests revealed that the advantage gained is marginal.

4. PARAMETRIZATION OF THE IMAGE CHARACTERISTICS

The characteristics of the image which we strive to parametrize are first of all those which are implicitly or explicitly assessed by the expert physician to form a diagnosis. In the next stage additional parameters are sought in order to enhance the reliability of the diagnosis. Following aspects are considered:

a) Not one parameter but a set (vector) of parameters is used to represent the image characteristics.
b) Fluctuations of tissue characteristics and their images thwart all efforts to find absolute parameters. One has to resort to *changes* in parameters or *relative* parameters.
c) By choosing homogeneous areas as regions of interest, one can characterize the region of interest with first order and second order statistics of the grey levels. These give reliable results, only if the statistics are related to those of a reference area.

The image features in the region of interest which are chosen to represent the state of health of the transplant pertain to the texture of the boundary region between the cortex and the surrounding tissue. In accordance with the features taken into account by an expert physician (Section 2), the following characteristics are quantified:

a) the *variations* of the mean grey level in the neighborhood of the boundary
b) the distinctness of the boundary
c) the variations of the coarseness of the texture in the neighborhood of the boundary.

4.1 First -order Statistics: Difference of Mean Grey Level in Original Image

First, the region of interest shown in Fig. 2 A1 is chosen. Next, the demarkation between the cortex and the surrounding tissue is identified and segmented. The parameter

p_1 = difference of the mean grey levels on either side of the demarkation boundary,

is calculated and taken to be the one of the characteristic parameters. Fig. 3 shows this parameter for a group of 21 patients.

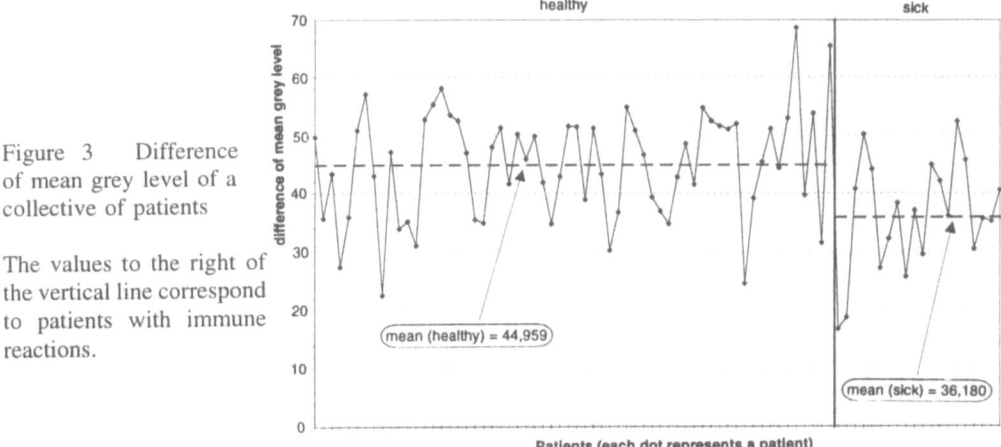

Figure 3 Difference of mean grey level of a collective of patients

The values to the right of the vertical line correspond to patients with immune reactions.

4.2 Parametrizing Second-order Statistics by Means of the Hurst coefficient

One parameter often used in signal processing to capture information of the second order statistics of a stochastic signal is the width of its autocorrelation function, generally represented by twice the value of the argument for which the autocorrelation function decreases to half its maximum [Albert et al, 1993, Greiner, 1994]. As this parameter is not independent of the scaling of the image, we exploit the fact that the intensity of the points of the image can be approximately represented by a fractional Brownian motion process $B_H[k,l]$. In such a case the variance of the difference of intensity at two points $[k,l]$ and $[n,m]$, is given by [Höfer, 1994]

$$E\{|B_H[k,l] - B_H[n,m]|^2\} \propto \sqrt{(k-n)^2+(l-m)^2}^{\,H} = \Delta^H \qquad (1)$$

where H denotes the Hurst coefficient, a parameter which is independent of scale. To determine H, the mean of the squares of the intensity increments for all pairs of points at the distance Δ from each other for various values Δ in the region of interest is calculated. The ordered set of values for increasing Δ is termed as the Brownian feature vector. The second order statistics of a fractional Brownian process are completely characterized by the Hurst coefficient H, which is given by the slope of the Brownian feature vector. The non-isotropic nature of the scan is taken into account by resolving the region of interest (which represents a two-dimensional process) into two one dimensional processes in the axial and the lateral directions. Fig. 4 shows plots of the Brownian feature vector of the axial stochastic process in the regions of interest of the ultrasound scan of the scans A1 and A2 in Fig. 2.

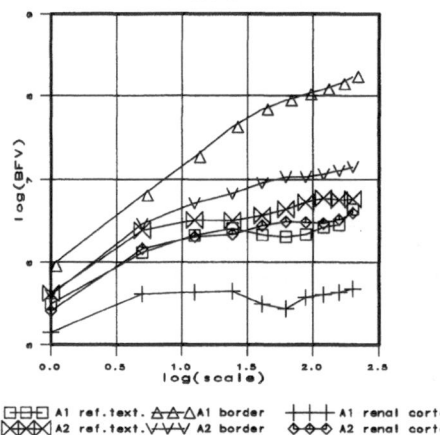

Figure 4   Brownian feature vector of scans A1 and A2

Each of the two regions of interest can be subdivided into three areas: the cortex, the transition area and the surrounding tissue. For both regions of interest, the Brownian feature vector of the transition area is steeper than that of the other two areas. *A quantitative measure for the distinctness of the boundary is the variation of the variance of increments over the range of scales considered, which can be directly expressed by means of the Hurst coefficient.*

The plots of the Brownian feature vectors in Fig. 4 exhibit that they have a different behavior for scale factors below 3, which corresponds to the distance covered by one impulse of the ultrasound pulse train in the tissue. In this region, the process deviates strongly from fractional Brownian motion.

4.3 Parameter Imaging with the Hurst coefficient

The parameter image of the region of interest with the local Hurst coefficient as parameter offers a scale independent representation of texture [Chen, 1989]. The Figs. 5a-d show such parameter images of the 4 regions of interest presented earlier. An abrupt transition in the original image yields a transition area which consists of a bright strip in the parameter image. The boundary area is marked off with two regression straight lines enclosing the strip in the parameter image.

Original
image of
ROI

Parameter
image of
ROI

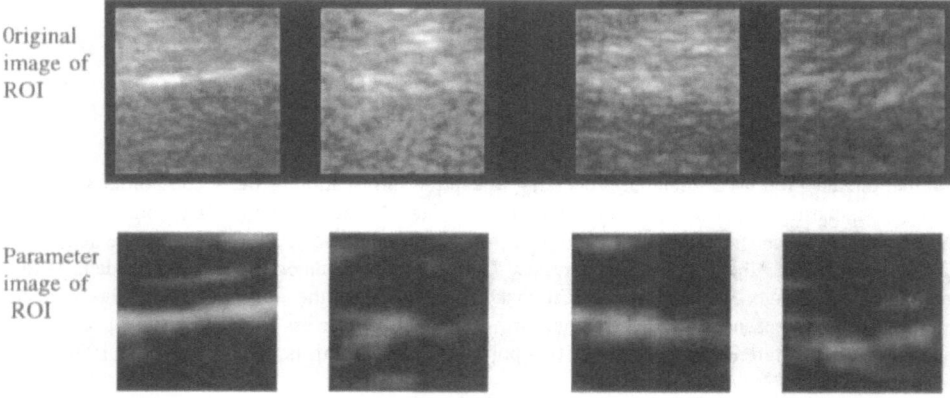

Figure 5 a - d   Original and parameter images of 4 regions of interest

The parameter

p_2 = Hurst coefficient of the original image in the boundary region

is a measure of the abruptness of the transition and is taken to be a further parameter for classifying an image.

Fig. 6 shows the value of parameter p_2 for the images of the patient collective.

Figure 6   Hurst coefficient in the boundary region

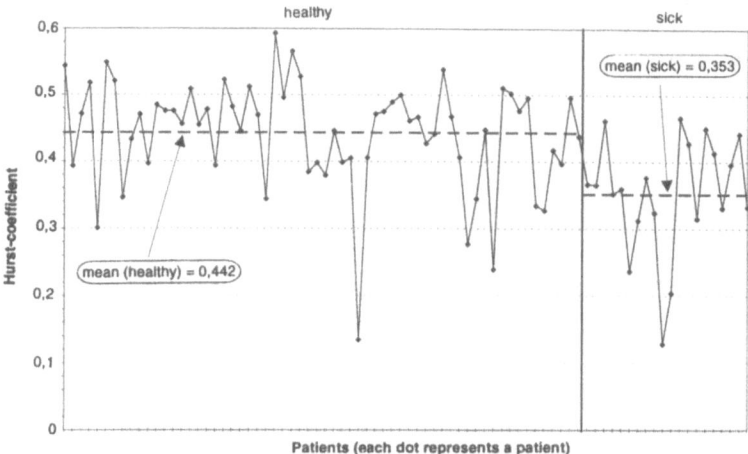

An alternative for quantifying the abruptness of the transition is based on the Brownian feature vector of *the parameter image*. The maxima of the increments in the two directions represent parameters for the quality of the transition region. The parameter

p_3 = product of the maxima is taken as a further measure

for classifyinging the images. Fig. 7 shows the value of p_4 obtained for the scans of the patient collective.

Figure 7   Product of maxima of variances

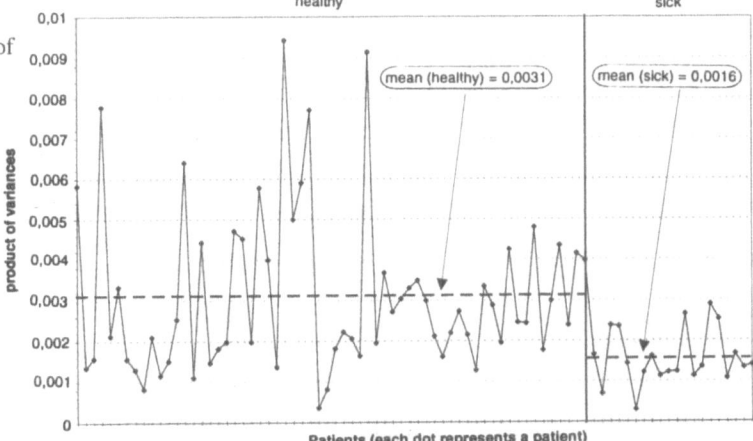

5. A KNOWLEDGE BASED LEARNING CLASSIFICATION SET-UP

For using the image parameters to arrive at a diagnosis, the following classification scheme is employed. First, a knowledge base is set up. This consists in locating the normalized representant parameter vectors $\mathbf{P_H} = [p_{H1}, p_{H1}, ..., p_{HN}]^T$ and $\mathbf{P_S} = [p_{S1}, p_{S1}, ..., p_{HN}]^T$ of the classes of vectors corresponding to the healthy and sick cases and the spread of the parameters about the representant. The representants are formed using the components p_{Hi} and p_{Si} obtained by averaging over the available data of n_H healthy n_S sick transplants respectively. The spread of the parameters are used

255

as weighting indices w_{Hi} and w_{Si} This is the knowledge base with which a diagnosis is formed for a measured parameter vector $p= [p_1, p_2,.., p_N]$. The distances d_h and d_s of the parameter vector under consideration from the representant of the classes corresponding to the healthy and sick transplants respectively are calculated The class at a shorter distance is selected.

6. ANALYSIS OF EXPERIMENTAL DATA

The parameters of the ultrasound scan of 21 patients were calculated and plotted. Fig. 8 shows distance difference d_S - d_H for each of the image parameter vectors. Ideally the transplants corresponding to the distance differences below 0 should be classified as healthy and those corresponding to distance differences greater than 0 should be classified as sick.

Figure 8 Difference of distance from representants

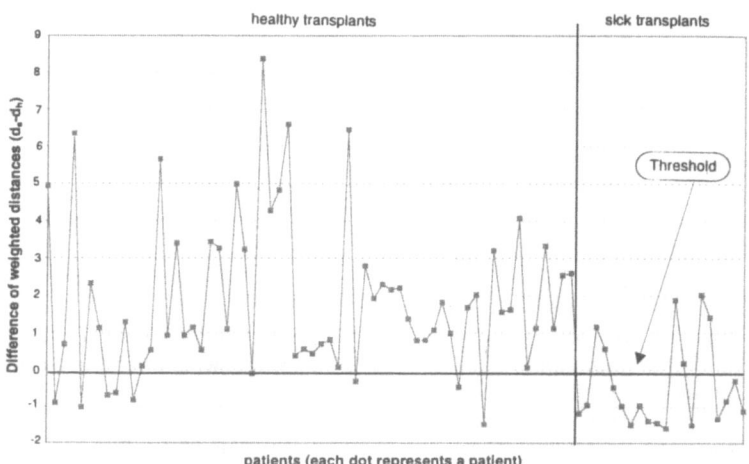

7. CONCLUSIONS

The computerised diagnosis of immune reactions in renal transplants using ultrasound scans seems to offer a fairly reliable non-invarsive means of diagnosis.

The classification method presented can be extended to take further parameters into account. Present work is directed towards incorporating the Doppler signal in the diagnosis.

8. REFERENCES

W. Albert, N. Berndt, T. Greiner, S. Höfer, J. Mauruschat, M. Pandit: Quantitative Ultraschallbildparameter als Indikator von Abstoßungsreaktionen in transplantierten Nieren. Symposium Quantitative Sonographie, Halle (Saale), November 1993.

C.C. Chen, J.S. Daponte, M.D. Fox: Fractal Feature Analysis and Classification in Medical Imaging. IEEE Trans. on medical imaging, Vol. 8, no. 2, pp. 133-142, 1989.

T. Greiner: Methoden der digitalen Bildverarbeitung zur computergestützten Gewebecharakterisierung mit Ultraschall unter besonderer Berücksichtigung der hierarchischen Texturanalyse. Verlag Shaker, Aachen, 1994.

S. Höfer, R. Kumaresan, M. Pandit, W.J. Ohley: Estimation of the Fractal Dimension of a Stochastic Fractal from Noise corrupted Data. AEÜ, Vol. 46, no. 1, pp. 13-21, 1992.

S. Höfer, M. Pandit: Two-dimensional Fractional Brownian Motion and its application in image analysis. Conference of the IPMU, Paris, 1994.

W. Sandritter, C. Thomas: Histopathologie. Schattaner Verlag, Stuttgart, 1983.

IMPROVED LESION DETECTION
BY LEVEL-DEPENDENT SPATIAL SUMMATION

Roland J. Collaris and Arnold P.G. Hoeks

Department of Biophysics
Cardiovascular Research Institute Maastricht
University of Limburg
P.O. Box 616
6200 MD Maastricht, The Netherlands

INTRODUCTION

The detection of focal lesions, such as tumors, is of utmost importance in medical diagnosis. In ultrasonic imaging, however, the perception of subtle differences in local mean gray level is impeded by the grainy character of the images. The speckle pattern exhibited is generally viewed as a form of acoustic noise, resulting from interference of ultrasound echoes coming from tiny structures too close to be resolved.

This paper describes a novel signal postprocessing method, which improves the detectability of slightly contrasting structures in B-mode ultrasound images. A signal distribution, directly related to the two-dimensional envelope distribution comprising a B-mode image, is convolved with a rectangular kernel of which both base area (i.e., window size) and height are modulated by the local signal level. The output includes information regarding the local mean brightness in the original image, which can be displayed as a gray scale image. A formal description of the process is given as well as a summary of the main properties of the output (in Collaris and Hoeks, 1994, referred to as the N distribution). Results are presented of the application of the algorithm to simulated hypoechoic (negative contrast) and hyperechoic (positive contrast) lesions. Thereby, a comparison is made with the optimal conventional low pass filter in lesion detection: the L_2-mean filter (Kotropoulos and Pitas, 1992; Verhoeven and Thijssen, 1993). As a measure of performance, the area under the Receiver Operating Characteristic is used (Swets and Pickett, 1982; Green and Swets, 1988). Also, one concrete example is given of the application of the algorithm to a positive contrast lesion.

METHOD

Input is a two-dimensional amplitude distribution A comprising a B-mode echo image. Each amplitude point $A(x,y)$ is raised to some nonzero power q and multiplied by a positive scaling constant c. Coordinates x and y denote lateral and axial positions, respectively. The output $N(x,y)$ is obtained by convolving the resulting distribution $cA^q(x,y)$ with a level-dependent kernel w_0 according to:

$$N(x,y) = \sum_{x'} \sum_{y'} cA^q(x',y') \cdot w_0\Big((x-x', y-y'); cA^q(x',y')\Big) \tag{1}$$

Kernel w_0 is defined by:

$$w_0\big((x,y); cA^q\big) = \begin{cases} \dfrac{1}{cA^q} & \text{for } |x| \le d_x \text{ and } |y| \le d_y \\ 0 & \text{otherwise} \end{cases} \tag{2}$$

where

$$d_x = \sqrt{\frac{kcA^q}{4}} \qquad \text{and} \qquad d_y = \sqrt{\frac{cA^q}{4k}} \tag{3}$$

are the lateral and axial radii of support, respectively. The factor $k=d_x/d_y$ defines the ratio between the lateral and axial dimensions of the kernel w_0, expressed in number of sample points. The rectangular area O covered by w_0 is dependent on the local signal level and equals $O=4d_xd_y=cA^q$. The volume enclosed by w_0, $O/(cA^q)$, is always equal to one. The convolution described by Eq. (1) can be interpreted as follows. Each amplitude point $A(x,y)$

Table 1. First-order statistics of the main variables involved in the process of level-dependent spatial summation. Γ denotes the gamma function, L_N is the correlation cell size as experienced by the algorithm. Depending on exponent q, L_N ranges between approximately 0.5 and 1 times the mean speckle size (in number of sample points), calculated as the area under the normalized autocovariance function of A.

mean window area	$\overline{O} = c \cdot \overline{A}^q \cdot (2/\sqrt{\pi})^q \cdot \Gamma(\frac{q}{2}+1)$	$(q>-2)$
variance in window area	$\mathrm{var}(O) = c^2 \cdot \overline{A}^{2q} \cdot (2/\sqrt{\pi})^{2q} \cdot \left\{ \Gamma(q+1) - \Gamma^2(\frac{q}{2}+1) \right\}$	$(q>-1)$
mean window length	$2\overline{d_y} = \sqrt{\dfrac{c}{k}} \cdot \overline{A}^{\frac{q}{2}} \cdot (2/\sqrt{\pi})^{\frac{q}{2}} \cdot \Gamma(\frac{q}{4}+1)$	
mean window width	$2\overline{d_x} = k \cdot 2\overline{d_y}$	
mean output level	$\overline{N} \approx \overline{O}$	$(q>-2)$
variance in output level	$\mathrm{var}(N) \approx L_N \cdot \left| 1 - 2^{-q/2} \right| \cdot \overline{N}$	

produces a rectangular window, centered around point *(x,y)*. The area *O(x,y)* covered by the window is proportional to $A^q(x,y)$, while the ratio $k=d_x(x,y)/d_y(x,y)$ of the lateral and axial dimensions of the window remains fixed. Afterwards, the *N* distribution includes the number of times each location was covered by a window. For a positive exponent *q*, high amplitude values produce large windows while low amplitudes produce small ones. Consequently, high *N* values indicate bright regions in the echo image while low *N* values correspond to dark regions. For a negative *q*, the opposite is true.

Table 1 summarizes the expected behavior of the algorithm in response to a stationary amplitude distribution exhibiting a fully developed speckle pattern at a mean level $\bar{A}$. Figure 1 shows the expected transient response along the line *x*=0 in case the algorithm is applied to the part of the stationary amplitude distribution *A(x,y)* for which *x*=0 and *y*≥0. Parameter *c* was adjusted for each *q* to produce a mean axial window length $2\bar{d_y}$ of 10 points. It is observed that the transient response is comparable to the response of a mean filter with a fixed axial window length of 10 points: the two algorithms produce the same slope at *y*=0, although for *q* close to -2 and *q*>5, the transition of the present algorithm exhibits long tails.

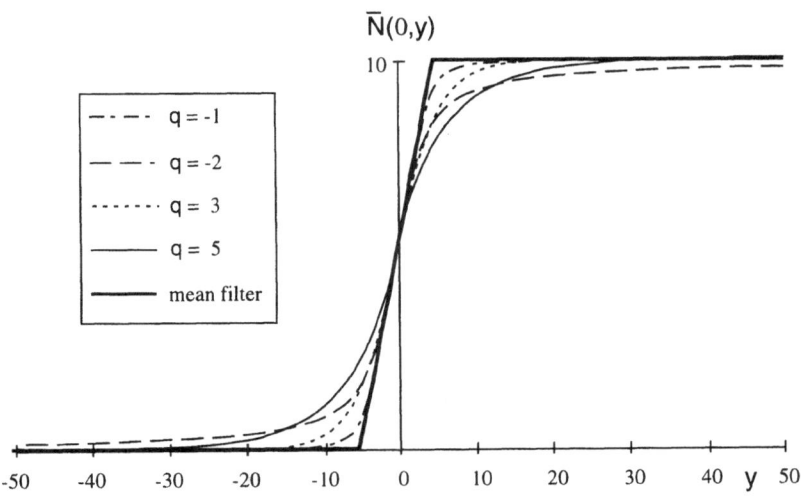

Figure 1. The expected transient response along the line *x*=0 for four different values of exponent *q*. The algorithm was adjusted to produce a mean axial window length of 10 points at (*x*=0, *y*≥0). The bold line shows the expected response of a mean filter with a fixed axial window length of 10 points.

LESION DETECTION

Two regions are identified in a lesion image as modeled by signal detection theory: one consisting of a signal plus noise (the lesion), the other only containing noise (the background). A complete description of how well the signal plus noise part can be distinguished from the noise only part is provided by the ROC curve, being a plot of the false positive fraction (*FPF*, the fraction of background points falsely recognized as lesion points) versus the true positive fraction (*TPF*, the fraction of correctly identified lesion points) as a function of a threshold applied to the image. The area A_{ROC} under the ROC curve has been recommended as a one-number summary of the curve (Swets and Pickett, 1982; Wagner, 1986; Green and Swets, 1988). It equals the fraction of correct answers in a two-alternative forced choice experiment and ranges from 0.5 in the pure guessing case to 1 in the perfect performance case. In the present paper, A_{ROC} is used to compare the performance of the L_2-

mean filter with the algorithm for level-dependent spatial summation in a lesion detection task.

Ten independent realizations of the same lesion were generated on a PC by convolving a 2-dimensional rf point spread function with spatial particle distributions (Bamber and Dickinson, 1980). The resulting images all contain a central, circular, hypoechoic (negative contrast) lesion with a diameter of 2 cm in a homogeneous background of 6x6 cm^2. Lesion scatter strength is 3 dB below background scatter strength. Particle density equals 26 per resolution cell. The number of parallel envelope lines is 113, the number of sample points per envelope line is 1024. The speckle size, measured as the area under the normalized autocovariance function, equals 3.6 sample points ($\approx$ 2 mm) in the lateral direction and 7.7 points ($\approx$ 0.5 mm) in the axial direction. In the same way, a similar set of ten independent hyperechoic (positive contrast) lesions was generated. In this set, lesion scatter strength is 3 dB above background scatter strength.

First, a 3 points lateral x 33 points axial L_2-mean filter was applied to the two sets of 10 spatial amplitude distributions. For each of the 20 filtered amplitude distributions obtained, an ROC curve was calculated: 100 different thresholds were applied to each distribution (corresponding to a single processed lesion image), ranging between the minimum and maximum signal levels present. Each threshold yielded a pair (*FPF*, *TPF*). All pairs together made up one ROC curve. Numerical integration of the curve yielded an estimate of A_{ROC}. This way, 10 values of A_{ROC} were obtained per set. The median values as well as the first and third quartiles are presented in Table 2 (first row).

Second, the algorithm for level-dependent spatial summation was successively applied to the 20 raw amplitude distributions. On the basis of the expression of $\overline{N}$ in Table 1, the algorithm could be adjusted to produce a mean N value (as measured over either the lesion area or the background area, whichever produces the largest $\overline{N}$) equal to the fixed window area of the L_2-mean filter. As a result, the number of summations involved in the convolution of Eq. (1) is comparable to the number of summations required by the L_2-mean filter. Windows were clipped at O_{max}=5000, which prevents incidental outliers to become extremely large. It is noted that the mean lateral and axial window dimensions, $2\overline{d_x}$ and $2\overline{d_y}$, respectively, were smaller than the fixed dimensions of the L_2-mean filter (which follows from the expressions in Table 1), indicating that edge blurring tended to be less in the N distributions than in the L_2-mean filtered amplitude distributions. The results of the ROC analysis are summarized in Table 2 for a number of different exponents q. A difference in the

Table 2. Area A_{ROC} under the Receiver Operating Characteristic as a measure of performance in lesion detection. The upper row shows, as a reference, the performance of a 3x33 points L_2-mean filter. The present algorithm was adjusted to produce a mean N level equal to 100. Parameter k equaled 0.1. Results are presented for various q 's.

	A_{ROC} over 10 negative contrast lesions			A_{ROC} over 10 positive contrast lesions		
	1st quartile	median	3rd quartile	1st quartile	median	3rd quartile
L_2-mean	0.8339	0.8480	0.8713	0.8442	0.8579	0.8682
q=-1.8	0.9316	0.9520	0.9679	0.9041	0.9158	0.9373
q=-1	0.8794	0.8989	0.9249	0.8749	0.8836	0.8949
q=-0.5	0.7905	0.8051	0.8337	0.7862	0.7980	0.8140
q= 0.5	0.7913	0.8005	0.8213	0.7988	0.8069	0.8146
q= 1	0.8584	0.8717	0.8892	0.8654	0.8771	0.8846
q= 2	0.8959	0.9082	0.9214	0.9080	0.9195	0.9225
q= 3	0.9225	0.9292	0.9379	0.9407	0.9424	0.9515
q= 5	0.9444	0.9516	0.9554	0.9722	0.9737	0.9821

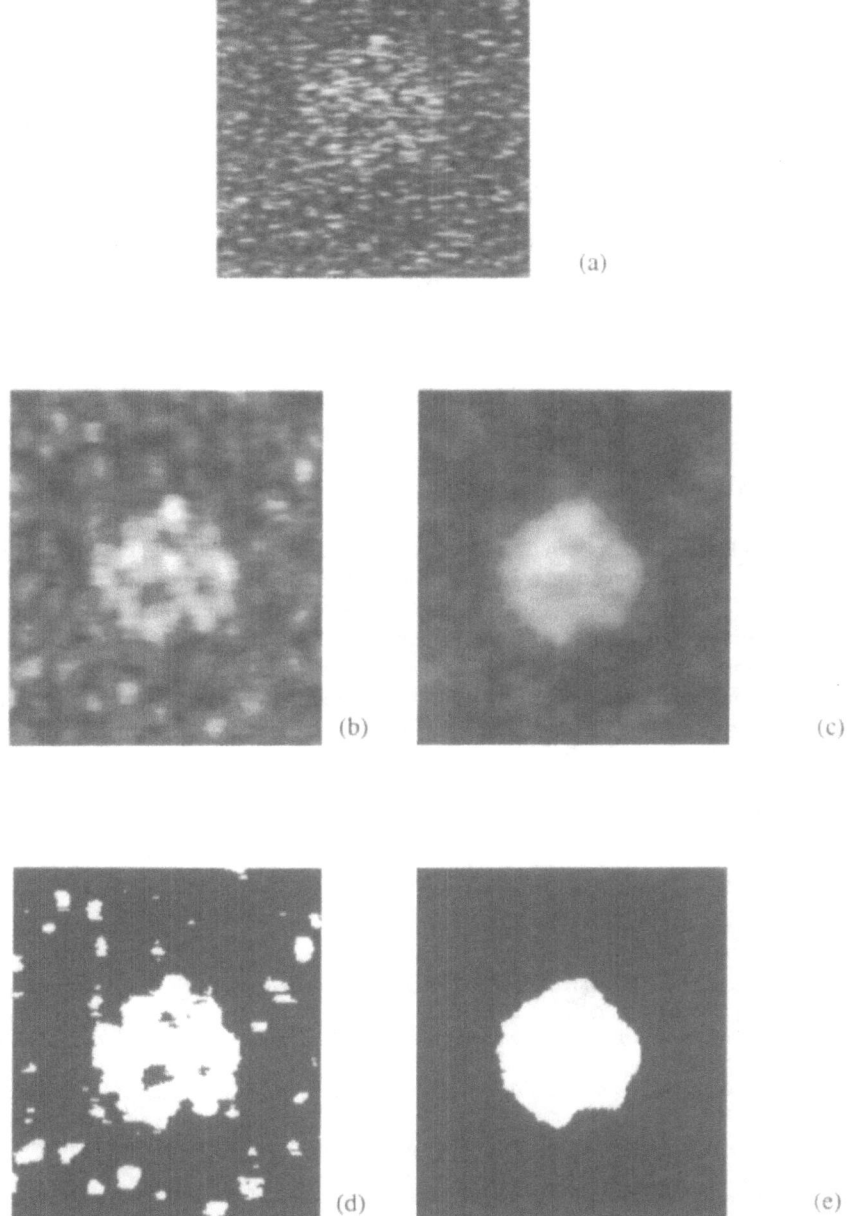

Figure 2. (a) Central part of one of the simulated positive contrast lesion images. (b) Response of a 5x55 points L$_2$-mean filter. (c) Response of the algorithm for level-dependent spatial summation, with q=5, k=0.1 and with $\overline{N}_{lesion} \approx 275$. (d) Binary image obtained by thresholding image (b). (e) Binary image obtained by thresholding image (c).

median value of a A_{ROC} of the order of 0.1 is noted for q=-1.8 and q>3 (negative contrast lesions) and q>3 (positive contrast lesions). Further, it is noted that for a negative q, the performance of the algorithm is best for negative contrast lesions, while for a positive q, the performance is best for positive lesions. For both negative and positive q 's, the deteriorated performance occurs if the background produces, on average, larger windows than the lesion. Especially for those q 's for which the transient response has long tails (q close to -2 and q close to 5) even relative distant background points may contribute to the interior of the lesion, this way increasing $\overline{N}_{lesion}$ and thus decreasing lesion contrast. This effect can be reduced by clipping the window area O, which causes the long tails in the edge response to disappear while the response in the vicinity of the edge remains unaffected.

Figure 2 gives a concrete example of lesion detection. Figure 2a shows one of the positive contrast lesion images simulated. Figure 2b displays the response of a 5x55 points L_2-mean filter and figure 5c shows the output of the present algorithm, adjusted with q=5, k=0.1 and a mean lesion window area $\overline{O}_{lesion}$ ($\approx \overline{N}_{lesion}$) equal to the fixed window area of the L_2-mean filter: 275. Note that the lesion boundary seems better defined in the N distribution than in the L_2-mean filtered amplitude distribution. The binary images of figures 5d and 5e were obtained by thresholding the images of figures 5b and 5c, respectively. The threshold applied to the output of the L_2-mean filter produced a TPF of 87 percent and a FPF of 7.5 percent. The threshold on the N distribution yielded TPF=94 percent and FPF=0.7 percent, demonstrating the superior performance of the present algorithm.

CONCLUSION

A novel technique has been developed for postprocessing spatial amplitude distributions. The output can be displayed as a gray scale image, exhibiting an improved detectability of focal lesions. An ROC analysis has shown that the present algorithm is superior to the L_2-mean filter in lesion detection: an improvement in the area under the ROC curve of the order of 0.1 was observed. The difference in performance is largest for exponent q between -2 and -1 and for q between 3 and 5. The algorithm is adjusted in the same way as a moving average filter: one parameter (c) determines the mean window area, and one parameter (k) determines the ratio of the lateral and axial window dimensions. If the algorithm is adjusted to produce a mean window area equal to the fixed window area of a moving average filter, the two methods use approximately the same number of summations in the processing of a frame, indicating that real-time implementation of the present algorithm is possible.

REFERENCES

Collaris, R.J., Hoeks, A.P.G., 1994, Ultrasonic B-mode image enhancement based on level-dependent spread functions, *Ultrasonic Imaging 16*: 205-230.
Bamber, J.C., Dickinson, R.J., 1980, Ultrasonic B-scanning: a computer simulation, *Phys. Med. Biol. 25*: 463-479.
Green, D.M., Swets, J.A., 1988, Signal Detection Theory and Psychophysics, Peninsula Publishing, Los Altos.
Kotropoulos, C., Pitas, I., 1992, Optimum nonlinear signal detection and estimation in the presence of ultrasonic speckle, *Ultrasonic Imaging 14*: 249-275.
Swets, J.A., Pickett, R.M., 1982, Evaluation of Diagnostic Systems, Academic Press, New York.
Verhoeven, J.T.M., Thijssen, J.M., 1993, Improvement of lesion detectability by speckle reduction filtering: a quantitative study, *Ultrasonic Imaging 15*: 181-204.
Wagner, R.F., 1986, Fundamentals and Applications of Signal Detection Theory in Medical Imaging, *in Proc. SPIE 626*, pp. 755-761.

THREE-DIMENSIONAL ECHOCARDIOGRAPHY: *IN VITRO* VALIDATION OF LEFT AND RIGHT HEART CAVITY VOLUMES

Riccardo Pini,[1] Giuseppe Giannazzo,[1] Mauro Di Bari,[1] Francesca Innocenti,[1] Leonardo Masotti,[2] and Richard B. Devereux [3]

[1]Institute of Gerontology and Geriatrics, University of Firenze, Firenze, Italy
[2]Department of Electronic Engineering, University of Firenze, Firenze, Italy
[3]Division of Cardiology, The New York Hospital-Cornell Medical Center, New York, NY, U.S.A.

INTRODUCTION

Two-dimensional echocardiography (2D-Echo) is currently the imaging modality most frequently used in cardiology due to its simplicity, lack of ionizing radiation, and relative low cost. However, 2D-Echo allows the visualization of only tomographic planar sections of the heart; thus, to obtain a complete evaluation of the heart anatomy and function, the physician must reassemble mentally a three-dimensional (3D) model from multiple two-dimensional (2D) images. Moreover, 2D-Echo relies on geometrical assumptions for the determination of heart chamber volumes and thus presents considerable measurement error, especially for the right ventricular and atrial volume determination[1].

Three-dimensional echocardiography (3D-Echo) may avoid the need for geometrical assumptions, thereby allowing accurate evaluation of chamber size and shape, even in the case of cavities with irregular or distorted geometry. We have previously described a new system for transthoracic 3D-Echo that allows the visualization of left ventricular wall and cavity in normal volunteers and in patients[2] as well as the accurate *in vitro* measurements of variably shaped test objects[3]. Thus, the present study was undertaken to evaluate the accuracy of *in vitro* measurements of left and right atrial and ventricular volumes obtained by our 3D-Echo.

METHODS

A total of 10 freshly excised sheep hearts were fixed in formalin for 48 hours, during pressure expansion to avoid chamber collapse. To obtain the ventricular and atrial casts, the heart cavities were filled with a heated 5% solution of agarose in water. After solidification of agarose with cooling, the excised hearts were suspended and immersed in a fish tank filled with water to acquire the echocardiographic images.

Acoustical Imaging, Vol. 22, Edited by P. Tortoli
and L. Masotti, Plenum Press, New York, 1996

To perform the 3D-Echo reconstruction of the hearts, we developed an echocardiographic system based on a 3.5 MHz dynamically focused anular array transducer rotating 180 degrees around its longitudinal axis with angular increments of 3.6 degrees. The ultrasonic beam profile exhibited a regular shape along the entire depth used to visualize the hearts, with a lateral resolution of 1.3 mm at -6 dB[4]. The transducer is connected to an echocardiographic system (SIM 5000, ESAOTE Biomedica) modified adding an electronic circuit that, with computer assistance (8086 personal computer), controls the acquisition process and displays the scanning plane number on the screen where the 2D-Echo images are displayed in real time; the same images are stored on 1/2 inch video tape at 25 frames/s.

For each heart cavity, the 50 videotaped 2D-Echo images are digitized using a 80386 personal computer (Compaq Computer Corporation) with 768x576 pixel resolution and 256 gray levels frame grabber (GTI/Freeland Medical Division). The 50 images acquired in cylindrical coordinates are processed by an algorithm for linear interpolation to reconstruct a 3D cone of information in Cartesian coordinates. Because the 2D-Echo fan-shaped images occupied only the central area of the screen, the computer uses the central 256x576 pixels of the original images to reduce both the computational time and the disk space without loosing important echocardiographic information. Thus, each 3D-Echo image has stored in matrices of 256x256x576 pixels and 256 gray levels preserving the spatial resolution of the original 2D-Echo images. From the 3D matrices stored in the computer, 2D-Echo images in any plane can be derived and visualized. In this *in vitro* validation, only one 3D-Echo matrix of data was reconstructed for each left and right heart cavity. The accuracy of the 3D reconstruction was tested with a standard ultrasound phantom (RMI, model 412A), and with variably shaped test objects immersed in a fish tank.

From the 3D-Echo matrix of each heart cavity, up to 48 parallel tomographic planes were derived beginning parallel to the mitral or tricuspid valve plane, and extending throughout the entire cavity in the short axis orientation with 256x576 pixel resolution. Slice thickness varied from 1.15 to 1.77 mm. The images were calibrated for distance and the volume in mm^3 of each voxel was calculated. For each short axis image, an observer, who did not know the actual volumes, traced manually the endocardial border using a mouse connected to the frame grabber. The cavity volumes were calculated summing the volumes of all the individual slices derived from the 3D-Echo reconstruction of each cavity.

After imaging the excised hearts, the myocardium was incised and carefully peeled off the cavity casts. The actual volume of the cavity casts were measured by water displacement in a graduated baker.

Data are presented as mean±standard deviation. Linear regression equations were calculated to verify the correlation between the measured and the actual volumes. The method of Bland and Altman[5] was used to assess whether 3D-Echo volumes systematically under- or overestimated the actual volumes. Mean biases were compared to zero value by one-sample Student's t test. The bias, or systematic error, was also expressed as a percentage of the anatomical volume [Percent error=(Measured volume-True volume)*100/True volume], and the mean percent bias with 1 SD was calculated for each heart cavity. The imprecision, or random error, was defined as the absolute difference between the anatomical volume derived from the regression line equation calculated for each heart cavity and the true anatomical volume; the imprecision was expressed as the percentage of the anatomical volume. A p value of <0.05 was accepted as significant.

RESULTS

The reconstruction of the heart cavities derived from the 3D-Echo data sets reproduced accurately the shape of the actual casts. The correlation between measured and true volumes was excellent for all four heart cavities (Table 1); the slopes and intercepts of the 4

Table 1. Linear correlations between 3D-Echo and true anatomic volumes for each heart cavity.

Heart Cavity	r	Regression Equation	SEE (ml)
Left Ventricle	0.99*	3D-Echo=-0.49+0.98*Anatomy	1.97
Right Ventricle	0.97*	3D-Echo=3.36+0.94*Anatomy	2.80
Left Atrium	0.99*	3D-Echo=-0.39+1.03*Anatomy	1.74
Right Atrium	0.98*	3D-Echo=-0.48+0.97*Anatomy	2.31

*$p<0.0001$

regression lines were statistically equal to the line of identity with a SEE ranging from 1.74 to 2.80 ml. 3D-Echo exhibited no systematic under or overestimation of the true volumes (Table 2), with a percent bias ranging from 0.6 to 5.2% and that remained constant over the entire range of explored anatomical volumes. The imprecision ranged from 4.1 to 6.5%.

Table 2. Bias and imprecision of 3D-Echo volumes for each heart cavity.

Heart Cavity	Bias (ml)	Imprecision (ml)
Left Ventricle	-0.93±1.86*	1.35±1.20
Right Ventricle	0.30±2.75*	2.05±1.69
Left Atrium	0.38±1.67*	1.16±0.99
Right Atrium	-1.49±2.21*	1.88±0.97

*$p=NS$

DISCUSSION

The present *in vitro* study demonstrated that our 3D-Echo system based on a transthoracic rotating transducer allows for excellent measurements of left and right ventricular and atrial volumes of excised hearts. Interestingly, no substantial over- or underestimate by 3D-Echo was detected, even for the trabeculated, irregularly shaped right ventricle. Indeed, the SEE was similarly low for all four cavities. Compared with probes connected to mechanical arms[6-8] or to an acoustic ranging device[9-12], our 3D system based on a rotating transducer presents the advantage of fully preserving both freedom of transducer movements and portability of the echographic instrumentation. A recently developed electromagnetic locator device allows for free movements of the transducer and maintains the echograph portability[13]; unfortunately, it requires the absence of metallic objects in the examining area. Moreover, the rotational probe we developed allows for the reconstruction of 3D matrices with the same spatial and temporal resolution as the original 2D images, while with the previously mentioned probes only wire frame models of the ventricular cavities can be obtained. Since our probe rotates with angular increments which are narrower than the transducer lateral resolution, the 2D tomographic images derived from the reconstructed 3D matrices are visually indistinguishable from the standard 2D images. The possibility to derive, from the 3D matrices, tomographic sections with any spatial orientation and with the same visual characteristics as the original 2D images, allows the selection of planes correctly oriented even in patients with non-standard acoustical windows. As demonstrated in a previous clinical series of 40 patients studied in the Echocardiography Laboratory of the New York Hospital-Cornell Medical Center[2], our system is easily applicable to a majority of subjects. Moreover, our 3D system requires a short acquisition time (less than 2 minutes) and allows for an immediate check of the technical quality of the study. In fact, immediately at the end of the acquisition, the operator can compare the first (0 degree rotation) and the last tomographic section (180 degree rotation) to verify the transducer stability before the reconstruction of the 3D matrices.

In conclusion, our 3D transthoracic echocardiographic system combines excellent measurement of atrial and ventricular volumes with the computed reconstruction of 3D volume images that maintain the same spatial and temporal resolution as the original 2D images without cumbersome external reference systems.

REFERENCES

1. W. Bommer, L. Weinert, A. Neumann, J. Neef, D.T. Mason, and A. DeMaria, Determination of right atrial and right ventricular size by two-dimensional echocardiography, *Circulation* 60:91 (1979).
2. A.S. Katz, D.C. Wallerson, R. Pini, and R.B. Devereux, Visually determined long- and short-axis parasternal views and four- and two-chamber apical echocardiographic views do not consistently represent paired orthogonal projections, *Am. J. Noninvas. Cardiol.* 7:65 (1993).
3. G.A. Mensah, R. Pini, E. Monnini, L. Masotti, K.L. Novins, D.P. Greenberg, B. Greppi, M. Cerofolini, and R.B. Devereux, Three dimensional echocardiographic reconstruction: experimental validation of volume measurement, *J. Am. Coll. Cardiol.* 17:291A (1991).
4. R. Pini, L. Ferrucci, M. Di Bari, B. Greppi, M. Cerofolini, L. Masotti, and R.B. Devereux, Two-dimensional echocardiographic imaging: *in vitro* comparison of conventional and dynamically focused annular array transducers, *Ultrasound. Med. Biol.* 13:643 (1987).
5. J.M. Bland, and D.G. Altman, Statistical methods for assessing agreement between two methods of clinical measurement, *Lancet* 1:307 (1986).
6. D.J. Skorton, K.B. Chandran, P.E. Nikravesh, N.G. Pandian, and R.E. Kerber, Three-dimensional finite element reconstructions from two-dimensional echocardiograms for estimation of myocardial elastic properties, *in:* "Computers in Cardioliology," IEEE Computer Society Press, Los Alamitos CA, U.S.A. (1981).
7. E.A. Geiser, L.G. Christie, D.A. Conetta, R. Conti, and G.S. Grossman, A mechanical arm for spatial registration of two-dimensional echocardiographic sections, *Cathet. Cardiovasc. Diagn.* 8:89 (1982).
8. H. Sawada, J. Fujii, K. Kato, M. Onoe, and Y. Kuno, Three dimensional reconstruction of the left ventricle from multiple cross sectional echocardiograms. Value for measuring left ventricular volume, *Br. Heart J.* 50:438 (1983).
9. W.E. Moritz, A.S. Pearlman, D.H. McCabe, D.K. Medema, M.E. Ainsworth, and M.S. Boles, An ultrasonic technique for imaging the ventricle in three dimensions and calculating its volume, *IEEE Trans. Biomed. Eng.* 30:482 (1983).
10. M.D. Handschumacher, J.P. Lethor, S.C. Siu, D. Mele, M. Rivera, M.H. Picard, A.E. Weyman, and R.A. Levine, A new integrated system for three-dimensional echocardiographic reconstruction: development and validation for ventricular volume with application in human subjects, *J. Am. Coll. Cardiol.* 21:743 (1993).
11. P.M. Sapin, K.D. Schroeder, M.D. Smith, A.N. DeMaria, and D.L. King, Three-dimensional echocardiographic measurement of left ventricular volume *in vitro*: comparison with two-dimensional echocardiography and cineventriculography, *J. Am. Coll. Cardiol.* 22:1530 (1993).
12. S.C. Siu, M. Rivera, J.L. Guerrero, M.D. Handschumacher, J.P. Lethor, A.E. Weyman, R.A. Levine, and M.H. Picard, Three-dimensional echocardiography. *In vivo* validation for left ventricular volume and function, *Circulation* 88:1715 (1993).
13. M.D. Handschumacher, L. Jiang, M.Y. Lee, M.J.A. Williams, T. Svizzero, and R.A. Levine, Accuracy of three-dimensional echocardiographic reconstruction by electromagnetic positional location: *in vivo* validation for right ventricular volume, *Circulation* 90:I-338 (1994).

ADVANCES IN TISSUE ELASTICITY RECONSTRUCTION USING LINEAR PERTURBATION METHOD

F. Kallel[1], M. Bertrand[1,2], J. Ophir[3] and I. Céspedes[3]

[1]Institut de génie biomédical, École Polytechnique, C.P./PO Box 6079, succ. Centre-ville, Montréal, Québec, Canada H3C 3A7

[2]Institut de Cardiologie de Montréal, 5000 Bélanger Est, Montréal, Québec, Canada H3T 1C8

[3]University of Texas Medical School, 6431 Fannin St., Houston, TX 77030, USA

INTRODUCTION

Elastography, recently proposed by Ophir et al.,[1] is a promising new acoustic imaging modality for tissue characterization. For example, it was proposed to use such technique for breast cancer detection on the basis of tissue hardness being indicative of the presence of tumor, as the practice of palpation examination indicates[2]. With this modality, tissue elastic properties are revealed through assessments of tissue displacements induced by a small external quasi static compression; specifically, the pre- and post-compression radio frequency (RF) ultrasound signals are used to estimate the axial tissue displacement and the associated strain component. Under the assumption of a constant stress-field, the strain-field can be interpreted as a relative measure of elasticity distribution; the strain being large in compliant (i.e. soft) tissue and small in a rigid (hard) one. This strain field visualized as a gray level image is termed an elastogram.[1]

Generally, the stress field would not be constant, being dependent on the boundary conditions[3] and, more importantly, on the elasticity distribution itself. From a practical stand-point, this leads to fundamental limitations on the strain/Young's modulus contrast-transfer efficiency (CTE), as shown by Ponnekanti et al.[4] and Kallel et al.[5] also, CTE itself is difficult to predict in situations where a complex arrangement of elastic inhomogeneities leads to a complex strain pattern. Hence it appears that elasticity imaging does require an approach that can go beyond strain imaging.

Indeed, a few groups including the authors recently proposed to compute tissue elasticity distribution within the framework of inverse problem (IP) solving.[6,7,8,9,10] The approach adopted by[6,8,9,10] is based on deriving a set of linear equations from a discrete model of the elasticity equations, where the unknowns are the spatial derivatives of the Young's modulus and the coefficients are the strains and their spatial derivatives. Under, the assumption of a plane-strain state[6,8] or a plane stress-state[10], both the axial and lateral tissue displacements (strains) are used for the elasticity reconstruction. The method we proposed is quite different in that it is based on a regularized linear perturbation approach (similar to a method used in electrical impedance tomography[11]) and can use only the axial tissue displacement field to solve the IP in elastography.[7]

In our earlier works, it was assumed that the contrast of the inclusions to be reconstructed was small.[7] Recently, we extended our method to reconstruct iteratively elastic inhomogeneties of arbitrary contrast[12], assuming on the other hand that the force distribution under the compressor was known. In practice, however it is generally the displacement of the compressor which is the known boundary condition. In the following paper, using a penalty technique we illustrate how it is possible to solve the inverse problem in elastography when it is the displacement rather than the force distribution under the compressor which is measured. Using computer simulations and clinical data we illustrate the potential of this proposed method.

SOLVING THE INVERSE PROBLEM IN ELASTOGRAPHY

The approach that we recently proposed to solve the inverse problem in elastography is based on the use of a linear perturbation method. It consists essentially of minimizing the least squares error between observed and predicted displacement field. The predicted displacement field is obtained using a theoretical model of the elasticity equations (constitutive equations). This model may be represented by a general mapping function "$\Im$" that maps a tissue elasticity distribution into a displacement field. Mathematically this may take the following form

$$\Im : \Re^{p+} \to \Re^3$$
$$E \mapsto \Im(E) = U = (u_1, u_2, u_3) \tag{1}$$

where "$\Re^{p+}$" represents the set of the real positive numbers where the tissue elasticity distribution is defined, "p" defines the number of parameters needed to fully characterize the tissue elasticity distribution; as will be seen below this number depends on the model. The quantity $U = (u_1, u_2, u_3)$ is the 3-D tissue displacement vector field. The mapping function itself $\Im$ is defined by the constitutive equations (stress/strain relationships) and the boundary conditions.

Defining general constitutive equations is a complex matter for soft tissues, these being nonlinear, viscoelastic and anisotropic materials. However under certain stress/strain conditions, soft tissues can be thought as being linear elastic materials for which the generalized Hooke's law is used as constitutive equations. In that case, 21 constants are needed to fully characterize the tissue elastic properties. As is often done, these are reduced to 9, 5, or 2 constants for orthotropic, transversely isotropic and isotropic material respectively.[13] When the tissue behaves as a non linear material the constitutive equation should be determined experimentally using a given strain-energy density function.[14]

In this paper numerous simplifications of the forward problem are made. Indeed, the tissue is assumed to behave as an inhomogeneous isotropic linear elastic medium.[9,10,12] Moreover, the ideal tissue is assumed to be incompressible. Hence, only one parameter is needed to define the tissue elastic properties. This parameter is known as the shear modulus (μ) or the Young's modulus (E). For an incompressible material for which the Poisson's ratio ν is equal to 0.5, the Young's modulus is equal to 3μ. In our work, in order to simplify the numeric implementation of the model we assumed a Poisson's ratio close to 0.5 (0.495) thus relaxing the incompressibility tissue condition. In addition, since only the components of the displacement field in the acoustical scanning plane can be measured, we assume a plane-strain state problem for which out of plane motion are neglected.[6,9,12]

After stating and fully defining a forward problem in elastography we now can solve the corresponding inverse problem. Solving this inverse problem may be considered as an optimization problem for which the solution is defined as that which best reproduces an observed set of internal displacements. For such optimization problem, we need to define an objective function to be minimized.[15] The least squares error (or the generalized least squares error) is often used as an objective function since it provides an explicit analytic formulation of the solution. For our specific problem the solution is given by

$$\hat{E} = arg \min_{E \in \mathfrak{R}^+} \left\| \mathfrak{I}(E) - U \right\|_W^2 \tag{2}$$

where $\hat{E}$ is the reconstructed elasticity distribution (Young's modulus distribution), $\mathfrak{I}(E)$ is the predicted displacement field, U is the observed displacement field and W is an operator used to add a constraint on the data (whitening matrix). Since $\mathfrak{I}(E)$ is a non-linear function of E, we solve equation 2 iteratively under the framework of the Newton-Raphson algorithm. Following this, we can show that at a given iteration k the elasticity distribution should be updated to

$$E_{k+1} = E_k + \Delta E_k$$
$$\Delta E_k = \left[S_k^T S_k \right]^{-1} S_k^T \Delta U_k \tag{3}$$

where $S_k = \mathfrak{I}'(E_k)$ is the Jacobian matrix that we also label the sensitivity matrix (to be discussed further below), T stands for matrix transpose, ΔE_k is the updating elasticity vector (elastic perturbation) and $\Delta U_k = U - U_k$ is the difference between observed (U) and predicted (U_k) displacement field, i.e. the displacement perturbation. At the first iteration the tissue is assumed to be homogeneous with known Young's modulus. At each iteration the elasticity distribution is updated according to equation 3 and used to compute a new sensitivity matrix. The iteration process may be stopped when the updating elasticity is less than a prescribed value.

In practice, we found the matrix to be inverted in equation 3 highly ill-conditioned which makes the solution unstable. For conditioning, a Tikhonov regularization scheme is used. Underlying this regularization is the introduction of a compromise between fidelity to the noisy data and to *a priori* information about the solution[16]. Following this, the regularized updating term of equation 3 is now given by[16,12]

$$\Delta E_k = \left[S_k^T W_k S_k + \lambda_k Q_k \right]^{-1} S_k^T W_k \Delta U_k \tag{4}$$

where W_k, Q_k are positive definite square matrices used to add constraint on the data and *a priori* information about the solution respectively, λ_k is the regularization parameter used to quantify the compromise between fidelity to the data and the *a priori* information. In this paper, the generalized cross-validation technique is used to find, in a least mean squares sense, the optimal regularization parameter[17,12] and the identity matrices are used for both W_k and Q_k. Using the identity matrix for Q_k is equivalent to constraining the elastic perturbation ΔE_k to have only small values.

Sensitivity matrix

The sensitivity matrix S_k in equation 3 is computed column wise using the FP. Each column of this matrix is analogous to a system impulse response (i.e. sample of a Green's function). In this system the input is a local small elastic inhomogeneity (elastic perturbation) and the output is the displacement field perturbation vector. It is computed as follows. For a plane-strain state problem the finite element (FE) formulation of the FP is reduced to the solution of a 2nx2n system of equations defined as

$$[K]\{U\} = \{F\} \tag{5}$$

where *[K]* is a *2nx2n* matrix known as the stiffness matrix which depends of the Young's modulus distribution, $\{U\}$ is a *2nx1* vector defining the nodal displacements (both axial and lateral) and $\{F\}$ is a *2nx1* vector defining the applied external forces and the body forces (in our model the body forces are neglected).[18] After taking the partial derivative of the left and right hand sides of equation 5 with respect to the Young's modulus E we have

$$\left[\frac{\partial K}{\partial E_j}\right]\{U\} + [K]\left\{\frac{\partial U}{\partial E_j}\right\} = \left\{\frac{\partial F}{\partial E_j}\right\} \tag{6}$$

where E_j is the Young's modulus at element number j (a given spatial position in the ROI). Assuming a known force distribution under the lower surface of the compressor means that forces should be used as boundary conditions; in that case the right hand side of equation 6 vanishes. Hence after some matrix manipulations we find that the "j^{th}" column from the sensitivity matrix is given by

$$\left\{\frac{\partial U}{\partial E_j}\right\} = -[K]^{-1}\left[\frac{\partial K}{\partial E_j}\right][K]^{-1}\{F\} = s^{(j)} \tag{7}$$

Equation 7 defines the sensitivity matrix for a *"force-driven"* elastography problem, which in practice would require force sensors under the compressor. A different formulation need to be developed for the more common *"displacement-driven"* elastography problem. For that purpose, we propose to use a penalty method, where the forward problem takes the following form

$$\left[K + K^p\right]\{U\} = \{F\} + \left[K^p\right]\{U^0\} \tag{8}$$

where $\left[K^p\right]$ is a *2nx2n* penalty stiffness matrix and $\left\{U^0\right\}$ is a vector containing either zeros or known displacements of the nodes under the compressor. The penalty matrix is a diagonal matrix that contains only a few identical non zero entries known as the penalty parameter γ_p. The positions of the non-zero elements of the matrix $\left[K^p\right]$ correspond to the position of the nodes where the non-zero displacement are prescribed. The value of the penalty parameter is arbitrary but should be such that the non-zero elements of vector *[K^P]U* are much larger than the corresponding elements of *[K]U* (the physical meaning of this will be discussed further below). Generally[18], it is set between $10^4\mu$ and $10^{12}\mu$. Now taking the partial derivative of both sides of equation 8 with respect to E_j, we obtain

$$\left[\frac{\partial K}{\partial E_j}\right]\{U\}+[K]\left\{\frac{\partial U}{\partial E_j}\right\}=\left\{\frac{\partial F}{\partial E_j}\right\}-\left[K^p\right]\left\{\frac{\partial U}{\partial E_j}\right\} \qquad (9)$$

Since the elements of the penalty matrix are large, the first term of the right hand side of equation 9 may be neglected. Hence, one column from the sensitivity matrix is given by

$$s^{(j)}=\frac{\partial U}{\partial E_j}=-\left[[K]+\left[K^p\right]\right]^{-1}\left[\frac{\partial K}{\partial E_j}\right]\left[[K]+\left[K^p\right]\right]^{-1}\left[K^p\right]\left\{U^0\right\} \qquad (10)$$

Equation 10 can be seen as a definition of a *"displacement-driven"* sensitivity matrix. Physically using such penalty method is equivalent to have very stiff springs connected between the compressor and the lower boundary. Therefore, the forces needed to provide a given displacement will be dictated mainly by the spring stiffness constant, and little by the elastic properties of the tissue underlying the compressor.

RESULTS AND DISCUSSION

In order to illustrate the potential of our linear perturbation method we use two simulation examples and one clinical application to breast cancer imaging.

Simulation example

For the computer simulation we consider a 100x100x100mm block of homogeneous elastic material embedding a cylindrical inclusion which could be set three times harder or three times softer than the background. We assume a plane-strain state problem. The entire region of interest (ROI) of 100x100mm is simulated with a total number of 1152 first order triangular elements, which leads to a total number of 625 global nodes. The Young's modulus of the background is set to 21 kPa with a Poisson's ratio of 0.495. The FP is solved to provide "ideal" displacement data with the following boundary conditions. The lower and upper surfaces of the tissue block are not allowed to move laterally (no slip conditions). In a first case we apply a <u>uniform pressure</u> of 210 Pa at the surface of the compressor which induces an average of 0.5mm of compressor axial displacement. In a second case we apply a <u>uniform</u> 0.5mm axial <u>displacement</u> of the compressor. A system based echographic image formation model is also used to generate the pre- and post-compression RF signals from which realistic noisy displacements data are computed using correlation;[19,12] in this paper, such displacement data are refered as "observed" displacements. The model parameters are: 5 MHz central frequency, 50% bandwidth, 2mm FWHM beamwidth and 32 MHz sampling frequency. Figures 1 and 2 summarize the results obtained for the hard and soft inclusion. These figures show contrast-transfer efficiency curves (CTE) as a function of iteration number. In Ponnekanti et al.[4] CTE was defined as the ratio of the measured contrast from the observed elastogram to the true contrast, the "observed" elastogram being the strain image corrected for stress decay. However, in this paper the elastogram is the reconstructed elasticity distribution. Hence, the contrast is defined as the ratio of the average reconstructed elasticity inside the inclusion to that of the background.

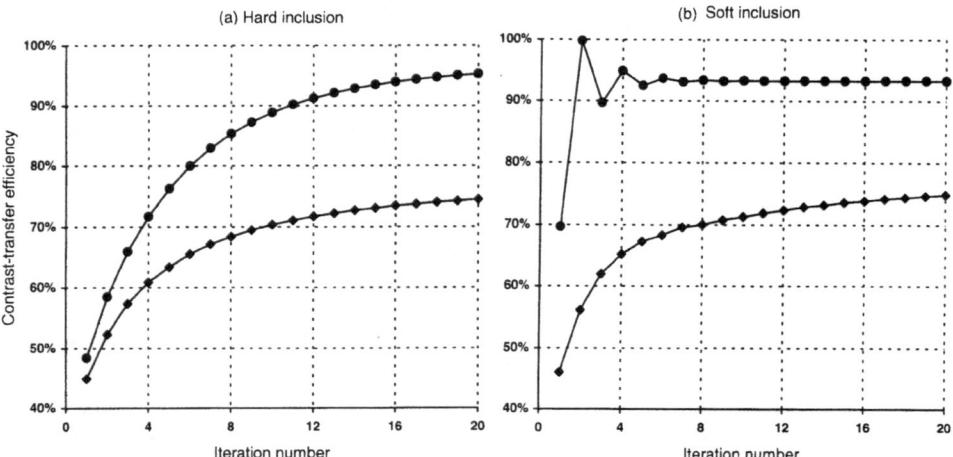

Figure 1. Contrast-transfer efficiency (CTE) versus iteration number for a uniform pressure boundary conditions situation. The CTE is defined as the ratio of the reconstructed elasticity contrast to the true contrast. (a) The inclusion is three times harder than the background. (b) The inclusion is three times softer than the background. ——•——: ideal displacements (noise free) are used for the elasticity reconstruction, : ——◆—— observed displacements (noisy) are used for the reconstruction. Notice a faster convergence rate for the reconstruction with "ideal" displacements data. Reconstruction using observed (noisy) displacements data requires a higher regularization, which also mean slower convergence.

Figure 1 shows the results obtained for both hard (Fig. 1a) and soft (Fig. 1b) inclusion and when a <u>uniform axial pressure</u> is used as boundary condition. In this case the sensitivity matrix is computed using equation 7. It is seen that when ideal[†] displacements data are used for the reconstruction (——•—— curves), after 20 iterations the CTE is of 95% for the hard inclusion (Fig. 1a) and of 93% for the soft inclusion (Fig. 1b). When the observed[ᵠ] displacements are used for the reconstruction (——◆—— curves), after 20 iterations the CTE is of 75% for both hard and soft inclusion.

[†] By ideal displacements, we mean the data computed by the FE model.
[ᵠ] By observed displacements, we mean data computed using correlation applied on simulated RF signals.

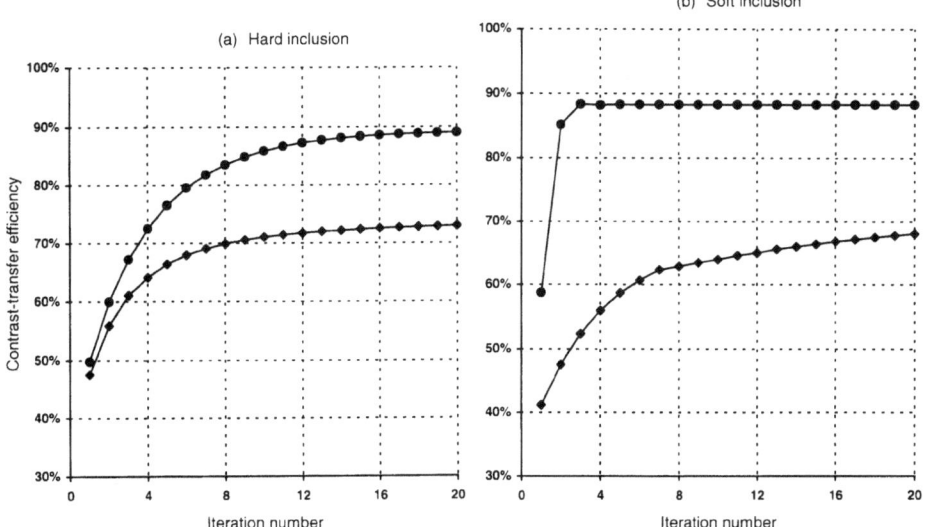

Figure 2. Contrast-transfer efficiency versus iteration number for a uniform displacement used as boundary conditions situation. (a) The inclusion is three times harder than the background. (b) The inclusion is three times softer than the background. ——●——: ideal displacements (noise free) are used for the elasticity reconstruction,: ——◆—— observed displacements (noisy) are used for the reconstruction.

Figure 2 shows the corresponding results when a <u>uniform axial displacement</u> is used as boundary condition, i.e. using equation 10 to express the sensitivity matrix. It is seen that after 20 iterations the CTE is of 89% and 88% for hard and soft inclusion respectively when ideal displacements data are used. CTE is 73% and 68% for hard and soft inclusion respectively when observed displacements data are used. These levels are a little lower than those obtained when the force distribution under the compressor is known (Fig. 1). Therefore, using a displacement driven system appears to lower the CTE by 1% to 6% in these examples.

Clinical application

We have applied our linear perturbation method to clinical data to reconstruct the elasticity distribution of a ROI from a cross-section of a volunteer breast cancer patient whose sonogram is shown in figure 3a. These data are obtained following the procedure described by Céspedes et al[2]. The ROI in the acoustical scanning plane is 40x60mm. A 0.7mm uniform compressor (ultrasound linear probe) axial displacement has been applied. For the specified depth a set of RF A-lines (100 A-lines) has been digitized before and after compression. The RF signals were digitized at 50 MHz (8 bits). Using correlation technique applied on the digitized RF data the displacement field in the ROI is computed. For the reconstruction method the ROI is divided into 1152 triangular elements resulting in 625 global nodes. Using a nearest interpolator the observed displacement field is resampled at the position of the global nodes for the reconstruction. The elasticity distribution is reconstructed iteratively using uniform axial displacements as boundary conditions. The sensitivity matrix is computed at each iteration using the penalty method (Eq. 10).

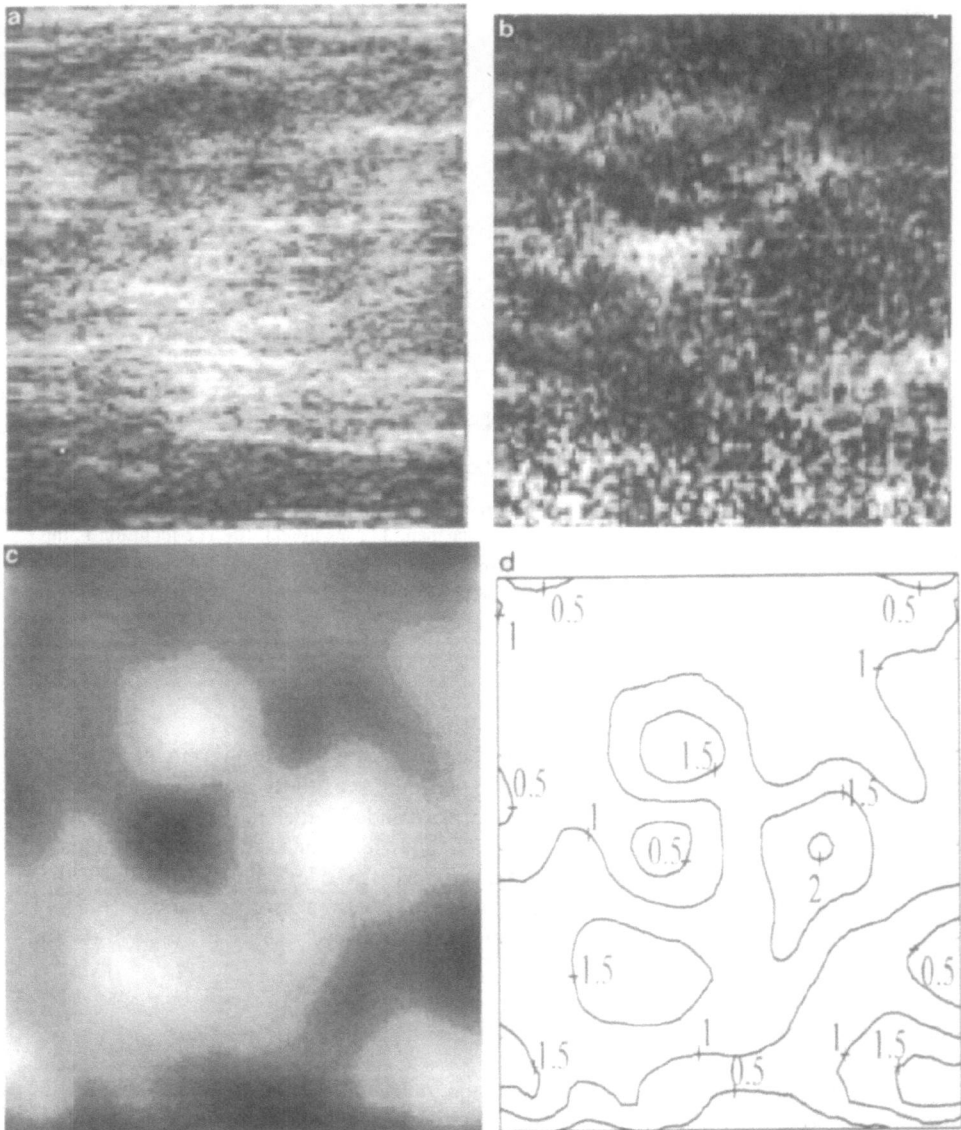

Figure 3. Clinical example of strain image and reconstructed elasticity distribution compared to sonogram. (a) Sonogram of the breast section. (b) Strain image of the breast section. On the gray scale hard tissue (low strain) appears black while soft tissue (high strain) appears white (c) Reconstructed elasticity distribution of the breast section obtained after 15 iterations of our linear perturbation method. On the gray scale hard tissue (high Young's modulus) appears white while soft tissue appears black (low Young's modulus); this convention is the reversal of that for strain image. (d) Contour plot of the reconstructed elasticity distribution. The numbers labeling the contours indicate the value of the relative elasticity. A small circular soft region near the center appears to be two times softer than the background. The hard inclusion on the right hand side is two times harder than the background. The two other hard inclusions are 1.5 harder than the background. The region of interest is 40x60mm, and the compression 0.7mm. The sonogram and strain image were generated by N. Maklad, MD from the University of Texas (1993).

Figure 3b shows the strain image of the ROI as processed on the elastographic machine at the University of Texas Medical school. Figure 3c shows the iteratively reconstructed elasticity distribution. These images display similar general features, i.e. on both images the tumor shows as a soft region surrounded by three hard "nodules". However, these seem to be delineated differently in the elasticity image; in particular the soft region (white on the strain image and black on the elasticity image) appears more circular in the elasticity image. Moreover, due to a better contrast-transfer efficiency the hardest nodules have a better contrast in the elasticity image than in the strain image. In addition, the strain image shows a softening at the bottom of the target which could be associated to a low accuracy of the estimation of the breast tissue displacements. As shown by the sonogram the ultrasound signal amplitude is low in this sub-region. The elasticity image shows that the regularization scheme takes into account such displacement noise and suggests that the presence of this soft region could be indeed real .

As shown by figure 3d elasticity reconstruction provides a quantitative measure of local tissue hardness. These measurements could be of interest for deriving discriminant features to determine the nature of the disease or to follow the progress of a treatment.

CONCLUSION

The proposed linear perturbation method provides a powerful framework to solve the new challenging inverse problem in elastography. It is based on the use of the Newton-Raphson technique combined with Tikhonov regularization. Our approach uses either force or displacement boundary conditions. When displacement conditions are used the Jacobian matrix (sensitivity matrix S) is computed using a penalty method.

The contrast transfer efficiency (CTE) has been used to study quantitatively the performance of the linear perturbation method with ideal and noisy displacements data. It was shown that for a 2 cm, 10 dB contrast (absolute value) inclusion, after 20 iterations, at least 88% of the CTE could ideally be recovered when using uniform displacement boundary conditions. Interestingly, when it is a uniform force under the compressor which is applied as boundary conditions, the CTE for the same inclusion reaches 95%. It therefore appears that a uniform pressure would, at least in this case, provides a better compressor drive than a uniform displacement drive. However, the benefits from using pressure rather than displacement boundaries appear smaller (but still significant) when noisy displacement data are used for reconstruction.

When noisy displacements are used for the reconstruction the optimal regularization parameter provided by the cross-validation technique is higher than that obtained when ideal displacements are used. Moreover, since higher regularization value means more fidelity to the constraint (small elasticity fluctuations with respect to the elastic modulus of the homogeneous medium), the CTE is lower. Indeed, when observed displacements are used for the elasticity reconstruction, only 68% to 75% of the CTE is recovered after 20 iterations (Fig. 1 and 2). Fortunately, this level could be improved by reducing the noise in the estimated axial tissue displacement field, and by adding *prior* information of the solution and constraint on the data. To improve the SNR of observed displacement it was proposed to stretch the post-compression RF signal prior to correlation.[20,21] The *prior* knowledge of the elasticity distribution variance (variance of the Young's modulus fluctuations in the homogeneous medium) could be used as *a priori* information of the solution. This *prior* information is added by an appropriate choice of a matrix Q in equation 4. The correlation coefficient may be used to generate a whitening matrix W adding a constraint on the data (Eq. 4); due to signal distortion the correlation coefficient will be lower for soft tissue than for hard tissue.

The results of application of our linear perturbation method to clinical data are very encouraging. The reconstructed elasticity seems better than strain imaging for showing the complex arrangement of masses inside the underlying breast cross-section with relatively high contrast (Fig. 3). However, we should consider these results as preliminary since we have a limited knowledge about the exact boundary conditions and the Young's modulus of the homogeneous medium. Controlled phantom experiments and direct measurement of elasticity constant from excised tumor are needed in order to confirm such results.

As stated before, many assumptions have been made to solve the inverse problem in elastography. Indeed, the tissue was assumed to behave as a linear isotropic material and being in a plane strain state. Such hypothesis also should be confirmed with appropriate controlled *in-vitro* experiments.

ACKNOWLEDGMENTS

The project upon which this publication is based was performed pursuant to the University of Texas grant No. P01-CA64597 from the National Institute of Health, Public Health Service (USA). The clinical images were generated by N. Maklad, MD from the University of Texas, under a grant from Diasonics Ultrasound, Milpitas, CA. A part of the project was also supported by the National Science Engineering Research Council of Canada and the Ministry of Education of Quebec (FCAR program).

REFERENCES

1. J. Ophir, I. Céspedes, H. Ponnekanti, Y. Yazdi, and X. Li, Elastography: a quantitative method for imaging the elasticity of biological tissues, Ultrasonic Imaging 14:111 (1991).
2. I. Céspedes, J. Ophir, H. Ponnekanti, and N.F. Maklad, Elastography: elasticity imaging using ultrasound with application to muscle and breast *in vivo*, Ultrasonic Imaging 15:73-88 (1994).
3. H. Ponnekanti, J. Ophir, and I. Céspedes, Ultrasonic imaging of the stress distribution in elastic media due to an external compressor, Ultras. Med. and Biol. 20:27 (1994).
4. H. Ponnekanti, J. Ophir, Y. Huang, and I. Céspedes, Fundamental mechanical limitations on the visualization of elasticity contrast in elastography, Ultras. Med. and Biol. 21:533 (1995).
5. F. Kallel, M. Bertrand, and J. Ophir, Fundamental limitations on the contrast-transfer efficiency in elastography: an analytic study, submitted for publication in Ultrasound in Medicine & Biology (1995).
6. S. Y. Emelianov, A.R. Skovoroda, M.A. Lubinski, and M. O'Donnell, Reconstructive elasticity imaging, to appear in Acoust. Imag. 21, J.P. Jones (ed), Plenum Press, NY (1994).
7. F. Kallel, M. Bertrand, J. Ophir, and I. Céspedes, Determination of elasticity distribution in tissue from spatio-temporal changes in ultrasound signals, to appear in Acoust. Imag. 21, J.P. Jones (ed), Plenum Press, NY (1994).
8. K.R. Raghavan, and A.E. Yagle, Forward and inverse problems in elasticity imaging of soft tissues, IEEE Trans. on Nuclear Science 41:1639 (1994).
9. A.R. Skovoroda, S.Y. Emelianov, M.A. Lubinski, A.P. Sarvazyan, and M. O'Donnell, Theoretical analysis and verification of ultrasound displacement and strain imaging, IEEE Trans. on UFFC 41:302 (1994).
10. C. Sumi, A. Suzuki, and K. Nakayama, Estimation of shear modulus distribution in soft tissue from strain distribution, IEEE Trans. on BME 42:193 (1995).
11. J.G. Webster. "Electrical Impedance Tomography," Adam Hilger, NY (1990).
12. F. Kallel and M. Bertrand, Tissue elasticity reconstruction using linear perturbation method, submitted for publication in IEEE Trans. on Medical Imaging (1995).
13. S. Saada. "Elasticity, Theory and Applications," Pergamon Press, New York, 1983.
14. N. C. Fung. "Biomechanics: Mechanical Properties of Living Tissues," Springer-Verlag, NY (1981).
15. Philip E. Gill, Waltre Murray, and Margaret H. Wright. "Practical optimization ," Academic Press, London (1981).

16. G. Demoment, Image reconstruction and restoration: Overview of common estimation structures and problems, IEEE Trans. Acoust., Speech, Signal Processing 37:2024 (1989).

17. S. Galatsanos and A. Katsaggelos, Methods for choosing the regularization parameter and estimating the noise variance in image restoration and their relation, IEEE Trans. Image Processing 1:322 (1992).

18. J.N. Reddy. "An Introduction to the Finite Element Method," McGraw-Hill, Inc., 1984.

19. J. Meunier, and M. Bertrand, Echographic image mean gray level change with tissue dynamics: a system-based model study, IEEE Trans. on BME 42:403 (1995).

20. I. Céspedes and J. Ophir, Reduction of image noise in elastography, Ultrasonic Imaging 15:89 (1993).

21. I. Céspedes, M. Insana, and J. Ophir, Theoretical bounds on strain estimation in elastography, IEEE Trans. on UFFC, in press, (1995).

VISUALIZING PROPAGATING TRANSVERSE MECHANICAL WAVES IN TISSUE-LIKE MEDIA USING MAGNETIC RESONANCE IMAGING

R. Muthupillai, D.J. Lomas, P.J. Rossman, J.F. Greenleaf, A. Manduca, S.J. Riederer and R.L. Ehman

Mayo Clinic and Foundation
Rochester, MN - 55905

INTRODUCTION

It is well known that mechanical properties of tissues like elasticity and viscosity depend on factors such as age and pathology. For example, physicians use palpation for assessing the mechanical properties of tissues to detect tumors and other abnormalities. However, palpation provides non-quantitative, subjective information and is limited to accessible regions of the body. Therefore, there has been a consistent interest in developing quantitative, non-invasive techniques that distinguish tissues based on their mechanical properties such as elastic modulus[1-11].

Elastic modulus can be measured by studying the response of materials to an externally controlled mechanical stress[12]. Several ultrsound based approaches have been suggested to measure the response of tissues subjected to external mechanical stress[1-4,6,7,13]. However, these ultrasound based approaches require appropriate acoustic windows of observation and are constrained with respect to signal to noise ratio and lateral resolution. Some of these techniques also require the knowledge of boundary conditions and appropriate mathematical models to quantitatively estimate the tissue mechanical properties[4,6,7,12].

Some MRI based techniques to measure response of tissue subjected to an external stress have also been reported[9-11]. Lewa et. al. have proposed MRI based methods, including a Stejskal-Tanner based approach to observe the effect of longitudinal mechanical waves on tissues and measure the mechanical properties of tissues. However, bulk modulus variations in soft tissues are only a few percent[5,14] and at low frequencies, the speed of longitudinal wave propagation is too high resulting in wavelengths longer than the objects that are imaged[1,14]. A phase contrast MRI technique using non-harmonic pressure transients, to visualize tissue compliance has also been suggested and like similar ultrasound based techniques[11], this technique may require the knowledge of boundary conditions or analytic solutions to quantitatively estimate the tissue mechanical properties.

We have proposed a MRI phase contrast based technique for directly imaging acoustic frequency propagating transverse acoustic strain waves in tissue like media[15]. Small cyclic displacements (sub-micron level) in the media caused by propagating mechanical waves, result in a measurable phase shift in the NMR signal, in the presence of appropriate motion-sensitizing gradients. Our preliminary experiments with tissue mimicking agarose gel phantoms confirm our hypothesis, that this method can sensitively image propagating mechanical waves.

THEORY

It is well known from the Larmor equation that the phase of transverse magnetization of a nuclear spin moving in the presence of a magnetic field gradient is:

$$\phi(t) = \gamma \int_0^\tau \bar{G}_r(t) \bullet \bar{r}(t) dt \qquad (1)$$

where $\bar{G}_r(t)$ is the time dependent magnetic field gradient, $\bar{r}(t)$ is the position vector of the moving spin and γ is the gyromagnetic ratio characteristic of the nuclear isochromat under investigation. If we consider the specific case of a propagating mechanical wave, the motion of the spin is given by:

$$\bar{r}(t) = \bar{r}_0(t) + \bar{\xi}_0 \exp(j(\bar{k} \cdot \bar{r} - \omega t + \varphi)) \qquad (2)$$

where $\bar{r}_0$ is the mean position of the spin, w is the angular frequency of the harmonic mechanical excitation , φ is the initial phase offset, $\bar{k}$ is the wave number and $\bar{\xi}_0$ is the peak displacement of the spin from its mean position. Under these circumstances, it may be useful to consider the gradient vector $\bar{G}_r(t)$ in Eq. (1) as one among a set of basis functions to estimate the harmonic components of the position vector $\bar{r}(t)$. Specifically, if we consider a rectangular gradient waveform of amplitude |G| Gauss/cm, switched in polarity at the same frequency as the motion vector (i.e., T = 2p/w), for N cycles then the phase shift in the received NMR signal can be shown as[15]:

$$\phi(\bar{r}, \varphi) = \text{Re}\left[\gamma \int_0^{\tau = NT < TE} \bar{G}_r(t) \bullet \bar{\xi}_0 \exp(j(\bar{k} \bullet \bar{r} - \omega t + \varphi)) dt \right],$$

$$(3)$$

$$= \frac{2\gamma NT(\bar{G} \bullet \bar{\xi}_0)}{\pi} \cos(\bar{k} \bullet \bar{r} + \varphi).$$

Sensitivity to small cyclic displacements caused by strain wave propagation can be increased by constructively accumulating phase shifts over multiple cycles of mechanical excitation and gradient waveform because of the dependence of the measured phase shift, on the period(T) and the number of gradient cycles(N). Equation 3 indicates that the measured phase shift is also a function of the initial phase offset φ between the gradient waveform and the mechanical excitation.

METHODS

Experimental system: The experimental system including the pulse sequence used for generating acoustic strain waves is shown in Fig. 1. A phase contrast based gradient recalled echo pulse sequence with provisions for cyclic motion-sensitizing gradients was developed (Fig. 1). The motion-sensitizing gradients can be trapezoidal or sinusoidal.

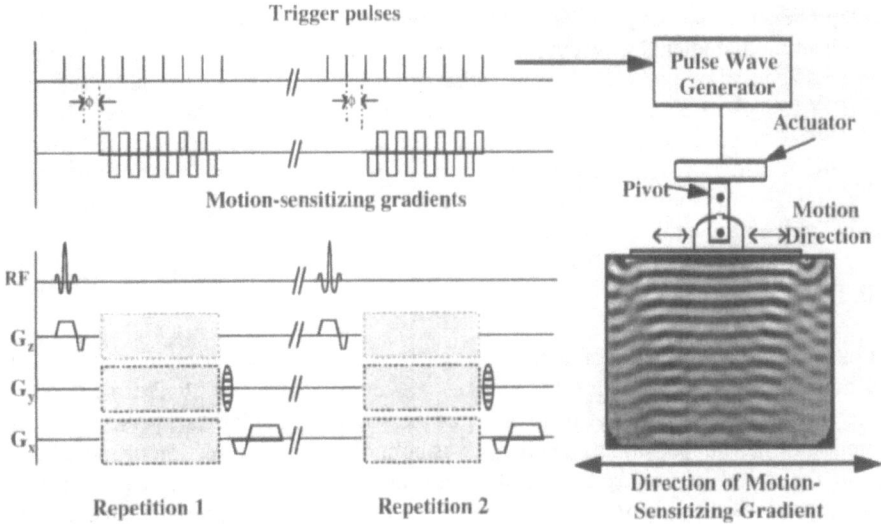

Figure 1 The actuator is a coil (4-6W) connected to a shaft oriented orthogonal to the main magnetic field. When an alternating current flows through the coil from a waveform generator, the resulting magnetic flux interacts with the main magnetic field moving the linkage cyclically. A flat plate coupled to the linkage acts as a plane source of mechanical waves. The waveform generator is triggered in synchrony with the motion-sensitizing gradients. The number and the frequency of motion-sensitizing gradients is variable and so is the phase relationship between the motion-sensitizing gradients and mechanical excitation. The motion-senistizing gradient waveforms can be applied along any axis, as shown by the shaded regions, in combination with imaging gradients.

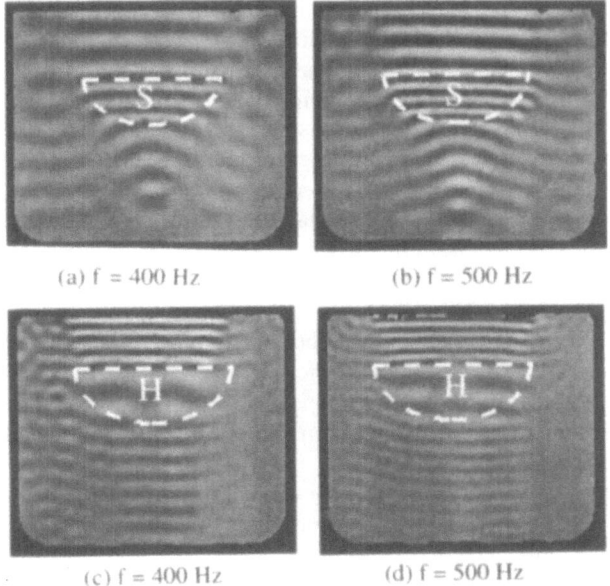

Figure 2 Focusing effect (Fig. 2a and 2b) due to a hemispherical lens-shaped soft (S) inclusion in a stiff gel. a) Frequency (f) of mechanical excitation=400Hz and number(N) of motion-sensitizing gradient cycles = 8. b) f = 500Hz and N = 10. Diverging effect (Fig. 2c and 2d) due to a hemispherical lens-shaped hard (H) inclusion in a soft gel. c) f = 400Hz and N = 8. d) f = 500Hz and N = 10 (white dotted lines indicate inclusions).

Data acquisition: Two acquisitons were made for each repetition. The polarity of the motion-sensitizing gradients was reversed between the two acquisitions, as shown in Fig. 1. Typical data acquisition parameters were: pulse repetition time(TR) ~ 50 - 300 msec, Echo time (TE) ~ 10 - 60 msec, acquisition matrix size ~ 128 x 256; acquisition time ~ 20 - 120 sec. The frequency of mechanical excitation and the number of gradient cycles (N) used are stated in the results. A displacement map of spins is obtained by reconstructing a phase difference image from these two acquisitions. These displacement maps can be viewed as a "snapshot" of the propagating transverse mechanical waves within the medium.

RESULTS

Phase difference images of a heterogenous agarose gel phantom in which a lens shaped hemispherical soft gel (1.25% w agar) was embedded in stiff agarose gel (2% w agar) are shown in Fig. 2. A flat plate acted as the source of planar transverse waves and the direction of motion-sensitizing gradients was colinear with the direction of motion. The change in the speed of propagation of the mechanical wave in the stiff and soft gel is evident and the lens shaped soft inclusion converges the wavefront within the phantom.

A snapshot of a propagating mechanical wave within a heterogenous tissue mimicking agarose gel phantom in which a lens shaped hemispherical hard gel (2% w agar) was embedded in a soft agarose gel (1.25% w agar), is shown in Fig. 3. As in Fig. 2, a flat plate was the source of planar mechanical excitation and the motion-sensitizing gradients were colinear with direction of motion.

DISCUSSION AND CONCLUSION

Our experimental results show that small cyclic displacements such as those caused by propagating mechanical waves in tissue-like media can be measured, using MRI. As shown in Figs. 2 and 3, one can observe various strain wave propagation characteristics such as attenuation, speed, reflection etc. Such wave propagation characteristics depend on the underlying biomechanical properties of the medium. Therefore, we speculate that this technique could be used to estimate the biomechanical properties of the medium. For example, the change in wavelength in Fig. 2 and 3 between the soft and stiff medium is evident and by quantitatively estimating the local spatial frequency it is possible to estimate the regional shear modulus of the material[15].

Our preliminary experiments indicate that it is possible to measure displacements as small as 200 nanometers. By acquiring a series of images at progressively varying phase offsets between the motion sensitizing gradients and the mechanical excitation and displaying these images in a "cine" fashion, it is also possible to observe dynamic strain wave propagation. Because the gradients can be superimposed along any axis, it is possible to estimate all components of the strain dyadic. This is a unique advantage of using MRI when compared to ultrasound based techniques. We speculate that the ability to study strain wave propagation in tissue-like media may find applications in medical imaging and materials science.

REFERENCES

1. T.A. Krouskop, D.R. Dougherty, F.S. Vinson, A pulsed Doppler ultrasonic system for making noninvasive measurements of the mechanical properties of soft tissue. J. Rehab. Res. Dev., 24(2), 1-8 (1987).
2. K.J. Parker, S.R. Huang, R.A. Musulin, R.M. Lerner, Tissue response to mechanical vibrations for sonoelasticity imaging. Ultrasound in Med. & Biol., 16(3), 241-246 (1990).

3. Y. Yamakoshi, J. Sato, T. Sato, Ultrasonic Imaging of Internal Vibration of Soft Tissue under Forced Vibration. IEEE Transactions on Ultrasonics, Ferroelectrics and Frequency Control, 37(2), 45-53 (1990).

4. J. Ophir, I. Cespedes, H. Ponnekanti, Y. Yazdi, X. Li, Elastography: a quantitative method for imaging the elasticity of biological tissues. Ultrasonic Imaging, 13(111-134) (1991).

5. A. Sarvazyan, et al. in "Proceeding of International workshop on interaction of ultrasound with Biological media," Valenciennes, France. 1994. p. 69-81.

6. M. O'Donnell, A.R. Skovoroda, B.M. Shapo, S.Y. Emelianov, Internal displacement and strain imaging using ultrasonic speckle tracking. IEEE Transactions on Ultrasonics, Ferroelectrics, and Frequency Control, 41(3), 314-325 (1994).

7. A.R. Skovoroda, S.Y. Emelianov, M.A. Lubinski, A.P. Sarvazyan, M. O'Donnell, Theoretical analysis and verification of ultrasound displacement and strain imaging. IEEE transactions on Ultrasonics, Ferroelectrics and Frequency Control, 41(3), 302-313 (1994).

8. M. Bertrand, J. Meunier, M. Doucet, G. Ferland. in "IEEE Ultrasonics Symposium," 1989. p. 859-863.

9. C.J. Lewa. in "Ultrasonics Symposium," Cannes. 1994. p. 691-694.

10. C.J. Lewa, J.D. de Certaines, MR Imaging of viscoelastic properties. Journal of Magnetic Resonance in Imaging, 5, 242-244 (1995).

11. D.B. Plewes, I. Betty, I. Soutar. in "Proc. of SMR," San Francisco. 1994. p. 410.

12. H. Kolsky, Stress waves in solids. Dover, New York, 1963.

13. A. Sarvazyan. in "125th Meeting: Acoustical Society of America," 1993. p. 2329.

14. S.A. Goss, R.L. Johnston, S.E. Shnol, Comprehensive compilation of empirical ultrasonic properties of mammalian tissues. J. Acoust. Soc. Amer., 64(2), 423-457 (1968).

15. R. Muthupillai, D.J. Lomas, P.J. Rossman, J.F. Greenleaf, A. Manduca, R.L. Ehman, Magnetic resonance Elastography by direct visualization of acoustic strain waves. Science, 269, 1854-1857 (1995).

FREEHAND ELASTICITY IMAGING USING
SPECKLE DECORRELATION RATE

Jeffrey C. Bamber and Nigel L. Bush

Joint Department of Physics
Institute of Cancer Research and Royal Marsden NHS Trust
Downs Road, Sutton
Surrey, SM2 5PT, UK

INTRODUCTION

Malignant tumours are often relatively stiff and immobile. These properties permit tumours such as breast cancers to be detected by palpation and have provided motivation for the development of new methods for imaging tissue elasticity, by ultrasonically estimating the displacement or strain distribution due to an externally applied stress. Various authors have now shown that such elasticity images may possess excellent spatial and contrast resolution[1,2,3].

Some elasticity imaging schemes use a carefully controlled stress (applied automatically with the tissue and transducer constrained to a fixed geometry) and relatively complex signal processing, with the aim of extracting the best possible estimate of strain distribution. In clinical ultrasound practice, however, elasticity information is already employed, subjectively, to assist in distinguishing between benign and malignant breast masses[4,5,6]. Judgements based on real-time observation of the dynamic behaviour of tissues in response to manual palpation yield a subjective assessment of stiffness and mobility. In an earlier study[4] it was demonstrated that breast masses which are composed of very soft tissue and are mobile are usually fibroadenomas and almost certainly never cancers, masses which are soft but immobile are probably cysts and masses which are composed of stiff tissue and are immobile are very likely cancers (Table I). It was also noted that the true tumour boundary, which for malignancies is frequently either echogenic or isoechoic relative to the glandular tissue of the breast, appeared to be more easily identified when observed during palpation.

Although it is necessary to pursue complex processing schemes if elasticity imaging is ever to achieve the desired goal of quantitative inverse reconstruction of elastic modulus, there may be immediate practical value to a simple free-hand, real-time elasticity imaging system which might enhance the visual appreciation of the dynamic features already found to be useful and is able to take advantage of the existing pool of knowledge on how best to carry out breast palpation under ultrasound observation. This paper reports on preliminary studies of two potential components of such a future system; firstly the use of speckle decorrelation rate to measure tissue internal displacement and, secondly, the use of hand-controlled transducer motion to apply the stress. Interest in the former is motivated by the fact that the computations are simple enough to be implemented in real-time with currently inexpensive hardware.

Table I. Visual assessment of dynamic tissue properties by real-time ultrasound observation during tissue palpation (after Ueno et al.[4])

| | No. Cases | Tissue stiffness | | | Tissue mobility | | |
		- (Soft)	0	+ (Hard)	- (Mobile)	0	+ (Immobile)
Cancer	37	0%	14%	86%	11%	16%	72%
Fibroadenoma	18	44%	17%	39%	67%	28%	6%
Cyst	15	80%	20%	0%	7%	7%	87%

SPECKLE DECORRELATION ESTIMATION OF TISSUE DISPLACEMENT

Under the commonly used assumption that echographic imaging can be modelled as a linear space-invariant convolutional process it can be shown[7] that the autocorrelation of B-mode image texture is given by:

$$R_{Texture}(\Delta x, \Delta y, \Delta z) = \tilde{a}(z)\,\tilde{b}(x)\,\tilde{c}(y) \otimes \tilde{T}(x,y,z) \tag{1}$$

where $R_{Texture}(\Delta x, \Delta y, \Delta z)$ is the (three dimensional) autocorrelation function of the B-scan image texture, $\tilde{a}(z), \tilde{b}(x)$ and $\tilde{c}(y)$ represent the autocorrelation functions of the ultrasound beam round-trip point spread function (PSF) in the axial, lateral and azimuthal directions respectively, $\tilde{T}(x,y,z)$ is the autocorrelation function of the tissue backscattering impulse response function[8] and $\otimes$ represents a convolution operation. From Equation 1 it may be seen that if the autocorrelation function of the tissue impulse response function is itself an impulse, and if the components of the PSF are both known and equal in x, y and z, then measurement of R, when the tissue is displaced from one position to another, will yield an estimate of the amount of displacement independent of direction. In practice the degree to which these requirements can be satisfied will determine the degree of artefact-free performance of such a method for measuring internal tissue displacement. For example, if $\tilde{T}(x,y,z)$ is not an impulse for a particular tissue region then it too must be known if measurements of R are to be successfully interpreted as tissue displacement. Another example arises from the spatial variance of the PSF which, if not corrected for, will cause a systematic under or overestimation of the displacement depending on distance from the transducer. In addition, R itself must be estimated in a statistical manner from the B-scan image. There will thus be a trade-off in the precision with which R can be estimated and the spatial precision with which the tissue displacement can be localized. This trade-off represents a compromise between contrast and spatial resolution in the final tissue displacement (or elasticity) image.

Phantoms of uniform open-cell reticulated foam, of various pore sizes (measured in pores per inch, or ppi), were used to study the trade-off between spatial resolution and signal-to-noise ratio of displacement when estimated from speckle decorrelation and to examine the behaviour of some artefacts associated with the method. For the convenience of this preliminary study lateral displacement only was considered, with the expectation that the trends observed will be applicable more generally.

SPATIAL/CONTRAST RESOLUTION TRADE-OFF

Using a custom mechanical scanner three dimensional (3-D) radio frequency (RF) backscatter data were acquired from a sponge fine enough (80 ppi) to produce fully developed acoustic speckle from the transducer being used (Aerotech™ 3 MHz, circular 20 mm diameter bowl focused at about 80 mm). The lateral distance between the RF lines in the 3-D scan was 0.4 mm. The RF echo signal was digitised to 12 bits at a sampling rate of 12.5 MHz and the echo envelope calculated by complex demodulation using the analytic signal. B-scans were then generated in parallel planes using groups of 128 lines. Correlation *images* were then computed between the first B-scan and all subsequent parallel B-scans by calculating, for a moving window, the cross-correlation coefficient between the B-mode data within the window in the first image and that for the same window in each of the other images. For studying the spatial/contrast resolution trade-off such correlation images were computed for a number of different moving window sizes and the data within a region of interest (ROI) of size 89 lines by 156 samples (about 225 speckle cells in area) situated beyond the focal region, where the beam is broadest, were used to compute the mean $\bar{R}$ and standard deviation σ_R of the correlation values.

Figure 1 shows a plot of one minus the mean of each correlation image $(1 - \bar{R})$ within the ROI as a function of the lateral distance separating the pair of B-scans from which the correlation image was computed, for 4 different window sizes used for computing the correlation coefficient. The following observations may be made from Figure 1. One lateral speckle cell size (full width at half maximum correlation) for this region of the beam appears to be about 6 samples, or 2.4 mm. There exists a region of the curves, between $(1 - \bar{R}) \sim 0.2$ and 0.8, where $(1 - \bar{R})$ is approximately linearly related to displacement. Therefore, in a practical freehand elasticity imaging system one would wish to employ a speed of hand motion in combination with an image sampling rate so as to be operating on the nearly linear portion of the equivalent to this curve. Displacements larger than a distance of one speckle cell can only be measured as being greater than or equal to this distance. There is a small bias associated with the size of the moving window over which the correlation coefficient is computed, causing the apparent displacement to be slightly reduced when one attempts to gain spatial resolution by decreasing the window size.

Figure 2 shows, for the same correlation images as used for Figure 1, an effective signal-to-noise ratio (SNR) for the displacement estimate, computed as the magnitude of $(1 - \bar{R})$ divided by the standard deviation of R within the same ROI. It may be seen that the

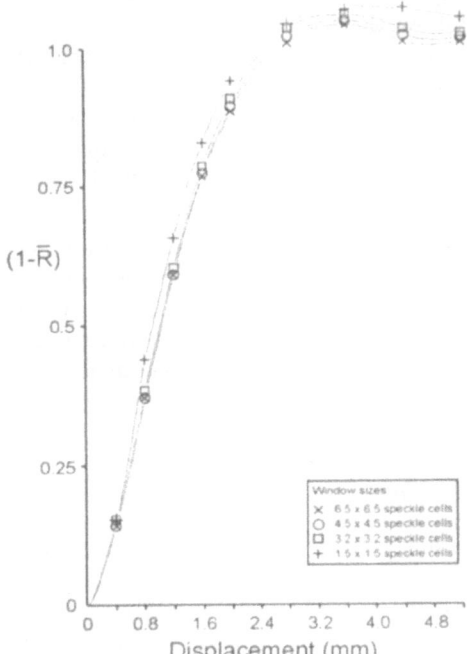

Figure 1. One minus the mean cross-correlation coefficient between pairs of parallel speckle B-scans as a function of the distance separating the image planes and for 4 different moving correlation window sizes.

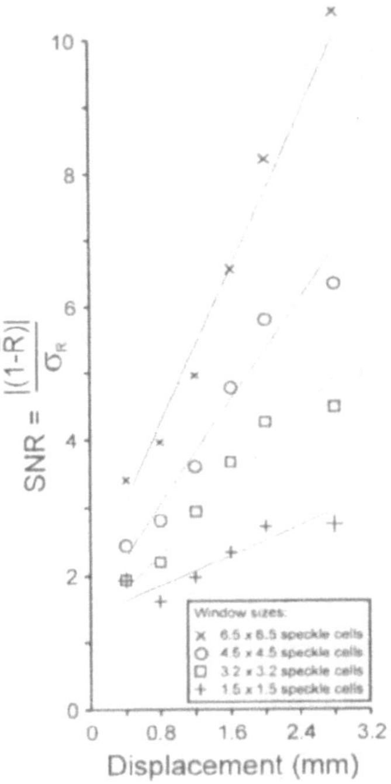

Figure 2. Signal-to-noise ratio computed as one minus the mean speckle correlation divided by the standard deviation of the correlation within the ROI versus lateral distance, as in Figure 1.

intrinsic SNR for this method of displacement imaging is itself displacement dependent and that, within the range for which displacement can be reasonably estimated, larger displacements should improve the SNR of the displacement image. Larger moving window sizes also improve the SNR. Using the correlation moving window size as a measure of spatial resolution it can be seen from Figure 2 that for a spatial resolution of similar order to that of the B-scan SNRs in excess of that for a speckle B-scan (where SNR ~ 2) are achieved for displacements greater than about 0.8 mm. Cespedes et al[1] have noted for a single-look tracking estimate of strain SNRs in excess of 4 can be achieved with spatial resolutions similar to those in B-mode imaging. The SNR for a $(1 - \bar{R})$ estimate of displacement may reach and exceed this value if, for example, the spatial resolution is about 3 times worse than B-mode and the displacement is greater than 2 mm. This, however, is close to the limit of displacement that can be measured (see Figure 1) and it may be better to work in the region of about 1 mm displacement with a larger moving correlation window corresponding to the same SNR. Usable SNRs as high as 8 may be obtained from this single-look correlation estimate if a greater than six times degradation in spatial resolution over B-mode imaging can be tolerated. Much greater SNRs than this are likely to be achievable if multi-look correlation estimates are combined, for example by averaging.

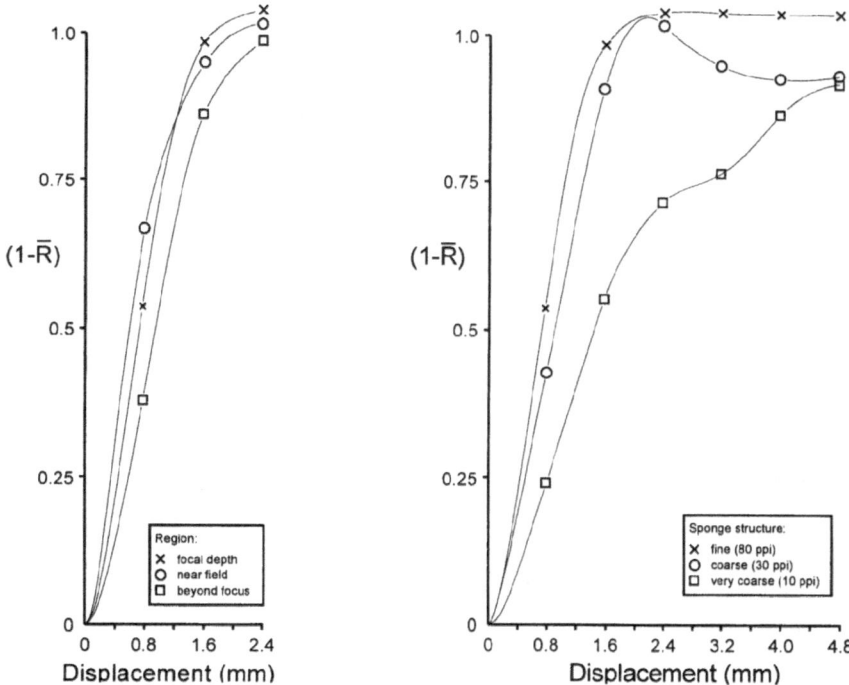

Figure 3. Results similar to those of Figure 1 for the largest window size repeated for 2 additional ROIs, to demonstrate the effect on the lateral displacement estimate of axial distance from the transducer.

Figure 4. Results similar to those in Figure 1 for the largest window size repeated using additional coarse and very coarse sponges, to demonstrate the effect on the lateral displacement estimate of resolved tissue structure.

ARTEFACTS

Potential artefacts of speckle decorrelation estimation of tissue displacement include the dependence of the displacement estimate on (a) distance from the transducer, associated with both the spatial variation in the system point response function and with the size of the transducer face, (b) direction and magnitude of the displacement, due to the asymmetrical and non-linear fall-off of speckle correlation with distance, (c) the degree to which resolvable structure is present in the object, and (d) the electronic signal-noise-ratio of the original B-mode images. Those associated with (a) and (c) were briefly illustrated by further experiments conducted exactly as described above but where, for (a), ROIs at the focal region and in the near field were also analysed and, for (c), other sponges with coarse (30 ppi) and very coarse (10 ppi) meshes were employed. The results are shown respectively in Figures 3 and 4. Lateral displacement may therefore easily be under or overestimated, depending on the distance from the transducer. This effect is related to the effect of diffraction on B-mode texture properties and it is likely that methods for diffraction correction, similar to those used in texture analysis, may be developed for displacement imaging. It may also be anticipated from previous work on texture analysis[9] that axial motion estimation is much less likely to be influenced by depth-dependent diffraction effects than lateral motion, although depth-related changes in speckle characteristics due to frequency dependent attenuation are likely to be important.

Figure 4 shows that when the tissue backscattering function is not an impulse the excess image correlation, due to the presence of resolvable tissue structure, would cause the displacement to be underestimated if curves such as those from Figure 1 were to be used to calibrate the system. At least two forms of adaptive processing may be conceived to overcome this problem. In one approach a filter, such as used previously for adaptive speckle reduction[10], might be used to select for use in displacement estimation only those B-scan regions which correspond to fully developed speckle. Alternatively, under the assumptions of small displacements, tissue structural isotropy and short-range stationarity of the statistical properties of B-mode image texture a preliminary regional autocorrelation analysis of one or more B-mode frames may be used to produce a spatially variant calibrating function which relates $(1 - \bar{R})$ to displacement for all tissue regions imaged. However, such approaches to the removal of these artefacts would detract substantially from the simplicity of the algorithm, which is one of its principle virtues.

HAND-INDUCED TRANSDUCER MOTION

Several potential problems raise doubts concerning the suitability of employing hand-induced transducer motion for elasticity imaging. Difficulty in avoiding rotational and out-of-plane translational probe motions may lead to loss of resolution, failure to estimate tissue displacement and artefacts in the displacement estimates. In addition, freehand palpation is likely to involve relatively large transducer displacements which, whatever signal processing method is used to estimate tissue displacement, will require correction if spatial and contrast resolution are not to be lost.

Gelatine phantoms were used to demonstrate that it is possible to maintain free-hand transducer motion within a single plane and without substantial rotational motion of the transducer, to a degree that is sufficient to produce high contrast displacement images of relatively inelastic inclusions, based on speckle decorrelation. Maximum axial transducer displacements were of the order of 1 cm and, as expected it was found necessary to compensate for the motion of the transducer. By way of an example, Figure 5 shows the B-mode scans at minimum and maximum transducer displacement, taken from a digitized real-time sequence of frames during freehand transducer palpation of one of the phantoms. The phantom was composed entirely of a gelatine-alginate complex[11] (ratio gelatine:alginate = 8:1) in water loaded with randomly distributed polyethylene granules (~119 μm mean diameter, concentration ~ 4 mm^{-3}). A "background gel" was made, consisting of a 9% gel:water mix, and set initially as a 3 cm layer in the base of a rectangular perspex box. An irregularly shaped hard inclusion was then produced from a 13.5% concentration of the same gel, containing the same number density of scatterers, set within a silicone rubber mould that was made using a beach pebble with an average diameter in the plane imaged of ~ 2 cm. This was placed on the first layer, over which more background gel was then poured and set to a depth of about 5 cm.

Figure 6 shows the "displacement image" (1-R) calculated for ROIs within the pair of images in Figure 5, both without and with a simple average axial displacement correction for the motion of the transducer (assuming that the average transducer-tissue distance changed equally for all depths within the ROI). In this case the approximate transducer motion correction was applied by displacing the axial position of the ROI in Figure 5 (right) to the point of maximum correlation between the data within this ROI and those in the ROI used for Figure 5 (left). No attempt was made to correct for rotational motion of the transducer or for the depth variation of the average transducer-tissue distance within the ROI.

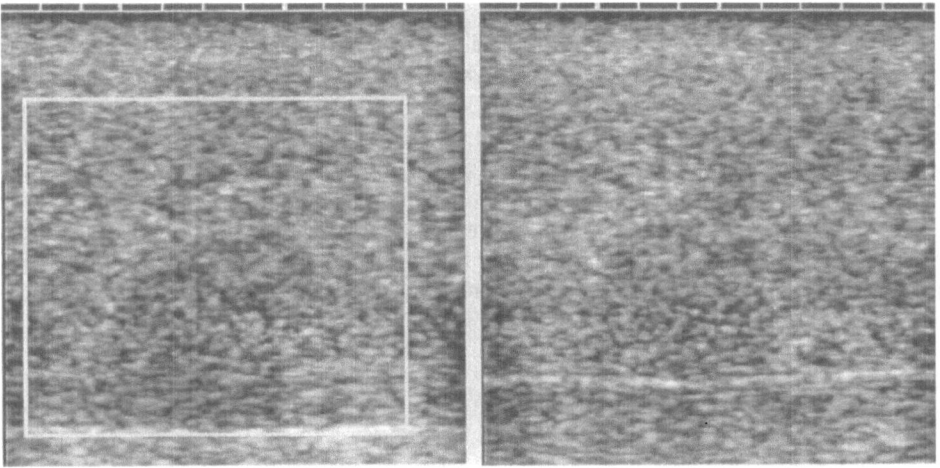

Figure 5. B-mode scans of a phantom containing an approximately isoechoic but relatively stiff inclusion during manual palpation with the transducer at minimum (left) and maximum (right) displacement.

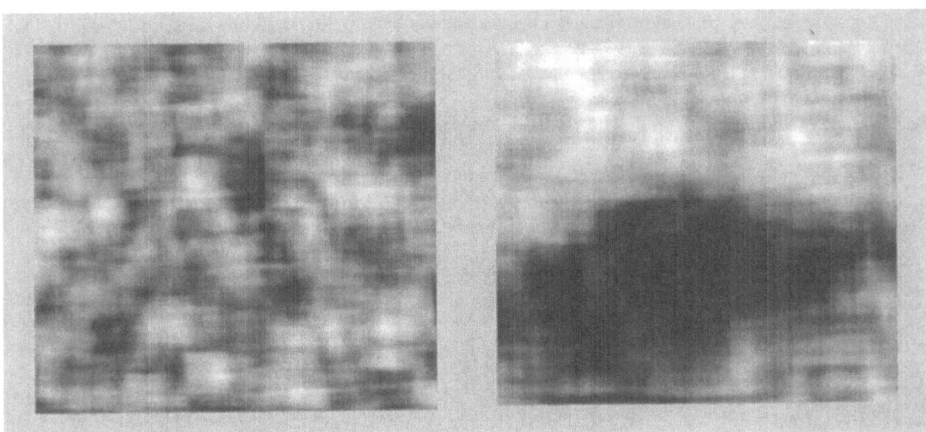

Figure 6. Relative internal tissue "displacement images" for the ROI indicated in Figure 5 computed as (1-R), where R is the cross-correlation coefficient between corresponding moving windows of echo data within the regions of interest shown in Figure 5, both without (left) and with (right) correction for the axial motion of the transducer. The largest of the moving window sizes shown in Figures 1 and 2 was used.

Figure 6 illustrates that without transducer motion correction substantial uncertainty exists with regard to the position of elastic structures, values for 1-R are principally a measure of transducer motion and the resulting image is little more than correlation noise. When even an approximate transducer axial motion correction is applied the relatively hard elastic inclusion within the phantom becomes plainly visible in the displacement image.

CONCLUSION

Freehand scanning and hand-induced transducer motion for elasticity imaging would appear to be feasible. Large transducer displacements may be expected but it should be possible to compensate for these. Speckle decorrelation then produces good images of relative tissue displacement. Speckle decorrelation imaging of internal tissue displacement performs reasonably well in terms of the signal-to-noise ratio versus spatial resolution compromise but the method is likely to suffer from a number of potentially serious artefacts. Solutions to eradicate these artefacts will require additional processing. Performance of the method in its present simple form may, however, be clinically valuable. It is therefore worthy of further development and evaluation.

REFERENCES

1. I. Cespedes, J. Ophir, H. Ponnekanti, and N. Maklad, Elastography: elasticity imaging using ultrasound with application to muscle and breast in vivo. *Ultrasonic Imaging.* 15:73 (1993).
2. K.J. Parker, S.R. Huang, R.A. Musulin, R.M. Lerner, Tissue response to mechanical vibrations for "Sonoelasticity Imaging". *Ultrasound Med Biol.* 6:241 (1990).
3. M. O'Donnell, A.R. Skovoroda, B.M. Shapo, S.Y. Emelianov, Internal displacement and strain imaging using ultrasonic speckle tracking. *IEEE Trans UFFC.* 41:314 (1994).
4. E. Ueno, E. Tohno, S. Soeda, Y. Asoaka, K. Itoh, J.C. Bamber, M. Blaszczyk, J. Davey, and J.A. McKinna, Dynamic tests in real-time breast echography. *Ultrasound Med Biol.* 14:53 (1988).
5. J.C. Bamber, L. de Gonzalez, D.O. Cosgrove, P. Simmons, J. Davey, and J.A. McKinna, Quantitative evaluation of real-time ultrasound features of the breast. *Ultrasound Med Biol.* 14:81 (1988).
6. M Walz, J. Teubner, M. Georgi, Elasticity as criterion of benignity for solid lesions of the breast. *Ultrasound Med Biol.* 20:S91 (1994).
7. R.J. Dickinson and C.R. Hill, Measurement of soft tissue motion using correlation between A-scans. *Ultrasound Med Biol.* 8:263 (1982).
8. J.C. Bamber and R.J. Dickinson, Ultrasonic B-scanning: a computer simulation. *Physics Med Biol.* 25:463 (1980).
9. B.J. Oosterveld, J.M Thijssen, W.A. Verhoef, Texture of B-mode echograms: 3-D simulations and experiments of the effects of diffraction and scatterer density. *Ultrasonic Imaging.* 7:142 (1985).
10. J.C. Bamber and C. Daft, Adaptive filtering for reduction of speckle in pulse-echo images. *Ultrasonics.* 24:41 (1986)
11. N.L. Bush, C.R. Hill, Gelatine-alginate complex gel: a new acoustically tissue-equivalent material. *Ultrasound Med Biol.* 2:25 (1983).

SHEAR ELASTIC PROPERTY MEASUREMENT IN VIVO USING ULTRASONIC IMAGING OF INTERNAL FORCED LOW FREQUENCY VIBRATION

Yoshiki Yamakoshi,[1] Masahiko Sanada,[2] Masaaki Ebara,[2] Keisuke Oomura,[1]
and Satoru Kobayashi[3]

[1]Faculty of Engineering, Gunma University, Kiryu-shi, Gunma 376 Japan
[2]Faculty of Medicine, Chiba University, Chiba-shi, Chiba 260 Japan
[3]Faculty of Engineering, Teikyo Kagaku-gijyutu University, Chiba 260 Japan

INTRODUCTION

Shear elastic properties of soft tissues, such as shear elasticity and shear viscosity, may give useful information in medical diagnoses, because these mechanical parameters relate closely to the tissue stiffness.[1,2] We have already proposed an ultrasonic imaging system which can reconstruct vibration maps inside soft tissue under forced mechanical vibration.[3] This system was applied to the experiments in vitro and internal vibration maps were reconstructed for uterine tissues.

In this paper, an internal vibration imaging system for the experiments in vivo is presented. One of the problems in internal vibration measurement in vivo is the existence of inherent ultra low frequency movements, which come from cardiovascular movements and/or breathing movements. To suppress these movements, we introduce a quantitative internal displacement measurement technique. In the method, first, the internal movements which are contaminated by the extra movements are measured by using Arc-tangent method. Then the forced vibration component is extracted by applying narrow pass band filtering around the vibration frequency. Another problem in the experiment in vivo is that the vibration wave propagates in a complicated way inside the soft tissues. The refraction and reflection of the vibration wave at internal tissue surface decrease the measurement accuracy of the vibration velocity. To this problem, we introduce two step data selection method. In this algorithm, an optimum section is selected on the vibration phase map by taking into account of both vibration wave propagation and vibration wave penetration, the vibration velocity is estimated on the section.

Since 1993, this system has been evaluated at Internal Department of Faculty of Medicine, Chiba University, in Japan in order to measure the vibration velocities of liver tissues in vivo. The vibration frequency used in the experiments is 40 Hz. More than 200 patients of chronic hepatitis, liver cirrhosis and normal liver are examined.

LOW FREQUENCY VIBRATION VELOCITY ESTIMATION USING ULTRASONIC PROBING WAVE

When ultrasonic probing waves are transmitted to the tissue under forced mechanical vibration, the internal tissue movement gives rise to the ultrasonic Doppler effect resulting in the frequency shift of the reflecting ultrasonic probing wave Δf as follows ;

$$\Delta f = \frac{2f_0}{c} \frac{\partial \xi}{\partial t} \tag{1}$$

where c, f_0 are the sound speed and center frequency of the ultrasonic wave, respectively. The ξ is displacement of internal tissue movement due to the forced vibration. Then the received ultrasonic probing wave is given by

$$I_r = I_0 \sin \left(2\pi \int (f_0 + \Delta f)dt + \phi\right)$$

$$= I_0 \sin (2\pi f_0 t + A(t) + \phi) \tag{2}$$

where

$$A(t) = \frac{4\pi f_0}{c} \xi(t) \tag{3}$$

and ϕ is phase shift due to the wave propagation. In order to detect the Doppler frequency shift components $A(t)$ from the received wave, we apply quadrature detection technique. Two quadratic output signals of the detector are written as follows ;

$$O_1 = K\cos\{A(t)+\phi\}$$

$$O_2 = K\sin\{A(t)+\phi\} \tag{4}$$

From eq.(4), the internal displacement $\xi(t)$ is derived by the following equation.

$$\xi(t) = \frac{c}{4\pi f_0}\{\arctan(\frac{O_2}{O_1}) + l\pi\} \tag{5}$$

where l is phase offset which compensates wrap round of the Arc-tangent function.

Since the internal displacement which is observed by the above algorithm is usually contaminated by ultra low frequency displacements, which come from inherent movements, such as cardiovascular movement and/or breathing movement, the $\xi(t)$ may be written as follows ;

$$\xi(t) = a \sin (\omega_b t + \varphi_b) + n(t) \tag{6}$$

where ω_b is angular vibration frequency, a and φ_b are vibration amplitude and vibration phase, respectively. The $n(t)$ is the extra noise. Since the dominant frequency range of the n

(t) is usually much lower than the vibration frequency, the vibration component is extracted by applying narrow pass band filtering around the vibration frequency. By repeating the algorithm until vibration displacement along the ultrasonic beam line is measured, the vibration displacement in depth direction (z direction) $\xi(z,t)$ by applying Fourier analysis is given by ;

$$\xi(z,t) = a(z)\sin\{\omega_b t + \varphi_b(z)\} \tag{7}$$

where $a(z)$ and $\varphi_b(z)$ are vibration amplitude and vibration phase in depth direction, respectively. From the vibration phase $\varphi_b(z)$ which can be derived from $\xi(z,t)$, the vibration velocity v_b is estimated as follows;

$$v_b = \omega_b \left(\frac{\partial \varphi_b(z)}{\partial z}\right)^{-1} \cos\theta \tag{8}$$

where θ is angle between vibration wave propagation direction and direction of ultrasonic probing wave propagation.

EXPERIMENTAL SET-UP

Figure 1 shows schematic diagram of the experimental set-up. Specification of the system is summarized in Table 1. The small mechanical vibrator with round surface vibration head is used to apply low frequency vibration to the soft tissue surface. The ultrasonic waves of center frequency of 3.5MHz are used as probing waves and they are transmitted and are received by ultrasonic transducer array. The Doppler frequency shift signals of the received ultrasonic probing waves are detected by using commercial base ultrasonic pulsed Doppler imaging instrument (TOSHIBA SSA-250) and the detector output signals(Q sig. and I sig.) are fed into micro computer through a pair of high speed A/D converters. In the micro computer, both internal vibration amplitude and vibration phase are estimated for all measureme-

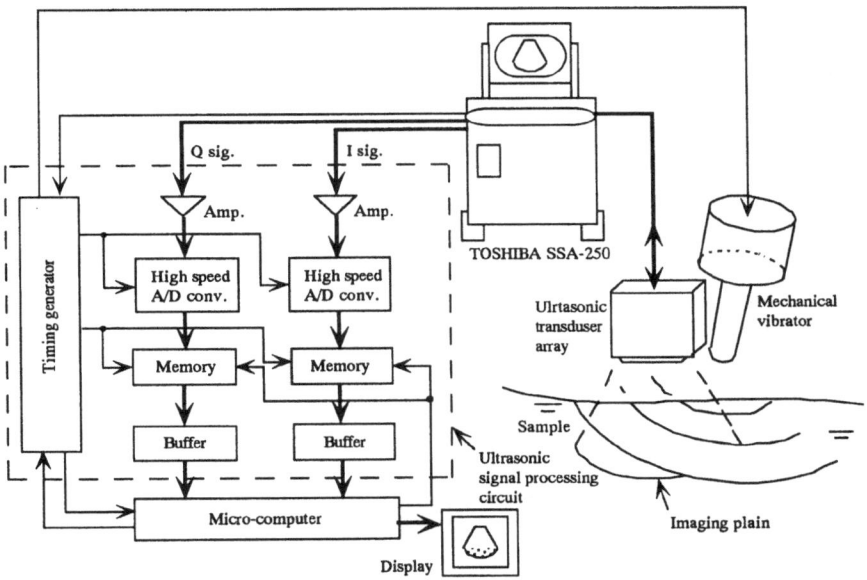

Figure 1. Experimental set up

Table 1. Specification of the constructed imaging system

Ultrasonic probing waves		Time required to get an image	
Center Frequency	3.5 MHz	B-mode type image	90 sec
		M-mode type image	in real time
Applied low frequency vibration		Reconstructed images	
Vibrator	Brüel & Kjaer type 4810	Vibration amplitude map	
Vibrator head	20mmφ half sphere	Vibration phase map	
Vibration amplitude	≈100 μm	Motion picture display	

nt points and the vibration maps are reconstructed. In this imaging system, both B-mode type vibration map and M-mode type vibration map are reconstructed. The B-mode type vibration map is suitable for observation of two dimensional internal vibration propagation, however it requires about 90 sec. for reconstructing an image. The M-mode type vibration map is suitable for vibration velocity estimation in the experiments in vivo, because this map is reconstructed in real time. From the vibration amplitude map, vibration wave penetration inside the tissue can be seen. The vibration phase advance due to the vibration wave propagation can be observed from the vibration phase map.

EXPERIMENTAL RESULTS

Two Step Data Selection In Vibration Velocity Estimation

The internal vibration maps which are observed by the constructed system are usually degraded by noises in the experiments in vivo. The additional ultra low frequency movements, such as cardiovascular movements and breathing movements, degrade the quality of the internal vibration maps. Besides these movements, large refraction and/or large reflection of the vibration wave at the internal tissue surface which give rise to standing wave also decrease the vibration velocity estimation accuracy. In order to suppress these noises, we introduce two step data selection method into the vibration velocity estimation. Figure 2 shows the schematic diagram of the method. Figure 2 (a) is an example of M-mode type vibration phase map $\varphi_b(z,t)$, which is obtained in liver tissue in vivo for vibration frequency of 40 Hz. This map shows the vibration phase advance due to the vibration wave propagation in depth direction (z direction). To this map, we apply first step data selection in order to select an optimum section where the vibration wave phase advances linearly in depth direction, because the vibration wave which propagates only in forward direction without producing any standing wave shows such a linear phase advance. Then, one dimensional vibration phase map $\varphi_{b,z}(z)$ is derived by using the following equation.

$$\varphi_{b,z}(z) = \frac{1}{\Delta T}\int_{\Delta T} \varphi_b(z,t)dt \tag{9}$$

Integration in eq.(9) is carried out over the section which is selected in the first step data selection. Figure 2(b) shows an example of the one dimensional vibration phase map $\varphi_{b,z}(z)$ calculated from the section A_2 shown in Figure 2(a). To this map, we introduce second step

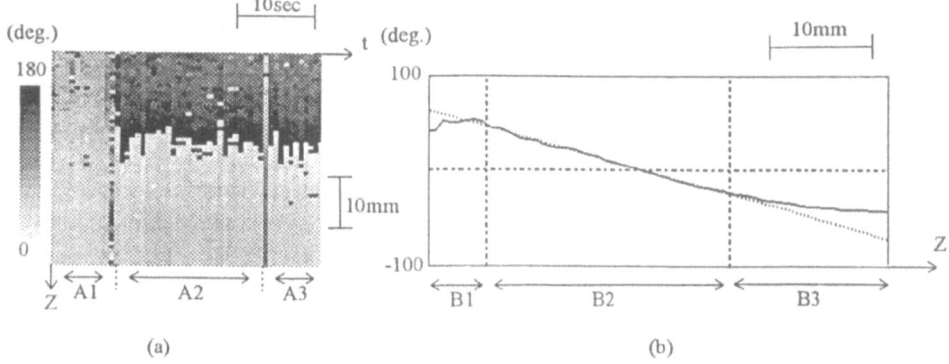

Figure 2. Two step data selection in vibration velocity estimation

(a): M-mode type vibration phase image. Almost linear vibration phase advance in depth direction is shown in both section A2 and A3.

(b): One dimensional vibration phase map in depth direction. Linear vibration phase advance is shown in section B2.

data selection in order to select an optimum section where the vibration wave phase advances linearly in depth direction. In this example, the section B_2 should be selected. From the selected section, the vibration phase differentiation $\delta\varphi_{b,z}(z)/\delta z$ is derived by applying least square error fitting method on the selected section. The vibration phase inclination calculated by the method is shown as dotted line in the figure. Then the vibration velocity is derived as follows ;

$$v_b = \omega_b \left(\frac{\partial\varphi_{b,z}(z)}{\partial z}\right)^{-1} \cos\theta \tag{10}$$

where the angle between the vibration propagation direction and the ultrasonic beam line θ is estimated beforehand from B-mode type vibration phase map.

Experimental Results In Vivo

In application of the two step data selection method to the experiments in vivo, we use the following four criteria in order to select an optimum section. i) Time width ΔT in eq.(9) is more than 2 sec. ii) Mean absolute error from linearly approximated line ε:

$$\varepsilon = \frac{1}{\Delta Z}\int_{\Delta Z} |\varphi_{b,z}(z) - \varphi_{b,z,L}(z)| dz \tag{11}$$

is less than 3 deg. in the second step data selection. In eq.(11), $\varphi_{b,z,L}$ is linearly approximated line estimated by least square error fitting method. iii) Section depth ΔZ is over 22.5mm. iv) Vibration amplitude average over the selected section is more than 10 μm. The criteria of ii) and iii) are taken so that a section where vibration wave propagates in almost forward direction (z direction) is selected. The criterion iv) is needed to select a section with relative large vibration wave penetration.

By using the constructed system, vibration velocities of liver tissues in vivo are measure-ed at Internal Department, Faculty of Medicine, Chiba University, Japan. Number of patients are 83 for chronic hepatitis, 153 for liver cirrhosis and 45 for normal patients. The frequency of low frequency vibration used in the experiments is 40 Hz. The result is shown in Figure 3. Low frequency vibration velocity of chronic hepatitis is 983 cm/sec (standard deviation S.D = 373cm/sec), liver cirrhosis is 1200 cm/sec (S.D = 416 cm/sec) and normal liver is 632 cm/sec (S.D = 148 cm/sec). For some of the patients with diseased liver, biopsy examinations are also carried out. The correlation between square root of low frequency vibration measured by the constructed system and the percentage of fibrotic area of the liver tissue measured biopsy examination is shown in Figure 4. From these results, we see that the diseased liver tissues show relative higher vibration velocities than those of the normal liver tissues. We also see that the vibration velocities have a correlation with the percentage of fibrotic area of the diseased tissues.

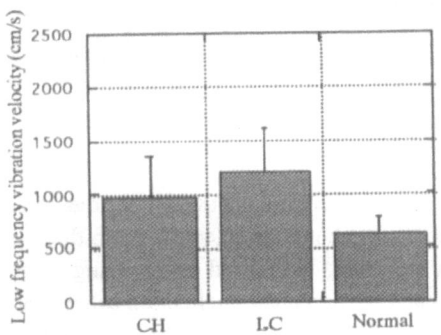

Figure 3. Experimental results in vivo

CH: Chronic hepatitis   LC: Liver cirrhosis

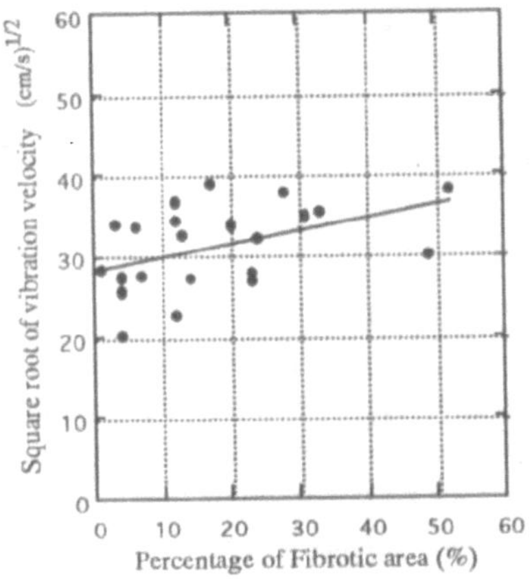

Figure 4. Correlation between vibration velocity and biopsy examination

CONCLUSION

An ultrasonic imaging system which can reconstruct internal low frequency vibration in the experiments in vivo is constructed. This system measures sinusoidal tissue motion due to the forced mechanical vibration by applying quantitative internal small displacement measurement based on ultrasonic wave pulsed Doppler technique. This system is applied to low frequency vibration velocity measurement of liver tissue in vivo and more than 200 patients have been evaluated.

REFERENCES

1. H.L.Oestreicher. "Field and Impedance of an Oscillating Sphere in a Viscoelastic Medium with an Applicaion to Biophysics" J. Acoust. Soc. Am. 27:707 (1951).
2. R.M.Lerner, S.R.Huang and K.J.Perker. "Sonoelasticity Images for Ultrasound Signals in Mechanically Vibrated Tissues" Ultrasound Med. Biol. 16:231 (1990).
3. Y.Yamakoshi, J.Sato and T.Sato. "Ultrasonic Imaging of Internal Vibration of Soft Tissue under Forced Vibration" IEEE trans. on UFFC. 37:45 (1990).

TEXTURE CLASSIFICATION OF ECHOGRAPHIC IMAGES BY MEANS OF THE COOCCURRENCE MATRIX

F.M.J. Valckx and J.M.Thijssen

Biophysics Laboratory, Department of Ophthalmology
University Hospital St-Radboud
6500 IIB Nijmegen, The Netherlands

INTRODUCTION

The cooccurrence matrix is used to analyse echographic video images. However, the size of the speckle, which depends on the transducer characteristics, depth, direction and underlying tissue,[1] influences the best choice for the displacement parameter d. Morris[2] suggested that an optimal value for d can be obtained in the following way: Firstly, a cooccurrence feature is calculated in the four directions and with different displacements. After this, the mean value of the four angles is calculated and the d at which this mean feature becomes independent of d is estimated. However, because of the different speckle dimensions in axial and lateral direction (expressed by the correlation length), the optimal d in fact is greatly different for these directions. Furthermore, the lateral speckle size is very much depth-dependent[1] (200% or more), so the optimal displacement d will also be depth dependent. Finally, it is not sure that all the features will have the same optimal d for a certain speckle size. In this paper the influence of the speckle size on the optimal d is investigated for five cooccurrence features. This investigation might lead to the optimal use of the cooccurrence matrix for detection or differentiation of tissue conditions wich are realted to the effective scattering number density.

COOCCURRENCE MATRIX

The cooccurrence matrix is a two-dimensional histogram which describes the occurrence of the gray levels of pairs of pixels which are separated by a certain distance d (Δx in x direction and Δy in y direction) and where the connecting straight line has a direction

θ (i.e. $\theta = arctan(\Delta y/\Delta x)$). Let $I(x, y)$ be a digital image with size $L_x \times L_y$ and M gray levels. So, for two pixels (x_i, y_i) and (x_j, y_j) with gray levels i and j, it follows that $\Delta x = x_j - x_i$ and $\Delta y = y_j - y_i$ and the cooccurrence matrix $C_{\theta,d}$ is defined as:

$$
\begin{aligned}
C_{\theta,d}(i,j) \; = \; & \#\{((x_i, y_i), (x_j, y_j)) \in |(L_x \times L_y) \times (L_x \times L_y)| \\
& \Delta x = d \sin\theta, |\Delta y| = d \cos\theta \\
& I(x_i, y_i) = i, I(x_j, y_j) = j\}
\end{aligned}
\tag{1}
$$

with # number of elements in the set. For $\theta = 45°$ or $135°$ it follows that the real displacement d' is $d\sqrt{2}$.

A great number of features can be calculated from the cooccurrence matrix.[3] In ultrasound images the most commonly used features are entropy (ENT), contrast (CON), correlation (COR) and angular second moment (ASM)[4,5] A parameter which has been suggested for detecting texture periodicity is the κ parameter.[6] This feature is added to the more commonly used features to see if it will give some extra information.

The entries in the cooccurrence matrix, and therefore, the features derived from it, depend on the displacement d and the angle θ. The size of the cooccurrence matrix equals the number of gray levels M. To avoid a large cooccurrence matrix, the number of gray levels is reduced after histogram equalization. This equalization is necessary to avoid the problem of losing information when reducing the number of gray levels. The method of equalisation employs the desired histogram (i.e. a uniform probability density function) to redistribute the bins of the original histogram.[7]

SIMULATIONS

The basic idea of the simulations is to create an rf-line by means of randomly distributed (in 1-D) scatterers. Since we want to change the speckle size of the images, the transducer characteristics are changed (Table 1). In all the simulations, the number of scatterers in the resolution cell is 10, so there is fully developed speckle.[8] The sampling frequency is set at 30 MHz and the sound velocity is 1540 m/s . From the above-mentioned parameters it follows that the size of one sample corresponds to 0.0255 mm. To obtain envelope signals, the rf-lines of the simulations are demodulated by means of the complex Hilbert transform. The number of graylevels is 256. The overall statistical sample comprises 64 lines which are 512 samples long. The cooccurrence matrix is calculated with d varied from 1 to 25 samples. The gray level reduction factor is eight, so the number of gray levels M used to calculate the cooccurrence matrix is 32.

Table 1. Transducer characteristics

f_c [MHz]	Δf [MHz]	$FWHM_{ax}$ [mm]	$FWHM_{ax}$ [samples]
2.5	1	0.46	18.3
3	1.66	0.28	11.4
3.75	2	0.23	9.2
5	2.66	0.17	6.9
7.5	4	0.115	4.7
10	5	0.09	3.7

RESULTS

In Figure 1 the features entropy and correlation are given with displacement d varied for different sizes of the speckle. The characteristics of the other features show similar dependence on the displacement d and speckle size. In Figure 2 the optimal displacement d is given for the different features versus the speckle sizes. The optimal d is defined as the displacement where the value of the feature surpasses 95% of the end value ($d = 25$ samples) of that feature. Since the end value is zero for the features COR and KAP, the boundary for these features is set at 5% of the value at $d = 0$ samples (i.e. 0.05).

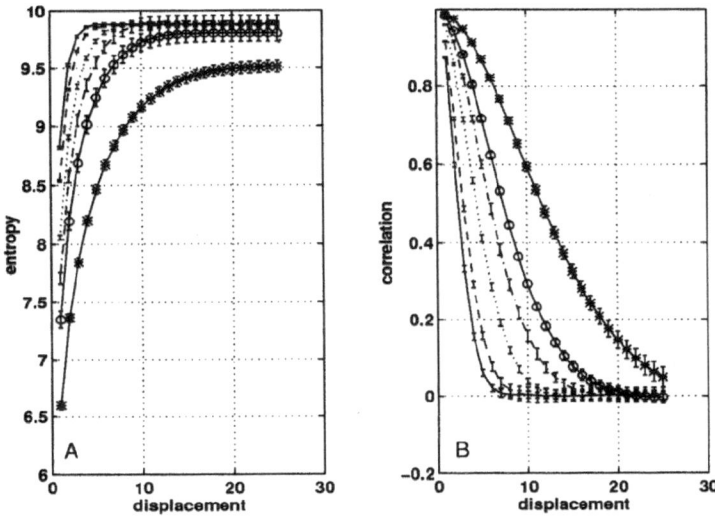

Figure 1. Features: entropy (A) and correlation (B) vs. displacement parameter d with different speckle size: asterix: speckle size 18.3 samples; circle: 11.4 ; dashed-dotted line: 9.2; dotted line: 6.9; dashed line: 4.7; solid line: 3.7. Errorbars indicate standard deviation of the mean.

DISCUSSION

As mentioned before, there is a clear dependence of the optimal d on the speckle size. Furthermore, it is obvious from Figure 2 that the features COR and CON have a higher optimal d than the other ones. Therefore, the notion of an optimal value of d as discussed by morris[2] has no practical importance. On the other hand, it follows from figure 1 that the value of the features depends strongly on the speckle size. At small values of the displacement parameter ($d = 1$), it is possible to distinguish between different sizes of the speckle when using ENT or ASM, whereas a medium value ($d \approx 4$) yields a good differentiation of speckle sizes for COR and CON. The speckle size depends both on the number density of the scatterers and on depth. Since the depth dependency is stronger compared to the number density dependency, it should be accounted if the cooccurrence matrix is used in texture classification. These (newly defined) optimal d-values correspond to 1/4 and 1 times the speckle size at fully developed speckle.

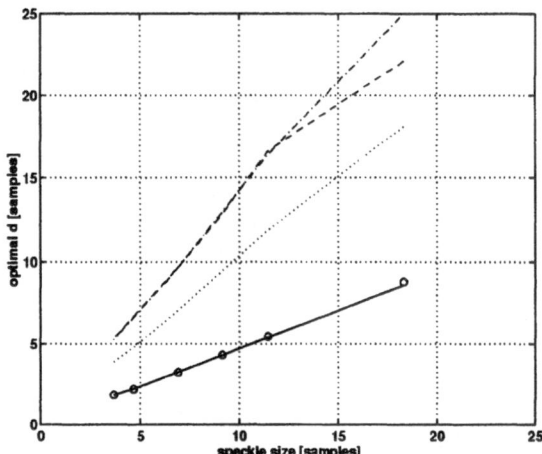

Figure 2. Optimal d versus speckle size for 5 features: solid line: KAP; dotted line: ASM; dash-dotted line: COR; dashed line: CON; circle: ENT. Number of scatterers in resolution cell: 10. Transducer properties are varied according to Table 1.

REFERENCES

1. B.J. Oosterveld, J.M. Thijssen, and W.A. Verhoef, Texture of B-mode echograms: 3-D simulations and experiments of the effects of diffraction and scatter density, *Ultrasonic Imag.* 7:142-160 (1985).

2. D.T. Morris, "An evaluation of the use of texture measurements for tissue characterization of ultrasonic images of in vivo human placentae," *Ultrasound Med. Biol.* 14:387-395 (1988).

3. R.M. Haralick, K. Shanmugan, and I. Dinstein, Textural features for image classification, *IEEE Trans. Syst. Man Cybbern.* 6:610-621 (1973).

4. J.M. Thijssen, B.J. Oosterveld, P.C. Hartman, and G.J.E. Rosenbusch, Correlations between acoustic and texture parameters from Rf and B-mode liver echograms, *Ultrasound Med. Biol.* 19:13-20 (1993).

5. U. Raeth, D. Schlaps, B. Limberg, et al. Diagnostic accuracy of computerized B-scan texture analysis and conventional ultrasonography in diffuse parenchymal and malignant liver disease, *J.Clin. Ultrasound.* 13:87-89 (1985).

6. J. Parkinnen, K. Selkäinaho, and E. Oja, Detecting texture periodicity from the cooccurrence matrix," *Pattern Recognition Letters.* 11:43-50 (1990).

7. R.C. Gonzales, and P. Wintz. "Digital Image Processing,"Addison-Wesley, Reading (1987).

8. J.W. Goodman, Statistical properties of laser speckle patterns, *in:* "Laser speckle and related phenomena," J.C. Dainty, ed., Springer, Berlin (1975).

POLAR COORDINATES TEXTURE ANALYSIS
METHODS FOR DIGITIZED INTRAVASCULAR IMAGES

I. Hardouin[1], E. Lieback[1], J. Armbruster[1], J. Boksch[2], M. Schartl[2], and R. Hetzer[1]

[1]German Heart Institute Berlin, Augustenburger Platz 1, 13353 Berlin
[2]University Clinic R. Virchow, Spandauer Damm 130, 14050 Berlin

INTRODUCTION

The image texture of ultrasound images correlates highly to the histological composition of the corresponding insonified tissue. Therefore, tissue identification can also be performed by analyzing textural features. Many methods of digital image texture analysis are based on measuring the relationship between pixels in a given environment. This environment is described in terms of pixel deviation in both horizontal and vertical directions. Intravascular images show different tissues such as thrombi, soft or calcified plaques, as well as the various wall layers arranged in concentric rings around the ultrasound catheter. Due to the circular geometry of blood vessels, regions of interest are better defined as circular portions of sectors than as quadratic windows as usually considered in image processing. Additionally, spatial resolution decreases with growing radii such that the normally computed texture spatial features, i.e. gray level run length or co-occurrence properties, should not be equally treated in all image regions. The texture appears to be finer with smaller radii, although for a histologically uniform region, tissue structure may not be different. Texture analysis methods were therefore computed in polar coordinates, as presented in the first section.

Once suitable texture features have been identified, i.e. once each particular texture has been described, automatic classification can be performed. In this report the result is presented as a segmented color encoded image.

METHODS

Conversion to Polar Coordinates Image

The first processing step consists of automatically determining the image center (catheter) and diameter. Although images were always acquired using the same ultrasound machine, i.e. the center could have been determined once for all samples, a computational method for finding the center was preferred. Such a method can be employed for other images or other imaging systems. After binarization at a user-defined gray level, one scans through rows and columns making note of the first and the last color transitions. The center is found at the crossing between row and column where the largest distance has been measured between the first and last transition. The average value of the largest horizontal and vertical distances measured is considered the imaging diameter.

The image is then considered as a disk and divided into n rings and m angles. The average gray value in each sector portion is stored in a two-dimensional buffer (n rows and m columns) which is considered a ring buffer in the angle direction. This means that first and last columns are considered adjacent. One should note that some angle values do not exist at small radii because of the discrete nature of digital images. For example, at a radius of 2 pixels the only angles which exist are those whose tangent has an absolute value of less than 2. Therefore, the $m \times n$ matrix of averaged sector values is not always full. This is important for computational reasons. The non-existing values must be noted in order to avoid being further considered in the matrix.

Texture Analyses

The gray level histogram, co-occurrence matrices[1], run lengths[2], and power spectrum were computed for texture analysis. Co-occurrence matrices were calculated using radial and tangential displacements of value one, two, and four for sectors and rings. Averaged gray value dependencies were measured within a greater portion of sector (covering 25 rings and 25 angles, see Fig. 1). Here the nature of the ring buffer was important as it permitted analysis of the continuous image properties between 359° and 0°. Run lengths were similarly computed over rings and angles. A ring run is an arc of the same gray value while an angle run is a portion of a diameter. Figure 1 illustrates both means of analyzing texture in polar coordinates. The 2D power spectrum was computed in the corresponding 25x25 table which is considered a standard image.

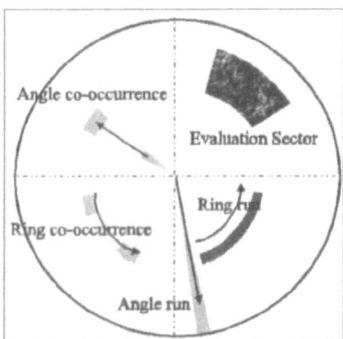

Figure 1. The principles of co-occurrence and run length analyses are extended in order to describe relationships between the parts of an image situated over the same ring at different angles or at the same angle among different rings.

The computed gray level histogram as well as co-occurrence, run length, and spectral matrices were described either using normal statistical moments or features proposed by other authors[1,2]. Initially 51 texture features found to be significant in a previous study[3] of echocardiographic image texture were retained.

Classification

Supervised learning classification methods were used for segmenting the intravascular images in this study. An IVUS specialist acquired the images and verbally described their contents thus allowing sets of feature vectors representing each class and possible variations within the class to be collected. The first classification was performed using statistical methods: maximum likelihood (Bayes) and Mahalanobis classifier. The feature vector to be classified was assigned by the statistically nearest class according to the classifier rule.

The learning sets were then taught to a back propagation neural network. Its architecture consisted of three layers. The size of input and output layer was given by features of the data: the number of texture parameters for the input layer, and the desired classification in three distinct classes. The size of the hidden layer was empirically set to 25 neurons.

STUDY DESIGN AND RESULTS

Intravascular images from patients undergoing P.T.C.A. were recorded on video tape. Complete standardization of the recordings was not possible since each examination had to be performed with a new sterile catheter. Variations in catheter performance did not seem negligible. The high quality recordings (SuperVHS) were replayed and acquired by a SONY 9500 video device and subsequently digitized into a 486 33MHz PC using the KONTRON system IMCO 10 (512x512, 256 gray levels).

Expert knowledge and learning sets

As learning and test sets were collected, the expert selected and named different image regions which were then analyzed for texture description. The final size of the learning sets was limited by the number of vectors representing calcified plaque, which was always smaller than that of soft plaque or thrombi, in order to acheive similar sets of all three. Regarding thrombi it should be mentioned that the expert considered that a particular texture (see figure 2, left image) must have represented thrombus because it disappeared after P.T.C.A. Fourty feature vectors of each class set were used for teaching the back propagation neural network as well as for creating the knowledge base while the remaining were used for test purposes.

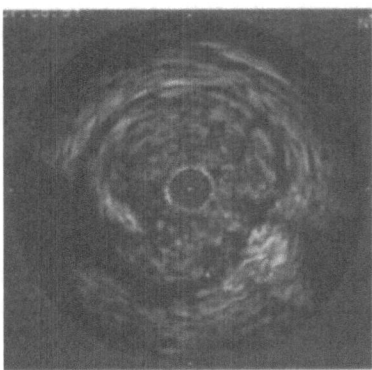 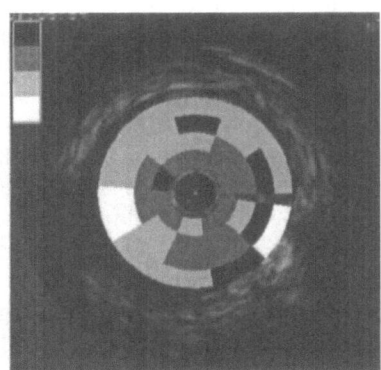

Figure 2. This IVUS image contains the three textures which were to be automatically recognized. The bright region represents calcified plaque while soft plaque is darker. The expert noticed thrombus around the catheter.

Classification results

Table 1 summarizes the classification results applying statistical and neural classification to the test sets. Both statistical classifier delivered the same results. A feature set was classified when the probabilty of belonging to a particular class was at least 50%

Table 1. Classification results when statistical base -learning sets- consisted of about 40 sets per class. Thrombi were not recognized using statistical classifiers.

	Calcified plaque (23)		Soft plaque (33)		Thrombi (33)	
	statistical	neural	statistical	neural	statistical	neural
Calcified plaque	87%	96%	0%	0%	3%	0%
Soft plaque	4%	4%	88%	85%	67%	6%
Thrombi	0%	0%	0%	12%	0%	76%
Not Classified	9%	0%	12%	0%	30%	18%

higher than the probability of belonging to another class. Note that thrombi were most often classified as soft plaque and never as thrombi.

The neural classification process (network recall) was therefore implemented as a visualization routine for on-line (but not realtime) PC-supported classification.

On-line recognition of image contents

After choosing the region to be classified, the 51 texture parameters were computed and sent to the recall routine which delivered the neural classification results. The set sector ROI was divided into evaluation windows (25 rings, 25 sectors) which were classified separately. The windows were then color encoded with red, blue, and green levels in order not to mask the texture. This allowed visual evaluation of the classification. If the net rejected any class affectation, the corresponding evaluation window remained unchanged (gray tones). Figure 2 (right image) illustrates the process with the exception of color encoding which has been replaced by uniform gray coding (white for calcified plaque, light gray for soft plaque and dark gray for thrombi).

CONCLUSION

Classification of texture features was performed by a back propagation neural network. This allowed for the characterization of three different tissues which may be present in intravascular images: calcified, soft plaque, and thrombi. Statistical evaluation of the specificity and sensitivity of this method would required knowledge about the histological composition of the inner vessel at the place where the image was acquired. Such evaluation was not possible as the images were made of patients during P.T.C.A. who are still alive. The classification reflected the expert knowledge very well since the image event, particularly for calcified plaque, which are small, fit well in the evaluation windows. This can be artificially achieved when one selects only the corresponding region.

Further topics in the domain will deal with in-vitro improvements of the method, analyzing images acquired in explanted hearts, and images of fresh thrombi. Improving computational speed, necessary for making clinical use of the system viable, is another goal which can be achieved by limiting hard disk accesses.

REFERENCES

1. R. Haralick, K. Shanmugan, and Its'hack, Textural features for image classification, *I.E.E.E. trans. on systems, man &cybernetics,* Vol. 3, n°6 (1973).
2. M. Galloway, Texture analysis using gray level run lengths, Computer Graphics and Image Processing, n°4 (1975).
3. E. Lieback, "Computer gestützte sonographische Gewebedifferenzierung des Myokards", Steinkopff Verlag, Darmstadt (1993).

CLINICAL VALUE OF ECHOCARDIOGRAPHIC TISSUE CHARACTERIZATION IN THE DIAGNOSIS OF MYOCARDITIS

E. Lieback[1], I. Hardouin[1], R. Meyer[2] , J. Bellach[2], J. Armbruster[1],
and R. Hetzer[1]

[1]German Heart Institute Berlin, Augustenburger Platz 1, 13353 Berlin
[2]Charité, Humboldt University Berlin, Schuhmann Str. 21-22, 10269 Berlin

INTRODUCTION

In international literature there are increasing indications that dilated cardiomyopathy is the result of an unrecognized episode of myocarditis (1). As no dependable clinical and praclinical diagnostic model exists for detecting florid myocarditis, endomyocardial biopsy remains the only sure means of diagnosing myocarditis. Although endomyocardial biopsy have a relatively low complication rate, they are nevertheless an invasive procedure. The consideration of using ultrasound for tissue characterization of the myocardium based upon the assumption that a reproducible connection exists between the anatomical structure of tissue and ist acoustical properties. Inflammatory changes of myocardium have been visually observed as alterations on echocardiograms. The goal of this study was to determine how these myocardial changes affect the texture of echocardiographic images, and how these could be described through quantitative texture analysis.

METHODS

One hundred six patients (41 women, 65 men) ranging in age from 19 to 57 years old (average age 40.41 ± 10.59 years) clinically suspected of having myocarditis underwent endomyocardial biopsies. In order to monitor their therapeutic course, some patients were biopsied several times thus yielding a total of 142 myocardial biopsies which were evaluated. The control group consisted of 24 patients (12 women, 12 men) ranging in age from 22 to 47 years old (average age 33.85 ± 7.51 years) who were healthy according to anamnestic, clinical, biochemical, electrocardiographic, and echocardiographic criteria. Additionally, biopsies of 8 of these patients revealed normal histological findings.

All of the patients underwent a standardized echocardiographic examination on the same day of their endomyocardial biopsy. When scanning the left parasternal short axis on the level of the papillary muscles, particular care was taken that the sonic beam's angle of incidence was approximately 90° in the area of the anteroseptal segment. The amplification settings used in their respective initial examinations were used again on patients who underwent repeated examinations. In general an amplification of 72-76 dB was used. The time gain compensation for proximity was set to 0. Other settings such as enhancement, dynamic range, reject, and gamma correction were kept constant for all examinations. A depth setting of 15 cm was always used. The ultrasound image was frozen at end diastolic time and digitized on-line to the image processing system. A Region Of Interest (ROI) was manually placed in the anteroseptal segment of the left ventricle. It was divided in windows of 25 x 25 pixels in which texture parameters were calculated. The ROI texture was analysed using measurements related to intensity and distribution of echo amplitudes (grey levels). In the present study, four major types of texture analysis were used: grey value histogram, co-occurrence matrix, run length and power spectrum analyses (2,3,4,).

The set of 117 originally calculated texture parameters were compared to the biopsy results. Since only a few of these texture parameters were essential for aimed discrimination (myocarditis - healthy control patients), the most expressive parameters were selected. The feature selection was performed iteratively, using the Fisher's discriminant analysis.

Table 1. Histological results of endomyocardial biopsies

myocarditis without fibrosis	33
myocarditis with fibrosis	19
persistent myocarditis	12
healed myocarditis without fibrosis	9
healed myocarditis with fibrosis	17
dilated cardiomyopathy	35
hypertrophy	5
ischemic heart disease and small vessel disease	4
no pathological changes	8
	142

RESULTS

The histological results, recorded from a total of 142 endomyocardial biopsies, are summarized in table 1. The co-ocurrence feature Angular Second Moment (ASM) was selected before proceeding with feature selection, since it already exhibited a high variation coefficient during the examination of beat-to-beat variability. The parameter reduction yielded the following important parameters: two histogram features -Mean Grey Level (MGL) and 90% percentile-, three undirected co-occurrence features -Inverse and Second Difference Moments (IDM and SDM), and entropy- computed for two distances -2 and 4 pixels-, four run length parameters -Grey Level and Run Length Non uniformity (GLN and RLN), both vertical and horizontal-, and two spectral features -Ring Sum R1-R6, Sector Sum 90°.

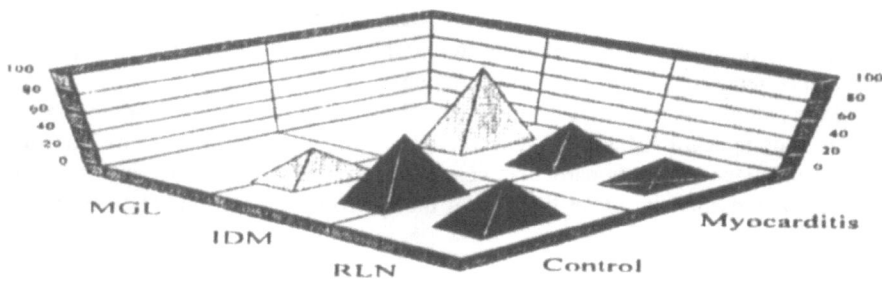

Figure 1. Average features values of MGL, IDM and RLN for two groups (myocarditis and healthy)

Figure 1 illustrates the variations of features MGL, IDM and RLN over two classes (myocarditis and control groups) whereas figure 2 shows the variations of the same features over six subgroups. The above-mentioned parameters formed the basis for further discriminance analysis. Before presenting the classification results, let us consider the results of feature selection.

The correlation matrix of the power spectrum's sector sum showed that they were highly correlated to each other. Therefore one sector was sufficient in describing the texture's directivity. By using the parameters of run length statistic, the parameters short run emphasis and long run emphasis could be dispensed with since their indicative value, as noticeable in the correlation matrix, was taken over by the other parameters of the run length statistic. Regarding the co-occurrence matrix, it was observed that a pixel interval of dx=6 did not deliver additive information.

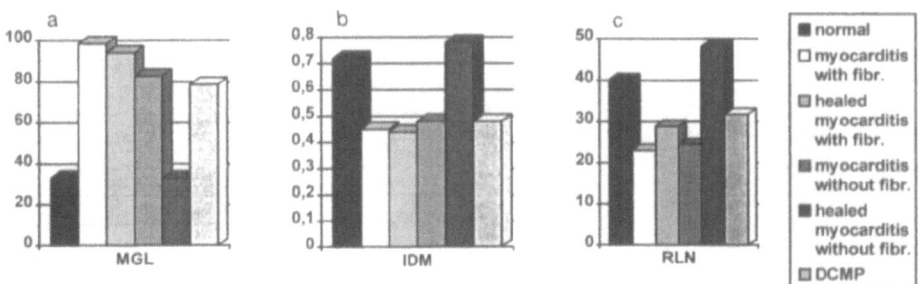

Figure 2 a,b,c . Average feature values of MGL,IDM and RLN for six groups

In order to differentiate between normal and pathological myocardium, myocarditis with or without fibrosis was defined as Class 1 and the normal group (normal with or without biopsy) as Class 2.

		Computer diagnosis		
		Myocarditis	Healthy	Sensitivity
Biopsie Results	Myocarditis	26	0	100%
	Healthy	3	29	91%

Figure 3 : Computer-generated attribution tables for myocardial and normal group

Performing discriminance analysis using upwards and downwards processing indicated that the texture parameters MGL, IDM -dx=2 undirected-, and RLN -horizontal- were optimal classification criteria. As indicated by the calculated attribution table (figure 3), these texture parameters made a very clear distinction between myocarditis and normal controls possible. The Bayes' risk for this classification, i.e. the probability of erroneous classification was then less than 5%. The attribution table shows that the sick patients, i.e. the patients with myocarditis, were always correctly identified whereas three of the control patients exhibited false positive results.

In order to further differentiate the various histomorphological states of the myocardium, a discriminance analysis was performed with the seven classifications given in table 2. In turn, performing Fischer's linear discriminance analysis using upwards processing indicated that the texture parameters MGL, IDM -dx=2 undirected-, and RLN -vertical- were optimal classification criteria.

309

A further increase in dimensions was not deemed justifiable by the computer. The attribution table calculated based upon these three texture parameters are illustrated in figure 4. The probability of erroneous classification was less than 10%.

Differentiating between florid myocarditis with and without fibrosis was not possible. Healed myocarditis without fibrosis could be clearly separated from florid myocarditis. Contrarily, healed myocarditis with heavy fibrosis could not be distinguished from florid myocarditis or dilated cardiomyopathy. Dilated cardiomyopathy could also not be distinguished from florid myocarditis or healed myocarditis with heavy fibrosis. There was a clear distinction from normal myocardium. The histological findings of hypertrophy, used as a control group, could be clearly distinguished from all pathological situations.

Computer Diagnosis

Biopsies	1	2	6	4	5	7	3	
1	11	7	2	0	0	0	0	20
2	4	2	0	0	0	0	0	6
6	8	4	1	1	0	0	0	14
4	1	0	0	4	2	0	1	8
5	0	2	0	9	6	0	7	24
7	0	1	0	4	1	0	1	7
3	0	0	0	1	1	0	3	5

Table 2. Histomorphological classes

Class 1	florid myocarditis without fibrosis
Class 2	florid myocarditis with fibrosis
Class 3	hypertrophy
Class 4	normal biopsy findings
Class 5	healthy patients (no biopsy)
Class 6	dilated cardiomyopathy
Class 7	healed, without fibrosis

Figure 4 . Computer-generated attribution table for various histological states of the myocardium

CONCLUSION

As the myocytolysis and cell infiltration associated with myocarditis change the acoustic properties of the myocardium, sonographic determination of these acoustic parameters appears to open a means of non-invasive diagnosis of myocarditis. The results of this study indicate that changes in echocardiographic texture occur in myocarditis and as well as during fibrosis of the myocardium. Already using a combination of only three texture features (mean grey value, inverse difference moment with a pixel interval of 2 undirected, and run length non uniformity horizontal) it was possible to separate the group with myocarditis (with or without fibrosis) from the control group.

As texture analysis of echocardiograms can not primarily differentiate if the changes in echocardiographic texture are caused by cell infiltration and oedema (florid myocarditis) or through fibrosis, control examinations are recommendable. Repeat examinations were able to detect regression of the changes in echocardiographic texture if they were caused more by inflammation, which was not the case with fibrosis. The echocardiographic texture of biopsy-proven cardiomyopathy was also similar to florid myocarditis in order to evaluate the course of the disease and the efficiency of therapy.

REFERENCES

1. J.B. O'Conell, The role of myocarditis in endstage dilated cardiomyopathy, Texas Heart Inst. 14: 268-275 (1987)
2. M.M. Galloway, Short note - Texture analysis using gray level run lengths, Comp Graph Imag Process. 4: 172-175 (1975)
3. R.M. Haralick, Statistical and structural approaches to texture, Proc IEEE. 67:785-804 (1979)
4. J.S. Weszka, Ch.R. Dyer, and A. Rosenfeld, A comparative study of texture measures for terrain classification, IEEE Trans Sys Man Cybern. 6: 269-285 (1976)

IN VITRO AND IN VIVO ASSESSMENT OF ULTRASOUND PARAMETRIC IMAGES OF BONE

Pascal Laugier, Pascal Droin, Bruno Fournier and Geneviève Berger

Laboratoire d'Imagerie Paramétrique, URA CNRS 1458
15 Rue de l'École de Médecine
75006 Paris, France

INTRODUCTION

Quantitative ultrasound techniques were introduced 10 years[1] ago as a new tool for the assessment of skeletal status in osteoporosis. Bone densitometry based on X-ray (Dual X-ray Absorptiometry [DEXA] and Quantitative Computed Tomography [QCT]) currently represents the state of the art approach for non-invasive assessment of mineralization of bone. A significant association between bone mineral density (BMD) and osteoporotic fracture risk has been established[2]. However, the prediction of fracture risk could potentially be improved further by additional assessment of bone's internal structure and bone quality (i.e. material properties of the mineralized tissue itself and of individual trabeculae). It appeared to be possible to obtain additional information unrelated to BMD from the ultrasound measurement of the frequency dependent attenuation (Broadband Ultrasonic Attenuation or BUA) and the speed of sound (SOS)[3].

Ultrasound is now commercially available from different companies. Most of the devices measure ultrasound in the calcaneus (heel bone) at a single location without accurate control of the transducer positioning with respect to the interindividual anatomical variations. Our group has previously shown that ultrasound parametric images of the calcaneus could enhance the technique by allowing reproducible repetitive measurements[4-5]. The purpose of the present study was to assess the imaging technique *in vitro* and *in vivo* . *In vitro*, the study aimed at adding new experimental evidence to clarify the relation between ultrasonic properties and BMD. To assess the clinical interest of this technique a new prototype was developed. To date, the first *in vivo* ultrasound parametric images of the calcaneus are presented and compared to X-ray radiography.

IN VITRO MEASUREMENTS

Material and method

Samples: Ultrasound images of bone were obtained in transmit mode by using a pair of focused transducers of center frequency 500 kHz. The experimental set-up has been previously described[6]. The present study was based on 15 heel bones (calcaneus) removed from cadavers (6 males and 9 females, ages ranging between 69 and 89). Samples of pure trabecular bone, with parallel sides and 12 mm of thickness approximately were obtained by slicing off the cortical lateral faces. The samples were then defatted using ethanol and ether and finally refilled with water under vacuum. A standard QCT was achieved for each sample using a 10 mm thick slice. With this procedure, the same volume of pure trabecular bone could be imaged with both ultrasound and X-ray.

Parameter measurement: The frequency dependent attenuation $\alpha(x,y,f)$ in decibels was found from the ratio of the magnitude spectrum $A(x,y,f)$ of the signal transmitted through the heel and a reference spectrum $A_r(x,y,f)$ of the signal transmitted through water according to:

$$\alpha(x,y,f) = 20 \log_{10}\left(\frac{A(x,y,f)}{A_r(x,y,f)} \right)$$

where x,y indicates the measurement position in the scan. In cancellous bone, the attenuation displays a nearly linear frequency dependence. The slope of attenuation or BUA(x,y) is the slope of the linear regression line fitted to $\alpha(x,y,f)$ between 200 and 600 kHz. The velocity v(x,y) was deduced from the difference of the transit time (measured from the first zero-crossing of the time signals) of both pulses between the transducers. Parametric image of the slope of attenuation (BUA) and of ultrasound velocity were derived in a few seconds from the calculation of the parameters value for each measurement site in the scan.

A software was developed, capable of reading both ultrasound and QCT images and of selecting site matched regions of interest (ROI) on the three images. The size, form, location of the ROI were adjustable by the operator. The average value of each parameter was then calculated for each ROI. An average Hounsfield Number (HN), which is linearly related to BMD, was derived from the QCT. We investigated the intra-sample variability by selecting several non-overlapping ROIs for each sample. Circular ROIs (7 mm in diameter) were chosen. According to the size of the specimen 12 to 20 non overlapping ROIs could be selected.

Statistical analysis: Analysis was performed using SYSTAT software (SYSTAT Inc, Evanston, Illinois). Simple linear regressions were used to model the intra-sample relationships between acoustic parameters and BMD. Then, all the data were pooled and multiple regression was used to model relationships between acoustic parameters and BMD. A categorical variable, representing the specimen (donor) was introduced in the model. The effect of introducing this new variable in the model was analysed with regard to the R^2 change.

Results

Images: Figure 1 shows the comparison between ultrasound images of BUA and SOS and the QCT image of the same sample. All the three images look alike and although the resolution achieved with ultrasound (the beam width at the focus was 4 mm approximately) is somewhat lower than with QCT, most of the macroscopic details (compared to the ultrasonic wavelength) can be identified on the three images. The regions of low BMD appearing in dark on the QCT image correspond to regions of low BUA or low SOS (also appearing in dark).

Correlation between ultrasound parameters and bone mass density: Figure 2 shows the data from a representative sample. The values of ultrasound parameters are plotted against the values of BMD (unit: CT Hounsfield number). For this sample, the correlation coefficients between BUA/BMD, SOS/BMD and SOS/BUA, respectively r=0.87 (r^2=0.76), r= 0.85 (r^2=0.73) and r= 0.98 (r^2=0.96) , were highly significant ($p<10^{-5}$). The average intra-sample correlation coefficient between BUA and BMD was 0.89 (r^2=0.79) ($p<10^{-4}$). The correlation coefficients for the 15 specimen were distributed between 0.52 and 0.97. The average intra-sample correlation coefficient between SOS and BMD was 0.93 (r^2=0.85) $p<10^{-4}$). The correlation coefficients for the 15 specimen were distributed between 0.82 and 0.97. The average intra-sample correlation coefficient between SOS and BUA was 0.93 (r^2=0.85) $p<10^{-4}$). The correlation coefficients for the 15 specimen were distributed between 0.50 and 0.99. An analysis of the constant and coefficient terms of the linear regression showed that these equations where specimen-dependent.

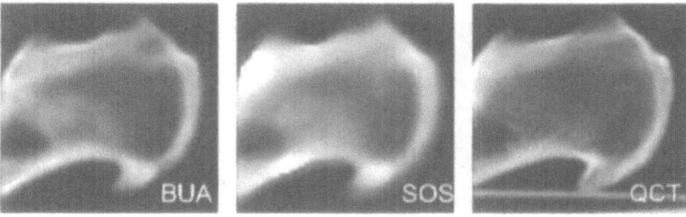

Figure. 1. BUA, SOS and QCT images of a slice of cancellous bone

Figure 2 also displays the data pooled from the fifteen samples and represents both the inter-sample and intra-sample variability of the ultrasound measurement. BUA values were ranging between 3 dB/cm.MHz and 38 dB/cm.MHz. SOS values were ranging between 1480m/s and 1685 m/s. When specimen is introduced in the model, the multiple correlation between BUA and BMD changes from 0.85 to 0.93, between SOS and BMD from 0.88 to 0.95 and between SOS/BUA from 0.94 to 0.99. This result support the idea that the relationships between acoustical parameters and BMD are specimen-dependent.

The strong correlation between BUA and SOS shows that they measure the same information. These results suggest that on the average 20% of the variability of attenuation and 15% of speed of sound remains unexplained by the variation of BMD. This not surprising since BMD is only one of the many parameters determining bone strength. Other parameters such as trabecular spacing, trabecular thickness and connectivity may also play an important role. However, it suggests also that the degree to which ultrasound measurement in transmission would provide information independent of BMD would be limited. This is corroborated by recent experimental evidences[7] showing that poor information on microarchitecture and connectivity of bone is given by acoustic parameters measured by ultrasound transmission.

IN VIVO MEASUREMENTS

Description of the device: An ultrasound transmission scanning system was constructed to make *in vivo* parametric images of the acoustic properties of the heel. Ultrasound images were obtained in transmit mode by using a pair of broadband focused transducers (center frequency 0.5 MHz, diameter 29 mm, focus 50 mm) immersed in a water bath at room temperature. With these characteristics, the theoretical beam width at the focus was approximately 5 mm. For the measurement in transmission, one of the transducers is the ultrasonic transmitter, the other one acts as a receiver. Both transducers were coaxially and confocally aligned. The ultrasound beam was moved (in a raster pattern) through the heel using a 2-axis scanning device. Both transducers were moved simultaneously. Individual measurement of the received signals were made and recorded as a function of position at each point in the scan. The space between measurement points was adjustable. In this study, the measurements were taken every 1 mm over a scale length of 60 mm in both XY directions. The transducers were excited with a monocycle pulse produced by a pulse generator (Sofratest, Ecquevilly, France). The received signals were amplified and digitized at a rate of 8 MHz with specially designed electronics (Sofratest, Ecquevilly, France). Data acquisition and collection of signals was controlled by an IBM compatible Pentium computer. Data acquisition was co-ordinated to motion position (Ultrasoft 2000 software, Sofratest, Ecquevilly, France). After digitization, the signals were transferred and stored on a hard disk of the computer for software processing. All further signal processing and image analysis were performed with proprietary softwares. The total duration of the acquisition period was 3 minutes.

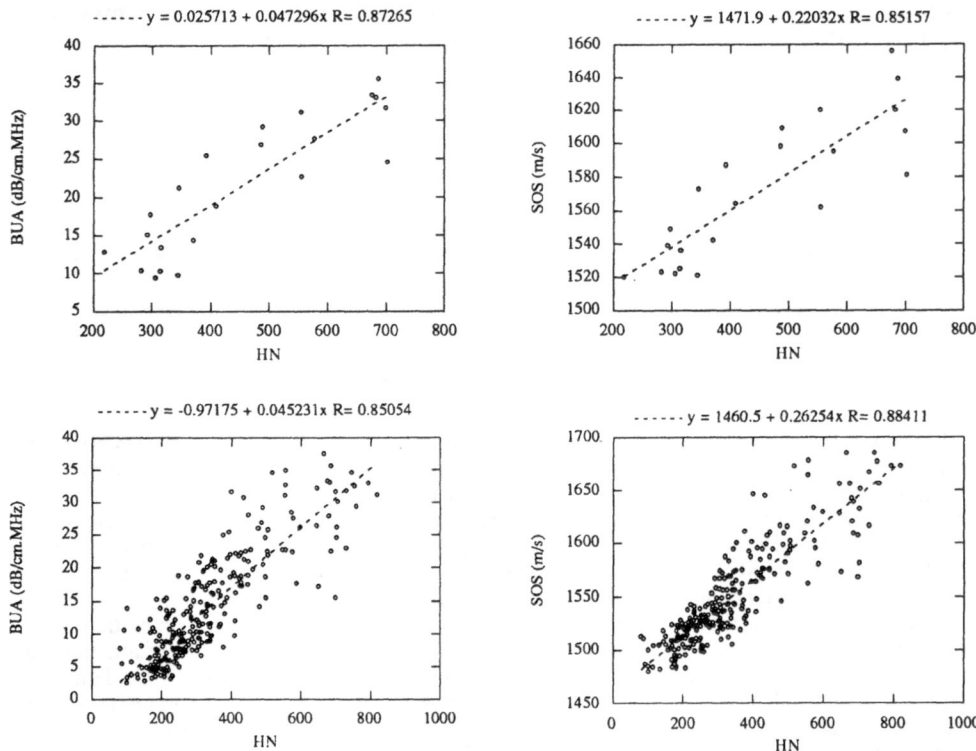

Figure.2. Upper curves: ultrasound parameters are plotted against BMD for a representative sample. Lower curves: data pooled for 15 specimen.

BUA was measured as explained in the previous section. As the calculation of the velocity requires the thickness of the bone, velocity images were obtained for *in vitro* samples only. We are currently developing a pulse-echo technique for the measurement of the thickness of bone at each point of the scan. The pulse echo will be combined to the *in vivo* transmission measurements.

Subjects: The technique was tested on a homogeneous cylinder that was constructed in our laboratory with a material (resin) having acoustic properties close to those measured in cancellous bone. *In vivo* images were obtained on healthy subjects from the regular staff of the laboratory. For one of the subjects, who previously had a radiograph of the ankle, the ultrasound image of the calcaneus could be compared to the X-ray image. Preliminary results on precision were obtained by repetitive measurements with intermediate repositioning on the phantom. Twenty measurements were taken in a period of one month and the reproducibility was assessed on a circular ROI (20mm in diameter) in the center of the phantom.

Results

Images: Figure 3 shows the BUA image of the homogeneous phantom. The measurement resulted in a homogeneous image except at the edges of the cylinder. The values of BUA and ultrasound velocity in a circular ROI (20 mm in diameter) in the center of the phantom were respectively 49 dB/MHz (standard deviation 0.8 dB/MHz) and 1495 m/s (standard deviation 1 m/s). The coefficient of variation for repetitive measurements was found to be 0.5% for the phantom. On the edges the BUA value was approximately 20 dB/MHz higher than in the middle of the phantom. An edge artefact was produced by the distortion of the wave front transmitted partly through water and partly through the cylinder.

Figure 4 compares the BUA images from three subjects (2 men 25 and 39 years old and one woman 59 years). The linear grey scale was the same for the three images (black: 0 dB/MHz and white: 165 dB/MHz). These images suggest that the bone of the woman is less attenuating than those of men. The average value of the BUA was measured in a circular ROI (20 mm in diameter) centered in the posterior part of the calcaneus. The average values were respectively: man (25 years): 90 dB/MHz, man (39 years): 84 dB/MHz and woman (59 years): 37 dB/MHz. Table 1 summarizes the results of the measurement ROI.

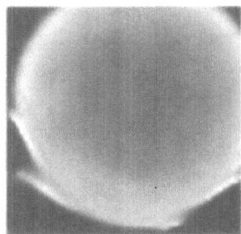

Figure. 3. BUA image of a homogeneous cylindrical phantom showing the edge artefact.

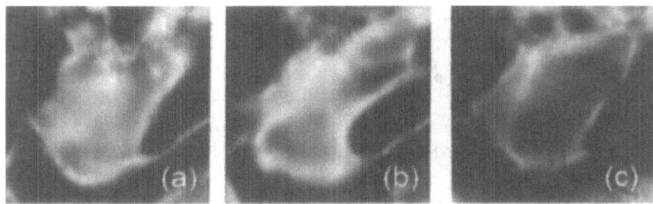

Figure.4. BUA images of three healthy subjects ([a] man 25 years old, [b] man 39 years old and woman 59 years old).

Table1. Average, standard deviation, maximum and minimum values of BUA measured in a circular ROI (20 mm in diameter) in the center of the posterior tuberosity.

| Subject (Age) | BUA (dB/MHz) | | | |
	Average	Standard deviation	Maximum	Minimum
Man (25)	90	9	110	68
Man (39)	84	16	131	54
Woman (59)	37	8	67	24

These results suggest that the calcaneous is acoustically extraordinarily heterogeneous and that the accurate control of the position of the measurement site is of the utmost importance for between subject comparison and for repetitive measurements.

Figure 5 shows the comparison between the BUA image and the radiograph of the calcaneus (man 25 years old). The ultrasound image compares favourably to the radiograph although the resolution of ultrasound is not sufficient to visualize bone trabecular structure.

The gross anatomical details can be identified in both images, however, somewhat different fine details are depicted on both images. This should be expected, since different physical means are involved in the formation of the ultrasound and X-ray images. Note also that the superficial plantar aponeurosis is visible on the ultrasound image (white arrow). It was found in all the subjects of our study.

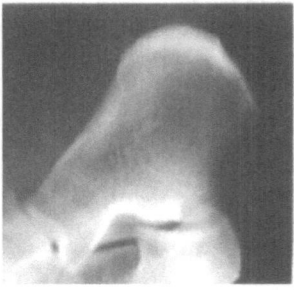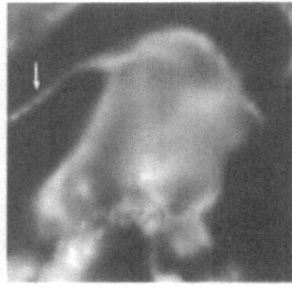

Figure 5. Comparison between the BUA and radiograph images of the calcaneus. The white arrow indicate the plantar aponeurosis.

Artefacts: The reconstructed ultrasound images are affected by refraction and waveform distortion, two phenomena which are negligible with X-ray sources. This may explain some of the differences between our ultrasonic image and X-ray image. These two phenomena, which are neglected in ultrasound conventional pulse-echo imaging where only qualitative information is expected, are of great importance when truly quantitative images are required. The apparent BUA values deviates from the true BUA values in regions where large variations in ultrasonic velocity are encountered. Such variations occur in regions containing several different tissues (bone edges, upper joint and aponeurosis). In transmission, variations of speed of sound encountered at the edges of bone lead to incorrect estimates of attenuation. The edge artefact on *in vitro* samples of bone was previously reported[6]. This effect is illustrated again in this work by the image of a homogeneous phantom (Figure 3).

Another consequence of the propagation through a heterogeneous medium is the unexpected high BUA values (approximately 50 dB/MHz) observed in each of the subjects in the region of the plantar aponeurosis. When the aponeurosis intersects the ultrasound beam, the wave front is partly transmitted through muscle tissue (approximate ultrasound velocity of 1540 m/s) and partly trough the adipose subcutaneous layer (approximate ultrasound velocity of 1460 m/s). This results in a complex waveform arriving at the receiving transducer (Figure 6).

The waveform is composed of two interfering waves propagating with slightly different velocities. The calculation of the frequency dependent attenuation is strictly valid for undisturbed pulses propagating through homogeneous media. In the case of interfering waves, only an apparent attenuation can be computed. In our example, the frequency dependence of the attenuation curve shows a sharp maximum and deviates significantly from a linear variation (Figure 6). Destructive interference (i.e. maximum attenuation) is possible between two waves when the time delay between these two waves is an odd number times the half-period. The sharp maximum of the attenuation curve displayed in figure 6 is occurring at a frequency in the bandwidth for which the condition for destructive interference is verified. We proposed the likelihood image as an efficient way of highlighting the regions of the image suspected to be subject to waveform distortion. A likelihood image was obtained by calculating the correlation coefficient between attenuation and frequency at each point of the scan. The results are then colour-coded and superposed to the grey-scale BUA image. It could be used to guide the selection of the optimal measurement site.

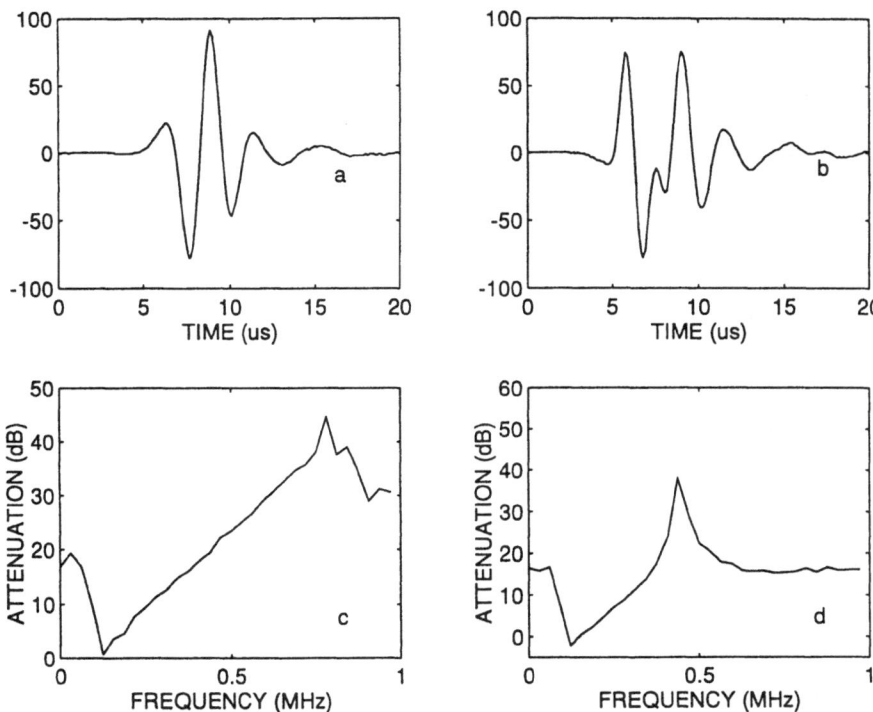

Figure 6. Comparison between (a) the transmitted waveform through the center of the calcaneus and (b) the waveform arriving at the receiving transducer when the plantar aponeurosis intersects the ultrasound beam. The waveform is composed of two interfering waves propagating with slightly different velocities. (c) Frequency dependent attenuation corresponding to wavefrom (a). (d) Frequency dependent attenuation corresponding to wavefrom (b) showing a sharp maximum.

DISCUSSION-CONCLUSION

We report the feasibility of ultrasound parametric imaging of the calcaneus. Large spatial variations of BUA values were previously reported. The images shown here, demonstrate the wide range of BUA found both in the whole bone and within the ROI thus reinforcing the idea of tremendous heterogeneity of the acoustic properties of bone. To make the between subjects comparison possible, a standard ROI must be defined. This is not possible with devices which deliver only a single point measurement without considering interindividual anatomical variations (amount of soft tissues surrounding the heel and concavity of the sole). Variations in the measurement site obtained with devices measuring a single point might be one of the various causes for the observed inter-individual variability. ROI location should be controlled to make comparisons, in particular between different techniques (ultrasound and DEXA) more conclusive. An edge detection algorithm that clearly defined the edges of the bone could help standardize the measurement site. Criteria for selecting an optimal standard measurement site should include the optimum precision and accuracy as well as the optimum ability to predict fracture risk. In these respects, the potential of ultrasound imaging is distinctly encouraging.

Our results suggest that ultrasound parametric imaging has the potential for enhancing the current ultrasound technique by (a) allowing reproducible, repetitive measurements, (b) permitting the selection of similar optimal measurement sites in all subjects and (c) avoiding accuracy errors due to waveform distortion.

The in vitro results support the idea that the ultrasound parameters measured in transmission appear to reflect bone quantity rather than bone microarchitecture. Reflection techniques such as the measurement of frequency backscattered signals might be helpful for the assessment of microarchitecture.

REFERENCES

1. Langton CM, Palmer SB Porter SW (1984) The measurement of broadband ultrasonic attenuation in cancellous bone. *Eng Med* 13:89-91.
2. Kanis JA, Melton LJ, Christiansen C, Johnston CC Khaltaev N (1994) The diagnostic of osteoporosis. *J Bone Miner Res* 9:1137-1141.
3. Glüer C, Wu C, Jergas M, Goldstein S Genant H (1994) Three quantitative ultrasound parameters reflect bone structure. *Calcif Tissue Int* 55:46-52.
4. Laugier P, Giat P Berger G (1994) Broadband ultrasonic attenuation imaging : a new imaging technique of the os calcis. *Calcif Tissue Int* 54:83-86.
5. Laugier P, Giat P Berger G (1994) New ultrasonic methods of quantitative assessment of bone status. *Eur J Ultrasound* 1: 23-38.
6. Laugier P, Berger G, Giat P, Bonnin-Fayet P Laval-Jeantet M (1994) Ultrasound attenuation imaging in the os calcis: an improved method. *Ultrasonic Imaging* 16:65-76.
7. Hans D, Arlot ME, Schott AM, Roux JP, Kotzki PO, Meunier PJ (1995) Do ultrasound measurements on the os calcis reflect more the bone architecture than the bone mass?:a two dimensional histomorphometric study. *Bone* 16:295-300.

MAPPING THE CONTINUOUS DISTRIBUTION OF SONIC VELOCITY AND ELASTIC MODULUS IN A CROSS-SECTION OF BONE

Sidney Lees, Douglas B. Hanson and Elizabeth A. Page

Bioengineering Department
Forsyth Dental Center
140 Fenway
Boston, MA 02115

INTRODUCTION

Ultrasonic scanning systems are now available which can be gated at several points in the A-scan, thereby locking on several echoes in the pulse train. Consider the situation of a flat plate where the gates are locked on the echoes from the front and rear surfaces. The time of flight through the flat plate can be measured by a digital clock that is turned on by the first pulse and turned off by the second. Since the two dimensional scanner locates each interrogation point in the plate, the variations of the transit time can be mapped. Any variations in the elastic modulus or the density associated with the variations in the time-of-flight will be detected.

METHODS AND MATERIALS

The Panametrics Multiscan Ultrasonic Imaging System (Panametrics Waltham MA) was used in this modality to study the variations in the sonic velocity and elastic modulus of three bone specimens, a human tibia, the metacarpal of a horse and the femur of an African elephant. The bone sections were ground flat and smooth with the aim of making the specimen uniformly thick. The average density was found by Archimedes Principle. Each time-of-flight map was evaluated by means of an image processor (4MIP, EPIX Inc, Buffalo Grove, IL) resulting in a map of the sonic velocity and a second map of the elastic modulus.

We experienced considerable difficulty when we tried to prepare a uniformly one mm. thick intact cross section of bone. It was necessary to segment the bone section and process each segment separately. As a consequence the segments have different thicknesses. We also found some taper in each segment despite our best efforts. The taper was due to the variations in the structure of the bone between the dense outer surface (the periosteum) and the spongy-like trabecular inner surface. The spongier structure grinds more readily.

The Multiscan System is programmed to determine the shortest and longest time-of-flight in the scan. The difference is divided into 128 levels (7 bits), so that a small range of values can be expanded. The time-of-flight map was stored on a floppy disc for further image processing by the 4MIP System. In the first step the image was broken down into the individual segment images and each segment in its proper geometric location was stored in one of the thirteen buffers on the 4MIP board and on the computer hard disk. The location of each segment was not disturbed, so that the ensemble could be reconstructed by adding the individual segment images.

Another command enabled the pixel value of the time-of-flight map to be converted into a pixel value representing the sonic velocity. The 4MIP employs 256 levels (8 bits). The command requires the thickness of the segment to be entered, hence it was necessary to repeat the process for each segment. The result was a set of sonic velocity maps, identical in geometrical shape to the first set of segment images, stored in the buffers and again on the hard disk. The segment images can be added to reassemble the bone section, this time showing a unified map of the sonic velocity as seen in the figures.

The elastic modulus is defined as the density times the square of the sonic velocity. Since the density varies around the bone section it is necessary to have a map of the density distribution for the highest accuracy. Lacking this informtion the nominal density of each point in the segment was taken to be average density of the segment. We would prefer to have a map of the density distribution within each segment but the method has not yet been developed. The pixel equivalent of the elastic modulus at each point was calculated from the segmental velocity map using the nominal density. Again, the set of segment maps was unified by adding the segmental elastic modulus images.

RESULTS

The data are shown in the three figures, one for each bone section. In Fig 1 a midshaft cross section of a human tibia (35 year female) is displayed. The upper left image is a photograph of the bone section. The upper right image is the time-of-flight image. The lower left image represents the sonic velocity distribution, while the lower right is that of the elastic modulus. The lower two images were processed to display three zones correponding to a range of values, so as to better show the distribution of the properties. The bone section was cut into two segments, each examined separately. The two separate images were merged, but the resulting image shows they overlapped accounting for the darker vertical strip. The anterior section (on the left) appears spotted, indicating localized areas where the elastic modulus as well as the sonic velocity are lower than for the major part of the bone. The elastic modulus of the posterior section (on the right) is less than in the anterior, suggesting that the load stress is unequally distributed.

Fig 2 is a similar display of the midshaft cross section of a horse metacarpal (20 year old male, 530 kgs). It was necessary to divide it into five segments. The segments were glued to a glass plate with epoxy. The time-of-flight image shows considerable variation because the segments are of different thicknesses. The velocity image, where the thickness is accounted for, shows a more consistent pattern. The two sides show the highest velocity, the top (anterior) has somewhat lower velocity. The bottom (posterior) has an even lower velocity. These findings are repeated in the modulus image even more

Figure 1

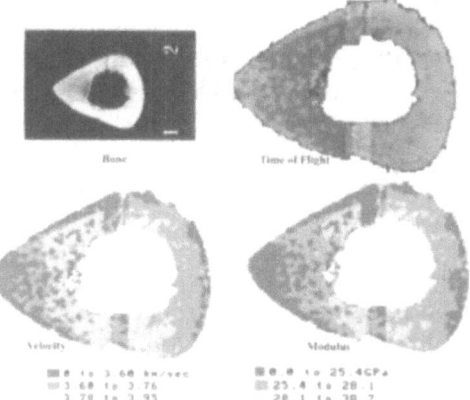

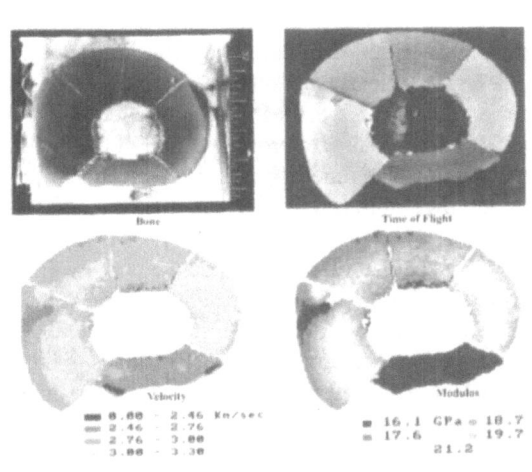

Figure 2

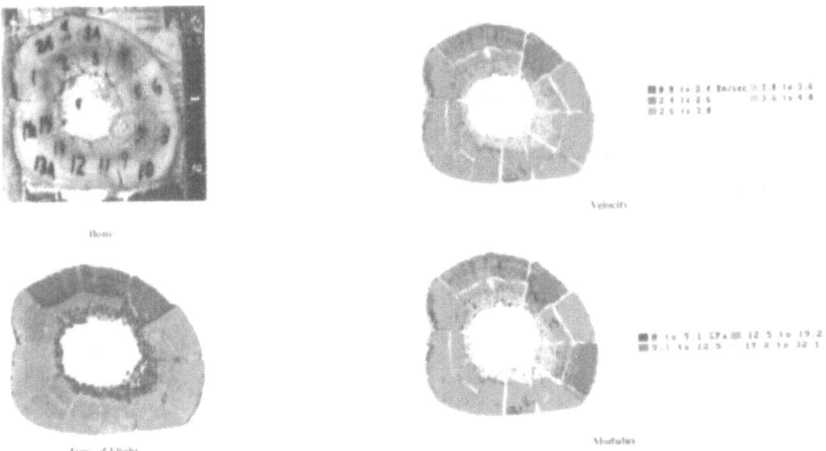

Figure 3A   **Figure 3B**

emphatically because of the variation in density among the segments. The posterior section had much lower density than did the other segments.

Finally Fig 3 shows the properties of the midshaft cross section of an African elephant femur (10-20 year male, 8300 kg). It was necessary to divide the bone into 18 segments. After grinding to uniform thickness the segments were glued to a glass plate with epoxy cement. The bone section seen in the upper left shows the central part to be rather porous. This part of the segment did not have the same thickness as the denser portion of the bone and the time-of-flight seen in the upper right is much less than for the rest of the bone. Also the the upper segments were thinner than the rest which also accounts for the darker aspect corresponding to a shorter time-of-flight. The properties are more uniformly distributed once the thickness and the segment density are taken into account as seen in the lower two images of Fig 3.

DISCUSSION

Despite a number of imperfections in these examples it has been demonstrated that it is feasible to obtain a continuous distribution of bone properties derived from ultrasonic time-of-flight measurements. It is clearly demonstrated that the properties of bone are not uniform. It does require refinement of the techniques, particularly to make the bone specimen uniformly thick. It will be useful to provide a map of the thickness variation as well as the density distribution. The time-of-flight values are in error since the corresponding sonic velocities are low. More careful calibration of the ultrasonic multiphase scanner is necessary.

Even with these limitations it is seen that these three specimens show different structural characteristics, apparently related to the stress distribution to which the bone was subjected. In particular the horse metacarpal shows higher modulus laterally and a much lower modulus in the posterior section. Presumably this corresponds to higher forces on the lateral aspects of the metacarpal when the horse runs. Since the anterior segment has a higher elastic modulus than the posterior, it is inferred that the anterior is more intensely loaded than the posterior but not as great as the lateral portions. The elastic modulus in the elephant femur appears to be more uniform, presumably corresponding to a uniform loading. The posterior portion of the human tibia exhibits a lower modulus than the anterior and the properties in the anterior are much less homogeneous.

Acknowledgements: The work was partially supported by National Institute on Aging Grant AGO 2325. The ultrasonic time-of-flight scanning was performed by Peg Danek under the supervision of Thomas J. Nelligan, Panametrics. We are indebted to Stephen White, Epix, Inc. for his assistance in processing the data on the 4MIP image processor.

ACOUSTIC MICROSCOPY APPLICATIONS FOR OBSERVING MICROSTRUCTURE OF BONES AND BONE - IMPLANT SYSTEM

Roman Gr. Maev[1], Vadim M. Levin[2], Robert M. Pilliar[3],
ElenaYu. Maeva[1], and Tamara A. Senjushkina[2]

[1,2]Center of Acoustic Microscopy , Institute of Biochemical Physics,
Russian Academy of Sciences, 4 Kosygin str, Moscow, 117977, Russia
[1]Current position,- University of Windsor, Ontario, N9B 3P4, Canada
[3]University of Toronto, 170 College str, Toronto, Ontario, MSS 1A1
Canada

INTRODUCTION

Bone tissue is a rigid material consisting of oriented collagen fibre layers and mineral microcrystals approximating calcium hydroxyapatite. It has a pronouced texture and a complicated microstructure with a nonuniform distribution of mineralized collagen lamellae encasing sites for cells and channels for blood vessels. For long bones, the struc-ture consists of an outer layer of dense cortical bone (with very few voids or pores) sur-rounding a more porous-structured bone (cancellous) and marrow space. Some ordering of this cancellous bone toward the ends of long bones results in a trabecular arrangements of plates and rods of bone tissue.

Acoustic microscopy may present an ideal method for investigating the structure of the bone as a result of the ability of acoustic images to display the microstructure of objects. The success or failure of a bone-interfacing implant system is dependent on the appropriate interface structure forming during the healing phase and being maintained for the long term.

Some studies of bone using acoustic microscopy have been presented but these studies have relied on rather low frequencies (25 - 50 MHz) by J.L. Katz et al, at 1988[1]. The choice of operating frequency is essential for obtaining informative acoustic images. For the characterization of microstructure of bone, low frequency acoustic imaging does not appear to provide sufficient microstructural detail. Therefore, we have utilized high-frequency acoustic microscopy with operating frequencies of 200 MHz to 1.3 GHz to study the structure of different regions of bone surrounding a metallic implant placed within the medulla of a long bone (femur) in a dog.

In this initial study, a specimen of bone containing a porous surface-coated (316L stainless steel core with a CoCrMo powder-made porous surface coat) was examined. This specimen had been retrieved from a much earlier study in which a porous-coated implant was used to bridge a purposely - created segmental gap in the femur of a dog by R. Pilliar et al, at 1981[2]. In that study, tissue ingrowth had been demonstrated with relatively large initial mopvements at the bone implant interface during the the implantation healing phase, resulting in fibrous tissue attachment of the implant within a compact shell of bone, which formed around the implant and within the surrounding cortical bone. This structure provided an interesting specimen for exploring the the usefulness of acoustic microscopy in identifying the different structures.

To examine the bone-implant sample, a block of bone containing the implant was retrieved following animal sacrifice (R.Pilliar, 1987)[3], and then dehydrated, while simultaneously being infiltrated with acrylic, to form an embedded block sample. A portion of this block, adjacent to a region in which a thin section had been taken for transmitted light microscopy, was provided for the acoustic microscopy study. It should be noted that because of the time of preparation and the methods used for specimen preparation, this specimen was far from ideal.

BACKGROUND

There are three basic problems in applying acoustic microscopy to bone investigations:

The first of them is the technique by which bone samples are prepared. Acoustic microscopy does not require the special fixing and straining of tissue samples, but it does imposes high demands upon the quality of the sample surface. Bone is a fragile material. It needs specific methods of preparing samples with a flat surface; the methods should allow minimal destruction and promote high quality of a section surface.

The second problem is choosing an appropriate state of the bone for study. Fresh bone contains a great deal of liquid. It is similar to two-phase porous Biot's medium. Dry bone is a heterogeneous solid.

Finally, the third problem is in choosing the parameters of a probing acoustic pulse. Acoustic images display the distribution of elastic or viscous parameters of a sample, which are averaged over the acoustic focus spot area on the sample surface.

The goal of this paper is to apply high-frequency acoustic microscopy for the visualiza-tion and characterization of the bone microstructure at the hystological and subhystological levels. The industrial multi-purpose reflection microscope Elsam (produced by Leitz, Germany) was used for the acoustic visualization of bone samples. The Elsam has the wide multi-level range of working frequencies: 100 MHz, 200 MHz, 400 MHz, 0.8 - 1.3 GHz, and 1.3 - 2.0 GHz; and a scanning field (≤ 1 mm^2) that compares with the field of view of optical microscopes. We have used the operation frequency of 200 MHz for the acoustic imaging of bone tissue structure (hystological level) and a high-resolution regime (f = 1.6 Ghz) for observing the fine details of the tissue microstructure (subhystological level).

Thus, we are coming to the essential role of ordered collagen in the formation of the mechanical microstructure of bone and, respectively, in displaying it through acoustical imaging.

INVESTIGATIONS OF THE ACOUSTIC PROPERTIES OF ORIENTED COLLAGEN

Single collagen fibers are 0.5 - 10 mkm in diameter. They are formed by microfibres, which, in turn, consist of very long oriented macromolecules. Models of tetragonal, hexagonal, cylindrical and helical packages of molecules in a microfibril were considered[4,5]. In the frame of each of these models a single fiber has hexagonal symmetry as well as a medium composed by identically oriented fibers as a whole. Cusak and Miller were able to measure the whole set of elastic modules $\{C_{11}, C_{12}, C_{33}, C_{13}, \tilde{N}_{44}\}$ of dry collagen in Brillouin light scattering experiments with pure collagen samples from a rat tail[7]. The elastic properties of collagen essentially depend on the water content in the tissue. As collagen is saturated by water a value of the longitudinal sound velocity is drastically reduced (by 30 - 40%) and the transverse sound disappears[8]. In oriented collagen all three acoustic modes are different. Two modes propagate with polarization vectors laying in the propagation plane: quasi-longitudinal (L) waves and fast quasi-transverse (TF) waves. For the quasi-longitudinal mode the angle dependence of velocity is described by a formula:

$$c_L = \frac{1}{\sqrt{2\rho}} \left\{ \left(C_{11} \sin^2 \theta + C_{33} \cos^2 \theta + C_{44} \right) + \right.$$

$$\left. \sqrt{\left[\left(C_{11} - C_{44} \right) \sin^2 \theta + \left(C_{44} - C_{33} \right) \cos^2 \theta \right]^2 + \left(C_{13} + C_{44} \right)^2 \sin^2 2\theta} \right\}^{\frac{1}{2}} . \tag{1}$$

The polarization vector e_L given by three of its components:

$$\mathbf{e}_L = \frac{1}{B} \left\{ \sin \theta, \quad 0, \quad \frac{C_{11} \sin^2 \theta + C_{44} \cos^2 \theta - \rho c_L^2}{\left(C_{13} + C_{44} \right) \cos \theta} \right\}, \quad \text{where}$$

$$B = \sqrt{\sin^2 2\theta + \left(\frac{C_{11} \sin^2 \theta + C_{44} \cos^2 \theta - \rho c_L^2}{\left(C_{13} + C_{44} \right) \cos \theta} \right)^2} \quad \text{is normalization factor;} \tag{2}$$

deviates from the wave vector direction. Analogous formulae describe the angle dependence of the phase velocity and polarization vector for the fast quasi-transverse mode. To obtain an expression for the angle dependence of phase velocity it is enough to change the sign in front of the internal root in (1):

$$c_{TF} = \frac{1}{\sqrt{2\rho}} \left\{ \left(C_{11} \sin^2 \theta + C_{33} \cos^2 \theta + C_{44} \right) - \right.$$

$$\left. \sqrt{\left[\left(C_{11} - C_{44} \right) \sin^2 \theta + \left(C_{44} - C_{33} \right) \cos^2 \theta \right]^2 + \left(C_{13} + C_{44} \right)^2 \sin^2 2\theta} \right\}^{\frac{1}{2}} , \tag{3}$$

The polarization vector for the TF mode is produced from (2) by a change of indices: $L \Rightarrow TF$. If we know the orientation of a polarization e_L we know a vector e_{TF} too; both the vectors are laying in the propagation plane and mutually perpendicular.

The third mode is a pure transverse (shear) TS mode with polarization perpendicular to the plane containing both the Z-axis and a propagation direction:

$$\mathbf{e}_{TS} = \{1, \quad 0, \quad 1\}. \tag{4}$$

The angle dependence of the phase velocity for the slow transverse mode is:

$$c_{TS} = \sqrt{\frac{1}{2\rho} \left\{ \left(C_{11} - C_{12} \right) \sin^2 \theta + C_{44} \cos^2 \theta \right\}} . \tag{5}$$

Calculated plots of the angle dependence for L and TF velocities are represented in Fig.1.

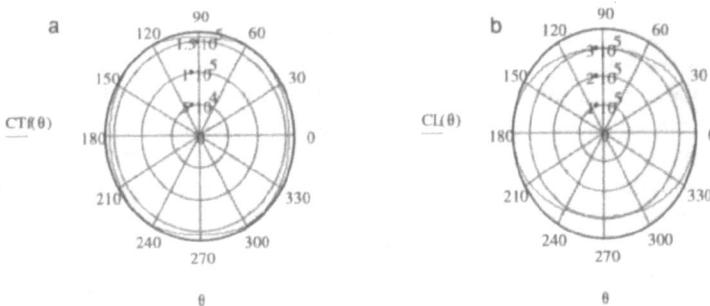

Figure 1. Angle dependence of phase velocities in dry oriented collagen: a). quasi-longitudinal waves (L mode);   b). fast quasi-transverse waves (TF mode).

The data for the maximum and minimum values of the velocities and for ratios c_{min}/c_{max} characterizing anisotropy for each acoustic modes are summarized in Tab. 2.

The longitudinal mode has the largest phase velocity along fibers ($c_L{}^{||}$=3.64 × 10^5 cm/s) and the smallest one across fibers ($c_L{}^{\perp}$=2.94 × 10^5 cm/s); the coefficient of anisotropy is 0.8. Angle β_L, characterizing a deviation of the vector e_L from a wave vector direction, reaches a maximum: $\varphi_{L,max}$ = 8.3°, near $\theta \approx 40°$. For quasi-transverse waves the maximum phase velocity arises near θ= 49° ($c_{TF}{}^{\theta=49°}$=1.64 × 10^5 cm/s), the smallest velocity corresponds to wave propagation across fibers as well as along them ($c_{TF}{}^{||}$= $c_{TF}{}^{\perp}$=1.56 × 10^5 cm/s). The phase velocity of the pure shear mode (TS mode) does not depend on propagation direction because of the equality $(C_{11}-C_{12})/2=C_{44}$ following from Cusak and Miller's experimental data (c_{TS} ≡ 1.56 × 10^5 cm/s).

Table 2a. Summary of properties of longitudinal waves in dry oriented collagen.

longitudinal waves		
maximum velocity, $c_{L,max}$	propagation along fibers	$c_L{}^{||}$=3.64 ×10^5 cm/s
minimum velocity, $c_{L,min}$	propagation across fibers	$c_L{}^{\perp}$=2.94 × 10^5 cm/s
anisotropy coefficient, K=$c_{L,min}$ / $\tilde{n}_{L,max}$	K=0.8	

Table 2b. Summary of properties of fast quasi-transverse waves mode in dry oriented collagen

transverse waves		
maximum velocity, $c_{TF,max}$	propagation at the angle of θ=49°	$c_{TF}{}^{||}$= 1.64 ×10^5 cm/s
minimum velocity, $c_{TF,min}$	propagation both along and across fibers	$c_{TF}{}^{\perp}$=1.56 × 10^5 cm/s
anisotropy coefficient, K=$c_{L,min}$ /$\tilde{n}_{L,max}$	K=0.95	

Material anisotropy is better characterized by slowness surfaces (surfaces of inverse sound velocities $1/c_L$, $1/c_{TF}$ and $1/c_{TS}$ or equivalent surfaces of wave vectors). The extent of energy concentration depends on the topology of the surfaces - their smoothness, and the existence of stationery points and flat regions on them [9].

Based on our calculations, we constructed the slowness surfaces for the L and TF modes:

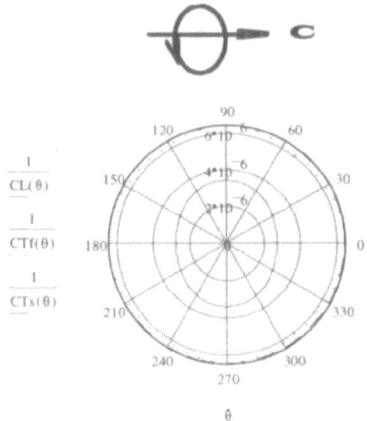

Figure 2. Acoustic slowness surfaces in collagen. Sections of the surfaces by a plane that goes through the symmetry axis C are shown. The surfaces themselves arise as a result of rotation of the curves about the C-axis.

Simultaneously, we calculate the angle dependence of the so-called energy coefficients of reflection V and transmission W_L and W_{TF}, that represent ratios of energy fluxes P_{refl}, and P_L, P_{TF}, which propagate in the liquid and collagen sample after the incident wave interaction with the interface, to a value of the incident wave energy flux:

$$V = \frac{P_{refl}}{P_{inc}} = \left|\frac{A_r}{A_0}\right|^2 = |R|^2 ; \qquad W_L = \frac{P_L}{P_{inc}} = \frac{\rho}{\rho_o} \cdot \frac{c_{gz}^L}{c_o \cdot \cos\theta} \cdot \left|\frac{A_L}{A_o}\right|^2 ;$$

$$W_{TF} = \frac{P_{TF}}{P_{inc}} = \frac{\rho}{\rho_o} \cdot \frac{c_{gz}^{TF}}{c_o \cdot \cos\theta} \cdot \left|\frac{A_{TF}}{A_o}\right|^2 . \qquad (6)$$

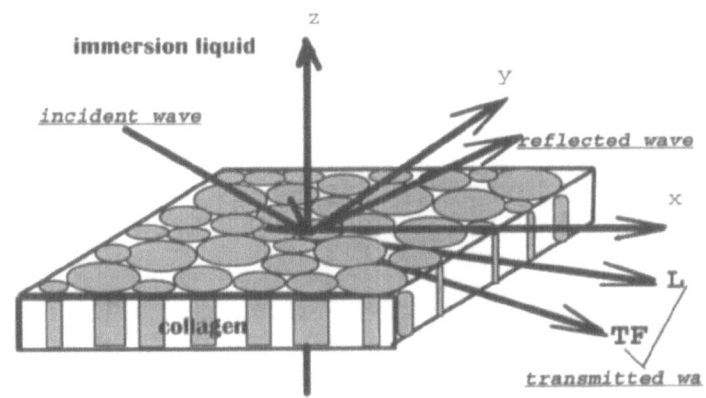

Figure 3. Scheme of reflection at liquid - collagen interface.

Here A_0 is amplitude of the incident wave; A_R - the amplitude of the reflected wave in liquid; A_L and A_{TF} - the amplitudes of refracted waves in collagen; ρ_0 and c_0 - density and sound velocity in liquid; ρ - collagen density; c_g^L and c_g^{TF} - the ray velocities of acoustic modes in collagen; and c_{gz}^α (α = L, TF) are their components normal to the interface. Energy coefficients are related by the energy conservation law:

$$V + W_L + W_{TF} = 1 \qquad (7)$$

We suppose our results confirm good prospects for acoustic microscopy methods in the study of bone and other hard tissue and for the applications of these methods in medicine and biology.

ACKNOWLEDGMENT

The authors acknowledge the financial support of the Russian National Fund of Fundamental Research.

REFERENCES

1. A.Meunier, J.L.Katz, P.Christel, L.Sedel, A reflection scanning acoustic microscope for bone and bone - biomaterials interface studies, J.Orthopaedic Research, 6: N5, 770 (1988).
2. R.Pilliar, H.Cameron,R.Welsh et al.,Radiographic and morphologic studies of load-bearing porous-surfaced structured implants, J.of Clinical Orthopedics,156:249 (1981).
3. R.M.Pilliar, Porous-surfaced metallic implants for orthopedic applications, J. of Biomedical material research: Applied biomaterials, 21: No A1,1 (1987).
4. R.D.Harkness, Collagen, Sci.Progress.Oxford,54:257 (1966).
5. A.Miller, Molecular packing in collagen fibrils, in "Biochemistry of collagen", eds. G.N.Ramachandran and A.H.Reddi, Plenum Press, NY: 85 (1976).
6. S.A.Goss, W.D.O'Brian,Jr, Direct ultrasonic velocity measurement of mammalian collagen threads J.Acoust.Soc.Amer.,65: 507 (1979).
7. S.Cusak, A.Miller, Determination of the elastic constants of collagen by Brillouin light scattering, J.Mol.Biol.,135: 39 (1979).
8. J.C.Bamber, C.R.Hill, J.A.King, F.Dunn, Ultrasonic propagation through fixed and unfixed tissues, Ultrasound in Med. Biol.,5: 159 (1979).
9. G.A.Northrop, J.P.Wolfe, Phys.Rev.Lett.,52: 2156 (1984).
10. R.G.Maev, Scanning acoustic microscopy of polymeric materials and biological substances, Tutorial, Archives of Acoustics,13, No 1-2: 12 (1988).
11. V.M.Levin, R.G.Maev, O.V.Kolosov, T.A.Senjushkina, M.A.Bukhny, Theoretical fundamentals of quantitative acoustic microscopy, Acta Phys.Slov.,40, N3: 171 (1990).

DENTAL DIAGNOSIS BY HIGH FREQUENCY ULTRASOUND

R. Wichard, J. Schlegel, R. Haak*, J.F. Roulet* and R.M. Schmitt

Fraunhofer-Institute for Biomedical Engineering, Dept. Ultrasound
St. Ingbert, Germany
* Charité, Centre for Dentistry, Berlin, Germany

INTRODUCTION

Caries is a progressive disease, which destroys teeth by demineralization of the dental hard tissues. Due to anatomic and technical circumstances, the diagnosis of caries is difficult. It is done by visual inspection or radiography. Different studies have clearly demonstrated that the sensitivity and specificity of these diagnostic methods are not satisfying (Espelid and Tveit, 1991; Lussi, 1991; Peers et al., 1993). Many carious lesions remain undetected or are diagnosed only after substantial destruction has already occured. From a clinical point of view, early diagnosis is very important because non- or minimal-invasive treatment becomes possible (Backer Dirks, 1966) .
The most frequent locations of undetected carious lesions are the surface between the teeth (approximal), in the fissures (occlusal), and at the margins of restorations (secondary caries). The visual inspection can only detect an inadequate number of lesions (Espelid and Tveit, 1991; Lussi, 1991; Peers et al., 1993) and cannot classify their penetration depth; especially in early stages. Radiologic imaging does not allow the detection of caries in front of or behind radioopaque materials. The ionizing radiation is considered a potential health risk and should be restricted to a minimum.
An ultrasound imaging system specialized for dentistry has a high potential for visualizing the layered structure of teeth and dental reconstructions for a non-invasive diagnosis of caries. The examination of teeth with ultrasound requires high frequencies due to the small structures and the high sound velocities of enamel and dentine. In earlier ultrasonic measurements in teeth the enamel-dentine and the dentine-pulp interfaces could be distinguished in the echograms, and the impedance matching for coupling was discussed (Barber et al.,1969). In this work the results of ultrasonic imaging in teeth are presented for the first time.
In the first phase of this project the sound velocities, acoustic impedances and attenuations of tooth sections and a selection of dental restorative materials (gold alloys, cements, composites, amalgam and ceramics) were measured at

high frequencies. The detailed results will be published elsewhere. It was shown that although there are large impedance mismatches at the layered structures of natural and restored teeth (e.g. enamel-dentine; gold alloy-cement-dentine; enamel-dentine-amalgam) even with water coupled immersion transducers significant echos from the inner interfaces can be measured. Also the occuring attenuation is tolerable for the thickness which is to be penetrated.

METHOD

Figure 1 shows the modular measurement system. It consists of a water tank with an adjustable cardanic sample holder, a focused transducer, which can be scanned in three dimensions and a high frequency pulser-receiver. The broadband transducer (30 MHz center frequency) has a 300 μm focal diameter and 10 mm depth of field. Together with a digitizing oscilloscope and a positioning control A-, B- and C-scans of teeth can be measured and visualized on a computer.

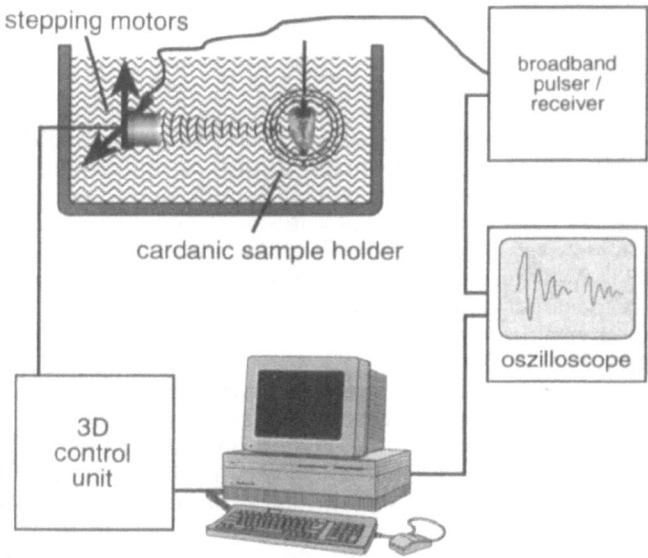

Figure 1. Experimental setup for dental ultrasound measurements

With this setup a selection of clinically relevant scenarios were examined at layered samples and extracted human teeth.

SIMULATIONS

Due to the geometric structure and the curvature of the interfaces, the propagation of ultrasonic waves in teeth is more complex than in parallel layered structures. By computer simulations of a quantitative ray tracing model and by

FEM simulations (figure 2) the influence of refraction, mode conversion and diffraction in teeth was studied. The simulations show that the propagation is not linear and that guided modes can occur in the enamel. In comparison to the experimental results it shows that only in plan-parallel sections the A-scans can directly be interpreted.

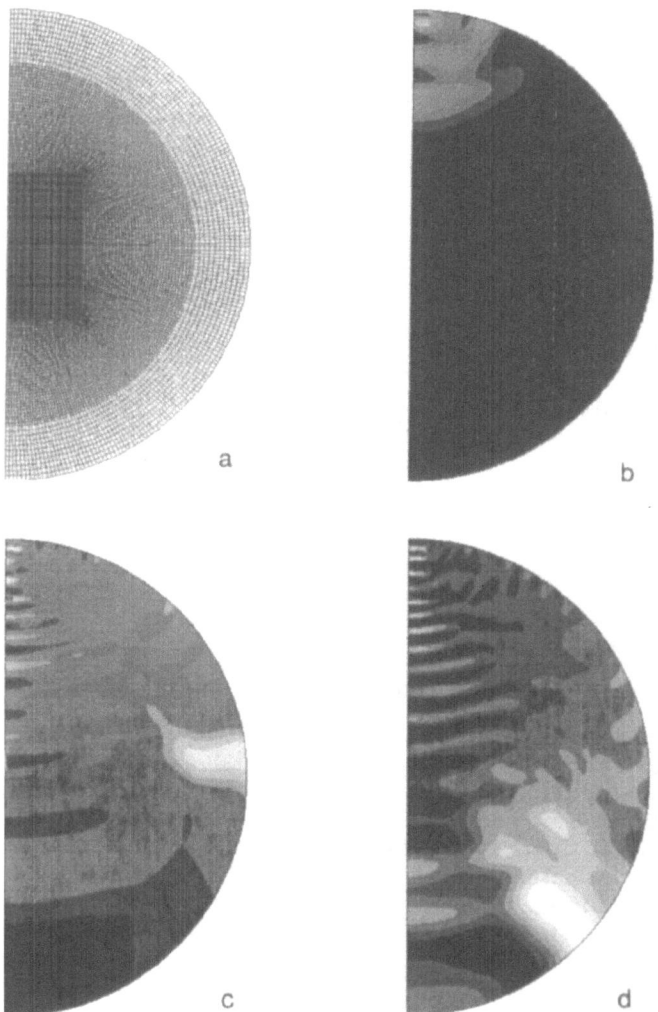

Figure 2. FEM simulation of the propagation of ultrasound in teeth. a) model for FEM calculation: enamel coated dentine cylinder, $\lambda/6$ mesh, two-fold symmetry assumed; b) 0.5 μsec after excitation from the top; c) 1.5 μsec: the direct wavefront reaches the bottom; d) 2.4 μsec: lamb mode guided in the enamel (bright area).

RESULTS AND DISCUSSION

The experimental measurements showed that the reflected signals from teeth are strongly dependent on the angle of incidence due to the refraction at the curved interfaces. Figure 3 shows an A-scan of a molar with an amalgam filling. The three significant echos correspond to the tooth surface, the enamel-dentine junction and the dentine-amalgam interface.

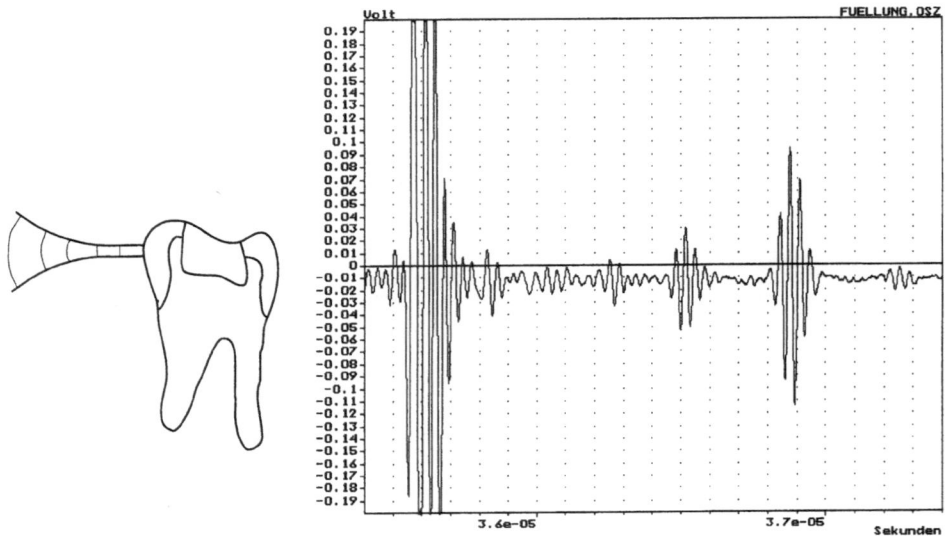

Figure 3. left: sketch of the molar with amalgam filling; incident beam from the left
right: A-scan signal (reflected amplitude vs. time).
time from first to second echo: 875 ns corresponds to 2.5 mm enamel
time from second to third echo: 395 ns corresponds to 0.75 mm dentine

The coupling at the interfaces between gold crowns, cement and hard tooth tissues was characterized with specially prepared samples. By analyzing the amplitudes of the echo sequences from tooth specimens with gold crowns it became possible to distinguish between cemented and uncemented interfaces. Such a diagnosis is not possible by X-ray due to the high absorption of radiation by metals. The clinical significance of this distinction is important because the first step in crown failure is the washing out of the cement between the restoration and the tooth leading to a secondary caries in the resulting gap.

For the very first time cross-sectional images of teeth were measured using ultrasound. Figure 4 shows the image of an approximal caries lesion. The amplitudes of the received echos were digitized with a sampling rate of 250 MHz and displayed in a grayscale. The surface of the molar is displayed in the lower third of the image. In areas with a perpendicular incidence of the ultrasonic beam a specular reflection is detected. The signals of other interfaces are detected by the transceiver due to their scattering. The depth of the conical shaped caries lesion, which has not penetrated the dentine in this case, is visualized quantitatively.

The following image (figure 5) shows a cross-section of a molar fissure. Visually, the surface of the lesion only appears as a dark line in the groove of the fissure. The water enamel interface and the stuctures behind it are displayed with a high resolution.

Figure 4. B-mode image of a caries lesion. 4.2 mm image width. Timescale 3.0 μsec from the bottom (corresponding to 2.25 mm image height) Insonification from the bottom.

CONCLUSIONS

The dental imaging method applied in this work has demonstrated the feasibility of a non-invasive, depth resolving diagnosis of caries and dental reconstrutions by using high frequency ultrasound. In B-mode images the caries lesions could be visualized for a quantitative diagnosis. For the prospective in-vivo system with a miniaturized scanner, which can be introduced into the oral cavity, micro-structured multi-element transducers in combination with high frequency phased-array techniques are required.

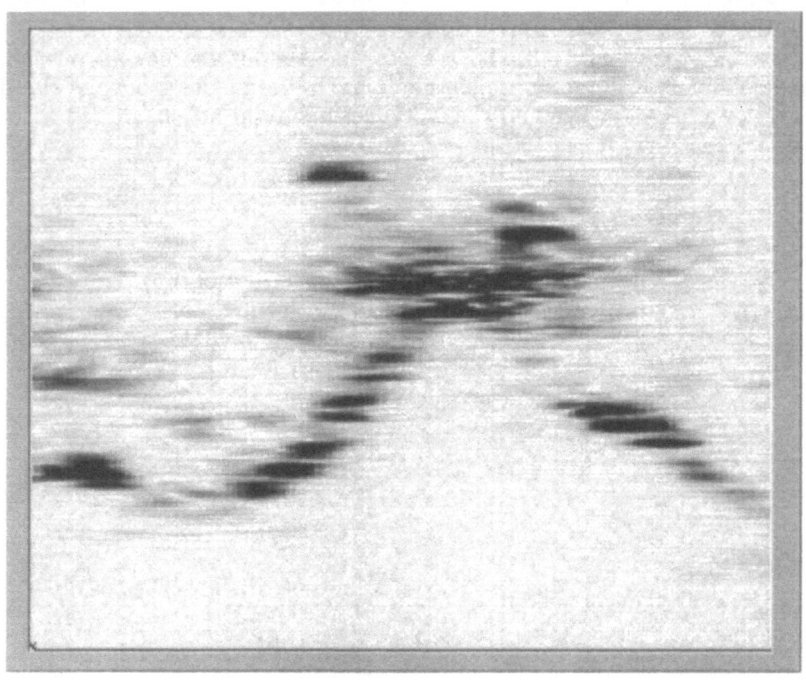

Figure 5. Molar fissure. (Image size: 4.2 mm width, 3.75 mm height)

ACKNOWLEDGEMENT

This work was supported by the Deutsche Forschungsgemeinschaft (DFG), Bonn, (grant Schm 754/6-1).

REFERENCES

Backer Dirks O.: Posteruptive changes in enamel. J. Dent. Res. Vol. 45 (supplement 3) pp. 503-511 (1966).

Barber, F.E., Lees, S. and Lobene, R.R.: Ultrasonic pulse-echo measurements in teeth, Archs oral. Biol. Vol. 14, pp. 745-760 (1969) Pergamon Press.

Espelid, I.; Tveit, A. B.: Diagnosis of secondary caries and crevices adjacent to amalgam, Int Dent J. Vol. 41, pp. 359-364 (1991).

Lussi, A.: Validity of diagnostic and treatment decisions of fissure caries, Caries Res. Vol. 25, pp. 296-303 (1991).

Peers, A.; Hill, F. J.; Mitropoulos, C. M.; Holloway, P. J.: Validity and reproducibility of clinical examinatin, fibre-optic transillumination, and bite-wing radiology for the diagnosis of small approximal carious lesions: an in vitro study, Caries Res. Vol. 27, pp. 307-311 (1993).

DEVEOLOPMENT OF ULTRASONIC SPECTROSCOPY FOR BIOMEDICAL USE

Yoshifumi Saijo, Hidehiko Sasaki, Hiroaki Okawai, and Motonao Tanaka

Department of Medical Engineering and Cardiology,
Institute of Development, Aging and Cancer, Tohoku University
4-1 Seiryo-machi, Aoba-ku, Sendai 980, Japan

INTRODUCTION

Ultrasonic tissue characterization of biological tissue has been widely treated. There are many interesting results. One of the most important results is the measurement of ultrasonic attenuation in tissue by analyzing the RF signal of the reflected ultrasound, because from the clinical point of view, quantitative values, provided by the method can be used for grading the disease state and for evaluating the efficiency of therapy.

Measurement of ultrasonic attenuation has been applied for evaluating diffuse liver disease, viz., fatty liver or liver cirrhosis[1-2]. In these studies, ultrasonic attenuation was considered to be affected with the chemical components in the tissue, viz., collagen or lipid.

On the other hand, much attention has been paid to the microstructure and its relationship to the acoustic properties of the tissue. For the assessment of the problems, we have developed a specially scanning acoustic microscope system operating in the frequency range 100–200MHz, and have used it to measure the attenuation constant and the sound speed of the liver tissues in the microscopic level[3-5].

Figure 1 is the acoustic microscopic images of liver cirrhosis obtained with our system. Two-dimensional distribution of the attenuation constant (left) and sound speed (right) are quantitatively demonstrated in the images. In practice, the values of attenuation constant and sound speed can be measured by comparing the optical and acoustical images.

Our data exhibits nearly the same values as that found in the literature for sound speed, but there are large difference between our data and that in the literature for the ultrasonic attenuation constant. Our data for the value of the attenuation constant of normal liver is 0.8 dB/mm/MHz between 100–200 MHz, while the literature values are around 1.0 dB/cm/MHz, i.e., our data are ten times greater.

One of the reason for high attenuation in our method was considered as that we have fixed the tissue by formalin. But our recent study has demonstrated that there are little change of attenuation or sound speed found in the fixation process.

Other reason of this confusing phenomenon is that the relation of ultrasonic attenuation and frequency was considered to be linear in overall frequency range. However, the measurement of ultrasonic attenuation in the biological tissue has not be done continuously, between 3 MHz and 100 MHz or over 100 MHz frequency range.

The purpose of the present study is to assess the frequency dependent characteristics of ultrasonic attenuation (FDA) in liver tissue in the frequency range between 100 MHz and 200 MHz ultrasound.

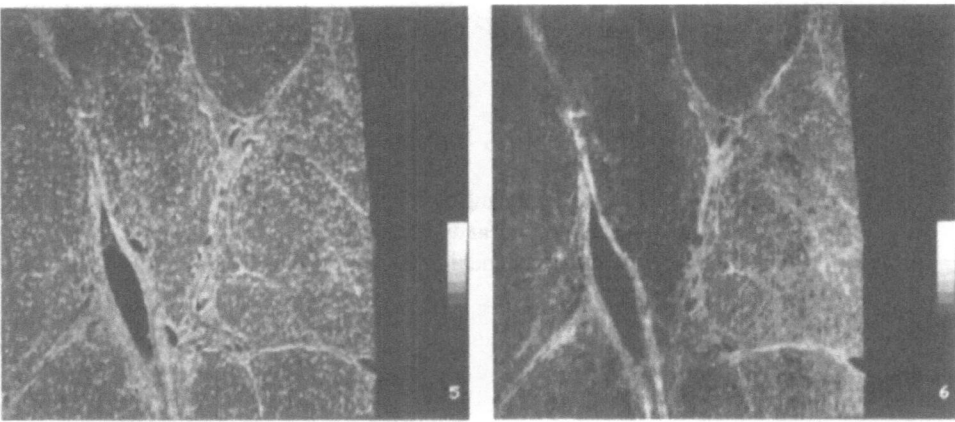

Figure 1. Acoustic microscopic images of liver cirrhosis. (left: attenuation image, right: sound speed image)

MATERIALS AND METHODS

Surgically excised liver tissues were frozen, cut at $10\,\mu$m in thickness, and mounted on glass slides for ultrasonic measurement.

Figure 2 is a block diagram of the ultrasonic spectroscopy system. The system is consisted of six parts; viz., (1)ultrasonic transducers, (2)mechanical scanner, (3)analogue signal processor, (4)frequency control unit, (5)image signal processor, and (6)display unit.

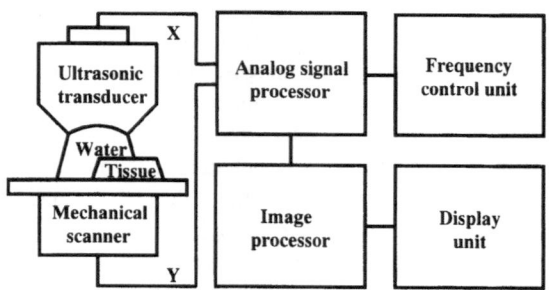

Figure 2. Block diagram of the ultrasonic spectroscopy system.

The distilled water coupling medium was maintained at 20 °C, as was the tissue specimen, during the measurement procedure.

The transducer is oscillated along x axis, and the mechanical scanner is scanned along y axis to obtain the C–mode image in the display unit, when this system is used as a scanning acoustic microscope to determine the region of interest (ROI).

The frequency control unit provides for change in frequency, amplitude and phase of the transmitted ultrasound in 1 MHz steps. In the present study, the x axis scan is done by varying the frequency from 100 MHz to 200 MHz. The y axis mechanical scan is not done.

Figure 3 shows the relationship between the A–mode, C–mode and frequency varying A–mode images of the ultrasonic spectroscopy. First, ROI (one line) was determined by the C–mode image. Next, changing the frequency, ultrasonic attenuation on the line of the ROI was measured and displayed in the frequency varying A–mode. Finally, the frequency dependent ultrasonic attenuation (FDA) of one point on the line of the ROI was obtained.

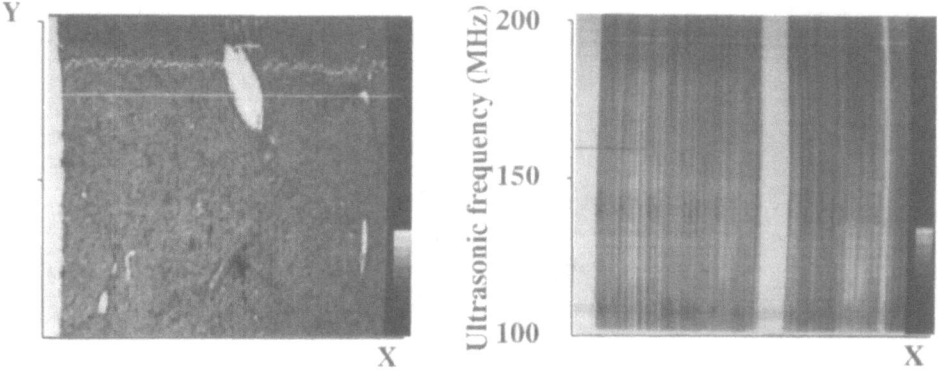

Figure 3. A–mode, C–mode (left) and frequency varying A–mode (right) images of liver cirrhosis.

RESULTS

Figure 4 shows the FDA of normal liver in the range 100 to 200 MHz. The attenuation increases with frequency. A sine wave–like component is also observed superposed on this frequency dependence. A straight line drawn through the data means the frequency rate of attenuation in this frequency. Figure 5 is the FDA of a sample of fibrosis. The pattern of the FDA is clearly different from that of normal liver. The slope is steeper than that of normal liver, and the sine–like curve superposed is more prominent.

By changing the frequency of continuous wave of the ultrasound, FDA of the normal liver and the fibrosis were obtained. And the pattern showed clearly different according to the tissue component.

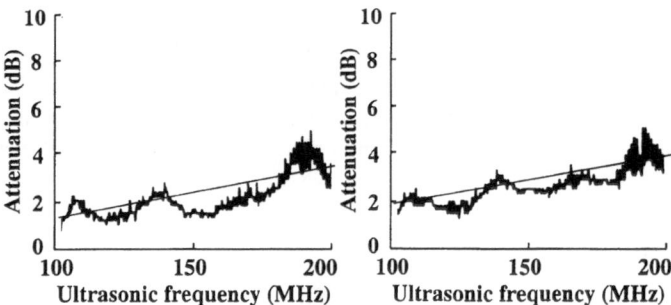

Figure 4. Frequency dependent characteristics of ultrasonic attenuation in normal liver.

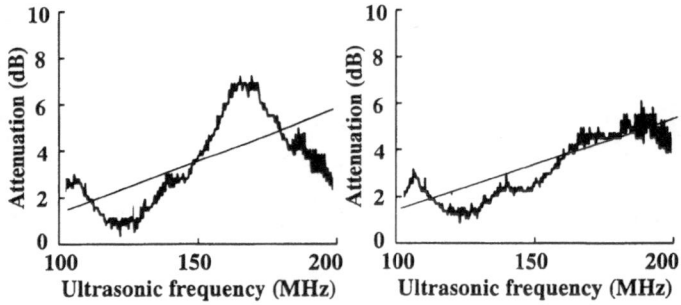

Figure 5. Frequency dependent characteristics of ultrasonic attenuation in a fibrotic lesion.

DISCUSSION

In this system, the received signal consists of two components. Figure 6 is the schematic illustration of the reflected ultrasound components. In the figure, u_1 is the component of reflected ultrasound from the nearer surface of the tissue, u_2 is the reflection from the interface between the tissue and the glass, u_3 is the received signal.

$$u_3 = u_1 + u_2$$

is satisfied. The thickness of the specimen is approximately $10\,\mu$m and the wave length of the ultrasound is $16\,\mu$m at 100 MHz and $8\,\mu$m at 200 MHz, when the velocity of the ultrasound is 1600 m/s in the specimen. The amplitude of the reflected ultrasound exhibits a high value at 160 MHz, where the wave length is 10 μm, u_1 and u_2 are in phase.

The reflected pressure wave y is given by

$$y = e^{-(a + j\frac{2\pi f}{c})yl}$$
$$= e^{-j\frac{2\pi f}{c}l}\, e^{-A_0 f^n l}$$

where f is the frequency, l is the thickness of the specimen, c is the sound speed in the specimen, $A_0 f^n$ is the attenuation function. In this equation, the first exponential component indicates sine−like curve, and the second exponential component indicates the increasing trend of attenuation.

For most soft tissues, it is commonly assumed that the value of n is very close to 1. Thus, the ultrasonic attenuation increases linearly with frequency in the tissue and a straight line is focused on the FDA. The slope of the line is assumed the attenuation coefficient of the liver tissue.

The attenuation coefficient thus obtained is 0.9 dB/mm/MHz in the normal liver, and 2.1 dB/mm/MHz in the fibrosis sample.

FDA pattern contains additional information for tissue. In the fibrotic sample, the wavy pattern is more pronounced than that of normal liver. This is taken to indicate that the amplitude of the component u1 is large in the fibrotic case. If the ideal reflection is occurred at the interface between the coupling water and the tissue, large value of u1 means that the difference of the specific acoustic impedance of the tissue is very large as compared with the water. In our previous study, the value of sound speed is 1580 m/s in the normal liver and 1670 m/s in fibrosis. It is not easy to measure the density of the small amount of tissue, but the sound speed data stress that the specific acoustic impedance showed large value in the fibrosis.

Thus, the FDA pattern is also considered to include the sound speed information of the tissue.

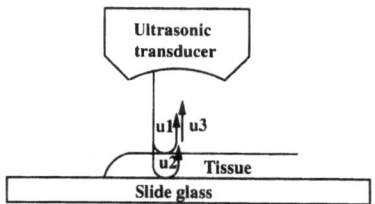

Figure 6. A schematic illustration of the received signal.

     (u_1: reflected ultrasound from the nearer surface, u_2: reflection from the interface between the tissue and the glass, u_3: received signal)

CONCLUSION

FDA of the normal liver considered to be nearly linear in all the specimens in this narrow frequency range. FDA of the fibrotic tissues showed two patterns, higher attenuation and lower attenuation than that of normal liver. In the optical microscopic observation, high attenuation areas were mainly consisted of collagen fibers, while the low attenuation areas contained much elastic fibers.

FDA may reflect chemical and structural features of the tissues. Quantitative analysis of FDA will provide new assessment opportunities of ultrasonic tissue characterization of liver tissues.

REFERENCES

1. Kuc R, Variation of acoustic attenuation coefficient slope estimates for *in vivo* liver, *Ultrasound in Med and Biol* 8:403 (1982).
2. Narayana PA, Ophir J, On the validity of the linear approximation in the parametric measurement of attenuation in tissues, *Ultrasound in Med and Biol* 9:357 (1983).
3. Okawai H, Tanaka M, Dunn F, Chubachi N, Honda K, Qualitative display of acoustic properties of the biological tissue elements, *Acoustical Imaging* 17:193 (1988).
4. Saijo Y, Tanaka M, Okawai H, Dunn F, The ultrasonic properties of gastric cancer tissues obtained with a scanning acoustic microscope system, *Ultrasound in Med and Biol* 17:709 (1991).
5. Saijo Y, Sasaki H, Okawai H, Tanaka M, Dunn F, Intravascular ultrasound and acoustic microscopy evaluation of aortic wall, *Acoustical Imaging* 21:423 (1994).

ULTRASOUND ATTENUATION ESTIMATION IN HIGHLY ATTENUATING MEDIA: APPLICATION TO SKIN CHARACTERIZATION

T. Baldeweck, P. Laugier, and G. Berger

Laboratoire d'Imagerie Paramétrique, URA CNRS 1458
15, rue de l'École de Médecine 75006 Paris, France

INTRODUCTION

The recent introduction of high frequency ultrasound in dermatology has opened new perspectives concerning skin architecture studies and skin diagnosis in dermatological disease. 25-MHz ultrasound technology is now widely available and has proven to be useful, particularly to define the dimensions of skin tumours [7, 8, 10]. In this case, ultrasound used in a traditional B-scan imaging provides a valuable clinical tool.

The measurement of acoustic attenuation in biological tissues has received much interest in the field of ultrasound tissue characterization. Over the past decade, several clinical applications have shown a correlation between attenuation values and pathological states. For *in vivo* applications, only echographic methods (i.e. radio-frequency signal analysis) could be used, the transmission mode being unavailable for skin examination.

Whatever the method used, attenuation estimation must be analyzed with respect to the particular characteristics of the backscattered signal. Firstly, the echo signal is a stochastic process because it results from interferences of randomly unresolved spaced scatterers (speckle noise). Consequently, an attenuation estimate is a stochastic variable, whose variance decreases when averaging information from uncorrelated echo lines. Secondly, this variance also depends on the depth of the explored tissue as the attenuation coefficient is estimated through a regression slope of the centroids as a function of depth. Finally, the spectral characteristics of the echographic signal not only depend on the properties of tissues versus depth but also on the local variations of the sound field (diffraction effect). The signal analysis discussed in this paper does not take into account this last effect and its possible correction.

Up to now, only a few quantitative studies on skin have been reported in the literature. Skin attenuation measurements [4-6, 12, 13, 15, 16] have been done in most cases in the transmission mode for characterization of wounds, aging and burns of the skin. A mean value of about 4 dB/cmMHz can be noted. The aim of this study was to estimate the attenuation slope in cases of highly attenuating media such as skin using high frequency (20 MHz).

We present here first a simulation study on attenuation measurements for the case of highly attenuating media, like skin. Experiments on two phantoms are also reported. Finally, *in vitro* results obtained on skin samples examined at 20 MHz are shown.

MATERIALS AND METHODS

Data simulation

Uncorrelated A-lines backscattered from a homogeneous attenuating medium are simulated using the method described in [14]. The scatterers are identical, randomly distributed in the medium and insonified by a transducer whose frequency response is a gaussian function.

Several data simulations were computed with the following parameters held constant : the sampling frequency F_S (100 MHz), the transducer central frequency F_0 (20 MHz), the scatterers density (400 cm^{-1}), the pulse duration (0.5 µs) (-6 dB bandwidth : 6.4 MHz).

Each simulation contains 256 A-lines of 1024 samples (which corresponds to 7.8 mm tissue depth with c=1530 m/s). The attenuation value is varied from 1 to 5 dB/cmMHz.

Acquisition procedure

Scattering phantom: We used a phantom developed by Computerized Imaging Reference Systems (CIRS, Norfolk, USA). The phantom is in the form of a cylinder which contains a gel with microspheres of 4 µm whose density provides an attenuation of the order of the attenuation of soft tissues (0.5 to 1 dB/cmMHz), and glass microspheres of 75 µm (10 scatterers per mm^3), which act as scatterers. This scattering phantom can be used in the transmission or in the reflection mode. The characteristics of this phantom given by the manufacturer are the following in a constant temperature (22°C) distilled water bath : c=1541 m/s and β=0.8 dB/cmMHz at F_0= 8 MHz.

This phantom was insonified with a focused 20 MHz center-frequency transducer. The sampling frequency was 200 MHz. A-lines were acquired over a scanning area of 2 cm by 2 cm with 1 mm step. This step size ensures the decorrelation between two adjacent rf-lines.

Layered foam phantom: In order to simulate a layered medium like skin, a phantom using different layers (1 cm thick) of different foams was constructed. The foams have been studied separately in the echographic mode, and also in the transmission mode at 5 MHz. Unfortunately, this phantom could not be studied at 20 MHz because of its low scatterer density.

This phantom was insonified with a focused 5 MHz center-frequency transducer with a -6 dB bandwidth of 4 MHz. The sampling frequency was 40 MHz. A-lines were acquired over a scanning area of 2 cm by 2 cm with 1 mm step.

Skin samples: Three female pigs (" Large-White "), 16 to 17 weeks old, were used in this experiment. Fresh normal skin samples of three sites (thigh, back and shoulder) on each pig were excised. In order to ensure the elasticity of the excised skin, it was stretched to its *in vivo* size and stapled to a wood frame. The samples were placed in a tank filled with normal saline and were examined within 10 hours after excision.

The skin samples were insonified with a focused 20 MHz center-frequency transducer. The sampling frequency was 200 MHz. A-lines were acquired over a scanning area of 2 cm by 2 cm with 1 mm step.

Throughout these experiments, the transducer is interfaced to a 486 PC computer with an analog-to-digital converter, which allows the acquisition of rf-lines. A software allows the reconstruction of the video signal (also called envelope) from the rf-lines. The mean envelope can also be estimated from the B-scan image.

Attenuation measurements

The frequency-dependent attenuation is an interesting acoustic parameter for tissue characterization. For most biological tissues, it varies linearly with frequency across a usable frequency bandwidth. One of the possible methods for the estimation of the attenuation of ultrasound by biological tissues is the spectral shift method, also called "centroid algorithm", which relies on the estimation of the local mean frequency of the echographic signal.

A rectangular window is shifted with a constant step along the digitized ultrasound signal. In each window, the spectral centroid f_C (first spectral moment) is calculated. Due to the random distribution of scatterers in the medium, the output of the estimator fluctuates. The variance is reduced by averaging the spectra of M independent echo signals for each window position before the centroid is calculated.

With assumption of a linear-with-frequency attenuation (β in dB/cmMHz) and a gaussian spectrum (the spectral variance σ^2 is a constant versus depth), the attenuation coefficient β can be calculated with the following relation :

$$\beta = \frac{8,68}{c\sigma^2} \cdot \frac{df_c}{d\tau} \qquad (1)$$

where c is the speed of sound in the tissue.

For soft tissue characterization in the reflection mode, the spectra can be estimated with the classical Short Time Fourier Analysis using Fast Fourier Transform (FFT) [3], and subsequently the centroid can be estimated. In the case of highly attenuating media, limitations due to the method (stationarity, window size) may prevent the use of Fourier analysis as this study shows.

Therefore, we have developed two new techniques which allow a quasi-local mean-frequency estimation for attenuation. The first technique is based on a second-order auto-regressive model (AR2) [2], which allows the use of short time windows without deterioration of the estimator performance. This study was done at low frequency (5 MHz). The second one relies on the first derivative of the autocorrelation function of the demodulated signal (adapted estimator AE) and was first developed for Doppler signal analysis [9]. These techniques were tested on simulated echographic signals and signals scattered by calibrated phantoms. The performance of each estimator was evaluated and compared to Fourier analysis.

RESULTS AND DISCUSSION

Simulation results

Throughout this study, the shift value of the time window is set to 50%. The window size ranges between 8 (0.08 μs) and 128 (1.28 μs) samples. In order to evaluate the performance of the different estimators (AR2, AE and FFT) for attenuation estimation, the relative error of attenuation (RE), was calculated with :

$$RE = \left| \frac{\beta_{th} - \beta_{est}}{\beta_{th}} \right| \qquad (2)$$

where β_{th} denotes the theoretical value of attenuation and β_{est} the estimated one. This parameter RE was evaluated for each estimator, each simulation and each window size and is shown on figure 1.

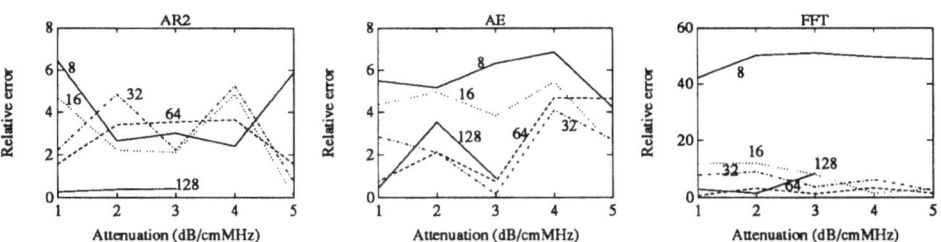

Figure 1. AR2, AE and FFT estimators on simulations. Relative error of attenuation RE (%) versus attenuation (dB/cmMHz). The numbers on the graphic correspond to the window size (samples).

The relative error of attenuation is always below 7% for windows duration of 128 to 8 samples for the AE and the AR2 estimators. The FFT estimator performance decreases when the window size decreases and the relative error reaches 40% with the 8 samples window.

It should be noted that the usable signal length decreases with attenuation due to limited penetration depth. Therefore, for attenuation value greater than 3 dB/cmMHz and a window size of 128 samples, the number of centroids is smaller than 4, which produces a high variance on the slope estimate. This is the reason why the curve 128 samples stops at 3 dB/cmMHz.

The conclusion of this simulation study is that the AR2 modeling and AE estimators are interesting alternatives to the FFT method, particularly when short windows are necessary.

Scattering phantom study

The phantom was located in the focal zone of the 20 MHz transducer. A depth of 3 mm around the focus was explored to minimize the diffraction effects.

Figure 2 shows the evolution of the centroid of the signal versus depth in the phantom for the three mean frequency estimators with a window of 64 samples and a shift of 50%. The AE and AR2 estimators exhibit the same behaviour, with a small shift between the two centroids evolution. On the other hand, the FFT curve is different and has a smaller frequency shift. Consequently, the attenuation estimate was smaller with the FFT.

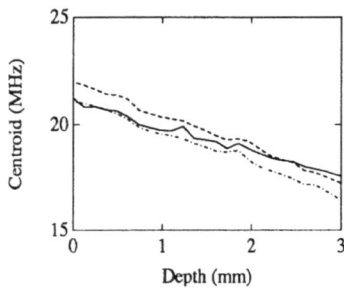

Figure 2. Scattering phantom (20 MHz). Centroid evolution (MHz) versus depth (mm) for the AR2 ('--'), AE ('..') and FFT ('-') mean frequency estimator.

In order to compare the behaviour of the different mean frequency estimators for attenuation measurements, we analyzed the results obtained for different window sizes. For window sizes ranging from 16 to 128 samples, the AE and AR2 estimators give constant and comparable estimates (0.69 dB/cmMHz for AE and 0.72 dB/cmMHz for AR2). The FFT estimator gives 0.5 dB/cmMHz for window size ranging from 128 to 32 samples. This value is lower than the value given by the manufacturer (0.8 dB/cmMHz).

A possible reason for this difference between the results obtained with the different techniques is that in addition to the dependence on the centroid as a function of time, there is a dependence on the spectral variance as a function of time. If the spectrum is a gaussian function, the theory shows that the spectral variance σ^2 at each depth is equal to the σ^2 estimated on the emitted pulse. In our study, we estimated the σ^2 on a reference pulse at the focus. Due to equation (1), accuracy of attenuation estimate depends on accuracy of σ^2 estimate. We are currently studying this problem in order to improve the estimation of the σ^2 using the FFT estimator, which allows the estimation of the σ^2 from each spectrum as a function of depth. Preliminary results on simulated data suggest that this method will improve the performance of the FFT technique and thus compensate the bias of the centroid estimation.

In order to verify this attenuation value, we also measured the attenuation in the phantom in the double transmission mode with the same transducer. We found 0,8 dB/cmMHz in the bandwidth 3-10 MHz, which corresponds to the linear zone in the spectral difference (the bandwidth is small because the total phantom thickness is 3 cm, which correspond to about 96 dB attenuation). However, the small difference between attenuation measured in transmission and in reflection with AR2 or AE could be due either to local heterogeneity of the phantom or to the σ^2 estimate.

On the other hand, one parameter of the method AE (frequency cut-off) [1] is very sensible, that is the reason why we chose to use only the AR2 modeling in the following experiments.

Layered foam phantom

The aim of this experiment is to show the feasibility of making attenuation profiles in a layered medium. This phantom was insonified with a 5 MHz central-frequency transducer and the centroids were calculated with a window of 64 samples. The results obtained with this phantom are shown on figure 3 (B-scan, mean envelope versus depth, and centroid versus depth).

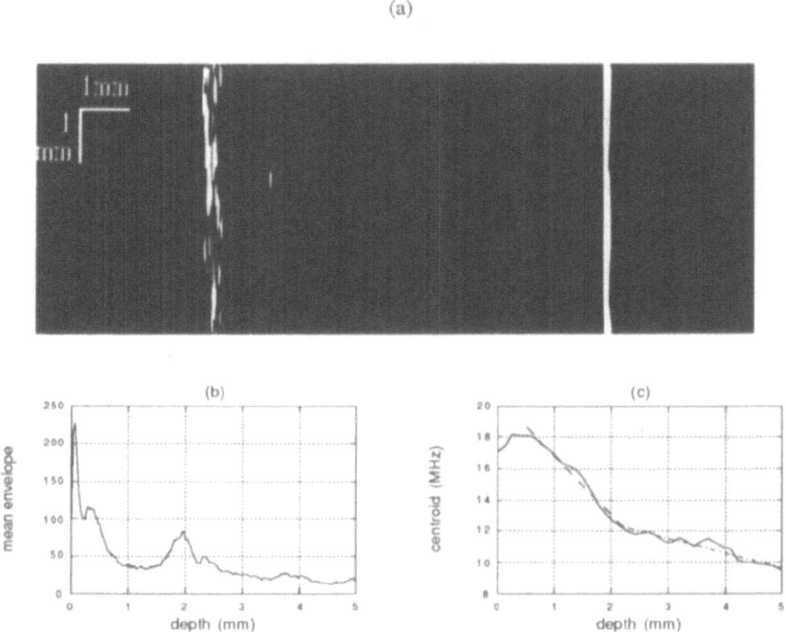

(a)

(b)

(c)

Figure 3. Layered foam phantom (5 MHz) : B-scan image(a), mean envelope (b) and centroid (MHz) versus depth (mm) and linear fits (c) using AR2 modeling.

On the traditional B-scan image, a difference can be observed between the two layers of foams. On the mean envelope, the exponential variation of the video signal can be appreciated and two different evolutions can be distinguished. Finally, on the centroid curve, attenuation profiles are shown with the two slopes in the two layers. Between the two peaks, there is a depth of about 10 mm, which corresponds to the thickness of the first foam. In the first foam, we found 0.9 dB/cmMHz, and 0.5 dB/cmMHz in the second one. These two values are in good agreement with the values measured in the transmission mode for each foam. The low value of centroid at 10 mm corresponds to specular echoes due to the interface between the two layers of foam.

The results obtained on this layered phantom with AR2 modeling allows us to distinguish different areas with different attenuation values, which correspond to different structures in the phantom.

Skin study

In vitro porcine skin was studied because it has been established as a human skin model [11]. Figure 4 shows an example of a B-scan image, the mean envelope versus depth and the evolution of the centroid (MHz) versus depth (mm) with the AR2 model. The regression slope in first area corresponds to the dermis, with an attenuation value equal to 2.7

dB/cmMHz. The second one corresponds to the hypodermis, with an attenuation value equal to 0.7 dB/cmMHz.

A difference can be seen on the B-scan image approximately 2 mm from the surface. On the mean envelope curve, a peak can also be seen at the same depth which corresponds to the interface of the two structures in the skin. Finally, this discontinuity is also observed at the same depth on the centroid curve. The results obtained on porcine skin with AR2 modeling allow us to distinguish two different areas with two different attenuation values corresponding to two different structures in the skin.

(a)

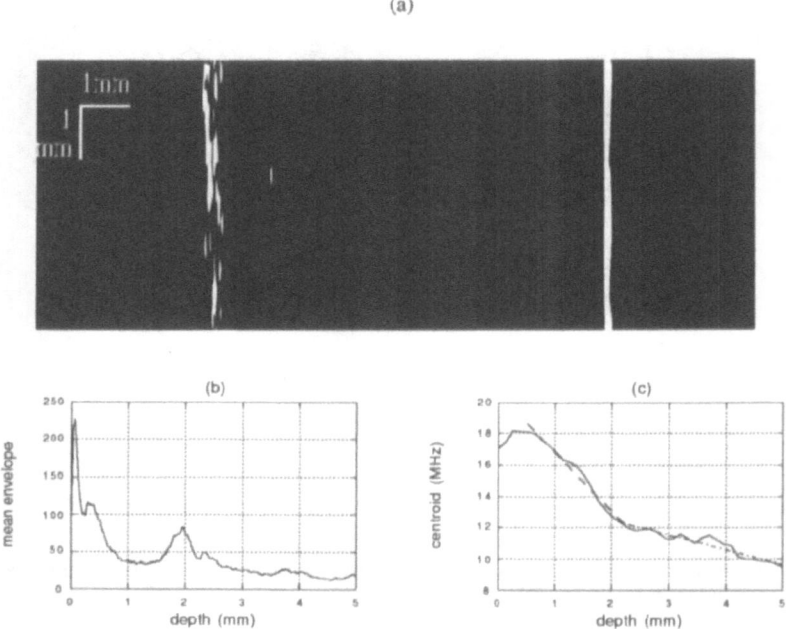

Figure 4. Porcine skin sample (20 MHz) : B-scan image, mean envelope and centroid (MHz) versus depth (mm) using AR2 modeling.

This result was reproducible on a set of samples. Table 1 shows the results obtained on nine skin samples. For the pig number 3, the thigh sample signal in the hypodermis does not have enough intensity to allow the centroid computation.

In the first area (dermis), the mean value of attenuation obtained from all the samples is 3.4 +/- 0.6 dB/cmMHz, which is in good agreement with the values obtained in the literature. In the second area (hypodermis), the mean value is 1.1 +/- 0.6 dB/cmMHz. We can note a significant difference between the hypodermis of the back and thigh and the hypodermis of the shoulder. This may be explained with the difference in the structure of the hypodermis in the shoulder (front body skin).

Table 1. Attenuation measurements on three sites on normal porcine skin (pig 1, 2 and 3).

Site	pig n°	Attenuation (dB/cmMHz)	
		area 1 (dermis)	area 2 (hypodermis)
thigh	1	3.8	0.7
	2	3.1	0.6
	3	4.8	/
back	1	2.7	0.7
	2	3.3	0.6
	3	2.7	0.7
shoulder	1	3.4	1.8
	2	3.4	1.6
	3	3.6	1.8

CONCLUSIONS

The frequency-dependent attenuation is an interesting acoustic parameter for tissue characterization. One of the possible methods for attenuation estimation from ultrasound scattered by biological tissues is the spectral shift method, also called the "centroid algorithm", which relies on the estimation of the first spectral moment of the echographic signal.

For soft tissue characterization in the reflection mode, the ultrasound attenuation can be estimated with the classical Short Time Fourier Analysis. In the case of highly attenuating media, limitations due to the method (stationarity, window size) may prevent the use of Fourier analysis. Therefore, we have developed two new techniques which allow a quasi-local mean-frequency estimation for attenuation estimation. The first technique is based on a second-order auto-regressive model (AR2) and the second one relies on the first derivative of the autocorrelation function of the demodulated signal (adapted estimator AE). These techniques were tested on simulated echographic signals and the performance of each estimator was evaluated and compared to that of Fourier analysis.

Experimental results obtained on phantoms with calibrated attenuation show the feasibility of determining the attenuation slope with these new estimators and characterizing layered media. In the first experimental study at 20 MHz on *in vitro* porcine skin, a continuous profile of centroids is obtained, which allows the determination of the attenuation slope in the different layers in the skin (dermis, hypodermis). This attenuation profile is confirmed by looking at the classical B-scan echographic image, which shows a difference between the different skin layers. These results are reproducible on a set of nine samples.

ACKNOWLEDGEMENTS

This work was supported by grants from INSERM (A9304194QAT) and from ARC (6446 and 6621).

REFERENCES

[1] Baldeweck T., Herment A., Laugier P. and Berger G., "Attenuation estimation in highly attenuating media using high frequencies : a comparison study between different mean frequency estimators", IEEE Ultrasonics symposium, 94CH3468-6, pp. 1783-1786, 1994.

[2] Baldeweck T., Laugier P., Herment A. and Berger G., "Application of autoregressive spectral analysis for ultrasound attenuation estimation : interest in highly attenuating medium", *IEEE Trans. Ultrason. Ferroelec. Freq. Contr.*, vol. 42 (1), pp. 99-110, 1995.

[3]   Berger G., Laugier P., Fink M. and Perrin J., "Optimal precision in ultrasound attenuation estimation and application to the detection of Duchenne muscular dystrophy carriers", *Ultrasonic Imaging*, vol. 9, pp. 1-17, 1987.

[4]   Bhagat P. K. and Kerrick W., "Ultrasonic characterization of aging in skin tissue", *Ultrasound in Med. & Biol.*, vol. 6, pp. 369-375, 1980.

[5]   Cantrell J. H., Goans R. E. and Roswell R. L., "Acoustic impedance variations at burn-nonburn interfaces in porcine skin", *J. Acoust. Soc. Am.*, vol. 64, pp. 731-735, 1978.

[6]   Forster F. K., Olerud J. E., Riederer-Henderson M. A. and Holmes A. W., "Ultrasonic assessment of skin and surgical wounds utilizing backscatter acoustic techniques to estimate attenuation", *Ultrasound in Med. & Biol.*, vol. 16, pp. 43-53, 1990.

[7]   Gassenmeier G., Kiesewetter F., Schell H. and Zinner M., "Wertigkeit der hochauflösenden Sonographie für die Bestimmung des vertikalen Tumordurchmessers beim malignen Melanom der Haut", *Hautarzt*, vol. 41, pp. 360-364, 1990.

[8]   Harland C. C., Bamber J. C., Gusterson B. A. and Mortimer P. S., "High frequency, high resolution B-scan ultrasound in the assessment of skin tumours", *British J. of Dermatol.*, vol. 128, pp. 525-532, 1993.

[9]   Herment A., Demoment G., Dumée P., Guglielmi J. P. and Delouche A., "A new mean adaptive mean frequency estimator : application to constant variance color flow mapping", *IEEE Trans. Ultrason., Ferroelec. Freq. Contr.*, vol. 40 (6), pp. 796-804, 1993.

[10]   Hoffmann K., Jung J., El Gammal S. and Altmeyer P., "Malignant Melanoma in 20-MHz B Scan Sonography", *Dermatology*, vol. 185, pp. 49-55, 1992.

[11]   Meyer W., Schwarz R. and Neurad K., "The skin of domestic mammals as a model for the human skin, with special reference to the domestic pig", *Curr. Probl. Dermatol.*, vol. 7, pp. 39-52, 1978.

[12]   Olerud J. E., O'Brien W., Riederer-Henderson M. A., Steiger D., Forster F. K., Daly C., Ketterer D. J. and Odland G. F., "Ultrasonic assessment of skin and wounds with the scanning laser acoustic microscope", *J. of Invest. Dermatol.*, vol. 88, pp. 615-623, 1987.

[13]   Olerud J. E., O'Brien W. D., Riederer-Henderson M. A., Steiger D. L., Debel J. R. and Odland G. F., "Correlation of tissue constituents with the acoustic properties of skin and wound", *Ultrasound in Med. & Biol.*, vol. 16, pp. 55-64, 1990.

[14]   Oosterveld B. J., Thijssen J. M. and Verhoef W. A., "Texture of B-mode echograms : 3-D simulations and experiments of the effects of diffraction and scattering density", *Ultrasonic Imaging*, vol. 7, pp. 142-160, 1985.

[15]   Querleux B., "Imagerie et caractérisation de la peau in vivo par échographie ultrasonore à haute résolution spatiale", *Journées d'étude : Interaction entre les ultrasons et les milieux biologiques, Valenciennes*, vol. , pp. 82-89, 1994.

[16]   Riederer-Henderson M. A., Olerud J. E., O'Brien W. D., Forster F. K., Steiger D. L., Ketterer D. J. and Odland G. F., "Biochemical and acoustical parameters of normal canine skin", *IEEE Transactions on Biomed. Eng.*, vol. 35 (11), pp. 967-972, 1988.

ON THE ECHOGRAPHIC IMAGE OF BIOLOGICAL TISSUE
WITH SELF-SIMILARITY STRUCTURE

Iwaki Akiyama[1], Nobuyuki Taniguchi[2] and Koichi Itoh[2]

[1]Department of Electrical Engineering, Shonan Institute of Technology
1-1-25 Tsujido-nishikaigan, fujisawa 251 Japan

[2]Department of Clinical Pathology, Jichi Medical School
3311-1 Yakushiji, Minamikawachi, Kawachi-gun 329-04 Japan

INTRODUCTION

Acoustical characteristics of biological tissue is dependent on pathological states of the tissue. Since an ultrasonic backscattering property of the tissue is dependent on tissue structure, it is clinically useful if a technique for characterization of those properties using an echographic image is established. Javanaud[1] implies a fractal model of medium structure like biological tissue for the angular dependence of the ultrasonic differential scattering cross section and the frequency dependence of the total scattering cross section. Chen, Daponte and Fox showed [2] that the fractal Brownian feature curves of echographic image of normal liver and abnormal liver are dependent on the histological states. Verhoeven and Thaijssen reported[3] that fractal analysis has a potential to identify the property of scatterer density of the tissue for lesion detection in echographic images. They also pointed out that Hurst coefficient offers no obvious advantage over statistical parameters such as a mean value and a standard deviation.

We have also reported [4,5]that Hurst coefficients which are experimentally obtained by fractal analysis of the echographic image are slightly dependent on the histological states of the diffuse liver disease. Further, the higher frequency used in the experiment the better dependence on the histological states. Since the tissue structure itself has a self-similarity characteristics[6,7,8], it is important to show how the echographic image of the tissue has the self-similarity characteristics and how it could be extracted from the image. In this paper we proposes a simple model for the simulation phantom with self-similarity using a fractal Brownian function. The Hurst coefficient of the echographic image of the phantom is compared with the initial Hurst coefficient over the range of 0.2 to 0.8. When Hurst coefficient is equal to 0.5, geometry is a nonfractal Brownian fluctuation. We found that the Hurst coefficient of the echographic image is dependent on the initial Hurst coefficient. Further we compute the echographic image of the phantom made from the optical macroscopic image of normal liver and cirrhosis liver tissue specimen. The Hurst coefficient of the echographic image is dependent on the histological state of the tissue. The results implies that the Hurst coefficient is expected to be one of the parameters for tissue characterization.

SIMULATION PHANTOM BASED ON FRACTAL BROWNIAN FUNCTION

We perform the computer simulation of the echographic image of the phantom with self-similarity structure for the investigation of the dependence of Hurst coefficient of the image on the initial Hurst coefficient. The simulation is made in a two-dimensional plane because of the elapsed time and memory capacity. We make the simple simulation model for computation. The reflection source is made of a square cell. We assume that ultrasonic waves are reflected at the center of the cell and the coefficient of the reflection is proportional to the area of the cell. Note that a large cell reflects ultrasound waves of large amplitude and occupies large area and a small cell reflects waves of small amplitude and occupies small area. The distribution of cell size is determined by the fractal Brownian function[9] $f(r)$;

$$Pr\left\{ \frac{|f(r + \Delta r) - f(r)|}{|\Delta r|^H} < t \right\} = F(t) , \tag{1}$$

where $Pr\{\}$ is a probability function, r is a position vector, and Δr is the distance between two points. Let the probability density function be a normal distribution, then

$$f(r + \Delta r) = f(r) + t|\Delta r|^H , \tag{2}$$

where t is a probability variable of the normal distribution with mean 0 and variance σ. Note that the difference in size between a cell at r and a cell at $r+\Delta r$ is statistically dependent on the distance $|\Delta r|^H$. If the distance is long, the statistical fluctuation is large. If the distance is short, the statistical fluctuation is small.

An algorithm for the generation of the phantom is as follows. First of all, we give the four initial cells; 200, 100, 20, and 4 pixels square in a restricted region of 4096 ×4096 pixel as shown in Fig.1(a). The cells are iteratively generated to satisfy eq.(2) and so as not to overlap the pre-generated cells until the 1500 cells are generated. Since t of eq.(2) is a random variable which satisfies normal distribution of mean 0 and standard deviation σ, the size of cell is determined by the random variables of the normal distribution of mean $f(r)$ and standard deviation of $\sigma|\Delta r|^H$. In this simulation we choose σ equal to one. Thus the size of the cell is determined by the random variable obeying the normal distribution of standard deviation σ and mean value m given by the eq.(3) and eq.(4).

$$\sigma = \sqrt{\frac{1}{\sigma_1^2 + \sigma_2^2 + \sigma_3^2 + \sigma_4^2}} \tag{3}$$

$$m = \frac{f(r_1)\sigma_2^2\sigma_3^2\sigma_4^2 + f(r_2)\sigma_1^2\sigma_3^2\sigma_4^2 + f(r_3)\sigma_1^2\sigma_2^2\sigma_4^2 + f(r_4)\sigma_1^2\sigma_2^2\sigma_3^2}{\sigma_2^2\sigma_3^2\sigma_4^2 + \sigma_1^2\sigma_3^2\sigma_4^2 + \sigma_1^2\sigma_2^2\sigma_4^2 + \sigma_1^2\sigma_2^2\sigma_3^2} \tag{4}$$

where σ_i (i=1,2,3,4) and r_i(i=1,2,3,4) are standard deviations of normal distribution and position vectors of the cells, respectively, and the subscript represents the number of the initial cell. The position and the size are determined by a computer-generated random value.

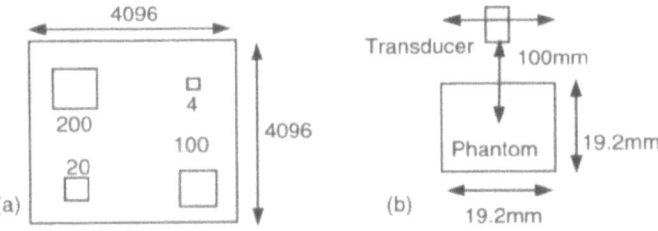

Figure 1 Initial cell distribution (a) and the arrangement of the transducer and the phantom (b).

Secondly, an echographic image is calculated from the simulation phantom which is generated by the above mentioned procedure. We consider the diffraction of ultrasonic waves by a concave transducer in the calculation, because the texture in an echographic image on which we concentrate is strongly dependent on the ultrasonic beam pattern. The aperture, the focal length, and the center frequency of a transducer are 10mm, 100mm and 3.5MHz, respectively. Those values are typical of commercially available equipment. The arrangement of the transducer and the phantom is illustrated in Fig.1(b).

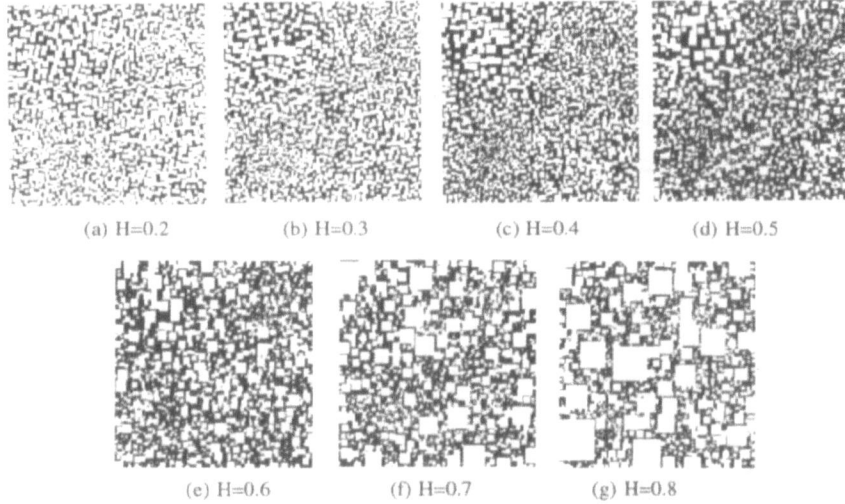

Figure 2 The series of the phantom which is computed by the proposed algorithm.

Figure 2 shows the series of computed phantom of Hurst coefficient range of 0.2 to 0.8. At lower Hurst coefficient cell distribution is clustered at the initial cell position, on the other hand at higher Hurst coefficient various sizes of the cell are scattered and the distribution is heterogeneous. The degree of complexity of the distribution is expressed as the value of Hurst coefficient. Figure 3 shows the echographic images of the phantom shown in Fig.2 where a typical speckle pattern is observed.

We calculate the Hurst coefficient of the echographic image at two different scales. Figure 5 shows the relation of Hurst coefficient of the image to the initial Hurst coefficient. We can find the dependence of Hurst coefficient of the echographic image on the initial coefficient. In the range of 10 to 60 pixel, at lower initial Hurst coefficient Hurst coefficient of the echographic image is increased as the initial coefficient is increased, on the other hand, at higher initial Hurst coefficient the Hurst coefficient of the echographic image is decreased as it is increased. In the range of 1 to 8 or 10 pixels, the Hurst coefficient of the echographic image is increased as the initial coefficient is increased in the range over 0.5. One pixel is correspondent to the 0.02mm. It is shown that we can estimate the difference of structure by Hurst coefficient of the echographic image.

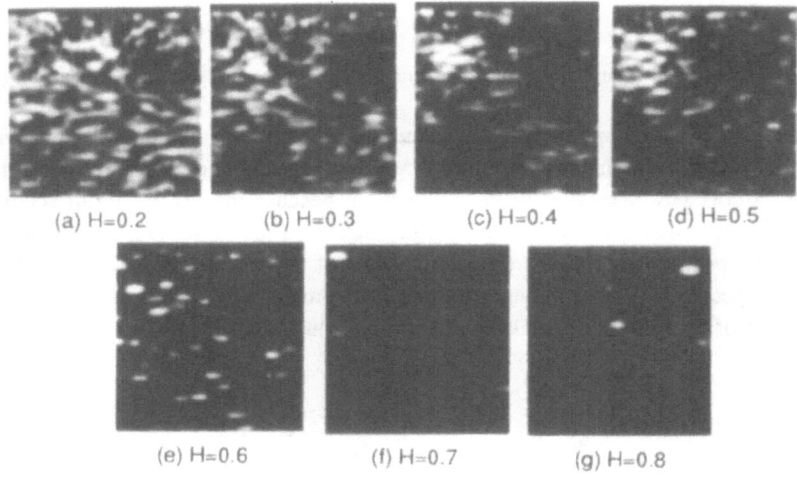

(a) H=0.2   (b) H=0.3   (c) H=0.4   (d) H=0.5

(e) H=0.6   (f) H=0.7   (g) H=0.8

Figure 3 The series of echographic image which are computed from the phantom shown in Fig.2. Frequency of ultrasound is 3.5MHz and the aperture and the focal length of transducer is 10mm and 100mm, respectively.

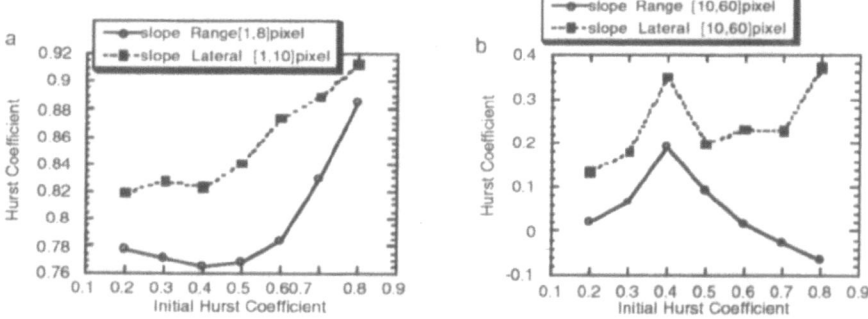

(a) Short scale range of [1,10] in pixel (b) Long scale range of [10,60] in pixel

Figure 4 Hurst coefficient of the echographic images vs. initial Hurst coefficient

SIMULATION OF ECHOGRAPHIC IMAGE OF LIVER SPECIMEN USING OPTICAL MACROSCOPIC IMAGE

The simulation described in the previous section is based on the phantom which is produced by calculating eq.(2). It means that the phantom is not based on biological tissue. It is important to show how self-similarity structure of the biological tissue affects the echographic image of the tissue. In this section we use liver tissue specimen of three different histological states; normal and two different level of cirrhosis. The optical macroscopic images of those tissue specimen are shown in Fig.5. Dimension of the specimen is approximately 2cm ×2cm. Obviously you can differentiate the structure of those tissues.

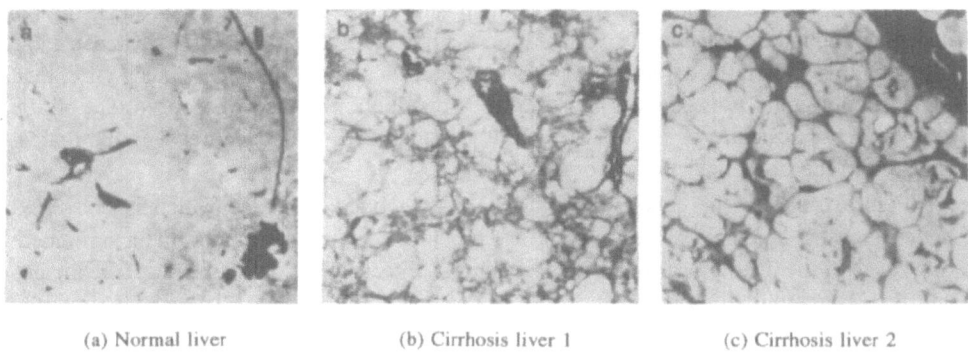

(a) Normal liver   (b) Cirrhosis liver 1   (c) Cirrhosis liver 2

Figure 5 Optical macroscopic image of liver specimen. Blue color component of the image by Azan stain.

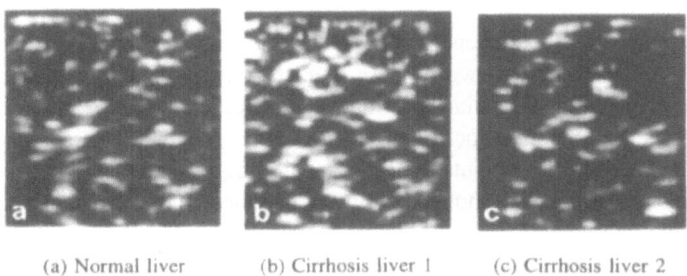

(a) Normal liver   (b) Cirrhosis liver 1   (c) Cirrhosis liver 2

Figure 6 Echographic images of the boundary distribution
which is extracted from macroscopic image shown in Fig.5.

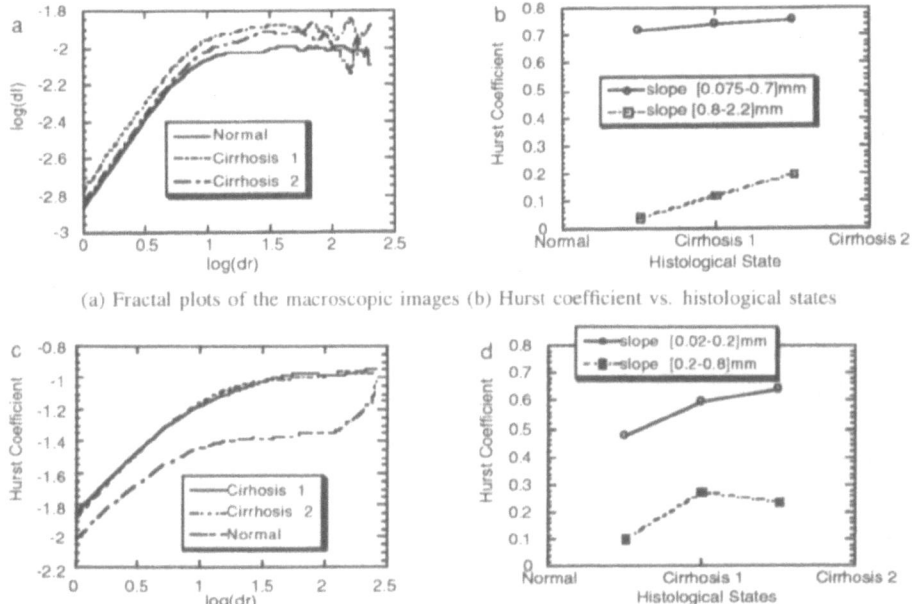

(a) Fractal plots of the macroscopic images (b) Hurst coefficient vs. histological states

(c) Fractal plots of the echograhic images. (d) Hurst coefficient vs. histological states

Figure 7 Fractal plots (a) and the Hurst coefficient (b) of the macroscopic images shown in
Fig.5 and fractal plots (c) and Hurst coefficient (d) of the echographic images shown in Fig.6.

353

To compute the echographic image of the phantom based on biological tissue using the specimen, boundary detection is operated for the optical image shown in Fig.5. because the ultrasonic wave is reflected at the boundary of acoustical properties. Obviously an optical image is not equal to acoustical characteristics distribution. Since we concentrate on the self-similarity structure of the tissue, there is no significant difference between the optical image and the acoustical image for this purpose.

Figure 6 shows the echographic image of the boundary image which is computed under the same condition as described in the previous section. We cannot find a significant difference of the histological states in those images. Fractal plots of the macroscopic images and the echographic images are shown in Fig.7(a) and (c). Figure 7 (b) and (d) show the relation of Hurst coefficient to the histological states. We can estimate the difference of the histological state by using Hurst coefficient of the echographic image.

CONCLUSION

We investigate Hurst coefficient of the echographic images of the media with self-similarity structure. First of all we propose a method for producing the phantom with self-similarity using fractal Brownian function. Hurst coefficient of the echographic image of the phantom is dependent on the initial Hurst coefficient which is used for production of the phantom. Second, we compute the echographic image of optical macroscopic image of liver specimen. The results show that Hurst coefficient of the echographic image is dependent on the histological states of the specimen. Consequently, Hurst coefficient is expected to be one of the parameters for tissue characterization using the degree of self-similarity of tissue structure.

REFERENCES

1. C.Javanaud, The application of a fractal model to the scattering of ultrasound in biological media, J.Acoust.Soc.Am. 86:493 (1989)

2.C-C. Chen, J.S. Danponte, and M.D. Fox, IEEE Trans. on Medical Imaging, 8:133 (1989)

3.J.T.M. Verhoeven and J.M. Thijssen, Potential of fractal analysis for lesion detection in echographic images, Ultrasonic Imaging, 15:304

4.I.Akiyama, T.Saito, M.Nakamura, N.Taniguchi, and K. Itoh, Tissue characterization using fractal dimension of B-scan images, IEEE Ultrasonics Symposium Proceedings, 1353 (1991)

5. I.Akiyama, A. Ohya, M.Nakamura, N.Taniguchi, and K.Itoh, Fractal analysis of ultrasonic images for tissue characterization, Jpn. J. Med.Ultrasonics, 20: 643 (1993)

6.E.B. Cargill, H.H. Barrett, R.D. Fiete, M. Ker, D.D. Patton, and G.W. Seeley, Fractal physiology and nuclear medicine scans, SPIE 914, Medical Imaging II, 355 (1988)

7.T.R. Nelson, Morphological modeling using fractal geometries, SPIE 914, Medical Imaging II, 326 (1988)

8. A. Chandra and C. Thompson, Ultrasonic characterization of fractal media, Proc. of the IEEE, 81:1523(1993)

9.N.Yokoya, K.Yamamoto and N.Funakubo, Denshi Joho Tsusin Gakkai Ronbunshi, 70D:2605(1987)

APPLICATION OF ARTIFICIAL NEURAL NETWORKS IN ULTRASONIC TISSUE CHARACTERIZATION

H.J. Huisman and J.M. Thijssen

Biophysics Laboratory, Department of Ophthalmology
University Hospital Nijmegen, The Netherlands

INTRODUCTION

We have shown[1] that Artificial Neural Networks (ANN) can be applied successfully at the classification stage (see Fig. 1) in Ultrasonic Tissue Characterization (UTC). The 'classifier' ANN was presented with a set of pre-chosen features (with a fixed calculation method). Only the number of features was reduced during learning. The classifier performance (compared to linear and quadratic discriminant rules) was improved.

In a fixed feature calculator the features themselves cannot be trained, only the number of features can be reduced. The reduction is optimized during training. A problem that remains is that the processing performance is, of course, also influenced by the chosen features themselves: a poor choice of features will lead to poor results; secondly, a high number of features (relative to the number of samples in the training set) may decrease classifier performance due to the peaking phenomenon.[2]

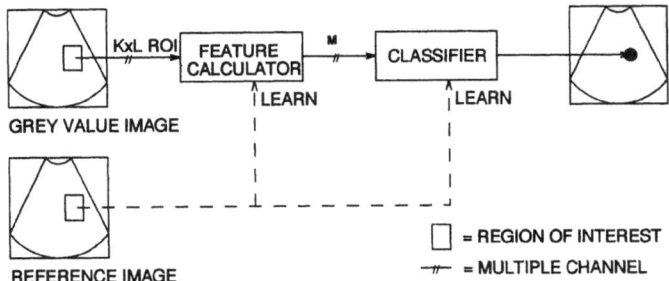

Figure 1. General UTC image processing setup

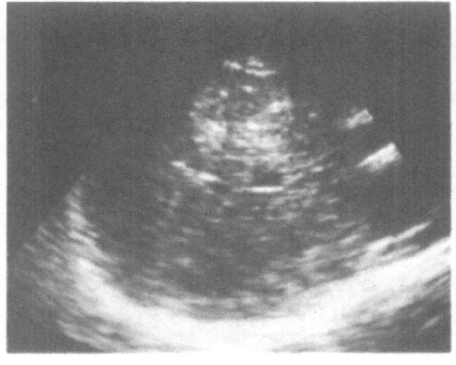 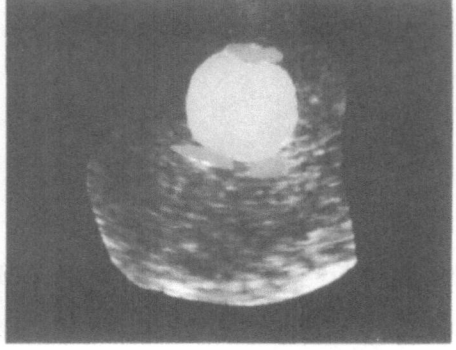

(a) Grey-value image   (b) Reference image (overlay)

Figure 2. Example input to the image processing system: a) B-mode echogram displaying transversal section through the liver, b) reference image is shown overlayed on the B-mode image. Non-liver region, blood vessels, and tumor are manually outlined and filled with a proper label, shown in the overlay. Parenchymal liver tissue is unlabeled (label=0)

The application of an ANN at the UTC feature calculation stage will improve the feature calculator. Firstly, due to the ANN's learning properties, the features are tuned to the problem at hand and need not be chosen. Secondly, the number of features is related to the number of classes in the image data (the ANN trains to discriminate classes).

Before explaining the two types of feature calculation methods, first an overview is given of the image processing setup in which a feature calculator is embedded. Figure 1 shows that during operational phase (solid lines) the grey values from a K×L region of interest (ROI) are fed into a feature calculator which outputs an M-dimensional feature vector used by the classifier to characterize the region of interest. During the learning phase (solid & dashed lines) a reference image is available that fully characterizes the grey value image (see Fig. 2 for example images). The reference image is used to adapt both the feature calculator and classifier such that the image processor will predict correct characterizations.

Using fixed features requires a two-stage feature calculator (see Fig. 3a): 1. calculation of N fixed features, 2. reduction to M features in a reduction stage. A large number of fixed features is known in literature (e.g. attenuation, average grey value, signal to noise ratio (SNR)). The classifier requires a low number of features (peaking phenomenon). Therefore, feature reduction (e.g. principle components) is often necessary. Two problems then arise. First, are there any other than the N fixed features

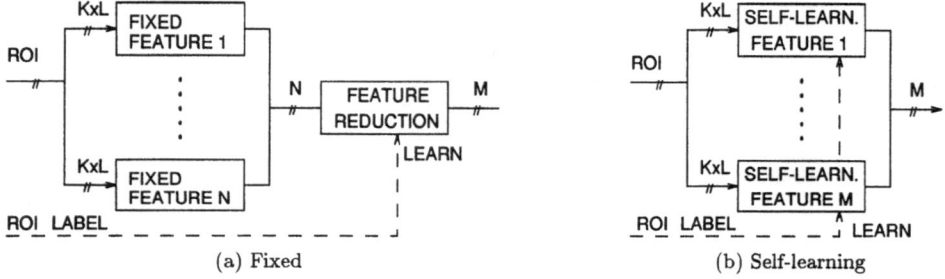

(a) Fixed   (b) Self-learning

Figure 3. Two types of feature calculators

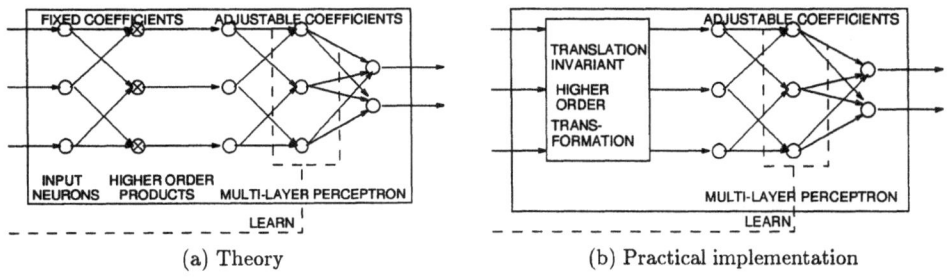

(a) Theory   (b) Practical implementation

Figure 4. Higher Order Neural Network

that could contribute to the discrimination process? Second, does not the reduction of N to M features also reduce the discriminatory information?

In the self-learning feature calculation method (see Fig. 3b) each feature is calculated with the same method. This method has adjustable coefficients which are adjusted (LEARN) to produce maximally discriminant features. As it will directly learn relevant and discriminative features, no reduction is necessary.

METHOD

To be a good self-learning feature calculator, the method should be capable of learning a wide variety of functions. ANNs are well-known to be capable of learning linear, as well as higher order functions.[3]

Direct application of an ANN to a ROI leads to an impractical network. Two reasons: the number of elements in a ROI is large (e.g. 10 lines, 100 samples = 1000 elements). This leads to a large number of inputs. Secondly, to learn higher order relations one or more hidden layers are necessary. The result is a very large network, which is difficult, if not impossible to train.

In a Higher Order Neural Network (HONN) (see Fig. 4a) the higher order terms are calculated in advance, and need not be learned. This reduces the number of hidden layers and therefore the complexity of the ANN.[4]

A direct implementation of a HONN is still impractical, because of a combinatoric explosion of higher order products. Incorporating invariances leads to translation invariant higher order transformations[5] (see Fig. 4b). For example: a translation invariant, second order transformation leads to the autocovariance function. Other examples are: histogram, spectrum.

RESULTS

1D RF data were simulated and the envelope signal was calculated. Two sets of ROIs were discriminated: one with scatterer number density 10 (in a resolution volume), the other with a lower level. All ROIs were normalized to zero average grey value. Calculated fixed features were: sigma, skewness, kurtosis. Setup of the HONN: histogram, single hidden layer perceptron. Train HONN with a learning set. All three fixed and two self-learning features were classified with a test set (independent from the learning set). Figure 5 shows the correct classification percentages at varying densities. As expected, the lower densities are easily discriminated from the high density (10) case. With increasing density the discrimination performance worsens. The self-learning HONN features are slightly better in discriminating low-contrast densities.

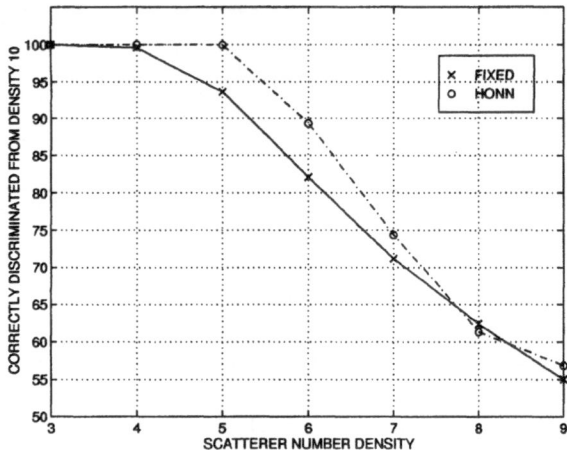

Figure 5. Discriminating high (10) from a varying level of low scatterer number density rf images using fixed and HONN features.

CONCLUSIONS

- Artificial neural networks can be applied as self-learning image features, when using HONN.

- A practical HONN implementation requires the choice of a higher order transformation

- A properly chosen HONN implementation operates as well as, or even better than, a properly chosen set of fixed parameters.

REFERENCES

1. M.S. klein Gebbink, J.T.M.Verhoeven, J.M.Thijssen, and Th.E.Schouten, Application of neural networks for the classification of diffuse liver disease by quantitative echography, *Ultrasonic Imaging* 15:205-217 (1993).

2. G.J.McLachlan, "Discriminant Analysis and Statistical Pattern Recognition," John Wiley & Sons, New York, 1992.

3. J.Hertz, R.G.Palmer, A.S.Krogh, "Introduction to the Theory of Neural Computation", Addison-Wesley, Redwood City, 1993.

4. Y.C.Lee, G.Doolen, H.H.Chen, G.Z.Sun, T.Maxwell, H.Y.Lee, C.L.Giles, Machine learning using a higher order correlation network, *Physica D* 22:276-306 (1986).

5. C.L.Giles, T.Maxwell, Learning, invariance, and generalization in higher-order neural networks, *Applied Optics* 26(23):4972-4978 (1987).

COLOR-CODED TISSUE CHARACTERIZATION IMAGES OF THE PROSTATE*

Georg Schmitz[1], Helmut Ermert[1] and Theodor Senge[2]

[1] Institut fuer Hochfrequenztechnik
[2] Urologische Universitaetsklinik
Ruhr-Universitaet Bochum
D-44780 Bochum, Germany

INTRODUCTION

In the diagnosis and early detection of prostatic cancer transrectal ultrasound is the most important imaging modality. However, the sensitivity and specificity of the standard sonographic methods are still not satisfactory. In this paper we describe a method which provides the clinician with additional information in form of color-coded tissue characterization images based on an estimation of the tissue malignancy from spectral and texture parameters.

In contrast to most approaches which are based on video images[1,2] or do not take into account the depth dependent diffraction and attenuation[2], we sampled the original rf-data and compensated for the depth dependent characteristics of the system and tissue.

Based on the preprocessed images several tissue characterizing parameters, which performed well in earlier works on tissue characterization, were computed. In order to obtain spatially resolved tissue characterization results the images were automatically segmented into rather small regions of interest (ROI) of 3×3.5 mm.

During a first clinical study we analysed over 200 sonographic images of 33 patients with localized prostatic carcinoma. We were able to histologically investigate all specimens after radical prostatectomy. From the histological findings we could select a large number of segments with known pathology. Using this data set we chose an optimal subset of tissue characterization parameters and investigated the linear and quadratic Bayes classifiers as well as several classifiers based on Kohonen-maps. For all classifiers not the optimistically biased apparent error-rate but the more realistic cross-validated error-rate was computed.

Furthermore, all classifers give the probability of malignancy, which is color-coded and thus leads to images providing additional information on tissue characteristics.

*This work was supported by the Bundesministerium für Forschung und Technologie through grant 01 KF 8903/2

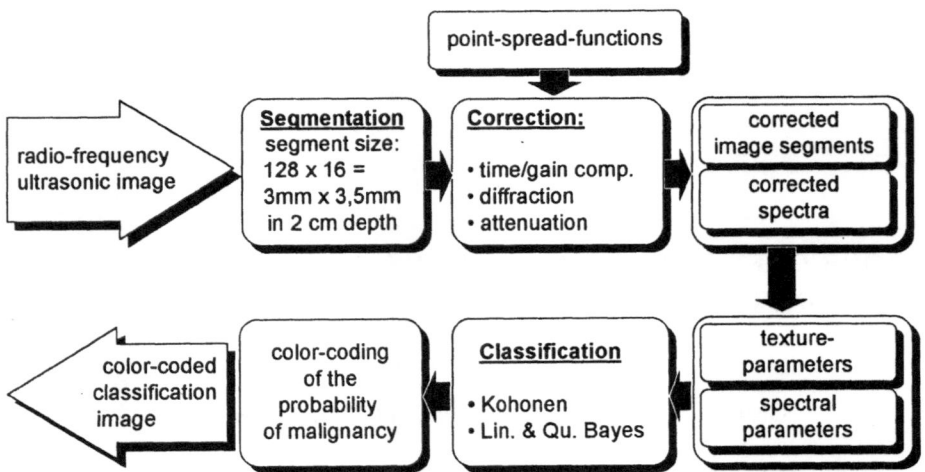

Figure 1. Block diagram of the image processing algorithm

MEASUREMENT SYSTEM

The tissue characterization system is realised as a workstation which is to be added to a conventional ultrasonic system. The rf-data is taken out of a conventional ultrasonic system immediately behind the preamplifier of a single crystal 7.5 MHz focused endoscopic sector-probe. A calibrated and software-controlled TGC amplifier is used to enable an 8-bit digitizer to sample the ultrasound signal in the desired dynamic range with a sampling rate of 33 MHz. A whole scan of 538 lines with 3584 samples each is digitized and stored in real-time. Transfering the image to an HP 9000/433s tissue characterization workstation and saving it on harddisk takes less than one minute.

SIGNAL PROCESSING

The entire image processing algorithm we used is shown in the block diagram of Figure 1. In order to calculate tissue characterizing parameters the rf-data have to be carefully preprocessed to compensate for any system or depth dependent effects. After the amplification due to the software-controlled TGC is compensated for, all images are segmented into overlapping segments of 16 lines and 128 sample points. The overlap of those segments is 50% and 75%, respectively. To characterize the system dependent effects due to focussing and electromechanical characteristics of the transducer we measured the system's point spread function. Using an inverse filter within the effective bandwidth of the system we compensated for these effects in the axial direction. After this step it was possible to estimate the average frequency dependent attenuation of the investigated tissue by the multi-narrowband (MNB) method[3] and correct for it.

For higher accuracy image segments with overflows, underflows or severe inhomogeneities are excluded from this estimation. The result of these operations are corrected spectra. Using an analog correction algorithm with a different segmentation corrected time-domain segments may be obtained by inverse Fourier-transformation. By an overlap-add algorithm the corrected segments form a corrected rf-image which is demodulated by the Hilbert-transform method. This image is segmented again into overlapping segments.

For each segment 16 tissue characterization parameters are computed. These build the feature set for the different classifiers used. The classification result in form of the probabilty of malignancy for the segment is color-coded and presented in a tissue characterization image.

PARAMETERS

In our study we included most of the major tissue characterization parameters which were proposed in recent years and reported to succeed in certain clinical applications. They can be divided into spectral parameters which are derived from the corrected spectra of the segments and texture parameters based on the corrected time domain data.

Spectral Parameters

The most important spectral characteristic of the tissue is its frequency dependent backscatter coefficient. We fitted a straight line to the backscatter spectrum and took its value at the center frequency of the usable bandwidth (BSM) as well as its slope (BSS) as parameters. Additionally, we included the deviation from the straight line fit as a parameter (DEV).

Another important property of tissues is the frequency dependent attenuation. Using data which is not corrected for the tissue's attenuation it is possible to calculate local attenuation estimates. By means of the MNB-method we computed the mid-band value (ATM) and the slope (ATS) of the local frequency dependent attenuation estimated using 21 overlapping segments in axial and 5 in lateral direction. It is obvious that local attenuation estimates cannot have the spatial resolution as other parameters which can be computed using only one segment.

Texture Parameters

To the texture parameters belong as a first group the mean (MEAN) and the signal-to-noise ratio (SNR) of the demodulated signal, the kurtosis (KUR) of the rf-data, the full width at half maximum (FWHM) of the demodulated signal's autocovariance function and the structural parameters proposed by Wagner and Insana[4]. In these authors' notation the parameters are the ratio of structural to the mean diffuse scattering intensity $I_s/\bar{I}_d$ (RSD), the ratio of structural variance to the mean diffuse intensity $\Sigma_s/\bar{I}_d$ (RVD) and the ratio of the structural to the Rician variance Σ_s/σ_R (RSR).

Sleefe and Lele[5] propose to use their scatterer number density estimator (SND) instead of the kurtosis which may be depth dependent. We found that in prostatic tissue the model of Slefe and Lele is not adequate and leads to a highly unstable SND estimator. On the other hand we used corrected rf-data which should lead to depth independent values of the kurtosis.

As a second group we included general texture analysis parameters based on co-occurrence matrices as proposed by Haralick[6] in our study. From the large amount of co-occurrence parameters we chose four parameters which have already been used in tissue characterization: contrast (CON), angular second moment (ASM), entropy

(ENT) and correlation (COR). Since the images are corrected in depth direction only axial co-occurrence parameters are computed with an axial distance of 0.5 FWHM which decorrelates the parameters.

STATISTICAL RESULTS

In a first evaluation step 3500 histologically as benign or malign classified segments from over 200 ultrasonic images of 33 patients were used to select those parameters which contribute significantly to the separation of the groups of benign and malign tissue. For the selection procedure a stepwise parameter selection based on Wilks Λ was carried out[7]. The following 10 parameters of the original set of 16 parameters were selected: BSM, DEV, ATM, ATS, SNR, RSR, RSD, FWHM, CON, COR.

The 3500 datasets of 10 parameters and their corresponding group membership make up the training set for the computation of the different classifiers. The different classification schemes were evaluated using the leave-one-out method also called the round-robin method.

Several different classification methods have been investigated. The results for the linear and quadratic Bayes classifiers are shown in Table 1.

Both classifiers are based on the assumption of normally distributed feature vectors. However, this assumption is not reasonable for the parameters we used. Therefore, we evaluated the performance of different classifiers based on the work of Kohonen[8]. We used unsupervised learning self organizing maps as well as supervised learning vector quantization. We confined ourselves to a two-dimensional hexagonal topology of the underlying map, but varied the number of neurons ($2 \times 2, 3 \times 3, 5 \times 5, 10 \times 10$) and the distance measure used in learning and classification (Euclidean distance, Mahalanobis distance, weighted Euclidean distances). The optimal performance was achieved by the 3×3 self organizing map using the Mahalanobis distance. It showed that the applied distance measure is crucial for the performance of the classifier. The results of the best Kohonen classifier are also shown in Table 1.

Table 1. Classification results of the linear and quadratic Bayes classifier as well as the selected Kohonen classifier. The last row gives the total percentage of correct classifications which is weighted by the group size.

	Linear Bayes		Quadratic Bayes		Kohonen	
	Benign	Malign	Benign	Malign	Benign	Malign
Actual class						
Benign	80.4%	19.6%	62.9%	37.1%	88.0%	12.0%
Malign	13.9%	86.1%	9.6%	90.4%	17.9%	82.1%
Total:	83.8%		76.8%		84.6%	

The results of the linear Bayes classifier and the Kohonen classifier are comparable although the significantly higher specifity of the latter makes it the method of choice which is used for the generation of color-coded tissue characterization images.

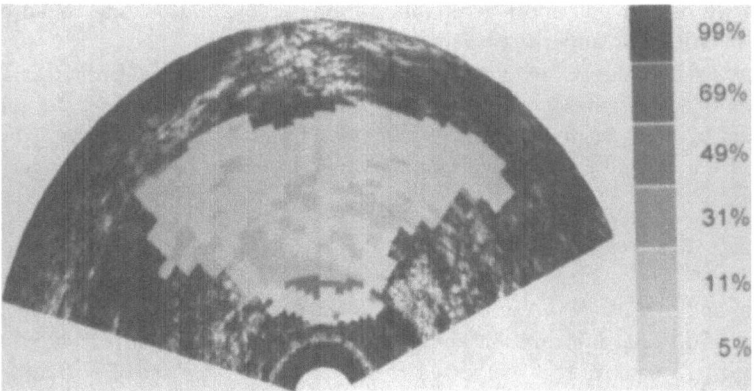

Figure 2. Tissue characterization image of a normal prostate. The bright gray-tones in the segments show the low level of estimated malignancy.

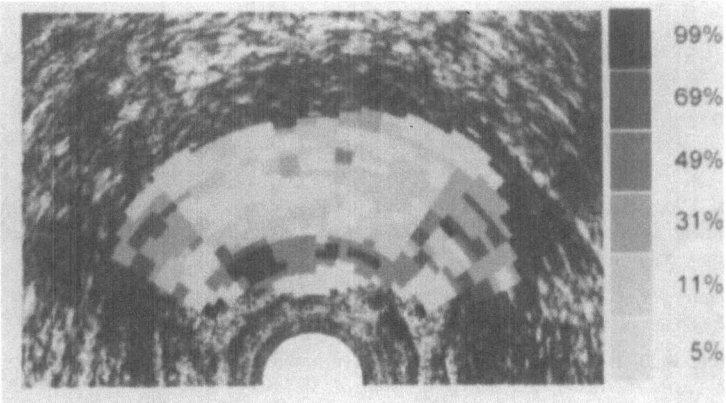

Figure 3. Tissue characterization image of a localized carcinoma in the right lobe of the prostate (left side of the image). The cumulation of black segments at the position of the actual tumor indicates the high probability of a malignancy.

COLOR-CODED IMAGES

Several images not belonging to the training set were analysed by the image processing algorithm described above. Typical results are shown in Figures 2 and 3. Instead of the color-coding used on the screen from green for benign to red for malign a gray scaling from light gray for benign and black for malign was applied. In all images segments which did not allow a reliable parameter estimation due to automatically detected overflows or underflows were omitted.

Figure 2 shows the tissue characterization results of a normal prostate. The bright gray tones of the segments show the low probability of malignancy. In contrast the cumulation of black segments on the left side of the image in Figure 3 indicate the position of a localized prostatic carcinoma which has been histologically verified.

CONCLUSIONS

Statistical and first clinical results of the application of our image processing algorithm show that additional information about the malignancy of the investigated tissue can be assessed. To achieve this, corrected rf-data, a reasonable set of parameters as well as an adequate classifier have to be used. In the group of classifiers we tested the self organizing Kohonen-map utilizing the Mahalanobis distance performed best. The optimal number of neurons is likely to depend on the number of training vectors and thus has to be optimized for each training set. It should be mentioned here that although more information about tissue properties is available an important fact must be kept in mind: Since no a-priori probabilities of malignancy are available the threshold for the diagnosis benign/malign cannot be given and has to be defined by the experience of the clinician using the additional information provided by the system.

REFERENCES

1. O. Basset, Z. Sun, J.L. Mestas, and G. Gimenez, Texture analysis of ultrasonic images of the prostate by means of co-occurrence matrices, *Ultrasonic Imaging.* 15:218 (1993).

2. A.L. Huynen, R.J.B. Giesen, J.J.M.C.H. de la Rosette, R.G. Aarnink, F.M.J. Debruyne, and H. Wijkstra, Analysis of ultrasonographic prostate images for the detection of prostatic carcinoma: The automated urologic diagnostic expert system, *Ultr. Med. Biol.*. 20:1 (1994).

3. M.J.T.M. Cloostermans and J.M. Thijssen, A beam corrected estimation of the frequency dependent attenuation of biological tissues from backscattered ultrasound, *Ultrasonic Imaging.* 5:136 (1983).

4. R.F. Wagner, M.F. Insana, and D.G. Brown, Unified approach to the detection and classification of speckle texture in diagnostic ultrasound, *Opt. Eng.*. 25:738 (1986).

5. G.E. Sleefe and P.P. Lele, Tissue characterization based on scatterer number density estimation, *IEEE Trans. Ultrason. Ferroel. Freq. Control.* 35:749 (1988).

6. R.M. Haralick, K. Shanmugam, and I. Dinstein, Textural features for image classification, *IEEE Trans. Syst. Man Cybern.*, 3:786 (1973).

7. D.J. Hand. "Discrimination and Classification", John Wiley and Sons, New York (1981).

8. T. Kohonen. "Self-Organization and Associative Memory", Springer, Berlin (1988).

TISSUE CHARACTERIZATION BY IMAGING OF ACOUSTICAL PARAMETERS

T. Gaertner,[1] K.-V. Jenderka,[1] H. Schneider,[1] and H. Heynemann [2]

[1] Institute for Medical Physics and Biophysics
[2] Urological Clinic
Martin-Luther-University, Halle, Germany 06097

INTRODUCTION

The main objective of the work of our group is to assist the diagnosis by ultrasonic imaging of cancer in human testis. For this we use a modern b-mode-device (ATL-HDI 9). In addition to the b-mode-image we estimate mean acoustic parameters like attenuation and backscattering by applying the technique of spectral analysis. Normally we find a higher attenuation in cancerous tissue. However the estimation of attenuation becomes difficult for small inhomogeneous areas in the image. So we use a greater region of interest to calculate mean values for the parameters. After a correction according to this values we are able to synthesize a new image what especially shows local deviations from the characteristics of the surrounding tissue.

MATERIALS AND MEASUREMENT METHODS

The rf-data (before any kind of signal processing except from the time gain compensation) from the ATL-HDI 9 are prestored in real time in a special FIFO-Module and then transferred to the parallel i/o-port of a computer (486/dx50). Depending on the type of transducer and the number of the selected focal zones it takes between 5 and 15 seconds to get one image stored (Figure 1).

The analysis of the parameters works off line. At first a region of interest is to be selected in the b-mode-like image on the computer monitor. This roi is a set of rf-scanlines. For further calculation a correction of the time gain compensation is carried out automatically. To get more information about the acoustical properties of the tissue we use a model to describe the ultrasonic echo process.

We assume that for the echo squared-amplitude $A(f)$ of a scanline for a time window $G(f)$ can be written:

$$A(f) = A_0(f) \ H(s,f) \ D(s,f) \ e^{-2\alpha_0(f) \ s_0} * G(f)$$

where $A_0(f)$ is the output spectrum of the transducer, $H(s,f)$ is the sample function, $D(s,f)$ is the distribution of the acoustic intensity in the soundfield and the exponential term is for the attenuation in the medium in front of the roi. s stands for the depth and f for the frequency.

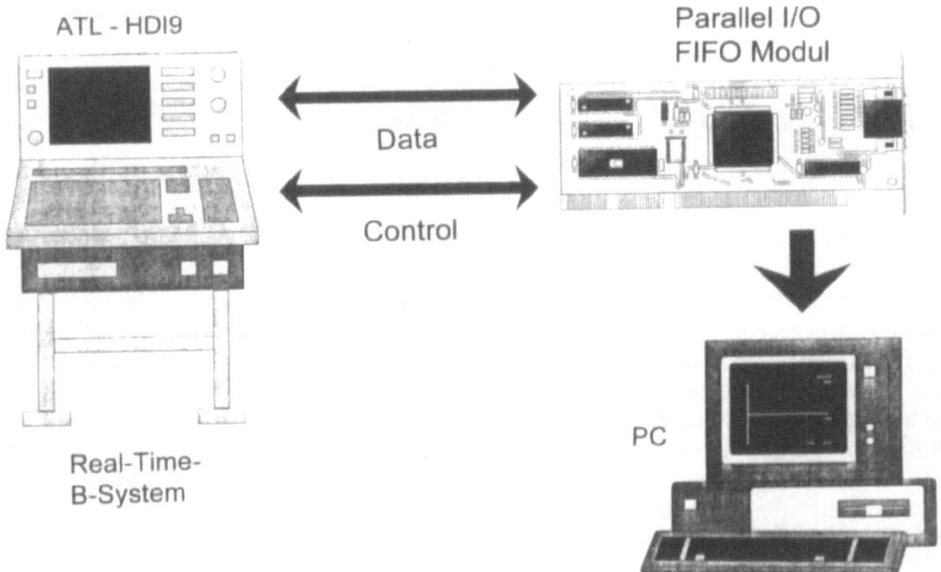

Figure 1. Rf ultrasound data acquisition system

Then we separate H as follows:

$$H(s,f) = Z(s,f) \ R(s,f) \ e^{-2\alpha(s,f) \ s}$$

Z is a structural function, R describes the backscattering properties and the exponential function should describe all losses of energy in the sample due to scattering, absorption and reflection. Z can be removed by the technique of cepstral smoothing. Therefore the data from the time window are transformed by a FFT-Algorithm. After a second FFT we remove the structural information in the so called cepstrum. With an inverse FFT we now get the smoothed spectrum and the basic data for the further calculations. If we move the time window along the scanline we get a new line what shows the log. squared-amplitude versus the time of flight (is equivalent to the depth in the roi) and this is possible for several frequencies within the bandwidth of the transducer.

EFFECT OF FOCAL ZONES ON ECHO-AMPLITUDE

ATL HDI 9 , agar based phantom

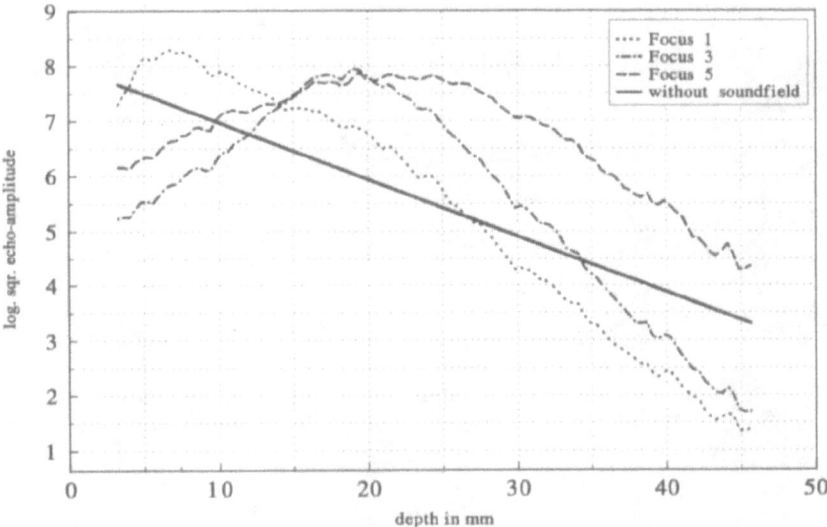

Figure 2. Measurement results for the selection of different focal zones and the theoretical shape without any influence of the pressure distribution in the soundfield

Now it is necessary to correct the influence of the soundfield of the transducer. Due to the possibility of focusing with a modern b-mode-device there is a real structure of the distribution of acoustic pressure in dependence on the distance from the transducer. Normally the sound pressure reaches its maximum in the region of the selected focus. Those effects can be corrected by applying special correction functions. Sometimes it is possible to get the correction by a calculation of the soundfield, based on the known or measured transducer characteristics. However for arrays this is nearly impossible. So we found the correction as we used several agar based phantoms with a different but known attenuation (measured with a substitution technique in a two transducer system) and all possible selections of focal zones (Figure 2).

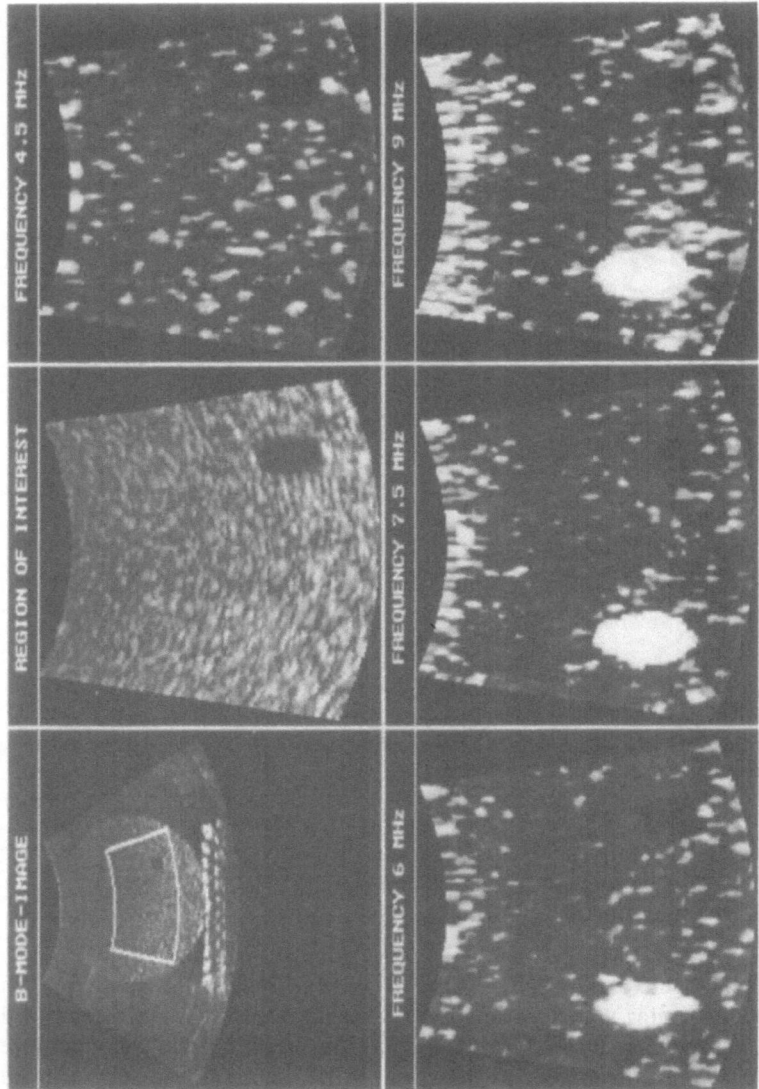

Figure 3. Here we find the results of a tissue mimicking agar-phantom (including graphite for attenuation and scattering). There are two inclusions one of pure agar and the other of the basic material with added polystyrol beads (100 μm). With a curved array scanner 7·5 MHz the second inclusion is not visible in the b-image and the images for the lower frequencies, but for the higher frequencies.

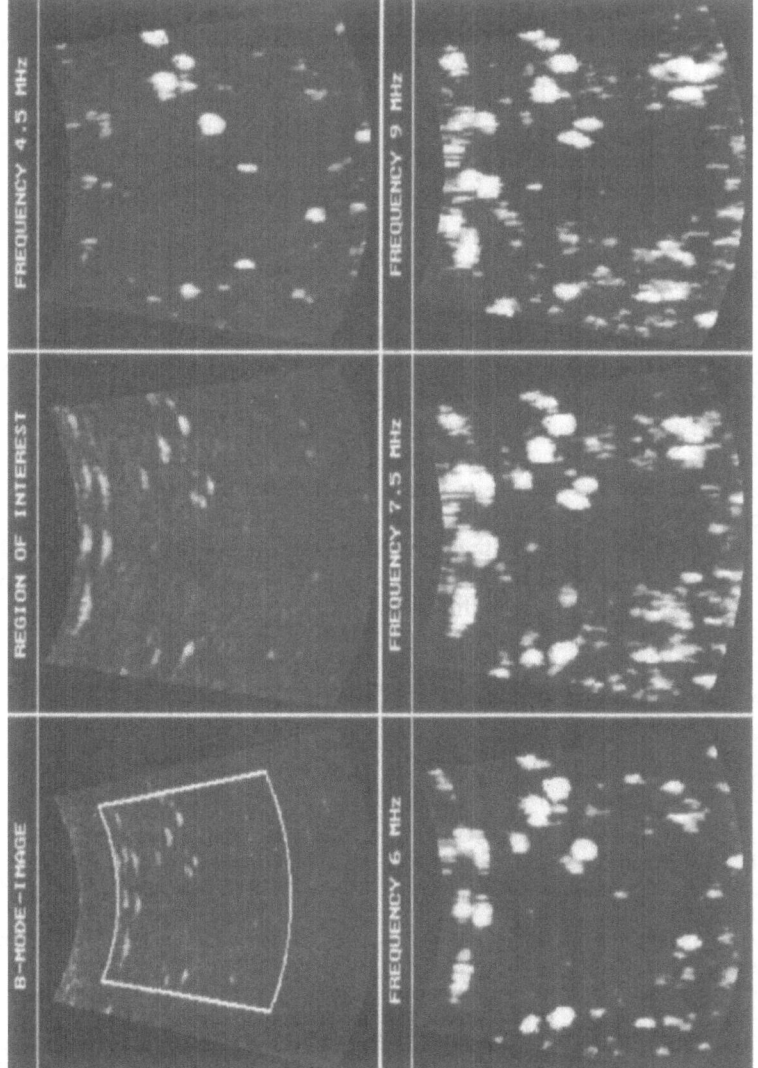

Figure 4. There is an example of cancerous testis tissue with typical inhomogenities. The differences between the frequency images is not as strong as for the phantom but also visible.

Now we estimate the attenuation and the backscatter parameter from the so calculated log. data. We do a lin. regression according to our model:

$$\ln A_c(s,f) = -2\alpha(s,f)s + \ln A_0(f) + \ln K_v(f) + \ln R(s,f)$$

The cepstral smoothed data A_c so only depend explicit on s with the factor α what is the attenuation. The transducer spectrum A_0 and effects in front of the roi K_v don't depend on s and should be constant for the same measurement, the backscatter parameter then is given by the lin. term of the regression. It has to be considered that attenuation and backscattering are not really independent. Stronger scattering effects will cause an increased attenuation. In human testis we find an attenuation what is a lin. function of the frequency (0.3 .. 0.5 dB/(cm MHz) for normal tissue) and a frequency dependence on the scattering parameter what is in the shape very similar to the transducer spectrum.

IMAGING PROCESS AND RESULTS

Because of that the estimation of acoustical parameters in a small region is strongly influenced by statistical uncertainties of the time windowing and of the properties of the sample itself we now synthesize a frequency image. With the help of the mean values for the attenuation and the backscatter parameter a new image is constructed. This can be understood as an picture of local differences from the main backscattered ultrasonic power. Especially the frequency dependence on the backscattering in inhomogeneous areas becomes visible.

Figure 3 and figure 4 show some examples for frequency images in comparison with the reconstructed b-mode-image. We hope that there will be a possibility to differ between several kinds of cancer in the future in this way.

Supported by the "Deutsche Forschungsgemeinschaft", Project He 2156/2-1.

REFERENCES

Boote, E.J., Hall, T.J., Madsen, E.L., and Zagzebski, J.A.,1991, "Improved resolution backscatter coefficient imaging", Ultrasonic Imaging, 13, 347-359

Boote, E.J., Zagzebski, J.A., and Madsen, E.L., 1992, "Backscatter coefficient imaging using a clinical scanner", Med. Phys., 19, 1145-1152

Parker, K.J., Lerner, R.M., and Waag, R.C, 1988, "Comparison of techniques for in vivo attenuation measurements" , IEEE Trans. on Biomed. Eng., 35, 12, 1064-1068

Walach, E., Shmulewitz, A., Itzchak, Y., and Heyman, Z., 1989, "Local tissue attenuation images based on pulsed-echo ultrasound scans", IEEE Trans. on Biomed. Eng., 36, 2, 211-221

Wear, K.A., Wagner, R.F., Insana, M.F., and Hall, T.J., 1993, "Application of autoregressive spectral analysis to cepstral estimation of mean scatterer spacing", IEEE Trans. on Ultr., Ferro. and Frequ. Control, 40, 1, 50-58

Yao, L.X., Zagzebski, J.A., Madsen, E.L., 1991, "Statistical uncertainty in ultrasonic backscatter and attenuation coefficients determined with a reference phantom", Ultrasound in Med. & Biol., 17, 2, 187-194

THE EFFECT ON ULTRASOUND INTENSITY OF REFRACTION AT A CURVED SURFACE

Rosemary S. Thompson,[1] Geoffrey K. Aldis,[2] and Laurence S. Wilson[3]

[1]School of Mathematics & Statistics, University of Sydney,
Sydney NSW 2006 [2]Department of Mathematics, ADFA,
Canberra ACT 2600 [3]Ultrasonics Laboratory, CSIRO
Division of Radiophysics, Chatswood NSW 2067 Australia

INTRODUCTION

Refraction occurs at an interface between two media of different acoustic impedance. If the interface is a curved surface such as a blood vessel wall, a uniform and planar insonating beam will be distorted. In the second medium the beam will not necessarily be uniform or planar, and this will affect both image intensity and the Doppler spectra of flows in the vessel.

In this paper a 3D model for the intensity distortion produced by a curved refracting interface is developed using a ray approximation. The results are applied to consideration of (i) insonation of flow in a uniform cylindrical vessel, with the transducer external to the vessel (ii) intravascular imaging of a blood vessel wall, where the lumen may not be circular.

MODEL

Assumptions

Figure 1 shows a uniform plane wave which encounters a curved acoustic impedance interface. Initially the beam propagates in medium 1, which has density ρ_1 and speed of sound c_1. In this medium the beam is assumed to have wavefronts perpendicular to its axis, and uniform intensity I_0. After encountering the interface the transmitted ultrasound propagates in medium 2, which has density ρ_2 and speed of sound c_2. There is no attenuation in the model; the ultrasound intensity is changed only by the acoustic impedance interface. It is assumed that refraction can be described using a ray approximation. This requires the wavelength of the ultrasound to be small compared to the radius of curvature of the interface. For example, the wavelength of 10MHz ultrasound in tissue is approximately 0.15mm, so at this frequency the ray approximation would require the vessel radius to be greater than about 1mm.

Refraction at a curved interface

Figure 1 shows an arbitrary ray which originates at the point P_0 in medium 1. At P_1, just inside medium 2, the magnitude of the intensity is I_1. At each point on the curved acoustic impedance interface the incident ray is assumed to be refracted as for a plane wave obliquely incident at a plane boundary. The ray changes direction, and only part of the energy is transmitted. The ratio of I_1 to I_0 is given by the intensity transmission coefficient τ. The four neighbouring rays shown in Figure 1 are refracted similarly but, because of the curvature of the interface, the angles of incidence and the refracted ray directions are all different. The uniform and planar incident beam is therefore distorted so that it is neither uniform nor planar in the second medium.

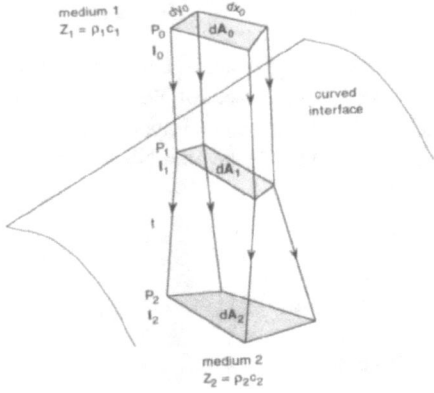

Figure 1. Incident ray $\overrightarrow{P_0 P_1}$ in medium 1 and refracted ray $\overrightarrow{P_1 P_2}$ in medium 2. The area elements formed at P_0, P_1 and P_2 by the neighbouring rays with small increments in x_0 and y_0 are also shown.

A refractive index for the two media is defined by $N = c_2 / c_1$. At P_1 the incident ray, the refracted ray, and the normal to the surface are coplanar, and $\sin \varphi_2 = N \sin \varphi_1$, where φ_1 is the angle of incidence and φ_2 the angle of refraction (Snell's law). The ratio of the transmitted intensity to the incident intensity is given by[1]

$$\tau = \frac{I_1}{I_0} = \frac{4 c_1 \rho_1 c_2 \rho_2 \cos^2 \varphi_1}{(c_2 \rho_2 \cos \varphi_1 + c_1 \rho_1 \cos \varphi_2)^2} \ . \tag{1}$$

At the point P_2 in medium 2, which is distance t from the interface, the magnitude of the intensity is I_2. If $\mathbf{r}_1$ is the position vector at P_1, the position vector at P_2 is

$$\mathbf{r} = \mathbf{r}_1 + \hat{\mathbf{R}} t \ , \tag{2}$$

where $\hat{\mathbf{R}}$ is the unit vector in the direction of $\overrightarrow{P_1 P_2}$. Let dx_0 and dy_0 be small increments in x_0 and y_0 which define an area element perpendicular to the beam at P_0 (Figure 1). The equations for the refracted rays can be expressed as a two parameter family of lines, with parameters x_0 and y_0, and the area elements $d\mathbf{A}_1$ and $d\mathbf{A}_2$ are then

$$d\mathbf{A}_1 = \left(\frac{\partial \mathbf{r}}{\partial x_0} \times \frac{\partial \mathbf{r}}{\partial y_0} \right)_{t=0} dx_0 \, dy_0 \quad \text{and} \quad d\mathbf{A}_2 = \left(\frac{\partial \mathbf{r}}{\partial x_0} \times \frac{\partial \mathbf{r}}{\partial y_0} \right)_t dx_0 \, dy_0 \ . \tag{3}$$

For the volume shown in Figure 1 which is bounded by the area elements dA_1, dA_2, and the four neighbouring rays, the intensity flux is zero since there is no energy loss between P_1 and P_2. The ratio of the intensity magnitude at P_2 to that at P_0 is therefore

$$\frac{I_2}{I_0} = \frac{I_2}{I_1} \frac{I_1}{I_0} = \tau \left| \frac{\hat{R} \cdot dA_1}{\hat{R} \cdot dA_2} \right| .$$

(4)

Uniform cylindrical vessel, external ultrasound transducer at Doppler angle θ_D

The geometry for this case is shown in Figure 2a. The incident ultrasound beam in medium 1 propagates in the $-z$ direction, and the vessel axis is along the z' direction. The parameters x_0 and y_0 are the beam coordinates of P_0. If the coordinates of the point P_2 in the tube based system are $x_2' = x_2$ and y_2', it has been shown[2] that the intensity at P_2 is given by

$$\frac{I_2}{I_0} = \begin{cases} \tau \left| \dfrac{x_0^3 N^2 \sin^2 \theta_D - x_0 \left(1 - N^2 \cos^2 \theta_D \right)}{x_0^3 N^2 \sin^2 \theta_D - x_2 \left(1 - N^2 \cos^2 \theta_D \right)} \right| & , \quad x_2 \neq x_0 \\[4mm] \tau \left| \dfrac{\sqrt{1 - N^2 \cos^2 \theta_D}}{N \sin \theta_D \left(1 + y_2' \right) - y_2' \sqrt{1 - N^2 \cos^2 \theta_D}} \right| & , \quad x_2 = x_0 = 0 \end{cases}$$

(5)

where x_0 is a solution of a non linear equation of the form $H(x_0, x_2, y_2') = 0$. The ray(s) passing through any point in the vessel are completely specified by finding these x_0 values. The coefficient τ can also be expressed in terms of x_0 and $\sigma = N(\rho_2 / \rho_1)$.[2]

Intravascular insonation of the blood vessel wall without circular symmetry

The transducer is located within the vessel lumen, at $T = (a, 0, 0)$, and the angular position of the rotating beam is given by ω (Figure 2b). The vessel is assumed uniform in the axial (z) direction, with cross section given by $r_L(\theta)$. The catheter and vessel axes are aligned. Again, the incident ray encounters the vessel wall at P_1, and P_2 on the refracted ray is distance t from the interface. The point P_1' is the projection of P_1 onto the $x - y$ plane, and similarly for the other primed quantities. The beam makes a constant angle Φ with the z axis; for a side viewing transducer $\Phi = \pi/2$.

To describe the ray paths in the intravascular case, where the beam direction is not constant, it is convenient to use (r, θ, z) coordinates rather than (x_0, y_0, z_0). The increments dx_0 and dy_0 define the area element perpendicular to the beam at P_0. In the absence of refraction this area element is assumed to remain the same at P_1 and P_2 (i.e. there is no divergence of the beam between these points). The increment $d\theta$ therefore depends on distance along the beam. If a constant incremental distance perpendicular to the unrefracted beam is projected onto the $\hat{\theta}$ direction, it is found that $s \, d\theta_1 \approx (s + t) \, d\theta_2$. Therefore

$$\frac{I_2}{I_0} \approx \tau \left(\frac{s+t}{s} \right) \left| \hat{R} \cdot \left(\frac{\partial r}{\partial \theta} \times \frac{\partial r}{\partial z} \right)_{t=0} \bigg/ \hat{R} \cdot \left(\frac{\partial r}{\partial \theta} \times \frac{\partial r}{\partial z} \right)_t \right| .$$

(6)

A general expression for the intensity ratio (6) has been obtained for a vessel lumen of the form $r_L(\theta) = 1 - k \cos\theta$, where k is a constant between 0 and 1.[3] A model for the lumen shape in vascular disease can be generated using a semicircle together with the $k > 0$ curve in the remaining two quadrants. The parameter k gives the degree to which the lumen shape has been altered by vascular disease.

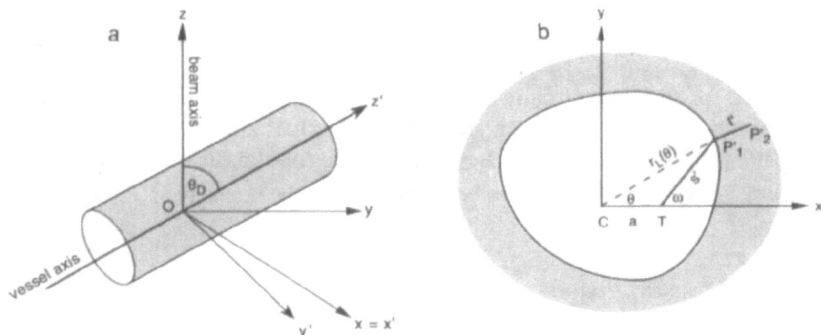

Figure 2. (a) Cartesian coordinate systems for the tube and the beam, which is at Doppler angle $\theta_D < 90°$ to the vessel axis. The tube based system (x', y', z') is obtained by rotating the beam based system (x, y, z) about the common $x' = x$ axis, through the Doppler angle. (b) Geometry for intravascular insonation of a lumen of arbitrary cross section $r_L(\theta)$. The transducer is located at T, and its rotational position is given by ω. TP_1' and $P_1'P_2'$ are the projections onto the $x - y$ plane of the incident and refracted rays.

RESULTS

At any interface part of the energy of the ultrasound beam is reflected and part is transmitted. For a plane boundary the transmitted power and intensity are uniform and so can be characterised by the usual transmission coefficients. This is not the case for a curved interface. In (4) the intensity change arising from deformation of the beam in medium 2 is multiplied by the value of τ for the appropriate angle of incidence. The density ratio affects τ, but not the beam deformation. Both the density ratio and I_0 are set to one in the following examples.

Insonation of flow in a uniform cylindrical vessel

The case of a single acoustic impedance interface between medium 1 and medium 2, the fluid, is considered for $N = c_2/c_1 = 1.2$ (diverging interface), and $N = 0.8$ (converging interface). Interfaces such as these may be encountered *in vitro* with flow phantoms, where the speed of sound in the blood mimicking fluid can be quite different to that in the surrounding tissue mimicking material.

With a diverging interface every point in the vessel is insonated by exactly one ray, given a sufficiently large incident beam. With a converging interface however there are points which cannot be insonated at all. For small enough N and θ_D there are also some points insonated by 2 or 3 (but not more) rays. In both diverging and converging cases the intensity depends on the Doppler angle; it is not uniform over the vessel lumen and the regions of maximum and minimum intensity are immediately adjacent. Figure 3 shows the intensity distribution across a tube slice at two different Doppler angles in the converging case. Theoretical CW Doppler spectra can be obtained by integration once the intensity distribution over a cross section of the vessel is known.[2] In the diverging case the intensity extremes are confined to small regions near the upper tube wall. With a converging interface the intensity extremes occupy a larger area, and have a greater effect on the spectrum. This is shown by the results for Poiseuille flow insonated by a circular ultrasound beam (Figure 4). The spectrum is very distorted with the converging interface, but in the diverging case it is closer to the theoretical rectangular result than when there is no refracting interface. This is because the divergence of the beam tends to counteract the incomplete insonation which near wall (lower velocity) regions experience as a result of the 3D geometry of the beam and tube.

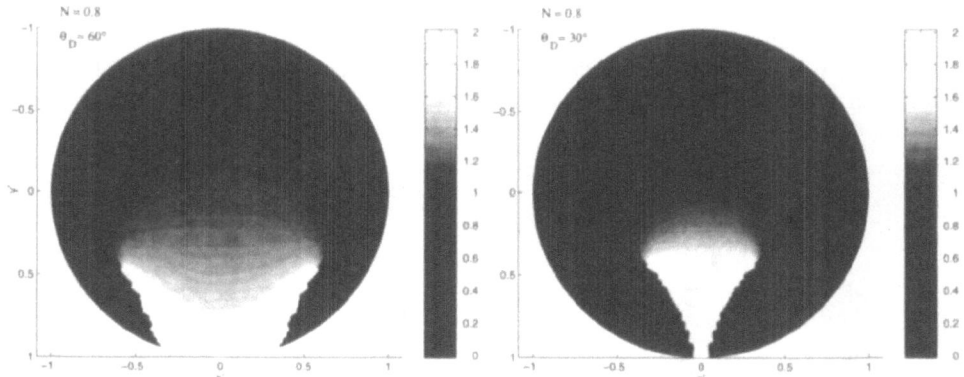

Figure 3. Intensity across a tube slice with the beam at different Doppler angles. The converging interface is the same in both cases $(N = 0.8)$, and the incident beam is assumed infinite.

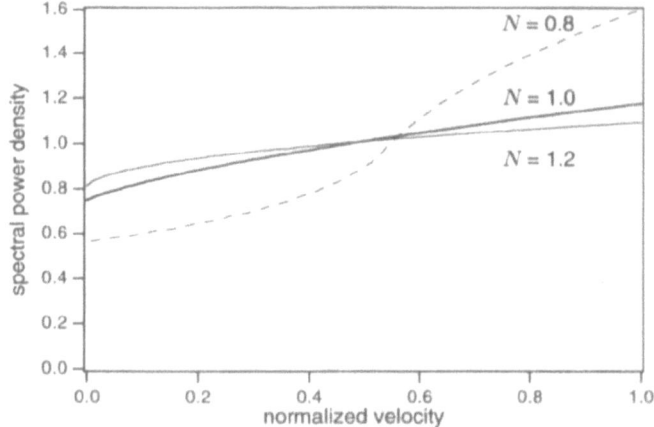

Figure 4. Doppler spectrum for Poiseuille flow insonated by a circular ultrasound beam with radius equal to the tube radius, and $\theta_D = 45°$. $N = 1.2$ diverging interface; $N = 1.0$ no refraction; $N = 0.8$ converging interface.

Intravascular insonation of the blood vessel wall

The speed of sound in blood at 37°C is 1580ms⁻¹. In vascular disease acoustic impedance interfaces will be encountered between blood and the vessel wall, since in fat $c \approx 1450 \, \text{ms}^{-1}$, whereas values of $c \geq 1700 \, \text{ms}^{-1}$ are found in fibrous and calcified plaques.

The intensity over the vessel wall, as calculated from (6), could be displayed using an (r, θ) grid, but this would not be the same as the image generated by an ultrasound system. In practice the returned echoes are mapped as a function of elapsed time for each rotational position ω of the transducer. Thus the coordinates of the image point corresponding to P_2 are $x_{P_2} = a + s' \cos \omega + (t'/N) \cos \omega$ and $y_{P_2} = s' \sin \omega + (t'/N) \sin \omega$. This image mapping gives the correct shape of the inner lumen with the transducer anywhere within an arbitrary lumen, and at any forward viewing angle, provided the point of origin of the beam is stationary.[3]

The cardioid-like curve $r_L(\theta) = 1 - k \cos \theta$ and a single impedance interface was used as a simplified model representation of an atherosclerotic vessel lumen, and results were obtained for a diverging interface (fibrous plaque), and for a converging interface (fatty plaque).[3] Figure 5 shows the intensity for such a lumen, displayed as is would be by the transducer, using ω and $s + t / N$, with the transducer at two different locations. The incident intensity is not uniform around the vessel wall, and depends on the position of the transducer in the lumen.

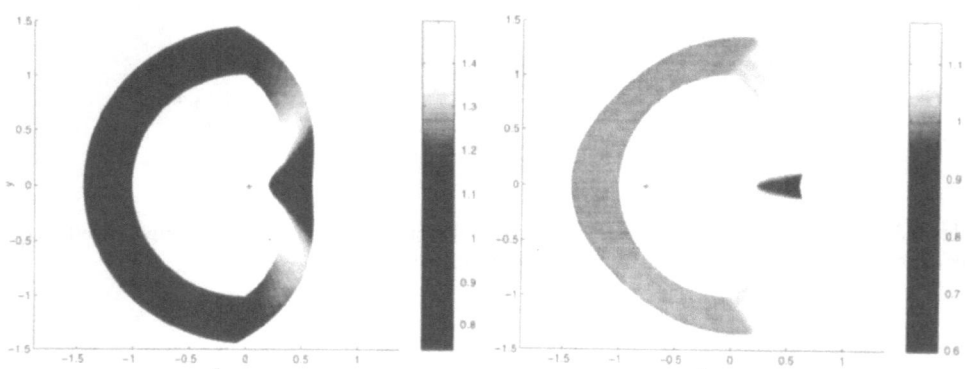

Figure 5. Intravascular insonation of a non circular vessel with the transducer at different locations in the lumen. There is a diverging interface ($k = 0.8$, $c_2 / c_1 = 1.1$) in first and fourth quadrants, and $\Phi = \pi / 3$.

SUMMARY

The intensity distribution across a slice of a cylindrical vessel has been calculated with an acoustic impedance interface between the fluid in the tube and the surrounding material. The intensity is not uniform and depends on the Doppler angle. A converging interface leads to greater distortion of the Doppler spectrum than a comparable diverging interface. Similarly with intravascular insonation, calculations using a non circular atherosclerotic lumen shape show that the intensity over the vessel wall is in general non uniform, and dependent on the transducer position. The exception is the special case of a central transducer in a circular lumen.

Acknowledgment: The National Health and Medical Research Council of Australia supported this work.

REFERENCES

1.   L. E. Kinsler and A. R. Frey. "Fundamentals of Acoustics", Second Edition, Wiley, New York (1962).
2.   R. S. Thompson and G. K. Aldis, Effect of a cylindrical refracting interface on ultrasound intensity and the CW Doppler spectrum, *IEEE Trans Biomedical Eng.* in press (1995).
3.   R. S. Thompson and L. S. Wilson, The effect of variations in transducer position and sound speed in intravascular ultrasound: A theoretical study, *Ultrasound Med Biol.* in press (1995).

AN ANALYSIS OF PULSED WAVE ULTRASOUND SYSTEMS FOR BLOOD VELOCITY ESTIMATION

Jørgen Arendt Jensen

Electronics Institute, Build. 349
Technical University of Denmark
2800 Lyngby, Denmark

ABSTRACT

Pulsed wave ultrasound systems can be used for determining blood's velocity non-invasively in the body. A region of interest is selected, and the received signal is range gated to measure data from the region. One complex sample value is acquired for each pulse emission after complex demodulation of the received signal. The time evolution and distribution of velocity can then be found by using samples from a number of pulse-echo lines. Making a short-time Fourier transform of the data reveals the velocity distribution in the range gate over time.

Such systems are called Doppler ultrasound systems implying that they use the classical Doppler effect. The velocity is typically on the order of 0.5 to 1 m/s giving a relative shift of 2 to 4 kHz of the center frequency of the received spectrum for a 3 MHz transducer. Finding such a shift is impossible since the unknown frequency shift from attenuation in tissue can be tens of kilohertz. Some recent reviews and articles state that the Doppler effect is used, and contradictory and wrong results and erroneous system diagrams arise from this assumption. Research done in the last fifteen years has revealed that it is the movement of the scatterers between pulse emissions, that is used for finding the velocity. This finding gives new insight into the role of the complex demodulation stage, and shows that this can be replaced by a matched filter and quadrature RF sampling. A derivation of this result is presented in this paper, and it reveals how the bandwidth of the pulse and the number of pulse emissions affect the result. The final equation for the received signal is quite complicated, and a simplified interpretation is therefore also given. This readily reveals the influence from transducer bandwidth, attenuation, non-linear effects, classical Doppler effect, and scattering.

1 INTRODUCTION

Pulsed ultrasound systems for the investigation of blood velocity emerged in the early 1970s. A seminal paper was that by Baker (1970), which detailed the sampling operation of the received backscattered signal from the blood. A single sample was acquired for each pulse emitted and Baker stated that the system tracked the slow movement of the scatterers through the range gate. Unfortunately Baker also calls this the Doppler effect, and the name has since then stuck to these systems, and has produced a number of mistakes in interpreting the function of and signal processing in these systems. It is here important to differentiate between the interaction between the moving scatterers and the pulse, yielding the Doppler shift in frequency, and the inter-pulse movement of the scatterers. The latter is the effect detected in these systems as

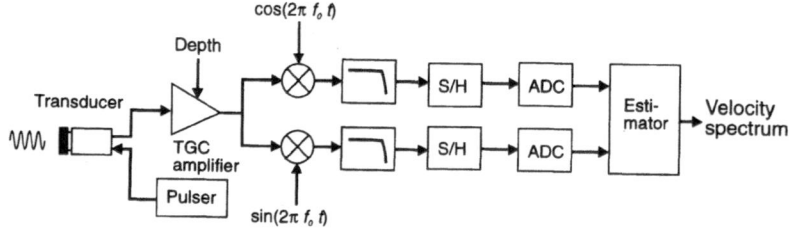

Figure 1. Block diagram of pulsed wave system for velocity estimation.

shown in Section 2. A model for the inter-pulse movement is derived in Section 3 and used in Section 4 for calculating the spectrum of the received signal. It is shown that sampling can be performed at a frequency in the kHz range although the received signal is in the MHz range. The derived model is somewhat complicated, and a simpler interpretation using frequency scaling is given in Section 5. From this the influence of, *e.g.*, attenuation and non-linear propagation is readily revealed and the role of the various parts of a PW system explained.

This paper shows that the frequency axis for the received signal is scaled by the factor $2v/c$, where c is speed of sound and v is blood velocity. This time dilation was first explicitly noted by Newhouse and Amir (1983). They gave a model for the received signal that demonstrated the stretched time factor. Experiments were also shown proving the model correct. The model was derived earlier and used for evaluating transit time effects (Newhouse et al. 1976) and the effects from scattering and absorption on the power density spectrum (Newhouse et al. 1977), although the time dilation effect was not noted explicitly. The influence from attenuation was measured by Holland et al. (1984) and found in agreement with the model. The scaling of the time axis was also used by Magnin (1986). He presented the signal processing fundamentals for the first commercial system to apply direct RF sampling of the received signal, thus skipping the analog demodulation scheme. The exact same scheme was presented by Forsberg and Jørgensen (1989). Magnin (1986) derived a model for the received signal using the time shift effect for a Gaussian pulse and infinitively many pulse emissions. The commercial implementation of this system is given by Halberg and Thiele (1986). Thomas and Leeman (1991) have also considered the time shift model and point out the difference between this and the Doppler effect, and elucidated the time scaling involved in forming the received signal. The difference was also experimentally demonstrated by Thomas and Leeman (1993).

This paper derives the received signal for a direct RF sampled signal and extends the model to include a limited number of pulses. It is also shown that the direct sampled system and the more conventional ones using analog demodulation actually are equivalent. The only difference is the implementation of the matched filter in the RF receiver.

2 DOPPLER SYSTEM

A traditional Doppler system is depicted in Fig. 1. A pulse is emitted and interacts with the moving blood. A Doppler shift of the pulse spectrum is introduced as

$$f_d = -\frac{2|\vec{v}|}{c}\cos(\theta)f_0 \tag{1}$$

as a result of the blood velocity. Here $|\vec{v}|\cos\theta$ is the blood velocity along the direction of the ultrasound beam, c is the speed of sound, and f_0 is the center frequency of the emitted pulse. The received signal is then quadrature demodulated. This is done by moving the spectrum down to around $f = 0$ by multiplying the signal with $\exp(j2\pi f_0 t)$ and then remove the sum component through a set of low-pass filters[1]. The resulting signal is sampled and processed to yield the Doppler shift, and thereby the blood velocity and its sign.

---

[1]These are often, falsely, perceived as being anti-aliasing filters before the sampling process. They actually act as matched filters. See Section 6.

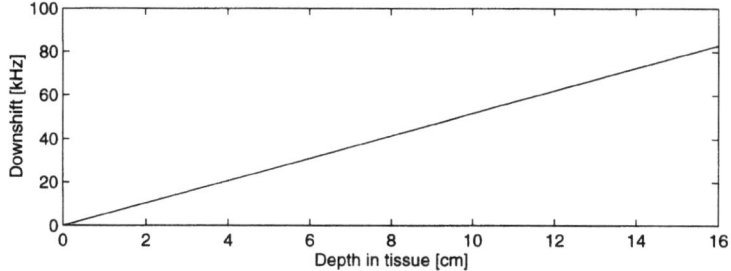

Figure 2. Shift in center frequency for a Gaussian pulse when propagating in tissue with an attenuation of 0.5 dB/[MHz cm]. The center frequency is 3 MHz and the relative bandwidth is 0.1.

This standard explanation of the function of the Doppler system has problems when attenuation of the ultrasound field by the tissue is included. The Doppler frequency is typically on the order of a few kHz, whereas the shift due to attenuation is considerably higher as can be seen from Fig. 2. The downshift in center frequency as a function of depth is depicted for a 3 MHz Gaussian pulse with a relative bandwidth of 0.1. The unknown attenuation shift can easily be ten times the Doppler shift, even for moderate depths. It is, thus, not possible to estimate the Doppler shift with any accuracy, and these systems do not employ the Doppler effect.

3 TIME SHIFT MODEL

A pulsed wave system for velocity estimation emits a number of pulses and extracts a single sample at the appropriate depth for each pulse. This sampled signal is then used for estimating the velocity or its spectrum. The estimation, thus, relies on the relation *between* samples in the sampled signal. This relation can be derived from the time shift of the signal from pulse to pulse. For a single scatterer $\vec{p}_1$ denotes the initial position of the scatterer and $\vec{p}_2$ the position after the next pulse emission T_{prf} seconds later. The distance between the two positions is given by

$$\vec{p}_2 - \vec{p}_1 = \vec{v}\, T_{prf} \tag{2}$$

The movement in the z-direction is perceived as a time shift in the received RF signal and is given by

$$t_s = \frac{|\vec{v}|\cos(\theta)}{c/2} T_{prf} = \frac{2|\vec{v}|\cos(\theta)}{c} T_{prf} = \frac{2v_z}{c} T_{prf}. \tag{3}$$

Two received signals are related by:

$$r_2(t_r) = r_1(t_r - t_s), \tag{4}$$

when denoting time since pulse emission as t_r. Emitting a sinusoidal pulse, one received response is given by

$$r_r(t_r) = a\sin(2\pi f_0(t_r - \frac{2d}{c})) \tag{5}$$

where d is the initial depth of the scatterer. Receiving the response from a number of pulses gives

$$r_r^{(i)}(t_r) = a\sin(2\pi f_0(t_r - \frac{2d}{c} - it_s)) \tag{6}$$

Taking one sample at a fixed time t_x relative to pulse emission gives a sampled signal

$$r_s(i) = -a\sin(2\pi f_0 \frac{2v_z}{c} T_{prf} i - \phi_c) \tag{7}$$

$$\phi_c = 2\pi f_0(\frac{2d}{c} - t_x)$$

where ϕ_c is a fixed phase factor. The received and sampled signal, thus, has a frequency of

$$f_p = -\frac{2|\vec{v}|\cos(\theta)}{c}f_0 = -\frac{2v_z}{c}f_0 \tag{8}$$

which has the same magnitude as the Doppler frequency, but is generated by the inter-pulse movement of the scatterers.

4 SPECTRUM OF RECEIVED SIGNAL

The model derived in Section 3 assumed that the emitted and received signals are essentially monochromatic. In reality a short pulse (8-16 cycles) are employed, and that affects the spectrum of the received signal. The influence of the pulse shape and duration is analyzed in this section. The analysis is, with no loss of generality, done for an RF sampled system in which the demodulation is replaced by a matched filter (see Section 6).

RF samples for N emitted pulses are received and the signal after the transducer is given by

$$r_s(t) = a\sum_{i=0}^{N-1} e(t - iT_{prf} - it_s - \frac{2d}{c}), \tag{9}$$

where t is absolute time since the first pulse emission ($i = 0$) and $e()$ is the signal emitted for each pulse. Transforming to the frequency domain gives:

$$
\begin{aligned}
R_s(f) &= aE(f)\sum_{i=0}^{N-1} \exp(-j2\pi f(i(T_{prf} + t_s) + \frac{2d}{c})) \\
&= aE(f)\sum_{i=0}^{N-1} (\exp(-j2\pi f(T_{prf} + t_s)))^i \exp(-j2\pi f\frac{2d}{c}) \\
&= aE(f)\frac{\sin(\pi f(T_{prf} + t_s)N)}{\sin(\pi f(T_{prf} + t_s))} \exp(-j\pi f((N-1)(T_{prf} + t_s) + \frac{4d}{c})), \tag{10}
\end{aligned}
$$

using that $\sum_{i=0}^{N-1} x^i = \frac{1-x^N}{1-x}$ and the Euler identities. This is the spectrum of the signal received by the transducer for a number of pulse emissions. An example is shown in Fig. 3 for a rectangular pulse with 8 sine periods at 3 MHz and for 10 pulse-echo lines. The local maxima are at

$$f = \frac{n}{T_{prf} + t_s} = \frac{nf_{prf}}{1 + \frac{2v_z}{c}} \approx nf_{prf}(1 - \frac{2v_z}{c}) \tag{11}$$

and their amplitudes are modified by the spectral amplitude of the pulse.

The received signal is then sampled with a frequency of $f_{prf} = 1/T_{prf}$, which can be described by a multiplication with a series of δ-functions as

$$r_f(t) = r_s(t)\sum_{n=-\infty}^{\infty} \delta(t - nT_{prf}). \tag{12}$$

The Fourier transform of this signal is

$$
\begin{aligned}
R_f(f) &= R_s(f) * \frac{1}{T_{prf}}\sum_{m=-\infty}^{\infty} \delta(f - \frac{m}{T_{prf}}) \tag{13} \\
&= \int_{-\infty}^{+\infty} \frac{1}{T_{prf}}\sum_{m=-\infty}^{\infty} R_s(s)\delta(f - \frac{m}{T_{prf}} - s)ds = \frac{1}{T_{prf}}\sum_{m=-\infty}^{\infty} R_s(f - \frac{m}{T_{prf}}).
\end{aligned}
$$

Combining with the previous equation gives

$$
\begin{aligned}
R_f(f) &= af_{prf}\sum_{m=-\infty}^{\infty} E(f - mf_{prf})\frac{\sin(\pi(f - mf_{prf})(T_{prf} + t_s)N)}{\sin(\pi(f - mf_{prf})(T_{prf} + t_s))} \tag{14} \\
&\quad \times \exp(-j\pi(f - mf_{prf})((N-1)(T_{prf} + t_s) + \frac{4d}{c})).
\end{aligned}
$$

380

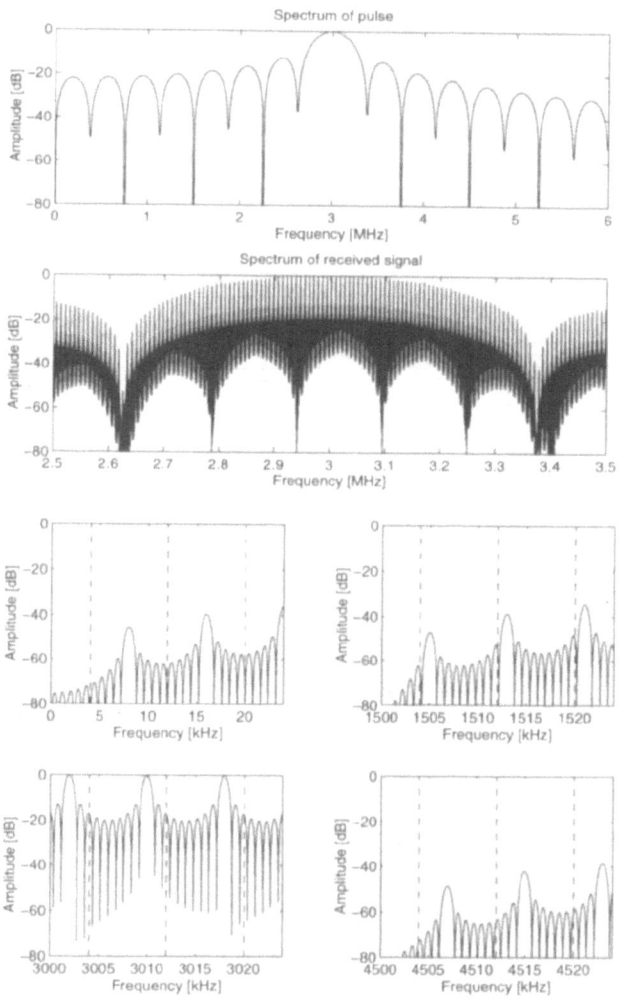

Figure 3. Spectrum of received signal at a velocity of 0.5 m/s. $M = 8, N = 10, f_{prf} = 8$ kHz (see Eq. (10)). The four small graphs are from different parts of the second graph.

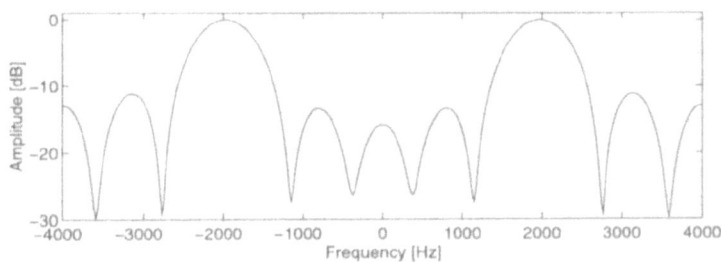

Figure 4. Resulting spectrum of received signal after sampling of signal with spectrum shown in Fig. 3.

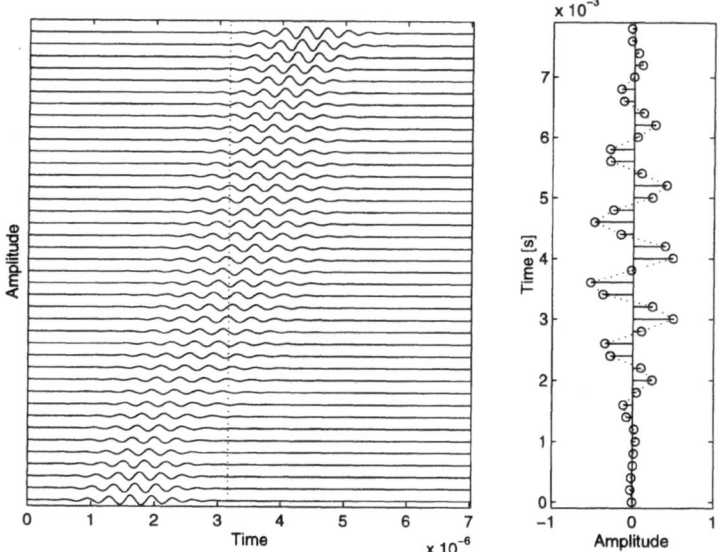

Figure 5: Signal from a single moving scatterer crossing a beam from a concave transducer.

The sampling operation moves the components of the spectrum down to the range from $-f_{prf}/2$ to $f_{prf}/2$. The maximum of the pulse spectrum is at f_0. The sampling operation is essentially a modulus f_{prf} operation, so the maximum in the resulting spectrum is at

$$f = nf_{prf}(1 - \frac{2v_z}{c}) \bmod f_{prf} = -\frac{2v_z}{c}nf_{prf} \tag{15}$$

and the dominant excursion is at $nf_{prf} \approx f_0$, so the maximum is at

$$f = -\frac{2v_z}{c}f_0 \tag{16}$$

An example of this is seen in Fig. 4. The velocity is 0.5 m/s and f_0 equals 3 MHz, giving a combined maximum at $f = \frac{2 \cdot 0.5}{1540}3 \cdot 10^6 = 1948$ Hz. Two peaks are seen, since quadrature sampling is not employed.

5   A SIMPLE INTERPRETATION

Equation (14) reveals the effects from using different number of pulses and pulse shapes, but the consequence is not directly obvious from the equation. A simpler model would thus be beneficial. This can be derived by keeping in mind that the shift in position is responsible for generating the signal. An example is shown in Fig. 5. A single scatterer crosses the beam from a pulsed concave transducer. The resulting sampled signal has the same shape as the pulse, since the RF pulse slowly moves past the sampling point indicated by the dashed line. The consequence in the frequency domain can be seen from Eq. (7). All frequency components are multiplied with $2v_z/c$, and are thus scaled to lie in the audio range. This scaling is indicated in Fig. 6. The RF frequency axis is replaced by the frequency axis for the sampled signal. The effect by a number of factors can now easily be deduced with this simple model in mind.

The pulse bandwidth is directly responsible for the bandwidth of the resulting spectrum. A longer pulse, thus, gives a more narrow spectrum. The influence from the number of pulse-echo lines can also be deduced. Only part of the pulse is sampled if too few lines are used, and this amounts to multiplying a rectangular window onto the data. A smearing of the spectrum,

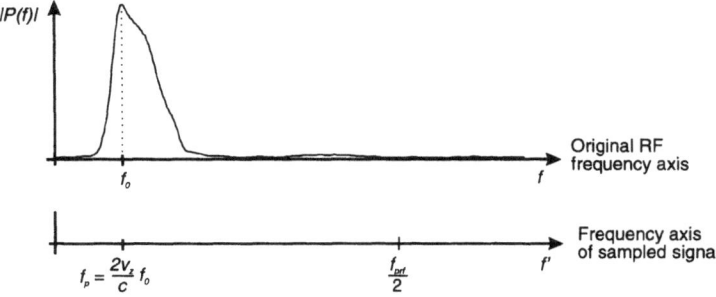

Figure 6. Frequency axis scaling for a sampled signal for a scatterer moving at a velocity of v_z.

thus, results. The sampling covers the full pulse length if

$$\frac{2v_z}{c}T_{prf}N \geq \frac{M}{f_0} \tag{17}$$

where M is the number of sine periods in the pulse. The lowest velocity possible to determine is therefore dependent on the number of lines used.

Attenuation will shift the center frequency of the pulse spectrum downwards, but this attenuation downshift is also multiplied with $2v_z/c$. So it is the downshift relative to f_0 that determines the bias of the estimate.

The instantaneous Doppler shift during the pulse's interaction with the moving scatterer gives a frequency shift of $\frac{2v_z}{c}f_0$. This shift is also multiplied with $2v_z/c$, and the resulting frequency deviation is

$$\Delta f_d = \frac{2v_z}{c}\frac{2v_z}{c}f_0 \tag{18}$$

An artifact, albeit with a small effect.

Non-linear effects during either propagation or scattering introduces higher harmonics into the spectrum. They alias around $\pm f_{prf}/2$ and introduce additional peaks in the spectrum and bias the result. The mean frequency is biased towards higher frequencies, if no aliasing takes place.

6 DEMODULATION

The model also makes the role of the demodulation process clear. The main purpose is to generate a signal that can detect the direction of the signal, and to reduce noise in the signal.

The signal after the demodulation can be written as

$$
\begin{aligned}
r_d(t_r) &= h_{lp}(t_r) * [r(t_r)\exp(j2\pi f_0 t_r)] \\
&= r(t_r) * [h_{lp}(t_r)\exp(j2\pi f_0 t_r)]
\end{aligned} \tag{19}
$$

where $h_{lp}(t_r)$ is a low-pass filter usually implemented as an integration over the pulse duration. It is, thus, possible to use a set of frequency modulated filters, and then apply the sampling operation. For a sinus burst the frequency modulated filter equals the pulse, and is equivalent to a matched filter (Skolnik 1980, Kristoffersen 1986), and can be replaced by such a filter. The demodulation can also be rewritten as

$$
\begin{aligned}
r_d(t_r) &= r(t_r) * [h_{lp}(t_r)(\cos(2\pi f_0 t_r) + j\sin(2\pi f_0 t_r))] \\
&= r(t_r) * [h_{lp}(t_r)(\cos(2\pi f_0 t_r) + j\cos(2\pi f_0 t_r - \frac{\pi}{2}))] \\
&= r(t_r) * [h_{lp}(t_r)(\cos(2\pi f_0 t_r) + j\cos(2\pi f_0 t_r - \frac{1}{4f_0}))]
\end{aligned} \tag{20}
$$

383

The imaginary part of the filter is just a delayed version of the real part. Measuring the signal a quarter period after the real signal, thus, gives a suitable signal (Halberg and Thiele 1986). An alternative is to perform a Hilbert transform on the RF signal to make the 90° phase shift for all frequencies.

The low-pass filters in the receiver are matched to the bandwidth of the pulse (Kristoffersen, 1986). This is done to reduce noise, and an alternative demodulation is to do a matched filtration on the received signal, and then a Hilbert transform. A complex sample set is then extracted and processed to yield the spectrum. Further information on direct sampling and its consequences can be found in Jensen (1996).

REFERENCES

Baker, D. W. Pulsed ultrasonic Doppler blood-flow sensing. *IEEE Trans. Son. Ultrason.*, SU-17:170–185, 1970.

Forsberg, F., and Jørgensen, M. Ø. Sampling technique for an ultrasound Doppler system. *Med. Biol. Eng. Comp.*, 27:207–210, 1989.

Halberg, L. I., and Thiele, K. E. Extraction of blood flow information using Doppler-shifted ultrasound. *HP-Journal*, 37:35–40, 1986.

Holland, S. K., Orphanoudakis, S. C., and Jaffe, C. C. Frequency-dependent attenuation effects in pulsed Doppler ultrasound: Experimental results. *IEEE Trans. Biomed. Eng.*, BME-31:626–631, 1984.

Jensen, J. A. *Estimation of Blood Velocities using Ultrasound: A Signal Processing Approach*. Cambridge University Press, New York, 1996.

Kristoffersen, K. Optimal receiver filtering in pulsed Doppler ultrasound blood velocity measurements. *IEEE Trans. Ultrason., Ferroelec., Freq. Contr.*, 33:51–58, 1986.

Magnin, P. A. Doppler effect: History and theory. *HP-Journal*, 37:26–31, 1986.

Newhouse, V. L., and Amir, L. Time dilation and inversion properties and the output spectrum of pulsed Doppler flowmeters. *IEEE Trans. Son. Ultrason.*, 30:174–179, 1983.

Newhouse, V. L., Bendick, P. J., and Varner, L. W. Analysis of transit time effects on Doppler flow measurement. *IEEE Trans. Biomed. Eng.*, BME-23:381–387, 1976.

Newhouse, V. L., Ehrenwald, A. R., and Johnson, G. F. The effect of Rayleigh scattering and frequency dependent absorption on the output spectrum of Doppler blood flowmeters. *Ultrasound Med.*, 3B:1181–1191, 1977.

Skolnik, M. *Introduction to Radar Systems*. McGraw-Hill, New York, 1980.

Thomas, N., and Leeman, S. Mean frequency via zero crossings. In *Proc. IEEE Ultrason. Symp.*, pages 1297–1300, 1991.

Thomas, N., and Leeman, S. The double Doppler effect. In *Acoust. Sensing and Imag.*, volume 369, pages 164–168. IEE, 1993.

TOWARDS AUTOMATED EXTRACTION OF VESSEL BOUNDARIES IN ULTRASONIC BLOOD FLOW IMAGES, BY MEANS OF EDGE DETECTION

T. Loupas

Ultrasonics Laboratory
CSIRO Division of Radiophysics
126 Greville Street, Chatswood NSW 2067
Australia

INTRODUCTION

Because of its non-invasive and real-time nature, ultrasonic colour Doppler blood flow imaging has become an important clinical tool in the assessment of vascular disease[1]. Until now, the morphological and haemodynamic information provided by this modality is used purely for qualitative visual interpretation. The key to the development and widespread acceptance of quantitative applications of ultrasonic blood flow imaging lies in the ability to extract vessel boundaries in a robust and automated manner. The automated extraction of vessel boundaries is a prerequisite for a variety of geometrical analysis (diameter, area, volume, and beam-vessel angle measurements) and advanced image processing operations (3D reconstruction of complex vascular structures, registration of image sequences, vessel identification and segmentation). In direct analogy with quantitative X-ray angiography[2], these operations are primarily applicable in the objective assessment of vascular morphology, but they are also relevant in the context of emerging techniques such as quantitation of tumour/organ vascularity[3] and contrast-based Doppler intensitometry[4]. Colour Doppler scanners carry out a simple form of blood/tissue segmentation, so that the colour-coded flow map and the grey-scale anatomical image can be combined in the same display, which is primarily based on Doppler power thresholding[5]. Briefly, a pixel is classified as flow or tissue depending on whether its corresponding Doppler power is greater or lower, respectively, than a threshold value which is user-adjusted by means of the colour Doppler "gain" (or "level") control. This criterion relies on the premise that, after clutter suppression, the Doppler signal associated with slowly moving or stationary soft tissue consists predominantly of electronic noise whose power is normally weaker than the power of the Doppler signal from moving blood. Although blood/tissue segmentation by means of power-thresholding is adequate for display purposes, it tends to be unreliable for quantitative applications. For example, the finite extent of the sample volume causes the Doppler power of pixels lying between adjacent vessels to be well above the noise floor, resulting in poor vessel separation. Also, the blurred nature of the flow-noise transitions, due to a combination of the transmitted pulse's shape and clutter filter, implies that the flow-region dimensions are threshold-dependent (see Figure 1) and hence inappropriate from a quantitative point of view.

The aim of this work was to investigate the feasibility of automated extraction of vessel boundaries by means of edge detection, i.e. identification of abrupt spatially consistent changes in the local attributes (Doppler power and/or velocity) of ultrasonic blood flow images. Given the obvious expectation that quantitative analysis should lead to accurate and reproducible results if it is to be of practical relevance, the main focus of the work was on edge detector validation. Performance evaluation based on a set of objective criteria (probability of true-positive v. false-positive edge detection, accuracy and reproducibility of edge localisation) and the effect and trade-offs associated with a number of acquisition and processing factors (SNR, intra- and inter-frame smoothing, compression, edge-detection window-size) were documented.

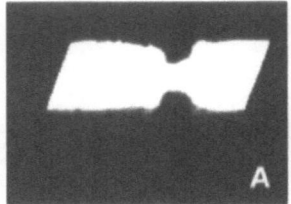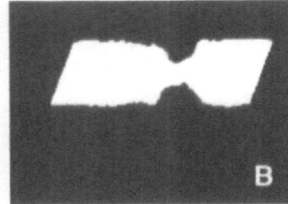

Figure 1. (A): Doppler power map of a stenotic flow model, after 15-frame averaging. **(B):** Two segmented versions (white region superimposed on mid-grey region) of Figure 1A, obtained using different threshold values to illustrate the threshold-dependence of the flow-region dimensions (particularly evident near the distal vessel walls, due to the skewed shape of the transmitted pulse).

PROCESSING AND ACQUISITION DETAILS

Edge detection was performed using two standard algorithms: the zero-crossings of the Laplacian of Gaussian proposed by Marr and Hildreth[6], and Canny's edge detector[7]. Both algorithms are based on the rotationally symmetric 2D Gaussian function

$$G(x,y) = \exp[-(x^2 + y^2)/2\sigma^2]$$

The first edge detector involves convolving the input image with the Laplacian of Gaussian

$$\nabla^2 G(x,y) = \frac{\partial^2 G(x,y)}{\partial x^2} + \frac{\partial^2 G(x,y)}{\partial y^2} = (\frac{x^2+y^2}{\sigma^2} - 2)\exp[-\frac{x^2+y^2}{2\sigma^2}]$$

and subsequent identification of the zero-crossing points on the convolution output. This approach has the desirable property of resulting in continuous one-pixel-wide edge contours but does not provide any information on edge strength, which is required to suppress zero-crossings due to negligible image discontinuities and/or noise. To overcome this drawback, an edge-strength map was generated by convolving the input image with the Gradient of Gaussian vector

$$\nabla G(x,y) = \frac{\partial G(x,y)}{\partial x}\vec{i}_x + \frac{\partial G(x,y)}{\partial y}\vec{i}_y = -(x\vec{i}_x + y\vec{i}_y)\exp[-\frac{x^2+y^2}{2\sigma^2}]$$

and, after calculating the magnitude of the gradient image, retaining only those points of the gradient-magnitude image previously identified as zero-crossings, while setting the rest to zero.

The starting point for Canny's edge detector is again convolution of the input image with $\nabla G(x,y)$, followed by calculating the magnitude of the gradient image. As is the case with all first-derivative operators, the edges generated by this approach are more than one pixel in width and, consequently, must undergo some form of thinning. Canny's algorithm achieves this task by estimating the edge orientation from the x- and y-components of the gradient image, scanning the gradient-magnitude image perpendicularly to the edge orientation searching for local maxima, and creating an edge-strength map which is equal to the gradient-magnitude values for local-maximum points and zero elsewhere.

Both edge detectors were implemented using a convolution window of $9\sigma \times 9\sigma$ pixels, to ensure an accurate representation of the $\nabla G(x,y)$ and $\nabla^2 G(x,y)$ functions. The choice of the standard deviation σ of the Gaussian represents a trade-off between accurate edge localisation and suppression of noise-induced false edges. Assuming for simplicity a 1D pulse edge $f(x)$ of width d and a 1D $\nabla^2 G(x)$ function

$$f(x) = \begin{cases} 1 & 0 \le x \le d \\ 0 & \text{elsewhere} \end{cases} \qquad \nabla^2 G(x) = (\frac{x^2}{\sigma^2} - 1)\exp[-\frac{x^2}{2\sigma^2}]$$

it can be proven analytically that their convolution is equal to

$$f(x) * \nabla^2 G(x) = x\exp[-\frac{x^2}{2\sigma^2}] - (x-d)\exp[-\frac{(x-d)^2}{2\sigma^2}]$$

Calculation of the roots of the above expression reveals that when σ is chosen so that the ratio d/σ becomes equal to 4, 3, 2, and 1, the zero-crossings overestimate the true edge width by 0.07%, 2%,

20% and 110%, respectively. These numbers imply that, in order to achieve accurate edge localisation, σ should be set to no more than one third of the size of the smallest feature of interest. Different edge models from the one considered here, as well as the presence of neighbouring edges, affect the zero-crossings in a complex manner, as has been documented in [8], introducing additional artefacts and errors for increasing values of σ.

From a practical point of view, the preceding arguments suggest that the value of σ should be kept relatively low. Since alternative forms of noise suppression are, consequently, required before edge detection to reduce the occurrence of false edges, the effectiveness of the following pre-processing operators was evaluated: median filtering within frames, which combines adequate noise suppression with satisfactory edge preservation; another nonlinear detail-preserving technique for intra-frame smoothing, known as "anisotropic diffusion"[9]; and straightforward averaging of multiple frames. However, the evaluation results of the anisotropic diffusion operator were comparable to those of median filtering and have not been included in the following section.

The input data used for the quantitative evaluation were multiple frames of raw I & Q sets acquired from an in-vitro flow model under low- and high-noise conditions (SNRs of 30 and 3 dB). The I & Q sets underwent clutter suppression by means of step-initialised IIR filtering, followed by mean axial velocity and Doppler power estimation based on the lag-one phase[10] and lag-zero squared-magnitude[11] of the complex autocorrelation function, respectively. However, because of space limitations, only results related to Doppler power images are reported here, whereas the velocity case is briefly discussed in the last section. Prior to edge detection, the Doppler power images underwent scan conversion with lateral interpolation to match the axial pixel size (0.1925 mm), and two compression schemes were used to reduce their dynamic range: a square-root operation, and logarithmic transformation (with clipping below the mean noise level, or more than 30 dB above it).

RESULTS

Figure 2 documents the effect of σ and intra-frame smoothing on the trade-off between missing true edges and accepting false edges. The input data for this figure were low-noise (SNR = 30 dB) scans of a straight vessel surrounded by tissue-mimicking material. The true-positive v. false-positive curves were obtained by applying different threshold levels to the edge-strength maps produced by the Laplacian of Gaussian detector, with a true-positive event defined as the existence of an edge point at a distance of less than 4 pixels (0.77 mm) from its correct position (determined by locating the vessel walls on grey-scale imaging).

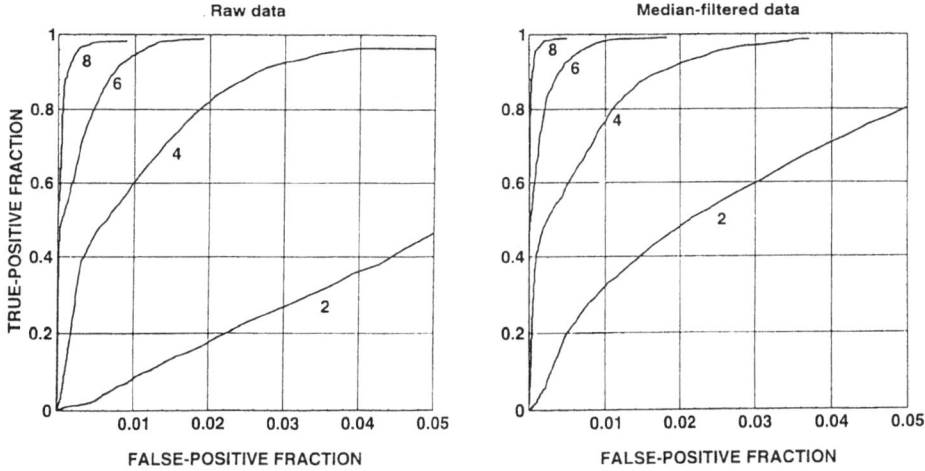

Figure 2. Performance of the Laplacian of Gaussian edge detector (σ = 2, 4, 6 and 8 pixels) applied to non-smoothed and median filtered (window of 5 X 5 pixels) Doppler power images with square-root compression.

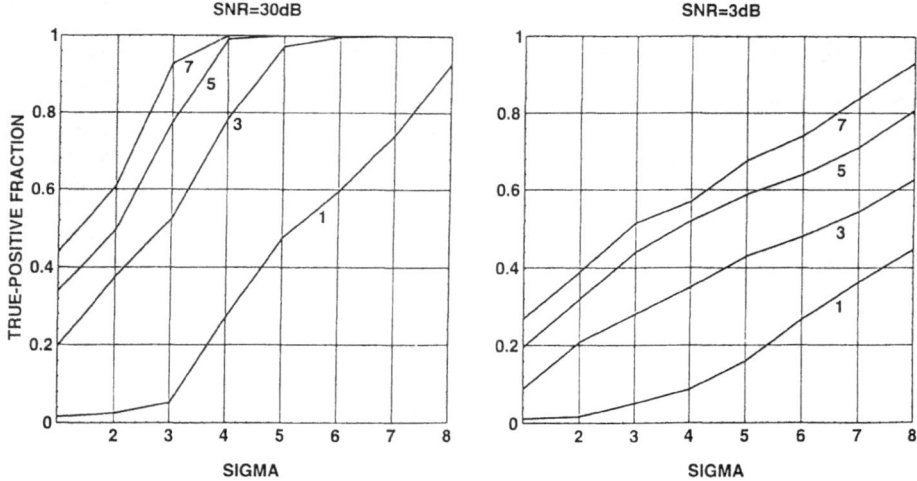

Figure 3. Effect of temporal averaging (1, 3, 5, and 7 frames) and SNR (30 and 3 dB) on the true-positive edge detection rates associated with the Laplacian of Gaussian detector, when the edge-strength maps are thresholded to maintain a constant false-positive fraction of 0.001.

Based on the median-filtered version of the input data used for Figure 2, as well as their high-noise (SNR = 3 dB) counterparts, and following the same methodology, Figure 3 demonstrates the effectiveness of inter-frame smoothing in achieving high rates of correct edge classification.

While the previous graphs quantify the detectability of edges, Table 1 provides data on the accuracy and reproducibility of edge localisation as a function of σ, SNR and compression scheme. The values of this table were obtained by: (i) applying Canny's edge detector to the straight-vessel (diameter of 8 mm) Doppler power images mentioned above, which had been pre-processed with median filtering and 7-frame temporal averaging; (ii) thresholding the edge-strength maps for a false-positive fraction of 0.001; (iii) scanning the thresholded maps along lines perpendicular to the vessel axis; (iv) and, for each line, measuring the distance between the proximal- and distal-wall edge points only if both of them were classified as true-positive events.

Table 1. Mean ± standard deviation of diameter measurements (mm), based on edge detection.

| σ | SNR = 30 dB | | SNR = 3 dB | |
	$\sqrt{}$ compression	log compression	$\sqrt{}$ compression	log compression
2	8.16 ± 0.17	8.94 ± 0.11	7.81 ± 0.31	7.97 ± 0.26
4	8.07 ± 0.13	8.92 ± 0.09	7.76 ± 0.23	7.88 ± 0.21
6	8.05 ± 0.14	8.92 ± 0.10	7.74 ± 0.25	7.88 ± 0.23
8	8.03 ± 0.15	8.93 ± 0.10	7.70 ± 0.30	7.83 ± 0.25

It is worth noting that the values of Table 1 do not exhibit any evidence of edge overestimation as σ becomes increasingly higher. This is due to the fact that the vessel diameter size – 8 mm over 0.1925 mm/pixel = 41.6 pixels – is well above the largest value of σ. However, edge overestimation is clearly evident in Figure 4 for $\sigma = 10$ pixels, which is comparable to the diameter size of the stenotic region (14 pixels), as well as the in vivo results of Figure 5, again for $\sigma = 10$ pixels.

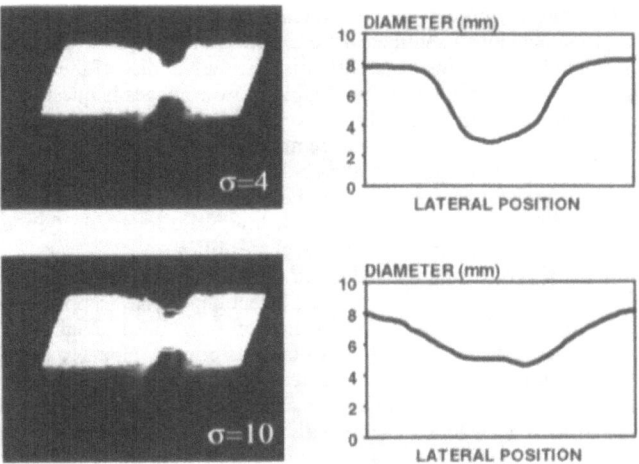

Figure 4. Left: The in vitro stenotic flow image of Figure 1, with the vessel boundaries obtained from Canny's edge detector (σ = 4 and 10 pixels) superimposed as white contours. **Right:** Automated diameter measurements, based on the edge detection outputs from the central part of the stenotic vessel.

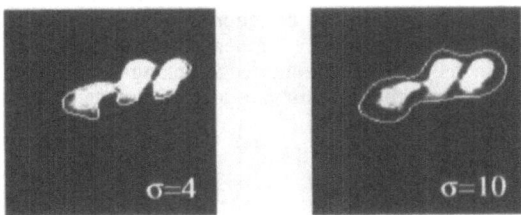

Figure 5. 15-frame averaged Doppler power image of the umbilical cord, with the vessel boundaries obtained from Canny's edge detector (σ = 4 and 10 pixels) superimposed as white contours.

DISCUSSION

Although space limitations did not permit a more extensive analysis, an attempt was made to include in the previous section a set of representative results which flag some important issues.

Figures 2 and 3 show that, in order to achieve satisfactory detectability of edges in Doppler power flow maps, some form of noise suppression must be employed through the use of a coarse-scale (i.e., large σ) edge detector and/or intra- and inter-frame smoothing during pre-processing. The effectiveness of the former option is limited by the fact that, as has been briefly addressed in the second section and illustrated by Figures 4 and 5, significant edge overestimation and distortion occurs when σ is set to values comparable to the size of the smallest feature of interest. The results of Figure 2 indicate that intra-frame smoothing by means of median filtering is indeed a useful pre-processing stage, particularly for small values of σ. However, the real key in achieving a high rate of true-edge detection while maintaining a low rate of false-edge occurrence[*] is inter-frame smoothing. As Figure 3 shows, modest amounts of frame averaging offer almost ideal performance under high-SNR conditions, even for very small values of σ. Nevertheless, it is also clear that a substantial number of frames would be needed to deal successfully with low SNRs.

The results of Table 1 demonstrate that accurate and reproducible dimensional measurements can be obtained by means of edge detection. At the same time, they emphasise the

---

[*]Note that the false-positive fraction of 0.001 used to form Figure 3 may appear very strict, but it is essential to avoid serious errors in the processing stages that follow edge detection.

importance of standardising the processing performed prior to edge detection, since it is evident that the two compression schemes examined have a noticeable effect on the obtained diameter values. A secondary implication of this observation is that the validity of measurements obtained by performing edge detection of frame-grabbed Doppler power images is questionable, because in many scanners the amount of compression is linked to the colour gain control. Another observation from Table 1 is the dependence of the measurement values on the SNR. This is simply a manifestation of an intrinsic characteristic of ultrasonic flow imaging – i.e. the gradual shrinkage of the flow region as the signal power approaches the noise floor – and its effect is modest (less than 5% for square-root compression) provided that a low pulse repetition frequency is used to avoid excessive attenuation of slow-flow areas by the clutter filter.

One issue which was not explicitly addressed in the previous section is the relevance of mean axial velocity flow maps in the context of edge detection. Unlike Doppler power data which represent one of the most invariant ultrasonic features of flow, mean axial velocity values are affected by a number of factors that make them unsuitable as primary inputs to edge detection. One such factor is aliasing although, to a certain extent, aliasing-induced discontinuities can be overcome by operating on the circular domain, i.e. mapping each velocity value v to a unit-length vector $\{\cos[\pi v / V], \sin[\pi v / V]\}$ where V denotes the Nyquist velocity limit. A second factor is turbulence, which makes flow velocity values indistinguishable from the random mosaic pattern of velocity noise in non-flow regions. Perhaps most important is the "colour bleeding" artefact which causes gross overestimation of the flow-region boundaries, due to the property of velocity estimators (such as the autocorrelator[10]) to register flow and produce coherent velocity values even for points well outside the vessel lumen with Doppler power values just above the noise floor. It seems, therefore, that the main role of velocity information in edge detection is to provide secondary evidence to subsequent processing stages (false-edge elimination and edge linking).

In summary, the results of this work establish the feasibility of, as well as the requirements for, performing accurate and reproducible edge detection in ultrasonic flow images. Although, in most cases, the edge detection outputs require further processing, the availability of reliable candidate points is critical for the success of either edge-linking[12] or active-contour[13] processes, which represent the next stage towards the ultimate goal of automated extraction of complete vessel boundaries.

REFERENCES

1. P. Lanzer and A. P. Yoganathan. "Vascular Imaging by Color Doppler and Magnetic Resonance," Springer-Verlag, Berlin (1991).
2. S. R. Fleagle, M. R. Johnson, C. J. Wilbricht, D. J. Skorton, R. W. Wilson, C. W. White, M. L. Marcus, and S. M. Collins, Automated analysis of coronary arterial morphology in cineangiograms: geometric and physiologic validation in humans, *IEEE Trans. Med. Imaging* 8:387 (1989).
3. D. O. Cosgrove, J. C. Bamber, J. B. Davey, J. A. McKinna, and H. D. Sinnett, Color Doppler signals from breast tumors, *Radiology* 176:175 (1990).
4. N. Nanda and R. Schlief. "Advances in Echo Imaging Using Contrast Enhancement," Kluwer Academic, Amsterdam (1993).
5. T. Loupas, R. B. Peterson, and R. W. Gill, Experimental evaluation of velocity and power estimation for ultrasound blood flow imaging, by means of a two-dimensional autocorrelation approach, *IEEE Trans. Ultrason. Ferroelec. Freq. Control* 42:689 (1995).
6. D. Marr, and E. Hildreth, Theory of edge detection, *Proc. Royal Soc. London* B207:187 (1980).
7. J. Canny, A computational approach to edge detection, *IEEE Trans. Pattern Anal. Machine Intell.* 8:679 (1986).
8. M. Shah, A. Sood, and R. Jain, Pulse and staircase edge models, *Comput. Vision Graphics Image Process.* 34:321 (1986).
9. P. Perona, and J. Malik, Scale-space and edge detection using anisotropic diffusion, *IEEE Trans. Pattern Anal. Machine Intell.* 12:629 (1990).
10. C. Kasai, K. Namekawa, A. Koyano, and R. Omoto, Real-time two-dimensional blood flow imaging using an autocorrelation technique, *IEEE Trans. Sonics Ultrason.* 32:458 (1985).
11. J. M. Rubin, R. O. Bude, P. L. Carson, R. L. Bree, and R. S. Adler, Power Doppler US: a potentially useful alternative to mean frequency-based color Doppler US, *Radiology* 190:853 (1994).
12. A. A. Farag, and E. J. Delp, Edge linking by sequential search, *Pattern Recogn.* 28:611 (1995).
13. M. Kas, A. Witkin, and D. Terzopoulos, Snakes: active contour models, *Int. J. Comput. Vision* 1:321 (1988).

THE PERFORMANCE OF AN ADAPTIVE RF-DOMAIN CLUTTER REMOVAL FILTER

Peter J. Brands, Arnold P.G. Hoeks, and Robert S. Reneman*

Department of Biophysics and *Physiology
Cardiovascular Research Institute Maastricht
University of Limburg
Maastricht, The Netherlands

INTRODUCTION

Ultrasound systems are widely used to visualize in real-time internal structures and blood flow velocity distributions. The latter are estimated from the received ultrasound radio frequency (RF) signals backscattered from moving red blood cells, using a mean frequency estimator. Since RF-signals contain not only scattering, but also reflections, reverberations and noise it is necessary to suppress the power of the reflections and/or reverberations (clutter removal) to estimate the temporal mean frequency of the signal component induced by scattering. Normally this is done with a static high-pass filter acting in the temporal direction with a fixed cut-off frequency. However, by using such a filter the time dependent aspect of the reflections is ignored. A more selective way is to use a band stop filter which adapts its rejection range to the mean frequency of the clutter. A great advantage of this adaptive filter (ADP-f) method is that the rejection range can be kept small.

This paper presents a comparison in performance of an ADP-f and the following three static filters: (1) a moving target indicator (MTI-f), (2) a second order IIR filter (IIR-f), and (3) a standard linear regression filter (SLR-f) (Hoeks et al. 1991). All these filter algorithms are used in combination with the same RF-domain cross-correlation model (CCM) mean frequency estimator to evaluate the influence of the filter algorithms on the temporal mean frequency estimation range of the CCM estimator (De Jong et al. 1990, Hoeks et al. 1993).

For different signal conditions (like noise level, signal bandwidth, clutter level) and a different filter algorithm the standard deviation of the estimate and the minimally detectable temporal mean frequency will fluctuate. To allow for a direct comparison in performance, as far as the temporal mean frequency estimation range is concerned, the different combinations of filter (MTI-, IIR-, SLR, ADP-f) and estimator (CCM) are evaluated for a common simulated RF-matrix (RF-signals in time and depth).

The different filters in combination with the CCM estimator are evaluated repetitively, using 100 estimates based upon independent estimation windows in the simulated RF-matrix, resulting in an estimate for the bias (systematic error) and standard deviation of a filter estimator combination. The bias b_1 and standard deviation s_1 of the CCM estimator with an clutter filter are tested statistically (5% confidence level) against the bias b_0 and standard deviation s_0 of the estimates of the CCM estimator without an clutter filter to establish whether the filter modifies the velocity estimate.

CLUTTER FILTERS

The amplitude of the clutter component is on the order of 40 dB higher than that of the scattered signal induced by flowing blood. For simulation a value of 25 Hz and 12.5 Hz is used as maximum temporal mean frequency and bandwidth, respectively. In combination with 2.5 kHz as maximum temporal mean frequency of the scattered signal (PRF of 5 kHz) this will give a normalized temporal mean frequency of 0.005 and a bandwidth of 0.0025 ($\bar{f}_r$=0.005, B_r=0.0025). Since the clutter has a single velocity the bandwidth depend only on the spectral broadening, caused by the transit time effect and acceleration. To realize a suppression of 40 dB for a given temporal mean frequency of the clutter component the cut-off frequency f_{cut} should be chosen in accordance with the roll-off of the filter used (Fig. 1).

Moving target indicator filter (MTI-f)

The *MTI-f* is a static first order FIR high-pass filter. The cut-off frequency is fixed at 0.25 times the *PRF* and the roll-off is 6 dB per octave, giving a suppression of 40 dB for a temporal mean frequency of 0.004 (Fig. 1). The settling time is only one sample point. The effect of including this clutter filter prior to a mean velocity estimation based on RF cross-correlation was investigated by Jensen (1993).

Infinite impulse response filter (IIR-f)

The IIR-f is a static second order IIR high-pass filter. The cut-off frequency is set at 0.04 times the PRF while the roll-off is 12 dB per octave, giving a suppression of 40 dB for a temporal mean frequency of 0.005 (Fig. 1).

Standard linear regression filter (SLR-f)

The *SLR-f* was specially developed for the use in color coded Doppler systems by Hoeks et al. (1991). The settling time of the SLR-f is zero, thus all N sample points within the temporal estimation window can be used for velocity estimation. The expected roll-off of this high-pass filter is 12 dB per octave. The latter is directly related to the temporal estimation window length. The normalized cut-off frequency for an estimation window length of 17 sample points is 0.038 giving a suppression of 40 dB for a temporal mean frequency of 0.006 times the *PRF* (Fig. 1).

Adaptive clutter filter (ADP-f)

A more selective way to suppress the clutter is to shift the temporal frequency distribution towards zero frequency where the shift is given by the estimated temporal mean frequency of the reflected signals. Subsequently, the reflections, then centered at zero frequency, are suppressed by a high-pass filter with a low cut-off frequency (Brands et al. 1995). The cut-off frequency is set at 0.02 times the PRF while the roll-off is 12 dB per octave, giving a suppression of 40 dB for a temporal mean frequency of 0.0025 (Fig. 1).

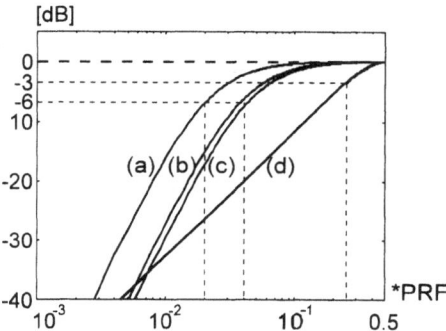

Figure 1. (A) The transfer functions of the different clutter filters: (a) adaptive clutter filter, (b) standard linear regression filter, (c) second order IIR clutter filter, and (d) moving target indicator.

STATISTICAL EVALUATION

To investigate the difference in temporal mean frequency estimation range, the different combinations of filter and estimator are evaluated for a common simulated RF-matrix with properties described by the parameters in signal space *SP*, subtended by 6 temporal and 2 spatial signal parameters:

$$SP(p, \bar{f}_s) = \{A_r, B_r, \bar{f}_r, A_s, B_s, A_n, \bar{f}_{rf}, Q_{rf}\}$$
$$SP(0, \bar{f}_s) = \{-60, 0.25, 0.005, 10, 5, 0, 0.25, 2\}$$
$$SP(1, \bar{f}_s) = \{10, 0.25, 0.005, 10, 5, 0, 0.25, 2\} \quad (1)$$
$$SP(2, \bar{f}_s) = \{20, 0.25, 0.005, 10, 5, 0, 0.25, 2\}$$
$$SP(3, \bar{f}_s) = (30, 0.25, 0.005, 10, 5, 0, 0.25, 2\}$$

where A_r, B_r, $\bar{f}_r$ are the amplitude, bandwidth and temporal mean frequency of the clutter component, A_s and B_s the amplitude and velocity bandwidth of the blood signal (red blood cells), A_n the noise amplitude, $\bar{f}_{rf}$ and Q_{rf} the mean frequency and quality factor ($Q_{rf} = \bar{f}_{rf}/B$) of the spatial RF-signals, p is a pointer to a given set of signal parameter values (0,1,2,3) and $\bar{f}_s$ the imposed temporal mean frequency of the signal component induced by the moving blood. The simulated RF-matrix was synthesized using the signal generation model described by Hoeks et al. (1993). For all simulations the noise bandwidth B_n is equal to the RF bandwidth ($Q_n=Q_{rf}$). For each set of signal properties the imposed temporal mean frequency $\bar{f}_s$ is logarithmically varied between 0.01 and 0.5 times the *PRF*. This variation is 98% of the total temporal mean frequency estimation range.

Each mean frequency estimate $\hat{\bar{f}}_s$, based on successive estimation windows, will have a bias $b_i(p, \bar{f}_s, w_t, w_s)$ and a standard deviation $s_i(p, \bar{f}_s, w_t, w_s)$. Here w_t is the temporal length of the estimation window and w_s the spatial length of the estimation window (both given in sample points) in a simulated RF-matrix with signal properties p and the true temporal mean frequency $\bar{f}_s$. The index i concerns the method of signal processing: (i=0 CCM estimator **without** clutter filter) and (i=1 CCM estimator **with** clutter filter).

For the CCM estimator without clutter filter (*i*=0) signal processing is executed in an RF-matrix with signal property *p*=0 (1). The CCM estimator with clutter filter (*i*=1) employes an RF-matrix with signal properties *p*=0, 1, 2 and 3 (1).

The estimation window for each estimate $\bar{f}_s$ is based upon a temporal length w_t of 16 or 32 sample points and a spatial length w_s of 8 sample points (*Q*=2). The temporal estimation lengths 16 and 32 sample points is chosen to achieve a normalized cut-off frequency of 0.04 and 0.02, respectively, for the SLR-f (Fig. 1). Note that the temporal estimation window for the SLR-f should be odd, so w_t for the SLR-f is set at 17 or 33 sample points.

SIMULATION RESULTS

Every filter estimator combination (MTI-, IIR, SLR-, ADP-CCM) is statistically tested and the simulation results are presented as $s_i(p, \bar{f}_s, w_t, w_s)$ for the standard deviation and as $|b_i(p, \bar{f}_s, w_t, w_s)|$ for the bias. Note that the latter is an absolute value.

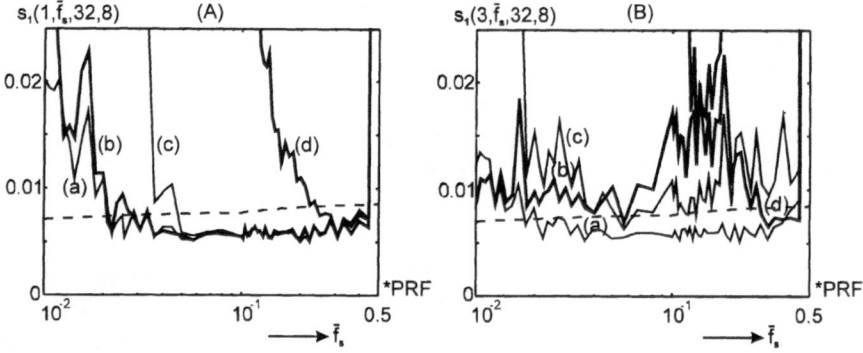

Figure 2. The estimates of standard deviation (A) $s_1(1, \bar{f}_s, 32, 8)$ and (B) $s_1(3, \bar{f}_s, 32, 8)$ as function of the temporal mean frequency for the CCM estimator in combination with (a) an ADP-f, (b) an SLR-f, (c) an IIR-f, (d) an MTI-f. The dotted line represents the 5% confidence level.

In the stop band of each clutter filter the mean frequency of noise is estimated, causing a corresponding standard deviation and bias. The mean frequency estimation range (5% confidence level) is different for each combination of filter and estimator (Fig. 2).

The temporal mean frequency $\bar{f}_r$ of the clutter component is set at 0.005, indicating that the reflections are moving with a low velocity. The adaptive filter will always suppress maximally the clutter component, independent of the temporal mean frequency of this signal component. On the other hand a static filter will only suppress maximally if the signal component has a zero temporal mean frequency (no movement). This difference in filter property is clearly recognized in figure 2B, where the amplitude of the clutter component is 30 dB (A_r=30 dB). In figure 2B the temporal mean frequency estimation range for the adaptive filter CCM estimator combination (a) is 96%, which is the same as in figure 2A for an amplitude of A_r=10 dB. However, the IIR filter CCM estimator combination (c) has a temporal mean frequency estimation range of 0% in figure 2B and 90% in figure 2A.

The overall performance of the different filter estimator combinations is shown in figure 3. A low root mean square can be associated with a "good" robustness (independence of signal properties), while a high mean value can be associated with a "good" quality in the sense of a "good" temporal mean frequency estimation range.

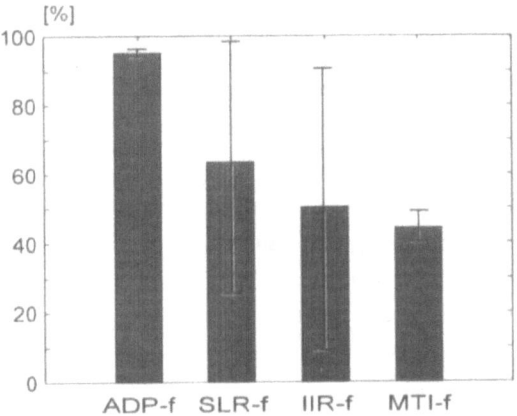

Figure 3. The mean and root mean square in percentage of the temporal mean frequency estimation range for the four filter estimator combinations (MTI-, IIR-, SLR-, ADP-CCM) for the simulation signal properties p.

Figure 3 confirms the excellent estimation range of the ADP-CCM estimator combination.

With the use of an IIR-f every cut-off frequency within the Nyquist frequency range can be realized at the expense of the response time. Model based filters such as the SLR-f have a limitation in the minimal cut-off frequency because of the finite length of the filtration window. Therefore, the length of the filtration window should be chosen in accordance with the PRF and with the cut-off frequency necessary for an effective filtration. Moreover, the temporal charateristics of the signal (non-stationarity) should be taken into account.

CONCLUSION

The influence of clutter suppression with different clutter filters on the mean frequency estimation range of the CCM RF cross-correlation model estimator was investigated. The clutter suppression from RF signals, using an clutter filter prior to the CCM mean frequency estimator, will restrict the measurement of low blood flow velocities. It has been shown that different clutter filters in combination with the CCM mean frequency estimator provide a different mean frequency estimation range. For the RF signal conditions considered the combination of the adaptive clutter filter and CCM mean frequency estimator shows the best overall mean frequency estimation range of 95%, which is almost independent of the signal properties reflected in an overall root mean square of 0.9%. The overall mean frequency estimation range for the standard linear regression and the infinite impulse response clutter filter in combination with the CCM mean frequency estimator is 63.4% and 50.1%, respectively. This range is sensitive to the variation in signal properties, reflected in an overall root mean square of 36.6% and 42.7%, respectively. The combination of moving target indicator clutter filter and CCM estimator shows an overall mean frequency estimation range of 46.3 % and an overall root mean square of 5.6%. These values show that the traditional static clutter filters give a large reduction in the mean frequency estimation range, confirming the findings of Jensen (1993). For the adaptive clutter filter the mean frequency estimation range is considerably greater because the reflections are suppressed by a small band stop filter that apparently adapts its central frequency to the temporal mean frequency of the slowly moving reflections or reverberations. In general we may conclude that the ADP-f provides a greater temporal mean frequency estimation range than a static filter. The extent of this increase may vary with the type of estimator.

REFERENCES

Brands, P.J., Hoeks, A.P.G., Hofstra, L., and Reneman, R.S., 1995, A non-invasive method to estimate wall shear rate using ultrasound. Ultrasound Med. Biol. 21:171-185.

De Jong, P.G.M., Arts, T., Hoeks, A.P.G., and Reneman, R.S., 1990, Determination of tissue motion velocity by correlation interpolation of pulsed ultrasonic echo signals. Ultrasonic Imaging. 12:84-98.

Hoeks, A.P.G., Arts, G.J., Brands, P.J., and Reneman, R.S., 1993, Comparison of the performance of the RF cross-correlation and Doppler autocorrelation technique to estimate the mean velocity of simulated ultrasound signals. Ultrasound in Med. & Biol. 19:727-740.

Hoeks, A.P.G., Van de Vorst, J.J.W., Dabekaussen, A., Brands, P.J., and Reneman R.S., 1991, An efficient algorithm to remove low frequency Doppler signals in digital Doppler systems. Ultrasonic Imaging. 13:135-144.

Jensen, J.A., 1993, Stationary echo canceling in velocity estimation by time-domain cross-correlation. IEEE Trans. on Med. Imaging. 12:471-477.

A FLOW RATE PROFILER AIMED AT SHEAR STRESS EVALUATION IN ARTERIES*

Moreno Bardelli,[1] Renzo Carretta,[1] Andrea Carletti,[2] Domenico Dotti,[2] Remo Lombardi,[2] and Bonaria Murtas[2]

[1]Cattinara Hospital, Trieste, Italy
[2]Dep. Informatica e Sistemistica, University of Pavia, Pavia, Italy

* Work partially supported by the Italian "Comitato Nazionale Ricerche"

INTRODUCTION

The shear stress of blood in arteries is an important parameter for the diagnosis of hypertension risks, even if its ultimate implications are not yet fully understood.

The evaluation of this parameter requires the contemporaneous measurements of viscosity and shear rate, the latter being the gradient of the blood velocity near the vessel wall; remark that the shear stress should be measured during at least one complete heart beat, in order to provide useful diagnostic information.

The viscosity can be measured through laboratory analysis, while the shear rate requires the acquisition of the blood velocity over a suitable space and time domain.

Our group has recently developed an instrument (Flow Rate Profiler, FRP II)[1] based on ultrasound techniques, capable of measuring the longitudinal component of the blood velocity along the path of the ultrasound; proper space and time periods can be selected by the operator. Unfortunately, the measurements obtained at the periphery of the vessel are substantially affected by echoes reflected by the walls; indeed, even if they originate at distances from the probe different from those at which the blood particles scatter the ultrasound, the two contributions partially overlap because of the limited resolution of the ultrasound apparatus. Furthermore the wall echo is not still, as a consequence of vessel and probe unavoidable displacements.

Since the wall motion is slower than the blood velocity, and hence the relevant Doppler frequency is proportionally lower, this interference is reduced by means of a proper high pass filter; however the much higher amplitude of the wall echo prevents its complete rejection, in particular for the most external blood layers. To overcome this difficulty, our acquisition system has been modified by introducing means for the stabilisation in space of the wall echo; so they can be easily cancelled by the high pass filter. In turn, wall echo stabilisation has been obtained by delaying the ultrasound pulse emitted in such a way to hold in place the first portion of the velocity profile. The underlying theory, the experimental apparatus and preliminary experimental results are described in the following sections.

ECHO STABILISATION METHOD

The closed loop control for the stabilisation of the wall echo is illustrated in Fig. 1.

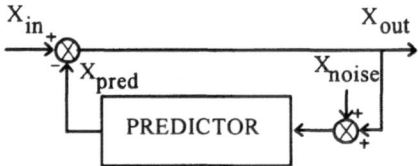

Fig. 1 - Block scheme of the closed loop stabilisation control

The actual wall position X_{in} is partially compensated by the predicted value X_{pred} so giving the resulting apparent wall position X_{out}; when the fluctuations of X_{out} are reduced to a minimum, the blood velocity profile near the wall is evaluated with the maximum accuracy, so yielding the best estimate of the shear rate.

However the predictor cannot operate neither on the input variable X_{in} nor on the corrected output X_{out}, but on an approximate estimate of the latter, obtained from the analysis of the velocity profile. The chosen algorithm is the linear regression of the velocity measures pertaining to the first quarter of the vessel diameter. Since the evaluated wall position is affected by the instrumental noise X_{noise}, the stability of X_{out} is improved as regards the fluctuations of X_{in},, but is worsened by X_{noise}. Eq. 1 describes the dependency of X_{out} on X_{in} and X_{noise}, as a function of the frequency:

$$X_{out}(j\omega) = \frac{X_{in}(j\omega) - P(j\omega)X_{noise}(j\omega)}{1 + P(j\omega)} \tag{1}$$

where $P(j\omega)$ is the transfer function of the predictor.

The optimal trade-off is achieved when Eq. 2 holds true for every frequency:

$$P(j\omega) = <X_{in}(j\omega)^2> / <X_{noise}(j\omega)^2> \tag{2}$$

where the notation $<X>$ specifies the ensemble average of X, and $P(j\omega)$ turns out to be real. Unfortunately the transfer function of Eq. 2 is not physically realisable; furthermore, a realistic assumption for $P(j\omega)$ must make sure that the closed loop is stable, matching for instance the conditions of the Bode criterion. Since the open loop chain does not include non minimal phase delay, the Bode diagram of gain has been adopted as a design instrument.

Eq. 2, even if not directly applicable, gives some hints for identifying the transfer function $P(j\omega)$; since the noise power spectrum can be considered white (because of the statistical independence of each realisation), while the contributions of the probe to vessel motion are confined to low frequencies, an appropriate choice consists of a Bode diagram rapidly decreasing with the frequency.

Furthermore, recalling that the disturbances on the velocity measurements depend on the derivative of X_{out}, it can be easily verified that, in order to put a bound to the contribution of the noise, the plot at the highest frequency must present a down slope of at least twelve dB per frequency octave (-2 in logarithmic scale).

In order to identify the best suited transfer function P(jω), a few implementation instances have been evaluated through computer simulation. The basic criterion was the minimisation of the derivative of X_{out} for the most common combinations of X_{in} and X_{noise}; in any case the r.m.s of the variables was considered.

After a few attempts, a satisfactory dependence on the frequency for P(jω) has been identified; the relevant Bode plot is illustrated in Fig. 2a. Approximately, it consists of a segment, having -1 slope, bounded by two rays with -2 slope; the segment is centered around the cross-over frequency, and extends over 2.54 octaves. As a consequence, the phase margin, evaluated at the cross-frequency, is 45°.

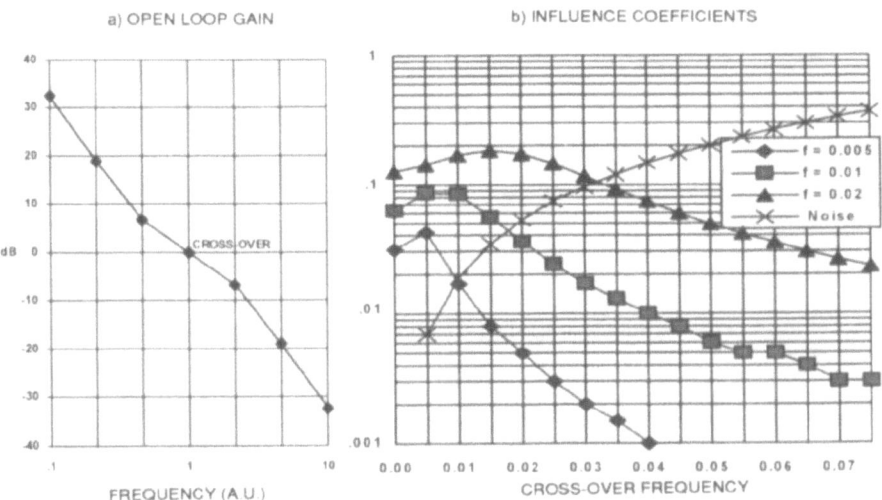

Fig. 2 - a) Bode plot of the selected frequency dependency of P (jω); b) the corresponding influence coefficients for the derivative of X_{out}, with inputs X_{in} and X_{noise}; the r.m.s. of each variable is considered.

Fig. 2b illustrates the dependency of the derivative of X_{out} on the input variables as a function of the cross-over frequency; inputs are three frequency components of X_{in}, or X_{noise}. Note that the linearity of Eq. 1 allows to superpose the effects of X_{in} and of X_{noise}, quadratically summing the contributions calculated from the plots of Fig. 2b.

Remark the following:

- It is assumed that the velocity profile are generated with a nominal frequency of 1.
- The point at zero cross-over frequency of every diagram shows the influence coefficient for no wall stabilisation.
- The closed loop control is profitable only when the cross-over frequency exceeds the frequency of X_{in}.
- The most appropriate cross-over frequency can be selected when the amplitude and the frequency of the main component of X_{in} and the amplitude of X_{noise} have been estimated.

EXPERIMENTAL FACILITY AND RESULTS

Fig. 3 shows the block scheme of the flow rate profiler, including the pulse delay controller.

The Personal Computer (PC) contains a special AT bus card, dedicated to the pre-processing of the ultrasound signal, for the evaluation of the blood velocity. The AT card is interfaced to the ultrasound probe by an external module, which emits the power pulses, when driven by the card, and amplifies the received echo signal, before giving it back to the card.

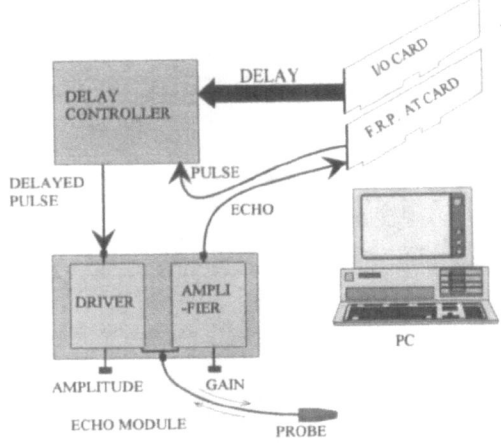

Fig. 3 - Block scheme of the acquisition and processing system.

The pulse delay controller is inserted into the connection that carries the signal driving the external module. The actual delay is controlled by the PC through a commercial parallel I/O card. Actually, the programmed delay is not immediately transferred to the exciting pulse, but through a low pass filter that smoothes the delay changes, so gradually compensating the relatively slow variations in the probe-vessel distance. The transfer function of the filter has been taken into account when implementing the wall position predictor. In order to establish the feasibility of the proposed wall stabilisation technique, an experimental apparatus, mimicking a circulation system, has been built,as shown in Fig. 4.

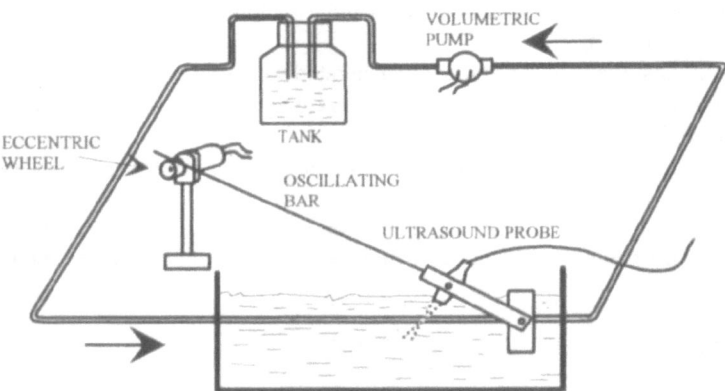

Fig. 4 - Circulation phantom, including the probe moving device

The liquid, the anti-freeze STILL FLU from MA-FRA, is made to circulate through a closed loop plexiglas piping, having internal diameter 13.5 mm, by a volumetric pump. A tank along the piping makes the flow rate more regular, so as to obtain a laminar regime, with minimal fluctuations. The ultrasound probe is tightened to an oscillating bar, moved by a rotating eccentric wheel; frequency and amplitude of the oscillation can be changed at will.

The following operating parameters have been adopted.

- acquisition range: 40 mm; the corresponding frequency of profile generation is about 150 Hz, for the adopted instrument (FRP II).
- range step: 0.2 mm.
- probe frequency: 7.5 MHz.
- on-axis velocity: about 55 cm/s.
- probe-tubing motion amplitude/frequency: 0.42 mm rms/1.3 Hz.
- cross-over frequency: 4 Hz.

The last parameter has been selected using the diagram of Fig. 2b, having estimated the noise level to approximately 0.3 mm rms. Fig. 5 illustrates the results.

Time diagrams of the wall position, without and with stabilisation, are plotted in Fig. 5a. The fluctuations of the latter are remarkably lower, even if more noisy, due to contributions of X_{noise}. The overall effect on the shear-stress can be appreciated from Fig. 5b, where, for both situations, the averages of 5 consecutive insets of the velocity profile are shown; measurements refer to the time abscissa 0.5 s in Fig. 5a. The profiles obtained with wall stabilisation are smoother in the neighborhood of wall, so making feasible the shear-rate evaluation in such a region.

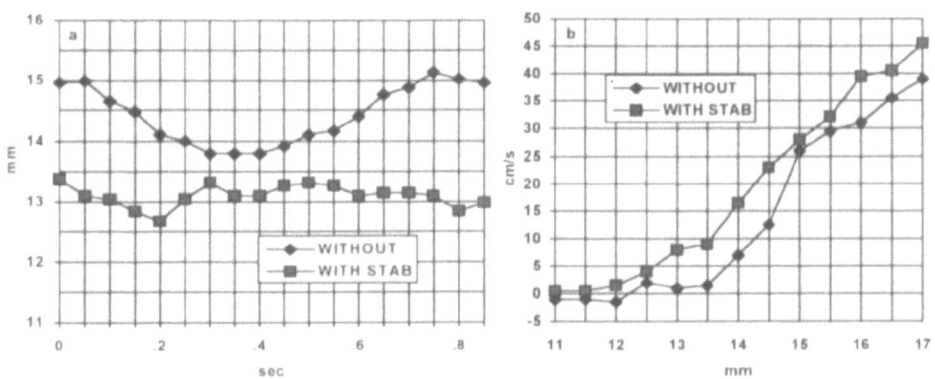

Fig. 5 a) time plots of the wall position; b) averaged velocity profile insets

CONCLUSIONS

The measurement of the shear stress near the wall of a blood vessel, by means of ultrasound, is frequently disturbed by even slow changes in the distance probe-wall. To overcome this difficulty, these changes are compensated by regulating the delay with which the ultrasound probe is fed with excitation pulses; the apparent wall velocity is so minimised.

This work describes the operating principles, the algorithms and the practical implementation of flow rate profiler that stabilises the wall position. Preliminary experimental results, obtained from an artificial circulation facility are also given. They confirm the applicability of the method and allow to identify the conditions for which the improvements are remarkable. New developments are under consideration, aiming at refining the methodology, in particular through the optimisation of the signal to noise ratio. The application of this technique to the above mentioned Flow Rate Profiler, and the examination of a large number of subjects, both healthy and suffering from hyper-tension, will allow to clarify the connection between shear-stress and artery damage.

REFERENCE

1. M.Bardelli, R.Carretta, G. Di Paolo, D. Dotti, F. Dotti, R. Lombardi, FRP II wide range high resolution .., Proc. Medicon 92, M.Bracale & F.Denoth ed., Capri, Italy (1992).

A NEW ULTRASONOGRAPHIC INSTRUMENT FOR MEASURING VESSEL WALL SHEAR STRESS: COMPARISON TO A MATHEMATICAL MODEL IN ESSENTIAL HYPERTENSIVES

Moreno Bardelli,[1] Renzo Carretta,[1] Domenico Dotti,[2] Bruno Fabris,[1], Fabio Fischetti,[1] Franco Cominotto,[1] Donatella Ussi,[1] Mario Calci,[1] Riccardo Candido[1] and Luciano Campanacci[1]

[1]Istituto di Medicina Clinica Università di Trieste
[2]Dipartimento di Informatica Università di Pavia
Italy

INTRODUCTION

Blood flow induces shear forces acting on the vessel wall[1,2]. These forces influence endothelium structure and metabolism[3-6] and the mechanical behaviour of the arterial wall.[2] The arterial wall shear stress[7] is representative of these shear forces; it is the product of blood viscosity and the velocity gradient near the endothelium (shear rate).

Low or oscillating shear stress has been linked to the development and progression of atherosclerotic lesions.[8-11] This correlation is based on "in vitro" studies or on theoretical mathematical models, since direct methods for "in vivo" studies have not been available to describe the velocity profile of blood flow.

We developed a new ultrasonographic instrument (Profilmeter FRP II) able to directly measure the velocity profile of blood flow inside the vessels.[12,13]

MATERIALS AND METHODS

The Profilmeter FRP II (Figure 1) is an instrument based on ultrasonographic multigate technique. Local blood velocities are simultaneously measured every 5 msec across the vessel transverse axis, in contiguous 0.2 mm sample volumes for the entire duration of the cardiac cycle.

Blood velocity is determined by a cross-correlation technique, which analyzes the time shift between correlated subsequent echo waves, instead of the frequency shift characteristic of the Doppler technique. It allows higher precision and less complicated electronic and computing networks in comparison to the Doppler method.

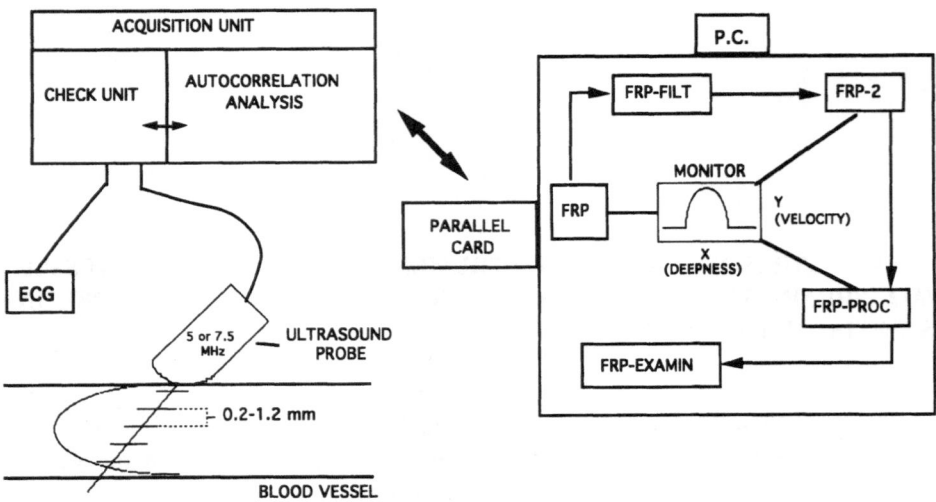

Figure 1. FRP II operative scheme

By means of a software package the autocorrelation functions are filtered (FRP-FILT program) and transformed into velocity functions (FRP-2 program).

Hence, the FRP-PROC program compares real data to a simplified model of laminar flow, resulting in the description of the instantaneous flow profile, by which is derived: the velocity gradient proximal to the vessel wall, the vessel diameter, the mean velocity and the blood flow (Figure 2). Shear stress is calculated multiplying the wall shear rate by the relative blood viscosity, measured by a rotating viscometer (Haake CV100).

We tested the precision and the sensitivity of the Profilmeter FRP II in measuring the shear stress in human subjects.

In a group of 9 elderly hypertensive subjects (age: 74.6 ± 6.7 years; mean arterial pressure: 120 ± 11 mmHg) we performed a series of 36 measurements of the peak systolic wall shear stress of the common carotid artery in different hemodynamic conditions, induced by the infusion of anti-hypertensive drugs (isradipine or clonidine), with the end point of reducing systolic blood pressure of 20 mmHg. Peak systolic shear stress was calculated as the average in the time interval between -15 and +15 msec surrounding the peak systolic flow.

The FRP II measured shear stress was compared to a calculated shear stress, based on the Womersley's mathematical model:[14-16]

$$Y = 2K \, (Vcl/d)$$

where: Y = vessel wall shear rate (sec^{-1}), Vcl = vessel center line blood velocity (cm/sec), D = vessel diameter (cm) and K = a complex parameter including heart rate, vessel diameter and blood viscosity.[14]

In addition the FRP II measurements of the vessel diameter and mean blood velocity were compared to those contemporarily performed by standard techniques: echo-tracking (Diamove, Teltec, Lund, Sweden) and echo-CW Doppler (TS Sistemi, Trieste, Italy).

RESULTS

In agreement with a previous mathematical analysis of blood flow profiles of the carotid vascular bed, performed by Pertkold et al.,[11] we observed two different morphologies during the systolic phase (Figure 2): a parabolic profile during the early and the peak systolic phase and the paramedian double parabolic profile "camel's humps", at the late decremental systolic phase.

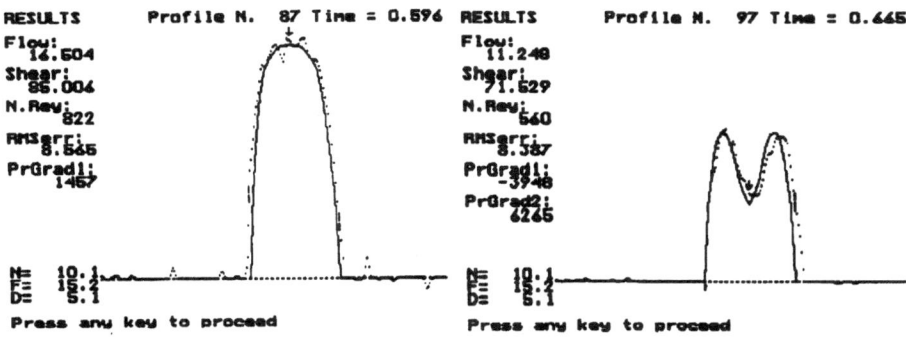

Figure 2. "In vivo" flow profile patterns described by FRP II in the common carotid artery: peak systolic parabolic (left) and late systolic double parabolic i.e.: camel's humps (right);

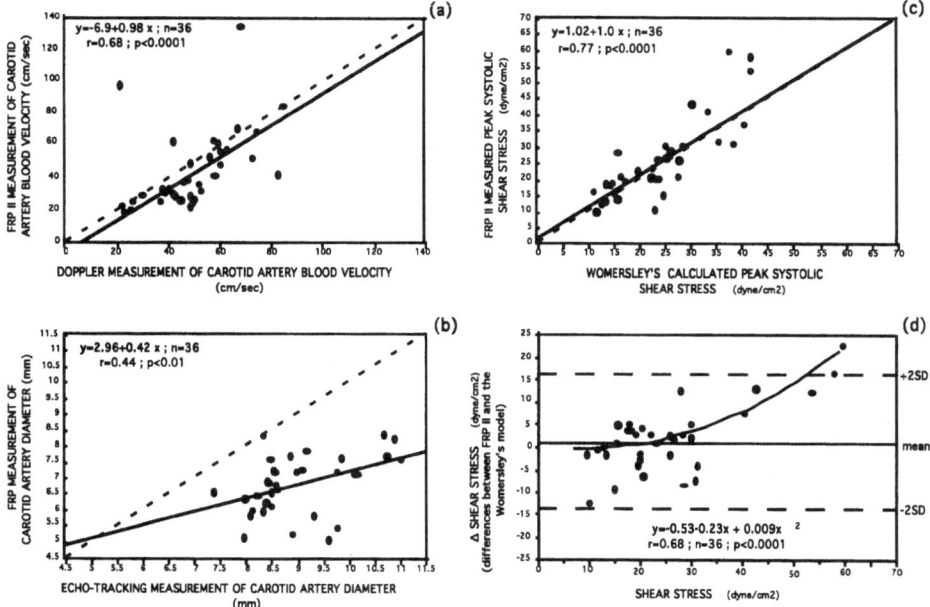

Figure 3. Correlations (continuous lines) between: mean blood velocities measured with FRP II and the echo-Doppler technique (Fig. 3a), vessel diameter measured with FRP II and the echo-tracking technique (Fig. 3b), peak systolic shear stress measured with FRP II and that calculated by the Womersley model (fig. 3c); Bland-Altmann analysis of the differences between the two methods in measuring the vessel wall shear stress (Fig. 3d).
Dotted lines=Identity lines. All measurements performed at the common carotid artery level.

Good correlation (r=0.68, p<0.001) was found between mean blood velocities measured with FRP II and echo-Doppler techniques (Figure 3a). A less significant correlation (r=0.44, p<0.01) was found for the vessel diameter measurement done by FRPII in respect to the echo-tracking technique (Fig. 3b).

Peak systolic shear stress measured by FRP II correlated (r=0.77, p<0.0001) with that calculated by the Womersley's mathematical model (Figure 3c), although a further Bland-Altmann analysis revealed a clear underestimation, by the Womersley model, for the highest values of shear stress (Figure 4d).

Intraindividual variability on repeated measures was 3.9 ± 2.4% and 6.7 ± 4.1% for the Doppler and FRP II blood velocity measurements respectively, and 2.4 ± 1.6% and 4.6 ± 4.1% for echo-tracking and FRP II vessel diameter measurements, respectively. The variability of the FRP II measured shear stress was 11.2 ± 5.6%.

DISCUSSION

The FRP II direct description of the carotid artery blood flow profile confirmed Pertkold's theoretical analysis of the pattern of distribution of velocities at that level.[11]

During peak systolic flow and late systolic phase two different patterns could be discerned: a parabolic pattern, as an expression of Newtonian laminar flow, and a complex profile with a double eccentric parabola and a central depression, i.e. a "camel's hump" profile. The latter could be determined by the combination of the longitudinal pressure gradient with that transverse, induced by vessel diameter variations during the cardiac cycle. This behaviour is in contrast with the simplified mathematical models of Newtonian laminar flow, such as that of Womersley, which forecast simple parabolic profiles of flow. The highest values measured by FRP II in the upper range of shear stress, in respect of the Womersley model, may depend on the "camel's hump" profile which could be detected by FRP II but not by the Womersley model; for a similar Vcl a parabolic profile produces a lower wall shear rate in respect to a camel's hump profile.

In spite of these limitations a good correlation between FRP II-measured shear stress and the Womersley's calculated shear stress was found. The relatively high intraindividual variability of the FRP II measured shear stress could be explained by the lack of echo guidance of the beam position, with a consequent blind angle beam.

The same problem may contribute to the lower correlation between the two methods for the diameter measurement. A further limitation of the FRP II is poor resolution, caused by the low noise, in detecting the lowest velocities near the arterial wall, with consequent underestimation of the vessel diameter.

To overcome the main limitations of the technique an echographic guidance is needed.

CONCLUSIONS

Although FRP II is still a prototype based on the autocorrelation analysis of the ultrasonographic signal, it is a promising instrument in noninvasive measurement of the true vessel wall shear stress.

REFERENCES

1. M. Okano and Y. Yoshida, Endotelial cell morphometry of atherosclerotic lesions and flow profiles at aortic bifurcations in cholesterol fed rabbits, *J. Biomech. Eng.* 114:301 (1992).
2. S. Glagov, C. Zarins, and D. Giddens, Fattori emodinamici e reattività tissutali durante l'aterogenesi nell'uomo, *G. Ital. Cardiol.* 20:1070 (1990).
3. D.L. Fry, Acute vascular endothelial changes associated with increased blood velocity gradients, *Circ. Res.* 22:165 (1968).
4. D.L. Fry, response of the arterial wall to certain physical factors, *Ciba Found. Symp.* 12:93 (1972).
5. A.M. Melkumyants, S. Balashov, and V.M. Khayutin, Endothelium dependent control of arterial diameter by blood viscosity, *Cardiovasc. Res.* 23:741 (1989).
6. A.M. Melkumyants and S. Balashov, Effect of blood viscosity on arterial flow induced dilator response, *Cardiovasc. Res.* 24:165 (1990).
7. R.M. Berne and M.N. Levy, Hemodynamics p. 53, *in:* "Cardiovascular Physiology," C.V. Mosby Company, Saint Louis (1977).
8. K. Morinaga, K. Okadone, M. Kuroki, T. Miyazaki, Y. Muto, and K. Inokuchi, Effect of wall shear stress on intimal tickening of arterially transplanted autogenous veins in dogs, *J. Vasc. Surg.* 2:430 (1985).
9. K. Okadone, T. Yikizane, S. Mii, and K. Sugikachi, Ultrastructural evidence of the effects of shear stress variations on intimal tickening in dogs with arterially transplanted autologous vein grafts, *J. Cardiovasc. Surg.* 31:719 (1990).
10. N.D. Nguyen and A.K. Haque, Effect of hemodunamic factors on atherosclerosis in the abdominal aorta, *Atherosclerosis* 84:33 (1990).
11. K. Perktold and M. Resch, Numerical flow studies in human carotid artery bifurcations: basic discussion on the geometric factor in atherogenesis, *J. Biomed. Eng.* 12:111 (1990).
12. M. Bassini, E. Gatti, T. Longo, G. Martinis, P. Pignoli, and P.L. Pizzolati, In vivo recording of flow velocity profiles and studies in vitro of profile alterations produced by known stenoses, *Texas Heart Institute Journal* 9:185 (1982).
13. M. Bardelli, R. Carretta, G. Di Paolo, D. Dotti, F. Dotti, and R. Lombardi, FRP II wide range low cost blood velocity profiler, *VI Mediterranean Conference on Medical and Biological Engineering*, Capri, July 5-10, 1992.
14. A.C. Simon, J. Levenson, and P. Flaud, Pulsatile flow and oscillating shear stress in the brachial artery of normotensive and hypertensive subjects, *Cardiovasc. Res.* 24:129 (1990).
15. J.R. Womersley, Oscillatory motion of a viscous fluid in a thin-walled elastic tube. The linear approssimation for long waves, *Phil. Mag.* 46:199 (1955).
16. S. Uchida, The pulsating viscous flow superposed on steady laminar motion of ancompressible fluid in a circular pipe, *F. Appl. Math. Phys.* 7:403 (1956).

REAL-TIME EVALUATION OF FLOW BEHAVIOR
BY A NOVEL ULTRASOUND MULTIGATE SYSTEM

P.Tortoli, G.Guidi, F.Guidi, L.Bessi and C.Atzeni

Electronics Engineering Department, University of Florence
Via S. Marta 3, Florence, 50139 - Italy

INTRODUCTION

As known, a pulsed emission of ultrasound energy allows discrimination of both range and velocity of moving targets such as the blood red cells in human arteries and veins. In multigate systems employed in Doppler flow-imaging equipment, the velocities of a number of targets located at different ranges along the ultrasound beam axis are simultaneously detected. The signals related to the various ranges must therefore be independently processed, either in the time or in the frequency domain, at high speed. Typically, in order to reduce the computation efforts, this processing is aimed at extracting a single parameter (e.g., the *mean* velocity) from each signal, even if this signal is known to be quite complex, being the result of the combination of all echoes returning simultaneously from a region in space called *range cell*, and therefore strongly depending on the range cell actual dimensions and position[1].

In this paper, a novel ultrasound multigate instrument is proven capable of providing a detailed analysis of Doppler signals originated from 64 different range cells. By means of a dedicated high-speed FFT processor, these signals are in fact transformed into the frequency domain without loss of information. The processed data are displayed in the form of *spectral* velocity profiles, with velocity on horizontal axis, range on vertical axis and power spectral densities computed from different ranges used to modulate the intensity of pixels in each horizontal line. This display gives immediate information about the spectral composition of all Doppler signals detected from a large region of interest. Experimental results shown here also point out that in general the results of multigate Doppler analysis may be heavily affected by the characteristics of the ultrasound beam transducer.

EXPERIMENTAL SYSTEM

The experimental multigate system accepts at its input the In-phase and Quadrature baseband components of the signal received from a conventional Pulsed Wave Doppler system. For each Pulse transmitted at a PRF rate, 64 samples are digitized with 12-bit resolution. The time distance between these samples, corresponding to the spacing between the range cells, can be changed by the operator in order to match the analysed range to the actual region of interest.

Acoustical Imaging, Vol. 22, Edited by P. Tortoli
and L. Masotti, Plenum Press, New York, 1996

Under control of a TMS320C50 Digital Signal Processor (by Texas Instruments), the digital data are stored in buffer RAMs and processed at high speed in order to provide in real time the complete spectrum of the Doppler signal related to each of the 64 range cells. The computed power spectral densities are sent to a host Personal Computer for real-time display or storage in a file.

INTERPRETATION OF SPECTRAL PROFILES

Fig.1a shows the Doppler spectra obtained by interrogating a steady flow in a 10 mm diameter plexiglass tube. Each horizontal line reports the power spectral density computed from the Doppler signal collected from one of the 64 range cells: since the range cell spacing was here 0.3 mm, a region of about 20 mm starting from a distance of 16 mm with respect to the transducer was analysed. An alternate display mode, used in real-time applications, is shown in Fig.1b. Like in spectrograms shown in commercial Doppler equipment, the power spectral densities are here coded in grey levels according to the look-up table shown on the right, where black corresponds to the maximum spectral power.

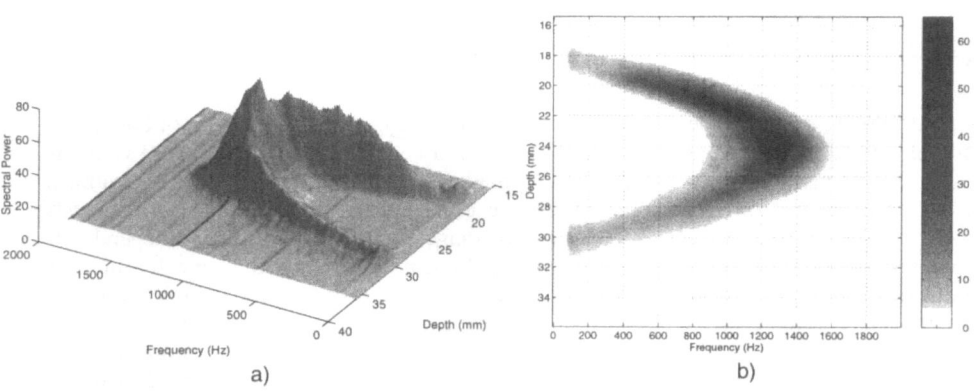

Figure 1 - Spectral profile obtained by the novel multigate Doppler velocimeter. a) 3-D representation. b) grey level representation. Note that in a) the frequency axis has been reversed with respect to b).

This spectral profile was obtained by exciting with short bursts of energy an 8-MHz ultrasound transducer having a 1.2 mm beamwidth approximately constant from 20 mm to 30 mm. The corresponding sample volume dimensions were so small that velocity gradients were limited in all the range cells, and the spectral profile turns out to be narrow at all depths. In particular, its width in correspondence of the maximum velocities (where the velocity gradient is minimum) may be almost completely attributed to geometrical broadening effects while near the walls, it is related to the large velocity gradient[2].

It has been demonstrated that the velocity distribution in the tube can be estimated from the distribution of maximum frequencies in the various range cells. In Fig.1b, the maximum frequency profile, corresponding to the right contour of the spectral plot, is well parabolic, consistent with the laminar flow conditions established with a Reynolds number <800. In particular, in the regions close to the tube walls, it may be appreciated the absence of those blunting effects which are often encountered in profilometers detecting only the mean Doppler frequencies .

However, it has to be pointed out that even in well controlled laminar flow conditions, an unsuitable choice of the ultrasound transducer may lead to results less easily interpretable. For example the distribution of detected power spectral densities shown in Fig. 2 appears quite asymmetrical with respect to the flow axis (located at a depth of approximately 24 mm). To understand the reason of this behavior, it is necessary to take into consideration the ultrasound beam of the transducer which was used in this experiment. Fig.3 shows the beam plots obtained by detecting the power reflected from a cylindrical 0.1 mm diameter graphite reflector located at different distances and lateral positions with respect to the transducer.

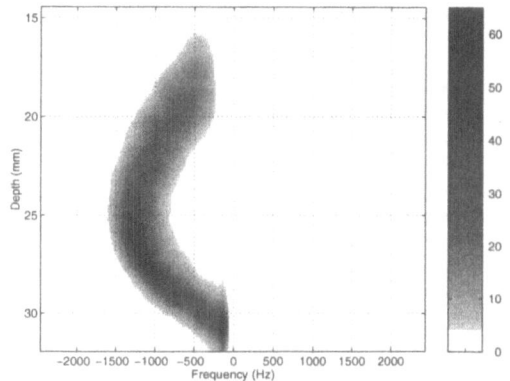

Figure 2 - Spectral profile obtained by insonating a flow receding from the transducer whose beam plots are shown in fig. 3.

These plots show that: a) the maximum energy is detected from the focal depth of about 30 mm (which is consistent with the fact that the spectral profile in fig. 2 is more "black" in the region corresponding to a 30-mm depth); b) the range cell lateral dimensions appreciably change over the region analysed in Fig. 2. In particular, for depths lower than 20 mm the ultrasound beamwidth is so large that the sample volume, while travelling back to the transducer, insists on a same flow line for several millimeters.

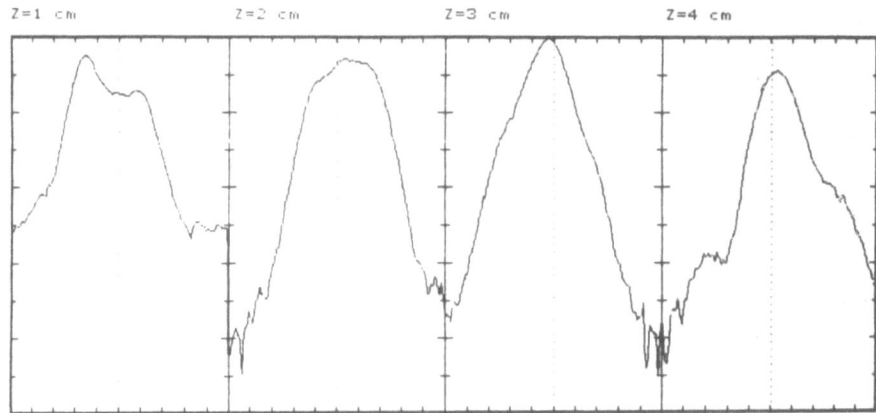

Figure 3 - Beam plots of a 4 MHz transducer, obtained by detecting the power reflected from a cylindrical 0.3 mm diameter graphite reflector located at different positions.

In regions where the ultrasound beam has a width equal to b, the distance between the extremal positions where the range cell intercepts the same flow line, is given by

$\{l+b/\tan\theta\}$, where l represents the sample volume length and θ is the beam-to-flow angle. This is consistent with the result shown in Fig.2, where the same Doppler frequency (i.e. ≈ 300 Hz) is detected from all signals collected from 16 mm to about 20 mm. As shown in figure 4, the mean velocity profile (which is here obtained by computing the mean frequency of each Doppler spectrum) turns out to be quite "distorted" with respect to the expected parabolic behavior, while the profile obtained from maximum frequencies better describes the actual laminar flow conditions.

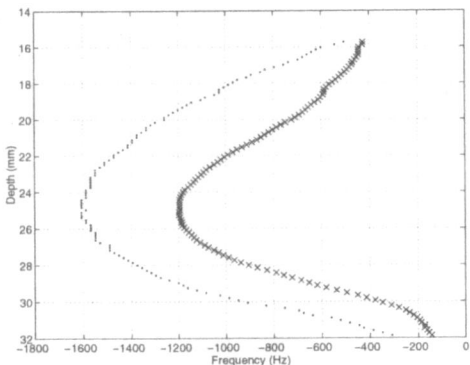

Figure 4 - Mean (crosses) and maximum frequency (dots) profiles detected from the spectral profile shown in Figure 2.

CONCLUSION AND DISCUSSION

The use of spectral analysis for computation of Doppler frequencies at all depths in a multigate instrument has here been demonstrated an useful means to get comprehensive information about the flow under investigation. The actual spectral distribution at different depths has been proven to depend on the flow velocity profile as well as on the characteristics of the transducer ultrasound beam. To avoid possible misleading results, the use of transducers with uniform, narrow bandwidth over the full range of interest is recommended.

Differently with respect to systems computing only the mean Doppler shift produced in the range cells (like, e.g., the ones employed in most color Doppler equipment) our system does not require the use of extremely short range cells. The profiles obtained from maximum frequencies are in fact found to be fairly unaffected by this parameter [3]. The use of relatively long range cells may allow the bandwidth in the Transmitter/Receiver section of the ultrasound system to be correspondingly limited, thus improving the attainable Signal-to-Noise ratio.

REFERENCES

1. D.H. Evans, W.N. Mc Dicken, R. Skidmore, J.P. Woodcock, *Doppler Ultrasound - Physics, Instrumentation, and Clinical Application*, pp. 131-133, John Wiley and Sons, 1989.

2. G. Guidi, V.L. Newhouse and P. Tortoli, "Doppler Spectrum Shape Analysis based on the Summation of Flow-Line Spectra", *IEEE Transactions on Ultrasonic Ferroelectrics and Frequency Control*, Vol. 42, No. 5, pp. 512-522, 1995.

3. P.Tortoli, F.Guidi, G.Guidi and C.Atzeni, "Spectral velocity profiles for detailed ultrasound flow analysis", submitted to *IEEE Transactions on Ultrasonic Ferroelectrics and Frequency Control*.

EFFECTS OF THE SIMULTANEOUS PRESENCE OF GEOMETRICAL AND VELOCITY BROADENING ON THE PULSED DOPPLER SPECTRUM

G. Guidi[1], P. Tortoli[1], and V.L. Newhouse[2]

[1] Electronics Engineering Department, University of Florence
     Via S. Marta 3, Florence, 50139 - Italy
[2] Biomedical Engineering and Science Inst., Drexel University
     Philadelphia, PA 19104 - USA

INTRODUCTION

This paper presents a study of the power spectrum produced by Pulsed Wave (PW) Doppler flowmeters which is based on a novel computational method[1]. With this approach, the spectrum is modelled by summing the single power spectra associated with each flow-line passing through a particular range cell. For a given velocity profile and a known sample volume position, it is possible to take into account the simultaneous presence of velocity gradient spectral broadening, due to the finite dimension of the sample volume, and geometrical spectral broadening related to the focussing properties of the transducer employed. Numerical results corresponding to a number of different operating conditions are presented and compared with the corresponding experimental results obtained by measuring laminar flow produced in a flow phantom.

THEORY

Theoretical studies show that a single line-flow insonified by an ultrasound focussed transducer, produces a triangle like Doppler amplitude spectrum, having mean frequency and base width proportional to the axial and transverse velocity components, v_a and v_t respectively[2].

The production of a triangular spectrum by a single flow line is a phenomenon usually called geometrical broadening. By increasing the velocity of the line flow, the related triangular spectrum moves towards higher frequencies while the bandwidth widens, and the maximum amplitude decreases to keep the total energy unchanged.

In flow analysis performed by means of PW Doppler systems, the detected Doppler signal is generated by all echoes backscattered from a limited Sample Volume (SV), whose length is proportional to the duration of the transmitted pulse, while its width is determined by the ultrasound beamwidth corresponding to that range. The actual distribution of spectral components is therefore affected by the contributions from different velocities within the range cell (velocity gradient broadening), each one intrinsically affected by the geometric broadening.

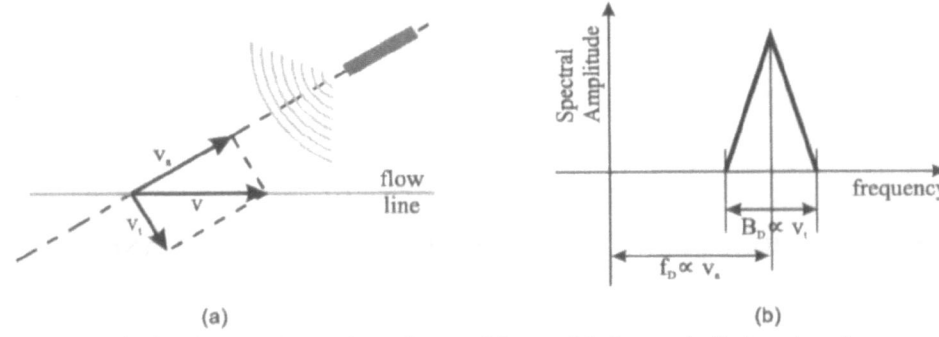

Figure 1 - (a) Line flow moving under a focussed beam. (b) Geometrically broadened spectrum produced by the line-flow.

The problem of evaluating spectral changes due to the beam geometry and to the velocity profile shape, have been previously addressed by Bascom et al. using an analytical approach[3]. In this paper, we present the application of a simple model for evaluating, directly in the frequency domain, the shape of the Doppler power spectrum produced in PW Doppler flowmeters. For a known range cell geometry, the PW spectrum is obtained by adding all the spectral contributions due to the flow lines contained in it. Hence, the broadening of the whole spectrum is seen as the summation of geometrically broadened line spectra associated with each velocity. According to this approach, geometrical broadening, originating from focussing properties of the ultrasound transducer, and velocity broadening, due to the presence of scatterers moving at different velocities within the range cell, are simultaneously considered.

By means of a computer program, we have implemented this simulation model which takes into account flow velocity profile, insonation angle, range cell dimensions, transducer geometry and focussing properties. Since each flow line (i.e. the source of the single element of the Doppler power spectrum) has been modelled as a continuous sequence of ultrasound scatterers, each one on the same line parallel to the flow direction, the influence of phase difference on the spectrum has been ignored.

By changing some controllable parameters, we have simulated the effects these exert on the final spectrum. Correspondingly, real flow experiments were made to obtain comparable measured spectra.

MATERIAL & METHODS

To evaluate the behavior of the Doppler power spectrum, a PW flowmeter involving an 8 MHz circular transducer with known beam pattern (aperture width over focal length ratio W/F=0.18) was used on a flow phantom. The flow phantom was made with an upper reservoir from which a test fluid flows to a lower reservoir through a plexiglass tube of 10 mm internal diameter, immersed in a tank filled with water. The test fluid in the phantom was a mixture of 70% water and 30% glycerol, with Sephadex particles to produce scattering. The fluid velocity was set at 0.4 m/s; a value low enough to have a Reynolds number lower than 2000, which produces laminar flow with a parabolic velocity profile[4]. The beam-to-flow angle was mechanically set at an arbitrary value of 65°.

During the experiments the range cell was positioned in several different locations and was changed in length. For each measurement condition the Doppler signal was digitized at 8 Ksample/sec using a DSP acquisition board with 8 bit A/D converters, installed in a PC compatible equipped with a 486 DX2 running at 66 MHz. It was then processed with a specifically developed C++ program running on the same PC. This program moves a sliding window of 512 samples over the entire acquisition period of about 65000 complex samples, with a 50% overlap factor. On each time window it gives a Hamming weighting to the signal and performs an FFT. The final spectral estimation of the measured signal is obtained by averaging 200 subsequent FFT analyses[5]. A simulation was run, corresponding to each experimental spectrum. The number of flow lines contributing to the spectrum formation, were chosen equal to 400. The computing time on the PC, varied from few minutes to one hour dependingly on the number of flow lines intercepted by the sample volume.

EXPERIMENTAL RESULTS AND DISCUSSION

To qualitatively check the validity of the approach, pairs of measured spectra and simulations have been normalized and superimposed. In each measurement, by ignoring geometrical broadening effects, one would expect a rectangular spectrum where each velocity component gives the same contribution. The expected rectangular shape is modified by the amplitude and shape of each spectrum associated with the flow-lines intercepted by the SV. Since the energy reflected by each flow-line remains constant independently of its velocity, higher velocities produce wider and lower power spectra.

In the first case the Doppler signal was produced by a 2 mm SV overlapped to a wall of the phantom tube, as sketched in fig. 2a. The above mentioned behavior of each spectral contribution is here exemplified for three flow lines intercepting the sample volume at slightly different depths. Due to the single flow-lines spectra behavior, the whole computed spectrum, illustrated by the dotted line in figure 2b, starts with a maximum at zero frequency, falling gradually as the frequency increases. This is because, having a roughly constant velocity gradient inside the SV, the spectral contributions are more or less uniformly distributed from zero up to the maximum velocity present in the sample volume, and their summation gives a spectrum approximately equal to the envelope of the different contributions shown here by dashed lines.

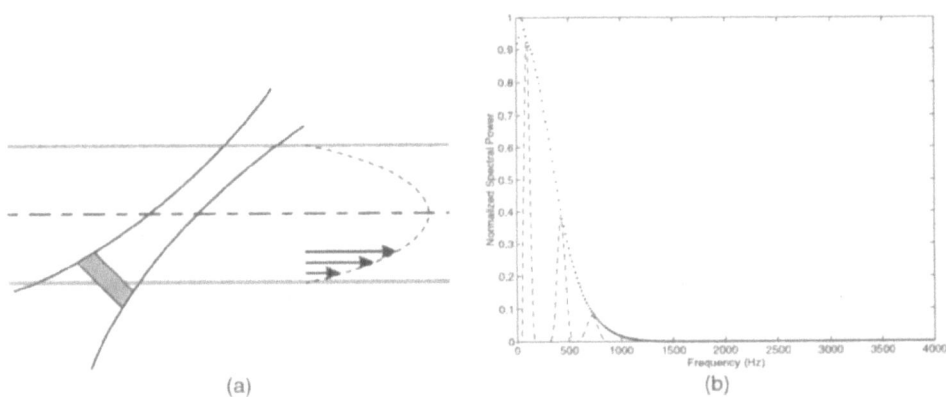

Figure 2 - Range cell position (a) giving the simulated spectrum shown in (b) with a dotted line. The dashed lines in (b) evidence three of the elementary spectral contributions produced by equally spaced flow lines.

On the other hand, in a SV intercepting the tube axis, as shown in figure 3a, the situation is reversed. For a sample volume of the same length as the one considered in fig. 2a, most of the intercepted flow lines, travel at a velocity close to the maximum, while a small number of them travel at slower velocities. It is due to the presence of a linear velocity gradient into the same sample volume. Therefore most of the spectral energy is produced by fast flowing lines, giving a spectrum with a gradual rise and a steep fall. This factor engenders an asymmetrical spectrum, wider than the symmetrical one expected of a single flow line (shown in figure 3b by the dashed line), and characterized by a "tail" on the lower frequencies side. The comparison of measured and computed spectra highlights similarities between the two behaviors, excluding the presence of noise and spectral variance that have not been taken into account during the simulation.

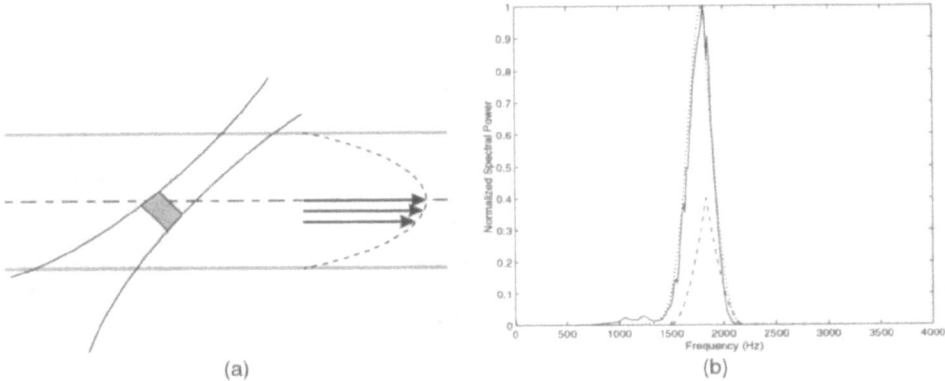

(a)   (b)

Figure 3 - Range cell position (a) giving the Doppler power spectra shown in (b). The solid line represents the measured spectrum, the dotted line shows the computed one. The dashed line indicates the symmetrical spectrum theoretically expected of the maximum velocity flow line (no velocity gradient).

As a result, for a given SV length, the enlargement of the Doppler spectrum near the tube wall can be mainly attributed to the velocity gradient, while in the center of the tube it is mostly (but not exclusively) due to geometrical effects.

In figure 4 are reported comparisons between simulated and measured spectra, seen to occur during an "in-vitro" experiment, which confirm the the previous considerations. A small SV was first located on the far wall of the tube, and afterwards its length was extended progressively, up to cover the lower half of the tube. Going from fig. 4a through to fig. 4f it is possible to observe the step-by-step movement of the spectrum barycentre from the left (lower) end to the right (upper) end, as the sample volume is enlarged. This is due to the gradual increase of the faster velocities in the sample volume boosting the weight of the upper frequencies' components.

To avoid the influence of long term flow phantom instability, the spectrum corresponding to each SV length was obtained averaging only 200 instantaneous spectra, therefore a relatively strong residual spectral variance is present in all the six diagrams composing figure 4. Also, the measured spectra have a lack of information that corresponds to the 0÷125 Hz clutter filters' rejection span. However, despite the above mentioned effects, in the remainder of the analyzed frequency range, the computed spectra are in accordance with the experimental ones.

CONCLUSION

The most advanced ultrasound flow imaging systems make use of pulsed emission of energy in order to obtain spectral sonograms of a particular range cell of an explored area (duplex scanner), or a color coded map of the flows present in the whole B-mode area (color Doppler).

The modifications in the Doppler spectrum shape intrinsically involved in the PW flowmeters working principle, bias the estimation of the diagnostically useful spectral parameters like mean or peak frequency. Furthermore, the Doppler spectrum broadening is

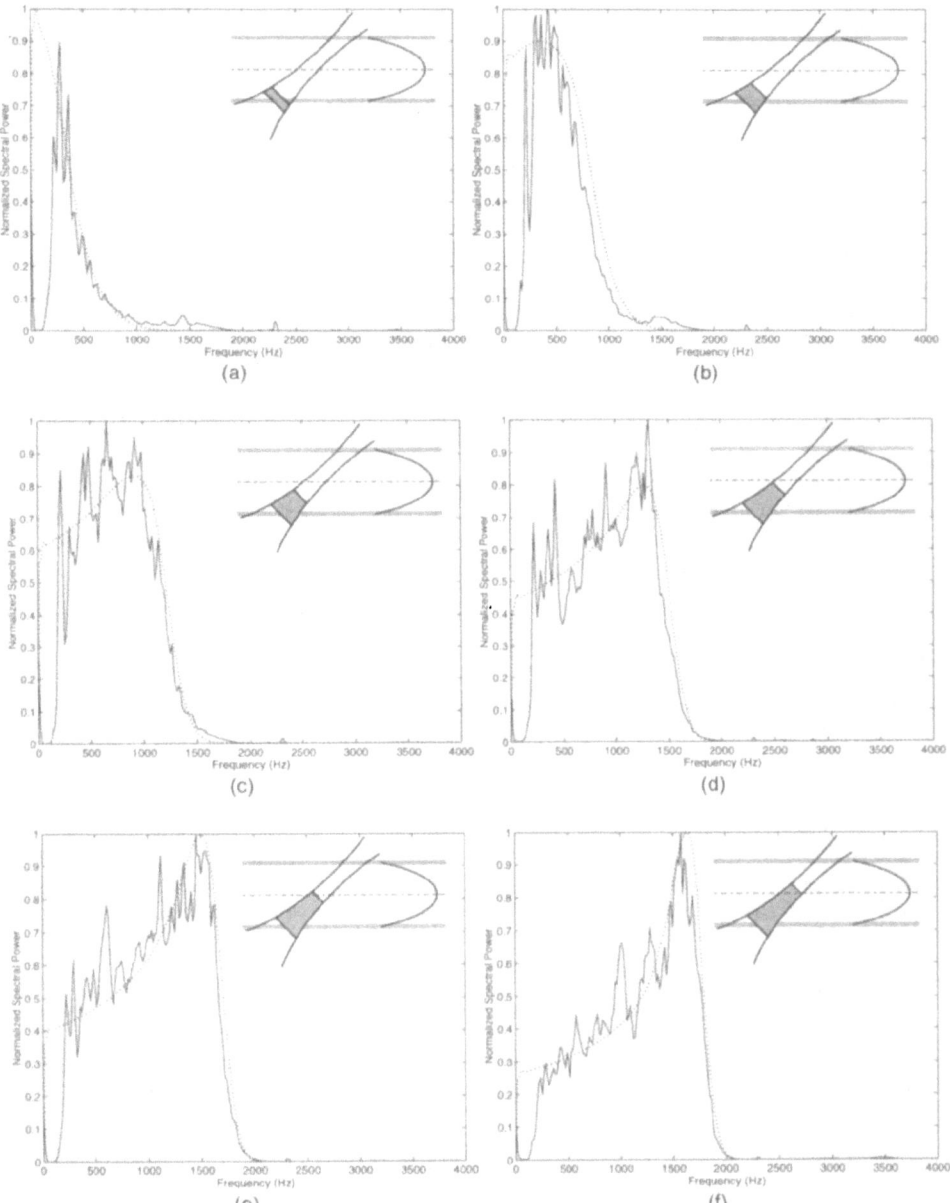

Figure 4 - Comparison between simulated and measured spectra obtained locating a 1mm range cell on the far wall of a flow phantom (a), and then enlarging the sample volume in steps of 1mm up to encompass all the lower half of the velocity profile (f). The dotted lines represent the computed spectra and the continuous lines the measured ones.

a parameter usually employed in clinical applications to evaluate the turbulence of a flow, from which the width of a stenosis can be estimated.

Hence, it would be extremely useful to discern between the unavoidable spectral broadening inducted by the PW measurement technique and the "informative" spectral broadening produced by the velocity gradient, when these are mixed.

The purpose of this work was to study, limited to the case of laminar flow, the concurrent effect on the shape of the whole PW Doppler spectrum of the two main broadening factors, usually addressed separately; the velocity gradient in the sample volume and the convergence of a focussed ultrasound beam.

The experimental results shown demonstrate that, once the focussing properties of the utilized transducer (i.e. the individual flow-line spectra), and the flow velocity profile are known, it is possible to predict qualitatively the shape of the Doppler spectrum, and its enlargement, due to the simultaneous presence of geometrical spectral broadening and velocity gradient spectral broadening.

Despite of the incidence of the Doppler signal phase due to different scatterer positions was here completely neglected, the good agreement between experimental measurements and associated numerical simulations, suggests a practical independence from the phase of the ensemble averaged spectra, used in our experiments to estimate the Doppler spectrum. However, a deeper study of the contribution of the phase to spectral broadening, also considering the work carried out by other authors[6], is planned.

A possible future development of this study is the extraction from the Doppler spectrum of a velocity histogram by eliminating the geometrical part of the whole spectral broadening, allowing in such way the detection of real flow disturbances involving an increase in the velocity gradient, rather than geometrical broadening artifacts.

REFERENCES

1. G. Guidi, V.L. Newhouse and P. Tortoli, "Doppler Spectrum Shape Analysis based on the Summation of Flow-Line Spectra", *IEEE Transactions on Ultrasonic Ferroelectrics and Frequency Control*, Vol. 42, No. 5, pp. 512-522, 1995.

2. D. Censor, V.L. Newhouse and T. Vontz, "Theory of ultrasound Doppler-spectra velocimetry for arbitrary beam and flow configurations" *IEEE Transactions on Biomedical Engineering"*, vol. 35, pp. 740-751, 1988.

3. P.A.J. Bascom and R.S.C. Cobbold, "Effects of transducer beam geometry and flow velocity profile on the Doppler power spectrum: a theoretical study", *Ultrasound in Medicine and Biology*, Vol. 16, No. 3, pp. 279-295, 1990.

4. D.H. Evans, W.N. Mc Dicken, R. Skidmore, J.P. Woodcock, *Doppler Ultrasound: physics, instrumentation and, clinical applications*, pag. 10, John Wiley and Sons Ltd., 1989.

5. P.D. Welch, "The use of fast Fourier transform for the estimation of power spectra: a method based on time averaging over short, modified periodograms", *IEEE Transactions on Audio Electroacoustics*, vol. AU-15, pp. 70-73, 1970.

6. D. Censor, M.D. Fox, E. Sonnenschein, "Microturbulences as the driving force for ultrasound Doppler signals and spectra irregularities: a computer simulation", *22nd International Symposium on Acoustical Imaging*, 1995.

MICROTURBULENCES AS THE DRIVING FORCE FOR ULTRASOUND DOPPLER SIGNALS AND SPECTRA IRREGULARITIES: A COMPUTER SIMULATION

Dan Censor[1], Martin D. Fox[2], and Elazar Sonnenschein[1]

[1]Department of Electrical and Computer Engineering
Ben Gurion University of the Negev
Beer Sheva, Israel, 84105
[2]Department of Electrical and Systems Engineering
University of Connecticut
Storrs, Connecticut, 06269

INTRODUCTION AND STATEMENT OF THE PROBLEM

Ultrasound Doppler velocimetry for medical diagnostics is based on displays of spectra derived from time signals due to acoustical waves backscattered by random collections of moving particles. Those spectra are notoriously very noisy, and accurate velocity measurements are usually difficult if not outright impossible. Models describing the stochastic signals have been suggested in the literature, e.g., by Mo and Cobbold (1986). Such models account for the interference produced by the random phased time signals. However, it is very well known that the often encountered non-steady, nonuniform, sometimes turbulent flows are conducive to randomly fluctuating spectra as well, causing also spectrum broadening. While the presence of a superposition of randomly phased time signals, originating from randomly located scatterers, adequately explains the fluctuations in the spectrum, giving it its typical "jagged" or "ridged" appearance, broadening is mainly due to Doppler frequency shifting, introduced by the motional modes mentioned above, whose analysis requires additional considerations. These broadening phenomena are manifested whenever additional spurious Doppler type frequency shifts occur, even without the presence of a fully developed turbulent flow regime. We therefore refer to them generically as microturbulences. While the statistical nature of the fluctuating phase signals and the associated spectra can be satisfactorily modeled, there is only a cursory understanding as to how these typically ridged spectra are related to the characteristics of the microturbulent flow in question. This problem might be very important for medical diagnostics, because the resulting displays might look similar in cases of turbulence due to stenoses, and cases of regurgitation or retrograding flows.

For a single, uniformly moving particle, the relation of the spectrum to the configuration parameters is well known, e.g., see Newhouse et al. (1987), and Censor (1988a). On the other hand, analyses involving other motional modalities usually ignore the beam structure. See for example Gimenez et al. (1990), also addressing a recent controversy (Censor, 1972;

Rogers, 1973; Censor, 1988b; Piquette et al., 1984; Piquette et al., 1988). In order to resolve some of the controversial issues, and provide a tool for compounding the effects of arbitrary transducer and flow configurations, Sonnenschein (1991) created an adequate computer simulator. The program turned out to be too slow to allow for simulation of an ensemble of many particles. Hence in order to test ensembles of particles, the transducer model must be kept as simple as possible.

One aspect of paramount importance seems to be missing in existing theories: It is well known that when a large ensemble of scattering particles is present, then the scattering is dominated either by a coherent intensity, proportional to the square of the number of scatterers, or, if the coherent intensity vanishes, by incoherent intensity proportional to the number of scatterers (Yu et al., 1988; Censor and Newhouse, 1988), the phase information gradually disappearing, finally leading to a power spectrum based on the sum of the contributions of the individual particles. This simple observation is the key to getting ever increasing incoherent signals when the particle density increases to the point where the scattering medium becomes uniform. The simulations we performed suggest that the jagged spectra are due not only to the simple location statistics of the particles, but to additional irregularities induced by microturbulences.

The present investigation uses the powerful "Mathematica" application for simulating various flow patterns, in an attempt to give better answers, which might lead to design improvement of future instrumentation.

A gallery of computed simulations will be presented and discussed. The results support the idea that microturbulence activity is important for the understanding of Doppler ultrasound spectra.

CONFIGURATIONS AND VARIOUS EXPERIMENTS

To save computation time, the transducer configuration has been assumed to be a flat, uniformly excited, focused, long strip transducer, as described for example by Newhouse et al. (1987). A Sinc function insonified region exists in the focal plane. A boxcar function was superimposed, in order to retain three sidelobes on each side of the main lobe.

In the simplest flow model (henceforth "random phase"), the individual particles are introduced into the insonified region at a typical random delay time on the order of 3 transit time durations. In addition, a random phase factor is assigned to each particle, modeling a random offset distance from the focal plane. The operating frequency, particle velocity and transducer dimensions are typical of existing systems.

In the next model ("phase wobbling") the phase of the signal is randomly changed at each sampling point. This is of course arbitrary, depending on how many sampling points one chooses, but it brings out the essentials of a situation, where the offset distance of the particle from the focal plane is perturbed as the particle moves. Note carefully that in this model the particle, so to speak, "jumps" from one distance to another without changing velocity. Results are displayed for various parameters of the phase wobbling range.

In order to simulate different velocities along the path of motion, transversally with respect to the main beam axis, the amplitude of the scattered signal is randomly perturbed at each sampling point. This model ("random bunching") corresponds to the bunching and unbunching effects due to different velocities, with particles converging and diverging along their path. Once again the bunching is attained without changing the velocity, hence as for the phase wobbling case, no genuine Doppler frequency changes are involved. Results are displayed for various values of bunching efficacy.

Finally, a microturbulence model was simulated, in which the velocity component along the main beam axis was perturbed. This effect, occurring in the frequency domain, gave rise to true Doppler frequency shifts and ensuing spectrum broadening. Here again results are displayed for various values the velocity perturbation along the main beam axis.

In all cases, we are showing spectra of ensembles of 1, 3, and 300 particles. The displays depict the absolute value, i.e., the square root, or the amplitude, of the power spectrum in question, subsequently referred to as the coherent amplitude spectrum.

The Random Phase Model

Figures 1-3 show spectra for ensembles of 1, 3, and 300 particles. Typically, the increasing number of particles does not lead to any spectral broadening. The time signals interfere, thus leading to characteristically ridged spectra. Very roughly speaking, the maximum amplitude is on the order of the square root of the number of particles, supporting the idea that we are dealing with coherent radiation. The one particle case, Fig. 1, is the trivial case for which a triangular spectrum is obtained (Newhouse et al., 1987).

The Phase Wobbling Model

Fig. 4-6 show spectra for ensembles of 1, 3, and 300 particles for severe phase wobbling, the phase changing in the range 0-π. The severe wobbling creates a noisy spectrum, but the "triangular" Doppler spectrum of Fig. 1 can be picked out of the noise. Figures 7-9 depict spectra for slight phase wobbling in the range 0-0.2π. Clearly there exists no broadening effect, only a noisy spectrum due to the random discontinuities introduced in the time signal. In the case of slight phase wobbling the noise is smaller, but it is also observed that the relative noise amplitude off the "triangular" Doppler spectrum is reduced as the number of particles increases. A more severe case (not shown here), where the range of phase wobbling is 0-2π, leads to a spectrum in which the Doppler spectrum is completely drowned in the noise.

The Random Bunching Model

Figures 10-12 depict spectra for many discrete levels of the amplitude corresponding to bunching events. This simulates particles whose velocity along their path is perturbed. Ensembles of 1, 3, and 300 objects are shown. For one object this is somewhat artificial, but shows how the spectrum deteriorates. For all ensembles it is clear that the central part of the Doppler spectrum stands out of the noise. Also note how the signal/noise improves as the number of particles increases. Figures 13-15 show the same, but here the number of discrete amplitude levels is smaller, and the noise is therefore higher, and does not smooth out as much as before.

The Microturbulences (Random Velocity) Model

Finally, we show ensembles of 1, 3, and 300 particles with velocity perturbations along the axis, i.e., normal to the focal plane. These are real Doppler broadening phenomena, and depend on the magnitude of the perturbations. Figures 16-18, for low velocity show almost no broadening. For moderate velocities, figures 19-21, and for high velocities, figures 22-24, it is clearly seen how spectrum broadening is effected. It is also noted that in the absence of phase wobbling and bunching, the broadening is independent of the number of particles.

Figures

Figure numbers correspond to the text above.

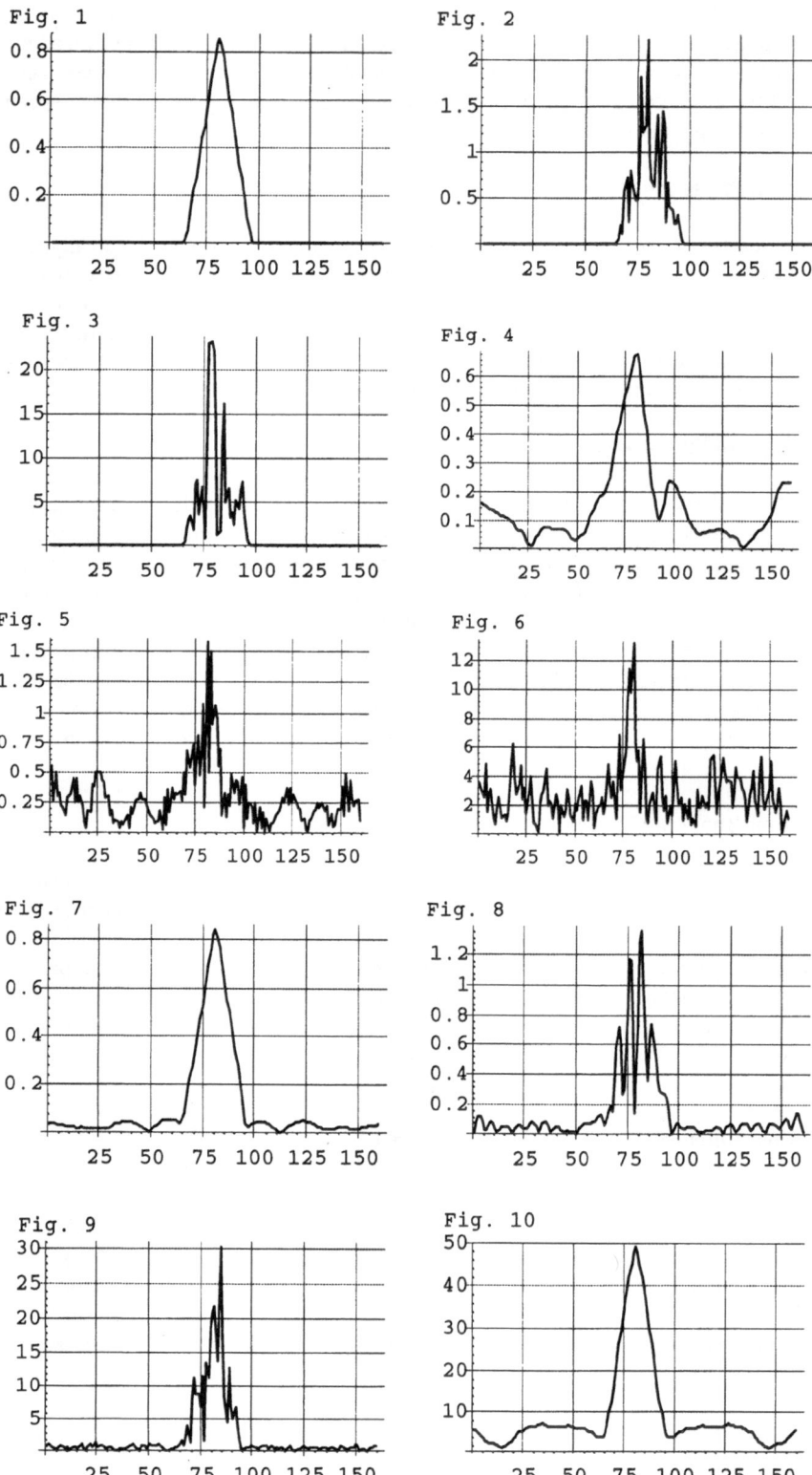

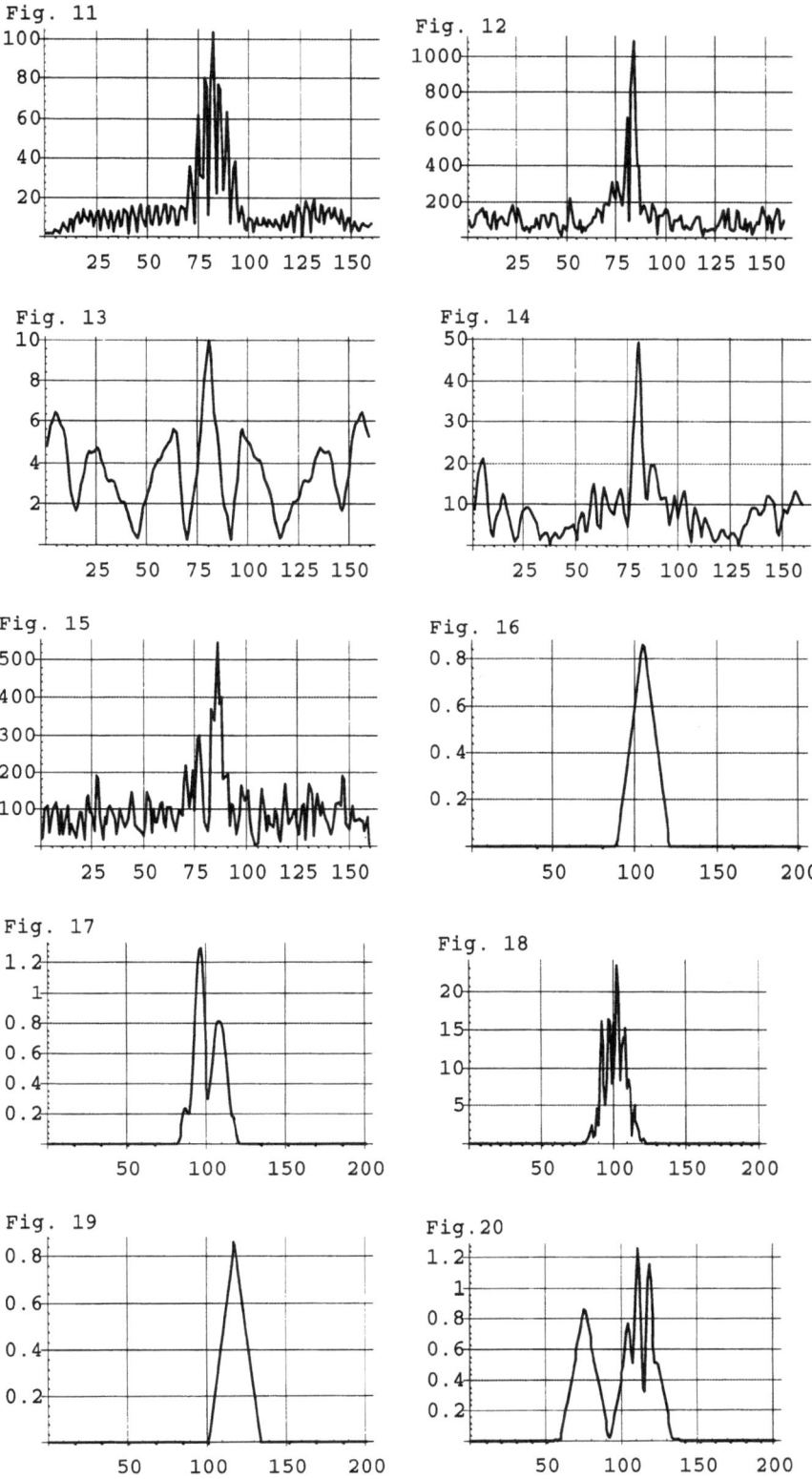

Fig. 11
Fig. 12
Fig. 13
Fig. 14
Fig. 15
Fig. 16
Fig. 17
Fig. 18
Fig. 19
Fig. 20

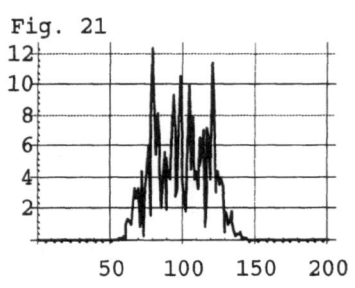

Fig. 21

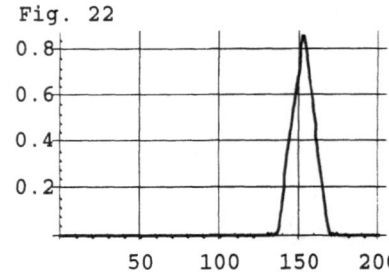

Fig. 22

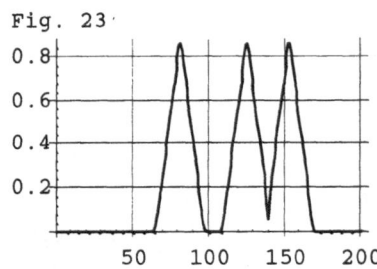

Fig. 23

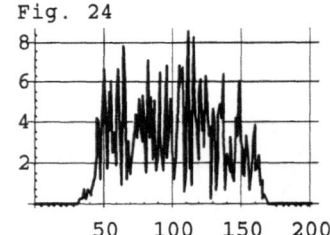

Fig. 24

REFERENCES

Censor, D., "Scattering by time varying obstacles", J. of Sound and Vibration, *25*, 101-110, 1972.

Censor, D., Newhouse, V.L., Vontz, T., and Ortega, H.V., "Theory of ultrasound Doppler spectra velocimetry for arbitrary beam and flow configurations", IEEE Transactions on Biomedical Engineering, *BME-35*, 740-751, 1988a.

Censor, D., "Acoustical Doppler effect analysis-is it a valid method?", J. Acoust. Soc. Am., *83*, 1223-1230, 1988b.

Censor, D., and Newhouse, V.L., "Generalized Doppler effect: Coherent and incoherent spectra", J. Acoust. Soc. Am., *83*, 2012-2019, 1988.

Mo, L.Y.L., and Cobbold, R.S.C., "A stochastic model of the backscattered Doppler ultrasound from blood", IEEE Trans. *BME-33*, 20-27, 1986.

Newhouse, V.L, Censor, D., Vontz, T, Cisneros, J.A., and B. Goldberg, "Ultrasound Doppler probing of flows transverse with respect to beam axis", IEEE Transactions on Biomedical Engineering, *BME-34*, 779-789,1987.

Piquette, J.C., and Van Buren, A.L., "Nonlinear scattering of acoustic waves by vibrating surfaces", J. Acoust. Soc. Am., *76*, 880-889, 1984.

Piquette, J.C., and Van Buren, A.L., and Rogers, P.H., "Censor's acoustical Doppler effect analysis - is it a valid method?" (also citing further communications), J. Acoust. Soc. Am., *83*, 1681-1682, 1988.

Rogers, P.H., "Comments on scattering by time varying obstacles" (and reply), J. of Sound and Vibration, *28*, 764-768, 1973.

Sonnenschein, E., "Computer simulations of ultrasound pulsed Doppler spectra for various source configurations and arbitrary velocity fields", Thesis for the Degree of M.Sc., Ben Gurion University of the Negev, Beer Sheva, Israel, 1991.

Yu, G., Newhouse, V.L., and Censor, D., "On coherent radiation scattered by random ensembles", J. of Sound and Vibration, *122*, 399-412, 1988.

DEVELOPMENT AND IN-VITRO EVALUATION
OF A DOPPLER VOLUME FLOW METER

J. Steck, J.P. Rupp and R.M. Schmitt

Fraunhofer Institute Biomedical Engineering
Dept. of Ultrasound

INTRODUCTION

Determination of blood volume flow (e.g. cardiac output) is very useful in the treatment of low birthweight neonates (< 1500 g). The clinically accepted method employs thermodilution. This, however requires central catheterisation. This treatment is invasive and therefore only acceptable in critical situations. For early detection of impending circulatory failures a non-invasive, accurate, reproducible monitoring method is required. Quantitative measurements of volume flow by means of Doppler ultrasound is a possible solution. To determine blood volume flow by ultrasound Doppler technique the following parameters are required: Doppler angle, beam profile, shape of the velocity profile and vessel size. Since this information is not available or inaccurate volume flow measurements are still not satisfactory. Hottinger and Meindl presented a new method in 1974, the attenuation-compensated volume flow meter[1]. A few papers dealing with this technique were presented so far[2-4]. A commercial system also has been developed but is no longer on the market probably because of unsatisfactory results. Since from the physical point of view the method is attractive it has been illucidated by a new development.

BASIC PRINCIPLE

Two different Doppler sample volumes (SV) have to be generated. The first SV (A_p) is the large one covering the entire lumen of the vessel and it is called the wide beam. The illumination of the vessel should be uniform. The second sample volume A_n is a small disc which must lie totally within the lumen. This one is called the narrow beam. Figure 1 shows the two sample volumes.

The following equations show the physical properties of the model. From the basic formulas

$$\dot{Q}(t) = \int \vec{v}(t, \vec{r}) \, \vec{n} dA,$$

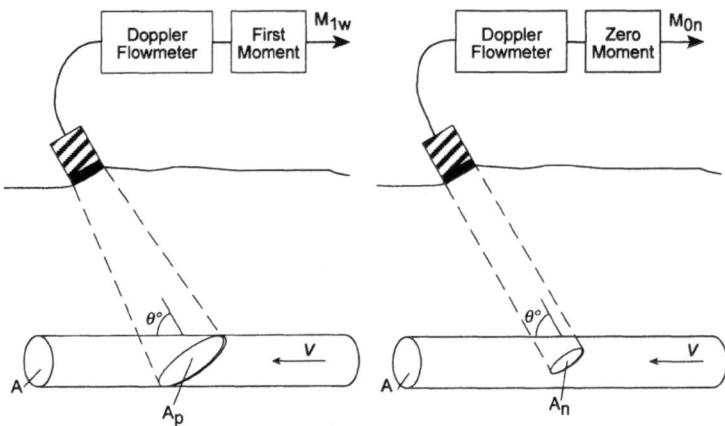

Figure 1. Required Doppler sample volumes

$$M_{0w} = \int S_w(f)\,df = T(z)\,\eta\,I_w(z)\,A_p,$$

$$M_{0n} = \int S_n(f)\,df = T(z)\,\eta\,I_n(z)\,A_n \text{ and}$$

$$M_{1w} = \int f S_w(f)\,df = T(z)\,\eta\,I_w(z)\,\frac{2}{\lambda}\dot{Q},$$

the following expressions are derived:

$$\dot{Q} = K(z)\,\frac{\int f S_w(f)\,df}{\int S_n(f)\,df} \quad \text{with}$$

$$K(z) = \frac{I_n(z)}{I_w(z)}\,A_n(z)\,\frac{\lambda}{2}$$

M_{ij} denotes the i-th Moment of either the wide (j=w) or narrow (j=n) beam, S_j the power spectrum of the beam, f the frequency, λ the wavelength in the medium, T the damping in tissue, η the backscatter coefficient of blood, I_j the sensitivity of the transducer for the corresponding beam and A_j the section of the beam with the vessel.

The fraction of the first moment of the wide beam and the zero moment of the narrow beam is proportional to the flow rate. The coefficient of proportionality is only dependent on system parameters like intensities of the beams, sample volume area of the narrow beam and the wavelength in the medium. It can be computed if these parameters are known or determined experimentally.

TECHNICAL REALISATION

The front end of the system is a four element concentric annular array with a centerfrequency of 4 MHz. Figure 2 shows on the left side the aperture of the single elements of the annular array and on the right side the application to the measurement of cardiac output through the suprasternal notch. The hardware for weighting and phase correction of the single elements is followed by a bidirectional pulsed Doppler System. The whole system is controlled by a PC. Figure 3 shows the block diagram of the system.

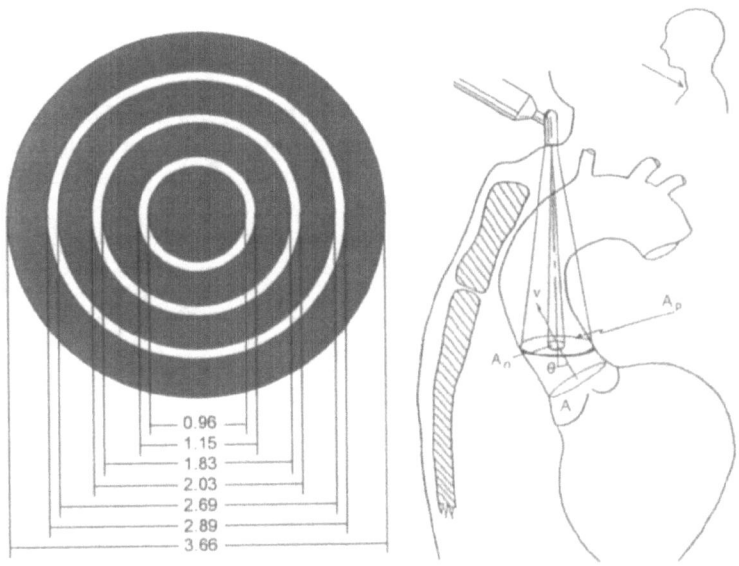

Figure 2. Single element diameters (in mm) of the annular array (left side) and application of the system (right side)

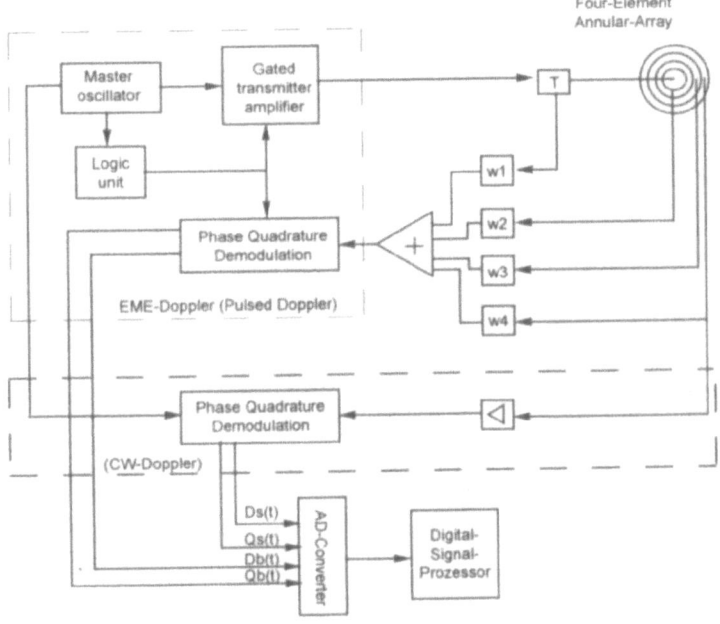

Figure 3. Block diagramm of the system

427

EXPERIMENTAL RESULTS

Calculated and measured beam patterns

Figure 4 shows the calculated and measured beam patterns for the narrow and the wide beam. The beam width of the wide beam is 14 degrees in both cases whereas the beamprofile is more uniform in the measured case. The beam width of the narrow beam is 2 degrees.

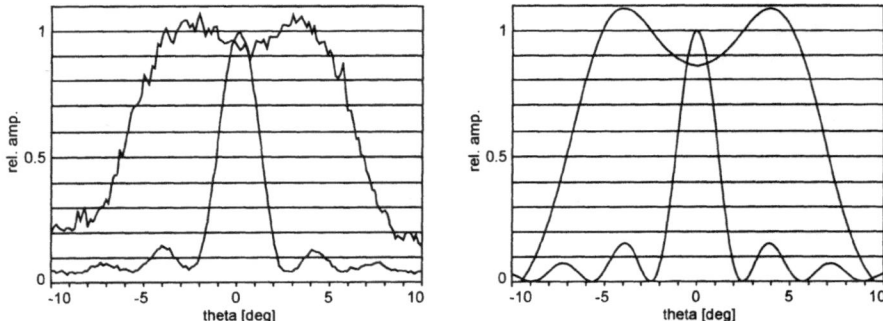

Figure 4. Measured (left) and calculated(right) beam patterns for the narrow and the wide beam

Doppler spectra of steady flow

According to the theory of steady laminar flow the spectrum of the wide beam (left side of figure 5) is rectangular shaped in the interesting range of frequency. The spectrum of the narrow beam (figure 5 right) exhibits higher amplitudes in the upper frequency range, which is due to the position of the sample volume in the center of the vessel.

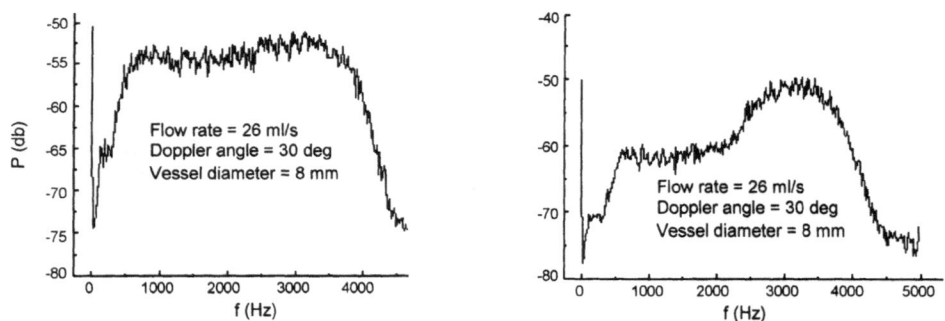

Figure 5. Doppler spectra of the wide (left) and narrow beam (right)

Measurement of steady flow

The measurements with steady flow shows encouraging results in determining real flow with the system developed. Figure 6 shows the results with a steady flow variing in the range from 5 to 40 ml/s. The regression coefficient r and the slope m, which is equal to 1/K, is also shown.

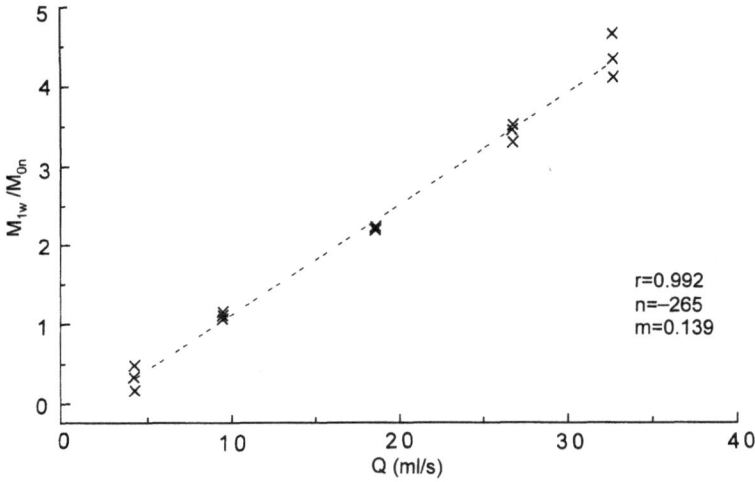

Figure 6. Measurement of steady flow

Angle independance

The angle independance of the system was verified with the previously determined proportionality factor K. The difference between measured and real flow was less then 5% in the range from 20 to 40 degrees. Presently the experimental setup does not allow angles less than 20 degrees. The variance at 45 degrees is due to the fact, that the wide beam does not cover the whole lumen of the vessel in this position. Therefore the volume flow is underestimated.

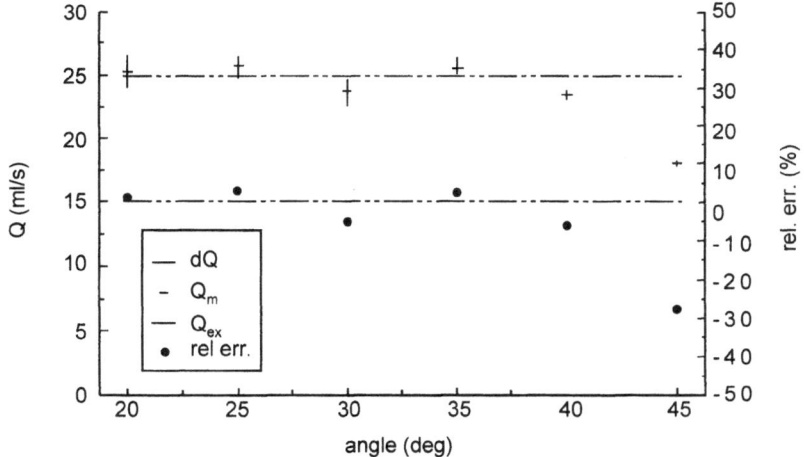

Figure 7. Doppler angle independance

CONCLUSIONS

- the transducer developed in combination with an experimental phased array electronic system is able to produce the required beam patterns in an excellent manner
- in vitro steady flow experiments show anticipated results in an angular range from 20 up to 40 degrees
- the comparison with a commercial pulsed Doppler system has shown an encouraging improvement in quantification of volume flow
- further experimental evaluation will focus on pulsed and turbulent flow

REFERENCES

1. C.F. Hottinger and J.D. Meindl, Blood flow measurement using the attenuation-compensated volume flow-meter, Ultrasonic Imaging 1, pp. 1-15 (1979).
2. C.F. Wippermann et al., Determination of cardiac output by an angle and diameter independant dual beam Doppler technique in critically ill infants, British Heart Journal 67, pp.180-4 (1992).
3. J.M. Evans, R. Skidmore, N.P. Luckman and P.N.T. Wells, A new approach to the noninvasive measurement of cardiac output using an annular array Doppler technique–I. Theoretical considerations and ultrasonic fields, Ultrasound in Medizine and Biology 15, pp.169-78 (1989).
4. J.M. Evans, R. Skidmore, J.D. Baker and P.N.T. Wells, A new approach to the noninvasive measurement of cardiac output using an annular array Doppler technique–II. Practical implementation and results, Ultrasound in Medizine and Biology 15, pp.179-87 (1989).

MEASUREMENT AND IMAGING OF A VELOCITY VECTOR FIELD BASED ON A THREE TRANSDUCERS DOPPLER SYSTEM

G. Bruni[1], M. Calzolai[1], L. Capineri[1], A. Fort[1], L. Masotti[1],
S. Rocchi[2], and M. Scabia[1]

[1]Dipartimento di Ingegneria Elettronica, Università di Firenze
 Via di S. Marta 3, 50139, Firenze, Italy
[2]Facoltà di Ingegneria, Università di Siena
 Via Roma 77, 53100, Siena, Italy

INTRODUCTION

A strong interest is growing in improving the diagnostic capabilities of current ultrasonic Doppler systems. In particular a consistent research effort is directed toward the development of new Doppler techniques capable to estimate the three-dimensional (3-D) velocity of blood flow.

In principle a pulsed (PW) 3-D Doppler system can estimate the magnitude and direction of the real velocity within a sample volume: therefore it is inherently more accurate than monodimensional Doppler techniques where only the projection of the velocity on the ultrasound beam axis can be measured. Thus with monodimensional techniques, the assessment of velocity requires the knowledge of the angle between the investigating beam and the flow direction. Obviously such angle can be only indirectly determined and with low accuracy, leading to errors in the velocity estimate up to 20÷30%.

In this context we developed a prototype system for the assessment of 3-D velocity profiles with PW Doppler technique employing three confocal transducers. The underlying idea is the straightforward application of the vector theory, by which an unknown vector can be completely determined by its projections along three independent directions, represented here by the three transducers beam axes. The work presents the basic theory and preliminary experimental results. The novelty of this approach, with respect to others reported in the literature [1, 2, 3], is the realization of a 3-D acquisition system working in PW, that it is essential to scan volumes of vessels and to obtain the 3-D morphology of blood flow.

Finally a tool for imaging a 3-D velocity vector field is proposed. Quantitative 3-D data are presented on different parallel sections of the vessel, by using color codes and arrows to represent respectively the magnitude and the orientation of the velocity vectors. User interactive software has been developed for this purpose in a MS-Windows environment.

THEORY

In this paragraph the theory underlying the quantitative estimation of a 3-D velocity field is briefly described. We assume an elementary scatterer with a velocity vector $\underline{v} = (v_x, v_y, v_z)$ (see Fig. 1) and a probe with four confocal transducers whose beams are directed along fixed directions $\underline{i}_0$, $\underline{i}_1$, $\underline{i}_2$, $\underline{i}_3$. As indicated in Fig. 1, the central transducer operates as a transmitter (Tx), while the lateral ones work as receivers (Rx). In theory, the 3-D velocity estimation could be carried out using only three transducers; the advantage of using a fourth transducer is to minimize the errors due to refraction effect [3].

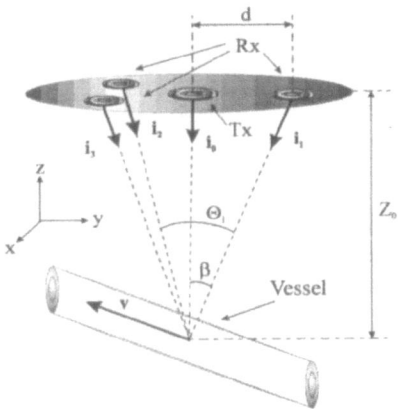

Figure 1. 3-D Doppler using four confocal transducers.

The distance between each transducer and the central axis $\underline{i}_0$ is d, the angular separation on the xy plane is 120° and Z_0 is the probe focus depth. The angle Θ_i between each lateral transducer pair and the angle β between each beam direction and the central axis $\underline{i}_0$ is completely determined by their geometrical relationships with Z_0 and d. For example in the here presented prototype system with $Z_0 = 9$ cm and $d = 3.55$ cm we obtain $\Theta_i = 37°$ and $\beta = 20°$.

The projections of $\underline{v}$ along the three directions $\underline{i}_0$, $\underline{i}_1$, $\underline{i}_2$, $\underline{i}_3$ can be written as follows:

$$
\begin{cases}
\underline{v} \cdot \underline{i}_0 = - v_z \\[4pt]
\underline{v} \cdot \underline{i}_1 = - v_y \, sen(\beta) - v_z \, cos(\beta) \\[4pt]
\underline{v} \cdot \underline{i}_2 = v_x \, \dfrac{\sqrt{3}}{2} \, sen(\beta) + (v_y/2) \, sen(\beta) - v_z \, cos(\beta) \\[4pt]
\underline{v} \cdot \underline{i}_3 = - v_x \, \dfrac{\sqrt{3}}{2} \, sen(\beta) + (v_y/2) \, sen(\beta) - v_z \, cos(\beta)
\end{cases}
\tag{1}
$$

The unknown values v_x, v_y, v_z can be derived by measuring the Doppler shifts with the three lateral transducers. Assuming that the scatterer is insonified by the central transducer (which corresponds to direction $\underline{i}_0$) with a burst at frequency f_t, the Doppler shifts detected by the three receiving transducers are respectively:

$$\begin{cases} f_1 = -(f_t/c)(\underline{v} \cdot \underline{i}_0 + \underline{v} \cdot \underline{i}_1) \\ f_2 = -(f_t/c)(\underline{v} \cdot \underline{i}_0 + \underline{v} \cdot \underline{i}_2) \\ f_3 = -(f_t/c)(\underline{v} \cdot \underline{i}_0 + \underline{v} \cdot \underline{i}_3) \end{cases} \qquad (2)$$

By substituting (1) in (2) and solving the linear system for v_x, v_y, v_z, the velocity $\underline{v}$ can be written as follows:

$$\begin{cases} v_x = -(c/f_t)\dfrac{(f_2 - f_3)}{\sqrt{3}\, sen(\beta)} \\ v_y = -(c/f_t)\dfrac{(f_2 + f_3 - 2f_1)}{3\, sen(\beta)} \\ v_z = (c/f_t)\dfrac{(f_1 + f_2 + f_3)}{3(1 + cos(\beta))} \end{cases} \qquad (3)$$

EXPERIMENTAL SETUP

The three-dimensional Doppler probe was built using four 2.5 MHz piezo-composite transducers provided by ESAOTE (Florence, Italy).

Such probe is driven by an experimental prototype of a 3-D PW Doppler system, which consists of three analog PW Doppler channels, a programmable logic unit that provides the timing signals to the analog channels, and a synchronous multi-channel acquisition unit. Analog Doppler channels have high dynamic range (80 dB at fixed gain) and use lower cost electronics than digital ones. Other characteristics of the developed Doppler channels are: PRF = 8 kHz, central frequency = 2.5 MHz, programmable number of cycles and voltage of the excitation burst, controllable gain in the range 24÷64 dB.

The six quadrature outputs from the three Doppler channels are sampled and processed using a periodogram algorithm based on 50% overlapped FFTs to evaluate the Doppler power spectrum density (PSD); the length of the acquisitions and the number of FFTs are also programmable. Mean velocity estimation in the voxel is carried out by PSD's centroid calculation. All these parameters can be varied from a virtual instrument developed with Labview.

A Doppler thread phantom was built in laboratory, to calibrate the system with a constant velocity elementary flow. The phantom uses a thin cotton wire whose velocity can be controlled over a range of 5÷150 cm/s with a speed accuracy of ±1%.

RESULTS

The 3-D experiments were performed using 20 ms acquisitions, with a periodogram algorithm based on three, 50% overlapped, FFTs.

The first three dimensional measurements were made at constant velocity $v = 93.1$ cm/s and constant angle $\Theta = 50°$, with four different values of ϕ, where Θ and ϕ are defined in the reference system fixed to the transducers assembly shown in Fig. 2b. The projections on the xy plane of the estimated vectors are shown in Fig. 2a.

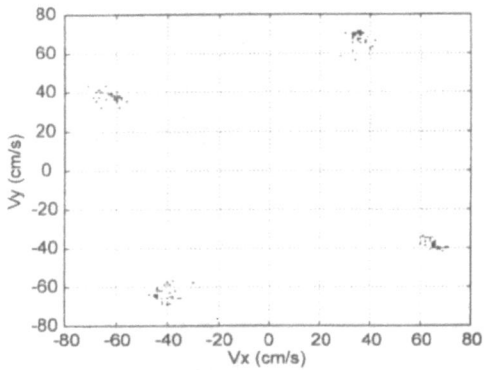

Figure 2a. Projection in the xy plane of the measured velocity at constant $\Theta = 50°$ and $\phi = -30°$, $60°$, $150°$, $-120°$. The speed of the wire was 93.1 cm/s.

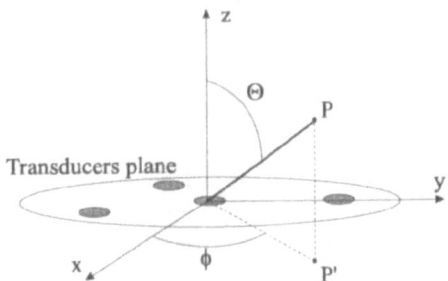

Figure 2b. Reference system fixed to the 3-D probe.

The uncertainty on the measured velocity module in these experiments was found to be 3÷7%.

The second series of measurements was performed at constant velocity $v = 79.8$ cm/s and constant angle $\phi = 35°$, by varying Θ in the range 45÷75°. The absolute velocities estimated in these experiments are shown in Fig. 3a, and the estimated angles Θ and ϕ are shown in Fig. 3b.

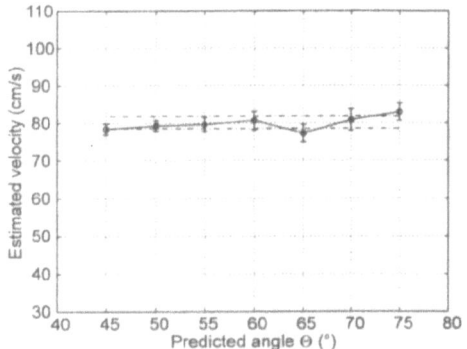

Figure 3a. Measured velocity module at constant angle ϕ as a function of predicted angle Θ. The predicted velocity falls in the range indicated by the dash-dotted lines.

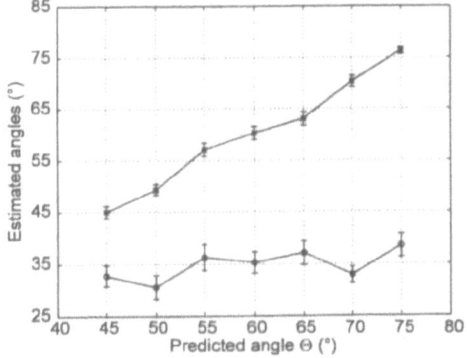

Figure 3b. Measured angles Θ and ϕ as a function of predicted angle Θ.

The uncertainty on the measured velocity found in these experiments was 4÷8%. Such errors are comparable with those found in the literature for other 2D and 3D Doppler techniques [1, 2, 3].

IMAGING OF A THREE DIMENSIONAL VELOCITY FIELD

The described technique provides three-dimensional Doppler data as a result of a volumetric sampling of the velocity vector field. The presentation of this kind of data is still an open problem: in fact it implies the handling of a large amount of information and can lead to a visualization of difficult interpretation. In the case of blood flow investigation, it is advisable to use a presentation which could be seen as an extension of the bidimensional color Doppler maps already routinely used in clinical exams. In this way the physicians' experience can be easily turned to the interpretation of 3-D Doppler data.

On the basis of these considerations the following presentation technique is proposed, developed in MS-Windows environment.

A Doppler 3-D data set corresponds to a set of voxel (resolution cells) in the 3-D space; as a result of the volume scanning, a vector is associated to each voxel.

To present data in a clear manner, the volume is exploded and presented as a collection of single images each corresponding to a volume slice made up of a voxel set. Different slices are determined by parallel planes. In this situation each voxel of the volume is visible. The velocity vector associated to each voxel is presented by separating the information corresponding to its module, orientation and direction. The voxel color represents the velocity module value in accordance to the color scale normally used for traditional color Doppler maps. Red tones are used for velocity flow crossing the slice from its bottom to its top, while blue tones are used for flows with opposite orientation.

To show the direction associated to each voxel, versors are drawn with the same direction as the corresponding velocity. To make more evident the versor directions, the presence of a light source is simulated above the considered volume. In this manner each versor has a shadow; this underlines the actual vector position by giving more information on its spatial location.

In Fig. 4 an example of the proposed presentation system is given for a simulated

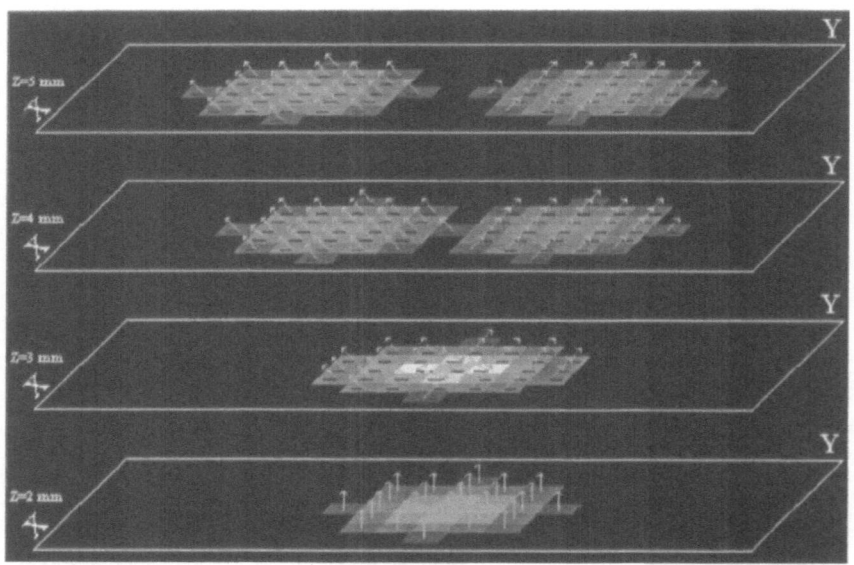

Figure 4. Presentation of the simulated flow in the bifurcation of a carotid artery.

blood flow in the carotid artery; a gray-level version of the color coding is shown.

The orientation of the parallel planes which cut the volume slice can be interactively selected orthogonal to one of the reference system axes (x, y, z). The number of slices presented together on the screen is limited, but below this limit it can be interactively selected.

The presentation program allows also to define and show 'green tags'.

To complete the presentation of the vector field a perspective view is provided which can be simultaneously displayed with the exploded sight. This is intended to give a qualitative global description of the 3-D flow. The perspective view presents the measured velocity field by means of color-coded flow lines with fixed density. Color scale is selected as previously described. The velocity vector direction coincides in each point with the flow line direction. This view derives from the classical Faraday's force line representation of a vector field, the difference being in the coding of the field intensity by color instead of line density variation. The perspective point of view can be interactively selected.

CONCLUSIONS

A prototype 3-D Doppler system based on three PW analog Doppler channels was developed. Such system has the capability to provide 3-D vector velocity maps.

The prototype performance was tested on a Doppler thread phantom which simulates a constant velocity elementary flow. These laboratory measurements revealed uncertainties on the velocity module estimate less than 8% for a range of 20÷100 cm/s.

BIBLIOGRAPHY

1. M. D. Fox, W. M. Gardiner, Three-dimensional doppler velocimetry of flow jets, *IEEE Trans. Biomed. Eng.*, vol. BME-35, pp. 834-841, Oct. 1988.
2. V. L. Newhouse, K. S. Dickerson, D. Cathignol, J. Y. Chapelon, Three-dimensional vector flow estimation using two transducers and spectral width, *IEEE Trans. Ultrason. Ferroelec. Freq. Contr.*, vol. 41, pp. 90-95, Jan. 1994.
3. J. R. Overbeck, K. W. Beach, Vector Doppler: accurate measurement of blood velocity in two dimensions, *Ultrasound Med. Biol.*, vol. 18, No. 1, pp. 19-31, 1992.

PATTERN - AND AMPLITUDE - OPTIMIZATION OF RANDOM SPARSE 2-D TRANSDUCER ARRAYS FOR 3-D ELECTRONIC BEAM STEERING AND FOCUSING

P.K.Weber and R.M. Schmitt

Fraunhofer Insitute Biomedical Engineering, Dept. of Ultrasound
Ensheimer Strasse 48, D-66386 St.Ingbert, Germany

INTRODUCTION

Two dimensional (2D) ultrasonic arrays provide the possibility of three dimensional (3D) electronic focussing and beam-steering and thus 3D imaging. To prevent grating lobes in the array response, it is necessary to fulfill the $\lambda/2$-condition in element spacing.[1] The focusing depth of a transducer is determined by the relation (D/λ) of aperture size D and the wavelength λ. The axial and lateral resolution is determined by the inverse of this factor and the beamwidth of the ultrasonic signal.[2] For focusing and beam steering in medical imaging in the lower frequency range a (D/λ)-factor of 30 up to 50 is necessary .For a fully sampled 2D-array this will result in a number of 3,600 up to 10,000 elements and electronical channels. Using existing technologies, the fabrication and implementation of such arrays and the beamforming electronics is not possible.

One possibility to reduce the number of array elements is the use of the synthetic aperture technique.[3] Another reduction technique bases on the principle of aperiodic statistical array.[1] The fully sampled aperture is thinned by randomly removing elements until a given order of reduction is reached. Grating lobes are prevented by avoiding periodicities in the element locations. The resulting array is called 'sparse array of order n' if the number of elements is reduced to 1/n-th. For 2D ultrasonic arrays Turnbull et al. have shown that the mainlobe is mostly unaffected by the random removal of elements but the average sidelobe level increases as the number of elements decreases.[4] A reduction of the sidelobe level can be obtained by aperture apodization, e.g. according to the Blackman or Tschebycheff approach.[1]

In this paper simulation studies are presented which show that the imaging quality of a sparse array can be improved by optimizing the random choice of elements. The optimization is done by the use of a genetic algorithm based on the principles of evolution.

METHOD

From a given fully sampled 2D-array with k elements the number s of sparse arrays of order n that can be obtained by random choice is

$$s = \binom{k}{r} = \frac{k!}{r!(k-r)!} \quad , \quad r = \frac{k}{n} \quad ,$$

e.g. for a 2D-array with 21x21 elements the number of randomly generated sparse arrays of order 8 yields $6.5 \ 10^{79}$. The question that arises with this great number of arrays is whether there is one of them with better sound field characteristics (sidelobe level and beamwidth) than others. There is no closed or unified theory to find the optimal sparse array but there are several optimization-procedures that can be used to get an optimum by numerical iteration.[5] Common to all these procedures is the evaluation of the quality resulting from an iteration step. The quality is described by the fitness function or quality-factor. Usually the fitness function is given by the least-square method. Aim of the optimization process is to find the best solution for a given problem, i.e. to find the minimum of the fitness-function. Compared to other optimization methods the evolution strategy gives good results for complex problems in short CPU-time.[5] In the case of sparse array optimization we implemented an optimization process based on the principles of evolution according to the following steps: 1. initialization, 2. selection, 3. recombination and 4. mutation.[6] The method will be presented for a 10x10mm matrix array with a center-frequency of 4MHz. The dense array contained 2916 elements with $\lambda/2$-spacing. Aim of the optimization process is to get a sparse array of the order 8 with a beam pattern that yields acceptable imaging quality. Starting from the dense array a set of 20 sparse arrays of order 8 was generated by a random choice of 364 elements out of the 2916 (initialization). For each array of this first set or first generation the cw-beam pattern was calculated using the method of point-source synthesis.[1] The fitness-function or quality factor F to be minimized was:

$$F = \sum_{i=1}^{N} [p_o(r_i) - p_k(r_i)]^2 .$$

$p_o(r_i)$ is the normalized pressure amplitude at a discrete point r_i in the beam pattern of the dense array, $p_k(r_i)$ the normalized amplitude of the k-th sparse array at the same point. N is the number of computation points for the beam pattern. The sparse arrays of this first generation were ranked according to their quality factor (selection). At the top position was the array with the smallest quality factor. Figure 1 shows a diagram of the ranking. A new set of 20 arrays (second generation) was generated according to the following procedure: the first five arrays of the ranking were copied to the new generation without any changes. Six new arrays were generated by recombination. This was done by randomly combining parts of the apertures of the first five arrays of the ranking and replacing the last six ones. The remaining nine arrays were mutated by randomly avtivating and deactivating single elements (mutation). For each array of the new generation the number of elements was controlled (364 as required by the order of reduction). Then the quality factors were calculated and selection, recombination and mutation were performed to get the next set of arrays. This procedure was continued until a given value of the quality factor has been reached. For the first five arrays of each generation a counter was increased with each transformation to a new set. This counter represents the 'duration of life'. If it reached a value of 20, the appropiate array was placed at position 20 of the old generation so that no transformation to the new set was performed. By doing this, the domination of the pattern of one or more arrays to the optimization process was prevented.

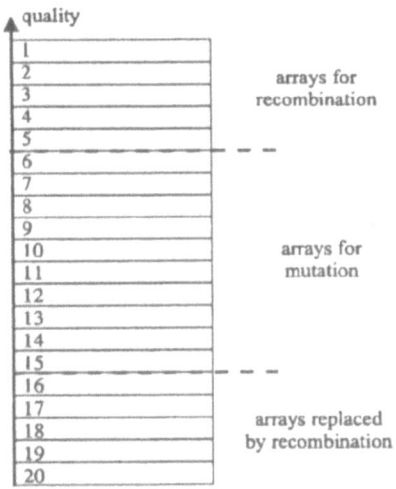

Figure 1. diagram of ranking

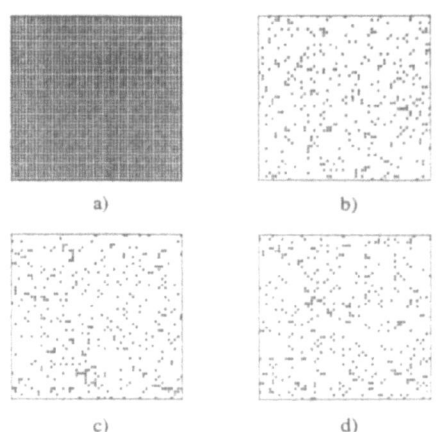

Figure 2. Element distribution for the dense array and the 8th order sparse arrays: (a) dense 2916 elements, (b) radom 364 elements, (c) sparse after 100 iterations, (d) sparse after 400 iterations

RESULTS

Figure 2 shows the element distribution of the fully sampled array together with those of the non optimized sparse array of 8th order and the results after 100 and 400 iterations. The transducer was focused at 60 mm without steering. Figure 3 shows the contour plots for the element patterns of the dense, the random and the otimized array after 400 iterations at a distance of 60 mm as a function of u and v according to figure 4.

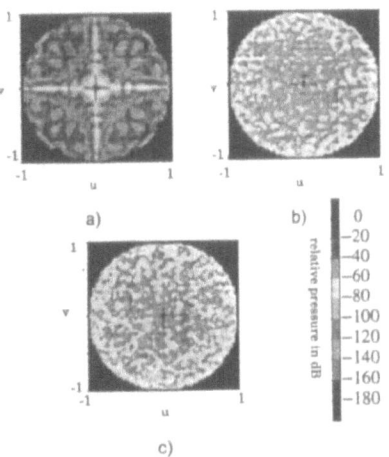

Figure 3. Contour plots at a distance of 60mm for (a) the dense , (b) the random and (c) the optimized array after 400 iterations

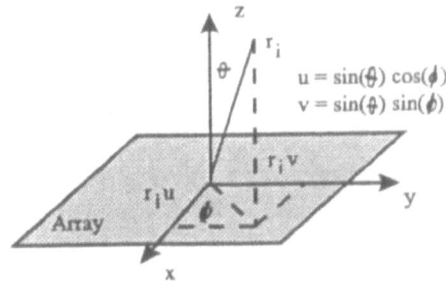

Figure 4: Coordinate system for the calculation of the beam pattern

The beam profiles of the three arrays are given in figure 5. The main lobe of the random array is unaffected by the reduction. Due to the effect of aperiodic element distribution in the sparse arrays the diffraction pattern of the random array is smoothed. The sidelobe level is increased by 40dB over a large area. The effect of optimization can be seen from figure 3. The average sidelobe level of the optimized array is clearly reduced, more energy is radiated to the mainlobe in the focal area. Figure 6 shows the quality factor F as a function of the number of iterations. Almost independent from the initial pattern the otimized pattern is obtained after 400 iteration steps as further simulation studies have shown.

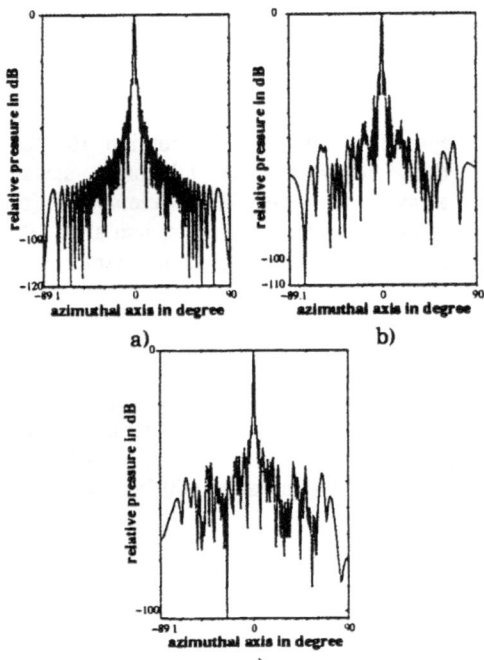

Figure 5. Beam profiles at v=0 for (a) the dense, (b) the random and (c) the optimized array after 400 iterations.

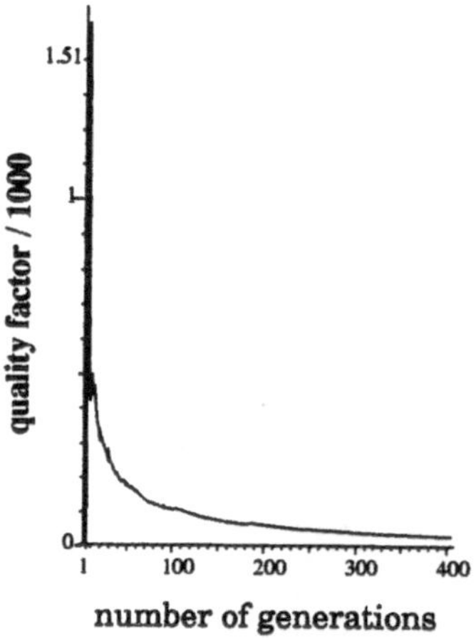

Figure 6. Quality factor as a function of the number of iterations.

For a steering angle of $(\theta, \phi) = (45^\circ, 45^\circ)$ the effect of optimization is reduced as can be seen from figure 7. The average sidelobe level of the random array is nearly the same as in the case of on-axis focusing but there is only slight improvement. This is due to the fact that the optimization was performed only for the beam pattern without steering, i.e. the quality function F was a function of one parameter.

442

The optimization procedure also can be used for quality functions of higher order as shown in the following example. The optimization was performed for an 2D-transducer array at 4 MHz driven by a digital phased array system with 128 channels. The fully sampled array contained 489 elements with a spacing of 1.1λ. The diameter of the array was 10 mm. It was necessary to reduce the number of elements by a factor of 4 to adopt it to our ectronic phased array system with 128 channels. Figure 8 shows the element distribution of the fully

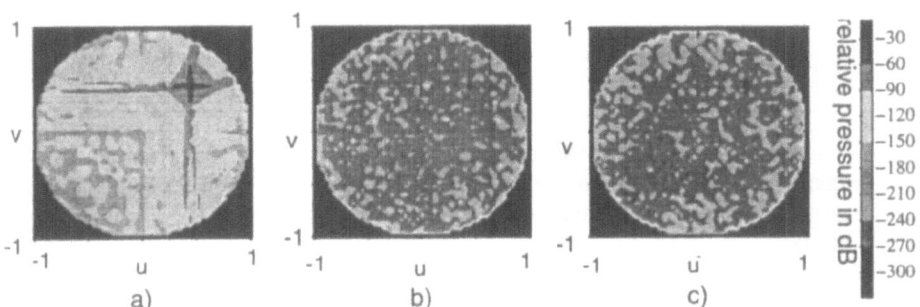

Figure 7. Contour plots at a distance of 60mm for (a) the dense , (b) the random and (c) the optimized array after 400 iterations. Arrays are steered at $(45^\circ, 45^\circ)$

sampled, the random sparse and the optimized sparse array. The beam patterns in the xz-plane are given in figure 9 for an off axis angle of $(0^\circ, 15^\circ)$. The optimization was done for both cases (no steering and off axis steering) in one procedure. Now also for the steering angle the average sidelobe level is reduced by a factor of 40 dB and more energy is radiated into the main lobe. A further improvement of the sidelobelevel can be expected from indroducing additional parameters to the fitness-function.

CONCLUSION

The optimization process presented bases on the principles of evolution. It significantly reduces the average sidelobe level of a random sparse array. The imaging quality of the array is improved because more energy is radiated to the focal area. The otimized result is obtained after 400 iteration steps almost independent from the initial array pattern. Further investigations will include the extension of the method to aperture apodization to obtain an additional improvement.

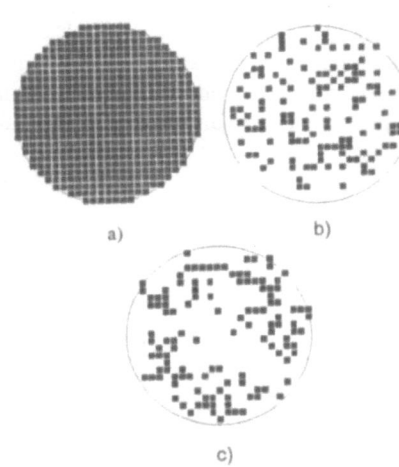

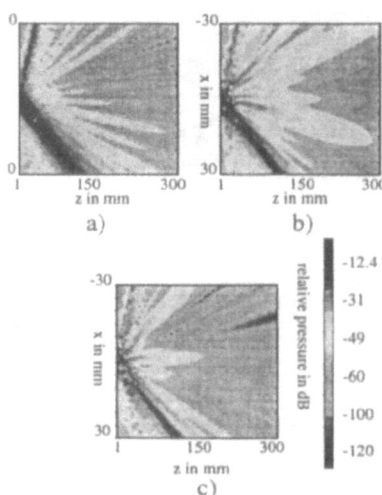

Figure 8. Element distribution for the fully sampled array and the 4th order sparse arrays: (a) dense 489 elements, (b) radom 122 elements, (c) sparse after 100 iterations.

Figure 9. Contour plots in the xz-plane for (a) the dense , (b) the random and (c) the optimized array after 100 iterations. Arrays are steered at $(0^0, 15^0)$

REFERENCES

1. B.D. Steinberg, Principles of Aperture and Array System Design, John Wiley and Sons, New York (1976)
2. M. Arditti, W.B. Taylor, F.S. Foster, J.W. Hunt, An Annular Array System for High Resoluiton Breast Echography, Ultrasonic Imaging 4, 1-31 (1982)
3. S.W. Smith, G.E. Trahey, O.T. von Ramm, Two Dimensional Arrays for Medical Ultrasound, Ultrasonic Imaging 14, 213-233 (1992)
4. D.H. Turnbull, F.S. Foster, Beam Steering with Pulsed Two Dimensional Transducer Arrays, UFFC, 38 4, 320-333 (1991)
5. T. Blümecke, Optimierungsverfahren in der Gegenüberstellung, c't, 12, 230 (1991)
6. I. Rechenberg, Evolutionsstrategie - Optimierung Technischer Systeme nach Prinzipien der Biologischen Evolution, frommann-holzboog-Verlag, Stuttgart (1973).

HIGH FREQUENCY INTEGRATED ULTRASOUND ARRAYS

A.D. Armitage[1], Q.X. Chen[2], J.V. Hatfield[1], P.J. Hicks[1] and P.A. Payne[2]

Departments of [1]Electrical Engineering and Electronics, and [2]Instrumentation and Analytical Science, University of Manchester Institute of Science and Technology (UMIST), P O Box 88, Manchester M60 1QD, UK

ABSTRACT

There is increasing interest in the use of ultrasound imaging techniques at frequencies in excess of 20 MHz. Examples may be found in both medicine and industry. This places a great demand on the imaging systems employed, particularly if the aim is to produce real time images using multi-element arrays. The Ultrasound Research Group at UMIST have been working along these lines for a number of years and this paper represents a report on work in progress covering beam plotting experiments to investigate the characteristics of the transmitted ultrasound field. We also report on preliminary results from the receive system which is based on phased array techniques.

INTRODUCTION

Ultrasound imaging applications at higher frequencies are increasingly numerous and can be found in both medicine and industry.

In medicine a good example is provided by the use of ultrasound in dermatology to provide both two- and three-dimensional images of skin lesions. Commercial systems designed to achieve such images are now available[1] and these operate at centre frequencies of up to 50 MHz. The disadvantage in the approach adopted is that the ultrasonic transducer is a single element device and the imaging modality is based on mechanical translation of the transducer across the surface of the skin with a small water bath acting as a stand-off between them. Three-dimensional (3-D) images are simply obtained by storing a series of B-scans and then using computer processing to assemble the 3-D version. The results obtained are extremely promising, but the time taken to acquire the image data is very long. The Ultrasound Research Group at UMIST has been working towards similar imaging capability, but in real time and based upon 2-D or 3-D piezoelectric transducer arrays[2-5].

An example of the industrial requirements for very high frequency ultrasound imaging is the increasing awareness of the effect of micrometre and sub-micrometre defects in engineering materials subject to high mechanical stress, such as an aero-engine turbine blade. In addition, there is increasing interest in micromachined parts for miniature transducers and actuators and methods of non-destructively examining such products for sub-micrometre defects will also become important.

If we examine the current state-of-the-art in ultrasonic medical imaging, there has been an increasing trend in the past ten years to provide ever more powerful computing facilities to not only store images produced, but also to enable the user to implement a variety of image processing techniques on the acquired data. Little has changed with respect to the transducer configuration employed. In the main for linear arrays this is still based on either piezoelectric ceramic materials or, more recently, piezo-composite materials[6,7]. The centre frequencies employed are mostly governed by the depth of penetration required for the particular diagnosis being performed. These can vary from around 2.5 MHz up to in excess of 10 MHz. However, if we wish to produce useful diagnostic images of organs such as the skin resolutions of better than 100 μm are demanded and this requires transducer centre frequencies to be increased to 15 MHz or higher. This brings with it problems associated with the connection between the transducer array and the imaging system. Even if some front-end electronics is built into the hand-held transducer case, there is still a need for a multi-wire cable to connect this back to the imaging system. This is a bulky and inconvenient arrangement, as most radiographers will confirm. Therefore, we have concentrated our endeavours on producing an integrated transducer which contains the transmitting and receiving electronic systems to as large an extent as possible. Once successful, this could result in a hand-held transducer with a connecting cable back to the imaging system with as few as four or five wires. The next step could even be a wireless connection to the imaging system.

PROTOTYPE TRANSDUCER DESIGNS

At the outset we made a decision to exploit the existing know-how within the Ultrasound Research Group on synthesis and applications of piezoelectric polymer materials. These are now widely available from commercial sources in convenient thicknesses either poled or unpoled and with or without metallic coatings. In addition, for a number of years, we have produced our own piezoelectric polymers based on the copolymer of polyvinylidene fluoride (PVDF) and trifluoroethylene (TrFE)[8]. We have also developed techniques for interconnecting between the elements of an ultrasonic array based on such materials to, for example, interconnect expansion pads that enable us to bring the transmit and receive circuits very close indeed to the transducers elements. The manner in which this was done has been described previously[3,5] and prototype hand-held 32- and 64-element packages have been constructed.

The arrangements adopted ensure that adjacent electrodes are not connected by adjacent interconnect expander leads. Nevertheless, there is a need to ensure that crosstalk between elements is of an acceptable level and we have reported on this previously[2]. The worst case measured crosstalk was some -32 dB, an acceptable figure.

TRANSMIT ELECTRONICS

We decided at the outset to work with transducers having a centre frequency of about 20 MHz. To some extent, the bandwidth that we operate with is under our control. We have conducted an extensive series of modelling and physical measurements over the years[8,9] which enable us to choose bandwidths from about 40% of the centre frequency up to approaching 100%. Having made the decision to operate at 20 MHz this places demands on the transmit and receive circuitry in terms of the time resolution required in generating the delays between transmitted pulses if we are to focus and scan on transmit and similarly for the receive circuitry which we would ideally also like to focus. For the transmit circuitry it turns out that a 1 ns timing resolution is required and we have turned to application specific integrated circuit (ASIC) technology to achieve this. The ASIC is based on 1.5 μm CMOS technology which has been developed under the EUROCHIP initiative. The modular design is capable of addressing groups of 16 elements of an array at a time and the concept is that it will be of a generic nature capable of operating with ultrasonic arrays over a wide range of frequencies and in a number of modes, ie, linear or phased array scanning. It proves impossible to obtain technology on the CMOS chip capable of providing the high drive voltages required to operate piezoelectric polymer transducers and we have therefore employed surface mount avalanche transistors for a solution[4]. These are capable of providing the necessary fast high voltage pulses needed to stimulate the transducer element and yet they occupy under 10% of the circuit area of conventional MOSFET pulsers. Examples of the pulse echo response from individual elements of prototype arrays that have been constructed from 28 μm thick Kynar™ film are shown in Figures 1 and 2. These also show the frequency spectra of the pulses and indicate the ease with which we can vary the centre frequency and bandwidth by adjusting array element dimensions and the acoustic impedance of the backing materials.

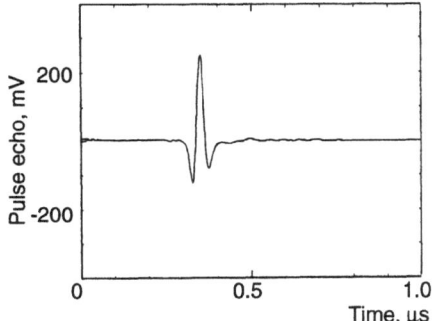 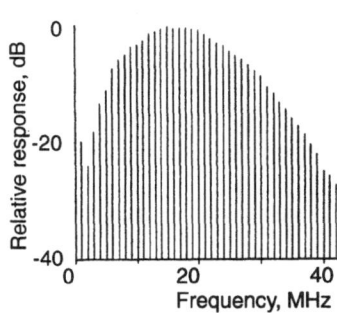

Figure 1. Typical pulse echo response and spectrum for a single element from a 32-element array (dimensions 0.2 x 7 mm with 0.05 mm gap), centre frequency 18 MHz, bandwidth 12 MHz (3 dB).

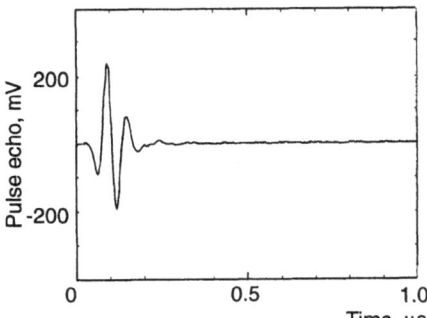 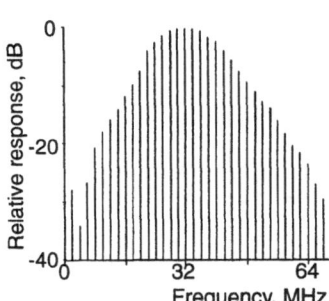

Figure 2. Typical pulse echo response and spectrum for a single element from a 64-element array (dimensions 0.05 x 5 mm with 0.025 mm gap), centre frequency 32 MHz, bandwidth 20 MHz (3 dB).

BEAM PROFILE MEASUREMENTS

We have conducted numerous beam profile measurements of transducer arrays constructed in the recent past. One example of these will be described. For this experiment we employed a 0.25 mm diameter PVDF based needle type hydrophone (Precision Acoustics Ltd, Model 348). This was connected to a submersible preamplifier having a wide bandwidth (up to 100 MHz) 10 dB gain and 50 Ω output impedance (Precision Acoustics Ltd, Model 824). The preamplifier was powered by a dc coupler unit (Precision Acoustics Ltd, Model F903) and the received signal was fed to a Panametric 5052PRX pulser-receiver which was then connected to a Hewlett Packard 54520A digital oscilloscope capable of downloading data to a PC.

For these measurements we used five delay times fixed at the values calculated to bring the beam to a focus at 5 mm in front of the array surface. The pulsers were connected to ten elements at the centre of a 32-element array and the hydrophone was then scanned in a raster fashion across the ultrasound field that was generated. The scanning resolution was 0.1 mm and we collected a total of 192 (16 x 12) data points for each experiment, 16 points at 1 mm per step along the transducer and 12 points at 0.2 mm step across the transducer. Figure 3 shows the form of the received pulse at the focal point, providing a centre frequency of something in excess of 20 MHz, but with a relatively low bandwidth. However, this could be

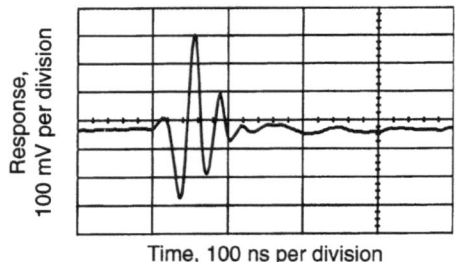

Figure 3. Pulse echo response produced from ten elements of a prototype 32-element transducer obtained at the focal point (5 mm from the array surface).

influenced by the hydrophone and preamplifier characteristics. Figures 4 and 5 show two views of the ultrasonic field produced by the transducer. There is evidence of considerable secondary lobe formation and this is due to the relationship between the centre frequency and

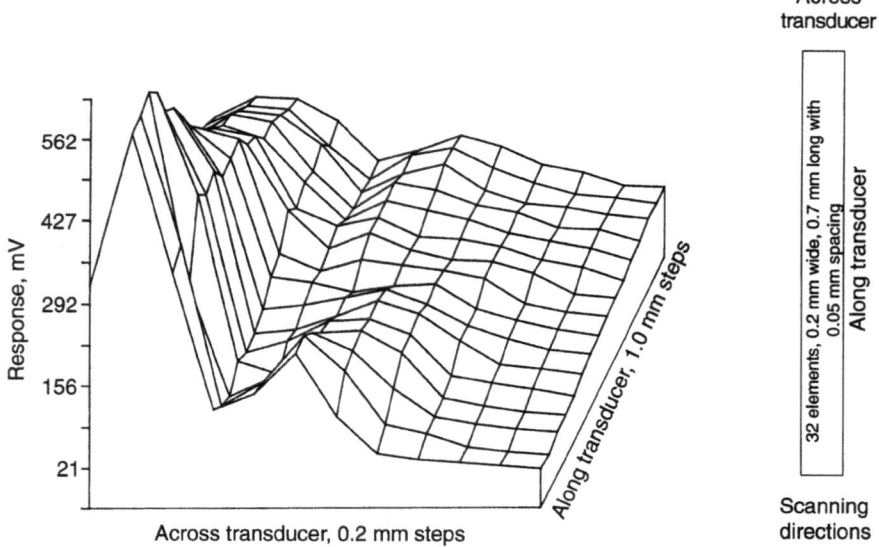

Figure 4. View 1 of the beam profile for prototype 32-element transducer (ten elements used for focusing as described in text).

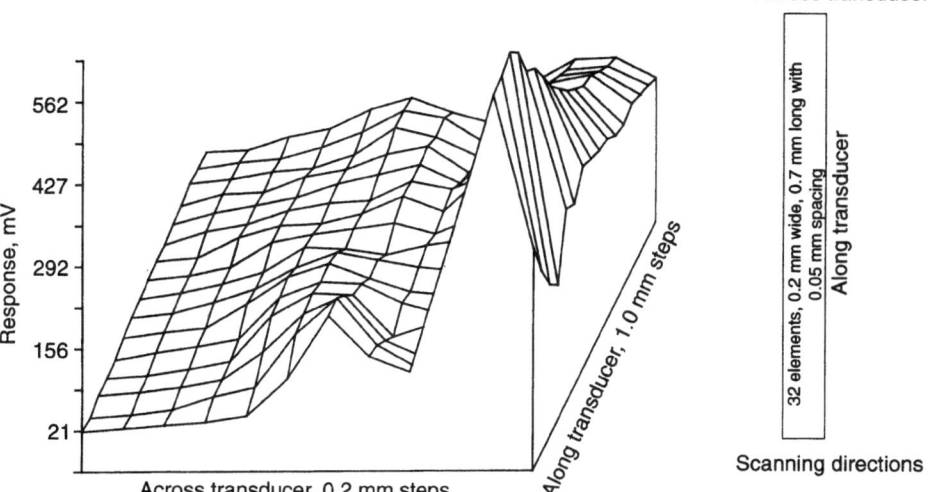

Figure 5. View 2 of the beam profile for prototype 32-element transducer (ten elements used for focusing as

the pitch of the transducer elements which were 0.2 mm wide with a 0.05 mm spacing. The transducer employed for these measurements was only lightly damped (Figure 3). With heavier damping the centre frequency will reduce, the bandwidth will increase and the secondary lobe energy will diminish. It should also be noted that the prototype array which was used as the basis for these measurements was a flat face design. A more pronounced focusing effect will result from using a concave inwards face which is perfectly compatible with the constructional techniques that we have developed.

RECEIVE SYSTEM

To date the transmit and receive electronic systems have been developed separately and work on integrating these as well as interfacing to appropriate display systems is yet to be done. However, some tests have been carried out to establish whether the principles that were described by Hatfield et al[4] and Armitage et al[10] will produce suitable images. These tests were based on scanning a piece of porcine skin mounted in a water tank and using a single element relatively low frequency transducer having a centre frequency of 5 MHz. The results are shown in Figure 6 and do indeed indicate that the system is feasible.

CONCLUSIONS

We have described progress being made towards an integrated ultrasound array containing a major proportion of the transmit and receive electronics. Further work to fully combine the transmit and receive systems and to interface these to suitable display technology will be reported in the near future.

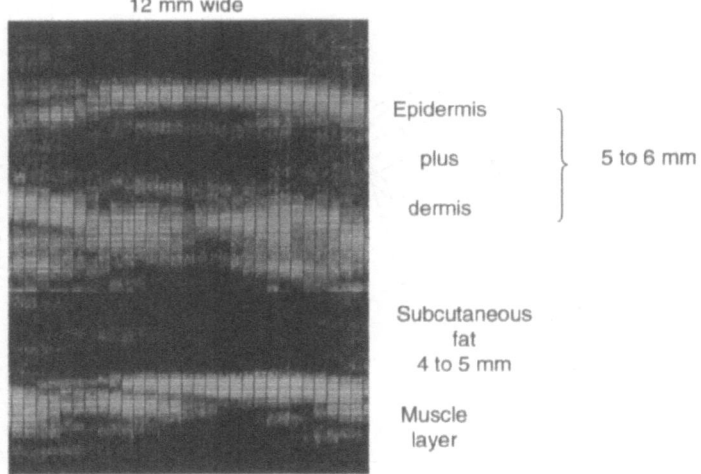

12 mm wide

Epidermis
plus
dermis
} 5 to 6 mm

Subcutaneous
fat
4 to 5 mm

Muscle
layer

Figure 6. B-scan of porcine skin.

ACKNOWLEDGEMENTS

Funding for this work is provided by the UK Engineering and Physical Sciences Research Council.

REFERENCES

1. P. Altmeyer, S. el-Gammal and K. Hoffmann (eds). "Ultrasound in Dermatology," Springer-Verlag, Berlin (1992).
2. P.A. Payne, J.V. Hatfield, P.J. Hicks, N.R. Scales, Q.X. Chen, A.D. Armitage and D.G. Lomas, Towards an integrated hand-held multi-element array transducer for ultrasound imaging, *in*: "Acoustic Sensing and Imaging", IEE Conf. Proc. No. 369, London (1993).
3. J.V. Hatfield, P.J. Hicks, P.A. Payne and N.R. Scales, High resolution multichannel pulse generator CMOS chip to drive ultrasonic array transducers, *Electronics Letters* 29:1632 (1993).
4. J.V. Hatfield, N.R. Scales, A.D. Armitage, P.J. Hicks, Q.X. Chen and P.A. Payne, An integrated multi-element array transducer for ultrasound imaging, *Sensors & Actuators A* 41-42:167 (1994).
5. N.R. Scales, P.J. Hicks, A.D. Armitage, P.A. Payne, Q.X. Chen and J.V. Hatfield, A programmable multi-channel CMOS pulser chip to drive ultrasonic array transducers, *IEEE J. Solid State Circuits* 29:992 (1994).
6. W.A. Smith, Piezocomposite materials for acoustical imaging transducers, *in*: "Acoustical Imaging, Volume 21," J.P. Jones, ed., Plenum Press, New York (in press).
7. W.A. Smith, A.A. Shaulov and B.A. Auld, Design of piezocomposites for ultrasonic transducers, *Ferroelectrics* 91:155 (1989).
8. Q.X. Chen and P.A. Payne, Industrial applications of piezoelectric polymer transducers. *Meas. Sci. Technol.* 6:249 (1995).
9. Q.X. Chen. "On Ferroelectric Polymer Transducers and Imaging Arrays," PhD Thesis, University of Manchester (UMIST) (1989).
10. A.D. Armitage, N.R. Scales, P.J. Hicks, P.A. Payne, Q.X. Chen and J.V. Hatfield, An integrated array transducer receiver for ultrasound imaging, *Sensors & Actuators A* 46-47:542 (1995).

EXPERIMENTAL VALIDATION OF A PIEZOCERAMIC ANNULAR PLATE THEORETICAL MODEL

A. Iula[1], F. R. Montero de Espinosa[2], N. Lamberti[1], and M. Pappalardo[3]

[1] Dip. di Ingegneria dell'Informazione ed Ingegneria Elettrica
Università di Salerno
Via Ponte Don Melillo, I-84084 Fisciano (SA), Italy

[2] Instituto de Acustica - C.S.I.C.
Serrano 144, 28006 Madrid - Spain

[3] Dipartimento di Ingegneria Elettronica - Univeristà di Roma III
Via C. Segre 2 - 00146 Roma - Italy

INTRODUCTION

The vibrational behavior of a piezoceramic ring is quite different from that of a solid disk with the same radius and thickness and therefore an "ad hoc" model is desirable. Berlincourt et al. proposed a simple lumped mode model[1] which is only valid when the inner radius approaches the outer one. More recently, Brissaud developed a three-dimensional stress free model with the purpose of characterizing piezoceramic materials[2]. In recent works we described a matrix model of the radial mode of a thin piezoceramic ring capable to take the interaction of the lateral surfaces with the surrounding medium into account[3,4]. The ring is modelled as a three port system with two mechanical and one electrical ports. In this work we measured the frequency spectrum and the lateral displacement of the ring and we compared with the computed results finding a good agreement. Further we compared the measured effective coupling factor k_{eff} with the computed one finding, also in this case, a good agreement and therefore validating the present model.

THE THEORETICAL MODEL

Fig. 1 shows a piezoceramic annular plate with external radius a, internal radius a' and thickness $2 \cdot b$. Starting from the classical constitutive equations of piezoceramic materials and from the wave differential equation describing the element vibration in the radial direction[5], we get the expression of the radial displacement[3]:

$$u_r = \left[AJ_1(kr) + BY_1(kr) \right]e^{j\omega t} \tag{1}$$

where k is the wave number, J_1 is the Bessel function of first kind and first order and Y_1 is the Bessel function of second kind and first order. By imposing continuity between the stresses and the forces on the external lateral surfaces of the ring, we obtain a matrix model able to describe the behavior of the ring as a transducer of radial waves, for any surrounding medium:

$$F_1 = -4\pi c_{11}^P\big(A_1 F_J(a') + B_1 F_Y(a')\big)v_1 - 4\pi c_{11}^P\big(A_2 F_J(a') + B_2 F_Y(a')\big)v_2 - 2\pi a' e_{31}^P V$$

$$F_2 = -4\pi c_{11}^P\big(A_1 F_J(a) + B_1 F_Y(a)\big)v_1 - 4\pi c_{11}^P\big(A_2 F_J(a) + B_2 F_Y(a)\big)v_2 - 2\pi a\, e_{31}^P V \qquad (2)$$

$$I = -2\pi j\omega e_{31}^P\big(A_1\Delta J_1 + B_1\Delta Y_1\big)v_1 - 2\pi j\omega e_{31}^P\big(A_2\Delta J_1 + B_2\Delta Y_1\big)v_2 + j\omega\pi\frac{(a-a')}{2b}e_{33}^P V$$

where:

$$\Delta J_1 = aJ_1(ka) - a'J_1(ka') \qquad \Delta Y_1 = aY_1(ka) - a'Y_1(ka')$$

$$F_J(r) = krJ_0(kr) - J_1(kr)\big(1 - \sigma^p\big) \qquad F_Y(r) = krY_0(kr) - Y_1(kr)\big(1 - \sigma^p\big)$$

$$A_1 = \frac{1}{j\omega J_1(ka')}\left(1 + \frac{Y_1(ka')J_1(ka)}{J_1(ka')Y_1(ka) - J_1(ka)Y_1(ka')}\right)$$

$$A_2 = -\frac{1}{j\omega}\left(\frac{Y_1(ka')}{J_1(ka')Y_1(ka) - J_1(ka)Y_1(ka')}\right)$$

$$B_1 = -\frac{1}{j\omega}\left(\frac{J_1(ka)}{J_1(ka')Y_1(ka) - J_1(ka)Y_1(ka')}\right) \qquad B_2 = \frac{1}{j\omega}\left(\frac{J_1(ka')}{J_1(ka')Y_1(ka) - J_1(ka)Y_1(ka')}\right)$$

σ^p is the planar Poisson's ratio; v_1 and v_2, F_1 and F_2 are the velocities and the forces in the external media at the boundary respectively. Referring to eqs. (2), the ring can be seen as a three-port system with one electric and two mechanical ports. With this approach it is possible to take the interaction of the lateral surface of the ring with the surroundings into account, by loading the mechanical ports with the acoustical impedances of the media. Connecting the electric port to an a.c. voltage $V = V_0\, e^{j\omega t}$, all the transfer functions of the ring, as well as the input impedance, can be computed. Finally, with this model, the dielectric and mechanical losses can be also taken into account by connecting an appropriate acoustical impedance in series with the load or by using complex mechanical and electrical parameters in eqs. (2).

ELECTRICAL MEASUREMENTS

The electrical measurements were carried out with an HP4194A impedance analyzer. A piezoceramic disk *Ferroperm 27* with diameter $2a = 24\ mm$ and thickness $2b = 0.68\ mm$ has been chosen as sample. It has been worked, making axial holes, increasing the diameter step by step. Following this procedure, we obtained a set of rings with different values of the ratio $G = a'/a$, using the same ceramic sample. For each obtained ring, the electrical input impedance was measured in order to obtain the resonance frequency spectrum. Fig. 2 shows the comparison between the experimental frequency spectrum and the computed one by using the above described model. The spectrum is plotted versus G. As it can be seen, the agreement between measured and computed values is excellent in all the range for the first mode. For higher order modes, and for $G > 0.5$, the measured values are slightly higher than the computed ones, however the shape of the computed and measured modes are the same. Some observations about the dynamic behavior of the ring can be done: as expected, when

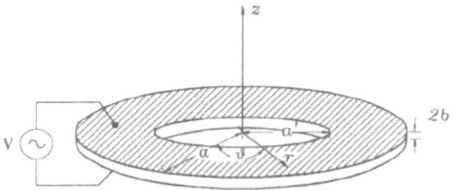

Figure 1. Coordinate system of the piezoceramic ring

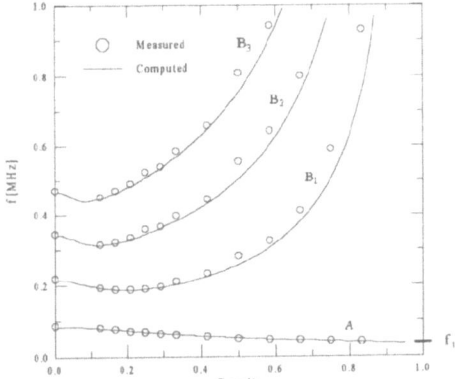

Figure 2. Computed and measured frequency spectrum of the piezoceramic ring (a = 12 mm, 2b = 0.68 mm).

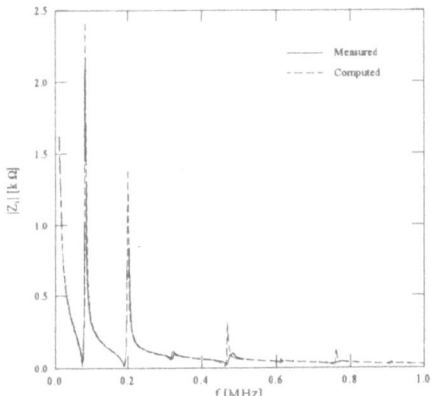

Figure 3. Computed and measured electrical input impedance of a piezoceramic ring (a = 12 mm,G = 0.167, 2b = 0.68 mm).

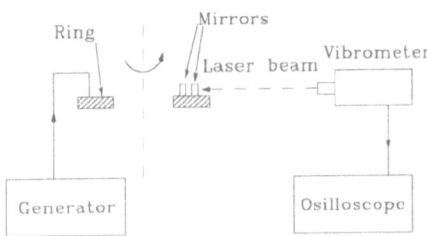

Figure 4. Experimental set up for radial displacement measurement.

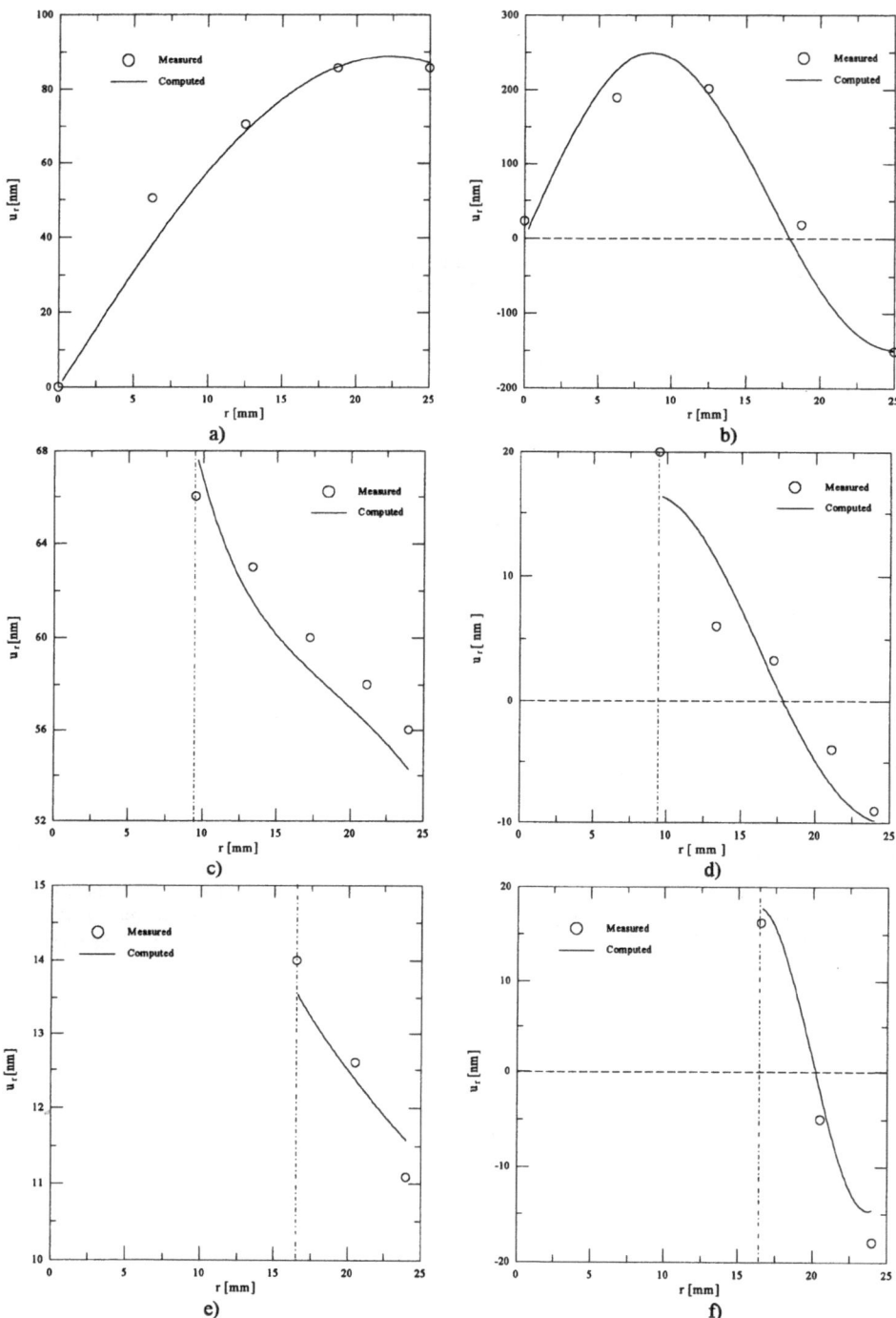

Figure 5. Computed and measured radial displacement: a) and b) 1th and 2nd mode of a thin disk ($G=0$); c) and d) 1th and 2nd mode of a thin ring with $G=0.4$; e) and f) 1th and 2nd mode of a thin ring with $G=0.69$;

$G \rightarrow 0$, the resonance frequencies of the different modes approach those of a thin piezoceramic disk with the same radius, because the ring degenerates into a plate; when G is close to I, as it was demonstrated by Berlincourt et al.[1], the ring approximates a lumped mode system. Our simulation and the experimental data confirm this behavior, in fact for $G \rightarrow I$ the first mode resonance frequency approaches the one computed by the Berlincourt model:

$$f_1 = \frac{1}{2\pi a \sqrt{\rho s_{11}^E}} = 37.2 \; kHz$$

while the higher order modes have so high frequencies that can be considered at infinity, justifying the lumped mode approximation. When G increases from 0 to I, the first mode resonance frequency decreases, while the others reach a minimum and then quickly increase to infinity. We can therefore consider two groups of resonance modes: to the A group belongs only the first mode, while the B group includes all the higher order modes. Since the A and B groups have different trends they should be related to different physical dimensions; B should be related to the radius of the annulus $(a-a')$ which decreases as G increases, while A should be related to the barycentric radius of the ring which increases with G. In fig. 3 it is shown the comparison between the measured and the computed input impedance for a sample with $G = 0.167$ in air. We can observe that we have a good agreement both for resonance frequencies and for impedance values. The small error arise because in the model the material losses are taken only in an approximate way into account.

OPTICAL MEASUREMENTS

To further investigate the nature of the radial mode, we also measured the radial displacement for rings with different inner radii. The experimental set-up used to this end is shown in fig. 4. A set of small mirrors - $2 \times 1 \; mm$ - were bonded on the ring major surface at different radii and equally distributed all around the surface. The mass and size of these mirrors must be small to prevent loading of the ring. The rings were placed onto a rotation table, to face the different mirrors to the vibrometer laser beam. The system permits an accurate positioning in order to avoid inaccuracies in the vibrometer measurement. The vibrometer is a BMI SH-120 probe, based on heterodyne interferometry. In fig. 5 the comparison between theoretical and measured radial displacement of the 1th (5a) and 2nd (5b) mode is shown for a full disk with a radius of $25 \; mm$ and for rings with the same external radius and inner radii of $9.5 \; mm$ (5c, 5d) and $16.5 \; mm$ (5e, 5f) respectively. For the disk the measured values well match the computed ones, confirming the classical theoretic Bessel function shape. For the rings, due to the small annulus, we were able to insert only few mirrors, so that, in this case, we verified only that, for the 1th mode the two lateral surfaces of the ring move in the same directions, while for the 2nd mode they move in opposite directions.

THE ELECTROMECHANICAL COUPLING FACTOR

The effective electromechanical coupling factor is the most simple parameter to be measured in order to estimate the capability of the piezoceramic to convert electrical into mechanical energy and vice-versa. It is defined as:

$$k_{eff}^2 = \frac{f_p^2 - f_s^2}{f_p^2} \tag{3}$$

where f_p is the frequency at which the electric input impedance of the ring reaches its maximum and f_s is the frequency of maximum admittance, for the fundamental mode. Fig. 6 shows the comparison between k_{eff} measured and computed with the model. The good agreement is a further confirmation of the validity of the model. From fig. 6 it can be seen that k_{eff} decrease when G increase from 0 to 1. For $G \cong 1$ the ring approaches a lumped-constant system; this means that the energy is uniformly distributed in the ring and k_{eff} coincides with the appropriate static one:

$$k_{31} = \frac{e'_{31}}{\sqrt{c_{11}^{E'} \varepsilon_{33}^{S'}}} = -0.344 \qquad (4)$$

where

$$e'_{31} = \frac{\left(e_{33}c_{13}^E - e_{31}c_{33}^E\right)\left(c_{12}^E - c_{11}^E\right)}{c_{11}^E c_{33}^E - c_{13}^{E\,2}}$$

$$c_{11}^{E'} = \frac{c_{33}^E\left(c_{11}^{E\,2} - c_{12}^{E\,2}\right) + 2c_{13}^{E\,2}\left(c_{12}^E - c_{11}^E\right)}{c_{11}^E c_{33}^E - c_{13}^{E\,2}} \qquad \varepsilon_{33}^{S'} = \varepsilon_{33}^S + \frac{e_{33}^2 c_{11}^E + e_{31}\left(e_{31}c_{33}^E - 2e_{33}c_{13}^E\right)}{c_{11}^E c_{33}^E - c_{13}^{E\,2}}$$

On the contrary, when $G \cong 0$ the ring degenerates into a disk and the static planar factor

$$k_p = k_{31}\sqrt{\frac{2}{1-\sigma^p}} = -0.6 \qquad (5)$$

is the appropriate for this radial vibration mode[5]; it differs from the dynamic one by a factor that can be recognized to be $\sqrt{8}/\pi$. Finally, in fig. 6 we can also observe that for $G > 0.6$ the k_{eff} becomes G independent, confirming, for this range, the approximation of the ring with a lumped mode system.

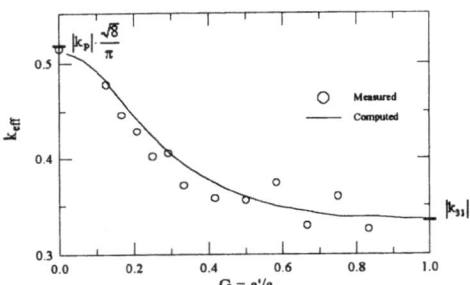

Figure 6. Computed and measured effective coupling factor for a piezoceramic ring.

CONCLUSIONS

In this work we compare the results obtained by a matrix model of the radial mode of a thin piezoceramic ring with the measured ones, finding a good agreement. The computed and measured frequency spectra show that, when the inner radius vanishes, the resonances of the ring coincide with those of a disk; furthermore, increasing G up to one, the first mode frequencies decrease, approaching the value obtained with the lumped mode model, while the higher mode frequencies increase up to infinity and therefore can be neglected. A good agreement is also obtained between the computed and measured spatial distribution of the radial displacement. Finally, investigating the effective coupling factor as a function of G, we find that for a thin annulus k_{eff} approaches k_{31}, while for a full disk k_{eff} is proportional to k_p by a constant that takes the energy distribution along the disk radius into account. Examining the behavior of k_{eff} with G we observe that the approximation of the thin ring using a lumped mode system is acceptable starting from $G > 0.6$. The authors intend to extend this model in three dimensions, which means to take the coupling with the thickness vibration of the ring into account, by means of the same approximated approach that they used for rectangular bars and circular plates, described in recent works[6,7].

REFERENCES

1. D.A. Berlincourt, D. R. Curran and H. Jaffe: "Piezoelectric and Piezomagnetic Material and their Function in Transducers", *in Physical Acoustic, vol. 1*, W. P. Mason Ed., New York: Academic Press, 1964, pp. 169-270.

2. M. Brissaud: "Characterization of Piezoceramics", *IEEE Transaction on Ultrasonics, Ferroelectrics and Frequency Control*, vol. 38, no. 6, pp. 603-617, November 1991.

3. A. Iula, N. Lamberti, G. Caliano, M. Pappalardo: "A Matrix Model of the Thin Piezoceramic Ring"; *1994 IEEE Ultrasonics Symp. Proc.*, vol. 2, pp. 921-924, 1994.

4. A. Iula, N. Lamberti, G. Caliano, M. Pappalardo: "A Model for the Theoretical Characterization of Thin Piezoceramic Rings"; *Accepted for Publication to IEEE Trans. on Ultrasonics, Ferroelectrics and Frequency Control.*

5. A. H. Meitzler, H. M. O'Brian and H.F. Tiersten: "Definition and Measurement of Radial Mode Coupling Factors in Piezoelectric Ceramic Materials with Large Variation in Poisson's Ratio", *IEEE Transaction on Sonic and Ultrasonic*, vol. SU-20, no. 3, pp. 233-239, JULY 1973.

6. N. Lamberti, M. Pappalardo: "A General Approximated Two Dimensional Model for Piezoelectric Array Elements", *IEEE Transaction on Ultrasonics, Ferroelectrics and Frequency Control*, vol. 42, no. 2, March 1995.

7. A. Iula, N. Lamberti, G. Caliano, M. Pappalardo: "A New Tridimensional Model for Circular Piezoelectric Transducers", *Proc. of 21st International Symposium on Acoustical Imaging*, 1994, Laguna Beach, California.

ACOUSTIC FIELDS OF ARBITRARY PLANE OR SPHERICAL TRANSDUCERS

Dominique Cathignol and Philippe Faure

INSERM, Unité 281, 151 Cours Albert Thomas, 69424 Lyon, Cedex 03
France

ABSTRACT

This work presents a method for calculating pressure fields radiated by arbitrary plane or spherical transducers vibrating in an infinite rigid baffle with a uniformly distributed velocity $v_n(t)$. Up to date the diffraction impulse response method is widely used, the pressure is calculated as a convolution process between the diffraction impulse response $h(M,t)$ and the time derivative function of $v_n(t)$. This method has two important limits: firstly, the $h(M,t)$ functions are known for only few typical shapes, and secondly, when $h(M,t)$ is analytically determined, it must be sampled in order to make the convolution process. We propose a new method which enables a faster calculation of the pressure field of any arbitrary plane or spherical transducer in the particular case of a uniformly distributed velocity. For this purpose, the following equation has to be solved:

$$p(M,t) = \frac{\partial\, h(M,t)}{\partial\, t} * v_n(t)$$

This equation leads to a simple line integral which can be interpreted as the sum of a plane or spherical wave, and a boundary diffraction wave. As a consequence, the computation time is decreased, because the calculation of $h(M,t)$ and the convolution process are no longer necessary. Pressure field distributions radiated by typical shapes were compared to those computed by other methods. In all cases, the agreement was excellent.

1. INTRODUCTION

This paper presents a method for calculating pressure fields radiated by arbitrary plane or spherical transducers vibrating in an infinite rigid baffle with a uniform distributed velocity. The motivation for this work was the need for an efficient, fast, and accurate algorithm to study specific transducers designed for medical applications such as diagnosis or therapy. A thorough knowledge of the radiation patterns of such transducers is essential to optimize the final apparatus. The diffraction impulse response h(M,t) is now widely used[1, 2, 3, 4], but suffers of two important limits: firstly, the h(M,t) functions are known for only few typical shapes, and secondly, when h(M,t) is analytically determined, it must be sampled in order to make the convolution process. We propose a new method, making it possible to calculate the transient pressure field of any plane or spherical radiator embedded in a plane or a sphere, respectively. Furthermore, in all cases, the computation time is lower than that obtained using the h(M,t) method. Because of the reciprocity of the wave propagation, the method described in the following sections is applicable both to emitting and to receiving transducers. For simplicity, however, only the emiting case is considered here. Further, the pressure amplitude profiles are calculated for continuous wave (CW) mode only.

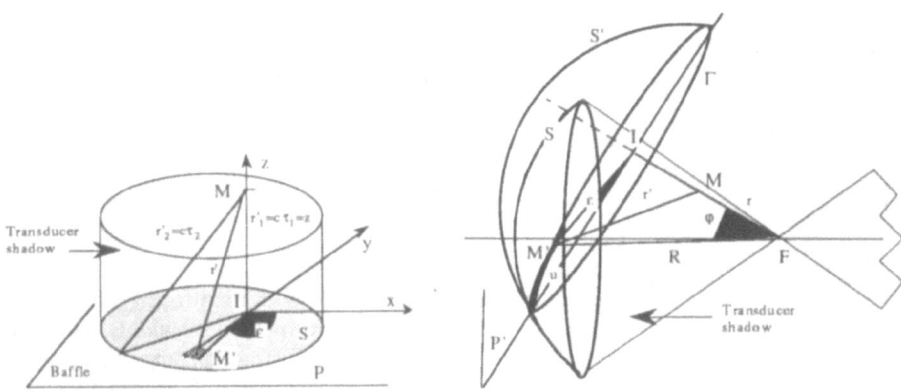

Figure 1   Figure 2

2. SUMMARY OF THEORY

Let us consider a CW radiator (wave number k) with face S embedded in a baffle B. In the limiting case where S is a plane radiator with an infinite rigid plane baffle B around it, an exact solution of the velocity potential φ in the time domain is:

$$\varphi(M,t) = \frac{1}{2\pi} \iint_S \frac{v_n\left(t - \frac{r'}{c}\right)}{r'} \, dS \tag{1}$$

and the pressure is given by

$$p(M,t) = \frac{\rho}{2\pi} \frac{\partial}{\partial t} \iint_S \frac{v_n\left(t - \frac{r'}{c}\right)}{r'} \, dS \tag{2}$$

Here r' is the distance MM', and dS is the elementary surface associated to the point M' (Figure 1). If the radiator surface is curved, equation (1) does not apply. The new method presented in this paper is based on equation (1). Accordingly, the results are exact for plane transducers and approximated for slightly curved radiators surrounded by an infinite and rigid baffle. The approximations are currently named the O'Neil approximations[5].

3. PRESSURE FIELD CALCULATION USING THE LINEAR INTEGRAL METHOD

The new method introduced in this section may be considered as a generalisation of the procedure which SCHOCH formulated for the special case of a plane piston in continuous wave (CW) mode[6]. For this, the formulation of the elementary surface dS has to be determined as a function of the distance MM'=r' in the two cases, i.e., the plane and the sphere. For a plane radiator dS = r'dεdr' where ε is the angle between IM' and a fixed direction, and I the projection of the observation point M on the plane of the radiator (Figure 1). For a spherical radiator dS = r'dεdr'(R/r). S' is the sphere centered in M and of radius MM', Γ the sphere intersection of S and S', and I the center of Γ, and ε is the angle between IM' and a fixed direction (Figure 2). It can be noticed that the case of a plane radiator is a limiting case of a sphere for which the ratio R/r equals 1. To solve simultaneously the two cases, plane and spherical radiator, in the following equations q equals 1 for the sphere, and 0 for the plane. Combining equation (2) and the analytical expression of dS, the pressure is:

$$p(M,t) = \frac{\rho}{2\pi} \left(\frac{R}{r}\right)^q \iint_S \frac{\partial}{\partial t} v_n\left(t - \frac{r'}{c}\right) dr' d\varepsilon \tag{3}$$

r' and ε being two independant variables , we have:

$$p(M,t) = -\rho c \frac{R}{2\pi r} \int_0^{2\pi} d\varepsilon \int_{r_1(\varepsilon)}^{r_2(\varepsilon)} \frac{df}{dr'} dr' \text{ , with } f(r') = v_n(t - \frac{r'}{c}) \tag{4}$$

$$p(M,t) = \frac{\rho c}{2\pi} \left(\frac{R}{r}\right)^q \int_0^{2\pi} \left[v_n\left(t - \frac{r_1'}{c}\right) - v_n\left(t - \frac{r_2'}{c}\right) \right] d\varepsilon. \tag{5}$$

Two cases have to be distinguished depending on whether M is in the geometric shadow of the transducer or not. In the case of a plane, we may say that point M is in the geometric shadow if the field point projection I lies inside the transducer area (Figure 1). In the case of a sphere, we may say that point M is in the geometric shadow if the line IM lies inside the transducer area (Figure 2).

The Point M Is In The Geometric Shadow Of The Transducer

The first time of flight $\tau_1 = r'_1 / c$ is constant and is equal to z/c. For all the points M located in the transducer shadow, equation (5) then becomes:

$$p(M,t) = \left(\rho c v_n\left(t - \frac{z}{c}\right) - \frac{\rho c}{2\pi} \int_0^{2\pi} v_n\left(t - \frac{r_2'}{c}\right) d\varepsilon \right) \left(\frac{R}{r}\right)^q \tag{6}$$

This shows that the pressure on any field point M located in the transducer shadow, can be described as the sum of a plane wave respectively a spherical wave (first term of equation (6)) and a second wave derived from the transducer boundary .

The Point M Is Not In The Geometric Shadow Of The Transducer

Let us now consider a point M in the second zone. In the case of a plane radiator, as illustrated in Figure 3, the pressure field can be written as the difference from the largest transducer T_2 and the smallest transducer T_1. The transducer boundary B can be divided into two distinct zones, B_1 and B_2, which are the active boundaries of T_1 and T_2, respectively. The times of flight $\tau_1 = r'_1 / c$ and $\tau_2 = r'_2 / c$ in equation (5), derive from B_1 and B_2, respectively, and the pressure may be expressed as:

$$p(M,t) = \frac{\rho c}{2\pi} \left[\int_0^\pi v_n(t - \tau_1(\varepsilon)) d\varepsilon - \int_0^\pi v_n(t - \tau_2(\varepsilon)) d\varepsilon \right] \left(\frac{R}{r} \right)^q \qquad (7)$$

In the case of a spherical radiator, an analogous demonstration may be made by replacing the plane P by the plane P' perpendicular to FM, and the radiator by its projection on P' following the direction FM. The pressure is given by the equations (6) and (7) where q=1.

In conclusion, this new formulation of the pressure field makes it possible to determine the acoustic field of any plane or spherical transducer by means of a line integral, instead of the conventional surface integral.

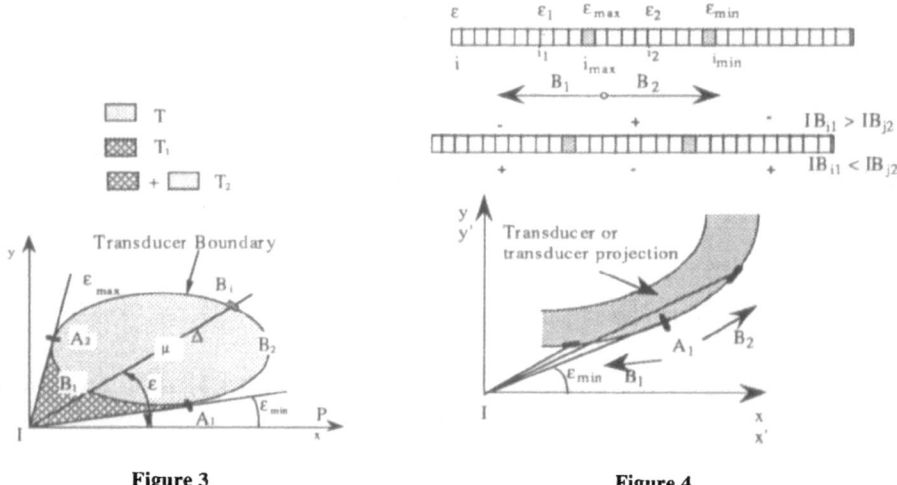

Figure 3   Figure 4

5. NUMERICAL IMPLEMENTATION

Numerical Equations

Let us consider first an arbitrary plane piston. The boundary of the transducer is described in the plane P with N points B_i, where ε is the polar co-ordinate of the boundary point B_i in the new co-ordinate system (Ix, Iy). For points M(x,y,z) located in the second zone, the numerical form of equation (7) is:

$$p(M,t) = \frac{\rho c}{2\pi} \left[\sum_{i=1}^{N_1} v_n \left(t - \frac{MB_i}{c} \right) \Delta\varepsilon - \sum_{i=N_1}^{N} v_n \left(t - \frac{MB_i}{c} \right) \Delta\varepsilon \right] \left(\frac{R}{r} \right)^q \qquad (8)$$

with $\qquad \Delta\varepsilon = \varepsilon(i) - \varepsilon(i-1) \quad and \quad MB_i = \left| \overrightarrow{MB_i} \right| \quad , q = 0.$

The N_1 points and the $(N-N_1)$ points describe the boundary parts B_1 and B_2 respectively (Figure 3). For points M(x,y,z) located in the first zone, equation (6) becomes:

$$p(M,t) = \left[\rho c\, v_n \left(t - \frac{z}{c} \right) - \frac{\rho c}{2\pi} \sum_{i=1}^{N} v_n \left(t - \frac{MB_i}{c} \right) \Delta\varepsilon \right] \left(\frac{R}{r} \right)^q \tag{9}$$

The new method may be applied in a similar way to the case of a spherical transducer. In this case, the corresponding plane P is the plane P' perpendicular to the line (FM), and the boundary B is the curve B' projection of B on P' parallel to (FM) and with q = 1. It can be noticed that the length of MB_i depends now on the co-ordinate z of B_i.

Determination Of The Sign Of The Contribution From Each Boundary Element

An important problem is the numerical determination of the two boundary parts B_1 and B_2, which make a positive and a negative contribution respectively (Figure 4). Let consider a one dimensional array $\varepsilon(i)$ including all the different values ε of all points B_i, in the co-ordinate system (Ix, Iy). It is easy to determine the two values ε_{min} and ε_{max} and their corresponding array co-ordinates i_{min} and i_{max} (Fig. 4). These two values are the ε co-ordinates of the two particular points A_1 and A_2 (Fig. 3). Let us consider now two points located on both sides of i_{min}. The comparison of the lengths IB_{i1} and IB_{i2} (Fig. 4) makes it possible to determine the sign of the contribution of the total array. If $IB_{i1} > IB_{j2}$, then, the inner interval makes a positive contribution; if not, a negative one. The acoustic field is the sum of each boundary element contribution multiplied by the value 1 or -1 according to the sign of the contribution as previously obtained. In the case of a spherical transducer, the same algorithm is used but the co-ordinates of the boundary and of the point M have to be expressed in the new cylindrical co-ordinate system (Ix', Iy') in P'.

The algorithm was implemented on a Sparc Station 10. The boundary of the transducer was generally divided into N>200 points. For the case of a polygonal transducer, we took care to place the points B_i at the corners of the polygon. Good results were obtained without any problem of sampling.

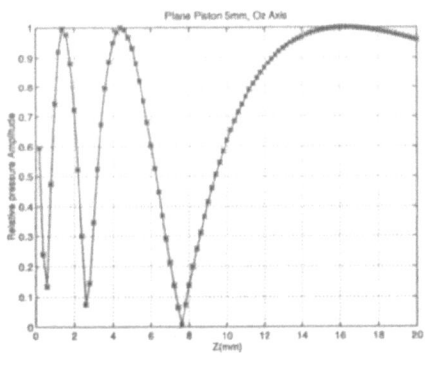

Figure 5

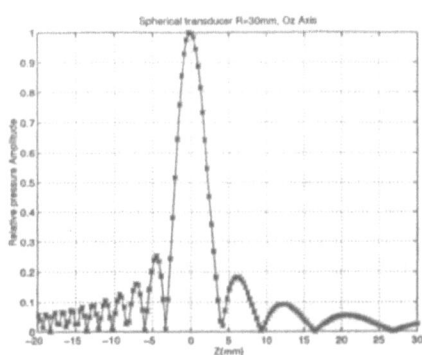

Figure 6

6. RESULTS

A comparison with the usual diffraction impulse response calculation was used to validate the new method. Curves given on Figure 5 show the comparison between the on-axis pressure field distribution for a plane piston (10 mm of diameter) calculated with both methods (solid line our method, stars h(M,t) method). The calculated pressure field distributions are presented for continuous wave (CW) mode at 1 MHz frequency. Curves given in Figure 6 show the on-axis pressure field distribution for a concave spherical transducer (R=30 mm in curvature radius and 17.5 mm in diameter) calculated with both methods. The calculated pressure field distributions are presented for CW mode at 2.2 MHz frequency. It is seen that the agreement between the classical diffraction impulse response method and the line integral method is excellent, in the both cases.

7. CONCLUSION

A new method to calculate the pressure field from a plane or spherical transducer of arbitrary shape has been presented. This method avoids the determination of the analytical solution h(M,t), simplifies the software implementation, and reduces the computation time. In contrast to the h(M,t) method, the line integral method is suitable for any geometric shape and does not require any convolution process between the time derivative of the normal velocity applied on the transducer and the diffraction impulse response h(M,t). The method is suitable for plane or spherical radiators of any shape provided the boundary can be described with N points. The pressure field is obtained as the sum of the excitation time signal delayed by the time of flight corresponding to the ultrasonic wave to go from M to the boundary elements. This value is weighed by the angle view from I, the projection of M in the P plane respectively in P'. Thus, the sampling rate is only limited by the highest frequency components of the excitation signal, and the computation time is considerably reduced. Comparaisons of pressure field distribution using standard methods and this new method show excellent agreement.

This model will be useful for the determination of the pressure field of transducers with complex shapes, which would require a long calculation time if the usual Rayleigh integral was used.

REFERENCES

1. F. Oberhettinger, On transient solutions of the baffled piston problem, *J. Res. Natl. Bur. Stand. NBS* 65:1-6 (1961)
2. P. R. Stepanishen, Transient radiation from pistons in an infinite planar baffled, *J. Acoust. Soc. Am.*, 49:841-849 (1971)
3. M. Arditi, F.S. Foster, and J.W. Hunt, Transient fields of concave annular arrays, *Ultrasonics Imaging* 3:37-61 (1981)
4. P. Faure, D. Cathignol, and J.Y. Chapelon, Diffraction impulse response of arbitrary polygonal plane transducers, *Acta acustica* 2:257-263 (1993)
5. H.T. O'Neil, Theory of focusing radiators, *J. Acoust. Soc. Am.* 21:516-526 (1949)
6. G. R. Harris, Review of transient field theory for a baffled planar piston, *J. Acoust. Soc. Am.* 70(1):10-20 (1981)

A 3D-SIMULATION TOOL FOR PREDICTING IMAGE QUALITY OF CONVEX 1D- AND 2D-ARRAYS

Machteld de Kroon and Frank Driessen

TNO Institute of Applied Physics
Inspection Technology Department
P.O. Box 155, NL-2600 AD Delft, The Netherlands
phone: (31) 15692302   fax: (31) 15692111
email: dekroon@tpd.tno.nl and fdriesse@tpd.tno.nl

ABSTRACT

This study was carried out within the framework of the Community Research Programme funded under the Commission's 3rd Framework Programme for R&D (Ref: ESPRIT Ultima project 008122).

A simulation tool was developed, which can be used for:

- The prediction of the performance of array prototypes
- Better understanding of the influence of the various mechanical and electronic probe parameters
- Debugging purposes if the simulated behaviour does not resemble the behaviour of probe and electronics in practice

The functionality of the developed simulation tool is twofold:

- Image simulation: Calculation of the 2D broadband image from a restricted number of objects,
- Beampattern simulation: Calculation of the one-way 3D monochromatic pressure field.

Experimental verification is performed by comparing results from the *image simulation* tool with results from experiments with a conventional convex array. Good correspondence between these results was found.

From the calculated *beampattern* several beam characteristics as a function of the axial distance (Figures-of-Merit) are deduced. These Figures-of-Merit, which are a measure of image quality, are the conventionally used *beam diameter* and *grating lobe level* and a Figure-of-Merit which is new in the field of medical imaging: *the object contrast parameter*. Their behaviour as a function of depth depends on the physical array properties and the electronic settings (e.g. delay settings). The Figures-of-Merit can be very helpful for finding the optimal design of the array and the optimal settings for electronic focusing .

Because full 3D computations are performed, the simulation tool is appropriate for 1D and 2D arrays.The Figures-of-Merit have been used to predict the benefits of applying transversal electronic focusing on a 2D convex array. Since the number of design parameters is higher for a 2D- array than for a 1D-array, performing simulations, which show the effect of all design parameters (e.g. element sizes, aperture, lens focus and electronic settings) on the Figures-of-Merit, is even more beneficial for a 2D than a 1D array.

INTRODUCTION

Although the quality and performance of medical diagnostic ultrasonic imaging equipment is quite reasonable, there is still a need to improve image quality in terms of resolution and contrast. The key component of the ultrasonic imaging equipment is the acoustic probe.

The performance of the probe mainly determines the performance and quality of the total imaging chain, therefore it is expected that most image quality improvement can be accomplished by improving the probe.

For imaging purposes, the ideal focus has a small lateral beamwidth (at image axis), a small transversal beamwidth (at elevation axis), a large depth of field, and an intensity and sensitivity as small as possible outside the focal region.

For these demands solutions can be found. However, the situation is complex since arrangements to improve a certain qualification can and usually will deteriorate others. So, compromises have to be made. The general side effects of actions to improve a specific aspect of the focus are commonly well known. However, it is hard to assess the exact consequences. With the help of numerical simulation tools it is possible to gain a better and more objective insight in the overall image quality improvement when changing a specific mechanical or electrical parameter of a probe.

With proper simulation tools, the probe and console manufacturers can anticipate on expected image improvements. In particular, in the designing of probes at lot of time and money can be saved if the effects of modifying mechanical probe properties (such as element sizes and radius of curvature) can be predicted, since probe prototyping is a time and money consuming process. The simulation tool can be used for determining a useful set of electrical parameters. A proper simulation tool can also be used as a debugging tool for probe and console designers in the case that test results of prototypes do not resemble the expected simulated behaviour.

SIMULATION METHOD

The simulation tool is written for a 2D convex array, but can be used without modification for 1D and linear arrays. With some minor modifications the tool can be used for phased arrays and annular arrays as well.

Probe element responses are assumed to be ideal. Simulation of single element responses and cross talk behaviour, for instance calculated by means of finite element computations, is not included.

Calculations are performed for homogeneous media with fixed density and propagation velocity and frequency dependent attenuation. The influence of a mechanical lens used for pre-focusing, fixed directly on the probe is taken into account.

The simulation tool has a twofold functionality:
- Beampattern simulation. Calculation of the one-way three dimensional monochromatic pressure field.
- Image simulation. Calculation of a (two dimensional) image from a restricted number of objects. For this broadband simulation, calculations are performed from the probe elements to one or more point scatterers, and from the point scatterers, acting as secondary sources, to the probe elements. This results in time domain receiving signals that can be represented in an image.

In the developing the tool a lot of attention has been paid to visualise the results. Full color cuts at different positions and orientations are possible of the beampattern results together with line graphs at different positions of the plot itself.

Beampattern Simulation

The calculation of the incident pressure field at any point of observation in a homogeneous medium radiated by a plane source can be evaluated by the Rayleigh-I integral[1], in which the Greens function describes the propagation from a monopole impulse source to the point of observation. It can be shown that this is also applicable for the pressure field of a slightly curved transducer with surface vibration normal to the transducer aperture.

After modifying the Greens function for attenuating media the pressure field of a single probe element can be calculated by the summation of the contributions from elementary monopole sources distributed over the element. The influence of a mechanical lens is approximated by a local time delay which depends on the focal distance of the lens and the position on the array aperture. Next, electronic focusing can be implemented by adding time delays to the pressure field of each probe element. These time delays can be discretised as is the case in real systems.

For beam pattern calculations the (three dimensional) region of interest is discretised into a grid. For each grid point the pressure field is to be calculated by adding the (complex) pressures at that point originated from all probe elements in the aperture of interest. For linear and curved arrays with a circular radius, symmetry can be used to reduce substantially the set of element to grid point pressure transfer functions that have to be calculated.

Image Simulations

For the image simulations the scattered pressure field at the probe elements have to be calculated. The scattered wave field is caused by inhomogeneities in the tissue. The inhomogeneities are characterised as spatial functions in density and propagation velocity about their mean values. If these fluctuations are small compared to their mean values and if multiple scattering can be neglected (Born approximation), the scattered wavefield can be regarded as caused by so-called secondary sources[2].

In order to calculate the image of a single point scatterer, the pressure field at the scatterer position for all frequencies and array elements of interest is calculated. Next the scattered pressure field at the array elements can be calculated with the same transfer functions. Then after adding appropriate time delays to simulate receive focusing, and adding the signals of the elements one image line can be calculated. This process has to be repeated as many times as there are image lines in the image. Symmetry of the probe can reduce the number of computations enormously.

FIGURE-OF-MERITS

Once the pressure field of an array is calculated, some Figure-of-Merits (FOM) can be deduced from the pressure field, which may help to qualify the performance of the array. The simulation program calculates six FOM's, five of which are well known and one, the object contrast parameter, is new.

The FOM's are:

1. Intensity on the main axis as a function of axial depth.
2. Level of grating lobes in the lateral plane relative to the intensity on the main axis as a function of axial depth.
3. Level of grating lobes in the transversal plane relative to the intensity on the main axis as a function of axial depth.
4. Lateral beamwidth as a function of axial depth. The lateral beamwidth is the full width of the beam at a fixed level (e.g. at -10 dB) below the maximum intensity on the main axis in the lateral plane.
5. Transversal beamwidth as a function of axial depth. The transversal beamwidth is the full width of the beam at a fixed level (e.g. at -10 dB) below the maximum intensity on the main axis in the transversal plane.
6. The object contrast parameter as a function of axial depth. This parameter is defined as the ratio of energy caught by a square object, which is centered at the beam axis, oriented perpendicular to the beam axis, and the energy in the pressure field outside this object. The size of the object is fixed (not a function of axial depth or beamwidth).

$$object\ contrast(z) = \frac{E_{object}(z)}{E_{total\ 3D}(z) - E_{object}(z)} \tag{1}$$

EXPERIMENTS

Two types of experiments have been performed: for validation of the simulation and for interpretation of the object contrast parameter.

Comparing the results of phantom experiments and image simulation of a linear 1 D probe, a high degree of similarity was found. A less than 0.1 mm difference in the -10 dB and -20 dB beamwidths of simulation and experiment was found. The level and position of the grating lobes showed a difference of respectively less than 0.5 dB and 1 mm.

After comparing some real images with calculated object contrast parameters it was found that the object contrast parameter can be interpreted as follows: less than -10 dB means a bad contrast, better than -5 dB means a good contrast.

EXAMPLES

Figure 1 shows an example of a beampattern simulation for a curved two dimensional array of 44 x 7 elements and a radius of curvature (ROC) of 40 mm. The figure shows the beam pattern normalised on the maximum per depth in a lateral cut in the region of interest. The grating lobe can be seen very clearly in the figure. Moreover figure 1 shows the pressure intensity at electronic focus (which is 79 mm in this example), the intensity on the main axis and the -3, -6, -10 and -20 dB (lateral) beamwidths. Physical parameters that can be changed in the simulation are: transversal and lateral element sampling, the ROC, the lens focus distance, the frequency, the sound velocity, and the attenuation. Electronic settings are: the electrical focus, the delay discretisation, and the lateral and transversal aperture (in number of elements).

Figure 2 shows an example of an image simulation of a single point scatterer. The responses at several depths are depicted as well, showing the broadband beamwidth and grating lobes.

An example of some calculated FOM's for two array types are shown in figure 3. For both arrays at eight (electronic) focal depths the following is calculated:

- relative intensity of the grating lobes
- the -10 dB lateral beamwidth
- the -10 dB transversal beamwidth
- the object contrast parameter for an object with a 2 x 2 mm² reflecting area.

The simulation tool is written at the moment in the numeric computation and visualization software Matlab running on a SUN SPARC 10 station. The examples shown, require a computation time of a few hours.

CONCLUSIONS

A software tool is developed with which it is possible to simulate the beampattern and imaging performance of 1D and 2D linear and curved ultrasonic medical probes with good accuracy. With the help of the beampattern and image plots, and some figure-of-merits, amongst which is a newly defined 'object contrast parameter', it is possible to predict the performance of a probe. Since a number of electrical and mechanical parameters can be changed, the simulation tool can be a great help in designing probes and consoles, reducing design and development time and costs.

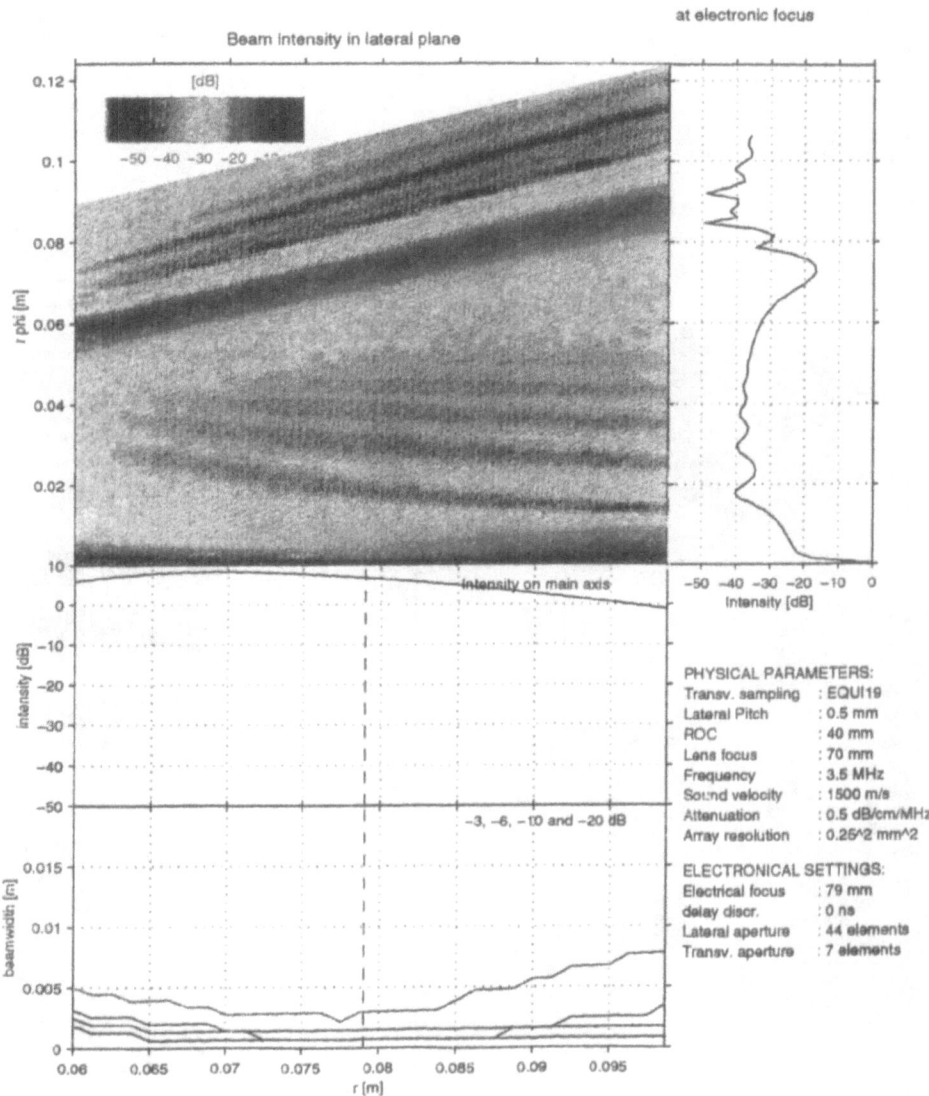

Figure 1. Example of full color simulated beampattern.

469

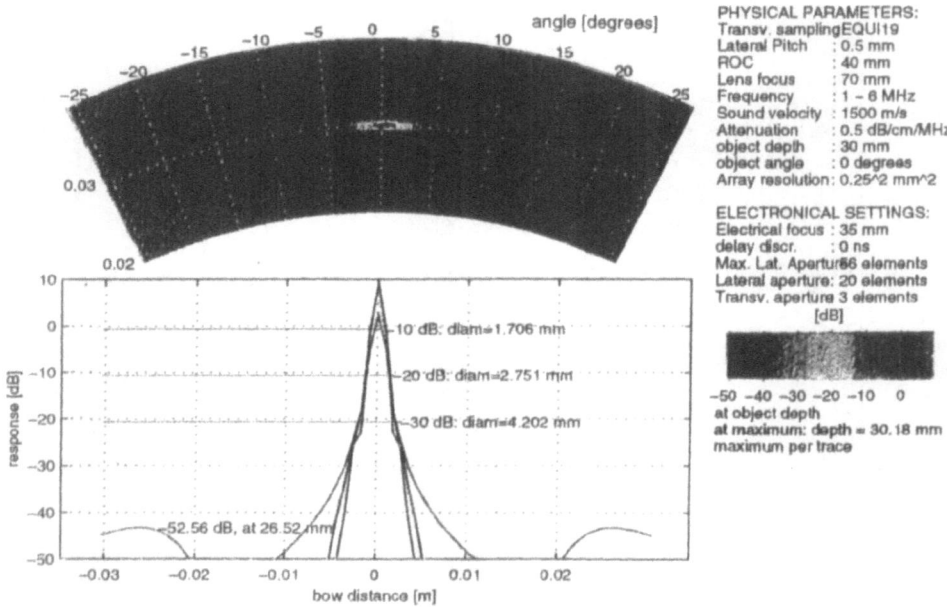

Figure 2. Example of full color image simulation.

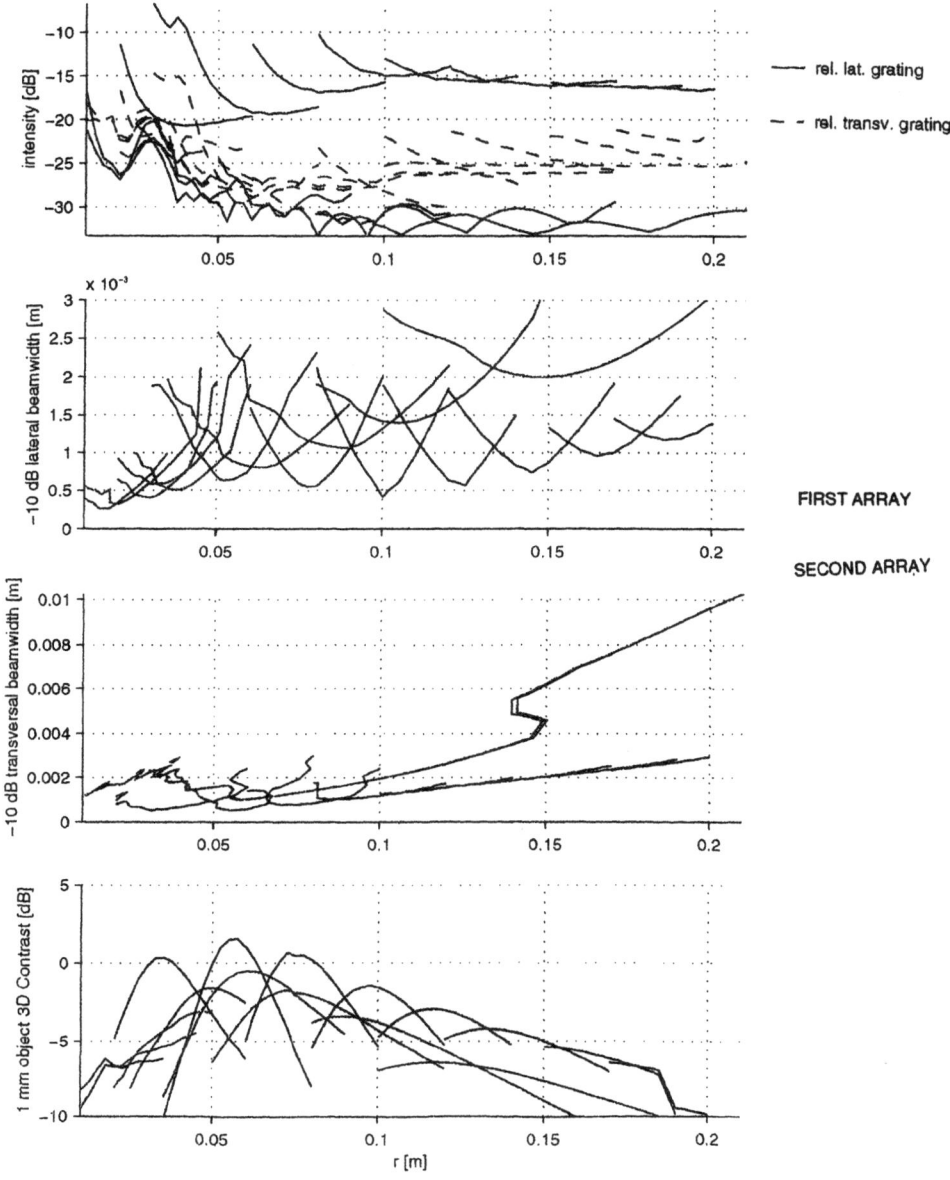

Figure 3. Example of Figure-of-Merits

REFERENCES

1. A.J. Berkhout, "Applied Seismic Wave Theory", Elsevier, Amsterdam (1987)
2. M.G.M. de Kroon and F.P.G. Driessen. "Deliverable D4 of Ultima project (no. 008122): Beam pattern and image simulations on convex array", TNO-report FSP-RPT-950031, Delft (12 May 1995)

SPATIAL SAMPLING OPTIMIZATION OF 2-D ARRAY FOR MEDICAL IMAGING WITH ELEVATION FOCUSING[1]

Y.Doisy*, C.Debaillon-Vesque*, and J.F.Gelly**

Thomson-Sintra ASM*, 06903, Sophia-Antipolis, France
Thomson Microsonics**, Sophia-Antipolis, France

INTRODUCTION

Imaging capability of an active acoustic imaging system in an incoherent homogeneous medium is characterized by a Contrast Figure of Merit (CFM) defined as the incoherent energy entering the mainlobe of the beampattern over the incoherent energy entering the sidelobes over all space. If this contrast is negative, the intensity at beam output is governed by the sidelobes and is not related to the steering direction, meaning that imaging capability is lost. This contrast is also related to the quality of the shadow formed behind an object of angular size larger than the mainlobe. The effect of spatial sampling on the CFM is analyzed in the case of 2D probes assumed to be oversampled in one dimension and undersampled in the other. First, the expression for the total incoherent intensity at beam output is derived for directive transmission and reception on a continuous rectangular focused array. Correction term due to undersampling is evaluated. In the following section, the expression for the incoherent intensity entering the mailobe is derived, for a rectangular array, in the same conditions. Corrections due to sampling and defocusing are given. The last section shows the behavior of the CFM on a specific example.

TOTAL INCOHERENT INTENSITY AT BEAM OUTPUT

Continuous Array

We consider in this section a continuous array, transmitting a short pulse e(t) focused at point F_t at distance x_t and using an area S_t of the array at transmission. At reception , the array focuses at point F_r at distance x_r using an area S_r. Then, assuming that:

1)the transmitted bandwidth B verify: $c/B \ll r = ct/2$, where c is the velocity of sound in the medium, t the time elapsed since transmission and r the observation distance,

2)the observation point is at a distance $r = ct/2$ much larger than array dimensions,

then the total incoherent intensity at beam output, assuming an homogenous scattering medium and neglecting absorption is:

$$I(r) = \frac{c}{2} \int_\omega \frac{|E(\omega)|^2}{r^2} \{ \int_\Omega | \frac{1}{S_t} \int_{S_t} e^{i\omega[\tau(M,\vec{r})-\tau(M,F_t)]} d^2M |^2 | \frac{1}{S_r} \int_{S_r} e^{i\omega[\tau(N,\vec{r})-\tau(N,F_r)]} d^2N |^2 \frac{d\Omega}{4\pi} \} \frac{d\omega}{2\pi}$$

where $E(\omega)$ is the spectrum of the transmitted pulse and $\tau(M,\vec{r})$ is the delay between point M on the array and the scatterer at $\vec{r}$ on the sphere $r = ct/2$ ("isochronous surface" of reference [1]). This expression is scaled with respect to the transmitted pressure of each element (assumed to be identical

[1] This work was performed with financial contribution of the European Community Research Programme ESPRIT, Ultima Project 008122.

for all elements) and with respect to the scattering strength of the medium. The intensity at a given frequency is therefore:

$$I_\omega(r) = \int_\Omega \mid \frac{1}{S_t} \int_{S_t} e^{i\omega[\tau(M,\vec{r}) - \tau(M,F_t)]} d^2 M \mid^2 \mid \frac{1}{S_r} \int_{S_r} e^{i\omega[\tau(N,\vec{r}) - \tau(N,F_r)]} d^2 N \mid^2 \frac{d\Omega}{4\pi} \tag{1}$$

showing the classical result that the incoherent intensity is the area under the product of the directivity patterns. Developping the squared magnitudes, performing the following change of variables:

$$\begin{cases} \vec{\sigma_t} = \frac{\vec{OM} + \vec{OM'}}{2} \\ \vec{\Delta_t} = \vec{OM'} - \vec{OM} \end{cases} \qquad \begin{cases} \vec{\sigma_r} = \frac{\vec{ON} + \vec{ON'}}{2} \\ \vec{\Delta_r} = \vec{ON'} - \vec{ON} \end{cases}$$

and noting that:

$$\begin{cases} c\tau(\vec{\sigma} + \frac{\vec{\Delta}}{2}, \vec{r}) - c\tau(\vec{\sigma} - \frac{\vec{\Delta}}{2}, \vec{r}) = \frac{\vec{\sigma}.\vec{\Delta}}{r} - (\vec{u}.\vec{\Delta})(1 + \frac{\vec{u}.\vec{\Delta}}{r}) \\ c\tau(\vec{\sigma} + \frac{\vec{\Delta}}{2}, F) - c\tau(\vec{\sigma} - \frac{\vec{\Delta}}{2}, F) = \frac{\vec{\sigma}.\vec{\Delta}}{x_F} - (\vec{u_0}.\vec{\Delta})(1 + \frac{\vec{u_0}.\vec{\Delta}}{x_F}) \end{cases}$$

where $\vec{u}, \vec{u_0}$ are the unit vectors pointing towards the scatterer at $\vec{r}$ and towards the probe focus axis (figure 1), we get after integrating over all the scatterer directions Ω:

$$I_\omega(r) = \int_{\vec{\Delta_t}, \vec{\sigma_t}, \vec{\Delta_r}, \vec{\sigma_r}} sinc(\frac{2\pi}{\lambda} \mid \vec{\Delta_t} - \vec{\Delta_r} \mid) e^{ik[(\frac{1}{x_t} - \frac{1}{r})\vec{\sigma_t}.\vec{\Delta_t} - (\frac{1}{x_r} - \frac{1}{r})\vec{\sigma_r}.\vec{\Delta_r} - \vec{u_0}.(\vec{\Delta_t} - \vec{\Delta_r})]}$$
$$K_t(\vec{\sigma_t} + \frac{\vec{\Delta_t}}{2})K_t(\vec{\sigma_t} - \frac{\vec{\Delta_t}}{2})K_r(\vec{\sigma_r} + \frac{\vec{\Delta_r}}{2})K_r(\vec{\sigma_r} - \frac{\vec{\Delta_r}}{2}) \frac{d^2\Delta_t d^2\sigma_t}{S_t^2} \frac{d^2\Delta_r d^2\sigma_r}{S_r^2}$$

where terms of order L/r (L is the larger array dimension) have been neglected with respect to 1, and where K_t ($/K_r$) are the transmitting (receiving) aperture functions. k is the wavenumber. We will consider in the following a plane array focusing on its normal axis so that $\vec{u_0}.(\vec{\Delta_t} - \vec{\Delta_r}) = 0$. Now, due to the sinc function (sinc(x)= sin(x)/x) whose range is $\lambda/2$, the order of magnitude of the argument of the exponential function can be evaluated by setting $\Delta_t = \Delta_r = \Delta$, and reads:

$$k\Delta[(\frac{1}{x_t} - \frac{1}{r})\sigma_t - (\frac{1}{x_r} - \frac{1}{r})\sigma_r] \sim \pi(\mid \frac{1}{x_t} - \frac{1}{r} \mid \frac{L_t}{2} + \mid \frac{1}{x_r} - \frac{1}{r} \mid \frac{L_r}{2})$$

Therefore, if the following conditions are met:

$$\mid \frac{1}{x_t} - \frac{1}{r} \mid \frac{L_t}{2} \ll 1, \text{and} \mid \frac{1}{x_r} - \frac{1}{r} \mid \frac{L_r}{2} \ll 1 \tag{2}$$

these terms can be neglected and the total incoherent intensity reads:

$$I_\omega = \frac{1}{S_t S_r} \int_{\vec{\Delta_t}, \vec{\Delta_r}} \Theta_t(\vec{\Delta_t})\Theta_r(\vec{\Delta_r}) sinc(\frac{2\pi}{\lambda} \mid \vec{\Delta_t} - \vec{\Delta_r} \mid) d^2\Delta_t d^2\Delta_r \tag{3}$$

where Θ is the aperture correlation function defined as:

$$\Theta(\vec{\Delta}) = \frac{1}{S} \int_{\vec{\sigma}} K(\vec{\sigma} + \frac{\vec{\Delta}}{2})K(\vec{\sigma} - \frac{\vec{\Delta}}{2}) d^2\sigma$$

Expression (3) means that provided:

1)the observation range is larger than the array dimensions,

2)the point of observation is close enough (conditions (2)) to the transmit and receive focus points, the total incoherent intensity is the same as it would be in the far field (except for the propagation losses). Expression (3) can be further simplified if we assume that array dimensions are much larger than a wavelength. Then, making the following change of variable:

$$\begin{cases} \vec{\Sigma} = \frac{\vec{\Delta_t} + \vec{\Delta_r}}{2} \\ \vec{\Delta} = \vec{\Delta_t} - \vec{\Delta_r} \end{cases}$$

where components of $\vec{\Sigma}$ vary from $-(L_t + L_r)/2$ to $(L_t + L_r)/2$ and components of $\vec{\Delta}$ vary from $-(L_t + L_r)$ to $(L_t + L_r)$, we get:

$$I_\omega = \frac{1}{S_t S_r} \int_{\vec{\Delta}, \vec{\Sigma}} \Theta_t(\vec{\Sigma} + \frac{\vec{\Delta}}{2})\Theta_r(\vec{\Sigma} - \frac{\vec{\Delta}}{2}) sinc(\frac{2\pi}{\lambda}\Delta) d^2\Delta d^2\Sigma$$

Since the sinc function has a range of $\lambda/2$, wich is much smaller than the characteristic scale of the aperture functions, we can neglect the dependence over $\vec{\Delta}$ in the aperture correlation functions. This is equivalent to neglect terms of order $(\lambda/L)^2$. Then, the integrals over $\vec{\Sigma}$ and $\vec{\Delta}$ split and each of them can be worked out. After some calculation we get the following expression:

$$I_\omega = \frac{\lambda^2}{4\pi} \frac{1}{H_M}(1 - \frac{H_m}{3H_M}) \frac{1}{L_M}(1 - \frac{L_m}{3L_M}) \tag{4}$$

where $H_t(/H_r)$ is the height of the transmitting (/receiving) array aperture, $L_t(/L_r)$ is the length of the transmitting (/receiving) array aperture, $H_M(/H_m)$ is the larger (/smaller) of transmitting and receiving aperture height, and $L_M(/L_m)$ is the larger (/smaller) of transmitting and receiving aperture length.

Spatial Sampling

When proceeding to the sampling of the array, we will assume that all the surface is covered by the transducers, neglecting the small spacing between adjacent elements. Then, the effect of the sampling is to apply quantized delays instead of continuous delays, resulting in an elementary directivity pattern of each transducer. Then the behavior of the system shows two regimes:

1) if the elements are sampled according to a non-aliasing condition, the beam patterns in (1) show no grating lobes and the array behaves as a continuous array. The reason for that is that in this case, the total incoherent intensity arises mainly from the mainlobe of the directivity pattern, and the attenuation due to elementary directivity patterns is the same for the signal as for the incoherent intensity, and spatial sampling has no effect on the CFM. The non-aliasing condition for a focused 1D curved array depends on the its radius of curvature R, its half angular aperture Θ, and the point of focus x_F and can be found for example from [2]:

$$p \leq \frac{\lambda}{1 + R\sin(\Theta)[\frac{1}{R} + \frac{\cos(2\Theta)}{x_F}]} \tag{5}$$

where p is the inter elements spacing, or pitch.

2) if the elements are strongly under-sampled, according to a condition which will be developed below, the total incoherent intensity becomes insensitive to the delays applied to the elements. The reason for that is that attenuation of the mainlobe due to elementary directivities is compensated by the incoherent intensity entering through the grating lobes.

The systems under investigation are rectangular arrays in the (y,z) plane fitted with a prefocused lens in the elevation direction (z-direction on fig.1) focusing at point x_0 on probe axis. The lens provides continuous delays along the z-axis, and the array is sampled in the z-direction in order to allow electronic focusing at other points in the vicinity of x_0 on x-axis. The elevation sampling has to be optimized to provide good imaging capability of the system in a given range about x_0. We will assume that the array is correctly sampled in the y-direction (condition (5) is fulfilled) and therefore consider only spatial sampling in the z-direction.

In deriving the total incoherent intensity, we will first consider a single line array along z-axis, consisting of N elements, N being odd, the array being symetric with respect to its center O (figure 2). Considering the upper half array, the edges of successive elements are noted z_n with n ranging from 1 to $(N+1)/2$ and the total length is $H = 2z_{(N+1)/2}$. For the lower half array, we have $z_n = -z_n$ n ranging from -1 to $-(N+1)/2$. The phase center of element n is noted ζ_n n ranging from $-(N-1)/2$ to $(N-1)/2$ and $\zeta_{-n} = -\zeta_n$. The array is fitted with the lens, so that to focus at point F_r the delay to apply to element n is:

$$c\tau(\zeta_n, F_r) = \frac{\zeta_n^2}{2}(\frac{1}{x_r} - \frac{1}{x_0}) = c\tau_n \text{ with } -(N-1)/2 \leq n \leq (N-1)/2$$

The lens provides a continuous focusing and we have seen in the previous section that we can discard its effect on the total incoherent intensity as long as conditions (2) are fulfilled. Then, with an omnidirectionnal transmission, the total incoherent intensity for the sampled line array reads:

$$I_\omega = \frac{1}{H^2}\sum_n\sum_{n'} e^{i\omega\tau_n} e^{-i\omega\tau_{n'}} \int_{z_n}^{z_{n+1}}\int_{z'_n}^{z'_{n'+1}} sinc[\frac{2\pi}{\lambda}\mid z - z' \mid]dzdz' = \frac{1}{H^2}\sum_n\sum_{n'} e^{i\omega\tau_n} e^{-i\omega\tau'_{n'}} J_{n,n'}$$

This expression can be figured out, keeping only terms involving the same and neighboring elements. Then we get after some calculation:

475

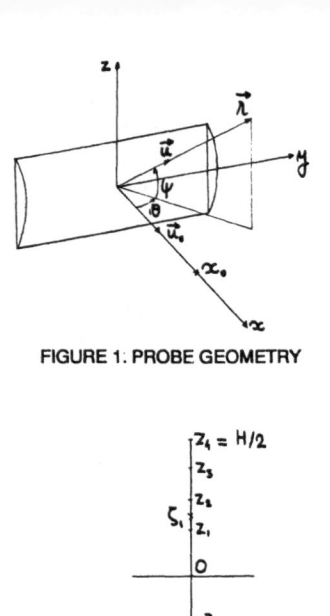

FIGURE 1. PROBE GEOMETRY

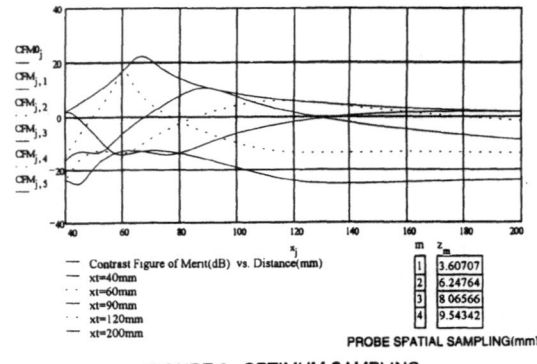

Contrast Figure of Merit(dB) vs. Distance(mm)
— xt=40mm
— xt=60mm
— xt=90mm
— xt=120mm
— xt=200mm

m	z_m
1	3.60707
2	6.24764
3	8.06566
4	9.54342

PROBE SPATIAL SAMPLING(mm):

FIGURE 3. OPTIMUM SAMPLING

$z_4 = H/2$
z_3
z_2
$\zeta_i \{ z_1$
O
$-z_1$
$-z_2$
$-z_3$
$-z_4$

FIGURE 2. ELEVATION SAMPLING

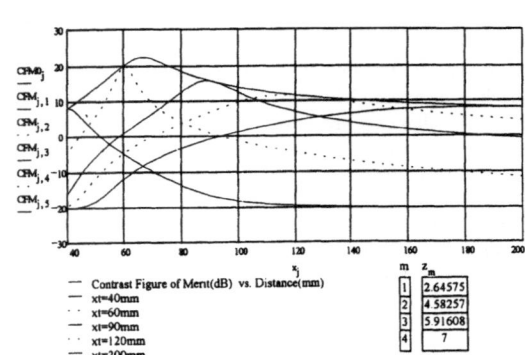

Contrast Figure of Merit(dB) vs. Distance(mm)
— xt=40mm
— xt=60mm
— xt=90mm
— xt=120mm
— xt=200mm

m	z_m
1	2.64575
2	4.58257
3	5.91608
4	7

PROBE SPATIAL SAMPLING(mm):

FIGURE 4. OPTIMUM SAMPLING

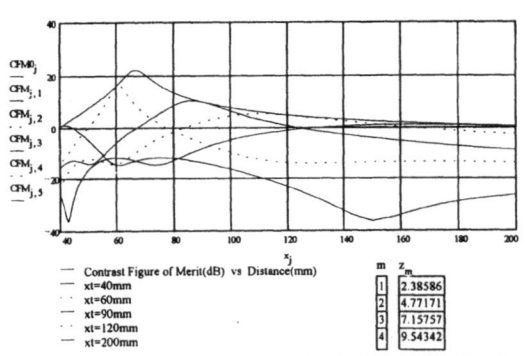

Contrast Figure of Merit(dB) vs. Distance(mm)
— xt=40mm
— xt=60mm
— xt=90mm
— xt=120mm
— xt=200mm

m	z_m
1	2.38586
2	4.77171
3	7.15757
4	9.54342

PROBE SPATIAL SAMPLING(m·r₁):

FIGURE 5. EQUI-IMPEDANCE

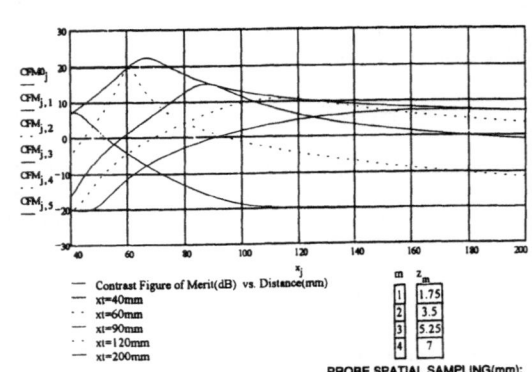

Contrast Figure of Merit(dB) vs. Distance(mm)
— xt=40mm
— xt=60mm
— xt=90mm
— xt=120mm
— xt=200mm

m	z_m
1	1.75
2	3.5
3	5.25
4	7

PROBE SPATIAL SAMPLING(mm):

FIGURE 6. EQUI-IMPEDANCE

$$J_{n,n} = \lambda(z_{n+1} - z_n)\{\frac{1}{2} - \frac{\lambda}{2\pi^2(z_{n+1} - z_n)} + O[\frac{\lambda^2}{(z_{n+1} - z_n)^2}]\} \text{ and } J_{n,n-1} = (\frac{\lambda}{2\pi})^2$$

$$I_\omega = \frac{\lambda}{2H}\left\{1 - \frac{\lambda}{\pi^2 H} - \frac{2\lambda}{\pi^2 H}\sum_{n=1}^{(N-1)/2}[1 - \cos(\omega\tau_n - \omega\tau_{n-1})]\right\} \tag{6}$$

In expression (6), the first corrective term is a second order correction to the continuous case expression, and the effect of sampling is associated to the last term only. If the sampling is properly optimized, the maximum delay differences between adjacent transducers are approximatively the same. Moreover, the attenuation of each element should not exceed 3dB, otherwise imaging capability of the system will be lost. Then the order of magnitude of the corrective term is:

$$\frac{2\lambda}{\pi^2 H}\sum_{n=1}^{(N-1)/2}[1 - \cos(\omega\tau_n - \omega\tau_{n-1})] \sim \frac{1}{\pi^2}(1 - \frac{1}{\sqrt{2}})\frac{(N-1)\lambda}{H} = 0.03\frac{\lambda}{\Delta z} \tag{7}$$

where Δz is the mean element spacing. We see therefore that the corrective term is only 3% for a mean element size of λ, and will be of order 0.5% for arrays considered in the following sections.

For two-dimensionnal array with continuous sampling in the y-direction, and under-sampling in the z-direction, exact evaluation of the total incoherent intensity has not been performed, but dimensional analysis shows that the corrective term has the same behavior as (7), with a slightly different coefficient.

We therefore conclude this section with the result that to an accuracy given by (7), and assuming that all the surface of the array is covered by transducers, the total incoherent intensity at the output of an array undersampled in one dimension, is insensitive to the delays applied to the elements, and is the same as at the output of a continuous focused array (expression(4)).

INCOHERENT INTENSITY IN THE MAINLOBE

Continuous Array

We derive in this section the incoherent intensity entering the mainlobe of the directivity pattern in case of directive transmission with the hypotheses that the array is rectangular, focusing at the same point at transmission and reception and at a distance equal to the focal point: $r = x_t = x_r$. The incoherent intensity is given by expression (1), but with integration over a solid angle $\delta\Omega$ given by:

$$\delta\Omega = \left\{ \begin{array}{l} -\frac{\lambda}{L_M} \leq \cos\psi\sin\theta \simeq \sin\theta \leq \frac{\lambda}{L_M} \\ -\frac{\lambda}{H_M} \leq \sin\psi \leq \frac{\lambda}{H_M} \end{array} \right.$$

instead of 4π because only the scatterers in this solid angle contribute to the intensity in the mainlobe. Proceeding as in section 2.1, we get the following expression:

$$I_\omega^{ml} = \frac{\lambda^2}{\pi S_M}\frac{1}{S_t S_r}\int_{\vec{\Delta},\vec{\Sigma}}\Theta_t(\vec{\Sigma} + \frac{\vec{\Delta}}{2})\Theta_r(\vec{\Sigma} - \frac{\vec{\Delta}}{2})sinc(\frac{2\pi}{L_M}\Delta_y)sinc(\frac{2\pi}{H_M}\Delta_z)d^2\Delta d^2\Sigma \tag{8}$$

where $S_M = L_M H_M$ and (Δ_y, Δ_z) are the components of $\vec{\Delta}$. This expression is similar to expression (3) except the fact that now the sinc functions (which represent the spatial correlation of incoherent scattering) have much larger ranges (L_M and H_M instead of $\lambda/2$) according to the fact that the concerned scatterers span a small solid angle [2]. The calculation of expression (8) is tedious but can be worked out. One finally gets for the intensity entering the mainlobe:

$$I_\omega^{ml} = \frac{\lambda^2}{4\pi S_M}\cdot\frac{2F(L_m/L_M)}{\pi(L_m/L_M)^2}\cdot\frac{2F(H_m/H_M)}{\pi(H_m/H_M)^2} \tag{9}$$

with:

$$F(x) = \frac{(1+x)^3}{6}Si[2\pi(1+x)] + \frac{(1-x)^3}{6}Si[2\pi(1-x)] - \frac{1}{3}Si(2\pi) \quad -\frac{x^3}{3}Si(2\pi x) - \frac{\sin^2(\pi x)}{3\pi}$$

Spatial Sampling

We assume the array is sampled in the z-direction as described in section 2.2.Then, taking into account the fact that the elementary transducers are much smaller than the array, the directivity function in the vicinity of the mainlobe will not be affected if the elementary directivity of each element is set constant and equal to its value in the direction of the mainlobe. The resultant lobe corresponds to the aperture function of the array weighted by the values of the elementary directivity patterns in the steering direction. The main effect is to attenuate the mainlobe of the following amplitude:

$$D = \frac{2}{H} \sum_{n=0}^{(N-1)/2} \int_{z_n}^{z_{n+1}} e^{i\frac{k}{2}(z^2 - \zeta_n^2)(\frac{1}{r} - \frac{1}{x_0})} dz$$

there is also a slight modification of the shape of the mainlobe due to the different values of the weighting coefficients. However, if the sampling is well designed, all the elements have approximatively the same attenuation in the steering direction and the preceeding correction is exact. This correction occurs at transmission and at reception(from expression (1)). So far, we have assumed that the transmit focus was the same as the receive focus and equal to the observation distance r. We will assume dynamic focusing at reception so that $x_r = r$ always holds true. We will take into account the fact that $x_t \neq r$ by the corresponding attenuation of the transmission mainlobe (assuming $L_t(/H_t) < L_r(/Hr)$), so that the amplitude corrections at transmission ($D_t(r)$) and at reception ($D_r(r)$) read:

$$D_t(r) = \{\frac{2}{H_t} \sum_{n=0}^{(N-1)/2} \int_{z_n}^{z_{n+1}} e^{i\frac{k}{2}[z^2(\frac{1}{r} - \frac{1}{x_0}) - \zeta_n^2(\frac{1}{x_t} - \frac{1}{x_0})]dz}\}\{\frac{2}{L_t} \int_0^{L_t/2} e^{i\frac{ky^2}{2}(\frac{1}{r} - \frac{1}{x_t})} dy\} \tag{10}$$

$$D_r(r) = \frac{2}{H_r} \sum_{n=0}^{(N-1)/2} \int_{z_n}^{z_{n+1}} e^{i\frac{k}{2}(z^2 - \zeta_n^2)(\frac{1}{r} - \frac{1}{x_0})} dz \tag{11}$$

The contrast figure of merit for a given transmit/receive strategy reads finally:

$$CFM = 10.\log[\frac{| D_t(r)D_r(r) |^2 I_\omega^{ml}}{I_\omega - | D_t(r)D_r(r) |^2 I_\omega^{ml}}] \tag{12}$$

where I_ω is the total incoherent intensity given by (4),I_ω^{ml} is the incoherent intensity entering the mainlobe for a continuous array given by (9), $D_t(r), D_r(r)$ are the correcting factors accounting for spatial sampling and transmission focusing, given by (10) and (11).

EXAMPLES AND DISCUSSION

The previous results are used to optimize the elevation sampling of a rectangular array of $N = 7$ elements in the z-direction, working at $3.5MHz$. The probe is designed to focus between $x_1 = 40mm$ and $x_2 = 200mm$, the lens focal point being set to $x_0 = x_1x_2/2(x_1 + x_2)$ .The lengths in y-direction are $L_t = 3mm$ and $L_r = 6mm$.Two heights are considered: $H_t = H_r = 2z_4 = 19.8mm$ and $14mm$, corresponding to mean element spacings of 47λ and 33λ, so that expression (4) hold. Figures 3 to 6 show plots of the CFM versus the distance to the probe for the transmission focus distances of $x_t = 40/60/90/120/200mm$, assuming continuous dynamic focusing at reception.The enveloppe curve of each plot, designated as CFM_0, corresponds to a transmission focused at the observation distance $x_t = r$ and isolates the effect of spatial sampling.

Figures 3 and 4 correspond to the sampling defined as $z_n \sim \sqrt{(n - 0.5)\lambda/(1/x_1 - 1/x_2)}$ designed to minimize the losses on axis at the edges of the field (x_1 and x_2).It is designated as "optimum"sampling.

Figures 5 and 6 correspond to an equi-impedance sampling. It can be seen from the curves that the behavior of the CFM is governed by signal attenuation on axis, a too large array in z-direction leading to low values of the contrast at the edges of the field. The loss associated to equi-impedance sampling instead of optimum sampling is about 1.5dB at the edges.

References

[1] R.MALLART and M.FINK The van Cittert-Zernike theorem in pulse echo measurements. J. Acoust. Soc. Am.,90(5),November 1991,p.2718-2727

[2] S.UMEMURA and K.KATAKURA Theoretical Analysis of Grating Lobe Intensity.IEEE Ultrasonic Symposium,1986,p.659-662

THE EFFECT OF CROSS-COUPLING IN THE ACOUSTIC FIELD
GENERATED BY A PHASED ARRAY TRANSDUCER

A. Gubbini[1], C. Lamberti[2], P. Palchetti[1], and A. Sarti[2]

[1]ESAOTE Biomedica. Via di Caciolle 15, 50127 Firenze, Italy
[2]DEIS, Università di Bologna. Viale Risorgimento 2, 40136 Bologna, Italy

INTRODUCTION

Focusing and steering in the phased array transducers are obtained by driving all sources with appropriate delay. To increase spatial resolution and sensitivity in the imaging plane, the main lobe of each source has to be as large as possible, and the emitting source dimension has therefore to be smaller with respect to the wave length.

The cross-coupling between the array elementary sources increases the emission area and decreases the beam performance.

A method for numerical simulation of acoustic field with an equivalent emission area is presented; moreover simulations relative to the prototipes of phased array transducers have been shown to study the cross-coupling effects in the acoustic beam.

METHOD

A phased-array transducer consists in an array of elementary rectangular sources independently and electronically driven to obtain a focused and steered beam. The computation of the pressure field radiated by transducer is based on the Huygen's principle, and each element has been assumed as a flat piston vibrating in an infinite rigid baffle[1,2]. The short burst wide band excitation and lossy media have been assumed since the lossy effects on the beam shape are not negligible[3,4]. The total pressure p from the whole array at a point in the field is given adding the contribute of all N sources, each with its proper delay[5]:

$$p(t,\bar{r}) = \sum_{i=1}^{N} \left\{ \iint_{S_o} F^{-1}\left[S_i(\omega) \cdot G(\omega,\bar{r},\bar{r}_o) \right] ds_o \right\} \tag{1}$$

where S is the driving function Fourier transform obtained measuring the acoustic pressure at the focus and assuming it as the pressure generated at the interface S_o between media and source:

$$S_i(\omega) = V(\omega) \cdot e^{-j(\varphi(\omega) + \omega\tau_{dfi})} \tag{2}$$

$V(\omega)$ and $\varphi(\omega)$ are, respectively, module and phase of excitation, while τ_{dfi} is the relative delay of the i-th source. G is the Green function and corrisponds to the lossy media transfer function:

$$G(\omega, r) = \frac{1}{2\pi} \cdot \frac{e^{-\alpha\omega r}}{r} \cdot e^{-j\omega r/c(\omega)} \tag{3}$$

where α is the attenuation coefficient, c is the phase velocity of the sound waves and r is the distance between the field point determined by the vector $\bar{r}$ and the elemental area of the source determined by $\bar{r}_o$.

To take into account the cross-coupling effect under the hipothesis that each array source is considered like a flat piston vibrating in an infinite rigid baffle, an equivalent element source's area has been considered. A single transducer of array has lateral dimension lower than $\lambda/2$ and elevation dimension between 50 and 100 times higher so that the longitudinal section field profiles in the two planes parallel to the sides are largely indipendent[6]. The cross-coupling effects on the single source lateral dimension can be found from a measurement of its directional response $D(f, \theta)$, comparing it to the theoretical directional response of a uniform circoular source; i.e., $2J_1(x)/x$, where J_1 is a first order Bessel function of first kind, and $x = \pi d_{eq} f \sin\theta/c$, where d_{eq} is the effective source dimension and c is the speed of sound in the propagation medium. If, for some f and θ, $D(f, \theta)$ have been measured and assumed to equalize the uniform source model relationship $2J_1(x)/x$, then a value for x can be computed and d_{eq} can be obtained. A driving function central frequency, and -3dB and -6dB deflection have been assigned to f and θ respectively.

RESULTS

Some fine pitch 128 elements phased-array have been build to investigate the cross-coupling causes and effects. Particularly, different element fillers and different matching layers have been compared. In this paper, the results concerning two phased-array transducers with different fillers, with or without $\lambda/4$ matching layers are presented . Moreover a special array with an isolated single element has been build as example of non-coupled source.

The phased-array transducers that have been built are cardiological devices with $2.5MHz$ resonance frequency, and the single element dimension is $20 \times 0.17 mm$.

In accord to the presented method the single element directional response has been measured and the equivalent lateral dimension has been calculated.

TABLE I
Equivalent lateral dimension

Transducers	parameters		
	θ_{-3dB}	θ_{-6dB}	deq(mm)
non-coupled element	26	40	0.7
coupled element without matching layers	18	25	1
coupled element with matching layers	15	22	1.2

The results are shown in Table I. The equivalent dimensions have been used for the numerical simulation in place of the real one.

The figures that follow show the simulation with three different equivalent dimension. Each figure show an isocronous image of B-mode acoustic pulse computed at the focal for a 12.8×12.8mm area in the imaging plane. Beam attenuation with respect to the no steering beam and -6dB lateral dimension with respect to the peak to peak maximum have been calculated to estimate the sensitivity and resolution loss steering the beam from 0 to 30 degree.

Figure 1 shows a simulation with equivalent dimension of single source equal to the effective; at 30°, -0.3dB of sensitivity and -30% of resolution have been calculated.

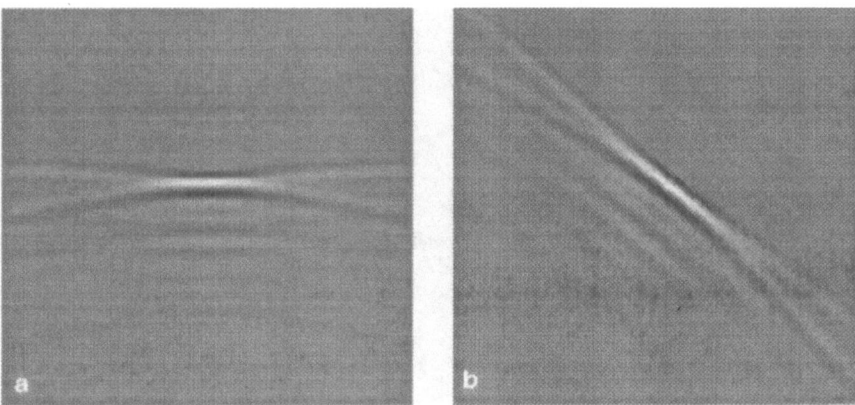

Figure 1. *Ultrasonic field in the imaging plane computed at the focus distance of 8cm with equivalent lateral dimension equal to the effective dimension ($d_{eq}=d_{eff}=0.17mm$): (a) 0° steering beam, (b) 30° steering beam.*

Figure 2 shows a simulation assuming the equivalent lateral dimension computed from the non-coupled element; at 30°, -4.6dB of sensitivity and -46% of resolution were been calculated.

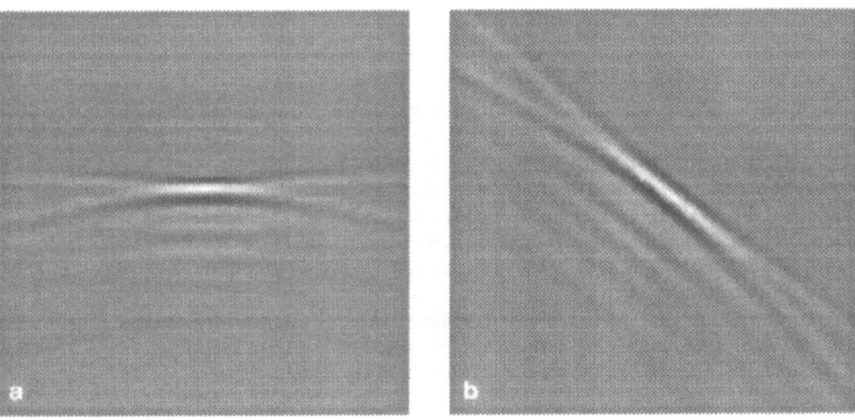

Figure 2. *Ultrasonic field in the imaging plane computed at the focus distance of 8cm with $d_{eq}=0.7mm$: (a) 0° steering beam, (b) 30° steering beam.*

Finally, Figure 3 shows a simulation assuming the equivalent lateral dimension computed from coupled element from the phased-array with two λ/4 matching layers; at 30° the sensitivity and resolution lossy raise to -17dB and to -94% respectively.

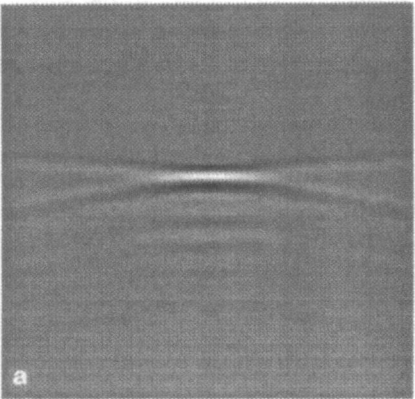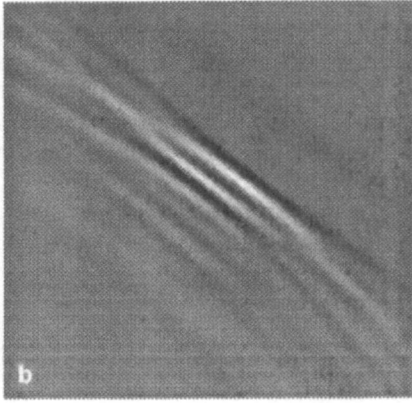

Fig. 3. Ultrasonic field in the imaging plane computed at the focus distance of 8cm with $d_{eq}=1.2mm$: (a) 0° steering beam, (b) 30° steering beam.

CONCLUSION

This work present a general method for modelling the acoustic field generated by phased-array transducers, taking into account the cross-coupling effects among elements. Simulations, by referring to specific transducers, have been made and -6dB lateral resolution and sensitivity have also been computed to quantitatively describe the effects on the acoustic beam. The results show as the cross-coupling can't be neglected for a most realistic simulation.

Numerical simulation can be adopted to study and evaluate the benefits that can be obtained trough different tecnological solutions.

REFERENCES

1. G.R.Harris, "Review of transient field theory for a buffled piston", J.Acoustic.Soc.Am., vol.70, pp.10-20, 1981.
2. J.W.Hunt, M.Arditi, F.S.Foster, " Ultrasound Transducers for Pulse-Echo Medical Imaging ", IEEE Trans.Biom.Eng., Vol.30, n.8, pp 1872-1884, 1989.
3. M.M.Goodsitt,E.L.Madsen, J.A.Zagzebski, "Field patterns of pulsed, focused, ultrasonic radiators in attenuating and not attenuating media." J.Acoustic.Soc.Am., vol.71, pp.318-329, 1982.
4. R.Learch, W.Friedrich, " Ultrasound fields in attenuating media ", J.Acoust.Soc.Am., Vol.80, n.4, pp.1140-1147, 1986.
5. A.Sarti, P.Bassi, C.Lamberti, " 3-D Modelling of Phased-Array Generated Ultrasounds in Lossy Media ", Computerized Medical Imaging and Graphics, Vol. 17, n. 4-5, pp.339-343,1993.
6. K.B.Ocheltree, L.A.Frizzel, " Sound Field Calculatio for Rectaungoular Sources ", IEEE Trans.Ultr.Ferr.Freq., Vol.36, n.2, pp.242-248, 1989.

MULTIBEAM PHASED ARRAY WITH DYNAMIC
FOCUSING AND IMPROVED IMAGE QUALITY

S.G. Mesochorjanakis[1], J.I. Burov[1], V.L. Strashilov[1], and
M.G. Christov[2]

[1]Sofia University, Faculty of Physics
Department of Solid State Physics and Microelectronics
5 James Bourchier Blvd.
Sofia BG-1126, Bulgaria

[2]Technical university
Faculty of Electronics and Electronic Technology
Durvenitza, Sofia BG-1000, Bulgaria

INTRODUCTION

Phased arrays of acoustic transducers have widely been used for obtaining 2D imaging [1]. For the last two decades a rapid growth of the application of these techniques in ultrasound diagnosis has been observed [2,3]. The principles of the linear phased array have been well established [4] and there is no use to go through them in detail. It has been acknowledged that, although these systems have the disadvantage of being complex and often costly, in most cases serious advantages come to outweight the limitations. Therefore, the interest to these arrays has remained constant, and the efforts to improve their performance are continuous. In this contribution we report on one such aspect, which, in our opinion, has been practically disregarded in the literature. It relates to the possibility to operate a phased array at several carrier frequencies and to process the information parallelly on receipt. In this way a number of field directions could be simultaneously interrogated with corresponding increase in the scan rate at a given lateral resolution or vice-versa.

The principles of multifrequency operation.

Frequency bandwidth has been known to be a key factor in signal processing. To make the received information as adequate to the real one in the studied object as possible, the individual transducer bandwidth should be broad enough to transmit and receive all Fourier components of interest. For this purpose the bandwidth is usually extended to its achievable maximum by impedance matching to both sides of the transducer. Under conditions of perfect matching the Q-factor of the loaded transducer becomes close to unity, i.e. the -3dB bandwidth equals the resonant frequency. It is interesting to compare the frequency band of a typical image pulse to that of a typical transducer. According to data in the literature a pulsed information on receipt usually has a Fourier spectrum at -3dB width not exceeding 2.5MHz for 5MHz carrier frequency [5]. On the other hand, the allowed frequency range is only restricted by the increased attenuation at higher frequencies. For the human body, this limit is established somewhere about 7MHz, i.e. we dispose of a frequency band which is too broad for the signals to process. There appears to be an excess in the system capabilities which remains unused. This can be reduced and more information obtained if the transducer band is devided into a number of channels for parallel image processing. In such a mode the transducer assembly is oscillated at several carriers simultaneously, each carrier being provided with its own packet of phase delays to focus the array along a definite direction. On receipt, the information from different carriers is frequency selected by filtering through a cascade of bandpass filters with appropriately designed characteristics. The applicability of this approach is subject to two major questions concerning: i) the geometry of the array and its optimization for multifrequency operation, ii) the information loss due to crosstalk between adjacent channels.

i) The array directivity. The diffraction limited resolution and the dynamic range of an array are known to be determined by its characteristic dimensions (full aperture D and interelectrode spacing d, respectively). An optimum relation for effective suppression of sidelobes is [6,7] $d \div \lambda = 0.65$. Since this relation can only be satisfied at a given frequency, a sound beam emitted at another frequency would necessarily have a worsened directivity. Let us consider a linear array with 33 transducer elements arranged as shown in Fig. 1.

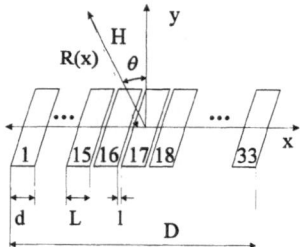

Fig. 1. A 33 element linear phased array.

The classical theory of array emission into a liquid medium of sound velocity c leads to the following expression for the directivity pattern in the xy plane [8] :

$$D(\theta) = \sqrt{[\text{Re}(\theta)]^2 + [\text{Im}(\theta)]^2} ,\qquad (1)$$

where

$$\text{Re}(\theta) = \sum_{n=-m}^{m} \int_{\frac{2 \cdot n-1}{2} \cdot L + n \cdot l}^{\frac{2 \cdot n+1}{2} \cdot L + n \cdot l} \frac{1}{R(x)} \cdot \cos\left[\omega \cdot \left(\frac{[DF(n)]}{c} + \frac{[DS(n)]}{c}\right) - \kappa \cdot R(x)\right] dx \qquad (2)$$

$$\mathrm{Im}(\theta) = \sum_{n=-m}^{m} \int_{\frac{2\cdot n-1}{2}\cdot L+n\cdot l}^{\frac{2\cdot n+1}{2}\cdot L+n\cdot l} \frac{1}{R(x)} \cdot \sin\left[\omega \cdot \left(\frac{[DF(n)]}{c} + \frac{[DS(n)]}{c}\right) - \kappa \cdot R(x)\right] dx \qquad (3)$$

We have introduced a linear phased delay variation for steering the main lobe at angle φ

$$DS(n) = n \cdot (L + l) \cdot \sin(\varphi), \qquad (4)$$

a quadratic phase delay variation for focusing at range r

$$DF(n) = \sqrt{[n \cdot (L + l)]^2 + r^2} - r \qquad (5)$$

and $R(x) = \sqrt{x^2 + H^2 - 2 \cdot x \cdot H \cdot \sin(\theta)}$ is the distance to the imaged point.

In Fig. 2 a, b we present two theoretical directivity patterns obtained at frequencies $f_1 = 2.5\text{MHz}$ and $f_2 = 5\text{MHz}$, if the condition for effective sidelobe suppression is satisfied at the lower frequency (a) or the higher frequency (b). In these figures the focusing range is normalized to the wavelength, i.e. both frequencies are focused at the same number of wavelengths. The degradation in the directivity at the other (unoptimized) frequency is considerable, which makes the case inapplicable because the two modes of operation are not comparable in efficiency. Since the geometry of the array is established apriori, the only means to affect the geometrical relationships is the phased delay distribution. In Fig. 3 we show the directivity of the same type of array as in Fig. 2b (geometry optimized at the hig-

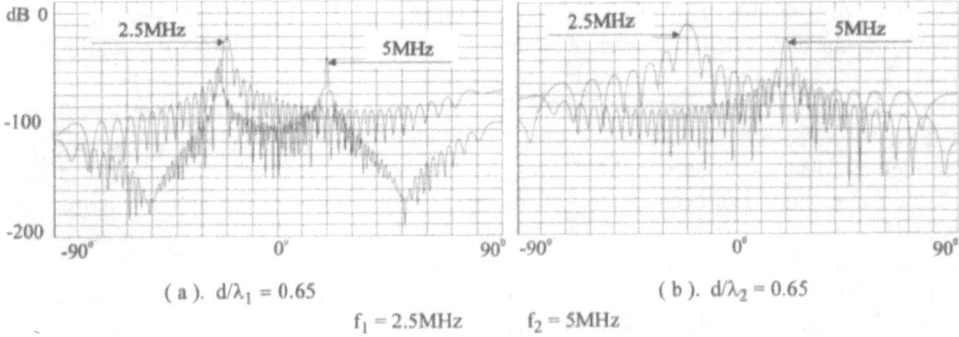

(a). $d/\lambda_1 = 0.65$ (b). $d/\lambda_2 = 0.65$

$f_1 = 2.5\text{MHz}$ $f_2 = 5\text{MHz}$

Fig. 2. Theoretical directivity pattern resulting from two frequency array excitation.

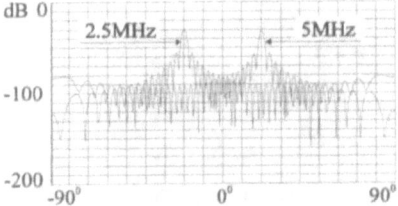

Fig. 3. Theoretical directivity pattern for geometry as in Fig. 2b with $d_1 = 2d_2$ (effective)

her frequency), but the phased delay is quantized in a different manner. While jumping steplike from transducer to transducer at f_2, it is arranged to have one and the same value on each two adjacent transducers at f_1. In this way an effective doubling of the interelement spacing is approximated which comes to compensate for the doubling in wavelength. If more carriers are used, we should compromise between array geometry and phase delay distribution, to obtain comparable mode efficiency at all carriers.

ii) Experimental Fourier analysis. The problem of interchannel crosstalk arises because the reflected pulses are known to have a close to Gaussian envelope, i.e. their frequency spectrum is also of similar shape. Accordingly, the channel selection filters should be broad enough and not much selective at the outskirts to transmit all frequency components. In this situation it is not possible to arrange a sequence of adjacent channels without a certain crossover of the characteristics, since the whole bandwidth is limited to about 4MHz. For example if we wish to open four channels, 2MHz each, the relative crossover will be larger than that in case of two channels 2MHz each. The question is what the acceptable limit is to compromise without destroying the spatial resolution of the array. In an attempt to answer this question we carried out the experimental study illustrated in Fig. 4. Three ceramic transducers with 2.1MHz resonant frequency are specifically located in a water tank. Two of them are arranged to operate as transmitter and receiver, the sound wave being reflected from the surface of a fused silica cube submerged in front of the transducers. The frequency band of this system was measured in CW regime and is shown in Fig.5. It is about 0.6MHz at -3dB and is somewhat narrower than that of the individual transducers. Then the surface of the cube was provided with an additional reflecting structure, modeling a periodic density fluctuation in the body with spatial period 1.8mm. It represents a sequence of two polymer foil sheets spaced at 1.8mm between themselves and from the glass surface also. This structure was irradiated with a pulsed sound beam from the input transducer at 2.1MHz carrier frequency, 0.95µs pulse width and 400µs repetition period. A typical waveform received by the output transducer and its Fourier spectrum are shown in Fig. 6a,b. The signals from the three reflecting surfaces are excellently resolved and we have a resolution well exceeding 1.8mm, although the information has been filtered through a -3dB band of only 0.6MHz. Now let us speculate that another frequency channel of the same shape has been introduced in the reception mode at the doubled frequency 4.2MHz (Fig. 5). With respect to the operating low-frequency channel, this one plays the role of a noise factor, the noise level depending on the frequency of operation. At 2.1MHz it is -6.3dB and is obviously exaggerated compared to a " real " situation because of the artificially produced bump of the characteristics in this range. We have simulated the noise action of this fictious channel by directly irradiating the receiving transducer with a CW

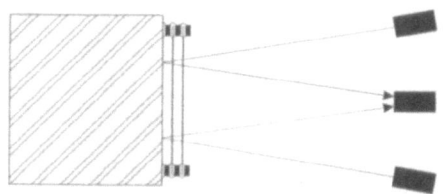

Fig. 4. Three transducer experimental arrangement
for interchannel crosstalk analysis.

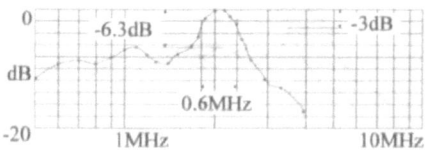

Fig. 5 Frequency characteristics on receipt.
Dashed: imaginary channels at 3.15
and 4.2MHz

sound wave from a third transducer of the same type. This wave models the most contributory Fourier component of the noise spectrum at the operating frequency 2.1MHz. By controlling the output of the sine-generator its level has been adjusted 6.3dB below the useful signal level, and we have also altered its phase in the range from 0^0 to 90^0. The signal from the receiving transducer at $\phi = 0^0$ is shown in Fig. 7. Its envelope retains the characteristic triple-step form of Fig. 6a and the smallest level difference in a step (denoted by horizontal lines) is much larger than the inherent system noise observed in Fig. 6. We therefore conclude, that the 1.8mm resolution is not affected by the spurious signal in this experiment. The same result has been found at other values of the phase, although the form of the envelope changes due to interference effects.

In another experiment the useful signal at 2.1MHz has been mixed with -3dB spurious CW signal at 2.4MHz to model the effect of a fictious channel of the same type inserted between those at 2.1 and 4.2 MHz. The resolution shows a clear tendency to worsen, although a 1.8mm field point spacing is still distinguishable (Fig. 8). It should be kept in mind, however, that these results hold for a situation in which the two frequency main lobes are totally overlapped. In a real scanning situation there is an angular separation between the lobes resulting in another -3dB decrease of the crosstalk between the adjacent channels, which restores the 1.8mm resolution observed in the previous experiment.

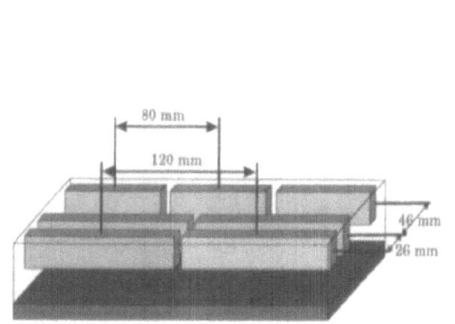

Fig. 6a. Received echoes from model reflecting surfaces in water.

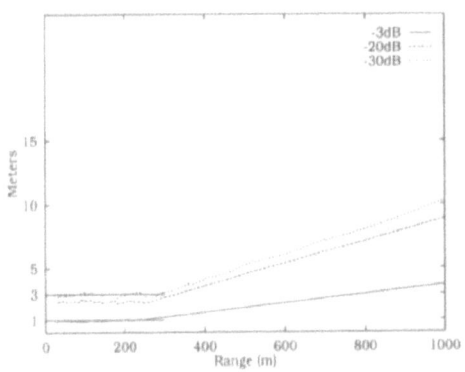

Fig. 6b. Fourier spectrum of the received echoes.

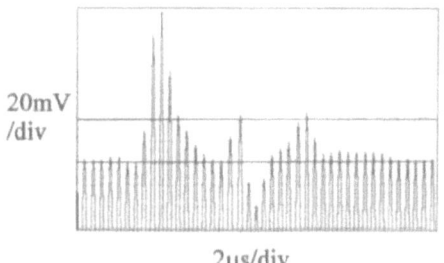

2µs/div

Fig. 7. Influence of noise on depth resolution at frequency 2.1MHz, level -6.3dB and phase 0^0

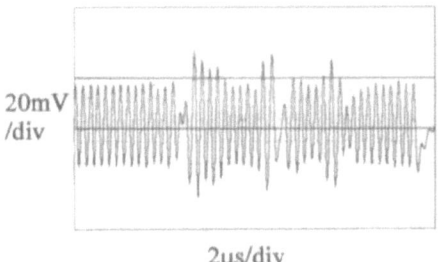

2µs/div

Fig. 8. Influence of noise on depth resolution at frequency 2.4MHz, level -3dB and phase 0^0

DISCUSSION AND FURTHER ACTIVITIES.

Since the multifrequency mode of operating an array requires parallel information processing on receipt, each array transducer should be provided with a series of parallelly connected delay lines. The phased delays are so distributed that the array focuses at several directions virtually simultaneously. In fact, the use of several carriers is not a prerequisite for making parallel processing. This can equally be done with only one carrier, with the limitation that the main lobes in operation should always be set well apart angularly to prevent the information from being spatially superimposed. This so called interlacing technique requires that, if a pulse is sent along a given line and other consecutive pulses are sent along other lines for parallel receipt, those lines which are adjacent to the first ones can be excited only after the whole previous line information has been received. This type of parallel processing is obviously faster than the conventional series processing, but is still slower than what can be achieved by multifrequency operation. Since the channels are selective in frequency, they can isolate two directions. Thus the limitation in the time of excitation is lifted, provided the consecutive excitation pulses do not overlap which might cause nonlinear effects. In this way we can either improve the spatial resolution, reaching the limiting values predicted by the diffraction theory, especially at large steering angles, or increase the scan rate by exploiting the full capability of the electronics.

From the results of this study it appears that three carriers is a critical number over which the depth resolution will be degraded beyond the 1.8mm limit because of interchannel crosstalk. In our opinion, a configuration which can readily be realized without the need of compromising and approaching performance limits is that employing two frequencies.

For the time being a prototype of such a system has been designed and is now being studied. The array consists of 64 PZT transducers 12mm wide and spaced at 0.2mm. The transducers have central frequency of about 3.75MHz and frequency bandwidth not narrower than 3MHz. The excitation is performed at 2.5 and 5.0MHz and all information collected on receipt is passed through two filters centered at these frequencies. The filters represent ceramic transducers of the same type loaded by air and backed according to the desired bandwidths allowing maximum crosstalk of -6dB. The angular sector is scanned line by line and each line is imaged by dynamic focusing. Since the lateral resolution at the higher frequency is twice better, it is not advisable to scan in two subsectors because they will be imaged with different quality. Instead, we suggest the interlacing mode in which the two carrier beam interchange their positions periodically so that the image is built of a sequence of lines of repeatedly oscillating quality, which will be averaged upon observation.

References

1. T. A. Shoup, J. Hart. Ultrasonic imaging systems, in: IEEE ULTRASONICS SYMPOSIUM (1988).
2. T. Mochizuki, C. Kasai. High speed acquisition scanner and 3-D display system using ultrasonic echo data, in: IEEE ULTRASONICS SYMPOSIUM (1990).
3. C. Lamberti, F. Scallari. A workstation-based system for 2-D echocardiography visualization and image processing, in: IEEE TRANSACTIONS ON BIOMEDICAL ENGINEERING vol. 37, N° 8, August 1990.
4. F. L. Thurston, O. T. von Ramm. Acoustical imaging with phased array, in: IEEE ULTRASONICS SYMPOSIUM (1975).
5. UNITY CORPORATION Products catalog (1986).
6. B. D. Steinberg. Principles of aperture and arrays system design. John Wiley & Sons Inc. 1976.
7. O. T. von Ramm, S. W. Smith. Beam steering with linear arrays, in: IEEE TRANSACTIONS ON BIOMEDICAL ENGINEERING vol. 30, N° 8, August 1983.
8. L. E. Kinsler, A. R. Frey, A. B. Coppens, J. V. Sanders. Fundamentals Of Acoustics. 3rd edition. John Wiley & Sons Inc. 1982.

ACOUSTIC IMAGES DETECTED BY SLAM GENERATED BY NEW BULK -TO - SURFACE WAVE TRANSDUCERS

V. I. Anisimkin*, V.A.M. Luprano, G. Montagna, and A. Quirini

* Institute of Radioengineering and Electronics Russian Academy of Science
Moscow, Russia
Pastis-CNRSM - S.S. 7 'Appia' km. 714 - 72100 Brindisi, Italy

INTRODUCTION

Development of new kinds of surface acoustic wave transducers is an important task because of numerous possible applications.

New kind of transducer is used to generate Rayleigh waves. The acoustic field generated by the transducer is tested using Scanning Laser Acoustic Microscopy (SLAM).

Images of the acoustic fields on fused quartz and Y, X-SiO$_2$ single crystal are detected by the scanned laser beam on the free surface in real time. The relative magnitudes of the radiated waves at different distances along the surface are measured to estimate the efficiencies of the transduction. The fields with and without acoustic absorber along the perimeter of the substrate surface are measured and compared with one another to estimate the contribution of reflected waves. The obtained results demonstrate new abilities of SLAM technique in non destructive testing. Possible applications of new transducers are discussed.

At present time, a large variety of different techniques for Surface Acoustic Wave (SAW) excitation are exploited in signal processing, material characterisation, non destructive testing and other applications. These techniques include piezoelectric driven, wedge-shaped, comb-type and edge-bonded transducers together with those based on transduction at liquid-solid boundary and by bulk-to-surface mode conversion [1,2]. Among different possible transducers the object of considerable interest is the SAW generation on non piezoelectric substrates. This type generation makes it possible to get appropriate insertion loss combined, either with the advantages of an isotropic medium, or of large or low propagation velocities, or of very low or high temperature coefficients.

In our paper new type of SAW transducer applicable both for piezoelectric and non piezoelectric, isotropic and anysotropic substrates is presented. Performance and main geometrical parameters of the transducer are described in experimental cases of fused and crystalline quartz.

In the past Scanning Laser Acoustic Microscope (SLAM) system was used to detect SAW travelling in the same plane of the interdigital transducers[3]. In our experiments we generate the SAW on the opposite surface of the transducer position. The acoustic fields generated are visualised and separately distinguished for surface and bulk acoustic components using SLAM.

EXPERIMENTAL

Ultrasonic images were obtained by SLAM system 2140 made by Sonoscan-Chicago. The SLAM usually uses a piezoelectric transducers to produce plane continuous ultrasonic waves at frequencies of 10, 30, 100 MHz and a scanning laser beam to detect the amplitude of the ultrasonic waves after their propagation through the sample [4]. Absorption, scattering or reflection of the ultrasonic waves crossing the sample are due to the changes of the elastic properties and/or density inside the sample. For our experiments we used SLAM only to detect with the laser the acoustic field generated by the Bulk Acoustic Waves - Surface Acoustic Wave (BAW-SAW) transducers (Fig. 1). To generate BAW a home made ultrasonic transducer for longitudinal wave at 30 MHz was used. Aluminium film (1500 Å) was spattered on the operating surface to obtain a surface reflecting the laser beam. The size of the scanning laser spot on the sample surface is about 25 μm [4]. After digitalisation of the acoustic image, the image analysis was made with a Data Traslation software.

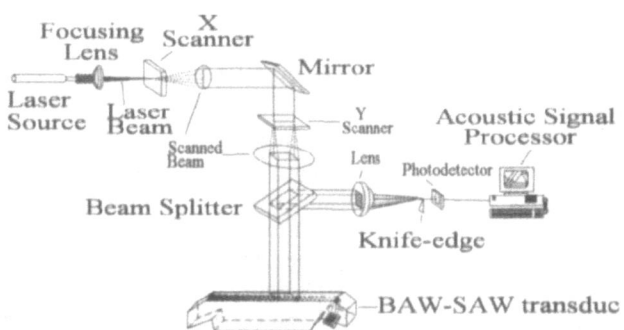

FIG. 1 SLAM operating system

The method of detection causes a Doppler shift, the magnitude of which depends on the relative velocity of the surface ripple and the laser scanning speed. In our case the value of the shift is estimated as of ±0.5 MHz. Since our photodetector response band has sharp edge at lower frequencies, we were not able to detect the signal when the direction of the SAW propagation was the same of the scanned laser beam direction and this is the reason that we used the opposite SAW propagation.

RESULTS

SAW-BAW TRANSDUCERS PERFORMANCE

Two new different kind of BAW-SAW transducers (**A** and **B**) were analysed by SLAM.

- Geometry of transducer **A** is shown on Fig. 2.a and b. It consists of the substrate 1 which has free surface 2 and two inclined planes 3,4. On the plane 4 there is a conventional bulk wave transducer 5. The longitudinal (P) bulk acoustic wave launched by transducer 5 propagates along the sample, reflects from the plane 3 and converts to the shear-vertical (SV) bulk wave. The reflected wave propagates at the vicinity parallel the free surface 2 and transforms gradually to Rayleigh-type surface acoustic wave owing to coincidence of mechanical displacement of SV-wave with mechanical (vertical) displacement of Rayleigh wave. The efficiency of SAW generation by proposed transducer depends upon the efficiency of transducer 5, its location on the plane 4, the distance between planes 4 and 3 (i.e. substrate thickness) and the values of φ and ψ angles. These angles determine the direction of propagation of reflected energy flow and the efficiency of P/SV mode conversion. The location of transducer 5 is adjusted to provide the maximum P-wave energy incidence on plane 3 at the neighbourhood of free surface 2. The substrate thickness determines diffraction loss and is taken minimal. The φ and ψ angles are the function of incident θ_{inc} and reflected θ_{ref} angles:

$$\varphi = 90° - \theta_{ref} - \gamma_{ref}, \quad \psi = \varphi - \theta_{inc},$$

where γ_{ref} is the angle of the energy flow direction of the reflected wave. In isotropic materials $\gamma_{ref} = 0°$, so φ and ψ angles can be calculated using the well-known P/SV mode conversion conditions [5] as simple functions of Poisson's module σ.

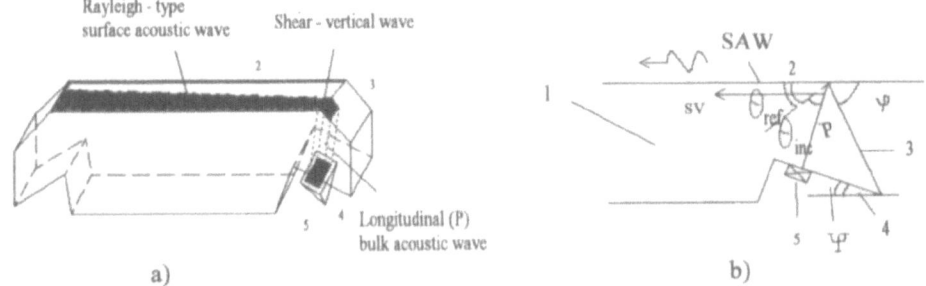

FIG. 2 a)3-D configuration b) 2-D configuration of the new BAW-SAW transducer (A)

- In anysotropic crystals (transducer **B**) φ and ψ angles are determined by the slowness curves in incidence/reflection plane (Fig. 3.a) [1]. Fig. 3.b shows the configuration of SAW transducer in the case $\{X\}$, $<Y> - SiO_2$ as an example.
Here $\psi = 0°$ for simplicity. In such transducer $\gamma_{ref} = 0°$, $\theta_{inc} = \varphi = 90° - \theta_{ref}$ and

$$\text{arc tg } \theta_{ref} = \frac{V_{SV}^{X}}{V_{L}^{Y}} = \left[\frac{C_{44} + C_{66} - \left\{ (C_{66} - C_{44})^2 + 4C_{14}^2 \right\}^{\frac{1}{2}}}{C_{11} + C_{44} + \left\{ (C_{11} - C_{44})^2 + 4C_{14}^2 \right\}^{\frac{1}{2}}} \right]^{\frac{1}{2}} = 28.8° \quad (1)$$

where C_{ij} are the elastic constant for quartz (trigonal system 32), as in our case.

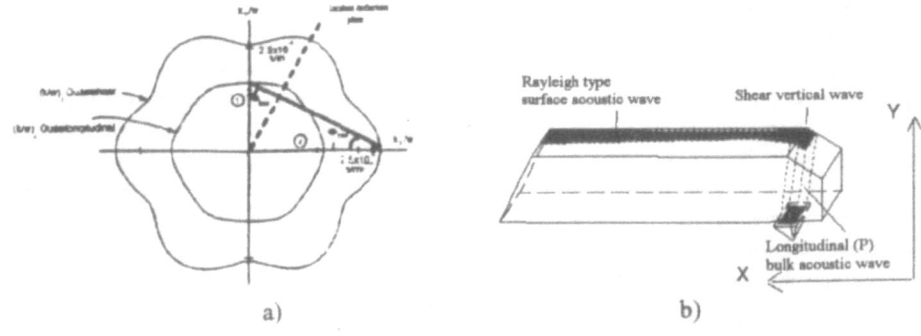

FIG. 3 a)Determination of θ_{inc} and θ_{ref} angles from the slowness curve of Y, X - SiO$_2$.
b) Configuration of new SAW transducer (B) on {Y}, <X> - SiO$_2$.

Fabrication and measurement of several test delay lines, using new SAW transducer, was begun, and initial results will be reported elsewhere, to demonstrate some basic features of such devices and to gain insight that will lead to improved design in subsequent devices. This result indicates that the more penetration depth and more relative vertical component of Rayleigh wave at the surface produces more effective SV/SAW mode conversion and hence SAW generation.

SLAM PERFORMANCE

SLAM images were detected on both BAW - SAW transducers with and without absorber on the corner, between surface 2 and 3, that produces SAW.
- Transducer **A**. As shown in Fig. 4 (at brighter zone corresponds lower acoustic attenuation) the surface acoustic field is radiated without preferred orientation. In this case no absorber was deposited along the perimeter of the substrate surface, in fact crossed lines produced by multireflections from the edges can be observed (Fig. 4, rows). The presence of the SAW is evident also from the diffracted acoustic field by a defect on the surface of the fused quartz (Fig. 4, square). When an absorber is placed on the corner between surface 2 and 3, the acoustic field does not present the SAW component, as can be seen from the absence of the diffracted acoustic field near the same defect (Fig. 5). On the other hand, some acoustic field is still presented in the substrate and can be related with the contribution of multireflected bulk waves in the bar substrate. The investigation of the nature of these bulk waves is in progress now in our laboratory.

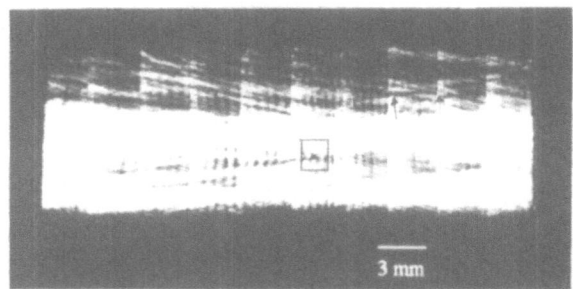

FIG. 4 Acoustic micrography of the acoustic field produced by transducer A.

- Transducer **B**. As shown in Fig. 6 the surface acoustic field is propagating along the side of the surface which is opposite to the location of the longitudinal transducer (see also Fig. 3). It is due to the quasilongitudinal wave in the Y, X-SiO$_2$ along the Y axis which exhibit energy-flux deviation from the wave normal [1,6]. Also in this case, when an absorber is placed along the corner, it is not possible to detect SAW through the sample.

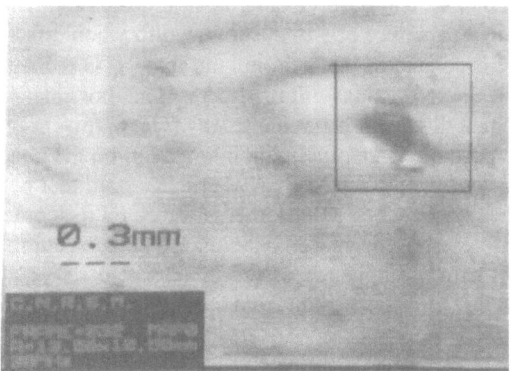

FIG. 5 Acoustic micrography of the acoustic field produced by transducer A with absorber at the corner (a selected zone)

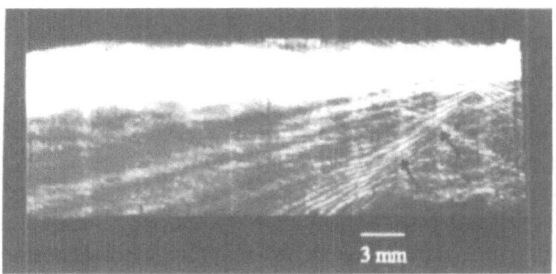

FIG. 6 Acoustic micrography of the acoustic field produced by transducer B.

DISCUSSION ON ACOUSTIC FIELD

Using the SLAM important information on the diffracted beam of the surface waves from the aperture were obtained. If we consider an oscillator, in a rigid wall, it produces the same effect as a diaphragm in a flat wave. The ratio of oscillator diameter D to wavelength λ determines the spread of the interference field and the number of maxima and minima. More precisely:

$$N = \frac{D^2 - \lambda^2}{4\lambda} \tag{2}$$

where N is the exact length of the near-field for a circular oscillator. Approximately at the position N/2, only a minimum remains on the axis. After the N distance begin the far-field. Furthermore it is shown that the spread of the beam has a definite angle γ_0 (angle of divergence). This angle is calculated according to the theory of diffraction

$$\gamma_0 = \sin^{-1}\left(1.2\frac{\lambda}{D}\right) \qquad (3)$$

A diffraction effect is present in both transducers considered. Transducer **A**, considering the aperture of 5 mm and the wavelength 0.1 mm, has a N=71.5 mm and a γ_0=1.3°. Due to the lower length of our sample, we can observe only the near diffraction field. The acoustic field of the traducer **B**, having an aperture of 3 mm, shows a diffraction effect also in the far-field. In this case the N value is calculated as 22.5 mm, the angle γ_0 is estimated as 2.3°, that is in agreement with the measured values. Furthermore it is possible to note a preferred orientation of the beam that diverge in a particular way (Fig. 6, rows).

By SLAM it also possible to get important information on the bulk wave generated by the new transducers to improve their BAW-SAW efficiency. In fact both transducers show more or less the same behaviour when an absorber is placed on the corner between surface 2 and 3. In particular an interesting decrease of the ultrasonic attenuation is detected after 15 mm from the corner. It can be explained as the impinging of the bulk waves with more intensity in this position due to the reflection from the surface 3.

CONCLUSIONS

SLAM is able to characterise BAW-SAW devices by real time imaging of the propagating acoustic field. No special sample preparation should be needed. This way it is possible to obtain important information on the mechanical or acoustic properties of the surface and subsurface of the substrate.

We believe the proposed conception of efficient mode conversion is most useful for acoustic wave based device and problems like: **a)** generation and receiving of Rayleigh-type SAW's having large penetration depth as well as Gulyaev-Bleustein waves and surface-skimming bulk waves; **b)** broad band acoustoelectronic devices, using proposed SAW transducer together with the ordinary interdigital transducer, forming desirable frequency response with essential out-of-band rejection; **c)** broad-band acoustooptic devices; **d)** nondestructive testing; **e)** liquid sensors (in the case of shear horizontal waves).

ACKNOWLEDGEMENTS

One of the authors (V.I.A.) would like to thank INEM, Genova for support in occasional collaboration and PASTIS-CNRSM, Brindisi for fruitful period spent for this paper.

REFERENCES

1. Auld B.A., 'Acoustic Fields and Waves in Solids', Wiley-Interscience Publication, 1973, Vol.2
2. C. Lardat, 'Surface Wave Edge Bonded Transducers and Applications'. *IEEE Ultrasonic Symposium Proceedings*, pp 433-436, 1974
3. W.P. Robbins, R. Mueller and E. Rudd, 'Thin film characterization using SLAM with surface acoustic waves', *IEEE Ultr. Ferr. Freq. Contr.*, Vol. 35, No. 4, (1988), pp, 477-483
4. L.W. Kessler, 'Imaging with dynamic ripple diffraction', *Acoustical Images*, Vol. 16, Plenum, New York, 1976, Chap. 10, pp 229-239
5. Brekhovskyh L.M., 'Waves in Layered media', *"Science"* (Soviet Ed.), 1973, Vol. 2.
6. L. G. Merkulov and L. A Yakovlev, *Sov. Phys.-Acoust.*, **8**, 72 (1962)

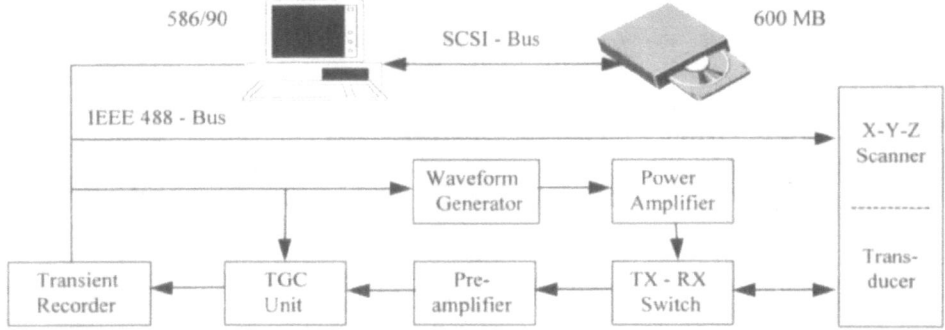

Figure 1. Block diagram of the imaging system

Currently, two quartz-lens-focused PZT-transducers with different imaging characteristics are available made by Ultran* and Panametrics†, respectively. The transducer geometry and formulas for the calculation of the axial resolution δ_{ax} and the lateral resolution δ_{lat} are shown in Figure 2 left. Echo spectra from a polished glass-plate, which was located in the focus of the transducers, are shown in Figure 2 right. Characteristic data of the transducers is presented in table 1.

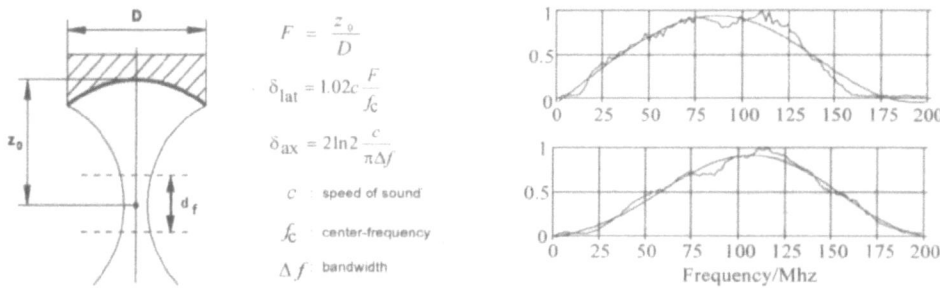

Figure 2. left: Transducer geometry, right: Spectra of the transducers – top: Ultran, bottom: Panametrics

Table 1. Characteristic transducer data

	$\delta_{ax}/\mu m$	$\delta_{lat}/\mu m$	f_c/MHz	Δf/MHz	F	$d_f/\mu m$
Ultran	8.5	27	86	107	1.34	400
Panametrics	6.5	11	106	103	0.85	110

IMAGING CONCEPTION

For medical applications a penetration depth of about 1 – 3 mm is required. This range can be covered despite the short focal zone d_f of the transducers using the outlined imaging conception:

A couple of "short" B-scans is acquired at different depths of the tissue by mechanical movement of the transducer in axial direction between each acquisition. Finally all scans are composed to get an image with a larger size. We call this additional axial

---

*Ultran Laboratories, Inc., Boalsburg, PA, USA
†Panametrics, Waltham, MA, USA

motion "Z-scan" and thus, the whole procedure "B/Z-scanning conception", which is illustrated in Figure 3. A modified B/Z-scan could also be obtained in real time using

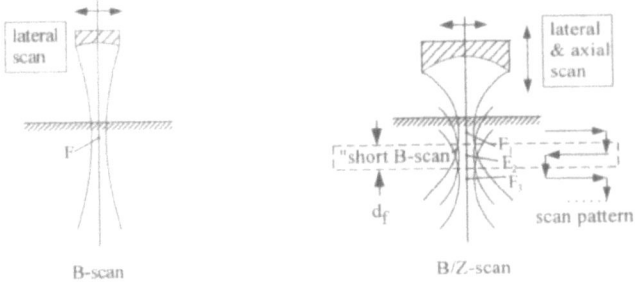

Figure 3. Comparisson of B-scan and B/Z-scan technique

anular arrays, but these are not yet available for the 150 MHz region. An example of a B/Z-scan is shown in Figure 4. It demonstrates the feasibility of the outlined scanning conception.

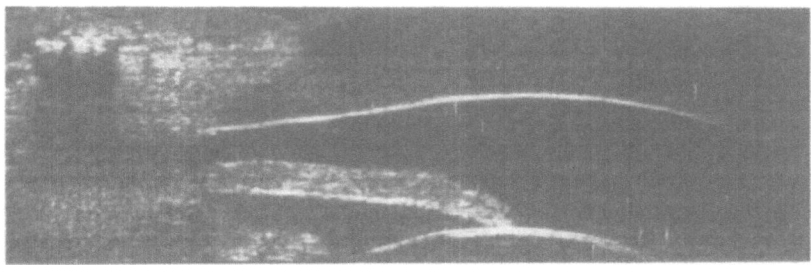

Figure 4. Angle of a pig eye in vitro – $10 \times 3.2 \ mm^2$

SIGNAL PROCESSING

Standard B-mode images usually suffer from inhomogeneous resolution as the exponentially growing attenuation of the high frequency spectral components of the signal remains uncorrected.

$$H_{att}(f) = 10^{-2 d_t \frac{\alpha_0}{20} f^n} \tag{1}$$

$$\alpha_0 = 1 \text{dB}/(\text{MHz cm}) \qquad n = 1.1 \tag{2}$$

Therefore, the bandwidth and center-frequency and consequently, the lateral and axial resolution are considerably lowered even in diffraction corrected images in the deeper parts of the image, which is shown for the Ultran transducer in Figure 5. A possible solution to that problem could be depth dependent pseudo-inverse filtering of the image data, but this reduces the signal to noise ratio (SNR) considerably, as in the frequency bands, which amplitudes have to be raised, the noise portion is high. This effect can be reduced by applying prefiltered transmitter signals $T_i(f)$ (Figure 6) instead of a postfiltering procedure, which can be accomplished without additional effort using the

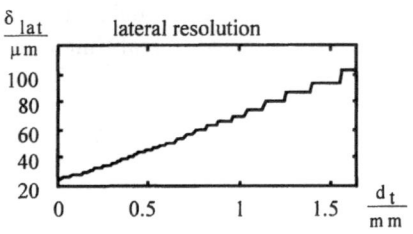
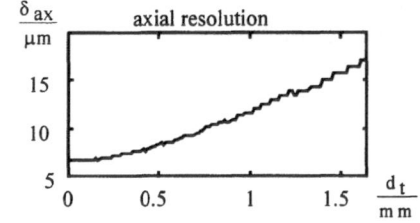

Figure 5. Resolution as a function of penetration depth in tissue d_t

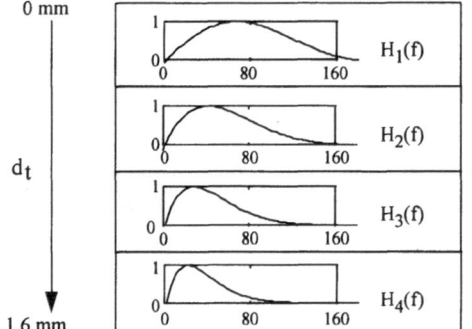
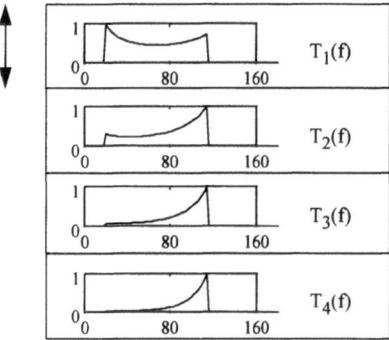

Figure 6. Attenuation Compensation, left: Transfer functions of skin in 4 different depths, right: Depth-adapted pseudoinversely prefiltered transmitter signals

B/Z-scanning procedure, as independent scans for different depth areas are carried out in any case. Even with the prefiltering procedure, the SNR is degraded, because the available transducer frequency band is not optimally utilized. Therefore, a signal energy increasing technique must be employed. We propose the well-known pulse compression technique in combination with nonlinearly-frequency-modulated (NLFM) chirp signals for that purpose. The basic principle of pulse compression[7] is illustrated in Figure 7 left: Instead of a short pulse (top), a frequency modulated (FM) chirp signal (bottom) with the same bandwidth as the pulse is transmitted. Due to the larger signal duration the signal energy is increased. The received echo is compressed by a digital or an analog allpass filter yielding a higher amplitude.

Prefiltering using the pulse compression method can be done in different ways, as shown in Figure 7 right: If a particular spectrum of the chirp signal is desired for a

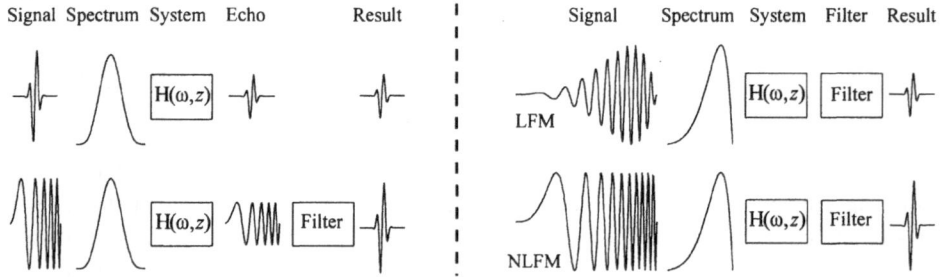

Figure 7. Pulse compression, left: Basic principle, right: Prefiltering

certain depth, it can be generated by a linear frequency modulated (LFM) chirp, which is in addition amplitude modulated, or by a nonlinear modulated (NLFM) chirp with constant amplitude. The latter transmits the higher frequency spectral components for a longer time, thus utilizing the full amplifier power for the complete transmission time. Therefore, the signal energy of the prefiltered NLFM-chirp is higher than of the LFM-chirp. We use NLFM-signals of $0.5\mu s$ length and with a bandwidth of 100 MHz yielding a gain of 12 dB.

As the pulse compression method relies on the assumption of a linear system, a problem arises in tissue regions with large inhomogeneities or strong scatterers, as these cause echos, which exceed the dynamic range of the imaging system. In this case, the system behaves nonlinearly and consequently large sidelobes are generated during the compression procedure, which is shown in Figure 8.

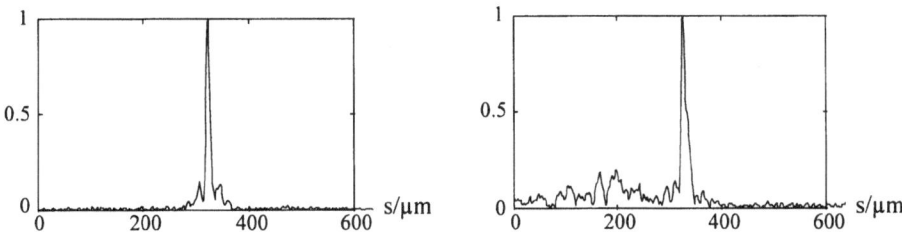

Figure 8. Pulse compressed echo of a weak scatterer (left) and of a strong scatterer (right)

To solve that problem, a depth-adaptive imaging procedure is proposed: It consists in (1) acquisition of a pulse and a chirp image at the same time, (2) detection of regions causing strong backscatter, (3) composing a combined image out of pulse data for the regions with strong backscatter and out of compressed chirp data for all other regions.

To further improve the detectability of small lesions, we apply an adaptive filter for speckle suppression, which has already been reported by Loupas[8]. We modified this filter for high frequency ultrasound applications by an adaptive estimation of the resolution cell size.

A chirp and pulse image, which were taken at the same time, are shown in Figure 9. The chirp image shows homogeneous resolution over the full depth area, but also

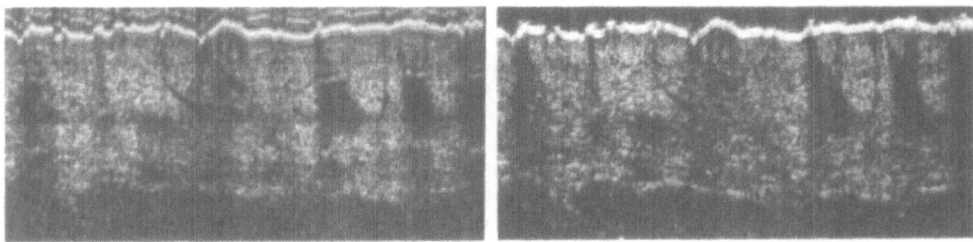

Figure 9. left: Chirp-mode skin image, right: Pulse-mode skin image – both: 10mm × 1.6mm

large sidelobes can be observed in the skin-entry-echo-region. The pulse image is free of artifacts, but shows degrading resolution in lower image regions.

Figure 10 shows the combined image and a speckle reduced version of it. The combined image shows homogeneous resolution over the full depth area and is free of artifacts. Therefore, the diagnostic value of the image is considerably improved. The speckle suppression processing allows easier detection of small image details.

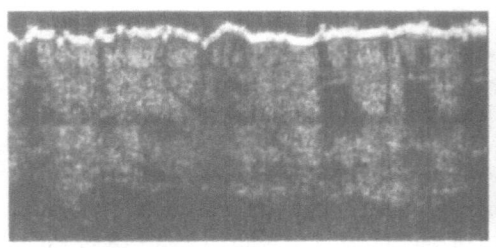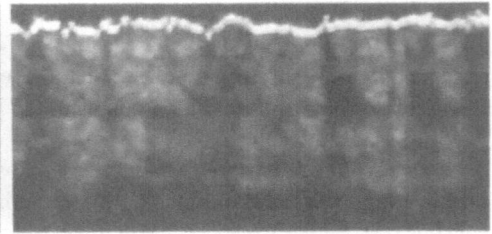

Figure 10. left: Combined skin image, right: Speckle reduced combined image – both: 10mm × 1.6mm

CONCLUSIONS

1. high frequency ultrasound provides image quality sufficient for skin and eye imaging

2. strongly focused transducers employing the B/Z-scan conception provide more isotropic images

3. attenuation compensation provides more homogeneous images

4. pulse compression provides signal energy required for attenuation compensation

5. adaptive speckle processing provides easier detectability of image details

The imaging capabilities of the system using the outlined conceptions have already provided valuable diagnostic information during clinical examinations of more than 60 patients suffering on different kinds of skin diseases, like cancer, psoriasis, lichen ruber, etc. In vitro experiments on eyes have been recently succesfully terminated. They indicate a very promising application of the system in vivo for glaucoma research.

REFERENCES

1. A. Höss, H. Ermert, S. el Gammal, and P. Altmeyer. A 50 MHz ultrasonic imaging system for dermatologic application. In *IEEE Ultrasonics Symposium Proceedings*, pages 849–852, 1989.
2. F.S. Foster, C.J. Pavlin, G.R. Lockwood, L.K. Ryan, K.A. Harasiewicz, L. Berube, and A.M. Rauth. High frequency ultrasound backscatter imaging. In *IEEE Ultrasonics Symposium Proceedings*, pages 1161–1169, 1991.
3. D.H. Turnbull, B.G. Starkoski, K.A. Harasiewicz, G.R. Lockwood, and F.S. Foster. Ultrasound backscatter microscope for skin imaging. In *IEEE Ultrasonics Symposium Proceedings*, pages 985–988, 1993.
4. C. Passmann, H. Ermert, T. Auer, K. Kaspar, S. el Gammal, and P. Altmeyer. In vivo ultrasound biomicroscopy. In *IEEE Ultrasonics Symposium Proceedings*, pages 1015–1018, 1993.
5. C. Passmann and H. Ermert. 150 MHz in vivo ultrasound of the skin: Imaging techniques and signal processing procedures targeting homogeneous resolution. In *IEEE Ultrasonics Symposium Proceedings*, pages 1661–1664, 1994.
6. C. Passmann and H. Ermert. A 150 MHz ultrasound imaging system for dermatologic and ophthalmologic diagnostics. *IEEE Trans. Ultrason. Ferroel. Freq. Control*, submitted for publication.
7. H. Ermert, M. Pollakowski, C. Passmann, and L. von Bernus. Acoustical imaging using an optimal combination of signal prefiltering and pulse compression. In J. P. Jones, editor, *Acoustical Imaging*, volume 21, in press, New York, 1994. Plenum Press.
8. T. Loupas, W.N. McDicken, and P.L. Allan. An adaptive weighted median filter for speckle suppression in medical ultrasonic images. *IEEE Trans. Ultrason. Ferroel. Freq. Control*, UFFC-36(1), pages 129–135, January 1989.

SUPERFICIAL TISSUE MICROSONOGRAPHY

Andrzej Nowicki, Jerzy Litniewski, Jacek Liwski, Wojciech
Secomski, Paweł Karłowicz, and Marcin Lewandowski

Ultrasound Department, Institute of Fundamental Technological
Research,
Polish Academy of Sciences, Świętokrzyska 21, 00-049
Warsaw, Poland

INTRODUCTION

A number of recent years have seen a growing interest among biologists and clinicians in surface tissue imaging and vessel wall examinations performed in the course of operations. Yano *et al.*[1987] were among the first to describe a system permitting sector skin imaging using a 40 MHz lithium niobate transducer. These authors achieved a beam width (-6dB) in the focus zone equal to about 0.1 mm. The transverse resolution as determined from measurements of the length of a pulse reflected from an ideal reflector was equal to 0.1 mm. Hoss, Emert *et al*. [1992] developed a system with a 40 MHz centre frequency with a very wide band [- 6 dB], close to 50 MHz. Such a wide band made it possible to use on the transmission side chirp modulation pulses and analog signal compression of a signal received using all-pass filters. Berson *et al*. [1992] described a scanner for skin imaging, working at 17 MHz, with 0.1 mm longitudinal resolution and 0.3 mm transverse resolution. Feuillard *et al*. [1994] improved the Berson system, replacing a ceramic PZT transducer by a 45 MHz wide-band transducer of P(VDF-TrFe). The development of high-frequency ultrasound diagnosis turns to completely new areas of application in dermatology and diagnosis of skin diseases, with consideration given to neoplastic lesions and watching progress in the treatment of them. Ophthalmologic applications seem to be very essential in terms of imaging cornea, sclera, iris and ciliary body. The purpose of our study was to develop a real time device for skin and eye imaging in 2D mode and image reconstruction in C mode with lateral and axial resolution better than 0.1 mm. The high resolution scanning acoustic microscope (SAM) for tissue structure study was also developed.

MODEL OF A SYSTEM FOR TISSUE IMAGING

In the system, several independent blocks can be distinguished: a mechanical sector probe, controllers, a transmitter of short high-frequency pulses, a broad-band receiver with an envelope demodulator, digital echo acquisition systems and a monitor control.

Transducer

Theoretical calculations were performed and then the optimum selection of a transducer and a focusing lens was checked on an experimental basis. For a 32 MHz transmitted wave the length of transmitted pulses was about 30 ns. It was assumed that the imaging range should be about 5 mm under the skin surface, and +/- 2.5 mm round the focus. Transducers of lithium niobate $LiNbO_3$ were applied; they were fixed on quartz glass lenses. The aperture of the lenses was 4 mm. For the two curvature radii, 5.8 and 9.8 mm, the lateral resolution in the focus zone (-3 dB) was about 0.06 mm.

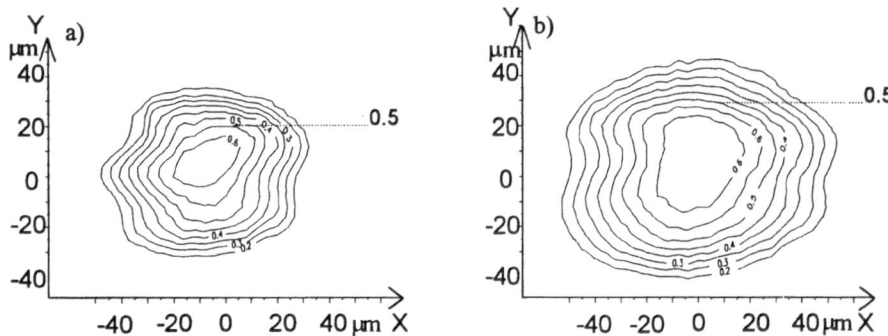

Figure 1. Ultrasound field radiated by a lithium niobate transducer, with 4 mm diameter and lens curvature radius of 5.8 mm, a) 5.8 mm and b) 6.4 mm from the transducer

Pulsed transmitter and receiver

The pulse forming system generates burst consisting one or two 32 MHz wave periods with a repetition frequency of 1.95 kHz. The amplifier ensures 34 dBm power output in the band 1-50 MHz. The receiver with the sensitivity higher than 10 µV consists two independent amplification circuits: a linear one with 65 dB amplification in the band 25.5-44 MHz and a logarithmic one with 60 dB amplification and 50 dB dynamics in the band 25-46 MHz. The output voltages from the two amplifiers can be summed up with each other using an appropriate weight; it permits smooth adjustment of echo levels from tissue boundaries (outline, linear amplifier) against the background of echoes from inside the tissues with a homogeneous structure (background, logarithmic amplifier). Apart from the adjustment of logarithmic and linear amplification as well as the sum of the two amplification modes, time gain control TGC can be applied. TGC is required because of large attenuation of ultrasound in skin, which was about 2 dB/cm/MHz, [Frew and Giblin, 1985], leading to larger attenuation than 60 dB at the 32 MHz transmitted frequency and the projected imaging depth of 5 mm.

Data acquisition

The data acquisition circuit consists of an 8-bit A/D converter with a maximum sampling frequency of 100 MHz and a 64 kB fast access memory. After the B-scan acquisition is switched on, the A/D converter is being initialized, and then data are collected from one 2D sector, 215 lines with 216 points each, into the internal memory of the system. The received signal is sampled with some delay to the transmitted pulse so that data could be collected from the focus zone of the probe. 2D presentation resembles that applied in standard ultrasonography. The echoes are sampled, coded in a 64 levels of grey scale and next displayed on the screen, to form one vertical line of the image. An image in C mode is composed of a predetermined number of successive sectors (64, 128 or 256). A C-scan consists of lines which lie in the XY plane - as selected from successive B-scans at given depth. In the current version of the programme, the C-scan image is built as data are acquired: only one line from a predetermined depth is selected and filed from each successive sector.

Acoustic microscope

The scanning acoustic microscope (SAM) developed in our laboratory operates in two modes , C-scan mode and B-scan mode. The sample is investigated with a well-focused sound beam and the amplitude or a phase of the waves reflected at the focal plane is used to control the brightness of the point on a monitor screen. The acoustic lens scans over the surface of the sample and image is build point by point . C-scan mode is used to visualise a surface or an interior of samples. The microscope can operate at a frequency of 32, 100 and 200 MHz and a lateral resolution of 45, 15 and 7 μm can be achieved respectively.

RESULTS

Sector scan

Skin was examined in 20 healthy volunteers aged from 27 to 56 years In each case, very distinct images of successive skin layers were obtained: of echogeneous epidermis, the intermediate layer between epidermis and dermis, dermis, subcutaneous tissue and adiposus layer. Spreading and depth of penetration of different skin melanoma (superficial spreading melanoma, nodular melanoma and lentigo malignant melanoma) were examined in 19 patients. After ultrasonic examination neoplastic changes were removed surgically and next their thickness were measured with optical microscope. In all cases the difference between histological and microsonographic measurements of melanoma thickness was below 0.12 mm, (depth estimated with ultrasound was always larger). It proved that the pre-operation microsonographic estimation of the depth of neoplastic changes is of a great clinical value in planning the safe width (according to the Breslaw scale) of the surgical intervention

Sector skin sections with "benign" naevi were also studied, e.g., discoloured and hypoplastic ones which could turn into neoplastic changes. Pigmented and thickened "benign" naevi showed a distinct boundary in skin. Naevi without this boundary infiltrated deeper.

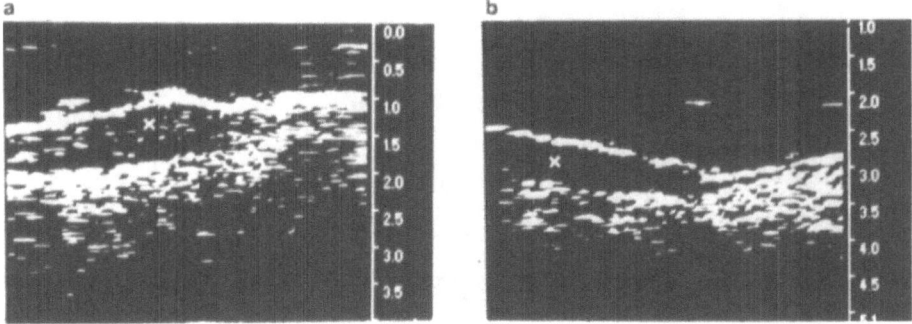

Figure 2. Sector scans of two different neoplastic changes (marked with cross), a) superficial spreading melanoma, ultrasonically measured thickness 0.8 mm and b) edge view of the nodular melanoma, vertical scale in mm.

The microsonograph made also possible to gain very detailed sections across the upper eye layers: cornea, sclera, anterior chamber, iris, ciliary body and lens

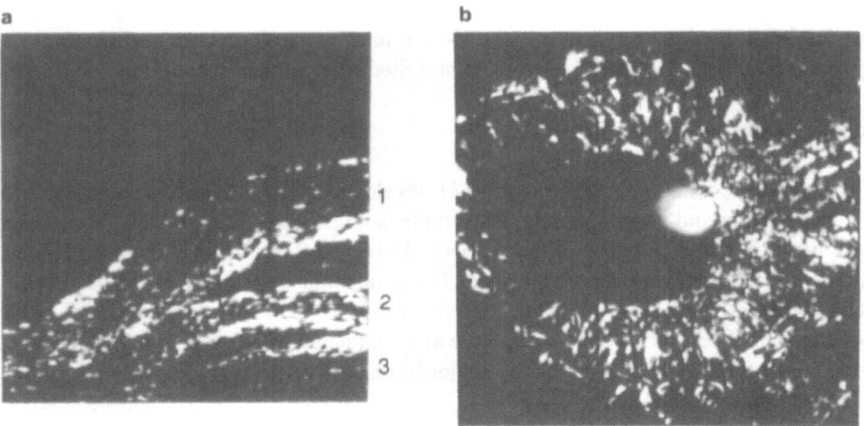

Figure 3. Sector scan of the eye section, (a) 1) cornea, 2) iris and 3) lens, scan area 6 x 6 mm (b) C scan SAM image of an interior of a man eyeball showing an iris and a pupil in a horizontal cross-section, scan area 6.4 x 6.4 mm

C-scan mode

The examined samples were approximately 20 μm thick slices of an eye fixed to the glass substrate. The images reflect distribution of a local acoustic attenuation and reflectivity in sections. The following acoustic images of eye structures (Fig. 4) are very similar to optical images. Different tissue types and structures are easily resolved and differentiated and the acoustic histology of an unstained tissue can be performed with SAM.

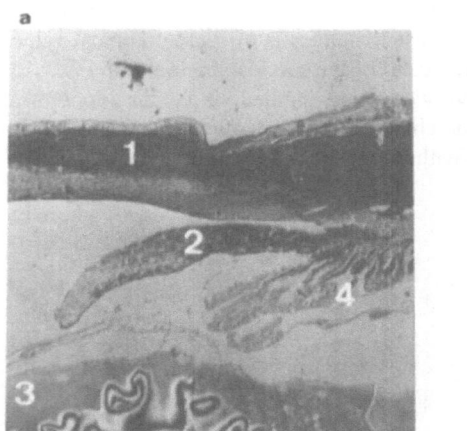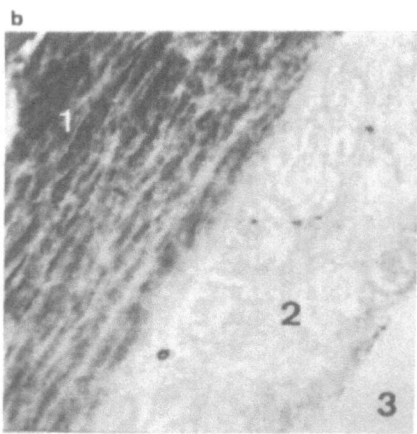

Figure 4. (a) C-scan SAM images of 20 μm thick, eye sections taken at 100 MHz. The image shows structure of an eye 1) cornea, 2) iris 3) lens 4) ciliary body, (b) SAM image of pathological eye tissue section. 1) sclera, 2) cancer tissue - adenocarcinoma infiltras penetrating into 3) anterior chamber, scan area 1.6 x 1.6 mm

The microscope can visualise samples of pathological eye structures. Fig.4 b shows the image of a tumour tissue (adenocarcinoma infiltras) adjacent to the sclera. Acoustic images of cancer tissues vary in contrast, density and shape spreading and can be used for tissue characterisation. The differences are so evident that we hope to recognise tissue types also when imaging "in vivo" with high frequency ultrasonography.

<u>Subsurface imaging in C-scan mode</u>

The microscope can also operate in the C scan mode. The images are obtained by focusing the waves under the surface of an eye globe (or other structure) and moving the time gate of the microscope to the focus position. Thus only acoustic signals from the focal plane inside the examined structure are accepted when the transducer is scanning over it in the x-y raster pattern. The image obtained in this way corresponds to the horizontal cross-section of an eyeball (Fig. 3 b).

CONCLUSION

A prototype of microsonograph was designed for diagnostic applications in dermatology. Better lateral and axial resolution than 0.1 mm was obtained and confirmed by investigating the radiation field of a focused LiNbO3 transducer. Three dimensional scanning permits visualization in 2D and C modes. Preliminary clinical studies showed that the system made it possible to obtain images of small skin structures (epidermis, dermis, subcutaneous ones and small vessels) and to distinguish such changes as malignant melanoma, psoriasis and other skin disorders. The oncological application proved to be very promising in determining the minimal margin of the cancerous tissue to be removed predicted on the basis of the tumor thickness measurements due to the Breslow scale The scanner can also be useful in eye examinations, particularly regarding changes in cornea, sclera, anterior chamber and iris.

Table 1. Comparison of infusion thickness measured with ultrasound and post operative

number of patients[*]	type of skin melanoma	thickness his-pat [mm] Breslow	thickness usg [mm] Breslow	max. difference [mm]
2	superficial spreading melanoma	0.7 - 2.0	0.82 - 1.8	0.12
7	nodular melanoma	1.0 - 8.0	1.0 - >4.0	for usg < 4.0, diff<0.11
1	lentigo malignant melanoma	1.0	1.0	0

* examination and surgery performed in Center of Oncology, Warsaw - Soft Tissue and Bone Cancer Clinic, head: Prof. W. Ruka M.D.

<u>References</u>

Berson M., Vaillant L., Patat F., Pourcelot L., 1992, High resolution real time ultrasonic scanner, Ultrasound Med.and Biol., 18, 5, 471-478.

Feuillard G., Lethiecq M., Tessier L., Patat F., Berson M., 1994, High resolution B scan imaging of the skin using a 50 MHz P(VDF-TrFe)- based ultrasonic transducer, European J.Ultrasound, 1, 2, 183-189.

Frew H.S., Giblin R.A., 1985, The choice of ultrasound frequency for skin blood flow investigation, Bioeng.Skin, 1, 193-205.

Hoss A., Ermert H., el-Gammal S., Altmeyer P., 1992, Signal processing ion high-frequency broadband imaging systems for dermatologic application, in: "Acoustical Imaging", vol.19, 243-249, H.Ermert and H.P. Harjes, eds., Plenum Press, New York.

Yano T., Fukuita H., Ueno S., Fukumoto A., 40 MHz ultrasound diagnostic system for dermatologic examination. IEEE 1987 Ultrasonic Symposium Proceeding ,1987; 857-878.

MODELLING OF ULTRASONIC FLOW METERS BY ACOUSTICAL RAYTRACING

H.-G. von Garßen and V. Mágori

Siemens AG, Central Research and Development, D-81730 München

INTRODUCTION

Ultrasonic flow meters based on time of flight measurements gain more and more importance in various areas of application. Examples are heat meters for the consumption of hot water supplied by district heating companies, and household gas meters[1]. The basic principle of such a flowmeter is shown in Fig. 1. From the left transducer, an ultrasonic pulse is sent into the direction of the flow. Its time of flight to the other transducer t_{down} is measured. For comparison, the time of flight t_{up} into the opposite direction is also determined. The effective flow velocity v_F defined by the ratio F/S of total flow rate F and nominal cross sectional area S of the tube, is calculated according to the formula

$$v_F = a\ \frac{t_{up} - t_{down}}{t_{up} \cdot t_{down}}. \tag{1}$$

To obtain high precision results, the calibration factor a should only depend on fixed tube properties like geometry, transducer design and electrical signal processing. Especially, it should not depend on fluid properties like temperature, speed of sound, density, viscosity and flow profile, which can vary considerably even within a fixed area of application. To minimize influences of unknown quantities, models for the complete signal path in the measurement tube are required. Since electric circuitry and finite element models of ultrasonic transducers can be treated successfully by commercial programs, we concentrate on the sound propagation inside the tube.

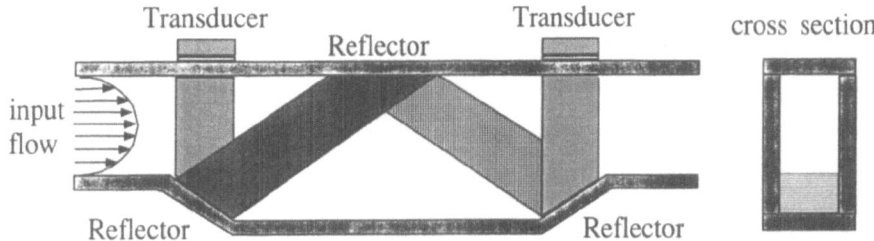

Figure 1. Basic principle of a flowmeter showing the basic soundpath provided by geometric design.

WAVE EQUATION AND GEOMETRICAL ACOUSTICS

Consider an incompressible fluid with density ρ. Let v and p be the velocity and pressure in the absence of the acoustic wave and v', p' the disturbances of these properties caused by the wave.

The sums $v+v'$ and $p+p'$ obey the Euler and continuity equations

$$\frac{\partial\,(v+v')}{\partial\,t}+\left[(v+v')\cdot\nabla\right](v+v')=-\frac{1}{\rho}\nabla(\,p+p') \qquad \nabla\cdot(v+v')=0. \qquad (2)$$

To describe the propagation of the acoustic wave, we separate the first-order terms with respect to the acoustic wave disturbation and get

$$\frac{\partial\,v'}{\partial\,t}+(v\cdot\nabla)v'+(v'\cdot\nabla)v=-\frac{1}{\rho}\nabla p' \qquad \nabla\cdot v'=0. \qquad (3)$$

Using the same method for adiabatic disturbances in compressible fluids, a similiar but more intricate set of three equations is obtained[2].

A direct numerical solution of the partial differential equations (3) requires a spatial discretisation small compared to the acoustical wavelength λ of the acoustical disturbance. Present flow meters typically use a wavelength of 1 mm, which must be compared to typical transducer diameters of 10 mm and a typical tube length of 100 mm. So this task is hardly solvable on present day computers. To avoid these difficulties and to get a better understanding of the physical principles of wave propagation, we employ a suitable variant of geometrical acoustics. Geometrical acoustics is an asymptotic method known to work very well especially in those cases like ours, where the typical spatial dimensions of wave propagation are large compared to the wavelength. The velocity components, the pressure disturbation and the velocity potential of the acoustic wave are represented by expressions of the form

$$A(x,\iota)\,e^{i\,\Psi(x,t)}, \qquad (4)$$

where amplitude factor $A(x,t)$ and wave number $k = i\nabla\cdot\Psi(x,t)$ are varying slowly within one wavelength as well as within one oscillation period of the acoustic field. Inserting (4) into equations (3) or its analogue for compressible fluids yields a set of first order partial differential equations, namely the so called eikonal equation for $\Psi(r,t)$ together with an asymptotic series of transport equations[3] for the amplitude $A(x,t)$. The main advantage of this description is that $\Psi(x,t)$ and $A(x,t)$ propagate along rays described by a set of ordinary differential equations[4] for the three spatial components of k and x

$$\frac{dk_i}{dt} = -|k|\frac{\partial\,c}{\partial\,x_i}-\sum_{j=1}^{3}k_j\frac{\partial\,v_j}{\partial\,x_i}, \qquad (5)$$

$$\frac{dx_i}{dt} = c\frac{k_i}{|k|} + v_i. \qquad (6)$$

While the propagarion of phase $\Psi(x,t)$ is given directly by (5) and (6), the propagation of the amplitude factor $A(x,t)$ can be derived easily from energy conservation applied to ray tubes formed by adjacent rays[3,5]. An example of such rays leaving a point source is shown in Fig. 2.

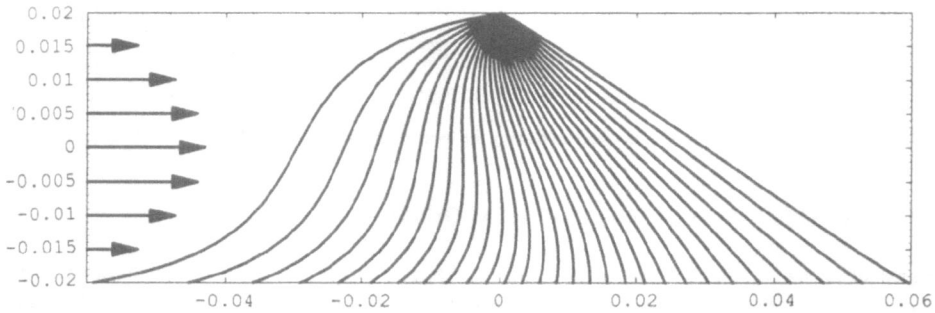

Figure 2. Ray propagation in a quadratic flow profile with a maximum mach number $v/c=0.5$. The axes are labelled in m.

Normally, the fluids in ultrasonic flow meters are characterized by sound velocities c nearly constant over the whole tube volume and by small mach numbers v/c of less than 1/10. Under these conditions, the rays can be approximated very well by straight lines. Up to the first order of v/c, the time of flight is given by:

$$t = \frac{1}{c}\left[L - \frac{1}{c}v_F I\right],$$
(7)

using the abbreviations

$$L = |x - x_s|, \qquad I = \frac{1}{v_F}\int_{x_s}^{x} v(x)\,dx$$
(8)

for the acoustical pathlength L and the effective influence I of fluid flow. The latter contains a path integral over the flow velocity component along the straight line from the source point x_S to the point x where the acoustic field is to be calculated.

TREATMENT OF ANGULAR DIVERGENCE, REFLECTION AND DIFFRACTION

The results of raytracing strongly depend on the boundary conditions imposed near the transducers and at the reflecting parts of the walls. For extended wall sections with typical dimensions up to about 100λ, a geometrical acoustics description is used, which avoids extreme computational problems with respect to spatial discretisation and usually promises to yield results of sufficient accuracy. However, small cross sections of the transducers (diameter $d\approx10\lambda$) as well as small reflecting surfaces should be modelled more exactly taking into account their wave theoretical diffraction pattern.

To illustrate this point, let the wall excitation of the sending transducer be described by a Gaussian amplitude distribution with constant phase. Within standard geometrical acoustics, only rays perpendicular to this surface would emerge. This would neglect the effects of angular divergence of the ultrasonic pulse. All rays would remain parallel on the idealized path shown in Fig. 1 and all would share the same pathlength L. To include angular divergence, the sending surface can be decomposed into a collection of point sources. Each point source generates rays in all directions and only the superposition of all rays from all point sources yields the total acoustic field.

The numerical evaluation is implemented on a workstation using an adapted type of raytracing. The primary rays start at all points of the sending transducer into all directions using a proper spatial discretisation. At extended sections of the wall, the rays are reflected according to the ordinary law of reflection. At small diffracting reflectors, secondary rays can emerge. There are two possibilities to generate these secondary rays. One uses the Kirchhoff approximation. Here each point of the reflector acts as a point source for the whole amplitude of the reflected field. The other possibility is based on Young´s principle and Keller´s geometrical theory of diffraction[6]. Here only the points at the boundary of the reflector generate additional rays needed to describe diffraction.

All rays arriving at the receiving transducer are classified with respect to both their different pathlength L and their effective flow influence I. This classification (represented as a 2-dimensional histogram) allows a concise insight into various sources of measurement error. Summing up all rays yields the output signal of the receiving transducer, taking into account the appropriate excitation function of the sending transducer and the sensitivity function of the receiving transducer as well as the correct amplitude factors and phase angles caused by sound propagation. The sensitivity of this output signal with respect to flow velocity and other quantitities can be used for a detailed error analysis.

SIGNAL PROPAGATION IN A THREE-DIMENSIONAL FLOW METER TUBE

As an example, the measurement tube of Fig. 1 is analyzed by the raytracing algorithm. For a first discussion illustrated by Fig. 3, only the center of the transducer surface is used as starting point for the rays. All starting directions of the rays are characterized by the spatial polar angles θ and φ. Here θ denotes the angle between the ray and the surface normal, while φ denotes the turning angle around the surface normal, measured with respect to the direction to the receiving transducer. All wall reflections are considered neglecting additional diffraction. Rays with a very long pathlength L do not contribute to the evaluation of pulse delay. Therefore they are cut off here.

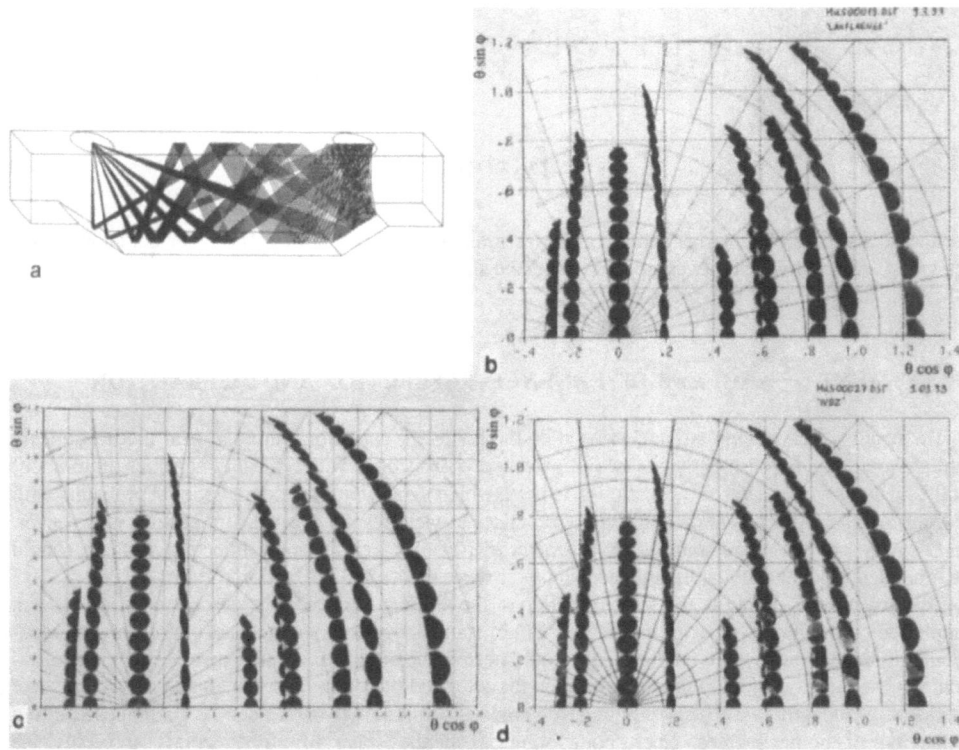

Figure 3. Ray propagation in the flowmeter tube shown in Fig. 1. (a) Perspective view. The left circle illustrates the surface of the sending transducer, the right circle the surface of the receiving transducer. To avoid overcharging the figure, only the angle θ is varied. (b) Length L of flight for all directions. Small L are shown in light tone, large L in dark tone. The filled circles on the positive part of the abscissa represent images of the receiving transducer, seen from the sending transducer in the directions shown in Fig. 3a. The circles following up in the ordinate direction correspond to increasing orders of additional sidewall reflections. (c) Phase factor $cos(kL)$. Negative values are shown in light tone, positive values in dark tone. (d) Flow influence I for all directions assuming a quadratic flow profile as an approximation for a laminar flow. Small I are shown in light tone, large I in dark tone.

The fringe pattern in Fig. 3c shows rapidly changing phases for most directions outside the origin of the coordinate system. So the complex amplitude contributions to the output signal will mostly be canceled by destructive interference. Therefore the basic path shown in Fig. 1 indeed should give the largest contribution to the resulting amplitude. Fig. 3d shows a very high flow influence I for the rays running near the symmetry plane of the tube, compared to those suffering many sidewall reflections.

Because of space restrictions, we cannot discuss in detail the rays starting from all the other points of the sending transducer surface. Instead, we will present some results for the integral over the whole area of the sending surface. To simplify the discussion, only rays of the central bundle are considered. That are those rays starting at all points of the sending transducer into all directions, which still follow the sequence (see Fig. 1) left transducer → lower left reflector → upper reflector → lower right reflector → right transducer. These rays are classified with respect to both their pathlength L and their effective flow influence I. For the excitation function of the sending transducer and the sensitivity function of the receiving transducer, a Gaussian spatial distribution around the center of the transducer surfaces was assumed. Fig. 4, (a) and (b), depicts the resulting histograms for a quadratic and for a rectangular flow profile used to model laminar and turbulent flow regimes within the tube, respectively. Both profiles correspond to the same effective flow velocity v_F.

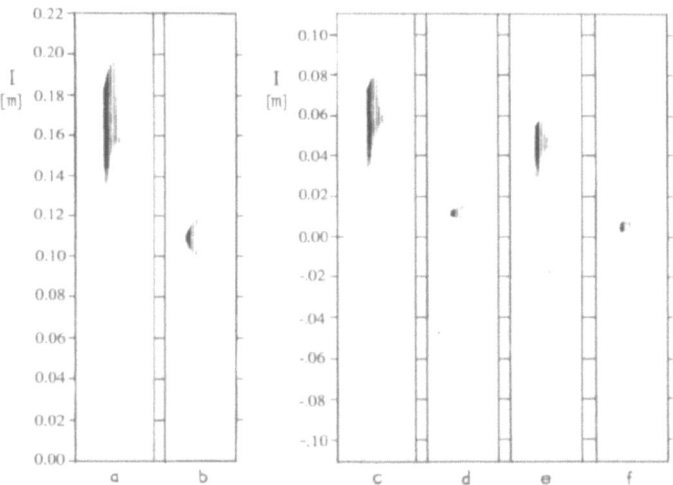

Figure 4. Two-dimensional histograms for the central ray bundle of the flowmeter. (a) quadratic flow profile approximating a laminar flow, (b) rectangular profile approximating a turbulent flow, same effective flow velocity v_F as in Fig. 4a. (c) Difference of flow 4a and 4b, equal to the sum of the last three Terms in eqn. (10). (d), (e), (f) The last three tems of eqn. (10) in the same order from left to right, exactly corresponding to the three flow fields in Fig. 5.

Obviously, for the quadratic profile the classified rays concentrate around much higher flow influences I compared to those of the rectangular profile. According to equation (8), this means a higher effect on the time of flight. In the signal processing algorithm (1) this requires a lowering of the calibration factor a. Such compensation is not feasible in a flowmeter, because the transition between the turbulent and the laminar flow regime depends on the viscosity of the fluid, which normally is neither known in advance nor measurable in a simple way.

To analyze the physical reason of the high flow influences I in a quadratic flow, this flow is decomposed into a rectangular flow and several disturbing flows. To demonstrate this method, we consider an unidirectional flow between two parallel infinite planes. The flow velocity is assumed to depend only upon the relative position $\tilde{x}$ between the two planes $\tilde{x} = -1$ and $\tilde{x} = +1$. Now we chose the velocity field $v_R(\tilde{x}) = 1$ to represent the rectangular profile and the field $v_D(\tilde{x}) = [1 - 3\tilde{x}^2]/2$ to represent a quadratic disturbation. The disturbation is chosen orthogonal to the rectangular flow and thus gives no net flow

$$\int_{-1}^{1} v_R(\tilde{x}) v_D(\tilde{x}) d\tilde{x} = 0 \qquad \int_{-1}^{1} v_D(\tilde{x}) d\tilde{x} = 0. \tag{9}$$

The complete quadratic flow profile expected to be valid in the laminar regime is given by the superposition of the rectangular flow and the quadratic disturbation $v_{1Q}(\tilde{x}) = v_R(\tilde{x}) + v_D(\tilde{x})$ $= (3/2)[1 - \tilde{x}^2]$. The objective of a profile independent flow measurement can now be formulated as eliminating the influence of the disturbation v_D upon the measurement results.

This point of view can be applied sucessfully to the flowmeter shown in Fig. 1. Let $\tilde{x}$ denote the relative position between the sidewalls and $\tilde{y}$ denote the relative position between the upper and the lower boundary of the central part of the tube volume. In the laminar regime, the flow can be approximated quite well by a product of two quadratic functions of the type defined above

$$v_{2Q}(\tilde{x}, \tilde{y}) = v_{1Q}(\tilde{x}) \, v_{1Q}(\tilde{y}) = 1 + v_D(\tilde{x}) + v_D(\tilde{y}) + v_D(\tilde{x}) \, v_D(\tilde{y}). \tag{10}$$

The first term on the right side of this equation represents a rectangular flow profile approximating a turbulent flow. The following three terms characterize the deviations between the laminar and the turbulent regime. They are illustrated in Fig. 5. For the measurement tube in Fig. 1, a comparison of histograms (d), (e) and (f) in Fig. 4 shows, that the term $v_D(\tilde{x})$ gives the main contribution to the

511

above mentioned discrepancy with respect to the flow influence I. This corresponds to an acoustic field mainly concentrated at low values of $|\tilde{x}|$ near the symmetry plane of the tube.

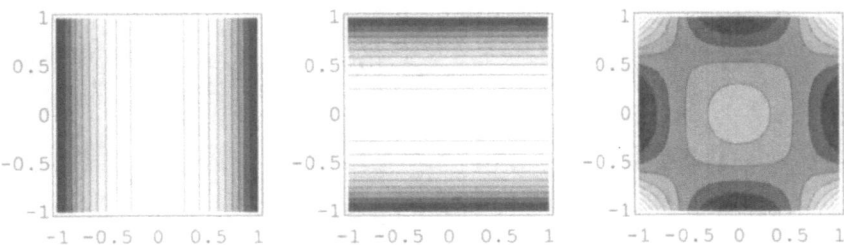

Figure 5. Contour plots of the last three terms in eqn. (10), shown in the order from left to right. The horizontal axes denote $\tilde{x}$, the vertical axes denote $\bar{y}$. Negative velocities are shown in dark tone, positive velocities in light tone.

The approach described above can be used to improve the design of the measurement tube attempting to minimize the influence of all the last three terms in (10). Our experience shows that there are good chances to achieve an accuracy of about 1%.

For each value of v_F , the histograms shown in Fig. 4 can be summed up over all values of I using eqn. (7). So the temporal response function of sound propagation is obtained. Folding this response function by suitable burst functions modelling transducer response, an estimate of the output signal of the flowmeter tube can be computed.

DISCUSSION

To facilitate the development of high precision ultrasonic flow meters, quantitative analytical and numerical models of the measurement process are required. In this paper, sound propagation in the moving medium is examined using a novel type of raytracing. Propagation and reflection at extended wall sections of the measurement tube is described by geometrical acoustics, while small cross sections of transducers and small reflecting surfaces are modelled more exactly taking into account effects of beam divergence and diffraction.

An alternative approach to the problem was developed using a simplified boundary element method for subsequent sections of the measurement tube[7]. This allows to consider diffraction occuring at the boundaries of these sections straightforwardly. On the other hand, our raytracing method enables a transparent analysis of acoustical effects across the whole measurement tube. It facilitates the use of intuitive reasoning and offers a high flexibility to model complex geometries with tolerable computational effort.

As an example, our method is applied to a typical flowmeter configuration. The classification of rays according to pathlength and flow influence is shown to provide a valuable tool for the analysis of measurement errors. Future work will include all rays outside the central bundle. This will require advanced summation methods to handle correctly terms with rapidly oscillating phases.

REFERENCES

1. Jena, A.v.; Mágori, V.; Rußwurm, W.; Ultrasound gas flow meter for household applications, Sensors and Actuators A Vol. 37-38 (1993), p. 135-140.
2. Brekhovskikh, L.M.; Godin, O.A.; Acoustics of Layered Media I, Springer, Berlin (1990).
3. Kravtsov,Yu. A.; Orlov, Yu.I. Geometrical Optics of Inhomogeneous Media, Springer, Berlin (1990).
4. Lighthill, J.; Waves in Fluids, Cambridge University Press, Cambridge (1978).
5. Pierce, A.D.; Acoustics, McGraw-Hill, New York (1981).
6. Keller, J.B.; Diffraction by an aperture, J. Appl. Phys., Vol. 28, p. 426 (1957).
7. Eccardt, P.C.; Landes, H.; Lerch, R.; Mágori, V., Simulation of acoustic wave propagation within flowing media using simplified boundary element techniques, Acoustical Imaging, Florence 1995.

ULTRASONIC TIME REVERSAL PROCESSING IN NON DESTRUCTIVE TESTING

Véronique Miette, Mathias Fink, and François Wu

Laboratoire Ondes et Acoustique
ESPCI, Université Paris VII
U.R.A C.N.R.S 1503
10 rue Vauquelin, 75005 Paris
France

INTRODUCTION

Non Destructive Inspection plays an important role in a number of high-tech industries such as aerospace, nuclear, petroleum. It is used to control an installation safety, to determine the component life times or to improve the product quality. These purposes can be summarized in only one aim: to detect without doubt the smallest defect whatever the material microstructure and the shape of the sample. Moreover, inspection time and cost have to be optimized.

To reach these objectives, we propose the use of the time reversal process[1,2,3,4] associated with several processing techniques. In the time reversal process, the acoustic pressure field reflected by a defect is detected with a set of transducer elements and is digitized and stored during a time interval T. The recorded acoustic signals are then retransmitted by the same transducers in a time reversed chronology (last in, first out). Such a time reversal mirror (TRM) allows conversion of a divergent wave issuing from a defect in a convergent one focusing on it. It is a natural focusing technique matched to the defect shape and to the geometry of the liquid-solid interface.

In collaboration with a French aircraft industry, S.N.E.C.M.A, we experiment the TRM process on titanium alloys. Indeed, during the titanium elaboration process, metallurgical inclusion defects called « hard-alpha » can appear. If not detected, hard-alpha inclusions can become crack initiation sites and lead to component failures in rotating component of a jet engine. The detection sensivity of this kind of defect is limited because a strong ultrasonic speckle noise is induced by the polycrystalline microstructure and because hard-alpha inclusion has a low reflectivity due to a small acoustic impedance. Moreover, a commonly used criterium of rejection is the amplitude threshold which is the ratio in dB between the maxima of the defect and the speckle noise signals. This criterion is the simplest one but it induces limitations for the classical ultrasonic techniques which become very sensitive to the following parameters: surface roughness, complex interface and speckle noise due to the material microstructure. Then a small defect echo can be confused with a signal resulting from grains configurations. On the other hand, false alarms resulting from constructive interferences of grains echoes can cause the rejection of a safe billet.

In this paper, we describe the experimental procedure and we present a new processing technique based on the matched filter properties of TRMs. The sample is a part of a titanium billet of 250 mm diameter containing artificial defects. These defects are flat bottom holes (FBH) because of the difficulty to fabricate synthetic « hard-alpha » inclusions. The theory will be illustrated by experimental results which demonstrate speckle noise reduction and improvement in defect detection.

EXPERIMENTAL TIME REVERSAL PROCEDURE

The TRM prototype

The real time electronic prototype is made of 128 channels working in transmit-receive mode. Each programmable transmitter is made of a 12 bits D/A converter followed by a linear power amplifier. In the receive mode, we use a set of 16 A/D converters through a multiplexer. The sampling frequency is 40 MHz and the central frequency of the tranducers array is 5 MHz. Different 2D arrays have been used in our study. In this paper we shall present results obtained with an optimal array designed to control specifically the center of large diameter billet. In order to improve the focusing capability of the time reversal technique in this zone, the 2D array geometry is built on a « Fermat surface » geometry designed to make all the propagation times between the transducer surface and the center of the billet equal, thus providing optimum focusing at this point. The array is prefocused through the water-titanium cylindrical interface (100 mm of propagation in water and 125 mm of propagation in titanium). It contains 121 elements distributed according to a six annuli structure of respectively 1, 8, 16, 24, 32 and 40 elements.

The experimental procedure

The basic theory [1] of the time reversal process is based on the time reversal invariance of wave equation which tell us that, if the acoustic pressure field $p(r, t)$ satisfies the wave equation then $p(r,-t)$ or $p(r, T-t)$ is also solution of the same equation. We take advantage of this property to achieve optimal focusing of a pulsed wave on a point-like reflector located in a solid sample immersed in fluid medium. The basic time-reversal process used requires four steps.

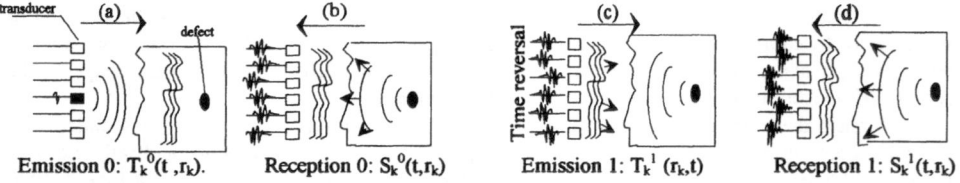

Emission 0: $T_k^0(t ,r_k)$.   Reception 0: $S_k^0(t,r_k)$ Emission 1: $T_k^1(r_k,t)$ Reception 1: $S_k^1(t,r_k)$

Figure 1. *Iterations number 0 and 1 of the time-reversed process. (a): first emission of a wide beam - (b reception on the 121 transducers array; the 121 signals $S_k^0(t,r_k)$ are then time reversed (c) Each $S_k^0(T-t,r_k)$ is emitted by the 121 elements and focusses on the defect- (d) the amplitude of the signal $S_k^1(t,r_k)$ reflected by the flaw is optimized.*

(a) A first incident pulsed wave is transmitted from the liquid towards the solid by one (or more) element(s) of the array. In this first transmission, the array sends unfocused acoustic energy into the material. For each transducer k of the array, located at position r_k the transmitted signals are $T_k^0(t, ,r_k)$ (Figure 1 (a)).

(b) The echoes coming from the block are measured by the same 2-D array on the N transducers. For each transducer k, the received signal $S_k^0(t,r_k)$ are recorded in storage memories. If a defect is present in the illuminated volume, it behaves as an acoustic source and reflects a small amount of the energy transmitted in the previous step (Figure 1 (b))..

(c) During this step, we choose the origin and the duration of the signals to be time-reversed. This is achieved through the definition of a temporal window identical for all the transducers corresponding to a given depth of inspection in the material. The depth of inspection is known by measuring the transit time of the acoustic pulse as in conventional ultrasonic inspection. For each $T_k^1(r_k,t) = S_k^0(r_k T-t)$ (Figure 1 (c))..

(d) The new echoes $S_k^1(r_k,t)$ coming from the titanium sample are recorded. If the time reversal window previously selected contains some echoes from a defect, the resulting time reversed wave refocuses naturally on it, the new signals $S_k^1(r_k,t)$ show a high level amplitude and the flaw is easily detected (Figure 1 (d)).

In some applications, a new iteration of the time reversal process can be done to improve the focusing on the defect and the signals $S_k^1(r_k,t)$ are windowed, time-reversed, and retransmitted by the array as $T_k^2(r_k,t) = S_k^1(r_k,T-t)$. The new echoes $S_k^2(r_k,t)$ are then recorded for signal processing.

Figure 2 shows a time reversal experiment conducted on a defect zone located at a depth 120 mm in a titanium billet. Three Bscans images of $S_k^0(r_k,t)$, $S_k^1(r_k,t)$ and $S_k^2(r_k,t)$ where k varies from 1 to 121 are presented. On each of them, the horizontal axis represents the time and the vertical axis the transducer number k. The amplitude is coded in gray level and it represents the logaritmic envelope of the echoes. On figure 2 (a) which shows $S_k^0(r_k,t)$ we can see the echoes coming from the front and the back faces of the titanium sample, and between them we can only notice the noise induced by microstructural inhomogeneities. The echo from the defect is superimposed upon a strong grain noise background. A time reversal window of 2 μs is chosen and its origin is located at a depth of 120 mm, where is located an artificial flaw. After one or two time reversal process, the signals $S_k^1(r_k,t)$ (Fig. 2 (b)) and $S_k^2(r_k,t)$ (Fig. 2 (c)) contain a strong echo coming from the flaw which has been amplified by the time reversal process. A wavefront appears on the data and we can notice the similarity between signals $S_k^1(r_k,t)$ and $S_k^2(r_k,t)$. The flaw signals are then differentiated from the backgroud noise. On each individual element, the signal to noise ratio is improved by more than 20 dB. Once the defect has been detected, the autofocus procedure remains focused on the flaw and the two sets of signals $S_k^1(r_k,t)$ and $S_k^2(r_k,t)$ are very similar. Note that in this experiment the wavefront appears as an oscillating line, which is due to the fact that the defect is located off axis of the transducer array.

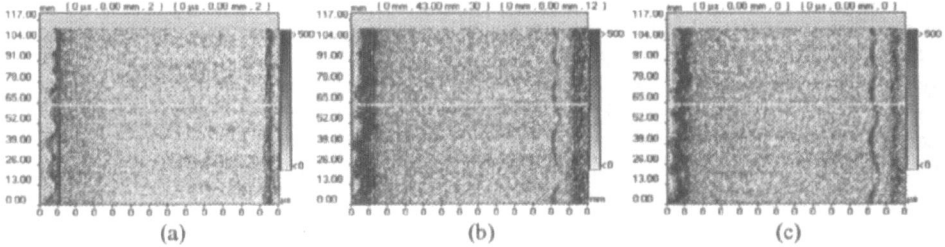

Figure 2. *Bscans of a cylindrical sample of 250 mm diameter and 120 mm depth with a 0.8 mm flat bottom hole*

In an another experiment conducted in a zone containing only the basic titanium microstructure, Fig 3 shows, that in this case, the signal behavior does not change between $S_k^0(r_k,t)$ and $S_k^1(r_k,t)$. We do not observe any wavefront appearing in the data. Another important point to be noticed is that if you look carefully to data $S_k^1(r_k,t)$ and $S_k^2(r_k,t)$, they are no more similar. The received data are quite different and no regular structure appears in the windowed signals. This is related to the fact that the titanium microstrucure had a mean dimension (a few μm) small compared to the wavelength (1.2mm for longitudinal waves). The time reversal process cannot exactly refocus the energy on the microstructure details due to the loss of information on these details during propagation of the backscattered wave with a 5-Mhz center frequency.

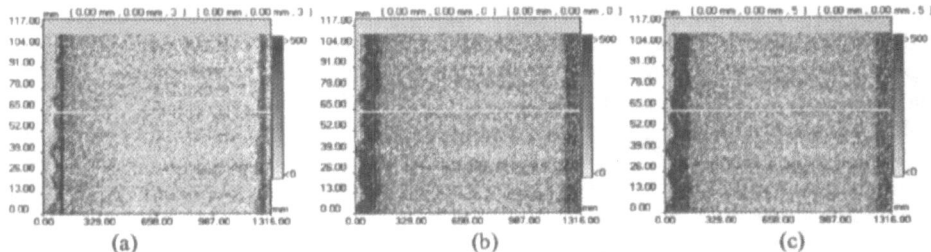

Figure 3. *Bscans of a cylindrical sample of 250 mm diameter and 120 mm depth in a noise zone.*

Figures 2 and 3 represent the complete Bscan informations on the 121 transducers of the array. A more compact presentation has to be implemented in order to reduce the 121 signals to one unique signal. Different processing of the data have been studied in our laboratory to reach this goal.

Incoherent summation

The simplest processing that we have implemented is an incoherent processing; where we compute the summation of the 121 logarithmic signals Log $S_k^1(r_k,t)$ (or Log $S_k^2(r_k,t)$).

$$\text{Inc}^i(t) = \sum_{k=1}^{121} \text{Log}(S_k^i(t,r_k)), \text{ where } i = 1 \text{ or } 2$$

Coherent summation

The incoherent processing is not optimal for defects located off axis of the probe. In this case, all the individual reflected signals, from the defect, have different arrival times on the array. The incoherent summation is not very efficient in this case. The total output signal can be improved significantly by correcting the differences in arrival times before the summation. Such a time compensating process corresponds to a *coherent summation* and allows to the echo level to be increased. A cross-correlation algorithm can be used to determine the time shifts between the individual echoes. This algorithm presents two drawbacks : it requires that all of the individual signals be stored and it is time consuming. We have developed a simpler and quicker algorithm to measure the time delays. This algorithm needs a complete time reversal sequence and uses the symmetrization property of the time reversal process. This symmetrization property is due to the fact that a time reversal process is equivalent to a matched filter[7]. From this property, the time delays are determined by finding on each element k, the position of the peak amplitude τ^{kl} of the symmetric received signal $S_k^1(r_k,t)$. As the signal becomes symmetric after one time reversal process, the location of the maximum pressure becomes precise and easy. This algorithm only requires to store the positions of the maximum pressure of $S_k^1(r_k,t)$. The coherent summation is determined according to the discrete summation:

$$\text{Coh}^1(t) = \sum_{k=1}^{121} S_k^1(r_k, t - \tau^{k1})$$

Then, the amplitude peak is optimized if there is a defect in the zone of interest. However, such a coherent processing can induced false echo in the case of a speckle noise zone. Indeed, in such a case 121 values for τ^k are computed when looking to the position of the maxima of signals $S_k^1(r_k,t)$. In the case of echoes from a speckle noise the values τ^k are randomly distributed and the coherent summation will give rise to some peack amplitude that can be considered as an artifact resulting from an alignment of the maximum of the individual speckle noise signals. In summary, the coherent summation is optimal if there is a defect in the zone of interest but it induces some false alarm from a spekle zone.

Iterative coherent summation

To solve this problem, we have developed a new coherent summation processing, which take into account the informations recorded on two consecutive iterations of the time reversal process. In the preceeding paragraphs, we have noted that when the time reversal window is located on a defect, the signals $S_k^1(r_k,t)$ and $S_k^2(r_k,t)$ are similar. The same wavefront appear at each iteration. This means that the positions τ^{k1} and τ^{k2} are aligned on the same curve. Besides, if the time reversal window is located on a pure speckle zone, we have seen that the signals $S_k^1(r_k,t)$ and $S_k^2(r_k,t)$ are very different and consequently the positions τ^{k1} and τ^{k2} are aligned on completely different curves. There is no correlation between these two curves. From these two remarks, a better algorithm can be implemented to increase the echo of a defect, while the echo of the speckle noise is reduced. It consists on the summation of signals $S_k^2(r_k,t)$ observed at iteration 2 delayed by the arrival times τ^{k1} observed at iteration 1.

$$\text{Coh}^2(t) = \sum_{k=1}^{121} S_k^2(r_k, t - \tau^{k1})$$

In the presence of echoes coming from a pure speckle noise such a procedure is equivalent to the one of a random phase transducer[6] which strongly reduces the speckle noise.

516

The iterative matched filter technique

The two preceeding remarks shows the interest to take into account the degree of similarity between the echoes $S_k^1(r_k,t)$ and $S_k^2(r_k,t)$. An even better algorithm that can be implemented to measure this degree of similarity between the two iterations consists in the computation of the cross correlation between the two sets of signals. A very classical theorem of signal theory tell us that it is equivalent to the computation of the convolution product of $S_k^1(r_k,t)$ with the time reversed signals $S_k^2(r_k,-t)$. Thus the resulting signal is

$$\text{Match (t)} = \sum_{k=1}^{121} S_k^1(r_k,t) \otimes_t S_k^2(r_k,-t)$$

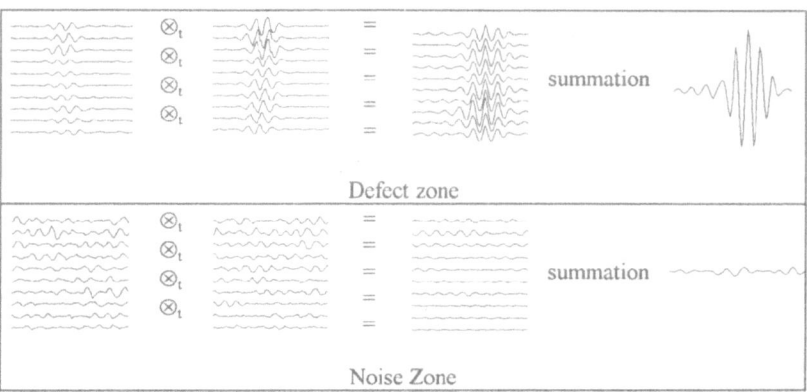

Figure 4: *Convolution of $S_k^1(r_k,t)$ and $S_k^2(r_k,-t)$: resulrs for a defect zone and for a noise zone.*

Such a signal becomes very high when the echo comes from a defect and remains very law when the echo comes from a pure speckle noise zone (Figure 4). This processing is in fact optimal to focus both in the transmit and in the receive mode on the defect[7]. However this processing is time consuming compared to the iterative coherent summation, but it allows a better detection of the defect. This is essentially due to the fact that the determination of the time delay law τ^{k1} can be noisy if the signal to noise ratio is poor after one iteration.

EXPERIMENTAL RESULTS

In following experiments, we have used the TRM prototype described above. The sample is part of a titanium TiA6V billet of 250 mm diameter. Its interface is cylindrical and it is 120 mm deep which is the most critical depth for the NDI techniques in account of the titanium microstructure and in account of the attenuation.

First, we present two CSCANs images obtained by scanning the 2D array in front of the titanium billet to demonstrate the ability of the time reversal process to improve the flaw detection in comparison with the classical techniques. Data coming from a window located around 120 mm depth in the billet are presented. The first one is the result of the classical focusing method : there is only one emission step and one reception step where the 121 transducer elements transmit a high focused beam which is reflected by a flat bottom hole of 0.8 mm and the field is received by the probe (figure 5 (a)).

The second one is the result of two time reversal iterations (figure 5 (b)). The experimental conditions are the same and the CSCAN are the results of the incoherent summation. For the classical focusing technique, the signal to noise ratio reaches 5 dB and the defect is detected only in a very small area. The time reversal technique allows to detect defect in a larger zone with a higher signal to noise ratio (24 dB). The time reversal technique detect the defect even if it is located off axis of the probe.

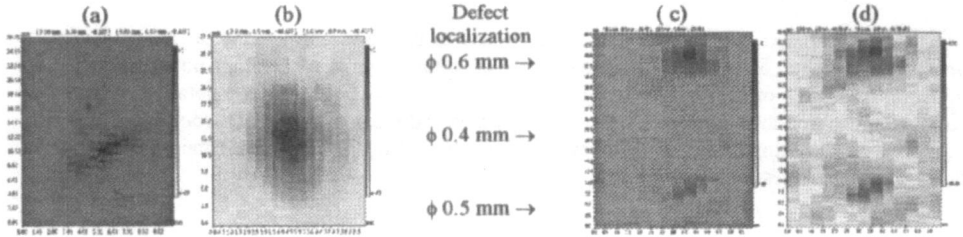

Figure 5. *Sample of 120 mm depth. The amplitude is presented in dB: the signal maximum is coded in black color.*

Figure 5 (c and d) illustrates the effectiveness of the matched filter algorithm. It shows two CSCANs examples which present the same sample of 120 mm depth with three flat bottom holes of 0.4, 0.5, 0.6 mm diameter. We can notice the capability of the matched filter technique to reduce the speckle noise. The signal to noise ratio is improved by 11 dB for the 0.6 mm diameter defect and by 13 dB for the 0.5 mm one. The defect of 0.4 mm diameter is not detected in this case.

CONCLUSION

We have demonstrated the ability of the time reversal process to focus on defect through complex interface in deep samples. This method overcomes the usual problems of the NDI method which are sensitive to the arrays alignment, to the interface shape and to the speckle noise. Moreover, NDI techniques are limited by the array precision focusing. The time reversal process allows a greater flexibility in reason of its capability to detect defect out of the probe axis: this system doesn't require a perfect quality or geometry of the titanium billet and no accurate adjustments.

The time reversal process reduced naturally the amplitude of the speckle noise. Due to the loss of information during the propagation this method cannot focus on microstructure whose dimension is small with respect to the wavelength.

In case of a very noisy microstructure, we can associated a new data processing to the time reversal technique. Indeed, we see that the matched filter in receive mode allows the reduction of false alarms detection and improves the small defects one. This technique present one drawback: it requires to store and to convolve the 121 signals of two iterations. However, with the used of a Digitized Signal Processor (DSP), these calculations can be implemented rapidly and the industrial time inspection of a billet can be optimized.

In conclusion, theses performances show the potential of the time reversal process to be adapted and used for industrial problems.

REFERENCES

1. M. Fink, « Time reversal of ultrasonic fields- Part 1 : Basic principles », IEEE Trans. Ultrason. Ferroelc, Freq. Contr., vol. 39, n°5, September 1992, p 555-566
2. F. Wu, J.L Thomas, M. Fink, « Time reversal of ultrasonic fields- Part II : Experimental results», IEEE Trans. Ultrason. Ferroelc, Freq. Contr., vol. 39, n°5, September 1992, p 567-578
3. N. Chakroun, M. Fink, F. WU « Basic principles of ultrasonic time reversal processing in non destructive testing», Review of QNDE, vol. 14, 1995, pp. 937-943
4. N. Chakroun, V.Miette, M. Fink, F. WU « Ultrasonic inspection of titanium alloy with a time reversal mirror», Review of QNDE, vol. 14, 1995, pp. 2105-2111
5. M.Fink, R. Mallart, F. Cancre, « The Random Phase Transducer : A New Technique For ncoherent Processing. :Basic Principles and Theory" IEEE Trans. Ultrason. Ferroelc, Freq. Contr, Vol 37,N°2, Mars 1990,p 54-69, 63
6. P.Laugier, M. Fink, S. Aboelkaram, « The Random Phase Transducer : A New Technique For Incoherent Processing : Experimental Results" IEEE Trans. Ultrason. Ferroelc, Freq. Contr, Vol 37, N°2, Mars 1990, p 70-78
7. C.Dorme, M. Fink « Focusing in transmit-receive mode through inhomogeneous media: the time reversal matched filter approach» J. Acout. Soc. Am., August 95, 98 (2), p 1155-1162

HIGH PRECISION ACOUSTIC RADAR
BY UNEQUALLY SPACED FREQUENCY ARRAY

Toyokatsu Miyashita

Dept. Electronics and Informatics
Ryukoku University
Seta Ohe-cho, Ohtsu, 520-21 Japan

INTRODUCTION

With an ideal impulse acoustic irradiation, a high precision measurement of the echo signal in the time domain will give directly the impulse response of the objects. This is a straight idea of the impulse response measurement by the pulse echo method. Any ideal impulse acoustic irradiation is not available, and a predictive extrapolation of the missing information, i.e. frequency components, were inevitable in the deconvolution processing[2]. The deconvolution with the aid of a priori knowledge gave an impulse response whose resolution was up to the fundamental sampling spacing in the time domain. However a limitation caused by an interference between the point spread function and the sampling points which is absolutely essential in the digital signal processing was left to be overcome.

In this paper, a measurement of the impulse response of the objects in the frequency domain, i.e. by a frequency array is considered. Here the predictive extrapolation, which corresponds to the extrapolation of the missing frequency components in the pulse echo method, is included naturally in the multifrequency measurement by a frequency array. The main consideration is how to overcome two essential limitations of the digital signal processing in order to get the highest precise location of the reflections with a given frequency aperture width.

(1) In general, the echo location of an object is not exactly on the fundamental sampling points, and a wide spreading response in the form of a sinc function is reconstructed in the spatial domain.

(2) The reconstructed response of an object is repeated infinitely in every fundamental section in the spatial domain whose length is determined by the element spacing of the frequency array.

This paper discusses an acoustic radar by an unequally spaced frequency array which overcomes these limitations and gives a superresolution up to 1/100 times the fundamental sampling spacing, i.e., 1/400 times the wavelength, using simulated complex multifrequency acoustic echo signals from an one-dimensional layered medium with multiple reflections.

A MODEL OF AN ACOUSTIC RADAR BY A FREQUENCY ARRAY

The measurement system is simply composed of a sinusoidal wave oscillator, cw transmitting and receiving transducers and a computer which controls the measurement system and makes data processings. The structure and the basic quantities of the reconstruction space and the frequency array are shown in Fig.1. The frequency array has K frequency elements f_k's and

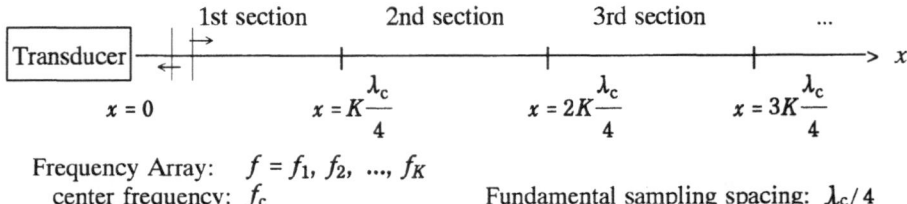

Frequency Array: $f = f_1, f_2, ..., f_K$
   center frequency: f_c Fundamental sampling spacing: $\lambda_c/4$
   its wavelength: λ_c Fundamental section: $K\lambda_c/4$

Figure 1. Reconstruction space of an acoustic radar by a frequency array.

the aperture width of $2f_c$ with the center frequency f_c. These quantities are not necessarily determined and fixed in the actual measurement, but possibly they are the results of a predictive extrapolation of the measured multi-frequency data. The fundamental resolution in the reconstruction space is equal to the fundamental sampling spacing $\lambda_c/4$, where λ_c is the wavelength of the acoustic wave in the medium at the center frequency f_c. The digital space does not have any intermediate point between the fundamental sampling points. The actual response of an object on an intermediate point would be reconstructed abnormally. The length of the fundamental section of the reconstruction space is $K\lambda_c/4$. The conventional measurement with an equally spaced frequency array gives an infinite periodic repetition of the reconstructed signal with an interval of the fundamental section. This requires that the object should locate only in a fundamental section, and that the all other sections should be empty.

The object is simulated by a one-dimensional $(N+1)$ layered medium. The width of the nth layer is d_n, and its complex reflection coefficient is r_n. The measured complex amplitude $A(\bar{f}_k)$ $(k = 1, 2, \ldots, K)$ is calculated by a recursive formula(1) which includes multiple reflections.

$$A(\bar{f}_k) = r_{e,1}, \quad z_{kn}^2 = e^{2\pi i \frac{\bar{f}_k \bar{d}_n}{K}}, \quad \bar{f}_k = K f_k/(2f_c), \quad \bar{d}_n \doteq d_n/(\lambda_c/4),$$

$$r_{e,n} = z_{kn}^{-2}\left[r_n + \frac{r_{e,n+1}(1 - r_n^2)}{1 + r_{e,n+1}r_n}\right], \quad \text{for } n = 1, 2, \ldots, N, \quad r_{e,N+1} = 0. \tag{1}$$

In the following theoretical considerations, this recursive formula is expanded in terms of z_{kn}^{-2} and represented as

$$A(\bar{f}_k) = \sum_{n=1}^{N_e} r_n z_{kn}^{-2}, \quad z_{kn}^2 = e^{2\pi i \frac{\bar{f}_k \bar{d}_n}{K}}, \tag{2}$$

where r_n $(n = 1, 2, \ldots, N_e)$ is the effective and equivalent complex reflection coefficient including multiple reflections and $\bar{d}_n$ in this formula is its effective distance from the transducers and equal to an integer on a fundamental sampling point.

EQUALLY SPACED FREQUENCY ARRAY AND
UNEQUALLY SPACED FREQUENCY ARRAY

The amplitude gain of the frequency array is usually tapered by a weighting function to suppress its side lobes below the desired level allowing a slightly larger main lobe. This tapering is effectively applied also to the modulation of the element spacing of the frequency array. As shown in

[1], an unequally spaced frequency array obtained by this tapering has not only the suppressed side lobes with slightly larger main lobe but also a very effective suppression of the periodic repetition in the reconstruction signal space which occurs with any equally spaced frequency array as illustrated below. These two types of frequency arrays, an amplitude tapered and a frequency tapered, are designed as shown in Fig.2 from a Taylor illumination function of -30dB side lobe level. These arrays are designed with a common

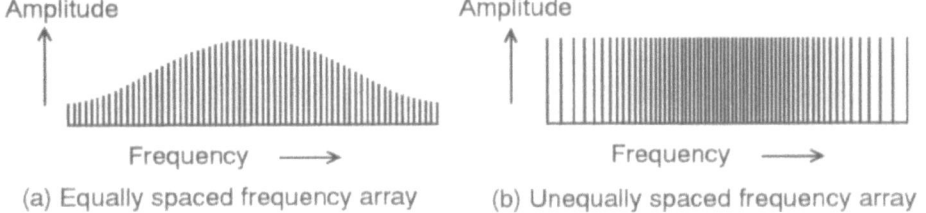

(a) Equally spaced frequency array   (b) Unequally spaced frequency array

Figure 2. Equally spaced frequency array and unequally spaced frequency arrays.

symmetric property $\bar{f}_{\frac{K}{2}+l} + \bar{f}_{\frac{K}{2}-l+1} = K$ $(l = 1, 2, \ldots, K/2)$ in order to have a simple linear phase dependence of their complex point spread functions as shown in the later sections. Here, $\bar{f}_k$ is equal to $k - \frac{1}{2}$ for the equally spaced frequency array, but it has, in general, no half integer value for the unequally spaced frequency array. The optimum frequency array in which both amplitude and frequency are tapered will exist for the suppression of the periodic repetition. However it is not pursued in this paper.

CONQUEST OF THE LIMITATIONS IMPOSED ON THE NORMAL DIGITAL SIGNAL PROCESSING

Point Spread Functions in the Digital Space

When the resolution in the signal space goes to the extreme of the fundamental sampling spacing determined by the frequency aperture width as in this paper, a well known but peculiar phenomenon occurs as shown in Fig.3. Any band-limited impulse response appears in the shape of a sinc

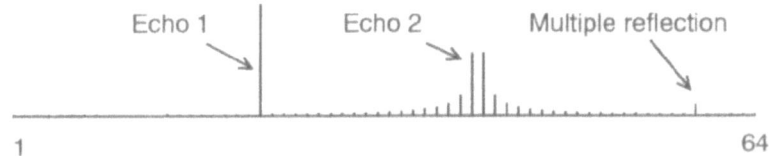

Figure 3. Reconstruction of sharp responses.

function, i.e. an interpolation function, on the continuous analog axis. Also on the digital axis, this shape appears in general. Only when its center is accidentally on a fundamental sampling point, the impulse response has a single nonzero value on the sampling points, because all other sampling points coincide with the null points of the sinc function. An application of an adequate tapering function as Taylor illumination function on the amplitude gain or on the frequency element spacing makes the side lobes of the sinc function smaller and consequently makes interferences between the impulse responses small enough to be investigated almost independently. A precise location of an impulse response requires an analog investigation of the function between the fundamental sampling points.

Independent of the practical possibility of large K for the multifrequency measurement, the reconstructed object response is a multiply folded version of the half infinite space into the fundamental section of a length $S = K\lambda_c/4$. The response of the object which occurs actually in the lth section between $(l-1)S$ and lS $(l = 1, 2, \ldots)$ is multiply folded into and reconstructed in the first section between 0 and S. Further more this multiply folded image is repeated in every section of the reconstruction space as shown in Fig.4(a). This phenomenon is sufficiently suppressed by the unequally spaced fre-

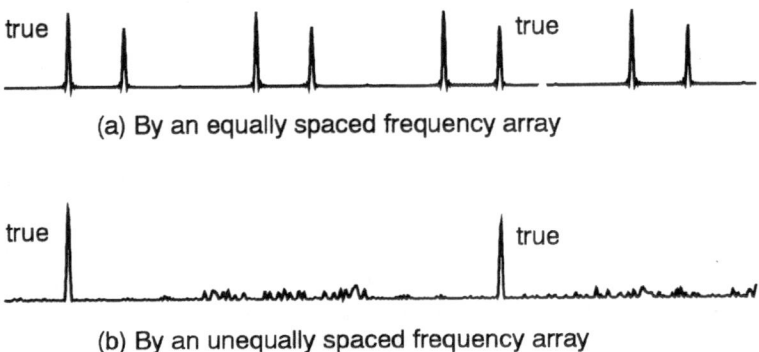

(a) By an equally spaced frequency array

(b) By an unequally spaced frequency array

Figure 4. Periodic repetition by an equally spaced frequency array and its suppression by an unequally spaced array.

quency array as shown in Fig.4(b). It is obvious that an unequally spaced frequency array should be used when the responses of the object are distributed also outside the first fundamental section.

PRECISE LOCATION OF THE REFLECTIONS

Phase Characteristics of the Reconstructed Impulse Response

The reconstructed signal $R(\bar{x}_m)$ is composed of a linear superposition of the impulse responses including multiple reflections as shown in Eq.(3).

$$R(\bar{x}_m) = \frac{1}{K} \sum_{k=1}^{K} A(\bar{f}_k) z_{km}^2 = \sum_{n=1}^{N_e} I_n(\bar{x}_m), \qquad z_{km}^2 = e^{2\pi i \frac{\bar{f}_k \bar{x}_m}{K}} \tag{3}$$

$$I_n(\bar{x}_m) = |I_n(\bar{x}_m)| e^{i[\pi(\bar{x}_m - \bar{d}_n) + \theta_n]}, \qquad \text{where} \quad r_n = |r_n| e^{i\theta_n}, \tag{4}$$

where $\bar{x}_m = x_m/(\lambda_c/4)$ $(m = 1, 2, \ldots)$ may be a general analog position on the signal space and it is equal to an integer on a fundamental sampling point. The impulse response $I_n(\bar{x}_m)$ changes its phase exactly by π between the fundamental sampling points across the true echo position as shown in Eq.(4). The phase characteristics is a result of the symmetric property imposed on the frequency array.

Numerical Examples

The phase characteristics are investigated on the numerical examples of two reflecting boundaries, i.e. targets, and their multiple reflections. The loci of the reconstructed complex amplitude $R(\bar{x}_m)$ are shown in Figs.5-7. The locations of the reflecting boundaries are very general, i.e., the first one is at $5.5\lambda_c$ in the first fundamental section, and the second one is at $32.0\lambda_c + 4.7\lambda_c$ in the third fundamental section. The length of the fundamental

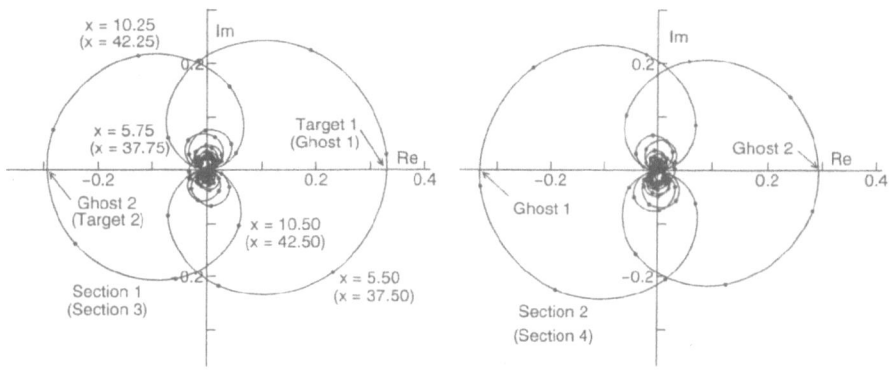

Figure 5. Locus of the response on the complex plain reconstructed with the equally spaced frequency array in the first section (left) and in the second section (right).

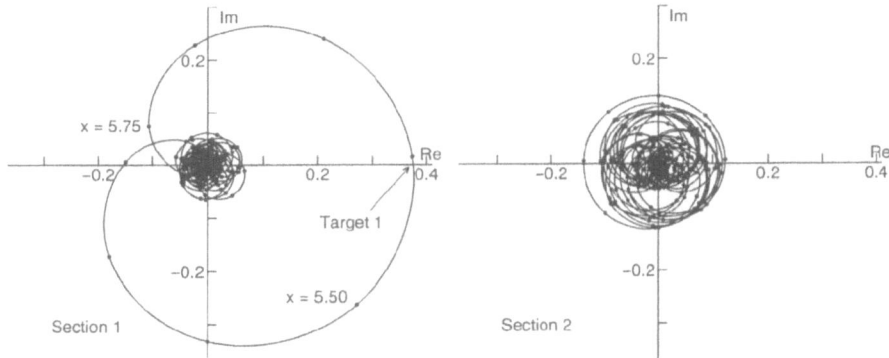

Figure 6. Locus of the response on the complex plain reconstructed with the unequally spaced frequency array in the first section (left) and in the second section (right).

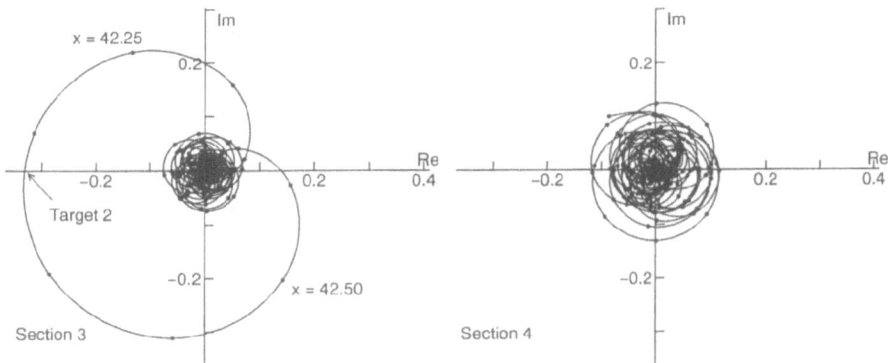

Figure 7. Locus of the response on the complex plain reconstructed with the unequally spaced frequency array in the third section (left) and in the fourth section (right).

Table 1. Precise location of the boundaries.

original boundaries	reconstructed positions			
	by equally spaced array		by unequally spaced array	
positions $[\lambda_c]$	positions $[\lambda_c]$	error $[\lambda_c/4]$	positions $[\lambda_c]$	error $[\lambda_c/4]$
5.55555555	5.55552987	0.000102	5.55812	0.010
42.33333333	10.33333516	0.000007	42.33025	0.012
	...	...		

section is $16.0\lambda_c$. The locus by an equally spaced frequency array repeats itself together with targets and their ghosts changing the phase by π in every section as shown in Figs.5. The position of the phase 0 or π where the amplitude is larger than the threshold value is exactly the location of the target or its ghosts in case of no phase shift by the reflections $\theta_n = 0$.

The locus by an unequally spaced frequency array has no such ambiguities as ghosts and the periodic repetition as shown in Figs.6-7. In a practical data processing, a computer program looks for the reconstructed responses whose amplitudes are larger than the threshold value, and exactly interpolates the phase 0 or π positions from the nearest neighbors. The reconstruction error i.e., precision of the locations is about $1/100$ times the fundamental sampling spacing, i.e. $\lambda_c/400$ by the unequally spaced frequency array. Table 1 shows the results and their precision.

CONCLUDING REMARKS

The phase 0 or π positions of the complex response function reconstructed with an unequally spaced frequency array have located precisely the reflection points even in the presence of a few reflections. This has a drawback that any nonzero phase shift θ_n by the reflection results in an effective displacement of the reconstructed location.

The maximum point of the response $|I_n(\bar{x}_m)|$ of Eq.(4) theoretically locates the reflection points $\bar{d}_n$ independent of the phase shift θ_n. However, the numerical interpolation of the maximum point has an accuracy of $1/10$ times the fundamental spacing even only a reflection exists. A differential filter applied simply on the multifrequency data gave no better accuracy.

The periodic repetition was successfully suppressed with an unequally spaced frequency array. An equally spaced frequency array has, however, an advantage of the possibility of the predictive extrapolation of a narrower frequency array. Then the lower frequency components subject to diffraction effects and the higher frequency components with short range attenuation, both sides of the frequency components are excluded from the actual measurement. This is a very practical requirement to be included in the multifrequency radar. AR model is not yet found for unequally sampled data, so a combination of equally spaced and unequally spaced frequency arrays will be a good solution of this radar.

REFERENCES

[1] T. Miyashita, T. Hamaguchi, H. Ogura, "Three-Dimensional Imaging with a Multifrequency Holographic Method," *Proc. IEEE ICASSP*, pp.2843-2846 (1986).

[2] T. Miyashita, H. Schwetlick and W. Kessel, "Recovery of Ultrasonic Impulse Response by Spectral Extrapolation," *Acoustical Imaging* (Plenum Press, New York, 1985) Eds. A. J. Berkhout *et al.*, vol. 14, pp. 247-257.

AN EFFICIENT ALGORITHM FOR COMPUTED ULTRASOUND IMAGING USING A CURVED SYNTHETIC APERTURE

Juha Ylitalo

Department of Electrical Engineering
University of Oulu
90570 Oulu, Finland

INTRODUCTION

Synthetic aperture imaging allows the design of a low-cost, high-quality ultrasonic scanner with simple hardware (Karaman et al., 1995). Such a scanner is capable of performing two-way synthetic focusing through software control. Computed ultrasound synthetic aperture imaging based on holography was introduced more than 20 years ago (Boyer et al., 1971). The method required fairly large two-dimensional linear apertures which suite well for nondestructive testing applications. In real-time medical applications, however, cross-sectional images of the target are desired. The imaging apertures are one-dimensional and they are often limited in size. For example, the acoustic window to the heart is limited by the rib structure. Therefore, especially in cardiology, the phased array method with digital beam forming has become the state-of-the-art technique.

Even though the phased array scanners are quite complex and expensive they produce only one-way dynamic focusing (in reception). With small aperture sizes this leads to rather modest lateral resolution. The main advantage of the present synthetic aperture approach is that two-way synthetic focusing is obtained with a single active channel. Moreover, it is possible to improve the lateral resolution if the monostatic mode with simultaneous scanning of the transmitter and the receiver is applied. In addition to a fast data acquisition sheme the proposed method has an efficient Fourier-domain image reconstruction algorithm which is capable of producing a sector scan image in real-time.

Fourier-domain synthetic aperture methods for the linear and enclosing circular apertures have been presented before (Ylitalo et al., 1989; Qin et al., 1989; Ylitalo and Ermert, 1994). This paper describes a novel reconstruction algorithm for a computed synthetic aperture imaging method employing a convex array. The algorithm is based on the linear-array model which has been modified according to the geometrical differences between the sector and the linear geometries. Water bath experiments with point sources were carried out to demonstrate the feasibility of the novel approach.

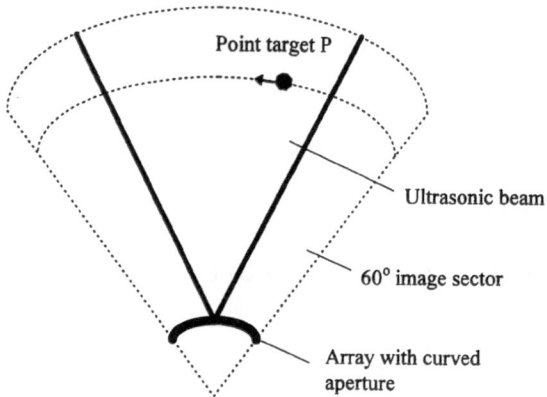

Figure 1. Geometry of the present imaging method.

METHOD

Figure 1 shows the geometry of the present imaging scheme. It is assumed that a 60° sector is scanned with 128 steps and that the radius of the curvature of the array is 15 mm. Actually, slightly larger apertures could be used. In experiments, however, the transducer was fixed in the central axis of the imaging sector and, instead, a point target was scanned mechanically across the 60° image sector. This equals the case where the transducer is scanned along the small imaging aperture and the target is kept steady. For a practical system, electronic scanning of a multielement curved array is needed.

In reconstruction, the backward propagation algorithm for a linear array (Ylitalo et al., 1989) was modified so that the geometrical differences between the linear and the curved approach were compensated for. In practise this means that the phase angle information of the complex hologram data has to be modified according to the differences in path lengths between the two geometries (Fig. 2). In the linear case the target P is seen by the transducer at a distance D1:

$$D1 = R / \cos(\beta) \qquad (1)$$

whereas in the curved geometry it is seen at a distance D2 which is determined by

$$D2 = (R^2 + r^2 - 2\,r\,R \cos (\alpha)) ^{1/2} \qquad (2).$$

If the curved geometry is modeled according to the linear one, the angle α can be expressed as

$$\alpha = (R - r) \tan (\beta) / R \qquad (3)$$

and the phase angle difference between the two geometries becomes

$$\Delta\phi = 4 \pi (D2 - D1) / \lambda \qquad (4).$$

Figure 2. Geometry of the linear aperture and the curved aperture approach.

The phase difference $\Delta\phi$ can be taken into account in the backpropagation algorithm as a multiplicative factor. Therefore, the reconstruction times for the linear and the curved geometries are essentially the same. As seen in the experimental part, this compensation factor is almost linear except in the very near field.

Reconstruction follows the principle of the wavefront backward propagation algorithm. First, Fourier-transform is applied to the complex data row by row. Next, the curvature compensation is performed in the spatial frequency domain according to (2). The differences between the linear and the curved geometries are taken into account as a multiplicative phase angle correction in the spectrum shift phase of the wavefront backward propagation procedure. Finally, the inverse Fourier-transform gives the numerically focused image (Ylitalo et al., 1989). Since only one-dimensional processing and FFT routines are required the image acquisition time is short. For example, the reconstruction of a 128*128 pixel image takes only about one second in an ordinary microcomputer.

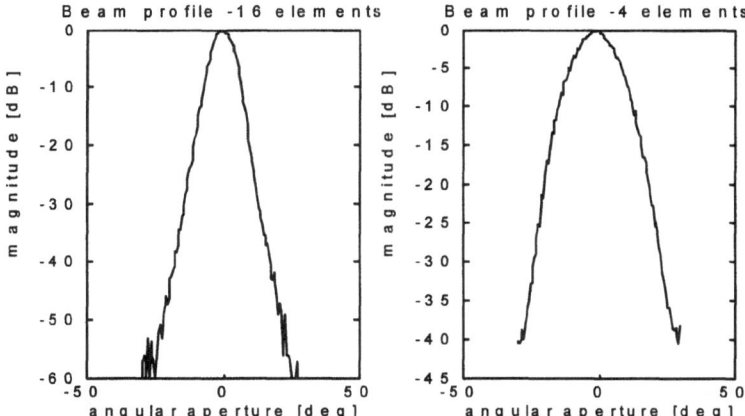

Figure 3. Beam profiles of the probe measured at a distance of 70 mm. 16 weighed elements operating in parallel give a two-way, -12 dB angular beam aperture of 15°. If only four unweighed elements are applied the -12 dB angular beam aperture is 22°. The pitch size is $\lambda/2$.

EXPERIMENT

Water bath experiments with point-like targets were carried out. In experiments an angular aperture of 60° was scanned and the radius of the curvature of the array was 15 mm. This corresponds to an aperture length of 16 mm, approximately. Data were collected by mechanical scanning using a 3.5 MHz multielement array (pitch 0.22 mm). Two different beam forms were employed (Figure 3). In the first case 16 weighed elements were operating in parallel. The weighing function had a shape of the Blackman-Harris window. In two-way measurement the corresponding -12 dB and -40 dB angular beam apertures were 15° and 35°, respectively. In the second case four elements were operating in parallel and no apodisation was applied. This resulted in the -12 dB and -40 dB angular beam apertures of 22° and 60°, respectively. The center frequency of the probe was actually 3.2 MHz and, accordingly, the wavelength was 0.48 mm. In the elevation plane the beam was focused by a lens.

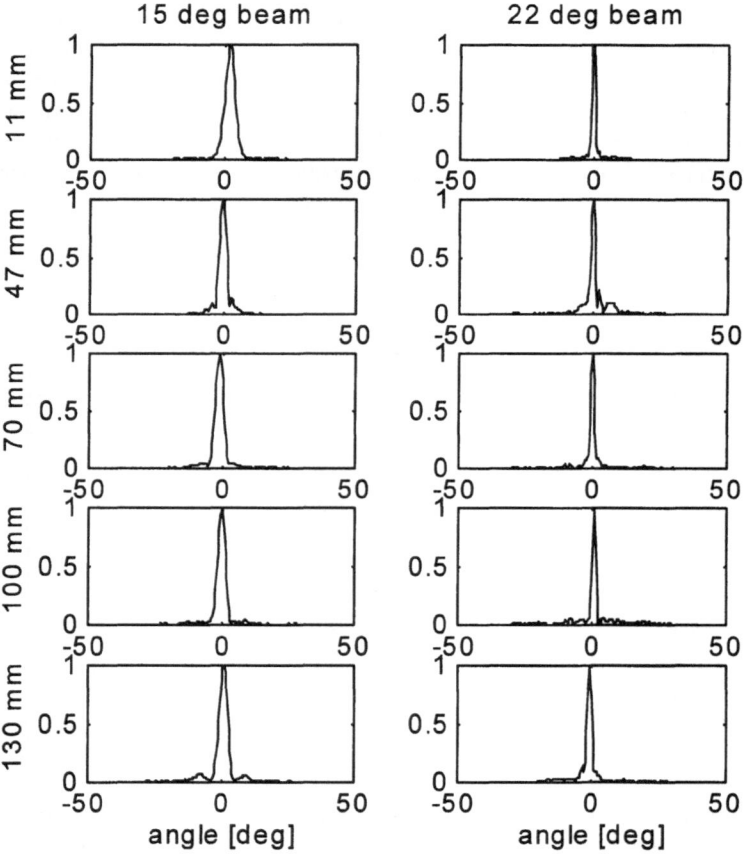

Figure 4. Lateral resolution of the present imaging method in the case of 15° and 22° beam profiles. Nearly uniform lateral resolution is achieved in a depth range of 11 mm - 130 mm.

Lateral resolution of the present imaging method was measured at depths of 11 mm, 47 mm, 70 mm, 100 mm, and 130 mm (Figure 4). Results show that the -6 dB and -20 dB lateral resolution is nearly uniform in a depth range of 11 mm- 130 mm. The lateral resolution was about 1.6° at -6 dB and 2.8° at -20 dB for the 22° beam. For the 15° beam the corresponding values were 3.3° and 6°, approximately. As expected, the lateral resolution is significantly better for the wider beam. Even though in a linear aperture case the 15° beam is adequate to give good lateral resolution (Ylitalo, 1994), the curved aperture approach requires a wider beam.

Figure 5 indicates that the side lobe level of the present imaging scheme is approximately at the level of -30 dB. From Figure 4 it is also evident that the 15° beam gives lower side lobes than the 22° beam. This is due to apodisation. Further improvement in the side lobe level could be obtained by increasing the temporal and the spatial sampling rates. In the present study the rf echo signal was sampled at four samples per wavelength and only one complex sample per wavelength was used in reconstruction.

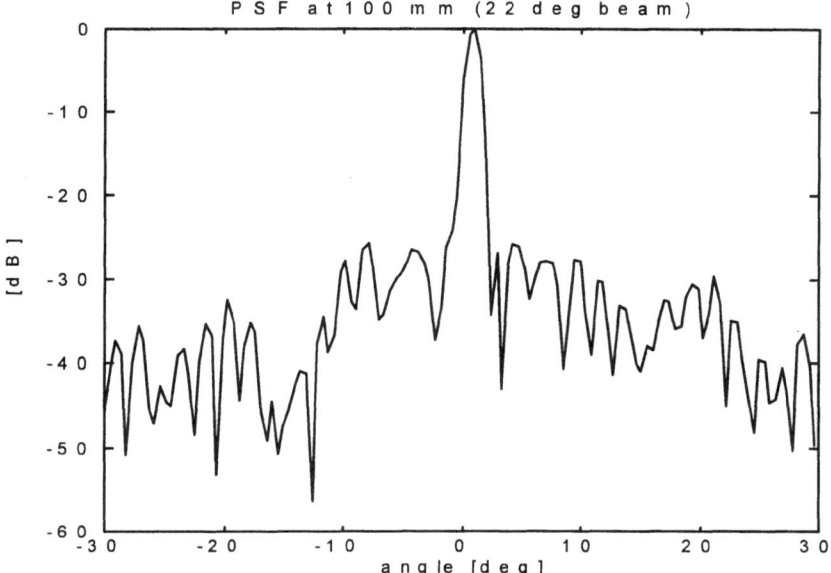

Figure 5. Lateral resolution at the depth of 100 mm when the 22° beam form was applied.

As discussed earlier, the reconstruction algorithm uses the linear aperture approach and a multiplicative factor is required to correct for the differences in the phase angle information. Figure 6 shows the compensation factor as a function of depth. The factor is inversely proportional to the depth. At this point the exact value of it is not known. It can be stated, however, that it is fairly linear in the depth range of 47 mm -130 mm but changes abruptly in the very near field. It also depends on the beam width. In these experiments it was quite difficult to locate the transducer and the rotating mechanics in a proper position with respect to each other. Possible disalignments affect, of course, to the compensation factor.

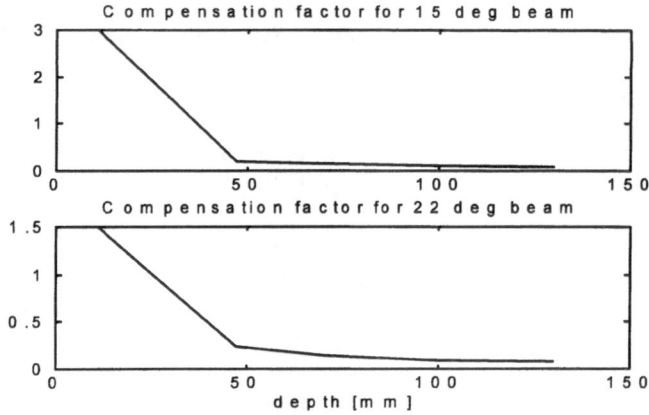

Figure 6. Compensation factor as a function of depth for the 15° and 22° beam profiles.

DISCUSSION

The present study shows that efficient Fourier-domain synthetic aperture imaging can be applied also for small convex apertures. Two-way synthetic focusing is achieved with a simple system which allows fast data acquisition and real-time numerical image reconstruction. Since only one active channel is needed the method is feasible for the design of a compact and a low-cost device. Potential application areas are in cardiological, intravascular and three-dimensional imaging.

ACKNOWLEDGEMENTS

The author acknowledges the assistance of Torgrim Lie in preparing the experiments. This work was supported by the Academy of Finland and the Nordic Academy for Advanced Study (NORFA).

REFERENCES

1. M. Karaman, P-C. Li, and M. O'Donnell, Synthetic aperture imaging for small scale systems. IEEE Trans. Ultrason., Ferroelec. and Freq. Contr., vol. 42, pp. 429-442, 1995.
2. A.L. Boyer, P.M. Hirsch, J.A. Jordan, Jr., L.B. Lesem, and D.L. Van Rooy, Reconstruction of ultrasonic images by backward propagation. Ac. Holography, vol. 3, pp. 333-348, 1971.
3. J. Ylitalo, E. Alasaarela, and J. Koivukangas, Ultrasound holographic B-scan imaging. IEEE Trans. Ultrason., Ferroelec., Freq. Contr., vol. 36, pp. 376-383, 1989.
4. Z.D. Qin, J. Ylitalo, J. Oksman, and W. Lü, Circular array ultrasound holography imaging using the linear array approach. IEEE Trans. Ultrason., Ferroelec., Freq. Contr., vol.36, pp. 485-493, 1989.
5. J. Ylitalo and H. Ermert, Ultrasound synthetic aperture imaging: monostatic approach. IEEE Trans. Ultrason., Ferroelec. and Freq. Contr., vol. 41, pp. 333-339, 1994.
6. J. Ylitalo, Optimisation of the beam profile for a digital computed ultrasonic hologhraphy imaging method. European J. of Ultrasound, vol. 1, pp. 327-335, 1994.

DYNAMICAL FOCUSING OF THE BOTH TRANSMITTED AND RECEIVED BEAMS VIA DIGITAL PROCESSING OF THE PULSED ACOUSTICAL SIGNALS, OBTAINED ON A SINGLE-ELEMENT SCANNING APERTURE

Z. M. Benenson and N. S. Kulberg

Scientific Council on Cybernetics
Russian Academy of Sciences
Moscow, Russia

INTRODUCTION

In the present paper there is represented the theoretical basis of the method, providing the dynamical focusing of both transmitted and received beams for the single-element scanning aperture. Also, there are included the experimental results that allow to verify the efficiency of the method.

The dynamical focusing by the method is possible providing that during the scanning neither reflecting nor scattering properties of the object can change. The theory of the method is based on the properties of the image as a convolution of the object scattering function and directional diagrams of the aperture. In the previous deconvolution methods (e. g., [1,2,3]) the problem statement is incorrect: to improve the image when the dynamical focusing is already implemented by beamforming on the aperture. It is known that these methods did not succeed significant image improvement.

The problem statement of the current paper is quite correct: to achieve the dynamical focusing by means of the processing of the received signals. In order to do it at first the 2-dimensional spectrum of the signals must be transformed. This transform does not change the signal-to-noise ratio. Obtained spectrum must be multiplied by the diagram correction factors, that corresponds to the convolution with some weight function. The inverse Fourier transform restores the image. Adduced simulation results corroborate that the dynamical focusing of both transmitted and received signals really can be achieved for the single scanning frame.

PRINCIPAL DEPENDENCIES

Let us consider the common method to build 2-dimension images that is based on the transmitting and receiving of the pulsed ultrasonic signals by a single transducer, scanning in the azimuth plane. Transducer aperture has a parabolic shape $Z(\xi,\eta)=\left(\xi^2+\eta^2\right)/2F$ that results in signals focusing to some point D on its axis $O\zeta$ at the distance F from its centre O (fig. 1). Let us enter the cylindrical coordinate system (R, β, y) to describe the signal in some point A. Both transmitted and received signals are multiplied on the aperture by some apodisation function $\psi(\xi,\eta)=\psi_1(\xi)\psi_2(\eta)$.

Transducer is rotating in a sector $(-\beta_0, \beta_0)$ of plane (R,y) with constant angular speed $\partial\beta/\partial t$.

To deduce the expression to determine the received signal let us use the results of paper[4]. Representing the transducer excitation function $V(t)$ as Fourier integral (t is time):

$$V(t) = \int V(\omega)\exp(i\omega\tau_1)d\omega, \quad \omega > 0 \tag{1}$$

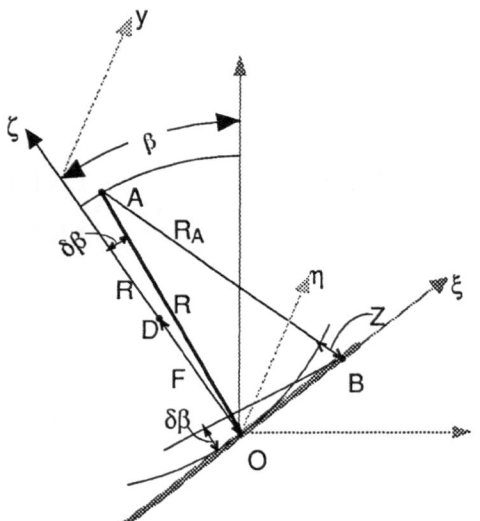

Fig. 1 Geometry

R_A —distance from point $B(\zeta,\eta)$ on the aperture to the scattering point A. In the aperture plane there is chosen the Cartesian system ξ, η (axis η is parallel to y axis). Rotation angle of the $O\zeta$ axis is β.

we obtain the expression for the disturbance in some point A:

$$P_i(R,\beta-\delta\beta,y,t) = \iint \frac{M_1(\omega,R)V(\omega)\exp\left(i\omega\left(t-\dfrac{R}{c}\right)\right)}{2\pi R} L_1(\omega,\delta\beta,R)L_2(\omega,y,R)d\omega dy, \tag{2}$$

$$L_1(\omega,\delta\beta,R) = \int \psi_1(\xi)\exp\left(\frac{i\omega\xi^2}{2Rc_0}\right)\exp\left(-\frac{i\omega\xi^2}{2Fc_0}\right)\exp\left(-\frac{i\omega\xi\sin\delta\beta}{c_0}\right)d\xi$$

$$L_2(\omega,y,R) = \int \psi_2(\eta)\exp\left(\frac{i\omega(\eta-y)^2}{2Rc_0}\right)\exp\left(-\frac{i\omega\eta^2}{2Fc_0}\right)d\eta \tag{3}$$

Expressions (2) and (3) are deduced by appropriate formula:

$$R_A = \sqrt{R^2 + \xi^2 - 2R\xi\sin\delta\beta + (\eta-y)^2} \approx R + \frac{1}{2}\frac{\xi^2}{R} - \xi\sin\delta\beta + \frac{(\eta-y)^2}{2R} \tag{4}$$

There is the common representation for the attenuation coefficient $M_1(A) = M_1(\omega,R)$.

Substituting (2) to (3) and taking into account (4) we obtain expression of the received signal:

$$P_r(\beta,t) = \iiint \frac{M_1^2(\omega,R)}{R^2}V(\omega)\gamma(R,\beta-\delta\beta,\omega)\exp\left(i\omega\left(t-\frac{2R}{c_0}\right)\right)L_1^2(\omega,\delta\beta,R)d\delta\beta dR d\omega, \tag{5}$$

$$\gamma(R,\beta-\delta\beta,\omega) = \int F_{op}(R,\beta-\delta\beta,y,\omega)dy, \tag{6}$$

$$F_{op}(R,\beta-\delta\beta,y,\omega) = \frac{\nabla[\delta\rho(R,\beta-\delta\beta,y)]\bullet\nabla[L_1(\omega,\delta\beta,R)L_2(\omega,y,R)]}{\rho_0 L_1(\omega,\delta\beta,R)} - \frac{\delta c}{c_0^3}\omega^2 L_2(\omega,y,R) \tag{7}$$

In these formulas $\delta\rho$ and δc are the variations of the density and velocity of sound, c_0 and ρ_0 are the mean values of the same parameters. The function

$$\Gamma_{im}(t,\beta-\delta\beta) = \iint \exp\left(i\omega\left(t-\frac{2R}{c_0}\right)\right)\gamma(R,\beta-\delta\beta,\omega)V(\omega)\frac{M_1^2(\omega,R)}{R^2}d\omega dR$$

is the ideal complex amplitude of the 2-dimension image at the distance $R=c_0t/2$ with infinite resolution for the angle β.

THE THEORY OF THE METHOD OF THE TRANSMITTING-RECEIVING DYNAMICAL FOCUSING

To formulate the theory of the method we shall use the expression (5), supposing that the functions of a real directional diagram $L_1(\omega,\delta\beta,R)$ and of the dynamical focusing diagram $L_0(\omega,\delta\beta)$ are known a priori:

$$L_0(\omega,\delta\beta) = \int_{-a}^{a} \psi_1(\xi)\exp\left(-i\xi\frac{\omega}{c}\sin\delta\beta\right)d\xi \tag{8}$$

where $2a$ is the aperture size for ξ coordinate.

Taking Fourier transform from (5) for spatial frequency Ω and using the convolution theorem, we obtain:

$$S_r(\Omega,t) = \int\exp(-i\Omega\beta)P_r(\beta,t)d\beta = \iint\Gamma(R,\Omega,\omega)V(\omega)M_1^2(\omega,R)\times$$
$$\times\frac{\exp\left(i\omega(t-2R/c_0)\right)}{R^2}L_3(\omega,\Omega,R)d\omega\,dR \tag{9}$$

where

$$L_3(\omega,\Omega,R) = \int\hat{L}_1(\omega,R,\Omega')\hat{L}_1(\omega,R,\Omega-\Omega')d\Omega'$$
$$\hat{L}_1(\omega,R,\Omega) = \int L_1(\omega,R,\delta\beta)\exp(-i\Omega\delta\beta)d\delta\beta \tag{9a}$$
$$\Gamma(R,\Omega,\omega) = \int\gamma(R,\beta,\omega)\exp(i\Omega\beta)d\beta$$

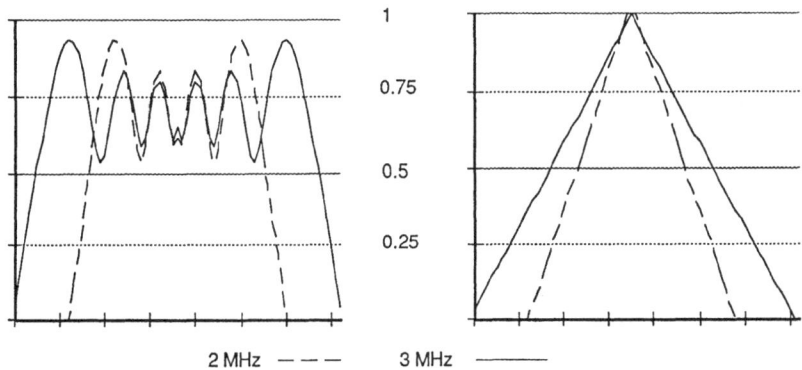

	1
	0.75
	0.5
	0.25

2 MHz – – –   3 MHz ———

Fig. 2. Spectra of real directional diagram obtained as Fresnel Integral (left) and of «ideal» directional diagram of the dynamically focused signal with the constant apodisation function, found as a convolution of receiving and transmitting diagram rectangular spectra (right). Spectra are shown at 2 different temporal frequencies.

It is evident that function

$$S_{r,1} = \int\frac{\Gamma(R,\Omega,\omega)V(\omega)M_1^2(\omega,R)\exp\left(-i\omega\frac{2R}{c_0}\right)}{R^2}L_3(\omega,\Omega,R_0)dR \tag{10}$$

is 2-dimension Fourier transform of the received signal.

For some arbitrary value $t_j = 2R_j/c_0$ we build the function:

$$S_d\left(\omega,\Omega,R_j\right) = \frac{\hat{L}_0(\omega,\Omega)}{L_3\left(\omega,\Omega,R_j\right)} S_{r,1}(\omega,\Omega), \quad \hat{L}_0(\omega,\Omega) = \int L_0^2(\omega,\delta\beta)\exp(-i\omega\delta\beta)d\delta\beta \tag{11}$$

Reverse Fourier transform of (11) for ω and Ω frequencies exactly coincides with the expression for the signal of the aperture focused at the depth R_j. Since R_j can be assigned any value, this formula restores the dynamical focusing of both transmitted and received signals. There is some insufficient change of the depth resolution in this case. Let us prove this assertion in the case of the Gaussian apodisation $f_1(\xi) = \exp\left(-\xi^2/a^2\right)$. Substituting this formula to (3), (8), (9a) we obtain:

$$L_3\left(\omega,R_j,\Omega\right) = \exp\left(-\alpha\Omega^2\right)\exp\left(i\zeta\Omega^2\right)$$

$$\hat{L}_0(\omega,\Omega) = \exp\left(-\alpha\Omega^2\right), \quad \alpha = \frac{c_0^2}{2a^2\omega^2}, \quad \zeta = \frac{c_0}{4\hat{R}_j\omega}, \quad \hat{R}_j = \frac{R_j F}{F - R_j} \tag{12}$$

Thus, (11) can be represented as

$$S_d\left(\omega,R_j,\Omega,\right) = \exp\left(-i\zeta\Omega^2\right)S_{r,1}(\omega,\Omega) \tag{13}$$

It is evident from (13) that in the considered case transformation (11) does not change signal-to-noise ratio, because $\left|\exp\left(i\zeta\Omega^2\right)\right| = 1$.

Let us implement reverse Fourier transform of (13) for ω:

$$S_f\left(t_j = \frac{2R_j}{c_0},\Omega\right) = \int\exp\left(i\omega\frac{2R_j}{c_0}\right)S_d\left(\omega,\Omega,R_j\right)d\omega = \iint\exp\left(-\alpha\Omega^2\right)\times$$

$$\times\Gamma(R_0,\Omega,\omega)V(\omega)M_1^2(\omega)\exp\left(i\omega\frac{2(R_j - R)}{c_0}\right)\exp\left(-i(\zeta - \zeta_1)\Omega^2\right)d\omega dR \tag{14}$$

where $\zeta_1 = \frac{c_0}{4R_{eq}\omega}, R_{eq} = \frac{RF}{F - R}$

Expression (14) differs from one for the dynamically focused signal by the factor $\exp\left(-i(\zeta - \zeta_1)\Omega^2\right)$. Representing $\omega = \delta\omega + \omega_0$, where ω_0 — is a carrier frequency and $|\delta\omega| < \Delta\omega$ ($\Delta\omega < \omega_0/2$), then the approximate formula is valid:

$$\exp\left(-i(\zeta - \zeta_1)\Omega^2\right) \approx \exp\left(\frac{ic_0(R_j - R)(F - R_j)\Omega^2}{4R_j^2\omega_0 F}\left(2 - \frac{\omega}{\omega_0}\right)\right) \tag{15}$$

In the worse case, if $F = \infty$, and $R_j = 2a$, the coefficient at ω in the exponent will be $\sim 1/4\left(R_j - R\right)/c_0$ and resolution degradation can be no more than 12%. If value of F is finite, at $R_j > F$ insufficient resolution improvement takes place.

Let us consider now the case with no apodisation: $\psi_1(\xi) = \text{rect}(\xi/a)$. Doing corresponding calculations, we obtain (fig. 2):

$$\hat{L}_0(\omega,\Omega) = 1 - \frac{|\Omega|c_0}{2a\omega}, \quad \left(\frac{|\Omega|c_0}{\omega} \leq 2a\right); \quad L_3\left(\omega,R_j,\Omega\right) = 2\int_0^{1 - \frac{|\Omega|c_0}{2a\omega}}\exp\left(\frac{i\omega u^2 a^2}{c_0\hat{R}_j}\right)du\exp\left(i\zeta\Omega^2\right) \tag{16}$$

At the maximum frequency Ω_{max} ratio $L_0/L_3 = 1/2$. For other Ω its value can exceed 1, that results from the defocusing of the wave. But the problem will remain correct because the ratio L_0/L_3 is always finite. Farther, substituting (16) to (11) we obtain the expression, analogous to (14) which will contain $\hat{L}_0(\omega,\Omega)$ instead of $\exp(-\alpha\Omega^2)$.

We have already mentioned that to achieve the dynamical focusing of the signal $S_f(t_j,\beta)$ for the function S_d 2-dimension Fourier transform must be performed. For Ω FFT algorithm can be used, but it is impossible for the variable ω, because S_d depends on R_j. For this reason we consider more reasonable algorithm.

ALGORITHM OF THE DYNAMICAL FOCUSING

The principal of the algorithm is that instead of the computation of the intermediate function $\tilde{S}_d(\omega,\Omega,R_j)$ and its Fourier transform we find immediately the function $S_f(t_j,\Omega)$ by convolution with some «short» weight function $W_j(\tau,\Omega)$:

$$S_f(t_j,\Omega) = \int W_j(\tau,\Omega)S_r(\Omega,t_j-\tau)d\tau \tag{17}$$

Let us construct the function $W_j(\tau,\Omega)$ as follows. Full frequency band of $2\Delta\omega$ of the receives signal must be divided by $1,...l,...L$ ranges (center frequency of $l+1$ range is ω_l). For each range $\omega_0 < \omega < \omega_{L-1}$ following expression is valid:

$$S_d(\omega,\Omega,R_j) = \sum_{l=1}^{L} \psi\left(\frac{\delta\omega}{\Delta\omega_0}\right)S_d^{(l)}(\delta\omega,\Omega,R_j) \tag{18}$$

where $\Delta\omega_0 = \dfrac{2\Delta\omega}{L}$, $\psi(x) = \begin{cases} 1-|x|, & |x| \le 1 \\ 0, & |x| > 1 \end{cases}$; $\delta\omega \in [-\Delta\omega_0, \Delta\omega_0]$

$$S_d^{(l)}(\delta\omega,\Omega,R_j) = H_{j,l}(\delta\omega,\Omega)S_{r,1}(\omega_l+\delta\omega,\Omega)$$

$$H_{j,l} = \frac{\hat{L}_0\left(\dfrac{\Omega c_0}{\omega_l(1+\delta\omega/\omega_l)}\right)}{L_3(\omega_l(1+\delta\omega/\omega_l),R_j,\Omega)}\exp\left(-\frac{i\Omega^2 c_0}{4\hat{R}_j\omega_l}\left(1-\frac{\delta\omega}{\omega_l}+\frac{\delta\omega^2}{\omega_l^2(1-\delta\omega/\omega_l)}\right)\right)$$

Having made the reverse Fourier transform $\int \exp(i\omega t_j)S_d(\omega,\Omega,R_j)d\omega$ and applying (18), we obtain the expression (17), and $W_j(\tau,\Omega)$ can be represented:

$$W_j(\tau,\Omega) = \sum_{l=1}^{L} \chi_{l,j}(\tau,\Omega)\exp(i\omega_l\tau),$$
$$\chi_{l,j}(\tau,\Omega) = \int \exp(i\delta\omega\tau)\rho_{l,j}(\delta\omega,\Omega)d\delta\omega, \quad \rho_{l,j}(\delta\omega,\Omega) = \psi\left(\frac{\delta\omega}{\Delta\omega_0}\right)H_{j,l}(\delta\omega,\Omega) \tag{19}$$

Supposing that $\delta\omega$ is small we can show that approximately

$$\chi_{l,j}(\tau,\Omega) = \exp\left(-\frac{i\Omega^2 c_0}{4\omega_l\hat{R}_j}\right)\mu_{l,j}\theta\left(\tau+\frac{\Omega^2 c_0}{4\omega_l\hat{R}_j}\right),$$

$$\mu_{l,j} = \frac{\hat{L}_0(\Omega c_0/\omega_l)}{L_3(\omega_l,R_j,\Omega)}, \quad \theta(\tau) = \int \psi\left(\frac{\delta\omega}{\Delta\omega_0}\right)\exp(i\delta\omega\,\tau)d\,\delta\omega = \frac{4\sin^2(\Delta\omega_0\tau/2)}{(\Delta\omega_0\tau/2)^2}$$

The width of the function $\theta(\tau)$ is about $\Delta\tau = 4\pi/\Delta\omega_0 = 2\pi L/\Delta\omega = L\tau_l$, where τ_l — is the initial pulse duration. The integral (18) can be changed by the finite sum on the interval $\Delta\tau$. After computation of $S(t_j,\Omega)$ the reverse Fourier transform must be done, and the result is the dynamically focused signal $s_f(t_j,\beta)$.

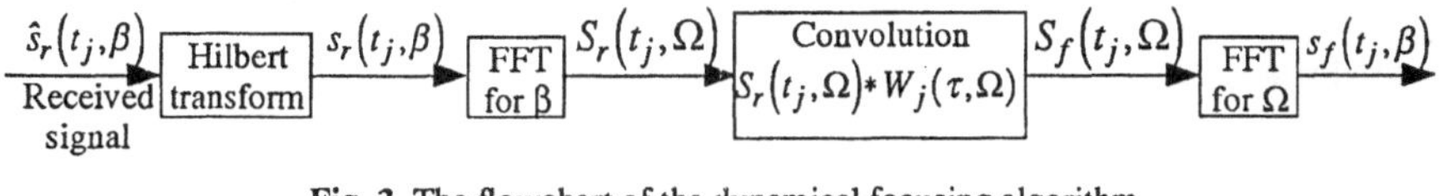

$\hat{s}_r(t_j,\beta)$ — Received signal → [Hilbert transform] → $s_r(t_j,\beta)$ → [FFT for β] → $S_r(t_j,\Omega)$ → [Convolution $S_r(t_j,\Omega)*W_j(\tau,\Omega)$] → $S_f(t_j,\Omega)$ → [FFT for Ω] → $s_f(t_j,\beta)$

Fig. 3. The flowchart of the dynamical focusing algorithm

535

The weight function $W(\tau)$ can be pre-computed and stored in the ROM. Full number of the operations needed for this algorithm is $\sim N_\beta \times N_R \times \left(2\log_2 N_\beta + n\right)$, where N_β is the samples number for the β coordinate, n is a number of terms in discrete representation of the integral (20), N_R is the samples number for the R coordinate.

THE EXPERIMENTAL RESULTS

There was done the computer simulation of the present algorithm. The input data were generated by the program to simulate the ultrasonic waves distribution. The «phantom» consisted of the several distinct reflectors. Function $V(t)$ was the Gaussian pulse with the carrier frequency $f_0=3$MHz and pulse duration $\tau_I=0.7\mu$s. The aperture of the transducer was a flat plate with size $2a=1.3$ cm. Also the dynamical focusing was simulated for the phased array with the same aperture size. Directional diagrams obtained as a result of the experiment are represented on the fig. 4.

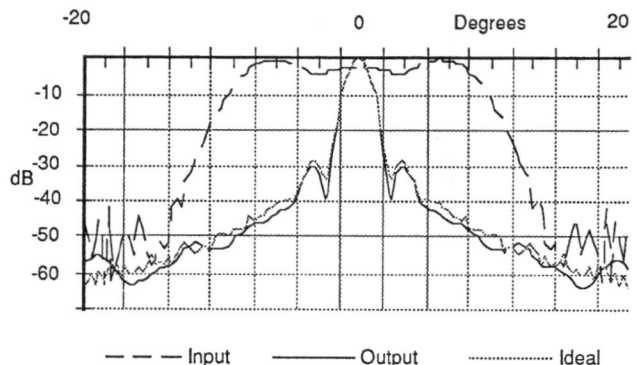

Fig. 4.

The directional diagrams of the non-focused aperture before (at input) and after processing by the algorithm. Signal at the output exactly coincides with one of dynamically focused phased array (ideal). Aperture size was 1.3 cm, distance to a target was 3 cm.

It is evident that the dynamical focusing really had been achieved. Also the algorithm was applied for the processing of the digital records of the signals obtained in clinic examinations. The characteristics of the processed signals also corresponded to ones that can be achieved only by using the dynamically focused phased array.

CONCLUSION

The algorithm of the signal processing had been developed that provides the dynamical focusing of the pulsed acoustical signals for the single-element scanning aperture. Focusing is made for each scanning frame. Directional diagram of the method coincides with one of the dynamical focusing for both transmitting and receiving. The algorithm has an insufficient effect to the depth resolution

REFERENCE

1. T. J. M. Jeurens, J. C. Somer, F. A. M. Smeets and A. P. G. Hoeks.
   The practical significance of two-dimensional deconvolution in echography. Ultrasonic imaging, 9, p. 106-116 (1987)

2. T. Loupas, S. D. Pye and W. N. McDicken.
   Deconvolution in medical ultrasonics. Practical considerations. Phys. Med. Biol., 34, 1691-1700 (1989)

3. Jorden Arendt Jensen.
   Deconvolution of ultrasonic images. Ultrasonic Imaging, 14, 1-15 (1992)

4 Jensen J. A.
   A model for the propagation and scattering of ultrasound in tissue. J. Acoust Soc. Amer. 89, 188-191 (1991)

SUPERRESOLUTION OF THE ACOUSTICAL BIOLOGICAL IMAGES VIA NON-LINEAR PROCESSING OF THE DYNAMICALLY FOCUSED TRANSMITTED/RECEIVED ULTRASONIC SIGNALS

Z. M. Benenson and N. S. Kulberg

Scientific Council on Cybernetics, Russia Academy of Sciences
Moscow, Russia

INTRODUCTION

In the modern digital signal processing there are 2 approaches to the improvement of the resolution of the visualization systems. First of them is based on the spectral estimations of the brief data sequences. Such sequences can be both temporal series and signals of the phased array elements. The cause of the improvement in these methods[1] (the auto-regressive spectral estimation, the maximum-entropy and minimum-dispersion methods) is based on the a priori information about the object.

The second approach is based on the signals approximation by the eigenfunctions of the directional diagram[2]. This approximation yields more efficient use of the full width of the spatial spectrum.

The algorithm of the present paper is to be applied to the pulsed acoustical signals that are dynamically focused for both receiving and transmitting. The non-linear processing results in the widening of the signal spatial spectrum and allows to improve the resolution of the acoustical imaging system. The signal after the non-linear processing has no imaginary part. Since the useful terms of the signal are always positive and side-lobes and noise can be negative, it is also used in the algorithm. The paper includes the simulation results that corroborate the resolution improvement.

THE THEORY OF THE METHOD OF THE SUPERRESOLUTION

For the dynamically focused signal main lobe width equals $\pi/1.4\Omega_{max}$ (-6 dB), where $\Omega_{max} = 2a\omega_{max}/c_0$ and ω_{max} is the maximum temporal frequency of the signal, c_0 is the velocity of sound and $2a$ is the aperture size. The spectrum of the diagram is represented on the figure 1.

The image forming is made by the intensity function. The intensity spectrum width is also represented on the figure 1. Though this spectrum is wider it does not yield resolution improvement. Let us construct the method that will provide more efficient use of the spectrum widening in the non-linear signal processing.

It is known that[3]:

$$\Phi_s(\Omega,\omega) = \Phi_d(\Omega,\omega)\Gamma(\Omega,\omega) \tag{1}$$

where $\Phi_s(\Omega,\omega)$ is a spatial-temporal spectrum of the received analytic signal $S_r(\beta,t)$ (β is the angle of the beam direction, t is time), $\Gamma(\Omega,\omega)$ is a spatial-temporal spectrum of the scattering function and $\Phi_d(\Omega,\omega)$ is a spectrum of the directional diagram.

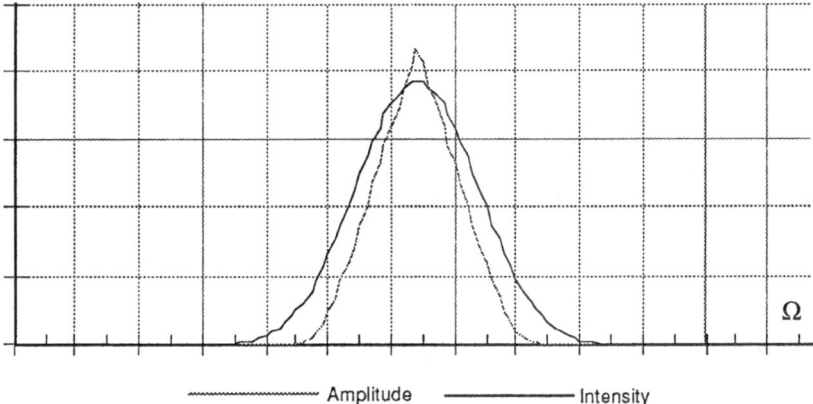

Figure 1. Spectrum of the directional diagram of the dynamically-focused signals. Spectra of both signal and of its intensity are included.

Let us multiply (1) by 2 functions $\Phi_1(\Omega,\omega) = \exp(\mu\Omega c_0/2a\omega)$ and $\Phi_2(\Omega,\omega) = \exp(-\mu\Omega c_0/2a\omega)$, where $\mu \sim 3$. It yields 2 spatial-temporal signal spectra:

$$\Phi_{s1}(\Omega,\omega) = \Phi_{d1}(\Omega,\omega)\Gamma(\Omega,\omega)$$
$$\Phi_{s2}(\Omega,\omega) = \Phi_{d2}(\Omega,\omega)\Gamma(\Omega,\omega)$$
(2)

Here the maximum of $\Phi_{d1}(\Omega,\omega) = \Phi_d(\Omega,\omega)\Phi_1(\Omega,\omega)$ is shifted to the positive Ω range and the maximum of $\Phi_{d2}(\Omega,\omega) = \Phi_d(\Omega,\omega)\Phi_2(\Omega,\omega)$ is shifted to the negative Ω range. These functions are real-valued and always positive. The relation is valid: $\Phi_{d1}(\Omega,\omega) = \Phi_{d2}(-\Omega,\omega)$. The reverse Fourier transform of (2) for both coordinates yields 2 signals:

$$S_1(\beta,t) = \int \hat{L}_1(\delta\beta,\tau)\gamma(\beta - \delta\beta, t - \tau)d\delta\beta$$
$$S_2(\beta,t) = \int \hat{L}_2(\delta\beta,\tau)\gamma(\beta - \delta\beta, t - \tau)d\delta\beta$$
(3)

where

$$\hat{L}_1(\delta\beta,\tau) = \iint \exp(i\Omega\delta\beta)\exp(i\omega\tau)\Phi_{d1}(\Omega,\omega)d\omega d\Omega$$
$$\hat{L}_2(\delta\beta,\tau) = \iint \exp(i\Omega\delta\beta)\exp(i\omega\tau)\Phi_{d2}(\Omega,\omega)d\omega d\Omega$$
$$\gamma(\delta\beta,t) = \iint \exp(i\Omega\beta)\exp(i\omega\tau)\Gamma(\Omega,\omega)d\omega d\Omega$$

To prove the algorithm let us represent (3) approximately as:

$$S_1(\beta,t) = \int L_1(\delta\beta)\gamma(\beta - \delta\beta, t)d\delta\beta$$
$$S_2(\beta,t) = \int L_2(\delta\beta)\gamma(\beta - \delta\beta, t)d\delta\beta$$
(4)

where $L_1(\delta\beta)$ and $L_2(\delta\beta)$ are equivalent to the direction diagrams of the formed signals. These functions are complex-valued and for them the relation is valid:

$$\int L_2^*(\delta\beta)\exp(-i\Omega\delta\beta)d\delta\beta = \int L_1(\delta\beta)\exp(-i\Omega\delta\beta)d\delta\beta$$
(5)

Let us enter the function

$$S_0(\beta,t) = S_1(\beta,t)S_2^*(\beta,t) + S_2(\beta,t)S_1^*(\beta,t) \tag{6}$$

where (*) means complex conjugation. The spatial spectrum of the entered function is 2 times wider than one of the received signal $S_r(\beta,t)$. It is obvious that $S_0(\beta,t)$ is a real value. For its spatial spectrum the formula is valid:

$$\Phi_{s0}(\Omega,t) = \int \Phi_\gamma(\Omega',t)\Phi_{\gamma^*}(\Omega-\Omega',t)\big(\Phi_{L_1}(\Omega-\Omega')\Phi_{L_1}(\Omega') + \Phi_{L_2}(\Omega-\Omega')\Phi_{L_2}(\Omega')\big)d\Omega' \tag{7}$$

where Φ is the Fourier transform of the corresponding function. To analyze the properties of the function (6) let us suppose that the scattering function is a superposition of some ideal reflectors:

$$\gamma(\beta,t) = \sum_n \gamma_n(t)\delta(\beta-\beta_n) \tag{8}$$

where β_n is the azimuth of the n-th reflector and $\delta(\cdot)$ is a Dirac function. Substituting (7) to (8) we obtain the expression $\Phi_{s0}(\Omega,t)$

$$\Phi_{s0}(\Omega,t) = \sum_n |\gamma_n(t)|^2 \exp(-i\Omega\beta_n)\hat{L}_I(\Omega) + \sum_{n\neq m}\big(\gamma_n\gamma_m^* + \gamma_n^*\gamma_m\big)\exp\left(-i\Omega\frac{\beta_n+\beta_m}{2}\right)\hat{L}_c(\Omega,\beta_n-\beta_m) \tag{9}$$

where $\hat{L}_I(\Omega)$ and $\hat{L}_c(\Omega,\beta_n-\beta_m)$ are some functions that were obtained from the convolution of the $\Phi_{L_1}(\Omega)$, $\Phi_{L_2}(\Omega)$ and $\exp(-i\Omega(\beta_n-\beta_m))$. The reverse Fourier transform of (9) yields the expression

$$S_0(\beta,t) = \sum_n |\gamma_n(t)|^2 L_I(\beta-\beta_n) + \sum_{n\neq m}\big(\gamma_n\gamma_m^* + \gamma_n^*\gamma_m\big)L_c\left(\beta-\frac{\beta_n+\beta_m}{2},\beta_n-\beta_m\right) \tag{10}$$

where $L_I(\beta)$ and $L_c(\beta,\beta_n-\beta_m)$ are the reversed Fourier transforms of $\hat{L}_I(\Omega)$ and $\hat{L}_c(\Omega,\beta_n-\beta_m)$.

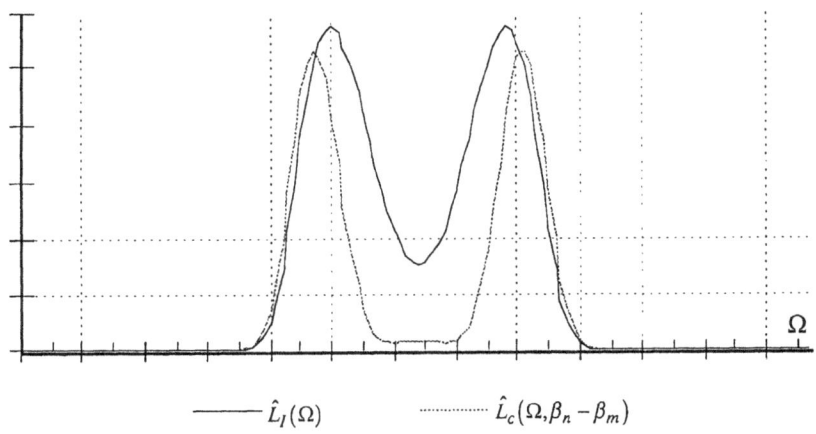

$$\text{———}\ \hat{L}_I(\Omega) \qquad \text{············}\ \hat{L}_c(\Omega,\beta_n-\beta_m)$$

Figure 2. The functions $\hat{L}_I(\Omega)$ and $\hat{L}_c(\Omega,\beta_n-\beta_m)$

First term in $S_0(\beta,t)$ yields the signal similar to one obtained by the aperture with the directional diagram $L_I(\beta)$; the effect of the second term looks like a speckle-noise, its level is proportional to the function $L_c(\beta,\beta_n-\beta_m)$. The farther are reflectors one from another, the less is $\max|L_c(\beta,\beta_n-\beta_m)|$. On the figure 2 there are represented functions $\hat{L}_I(\Omega)$ and

$\hat{L}_c(\Omega,\beta_n-\beta_m)$. It is obvious that $\hat{L}_c(\Omega,\beta_n-\beta_m)$ has its maximum at higher frequency than $\hat{L}_l(\Omega)$. For this reason the signal should be filtered by a low-pass filter. We have used the filter with the frequency response $\exp(-\Omega^2/\Delta_0^2)$, where Δ_0 is a heuristic value.

Graphs on the figure 3 represent functions $S_0(\beta,t)$ for single and two reflectors before and after the low-pass filtration. The width of the main lobe of $L_l(\beta)$ is ~2 times less than in the case of the dynamical focusing. But there are large negative side-lobes. To allow for them let us take into account the fact that the scattered signal is always positive.

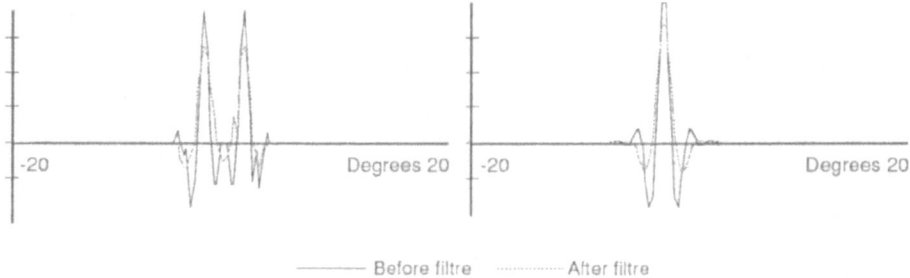

Figure 3. Graphs of the function $S_0(\beta,t)$ before and after the low-pass filter. There are plotted signals obtained from one (right) and two (left) reflectors.

THE ALGORITHM OF THE SIGNAL PROCESSING TO FORM THE IMAGE

On the figure 4 there is represented the flowchart of the algorithm. It consists of 4 principal stages. First 3 of them (signal's spectrum transform, 2-dimension Fourier transform in order to determine the signals $S_1(\beta,t)$ and $S_2(\beta,t)$, and computing of the signal $S_0(\beta,t)$) were described above.

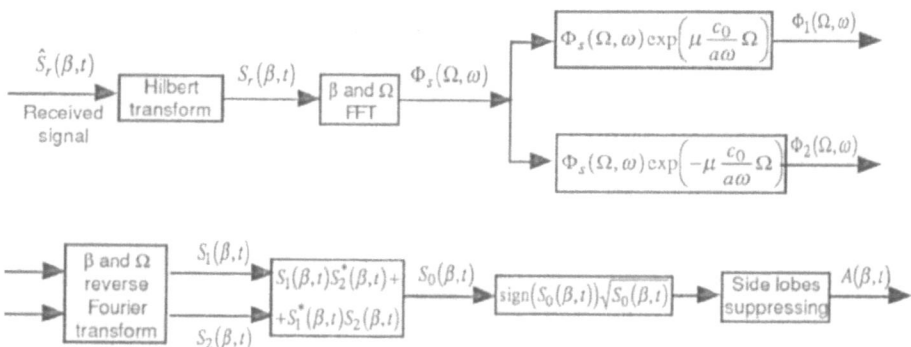

Figure 4. The flowchart of the algorithm.

To construct the algorithm to suppress the side lobes let us enter the function $A_0(\beta,t)=\text{sign}(S_0(\beta,t))\sqrt{S_0(\beta,t)}$. Its absolute value equals to the square root of $|S_0(\beta,t)|$ and its sign is the same as the sign of $S_0(\beta,t)$. The algorithm idea is the following. Let us choose the digital grid β,t: β_j,t_l $(j=1,...J,\ l=1,...L)$. If $A_0(\beta_j,t_l)\geq 0$ then resulting signal $A(\beta_j,t_l)=A_0(\beta_j,t_l)$. Else

$$A(\beta_j,t_l)=\left|A_0(\beta_j,t_l)+w(\beta_j,t_l)\times\max\{A_0(\beta_j+\Delta,t_l),\ A_0(\beta_j-\Delta,t_l)\}\right| \qquad (13)$$

In these expressions Δ is a first side lobe width, it is determined from the function $L_I(\beta)$; $w(\beta_j,t_l)$ is some weight function that is determined as follows. For all values $\beta' \in [\beta_j - 2\Delta, \beta_j + 2\Delta]$ the value $a(\beta_j,t_l)$ must be found as a mean of values $\max\{A_0(\beta'+\Delta,t_l),\ A_0(\beta'-\Delta,t_l)\} > 0$; and the value $b(\beta_j,t_l)$ as a mean value of values $A(\beta') < 0$. Let us assign

$$w(\beta_j,t_l) = -\alpha \frac{b(\beta_j,t_l)}{a(\beta_j,t_l)} \tag{14}$$

where α is some scale factor. The idea of the construction of the w function is in determining of the negative signal component in (β_j,t_l) point, which is caused by the negative side lobes of the diagram at the interval $\beta' \in [\beta_j - 2\Delta, \beta_j + 2\Delta]$

It is obvious, that the expression (13) can be represented as $\sqrt{y} - \sqrt{x}$, where

$$\sqrt{x} = \sqrt{|S_0(\beta_j,t_l)|};\ \sqrt{y} = \sqrt{w^2(\beta_j,t_l)S_0(\beta_j + \Delta,t_l)}$$

$$\text{or } \sqrt{y} = \sqrt{w^2(\beta_j,t_l)S_0(\beta_j - \Delta,t_l)} \tag{15}$$

The relation is valid: $|\sqrt{y} - \sqrt{x}| \le \sqrt{|y - x|}$ \tag{16}

Since $|y - x| = |w^2(\beta_j,t_l)S_0(\beta_j \pm \Delta,t_l) - |S_0(\beta_j,t_l)||$, it follows from (15) and (16) that the remainder value after the side lobes suppressing will be less when substituting $A_0(\beta,t)$ instead of $S_0(\beta,t)$ into the expression (13).

The present algorithm of the side lobes suppressing was chosen as a result of numerous digital experiments with different images, both simulated and resulting from the clinical examinations.

THE EXPERIMENTAL RESULTS

The algorithm was tested be simulating using the special program[3]. Figure 5 represents the directional diagram $L_I(\beta)$ after the processing by the algorithm. Also there is represented the diagram of the dynamically focused phased array. The width of the diagram main lobe after processing by the algorithm is ~2 times less than before processing. The 1-st side lobe level is about -20 — -30 dB

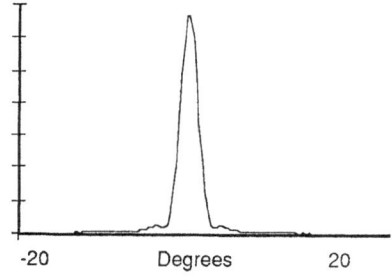 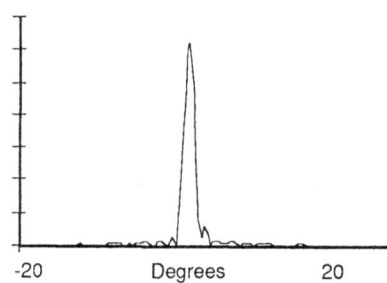

-20   Degrees   20   -20   Degrees   20

Figure 5. Directional diagrams of the dynamically focused phased array (left) and of the superresolution method (right).

On the figure 6 there is represented the amplitude distribution of the signals received from two reflectors that were placed at the distance of 2/3 Rayleigh resolution limit. The distributions were obtained by applying of the dynamical focusing and the superresolution algorithm. Also, the main lobe have become 2 times narrower; the gap between the objects has become sufficiently more; the side lobes level also is about 20-30 dB.

The algorithm was also applied to process the digital records of the ultrasonic signals that were obtained in clinical abdominal examinations. The result was the significant improvement of the images contrast; the sharpness of the image details has been improved; the characteristic size of the speckle-noise particles also reduced by ~2 times.

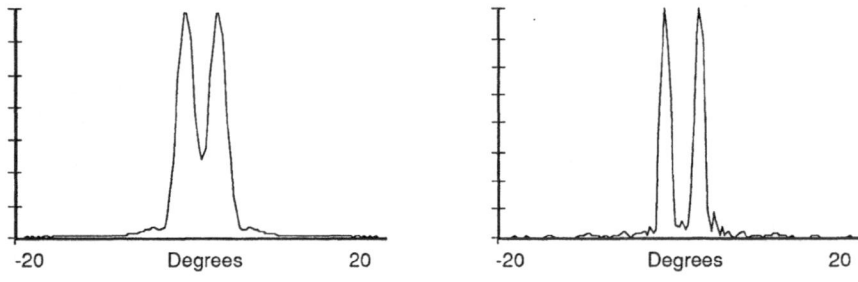

Figure 6. Distribution of the signals amplitudes obtained from 2 reflectors, before (left) and after (right) processing by the superresolution algorithm.

CONCLUSION

The developed superresolution algorithm yields the image improvement by ~2 times for the images, obtained using the dynamical focusing of both transmitted and received beams; for the images obtained by the dynamical focusing of the received beam only the resolution improvement is ~3 times.

The algorithm can be realized using modern high-performance digital signal processors in real-time for the full frame or for its several parts, depending on the processor complexity.

Comparing of the images obtained by the superresolution procedure with ones obtained by the traditional methods will allow to get more information and to differ reliably useful image details from the speckle-noise.

REFERENCE

1. S. Lawrence Marple jr.
    Digital Spectral Analysis with Applications. Prentice-Hall, Inc., Englewood Cliffs, New Jersey (1987)
2. M. Bertero, P. Branzi and E. R. Pike
    Superresolution in confocal scanning microscopy. Inverse Problems 3, 195-212 (1987)
3. Z. M. Benenson, N. S. Kulberg
    Dynamical focusing of the both transmitted and received beams via digital processing of pulsed acoustical signals obtained on a single-element scanning aperture. Acoustical Imaging 22 (1995)

A SIMPLE SOLUTION FOR REMOVING ECHO BARS
FOR URTURIP TECHNIQUE

Bruno Migeon, Pierre Vieyres, and Pierre Marché

Laboratoire Vision et Robotique
63, avenue de Lattre de Tassigny
18020 Bourges Cedex, France

INTRODUCTION

In standard echography[1], when a probe faces a hyperechogenic structure equivalent to an acoustical mirror, multiple reflections may be present. When using ultrasound tomography in reflection-mode for the exploration of the breast[2], the skull[3] and long bones[4,5], the same phenomenon may occur, especially in the case of Ultrasound Reflection-mode Tomography Using Radial Image Processing (URTURIP) which uses B-scan images obtained from the video signal of the echograph instead of projections[6].

When a planar acoustical mirror faces a probe, multiple echoes appear on the images as horizontal lines called *echo bars*. It is, however, more appropriate to talk about multiple reflections in the general case.

It is possible to reduce this echo bars phenomenon by using impedance matching materials or image processing. As an example, a frequency filter can be applied to each image when the frequency signature of the echo bars is known. However, this technique is time consuming and blurrs the image.

We show in this paper that it is possible to choose a distance between the probe and the depth-adjustable acoustical mirror to eliminate the echo bars phenomenon.

ECHO BARS PHENOMENON

Whatever acquisition system used, if an acoustical mirror is set at a distance H from the probe, it is equivalent to considering the simple case of a flat bottom tank (fig. 1) filled with water up to a level H, where a probe located at the surface of the water gives an image with echo bars as represented in figure 2.

Let us consider c', the ultrasound velocity in water, and P, the investigation depth of the probe; t_e, t_r and t_d represent the emission time, reception time and delay time respectively. $2H_{max}$ is the maximal distance at which the ultrasounds are no longer detected.

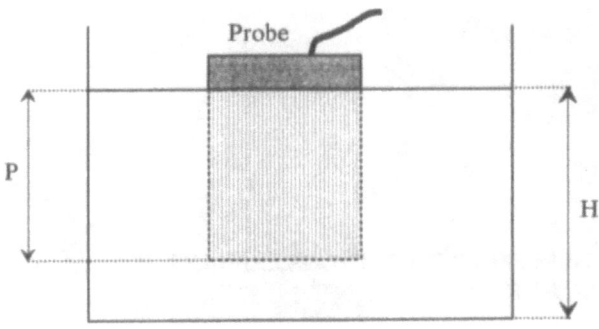

Figure 1. Investigation depth of the probe and depth of water

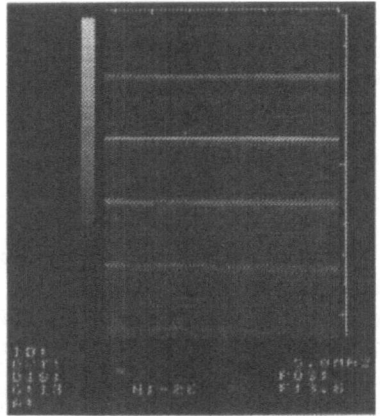

Figure 2. B-Scan image with several echo bars

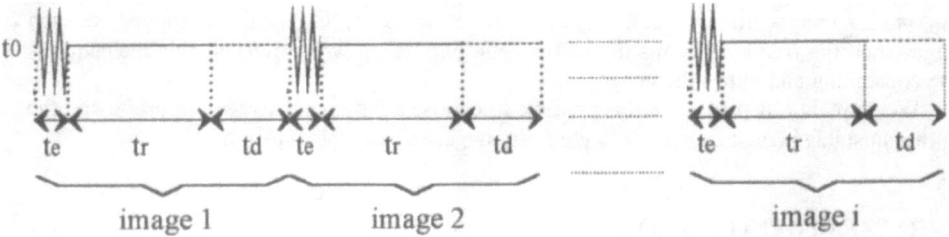

Figure 3. Emission time t_e, reception time t_r and delay time t_d

An ultrasound pulse propagates in water from the probe, reflects back onto the bottom, onto the probe and back onto the bottom and so on. Several ultrasound echoes may arrive at the probe during a reception time t_r to form an image. It is then possible to find, on the same image of interest, ultrasound echoes, having traveled several times the 2H distance, represented by echo bars.

In this paper, echo bars of order j, represent received echoes having traveled 2jH, that is to say j times back and forth between the probe and the bottom of the tank.

We define $T_e(i)$, $T_r(i)$ and $T_d(i)$ to be the beginning times of the emission, reception and delay respectively, associated with the creation of the image i. $F_j(i)$ is the arrival time of the signal corresponding to an order j echo coming from the emission to create the ith image; $T_t = 2H/c'$ is the time over which ultrasounds travel a round trip distance (2H). We can then write the following relations :

$$T_e(i) = t_0 + (i-1)(t_e + t_r + t_d) \tag{1}$$

$$T_r(i) = t_0 + it_e + (i-1)(t_r + t_d) \tag{2}$$

$$T_d(i) = t_0 + i(t_e + t_r) + (i-1)t_d \tag{3}$$

$$F_j(i) = T_e(i) + jT_t \tag{4}$$

MODELIZATION OF THE PHENOMENON

The eventual echo bars present on an image k are generated from signals arriving between $T_r(k)$, and $T_d(k)$. Among all these signals, only a few will be detected as most of them will be too attenuated. The number of detected signals is the number N corresponding to signals traveling less than $2H_{max}$. Knowing $2H_{max}$, it is possible to determine the maximal order J_{max} of the echoes by the relation $J_{max}=E(H_{max}/H)$ (where E(x) denotes the integer part of x). We can then write :

$$N = \sum_{i=1}^{k} \sum_{j=1}^{J_{max}} \delta_{[T_r(k), T_d(k)]}\big(F_j(i)\big) \quad \text{with} \quad \delta_{[a,b]}(x) = \begin{cases} 1 & \text{if } x \in [a,b] \\ 0 & \text{if } x \notin [a,b] \end{cases} \tag{5}$$

As the time interval between two successive emissions is $(t_e + t_r + t_d)$, t_r being the reception time, the probe can receive, at the most, a sole echo of order j for a k image. Under this consideration relation (5) can be simplified as i and j indices are linked together, and the presence of an echo bar of order j may be known without its index.

In order for a k image to represent an echo bar of order j whose signal comes from the emission i, the following relation should be satisfied :

$$T_r(k) \leq F_j(i) < T_d(k) \tag{6}$$

We can write $i = k - m$, where m is an integer, and by replacing (4) in (6) we obtain :

$$T_r(k) \leq T_e(k-m) + jT_t < T_d(k) \tag{7}$$

From (1), (2) and (3), we get the following relations :

$$T_r(k) = T_e(k) + t_e \tag{8}$$

$$T_e(k-m) = T_e(k) - m(t_e + t_r + t_d) \tag{9}$$

$$T_d(k) = T_e(k) + t_e + t_r \tag{10}$$

Taking (8), (9) and (10) in relation (7) we finally obtain :

$$\frac{jT_t - (t_e + t_r)}{t_e + t_r + t_d} < m \le \frac{jT_t - t_e}{t_e + t_r + t_d} \qquad (11)$$

If an integer m verifies condition (11), then a k image will eventually represent an echo bar of order j whose signal comes from the emitted signal for the formation of the (i=k-m) image.

Let's consider an echo of order j, $j \in [1,..,J_{max}]$, which verifies relation (11) and is present on the images. Let d_j be the distance represented by the echograph between the probe and the echo bar of jth order. d_j is associated with an obstacle, met by the ultrasounds traveling a distance $2d_j$ at velocity c taken into account by the echograph device. In reality, they have traveled 2jH at velocity c'. The traveling time for the distance 2jH is :

$$jT_t = q(t_e + t_r + t_d) + t_j \qquad (12)$$

where q is an integer. An analogous relation can be written in terms of distance :

$$2jH = 2qP' + 2d'_j \qquad (13)$$

with $\qquad P' = (t_e + t_r + t_d).c'/2 \qquad$ and $\qquad d'_j = c't_j/2$

which gives: $\qquad\qquad\qquad\qquad d'_j = jH \text{ modulo } P' \qquad (14)$

However, the echograph device considers a velocity c for the ultrasounds, hence the echo bar of jth order is not represented at a distance $d_j = c't_j/2$, but rather at a distance defined by $d_j = ct_j/2$. d_j and d'_j are related to each other by :

$$d'_j / d_j = c'/c \qquad (15)$$

In conclusion, knowing P, H, c, c', t_e, t_r, t_d and H_{max}, it is possible to determine the number N of echo bars present on the images as well as the distance d_j between the probe and the echo bar of jth order :

$$N = \sum_{j=1}^{J_{max}} \delta_{\left]t_j^-, t_j^+\right]}(m_j)$$

$$d_j = \delta_{\left]t_j^-, t_j^+\right]}(m_j).\left[\frac{c}{c'}\left(jH \text{ modulo } P'\right)\right] \qquad \forall\ j = 1...J_{max} \qquad (16)$$

with $\qquad\qquad\qquad\qquad J_{max} = E(H_{max}/H)$

$$t_j^- = \frac{jT_t - (t_e + t_r)}{t_e + t_r + t_d} \qquad\qquad t_j^+ = \frac{jT_t - t_e}{t_e + t_r + t_d}$$

$$m_j = E(t_j^+)$$

$$P' = \frac{c'}{2}(t_e + t_r + t_d)$$

$$T_t = 2H/c'$$

ECHO BARS CANCELLATION

For a set of known parameters $(P, c, t_e, t_r, t_d, H_{max})$ of a given device, what distance H between the probe and the acoustical mirror should be required to cancel all the echo bars ? Figure 4 represents the function d_j given by (16), for a given j as a function of H (solid line) and the graph of function : (jH modulo P').c/c' (dotted line).

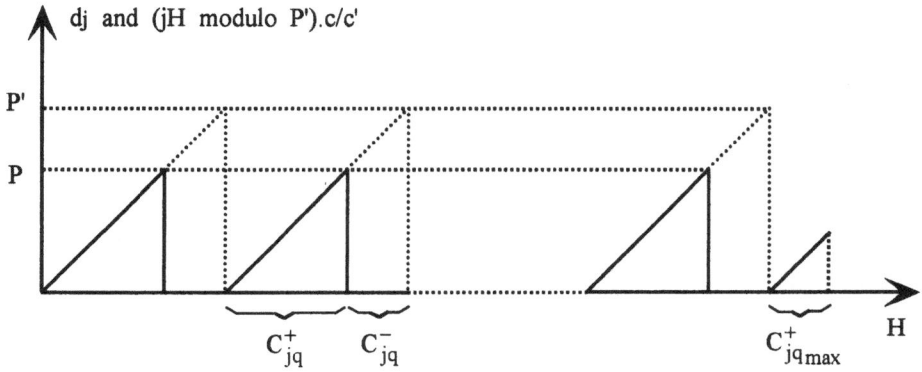

Figure 4. Graphs of the functions d_j and (jH modulo P').c/c'

C_{jq}^+ and C_{jq}^- are the intervals where condition (11) is valid or not valid repectively *i.e.* where the echo bar of jth order is or is not present on the images. Except for the last set of intervals $C_{jq_{max}}^+$ and $C_{jq_{max}}^-$, relations (14) and (15) define a set of intervals C_{jq}^+ and C_{jq}^- covering a distance P'/j by :

$$C_{jq}^+ = \left]\frac{1}{j}qP', \frac{1}{j}\left(qP' + \frac{c'}{c}P\right)\right] \quad \text{and} \quad C_{jq}^- = \left]\frac{1}{j}\left(qP' + \frac{c'}{c}P\right), \frac{1}{j}(q+1)P'\right] \quad (17)$$

Let s be the ratio between H_{max}/j, representing the distance after which echo bars of jth order are not received, over the distance covered by a set of parameters C_{jq}^+ and C_{jq}^-. $s = (H_{max}/j)/(P'/j) = H_{max}/P'$ and is independant of j. The maximum index of the interval couples C_{jq}^+ and C_{jq}^- is constant for all the echo orders and equals :

$$q_{max} = E\left(\frac{H_{max}}{P'}\right) + 1 \quad (18)$$

For $q = 1...(q_{max}-1)$ the intervals C_{jq}^+ and C_{jq}^- are given by (17) and for $q = q_{max}$:

if $\dfrac{c}{c'}(H_{max} \text{ modulo } P') \le \dfrac{c'P}{cj}$,

then $\quad C_{jq_{max}}^+ = \left]\dfrac{1}{j}q_{max}P', \dfrac{H_{max}}{j}\right] \quad$ and $\quad C_{jq_{max}}^- = \left]\dfrac{H_{max}}{j}, +\infty\right[$

or else $\quad C^+_{jq\,max} = \left]\dfrac{1}{j}q_{max}P', \dfrac{1}{j}\left(q_{max}P'+\dfrac{c'}{c}P\right)\right]$ and $\quad C^-_{jq\,max} = \left]\dfrac{1}{j}\left(q_{max}P'+\dfrac{c'}{c}P\right),+\infty\right[$

To eliminate the echo bar of jth order on the images, the distance H should be :

$$H \in I_j = \bigcup_{q=1}^{q_{max}} C^-_{jq} \qquad (19)$$

For a total cancellation of the bars, one must choose H such as :

$$H \in I = \bigcap_{j=1}^{+\infty}(I_j) = \bigcap_{j=1}^{+\infty}\left\{\bigcup_{q=1}^{q_{max}} C^-_{jq}\right\} = \overline{\bigcup_{j=1}^{+\infty}\left\{\bigcup_{q=1}^{q_{max}} C^+_{jq}\right\}} \qquad (20)$$

Remark : We have to consider an intersection on $+\infty$ as the maximum echo order j_{max} can be calculated only if H is given ($j_{max} = E(H_{max} / H)$). However, in this case, H is unknown.

CONCLUSION

In standard echography, as well as in Utrasound Reflection-mode Tomography Using Radial Image Processing (URTURIP), a phenomenon of multiple reflections between the probe and the hyperechogenic structures is encountered. When a planar acoustical mirror faces the probe, multiple reflections appear on the images as horizontal lines called *echo bars*.

We have presented in this paper a modelization of this phenomenon, and shown that it is possible to determine a set of intervals of possible positions for the mirror to suppress the echo bars. As part of our research program on the development of an ultrasound scanner based on the URTURIP Technique for 3D medical imaging, we have used an acoustical planar mirror to solve this echo bars problem.

As further studies for this solution, multiple reflections may be attenuated or cancelled when investigating abdominal tissues using the same model by considering the delay time as the adjustable parameter rather than the distance between the hyperechogenic structures and the probe.

REFERENCES

1. P.N.T. Wells. "Biomedical Ultrasonics", Academic, New York, (1977).
2. M. Friedrich, E. Hundt, and G. Maderlechner, Computerized ultrasound echo tomography of the breast, *Europ. J. Radiol.*, vol. 2, pp 78-87 (1982).
3. J. Ylitalo, J. Koivukangas, J. Oksman, Ultrasonic reflection mode computed tomography through a skullbone, *IEEE Transactions on Biomedical Engineering*, vol. 37, no 11, (1990).
4. C.M. Sehgal, D.G. Lewallen, R.A. Robb, and J.F. Greenleaf, Ultrasonic Imaging of Musculoskeletal System, *News in Physiological Sciences*, Vol. 6, pp. 16-20 (1991).
5. B. Migeon and P. Marché, 3D surfacic representation of limb long bones using ultrasound tomography, *Innov. Tech. Biol. Med.*, vol. 15, no 6 (1994).
6. B. Migeon, P. Marché, Ultrasound Tomography by Radial Image Processing, *Innov. Tech. Biol. Med.*, Vol 13, no 3, pp 292-304 (1992).

A NONLINEAR ULTRASONIC IMAGING METHOD
BASED ON THE MODIFIED INFORMATION CRITERION

Sumio Watanabe[1] and Masahide Yoneyama[2]

[1] Faculty of Engineering, Gifu University
1-1, Yanagi-do, Gifu, 501-11 Japan
[2] Faculty of Engineering, Toyo University
2100, Kujirai, Kawagoe, Saitama, 350 Japan

INTRODUCTION

It has been shown that acoustical imaging is useful for three-dimensional object recognition in robotic sensing and factory automation, and that it is refined by neural information processing[1,2]. The 3-D shape information which is extracted from the backscattered acoustic waves is precisely restored and identified by artificial neural networks. However, there remains a problem of the neuro-ultrasonic recognition system. It is still difficult to optimize several parameters automatically, resulting that a lot of trials and errors are needed to construct a reliable system for practical uses.

This paper proposes an automatic design method for the neuro-ultrasonic 3-D visual sensor based on the statistical properties of learning machines. The conventional design method minimizes the squared error of the training samples, which does not ensure the best performance for the unknown testing samples. The proposed method minimizes the averaged squared error of the testing samples, which means the automatic design of the neuro-ultrasonic 3-D sensor. We show the basic structure of the system, and illustrate the effectiveness of the proposed method by expermental results.

THREE DIMENSIONAL ACOUSTICAL IMAGING

Figure 1 shows an acoustic imaging system of a transmitter, a receiver array, and an object. An incident wave is emitted at $t = 0$ from the transmitter, whose sound pressure at location $\mathbf{r}$ and time t is given by

$$P_{in}(\mathbf{r}, t) = \Theta(ct - |\mathbf{r} - \mathbf{r}_0|) \exp(j\mathbf{k}_{in} \cdot (\mathbf{r} - \mathbf{r}_0) - j\omega t), \qquad (1)$$

where $\mathbf{r}_0$ is the location of the transmitter, $\Theta(x)$ is a Heaviside function, c is a velocity of the sound, ω is an angular frequency, $\mathbf{k}_{in} = (k \sin \theta, 0, -k \cos \theta)$ is a wavenumber vector, and θ is an illuminating angle.

It is assumed that the object's reflection coefficient is $\xi(x, y)$ and that its surface function is $z = \zeta(x, y)$. Let $g(x, y, z)$ be a three-dimensional shape of the object defined by

$$g(x, y, z) = \xi(x, y)\Theta(\zeta(x, y) - z). \tag{2}$$

Our purpose is to reconstruct $g(x, y, z)$ from the backscattered sound pressure $P(\mathbf{r}', t)$ at the receiver's place $\mathbf{r}'$ and time t.

We use an acoustic imaging method combining the holographic reconstruction in XY-axis with the time-flight measuring method in Z-axis. By using the Fourier transform, we have a reconstruction formula[2],

$$g(x, y, z) = \frac{(kz)^2}{\pi} |\int\int Q(\mathbf{r}', z) \exp(j\frac{k}{r}(xx' + yy'))dx'dy'|, \tag{3}$$

$$Q(\mathbf{r}', z) = \frac{P(\mathbf{r}', (r' + r_0 - 2z \cos \theta)/c)(H + z \cos \theta)}{\{(x' - r' \sin \theta)^2 + y'^2 + (H + r' \cos \theta)^2\}r'^3 \exp(jkr')}, \tag{4}$$

where $r = (x^2 + y^2 + z^2)^{1/2}$, $\mathbf{r}' = (x', y', H)$, and H is the height of the receiver array. By this formula, 3-D shape information $g(x, y, z)$ can, in principle, be calculated from the measured acoustic pressure $P(\mathbf{r}', t)$. However, because of the nonlinear response of the receiver, the limited number of receivers in the array, and noises of the system and the environment, the reconstructed image is not so fine for 3-D object recognition. This is the reason why we employ an artificial neural network for the nonlinear filter from the acoustic image to a precise image.

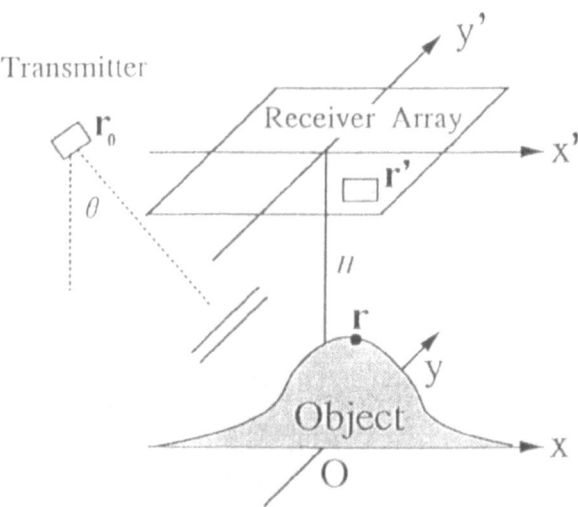

Figure.1 Three-Dimensional Acoustic Imaging
An image in XY-axis is reconstructed by the holographic method,
and that in Z-axis by the time-fight mesuring method.

THE MODIFIED INFORMATION CRITERION

In this section, we propose a method how to estimate the nonlinear imaging function $y = f(w; x)$, where w is a parameter to be estimated by using N training samples, $\{(x_i, y_i); i = 1, 2, ..., N\}$. An input x_i is a part of the acoustic image, and its desired output y_i is a corresponding part of the precise image. We define the training error $E_N(w)$ and the testing error $E(w)$ respectively by

$$E_N(w) = \frac{1}{N} \sum_{i=1}^{N} \|y_i - f(w; x_i)\|^2, \tag{5}$$

$$E(w) = \int \|y - f(w; x)\|^2 P(x, y) dx dy, \tag{6}$$

where $P(x, y)$ is the unknown probability density function from which training samples are independently taken. By applying the statistical asymptotic theory, we can derive a quantitative relation between the training error and the testing error[3],

$$< E(w^*) > = (1 + \frac{2F(w^*)}{NL}) < E_N(w^*) >, \tag{7}$$

where w^* is the parameter that minimizes $E_N(w)$, $< \cdot >$ is the expectation value for sets of training samples, $F(w^*)$ is the number of non-zero parameters in w^*, and L is the dimension of the output y. This relation indicates that the optimal parameter for the minimum testing error can be chosen by minimizing an criterion,

$$I(w) = (1 + \frac{AF(w)}{NL}) E_N(w), \tag{8}$$

which is called AIC when $A = 2$. If we set $A = \log(N)$, $I(w)$ is equal to the logarithm of the description length of the data and the model (MDL)[4], which is the best criterion for the universal data compression.

The usual training rule for the artificial neural network is given by the steepest descent for the training samples,

$$\frac{dw}{dt} = -\frac{\partial E_N(w)}{\partial w}. \tag{9}$$

We propose the new training rule given by the steepest descent for the averaged testing error,

$$\frac{dw}{dt} = -\frac{\partial I(w)}{\partial w} \approx -\frac{AE_N(w)}{NL} \frac{\partial F(w)}{\partial w} - (1 + \frac{AF(w)}{NL}) \frac{\partial E_N(w)}{\partial w}. \tag{10}$$

To make the function $F(w)$ differentiable, we modify and soften $F(w)$ into $F_\alpha(w)$,

$$F_\alpha(w) = \sum_{i,j} (1 - \exp(-\frac{w_{ij}^2}{2\alpha^2})), \tag{11}$$

where $\sum_{ij}$ is the sum for all parameters in $f(w; x)$, and $F_\alpha(w)$ is slowly controlled back to $F(w)$ during training by $\alpha \to 0$. The softened information criterion $I_\alpha(w)$ is called a modified information criterion[5]. Since artificial neural networks can be understood as statistical parametric models, and their learning is formulated as the maximum likelihood method, the modified information criterion can be applied to other neural network models [6].

EXPERIMENTAL RESULTS

Figure 2 shows a neuro-ultrasonic 3-D visual sensor. The transmitter was located at the center of the receiver array ($\theta = 0$). Objects were placed on the table, and the distance from the receiver array and the table was set as 400mm. From the transmitter, a 40KHz ultrasonic burst wave made of 5 cycles of sine waves was illuminated, and the scattered wave was measured by 8×8 receivers. A 3-D image was calculated by the formula, eq.(3). For the Z-axis, 12 pixels were calculated, and the length of each pixel was the half of the wavelength. For the XY-axis, an image of 32×32 pixels was reconstructed by extrapolating 0 as the pressure at the outside of the array.

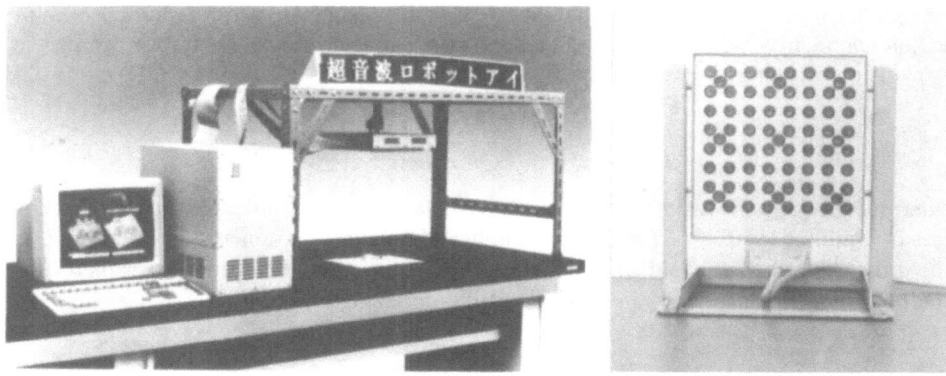

Figure.2 A Neuro-Ultrasonic 3-D Visual Sensor

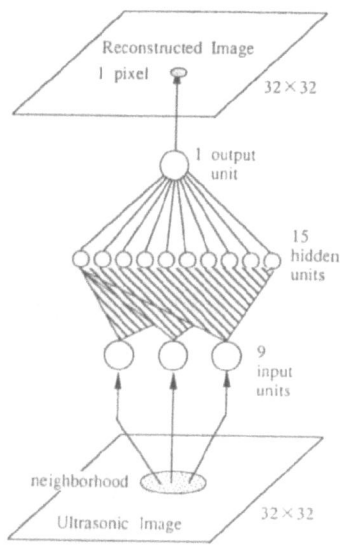

Figure.3 A Neural Network
for Image Reconstruction

Figure.4 Sample Objects

In image restoration, we used a three-layered perceptron from 3×3 neighborhood pixels in an acoustic image to one pixel in an reconstructed image. Figure 3 illustrates the neural network for image reconstruction. It is shown mathematically that a three-layered perceptron can approximate an arbitrary differentiable function more efficiently than any linear combinations of the fixed basis functions[7]. As is shown in eq.(7), a system with the smaller number of parameters can realize that with the better generalization, the three-layered perceptron is universally the better statistical model to estimate unknown relation between inputs and outputs than any models with the linear parameters.

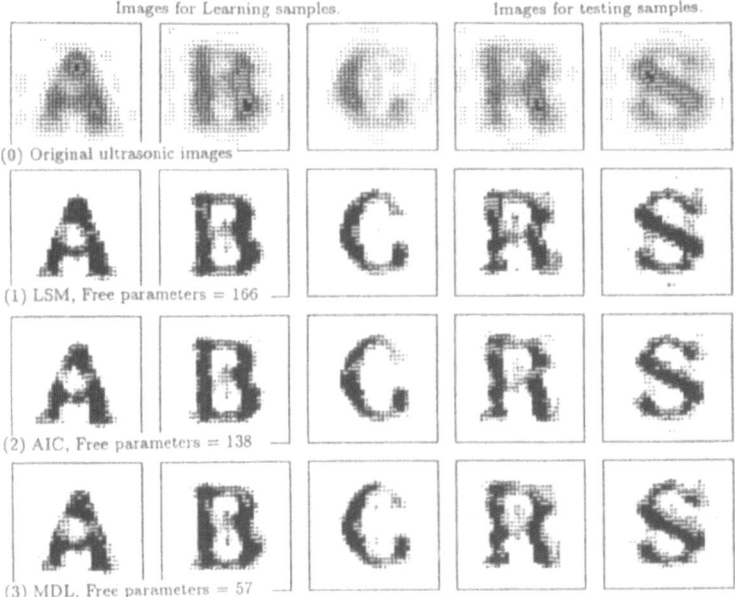

Figure.5 Reconstructed 2-D Images

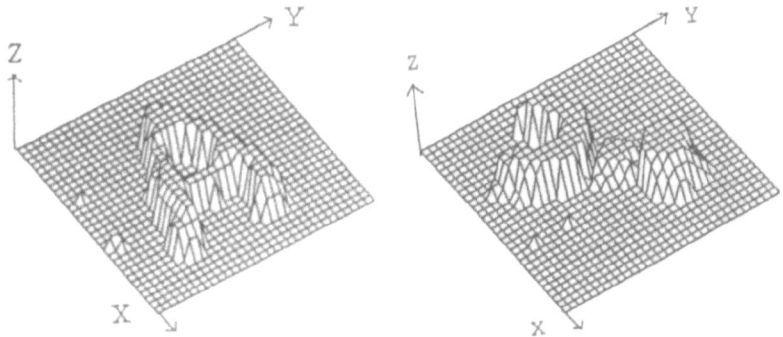

Figure.6 Reconstructed 3-D Images

Figure 4 is a photograph of sample objects. The size of 'A' is about 100mm×100mm × 20mm. Figure 5 shows experimental results by the acoustic imaging and neural network image reconstruction. Three images 'A', 'B', and 'C' were used for training the neural network. The number of the training samples were $3 \times 32 \times 32$. Two images 'R' and 'S' were used for testing. Three methods were compared, the conventional least squared error method (LSM), the minimum AIC method, and the minimum MDL method. For minimizing AIC and MDL, modified information criteria were used. The number of training cycles was 30000, and α was controlled as $\alpha = 3.0(1 - t/30000)$, where t is the training cycle. Note that, at the end of training, the modified information criteria were controlled back to the conventional information criteria.

The pairs of training errors and the testing errors per one image by the LSM, AIC, and MDL were respectively (12.573, 13.428), (12.688,12.879), and (12.895, 13.061). The parameters which were not zero at the end of training were 166, 138, and 57. Testing errors obtained by AIC and MDL were smaller than that by LSM. By comparing images, noises in the image by the AIC were smaller than those by LSM, and the 'tail' of 'R' could be more precisely resolved. Figure 6 shows 3-D images obtained by integrating 2-D images.

By eq.(7), the difference between the AIC or the MDL and the LSM is given by the $1/N$ or $\log(N)/N$ order term. In our experiment, the number of training samples was rather large, 3072, resulting in small improved testing errors. However, even in a such case, information criteria enable us to design the best size system automatically without trials and errors.

CONCLUSION

A nonlinear acoustic imaging system is optimized automatically based on the modified information criterion, and its effectiveness is shown by experimental results.

REFERENCES

1. S.Watanabe and M.Yoneyama, "Ultrasonic robot eyes using neural networks," *IEEE Trans. on UFFC*, Vol.37, No.3, pp.141-147,1990.

2. S.Watanabe and M.Yoneyama, "An ultrasonic 3-D visual sensor using neural networks," *IEEE Trans. on Robotics and Automation*, Vol.8, No.2, pp.240-249, 1992.

3. H.Akaike,"A new look at the statistical model identification," *IEEE Trans. on Automatic Control*, Vol.19, No.6, pp.716-723,1974.

4. J.Rissanen,"Universal coding, information, prediction, and estimation," *IEEE Trans. on Information Theory*, Vol.30, pp.629-636, 1984.

5. S.Watanabe, "An optimization method for layered neural networks based on the modified information criterion," *Advances in Neural Information Processing Systems*, Vol.6, pp.293-300, San Mateo, Morgan Kaufmann, 1994.

6. S.Watanabe and K.Fukumizu,"Probabilistic design of layered neural networks based on their unified framework," *IEEE Trans. on Neural Networks*, Vol.6, No.3, pp.691-702, 1995.

7. A.R.Barron, "Universal approximation bounds for superpositions of a sigmoidal function," *IEEE Trans. on Information Theory*, Vol.39, No.3, pp.930-945,1995.

EVALUATION OF ULTRASONIC SENSOR SIGNALS USING FUZZY LOGIC

M. Vossiek, P.-C. Eccardt, V. Mágori

Siemens AG, Corporate Research and Development
Otto-Hahn-Ring 6, 81730 München, Germany
email: vossiek@curry.zfe.siemens.de

ABSTRACT

Novel ultrasonic sensor systems are introduced which demonstrate the use and strength of fuzzy logic for evaluating ultrasonic signals. As one application the recognition of objects and situations by ultrasonic means will be discussed. To inspect printed circuit boards the correlation function of the actual signal from a board under test and the reference signal is evaluated using fuzzy logic. The sensor system does not only recognize assembly defects with high resolution and high reliability, but also classifies the kind of defects. Another application of fuzzy logic for ultrasonic sensor systems is the level measurement of liquids or bulk materials. Employing fuzzy logic, rather than Boolean, for discriminating the echo of the material surface, a high performance of the sensor in critical environment conditions has been achieved. Besides the basic principles of processing ultrasonic signals with fuzzy methods the paper presents implementations of the fuzzy algorithms on a fuzzy coprocessor (SAB 81C99A) or a digital signal processor.

INTRODUCTION

Ultrasonic sensors for applications in air can be used for many measuring tasks such as distance or presence measurements and object recognition[1]. As typical smart sensors, ultrasonic sensors evaluate the information impressed on the propagating ultrasonic signal by the quantity to be sensed. Hence, the task of a smart sensor is to assign the measured raw data to a scale or a class based on a set of suitable rules. Since ultrasonic sensor signals are affected by several physical variables, this transformation or decision process can be very complex and may require an intelligent combination of different signal characteristics using models of signal theory or heuristic approaches.

As proved by recent applications in the area of ultrasonic distance measurement[2,3], robotic sensing[4] and sonar[5], fuzzy logic is advantageously applicable for this kind of intelligent ultrasonic signal processing. Fuzzy logic allows to incorporate expert knowledge very easily and has the ability to process multiple varying inputs and to make comparison with a range of values. Another strength of fuzzy logic is its robustness, fault tolerance and the simplicity of implementing new evaluation rules[6]. In the area of signal processing, fuzzy logic is mainly used in the original sense proposed by Zadeh[7] namely as a tool to analyze complex systems and decision processes rather than for system control[8] being the other predominant application of fuzzy today.

ULTRASONIC SIGNAL PROCESSING USING FUZZY LOGIC

The basic principles of evaluating sensor signals using fuzzy are reviewed on the basis of a simple example. A typical signal processing problem for many fault diagnosis systems is the comparison of a known reference signal with a measured signal from an object or situation observed by the sensor. This signal could be the echo of an object 'illuminated' by an ultrasonic transducer. Fig. 1 illustrates such a measuring situation.

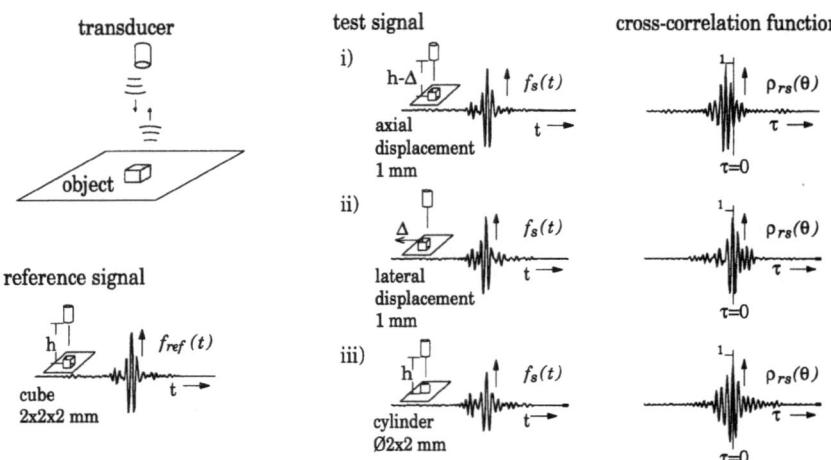

Fig. 1. Arrangement of transducer and inspected object with three measuring situations and the resulting cross-correlation functions of the corresponding test signals and the reference signal

There are several techniques to compare two signals; a common method is to cross-correlate the signals. For stationary real signals the cross-correlation function $\rho_{rs}(\theta)$ is given by:

$$\rho_{rs}(\theta) = \int_{-\infty}^{\infty} f_{ref}(t) \cdot f_s(t+\theta)\, dt \qquad and \qquad \rho_{rs_n}(\theta) = \frac{\rho_{rs}(\theta)}{\sqrt{\rho_{rr}(0) \cdot \rho_{ss}(0)}}, \tag{1}$$

where f_{ref} and f_s denote the reference signal and the test signal, respectively. An amplitude independent cross-correlation function $\rho_{rs_n}(\theta)$ can be obtained by normalizing $\rho_{rs}(\theta)$ with respect to the average power of the reference signal, that is given by the maximum of the auto-correlation function $\rho_{rr}(\theta=0)$, and the average power of the test signal $\rho_{ss}(\theta=0)$.

If $f_{ref}(t)$ and $f_s(t)$ are identical except for a constant amplitude factor and a time delay τ, the maximum of $\rho_{rs_n}(\theta)$ equals 1. The maximum is located at the point of time $\theta=\tau$. If $f_{ref}(t)$ and $f_s(t)$ are totally independent, $\rho_{rs_n}(\theta)$ is 0. Because the auto-correlation function is a symmetric even function, the asymmetry of the cross-correlation function around its maximum is another quantity indicating whether the reference equals the measurement. Fig. 1 shows some typical measuring situations, the pertaining signals and the resulting cross-correlation functions.

Due to the finite length of the measured ultrasonic signals and external disturbances such as temperature and flow fluctuations or noise effects, the normalized cross-correlation of two ultrasonic signals, in practice, will never be identical one or zero. Thus the user has to decide above which level the echo shape of reference and measurement are classified to be equal; as displayed in Fig. 2a, this crisp level may be chosen as 0.9. Nevertheless, if $\rho_{rs_n}(\theta)$ is 0.8, it is still much more likely that both signal shapes are equal than that they are totally different. In contrast to classical logic, fuzzy allows partial membership with arbitrary membership values in the range [0...1]. The degree of membership depends on the value of the input variable and a user defined membership function. With the membership function depicted in Fig. 2b and an input $\rho_{rs_n}(\theta)$ of 0.8, the degree of membership μ is 0.9. The assignment of the input data to linguistic variables is called fuzzification. A fuzzy variable can belong to one class or several classes with special degree of membership.

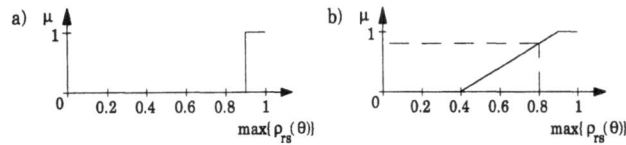

Fig. 2. a) crisp function $\mu=f(max\{\rho_{rs}(\theta)\})$ for the probability that f_{ref} equals f_s
b) fuzzy function for the probability that f_{ref} equals f_s

Usually there are many input variables and several fuzzy sets consisting of various fuzzy membership functions. To describe the asymmetry of the cross-correlation function, a fuzzy set partitioned into the membership functions {*negative big, negative medium, small, positive medium, positive big*} has been defined (Fig. 3a). Shape differences, differences in the signals' time of flight and differences in the reflected average power are represented by similar fuzzy sets. As fuzzy variables provide a direct image of a quantity in the range [0...1], different quantities can be combined and compared straightforwardly. Furthermore, by translating a wide input range into a small number of membership values the intervals and number of variables to be considered in the evaluation can be reduced significantly.

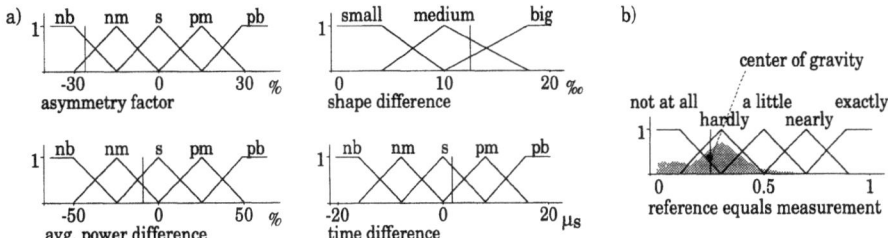

Fig. 3. a) Fuzzy input and b) fuzzy output variables used for evaluating the cross-correlation function

The central element of a fuzzy evaluation is the so called fuzzy inference. Fuzzy inference systems consist of a number of conditional "if-then" rules describing relations between fuzzy input and fuzzy output variables. In the example one simple rule may be: '*If the shape difference is big and the asymmetry factor is negative or positive big, then the measurement does not equal the reference at all*'. As it can be seen in Fig. 3b, this rule is met only to some degree because the premises '*shape difference is big*' and '*asymmetry factor is negative or positive big*' is given only partially by the membership factor of 0.8 and 0.3, respectively. As the two input variables are combined by the intersection operator '*and*', the output is limited by the minimum of both, accordingly 0.3. In addition to this intersection operator *(A and B <=> min{μ_A, μ_B})*, further operators are the union operator *(A or B <=> max{μ_A, μ_B})* or the complement operator *(not A <=> 1 - μ_A)*. Beside these common min-max relations, there are many other operators proposed in literature.

The union of all truncated areas of all rules is the output of the inference. This complex distribution has to be defuzzified to get a crisp output value. As depicted in Fig. 3b, a defuzzification can be performed by calculating the center of gravity of the area enclosed by the rule outputs. Obviously, at the end of the fuzzy evaluation a crisp level must still be defined, which classifies reference and measurement to be equal or not. However, up to that point a loss of information can be prevented by avoiding crisp values during the evaluation.

Fuzzy classifiers work quite similar to distance measuring classifiers that use a weighted distance measure. The advantage of fuzzy is the use of human language rules instead of algebraic weighting and operator functions. The fuzzy systems converts these user defined rules to their mathematical equivalents - nevertheless, normally at the expense of increased computation time. In our opinion, fuzzy is not a unique solution of tasks that could not be performed by other techniques. Foremost, fuzzy is a very elegant programming language, which allows to describe and rule complex processes in a well understandable and compact way. Variables are weighted and combined just by defining appropriate sets and rules. This approach enables the user who understands the given problem to create fuzzy systems straightforwardly. At the same time, it simplifies the design of systems and ensures easy system update and maintenance.

APPLICATIONS

Inspection of printed circuit boards

Based on the fuzzy evaluation of the cross-correlation function described above, an ultrasonic sensor system for inspecting SMD (surface mounted devices) printed circuit board has been set up. The board under test is scanned by an ultrasonic transducer moved by a stepper motor. The transducer has a center frequency of 250 kHz and a 3 dB bandwidth of 90 kHz. The lateral footprint of the transducer directivity pattern in the measuring plane is about 10 mm^2. The measured signal from each scan point is compared with the stored corresponding signal of a reference board. By utilizing several features of the cross-correlation function the fuzzy algorithm does not only detect assembly defects with high resolution and high reliability but it is also able to decide whether the defect is an axial or lateral displacement or whether a part is completely missing. Fig. 4 illustrates exemplary results of the fuzzy classification.

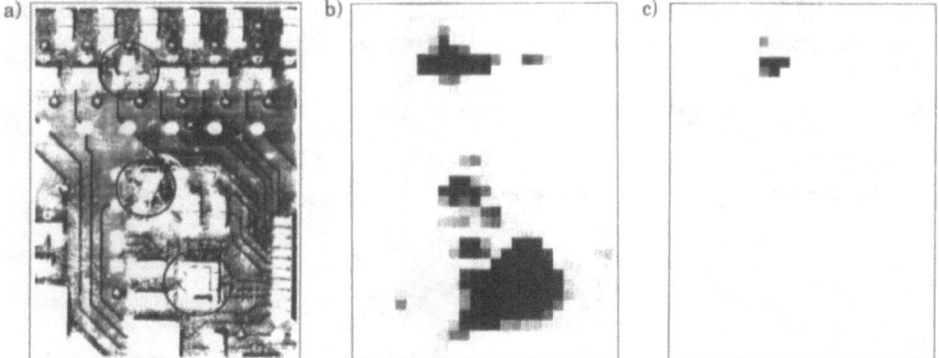

Fig. 4. Recognition of defects in a printed circuit by fuzzy evaluation of the correlation function
a) photo of the board under test,
b) error probability for an assembly defect, c) error probability that a part is missing

The selectivity of the system with respect to deviations of the object dimensions is about 0.2 mm. This is significantly better than the image resolution achievable with the given system. And hence, Figs. 4b-c should not be confused with images of the defects - the figures just illustrate error probabilities. The entire signal processing including correlation, feature extraction and fuzzy evaluation are performed by a digital signal processor (DSP 56001). This way it is possible to obtain a high measuring rate of about 20 Hz.

Level measurement in a complex environment

An other application is the level measurement of liquids or bulk materials in containers. A high performance level meter has been developed, which uses fuzzy logic, rather than Boolean, to detect the echo of the content level in a large number of misleading echoes caused by stationary or moving disturbing objects or by multiple reflections.

A typical problem of ultrasonic distance sensors occurs if good reflecting objects cause multiple echoes due to repeated reflections between sensor and object. These multiple echoes simulate the existence of further objects, which is very critical if the distance to more than one object is to be determined. The main characteristic of a multiple echo is, that it can be derived from one or several preceding echoes with shorter time of flight. By taking into account the features of previous echoes (e.g. time of flight and amplitude or shape), expected values of a multiple echo can be calculated by evaluating the echo propagation time, the spatial divergence of the sound beam and the attenuation along the measuring path. The probability that this echo is a multiple echo is the higher the more the calculated values of a theoretical multiple echo correspond with the values of a received echo. Because of disturbing influences on the sound propagation and unknown reflector features, the hypothesis of a multiple echo must always be vague and incomplete.

Therefore, fuzzy is an ideal tool for describing the previously mentioned relations avoiding "yes/no" decisions. During the fuzzy evaluation each partial echo is assigned by a probability value for being a multiple echo.

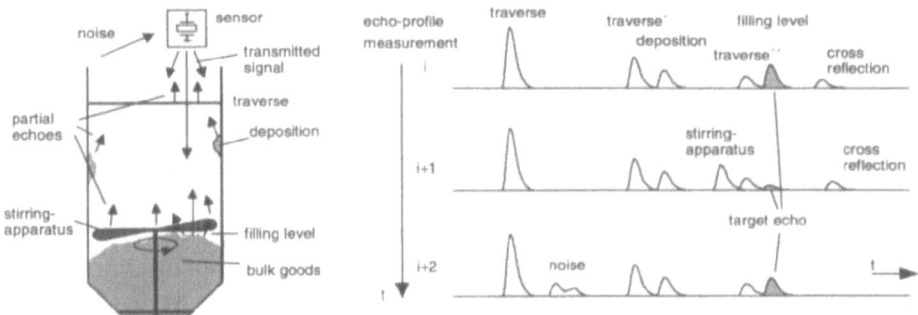

Fig. 5. A typical echo profile with a number disturbing echoes and their temporal development

For level measurements, the maximum velocity of filling level changes is a very useful a priori knowledge. Considering the time history of the echoes allows to suppress echo fluctuations, caused, for instance, by a stirring apparatus. Additionally, it is known that the target echo is the echo which has the longest propagation time but is not a multiple echo. In particular critical measurement situations, as in narrow silos with reflecting constructive elements in the measurement path or in silos containing bulk goods with rough surfaces, additional knowledge about fixed constructive elements must be involved. The suppression of the fixed target echoes is similar to the suppression of multiple echoes mentioned above. The main difference is that the current echo profile is not compared with a hypothetical signal but with a teach-in-profile of the empty tank. In order to compensate drift effects, the data from the teach-in-phase are updated from time to time during the measurement.

For an exemplary echo profile, the developed progressive fuzzy evaluation is demonstrated in Fig. 6. Apparently, the echo of the content level is separated from a large number of interfering echoes very efficiently. The fuzzy algorithms contain a total of 80 rules, that mostly have to be processed for several combinations of partial echoes. Thus, the number of rules to be processed for a complete evaluation of a complex echo profile can be rather big. For a high measuring rate, a dedicated hardware such as a fuzzy coprocessor is required.

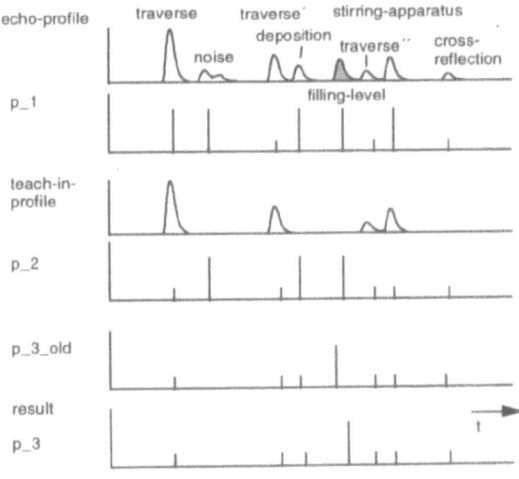

Fig. 6. Progressive fuzzy-evaluation of the echo-profile. The signal processing is divided in three modular steps.
Probability of target echoes after:
p_1 - suppression of multiple echoes
p_2 - suppression of fixed target echoes by a comparison of a teach-in-profile with the current measurement
p_3 - evaluation of the temporal echo development by a comparison of preceding results (p_3_old) with the current result.

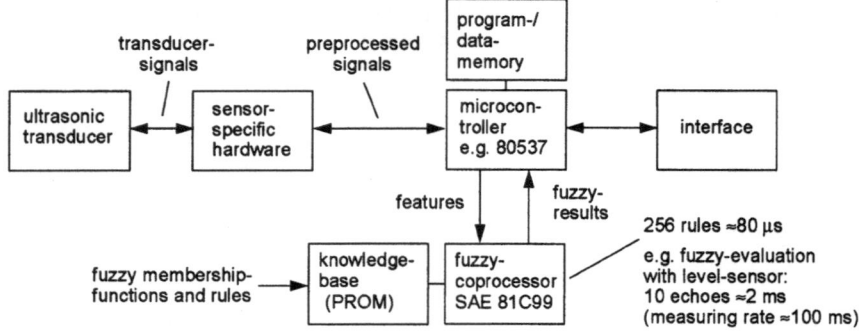

Fig. 7. Block diagram of the system

The setup of the developed sensor system is shown in Fig. 7. A microcontroller (SAB 80537) controls the measurement and performs the communication with the peripherals. The Siemens fuzzy coprocessor SAE 81C99A[9] combined with the microcontroller allows rapid and efficient signal processing at low cost. The microcontroller extracts the features for the fuzzy evaluation. The fuzzy coprocessor with the ability to process up to 8 million rules per second carries out the fuzzy algorithms and supplies the microcontroller with the defuzzified results. For the described level measurement, the signal processing for a typical measuring situation (echo profile with approximately 10 partial echoes) takes about 2 ms.

The fuzzy knowledge base is stored in a separate memory chip. The sensor specific hardware interfaced to the microcontroller performs several signal preprocessing steps such as signal demodulation, filtering and parts of the feature extraction. The modular hardware concept and the simplicity of changing the fuzzy rules result in a great flexibility of the systems.

REFERENCES

1. Mágori, V., "Ultrasonic Sensors in Air," in IEEE International Ultrasonics Symposium, Cannes, France, 1-4 Nov. 1994.

2. Kroemer, N., Vossiek, M., Eccardt, P.-C., Mágori, V., "Ultraschall-Distanzsensor mit Fuzzy-Auswertung," in Sensor 93, 6. Internationale Fachmesse mit Kongress für Sensorik & Systemtechnik, Nürnberg, Germany, Vol. II, pp. 113-122, 11.-14. Oct. 1993.

3. Burkard, R., "Fuzzy-Logik-Elemente am Beispiel der Füllstandsmessung mit Ultraschall," in Sensor 93, Nürnberg, Germany, Vol. II, pp. 123-129, 11.-14. Oct. 1993.

4. Foulloy, L., Mauris, G., "An Ultrasonic fuzzy sensor," in Advanced Sensor Technologie, Zürich, Switzerland, pp. 161-170, 2-4 Feb. 1988.

5. Kummert, A., "Fuzzy Technology Implemented in Sonar Systems," IEEE Journal of Oceanic Engineering, Vol. 18(4), pp. 483-490, Oct. 1993.

6. Self, K.: "Designing with fuzzy logic," IEEE Spectrum, Nov. 1990, S.42-44 and 105.

7. Zadeh, L. A.: "Outline of a New Approach to the Analysis of Complex Systems and Decision Processes," IEEE Trans. on System, Man, and Cybernetics, Vol. SMC-3, No 1, 1973, S.28-44.

8. Tong, R.M., "Fuzzy Control Systems: A Retrospective," in Proc. of the American Control Conference, San Francisco, CA, USA, Vol. 3, pp. 1224-1229, June 1983.

9. Eichfeld, H.P., Künemund, T.N., Klimke, M.G., "An 8bit Fuzzy Coprocessor for Fuzzy Control," in IEEE International Solid-State Circuits Conference, San Francisco, USA, pp. 180,181,286, Feb. 1993.

A PARALLEL PROCESSING ALGORITHM FOR ACOUSTIC TOMOGRAPHY

R. Kline and C. Sullivan[1], R.B. Mignogna [2], and P.P. Delsanto [3]

[1]School of Aerospace and Mechanical Engineering
University of Oklahoma
Norman, OK. 73019

[2]Mechanics of Materials Branch
Naval Research Laboratory
Washington, D.C. 20375-5000

[3]Dipartimento di Fisica
Politecnico di Torino
10129 Torino, Italy

INTRODUCTION

Acoustic tomography is a widely used technique to reconstruct material properties from global wave propagation data. Geophysical applications include pre-stack depth migration [1] and cross-borehole tomography. Acoustic tomography is also commonly used in nondestructive testing of materials. The utility of the method is restricted by intensive computational demands. In this work, we present a tomographic reconstruction algorithm, including ray tracing, which fully utilizes thesimultaneous computation capability of the Connection Machine. The advantage of the parallel program is the computation time remains relatively constant as the complexity or problem size increases, whereas the time required for a similar sequential program increases rapidly. Results are presented for several synthetic reconstructions, illustrating the utility of this approach for complex problems.

Two models were tested, a two-layer sample with a 50% velocity increase and an Epstein layer with a gradual 50 % velocity increase. The source-receiver configurations were varied to test the boundary resolution. The speeds of the parallel and the serial computation were compared as the array size increased.

Previous work in this area has concentrated on modeling the ultrasonic wave propagation through the model. Delsanto et al. [2] developed a parallel algorithm to model wave propagation in a two-dimensional model. Mignogna et al. [3] has used this technique for acoustic tomography, but the ray tracing was performed using a serial algorithm

PARALLEL PROCESSING REQUIREMENTS

The CM200 has 8192 independent processors capable of performing simultaneous operations on large data sets. The main difficulty in adapting current tomographic imaging algorithms for parallel computation was in implementing the ray tracing. The ray tracing procedure requires velocity information from an nxm matrix representing the spatial grid to be used in conjunction with kxl matrices that store the current ray path angle and spatial location. These matrices are usually unconformable in size. In addition, the current spatial location must be accounted for without sequentially progressing through the spatial grid to maintain a strictly parallel operation. These requirements can be met by using <u>vector-valued</u> <u>subscripts.</u> Vector-valued subscripts can be thought of as place-holders, used to associate the correct velocity to each raypath, dependent on its current location. For example, conformable velocity arrays can be created by

$$vray = v \ (\ ipost(k,l), jpost(k,l) \ ) \tag{1}$$

where $ipost(k,l)$ contains the current row location of the kl^{th} ray, $jpost(k,l)$ contains the column location and v is the spatial grid velocity array. The new velocity array, $vray$, is conformable with any kxl sized array containing raypath information.

ALGORITHM DESIGN

Three major components comprise the tomographic algorithm: time delay calculations, ray tracing and the tomographic inversion. The time delay calculations replace experimental measurements in this numerical simulation.

Time delay calculations

Time delay data are calculated using Snell's law. The ray path distances are divided by the cell velocities and summing over the entire ray for each source-receiver pair. A polynomial approximation to the time delay curve replaces the specific source-receiver time delay measurements and is used instead of the traditional shooting method. As a result, only one iteration of the ray tracing procedure is required for each tomographic iteration. Determination of specific launch angles is no longer required, since the measured time delay data can be calculated by the polynomial expression regardless of where the ray terminates.

Ray Tracing

As the complexity of a tomographic reconstruction increases, the number of raypaths increases dramatically. A simultaneous ray tracing calculation can reduce computation time significantly. The ray tracing procedure in this work is based on Snell's law. The angle at which a ray travels through a cell is dependent on the incident angle and the ratio of the velocities of the adjacent cells. Vector-valued subscripts are used to convert the velocity information into conformable sized arrays matching the current raypath cell locations.

The raypath length is found using straight-line calculations within each cell, and is stored in a four dimensional array corresponding to each source-receiver pair and cell location. This information is then used in the tomographic inversion.

Tomographic Inversion

The tomographic inversion algorithm adjusts the estimated slowness values in a systematic fashion until the estimated time delays, T_{kl}, match the measured time delays. This work uses the simultaneous iterative reconstruction technique (SIRT), one of many different procedures for tomographic inversion. Commonly used techniques such as ART (algebraic reconstruction technique) examine the data on a ray-by-ray basis when making the slowness adjustments. This is less then satisfactory for parallel reconstruction where all raypath data are available simultaneously. For our study, we have implemented a parallel version of SIRT. It makes use of the output from the parallel ray tracing procedure by evaluating the time delays for all rays simultaneously. In addition, adjustments to the slowness values are made for the entire grid at one time.

The parallel algorithm must account for <u>all</u> the raypaths at one time. The adjustment to the slowness values is calculated by

$$\Delta S_{ij} = \frac{\sum_{kl} (T^{kl} \Delta s_{ij})}{\sum_{kl} T^{kl}} \tag{2}$$

T^{kl} is the time delay that is calculated by

$$T^{kl} = \sum_{ij} s_{ij} a_{klij} \tag{3}$$

where s_{ij} is the slowness value for cell ij and a_{klij} is the length of ray kl through cell ij. The ART algorithm for the slowness correction is Δs_{ij} and is calculated by

$$\Delta s_{ij} = \frac{\Delta T^{kl} a_{klij}}{\sum_{ij} (a_{klij})^2} \tag{4}$$

where
ΔT^{kl} = the time difference between the actual travel time of ray kl and the travel time calculated by equation (3).
a_{klij} = the distance traveled by ray kl across cell ij.

Proportionally greater weight is accorded rays whose cell transit times are large in comparison to rays that spend only a short time in that particular cell. The slowness values of each cell are updated for each iteration until the calculated time delays match the actual time delays. At this point the slowness values are assumed to match the actual slowness values. The slowness or the velocity of each pixel can be used to determine the material properties of the specimen.

RESULTS

Two models were tested, a two-layer sample with a 50% velocity increase and an Epstein layer with a gradual 50 % velocity increase. The source-receiver configurations were varied to test the boundary resolution. Figure (1) shows the different source-receiver configurations and the corresponding tomographic inversions. The configuration with a minimal number of raypaths converged to a non-unique solution for the two layer problem. Excellent results were obtained for the Epstein layer problem. The error in the two layer problem decreased as the number of raypaths increased, as shown in the two middle configurations in figure (1). The bottom configuration shows an increase error with the increase in the number of raypaths. This is caused by a compounding of the error in the time delay estimates. By improving the order of the polynomial representing the time delay curve this error is reduced from 100 m/s or 2.0 % to 9.3 m/s or 0.2%.

The Epstein layer model converges in fewer iterations then the two layer sample. Figure (2) shows the decrease in velocity error with increasing iterations for both models.

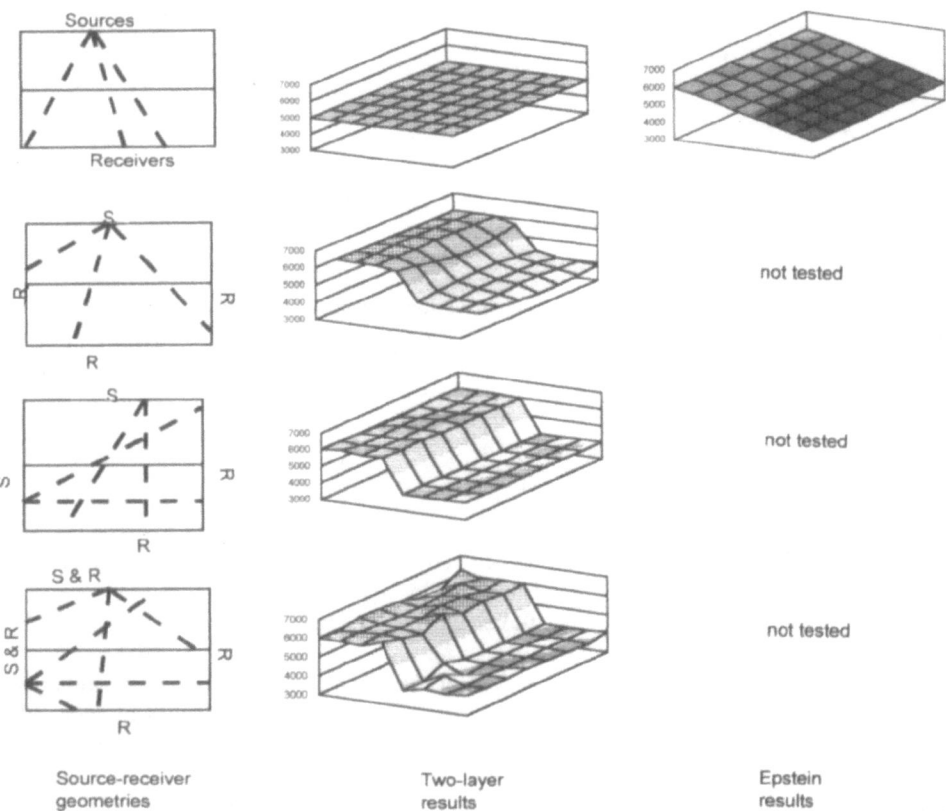

Figure 1. Different source-receiver geometries and the corresponding tomographic reconstructions.

564

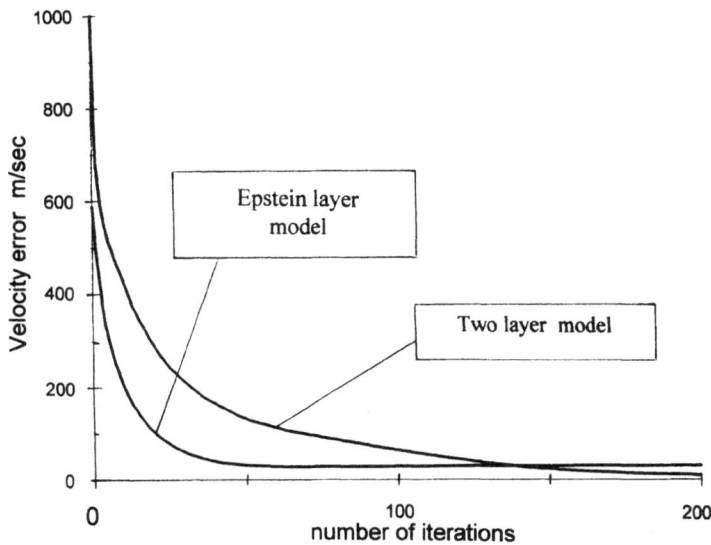

Figure 2. Convergence rates for Epstein layer and two-layer models.

The two-layer model was used for the timing study. The problem size was increased and the corresponding CPU time was recorded. A comparison was made between the parallel program on the CM200 and a similar serial program. The results are shown in figure (3). It can be seen that as the array size increases the CPU time remains relatively constant for the parallel algorithm, while the CPU time for the serial program increases significantly. This demonstrates the utility of this parallel algorithm for large tomographic reconstructions.

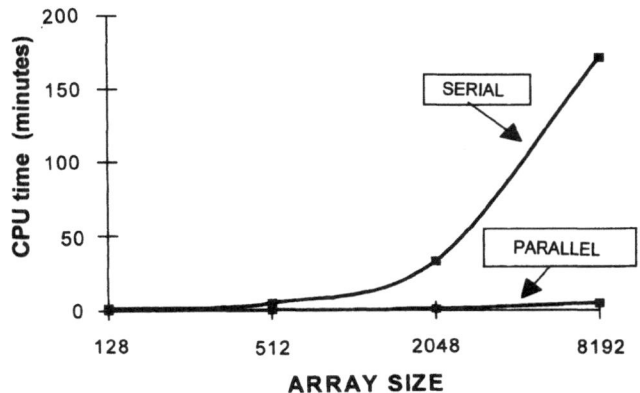

Figure 3. Comparison of CPU time for the serial and parallel tomographic reconstruction algorithm.

CONCLUSIONS

A parallel approach to acoustic tomography in nonhomogeneous media has been developed. The ray tracing procedure is based on Snell's law, and the tomographic inversion is a parallel version of SIRT. The use of a polynomial approximation to the time delay curve eliminates the need for a shooting method.

Different source-receiver configurations and different sized arrays of sources and receivers were tested to determine the accuracy and the speed of the parallel tomographic algorithm. Sources along the side of the sample were required to resolve the boundary in the two-layer model. The Epstein layer model was successful with minimal source-receiver pairs. The convergence rate was faster for the Epstein layer model.

The CPU time required for the solution of the tomographic inversion of the two layer model was compared for serial and parallel operations. The CPU time required for the solution of the serial operation increases rapidly as the array size increase, limiting the applicability of the technique. The CPU time required for the parallel solution remains relatively constant as the problem size increases, demonstrating the suitability of this technique for large problems.

REFERENCES

1. D.W. Ratcliff, Mahogany salt images show key role of velocity model, *Oil and Gas Journal,* 3-53, Oct. 1994
2. P.P. Delsanto, R.S. Schechter, H.H. Chaskelis, R.B. Mignogna, and R.A. Kline, Connection machine simulation of ultrasonic wave propagation in materials. II: The two-dimensional case, *Wave Motion, 20:295 (1994)*
3. R.B. Mignogna,, R.S. Schechter, H.H. Chaskelis, P.P. Delsanto, R.A. Kline and C. Sullivan, Parallel Processing technique for acoustic tomography of multilayers, in "Review of Progress in Quantitative NDE", Vol. 12, D.O. Thompson and D.E. Chimenti, ed., Plenum Press, New York, (1993)

RESTORATION OF IMAGES OF LINEAR
ULTRASONIC ARRAY BY NEURAL NETWORK

Shahram Shirani[1] and Vahid R. Nafisi[2]

[1] Electrical Eng. Department
Tehran University
P.O. Box 14155-6442
Tehran, Iran

[2] Iran Research Organization of Science & Technology
Forsat St., Tehran
Iran

INTRODUCTION

Image restoration techniques are used to improve the image quality. Inverse filtering and Wiener filters are classical image restoration techniques. In these techniques some simplifying assumptions are necessary (e.g. position invariancy and linearity). In 1988 Zhou[1] proposed using Hopfield neural network for image restoration. Advantage of this technique compared to previous techniques is that no assumption about the statistics of the image is necessary e.g. stationarity and position invariancey of imaging system. The main idea of this technique is to define a criterion function and equating it with the energy function of the Hopfield network, then determining weights and initial conditions of the network[1]. This technique has several shortages :
1. Imaging system must be linear.
2. The criterion function must be of second order so that it world be possible to make a correspondance between this and the energy function of the Hopfield network.

In this work we use a perceptron for image restoration. We apply our method on ultrasonic B-mode images. Our choice is based on two reasons. First, ultrasonic images generaly have poor resolution and any improvement is of interest in medical imaging community. Second, these images are nonstationary, also point response of ultrasound systems is position variant. These conditions limit the utility of classical restoration approaches to such images. The images that will be used are simulation results for a linear ultrasonic array.

METHODS

Simulations

The array that will be considered in this paper is shown in Figure 1. Ultrasonic pulses are sent to the objects by a source or sources. It will be assumed that the elements of the array are located along the x axis, and that the width of each element is smaller than a wavelength. Object is considered as a distribution of scatterers in a medium with constant sound velocity, $v_\bullet$.

Consider a single scatterer at position (x_s, x_s). The time which it takes the wave to travel from the transmitter element to the scatterer and then to the receiving element (at position $(x_n, 0)$) is :

$$t_n = (\sqrt{(x_s - x_t)^2 + z^2_s} + \sqrt{(x_s - x_n)^2 + z^2_s})/z_\bullet \qquad (1)$$

To compensate delays of a spherical wavefront detected over the elements of the transducer array, received wavefronts are delayed according to Eq. (1) and then added together.

Let $f_n(t)$ denote the rf signal recorded by the n-th array element. If $I(x_i, z_i)$ is the amplitude of the image at position (x_i, z_i), imaging algorithm can be written as:[2]

$$I(x_i, z_i) = \sum_{n=1}^{N} f_n [t_n(x_i, z_i)] \qquad (2)$$

N is the number of receiving elements. Let the transmitted pulse be p(t), then the signal detected by the n-th element of the arry will be :

$$f_n(t) = \sum_i \sigma_i \, p(t - t_n(x_i, z_i)) \qquad (3)$$

where σ_i is the strength of i-th scatterer.

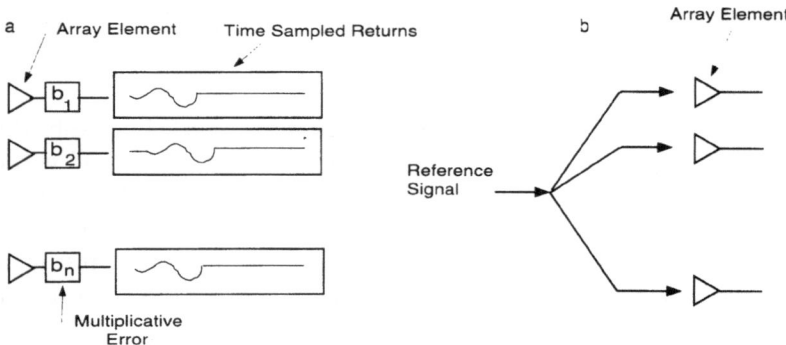

Figure 1. The linear array

Restoration

Let $\bar{g}$ denote the disturbed noisey image, $\bar{f}$ original image, h point spread function of the imaging system and $\bar{n}$ the noise. Therefore we can write :

$$\bar{g} = (\bar{f} ** h) + \bar{n} \tag{4}$$

since we will use digital images, Eq. (4) can be written in matrix form :

$$g = Hf + n \tag{5}$$

g, f and n are vectors formed by succession of the rows of $\bar{g}$, $\bar{f}$ and $\bar{n}$ matrixes respectively. H is system matrix and its elements are calculated from $h^{1,3}$.

In image restoration process the main object is to abtain the best estimation of f form g. Usually this is done by minimizing a criterion function. General structure of this criterion function is :

$$J = \frac{1}{p} \|g_i - H_i \hat{f}\|^p + \frac{\lambda}{q} \|D_i \hat{f}\|^q \tag{6}$$

H_i is the i-th row of H and g_i is the i-th element of vector g.

Since image restoration is an ill problem[3] (a small change in g produces large changes in $\hat{f}$) D matrix is used to regulate the noise[1]. Hopfield network can be used for minimizing Eq. (6) if p and q are selected to be 2^1. In this situation Eq. (6) can be written as :

$$J = \frac{1}{2} \sum_i \sum_j \sum_p h_{p,i} h_{p,j} \hat{f}_i \hat{f}_j + \frac{\lambda}{2} \sum_i \sum_j \sum_p d_{p,i} d_{p,j} \hat{f}_i \hat{f}_j - \sum_i \sum_p g_p h_{p,i} \hat{f}_i$$
$$+ \frac{1}{2} \sum_p g_p^2 \tag{7}$$

It is assumed that image is LxL and $h_{p,i}$ is the (p,i) th element of H. On the other hand Hopfield network which is a dynamic neural network with feedback, after adjusting its initial conditions, reduces its internal energy and approaches its equilibrium point. This energy function is as follow (the network has L^2 nodes)[1] :

$$E = \frac{-1}{2} \sum_{i=1}^{L^2} \sum_{j=1}^{L^2} T_{ij} v_i v_j - \sum_{i=1}^{L^2} I_i v_i \tag{8}$$

T_{ij} and I_i are weights and biases of the network respectively. Comparing Eqs. (7) and (8) T_{ij} and I_i are obtained as :

$$T_{ij} = - \sum_{p=1}^{L^2} h_{p,i} h_{p,j} - \lambda \sum_{p=1}^{L^2} d_{p,i} d_{p,j} \tag{9}$$

$$I_i = \sum_{p=1}^{L^2} g_p h_{p,i} \tag{10}$$

Therefore using g and h, weights and biases of the network are obtained and g is used as the initial condition of the output of the network. After several iterations, estimates of f are obtained at the output.

Although using Hopfield network, good results are obtained sometimes it is advantagous to use a norm 1 criterion function[4]. In this work we use a simulated annealing technique for criterion function. This means that when the output error is small and we are near to convergence point a second order (p=2) criterion function is used. If the error is above a threshold, first order (p=1) criterion function is used.

The perceptron which is used in image restoration is shown in Figure 2. Input and output training patterns are H_i and g_i both availble. In adjusting weights of the network a special kind of back propagation error is used :

$$\hat{f}(k+1) = \hat{f}(k) - \mu \frac{\delta J}{\delta f} \mid \hat{f}(k) \tag{11}$$

There isn´t a close mathematical equation about the way each of the parameters of Eqs. (6) and (7) affect the convergence. A trial and error approach must be used to answer this problem[5].

RESULTS

In simulations a linear array with following parameters is considered :
- 32 receiving elements and a single transmitter which is located at the center of the array. Spacing between elements is 5 mm.
- Center frequency of transmitted pulse is assumed to be 2.5 MHz and a duration of about 3.5 cycles. In other words p(t) is assumed to be :

$$p(t) = \begin{cases} 0 & t < 0 \\ e^{-t/\tau} \sin(2\pi f_0 t) & t > 0 \end{cases} \tag{12}$$

where $\tau = 0.5 \ \mu s$.
- Only a single scatterer is considered. In other words the image is the PSF (point spread function) of the system.
- Images are formed in a 80 × 80 square grid with 0.25 mm pixels.

Figure 3 shows the PSF of the imaging system. In Figure 4 white noise with relative power of 10% is added to the image of Figure 3. If the technique introduced in this paper is used to restore this image, Figure 5 results. It is seen that the quality of the restored image has increased. Number of nonzero pixels has reduced in both axial and lateral directions, also ratio of the first side lode to the main lobe has reduced.

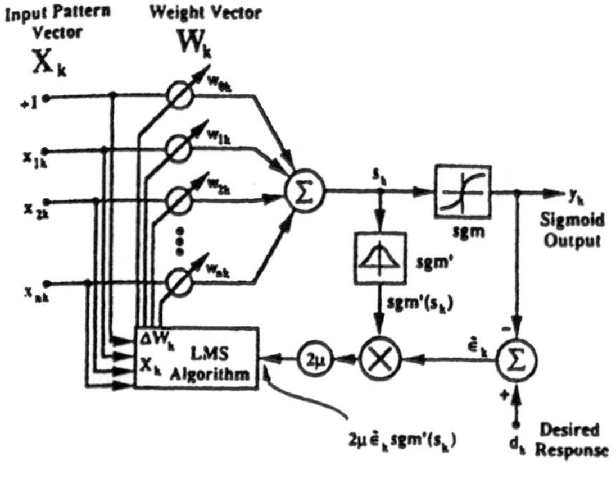

Figure 2. The perceptron

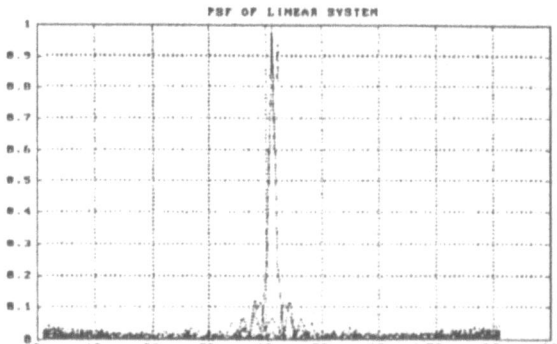

Figure 3. PSF of imaging system

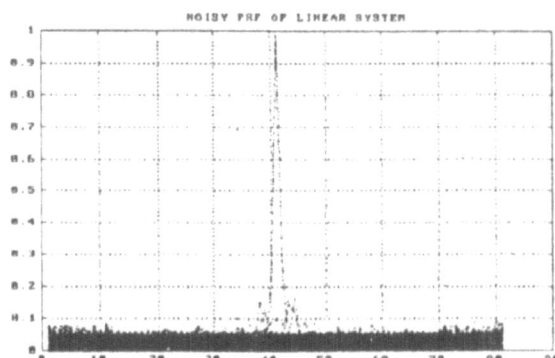

Figure 4. PSF after adding noise

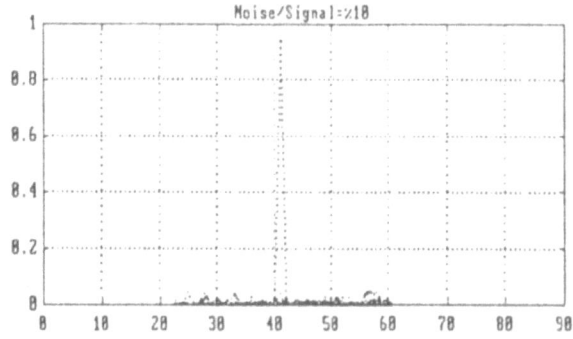

Figure 5. Restored image

Table 1 shows numerical values of these two criterions.

Table 1

	Number of nonzero pixels in axial direction	Ratio of first shde lobe to main lobe
Noisy image	80	0.78
Restored by perceptron	5	0.15

Table 2. Comparison of sensitivity to noise

Noise power	10%	20%	30%
Hopfield	0.626 (80)	0.28 (80)	0.75 (80)
Perceptron	0.15 (5)	0.16 (18)	0.37 (35)

Table 2 compares sensitivity of the results of perceptron and Hopfield network to the noise of the image. This table shows the ratio of the first side lobe to the main lobe and number of nonzero pixsels (numbers in parantheses) in axial direction. It is seen that technique proposed in this paper is less sensitive to noise.

Table 3 shows numerical values of lateral and axial resolutions for the noisy image and after restoration by Hopfield network and by perceptron. Numbers inside parentheses are resolution in axial direction. Resolution in each direction is computed as the width of a rectangle which its length is equal to the maximum amplitude of the image in that direction and its area is equal to the area under the cross section curve in that direction. This table shows that using neural processor image with higher quality and less sensitive to noise can be obtained.

Acknowledgments

The authors would like to express their sincere appreciation to Dr. M. Fatemi for reviewing the manuscript and making many constructive suggestions.

Table 3. Comparison of resolution

Noise power	10%	20%	30%
Noisy image	7.21 (5.15)	11.13 (8.89)	19.65 (17.81)
Hopfield	1.67 (2.13)	1.93 (2.48)	6.05 (9.15)
Perceptron	1.12 (1.07)	1.33 (1.3)	1.58 (1.7)

REFERENCES

1. Y.T. Zhou, R. Chellappa, S. Vaid and B. K. Jenkins, Image restoration using a neural network, IEEE Trans. on ASSP. 36 : 1141 (1988).

2. S.J. Norton, Adaptive imaging in aberrating media : a broadband algorithm, Ultrasonic Imaging. 14 : 300 (1992).

3. R.C. Gonzalez. "Digital Image Processing," Addison - Wesley (1989).

4. S.S. Kud and R. J. Mammone, Restoration by convex projections using adaptive constraints and the L1 norm, IEEE Trans. on Signal Processing. vol. 40, No. 1,(1992).

5. V.R. Nafisi, M. Fatemi and M. Hajivandi, Processing of images of limited diffraction beams by neural network, M.S. dissertation, Biomedical Eng. Dept., Amirkabir University of Technology, Nov. 1994, Tehran, Iran.

A HIGH RESOLUTION BATHYMETRIC SIDESCAN SONAR
USING DYNAMIC FOCUSING AND BEAM STEERING

François Ollivier, Pierre Alais, and Pierre Cervenka

Laboratoire de Mécanique Physique
Université Pierre et Marie Curie (Paris 6) - CNRS URA 879
2 Place de la Gare de Ceinture,
78210 Saint-Cyr-l'Ecole, France

INTRODUCTION

Investigating the sea-floor may be done through several different means:

- The simplest and oldest one is the sidescan sonar system which is made of two lateral transducers set on a fish towed by the surveying ship. These transducers deliver lateral beams whose shape is open in the vertical across-track plane and narrow in the axial direction, thus illuminating a narrow strip of the sea-floor. The reverberation information, properly decoded in distance from the echographic time of flight and the altitude of navigation, permits to obtain good images of the sea-floor when the local relief perturbations are well visualized due to the shadowing effect.

- The multibeam echosounder delivers the same kind of information using the same geometry for the illuminating beam and a transversal array for decoding the site angle associated to the incoming backscattered information, so that it is possible to obtain a topographic map of the sea-floor. After the *SeaBeam* developed in the mid 70's, several other systems have been conceived providing wider swaths comparable to the sidescan sonar potentiality. Most often, they have been conceived for large ranges up to 10 km and for the elaboration of topographic maps. However, some recent systems have been successful at shorter ranges (~ 1000 m). Although the along-track resolution remains poor, they deliver images similar to those of the sidescan system.[1]

- The bathymetric sidescan sonar system features an interferometric pair of receiving transducers on each side, allowing to derive the site angle of the incoming echoes through phase measurements. It competes with the multibeam echosounder. The essential objective remains access to a topographic map, with a relatively poor along-track resolution. Its advantages are cheapness and simplicity.[2]

- The synthetic aperture sonar, which is the underwater transposition of the synthetic aperture radar, has been investigated for more than 20 years. This system has very good potentialities but encounters a major difficulty, unknown with the radar version, caused by the erratic motion of the vehicle. New attempts have been made recently [3] and a bathymetric version should be studied.

But for the synthetic aperture sonar which is not yet operative, these systems are devoted either to sea-floor imaging with a good resolution or to the sea-floor topographic

Acoustical Imaging, Vol. 22, Edited by P. Tortoli
and L. Masotti, Plenum Press, New York, 1996

mapping with a lesser resolution. The system we propose here has been designed to provide the best echographic resolution and a topographic mapping capability. Instead of using unique transducers, as it is done for classical or bathymetric sidescan systems, we are using arrays which allow to control the focusing and the steering of the transmitted (or received) beam. The erratic perturbations in the attitude (essentially yaw and pitch) of the vehicle is compensated continuously. The along-track resolution is controlled and homogenized in an optimal way, hence solving the main problems encountered with imaging sidescan sonars. In addition, the use of two receiving arrays permits, through phase measurements of the focused information provided by each array, to obtain a bathymetric information that is more accurate than data delivered by classical bathymetric sidescan sonars. We had already shown the beginning of our work at the 20th ISAI [4]. Now the system has been completed and tested at sea.

THE ACOUSTICAL SYSTEM

We designed and built a 4 m long antenna whose operating frequency, 112 kHz, is well suited for the observation of the continental shelf. The maximal slant range is about 1000 m, and the far field angular resolution is 0.25°. Each of the two receiving arrays is made of 32 transducers at the pitch of 120 mm. It permits to steer the beam in a [-3°, +3°] angular interval which is large enough compared to the amplitude of the erratic attitude perturbations of a towed fish. The third array is used at transmit. It is made of 36 transducers at a pitch of 80 mm. This array is shorter (3 m long) than the receivers. It is not inconvenient because the footprint of the transmitted beam must overlap the beam pattern at receive. At transmit, the electronically steered array generates a beam whose conical shape illuminates a narrow strip on the seafloor, the mean line being an hyperbole. So, even if this curve is adjusted to come close the ideal straight line orthogonal to the fish track, the transmitted beam must be kept slightly diverging and much larger than the beam pattern at receive.

On the other hand, the electronic steering at receive may compensate exactly the yaw and pitch effects. The dynamic focusing which may be operated simultaneously insures a constant along-track resolution, at least in the near range image.

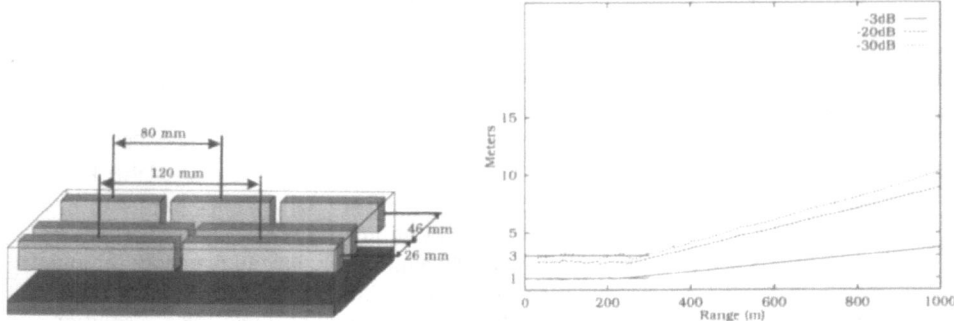

Figure 1. The elementary antenna module

Figure 2. Across-track resolution and side lobes level

Actually, the dynamic focusing may be correctly obtained by using 16 different focusing laws with adequate shaded apertures, as shown on Figure 2, where a constant -3 dB resolution of 1 m has been chosen up to a 300 m slant range. The -20 dB and -30 dB resolution performances are the result of carefully chosen weighted apertures. Of course, other choices for the constant resolution may be done, e.g. a 2 m resolution, up to 600 m in slant range.

It must be noticed that using different pitches for the transmitting array and the receiving arrays insures that grating lobes artifacts are canceled [4], and allows to operate in a large steering interval.

The antenna has been built by juxtaposition of 16 identical modules. Each module (Figure 1) includes 3 elementary emitters and 2 receivers for each receiving array. The spacing between the receiving arrays is equal to 26 mm, i.e. about 2 wavelengths.

SIGNAL PROCESSING

The 8 ms transmit signal is modulated both in amplitude and frequency. A linear frequency modulation is used with a bandwidth of 3 kHz centered at 112,5 kHz. At receive, a matched filter performs pulse compression. The amplitude of this signal is gaussian modulated in order to lower the temporal side lobes resulting from the pulse compression process. Due to the relatively low bandwidth of this signal, focusing and steering the beam, both at transmit and at receive, may be done just by phasing correctly the signals that drive each elementary transducer. This operation is digitally controlled with Programmable read-only Memories dedicated to each transmitting and receiving channel. The phasing of each receiving channel is obtained by heterodyning the amplified incoming echoes with adequately phased reference signals, at 108 kHz. The dedicated EPROMs permit in these conditions to steer the beam at transmit and at receive in 32 directions, with a pitch of .25°, i.e. in a [-4°, +4°] angular interval to control the divergence of the emitting beam by steps of 0.5° and to focus the receiving beam with 16 different focusing apertures in accordance with the theoretical predictions of.

The low frequency signal obtained after summing the 32 heterodyne signals is centered at 4 kHz and digitized at a 32 kHz sampling rate. Then, the matched filtering, equivalent to a correlation, is performed electronically in real time over 256 samples at 16 bits. After the compression step, the -3 dB across-track resolution is about 35 cm, i.e. significantly better than the expected along-track resolution. Hence, a higher compression ratio would be useless. Another reason for limiting the compression ratio is that the resulting signal is a 4 kHz amplitude modulated signal on which an ordinary phase measurement may be exerted to derive the interferometric function with an acceptable accuracy. A dedicated electronic circuitry provides in real time the mean amplitude and phase difference of the signals received by both arrays, averaged over 8 successive samples, hence deriving one complex data every 256 µs (i.e. a 4 kHz complex data stream). This information is sent (as 16 bits integers), together with a record of roll, to a PC for real time imaging and post-processing of the bathymetric information.

LABORATORY AND SEA TRIALS

In a first step, the system was tested with a 1 m long antenna (4 modules). The accuracy of the steering and focusing capabilities were checked in our 12 m long tank [4]. Then, the complete system was tested in the IFREMER tank (up to 40 m range). Figure 3 shows the theoretical and experimental echographic patterns obtained for 2 steering angles (0.25° and 1.25°). The selected aperture is one of the settings for obtaining the resolution shown on Figure 2 i.e. 1 m at -3 dB. It explains the rather poor angular resolution of the main lobe at this short range of 40 m. Theoretical and experimental shapes of the main lobe are remarkably matching, as well as the side lobes level.

Two sea trials have been led on board IFREMER ships. The first one, in November 93, permitted surveys of various areas in the "Rade de Brest". The antenna was fastened to the hull, oriented downwards 40° below the horizontal direction. Pitch and roll measurements were made by a 2 axis inclinometer, and yaw was obtained from an electronic magnetic compass. The steering angle was computed in real time with a dedicated electronics using read-only memories. The sea state was very rough so that the measurements of yaw were often not reliable because of the lack of stability of the compass.

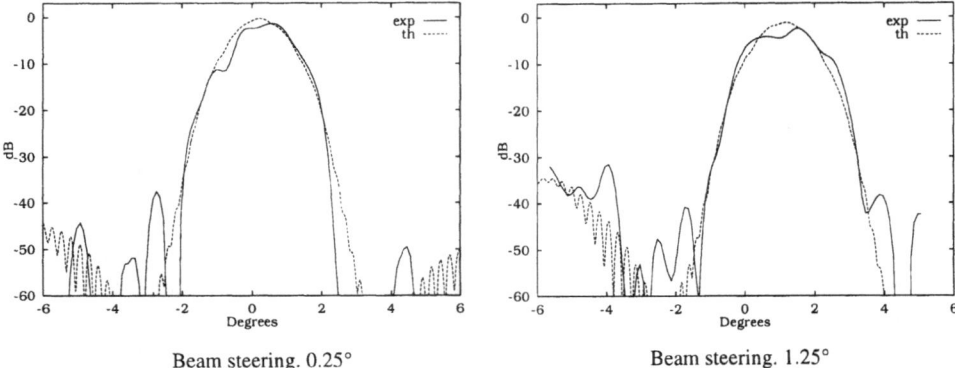

Beam steering. 0.25°   Beam steering. 1.25°

Figure 3. Experimental to theoretical echographic beam patterns at 40 m range.

Nevertheless, our one-sided sonar system succeeded in providing high resolution images for ranges exceeding 10 times the depth which remained less than 40 m for the different surveyed locations. Although this first trial was essentially dedicated to imaging, some inteferometric data were gathered and used to design an original on-line process. For now, this process provides, without loss of accuracy, amplitude and phase measurements with eight times less data to be recorded.

In June 94, a second sea trial took place off Toulon in the Mediterranean sea. There it was possible to investigate greater depths (up to more than 100 m) and more chaotic reliefs. Bathymetric capabilities were tested and high resolution imaging checked again. The antenna was again fastened to the hull with a 40° tilt angle, and yaw measurements were elaborated using a more accurate, reliable gyroscopic system than the previous magnetic compass. As will be pointed out in the next sections, this new bathymetric sidescan sonar system proved unmatched capabilities, despite the very unfair operating setup (e.g. hull support instead of a more stable platform, rough sailing conditions ...).

HIGH RESOLUTION IMAGES

The electronic system transmits on-line to a PC the samples of both the backscattered amplitudes and the complex samples derived from the interferometric process. The former are used on-line to display the scan line image in slant range coordinates. The latter are recorded for image post processing and for building the topographic maps.

An accurate bottom detection scheme is needed for processing the images, e.g. for building images with the flat bottom assumption. The operating frequency being around 110 kHz, the first bottom return has to be discriminated from parasitic echoes coming from fish shoals located at a closer distance of the antenna.. The processing technique is based on an analysis of the amplitudes only. We designed an adaptive algorithm which tracks the bottom, avoiding in most cases to be fooled by occasional clusters caused by the above mentioned artifacts. A median along-track filter is then applied in order to remove most of the remaining spikes. The sounding line is finally checked by superimposing the data and the image displayed in slant range coordinates.

Using the first bottom return information, images are corrected according to the classical flat across-track profile. During this process, a statistical gain compensation is performed to alleviate for the possible TVG deficiencies.

A few recorded pings were out of the range of the yaw-pitch compensation. The resulting images exhibit very dark lines that have to be corrected. For this purpose, a filtering technique based on the decomposition of each scan line amplitude with Chebyshev polynomials is applied [6].

Finally, a smooth histogram equalization is performed in order to get a better contrast, if needed.

Images 1 to 4 were obtained after such processing. The along-track versus across-track scale ratio is estimated from the navigation speed and the ping rate and knowing there is one slant-range sample every 0.19 m. In all these images, the ship is sailing upwards, its track being on the left side (the one-sided sonar system looks at right).

Image 1 shows somewhat ideal experimental targets, consisting of two concrete structures mounted on piles, known as the « Ducs d'Albe ». The water depth is maximum (18 m) at nadir. Isobath lines are running parallel to the ship-track, with the slope going upward from left to right. The horizontal size of the concrete blocks is about 20 m × 20 m. They emerge from the sea surface so that they induce infinite shadows behind their specular echoes. However, because of the basement mounted on piles, small areas located next to the right side of the blocks are insonified and can be seen. In the surroundings of these obvious targets, many small topographic details can be clearly noticed, such as a network of ripple marks at the top left, or sandy mounts and accretions that are formed near the piles by the strong currents prevailing in the area. Between the two block shadows, rocky structures are clearly visible in the far range where the grazing angle is very small. All these details confirm the expected high resolution of the system. In this image, the swath width is about 10 times the water depth at nadir.

In Image 2, the strip close to nadir is removed (about 100 m wide). The object is the wreck of a 70 m long coal cargo, the *Swansea*, laid on a flat, 25 m deep area, near the harbour of Brest. This image gives another evidence of the system high resolution. The thwarts of the ship are well outlined in both directions. The size of the shadows generated by the superstructures shows that their height is about 4 m over the bottom. Compared to the length of the ship, it gives an indication of either the bad state of the wreck preservation, or its deep stranding.

Image 3 has been collected in very shallow water (<15 m), near the Giens peninsula off Toulon. The area is well known for it has been already surveyed with various imaging or bathymetric systems. Therefore, this survey was intended to calibrate our interferometric setup. The backscatter image shows many qualities that have to be pointed out. It gives also clear evidences about the high resolution. For instance, the large black spot that is probably a posidonia bed, is crossed by thin trawling marks that are clearly visible. Many other small vegetal or sandy structures can be seen in other areas of this image. The most important remark concerns the swath width which is here more than 25 times the water depth. With a nearly flat bottom that is about 15 m deep, the backscattered amplitude is still relevant at a distance of 380 m, where the grazing angle is very small ($< 2°$).

Except for Image 2 that has been collected with a rough sea condition (heavy swell), it can be noticed that these images (including Image 4 that will be discussed later) bear no artifacts caused by yaw or pitch, thank to the efficiency of the attitude compensation.

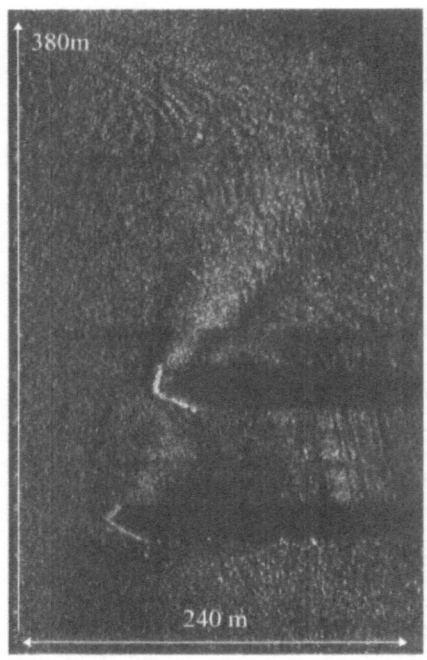

Image 1. Emerging concrete structures
built on piles in the " Rade de Brest "
Average depth at nadir around 18 m

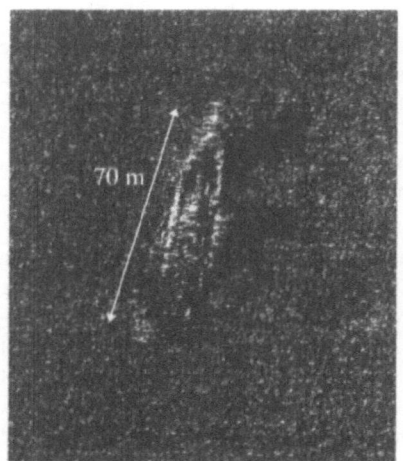

Image 2. Wreck of the coal cargo *Swansea*
in the " Rade de Brest "
Average depth at nadir around 25 m

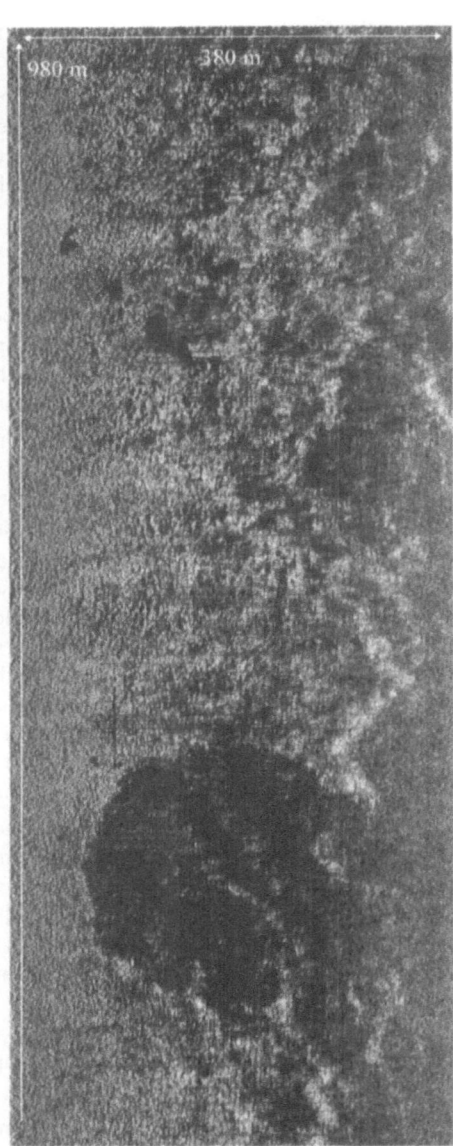

Image 3. Very shallow water area near
the Giens peninsula (off Toulon)
Average depth at nadir < 15 m

BATHYMETRY

The bathymetric capability is obtained by means of a classical doublet interferometer (i.e. the two receiving arrays). The original characteristic of this interferometer is the size of the baseline - two wavelengths. Other bathymetric sidescan sonars are usually based on doublets whose spacing is half a wavelength in order to avoid ambiguities in the interferometric phase measurement [5]. Some other systems use the vernier technique which requires a third receiver that is mounted at a larger distance of the doublet (about 10λ) [2]. Since the ambiguity problem can be solved, our system takes full advantage of the extended spacing of the interferometric doublet to produce accurate phase measurements.

The recorded complex interferometric signal undergoes first a low-pass filtering process : Each ping is convolved with a varying size shaded window. The length of the filter is proportional to the slant range. Figure 4 exhibits the phase of the signal that is derived after such a process.

Within the next step, phase angles are converted into geometric angles corresponding to the directions of the incoming echoes. This conversion requires to perform the calibration of the interferometer. For this purpose, we surveyed a large area whose relief is known to be reasonably flat, horizontal. From the few thousands of crossed records thus gathered with various headings, a statistical method enabled to build a calibration table.

The ambiguity problem arises since two or three directions lead to the same phase angle. We developed a reliable algorithm based on continuity property to unfold the phase values. The absolute direction is finally corrected by taking into account the roll angle.

A penultimate process consists of an along-track smoothing of these geometric angles using once more a shaded window whose length varies as a function of the slant range. Data are then selected according to the level of the associated backscatterred amplitude (threshold selection).

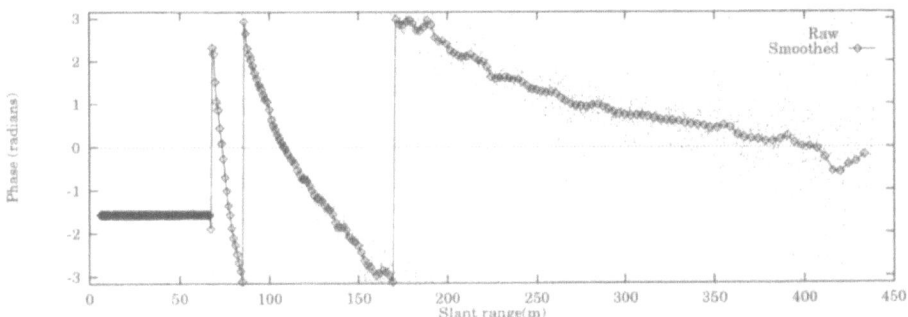

Figure 4. Example of a phase signal obtained from the interferometric signal related to one ping.

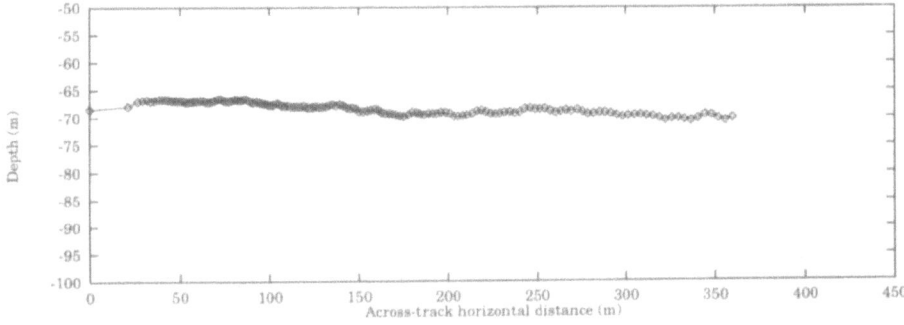

Figure 5. Depth profile derived from the phase signal shown on Figure 4.

Finally, soundings are derived by combining angles and slant ranges.

This processing scheme was applied to the smoothed phase signal shown on Figure 4.

The resulting depth profile is displayed in Figure 5 that calls for the following comments:

- The soundings in the nadir area are not numerous because of the fast variation of the incidence angle. All the existing bathymetric systems (side scan sonars systems as well as multibeam echosounders) encounter the same problem that occurs for imagery too.

- However, the most important remark concerns the across-track extent of the bathymetric capability. With an average depth around 70 m, consistent soundings are available beyond 350 m from nadir, i.e. five times the depth value with this single-sided system. The swath even reached 600% of the water depth in other shallower areas.

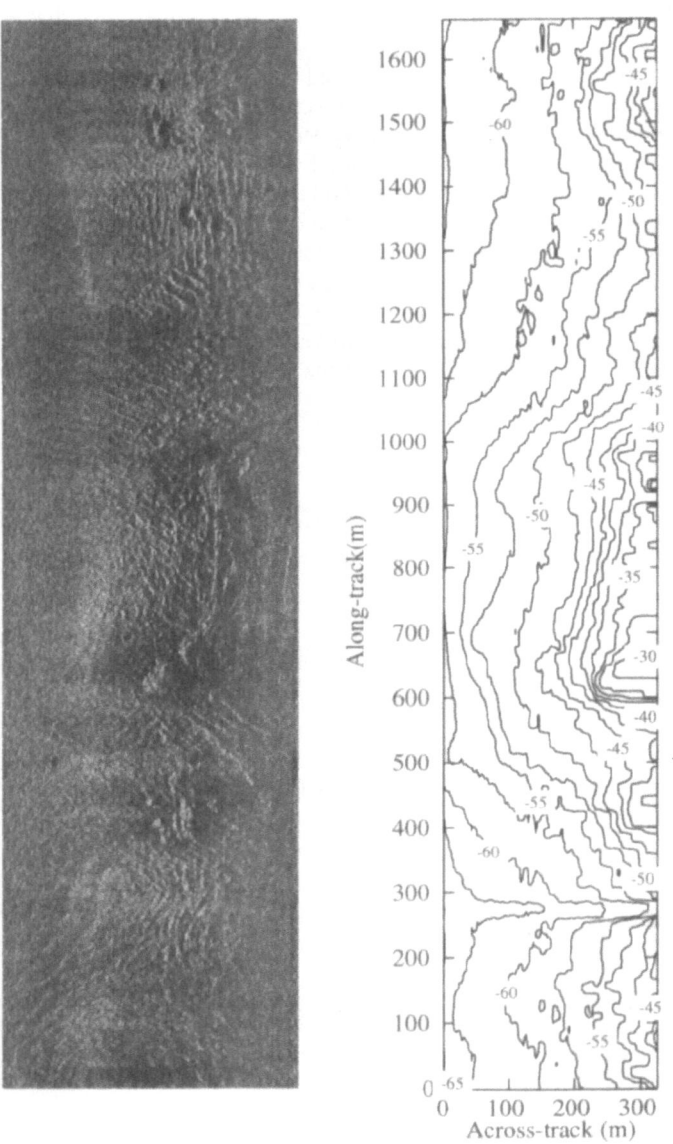

Image 4. Background surrounding the base of two islands, « Les Fourmigues », not far from Toulon.
Backscatter high resolution image (left) & Topographic contour map (right)

This is a very surprising performance and we do not know of any other bathymetric sidescan systems featuring such a large bathymetric swath [1][2].Valid data are usually gathered over an angular sector limited to 60° on each side. Starting from this angle and beyond, signals occurring from multipath specular reflections interfere with the backscatterred echoes so that phase measurements are no longer reliable. The outstanding capability of our system is probably linked to the combination of the high signal/noise ratio provided by the pulse compression technique, with the incoherent phase characteristics of the specular echoes that the large 2λ doublet spacing still amplifies.

Image 4 presents the high resolution image of the area running along the basement of two small islands, « les Fourmigues », off Toulon. Many details can be noticed and should be of interest for geologists. The rocky base of the islands is clearly outlined, as well as other rocky or sandy structures. The associated contour map covers a shorter distance (350 m instead of 450 m i.e. constantly more than 500% of the water depth). It uses an isobath line every 2.5 m for depth lying between -65 m and -30 m. It reveals topographic details that are not visible in the backscatter image, e.g. the across-track deep fissure that appears in the lower part of the image. The image and the contour map are well correlated with regard to the large scale features, i.e. the island bases.

CONCLUSION

The evidence has been brought that phased arrays increase in a significant way the performances of the side scan sonar imaging technique in terms of along track resolution as well as of parasitic motions compensation. The high quality of the images presented in this paper comes, on an other hand, from the use of the fast numerical adaptation of the pulse compression technique that gives access to high signal to noise ratios, thus leading to unusually wide swaths. Last but not least, the combined use of these techniques with a 2 wavelength spaced interferometric doublet constitutes a seemingly very wide swath echosounder. Nonetheless, these bathymetric results have to be compared to other systems results in order to be confirmed. We thank IFREMER for its support.

REFERENCES

[1] C. de Moustier « State of the art in swath bathymetry survey systems »
Int. Hydro. Rev **65**, 25-54, 1988.

[2] R.L. Cloet & C.R. Edwards « The Bathyscan precision swath sounder »
Proc MTS-IEEE Oceans 86 conf.1., 153-162, 1986.

[3] J. Chatillon, M.E. Bouhier & M.E. Zakharia « Synthetic aperture sonar for seabed imaging : Relative merits of narrow-band and wide-band approaches »
IEEE J. Ocean. Eng. **17** (1), 85-105, 1992.

[4] P. Alais, P. Challande, F. Ollivier & N. Cesbron « A new generation side scan sonar »
Proc. 20 th Int. Symp. Acoustical Imaging.

[5] P.N. Denbigh, « Swath bathymetry : principles of operation and analysis of errors »
IEEE J. Ocean. Eng. **14** (4), 289-298, 1989.

[6] P. Cervenka & C. de Moustier, « Sidescan sonar image processing techniques »
IEEE J. Ocean. Eng. **18** (2), 108-120, 1991.

THREE DIMENSIONAL INTERPRETATION OF SONAR IMAGE FOR FISHERIES RESEARCH

Kohji Iida[1], Tohru Mukai[1], Yoshinao Aoki[2] and
Tomoko Hayakawa[1]

[1]Department of Fisheries Science
Hokkaido University, Hakodate Japan

[2]Department of Information Engineering
Hokkaido University, Sapporo Japan

ABSTRACT

Three-dimensional interpretation of fish images from a sector scanning sonar to determine the shape and the distribution of fish schools was investigated. The sonar drove a half-circular cylindrical array transducer, which was mechanically rotatable on two axes. It emitted 2ms acoustic pulses of 160kHz in a half-circular plane omnidirectionally, then received echoes by scanning the insonified plane. In general use, the scanning plane is tilted a few degree downward from the surface, and the fish school echoes are indicated on a display as a PPI image around the ship. Therefore we can only observe fish echoes as a two-dimensional image, even if the scanning plane is tilted largely from the surface.

The new idea we discuss here is to reconstruct a three-dimensional image from hundreds of sectional sonar images using a computer image processing technique. In this method, the scanning plane is set vertically downward and perpendicular to the ship's course. Two hundred successive sectional images were digitized and the three-dimensional information was displayed as a top view projection and a side view projection along the ship's course. this method of data processing allows the shape and the distribution of fish schools and bottom features to be easily understood.

A fisheries survey using sonar was conducted near artificial fish reefs located at about one hundred meters depth. Many fish-school echoes were observed and the sonar images were analyzed. While the general fishery echo sounder provides information only for fish schools directly below a ship, this new method is a promising technique to obtain three-dimensional information from fisheries surveys.

INTRODUCTION

An acoustical wave is the most useful medium for underwater communication, because the propagation losses are much less than for any other media, including radio waves and optical waves, in water. Echo sounder have been used for a long time in order to find fish schools quickly and efficiently. Sonar which can help quickly locate fish schools near a ship is an indispensable tool for modern fishing.

Recently, for the purpose of conservation and effective utilization of fish resources, the scientific use of hydroacoustic equipment has become more important. Today, hundreds of scientific quantitative echo sounders are being used for fish abundance estimation and other scientific purposes throughout the world, because they can detect fish schools quickly with high resolution and over a wide range. Besides fish detection, sonar can provide other scientific information, e.g., the sea bottom features, the spatial distribution of fish, the school shape, and fish swimming behavior.[1]

In this paper, the acoustic tomography and the image processing techniques,[2] which are effectively used in medical diagnosis, are introduced to analyze the fish school echoes obtained from fishery sonar, and the potential for fisheries research is discussed.

PRINCIPLES OF FISHERY SONAR

Modern fishery sonar, called scanning sonar, uses an arrayed transducer to scan a beam electrically at high speed in a horizontal plane. There are two types of scanning sonar, namely, the omnidirectional scanning type, which fixes the cylindrically arrayed transducer mounted on the ship's hull, and the half-circular sector scanning type, which can control both the dip angle and the turning angle of the half-cylinder shaped arrayed transducer. Figure 1 explains the concept of a half-circular sector scanning sonar. In ordinary use, the scanning plane is kept near the horizontal plane, as in Fig. 1.a, and the forward half space is scanned and displayed as a half-circular PPI image. Also, for this type of sonar, the dip angle can range from 0 to 90 degrees. A cross-sectional image under the cruising course is obtained when the dip angle is set to 90 degrees as shown in Fig. 1b.

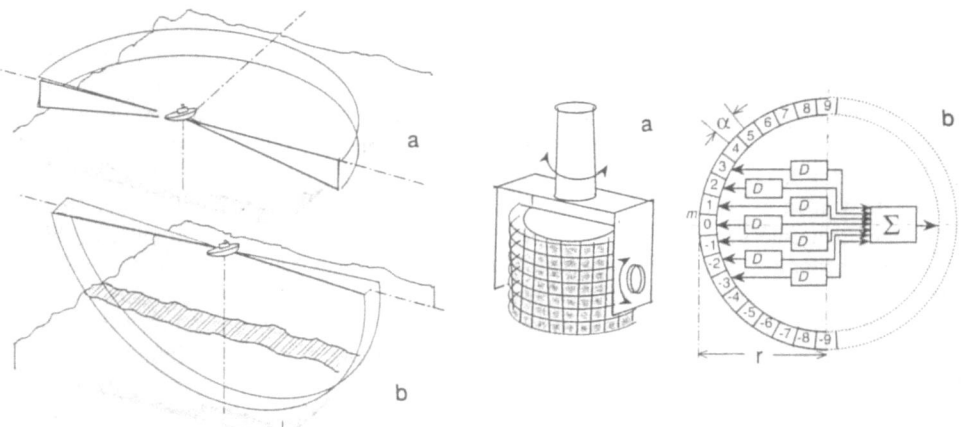

Fig.1 An omnidirectional scanning sonar and a half circular sector scanning sonar for fishing.

Fig.2 Construction of a transducer and a principle of beam forming of a half-circular sector scanning sonar.

Figure 2.a explains the principle of beam forming of a half-circular sector scanning sonar. The transducer is made by arraying the elements regularly on a half ring, and stacking the rings cylindrically. In the transmitting mode, all elements are driven simultaneously by a power amplifier, and the acoustic pulse is emitted omnidirectionally in a half plane. In the receiving mode, since each element has its own independent preamplifier, by composing the outputs of element with suitable phase and amplitude, a free directional pattern is formed. Figure 2.b shows a top view of an arrayed transducer. In order to form a beam pattern to the left direction of the figure, we can replace the elements electrically on the diameter of the ring perpendicularly to the beam direction. To realize such a replacement equivalently, delay lines are employed after each element and the outputs are summed by an adder. The delay time D_m is calculated from the following equation:

$$D_m = (r \cos m \alpha) / C \qquad (1)$$

$$(m=0, \pm 1, \pm 2, \pm 3, ..., \pm N/2),$$

where r is the radius of the cylinder, m is the order of the element originating on the beam axis, α is the sector angle between the neighboring elements, and c is the speed of sound in water. These delay line circuits can be slid along the circumference within a half circle of the transducer using analogue switches. An electrical beam scanning in a half-circular plane is then realized. A vertical beam pattern perpendicular to the scanning plane is determined by the height of the cylindrical transducer. Since both the tilt angle and turning angle are controllable mechanically, the acoustic beam can be directed to any direction in a half sphere space.

DIGITIZING AND PROCESSING OF SONAR IMAGE

The sonar observation was conducted with the R/V Ushio-maru of Hokkaido University near an area with artificial fish reefs at a depth of 100m. The ship's sonar was a half-circular sector scanning sonar, model KCH1827. The dip angle of the sonar was set to 90 degrees during the survey, and the cross-sectional images under the cruising course were displayed.

The vertical-section sonar images were digitized every 2 seconds with a resolution of 256 x 256 pixels with 8 bits using a video capture board of a personal computer. The digitized sonar images were processed to extract the edge of the fish school and the features of the bottom topography. The vertical-section image of a half-circular scanning sonar, originally is distorted depending on the directional resolving power. This distortion causes the sea bottom echo to thicken and bend with an increasing incident angle, as shown in Fig. 3. The true depth d is then obtained from following equation:

$$d = d' \cos(\phi - \theta/2) / \cos \phi, \qquad (2)$$

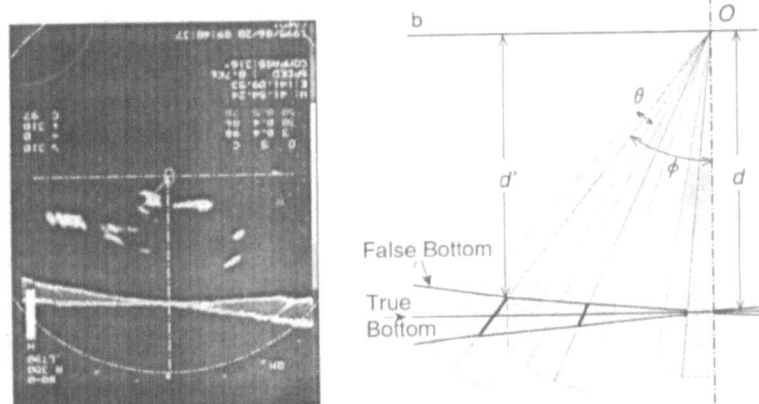

Fig. 3 Sectional image of sea bed and the distortion due to the directional resolution of sonar beam.

where d' is the false depth at the incident angle ϕ to the sea bed with a resolution angle θ of the sonar beam.

Next, successive sectional sonar images were stacked on a computer memory space as seen in Fig. 4, and a three-dimensional image was reconstructed. Each unit was composed of 100 to 200 frames, and was displayed as two perpendicular projection planes, namely the top view projection and the side view projection.

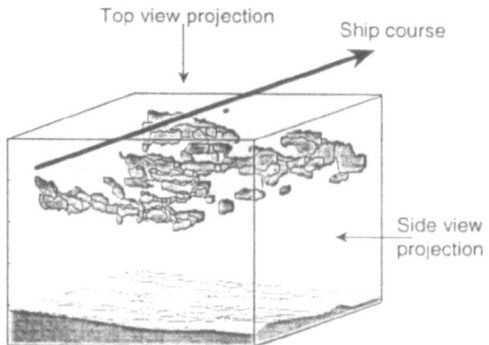

Fig. 4 Display method of three-dimensional sonar image by the two perpendicular projection planes.

SPATIAL DISTRIBUTION OF FISH SCHOOL

Figure 5 shows the spatial distribution of fish schools and sea bottom features from the top view projection and the side view projection. The upper chart of each figure indicates the top view projection, and the horizontal middle line represents the course of the ship. As the ship proceeds from left to right in the figure, the vertical axis over the track line indicates the horizontal distance to the left, and the lower vertical axis indicates the distance to the right.

The lower chart of each figure indicates the side view projection. The horizontal axis indicates the cruising distance, and the top horizontal frame of the figure represents the sea surface, and the vertical axis indicates the depth. These displays help us easily understand the spatial distribution of fish schools. For example, in Fig. 5 we can determine the fish school in the middle layer (A) is distributed to the left hand side of the course, and the fish school near the bottom (B) is located to the right side of the ship's course. Also, the fish school (C) is distributed symmetrically to the course, but the central part of school seems diffuse, suggesting that the fish school dispersed away from the wake of the ship.

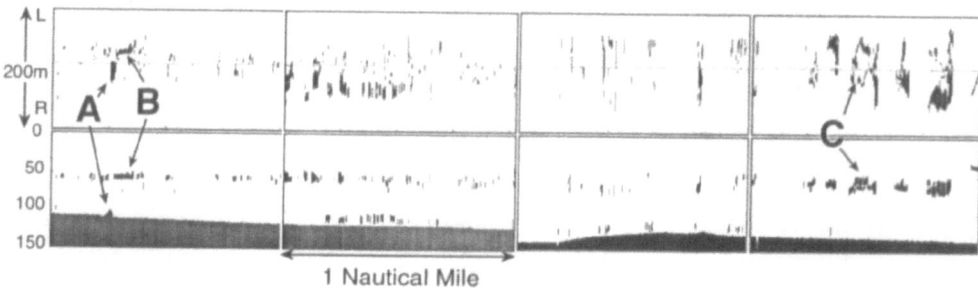

Fig. 5 Spatial distribution of fish schools by the top view projection and the side view projection.

THREE DIMENSIONAL SHAPE OF FISH SCHOOL

The three-dimensional analysis technique of sonar images is applicable to shape analysis of underwater objects. Figure 6 explains the concept of the solid-shape analysis method. As the vertical sectional images are sampled every 2 seconds, the distance between each section is 1 meter when the ship cruises at a speed of 10 knots. In order to determine the shape of a fish school, the outline of the fish school which was displayed on each sectional image was extracted and a scaled down replica of a fish school was made from polystyrene boards, as shown in Fig. 7. The same procedure was done using computer analysis, the projected shape of the fish school from various directions was analyzed.

In order to quantify the characteristics of fish school shape, the height (H), the length (L), the perimeter (P), and the area (A) of the fish school image were measured. The shape parameters, i.e., the elongation (E), the circularity (C), and the Fractal dimension (FD), were defined and calculated by the following equations:

$$E = L / H \qquad (3)$$
$$C = P^2 / 4 \pi A \qquad (4)$$
$$FD = 2 \ln(P/4) / \ln A \qquad (5)$$

Several types of fish schools and their shape parameters were shown in Fig. 8. The set of these three shape parameters of the fish schools had different values by season, by area, and by the depth at which the schools were observed.

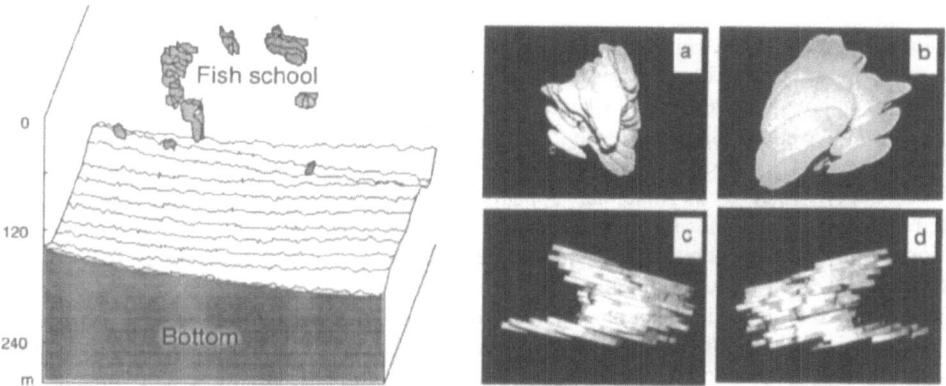

Fig. 6 Shape analysis method of fish school image.   Fig. 7 Model of fish school based on image processing.

MOVEMENT AND SWIMMING BEHAVIOR OF FISH SCHOOL

Three-dimensional analysis of sonar images is also applicable to analyze the swimming behavior of a fish school by tracking echoes. Scanning the sonar beam both electrically and mechanically around the target fish school, the swimming route of the school can be analyzed. Note that the movement of the fish school on the sonar display indicates the motion of the fish school relative to the movement of the ship. Hence the ship's motion is subtracted from the relative fish movement to determine the true movement of the fish school. This analysis is often delicate because the fish school moves much slower than the cruising ship.

Figure 9 shows the position changes of the target fish school on a sonar display when the ship approached the school. The three-dimensional coordinates were measured during the tracking by scanning the tilt angle. The tracked data show that the fish school moved across the ship's path, proceeding from left to right, and the school approached the surface when the ship approached closest to the school.

Type	Shape	Parameters	Type	Shape	Parameters
I		E : 1.300 C : 1.350 FD: 1.010	II		E : 2.222 C : 3.034 FD: 1.142
I		E : 0.808 C : 1.629 FD: 1.043	II		E : 1.800 C : 3.515 FD: 1.150
I		E : 1.885 C : 2.110 FD: 1.075	II		E : 1.545 C : 3.219 FD: 1.154
II		E : 1.211 C : 2.947 FD: 1.121	III		E : 1.389 C : 4.776 FD: 1.201
II		E : 1.033 C : 2.682 FD: 1.124	III		E : 0.857 C : 4.894 FD: 1.207

Fig. 8 Typical fish school shapes and the characteristic parameters.

Fig. 9 Swimming behavior of fish school by tracking the target school echo on a sonar display.

DISCUSSION

Today the most important application of acoustic techniques to fisheries research is for fish resource estimation. The scientific quantitative echo sounder, which was developed to measure the density of fish schools, quantifies the fish echo intensity in an acoustic beam directed downward vertically. However it only obtains density information from the narrow channels just under the ship's track lines. For more precise estimation of fish abundance, we have to set survey lines very close together. Unfortunately, there is often not enough time and money to survey enormous areas of the sea.

The three-dimensional shape of a fish school provides useful information for identifying the fish species. Some species form dens and patchy schools. A hydrodynamic discontinuity, such a thermocline, appears as a clear linear echo and zooplankton aggregations appear as large-scaled layer echoes. Therefore, besides the intensity information of a fish school echo, the characteristics of the shape parameters of fish echoes are powerful aids for identifying the object.

Lastly, the three-dimensional analysis of swimming behavior of a fish school is important, especially for fish resource surveys. If fish avoid the ship during the survey, the measured data may underestimate the actual stock abundance. There are several precautions can be taken to ensure more accurate estimates, e.g., slowing the cruising speed or selecting an appropriate survey season and survey time may help reduce fish abundance caused by underwater noise emitted by the ship. Also the knowledge of swimming direction and the speed of a fish school is indispensable for fish avoidance evaluation.

ACKNOWLEDGEMENTS

We thank to Dr. D.J. Hwang for his assistance in preparing the manuscript, and all the crew of the R/V Ushio-maru who helped conduct the experiment.

REFERENCES

1. O. A. Misund, A. Aglen, and E. Frønaes: Mapping the shape, size, and density of fish schools by echo integration and a high-resolution sonar. ICES J. mar. Sci., 52, 11-20(1995).
2. Y. Aoki, T. Sato, P. K. Zeng, and K. Iida. Three-Dimensional Display Technique for Fish-Finder with Fan-Shaped Multiple Beams., Acoustical Imaging, 18, 491-499(1991).

MUTUAL REFERENCE TECHNIQUE FOR
ARRAY CALIBRATION

Lewis Thompson and Hua Lee

Department of Electrical and Computer Engineering
University of California Santa Barbara, CA 93106–9560

ABSTRACT

The resolving capability of a sonar system is governed by several key parameters such as aperture size, operating frequency, bandwidth of transmitted pulse, and noise level. The characteristics of the receiving transducers of a multiple-element array are typically assumed to be ideal and identical. In practice, each receiving transducer is unique in terms of functions and responses. Accurate signal detection has become increasingly important as high-performance sonar imaging continues to emphasize resolving capability. As a result, array calibration now plays an important role in data acquisition as well as overall system performance.

In the past, array calibration was performed in a relatively primitive manner by using one of the channels as the reference signal. The error characteristics of the reference channel can propagate through the entire calibration process and result in degradation of overall system performance. In this paper, we present a systematic mutual reference calibration procedure technique for multiple-element sonar arrays. This method provides a simple and systematic calibration procedure with superior accuracy and computational efficiency. The accuracy of the calibration procedure is optimized by a singular value decomposition process and the complex correction vector is constructed from the most significant eigenvector.

The presentation of this paper includes a complete system modeling procedure, theoretical analysis, algorithm implementation, and results from full-scale experiments with a Sonatech SAS-10 sonar array.

Acoustical Imaging, Vol. 22, Edited by P. Tortoli
and L. Masotti, Plenum Press, New York, 1996

INTRODUCTION

Many modern imaging techniques for coherent sonar, such as synthetic aperture techniques, rely on well calibrated arrays capable of detecting very small phase differences across the array. Several techniques are presently available for calibration including the propagator operator[1], however this and related array shape estimation algorithms rely on the presence of strong targets or active sources and require a relatively high level of array calibration as a base. Other methods, such as the eigen structure approaches[2] try to simultaneously determine angle of approach and perform array calibration. This is optimized with respect to angle of approach algorithms and not the goals of coherent image formation. The calibration technique presented in this paper is designed to compliment the existing calibration techniques by improving the predeployment calibration of arrays. The motivation was to improve the acurracy of arrays used for image formation through techniques such as backward propagation[3] and phase or time-delay beamforming. These techniques are near field imaging algorithms that rely on the curvature of the wave front, making these techniques particularly sensitive to phase errors. Therefore, only data taken in a controlled environment will permit the calibration accuracy necessary for high resolution imaging.

A common model for the errors present in arrays is that each channel has a range and angle-independent delay and scaling error which is described by multiplication with a complex constant. This assumes a narrow band system where the delay can be modeled as a constant phase shift across the operating bandwidth. The wavefield is detected by the hydrophone and time samples are stored over some time segment of interest. The complete array model is shown in figure(1a). For pre-deployment calibration, a known signal is injected into each channel simultaneously, figure(1b). In the absence of errors, the time

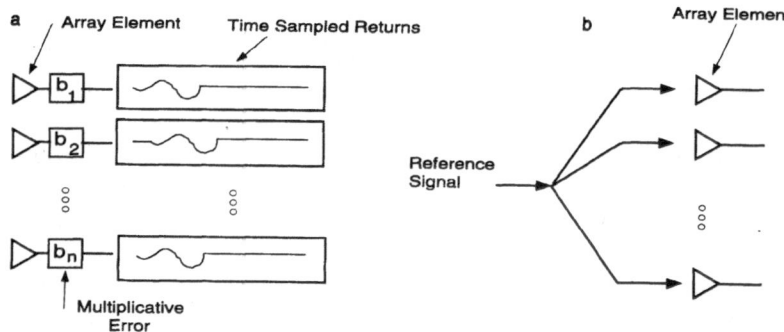

Figure 1 System Model (a) Complete Array, (b) Data Aquisition

sampled returns in each channel should be identical. The goal of all calibration techniques with this type of data is to identify the complex constants that modify the crosscorrelations such that they are as close as possible to the square of the signal energy. The techniques used today consider the crosscorrelations relative to only one arbitrary array element. This is susceptible to error caused by choosing a damaged channel. In addition, if only one channel is used as a reference, then the noise in that channel has a substantial effect on all calibration coefficients. By considering all channels simultaneously such complications can be avoided. The following section describes the conventional method[4], subsequently the Mutual Reference Technique is derived. Simulation and experimental results are then presented.

CONVENTIONAL SINGLE REFERENCE METHOD

Assuming the only difference between two channels in an array is a multiplicative complex constant of the form,

$$b_i = |b_i| e^{j \angle b_i} \tag{1}$$

the crosscorrelation of the time returns of each channel is used to estimate these constants. Let r_i be a vector of time returns from the i th channel of the array. The first step is to selected a reference channel in the array, typically one of the center-most channels is chosen because it has increased correlation to the other channels in the array. The auto correlation of the reference channel is calculated,

$$C_{ref,ref} = E\left[r_{ref}^H r_{ref}\right] \tag{2}$$

where ref refers to the position of the reference channel. The complex errors are relative, so setting $b_{ref} = 1$ and $r_{ref} = b_i r_i$ allows the crosscorrelation to be expressed in the form

$$
\begin{aligned}
C_{ref,i} &= E\left[r_{ref}^H r_i\right] \\
&= E\left[r_{ref}^H b_i^{-1} r_{ref}\right]
\end{aligned}
\tag{3}
$$

The multiplicative calibration error can be removed from the expectation,

$$
\begin{aligned}
C_{ref,i} &= b_i^{-1} E\left[r_{ref}^H r_{ref}\right] \\
&= b_i^{-1} C_{ref,ref} \Rightarrow b_i = \frac{C_{ref,ref}}{C_{ref,i}}
\end{aligned}
\tag{4}
$$

Assuming a high signal to noise ratio these multiplicative errors can be extracted.

MUTUAL REFERENCE TECHNIQUE THEORY

The Singular Value Decomposition(SVD) is a matrix factorization of the form

$$A = U\Sigma V^T \tag{5}$$

where $A \in C^{mxn}$, $U \in C^{mxm}$, $V \in C^{nxn}$, where C^{mxn} denotes the space of mxn complex valued matrices. The matrices U and V are mxm and nxn respectively, and Σ is a real, diagonal matrix with the square root of the eigenvalues of $A^H A$ in decreasing order on the diagonal. The columns of U and V are the eigenvectors of $A^H A$ and AA^H respectively. Note that if the matrix A is Hermitian (positive semi-definite) the SVD is simply the eigen decomposition of the matrix A.

The data from an n-element array can be represented in matrix form as

$$R = [r_1 \ r_2 \ .. \ r_n]^T \tag{6}$$

where r_i is an m length column vector. Assuming only the multiplicative complex error between array elements, the corrected data is expressed in matrix form as

$$\tilde{R} = B^{-1}R = \begin{bmatrix} b_1 & 0 & \cdots & 0 \\ 0 & b_2 & \ddots & \vdots \\ \vdots & \ddots & \ddots & 0 \\ 0 & \cdots & 0 & b_n \end{bmatrix}^{-1} R \qquad (7)$$

where b_i are the corresponding complex error coefficients. The expectation $E\left[\tilde{R}\tilde{R}^H\right]$ is then expanded in terms of the original data and the correction matrix B

$$\begin{aligned} E\left[\tilde{R}\tilde{R}^H\right] &= E\left[B^{-1}RR^H B^{-H}\right] \\ &= B^{-1}E\left[RR^H\right]B^{-H} \end{aligned} \qquad (8)$$

and the SVD of $E\left[RR^H\right]$ is then performed noting that in this case the matrix being decomposed is Hermitian, therefore U = V, leading to

$$E\left[\tilde{R}\tilde{R}^H\right] = B^{-1}U\Sigma U^H B^{-H} \qquad (9)$$

Under the assumption that the autocorrelation matrix of the original uncalibrated data is nearly rank one, or equivalently assuming that the ratio of the largest singular value to the second largest is much greater than one, we can approximate the above matrix by

$$E\left[\tilde{R}\tilde{R}^H\right] = B^{-1}[\mathbf{u_i} \, 0 \, 0 \, \cdots \, 0] \begin{bmatrix} \sigma_1 & 0 & \cdots & 0 \\ 0 & 0 & & \vdots \\ \vdots & & \ddots & \\ 0 & \cdots & & 0 \end{bmatrix} [\mathbf{u_i} \, 0 \, 0 \, \cdots \, 0]^H B^{-H} \qquad (10)$$

This is a reasonable assumption considering that the data from different channels of an uncalibrated array still share most of their characteristics. In addition, a further consequence of the data being corected, or equivalently perfectly calibrated, is that the normalized correlation matrix is

$$E\left[\tilde{R}\tilde{R}^H\right] = \begin{bmatrix} 1 & 1 & \cdots & 1 \\ 1 & 1 & & \vdots \\ \vdots & & \ddots & \\ 1 & \cdots & & 1 \end{bmatrix} = \begin{bmatrix} 1 & 0 & \cdots & 0 \\ 1 & 0 & & \vdots \\ \vdots & \vdots & \ddots & \\ 1 & 0 & \cdots & 0 \end{bmatrix} \begin{bmatrix} 1 & 1 & \cdots & 1 \\ 0 & 0 & \cdots & 0 \\ \vdots & & \ddots & \vdots \\ 0 & \cdots & & 0 \end{bmatrix} \qquad (11)$$

and by setting the factored version of Eq. (11) equal to Eq. (10) and it is possible to solve explicitly for B by setting the matrix square roots equal

$$\begin{bmatrix} \sqrt{\sigma_1} & 0 & \cdots & 0 \\ 0 & 0 & & \vdots \\ \vdots & & \ddots & \\ 0 & \cdots & & 0 \end{bmatrix} [\mathbf{u_1} \, 0 \, 0 \cdots 0]^H = \begin{bmatrix} b_1 & 0 & \cdots & 0 \\ 0 & b_2 & \ddots & \vdots \\ \vdots & \ddots & \ddots & 0 \\ 0 & \cdots & 0 & b_n \end{bmatrix} \begin{bmatrix} 1 & 1 & \cdots & 1 \\ 0 & 0 & \cdots & 0 \\ 0 & 0 & & \vdots \\ \vdots & & \ddots & \\ 0 & \cdots & & 0 \end{bmatrix} \qquad (12)$$

or equivalently in vector form

$$\sqrt{\sigma_1}\mathbf{u_1^H} = [b_1 \, b_2 \, .. \, b_n] \qquad (13)$$

which leads to the scalar equation

$$b_i = u_{1i}^* \sqrt{\sigma_1} \qquad (14)$$

These are then the corresponding complex multiplicative errors to be removed form each channel.

SIMULATIONS AND EXPERIMENTS

The effectiveness of the MRT and the conventional Single Reference Technique (SRT) were compared on both simulated and actual calibration data. The effectiveness of the algorithms is judged by their ability to properly identify the phase errors, a measure chosen to reflect the importance of phase calibration in nearfield processing.

This simulation was based on actual hardware calibration procedures where a single frequency was sent simultaneously into each channel of the array. Random phase and magnitude errors were added to each channel as was additive white gaussian noise. Figure 2 shows the variance of the difference between the actual phase error and the estimate in the presence of increasing noise power. The second demonstrates the same measure but instead of increasing the noise power the variance of the errors themselves is increased. In both cases the MRT performs significantly better than the SRT.

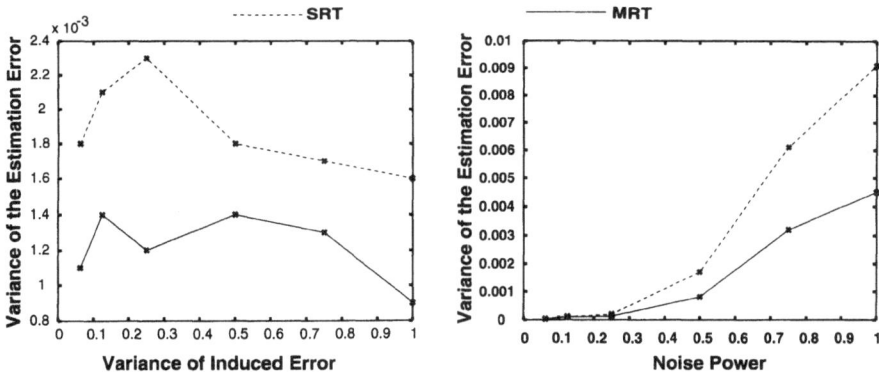

Figure 2 Results of simulations. Left: Variance of the estimation error vs. the variance of induced error with constant noise power, Right: Variance of the estimation error vs. power of additive white Guassian noise with constant induced error variance.

Before comparing the performance of the two calibration techniques with real data it is necessary to determine a reasonable measure of effectiveness. The measure used in this comparison is the variance of the cross correlations. This is a logical measure of the "flatness" of the cross correlation matrix that we wish to set equal to a constant. Once again only the variance of the phase of the cross-correlations is considered, although it is important to note that the magnitude performance was comparable.

The calibration data is from the Sonatech SLS10 sonar system, a ten element system with a 600 kHz operating frequency. These data were obtained by attaching leads directly to the array electronics and sending a 600 kHz sinusoid. Therefore these results do not include the effects of the acoustic transducers, however do compensate for the effects of samplers, time varying gain, and other non-linear effects in the electronics.

Table 1 Flatness measurements of several independent data sets.

Data Set	1	2	3	4	5	6
MRT Flatness	9.91	0.63	5.45	2.50	0.98	1.94
SRT Flatness	21.8	1.22	7.85	4.87	1.71	3.90
Flatness ratio	2.20	1.93	1.44	1.95	1.74	2.01

FUTURE WORK AND CONCLUSIONS

The Mutual Reference Technique is an effective new array calibration procedure. The performance of this algorithm has been verified by both simulation and experimentation. Future extensions of this work include range dependent calibration where the correction coefficients would be a function of array position and sample time. The final goal is wideband calibration where the correction factors would be channel dependent filters.

ACKNOWLEDGMENTS

This work is funded by the UC-MICRO program and Sonatech, Inc.

REFERENCES

1. S. Marcos, "Calibration of a distorted towed array using the propagation operator," *Journal Of the Acoustical Society of America*, vol.93:1987–94, April 1991.
2. G. C. Brown, J. H. McClellan, E. J. Holder, "Eigenstructure approach for array processing calibration with general phase and gain perturbations," *International Conference on Acoustics, Speech and Signal Processing*, vol.5:3037–40
3. Z. C. Lin, H. Lee, and G. Wade, "Back-and-forth propagation for diffraction tomography," *IEEE Transactions on Sonics and Ultrasonics*, SU-31:626–634, Mar. 1984.
4. B. L. Douglas, "A calibration technique for sidescan sonar systems based on the second order statistics of returned signals," *Proceedings of 1992 IEEE Oceans Conference*, vol–1:300–305, 1992.

SEA BOTTOM CHARACTERIZATION BY ACOUSTIC MEANS: OBLIQUE INCIDENCE

Panagiotis J. Papadakis and Bradley Carr

Foundation for Research and Technology-Hellas
Institute of Applied and Computational Mathematics
P.O.Box 1527, 711 10 Heraklion, Crete, Greece

INTRODUCTION

Sea experiments using acoustic measurements are an effective and feasible way for the characterization of the ocean bottom structure. Knowledge of the acoustic properties of the bottom (density and sound velocity profiles, absorption etc.) is necessary not only in the successful modeling of the direct problem of underwater acoustic propagation but to other sciences as well such as oceanography, geology and seismology.

In this work, the problem of the reconstruction of a bottom consisting from a sediment fluid layer over a semi-infinite fluid substrate using oblique plane waves incident upon the surface of the bottom, will be examined. The input data are the incident and reflected signal (in time) for two angles of incidence and the parameters recovered are the thickness of the sediment layer, the density and sound speed in each layer and the absorption (attenuation coefficient) of each material. In a previous work[1] the same problem was examined using normal incident. In that case only the impedance (the product of the sound speed by the density of each bottom layer) was recoverable. However normal incidence is not possible to have in case of a sea experiment with a towed array.

In the first part of the paper the model under examination and the mathematical formulation will be presented. Since the input data for the main part of the algorithm will be the reflection coefficient, the spectra of the incident and reflected signals obtained using Fast Fourier Transform are divided in order to provide the reflection coefficient as a complex number. In the second part the inverse method developed as well as postprocessing methods which improve the initial approximation to the acoustic parameters will be examined. They are based in the fact that the reflection coefficient is known for several frequency values which is feasible if the previous procedure is used.

Finally in the third part a test using synthetic data will be presented. In this test error free data were modified so that random errors were introduced in their values and the error induced values were used as input to the algorithm. Also a test with data obtained through a laboratory experiment is included. The results of such tests show that the inverse method is fast, reliable and accurate.

THE MATHEMATICAL FORMULATION

The model under consideration is shown in Figure 1. A plane wave is emitted incident on the bottom with incident angle θ_1. From the reflected signal the reflection coefficient is calculated for two different angles of incidence. The properties to be recovered are the densities (ρ_2, ρ_3), the sound speeds (c_2^*, c_3^*) and the attenuation coefficients (α_2, α_3) of the bottom layers, as well as the thickness d of the sediment layer. We assume that the density ρ_1 and sound speed c_1 of the water are known.

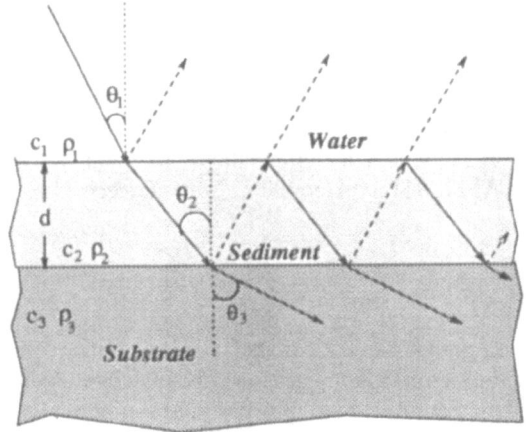

Figure 1. The environment

Actually in laboratory or sea experiments the emitted signal is a narrow or wide beam and not a plane wave. However if the source is far away from the bottom surface it can be considered as plane wave. A sample incident and reflected signal and their spectra obtained using Fast Fourier Transform are shown in Figures 2 and 3. Dividing then the reflected by the incident spectrum we obtain the reflection coefficient.

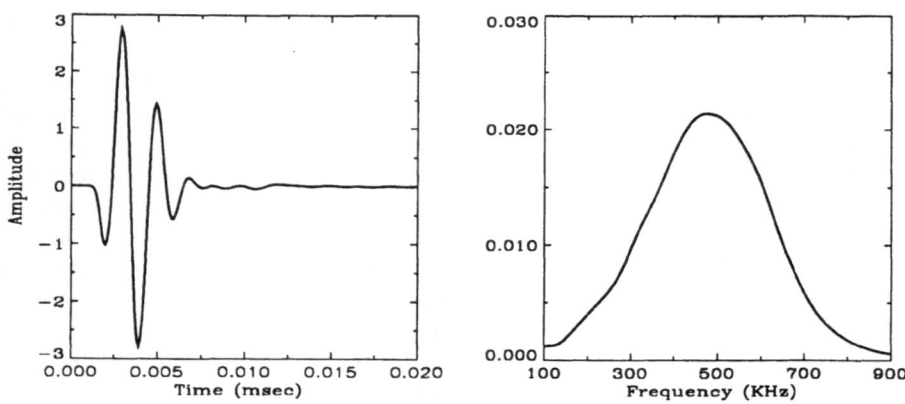

Figure 2. The incident signal and its spectrum

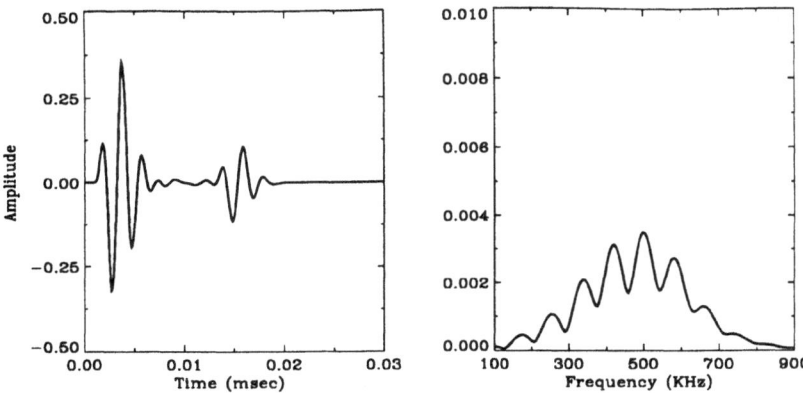

Figure 3. The reflected signal and its spectrum (angle $15°$)

REMARK: With this procedure the reflection coefficient is known for as many frequencies as we like by reducing the sampling rate for the Fast Fourier Transform. We are only limited by the width of the beam.

Under these assumptions the reflection coefficient is given by the following formula[2]:

$$R(f, \theta_1) = \frac{R_{12} + R_{23}e^{4\pi i z f}}{1 + R_{12} \cdot R_{23}e^{4\pi i z f}}, \qquad z = \frac{d}{c_2}\cos\theta_2 \tag{1}$$

where

$$R_{12}(\theta_1) = \frac{\rho_2 c_2 \cos\theta_1 - \rho_1 c_1 \cos\theta_2}{\rho_2 c_2 \cos\theta_1 + \rho_1 c_1 \cos\theta_2}, \qquad R_{23}(\theta_1) = \frac{\rho_3 c_3 \cos\theta_2 - \rho_2 c_2 \cos\theta_3}{\rho_3 c_3 \cos\theta_2 + \rho_2 c_2 \cos\theta_3} \tag{2}$$

and

$$c_2 = \frac{c_2^*}{1 + \alpha_2^2}(1 - i\alpha_2), \qquad c_3 = \frac{c_3^*}{1 + \alpha_3^2}(1 - i\alpha_3) \tag{3}$$

The angles θ_1, θ_2, θ_3 are those shown in Figure 1. Notice that the reflection coefficient is a complex number. This expression for the reflection coefficient represents the relation between the input for our algorithm and the quantities to be recovered and it will be used as the basis for our inverse procedure.

THE INVERSE METHOD

The main formulation

We will assume that the reflection coefficient $R(f, \theta_1)$ is given for the following frequencies:

$$I_j = R(f_0 + k\Delta f, \theta_1) \qquad j = 1, \ldots, 8 \tag{4}$$

$$k = j \text{ for } j = 1, \ldots, 5 \qquad k = j + 1 \text{ for } j = 6, 7 \qquad \text{and} \qquad k = j + 2 \text{ for } j = 8$$

We choose these 8 frequency values around the main frequency of the pulse. From equation (2) we notice that R_{12} and R_{23} do not depend on the frequency. Thus equating in pairs the 8 equations in (4) we eliminate the R_{12} and R_{23}. At the end of this algebraic procedure we have a third degree polynomial with complex coefficients of the form:

$$b_1 y^3 + b_2 y^2 + b_3 y + b_4 = 0 \qquad (5)$$

where b_j $j = 1, \ldots, 4$ are complex numbers which depend only on the known values I_j, Δf is the sampling rate of the Fast Fourier Transform, and y is the expression:

$$y = e^{4\pi i z \Delta f} \quad \text{where} \quad z = \frac{d}{c_2} \cos \theta_2$$

Thus it can be seen that the expression $z = \frac{d}{c_2} \cos \theta_2$ can be calculated, solving numerically the complex polynomial of equation (5).

REMARK: If the exact values of I_j were used in the inverse procedure then the roots of polynomial (5) are zero and two conjugate roots. Only one of the conjugate roots gives z in the first quadrant of the complex plane. This is the correct root. If the data I_j are error contaminated the roots are not any more conjugate but we choose the one that gives z in the first quadrant.

Knowing z we can then backtrack our steps to calculate R_{12} and R_{23}. Knowing these quantities for one angle of incidence, only the impedances (the product of the dencity and the sound velocity) of each bottom layer can then be calculated. So the same procedure is performed for another angle of incidence ϕ_1 in which case all the unknowns ρ, c, α and d can be computed separately.

Postprocessing

For exact input data the algorithm returns always the exact values for the unknowns. If the data are not error free, which will be the case in real experimental situations or during sea trials, the initial value of z may be incorrect. In this case the approximation is improved in the following way: An area around z in the complex plane is searched and for each search point the reflections coefficients for the frequency range under consideration is calculated and compared with the data in the following norm:

$$Norm = \frac{1}{2} \left[\frac{1}{N} \sqrt{\sum_{k=1}^{N} |I_k - J_k|^2} + \frac{1}{M} \sqrt{\sum_{l=1}^{M} |I_l^* - J_l^*|^2} \right]$$

where,

I_k, J_k $k = 1, \ldots, N$ are the data and the reflection coefficients calculated for a particular z for angle θ_1 and I_l^*, J_l^* $l = 1, \ldots, M$ are the same quantities for angle ϕ_1. Notice that N and M are greater than 8 since we know the reflection coefficient for many more frequencies than those used in the main procedure. The z that gives the minimum Norm is consider as the best z and is used subsequently for the calculation of the acoustic parameters. Several recursive calls as well as other safety measures ensure that the z obtained in this way is the best possible. The size of the area to be searched is either given by the user or the default size computed by the algorithm is used. This area is always included in the first quadrant of the complex plane, since z has positive real and imaginary parts.

REMARK: It can be seen that z is the time of flight of the plane wave inside the sediment and an accurate approximation can be obtained by observing the distance between the first and second return in the reflected signal. However this is not used in our method since there may be cases where the two returns partially overlap each other.

RESULTS

Synthetic Data

The algorithm was tested with several sets of data using synthetic data and performed satisfactory with all of them. One of these tests is presented here.

We assume a two-layered bottom with parameters given in the second column labeled *exact* in Table 1. The exact reflection coefficient I was calculated using the formula in equation (1) and for several frequency values in the range of 4-8kHz.

Then errors were introduced in the exact values I to obtain new values I^* in the following way

$$I^* = I + \frac{\alpha}{100} I \cdot [x]$$

where x is a random variable between -1 and 1.

Table 1. Results with synthetic data

	exact	3	6	12	24
			a		
ρ_2 (kgr/m^3)	1400	1394	1382	1368	1321
c_2 (m/sec)	1700	1691	1684	1671	1595
ρ_3 (kgr/m^3)	1950	1930	1918	1907	1870
c_3 (m/sec)	2100	2114	2055	2036	1940
d (m)	2.0	1.98	1.95	1.92	1.90
α_2 (dB/λ)	0.3	0.27	0.2	0.16	0.07
α_3 (dB/λ)	0.5	0.7	0.97	1.56	3.6
Norm		2×10^{-3}	6.6×10^{-3}	9.1×10^{-3}	7.6×10^{-2}

The values I^* were used in our inversion method and the results for several values of α are presented in Table 1. It can be seen that the results are acceptable with the exception of the results of the last column where the error was 24%. Also the approximation for the attenuation in the substrate is not good. This can be explained by the fact that the overall contribution of the attenuation in the substrate in the calculation of I is very small.

Experimental Data

A laboratory experiment using synthetic materials to model the ocean bottom was performed in Laboratoire de Mécanique et d'Acoustique CNRS France. The properties of the materials used are given in the first column of Table 2.

Table 2. Results with Experimental data

	Measured	Calculated
ρ_2 (kgr/m^3)	1200	1181
c_2 (m/sec)	1550	1619
ρ_3 (kgr/m^3)	1200	1275
c_3 (m/sec)	1760	1700
d (m)	0.009	0.0093
α_2 (dB/λ)	1.14	0.78
α_3 (dB/λ)	1.29	1.7
Norm	3.6×10^{-3}	1.4×10^{-3}

The incident and reflected signal used are presented in Figures 1 and 2. The resulting reflection coefficient is shown in Figure 4 as a solid line. The dotted line presents the theoretical reflection coefficient obtained using the values of column 1 of table 2. Our inverse procedure gave the results presented in column 2 of table 2 and shown as a dashed line in Figure 4. It can be seen that the calculated values give very accurate results.

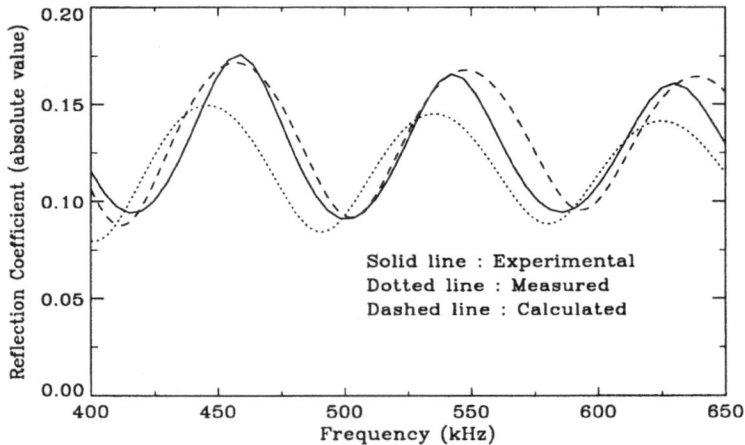

Figure 4. The experimental, measured, and computed reflection coefficient (angle $15°$)

CONCLUSIONS

The inverse method presented can be used if reflection coefficient data for two different angles of incidence are available. The properties of a two layered bottom consisting of fluid materials can be recovered as well as the thickness of the first layer.

Extensive testing showed that the algorithm performs satisfactory for the majority of situation even in the presence of errors in the data.

A new version of the algorithm where elastic properties of the bottom will be taken into consideration will be the next step to be examined.

REFERENCES

1. P.J. Papadakis, J.S. Papadakis, M.I. Taroudakis and M. Tsachnopoulou, Ocean bottom recognition from acoustic reflection data, *in*: "International Conference on Acoustic Sensing and Imaging, No 369" (1993)
2. L.Brekhovskikh "Waves In Layered Media", Springer Verlag (1982)

MATCHED FIELD OCEAN ACOUSTIC TOMOGRAPHY USING GENETIC ALGORITHMS

Michael I. Taroudakis[1,2] and Maria G. Markaki[2,3]

[1]Department of Mathematics,University of Crete
714 09 Heraklion, Crete, Greece
[2]Foundation for Research and Technology-Hellas
Institute of Applied and Computational Mathematics
P.O.Box 1527, 711 10 Heraklion, Crete, Greece
[3]Department of Oceanography, University of Athens

INTRODUCTION

One subject of Underwater Acoustics that is currently receiving much attention, is the solution of inverse problems, in which some (or all) of the ocean-acoustic parameters characterizing a channel (e.g., the sound speed profile in the water column, and the subbottom structure) are determined from measurements of the sound field at several points in the channel. Matched Field Processing (MFP) is an example of an inverse technique that relies heavily on numerical propagation codes. The matched field problem is not solved by direct inversion, but by exhaustively solving the forward problem (i.e. the wave equation with the appropriate boundary conditions) using a propagation model, and finding the set of parameter values which give the best match to the measured data, that is the complex pressures from a hydrophone array (Figure 1). Thus, MFP can be formulated as an optimization problem that seeks to maximize a correlation coefficient between the observed and predicted fields; this coefficient can be either linear, such as the Bartlett processor, or nonlinear, such as the Capon processor[1].

MFP has been most often used in Underwater Acoustics for determining the spatial coordinates of passive sources. Only recently MFP - for a known source location - has been applied to environmental inverse problems, where one seeks to determine environmental parameters. In this paper, we implement MFP on ocean acoustic tomography, the remote sensing problem of determining the sound speed in the water column by acoustic means. The recovered sound speed profiles are directly associated to the measurement or estimate of the physical parameters that affect the sound propagation in the water column, i.e., water temperature and density. The tomography methods that have been successfully applied so far, are based mainly on linear inversion and acoustic travel times measurements[2].

Travel-time tomography methods exploit just a fraction of the information available in the data (i.e., only ray or modal arrivals). Matched-field tomography exploits all the information in the acoustic field, using both phase and amplitude, in exchange for an intensive computer processing, since it is usually neccessary to calculate thousands of replica fields. Moreover, tomography methods based on linear inversion schemes require a good *a priori* estimate of the sound-speed distribution, which is often limited[2].

In this paper we will follow matched-field techniques established in Ocean Acoustic

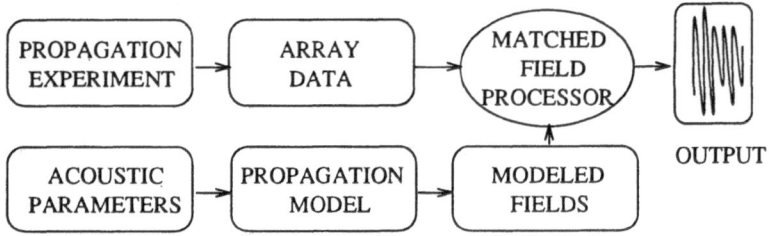

Figure 1. Basic components of ocean acoustic MFP.

Tomography by Tolstoy and others[3,4], to perform nonlinear inversion of a single sound speed profile (SSP), incorporating a genetic algorithm (GA) as an adaptive search technique to reduce the computational load[5,6,7].

GENETIC ALGORITHMS & MFP

A simple GA, as defined by Goldberg[7], proceeds in two stages: it starts with the current population of sambles from the search space; selection is applied, to create an intermediate population of above-average members; then crossover (that is, recombination) and mutation (random bit change) are applied to the intermediate population to create the next one. The process of going from the current population to the next, constitutes one generation in the execution of a GA.

GAs use performance-oriented feedback to bias the search process in the direction of better performance. This generally results in a more global search procedure than traditional methods but also means that the search time can be considerably longer. The overhead is quite evident on simpler problems but the payoff is clearly seen as the problem comblexity increases: because GAs maintain a population of sambles, they are much less sensitive to noise than other techniques and have been shown to be effective in the presence of hiqh noise levels without assuming any model of noise distribution prior to the learning. Also, the performance oriented GA search provides the ability to escape from local optima traps, resulting from limited environmental information[6,7].

In fact, matched field problem has been shown to be highly unstable to such effects as noise, errors in model parameters and the introduction of new data. The difference between the actual acoustic environment and the model used to compute replica fields is usually referred to as the mismatch problem. Generally, there have been two approaches to solving it: one is to construct a processor which is robust against a certain amount of mismatch; the other is to attempt to better estimate some (or all) of the uncertain parameters, by including them in the search space, along with any constraints known. The use of a propagation model that considers the 3-D complexity of the environment would provide additional uniqueness to the replicas and enable the discrimination between them[2].

APPROACH

Empirical Orthogonal Functions (EOFs) are used as the most efficient basis functions for an expansion of the sound-speed profiles[8]:

$$c(z) = \bar{c}(z) + \sum_{j=1}^{m} a_j f_j(z),$$ (1)

where the mean sound speed profile, $\bar{c}(z)$, and the EOFs, $f_j(z)$, are known functions of depth for the area considered (they are obtained by analysing hydrographic data from long

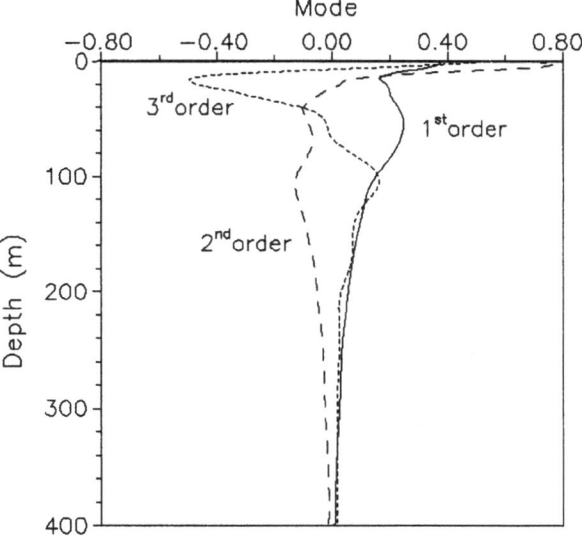

Figure 2. The Empirical Orthogonal Functions for the test case.

term observations). Therefore, only the EOFs' coefficients $a_j, j = 1, ..., m$ are to be determined. In well surveyed areas, a limited number of EOFs are adequate for an accurate representation of a sound speed profile. In the test case considered below, we use the first three of them, reducing the search space for the matched field processing (Figure 2).

A simple dot product, that is, Bartlett processor, is used to cross-correlate the measured with the modeled fields:

$$P(\vec{a}) = \mathbf{w}^+ \underline{C} \mathbf{w}, \tag{2}$$

where $\mathbf{w}(\vec{a})$ is a normalized prediction of the acoustic field on the array due to a sound speed profile with EOF coefficients $(a_1, a_2, a_3) = \vec{a}$ (the superscript $^+$ means conjugate transpose) and $\underline{C} = \langle \mathbf{FF}^+ \rangle$ is the cross-spectral data matrix. The brackets denote the averaged value of $\mathbf{FF}^+$, where $\mathbf{F}$ is the measured, normalized acoustic field at frequency ω and it contains noise as well as signal. For the simulation "data" considered here, noise is treated as Gaussian, white and uncorrelated (which can be interpreted as equipment noise) and enters the calculations by the addition of the quantity $N_o = 10^{-\frac{SNR}{10}}$ to the diagonal of $\underline{C}$, the estimated cross-spectral matrix (SNR is the average input signal-to-noise ratio). That is, for our simulations $\underline{C} = \mathbf{SS}^+ + N_o \underline{\mathbf{I}}_N$ where $\mathbf{S}$ is the simulated signal seen at the array (consisting of N hydrophones) and $\underline{\mathbf{I}}_N$ is the NxN identity matrix[1].

THE ALGORITHM

Usually, there are two main components of most GAs that are problem dependent: the problem encoding and the evaluation (object) function[7].

One widely used method of coding multiparameter optimization problems of real parameters, as it is here, is the concatenated, multiparameter, mapped binary coding. Each variable is mapped linearly from a specified interval $[Umin, Umax]$ to $[0, 2^l]$, and thus it is coded as a l-bit substring. The precision of the mapped coding is: $\pi = \frac{Umax - Umin}{2^l - 1}$ and it is assumed that the resolution provided is "visible" at the output.

The evaluation function must be relatively fast, since a GA works with a population of potential solutions and it incurs the cost of evaluating this population. The relative simplicity of the Bartlett processor was thus the main reason for its choice. We should point here that the normal mode model used, SNAP[9], can be parallelized for modes computed[10], reducing the evaluation time for each member of the population. GAs too, are intrinsically parallel; their implementation on parallel architectures that has recently received increased attention, could reduce the total evaluation time according to the population size and the number of processors used.

The selection process chosen here is *remainder stochastic sambling without replacement*, which minimizes stochastic errors in sampling. Selection pressure was introduced by linear fitness scaling, to prevent premature convergence to suboptimal solutions, as well as a meaningless wandering among mediocre solutions in the end[7].

Typically, population size, and crossover and mutation rates are interrelated. Their choice might be a difficult nonlinear problem by itself. We prefered to keep a small population size (50) to speed up the convergence rate, along with a 100% crossover rate, as recommended for the above mentioned selection procedure[7]; two-point crossover was performed. Since mutation is useful to keep a continuing "genetic" diversity but should be controlled too, we introduced its rate among parameters for iteration. The algorithm therefore modifies adaptively the mutation operator as the evolution proceeds, and the mutation propability varies between 0.1% and 4%. This would be interesting in the case of nonstationary object functions, where the fitness landscape changes over time.

In small populations, stochastic errors in sambling result in ultimate convergence on one "peak" or another without differential advantage, sometimes missing the true peak. To reduce the effects of these errors and enable stable subpopulations to form around each peak, we induce niche and "species" formation. A power-law sharing function, based on the relative bit difference, is used to define the neighborhood and degree of sharing for each string in the population. Similar strings derate one another's fitness values and limit the uncontrolled growth of particular species within a population. Speciation has been further exploited through methods of mating restriction. When "mating" occurs at random, high-performance but dissimilar parents tend to generate low-performance "offsprings" (lethals)[7]. To cause *similar* individuals (i.e., belonging to the same niche) to be recombined, various approaches have been examined; the algorithm used here chooses them according to their Eucleidian distance, i.e., it selects the "nearest-neighbours", as long as the average performance increases; otherwise, random recombination is performed temporarily.

THE TEST CASE

We apply the genetic algorithm to reconstruct a single sound speed profile in a range-independent, shallow water environment of 400m depth, with a semi-infinite bottom of density $1.5 g/cm^3$ and compressional speed of 1600 m/sec. Results of sound speed inversion using modal travel time tomography methods, that already existed[11], are used as a basis for the comparison of these methods to the matched field inversion.

We consider a vertical array with 20 sensors spanning the first 75m in depth, and a continuous wave (single frequency) source (200Hz) in a noisy environment. Since the sensors in the array are spaced half-wavelenth apart, to first order, noise may be modeled as being spatially white[1].

Inversion methods are generally formulated indepedently of forward modelling routines. Here the matching fields are computed using a range independent version of the normal mode model SNAP[9]. The same approach would be valid in a range-dependent environment for the reconstruction of the range averaged sound speed profile.

The convergence rate of the algorithm can be observed in Figures 3(a) and 3(b), showing the best-of-generation results versus number of generations for two different SNRs, 10 and -5dB respectively; they are the average of twenty independent runs using a different random seed. GA seems relatively insensitive to the decrease of SNR. Actually,

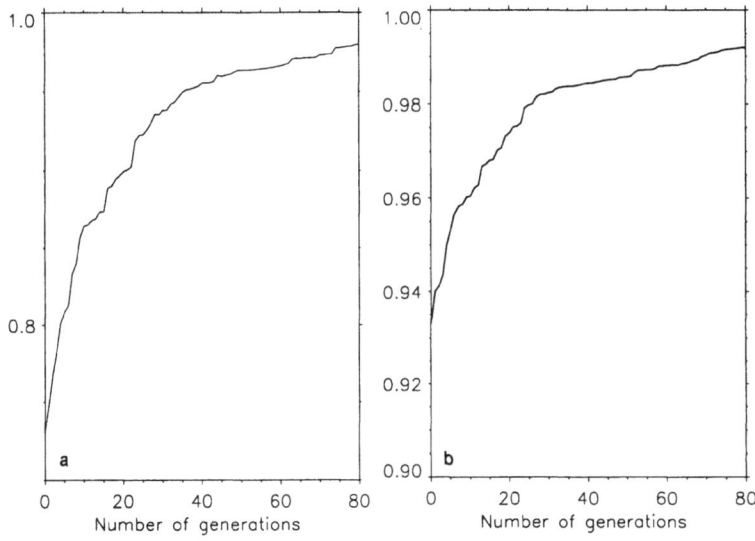

Figure 3. Bartlett optimization best-of-generation results averaged over 20 runs: (a) for SNR=10dB, (b) for SNR=-5dB.

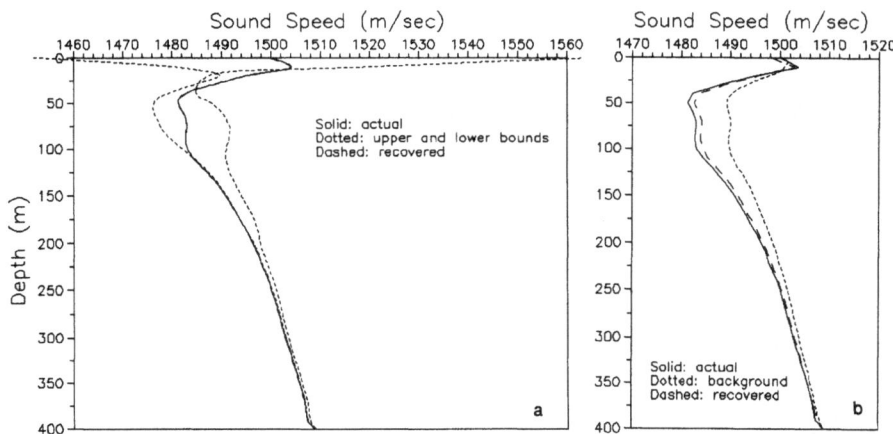

Figure 4. The actual and recovered sound speed profiles: (a) plus upper and lower bounds, for matched field inversion, (b) plus background profile, for modal travel time inversion.

in the case of white noise, the true peak always corresponds to the maximum value of Bartlett processor; increased noise levels simply produce large numbers of local optima that, instead of diverting the algorithm from further improvement, are exploited to improve performance, while the search for more global optima goes on[6]. Most of the CPU time is consumed in solving the forward problem for each model's evaluation; thus it is proportional to the total number of models evaluated, which can vary between 1000 and 4000 models, due to the different random seed used in every run of the algorithm.

Figure 4(a) shows actual and recovered SSPs, as well as upper and lower bounds of the search space, for the matched field inversion, where the recovered profile corresponds to the fittest of the models. Statistical analysis of the best models obtained from 20 trials gives a mean value of -44.859, 44.209 and -14.994, with a variance of 1.054, 1.167 and 0.842, corresponding to the EOF coefficients a_1, a_2, a_3, in the case of SNR=10dB, and a mean value of -44.555, 44.546 and -14.865, with a variance of 0.966, 1.331 and 1.352, respectively, for SNR=-5dB. For comparison, the true values of the EOF coefficients are -44.713, 44.438 and -14.892. Figure 4(b) shows modal inversion's results, and the background profile. Although the SSPs are well reconstructed in both cases, the improvement achieved by the MFP approach is evident if one takes into account that modal approach does not consider noisy simulated data, in this case; moreover, it requires a background profile that has to be a linear approximation of the actual one.

CONCLUSIONS

We have described a method for inverting sound speed profile in an ocean environment, using a matched-field processor and a genetic algorithm, in the context of noise and limited knowledge of the sound speed distribution. Implementation of GAs on parallel hardware and further investigation on the adaptive modification of the control parameters, could enhance their performance in the domain of ocean acoustics inverse problems, involving high dimensionality and multitudes of local optima. A further step, under consideration now, is the application of GAs in range-dependent problems of ocean acoustic tomography.

REFERENCES

1. A.Tolstoy."Matched Field Processing for Underwater Acoustics", World Scientific, Singapore (1993).
2. M.D.Collins and W.A.Kuperman, Inverse problems in ocean acoustics, *Inverse Problems* **10** 1023-1040 (1994).
3. A.Tolstoy, O.Diachoc, and L.N.Frazer, Acoustic tomography via matched field processing, *J.Acoust.Soc.Am.* **89** 1119-27 (1991).
4. A.Tolstoy, Linearization of the matched field processing approach to acoustic tomography, *J.Acoust.Soc.Am.* **91** 781-87 (1992).
5. P.Gerstoft, Inversion of seismoacoustic data using genetic algorithms and *a posteriori* propability distributions, *J.Acoust.Soc.Am* **95** 770-82 (1994).
6. J.Holland. "Adaptation in natural and artificial systems", M.I.T. Press, Massachusetts (1992).
7. D.E.Goldberg. "Genetic algorithms in Search, Optimization and Machine Learning", MA: Addison-Wesley, Reading (1989).
8. M.I.Taroudakis and J.S.Papadakis, A modal inversion scheme for ocean acoustic tomography, *J.Comput.Acous.* **4** 395-421 (1993).
9. F.B.Jensen and M.C.Ferla. "SNAP: The Saclantcen Normal-Mode Acoustic Propagation Model", Saclantcen Memorandum, SM-121 (1979).
10. E.Vavalis and C.Georgalakis. "Normal Mode Acoustic Propagation Models on Distributed/Parallel Computer Systems", F.O.R.T.H., I.A.C.M., No 94-9 (1994).
11. G.Athanassoulis, J.Papadakis, E.Skarsoulis, and M.Taroudakis, A comparative study of two wave-theoretic inversion schemes in ocean acoustic tomography, *in* : "Full Field Inversion Methods in Ocean and Seismo-Acoustics", O.Diachok et al., eds., Kluwer Academic Publishers, Dordrecht, pp. 127-132 (1994).

A 3D UNDERWATER ACOUSTIC CAMERA - PROPERTIES AND APPLICATIONS

R.K.Hansen and P.A.Andersen

OmniTech AS
Nedre Åsteveit 12
N-5083 Øvre Ervik
Bergen, Norway

INTRODUCTION

The offshore activities in the North Sea have motivated the development of a 3D acoustic camera. It is considered important to avoid the use of divers subsea and instead develop advanced remotely controlled machines and sensing devices. Also, water turbidity is high in many areas and the operation of conventional video is often restricted. Furthermore, conventional 2D sonars are not capable of providing a real time view of the volume in front of an underwater vehicle.

In order to provide satisfactory sensor information when conditions are near "zero visibility", OmniTech has, with the support of Statoil and the Research Council of Norway, developed a 3D acoustic camera called the EchoScope.

TECHNOLOGICAL BASIS

Image generation is carried out using spectral decomposition of the aperture field. This technique has been described in earlier papers by the authors[1,2] and is based upon theory from optics[3] and linear algebra[4,5]. The method is based on a finite number of hydrophones and detection of the quadrature components of the acoustic field at each hydrophone. The acoustic signal is a monofrequency burst. For a range of distances and a set of frequencies, the inverse matrices are computed in advance and stored in the processing unit. The images are generated by simple complex matrix multiplications.

The processing method do not require paraxial approximations and the inverse propagation generates an image on a curved surface with the camera in the centre.

Separate detectors and transmitters are used. The detectors are arranged in an array of 40 by 40 hydrophones.

Up to 150 kHz, no aliasing occurs since the detector to detector distance is half a wavelength. At reduced fields of view, imaging at 300 kHz and 600 kHz is also possible, and switching between these frequencies is done in software.

Data are acquired in a parallel set of detector channels, digitized and transferred to the surface based processor via optical fibre.

Figure 1 shows the underwater camera housing which is about 19 litres. The topside processing electronics is housed in a standard PC enclosure.

Figure 1. The EchoScope Underwater Housing.

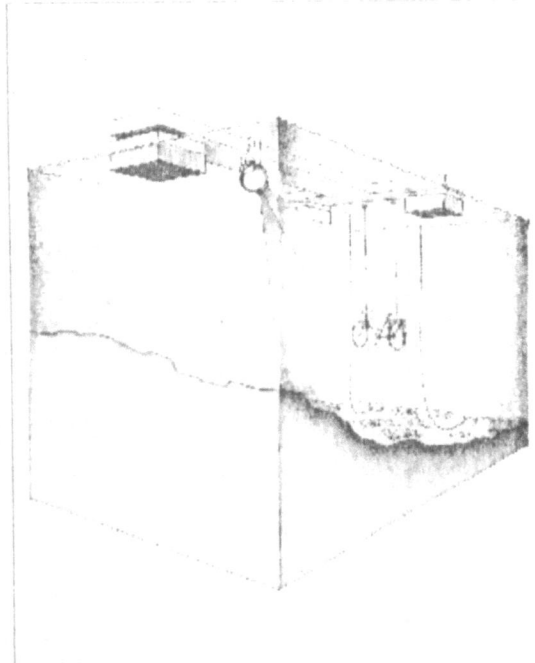

Figure 2. The Test Arrangement.

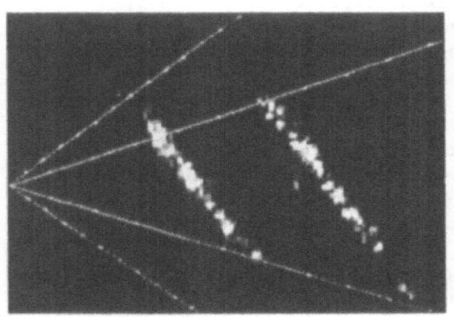

Figure 3. Image of the Anchor Chains at 600 kHz.

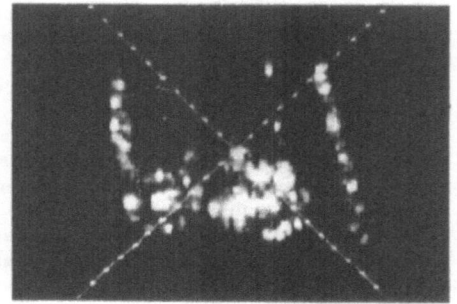

Figure 4. Image of a Bicycle Hanging between the Anchor Chains.

EXAMPLE

A prototype of the camera has been tested in shallow water and some images are presented here. The camera display shows real time colour images and black and white still photos attached do not provide nearly as good images. However, figures 2-7 give an indication of the type of information available from the camera in some typical application areas.

Figure 2 shows the arrangement for some demo tests which have been carried out. The camera is placed under a floating structure headed downwards in the direction of two anchor chains. Figure 3 shows a 600 kHz image of the anchor chains with a viewing angle of 25°. The straight lines enclose the viewing volume. In order to reveal the orientation of the anchor chains, the scene has been slightly rotated - this is one of the great advantages with 3D acoustics. Figure 4 shows the frontal image when a bicycle has been lowered between the anchor chains. The data has been processed with a rectangular window which maximizes resolution. If sidelobe suppression is required at the cost of resolution, software selectable windows are available as an option.

Underwater leakages from gas pipelines near offshore production platforms or near land are crucial threats not only to the environment, but also to personnel safety. As a detection system for underwater gas leakages, the underwater camera is able to provide images of the gas plumes indicating shape, orientation, leak rate and location. Other methods like 2D sonar can not reveal the shape of the plume, consequently, false alarms due to divers, underwater vehicles or fish occur easily. Figures 5 and 6 show some images of air escaping from a diver - again between the before mentioned anchor chains. The frequency is 300 kHz and the viewing angle is 50°.

Imaging of the sea bottom is relevant to the production of underwater maps. The acoustic camera is able to produce sea bottom images with a distance resolution of 5 cm at 600 kHz. Figure 7 shows the image of the sea bottom with 600 kHz, distance resolution 7 cm and distance 4.5 m. The viewing angle is 25°. Here, the distance has been colour coded and the closest pixels show up as nearly white reflectors in the black and white image. The slightly curved vertical object is a 35 mm cable laying on the sea bed. The image is based on a single acoustic pulse, thus movement of the camera is practically irrelevant to the internal structure of the image.

APPLICATIONS

3D acoustics has important application areas as an instrument for navigation and positioning in turbid water. Also, since the effective range is more than 100 m, it will be a useful tool even under normal conditions with respect to optical visibility. Thus it will de used both for seafloor mapping as well as for close range work operations.

3D acoustics also has interesting perspectives for monitoring of the underwater world - for example as a detector of underwater gas leakages.

The fisheries is another market segment for 3D acoustics and development is under way in order to make a camera available as a tool for fishing vessels in the detection and tracking of fish schools.

TECHNICAL PERSPECTIVES

3D acoustics benefits from the development of faster processors and better operating systems as well as effective software tools for 3D data presentation. The technical perspectives for the acoustic camera are :

Rate : 500 images per second. One image represents one distance segment. With 100 distance segments per image and 20 cm distance resolution, each 3D image covers a distance range of 20 m. The 3D image is the updated 5 times per second.

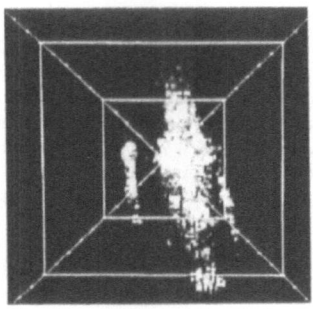

Figure 5. Image of Air Escaping from a Diver
Frequency 300 kHz.

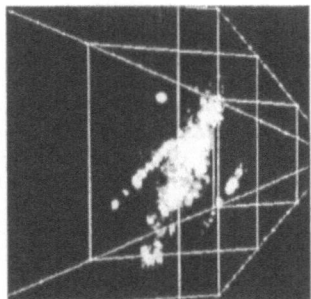

Figure 6. The Same Image as in Figure 5,
but the Observation Volume has
been Rotated.

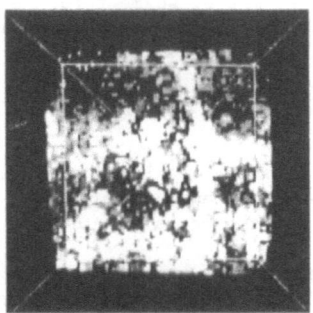

Figure 7. Image of the Sea Bottom at 4.5m Distance.
600 kHz, Viewing Anghle 25° and Distance
Resolution 5 cm.

Frequencies of operation :   150 kHz, 300 kHz or 600 kHz

Resolution :   150 kHz : 2.8°
                      300 kHz : 1.4°
                      600 kHz : 0.7°

Viewing angle :   150 kHz : 90° or 45°
                      300 kHz : 50° or 25°
                      600 kHz : 25° or 12.5°

Operating distance :   From 0.5 m to more than 100 m

Distance resolution :   5 cm (600 kHz), 10 cm, 20 cm or 40 cm

ACKNOWLEDGEMENTS

The authors would like to thank Statoil and the Norwegian Research Fund for supporting the development.

REFERENCE

1. R.K.Hansen. Acoustical imaging using spectral decomposition of the aperture field, *In* :
    Acoustical Imaging, Volume 19, H.Ermert and H.-P-Harjes, ed., New York (1992), pp.103-107.

2. R.K.Hansen, P.A.Andersen. 3D acoustic camera for underwater imaging. *In* :
    Acoustical Imaging, Volume 20, Y.Wei and B.Gu, ed., New York (1993), pp.723-727.

3. J.W.Goodman."Introduction to Fourier Optics", McGraw Hill, New York (1968).

4. D.G.Luenberger. "Optimization by Vector Space Methods", Wiley, New York (1969).

5. G.Strang. "Linear Algebra and its Applications", Harcourt Brace Jovanovich, Orlando (1980).

ACOUSTIC CAVITATION-BASED TRANSDUCER FOR SUB-SEA BOTTOM STRUCTURE IMAGING

Giovanni B. Cannelli, Enrico D'Ottavi, Luca Pitolli and Giorgio Pontuale

CNR-Istituto di Acustica "O. M. Corbino"
Via Cassia 1216, 00189 Roma, Italy

INTRODUCTION

Many acoustic-wave devices are currently in use to explore sea subbottom, especially in marine geology where a large number of researches aim at obtaining information on the nature and distribution of sediments. In this type of research, two requirements for the acoustic pulse are normally needed, which are often antithetic: high resolution to resolve also small-dimension subsurface anomalies and penetration capability for deeper prospecting. Current available devices, also including the most recent parametric systems, have not both these requirements. So, they don't allow utilization in a wide range of different experimental conditions and environmental situations.[1]

A compromise between resolution and penetration has been achieved in a prototype of broad-band, high-power electroacoustic paraboloidal transducer developed at Istituto di Acustica "O.M. Corbino"(IDAC).[2] This transducer is based on sparker source technology. In comparison with traditional sparker devices, the paraboloidal source is characterized by a more controllable electronics and mechanics which makes it more versatile. Generally, the sparker source generates a primary pulse followed by one or more cavitation bubble pulses which appear randomly in time in a commercial sparker-array. Instead, in the IDAC source the pulse can be electrically controlled in time and the secondary pulses can be minimized by means of a sophisticated paraboloidal-source tuned-array.[3]
As an alternative, an improved electromechanical technique can be used which allows the acoustic energy to be concentrated in a unique pulse, generated in only one transducer by the cavitation bubble phenomenon.[4,5] The presentation of this new tranducer is just the purpose of the present paper.

PRELIMINARY LABORATORY TESTS

The geometry and spark-gap distance of the electrodes have a funtamental role in determining the performance of sparker-based source, once the other electrical, physical and chemical parameters are fixed. So, preliminary measurements were made of voltage at bank-capacitor terminals, and acoustical pressure as a function of time, for different values of spark-gap, for a given geometry of the electrodes. Tests were carried out both for a simple sparker and a paraboloidal sparker-based source. The experimental apparatus is the same used in a previous work.[6] Namely, a laboratory tank having dimensions 1m x 0.64 m with a depth of 0.66 m, where the relative position and depth of the source and receiver are adjustable by means of suitable scanning and turning devices. When using a paraboloidal source, it is set in a horizontal position on the same axis of a hydrophone, and an electrode setting allows the spark gap to be varied around focus. The following experimental conditions were utilized for both transducers: initial voltage = 2250 V, capacitance = 360 μF (corresponding to a discharge energy of 911 Joule), temperature = 19 °C, salt content of the water = 3.4% by weight. Figure 1. shows the results for a simple sparker having hemispherical-shaped electrodes with radius of 5 mm. It can be noted that, after closing the electrical circuit, the voltage, V_C, starts decreasing exponentially and keeps fairly monotonic for large gaps (as far as 24 mm, for the first curve). Actually, for the first two cases, it seems that electrical current across the gap is essentially used up to heat and vaporize the water, rather than creating plasma. In any case, a bubble is generated which collapses after about 7 ms and causes a remarkable pressure pulse. The other curves corresponding to spark-gaps, d, having decreasing dimensions of 15, 12, 9.0, and 5 mm, show a voltage which abruptly decreases at the istant, t_B, of electrical breakdown and at the same time a primary pressure signal also appears. In these latter cases, the distance, d, of the electrodes is close enough that the electrical field becomes so intense as to generate a discharge and a consequent plasma. For spark-gaps smaller than 5.5 mm, the voltage, after the breakdown time, exhibits an oscillating damped behaviour which is typical of the discharge of a capacitor C in a circuit with resistance R and inductance L in series. For the purpose of the present work it is important to gain a better understanding of the genesis of the vapour bubble formation which affects the subsequent phase of bubble growth and collapse. So, a first step of this research is to study the vapour bubble by acoustic detection of the collapse and observing the sequence of pulses as a function of time.The behaviour of the acoustic pressure pulse as a function of time is shown in the sequence of signal signatures of Fig.2 corresponding to a paraboloidal sparker-based source having focal length of 3 cm and base diameter of 13 cm, in the same experimental conditions of the previous test.

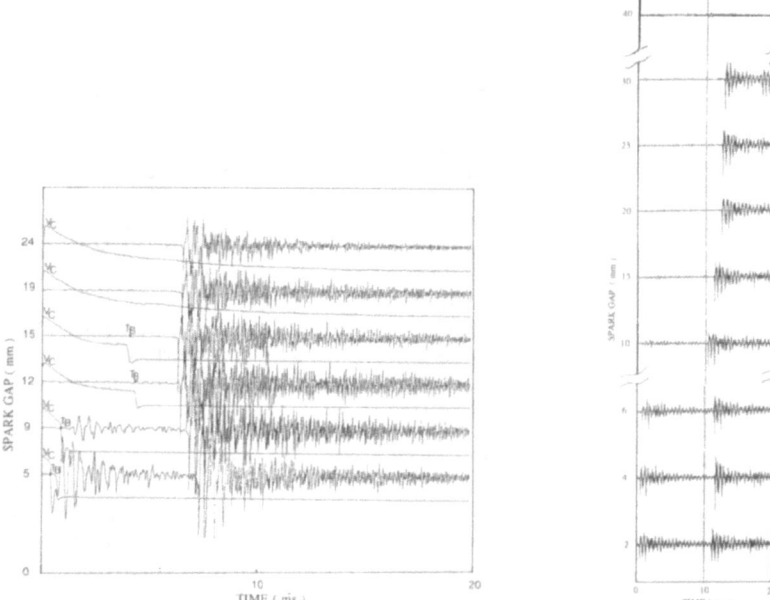

Figure 1. Behaviour of voltage at bank-capacitor terminals, and acoustical pressure as a function of time, for different values of spark-gap in a sparker

Figure 2 Behaviour of acoustical pressure pulse as a function of time, for different values of spark gap in the paraboloidal reflector

In these measurements, wider spark gap is considered, and no appreciable signal appears for d ≥40 mm. While increasing amplitude of pressure pulses can be seeen as the spark gap decreases, as far as a spark gap of 25 mm, for which a maximum is attained, then the amplitude decreases and reach a minimum for d = 10 mm. Also, it can be noted that a unique pulse is obtained only for spark gap distances not smaller 10 mm, and greater pulse amplitudes occur at longer times. Of course, an anomalous growth of the bubble occurs when breakdown is present, as it is evident for spark gap smaller than 6 mm, for which a regular growth of a unique vapour bubble is interrupted.

THEORETICAL CONSIDERATIONS

Some theoretical considerations are suggested by observation of the experimental curves of Figures 1 and 2, and existing interpretations on the origin of the bubble collapse phenomenon.

A possible mechanism of discharge in a conducting liquid, in which the process should be energy dependent rather than field dependent, has been recently proposed.[7] Experimental observations, currently under way also at IDAC, should indicate that there is a threshold energy for the generation of a vapour bubble in the spark gap, whose collapse causes the acoustic pulse. In particular, should exist a threshold, T_{vap}, of temperature, corresponding to a threshold, E_{vap}, of the electrostatic energy, above which the liquid begins to vaporize and a consequent vapour bubble is generated. On the contrary, if the maximum energy supplied by the system is lower than E_{vap}, no vaporization there should be, and a faint discharge will occur inside the liquid, without generation of any bubble cavitation.

In the light of the above considerations, the different curves of Fig. 2, all obtained in the same experimental condition of energy E = 720 Joule, obviously greater than E_{vap}, can be interpreted in terms of different discharge and vaporization mechanism into the liquid. According to whether the spark gap is greater or smaller than the value d = 10 mm, two cases can be distinquished. In the first case (for d values of 15, 20, 25 and 30 mm), there is the presence of a unique powerful pressure pulse that, indicates the collapse of a vapour bubble of relatively large size. This should be obtained by an efficient heating of the liquid and consequent optimum vaporization. It was observed in the same test that increasing the gap distance beyond a limiting value, the bubble pulse disappeares, thus indicating that the ability of the system to deliver energy to the liquid for its vaporization is greatly reduced.

On the other hand, for d smaller than 10 mm, the vapour bubbles, which could possibly imply two regions of vapour (bubbles) near the electrodes, start to expand towards one another, due to the combined effect of internal pressure difference and vaporization of the liquid surrounding them. Eventually, these two regions may join together, bridging the gap and starting a discharge in vapour rather than in the liquid. This is revealed by a sudden drop in the voltage and a rise in the current, and is commonly identified as breakdown; at this instant a primary pressure pulse also appears. Since the discharge takes place largely in vapour, little heat is dissipated into the liquid, due to the high electric conductivity. In this case, the energy left in the source is not used for further heating of the liquid, but mainly spread in the oscillating circuit according to the well-known RLC law correlating the capacitive and the inductive component of the load. Thus, further vaporization of the liquid and consequent regular growth of the bubble are interrupted. In this case, the performance of the acoustic source is not the best one, since the acoustic efficiency of the sparker-based transducer is directly related to the maximum size reached by the uninterrupted vapour bubble.

In order to evaluate the performance of the acoustic cavitation-based transducer, like that proposed in the present paper, and to compute the corresponding order of magnitude of the acoustic efficiency, one can start from the equation of Rayleigh,[8] which gives the collapse time of a vapour bubble imploding under a constant pressure difference, p - pv (T) from an initial maximum radius R to a negligible small size. The bubble internal pressure (neglecting the presence of any incondensible gas) is considered equal to pv (T), which is the saturation pressure of the vapor at the indisturbed temperature T of the liquid, of density ρ, throughout the stages of the bubble motion, while p is the ambient pressure. The maximum radius R of the bubble can be related to the energy of electrical discharge by computing the work that the bubble performs in collapsing from the maximum radius R to a zero radius under the constant pressure p - p_v . Other than for a conversion efficiency η of the electric energy into the mechanical energy stored in the bubble (which also accounts for the acoustic energy radiated), this work derives from the electrostatic energy, E, supplied by the system, so that the efficiency can be related to the collapse time, t, by the relation

$$\eta = 5.455 \ t^3 \frac{[p - p_v(T)]^{5/2}}{\rho^{3/2} E} \qquad (1)$$

from which it is evident that efficiency is strongly dependent on the collapse time of the vapour bubble.

Again considering Figures 2, one can observe that the acoustic pulse having the maximum peak amplitude (for d = 25 mm) corresponds to a bubble collapse occurring at the maximum time too, thus indicating, according to relation (1), that a maximum efficiency coefficient is attained for the following geometrical condition: radius of hemispherical electrodes = 5 mm; spark-gap distance = 25 mm. (Note that, although the time at which occurs the bubble collapse in each curves of Figures 2, is not coincident with the collapse time which figures in relation (1), however they are closely related to each other, and this fully justifies the above considerations).

The experimental observations so far obtained, together with the above theoretical considerations, indicate that investigation on the suitable conditions to optimize the process of energy deposition into the liquid, in order to maximize its vaporization before breakdown, is fundamental in designing the sound source The possible existence of an optimal value for the ratio of the spark-gap dimension to the electrode radius, suggests that the design of sparker-based sources can be improved according to this criterium, once the electrical parameters of the system are fixed.

DESIGN AND CHARACTERIZATION OF THE TRANSDUCER

On the basis of the above experiments and theoretical considerations, a paraboloidal sparker-based transducer was designed with the following geometrical dimensions: focal length = 10 cm, spark-gap distance = 25 mm, with hemispherical electrodes of radius 5 mm, inside base diameter = 20 cm. The height (h = 2.5 cm) of the paraboloidal reflector was chosen in such a way that its focus get a position far enough from the metallic wall, in order to avoid any possible obstacle to a regular growth of the vapour bubble. The backing of the metallic dome has a thickness of 11 cm; it was obtained from a unique block of massive aluminum cylinder (dimensions: 22 cm x 11 cm) inside which a paraboloidal dome-shaped surface was turned by lathe.

The acoustic characterization of the transducer was carried out in a tank larger than that used in the laboratory of the Institute, and having dimensions 25 m x 12m with a depth of 3.5 m. The experimental apparatus is similar to that already used in the above tests. Source level and amplitude spectrum were tested and compared with those corresponding to the simple sparker in the same experimental conditions. The electric parameters were the following: voltage = 2250 V, capacitance = 360 μF, corresponding to an electrostatic energy of 911 Joule. Figure 3 shows the acoustical signal of the paraboloidal source detected at 1 m, on the paraboloid axis, by a hydrophone with sensitivity of 12.6 mV/ Pa, and frequency band flat from 0 to 350 kHz. The two separate pulses represent a significant comparison between the signature of the simple sparker and that of the paraboloidal source. Infact, the first signal corresponds to the direct spherical wave generated by the sparker, without reflection on the metallic wall of the paraboloid, and the second signal represents the wave reflected by the paraboloidal reflector. The peak acoustic pressure of the paraboloidal source is about 700 kPa, while that of the simple sparker is about half this value. Moreover, this latter is strongly attenuated at a distance of a few meters from the source and becomes negligible in comparison with that of the paraboloidal source.

An interesting comparison is that of the respective spectra in the two cases, which are illustrated by Figures 4a and 4b. It can be noted that the presence of the paraboloidal reflector has a twofold effect: (i) it improves significantly the amplitude spectrum in the band of the lower frequencies, which exhibits very intense components as far as 50 kHz, (ii) and provides an almost flat frequency band, within a few dB, from 100 to about 400 kHz.

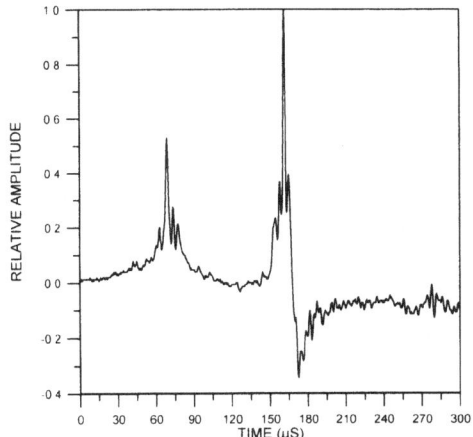

Figure 3. Signature of the paraboloidal sparker-based source.

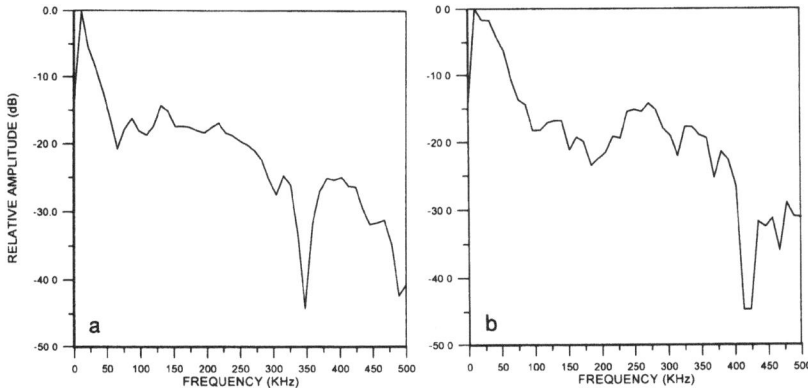

Figure 4. Frequency spectrum of the simple sparker (a), and of the paraboloidal sparker (b).

UNDERWATER ACOUSTICAL IMAGING

Some preliminary experiments, in order to image objects of various nature buried in material simulating sea bottom sediment, were carried out in the same water tank and experimental conditions used to characterize the transducer. Figure 5a shows the acoustical imaging of a little terra-cotta pot having the maximum size 12 cm, set down on a tufa laid over the tank floor, as illustrated by the photograph. An other object of different material (nickel silver) is the little jug (heigth = 10 cm) shown in Fig. 5b, where the acoustical imaging is also given. However, in this case the object, placed on the same tufa brick, was completely buried under the sand, before insonifying the tank bottom. To obtain the images, a linear scanning was made by insonifying the tank bottom by the transducer and using in reception a high-sensibility detector with a circular aperture of an inch diameter, and frequency response until 500 kHz. The images were obtained,without any elaboration on the signal, by employing 103 lines corresponding to a plane area of visualization of about 52 cm x 51 cm, and utilizing a computer system which allows 16 signal levels to be presented by means as many colors.

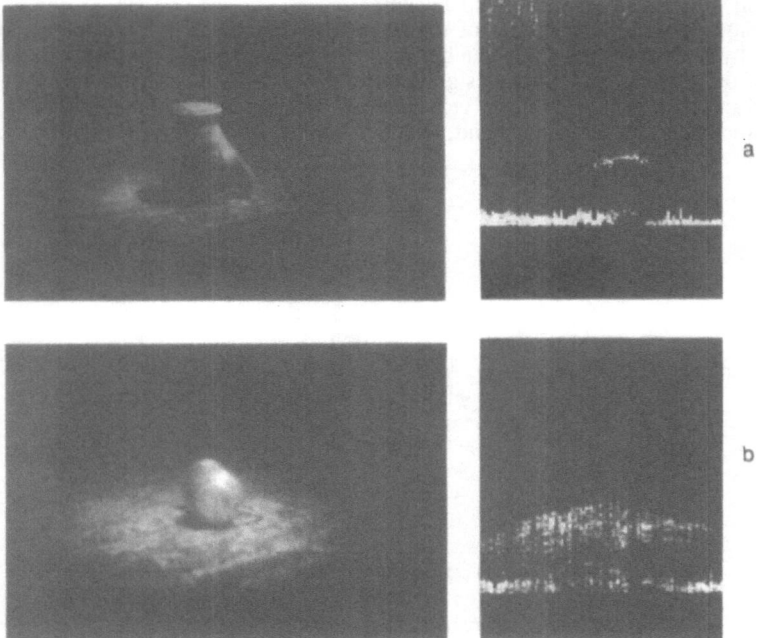

Figure 5. Acoustical imaging of two different objects lying on salt-water tank bottom, simulating sea floor with tufa and sand sediment. (a): terra-cotta pot, (b): nickel-silver jug.

REFERENCES

1. R.J. Urick. "Principles of Underwater Sound," McGraw-Hill Book Company, New York (1983).

2. G.B. Cannelli, E. D'Ottavi and S. Santoboni, "Electroacoustic pulse source for high-resolution seismic explorations",. U.S.A. Patent N. 4734804, (1988).

3. G.B.Cannelli and E.D'Ottavi, Tuned array of paraboloidal transducers for high resolution marine prospecting" *in* : "Acoustical Imaging," H. Lee and G. Wade, eds., Plenum, New York (1990).

4. G.B.Cannelli, E.D'Ottavi and A. Prosperetti, "Bubble activity induced by high-power marine sources" IEEE Oceans 90, pp.533-537 (1990).

5. M.S.Plesset,and A. Prosperetti, Bubble dynamics and cavitation. *Ann. Rev. Fluid Mech.* 7:185 (1977).

6. G.B.Cannelli and E. D'Ottavi, Physical phenomena involved in sea-water plasma based sound source, *in* : " European Conference on Underwater Acoustics," M.Weydert ed., Elsevier Applied Science, London (1992).

7. A. H. Olson andS. P. Sutton, The physical mechanism leading to electrical breakdown in underwater arc-sound sources, *J. Acoust. Soc. Am.* 94:2226 (1993).

8. Lord Rayleigh, On the pressure developed in a liquid during the collapse of a spherical cavity, *Phil. Mag.* 34:98 (1917).

A PORTABLE ACOUSTIC MAPPING/IMAGING SYSTEM
FOR ASSESSING AND MONITORING OF AQUATIC RESOURCES
AND ENVIRONMENTAL PARAMETERS

Andrzej Stepnowski[1], Massimo Azzali[2], Janusz Burczyński[3],
Jacek Lenkiewicz[1], Andrzej Partyka[1] and Marek Moszyński[1]

[1] Technical University of Gdańsk, Acoustic Dept. 80-952 Gdańsk, Poland
[2] CNR-IRPeM, Largo Fiera della Pesca, 60100 Ancona, Italy
[3] C-MAP Environmental/BioSonics, 54036 Marina di Carrara (MS), Italy

INTRODUCTION

Researchers who conduct acoustic, oceanographic and fish catch surveys are in real need of a computerized Data Base System, which can allow on-board processing of the acquired data along with their analysis and mapping in geographical context [1].

One such system was developed in 1985 by CNR/IRPEM in Ancona, for storage and cartographic imaging of data on pelagic fish resources in the Adriatic Sea, but the present technological and operational needs (mostly speed of data processing and variability and volume of data to be handled) called for an enhanced and more versatile system's version.

To meet these requirements a new computerized fishery research data base system called EchoBase was developed in 1994 at the Technical University of Gdańsk[2]. The system, which uses state of the art electronic technology, combines features of database software package with display of electronic cartography and imaging. With these applications in place it plays the role of a portable dynamic Geographic Information System (GIS). This provides on-board management, mapping and evaluation of data on marine fish resources from acoustic surveys and catch samples, along with environmental parameters acquired by a CTD probe.

ENVIRONMENT OF ECHOBASE

The environmental context of EchoBase is shown in the system's functional diagram in Figure 1.

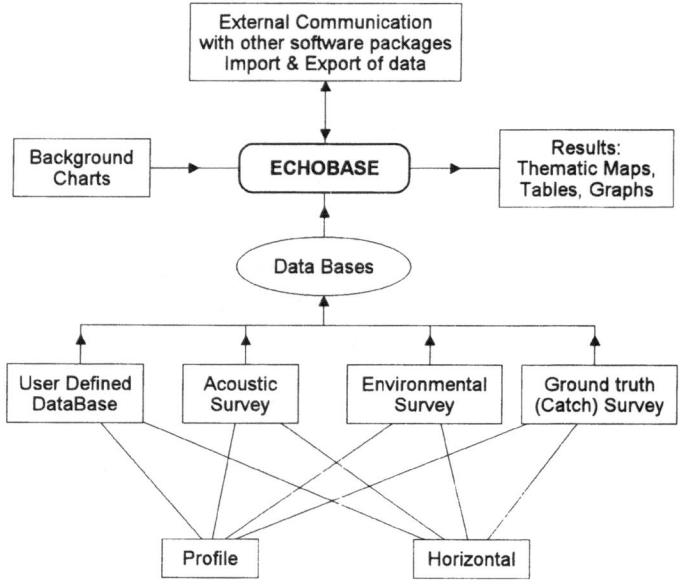

Figure 1. Schematic block diagram of EchoBase.

The following three data base formats are ready for use:
- Acoustic Surveys Data (aquatic resources data from echosounder and echo signal processor, such as fish or plankton density and bathymetry),
- Environmental Survey Data ('*in situ*' data on aquatic environment from CTD probe)
- Catch Data (set of data from sample catch by trawl or other ground truth gear)

The user can also define his own (customized) data base format which can be stored as a profile or horizontal record.

Results obtained by manipulation and processing of stored data can be presented in the form of:
- thematic maps (e.g. maps of fish density distribution, water temperature etc.),
- graphs (e.g. vertical profile of water salinity or temperature),
- tabular reports (e.g. fish density table).

Import/Export function allows communication of EchoBase with other software packages. Data can be imported from ASCII files and from private files of data acquisition instruments (e.g. Echo Signal Processor, CTD profiler). Data can also be exported to commercial packages such as dBase, spreadsheets, GIS or as ASCII files

Characteristics of Data

Two types of spatially distributed data can be handled by EchoBase i.e. horizontally distributed data and vertical profiles. Data are allocated in time and in space (georeferenced in Latitude & Longitude). Horizontally distributed data are stored as a set of reports from measurements representing point values on a surface (e.g. surface temperature of the water, catch data, bottom depth). Profile data are stored as an array from measurements representing vertical profiles (e.g. temperature profile of the water, fish distribution profile).

Interpolation of horizontally distributed data creates a horizontal thematic map of an interpolated variable. Profile data can be interpolated in two planes: in horizontal plane (horizontal map for arbitrary depth layers) and in vertical plane (vertical cross-sectional map)

The following data (from sensors) are acquired *on-line:*
- bathymetry, fish density and other data (from an echosounder and echo signal processor),
- oceanographic parameters of water - temperature, salinity etc. (from a CTD profiler),
- geographical location (from a GPS receiver),
- user input data.

The *off-line* data are gathered from:
- digitized chart data files (e.g. background charts in C-MAP public or MapViewer format),
- data bases and data files acquired and stored previously,
- various software packages such as spreadsheets, GIS, dBase III.

Basic Functions of EchoBase

EchoBase provides the following basic operations with data:
- collecting and storing of survey data from various instruments and sensors,
- reduction and conversion of survey data,
- processing of data base information (queries, filtering, searching, extraction etc.),
- calculation on system data bases (interpolation, extrapolation, statistical computation),
- storing, retrieving and exchange of data base with other data bases and software packages,
- graphical presentation (maps, graphs, tables, reports).

The output data or results can be produced in three basic forms:
- charts (various thematic charts).
- numerical tables (e.g. tables of fish density, water temperature etc.),
- graphs (2^D or 3^D presentation of measured and stored variables),

Functional Modules of EchoBase and Data Flow

There are three basic functional modules of EchoBase:
- Mapping and Presentation Module,
- Data Base Module,
- Computation and Processing Module.

The interacting of modules is shown on the diagram in Figure 2.

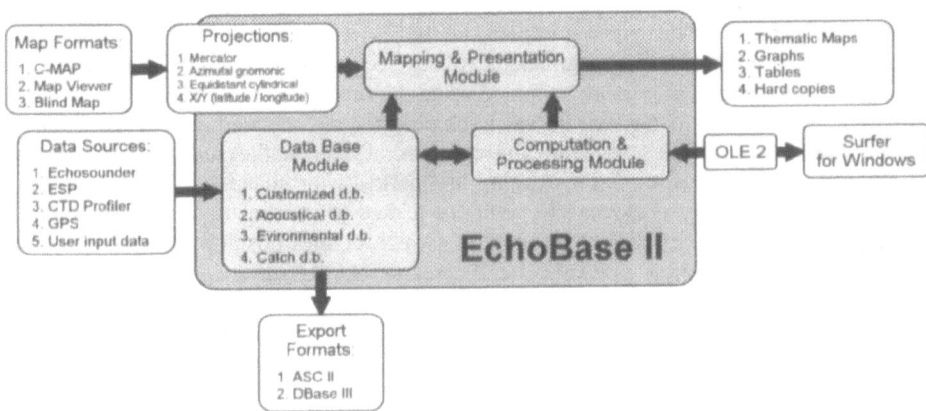

Figure 2. Functional modules of EchBase and data flow

Mapping and Presentation Module (MPM) creates background nautical charts, based on C-MAP public format files, containing digitized coastlines, depth contour lines and other navigational map elements. It provides an active on-screen map that the user can scroll, pan, zoom in and out and rotate. MPM supports four kinds of cartographic projections: the Mercator, the Azimuthal Gnomonic, the Equidistant Cylindrical and the Rectangular. MPM collects queries and extraction results from the Data Base Module along with results of calculations from the Processing Module and transforms these to maps.

Data Base Module (DBM) designed around the Paradox PC database standard, provides a relational database for acquiring, storing and managing acoustic, environmental and catch data. DBM translates data from various sensors, other databases and results of computations, to the form understood by EchoBase. The tasks carried out by DBM involve: creation, insertion and selection of records, browsing, queries and other database functions.

Computation and Processing Module (PM) consists of an interpolation engine and tools for data processing and evaluation. EchoBase builds a logic around the popular PC interpolation package SURFER, which performs interpolation in the background. Interpolation converts the data (randomly distributed) into a two-dimensional array of values having a uniform spatial step and stores this array in an image file. The latter is later displayed on the on-screen map showing distribution of the interpolated property. Further data processing includes estimation of biomass of fish, distribution of environmental parameters and statistical calculations. Storing and exporting capabilities are also provided.

SYSTEM IMPLEMENTATION

The basic hardware platform of the EchoBase system is IBM PC with a '486 processor. The minimum hardware requirements are 4 MB RAM, VGA card, and 5 MB of free disk space. The basic software environment of EchoBase is Windows™ 3.1. Data Base Module is designed around the Borland's Paradox Engine 3.0 package. The EchoBase software was written in C++ (Borland C++ 3.1) as Windows™ 3.1 application. It has implemented the OLE feature for communication with other Windows™ applications.

The system is user friendly and easy to operate. The user interacts with the system through menu bar commands, which are grouped as follows:

Maps menu provides access to map selection (OPEN command), map compilation from a C-MAP public format file (IMPORT command) and control over map attributes. Also rotation of maps, opening of map images and saving of maps in Widnows Meta files format are managed from this menu group.

Display menu manages map zooming, scrolling, redrawing and unzooming options. Any required portion of a map can be zoomed using the rubber rectangle tool.

Track menu brings to display a survey track obtained directly from an opened database or from a text file. EchoBase displays the contents of such a file superimposed on the currently selected map. Location of data points are marked with small red circles.

Interpolation menu manages all operations related to interpolation of survey data. It is initiated by the DEFINE REGION command, which defines position of an interpolated area (rectangle) on a map, its dimensions and a search ellipse size. The position of such a region on the map along with a pulled down INTERPOLATE menu is shown in Figure 3. Likewise the track selection option, the interpolation process may use external text data files, or access properties directly in an opened data base. Interpolation uses either the Surfer program which provides five basic $2D$ interpolation methods (inverse distance, kriging, minimum curvature, polynomial regression and triangulation) and works in the background, or alternatively, the internal interpolation program, using the inverse distance method.

Evaluate menu provides creation of images from datasets and associated operations. Immediately following the interpolation process its resulting array is stored to an image file. Then the user displays this file on the on-screen map as a color image showing the distribution of an interpolated property. The user subdivides the range of values in the image

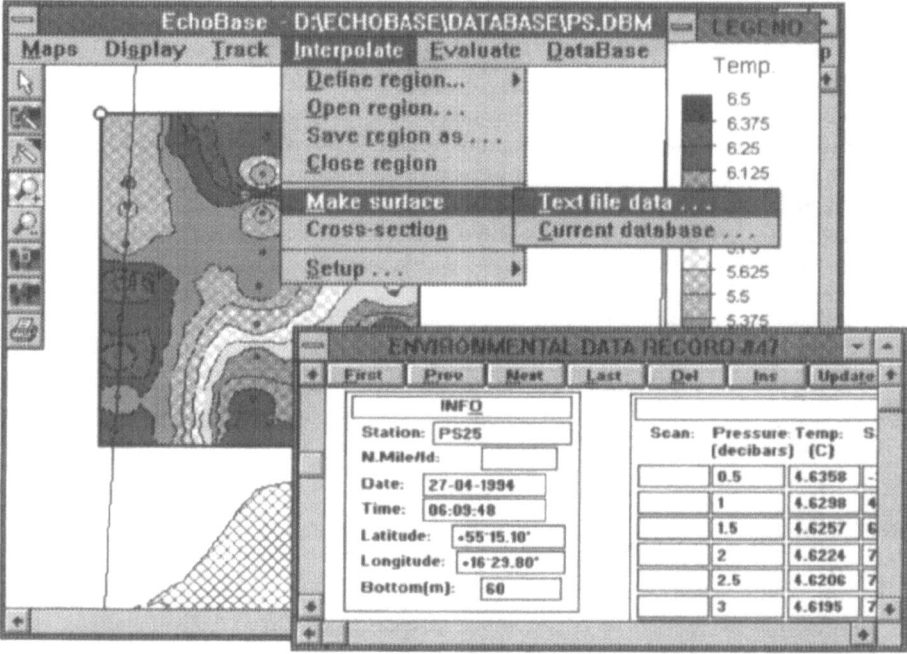

Figure 3. Sample data entry form from the environmental database, INTERPOLATE menu pulled down

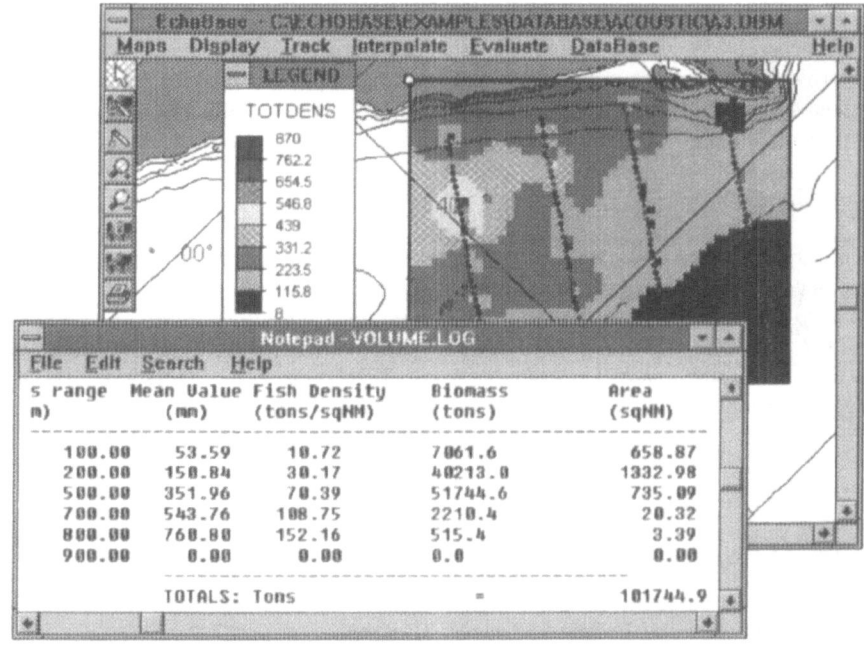

Figure 4. Sample interpolated image of the biomass distribution in the Adriatic Sea along with the assessment report

into discrete classes and assigns a distinct color to each class. Figure 4 shows a sample image depicting the results of interpolation of fish biomass distribution obtained from the acoustic survey at Adriatic Sea. The evaluate menu includes also the ASSESSMENT option for estimating surveyed fish population density or biomass in a selected strata. An example of such an assessment report extracting the last image data, is shown in Figure 4

DataBase menu manages all operations related to databases, i.e. opening, closing, import and export of databases, browsing of records and queries. EchoBase has a built-in internal database system for management of acoustic survey data, fish catch results and basic oceanographic parameters. Additionally, any other kind of data can be managed using the user defined database. Selection of a database type is made upon creation.

A sample dataset provided with EchoBase contains several prepared databases with results of surveys in the Mediterranean, Adriatic and Baltic Seas. The selection of the BROWSE command brings to the screen a data entry form relevant to the database type. Following the selection EchoBase scans a current database and sends to a file the latitude, longitude and a value of a selected property. This file may be later used during ship track plotting or interpolation by EchoBase, or it may be exported to other software packages.

CONCLUSION

EchoBase allows for a wide range of operations on vector maps e.g. different kinds of geographical projections, rotation, zooming, scrolling, multi-layered manipulations, etc. Background charts can be imported from C-MAP public format or GoldenSoftware™ MapViewer format. User-defined "grid maps" are also available.

Although the EchoBase system cannot aspire to a fully featured Geographic Information System, it can play the role of a compact portable GIS particularly suitable for mobile surveys. Users of the first EchoBase version have found it to be a quite useful, portable and yet powerful tool for on-board processing and mapping of data from acoustic and oceanographic surveys..

The prospective system should be enhanced with satellite image processing which would provides map/image comparison and utilizing standard software tools available in GIS systems.The new system will also include a larger volume of data and will provide more processing and imaging capabilities (e.g. fish schools imaging and bottom typing).

Acknowledgments

The authors would like to express their thanks for the support for the EchoBase development to the European Community Project AIR-1-3003 92 0314 (*Assessment of the relationship between oceanographic parameters and spatial/temporal fluctuations of pelagic fish population characteristics using satellite and acoustics, "T-ECHO"*) and to the Polish National Committee of Scientific Research (KBN) Project 6 PO4E 031 09 (*Design of the Hydroacoustics System for Monitoring and Mapping of Sea Bed*).

REFERENCES

1. Stepnowski A. and Burczynski J., Compact Geographic Fisheries Information System. *6th Interdisciplinary Conf. on Natural Resources Modeling and Analysis*, Sabaudia, Italy, art. 4.7. (1993).
2. Stepnowski A., Ostrowski M., Burczyński J., Geographic fishery research data base system; *Proc. of the Second European Conf. on Underwater Acoustics, Copenhagen*, vol. II, 847-853, (1994).
3. Stepnowski A., Burczyński J, Ostrowski M., Moszyński M., Partyka A., Lenkiewicz J., EchoBase-a portable Geographic Fishery Research Data Base System; *Marine Geodesy - an International Journal of Oceans Surveying, Mapping and Remote Sensing*, vol.18, No. 3, (1995).

AN ULTRASONIC 3D ROBOTIC VISION SYSTEM FOR A NUCLEAR ENVIRONMENT

Gerrit Blacquière, Frits van der Putten, and Hugo Vos

TNO Institute of Applied Physics
Inspection Technology Department
P.O.Box 155, NL-2600 AD Delft,
The Netherlands
e-mail: blacquie@tpd.tno.nl

INTRODUCTION

The hostile environment in nuclear power plants has considerably pushed forward the development of smart robot systems which are capable of carrying out 'human' tasks. Obviously, many of these tasks require the robot to interact properly with the outside world. For that purpose, it should be equipped with a vision system. In many cases, a solution based on video equipment will be sufficient. However, if 3D information is required, an ultrasonic technique is a good alternative, in particular under circumstances of poor visibility. Furthermore, ultrasonic transducers are known to sustain radiation better than CCD camera's, which is a major advantage in a radioactive environment.

An example of a typical robot task - which is encountered in the process of decommissioning nuclear power plants - is the removal of welded keys that are securing bolts and the successive unscrewing of these bolts. The actual removal of the keys is carried out by a robot-operated electro-erosion tool. An ultrasonic 3D vision system is used to provide the robot with the navigation data, in such a way that it is able to accurately position the electro-erosion tool. In addition, the result of the key removal action is checked afterwards with the same system. The vision system is based on a medical ultrasonic scanner that is operated by the robot (see Figure 1).

In this paper such an ultrasonic 3D robot vision system is presented.

ULTRASONIC 3D VISION SYSTEM

The ultrasonic 3D vision system consists of three parts that deal with data-acquisition, data processing and post-processing respectively. These parts will be discussed next.

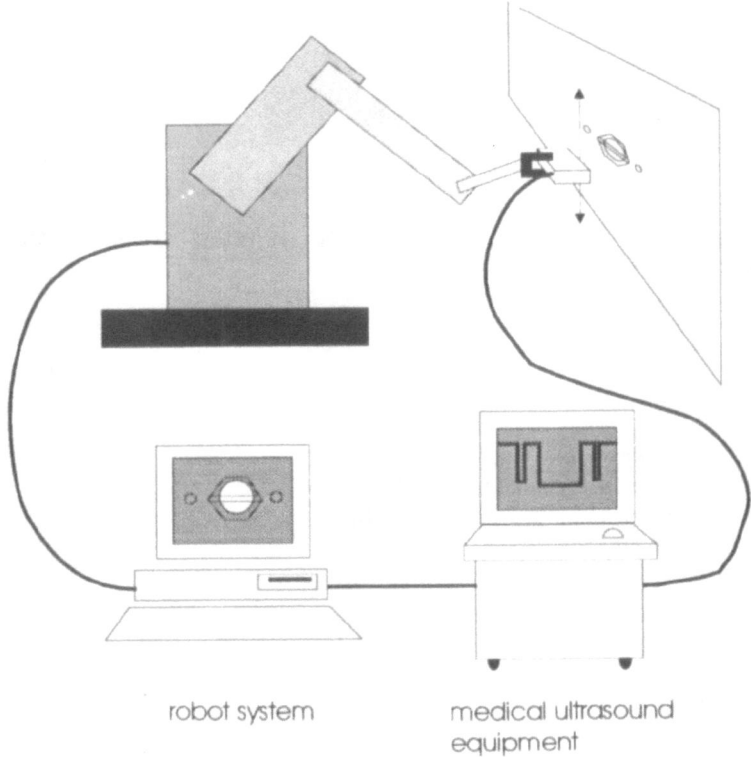

robot system   medical ultrasound
                      equipment

Figure 1. A robot scans the bolt using medical ultrasonic equipment. The data are processed automatically and the exact location of the bolt is computed. Later, the robot uses this information to unscrew the bolt.

Data Acquisition

The data acquisition part of the system is based on a commercially available, general purpose, medical ultrasonic scanner, equipped with a linear-array probe (Scanner 250 of the PIE Medical Corporation). The probe has a length of 11 cm and consists of 80 elements. The centre frequency of the system is 2.25 MHz and the bandwidth is appr. 3 Mhz.

One B-scan with this probe results in a 2D image. In order to obtain 3D images the probe is translated by the robot system. In this way a number of sequential B-scans are obtained that together form a 3D data space. The data acquisition is based on a fast 20 Mhz 8-bit A/D converter. The output of the ultrasonic scanner is transferred to a workstation for data processing, interpretation and visualisation. A complete 3D data set can be acquired in a few seconds.

Special software was written for the communication between the data acquisition system and the robot system. In particular, attention was paid to a proper handling of the information on the position of the robot arm while scanning and to the time stamping.

Data Processing

The data processing is relatively simple: it basically consists of peak detection, i.e., the maximum value and its traveltime are determined. The maximum value is considered to be representative for the surface reflectivity. The results are a reflectivity image (see Figure 2) and

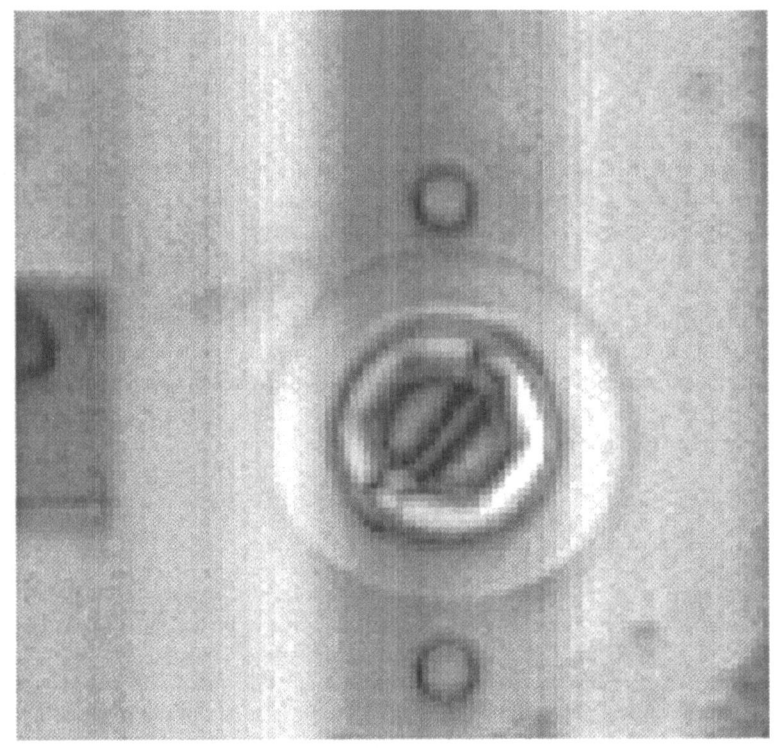

Figure 2. Reflectivity image (maximum amplitude) of the bolt.

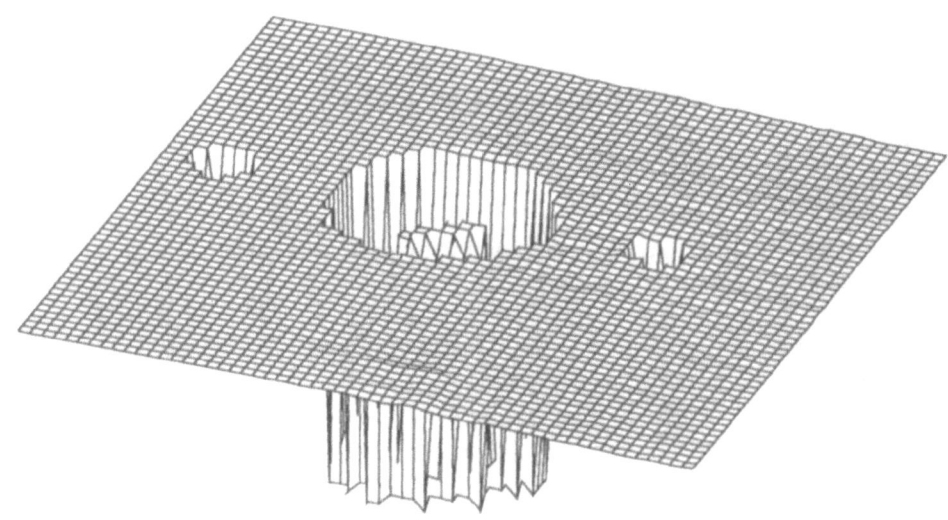

Figure 3. Traveltime image of the bolt.

a traveltime image (see Figure 3). Note that the traveltime is a measure for the distance to the object.

Post Processing

The post-processing however is more advanced. The system should be able to recognize the bolt in the ultrasonic image automatically and compute its exact location and orientation. The approach is to include as much *a-priori* information as available. E.g., the shape of the bolt is known exactly. This knowledge can be fed into an object matching technique. The applied technique has proven to work well for the case of laser-range data (Pentland, 1990). Its principle is as follows. The mathematical description of the objects to be matched is based on a composition of various 'primitive' elements like blocks, spheres and cylinders (superquadrics). The matching technique minimizes the error between the observed range values and the mathematical model in a least squares sense. After the post-processing, a nice image as well as the required quantitative information about this image - the bolt's location and orientation - are available.

RESULTS

The initial results, obtained from water-basin experiments, prove the feasibility of this concept, see Figure 2 and Figure 3. In both images the bolt can be recognized easily. The bolt has a diameter of appr. 3 cm. The reflectivity image (Figure 2) clearly shows the hexagonal shape of the bolt. Furthermore, the orientation can be seen. The quality of the image is almost photographic. From the traveltime image (Figure 3), the distance infomation can be deduced. One can imagine that the location and orientation of the bolt can be easily obtained automatically from these type of pictures by the object matching algorithm.

Apart from the good results, an attractive property of the system is that it is built from 'of-the-shelf-components'. It is a typical example of the successful translation of existing technology into a new application area with relatively little effort.

ACKNOWLEDGEMENT

This work was partly funded by the EC under contract no. F12D-0041.

REFERENCE

Pentland, A.P., 1990, Automatic extraction of deformable part models, *Intern. J. Comput. Vision*, 4, 107-126.

ULTRASONIC SYSTEM FOR 3D OBJECT PROFILE RECONSTRUCTION IN AIR

M. Calzolai[1], L. Capineri[1], A.S. Fiorillo[3], L. Masotti[1], J. Ping[4], and S. Rocchi[2]

[1] Dipartimento Ingegneria Elettronica, Università di Firenze, Firenze, Italy
[2] Facoltà di Ingegneria, Università degli Studi di Siena, Siena, Italy
[3] DIIIE, Università di Salerno, Fisciano, Italy
[4] Changchun Institute of Optics and Fine Mechanics, Chinese Academy of Science, Chang Chun, P.R. China

INTRODUCTION

Ultrasounds have been widely used in medical and NDT fields but recently they have gained importance also in industrial applications. For these applications the development of new type of low cost airborne ultrasonic transducers leads to consider the ultrasonic instrumentation as a valuable alternative to both optical and mechanical systems. This work investigates the feasibility of an ultrasonic 3D scanner in air based on ferroelectric polymer (PVDF) transducers. A laboratory system has been developed to extract 3D profiles of wood shoe models. The characteristics of accuracy, quickness and absence of mechanical movements make the developed system suitable for just in time fabrication. Common systems are based on 3D pantographs which suffer of low speed due to the mechanical movements and frequent maintenance due to the contact of the measuring head with the wooden model. On the other side advanced systems exploit optical techniques but they result expensive, bulky and sensitive to illumination, dust and dirty.

MATERIALS

A laboratory prototype of the proposed US scanner is shown in Fig. 1. The shoe model is scanned along X direction with a circular array made of hemicylindrical PVDF transducers, which resonance frequency can be easily changed because it is inverse proportional to the bending radius of the Ag metallized PVDF film[1]. The actual work frequency is 61 kHz, corresponding to a bending radius of 3.6 mm, and 85 dB two way insertion loss with a $\lambda/4$ back reflector[1]. The circular array has a radius equal to 23.6 cm and it lies on a plane perpendicular to the scanning direction X. A set of parallel 2D profile data are acquired during each scan at fixed positions along X axis; once the scanning procedure is completed, data are combined in order to obtain a 3D mathematical model of

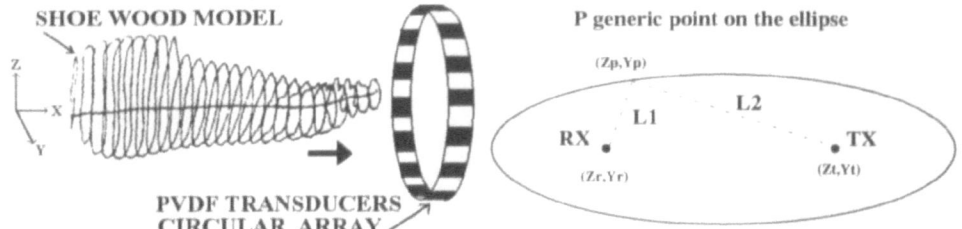

Figure 1 Laboratory prototype of the airborne US scanner based on a circular array.

Figure 2 Acquisition system of a single object cross-section.

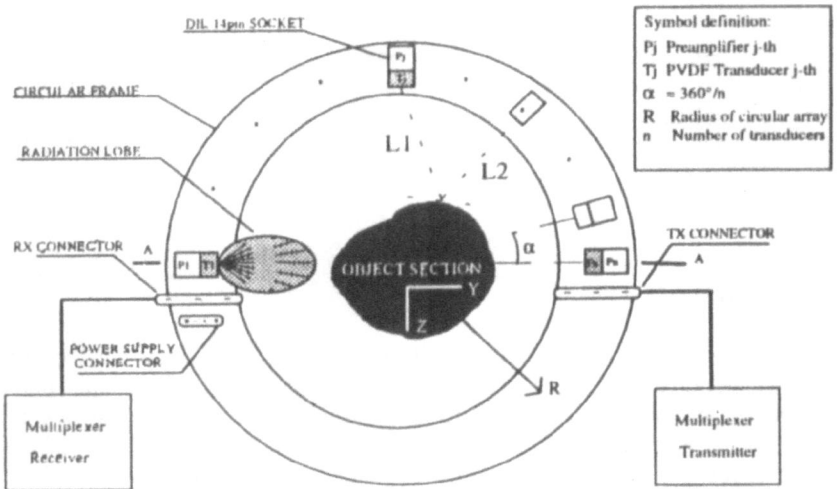

Figure 3 Profile reconstruction with ellipsis method.

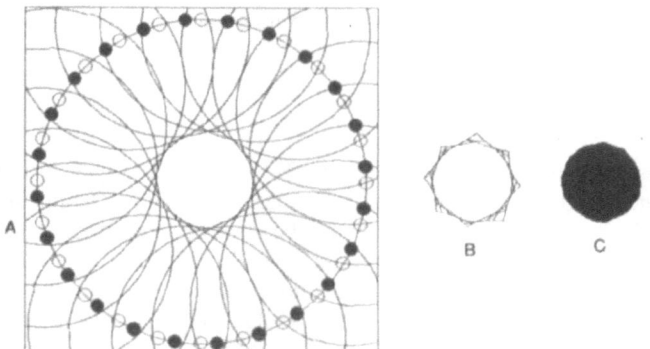

Figure 4 Experimental results of a cylinder cross-section with 24 RX-TX: (A) reconstruction with ellipsis method; (B) intermediate profile extraction; (C) final extracted inner contour.

the object that can be used as CAD system during manufacturing processes[3]. The PVDF transducers are used in the unimodal configuration as transmitter (TX) or receiver (RX) as shown in Fig. 2. A PC addresses, in a clockwise sequence, the transmitters along the circular frame and enables by a multiplexer the side receivers for the acquisition of the signal reflected from the target. The measurements are performed in a controlled environment since the US propagation is sensitive to air motions and to temperature gradients. In the final system design these error sources can be minimized by encasing the whole system and the circular array can be replaced by a cylindrical one for avoiding mechanical movements.

METHODS

The profile reconstruction of each section is obtained by measuring the Time Of Flight (TOF) between different TX-RX pairs sequentially selected along the circular frame. The TOF measurement is performed by transmitting a narrow band signal (burst of 20 pulses at 61 kHz) and by detecting the received signal with a digital coherent demodulation technique. This technique allows accurate ranging[4] even if it is used with low frequency transducers. The phase measurement is derived by processing the samples of in phase and in quadrature components of the received signals which are sorted from a sampling at four times the resonance frequency. The involved samples with 8 bit resolution are processed with a low cost ad-hoc electronic unit based on a TMS320C25 DSP[5]. The calculation of the TOF is carried out in two steps: at first a coarse evaluation of the echo arrival time is obtained through the envelope detection providing a resolution of $\lambda/4$ ($\lambda = 5.4$ mm at 61 kHz); in the second step a fine TOF calculation is carried out by evaluating the instantaneous phase in each cycle and the average phase within the central portion of the received burst. Experimentally the ranging unit has been tested with a plane reflector at 27 cm providing a max error of 0.13 mm (about $\lambda/50$) and r.m.s. error= 0.03 mm. The TOF information obtained with this ranging system is then converted into an ellipsis that is defined by the current positions of the TX and RX (representing the foci) and the sum of the distances given by the range value (sum of distances L1+L2) in Fig. 3. According to the defined acquisition method, all the ellipses are accumulated in the image memory of the computer, which is linked to the electronic board by the parallel port. The ellipses are tangent to the profile of the cross section and by suitable image processing the inner contour can be extracted. The assumption underlying this method is the validity of ultrasound propagation by ray tracing.

RESULTS

Experiments on 2D profiles are carried out in laboratory environment to test the accuracy of the method. A work in progress is dedicated to the calculation of a 3D mathematical model starting from 2D object profiles. A test object consisting in a solid cylinder with diameter 11.5 cm is scanned with 24 transducer equally spaced over 360°. Fig. 4A shows the experimental results of the profile reconstruction with the ellipsis method. The acquisition is performed by collecting the signals with single TX-RX pairs. A fast image processing program has been developed to extract the inner contour. The program finds cross points between every pair of ellipsis in a window containing the whole profile (see Fig. 4B). Only the part of the ellipsis between two cross points is retained and an intermediate picture formed by arcs of ellipsis is obtained. Because the circular scanning the arcs result tangent to the object and constitute a closed curve from which the inner contour can be easily extracted (see Fig. 4C). The obtained maximum rms error of the cylinder radius is equal to 0.08 mm that is in good agreement with the range rms error obtained with a simple plane reflector. The profile reconstruction algorithm doesn't enlarge the overall processing time, because it is time-overlapped with the acquisition process.

CONCLUSIONS

Preliminary results have pointed out the capabilities of a low cost US 3D scanner in air. The system will be tested with actual shoe models to investigate the validity of the US propagation according to the geometrical approximation that does not consider multiple reflections and diffractions from sharp points.

REFERENCES

1. A.S. Fiorillo, Design and characterization of a PVDF Ultrasonic Range sensor, *IEEE Transactions on UFFC*, Vol. 39, N.6, pp. 688-692 (1992)
2. L. Capineri, A.S. Fiorillo, L. Masotti, S. Rocchi, "Array of PVDF sensors for ultrasonic imaging in air", *IEEE Ultrason. Symp. Proc.*, Cannes, F, pp. 487-490, (1994)
3. Y. Kim and Ren C. Luo, "Validation of 3-D curved objects: CAD model and fabricated workpiece", *IEEE Trans. on Ind. El.*, Vol. 41(1), pp. 125-131, (1994)
4. J.D. Fox, B.T. Khury Yakub, G.S. Kino, "High frequency acoustic wave measurements in air", *IEEE Ultrason. Symp. Proc.*, Atlanta, GA, USA, pp. 581-584, (1983)
5. L. Capineri, M. Calzolai, A.S. Fiorillo, S. Rocchi,"A digital system for accurate ranging with airborne PVDF ultrasonic transducers", *IMEKO, Proc. 4th International Symposium on Measurement and Control in Robotics*, Smolenice Castle, Slovakia, pp. 71-74,(1995)

MODELING IN-AIR ULTRASONIC
TIME-OF-FLIGHT NOISE

Angelo M. Sabatini

Advanced Robotics Technology & Systems Lab (ARTS Lab)
Scuola Superiore Sant' Anna
Pisa, Italy

ABSTRACT

The statistical properties of the random fluctuations observed in the Time-of-Flight (TOF) measurements performed by in-air sonar sensors are analysed. The proposed model recognises the presence of three components in the TOF noise, e.g., the deterministic time-varying mean, the correlated random component and the uncorrelated random component.

The time-varying mean correlates with global thermal changes and drafts affecting the environment. The detrended noise samples are the sum of the correlated random component, due to inhomogeneities in the medium, such as temperature gradients and air turbulence, and the uncorrelated random component, mainly due to the broadband electronic noise generated at the receiver. The correlated and uncorrelated random components are described using autoregressive-moving average (ARMA) modeling and adaptive segmentation techniques. The experimental results confirm the assumptions underlying the modeling effort.

INTRODUCTION

In-air US sensors are widely used for performing range measurements in a cheap and simple manner. Most current sonar ranging systems are of the pulse-echo type, namely an US pulse is emitted in the medium and the time taken by the pulse to travel from the emitting transducer to the object's surface and back to the receiving transducer, e.g., the TOF, is measured. Multiplication of this time by the speed-of-sound value yields the distance between the sensor and the object.

In spite of their ease of implementation, sonar ranging systems suffer from several drawbacks, mainly due to the adverse properties of the typical propagating medium of interest in the robotic field, namely air, against the propagation of US waves[1,2]. In this paper we are interested in the development of analytical tools for analysing the statistical properties of random noise affecting the sensor information. Our method of analysis decomposes the TOF noise into three components with different physical origin and properties: the deterministic time-varying mean, the correlated random component, and the uncorrelated random component. The time-varying mean correlates with global thermal changes and drafts affecting the environment. The detrended data are assumed to result from the additive combination of the correlated random component, due to inhomogeneities in the medium, such as temperature gradients and air turbulence, and the uncorrelated random component, mainly due to the broadbandband electronic noise superimposed on the echo signal. A similar assumption regarding the presence of different random components in the TOF noise is also reported in[3]. However, no attempt is made there to analyse the correlation structure of the correlated random component; additionally, the time-varying mean is not even considered in the noise model. We propose to use parametric modeling techniques combined with a method of adaptive segmentation in order to capture the salient features of the piecewise stationary correlated random component.

SOURCES OF ERRORS IN THE TOF MEASUREMENTS

The ability of a sonar ranging system to perform TOF measurements is affected by several noise sources. These include electronic timing errors, acoustic amplitude fluctuations, as well as changes in properties of the medium that affect the propagation velocity[4].

How the noise superimposed on the echo signal affects the measured TOF and its statistical properties depends on the method for TOF estimation and the Signal-to-Noise ratio (SNR). Random variations of the properties of the propagating medium affect the SNR through their influence on the acoustic amplitude and, to some extent, the shape of the emitted pulses. However, these effects are negligible, compared to the influence the same variations have on the speed-of-sound value. The reflecting object, or target affects the SNR with its strength; in the sequel we are not interested in modeling the nonstationarities caused by the time-varying strengths of moving targets. For sufficiently high SNRs, it is commonplace to consider that the electronic timing errors resulting from the contamination of the echo signal are spread out uniformly in frequency and are Gaussian in amplitude as a function of time. Further, the electronic timing errors are commonly assumed to have a zero mean. For the digital correlator used in the experimental system of this paper, these assumptions are approximately valid[5,6].

The flight of the emitted pulse takes place in a medium with randomly varying propagation velocity which directly affects the measured arrival times. The local distribution of physical properties of the medium like temperature and humidity determine a space and time-variant field of sound velocities with a high degree of statistical correlation, both in spatial and temporal terms. The correlated variations are slow relative to the instantaneous fluctuations of the timing samples, such that the noise appears to have a fixed mean over a short enough time period. This would appear to indicate a certain degree of stationarity. However, the drifts or trends determined by global temperature changes and other macroscopic environmental changes imply that the mean value of the TOF noise varies over appropriately long observation intervals, calling for stochastic models of nonstationary physical processes.

A MATHEMATICAL MODEL OF THE TOF NOISE

The model we propose assumes that the TOF raw data $n(t)$ results from the additive combination of three different components:

$$n(t) = \mu(t) + n_c(t) + n_u(t) \tag{1}$$

$\mu(t)$, henceforth called the trend, is the time-varying mean. $n_c(t)$ is modeled as a short-term stationary, zero-mean, correlated Gaussian random process with variance σ_c^2 (colored noise component). $n_u(t)$ is modeled as a stationary, zero-mean, white Gaussian random process, with variance σ_u^2 (white noise component). In the sequel $n_c(t)$ and $n_u(t)$ are considered to be independent of one another[3].

We further assume that the colored noise component is described in terms of a Gauss-Markov random process with variance σ_c^2 and decorrelation time τ,[7]. $n_c(t)$ is produced from passing a zero-mean, unit variance, white noise process through a first-order, shaping filter with frequency response:

$$H_c(\omega) = \frac{\beta}{1 + j\,\omega\,\tau} \tag{2}$$

where σ_c^2 is given by:

$$\sigma_c^2 = \beta^2/2\tau \tag{3}$$

The white noise component $n_u(t)$ can be described in a similar fashion. In this case, however, the shaping filter is approximated over the system's bandwidth by a constant gain α:

$$H_u(\omega) = \alpha \tag{4}$$

Note that $\sigma_u^2 = \alpha^2$. Since $n_c(t)$ and $n_u(t)$ are independent, the spectral density of their additive combination is as follows:

$$N(\omega) = |H_c(\omega)|^2 + |H_u(\omega)|^2 \tag{5}$$

Applying the spectral factorization technique yields the expression of the shaping filter whose output is colored noise with the spectral density of Eq.(5) when driven by a zero-mean, unit variance, white noise. Also, since the TOF samples are available only at discrete times, the concept of an equivalent discrete-time model has to be considered. Let T be the time interval between successive measurements. The bilinear transform technique[8] is used to obtain the z-transform

$$H(z^{-1}) = H_o \frac{1 + b\,z^{-1}}{1 + a\,z^{-1}} \tag{6}$$

of the resulting digital shaping filter, where:

$$b = \frac{(1-2\tau_o/T)}{(1+2\tau_o/T)}$$
$$a = \frac{(1-2\tau/T)}{(1+2\tau/T)} \tag{7}$$

The time constant τ_o is given by:

$$\tau_o = \tau\,\alpha\,/\sqrt{\alpha^2 + \beta^2} \tag{8}$$

PARAMETRIC MODELING TECHNIQUES

The TOF data $n(kT)$ result from summing the trend $\mu(kT)$ to the output of a linear system driven by a white noise sequence $w(kT)$:

$$n(k) = \mu(k) - \sum_{i=1}^{p} a_i\,n(k-i) + \sum_{j=0}^{q} b_j\,w(k-j) \tag{9}$$

For the sake of generality, we extend the sums beyond the limits of simple first-order models (p=1, q=1). A recursive digital low-pass filter is applied to the data n(k) in order to produce the trend estimate[9]. The damped Gauss-Newton iterative method is used to perform the estimation of the model parameters for the detrended samples, according to the prediction-error approach[10]. The Akaike's final prediction-error (FPE) criterion is adopted for determining the model's order. The FPE criterion is also a function that measures some complexity of the model, since the dimensionality of the model is explicitly incorporated in its definition. The Gauss-Markov model for the colored component leads to an adequate representation of the underlying physics if the corresponding FPE measure turns out to be comparable to that of higher order models.

The assumption that the colored noise component is stationary in the sense of exhibiting constant spectral characteristics is generally valid only within a short-time interval. The method for adaptive segmentation of ARMA time series reported in[11] is applied to test the data for piecewise stationarity. The method uses constant length reference and sliding windows. The basic idea is to estimate an ARMA model for the time series into the reference window and to observe the variations of the spectrum in the sliding window. Segmentation is performed by placing a threshold on a purposely devised quantity, e.g., the Spectral Error Measure, as described in[10].

Once the model parameters are estimated, a Kalman based observer of the correlated component can be designed[12]. The correlated component is the system's state variable we intend to estimate, provided that a) its observations are available, although corrupted by white Gaussian noise, i.e. the uncorrelated random component, b) the first-order dynamics of the correlated component is known from the previously described modeling effort.

EXPERIMENTS

The experimental verification of the modeling methodology presented in this paper is performed using piezoelectric MASSA Mod. E-188 transducers, resonating at f_0=225 kHz. The sonar ranging system - a DSP-based digital correlator - has been thoroughly described elsewhere[6].

A number of measurements (M=2000) are acquired from a planar reflecting object at some distance relative to the sensor with a sampling interval T=1 sec. In Fig. 1, the TOF raw data are plotted. Figs. 2-3 report the estimated trend and the detrended data, respectively.

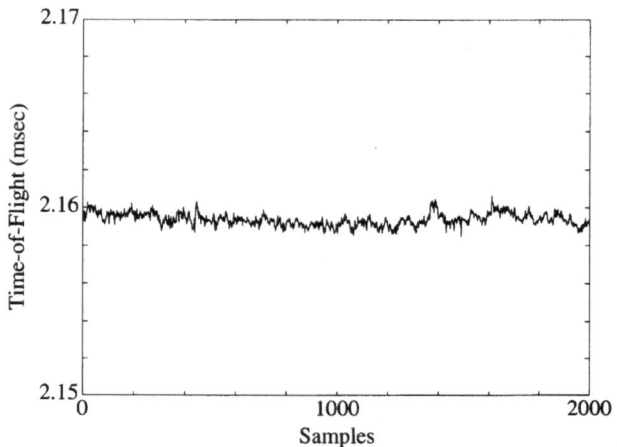

Figure1. Time history of the TOF measurements for the described experiment.

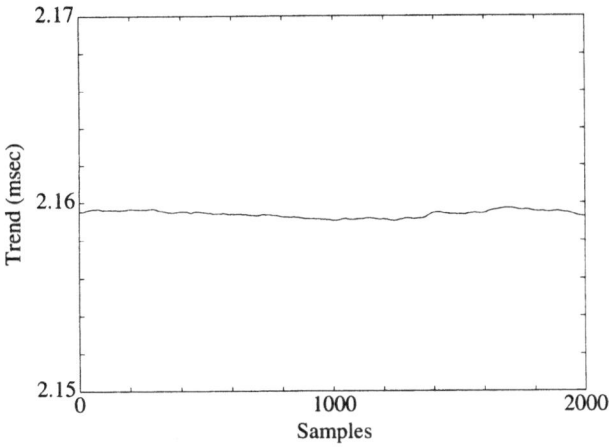

Figure 2. Time history of the time-varying mean of the TOF time series.

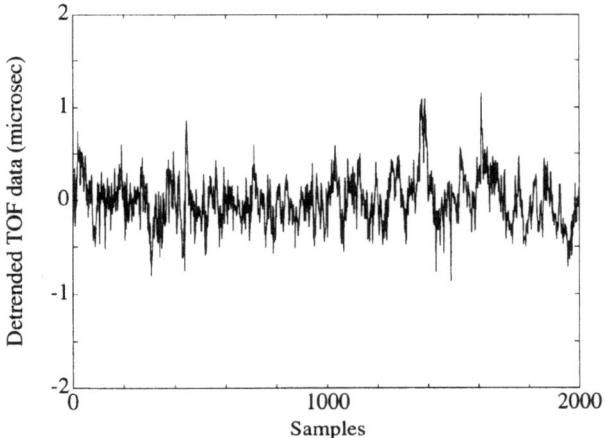

Figure 3. Time history of the detrended data subjected to ARMA modeling.

Short-term stationarity of the detrended data of Fig. 3 has been successfully tested. The quality of the fitting is only slightly affected by the order of the model. For the ARMA (1,1) model that is our preferred candidate, we obtain FPE = 2.167 10^{-14}; the minimum value is achieved for the ARMA (3,3) model (FPE = 2.152 10^{-14}). The whiteness of the prediction-error sequence has been confirmed for the selected ARMA (1,1) model by means of the Bartlett's test[9]. Using Eqs.(3)-(8) allows for deriving the desired estimates of the decorrelation time and the white and colored standard deviations ($\tau \approx 10$ sec, $\hat{\sigma}_u \approx 0.10$ μsec, $\hat{\sigma}_c \approx 0.26$ μsec). Finally, in Fig. 4 the correlation function of the innovations produced by the Kalman filter is plotted; the reported dashed lines are the bounds for the Bartlett's test at a level of significance p=95%. The whiteness of the innovation sequence proves the capability of the filter to properly separate the correlated component from the uncorrelated component.

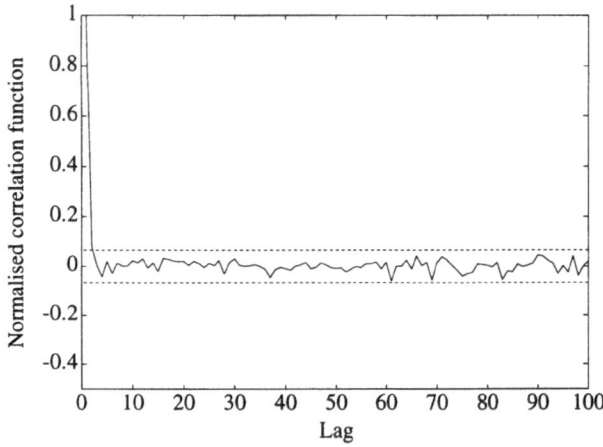

Figure 4. Correlation function of the innovations produced by the Kalman filter.

CONCLUDING REMARKS

The analytical tools described in this paper can be used to acoustically probe the environment where the sonar sensors are to be employed. Investigating how the model parameters change with the length of the acoustic path and quantitating the spatio-temporal correlation existing between the timing samples of multiple transducers have several practical implications, which include the development of a) signal processing techniques for improving the accuracy of differential TOF estimators in binaural sonar devices, b) averaging algorithms for improving the resulting SNR of echoes affected by random time shifts. These topics are currently inserted in our research agenda.

REFERENCES

1. R. Hickling and S. P. Marin, The use of ultrasonics for gauging and proximity sensing in air, *J. Acous. Soc. Am.*, 79(4):1151 (1986).
2. J. D. Fox, B. T. Khuri-Yakub and G. S. Kino, High-frequency acoustic wave measurements in air, *Proc. IEEE Ultrason. Symp.*, 581 (1983).
3. M. K. Brown, Feature extraction techniques for recognizing solid objects with an ultrasonic range sensor, *IEEE J. Robotics Automat.*, 1(4):191 (1985).
4. J. A. Kleppe. "Engineering Applications of Acoustics," Artech House, Inc., Norwood (1984).
5. H. L. Van Trees. "Detection, Estimation and Modulation Theory, Part I" J. Wiley & Sons, New York (1968).
6. A.M. Sabatini and E. Spinielli, Digital correlation techniques for time-of-flight measurements by airborne ultrasonic rangefinders, *Proc. IEEE/RSJ Int. Conf. Intell. Robots Syst.*, 2168 (1994).
7. P. S. Maybeck. "Stochastic Models, Estimation and Control," Academic Press, New York, San Francisco, London (1979).
8. A. Oppenheim. "Digital Signal Processing," Prentice Hall, Inc., Englewood Cliffs (1975).
9. G. E. P. Box and G. M. Jenkins. "Time Series Analysis," Holden-Day, Inc., Oakland (1976).
10. L. Ljung. "System Identification. Theory for the User," Prentice Hall, Inc., Englewood Cliffs (1987).
11. G. Bodenstein and H. M. Pretorius, Feature Extraction from electroencephalogram by adaptive segmentation, *Proc. IEEE*, 65(4):642 (1979).
12. J. S. Bendat and A. G. Piersol. "Random Data: Analysis and Measurement Procedures," Wiley Interscience, New York, London, Toronto (1971).

ULTRASONIC SETUP FOR FINGERPRINT PATTERNS DETECTION AND EVALUATION

Zbigniew Gumienny [1], Mieczysław Pluta [1], Wiesław Bicz [2], and Dariusz Kosz[2]

[1] Institute of Physics, Technical University of Wrocław
 Wyb. Wyspiańskiego 27, 50-370 Wrocław, Poland
[2] Research - Production Enterprise OPTEL Ltd.
 ul. Otwarta 10a , 50-212 Wrocław, Poland

INTRODUCTION

The scope of potential use of identification systems widens, including restricted area or confidential data access , credit card use etc. Only in 1993 banks lost more than three billions dollars due to credit cards forgery and misuse [1].

"No two fingerprints from different fingers have ever been found identical ..." [2]. That is why the fingerprint structure is successfully used in criminology [3] and is one of the properties that can be used for the person verification purpose [3]. Any fingerprint can be easily scanned and the verification process which goes along with it and lasts within pushing of a button seems natural and ergonomic. So far , optical sensors for the recognition of fingerprints, palm of the hand and the structure of eye blood vessels have been utilized and all of them have some disadvantages.

The authors have proposed and have been developing ultrasonic method of fingertip papillary lines representation acquisition [4]. We have stated that ultrasound is highly sensitive and brings out high contrast at the subsurface structure of the finger . One may expect that due to the uniqueness of the skin structure and its specific physical properties, the preparation of an artificial finger would be very complicated .

. At the early stage of the research we concentrated on the far field diffractive representation (or Fourier transform) of the fingerprint structure. Now we are able to measure pulse response representation of the finger-tip and perform reconstruction of papillary lines from the set of measured data. Some techniques of reconstruction quality improvements and results achieved are presented below.

DETECTION AND RECONSTRUCTION METHODS

In all of our setups finger-tip is applied to a window of ultrasonic head. The head contains one , two or a ringshaped matrix of electro-acoustic transducers (Figure 1a). Ultrasonic method of acquiring fingerprint representation is based on sending acoustic signals towards the finger and detecting the echo.

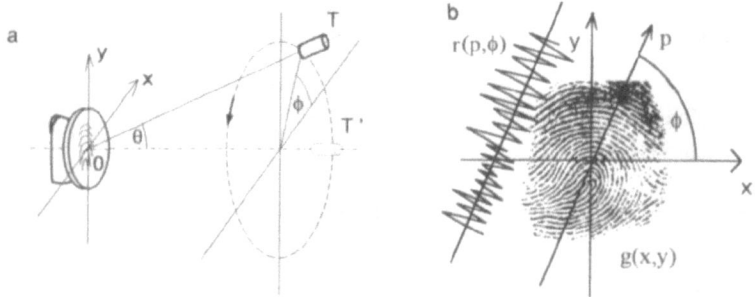

Figure 1. Scheme of the data acquisition head (a), Projection geometry (b),

Localization of the sender T 'and receiver T allows us to detect first diffractive order of the fingerprint structure :

$$\sin(\theta) = \lambda / d, \tag{1}$$

λ - ultrasound wave length   (0.25 mm in water by 6 MHz),
d - papillary line distance   (0.3 - 0.9 mm for most fingers) .
     what makes $\theta \approx 30°$.

Reflected pulse model

To get the data needed for the reconstruction of the finger-tip structure we apply the pulse method. The sender generates short acoustic pulses and the receiver , moving around a circle , detects the responses sampled then by a 50 MHz scope board. For 3D object observed by a flat receiver, tilted by angle Θ, space variables p, z and time t are related by the time of flight relation:

$$t = const - (p \cdot \sin(\theta) + z \cdot \cos(\theta)) / c, \tag{2}$$

c - sound velocity

The same relation holds for spherical cross-sections of the object and properly located point transducers.

By the fixed angle Θ, we can then measure time in millimeters and use the space variable p paralely with t . For 2-D flat object $g(x,y)$, the receiver rotated by the angle ϕ observes the scaled projection

$$r(p, \varphi) = \int_{-a}^{a} g(p \cdot \cos(\varphi) + s \cdot \sin(\varphi), -p \cdot \sin(\varphi) + s \cdot \cos(\varphi)) ds \tag{3}$$

In real case we have to take into account the shape of voltage spike $u_0(t)$, used to drive the transducer, and time depending transfer functions of the sender and receiver $t_S(t)$, $t_R(t)$. The echo $e(t)$ is then described by the multiple convolution

$$e(t,\varphi) = u_0(t) \otimes t_S(t) \otimes r_t(t,\varphi) \otimes t_R(t) = r_t(t,\varphi) \otimes h_0(t) \qquad (4)$$

where $h_0(t) = u_0(t) \otimes t_S(t) \otimes t_R(t)$ represents the pulse response of the setup and $r_t(t,\varphi)$ set of projections of the object scaled to the time domain. An example of the echo of a finger tip structure is presented in the Figure 2.

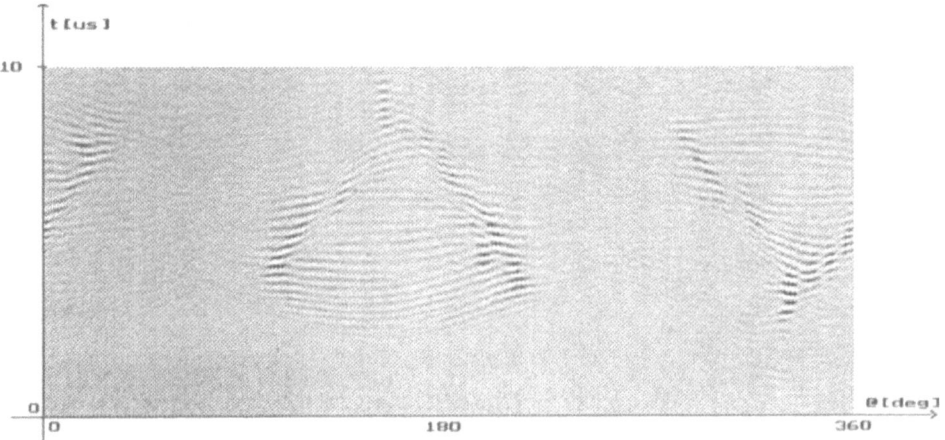

Figure 2. Visual presentation of measured echo matrix $e(t,\phi)$ of a fingerprint.

Point Spread Distribution of reconstruction

There exist well known methods of reconstruction of the 2D function $g'(x,y)$ from the set of projections $r(p,\phi)$. Back projection algorithm which we use is linear and stationary, which means that we can analyze the quality of reconstruction using 2D point spread distribution (PSD). The projection of a point $\delta(x0,y0)$ gives a trace in a form of a sinusoidal line, therefore during the reconstruction we find the value for every point by integration along such lines.

$$g'(x,y) = 1/\pi \cdot \int_0^{2\pi} e_p(z \cdot tg(\theta) - \rho \cdot \cos(\varphi - \varphi_0), \varphi)d\varphi, \qquad (5)$$

z - defocusing parameter, e_p - echo scaled to the space domain,

For a point object we get the 2D PSD function of the setup (including reconstruction algorithm). Due to the rotational symmetry of the PSD, it is enough to calculate and show its crooss- section. The shape of PSD depends on the defocusing z, and thus we get 3D PSD. Now we can consider transversal and axial resolution of reconstruction in Reileigh's sense (a distance from the maximum to the first zero of PSD in p and z direction).

641

There are three examples of pulse responses in Figure 3 .We have analyzed transducer with short pulse response,narrow band transducer (with many oscilations per pulse)   and   its signal after deconvolution.

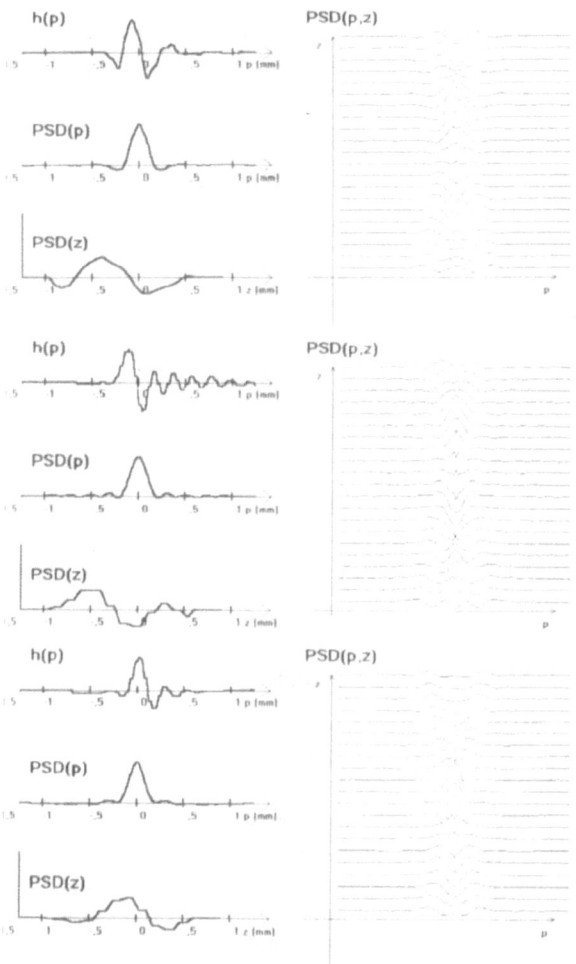

Figure 3. Examples of pulse responses h(p) , PSD and its transvwersal PSD(p) and axial PSD(z) cross-sections , for electronic setup and transducers of :
good quality (a), poor quality (b), poor quality transducer after deconvolution (c).

Figure 3. shows that even   using transducers of several periods per one pulse response we are able to obtain narrow central maximum of PSD, due to the suppression of side loops by the reconstruction integral (5). The disadvantage of such transducer is that we can get also a false reconstruction for another defocusing parameter .

Deconvolution dramatically improves the pulse response of the setup, or the amplitude and phase of of the signal spectrum specially at the higher frequencies (Figure3(c)). Sampling grain effects were not corrected.

ENHANCEMENTS OF THE RECONSTRUCTION

Due to the limited quality of transducers , electronic noises and interference with signals from deeper layers , we have to use some enhancement techniques to obtain satisfactorily reconstructed image. The first of them is deconvolution of the echo. Also three methods of 2D image processing are used by us:
smoothing , directional filtering in spectral domain and binearization.

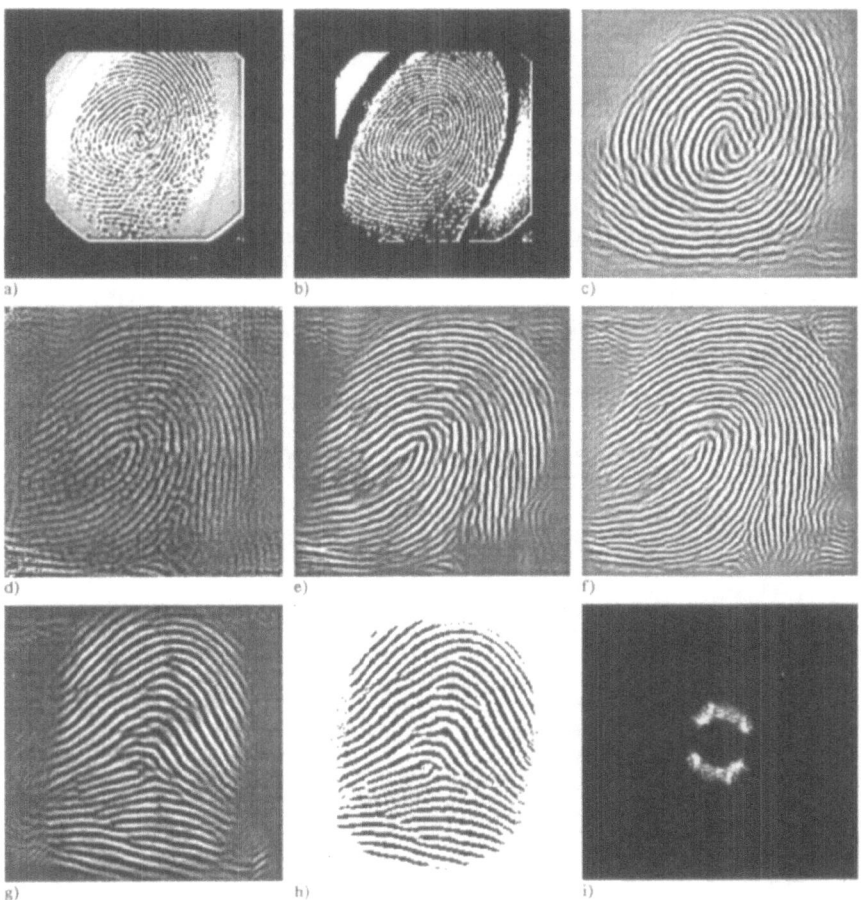

Figure 4.   (a) - optical picture of a fingerprint,
        (b) - optical picture after binearization,
        (c) - reconstructed an enhanced acoustic picture of the same finger,
        (d) - acoustic fingerprint reconstruction,
        (e) - acoustic fingerprint reconstruction after directional filtering,
        (f) - acoustic fingerprint reconstruction after deconvolution and directional filtering,
        (g) - another acoustic fingerprint reconstruction after deconvolution and directional filtering,
        (h) - binearization of (g),
        (i) - 2D Fourier spectrum of (g),

We see suppresion of the noises after directional filtering (Figure 4e), and appearance of fine detals after deconvolution (Figure 4f).

CONCLUSIONS

Measurements and reconstructions of finger-tips show that acoustic field diffraction occurs mainly on fingerprint lines. Quality of the reconstructed images depends on the bandwidth of the transducer , but may be improved by some signal and image enhancement methods. Reconstructed images are so similar in form to the optical ones that one can use the same classification and recognition methods. The advantage of the ultrasonic method is the uniqueness of the acoustic properties of the finger-tip , which means that preparing its dummy should be very difficult.

Presented setup may by also used for measuring of other objects and be treated as synthetic aperture microscope.

REFERENCES

1. Kriminalität , "Mafia nera" , Der Spiegel, No.11, p. 81, März 1994.
2. A.A. Moenssnens, "Fingerprint Techniques " , Chilton Company (1971).
3. Cz. Grzeszyk , "Daktyloskopia ", PWN, Warszawa, 1994,(in polish).
4. R.H. Andersen, P. Jürgensen , "Fingerprint Verification - For use in Identity Verification Systems" , Aalborg University , (1993).
5. W. Bicz, Z. Gumienny, M. Pluta, "Ultrasonic sensor for fingerprints recognition ", COE '94 Warszawa , Poland, (to be published by SPIE).

SCANNING TOMOGRAPHIC ACOUSTIC MICROSCOPY: A HISTORICAL OVERVIEW

Hua Lee

Department of Electrical and Computer Engineering
University of California, Santa Barbara

ABSTRACT

The development of the Scanning Tomographic Acoustic Microscope (STAM) spans over a period of over ten years and involves every facets of engineering ranging from mathematical and physical modeling, theoretical analysis and signal processing, hardware and software development, and overall system optimization. This paper provides a summary of the technical development, starting from the initial development of the concept, preliminary system prototyping, holographic and tomographic image formation, and system integration.

INTRODUCTION

The initial concept of a new imaging approach can be in various forms, ranging from data acquisition configurations to image formation techniques. Nevertheless, the successful development of the approach involves the integration of many technical components. In this paper, the development of the scanning tomographic acoustic microscope (STAM) is utilized as an example of system integration and optimization. The paper is an overview of the complete research program, starting with the formation of initial concept for subsurface imaging, development of image reconstruction algorithms, modification of data acquisition hardware, and various challenges during the process of holographic and tomographic image formation. It also illustrates the contributions, from related areas such as estimation and detection theory as well as computer vision, to the integration and optimization process. The results and experiences of this research program have also made profound contributions to the advancement of other imaging projects including synthetic-aperture sonar imaging, microwave subsurface NDE, radar RCS estimation, and image formation of time-varying objects.

CONCEPT OF TOMOGRAPHIC ACOUSTIC MICROSCOPY

The fundamental concept of scanning tomographic acoustic microscopy (STAM) was first established in 1980. At that time, it was in a form of theoretical development of an image formation technique without specific system configurations. An outline of the technique was first presented in the 11th International Symposium on Acoustical Imaging and published with the title *Ultrasound Planar Tomography*.

The algorithm is structured based on the plane-to-plane wavefield backward propagation model with a simple demodulation step for the removal of the effects due to the illumination wavefield. In terms of theoretical framework, it is equivalent to the technique developed by A. J. Devaney earlier for applications in inverse scattering.

The technique was known as planar ultrasound tomography is due to the signal filtering implementation which is built upon a simple transfer function associated with the plane-to-plane model. This particular transfer function was selected for implementation because of the characteristics. It is lowpass, lossless within the passband, and stable, which matches with the implementation criteria of stability, wavefield sampling, cascade realization format, and sensitivity level to noise.

The development was limited to an elegant piece of theoretical work until we decided to apply it to acoustic microscopy after a technical discussion with Larry Kessler of Sonoscan. The application was a very logical choice because of the planar wavefield detection scheme of *Scanning Laser Acoustic Microscope* (SLAM). In the mean time, the application to acoustic microscopy appeared most exciting and challenging due to the microscopic scale and dimension.

During the early period, the research program was supported by the UC MICRO Program with industrial partner of Santa Barbara Research Center of Hughes Aircraft Company. Z. C. Lin joined the research team as the first graduate assistant in 1982, and performed an in-depth theoretical study of system performance and resolution analysis.

The output signal of a conventional SLAM system is the intensity of the wavefield distribution. In order to achieve subsurface imaging, the phase information becomes critical. Thus, data acquisition system of SLAM needs to be modified to detect the complete complex wavefield.

HOLOGRAPHIC AND TOMOGRAPHIC IMAGE FORMATION

In 1984, propelled by a new NSF grant, the prototyping of the STAM system started at the University of Illinois on an early model of SLAM. Carlos Ricci then joined the research team as a graduate student responsible for the modification with a quadrature receiver for the detection of complete surface wavefield. The output signal was first down converted from the operating frequency of 100 MHz to 32.4 MHz. Then the quadrature receiver, operating at 32.4 MHz, detects the wavefield and outputs the real and imaginary components of the wavefield through the twin channels sequentially. In the same period, Z. C. Lin joined the Illinois team to implement data conversion, signal processing, and image formation procedures. In early 1986, the first holographic reconstruction of subsurface profiles was successfully performed, which was widely regarded as a critical step toward the full-scale realization of STAM.

Subsequent to the successful experiment of holographic subsurface imaging, the research effort turned its focus to a more ambitious objective of tomographic imaging. As expected, the research program was elevated to a level of extreme complexity especially for the component related to signal analysis and processing.

The tomographic imaging schemes include both the multiple-frequency and multiple-angle operating formats. During a long period of time, our effort toward tomographic imaging capability encountered various technical challenges. These difficult technical problems have been attacked and solved sequentially through mainly software development, which has become an excellent example of direct applications of mathematical modeling, estimation and detection theory, and high-performance signal processing algorithms.

The first task was the data acquisition error at the stage of quadrature-receiver operation. The error in the twin-channel outputs was first identified in a form of background fringe patterns in the holographic images. For error removal, Lin first proposed and implemented a side-band elimination technique based on the assumption of constant phase error. The applicability of this approach was limited to holographic reconstruction because the error was not totally removed and remained embedded in the other sideband. Richard Chiao approached the problem with a different model by relaxing the constant-phase error assumption and developed the self-checking single-sideband algorithm, which effectively removed the data acquisition error. Due to the effectiveness of this technique, it has not only been widely applied in acoustic systems, but also propagated into microwave data acquisition devices in various applications.

Subsequently to the successful removal of quadrature-receiver error, we encountered another layer of system error during the superposition process of multiple-frequency tomography. This is due to the inconsistent reference position, in time, of the acoustic illumination. Typically, the time-axis reference was implemented by hardware electronics by using signal mixers. Due to the inaccuracy of the electronic reference signal, largely because of the different operating frequencies and associated hardware filtering components, there exists a phase offset in each sub-image. This phase offset seriously degraded the effectiveness of the superposition process. This error was identified and successfully removed by a cross-correlation technique.

The most challenging task came as we engaged in the image formation procedure for multiple-angle tomography. The rotational scan was conducted manually so displacement error was not negligible. This is because the image resolution is extremely sensitive to displacement error at a microscopic level, which is in the order of wavelength. On the other hand, due to the lack of point features and correspondence information, conventional image registration methods were not adequate for either the accuracy or computational requirement. To overcome this difficult problem, a singular value decomposition (SVD) based approach was developed. The rotation component of the error was constructed from the matrix components of the singular value decompositions of the autocorrelation matrices before and after the rotation, and the translation error was obtained from the change of centroids. This technique provided an excellent balance of accuracy and computation time and, as a result, was responsible for the successful realization of multiple-angle acoustic tomographic microscopy in 1989.

The proof of concept as well as the first holographic and tomographic reconstructions were performed on an early version of SLAM at the campus of the University of Illinois. The simplicity of the hardware electronics of the early model SLAM system provided an excellent degree of flexibility for modification into a STAM prototype. The research staff, Z. C. Lin and Richard Chiao, were responsible for many important technical milestones and made enormous contributions to the overall success of this program.

FULL-SCALE SYSTEM INTEGRATION

In 1992, 3M made available a more recent model of SLAM to the Imaging Systems Laboratory at UC Santa Barbara, and the full-scale integration of STAM took place subsequently. Davis Kent was responsible for the complete system integration and optimization. Currently the STAM system is equipped with a dual-channel quadrature receiver, a computer-control rotating stage for tomographic scan, and a complete software support for data preprocessing, error reduction, holographic and tomographic image formation, and visualization display.

In the current system, the quadrature receiver is operating at the frequency of 36.4

Figure 1. Multiple-frequency tomographic image of a US Penny.

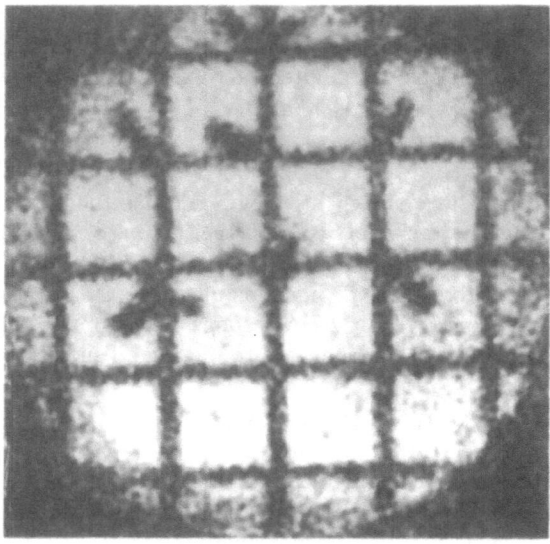

Figure 2. Multiple-angle tomographic image of a subsurface layer.

MHz, after down-shifting the detected signal from 100 MHz. The analog-to-digital conversion is performed on both channels simultaneously, instead of sequentially in the early prototype. The rotating stage is capable of performing the rotational scan with precision up to 1/1000 degree per increment.

Experiments have been performed extensively with various test specimens for system performance evaluation and limitation study. To demonstrate the capability, two experimental results are included here in this paper. One experiment is the multiple-frequency image formation of a US penny. The data acquisition is conducted on one side of the penny and then the detected wavefield is backward propagated to the other side of the penny, with a propagation distance of 630 microns, to form the image, shown as Figure 1. Another experiment is a multiple-angle tomographic reconstruction of a specimen containing a subsurface grid layer. Figure 2 shows the reconstruction of the grid layer reconstructed from 36 projections and the propagation distance is 690 microns.

CONCLUSION

The development of STAM has been regarded as an exciting engineering expedition. The research period spanned over more than ten years, and the research program involved two campuses, three generations of researchers, and every facet of engineering advancement. This paper provides an overview from the historical perspective, summarizing the development of concept, preliminary experiments, and final system integration.

ACKNOWLEDGMENT

This research program was supported by Santa Barbara Research Center, UC MICRO Program, National Science Foundation, and 3M.

REFERENCES

Hua Lee, Carl Schueler, Gail Flesher, and Glen Wade, "Ultrasonic Planar Scanned Tomography," *Acoustical Imaging,* vol. 11, J. Powers, Ed., Plenum Press, New York, pp. 309-323, 1982.

Hua Lee, Carl Schueler, Gail Flesher, and Glen Wade, "Ultrasonic Planar Scanned Tomography," *Acoustical Imaging,* vol. 11, J. Powers, Ed., Plenum Press, New York, pp. 309-323, 1982.

Zse-Cherng Lin, Hua Lee, and Glen Wade, "Back-and-Forth Propagation for Diffraction Tomography," *IEEE Transactions on Sonics and Ultrasonics,* vol. SU-31, no. 6, pp. 626-634, November 1984.

Zse-Cherng Lin, Hua Lee, and Glen Wade, "Scanning Tomographic Acoustic Microscope: A Review," *IEEE Transactions on Sonics and Ultrasonics,* vol. SU-32, no. 2, pp. 168-180, March 1985.

Hua Lee and Carlos Ricci, "Modification of the Scanning Laser Acoustic Microscope for Holographic and Tomographic Imaging," *Applied Physics Letters,* 49(20), pp. 1336-1338, November 1986.

Zse-Cherng Lin, Hua Lee, Glen Wade, Michael G. Oravecz, and Lawrence W. Kessler, "Holographic Image Reconstruction in Scanning Laser Acoustic Microscopy," *IEEE Transactions on Ultrasonics, Ferroelectrics, and Frequency Control,* vol. UFFC-34, no. 3, pp. 293-300, May 1987.

Hua Lee and Richard Y. Chiao, "Holographic Acoustic Microscopy for Quantitative Velocity Profile Imaging," *Journal of Acoustical Society of America,* 85(3), pp. 1375-1376, March 1989.

Hua Lee and Richard Chiao, "Phase Error Estimation and Correction in Acoustic Microscopy," *IEEE Transactions on Acoustics, Speech, and Signal Processing,* vol. 38, no. 1, pp. 171-173, January 1990.

Hua Lee and Richard Chiao, "Tomographic Reconstruction of Multiple-Layer Specimens with the Scanning Tomographic Acoustic Microscope," *International Journal of Imaging Systems and Technology,* Vol. 2, No. 3, pp. 200-202, 1990.

Richard Chiao and Hua Lee, "Recent Advances in Scanning Tomographic Acoustic Microscopy," *International Journal of Imaging Systems and Technology,* Vol. 3, No. 3, pp. 334-353, 1991

Bretton L. Douglas, S. Davis Kent, and Hua Lee, "The Use of Contrast and Entropy Criteria for the Estimation of Imaging Parameters in Acoustic Microscopy," *Journals of Acoustical Society of America,* Vol. 95, No. 1, pp. 306-312, January 1994.

Richard Y. Chiao and Hua Lee, "Scanning Tomographic Acoustic Microscopy," *IEEE Transactions on Image Processing,* 4(3), pp. 358-369, March 1995.

S. Davis Kent and Hua Lee, "Quantization Bit Rate Reduction in Scanning Tomographic Acoustic Microscopy," *IEEE Transactions on Ultrasonics, Ferroelectronics, and Frequency Control,* vol. 42, no. 3, pp. 464-477, May 1995.

Richard Y. Chiao and Hua Lee, "Estimation and Correction of Quadrature Signal Error for High-Resolution Acoustic Microscopy," *Journal of Acoustical Society of America,* 97(3), pp. 1987-1990, March 1995.

BIOMEDICAL APPLICATIONS OF ACOUSTICAL MICROSCOPY

Joie P. Jones

Department of Radiological Sciences
University of California Irvine
Irvine, CA 92717 USA

INTRODUCTION

Even from the beginnings of acoustical microscopy in the mid-1970's, biomedical problems were thought to be an important application area for this new technology. In fact, some of the first acoustical microscopy images produced by both Quate at Stanford and Kessler at Zenith were of biological structures. Unfortunately, over the past two decades the wide application of acoustical microscopy to biomedicine never developed as initially predicted. Recently, however, a new generation of biomedical scientists have rediscovered acoustical microscopy and are applying this tool to a myriad of applications in both the basic sciences as well as diagnostic medicine. Biologists are beginning to apply acoustical microscopy to the study of living cells, including cellular development and biochemical processes, in a way simply not possible with any other imaging modality. Pathologists are beginning to use acoustical microscopy to develop a new gold standard by which tissue changes and disease processes may be defined. Medical scientists are developing new tools, based on acoustical microscopy technology, for the noninvasive and minimally invasive diagnosis of tissue state.

In this brief review we outline the historical developments of acoustical microscopy, indicate the advantages and disadvantages of this modality, suggest reasons why this technology has not developed more extensive biological applications, and describe several biomedical applications which show considerable promise.

HISTORICAL DEVELOPMENTS

The basic idea to image structures at the microscopic level using acoustics seems to have been first proposed by Sokolov [1] in 1936. He noted [2] that the wavelength of a 3 GHz sound wave in water is equal to the wavelength of green light and proposed an imaging system which might compete with optical instruments. The proposed system was to read out the localized electric charge developed on a piezoelectric crystal in response to an acoustic input. For various technical reasons such systems are difficult to construct at high frequencies and today this concept is universally limited to low frequency ultrasonic imaging.

Serious attempts to develop an acoustical microscope had to wait until after the early 1960's and the development (largely at Bell Labs, UCLA, and UCB) of very-high frequency ultrasonic transducers. With this technology in hand, two groups in the U. S. developed two very different acoustical microscopy systems. At the Zenith Laboratory in Chicago, Adrian Korpel and Larry Kessler developed a system in which the acoustical wave traveled through the sample and deformed a plastic membrane which, when scanned with a laser beam, yielded an image [3]. This general type of system has become known as a Scanning Laser

Acoustical Microscope (or SLAM). Such systems available today are generally easy and fast to use, are capable of producing images in real time, but are limited in frequency to those below a few 100 MHz.

Almost in parallel with the developments at Zenith, Bertram Auld and his students at Stanford analyzed several possible acoustical imaging systems while Quate and Lemons at the same institution built a direct scanning system which did not require a laser [4]. This general type of system has become known as a Scanning Acoustical Microscope (or SAM). Such systems available today are generally more difficult to use than SLAM, do not produce real-time images, but can operate at much higher frequencies, well into the GHz range.

BIOMEDICAL APPLICATIONS

Some of the first applications with both SLAM and SAM were in biomedicine [4,5,6]. Such applications have continued to play a major role [7-14], but have not made the impact once expected and anticipated. There are strong physical arguments as to why acoustical microscopy should play a major role in biomedicine. For example, the optical properties of soft tissue vary by only 0.5% whereas the acoustical properties range over two orders of magnitude. Thus, optical microscopy can visualize tissue structures with even a modest degree of sensitivity only if some staining techniques are used which can emphasize particular tissue features of interest. Such staining techniques can be quite involved and elaborate; many can take several days to implement. Thus, with optical microscopy it is virtually impossibly to scan fresh or living tissue. Acoustical microscopy, on the other hand, has great inherent sensitivity and requires no staining to bring out the features of interest. Thus, acoustical methods permit us, in principle, to scan fresh as well as living tissues. Although, as will be seen, many similarities do exist between the optical and the acoustical images, many of the important differences (such as the differences between the optical and elastic properties of tissue) are not well known or fully appreciated. Part of the problem centers on the fact that many of the subtle features noted in the acoustical image are not understood or have not yet been correlated with a particular biological process. It is difficult to create a new gold standard when the new one is significantly more sensitive and reveals details as yet uncorrelated with other processes.

To illustrate several important biomedical applications of acoustical microscopy, I have chosen a sampling of acoustical images made in my laboratory at the University of California Irvine over the past six years using an Olympus UH3 Scanning Acoustical Microscope. The UH3 operates in the reflection mode; i.e., a fixed frequency piezoelectric transducer generates a pulse of ultrasonic energy which is focused by a sapphire lens through a drop of coupling liquid (in this case, water), through the sample (in most of these cases a 6 micron thick section of tissue), and onto the interface between the tissue and an underlying glass block. Transducers are available in a range of frequencies between 200 MHz and 1 GHz. The echo reflected from the focal point returns by the same path to the transducer where it is detected and the maximum returning signal displayed as a single point of some relative intensity on a video monitor. Scanning the sample in a rectilinear fashion and keying this motion to the spot position on the display yields an image of the sample which is parallel to the face of the sample but perpendicular to the scanning ultrasonic beam. Such an image is often termed a C-mode display. Note that with the transducer focused on the highly reflecting interface behind the sample the resulting image is, in effect, a relative mapping of attenuation in the sample.

Much of my work in acoustical microscopy over the past six years has been with my clinical colleagues in Dermatology at UCI where this instrument has become an important and useful clinical tool. Our studies began with a detailed comparison between optical and acoustical microscopy. A typical result is shown in Figure 1. Here a biopsy specimen of a malignant melanoma is viewed optically (on the left) and acoustically (on the right). The image on the left is a standard H & E section; i.e., the tissue has been fixed and stained to bring out those features which could lead to a diagnoses. The acoustical image on the right was made from the adjoining tissue block and was neither fixed nor stained. However, it was frozen to permit accurate slicing. Both samples were 6 microns thick. Since the optical images represented the standards used for making a pathological evaluation, we attempted to produce (at least initially) acoustical images which were similar to their optical counterpart. This led us to image at 600 MHz with a 2 mm field of view in most cases. Although the

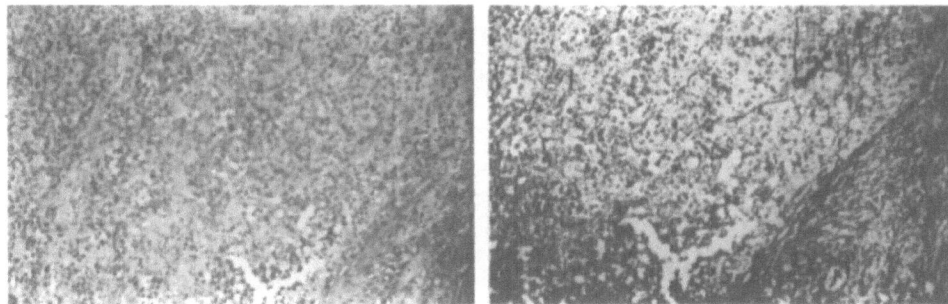

Figure 1. Left: Optical microscope image of malignant melanoma obtained at biopsy. A standard H & E section. **Right:** Acoustical microscope image (600 MHz) of adjacent section of same lesion. Here the tissue is neither fixed nor stained. Dimension across the bottom of each image is about 2 mm. Note the dermis in the lower right of each image. The remaining tissue is all tumor. The cells are arranged in confluent nests with some loss of cohesion. The acoustical image has greater contrast. Both images are diagnostic.

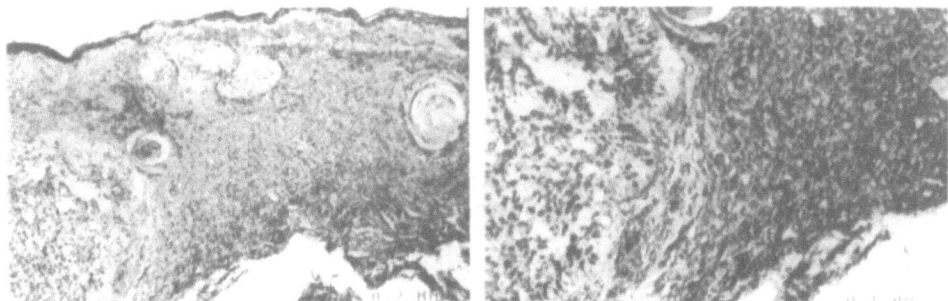

Figure 2. Left: Acoustical image (600 MHz) of biopsy section of basal cell carcinoma. **Right:** Detail of lower left hand section of image on the left. Note the individual basal cells (small solid black structures).

acoustical image has more contrast but less detail than the optical image (at least in this case) it proves to be equally diagnostic. Such has been the case for several hundred similar images over a wide range of pathologies. Thus, although the acoustical images are different than their optical counterparts, they are easily recognized by a trained pathologist.

Figures 2 and 3 are from an extensive retrospective study on basal cell carcinoma (BCC). Figure 2 is a very typical BCC. Note that the individual basal cells are easily seen, especially in the more detailed image on the right. Figure 3 provides a series of acoustical images of ever narrowing field of view of a particular subset of BCC known as keratotic BCC. The horn cysts characteristic of this subtype are easily seen in the image at the upper left which has the largest field of view. Moving clockwise around the four images, the field of view continues to narrow and the dark foci of keratin become more evident. The image at the lower left has the smallest field of view and provides the greatest detail of the keratin centers. The white structure at the center of several of these is calcium. Given the acoustical properties of calcium, it is not surprising that acoustical microscopy is far better than optical means, even with staining, in evaluating keratotic BCC. In fact, acoustical microscopy proved equal to optical means in the initial classification of BCC type. This is particularly impressive given the fact that acoustical methods require no staining or elaborate sample preparation.

The two images in Figure 4 are perhaps the most interesting ones we have acquired. On the left is an acoustical image of a biopsy sample from the heart of a heart transplant patient at USC Medical Center. The dark black circles on the right are regions of necrosis, representing early rejection of the organ. Since this process is seen much earlier with acoustical microscopy than with optical means, it provides a way of intervening before further complications arise.

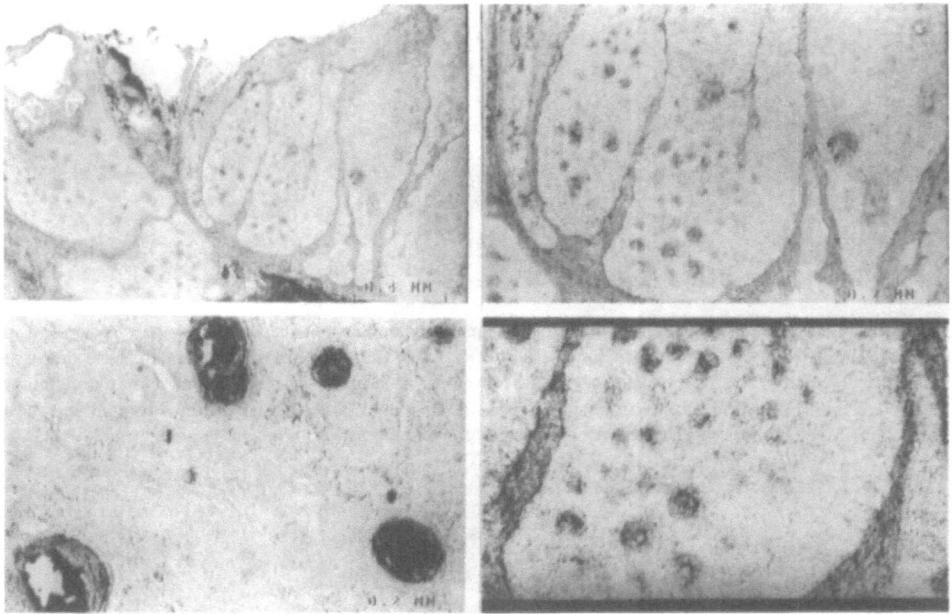

Figure 3. Acoustical images (600 MHz) of Keratotic Basal Cell Carcinoma. The field of view decreases moving from the upper left clockwise. Note the foci of keratin, especially in the lower left hand image.

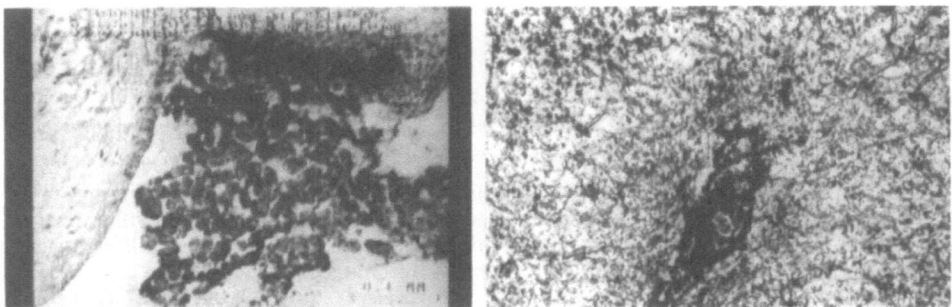

Figure 4. Acoustical images (600 MHz). **Left:** Biopsy from heart transplant patient showing early necrosis. **Right:** Section of human brain with Alzheimer's plaque.

On the right of Figure 4 is an acoustical image of a section of human brain with Alzheimer's disease. The plaque associated with this disease process is the large black structure in the lower center of this image. This plaque is very difficult to visualize with optical microscopy even with elaborate staining techniques. With acoustical microscopy, the plaque is immediately evident even with a fresh tissue sample that has been neither fixed nor stained. Clearly, acoustical microscopy offers a new tool to investigate Alzheimer's disease and other forms of dementia on a laboratory basis using various animal models as well as cell layers in culture media. By better understanding how the elastic properties of tissue are changed by this disease process we may be able to develop ultrasonic techniques for evaluating Alzheimer's disease on an in vivo basis.

At our institution acoustical microscopy is already playing a significant role in the clinic. My dermatology colleagues are using this technology on a regular basis to evaluate skin biopsy samples. They can obtain an immediate assessment of the pathology long before standard techniques have even begun. Our goal is to provide the same assessment of pathology without requiring a biopsy. We believe a specialized ultrasound scanner applied directly to the skin could provide acoustical microscopy on an in vivo basis.

In vivo acoustical microscopy may also be possible with a catheter based system in which a small millimeter sized transducer is threaded via an artery or some other small passageway to the region of interest. This would mean that the heart transplant patient of Figure 4 could be evaluated on a regular basis without a biopsy. Such in vivo systems are under development at a number of institutions, including ours. Coupled with the development of new prototype systems better suited for laboratory work in the biological sciences, acoustical microscopy may well be entering a new phase of development.

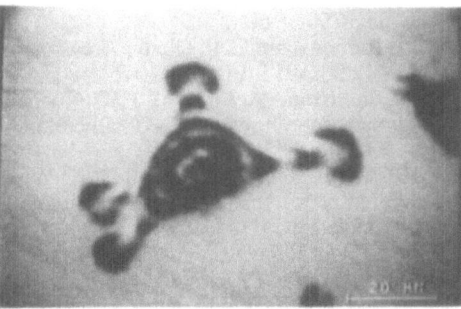

Figure 5. Acoustical images (600 MHz) of human HeLa cells in culture.

There are at least two reasons for the new interest in acoustical microscopy. First, a new generation of biomedical scientist seems open to new approaches and new technology. Twenty years ago it was virtually impossible to get a pathologist to even look at an acoustical image. Today a small but growing number of pathologists are using acoustical microscopy as a tool.

Secondly, recent work has used a range of frequencies which were previously unavailable and which seem particularly useful for pathological evaluation of tissues. The initial instrumentation operated either at a few hundred MHz, defined by the limits of attenuation, as with SLAM, or in the low GHz range, in an attempt to gain the highest resolution, as with SAM. Frequencies in the range 400 to 800 MHz were largely ignored, although it is precisely in this range that images can be made which are best compared with light microscopy. In this same range an important transition occurs between the wavelength of the sound and the scatterer size such that the rather fuzzy images produced between 1 and 200 MHz now become crisp and well defined at about 400 MHz.

Finally, Figure 5 provides an example of how acoustical microscopy can be used as a fundamental tool in the biological sciences. Here we have acoustical images of a particular human cell line growing in culture. The dark concentric rings are produced by diffraction and provide a means to determine the size of this three dimensional structure. The ability to observe and study living cells in this detail would be impossible with any optical technique since most of these structures are optically opaque and since their optical properties range over such a small value.

ACKNOWLEDGMENTS

My sincere thanks go to Dr. Jacqueline Gallet, formally a graduate student with me and now at the University of Toronto, who actually made most of the images in this presentation and to Dr. Ron Barr, a long standing clinical colleague who, over the years, has provided great insights and lively discussions.

REFERENCES

1. S. Sokolov, USSR Patent No. 49 (31 August, 1936); British Patent No. 477139 (1937); U. S. Patent No. 21 64 125 (1939).
2. S. Sokolov, The ultrasonic microscope, *Akademia Nauk SSSR, Doklady,* 64: 333-345(1949).
3. A. Korpel, L. W. Kessler, and P. R. Palermo, Acoustic microscope operating at 100 MHz, *Nature,* 232: 110-111 (9 July 1971).
4. L. W. Kessler, A. Korpel, and P. R. Palermo, Acoustic microscopy of biological specimens, presented at the 17th Annual Meeting of the American Institute of Ultrasound in Medicine, Philadelphia, PA (29 November 1972).
5. L. W. Kessler, VHF ultrasonic attenuation in mammalian tissue, *J. Acoust. Soc. Am.* 53:1759-1760 (1973).
6. R. A. Lemons and C. F. Quate, Acoustic microscopy: biomedical applications, *Science* 188:905-911 (30 May 1975).
7. R. N. Johnston, A. Atalar, J. Heiserman, V. Jipson, and C. F. Quate, Acoustic microscopy: resolution of subcellular detail, *Proc. Natl. Acad. Sci. USA* 76:3325-3329 (July 1979).
8. J. Heiserman, C. F. Quate, and H. K. Wickramasinghe, Acoustic microscopy in biophysics, *in*: "Advances in Biological and Medical Physics, Volume 17," 325-364, Academic Press (1980).
9. J. A. Hildebrand, D. Rugar, R. N. Johnston, and C. F. Quate, Acoustic microscopy of living cells, *Proc. Natl. Acad. Sci. USA* 78:1656-1660 (March 1981).
10. J. A. Hildebrand and D. Rugar, Measurement of cellular elastic properties by acoustic microscopy, *J. of Microscopy* 134:245-260 (June 1984).
11. J. E. Olerud, W. O'Brien, M. Riederer-Henderson, D. Steiger, F. K. Forster, C. Daly, D. J. Ketterer, and G. F. Odland, Ultrasonic assessment of skin and wounds with the scanning laser acoustic microscope, *J. Invest. Dermatol.* 88:615-623 (1987).
12. O. V. Kolosov, V. M. Levin, R. G. Mayev, and T, A. Senjushkina, The use of acoustic microscopy for biological tissue characterization, *Ultrasound in Med. & Bio.* 13:477-483 (1987).
13. N. Buhles and P. Altmeyer, Ultrasonic microscopy of skin sections, *Zeitschrift fur Hautkrankheiten* 63:926-934 (1988).
14. R. J. Barr, G. M. White, J. P. Jones, L. B. Shaw, and P. A. Ross, Scanning acoustic microscopy of neoplastic and inflammatory cutaneous tissue specimens, *J. Invest. Dermatol.* 96:38-42 91991).

MATERIALS CHARACTERIZATION BY SURFACE ACOUSTIC WAVES FROM 200 MHz TO 20 GHz

G.A.D. Briggs, O.V. Kolosov, and M.M. Puentes Heras

Oxford University
Department of Materials
Parks Road
Oxford OX1 3PH
England

INTRODUCTION

There are many occasions when it is desirable to be able to characterize the mechanical properties of engineering surfaces. The surface of a material may have been prepared mechanically, by grinding or polishing, or it may have been implanted with ions, or it may have had one or more layers or coatings grown or deposited on it. In such situations the surface can be characterized by measuring its elastic properties. Rayleigh waves are useful for this purpose, because they have longitudinal and shear components which decay exponentially with depth below the surface. Since their energy is confined to the surface in this way, Rayleigh waves are also known as surface acoustic waves. Their velocity and attenuation are determined by the elastic properties and structure of the depth of material to which they are confined.

Two methods of making such measurements are now available which together enable the depth to be sampled to be chosen from more than 20 μm to less than 0.3 μm. The techniques are quantitative acoustic microscopy (QAM) and surface Brillouin spectroscopy (SBS). The depth sampled is approximately equal to the Rayleigh wavelength, which can be chosen by selecting an appropriate frequency. The wave will be a pure Rayleigh wave only in the case of a material which is isotropic with properties that are uniform to some depth much greater than a Rayleigh wavelength. In most cases the structure will be more complicated than this, and then a whole wealth of mathematical and experimental knowledge about the propagation of waves in such structures can be brought to bear in order to extract as much information as possible about the materials from the measurements of the surface acoustic waves.

Acoustical Imaging, Vol. 22, Edited by P. Tortoli
and L. Masotti, Plenum Press, New York, 1996

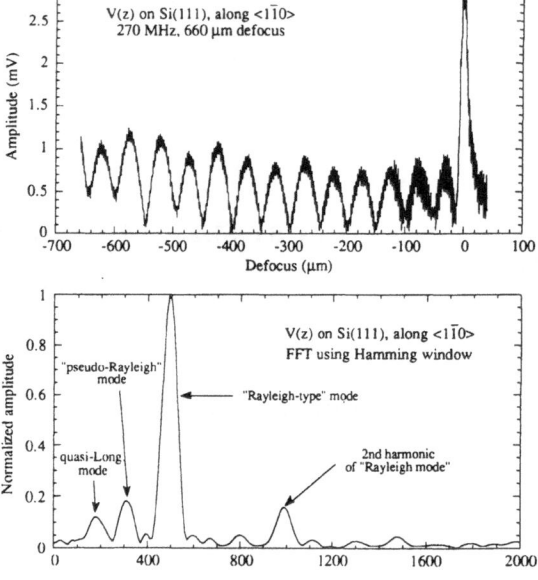

Figure 1. An analysis of 270 MHz $V(z)$ data in a $\langle 1\bar{1}0 \rangle$ propagation direction on Si(111): the upper figure contains the measured data; the lower figure shows the result of applying a fast Fourier transform with a Hamming window.[3]

QUANTITATIVE ACOUSTIC MICROSCOPY

Acoustic microscopy has remarkable sensitivity to factors which affect the propagation of Rayleigh waves in the surfaces of materials.[1] By defocussing the sample towards the lens interference can be obtained between specularly reflected rays and rays that excite Rayleigh waves in the sample. This interference can be optimised to give strong contrast in acoustic images from features that affect the propagation of Rayleigh waves. The same effect can be exploited to give quantitative measurements of Rayleigh wave propagation. For this purpose a cylindrical (line-focus beam) lens is often used instead of a spherical (point-focus beam lens).[2] The cylindrical lens excites Rayleigh waves in only one azimuthal direction in the surface at a time, and so enables Rayleigh velocity and attenuation to be measured as a function of direction in anisotropic surfaces. The use of cylindrical lenses also allows certain useful approximations to be made in the analysis of the raw data.

Quantitative measurements are made by moving the lens relative to the sample in a direction normal to the sample surface, instead of scanning a raster parallel to the surface as would be done for imaging. Data are shown in Fig. 1 from a measurement on a Si(111) wafer, with the cylindrical lens oriented to excite Rayleigh waves in a $\langle 1\bar{1}0 \rangle$ direction in the surface.[3] The horizontal axis indicates the distance z between the surface of the sample and the focal plane of the lens, with negative defocus corresponding to moving the lens towards the sample. The vertical axis is the video detected signal V from the transducer. Such data are known as $V(z)$ curves. It can be seen that there are oscillations in $V(z)$ of regular periodicity. The periodicity can be found by taking a Fourier transform of $V(z)$, as shown in the lower graph. Peaks corresponding to other acoustic modes in the surface are also indicated, but by far the strongest peak is the one corresponding to Rayleigh waves. The period Δz of the oscillations in $V(z)$ is related to the angle of incidence θ_R of the rays from the lens that excite Rayleigh waves in the sample by

$$\Delta z = \frac{\lambda_0}{2(1 - \cos\theta_R)},$$ [1]

where λ_0 is the wavelength of the acoustic waves in the coupling fluid between the lens and the sample. The velocity v_R of the Rayleigh waves in the sample is related to the Rayleigh angle by Snell's law,

$$\frac{v_0}{v_R} = \sin\theta_R,$$ [2]

where v_0 is the velocity in the coupling fluid. Both Equations [1] and [2] require this to be known, and since it is a function of temperature it is necessary to measure the temperature of the coupling fluid accurately. The Rayleigh wave attenuation can also be determined from the Fourier transform of $V(z)$.

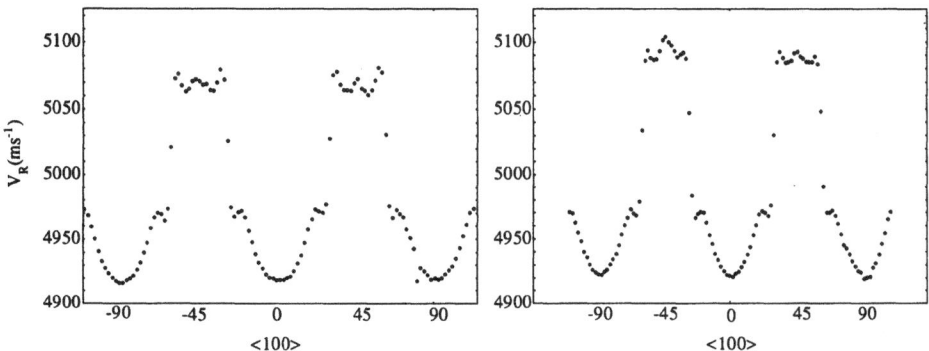

Figure 2. The effect of polishing damage on Si(001). Each figure shows the surface wave velocity deduced from $V(z)$ measurements of the type shown in Fig. 1, as a function of the propagation direction in the surface. The left figure is for a wafer which had been polished with 3 μm alumina powder; the right figure is for a wafer which had been polished with syton.

The effect of polishing damage on Rayleigh velocity is illustrated in Fig. 2, which shows two sets of measurements on Si(001) wafers. The vertical axis is the wave velocity deduced from $V(z)$ measurements using a line-focus beam cylindrical lens at 225 MHz. The horizontal axis is the propagation direction in the surface, measured from $\langle 100 \rangle$. The left plot presents measurements on a wafer that had been polished with 3 μm alumina powder, which would be expected to give considerable damage; the right plot presents measurements following polishing with Syton, which gives a surface which is almost free of damage. Around -90°, 0, and 90°, Rayleigh waves are excited, and the velocity of these is similar for both specimens. Around angles at 45° to these pseudosurface acoustic waves are excited. These occur in anisotropic materials when there is a slower bulk wave to which the surface waves can weakly couple. The pseudo-surface wave velocities differ for the two samples, in the alumina polished samples they are 10 - 20 m s^{-1} slower. This corresponds to greater subsurface damage introduced during the polishing, leading to locally reduced stiffness and hence lower wave velocity. There is also a reduction in wave velocity due to surface roughness, but this is a somewhat smaller effect.[4] Silicon has {111} cleavage planes, and it may be that the direction dependence of the change in velocity associated with polishing damage is related to preferred crack orientations.

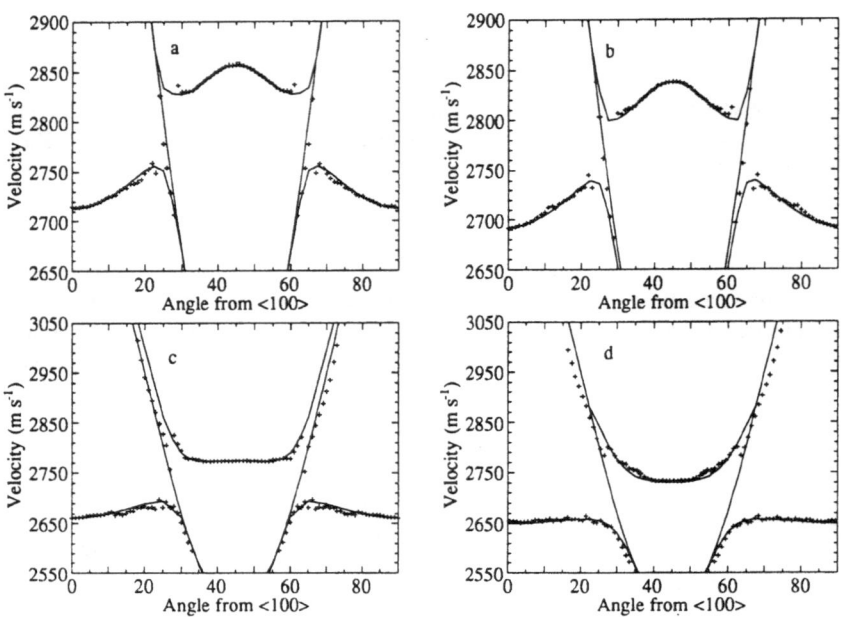

Figure 3. Surface acoustic wave velocities measured as a function of propagation direction for GaAs(001) wafers implanted with Si^+ ions of 1.5 MeV energy at 77 K, with doses: (a) 5×10^{12} cm^{-2}; (b) 5×10^{13} cm^{-2}; (c) 5×10^{14} cm^{-2}; (d) 5×10^{15} cm^{-2}; the theoretical curves were fitted using the parameters in Fig. 4.[3]

Sophisticated methods are now available for determining the elastic constants of surface layers from $V(z)$ data, by varying the materials parameters either to give the best fit between simulated and measured $V(z)$ curves,[5] or to give the best fit to measured surface wave velocities.[3] It is even possible to include the effect of surface stresses using third order elastic constants. One must not try to handle too many unknowns: if you start with zero knowledge you will end up with infinite ignorance. But it is possible to determine a limited number of parameters if the others are known. Figure 3 shows Rayleigh and pseudo-surface wave velocities measured at 225 MHz on four samples of GaAs(001) which had been implanted with increasing doses of Si^+ ions at 1.5 MeV and at 77 K.[6] A progressive flattening out of the anisotropy with increasing ion implantation is apparent. The elastic constants of the substrate material were determined directly by QAM, and the profiles of the implanted layers were measured by Rutherford backscattering. The solid curves in Fig. 3 show the calculated bulk and surface wave velocities calculated by fitting the elastic constants c_{11} (the longitudinal wave modulus) and c_{44} (the shear wave modulus) to the experimental data. The fitted values are shown in Fig. 4. Also shown in Fig. 4 are the values of the anisotropy factor $A = 2c_{44}/(c_{11}-c_{12})$ for the four doses. At the two lowest doses the anisotropy factor is scarcely changed from its bulk value, whereas for higher doses there is a substantial decrease in the anisotropy, associated with the disruption of the crystal lattice by the ions.

The highest frequency routinely used in acoustic microscopy is 2 GHz, though frequencies in the hundreds of MHz are increasingly popular for many industrial applications (e.g. the SAM 100 series[7]). For the kind of quantitative acoustic microscopy described in this section, 225 MHz has been established as the most usual frequency.[2] This corresponds to a Rayleigh wavelength of 10 - 20 μm, which means that a comparable depth is sampled. In many applications the significant depth can be considerably less than this, and greater sensitivity to elastic properties can be obtained using higher frequencies. This is possible using surface Brillouin spectroscopy.

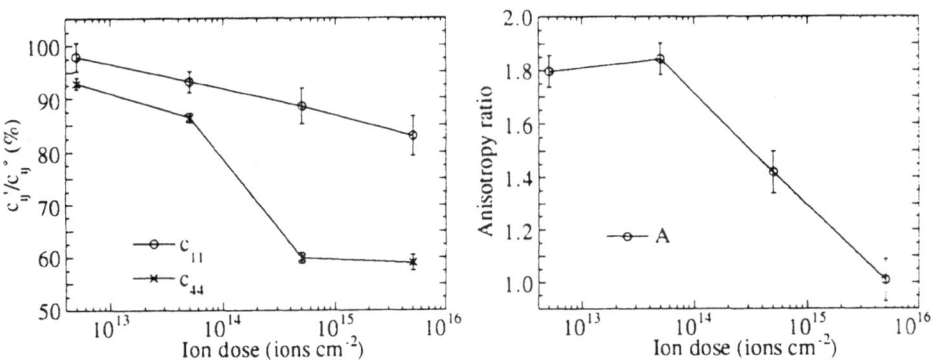

Figure 4. Elastic properties of the ion-implanted GaAs specimens obtained by fitting to the data presented in Fig. 4: (left) elastic constants c_{ij}' of the implanted layer normalized to the fitted constants of the substrate $c_{ij}{}^{o}$; (right) anisotropy ratio $A = 2c_{44}'/(c_{11}'-c_{12}')$ of the implanted layers.[3]

SURFACE BRILLOUIN SPECTROSCOPY

In Brillouin spectroscopy, light interacts with the elastic properties of a solid to produce a frequency shift in the light which is characteristic of an elastic wave in the solid that scatters the light.[8] In quantum terms, a photon of light either creates or absorbs a phonon (Stokes and anti-Stokes respectively). In this process energy and momentum are both conserved. Conservation of energy is described by a scalar equation,

$$\pm\Omega = \omega_2 - \omega_1, \qquad [3]$$

where Ω is the phonon frequency, ω_1 and ω_2 are the incident and scattered photon frequencies, and + or − refers to Stokes or anti-Stokes. Conservation of momentum is described by a vector equation,

$$\pm\mathbf{K} = \mathbf{k}_2 - \mathbf{k}_1, \qquad [4]$$

where $\mathbf{K}$ is the phonon wavevector and $\mathbf{k}_1$ and $\mathbf{k}_2$ are the incident and scattered photon wavevectors. In surface Brillouin spectroscopy a Rayleigh phonon is involved and the component of wavevector parallel to the surface is conserved, which is described by a scalar equation,

$$\pm K = k_2 \sin\theta_2 - k_1 \sin\theta_1, \qquad [5]$$

where θ_1 and θ_2 are the angles of the incident and scattered light beams. Equations [3] and [5] can also be derived by considering the light to be diffracted by a moving grating formed by the surface wave, with a consequent Doppler shift. The frequency of the phonon is usually about five orders of magnitude less than the frequency of the photons, and so to a good approximation $k_2 \approx k_1$. In the usual implementation of surface Brillouin spectroscopy the scattered light is detected back in the direction of the incident light, so that $\theta_2 \approx -\theta_1$. To a good approximation we may then write,

$$\pm K = 2k_1 \sin\theta_1. \qquad [6]$$

Thus if we know the wavelength of the incident light and the velocity of light, and the angle of incidence, then we can deduce the wavenumber of the surface phonon. In Brillouin spectroscopy the frequency shift of the inelastically scattered light is measured, which is the frequency of the phonon, and hence we can deduce the velocity of the surface phonons $v_R = \Omega/K$.

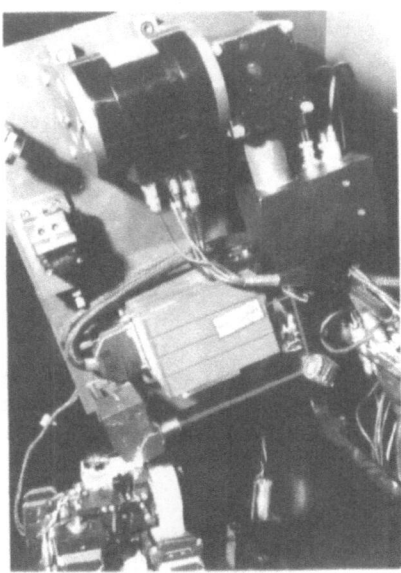

Figure 5. Bede BriSc surface Brillouin spectrometer.[9]

A commercially available surface Brillouin spectrometer is shown in Fig. 5.[9] The sample is mounted on a stage which permits scanning, rotation and tilt in six axes, for samples as large as 150 mm wafers. The complete optical system is mounted on a vertical plate, which can be rotated through 90° in its plane. Light of wavelength 532 nm is generated by a frequency-doubled Nd-YAG laser and is incident on the sample at an angle determined by the orientation of the plate. The scattered light is collected by a lens which is coaxial with the incident beam, and is then reflected to pass through a spatial filter and a Fabry-Perot interferometer. Brillouin scattering is a relatively weak process, so that the intensity of the inelastically scattered light is relatively weak. Moreover the frequency shift is typically 10 - 20 GHz, which corresponds to a rather small fractional change in the wavelength of the light. The interferometer is therefore of a demanding specification. In this instrument it is a seven-pass parallel plate interferometer, with active feedback control of the plate spacing; the high stability means that only that part of the spectrum which is required need be measured, enabling data to be collected very efficiently. The light passes to a detector with a dark count close to one photon per second, with full computer control of the whole instrument and data acquisition.

Surface phonon peaks from a GaAs(001) wafer on which 80 nm GaSb had been grown are shown in Fig. 6. The angle of incidence was 70°, giving a phonon wavelength of about 0.3 nm. Curve (a) is the peak associated with scattering from Rayleigh-type phonons propagating in the [100] direction, with a frequency shift $\Omega = 7.90$ GHz giving a velocity $v_R = 2,236$ m s^{-1}. Curve (b) is due to pseudosurface phonons propagating in the [110] direction, with a frequency shift 8.32 GHz, giving a velocity $v_{ps} = 2,355$ m s^{-1}. This demonstrates the ability to measure the difference in surface wave velocity in different directions in an isotropic surface with a thin layer grown on it.

A surface Brillouin spectrum from an Au/Co superlattice grown on a metal substrate is compared with the spectrum from the uncoated substrate in Fig. 7. The superlattice gives a velocity considerably slower than the Rayleigh velocity in the bare surface, and it also introduces an additional wave mode with a much higher phase velocity, possibly a Sezawa mode. This illustrates the ability to measure surface waves in a metal sample with a superlattice structure that leads to more complicated modes. The full armoury of techniques that have been developed for relating surface mode velocities measured by acoustic microscopy to the elastic structure and properties of surfaces is available for analysing this kind of data measured by surface Brillouin spectroscopy.

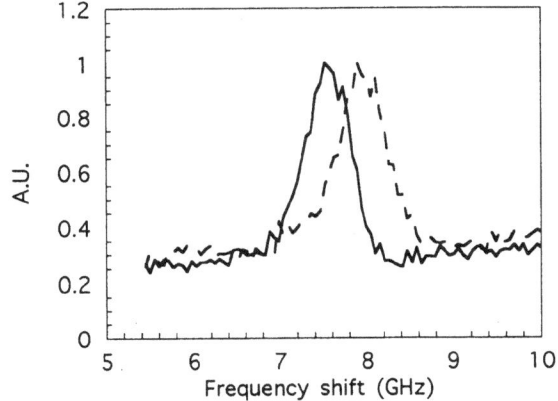

Figure 6. Surface Brillouin spectrum of a GaAs(001) wafer with a 80 nm GaSb layer: ———— [100] propagation; - - - - [110] propagation.

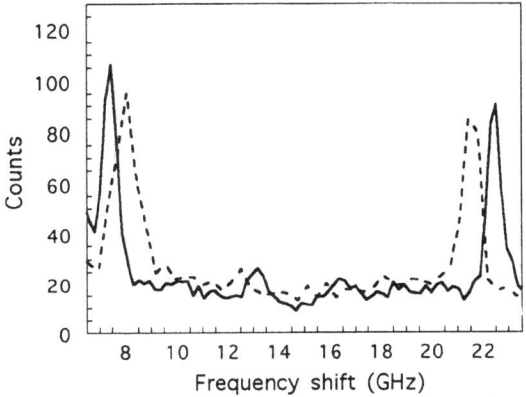

Figure 7. Surface Brillouin spectrum of a Au/Co metal superlattice: ———— the superlattice; - - - - the substrate.

CONCLUSION

By choosing the appropriate frequency it is possible to tailor the measurement of surface wave velocity to the depth of interest in a given sample. Quantitative acoustic microscopy is useful for sampling properties in the depth range extending to 10 μm or so. It can give the Rayleigh velocity with an accuracy which can routinely be better than 0.1%. Surface Brillouin spectroscopy samples the properties within a depth of less than 0.3 μm, with an accuracy in Rayleigh velocity which is generally better than 1%. The reduced accuracy is more than compensated by the increased sensitivity at shallower depths. Both techniques work best with surfaces that are flat and smooth to a tolerance considerably better than the wavelengths involved. The techniques are now established and are finding a growing range of applications in the characterisation of materials surfaces during and after processing.

ACKNOWLEDGEMENTS

The use of quantitative acoustic microscopy and surface Brillouin spectroscopy at Oxford are part of the LINK Nanotechnology Project *Characterization of Surface Topography and Subsurface Damage*. The GaSb on GaAs sample was provided by Dr N. Mason; the Au/Co superlattice sample was provided by Professor B.K. Tanner. MMPH is supported by a CEC HCM Fellowship.

REFERENCES

[1] G.A.D. Briggs, *Acoustic Microscopy*, Oxford: Clarendon Press (1992).

[2] J. Kushibiki and N. Chubachi, Material Characterisation by line-focus beam acoustic microscope, *IEEE Trans* **SU-32**, 189-212 (1985).

[3] Z. Sklar, P. Mutti, N.C. Stoodley, and G.A.D. Briggs, Measuring the elastic properties of stressed materials by quantitative acoustic microscopy, in *Advances in Acoustic Microscopy* 1, G.A.D. Briggs, ed., pp. 209-247. Plenum Press, New York (1995).

[4] C. Pecorari and G.A.D. Briggs, Acoustic microscopy and dispersion of leaky Rayleigh waves on randomly rough surfaces: a theoretical study (1995, submitted for publication).

[5] J.D. Achenbach, J.O. Kim, and Y.-C. Lee, Measuring thin-film elastic constants by line-focus acoustic microscopy, in *Advances in Acoustic Microscopy* 1, G.A.D. Briggs, ed., pp. 153-208. Plenum Press, New York (1995).

[6] P. Mutti, Z. Sklar, G.A.D. Briggs, and C. Jeynes, Elastic properties of GaAs during amorphization by ion implantation, *J. Appl. Phys.* **77**, 2388-2392 (1995).

[7] Krämer Scientific Instruments GmbH, PO 1922, D-35745 Herborn, Germany.

[8] P. Mutti, C.E. Bottani, G. Ghislotti, M. Beghi, G.A.D. Briggs, and J.R. Sandercock, Surface Brillouin spectroscopy—extending surface acoustic wave measurements to 20 GHz, in *Advances in Acoustic Microscopy* 1, G.A.D. Briggs, ed., 249-300. Plenum Press, New York (1995).

[9] Bede Scientific Instruments Ltd, Lindsey Park, Bowburn, Durham DH6 5PF, England.

NANOSCALE IMAGING OF MECHANICAL PROPERTIES
BY ULTRASONIC FORCE MICROSCOPY (UFM)

Oleg Kolosov[1], Andrew Briggs[1], Kazushi Yamanaka[2], and Walter Arnold[3]

[1]Department of Materials, University of Oxford, Oxford, OX1 3PH, UK
[2]Mechanical Engineering Laboratory, Namiki 1-2, Tsukuba, Ibaraki 305, Japan
[3]Fraunhofer Institute for Non-destructive testing, Bldg. 37, D- 66123, Saarbrucken, Germany

Acoustic microscopy[1] enables the elastic properties of materials to be imaged and measured with submicron spatial resolution limited only by the wavelength of the ultrasound. Nevertheless, there are practical reasons why this resolution cannot be increased indefinitely. Therefore, in order to improve the spatial resolution below the wavelengths presently available, a near field technique is needed. Atomic force microscopy (AFM) offers the resolution required[2], but by itself AFM does not measure material properties; in its usual mode it gives an image of the topography of a surface at constant normal force.

Recent advances in AFM, such as Force Modulation Mode (FMM)[3], provided the methods for discrimination of elastic properties from object topography, but they are unable to reveal the HF dynamics of elastic interaction. The highest frequencies achievable in such approaches are restricted by the resonant frequencies of the force sensitive cantilever or tip. These are of order of 10-100 KHz, and do not reach the interesting region of intermolecular vibration and relaxation frequencies which are in the MHz and GHz frequency range[4]. A direct relationship, connecting the resonant frequency of the cantilever ω_r and the cantilever effective mass m and rigidity k as $\omega_r^2 = k/m$ in AFM experiments, suggests that the higher the vibration frequency, the stiffer the cantilever must be in order to sense this vibration[3,5] provided that the mass m is constant. However, the stiffer the cantilever, the worse the force sensitivity, reducing the performance of the AFM system. A different approach for sensing a HF vibration is nonlinear detection using the well known extreme nonlinearity of the force-versus-separation dependence $F(z)$[2,5]. This approach has been realized for a cantilever of very high rigidity[6], leaving unresolved the trade-off between high force resolution and sensitivity to viscoelastic dynamic properties.

Rather than to tolerate this disappointing trade-off, we have resolved it using the technique described in reference 7. We recognized at first, that if the object surface is vibrated at the high frequency $\omega_v \gg \omega_r$, HF rigidity of the cantilever-tip system is defined by its inertial properties rather than the static cantilever rigidity k. This inertial rigidity of cantilever k_v increases drastically with the frequency increase (e.g., assuming the simplest point-mass model of cantilever, $k_v \sim k(\omega_v/\omega_r)^2$ and for a typical cantilever with ω_r of 10 KHz and ω_v=3 MHz, k_v exceeds the static rigidity k by a factor of 10^5 times !). This extreme HF rigidity of the cantilever prevents it from vibrating along with the surface. This results in the HF oscillation of the tip-surface separation z and, therefore, in nonlinear detection ("rectifying") of vibration resulting in an average cantilever shift[7]. At the same time, the force sensitivity of the system, specified by the low static rigidity k of cantilever, remains high, therefore providing the exciting possibility of studying the elastic properties of *rigid objects* using statically *soft*, highly force sensitive microfabricated cantilevers.

Subsequently it was demonstrated that this approach, Ultrasonic Force Microscopy, or UFM (which we call "ultrasonic" due to the frequency ω_v falling in the ultrasonic regime), can provide new information on the local elastic properties[8] and subsurface features[9] of objects. The direct measurement of the HF component of cantilever vibration by adding the HF optical knife-edge detector to AFM was recently reported[10]. The complex nonlinear dynamics effect of the tip bouncing on the surface at ultrasonic frequencies was also investigated[11].

The response of AFM to ultrasonic vibrations could be described by the introduction of a new force-versus-separation dependence $F_m(z)$, derived from the original dependence $F(z)$ by averaging over a vibration period. This reduces the problem of UFM response to the well-known for AFM force balance equation (3) $F_m(z)=k(z-z_s)$, where k is a cantilever rigidity, z_s is the undeformed position of the surface and z is the tip-surface separation (where $z_c=z-z_s$, the cantilever displacement, is measured in the experiment). The $F_m(z)$ is then defined as[7]

$$F_m(z,a) = \frac{1}{2\pi} \int_0^{2\pi} F(z - a\cos x)dx. \tag{1}$$

It can be seen from eq. 1, that the modified $F_m(z)$ does not directly depend on the vibration frequency ω_v and is defined solely by the amplitude of the ultrasonic vibration and the original force-versus-separation dependence $F(z)$. At the same time, $F(z)$ itself can vary with the frequency, manifesting the dynamic viscoelastic behavior of the object, which is related to molecular relaxation. Such dynamic behavior is described by the introduction of a frequency varying force-versus-separation dependence $F^*(\omega,z)$, with $F^*(0,z)$ standing for $F(z)$ used in low frequency AFM measurements. Thus, UFM provides AFM with a new viewpoint on the nature of the object, adding a new dimension to the experimental data - a frequency, and therefore providing a unique possibility for the study of molecular dynamics and the nature of interatomic forces with nanoscale resolution.

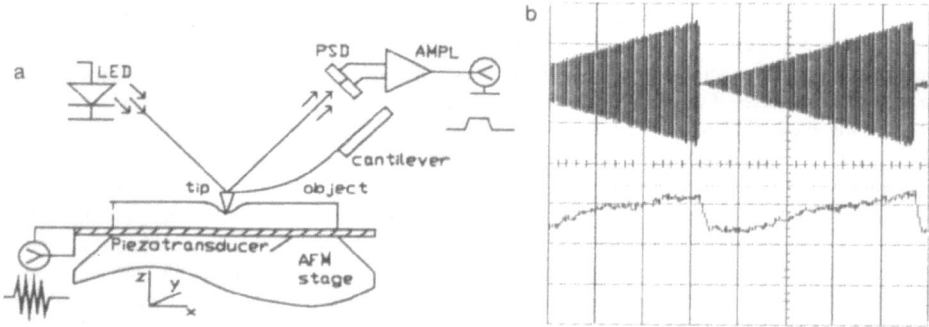

Figure 1. a) An experimental UFM setup. A microfabricated cantilever with a tip of about 50 nm radius with a resonant frequency of 40 kHz and spring constant k of 0.09 N/m was used. The studied object was bonded to PZT ceramic piezoelectric transducers driven at ultrasonic frequencies. The UFM images were formed by synchronous detection of LF vibration of the cantilever at the AM frequency using a lock-in amplifier. **b)** Cantilever deflection vs amplitude of the ultrasonic wave measured by interferometric AFM head - z(A) curves (upper trace - an envelope of the saw-tooth-modulated HF signal on the piezotransducer, frequency - 3.2 MHz, 0.5 V/div, lower trace- - cantilever deflection, 0.8 nm/div, horizontal scale - 0.2 ms/div).

The experimental UFM setup is described elsewhere[7] (Fig. 1). We used both the commercial AFM system (detecting a laser beam deflected by the cantilever) and a laboratory made Nomarski-type laser interferometer head providing us with absolute values of the cantilever displacement and also able to measure the HF component of the cantilever vibration. We compared UFM performance with standard AFM modes (e.g. topography, friction, etc.) and modes allowing investigation of elastic properties, such as FMM[3]. In Fig. 2 it is easy to see that the contrast of UFM to the grain structure of the magnetic floppy disk is much higher than the contrast of standard topography and FMM images.

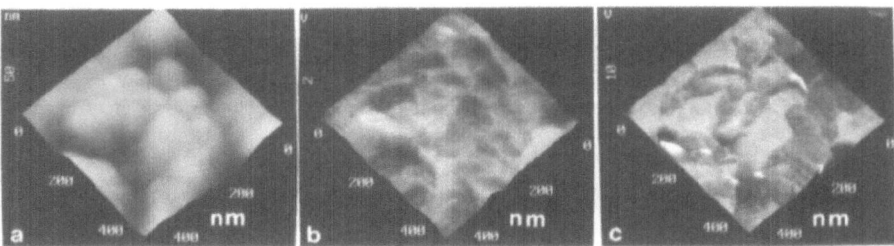

Figure 2. Standard topography a), and Force Modulation Mode (FMM) b) images of the magnetic floppy disk surface. A grain structure of the surface with grain size of about 100 nm is presented in the pictures. All grains have nearly equal brightness in both images, in the FMM image the topography features have been enhanced. The contrast and resolution in UFM images c) is much higher than those of the topography and FMM images showing the fine grain structure of 10-20 nm width, and the previously unobserved contrast (i.e. between grains (I) and (II)) is clearly visible.

The resolution of the grain structure seems to be better for the UFM mode and the fine grain structure of 10-20 nm width becomes clearly visible. Furthermore, the contrast of the different grains in UFM images is different, reflecting the different dynamic elasticities of ferrite particles and the polymer matrix of the magnetic disk structure. Hence, the UFM mode is sensitive to the local variation of the high frequency elastic properties and reveals features not visible by standard AFM methods.

To investigate the ability of UFM to detect mechanical damage on nanoscale, we imaged a Vicker's micro-indentation crack on the single crystal Si surface.

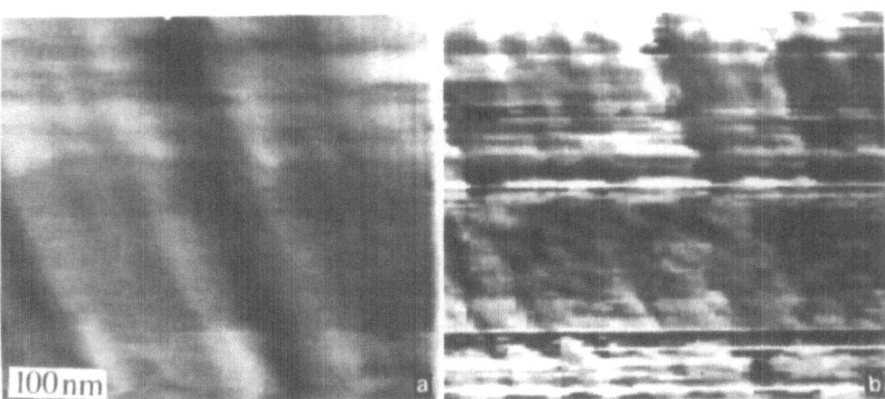

Figure 3. The AFM a) and UFM b) images of the same GaAs crystal polished with a chemical mechanical process (*chemlox*, Logitech, Glasgow, UK) with an abrasive particles of 0.3 um diameter. Unidirectional surface scratches revealed in topography image are associated with extensive mechanical damage (black areas and bands in UFM image). It is to be noted here that in the upper part of UFM image the darker bands of different orientation (probably, associated with subsurface damage from another polishing direction) are revealed, not visible in the topography image.

Comparing AFM and UFM images we found: i) the crack reduces the apparent surface elasticity in its vicinity, ii) UFM possess a high contrast to the crack with almost no contrast to the sample height, iii) the crack can be tracked down to the tip with resolution about 10 nm. A comparison of the AFM surface topography images with the UFM images reflecting elastic compliance of the surface and, thereby, material properties and mechanical damage of the crystal allowed us to receive information regarding crystal damage after polishing and about the polishing process itself. It was found, that surfaces polished using different techniques (chemical and mechanical) are of entirely different nanotopography and nanoscale elastic properties. In contrast, UFM pictures of chemical - mechanically polished GaAs (*chemlox*, Logitech Ltd., Glasgow, UK)) in addition to the surface scratches revealed extensive subsurface damage (Fig. 3) not revealed by AFM topography.

An important problem concerning the maximum achievable resolution is of great interest in UFM mode, where the sample and cantilever vibration likely to smear the sharp features of the image. Nevertheless, UFM images of an atomic lattice of freshly cleaved mica and of monatomic steps on a mica surface[12] clearly demonstrate that a sub-nanometer resolution is routinely achieved in UFM.

ACKNOWLEDGEMENTS

Authours acknowledge the support of the EPSRC-LINK Nanotechnology Project *Characterisation of Surface Topography and Subsurface Damage* and the CEC *Atomic Force Acoustic Microscopy* Network and also support of the Paul Instrument Fund (c/o The Royal Society).

REFERENCES

1. Lemons and C. F. Quate, Appl. Phys. Lett., **24** 163-165 (1974); Briggs, *Acoustic Microscopy*, Clarendon Press, Oxford (1992).
2. Binnig, C.F. Quate, Ch. Gerber, *Phys. Rev. Lett.* 56, 930 (1986); Martin, C. C. Williams, H. K. Wickramasinghe, *J. Appl. Phys.*, **61** 4723 (1987).
3. Maivald *et al.*, *Nanotechnology* 103 2 (1991); Radmacher, R.W. Tillmann, M. Fritz, H. E. Gaub, *Science*, 1990 257 (1992).
4. M. North, R.A. Petric, D. W. Phillips, *Macromolecules* 10, (1977). Kolosov, Suzuki M., Yamanaka K., *J. Appl. Phys.*, 6407 **74** (1993).
5. Takata, T. Hasegawa, S. Hosaka, S. Hosoku, T. Komoda, *Appl. Phys. Lett.* 17 **55** (1989); Cretin and F. Sthal, *Appl. Phys. Lett.* 829 **62** (1993).
6. Rohrbeck and E. Chilla, *Phys. Stat. Sol.(a)*, 69 **131** (1992).
7. Kolosov and K. Yamanaka, *Japanese J. Appl. Phys. B (Letters)* L1095 **32** (1993).
8. Kolosov, H. Ogiso, K. Yamanaka, *Proceedings of the 3rd Japan SAMPE Symposium*, Tokyo, JAPAN, 1993.
9. Yamanaka, H. Ogiso, O. Kolosov, *Appl. Phys. Lett.*, 178 **64** (1994); S. Myers, MRS Bull. 8 **XIX** (1994).
10. Rabe and W. Arnold, *Appl. Phys. Lett.*, **64** :1493 (1994); Rabe, W. Arnold, *Annalen Der Physik*, **3** :589, (1994).
11. Burnham, A.J. Kulik, G. Gremaud and G.A.D. Briggs, *Phys. Rev. Lett.*, **74** :5092 (1995).
12. O. Kolosov, H. Ogiso, H. Tokumoto, K. Yamanaka, in *Nanostructures & Quantum Effects*, ed: Sakaki H. & Noge H., Springer Series in Material Sciences, Springer-Verlag, vol. 31, pp.345-348, 1994.

ACOUSTIC MICROSCOPY WITH RESOLUTION IN THE NM-RANGE

U. Rabe, K. Janser and W. Arnold

Fraunhofer-Institute for Nondestructive Testing
University, Bldg. 37
D-66123 Saarbrücken, Germany

INTRODUCTION

In conventional Acoustic Microscopy[1] (SAM) a focused acoustic beam is scanned over the sample surface and the lateral resolution is given by the focal spot diameter. According to Abbe´s principle this is determined by the wavelength which in practice limits the resolution of SAM to about 1 µm because of the high absorption at high acoustic frequencies. The idea of near-field microscopy is to guide the acoustic wave towards the sample by a structure, e.g. a pin or a pin hole, with dimensions smaller than the acoustic wavelength. When the sample is placed in the near field radiated from the pin, a resolution can be achieved determined by the pin diameter and not by the acoustic wavelength. Fig. 1 shows an example of SAM working in transmission at a frequency of 35 MHz[2]. With the pin probe a lateral resolution of 10 µm was achieved, while with the lens, the resolution was only 50 µm. Other examples of near-field acoustic microscopes based on this principle can be found in the literature[3,4,5].

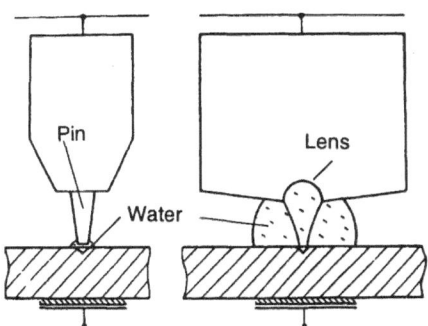

Figure 1. Ultrasonic microscope[2]. Left: arrangement with pin probe, right: ultrasonic lens.

In 1982 with Scanning Tunneling Microscopy[6] (STM) a near-field technique became public with a very high lateral resolution reaching atomic dimensions. In the STM a sharp electrically conducting tip is scanned over a conducting sample surface at a distance of less than 1 nm. A voltage is applied and the tunneling current is used for tip-sample distance control. As an acoustic surface vibration causes a modulation in the tip-sample distance and therefore a modulation of the tunneling current, acoustic waves can be detected with the STM. Several research groups[7,8,9,10,11] work on the combination of Acoustic Microscopy and STM. The main disadvantage of STM is, however, that many materials cannot be examined because electrically conducting surfaces are required.

A scanning tip microscope similar to the STM is the Atomic Force Microscope[12] (AFM), invented in 1986. Here, the sensor tip is mounted on a soft microfabricated cantilever of typically 100 μm to 200 μm length. Instead of the tunneling current, the deflection of the cantilever caused by the forces acting between the sensor tip and the sample surface is measured with a position detector and used for distance control. An example of a low-frequency setup working at frequencies below 1 MHz is the Scanning Microdeformation Microscope[13], where acoustic waves are generated by vibrating an AFM tip made of sapphire. The waves generated by this point source are detected by a piezoelectric tranducer after transmission through the sample. Reverse operation, e.g. piezoelectric generation and detection via the tip is also possible. In another technique, mixing of two waves[14] and the average cantilever deflection[15,16,17], both caused by the nonlinear tip-sample interaction, is detected as a response to sample surface vibration at frequencies of up to 100 MHz. Acoustical imaging based on detection and generation of high-frequency cantilever vibrations[18,19] is the subject of the following.

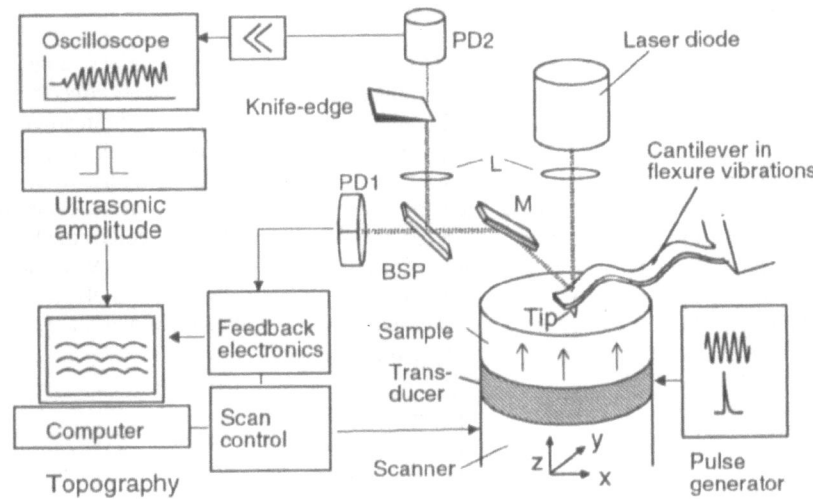

Figure 2. Principle of the AFAM working with high-frequency cantilever vibrations. An ultrasonic transducer is inserted between the scanner and the sample of the AFM. Sample surface vibrations are transmitted into the cantilever and measured with a fast optical knife-edge detector added to the AFM position detector. The signal of the knife-edge detector is amplified and displayed by a digital oscilloscope. For imaging, part of the ultrasonic amplitude is gated out and displayed as an ultrasonic image alongside to the topography image.

DESCRIPTION OF THE ATOMIC FORCE ACOUSTIC MICROSCOPE

Acoustical imaging (AFAM) (Fig. 2) is integrated into a commercial AFM (Nanoscope III, manufactured by Digital Instruments, Santa Barbara, Ca., USA). For generation of the acoustic waves, an ultrasonic transducer is inserted between the scanner of the AFM and the sample and bonded to it by a coupling gel. The piezoelectric transducer is excited with c.w. or spikes and emits longitudinal waves which cause out-of-plane vibrations of the sample surface. Because of the contact force between the sensor tip and the sample surface, these vibrations are transmitted into the AFM cantilever. The AFM is supplied with an optical position detector in which the beam of a laser diode is reflected at the cantilever and falls onto a position sensitive photodiode. For detection at ultrasonic frequencies part of the beam reflected from the cantilever is split and directed to a knife-edge detector[20], consisting of a razor blade shading half of the beam and a fast photodiode (rise time 1 ns). The signal of the knife-edge detector is amplified and displayed by a fast digital oscilloscope. Several processes such as averaging or Fourier transforming can be performed by the instrument. For imaging, the ultrasonic amplitude is detected and displayed as an ultrasonic image alongside the topography image.

The maximal scan size of the instrument is 150 µm × 150 µm and the ultrasonic frequencies at which we operate generally range from about 100 kHz to 10 MHz. In most cases the scan size is smaller than the acoustic wavelength in the sample, and the image contrast results from the changes in local tip-sample interaction and sound transmission. For an interpretation of such images, it is necessary to understand how the AFM cantilever reacts to a certain sample vibration amplitude.

FLEXURAL WAVE THEORY AND EXPERIMENT

The cantilever can be regarded as a small elastic beam which is excited to forced flexural vibrations by the vibrating sample surface. The equation of motion for flexural vibrations in an elastic bar with homogeneous crossection is[21]:

$$EI\frac{\partial^4 y}{\partial x^4} + \rho A\frac{\partial^2 y}{\partial t^2} = 0 \tag{1}$$

where E is the modulus of elasticity, ρ is the mass density, A is the cross section, and I is the area moment of inertia. x is the coordinate along the cantilever and y is the deflection of the length element at x from its rest position. This equation has to be solved under consideration of the boundary conditions of the beam (Fig. 3). If the cantilever is far from the sample surface, no forces act between the sensor tip and the sample surface and the boundary conditions of a beam clamped at one end and free at the other have to be applied. In the other extreme case, when the sensor tip is fixed to the sample surface, the beam is clamped at one end and hinged at the other. A very simple case of a tip-sample interaction force is a linear spring with stiffness k* fixed between the end of the cantilever and the surface. The free and the hinged cantilever can be regarded as extreme cases of the spring coupled cantilever with k* = 0 and k* = ∞, respectively. The solutions of the extreme cases can be found in acoustics textbooks[21,22] and also in the AFM - literature[23,24,25]. For every set of boundary conditions one obtains an infinite set of flexural vibration modes with discrete wave numbers k_n and frequencies f_n, where n is the mode number. The resonance frequency of the clamped/free mode with same mode number is always lower than the resonance frequency of the clamped/hinged mode. If one solves the equations for the spring coupled case, one can calculate that in this case the resonance frequencies lie always between the resonance frequency of the clamped/free and the clamped/hinged mode (Fig. 4)[26]. If the ratio of k^*/k_c increases from 0 to ∞, the frequencies and shapes of all modes shift from the clamped/free solution to the clamped/hinged solution, where k_c is the cantilever stiffness.

To test the flexural wave theory we excited forced vibrations of free rectangular-shaped cantilevers by vibrating the clamped end. We measured the local vibration amplitude by scanning the focal spot of an optical interferometer along the cantilever and compared the experimental mode shapes and resonance frequencies to the theoretical ones. We found very good

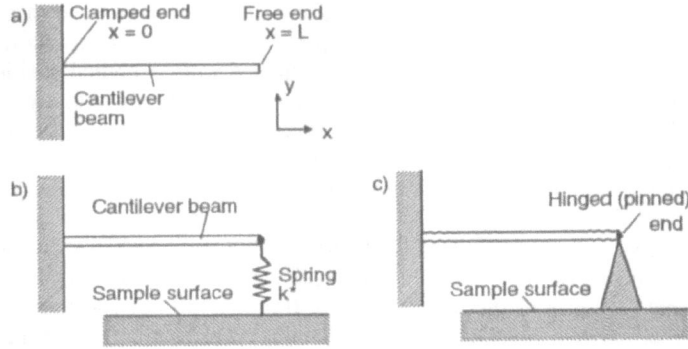

Figure 3. Boundary conditions in the elastic-beam model for the AFM cantilever. One end of the cantilever is clamped: a) when the sensor tip is so far from the sample surface that no tip-sample forces are present, the other end of the cantilever is free; b) a linear spring with spring constant k* fixed between the end of the cantilever and the sample surface is a simple model for a tip-sample interaction force; c) when k* is increased to infinity, the spring-coupled case converges towards the case of a hinged end.

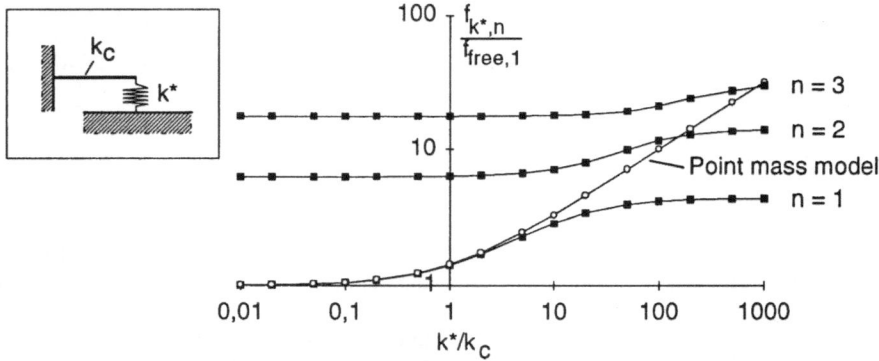

Figure 4. The solutions of the clamped/free cantilever and the clamped/hinged cantilever can be found in acoustics textbooks. As shown here for the first three modes, the resonance frequency of the n´th mode of the clamped/spring-coupled cantilever lies between the frequency of the n´th clamped/free mode and the n´th clamped/hinged mode. When $k*/k_c$ is increased from 0 to ∞ the resonance frequency and shape of every mode shifts from the free case to the clamped case. A comparison with the point-mass model for the cantilever shows that this model predicts too large frequency shifts at $k*/k_c > 1$. For silicon cantilevers with $k_c < 1$ N/m, $k*/k_c$ is larger than 100 for sample surfaces like for example glass.

agreement between theory and experiment. Fig. 5 shows the free vibration spectrum from the second to the ninth mode of a rectangular cantilever made of monocrystalline silicon with the dimensions 233 µm × 51 µm × 1.5 µm (length × width × thickness) and a stiffness k_c of 0.6 N/m. Mode n = 8 is missing because parametric coupling between a lower frequency flexural mode and a torsional vibration mode occured at its excitation frequency. The quality factor Q of the resonances increases for the first four modes to a maximal value of about 900 and then decreases.

When the sensor tip contacts the surface, forced cantilever vibrations are excited by the surface vibrations. To examine the spectra of surface-coupled cantilevers experimentally, we excited the transducer with an electrical spike, so that the surface movement was a short pulse containing a broad band of frequencies. The resulting cantilever vibration was measured with

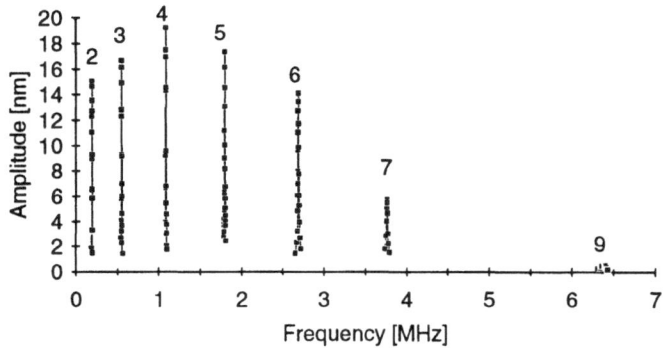

Figure 5. Spectrum of a free rectangular cantilever made of silicon measured experimentally with an optical interferometer. The qualitiy factor of the resonances has a maximum of 900 at the 4th mode. Mode number 8 could not be excited because of mode coupling to a torsional vibration mode.

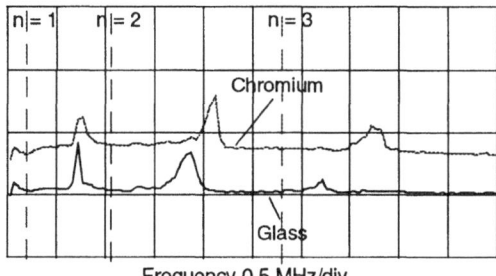

Frequency 0.5 MHz/div

Figure 6. Shifted spectra of a rectangular cantilever due to tip-sample contact. The broken lines show the free resonances. The sample was a thin glass plate partially covered by chromium. The position of the shifted resonance frequencies is material dependent.

the knife-edge detector and Fourier transformed. The sample was a thin glass plate partially covered with chromium. At high excitation voltages the cantilever vibration spectrum contained many signals, but at very low exciting voltage most of the peaks disappeared and the remaining resonances could be assigned to the theoretical modes. From the shifted resonance frequencies in the spectrum (Fig. 6), in principle k* can be calculated using the characteristic equation for the spring-coupled cantilever[26]:

$$\frac{\sinh(c_c\sqrt{f_n})\cos(c_c\sqrt{f_n}) - \sin(c_c\sqrt{f_n})\cosh(c_c\sqrt{f_n})}{1 + \cos(c_c\sqrt{f_n})\cosh(c_c\sqrt{f_n})} \cdot \frac{3}{(c_c\sqrt{f_n})^3} = \frac{k_c}{k^*} \qquad (2)$$

Here $c_c = L\sqrt{2\pi}\sqrt[4]{\rho A / EI}$ where L is the length of the cantilever. When the first three resonance frequencies f_n were inserted and k* was calculated for every individual resonance frequency, we obtained a different value of k* for each resonance frequency. k* varied from 864 N/m to 1618 N/m on glass and from 1295 N/m to 2704 N/m on chromium. One main reason for the disagreement with theory is the shape of the cantilever, for example the sensor tip is not exactly at the end of the beam. Frequency shifts caused by damping must also be considered. An additional reason is probably the nonlinearity of the tip-sample interaction force F(z).

NONLINEARITY OF THE TIP-SAMPLE FORCE F(Z)

When the tip is brought towards the sample surface the interaction force is first attractive, goes through a maximum and becomes repulsive, when the tip has reached a certain penetration depth. In this contact mode a small cantilever force is applied onto the sample surface. In the absence of vibrations, the tip is in an equilibrium position z_e, where the repulsive tip-sample force equals the cantilever force $F(z_e) = k_c (z_e - z_0)$. If the sample surface starts to vibrate with small amplitude a(t) and frequency ω, so that $a(t) = a_0 \cos(\omega t)$, the sensor tip is forced to vibrate as $d(t) = a_c \cos(\omega t - \varphi)$. A linear approximation for F(z), which enables one to use the clamped/spring-coupled cantilever-model, $F(z) \approx F(z_e) - k^*(z - z_e)$, with $k^* = - (\partial F/\partial z)_{z=z_e}$ is only possible if the amplitude of the surface vibration *and* the amplitude of the cantilver vibration are both smaller than about 0.1 nm. At higher vibration amplitudes, the setpoint z_e is shifted because of the nonlinearity of F(z)[15,16,27] and at very high acoustic amplitudes, the tip can jump out of the adhesion potential. Its vibration spectrum then contains other frequencies additional to the exciting frequency[28], like for example free cantilever vibration frequencies and subharmonics[29].

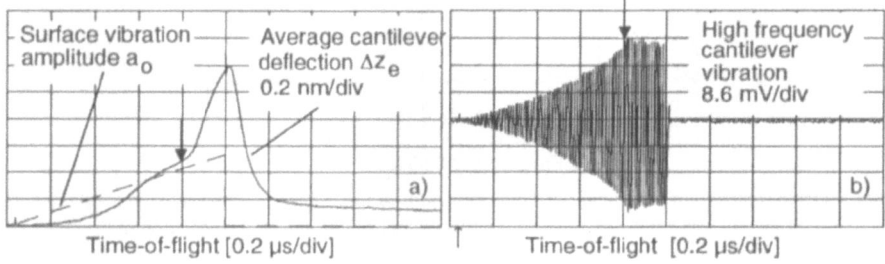

| Surface vibration amplitude a_0 | Average cantilever deflection Δz_e 0.2 nm/div | | High frequency cantilever vibration 8.6 mV/div |

a)

b)

Time-of-flight [0.2 µs/div]   Time-of-flight [0.2 µs/div]

Figure 7. a) When the surface vibration amplitude is ramped (here from 0 nm to 0.69 nm), the cantilever is pressed away from the sample surface. b) The cantilever vibrates at the excitation frequency, here 1.64 MHz, with increasing amplitude. When a certain tip-sample distance is reached (marked by the arrow in the figure), the transmission of ultrasound abruptly changes.

An example for the shift of the equilibrium position z_e caused by the nonlinear tip-sample forces is shown in Fig. 7. The vibration amplitude of a glass sample covered with a thin chromium layer was ramped from 0 to 0.69 nm (Fig. 7a). As a result, the cantilever is pressed away from the sample surface. Note that the cantilever continues to vibrate at the surface vibration frequency (Fig. 7b). Its vibration amplitude first increases, until at a certain tip-sample distance the high-frequency coupling from the surface to the cantilever seems to be interrupted, so that the cantilever vibration amplitude no longer grows (Fig. 7b, arrow). We made many similar experiments which showed that the cantilever vibration amplitude is comparable to the surface vibration amplitude and must be considered when the average shift of the setpoint is calculated, *even* if one does not work at one of the cantilever resonance frequencies.

In Fig. 8a the topography image taken on top of a SiO_2-sample coated with tantalum is shown. Notice that the sample is extremely flat. The gray scale covers only 3.9 nm. The ultrasonic images b) and c) were taken at two different excitation voltages, 30 V and 110 V, respectively. All other parameters such as excitation frequency (1.91 MHz), repetition rate (10 kHz) and number of cycles in the tone burst (50) were the same for both images. Both images show a lateral resolution comparable to the tip diameter of approximately 20 nm, much higher than the acoustic wavelength. In the ultrasonic image with lower excitation voltage the acoustic image is influenced by the sample topography. At higher excitation voltage the contrast changes, and we assign this to an additional image contrast generated by variation in local tip-sample adhesion.

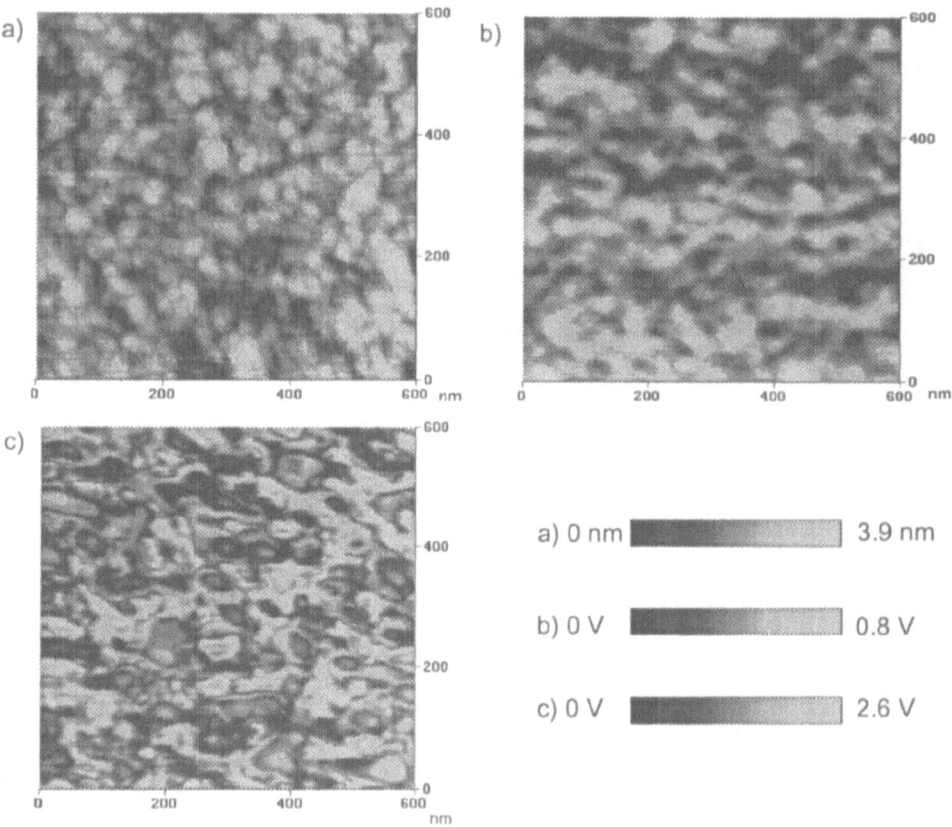

Figure 8. a) topography image of an SiO_2-substrate coated with tantalum; b) AFAM-image at low excitation amplitude; c) AFAM-image at high excitation amplitude (11 dB difference compared to b). Note the change in contrast between b) and c).

CONCLUSION

We have demonstrated that AFM cantilevers with low stiffness have resonances in the MHz range which can be excited with high amplitude, measured with high signal-to-noise ratio and therefore can be used for imaging purposes. The vibrations of cantilevers without surface contact can be well described by flexural wave theory of elastic beams. In the near-field technique demonstrated here, the lateral resolution is not given by the wavelength of the acoustic field in the sample but by the sensor tip radius. The image contrast stems from the local variation of sound transmission due to tip-sample interaction forces.

Atomic Force Acoustic Microscopy has the potential to determine different material constants of the sample surface. Surface stiffness can be derived from the shift of the cantilever resonance frequencies at low vibration amplitudes. Information about the shape of the $F(z)$-curve can be extracted from the shift of the equilibrium position z_e caused by ultrasonic vibrations. The adhesion energy can be determined from the vibration amplitude at which the cantilever jumps out of the adhesion potential. For *quantitative* modeling, however, the nonlinear tip-sample force $F(z)$ *and* flexural wave theory have to be considered. To compare theory with experiment, the cantilever vibration amplitude and the surface vibration amplitude below the sensor tip must be measured absolutely.

ACKNOWLEDGEMENT - Part of this work was carried out within a H.C.M. research project supported by the E.U. We thank B.Bendjus, FhG-EADQ Dresden for providing us the SiO_2-sample covered with Tantalum.

REFERENCES

1.   G.A.D. Briggs, "Acoustic Microscopy", Oxford University Press, Oxford (1992).
2.   J.K. Zieniuk and A. Latuszek, *Acoustical Imaging*, Vol. 17, Eds. H.Shimizu, N.Chubachi, and J. Kushibiki, Plenum Press, New York: 219 (1989).
3.   W. Dürr, D.A. Sinclair, and E.A. Ash, *Electr.Lett.* 16: 805 (1980).
4.   A. Kulik, J. Attal, and G. Gremaud, *Acoustical Imaging*, Vol. 20, Eds. Y. Wei and B. Gu, Plenum Press, New York: 241 (1993).
5.   B. T. Khuri-Yakub, C. Cinbis, C.H. Chou, and P.A. Reinholdtsen, *Proc. 1989 IEEE Ultrasonics Symp.*: 805 (1989).
6.   G. Binnig and H. Rohrer, *Sci. Am.* 253: 50 (1985).
7.   S. Akamine, B. Hadimioglu, B.T. Khuri-Yakub, H. Yamada, and C.F. Quate, *Transducers '91, IEEE Int. Conf. on Solid State Sensors and Actuators*: 857 (1991).
8.   K. Uozumi and K. Yamamuro, *Jpn. J. of Appl. Phys.* 28: L1297 (1989).
9.   K. Takata, T. Hasegawa, S. Hosaka, S. Hosoki, and T. Komoda, *Appl. Phys. Lett.* 55: 1718 (1989).
10. A. Moreau and J. B. Ketterson, *J. Appl. Phys.* 72: 861 (1992).
11. W. Rohrbeck, E. Chilla, H.-J. Fröhlich, and J. Riedel, *Appl. Phys.* A 52: 344 (1991).
12. G. Binnig, C.F. Quate, and Ch. Gerber, *Phys. Rev. Lett.* 56: 930 (1986).
13. B. Cretin and F. Sthal, *Appl. Phys. Lett.* 62: 829 (1993).
14. W. Rohrbeck and E. Chilla, *Phys. Stat. Sol.* 131: 69 (1992).
15. O. Kolosov and K. Yamanaka, *Jpn. J. Appl. Phys.* 32: 22 (1993).
16. K. Yamanaka, "New Approaches in Acoustic Microscopy for Noncontact Measurement and Ultra-High Resolution", in "Advances in Acoustic Microscopy", Vol. 1, Ed. A. Briggs, Plenum, N.Y., 1995.
17. O. Kolosov, *Acoustical Imaging*, Vol. 22, Proc. 22nd Int.Symp.Acoust.Img., Florence, 1995, Ed. P.Tortoli, to be published.
18. U. Rabe and W. Arnold, *Ann. Phys.* 3: 589 (1994).
19. U. Rabe and W. Arnold, *Appl. Phys. Lett.* 64: 1493.(1994).
20. See for example: J. W. Wagner, "Optical Detection of Ultrasound", p. 201 in "Physical Acoustics", Vol. 19, Eds. W.P. Mason and R.N. Thurston, Academic Press, New York, 1990.
21. P.M. Morse und K. U. Ingard, "Theoretical Acoustics", McGraw-Hill, New York (1968).
22. W. F. Stokey, "Vibration of Systems Having Distributed Mass and Elasticity", Chap. 7 in "Shock and Vibration Handbook", Eds. C.M. Harris und C.E. Crede, McGraw-Hill, New York (1976).
23. D. Sarid, "Scanning Force Microscopy", Oxford University Press, Oxford (1991).
24. M. Nonnenmacher, PhD-Thesis, Gesamthochschule Kassel, Fachbereich Physik (1990).
25   J. Colchero, PhD-Thesis, Universität Konstanz, Fakultät für Physik (1993).
26. U. Rabe, K. Janser, and W. Arnold, in preparation.
27. U. Rabe and W. Arnold, *Acoustical Imaging*, Vol. 21, Proc. 21th Int.Symp.Acoust.Img., Laguna Beach, 1994, Ed. J. Jones, Plenum Press, New York, to be published.
28. U. Rabe, M. Dvorak, and W. Arnold, *Thin Solid Films* 264: 165 (1995).
29. N.A. Burnham, A.J. Kulik, G. Gremaud and G.A.D. Briggs, *Phys.Rev.Lett.* 72: 5092 (1995).

MEASUREMENT OF SURFACE WAVE VELOCITY AND ANISOTROPY AT EDGES USING POINT-FOCUS ACOUSTIC MICROSCOPY

Klaus Kosbi, Wieland Weise, Ulla Scheer, Uwe Laun, and Siegfried Boseck

Institute of Materials Science and Structure Research
Physics Department, University of Bremen
28334 Bremen, Germany

ABSTRACT

A new technique for measurement of SAW velocities by lateral scanning in the vicinity of an elastic discontinuity is presented. The method uses a point-focus SAM with a tilted lens. The angular dispersion of SAW propagation on Si(100) is measured. The accuracy of this method is considered.

INTRODUCTION

Material characterization with the scanning acoustic microscope (SAM) is commonly performed by analysis of $V(z)$ curves. A theoretical understanding of the dependence of the oscillations in $V(z)$ curves on the surface wave (SAW) velocity has been developed on the basis of a ray model. But the analysis recovers difficulties because the pupil function affects the oscillations in $V(z)$.[1] Especially for high frequencies > 1 GHz $V(z)$ curves show only a small amount of oscillations because high frequency lenses offer only a small working distance due to the strong attenuation in the coupling fluid (water). This leads to considerable errors in the determination of SAW velocities. Moreover measurement of anisotropy is only possible with line-focus acoustic lenses.

Chizhik *et al.* presented a method to determine properties of discontinuities that makes use of the analysis of interference fringes in SAM images.[2] In this paper we use discontinuities for a new technique of SAW velocity measurement. Interference fringes that occur due to the reflection of the SAWs at discontinuities are imaged by a tilted lens. For a non-tilted lens the intensity of the SAW signal is weak compared with the specularly reflected signal so that fringes can only be seen near discontinuities where the influence of the halo contrast is very strong.[4] It will be shown that by tilting the lens the specular signal can be reduced relatively to the SAW signal.

EXPERIMENT

The experimental setup is shown in fig. 1: The acoustic lens is tilted by an angle θ_t that equals approximately the critical angle for stimulating SAWs of the analyzed material. This ensures that the SAW is stimulated with maximum intensity. The SAW which propagates a distance d towards the elastic discontinuity leaks energy into the water with wavevector $\vec{k}_{SAW}^f$. The reflected SAW leaks energy into the direction $\vec{k}_{SAW}^r$ that is measured by the lens along its optical axis where the sensitivity has a maximum. The transducer measures the interference of the reflected SAW $\vec{k}_{SAW}^r$ and the specularly reflected wave $\vec{k}_s$. As the lens scans along the x direction the distance d varies and consequently alternating destructive and constructive interference occurs. The wavelength of the oscillations in the image is equal to half the wavelength of the SAW. While fig. 1 shows the sample in-focus it is also possible to perform the

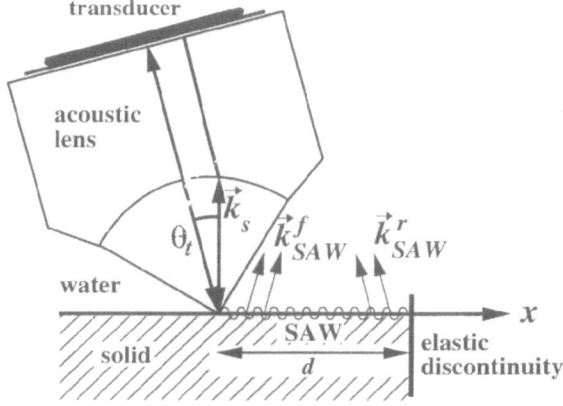

Figure 1. Experimental setup for imaging interference fringes due to SAWs

experiment in the over- and underfocus region. Especially for frequencies > 1 GHz it may be convenient to use the underfocus to obtain better contrasts although this leads to a bigger insonified spot (see fig. 5). As long as the insonified spot includes the discontinuity this gives rise to halo fringes around the discontinuity because the interference of the specularly reflected signal with a cylinder wave emitted from the discontinuity is measured. Additionally it can also be seen from fig. 1 that also low aperture lenses may be used that do not stimulate SAWs in a perpendicular setup. It is also possible that interference fringes due to the interference of internal lens echoes with $\vec{k}_{SAW}^r$ may occur. In this case the internal lens echoes might serve as the reference wave instead of the specular signal.

RESULTS

The investigated sample is a silicon wafer Si(100) that has a SAW velocity of 4921.2 m/s in the [010] direction corresponding to a SAW-angle of 17.8° in water at 23°C.[5] Fig. 2 shows images with a non-tilted lens (a) and with a lens 17° tilted (b), both taken at 970 MHz. It is obvious that image (b) shows much better contrast than (a). Image (a) shows mainly contrast due to the halo effect so that no SAW velocity can be derived. In image (b) many fringes are visible, therefore the area affected by the halo effect can be excluded from consideration. The distance between maxima/minima in image (b) corresponds to half a wavelength of the SAW so that the SAW velocity can easily be derived by measuring the spacings between the fringes. This is done by calculating the Fourier spectrum of a column in the image and determining the peak corresponding to the periodicity in the fringe pattern. In this way different frequencies - caused by different SAW modes or the halo effect around edges - in the image may be distinguished. Fig. 3 shows the sum of the spectra of all colums in fig. 2(b). The wavenumber of the main peak is $k_{SAW} = 0.1983$ and with an ultrasound frequency of 970 MHz this leads to a SAW velocity of $v_{SAW} = 4891$ m/s which differs by 0.6% from the literature value. The ultrasound frequency is determined by measuring the frequency of the water ripple oscillations in $V(z)$ curves with the same lens. To measure

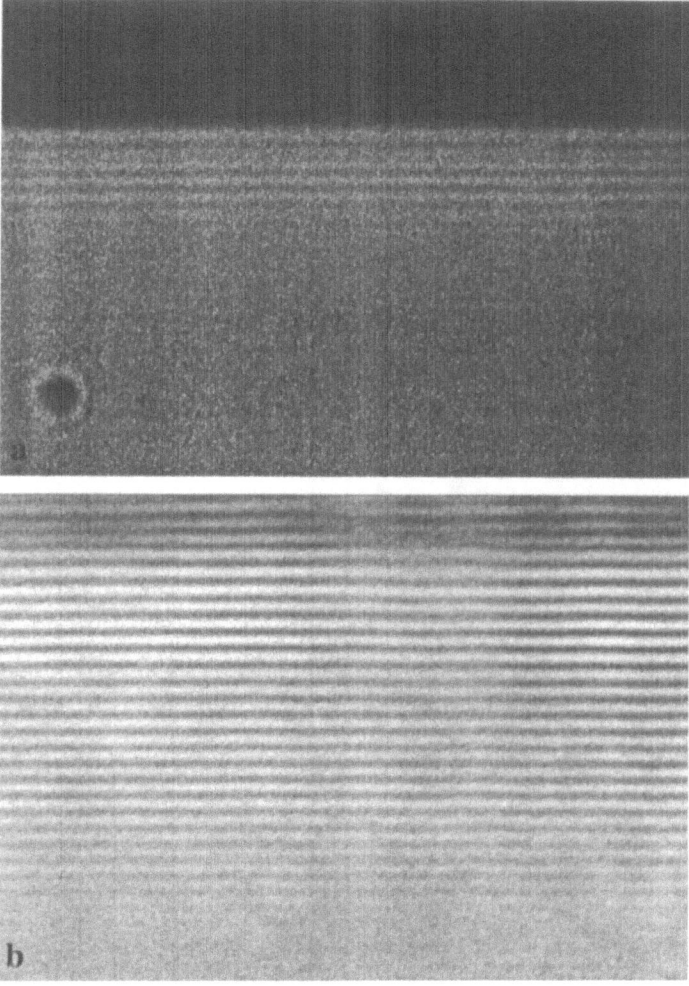

Figure 2. Interference fringes caused by a SAW travelling in [010] direction on a Si(100) wafer. (a) imaged with lens perpendicular to the sample's surface and 13 μm in the overfocus. (b) lens tilted 17° against the surface normal direction, 45 μm in the underfocus. The reflecting edge in (b) lies about 10 μm above the upper image border. For both images the frequency is 970 MHz, lens aperture $2 \times 50°$, image height 75 μm.

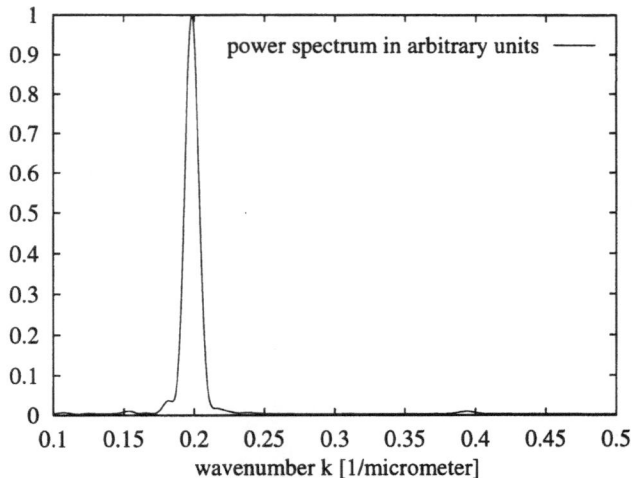

Figure 3. Power spectrum of the image in fig. 2(b): The maximum at $k = 0.1983$ corresponds to a SAW velocity of $v_{SAW} = 4891$ m/s

the anisotropy of SAW propagation on a Si(100) surface we used an etched silicon wafer with a circular ring structure. The depth and the width of the etched ring is 15 μm. The focus position is outside the ring and the lens is tilted by 17° towards the center of the ring. The sample is rotated in steps of 3.6° and thus the measured reflected SAWs are propating in different crystallographic directions. The angular dispersion between [010] and [011] direction on Si(100) is shown in fig. 4. The biggest deviation of the experimental data from the theoretical curve is 35 m/s or 0.7%. We attribute these deviations to instabilities in the ultrasound frequency of the used microscope.

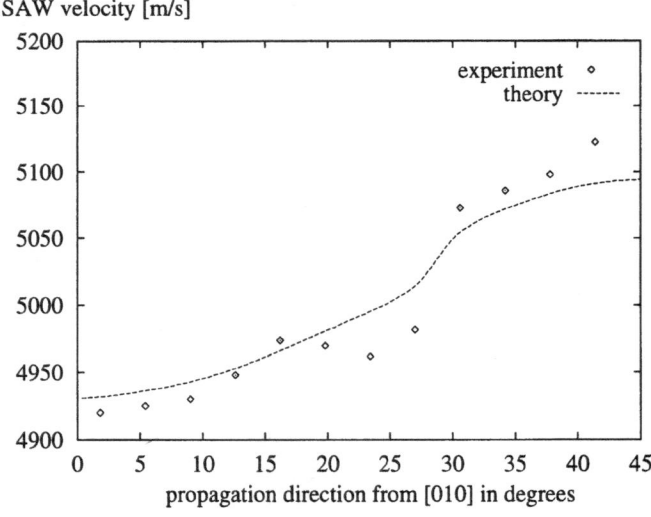

Figure 4. Angular dispersion of SAWs on Si(100) from [010] direction (0°) to [011] direction (45°). The theoretical curve is taken from Kushibiki et al.,[6] the experimental points are derived from analysis of images taken with tilted lens containing SAW interference fringes. The used frequency is 970 MHz.

ACCURACY

To measure undisturbed interference patterns it is necessary to use straight discontinuities so that no undesirable scattering takes place. With reflection at scratches no
satisfying results were achieved. The most efficient way of reflecting SAWs is using a
right-angled edge.

The main reasons for errors in the determination of precise SAW velocities are
uncertainties in the ultrasound frequency and misalignment of the sample relative to
the lens. By geometrical consideration the influence of a tilt by 1° results in an error
of 0.6% for a SAW angle of 17°.

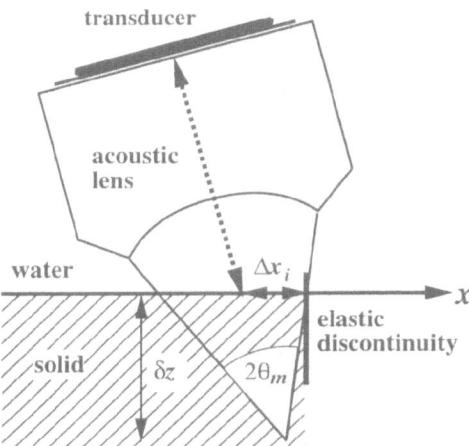

Figure 5. The insonified spot increases with increasing negative (or positive) defocus δz; θ_m
is the half-opening angle of the lens. As long as the discontinuity lies within the area of width
Δx_i halo effect and SAW fringes occur simultaneously.

The high values of defocus in fig. 2 lead to large insonified areas Δx_i in which halo
contrast and contrast due to SAWs occur simultaneously (see fig. 5). For the defoci
$\delta z = 13$ μm and $\delta z = -45$ μm in fig. 2(a) and (b) Δx_i is 27 μm and 43 μm, respectively.
Therefore, it is interesting to consider additionally the influence of the halo effect in
more detail. Assume that the output voltage V of the microscope in dependence of the
scanning position x is a superposition

$$V = V_0 + H(x) + A(x)\exp(-2ikx)$$

of the specularly reflected signal V_0, $H(x)$ is the signal due to diffraction at the edge
situated at $x = x_0$. $A(x)\exp(-2ikx)$ is the SAW signal where $A(x)$ is the exponentially
decaying amplitude of the SAW signal and k its wavenumber. For reasons of readability
the arguments of A and H are omitted in the following equations. Assuming that the
intensity in the image is the absolute value of V and that A and V_0 are real this leads
to

$$|V| = \left(V_0^2 + A^2 + |H|^2 + 2V_0\Re[H] + 2A(V_0 + \Re[H])\cos(2kx) - \right.$$
$$\left. 2A\Im[H]\sin(2kx)\right)^{1/2},$$

where $\Re$ denotes the real, $\Im$ the imaginary part. V_0 is the main contribution to the
measured signal $|V|$. Away from the discontinuity $H(x)$ is small compared with V_0.

Thus the root can be expanded taking into account terms up to order $1/V_0$:

$$|V| = V_0 + \underbrace{A\cos(2kx) + \frac{A^2}{2V_0}\sin^2(2kx)}_{\text{SAW}} + \underbrace{\Re[H] + \frac{\Im^2[H]}{2V_0}}_{\text{halo effect}} - \frac{A\Im[H]}{V_0}\sin(2kx)$$

The spectrum, given by the Fourier transform of $|V|$, contains terms which are only determined by the SAW, consisting of the peak at $2k$ and a small peak at twice this frequency. The next terms describe the pure halo effect which is spread over the range of the optical transfer function. The last term of order $1/V_0$ consists of the spectral components of the halo effect shifted by $2k$ in positive and negative direction due to the sine function. Because the halo effect is wide spread in the spectrum only the underground will be raised and the position of the SAW peak remains in good approximation unshifted. So we conclude that the halo effect does not affect the accuracy of the determined SAW velocity.

SUMMARY

A new technique for measurement of SAW velocities that uses a point-focus SAM with a tilted lens was presented. By lateral scanning in the vicinity of an elastic discontinuity it is possible to image interference fringes due to SAWs with a wavelength equal to half of the SAW wavelength. The tilting of the lens leads to a reduction of the strong specularly reflected signal while the weak SAW signal is enhanced. By chosing discontinuities in different crystallographic directions of an anisotropic material it was demonstrated that the anisotropy of SAW propagation can be measured.

REFERENCES

1. K.K. Liang, G.S. Kino, and B.T. Khuri-Yakub, Material characterization by the inversion of $V(z)$, *IEEE Trans. on SU* SU-32:213 (1985).
2. D. Chizhik, D.A. Davids, and H.L. Bertoni, A modified ray theory for predicting the $V(x,z)$ response of a point-focus acoustic microscope in the presence of a crack, *J. Acoust. Soc. Am.* 91:3285 (1992).
3. W. Weise, P. Zinin, and S. Boseck, Modelling of inclined and curved surfaces in the reflection scanning acoustic microscope, *J. Microscopy* 176:245 (1994).
4. H. Block, S. Boseck, and G. Heygster, Description of the defocus effects of an acoustic reflection scanning microscope (SAM), *Optik* 86:27 (1990).
5. B.A. Auld. "Acoustic Waves and Fields in Solids," Vol. II appendix 4, John Wiley and Sons, New York (1990).
6. J. Kushibiki, K. Horii, and N. Chubachi, Leaky SAW velocity on water/silicon boundary measured by acoustic line-focus beam, *Electron. Lett.* 18:732 (1982).

IMAGING AND QUANTITATIVE EVALUATION OF MICRO-CRACK DENSITY IN CHROMIUM COATINGS ON STEEL BY MEANS OF ACOUSTIC MICROSCOPY

C. Maurel[1], J.M. Saurel[1], L. Lelait[2], J.M. Soro[3] and J. von Stebut[3]

[1]Laboratoire d'Analyse des Interfaces et de Nanophysique, URA CNRS 1881, CC82, Place Eugène Bataillon, 34095 Montpellier, France
[2]EDF, Centre de Recherches Les Renardières
BP 1, 77250 Moret/Loing, France
[3]Laboratoire de Science et Génie des Surfaces, URA CNRS 1402, INPL, Ecole des Mines, 54042 Nancy, France

ABSTRACT

A non-destructive method of high frequency acoustic microscopy (**HFAM**) is applied to image and quantify micro-cracks in approximately 10 μm thick hard chromium coatings, especially in-depth where conventional reflected light and scanning electron microscopy (**RLOM** and **SEM**) are inadequate. Detailed information on precise in-depth micro-crack geometry and position is gained. The extension of this method leads to a study of adhesive failure (interface cracks).

In the present work cracks are generated by micro mechanical techniques (scratching tests). Their density as assessed by our HFAM method is compared with results by standard metallography. In addition it is shown that quantitative information about both coating elasticity and quality of adhesion can be obtained with microprobe precision in between visible cracks.

I. INTRODUCTION

In modern engineering two of the major objectives of surface treatments (thermochemical "transmutations", coatings) are to protect technical parts from corrosion and mechanical damage. Cracks in treated surface layers are especially detrimental to service life, especially when spreading across the entire layer thickness and/or when propagating along the interface between the base material and the treated surface layer (adhesive failure for coatings).

In the present study through thickness and hidden cracks perpendicular to the specimen surface as well as interface cracks are introduced deliberately into a coated system of particular technological interest ; hard chromium coatings on steel substrates. This is achieved by specific micro mechanical testing techniques like scratching in single pass and multipass operation mode[1-4]. For these testing procedures, the corresponding failure thresholds for cracking (critical normal load and/or number of cycles to failure) are vital criteria for coating quality characterisation.

Standard non-destructive metallography can only give information on sufficiently wide surface cracks. In-depth cracks that have not reached the surface, or even emerging cracks, too thin to be detected (e.g.: owing to closing-up after stress relaxation), require cross-sectional cuts likely to introduce additional brittle damage.

Acoustic microscopy (AM) has been presented in the past as an appropriate non-destructive, analytical tool[5,6]. In the present study we illustrate in detail the scientific potential of this technique.

II. EXPERIMENTAL

II.1. Specimens

Approximately 10 µm thick *hard chromium coatings* having dense columnar structure were deposited by means of a non-reactive magnetron sputtering technique from sintered chromium targets containing 1 wt.% of carbon[1]. The coatings had a surface hardness $H_{V,0.1}$= (15±1) GPa and contained from the beginning a population of pre-existing cracks (Fig. 1), probably owing to tensile type internal stress resulting from the surface treatment.

Substrates were 5 mm thick AISI 304L stainless steel flats with a surface roughness R_A=0.1µm after electrolytic polishing and a surface hardness $H_{V,0.1}$=(1.8±0.1) GPa.

II.2. Micro mechanical testing for crack generation

The multipass scratch rig retained for deliberate crack generation has been described in detail elsewhere[1,4]. In simple multipass operation a spherical pin (radius, R=0.79 mm) with a contact load of 10N slides N times in the same scratch track of the coated specimen. In between consecutive unidirectional passes the slider returns frictionless to its starting point. The distance per cycle of dry sliding at a speed of 1 mm/s is 4mm.

Three different scratches have been studied, corresponding to N=1, 10, and 20 passes.

II.3. Surface damage analysis

II.3.1. Acoustic imaging and elastic microanalysis: A spherical transducer operating at 600 MHz is excited with a pulse width of less than 10ns. This width does not allow separation of surface and interface echoes owing to too small a difference (only 1.5%) in acoustic impedance of chromium and stainless steel. Operating at a frequency of 600 MHz, for a layer thickness t = 10 µm, does not allow Lamb wave but rather Rayleigh mode generation. With these operation conditions layer defects can be imaged over a depth of about one Rayleigh wavelength, λ_R ($\lambda_R \approx$ 6µm in chromium at 600 MHz) and, for large defocus, defects close to the interface can be detected. The corresponding image contrasts are sensitive to various material parameters.

The acoustic signature is well known to all AM users for mechanical properties assessment of substrates or thin layer materials deposited on substrates[7,8,9]. While for surface layers with t >> λ_R, the material can be taken as a semi-infinite half-space, for surface layers with t ≤ λ_R the acoustic response is sensitive to intrinsic coating properties as well as the coating/substrate interface quality. Under these circumstances internal micro structures can be imaged and analysed by acoustical signatures V(z) and V(x). The latter signal is obtained by a line scan in the transverse direction of the micro crack. Fringe spacing is given by $\lambda_R/2sin\theta$[10] where θ is the micro crack inclination with respect to the surface normal.

III. RESULTS AND DISCUSSION

III.1. Crack analysis before micro mechanical testing

In order to understand contrast from acoustic images, correlation with conventional techniques like optical and scanning electron microscopy (RLOM and SEM) is necessary,

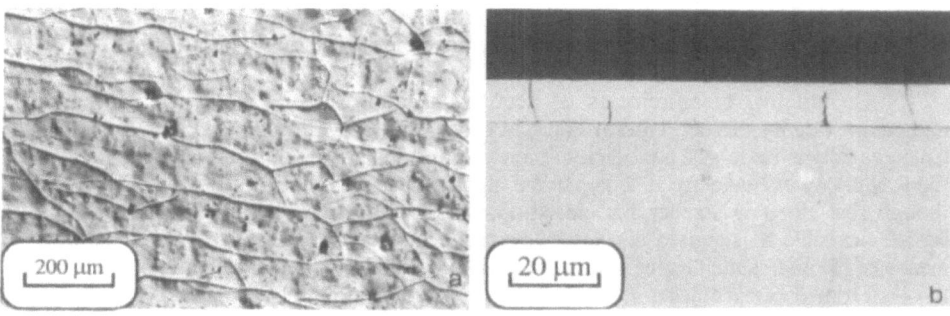

Figure 1. optical micrographs of the chromium coating. (a) Top view. (b) Transverse view after chemical attack (Nital and Murakami). Note the presence of a crack network.

even if these techniques only yield surface information and in-depth analysis requires destructive cross-sectional metallographic cuts. Such conventional RLOM micrographs of our as deposited coatings are shown in figure 1. On the surface areas without additional micro mechanically induced damage the above techniques reveal an array of pre-existing linear micro defects (lines parallel to a given direction). They are surrounded by numerous "point"-defects in the coating as would be expected for cracks linking up soft spots in the coating structure. These line defects appear as changes in topography (between 0.1 and 0.3 µm steps) possibly due to slight variations in chromium structure. However identification as cracks by means of standard RLOM and SEM often remains ambiguous.

When applying HFAM, such identification becomes possible. Indeed, inspection of the region outside of the scratch track in fig.2 shows interference fringes next to all of the above line defects giving clear evidence of their crack identity. This image taken at 40 µm defocus is typical of the majority of parallel unidirectional cracks (**PUCs**) in the as-received state. They are open micro-cracks with a density of (15 ± 1) mm^{-1}. Few cracks propagate in the transverse direction; most of the latter are closed-up, with a density of (3 ± 0.5) mm^{-1}.

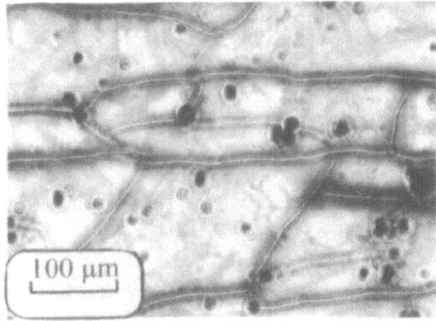

Figure 2. Acoustical image at 40 µm defocus of a free damage zone.

Open and closed-up cracks can be distinguished:
-For open micro-cracks (cracks n°1,3,4 in fig.3b) the V(x) profile shows important oscillations.

-For closed-up micro-cracks (crack n°2 in fig.3b), the number of fringes is smaller and amplitudes have less contrast, as these parameters depend on crack size and depth.

Quantitative measurements of acoustic signature V(z) have been realised in the immediate vicinity of each type of crack in order to improve the quality of analysis (fig.3a). Changes in amplitude and periodicity in inhomogeneous zones cause a systematic decrease in the Rayleigh wave velocity. The larger the crack dimensions the smaller is this velocity. Even though this effect is smaller for closed-up and/or in-depth micro cracks (as crack n°2 in fig.3a), invisible to standard surface analytical techniques, it can still be clearly resolved by means of HFAM. Thus this technique can give information on defect size and depth.

From V(x) profiles on figure 3b , crack inclinations less than 10° off the surface normal are computed.

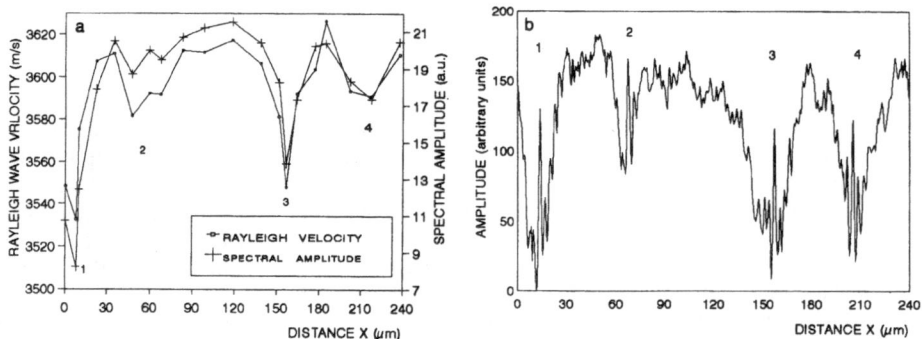

Figure 3. (a) V(z) measurements in the vicinity of microcracks. (b) V(x) profile in the same area.

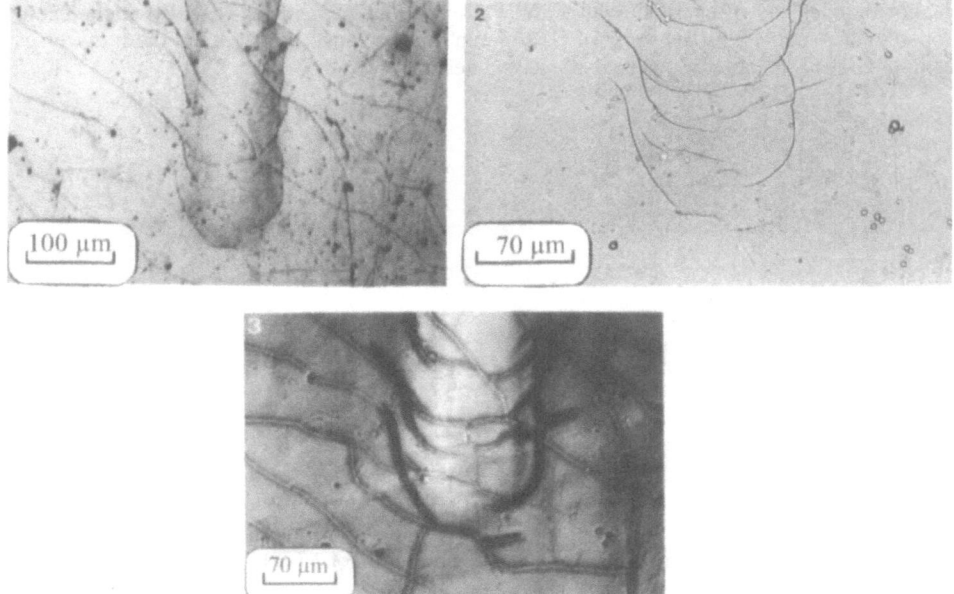

Figure 4. Analysis of surface damage around the beginning of a scratch track (b1: RLOM, b2: SEM, b3: HFAM).

III.2. Analysis of scratch-induced cracks

III.2.1. Single pass operation

Figures 4b$_1$, 4b$_2$, and 4b$_3$ show RLOM, SEM, and HFAM images of the same zone around the beginning of a single pass scratch track. Only HFAM allows unambiguous, strong contrast evidence of closed-up micro-cracks with the typical fringe pattern on either side. Crack density assessed within the scratch track itself is about (42 ± 1) mm^{-1} and includes both (18 ± 1) mm^{-1} pre-existing cracks, (15 ± 1) mm^{-1} diverging border cracks and (9 ± 0.5) mm^{-1} hidden in-depth cracks.

II.2.2. Multipass operation.
The multipass scratch track (N=20) has been studied in its centre with different defocusing depths. Images in figures 5a and 5b are respectively observed by RLOM and HFAM, the latter at 5µm defocus. Crack density in this scratch track is about (45 ± 1) mm^{-1} including a population of (18 ± 1) mm^{-1} pre-existing cracks, (16 ± 1) mm^{-1} diverging border cracks and (11 ± 1) mm^{-1} hidden in-depth cracks. As expected, crack density is higher after multipass than after single pass operation but only by 7% and does not seem to change any more between 10 and 20 passes. As to crack inclination, it can be assessed by analysing fringes width and symmetry on both sides of the crack.

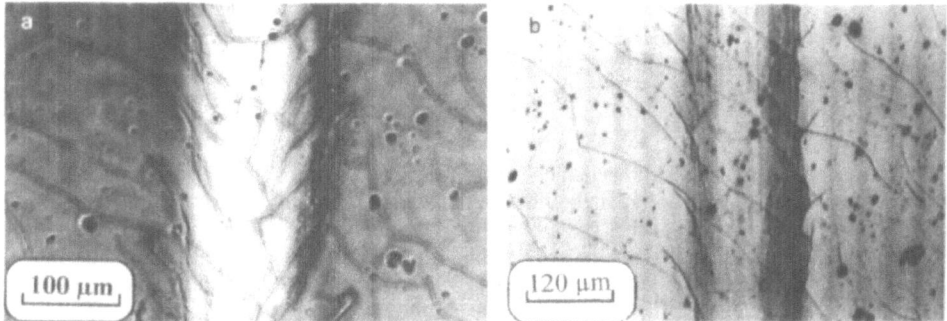

Figure 5. Analysis around a multipass scratch track (a) : RLOM. (b) : HFAM.

Acoustic image interpretation has led to identification of three types of cracks :

-Essentially *closed-up and in-depth cracks* in the zones free from scratch damage.

-*Diverging border cracks (DBCs)* propagating in a forward direction from the track borders. This propagation direction appears as essentially uncorrelated with respect to that of the pre-existing PUC array and is clearly specific of the scratch-induced strain field. However, most DBCs emerge from pre-existing PUCs that appear to be acting as crack initiators. DBCs can be observed both in RLOM and HFAM. SEM images yield a typical crack height of 0.3 µm. The number of HFAM fringes of DBCs is twice or three times larger than for PUCs.

-Numerous, *hidden in-depth cracks within the scratch track, parallel to the sliding direction* and generally perpendicular to the surface. To our knowledge the existence of such scratch-induced, longitudinal, hidden cracks has never been discussed in the literature.

Acoustical signature measurements and V(x) profiles have been made on a line across a diverging microcrack running along the single pass scratch as shown respectively on figures

6a and 6b. A strong increase in the Rayleigh wave velocity appears in this area as well as two different fringe spacings on V(x) profile. If, for some microns width cracks, a strong wave attenuation decrease is expected, an increase in Rayleigh velocity is difficult to interpret without taking into account a multiwave system of interference. In comparison with V(z) measurements made on pre-existing cracks, this effect shows that microelastic analysis is able to make a distinction between different types of cracks.

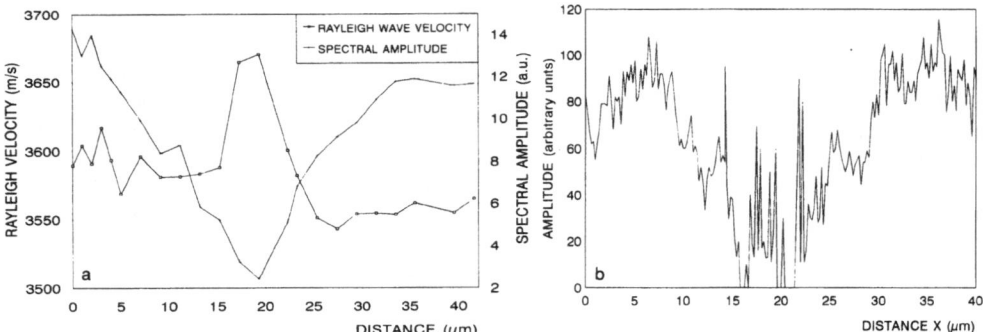

Figure 6. (a) V(z) measurements near a diverging crack track. (b) V(x) profile in identical area.

CONCLUSION

This work using micro mechanical tests and observations in optical, electron and especially acoustical microscopy has confirmed the scientific potential of the latter technique. In fact, acoustic microscopy offers the advantage over conventional techniques to distinguish different types of cracks and to image, with strong contrast, defects below the surface such as closed-up and/or hidden in-depth micro-cracks. Moreover, quantitative morphology analysis of such cracks ("invisible" with standard metallographic techniques like RLOM and SEM) can be achieved. For fatigue experiments it is perfectly conceivable to do HFAM on-line tracking of crack propagation for follow up of surface damage in a coating under contact mechanical ageing.

REFERENCES

1. A. Darbeïda, J. von Stebut, M. Barthole, P. Belliard, L. Lelait, G. Zacharie, Surface and Coatings Technology 68/69 (1994) 582-590.
2. S.J. Bull, Surface and Coatings Technology 50 (1991) 25-32.
3. H.E. Hintermann, Fresenius J. Anal. Chem. 346 (1993) 45-52.
4. A. Darbeïda, J.von Stebut, M. Assoul, J. Mignot, Int. J. Mach. Tools Manufact. 35 (1995) 177-181.
5. J.H. Cantrell, M. Qian, M.V. Ravichandran, K.M. Knowles, Appl. Phys. Lett. 57 (1990) 1870-1872.
6. R. Caplain, L. Ferdj-Allah, J.M. Saurel, J. Attal. "Aspects récents des signatures et de l'imagerie acoustiques à 50<F<2000 MHz- application à la métallurgie", Mémoires et Etudes Scientifiques. Revue de Métallurgie (Oct.1992). 643-651.
7. R.J.M. Da Fonseca, L. Ferdj-Allah, G. Despaux, A. Boudour, L. Robert and J. Attal. "Scanning acoustic microscopy - recent applications in materials science", Advanced Materials. Vol 5 (7/8). (1993). 508-519.
8. W. Parmon and H.L. Bertoni. "Ray interpretation of the material signature in the acoustic microscope", Electronics Letters Vol 11. (1979). 684-686.
9. R.G. Wilson and R.D. Weglein. "Acoustical material signatures using the reflection acoustic microscope", Proc. 1st International Symposium On Ultrasonic Materials Characterization. (June 1978).
10. A. Briggs. "Surface Cracks and Boundaries", *in:* "Acoustic Microscopy", Clarendon Press. Oxford (1992).

HIGH-FREQUENCY FOCUSING TRANSDUCER FOR ACOUSTIC MICROSCOPE

Konstantin I. Maslov[1], Roman Gr. Maev[1], Leonid M. Dorozhkin[2], and
Valery S.Doroshenko[2]

[1] Acoustic Microscopy Center, Russian Academy of Sciences
Kosygyn str. 4, Moscow, 117334, Russia
[2] Instit. of General and Inorganic Chemistry, Russian Academy of Sciences
Leninskiy prosp. 31, Moscow, 117907, Russia

INTRODUCTION

The characteristics of any type of acoustic microscope are determined mainly by the characteristics of the acoustical focusing system being used. At present, two types of acoustic microscope focusing systems are available. The first of them is the commonly used acoustic lens[1]. The piezoelectric transducers for acoustic lenses can be made of highly piezoactive materials such as LiNbO, LiTaO, LiIO, textured ZnO or AlN films. The natural drawback of the acoustic lens is the high sound reflection from the spherical convex surface inside the lens body, which gives rise to the strong coherent background noise.

Another type of focusing system is a thin film spherical piezoelectric transducer. At the 1-100 MHz frequency region, PVDF or P(VDF,TrFE) polymer piezofilms are used as a rule[2]. These film materials is feasible to obtain piezotransducers with sufficient bandwidth and good pulse response. However, their piezoelectric parameters of the polymer film materials are unstable and coupling coefficient K is less than 0.28[2]. The other drawback of this system is the complexity of its fabrication. It seems hard to obtain the high quality transducer for frequencies above 100 MHz.

Here we present a new type of spherical focusing piezoelectric transducer - a piezofilm directly deposited onto the convex spherical surface of the acoustic lens. The piezofilm material is the crystalline organic substance-based original textured piezoelectric film with high and stable piezoelectric parameters and low acoustical impedance. The high quality films of this material can be obtained by vapor deposition on the flat, concave or convex surfaces of any dielectric. In this way it is feasible to fabricate high frequency (up to 600 - 700 MHz) well-shaped focusing transducers for acoustic microscopy.

THE THIN-FILM PIEZOELECTRIC MATERIAL

The thin-film material we developed is a crystalline organic compound of the carbomide type designated as CA. This material is similar to the well known textured

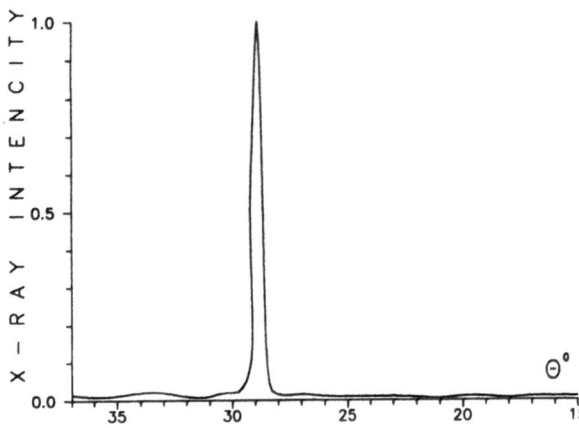

Figure 1. The X-ray pattern of CA textured film theta-2theta method, CuK as X-ray source. The only peak corresponds to [111] direction.

materials as CdS, ZnO and AlN[3]. X-ray analysis showed that [111] the crystallographic axis of CA micro crystals are directed mostly along the normal to the substrate surface with high degree of orientation and no other orientation is present (see figure 1).

Thus the CA thin-film piezomaterial has only one symmetry element - the infinite order axis which is normal to the substrate surface. This type of piezomaterial according to its piezoelectric tensor is capable to generate longitudinal acoustic waves in the substrate under the action of the electric field applied parallel to the substrate normal.

The CA textured films were prepared by the method of thermal vacuum evaporation. Depending on the bell jar size, several dozens or even several hundred of the CA-based transducers may be fabricated in one cycle. The transducer electrodes are produced by the common vacuum evaporation technology. The bottom electrode is a Cr-Au 0.3 mm thickness film (Cr-Ni film may be used); the top electrode is an Al 0.15 mm thick film. The CA thin-film piezoelectric material offers a number of advantages in comparison with similar materials (for example, ZnO and AlN). No special heating of the substrate is required here. So, high quality CA piezoelectric textured films can be deposited on the polished surface of various materials: fused and crystalline quartz, various types of glass, silica, sapphire, garnet, polymeric compounds (polymethylmethacrylate, polystyrene) and oxidized metals can be used as substrates. Flexible materials (polymeric or metallic) are also suitable as a substrates.

At present we can fabricate CA films of any sickness between 0.5 - 12 mm with a operation frequencies from 50 Mhz up to 1 Ghz. The main properties of CA textured piezofilms are listed in the following table:

Table 1. Piezofilm properties.

Effective piezoelectric constant	d_{33} = 18 pK/N
Effective electro-mechanical coupling	K = 0.4-0,6
Density	n = 1.5 g/cm
Longitudinal sound velocity	$v_1 = 2.2*10_5$ cm/s
Dielectric constant	e = 4

We have already used the CA material as a piezoactive element of various devices including wide-band transducers[4] and piezosorption chemical sensors[5]. In all previous applications the piezoelectric films were deposited on the flat surfaces and were used for the radiation of sound inside the substrate. Here we exploit the remarkable property of the CA film material to form a highly oriented texture on the convex surfaces.

TRANSDUCER PROPERTIES

The substrates we used were the fused quartz rods with a 3.2 mm diameter a spherical cavity with curvature radius 4 mm. The electrode area exceeds that of the cavity by 10%. The thickness of the piezoelectric film obtained was l = 8 mm. It was controlled during evaporation process by quartz microbalance. Several 'witness' transducers were fabricated in the same process on a flat substrate. The X-ray investigation showed a good quality of the film texture. The insertion losses of flat transducers were within 7 - 8 dB.

Various properties of focusing transducers that determine the quality of an acoustic microscope objective: the frequency dependence of insertion losses, signal to background noise ratio, and sound field distribution in focal region were investigated.

The frequency dependence of insertion losses of the transducer is shown in figure 2. At a frequency of maximum efficiency total insertion losses were within 30 db. The bandwidth of about one octave at the 3 dB level was implemented. Theoretical curves obtained for flat transducer of the same dimensions and materials[6] for different generator impedance are also shown in figure 2 by dashed lines. The transducer behavior corresponds to K of 0.3 instead of K = 0.5 for transducer material. It can be explained by the resistive losses in transducer electrodes and piezomaterial, the influence of finite electrode thickness and diffraction losses in immersion liquids.

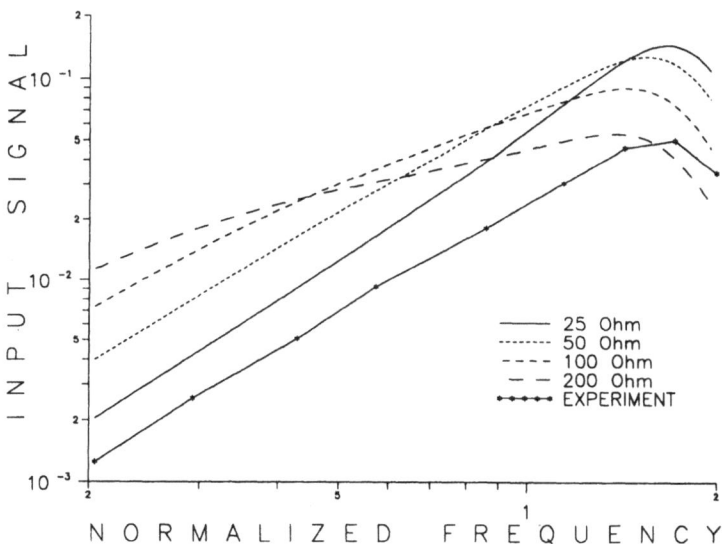

Figure 2. The frequency dependence of insertion losses of the transducer.

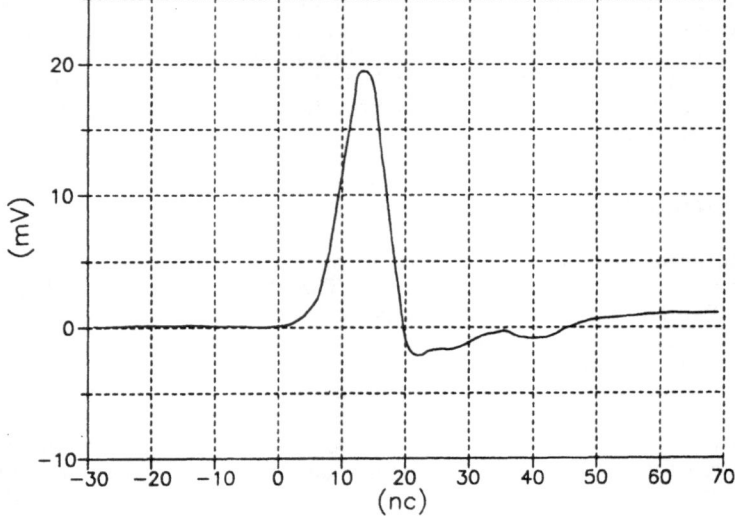

Figure 3. A typical transducer response to 5 ns fall time step voltage.

The acoustical pulse properties of the transducer where investigated. by exciting it with 5 ns fall time step voltage. The responded ultrasonic pulse was reflected by a perfect reflector (liquid/air interface) placed in the focal region of the transducer and received by the same transducer. A typical form of the response is shown in figure 3. The received pulse exhibits almost unipolar shape with no delayed oscillations.

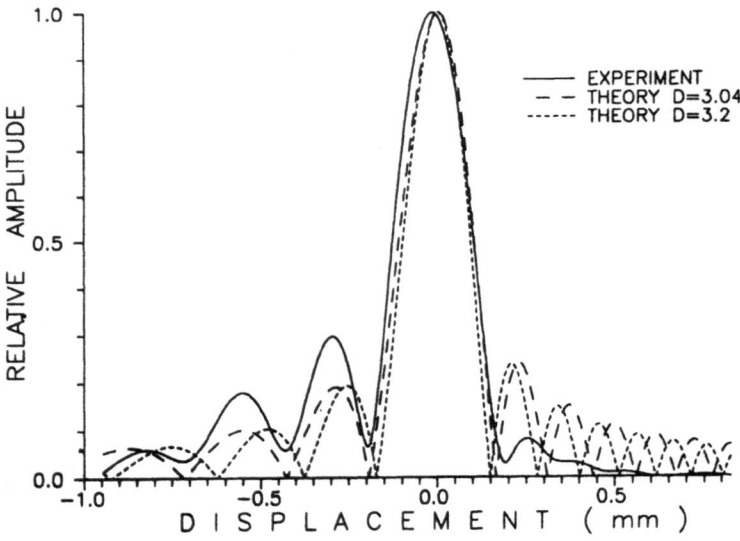

Figure 4. Calculated and measured V(z) curves for absolute reflector.

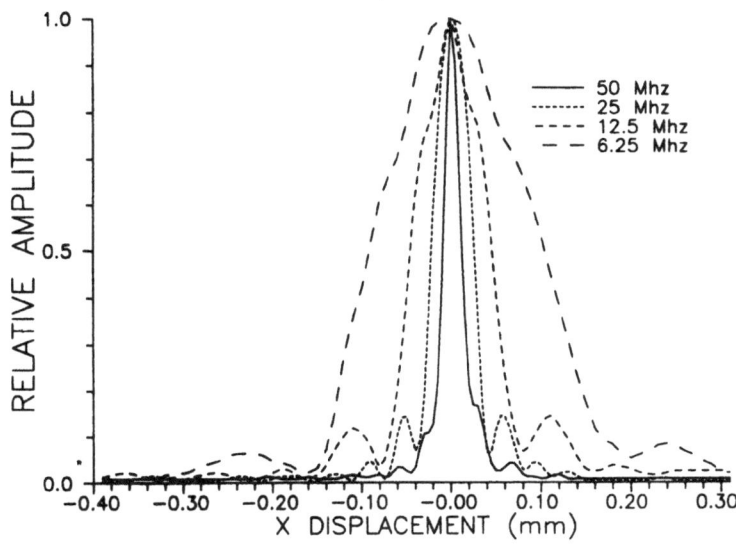

Figure 5. Intensity variations of the echoes from a 4 mm diameter steel ball.

The transducer was used and tested in 100 Mhz scanning acoustic microscope made in our laboratory[7]. The axial resolution of the transducer was estimated by V(z) curve measurement for a perfect reflector. The result is presented in figure 4. along with theoretical calculation[8]. One can see that the experimental minima and maxima coincide with those on the theoretical curve for the transducer with a slightly smaller aperture. The lack of agreement between the minima and maxima positions apparently demonstrate the unperfectness of the transducer edge region.

Lateral resolution of the transducer was estimated by using method based on the imaging of a large spherical particle placed in its focal plane[9]. In this case the image brightness distribution is proportional to the lateral acoustic potential distribution in the focal region of the transducer. The experimentally obtained values of the transducer voltage are shown in figure 5 versus the lateral displacement of a particle with respect to transducer focus. The excitation frequencies were 50 Mhz, 25 Mhz, 12.5 Mhz and 6.25 Mhz. The measured widths of the zero order peaks are in good agreement with their theoretical value.

CONCLUSIONS

The novel focusing ultrasonic transducer for acoustic microscope is developed. The piezoelectric textured film of a new organic-based material was used as its active element.. The highly piezoactive film of this material can be deposited by vacuum evaporation on the concave surface of practically any material. A set of new-type transducers were manufactured and studied.

The level of insertion loss obtained within 30 bB was achieved in octave frequency band from 50 to 100 Mhz. The insertion losses frequency dependence curve is smooth and slow within the frequency band from 5 to 120 Mhz. The transducer response to the step voltage with 5 ns fall time is close to an unipolar 15 ns duration pulse.

The lateral and axial resolution is close to their theoretical value even for the transducers with large aperture angle. An absence of reflective surface between the transducer and studied object dramatically improves the signal-to noise ratio with respect to the acoustic lens system.

The transducer is acoustically transparent. It can be formed on the concave surface of standard acoustic lens with flat piezoelectric transducer and work simultaneously with it. Resulting two focus system can be used for subsurface imaging using mixed shear and longitudinal wave or in continuous wave reflection microscope.

The main drawback of the piezofilm developed is its rather high solubility in water but drawback can be prevented by the use of the various kind of the appropriate protective film.

Thus, the transducer we developed exhibits the high parameters and is a promising element for application in acoustic microscopy.

REFERENCES

1. C. F. Quate, A. Atalar, and H. K. Wickramasinghe, "Acoustic microscope with mechanical scanning - a review", *IEEE Proc.,Ultras.,Ferroel.,&Freq.control,* 67:1092 (1979).
2. H. Ohigasi, K. Koga, and T. Nakanishi, "Ultrasonic transducer employing piezoelectric polymer material," U.S. patent No 4,577,132, May 18, (1986).
3. A. V. Shubnikobv, Piezoelectric textures, *(in Russian),* Moscow - Leningrad: AN USSR, (1946).
4. L. M. Dorozhkin, V. S. Doroshenko, A. N. Karabutov, V. V. Lasarev, E. D. Pol'skikh, and B. A. Chaianov, "Wide-band piezoelectric transducer based on new type of textured material, *" Sov.- Phys. acoustics",* 38: 463 (1992).
5. L.M.Dorozhkin, V.S.Doroshenko, A.N. Karabutov, V.A.Ketsko, N.T.Kouznetsov, New type of piezoelectric chemical sensor based on thin film high frequency transducer, *in:* Proc. Intern. Conf. of Sensor Systems. St. Petersburg (1993).
6. G. S. Kino, Acoustic Waves: Devices, Imaging, and Analog Signal Processing, Englewood Cliffs: Prentice-Hall, (1987).
7. K. I. Maslov, "Acoustic scanning microscope for investigation of subsurface defects," *in Acoustical Imaging,* 19: 645, Ed. by H. Ermert and H.-P. Harjes, NewYork: Plenum, (1992).
8. H. T. O'Neyl, "Theory of focusing radiators," *J. Acoust. Soc. Amer.,* 21: 516, (1949).
9. K.I.Maslov, L.M.Dorozhkin,V.S.Dorosenko, R.G.Maev, A new focusing ultrasonic transducer and the two foci acoustic lens for acoustic microscope on its basis, *IEEE Proc.,Ultras.,Ferroel.,&Freq.control (in press).*

USING WAVELETS TO IMPROVE ELASTIC MICROQUANTIFICATION

J. C. Schlegel[*], R. Wichard, and R. M. Schmitt

Fraunhofer Institute Biomedical Engineering
Dept. of Ultrasound
66386 St. Ingbert
Germany

INTRODUCTION

Conventional scanning acoustic microscopy (SAM) offers very high resolution in the lateral imaging plane. The major drawback of this method is that echos from different planes as well as different wave modes that can be produced at the specimen surface contribute to the detected signals. Thus the quantification of elastic properties is strongly restricted. Using $V(z)$ for instance has been proved to be not very accurate or even not suitable in many cases.

However the quantification of elastic parameters in microscopic resolution would be very useful, especially to completely understand the principles of the macroscopic behaviour of tissue in diagnostic ultrasound with respect to tissue characterisation. In order to estimate the elastic properties of biologic material and solid bodies the received RF-signals of a common SAM are digitized, stored and processed in a bypass unit. From time of flight (ToF) and amplitude measurements the speed of sound, the impedance and the attenuation can be determined in microscopic resolution (Briggs et al. 1993, Schlegel & Schmitt 1994).

One of the major drawbacks of this method is the low signal to noise ratio (SNR) due to the very high attenuation at higher frequencies and to the need of detecting several surfaces simultaneously. Digitizing A-scans further allows only a low pulse repetition frequency (PRF). Thus the necessary averaging leads to an unacceptable acquisition time. In our case a sufficient SNR could only be reached by 64 times averaging or more. Due to the digitizing with 4 GSamples the PRF could not be increased far beyond 10 Hz. Thus a B-scan image of e.g. 200 A-Scans would at least take more than 15 minutes, a 100*100 points volume scan would take nearly one day.

Wavelet analysis was introduced to overcome this problem. The idea was to perform an off line limitation of bandwidth because the noise energy is spread over the electronic bandwidth of the equipment while the signal energy is concentrated in the center of the transducer bandwidth. Besides the Wavelet transform provides an optimum resolution in time and frequency scale. Thus the resolution will not be suppressed. A Gabor Wavelet was designed that can be scaled to the shape of the reflection signals. Simulations show that a reflection can properly be detected without significant worsening of time resolution even if the noise level is in the range of the signal level. Consequently the SNR can be improved without time consumptive averaging.

Practical examples on living cells, caries lesions in tooth material and layered structures in solid bodies are showing the versatility of this quantification method.

---

[*] Correspondence to: J. C. Schlegel, Univ. Hospital Vienna, Div. Pediatrics, Dept. Neonatology, Waehringer Guertel 18-20, A-1090 Vienna, Austria

THEORY

Elastic Microquantification

It can be shown that any reflection signal $S(t,z)$ in dependence of time t and defocus z can be described using a defocus function $g(z)$ that scales the amplitude with regard to the placement of surfaces refering to the focal point z of the lens:

$$S(t,z) = R \cdot s(t - \tau) \cdot g(z) \tag{1}$$

where R is the reflection coefficient between coupling fluid and specimen and $s(t)$ is the lens response function which is a constant for a given set of specimen and coupling. In the case of a layered specimen each reflection from different surfaces can individually be described by formula (1). Superimposing all $S(t,z)$ leads to the complete reflection signal. In order to separate the single variables in (1) the transducer has to be calibrated by an initial defocus experiment using an ideal reflector. Then R gets equal to unity. If the focus of the transducer is placed on the surface of the reflector the reflection amplitude is in it´s maximum and $g(z)$ is set to unity. Thus the lens response function can be determined by

$$S_0(t, z = 0) = s(t - t_0) \cdot g(z = 0) = s(t - t_0) \tag{2}.$$

Now $g(z)$ can be evaluated from different echos $S(t, z \neq 0)$ by scanning in z-direction in a relevant range.

ToF and amplitude measurements now can lead to the following geometric and elastic properties in a quantitative way without any further assumptions (Briggs 1993). From the difference in ToF of reference and surface reflections the heigth of the layer can be determined at known coupling fluid sound velocity. The difference in ToF of top and bottom reflection yields to the layer´s sound velocity. The amplitude ratio of front and reference reflections is determined by the impedance of the layer at known impedance of the coupling. The attenuation of the layer is determined by the ratio of the amplitudes of top and bottom reflection with respect to the reflection coefficients of all surfaces determined by their impedance.

Wavelet Transform

The Wavelet transform (WT) can reveal instantaneous frequency contents of a signal and does not degrade time resolution. It is defined as follows

$$WT_{\Psi} f(a,b) = \frac{1}{\sqrt{a}} \int_{\Re} f(t) \Psi_{a,b}^{*}(t) dt \tag{3}$$

and reveals in fact the similarity between the signal $f(t)$ and the basis function $\Psi(t)$ called Wavelet (Rioul & Vetterli 1991). In order to preserve energy the mean of $\Psi(t)$ has to be zero:

$$\hat{\Psi}(0) = \frac{1}{\sqrt{2\pi}} \int_{\Re} \Psi(t) dt = 0 \tag{4}.$$

To achieve the optimum time and frequency resolution after the uncertainty principle and to compare both functions the origin of the Wavelet or "mother Wavelet" has to be scaled and translated accross the time function. This can be described by two operators D and T which can be seen as pupil functions:

$$\Psi_{a,b}(t) = T^{b}\left(D^{a}\Psi\right) = D^{a}\left(T^{b}\Psi\right) = \frac{1}{\sqrt{a}} \Psi\left(\frac{t-b}{a}\right) \tag{5}.$$

Thus the WT is able to create a set of similarity functions $f(a,b)$ from the original time function $f(t)$. A wavelet is needed which delivers results $f(a,b)$ that can be interpreted as frequencies and should be well localised in time and frequency representation. Partitioning a signal into sinusoids leads to frequencies and good localisation in spectrum but not in time. However the Gaussian function is well localised in both representations. Combining sinusoids and the Gaussian function leads to the Gabor Wavelet, a frequency modulated Gaussian window:

$$g_{\omega,\sigma,\tau}(t) = \frac{1}{\sigma\sqrt{2\pi}} e^{-\frac{(t-\tau)^2}{2\sigma^2}} e^{-i\omega t} \tag{6}.$$

The original Gabor transform (Chui 1992) does not scale the kernel function. But in fact there are two parameters to determine time and frequency resolution. With their help it is possible to adapt the wavelet to the real shape of the reflection signals. The frequency is directly related to ω and σ is responsible for the width of the Gaussian function. Both parameters can scale the wavelet in a way that time resolution is high at high frequencies and low at small frequencies. Moreover the shape of the Gabor Wavelet can be adapted to the shape of the real reflection signals. This provides maximum time and frequency resolution after the uncertainty principle at each frequency. τ is the translation parameter.

IMPLEMENTATION

In order to achieve a sufficient accuracy for estimation of geometry, velocity of sound, impedance and attenuation from ToF and amplitude measurements the signal chain of a common SAM was branched off, the transmitting pulse was shortened and the received RF signals were amplified, digitized, stored and processed in a bypass unit (see fig. 1). The bypass unit has an overall bandwidth of 1GHz and consists of an 80 dB amplifier, a HP54720 digitizing scope (4 GSamples single shot, 8 bit) and a Macintosh Quadra 700 for signal processing. To minimize cable reflections and to obtain an optimum SNR the transmitter/receiver-switch and a preamplifier are mounted directly at the transducer. To further suppress cable echos a time signal with no target in front of the lense is digitized and substracted from all echos. Inherent lens reflections can be erased using this method, too.

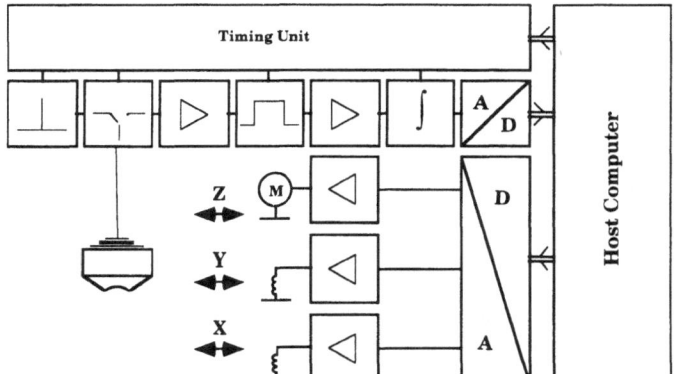

Fig. 1. Experimental setup for elastic microquantification conisting of a conventional SAM and an RF bypass acquisition and processing unit.

RESULTS

In order to calibrate the system at first the defocus function $g(z)$ of the time resolved acoustic microscope has to be obtained. Fig. 2 shows the z-scan of a glass slide in a defocus range of $z = [-50,200]$ µm (x-axis). The envelope of the time signals determined by Wavelet transform at the center frequency of the lens is shown in y-direction in the left image. Because the total ToF is decreasing with increasing defocus the reflections occur earlier with increasing defocus z. The total time in the y-direction is 1 µsec. The amplitudes are coded in grayscale. The maximum of each time signal is then mapped in dependence of the defocus z in the graph on the right side. This curve was smoothed by a moving average filter of rank five. The maximum was then scaled to unity leading to the defocus function $g(z)$.

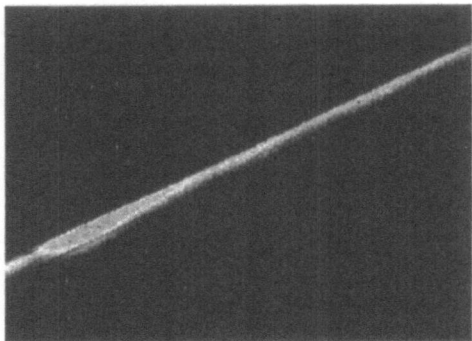 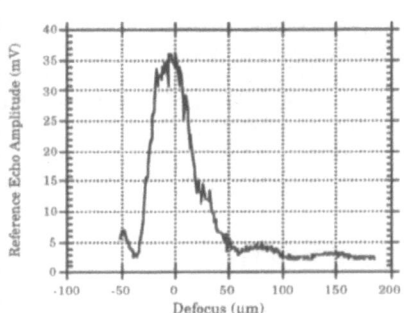

Fig. 2. left side: defocus scan taken at 400 MHz on a glass slide in order to reveal the reference defocus function, right side: maximum amplitude of the same scan

A simple experiment is shown in fig. 3 to demonstrate the measuring principle. In the left image the defocus scan of a 50 µm thick polypropylene layer is shown in the same defocus range of $z = [-50,200]$ µm. The upper line in the image is the top reflection of the layer. The lower line represents the reflection from the surface between the layer and the glass substrate. Because of the confocal arrangement the amplitude reaches it´s maximum if the focal point of the transducer is localised on the surface. However the need of simultaneous recognition of several surfaces requires defocussing and therefore loss of amplitude. Thus for quantification of desired parameters an optimum adjustment for all reflections has to be found. The graph on the right side of fig. 3 shows the maxima of both reflections in dependence of the defocus.

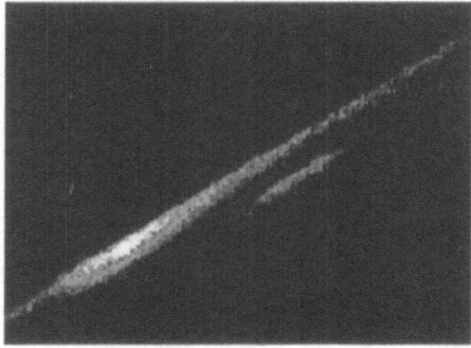 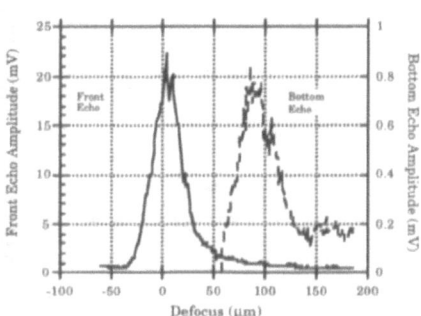

Fig. 3. left side: defocus scan of a 50 µm polypropylene layer on glass showing top and bottom reflections, right side: maximum amplitudes of the same scan (different amplitude scales of top and bottom echos)

As seen the amplitude of the bottom reflection is more than 25 times smaller than the amplitude of the top reflection and the optimum defocus for recognition of both surfaces is reached if the lens is focused on the bottom of the layer. But this means that the SNR is strongly decreased in the top echo. Using the Gabor WT both echos have been detected.

In order to measure the elastic properties of cells 3T3 fibroblasts were cultured on a glass slide. They adhere on glass very well. This is improtant because the scanning motion otherwise may change the position of the cell. During the measurement the samples were heated to 37°C and as the coupling fluid DMEM (Dulbecco´s modification of minimal essential medium) + 10% FCS (fetal calf serum) has been used. The upper left image in fig. 4 shows the high frequent B-scan-image of a single cell. The grayscaled unrectified A-scans are mapped therefore in x-direction in dependence of the transducer position with respect to the cell (y-direction). The total distance is 50 μm. The total ToF is 10 nsec. The grayscale range is [0,255] and 128 is set to zero. Every reflection is a burst of two cycles. The reflection from the glass slide is seen over the whole distance while the top reflection of the cell occurs only in the center of the image.

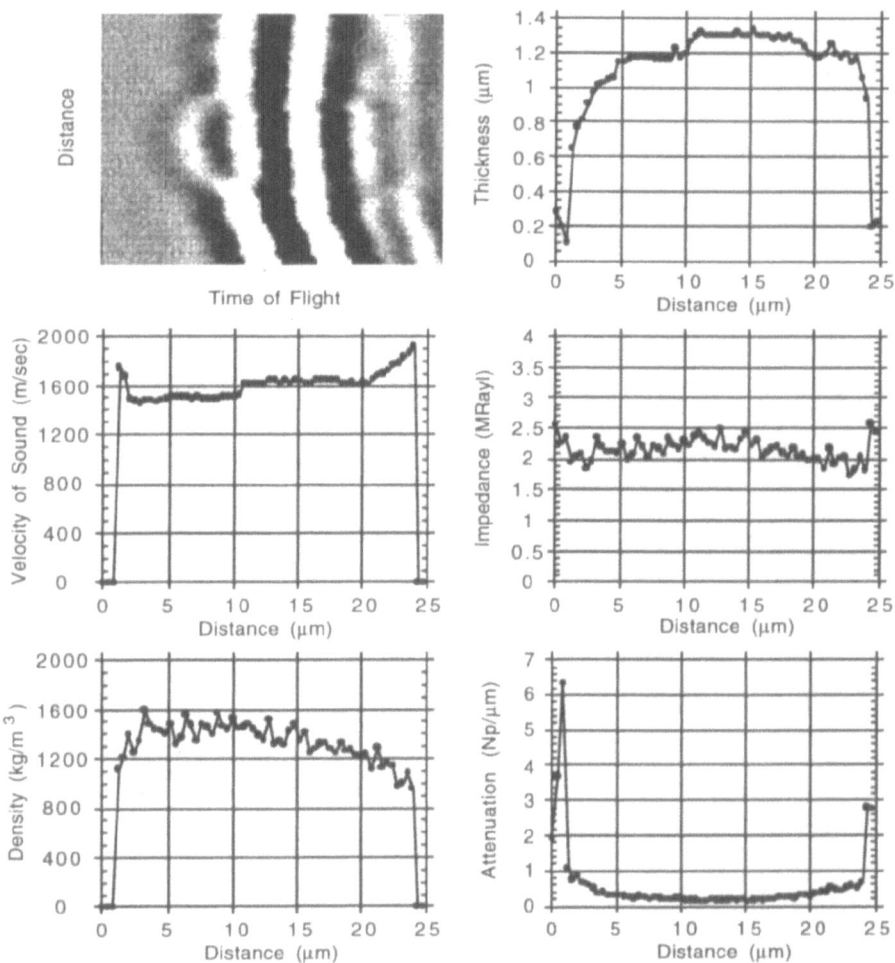

Fig. 4. B-scan image of a single cultured cell on a glass slide taken at 550 MHz (upper left side) and corresponding geometric and elastic properties calculated from ToF and reflection amplitudes

A region of interest (ROI) was then set to cut out the top and bottom reflections of the cell. A second ROI was set near the cell on the glass substrate to calculate a reference ToF and amplitude. The ToF and reflection amplitudes have been determined by the use of the Gabot WT. The geometric and elastic properties of the cell shown in the graphs in fig. 4 have been calculated after Briggs et al. 1993. The major advantage of using Wavelets besides the time saving was an increase in time resolution and therefore in thickness resolution.

CONCLUSIONS

The total error in the results of the cell experiments is unknown. From macroscopic behaviour of tissue the calculated sound velocity and impedance values are in the expected range. However, the very high values in the attenuation curve at the borders of the cell shows that below 1 μm layer thickness the amplitude error increases rapidly.

The effectiveness of the described quantification method has been evaluated on several objects like PZT composites and thick films at known mechanical properties and in imaging caries lesions in teeth. Using Gabor Wavelets has always shown sufficient confidence to long term averaging.

AKNOWLEDGEMENT

We would like to thank G. A. D. Briggs from Oxford for the possibility to use his time resolved acoustic microscope for doing the cell experiments.

REFERENCES

G. A. D. Briggs, J. Wang & R. Gundle, "Quantitative acoustic microscopy of individual human cells", J. Microsc., Vol. **172**, pp. 3-12 (1993).

K. C. Chui, "An Introductuion to Wavelets", Academic Press (1992).

O. Rioul & M. Vetterli, "Wavelets and signal processing", IEEE SP Magazine, Vol. **8** No. 4, pp. 14-38 (1991).

J. C. Schlegel & R. M. Schmitt, " Quantitative Time Resolved Acoustic Microscopy Using Wavelet Analysis", Proc. IEEE Ultras. Symposium, pp. 1393-1396 (1994).

SCANNING MICRODEFORMATION MICROSCOPY USING AN ELECTROMECHANICAL OSCILLATOR

Pascal Vairac and Bernard Cretin

Laboratoire de Physique et Metrologie des Oscillateurs
associé à l'Université de Franche-Comté-Besançon
32 Av. de l'Observatoire - 25044 Besançon Cedex - France

INTRODUCTION

Near-field techniques have for some years an important influence on development of microscopy. Their use, reported by many authors[1-9], enables super-resolution. In near-field microscopy, a small tip often generates or detects an acoustic field at the sample surface.

We have developed the Scanning Microdeformation Microscope[10] which uses a vibrating tip contacting the surface of the tested material. This near-field technique allows subsurface imaging and high resolution at low frequencies (typically about 1 micron at a few 10 kHz). In the first setup[10] the microscope was operated in transmission mode: dynamic microdeformation was generated by the vibrating tip and the corresponding mechanical strain was detected with a PZT transducer fixed on the opposite face of the sample.

The microscope has been improved and now operates in reflection mode: an electromechanical oscillator is used to detect variations in mechanical coupling related to elastic inhomogeneities of the sample. A small tip, a cantilever, a PZT transducer and an electronic amplifier make up this oscillator.

The paper reports the principle of the new setup and a first phenomenological approach is presented. Sensitivity of oscillation frequency to elastic coupling is discussed and a few images are showed. They demonstrate subsurface imaging in reflection mode and the high resolution related to near-field acoustics, already shown in transmission mode.

PHENOMENOLOGICAL APPROACH

Scanning Microdeformation Microscopy requires a contact pressure between tip and sample in order to obtain a high acoustic coupling. We have modelled the tip-sample interaction by a stiffness K_i and tip mass m, and the cantilever by two springs of stiffness K and mass M. Figure 1.a shows a schematic diagram of the model. This rudimentary model

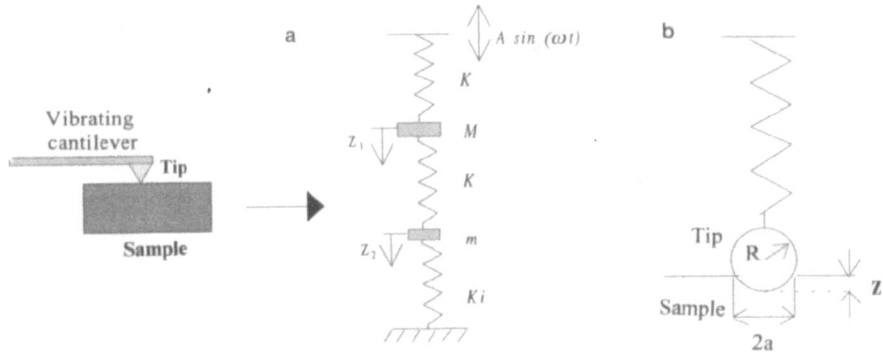

Figure 1. a) Schematic of the phenomenological model used for the theoretical approach. **b)** representation of the interaction corresponding to the spring of stiffness K_i.

represents the simplest method to account for different finite resonant frequencies in the limits $K_i=0$ and $K_i=\infty$.

When there is no interaction between tip and sample, the resonance angular frequency of the cantilever can be written $\omega_0=K/M$. In contact mode, the contact pressure level between tip and sample is modelled by $K_i=\partial F/\partial Z$ where the quantity $\partial F/\partial Z$ is the force gradient acting on the cantilever due to the interaction with the sample. A change in $\partial F/\partial Z$ induces a shift in the resonance angular frequency. To caculate the force F, we used a model based on Hertzian contact[11,12]. This theory gives the radius a of contact of a sphere and a plane, assuming the tip to be a sphere of radius R

$$a^3 = \frac{3}{4}\Pi\,(k_1+k_2)RF(Z) \tag{1}$$

where $F(z)$ is the bearing force, $k_i = [1-v_i]/\pi E_i$, v_i is the Poisson's ratio and E_i the Young's modulus (i=1 for sample and i=2 for tip). As shown in Fig. 1b, if a sphere with radius R deforms a plane, a small indentation is induced within a depth $Z<<R$ together with a radial indentation $a=\sqrt{RZ}$. Inserting this value in (1) yields to:

$$F(z)=\frac{4\sqrt{R}}{3\Pi(k_1+k_2)}(\sqrt{Z})^3 \tag{2}$$

and the stiffness interaction K_i is given by:

$$K_i=\frac{2\sqrt{R}}{\Pi(k_1+k_2)}\sqrt{Z} \tag{3}$$

The equations of motion of the oscillating system are:

$$M\frac{d^2 z_1(t)}{dt^2}+K(z_1-z_2)+K(z_1-A\sin\omega t)=0 \tag{4}$$

$$m\frac{d^2 z_2(t)}{dt^2}+K(z_2-z_1)+k_i z_2=0 \tag{5}$$

where displacements z_1 and z_2 are defined in Fig. 1a. Resolution of these equations gives a resonance frequency which depends on K_i. According to quantitative estimates, the results are in agreement with the general trend of the observations (see Experimental Results section). An example of a simulation result is shown in Fig. 2. Realistic values of the different parameters have been chosen. This simple model is a first approach to understand the system. Refinement of the model to produce the same frequency excursion as in experiment would be possible by using different stiffness for the cantilever springs. In a future paper a complete model will be presented including the effect of tip-sample interaction just before contact. The presented results describe small changes in the force gradient F due to changes in the elastic properties of the sample.

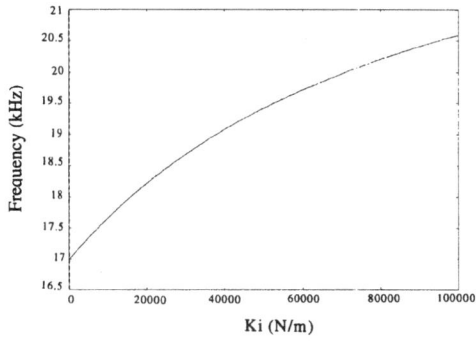

Figure 2. Simulation of the detected frequency versus the interaction stiffness K_i.

MICROSCOPE PRINCIPLE

The figure 3 shows the principle of the new set-up operating in reflection mode. A diamond or sapphire tip is bonded to a cantilever, in turn linked to a piezoelectric ceramic. This electromechanical resonator (tip, cantilever and piezoelectric ceramic) with the amplifier constitute an electromechanical oscillator of which the oscillation frequency depends on the mechanical coupling between sample and tip. Due to the presence of defects in the microdeformation volume, the change in the tip-sample coupling induces a variation of the oscillation frequency. Piezoelectric translation units allow sample scanning and force adjustment. The oscillation frequency is measured with a programmable counter.

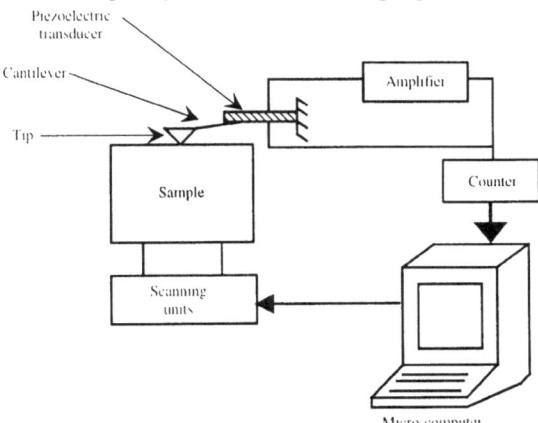

Figure 3. Principle of the Scanning Microdeformation Microscope(SMM) in reflection mode.

EXPERIMENTAL RESULTS

The electromechanical resonator used is made of a cantilever, a sapphire tip and a piezoelectric ceramic. Figure 4 shows the electromechanical transfer function of the resonator with the fundamental vibrational mode measured at 17,75 kHz.

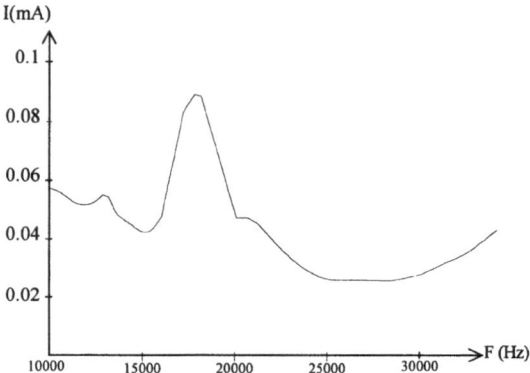

Figure 4. Transfer function of the electromechanical resonator versus frequency (without contact between tip and sample).

The low value of the Q-factor (about 8) allows high mechanical coupling without cancelling oscillation.

Figure 5 describes the effect of the tip contact force on the oscillation frequency. The distance between the tip and the sample is represented on the horizontal axis and the contact position by the crossing point of the horizontal axis and the curve.

The detected frequency increases after the contact with a quasi-linear evolution until plastic deformation of the sample.

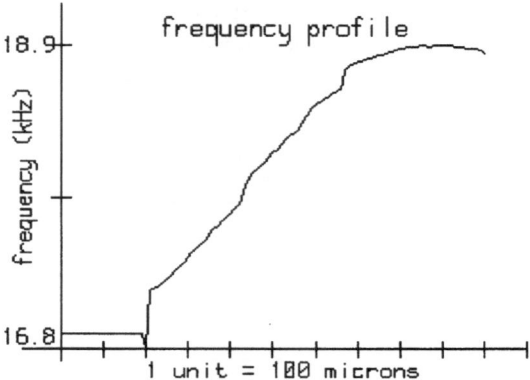

Figure 5. Detected frequency versus distance of probe head for a silicon sample. (cantilever beam stiffness~100N/m, tip radius : 15μm).

IMAGING OF SAMPLES

One of the samples examined is a silicon wafer (360 μm thickness, crystalline orientation [100]) with parallel grooves chemically etched on one face, the opposite face remaining polished. Figure 6a shows the geometry of the sample. On figure 6b subsurface grooves appear as parallel black stripes, after scanning the sample with a tip of 40 μm radius.

The grey scale gives the frequency and the white bar the spacial scale.

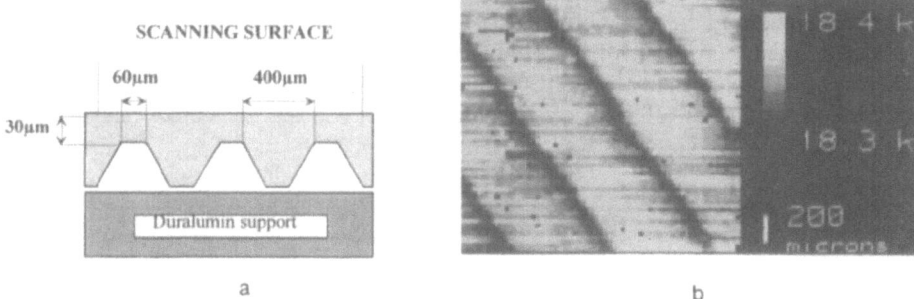

Figure 6. a) Geometry of the etched silicon sample. **b)** Microdeformation image of the subsurface pattern in reflection mode.

The second sample is made of duralumin (AU4G) and a 50 μm diameter tungsten wire was inserted in a diffusion bond. The sample was cut at an angle and polished progressively so that the tungsten wire just appeared on the sample side. The spacing between the tungsten wire and the surface is estimated to be in the range 25-35 μm.

Figure 7. Image of the tungsten wire inserted in a diffusion bond. Observation in reflection mode.

After scanning the sample, figure 7 reveals the presence of the subsurface simulated defect in the sample.

These results in reflection mode allow us to present the Scanning Microdeformation Microscope as a simple method for acoustic subsurface imaging. The system lateral resolution is essentially related to the tip diameter as in the other near-field microscopes.

CONCLUSION

Scanning Microdeformation Microscopy, in the new operating reflection mode enables subsurface imaging at low frequency. Due to the simplicity of operation and less constraint on sample geometry, the reflection mode seems to be more attractive and interesting.

The observation depth, according with the theory of Hertzian contact, seems limited to a few tip diameters in the samples tested (~100 µm). In subsurface defects imaging The contrast depends on the acoustic coupling between the tip and the sample. Coupling is indeed directly related to the contact area; coupling variations are due to the presence of subsurface features and variations in the local elastic constants of the tested material.

Such a microscope which enables high resolution at very low acoustic frequency could become a complementary instrument to other classical acoustic microscopes. Grain characterization, evaluation of local elastic properties and subsurface defect imaging are potential applications of this system.

REFERENCES

1.   W. Durr, D.A. Sinclair, E. A. Ash, "A high resolution acoustic probe", *Electron. Lett.*, vol. 21, p. 805. (1980).
2.   J.K. Zienuk, A. Latuszek, "Ultrasonic pin scanning microscope. A new approach to ultrasonic microscopy", in Proceedings IEEE Ultrasonics Symposium, 17-19 November 1986, Williamsburg,
pp.   1037-1039 ed. B.R. Mc Avoy (Pittsburgh).
3.   B.T. Khuri-Yakub, S. Akamine, B. Hadimioglu, H. Yamada and C.F. Quate, "Near field acoustic microscopy", SPIE, 1556, *Scanning Microscopy Instrumentation*, 30 (1991).
4.   K. Takata, T. Hasegawa, S. Hosaka, S. Hosaka, T. Komoda, "Tunneling acoustic microscope", *Appl. Phys. Lett.*, vol. 55, p. 1718 (1989).
5.   P. Güthner, U. Ch. Fisher, K. Dransfeld, "scanning nearfield acoustic microscopy", *Appl. Phys. B*,
vol.   48, p. 89 (1989).
6.   A. Kulik, J. Attal, G. Gremaud, "Nearfield scanning microscopy", Acoustical Imaging, Nanjing, China, Plenum Press (1992).
7.   B. T. Khuri-Yakub, C. Cindis, C.H. Chou, P.A. Reinholdtsen, "Nearfield scanning acoustic microscope", 1989 Ultrasonics Symposium Proc., IEEE cat. N°89CH2791-2, p.805.
8.   W. Rohvbeck, E. Chilla, "Detection of surface Acoustic waves by Scanning Force Microscopy",
Phys. *Stat. Sol. (A)* 131, 69 (1992).
9.   U. Rabe, W. Arnold, "Acoustic microscopy by atomic force microscopy", *Appl. Phys. Lett.*, vol. 64,
p.   1493 (1994).
10.   B. Cretin, F. Sthal, "Scanning Microdeformation Microscopy", *Appl. Phys. Lett.*, vol. 62, p. 829 (1993).
11.   K. Yamanaka, H. Ogiso, "Ultrasonic force microscopy for nanometer resolution subsurface imaging", *Appl. Phys. Lett.*, vol. 64, p. 178 (1994).
12.   A. Briggs, "Advances in acoustic microscopy", plenum press 1994.
13.   K. L. Johnson, "Contact mechanics", Cambridge university press 1987.

ANGULAR SPECTRUM APPROACH FOR IMAGING OF SPHERICAL PARTICLES IN REFLECTION AND TRANSMISSION SAM

Wieland Weise, Pavel Zinin, and Siegfried Boseck

Physics Department, University of Bremen
28334 Bremen, Germany

ABSTRACT

The output signal for imaging a spherical particle with reflection and transmission scanning acoustic microscope is derived nonparaxially with the angular spectrum approach, which is based on the reciprocity principle. 2-dimensional computed and experimental results for continuously scanning vertically and laterally are presented.

INTRODUCTION

The analytical solution of the problem of imaging spherical particles in a spherical transducer reflection SAM based on the Kirchhoff integral was described in (Lobkis, 1990). Here we present the derivation of the solution in terms of the spectrum approach for the reflection SAM and also for the transmission SAM. The advantage of the spectrum approach is that for its application it is sufficient to know the structure of the velocity potential field in a single arbitrary plane. It is not neccessary to consider the field propagation above this plane, for example inside of the microscope. Moreover, the spectrum approach offers an easy possibility to take into consideration additional obstacles like interfaces, which might be included step by step, when neglecting multiple reflection between them.

THEORY

The microscope images a sphere of radius a as depicted in fig. (1). In the case of a reflection microscope emitter and detector are identical, for the transmission microscope the detector is confocally facing the emitter. The centre of sphere is placed at distance Y of the acoustical axis and at distance Z from the focal plane.

The change in output voltage V of the reflection microscope due to the presence of a scatterer (Atalar, 1980) and the output voltage of the detector for the transmission microscope (Atalar, 1988) can be calculated with the help of the reciprocity principle.

An expression was obtained which is nonparaxial and describes both problems:

$$V(Y, Z)^{t/r} = B \iint\limits_{-\infty}^{+\infty} U^d(-k_x, -k_y) \, U^{t/r}(k_x, k_y) \sqrt{k^2 - k_x^2 - k_y^2} \, dk_x \, dk_y \qquad (1)$$

with U^d denoting the angular spectrum of the velocity potential field that would be generated if voltage was applied to the detector in absence of the obstacle, $U^{t/r}$ the transmitted or reflected spectrum, due to the scatterer in case of an incident spectrum $U^i(k_x, k_y)$, generated by the emitter. k is the wavenumber, B a proportionality constant. The above spectra are expressed for simplicity in the focal plane in a coordinate system centered at the focal point (see fig. (1)).

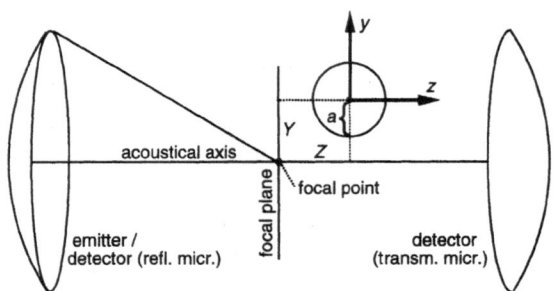

Figure 1. Geometry for imaging the spherical particle

To describe the scattering from the sphere a coordinate system $\vec{r} = (x, y, z)$ with origin at the centre of the particle is used, the z axis is directed away from the emitter. The field that is scattered from the sphere according to a particular incident spectral component is determined by decomposing this incident plane wave into spherical functions. By integrating over the incident spectrum, an expression for the total scattered field Ψ_p^s is obtained, which has been derived in (Gaunaurd and Ueberall, 1979):

$$\Psi_p^s(\vec{r}) = \frac{1}{\pi} \sum_{l=0}^{\infty} \sum_{m=-l}^{l} (i^l) \, A_l \, h_l^{(1)}(kr) \, Y_{l,m}(\vec{e}_r) \iint\limits_{-\infty}^{+\infty} U_p^i(k_x, k_y) \, Y_{l,m}^*(\vec{e}_k) \, dk_x \, dk_y \qquad (2)$$

p is denoting particle coordinates, $Y_{l,m}$ are the spherical functions with $*$ denoting the complex conjugate, $h_l^{(1)}(kr)$ are the spherical Hankel functions. A_l are the scattering coefficients which are independent of m because no transverse waves are present in the liquid (Prosperetti, 1980). $\vec{e}_r$ is the unit vector of the spacial coordinate. $\vec{e}_k$ is denoting the unit vector of the incident field wavevector $\vec{k} = (k_x, k_y, \sqrt{k^2 - k_x^2 - k_y^2})$. The incident spectrum in particle coordinates is:

$$U_p^i(k_x, k_y) = U^i(k_x, k_y) \exp\left(i(k_y Y + k_z Z)\right) \qquad (3)$$

The factor $\exp(i(k_y Y + k_z Z))$ accounts for the offset of the sphere from the focal point. For $Z > 0$ the focal point is situated left of the centre of sphere (see fig. (1)). In Debye approximation the spectra are restricted to real values of the wavevector: $k_x^2 + k_y^2 \leq k^2$.

The Cartesian coordinates in integral (2) are substituted by cylindrical coordinates $k_\rho = \sqrt{k_x^2 + k_y^2}$ and azimuthal angle φ_k. Eq.(2) then becomes in Debye approximation:

$$
\Psi_p^s(\vec{r}) = \sum_{l=0}^{\infty} \sum_{m=-l}^{l} A_l \frac{(i^l)}{\pi} h_l^{(1)}(kr) Y_{l,m}(\vec{e}_r) \cdot
$$

$$
\int_0^k \int_0^{2\pi} U^i(k_\rho, \varphi_k) \exp\left(i(k_\rho \sin \varphi_k Y + k_z Z)\right) Y_{l,m}^*(\vec{e}_k) k_\rho \, d\varphi_k \, dk_\rho \tag{4}
$$

Next eq. (4) is simplified. At first the spherical functions under the integral in (4) are decomposed:

$$
Y_{l,m}^*(\vec{e}_k) = (-1)^m \left[\frac{(2l+1)(l-m)!}{4\pi(l+m)!} \right]^{1/2} P_l^m\left(\frac{k_z}{k}\right) \exp(-im\varphi_k) \tag{5}
$$

P_l^m are the associated Legendre polynomials. By assuming from now on that U_i is independent of φ_k, the integral over φ_k in (4) can be solved with help of the defining integral for the cylindrical Bessel functions:

$$
J_m(k_\rho Y) = \frac{1}{2\pi} \int_0^{2\pi} \exp\left(\pm i(k_\rho \sin \varphi_k Y - m\varphi_k)\right) d\varphi_k \tag{6}
$$

where the positive sign must be used here. Furthermore an abbreviation is introduced:

$$
I_{l,m}(Y, Z) = 2\pi (-i)^m \left[\frac{(2l+1)(l-m)!}{4\pi(l+m)!} \right]^{1/2} \cdot
$$

$$
\int_0^k U^i(k_\rho) \exp\left(i\sqrt{k^2 - k_\rho^2}\, Z\right) P_l^m\left(\frac{\sqrt{k^2 - k_\rho^2}}{k}\right) J_m(k_\rho Y) k_\rho \, dk_\rho \tag{7}
$$

With eqs. (5) to (7) eq. (4) then finally leads to:

$$
\Psi_p^s(\vec{r}) = \sum_{l=0}^{\infty} \sum_{m=-l}^{l} A_l \frac{(-i)^m (i^l)}{\pi} h_l^{(1)}(kr) Y_{l,m}(\vec{e}_r) I_{l,m}(Y, Z) \tag{8}
$$

Now the angular spectrum of the scattered wave $\Psi_p^s(\vec{r})$ is to be calculated. Therefore an expression for the decomposition of a spherical wave in terms of plane waves given in (Devaney, and Wolf, 1974) is used:

$$
h_l^{(1)}(kr) Y_{l,m}(\vec{e}_r) = \frac{(-i)^l}{2\pi} \int_C \int_0^{2\pi} \exp\left(ikr\, \vec{e}_k \vec{e}_r\right) Y_{l,m}(\vec{e}_k) \sin \vartheta_k \, d\varphi_k \, d\vartheta_k \tag{9}
$$

The path of integration C of the polar angle ϑ_k is from 0 to $\pi/2$ and then to $\pi/2 - i\infty$ for $\vec{e}_k$ pointing into the positive z hemissphere, and otherwise from $\pi/2 + i\infty$ to $\pi/2$ and then to π. Introducing (9) into (8) and substituting the spherical integration variables by Cartesian via $\sin \vartheta_k \, d\varphi_k d\vartheta_k = dk_x dk_y/(k\sqrt{k^2 - k_x^2 - k_y^2})$ yields:

$$
\Psi_p^s(\vec{r}) = \sum_{l=0}^{\infty} \sum_{m=-l}^{l} \iint_{-\infty}^{\infty} A_l \frac{(-i)^m}{2\pi^2} \exp\left(i(k_x x + k_y y + k_z^{f/b} z)\right) \frac{Y_{l,m}(\vec{e}_k^{f/b}) I_{l,m}(Y, Z)}{k\sqrt{k^2 - k_x^2 - k_y^2}} dk_x dk_y \tag{10}
$$

If the foreward scattered part of the wave is considered (transmission microscope), $\vec{k}^f$ with z-component $k_z^f = +\sqrt{k^2 - k_x^2 - k_y^2}$ has to be choosen , and if the backward scattered part is considered $\vec{k}^b$, with $k_z^b = -\sqrt{k^2 - k_x^2 - k_y^2}$.

The integral in eq. (10) has the structure of a Fourier back transform. Hence by changing the order of integration and summation, and putting z to zero, the angular spectrum $U_p^{f/b}$ in the $z = 0$ plane of the forward-/ back-scattered wave is identified as:

$$U_p^{f/b}(k_x, k_y) = \sum_{l=0}^{\infty} \sum_{m=-l}^{l} 2(-i)^m A_l \frac{Y_{l,m}(\vec{e}_k^{f/b})}{k\sqrt{k^2 - k_x^2 - k_y^2}} I_{l,m}(Y, Z) \tag{11}$$

Comparing eq. (8) and eq. (11) we see that the angular dependence of the spectrum is similar to the far field angular dependence in the spatial domain.

The spectrum is shifted back from the particle coordinate system to the system centered in the focal point by including the factor $\exp\left(-i(k_y Y + k_z^{f/b} Z)\right)$. The reflected spectrum is given by $U^r(k_x, k_y) = U_p^b(k_x, k_y)\exp\left(-i(k_y Y + k_z^b Z)\right)$ whereas the transmitted spectrum $U^t(k_x, k_y) = U^i(k_x, k_y) + U_p^f(k_x, k_y)\exp\left(-i(k_y Y + k_z^f Z)\right)$ is the sum of the incident spectrum and the scattered spectrum.

From now on the calculation is to be restricted to the case of the reflection microscope. Because emitter and detector are equal, $U^d(k_x, k_y) = U^i(k_x, k_y)$ in eq. (1). Introducing eq. (11) and substituting with cylindrical integration variables, eq. (1) gives:

$$V^r(Y, Z) = \sum_{l=0}^{\infty} \sum_{m=-l}^{l} \frac{B\, 2(-i)^m}{k} A_l\, I_{l,m}(Y, Z) \cdot \tag{12}$$

$$\int_0^{\infty} \int_0^{2\pi} U^i(k_\rho) \exp\left(i(-k_\rho \sin\varphi_k Y + \sqrt{k^2 - k_\rho^2}\, Z)\right) Y_{l,m}(\vec{e}_k^b)\, k_\rho dk_\rho d\varphi_k$$

Eq. (12) has a similar structure as eq. (4). Introducing the complex conjugate of eq. (5) for $Y_{l,m}$, and again solving the integral over φ_k with help of the definition of the cylindrical Bessel functions eq. (6) (now using the negative sign in the exponent), eq. (12) becomes in Debye approximation:

$$V^r(Y, Z) = \sum_{l=0}^{\infty} \sum_{m=-l}^{l} \frac{B\, 4\pi\,(i)^m}{k} \left[\frac{(2l+1)(l-m)!}{4\pi(l+m)!}\right]^{1/2} A_l\, I_{l,m}(Y, Z) \cdot \tag{13}$$

$$\int_0^k U^i(k_\rho) \exp\left(i\sqrt{k^2 - k_\rho^2}\, Z\right) P_l^m\left(\frac{-\sqrt{k^2 - k_\rho^2}}{k}\right) J_m(k_\rho Y)\, k_\rho dk_\rho$$

Making use of the identity: $P_l^m(-\sqrt{k^2 - k_\rho}/k) = (-1)^{l+m} P_l^m(+\sqrt{k^2 - k_\rho}/k)$ we can again identify $I_{l,m}(Y, Z)$. Eq. (12) finally gives:

$$V^r(Y, Z) = \sum_{l=0}^{\infty} \sum_{m=-l}^{l} \frac{B\, 2(-1)^l}{k} A_l\, I_{l,m}^2(Y, Z) \tag{14}$$

In the case of the transmission microscope we assume that U^d is also independent of φ_k. Then eqs. (11) and (1) yield in Debye approximation instead of eq. (13):

$$V^t(Y, Z) = B \int_0^k \left\{ U^d(k_\rho)U^i(k_\rho)\sqrt{k^2 - k_\rho^2} + \sum_{l=0}^{\infty} \sum_{m=-l}^{l} \frac{2(i)^m}{k} \left[\frac{(2l+1)(l-m)!}{4\pi(l+m)!}\right]^{1/2} \cdot \tag{15}\right.$$

$$\left. A_l\, I_{l,m}(Y, Z) \exp\left(-i\sqrt{k^2 - k_\rho^2}\, Z\right) P_l^m\left(\frac{\sqrt{k^2 - k_\rho^2}}{k}\right) J_m(k_\rho Y) \right\} k_\rho dk_\rho$$

Assuming that the spectra of emitter and detector are equal and real (aberration free), we can identify in eq. (15) $I_{l,m}^*$ and finally obtain:

$$V^t(Y, Z) = B \left[\int_0^k \left(U^i(k_\rho) \right)^2 \sqrt{k^2 - k_\rho^2}\, k_\rho\, dk_\rho + \sum_{l=0}^{\infty} \sum_{m=-l}^{l} \frac{2}{k} A_l \, |I_{l,m}(Y, Z)|^2 \right] \qquad (16)$$

RESULTS

In fig. (2) images for a reflection microscope calculated with eq. (14) and for a transmission microscope calculated with eq. (16) are presented. The spectra are corresponding to a spherical transducer: $U_i(k_\rho) = 1/k_z$ up to the aperture at $k_\rho/k = \sin(30°)$, zero outside. The size of the considered steel sphere is $a = 100/k$. The coordinates denote the position of the focal point in units of the radius of the particle. The geometry is as depicted in fig. (1). For the reflection microscope high signal strength occurs when scanning at constant Z through the top or the centre of the sphere. The size of the image at the top is depending on the aperture size and is always smaller then the particle. The central spot is independent of the particle size. For the transmission microscope the image at constant Z is determined by the contour of the particle. The light spot exhibits a sharp edge when scanning through the centre of sphere.

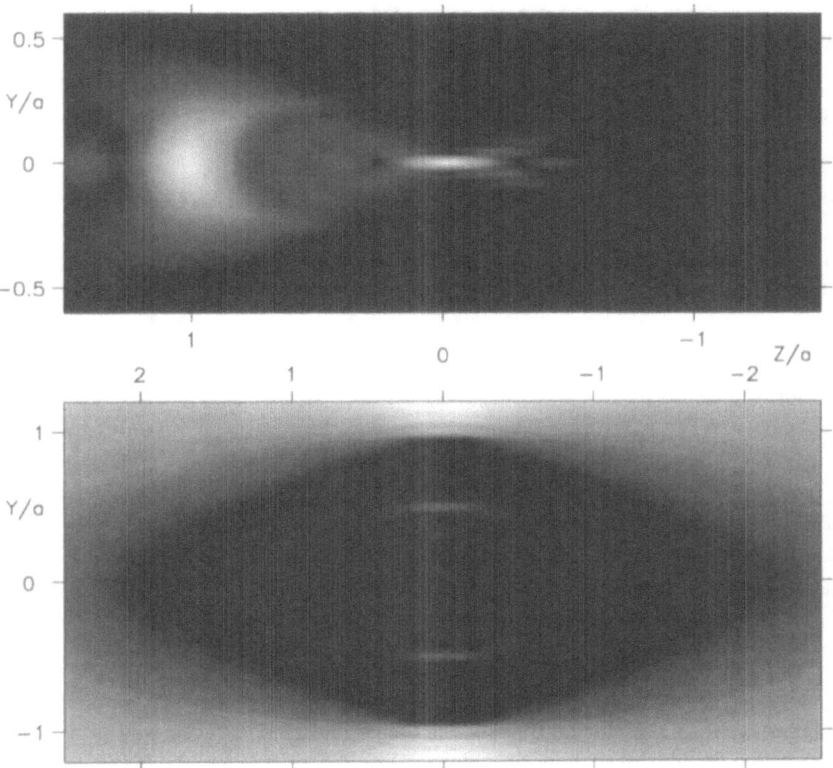

Figure 2. Calculated $y - z$ scan through a steel sphere for reflection (top) and transmission microscope (bottom). For $Z/a > 0$ the focus is on the left side of the particle centre.

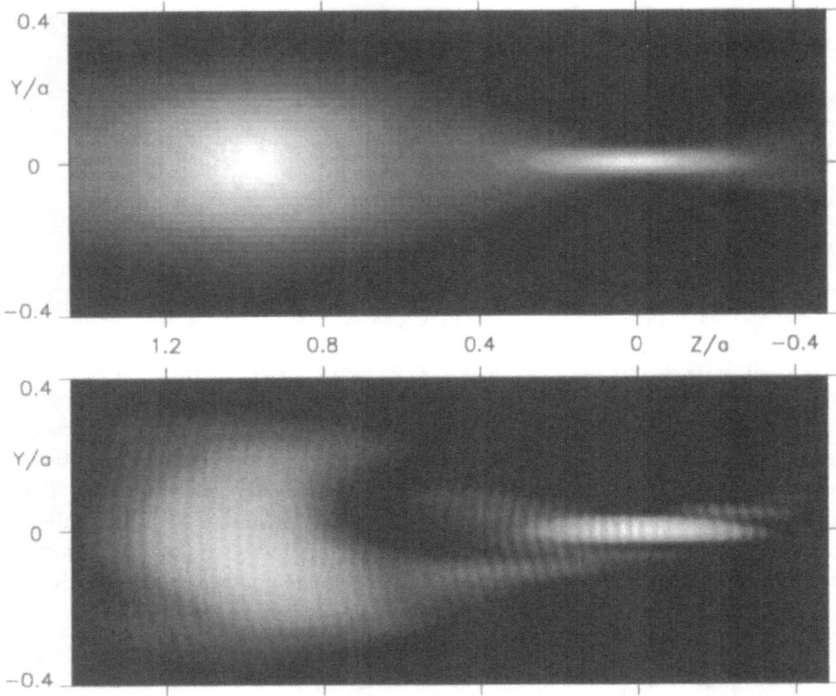

Figure 3. $y-z$ scan through a steel sphere for reflection microscope calculated with continuous spectrum (top) and meassured (bottom).

Fig. (3) shows reflection microscope scans with the same maximum aperture as in fig. (2) through a steel sphere with $a = 100/k$: above, calculated for a continuous spectrum as proposed by (Sasaki et.al., 1992), and below an experimental result. Experimentally asymmetry occurs because the transducer is not coaxial with the lens. Obviously the appearance is intermediate of the two calculated scans.

REFERENCES

A. Atalar: A backscattering formula for acoustic transducers. J. Appl. Phys. **51** (1980) 3093-3098.

A. Atalar: A fast method of calculating diffraction loss between two facing transducers. IEEE. Trans. on UFFC **35** (1988) 612-617.

A.J. Devaney, E. Wolf: Multipole expansions and plane wave representations of the electromagnetic field. J. Math. Phys. **15** (1974) 234-244.

G.C. Gaunaurd, H. Ueberall: Acoustics of finite beams. J. Acoust. Soc. Am. **63** (1979) 5-16.

O.I. Lobkis, P.V. Zinin: Acoustic microscopy of spherical objects. Theoretical approach. Acoust. Lett. **14** (1990) 168-172.

A. Prosperetti: Normal mode analysis for the oscillations of a viscous liquid drop in an immiscible liquid. Journal de Mecanique. **19** (1980) 149-182.

Y. Sasaki, T. Endo, T. Yamagishi, M. Sakai: Thickness measurement of a thin-film layer on an anisotropic substrate by phase-sensitive acoustic microscope. IEEE Trans. on UFFC **39** (1992) 638-642.

TWO KINDS OF CONTRAST
OF SURFACE IMAGING

Pavel Zinin[1], Wieland Weise[2], and Siegfried Boseck[2]

[1]Department of Materials, University of Oxford
Oxford OX1 3PH, UK
[2]Physics Department, University of Bremen
28334 Bremen, Germany

INTRODUCTION

Application of Scanning Acoustic Microscopy for nondestructive evaluation of mechanical properties of objects having non-planar surface reveal two kinds of contrast. Images of ceramic bearing balls (Chow *et al.*, 1991) and the images of steel spheres (Kolosov *et al.*, 1991) exhibit the shape of a light spot which diameter is proportional to the diameter of the spheres. In contrary, acoustic micrographs of steel bearing ball obtained by (Weglein, 1981) and cavities in alumina (Poirier *et al.*, 1984) exhibited a fringe pattern associated with a V(z) curve (Poirier *et al.*, 1984). In this paper we present the theoretical explanation of this phenomena using the approximation of the uneven surface by an inclined plane (Weise *et al.*, 1994). The dependence of the contrast of surface images on the wave length, radius of surface curvature and the aperture angle of the lens is investigated. The conditions for different kinds of contrast are obtained.

THEORY

Fig. 1 depicts the sketch of our model. The focused transducer with a semi aperture angle α irradiates a focussed sound wave and receives waves scattered by the uneven surface. As it was shown elsewhere (Weise *et al.*, 1994) the image of a non-planar surface can be modelled locally in every point by a plane inclined surface. The output signal of the microscope up to a constant within this model has the form:

$$V = 2\pi \int\limits_{0}^{\alpha-\beta} \exp\left(i2kL\cos\beta\cos\vartheta\right)\mathcal{R}\left(\vartheta\right)\sin\vartheta\ d\vartheta \quad +$$

$$\int\limits_{\alpha-\beta}^{\arccos\left(\frac{\cos\alpha}{\cos\beta}\right)} \exp\left(i2kL\cos\beta\cos\vartheta\right)\mathcal{R}\left(\vartheta\right)\sin\vartheta\left[2\pi - 4\arccos\left(\frac{\cos\vartheta\cos\beta-\cos\alpha}{\sin\vartheta\sin\beta}\right)\right]\ d\vartheta \quad (1)$$

Here k is the wave number, $\mathcal{R}(\vartheta)$ is the reflection function of the material, β is the angle of inclination of the plane tangentially to the surface (see fig. 1). Function (1) has a maximum at $\beta = 0$, and it drops to zero at $\beta = \alpha$. For a perfectly reflecting plane, the first integral can be solved analytically and is equal to the sinc-function

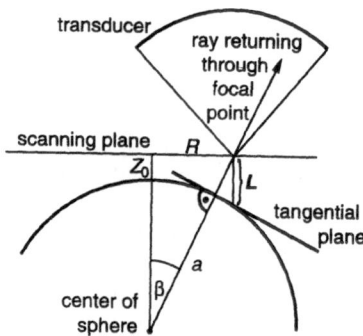

Figure 1. Model of the Reflection Acoustic Microscope scanning a non planar surface.

which is obtained as the $V(z)$ curve with $z = L \cos \beta$ of a circular transducer with semi-aperture angle $\alpha - \beta$. The second integral in eq. (1) is solved numerically. We will use this model to analyze qualitatively the features of image formation of non planar surfaces. The analysis is performed for a spherical surface since it exhibits all possible angles of inclination (Atalar, 1979). For numerical calculation the rigorous solution of imaging of the top surface of a spherical particle will be used (Lobkis and Zinin, 1990).

$$V(R, Z_o) = \sum_{n=0}^{\infty} \sum_{m=0}^{n} (-1)^n (2 - \delta_{0m}) A_n I_{nm}^2(R, Z_o) \tag{2}$$

$$I_{nm}(R, Z_o) = (-i)^m \int_0^{\alpha} \mathcal{P}(\theta) \exp\{ik(Z_o + a) \cos \theta\} J_m(kR \sin \theta) \overline{P_n^m}(\cos \theta) \sin \theta d\theta,$$

where a is the radius of the particle, R is the radial image coordinate and Z_0 is the defocus from the top of the sphere. A_n are the constants derived from the boundary conditions, δ_{0m} is the Kronecker delta symbol, $J_m(x)$ are the cylindrical Bessel functions, $\overline{P_n^m}(\cos \theta) = P_n^m(\cos \theta) N_{nm}$ are the normalized associated Legendre polynomials, $P_n^m(\cos \theta)$ the associated Legendre polynomials, $N_{nm} = \sqrt{(2n + 1)(n - m)!/[2(n + m)!]}$ are the normalizing coefficients. A normalising constant in (2) is omitted.

RESULTS

Two different kinds of contrast can be easily understood if we consider imaging of the surface of a rigid body ($\mathcal{R}(\theta) = 1$) when planarly scanning across the top of the surface. Comparison between the rigorous solution (2), and numerical calculations based on (1), made using paraxial approximation (Zinin et al., 1994) have shown that the first term in (1) can be considered as a good approximation to the rigorous solution. So for $Z_o = 0$ we will approximate the exact solution for a rigid spherical surface by

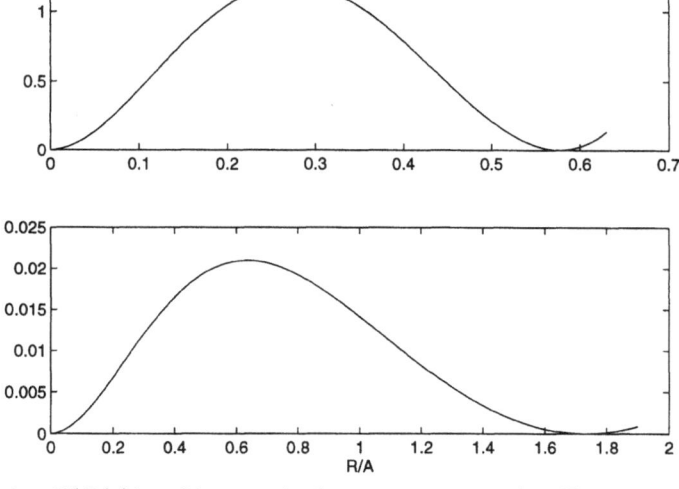

Figure 2. Function $N(R/a)$ in arbitrary units for two aperture angles. Top: $\alpha = 30^\circ$, bottom: $\alpha = 60^\circ$.

the formula:

$$V(Z_o) = \frac{\sin\left\{(ka)\left[\sqrt{1+(R/a)^2}-1\right]\left[1-\cos\left(\alpha-\arctan(R/a)\right)\right]\right\}}{(ka)\left[\sqrt{1+(R/a)^2}-1\right]\left[1-\cos\alpha\right]}. \tag{3}$$

The term $(ka)\left[\sqrt{1+(R/a)^2}-1\right]$ is the distance between the focal point and the sphere's surface. The value $[1-\cos(\alpha-\arctan(R/a)]$ or the second term in (3) is determined by the inclination of the tangential plane $\tan\beta = R/a$. According to (3) the image can be represented as a light spot which diameter is determined by the condition $\sin[kaN(R/a)] = 0$, where we introduce the notation

$$N(R/a) = \left[\sqrt{1+(R/a)^2}-1\right]\left[1-\cos\left(\alpha-\arctan(R/a)\right)\right]. \tag{4}$$

Obviously $\sin[kaN(R/a)] = 0$, when $kaN(R/a) = 0, \pi$. N is dependent only on the aperture angle of the lens (α) and is presented in fig. 2 for different aperture angles. Obviously $N = 0$ at the point $R = ka\tan\alpha$ for arbitrary ka. In this point the inclination of the tangent plane causes the reflected ray which is returning through the focal point, to miss the receiver. So in this case the inclination of the tangent plane determines the dimension of the image. However, function (3) also drops to zero when the product $ka\,N(R/a)$ is equal to π. For a fixed aperture angle α this is possible only when the radius of curvature of the surface is large. As it is obvious from fig. 2, the value of the product of the function N at the maximum (N_{max}) and ka must be greater than π:

$$kaN_{max} > \pi. \tag{5}$$

The maxima of the function N have different values for different aperture angles. The bigger the aperture angle, the bigger is the value of N_{max} and the smaller is the desired

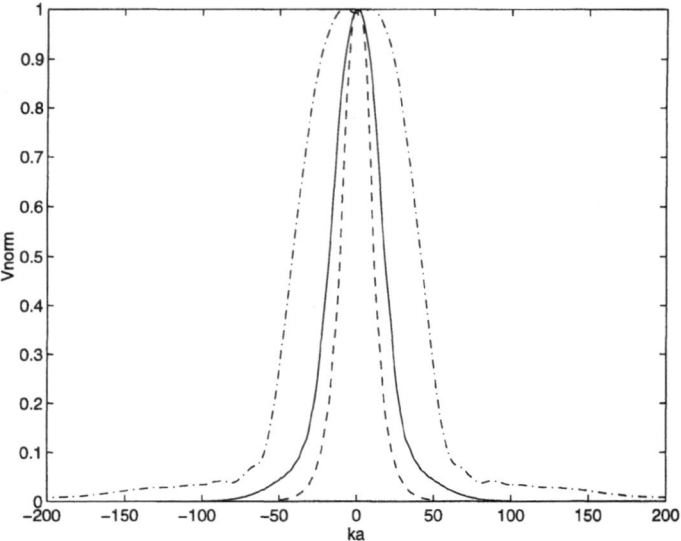

Figure 3. Images calculated for steel spheres with different radius. Solid line: $ka = 50$, dashed line: $ka = 25$, dotdashed line: $ka = 200$. $\alpha = 60°$.

radius of curvature to satisfy condition (5). This criterion is obviously rough, but it shows that the appearance of the top images is strongly dependent on the aperture angle and the radius of curvature of the surface. For half aperture angle equal to 30°, ka must be greater than 2500, for $\alpha = 60°$, ka must be greater than 150, and for $\alpha = 80°$, ka must be only greater than 40. The condition $kaN = \pi$ corresponds to the first minimum of the $V(Z)$ curve of a perfectly reflecting surface, when the central ray and the edge ray reach the lens with opposite phase. In this case the diameter of the central light spot will be determined mostly by the distance between the focal point and the tangent plane and the image will simply reflect the $V(z)$ curve, where z is the defocus of the curved surface.

Images of steel spheres with different radii calculated by eq. (2) are presented in fig. 3. The behavior of the images is in agreement with the above criteria. Up to $ka = 100$ the radius of the central light spot is proportional to $ka \tan \alpha$. For $ka = 200$ this is no more valid.

Since for a small aperture angle the changes in contrast due to condition (5) appear to occur only for very big particles ($ka > 2500$), we can conclude that for the low aperture lens the top images have the shape of a light spot, which radius is proportional to $ka \sin \alpha$ ($\tan \alpha \approx \sin \alpha$). In this case the contrast is determined mainly by the inclination of the surface under the focal point. Those kinds of images were experimentally obtained by (Kolosov et al., 1992) and (Chow et al., 1991).

For large aperture angle and for large radius of the surface curvature the top images have a qualitatively different character than for small aperture angle. In the first case the contrast is determined by the distance between the focal point and the surface of the object, hence by the $V(z)$ curve. Such kind of contrast was experimentally observed by (Weglein, 1984) and (Poirier et al., 1984). For an object with a smooth surface profile this kind of contrast plays a dominant role.

During defocusing (increasing Z_o in (3)) the contrast will decrease. With increasing

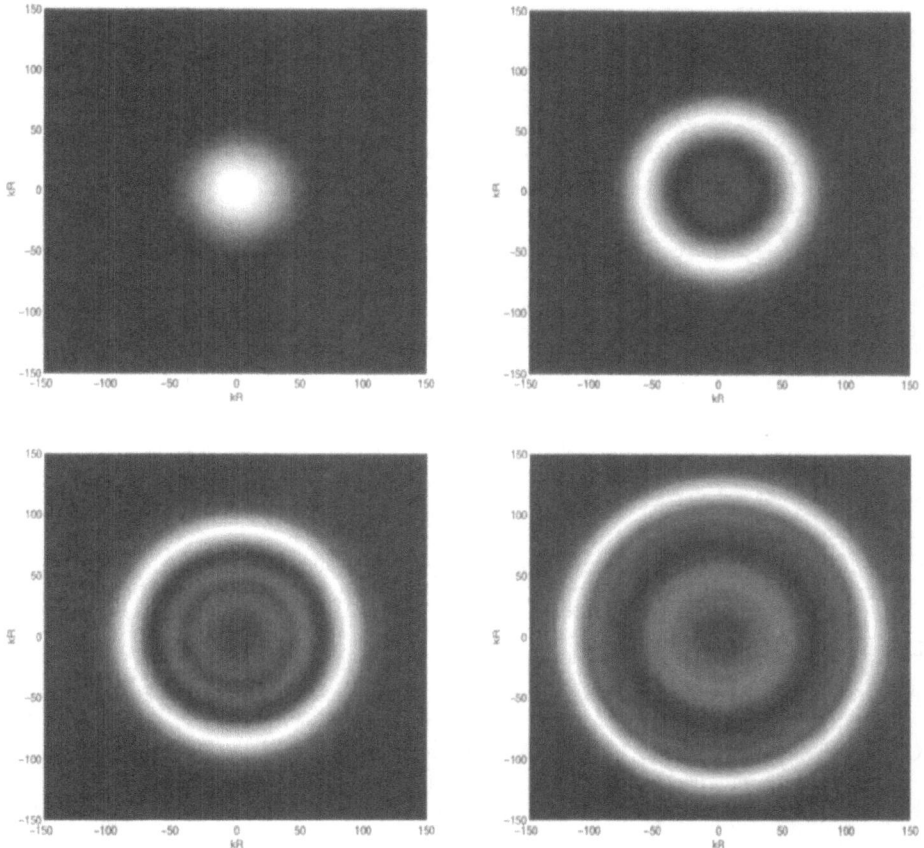

Figure 4: Images of a steel sphere (ka=200) surface calculated for reflection SAM $\alpha = 60°$ at different focal plane positions. Axes scaled in units of the sphere's radius.
Top left: $kZ_o = 0$, Top right: $kZ_o = -10$, Bottom left: $kZ_o = -20$, Bottom right: $kZ_o = -40$.

negative value of Z_o, the subsurface image becomes a ring (see fig. 4). The contrast maximum of the ring occurs at the radius R where the scanning plane crosses the sphere's surface. The radius of this ring in dependence on the defocusing Z_o can be described as:

$$R_a = a\left[1 - \sqrt{1 - (Z_o/a)^2}\right]$$

Inside the ring the images exhibit a fringe pattern (see fig. 5), which is simply the $V(Z)$ curve when the distance Z is changed as $(ka)\left[\sqrt{1 + (R/a)^2} - 1\right]$ during scanning.

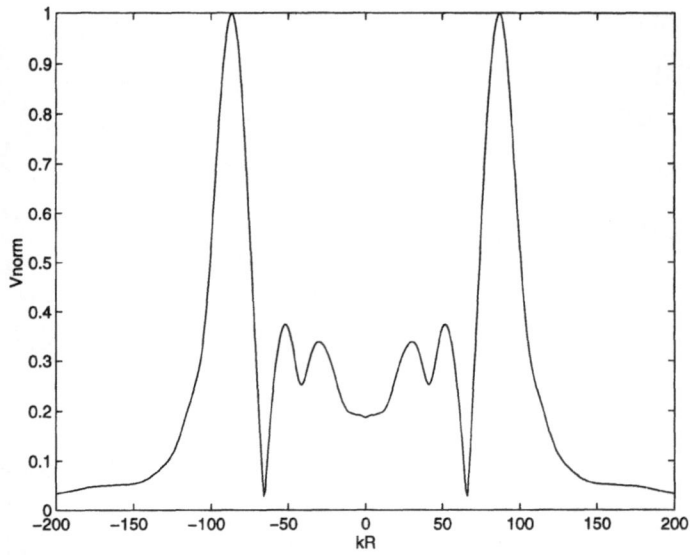

Figure 5. Defocused images of steel sphere. $ka = 200$, $\alpha = 60°$, $kZ_o = -20$.

REFERENCES

Atalar A., 1979, Modulation transfer function for the acoustic microscope, *Electron. Lett.* 15:321.

Chow C.H., Khuri-Yakub B.T., 1991, Acoustic microscopy of ceramic bearing balls, *in:* Acoustical Imaging, Vol.18," K.Y.Wang, ed., Plenum Press, New York.

Kolosov O.V., Lobkis O.I., Maslov K.I. and Zinin P.V., 1992, The effect of the focal plane position on the images of spherical objects in the reflection acoustic microscope, *Acoust. Lett.* 16:84.

Lobkis O.I. and P.V. Zinin, 1990, Acoustic microscopy of spherical objects. Theoretical approach, *Acoust. Lett.* 14:168.

Poirier M., Castonguay M., Neron C. and Cheeke J.D.N. 1984, Nonplanar surface characterization by acoustic microscopy, *J. Appl. Phys.* 55:89.

Weglein, R.D, 1981, Acoustic microscopy of curved surfaces, *Appl. Phys. Lett.* 38:516.

Weise W., Zinin P. and Boseck S., 1994, Modeling of inclined and curved surfaces in the reflection scanning acoustic microscope, *J. Microscopy.* 176:87.

Zinin P., Weise W., Lobkis O., Kolosov O. and Boseck S. 1994, Fourier optics analysis of spherical particles image formation in reflection acoustic microscopy. *Optik.* 98:45.

CRACKS AND FRACTURES IN ROCK: GEOLOGICAL ANALYSIS WITH THE ACOUSTIC MICROSCOPE

Nicholas E. Pingitore Jr.[1,2], Leon DuPlessis[2],
Laura Lopez[1], Kate C. Miller[1], and Lawrence E. Murr[2]

[1]Department of Geological Sciences
[2]Materials Research Institute
The University of Texas at El Paso
El Paso, TX 79968-0555

INTRODUCTION

Because of its ability to image in three dimensions, the acoustic microscope can become an important tool for the analysis of cracks and fractures in rock. Most rock exhibits fracture, indicating material failure at scales from the global fault systems of plate tectonics to minuscule shears visible only with a microscope. Fractures are of great interest to the geologist, whether they are still open or have been healed by precipitates of such natural cements as calcium carbonate or silica. Healed fractures record the effects of one or more generations of stress fields, from which the tectonic history of the rock may be reconstructed. Open fractures are potential reservoirs and conduits for the storage and migration of water and hydrocarbons, and thus have major economic and social significance.

Acoustic microscopy provides the opportunity to rapidly, inexpensively, and non-destructively acquire three-dimensional images of pore spaces and cracks in sedimentary materials. Other recent approaches to this problem include confocal optical microscopy of fluorescently impregnated serial sections (Fredrich et al., 1995), x-ray tomography by CAT scan (Carlson and Denison, 1992), and x-ray micro-tomography with synchrotron radiation (Flannery et al., 1987). Although each of these techniques provides useful and unique information, such factors as convenience, expense, and time make it unlikely that any of the latter will win wide acceptance by such potential geological users as the petroleum industry. In contrast, acoustic microscopy provides subsurface images in times ranging from a few minutes to perhaps half an hour, depending on the resolution desired, at an initial equipment expenditure of $ 75,000 to $ 150,000, with a robust, user-friendly, desk-sized instrument. Drill core or rock slabs can be examined with little or no preparation; a smooth, flat top surface provides the best three-dimensional images.

The Acoustic Microscope

Acoustic microscopy exploits a confocal source/receiver shaped into a hemispherical lens which is coupled to the solid specimen through a fluid, normally water. Pulsed acoustic waves of MHz to GHz frequency are focused on the surface of the solid, and the travel time and strength of the reflected wave is recorded back at the lens. Vertical motion of the lens provides focus; source pulsing creates a time window to receive the reflected signal. Images are created pixel by pixel by rastering the lens over the surface and processing the accumulated data. Acoustical images of surfaces can rival light microscopy in resolution. The gain in resolution with increasing frequency is accompanied, however, by a deterioration in penetration for subsurface imaging. Subsurface imaging can achieved by changes in the focus and in the time gating for the returning signal. An introductory description of acoustic microscopy is found in Briggs and Hoppe (1991) and an advanced treatment is available in Briggs (1992).

Purpose of Study

Acoustic microscopy has become widely accepted in the last decade in the fields of materials science and high-technology manufacturing, especially for non-destructive quality-control testing of microchip packaging. Commercial instruments designed and used chiefly to detect delaminations in layered electronic components are available from several vendors. Because such instruments are still unusual in universities, there have been few studies of geological materials utilizing acoustic microscopy. Published research includes the examination of individual mineral specimens by Bonner (1978) and Morales *et al.* (1991), of rocks by Dansburg and Yuhas (1978) and Rodriguez-Rey (1990), of fossils by Kustov *et al.* (1987) and Scott and Hemsley (1991), and of a variety of geologic materials by Pingitore *et al.* (1995).

The objective of the this work is to explore the possible application of acoustic microscopy to the specific problem of imaging cracks and fractures in three dimensions in rock. As the initial step to this goal, we seek to define the "envelope" of this application by answering questions regarding spatial resolution, depth of penetration, production of artifacts in images, ability to detect healed fractures, and ease of use.

TECHNIQUES

Instrumentation

We conducted our investigation utilizing a Hitachi AT 6000C scanning acoustic tomograph equipped with 10, 25, and 50 MHz acoustic sources. The proprietary Hitachi software was used for image collection and generation, and no additional image processing was done. The Hitachi AT 6000C instrument is well suited for working with bulky rock specimens, such as slabbed oil-well drill core. The relatively large water tray, approximately 25 x 25 x 10 cm, permits simple introduction and orientation of such samples.

The 10-to-50 MHz frequency range which we employed lies just below that of instruments traditionally termed acoustic microscopes (100 MHz-10 GHz). Instruments such as our Hitachi sometimes are referred to as Scanning Acoustic Tomographs. Our lateral resolution of about 10 μm is sufficient to address many geological questions which involve grains or pores of somewhat larger dimension. Signal scattering at grain boundaries is less severe at these frequencies and therefore penetration is enhanced.

Samples

Fractures and cracks in a variety of rock types were examined, including representatives of each of the three rock groups: igneous, sedimentary and metamorphic. Because of the potential for application to the petroleum industry, we focused our attention on the sedimentary rocks, the group which hosts virtually all of the world's hydrocarbons. The varieties studied included sandstones, carbonates, shales, and chert. Both open and naturally cemented fractures were examined.

In this report we present images of a healed fracture in limestone. Limestone designates a rock composed principally of the mineral calcite ($CaCO_3$), the rhombohedral polymorph of calcium carbonate. Many of the calcite grains were below 10 micrometers in maximum dimension, but fossils, up to a few mm in diameter, were interspersed throughout the material. The fractures in this specimen were cemented nearly completely by calcite. Hand specimens of the rock exhibited no porosity.

Sample preparation

The calcite specimen was cut into a rough cube, approximately 1.5 cm on a side. The surface to be examined was ground and polished to minimize scattering of the incoming acoustic wave at the water-rock interface. The orientations in the third dimension of the larger cracks could be approximated by inspection of the four sides of the cube bordering the examination surface. Other selected samples (not shown) were serially cut subsequent to insonification in order to permit direct visual comparison of the crack orientations with those on the acoustic images.

Generation of Images

Subsurface tomographic images can be composed from multiple B-scans, in which the lens scans a line and focuses into the sample at multiple depths at every pixel, or C-scans, where the lens repeats areal scans at several different depths. We present figures generated from multiple C-scans, along with synthetic B-scans derived from them by computer tomography.

Subsurface focus and gating of the return signal influenced image quality. The time to the gate (depth in sample), width of the gate (thickness of sampled region), and gate interval (depths examined) all affect the final image.

RESULTS

Images

Figure 1 shows areal (C-scan) images of a crack in limestone cemented by calcite, taken from several depths with the 25 MHz lens. The reflected signal was gated at 50 ns, and the vertical distance between images is approximately 1 mm.

The intersection of the major crack with the imaging plane is clearly seen to migrate toward the upper right corner in images at successive depths, revealing its true orientation ("strike and dip" to the geologist) in three dimensions. From the surface to the deepest image, the fracture appears close to planar. The intersections are nearly straight lines and are close to parallel. Thus both the orientation and the shape of the crack can be discerned without difficulty.

A second, nearly vertical, cemented crack intersects the main crack at the upper left of the image. At depth there is no reflected signal from this region, yielding a dark area in the images. Increased scattering of the acoustic signal by this zone yields little energy for reflections from deeper levels.

Several features unrelated to the cracks appear as bright spots in the images. The most prominent of these is seen toward the upper right corner of Figure 1. These are fossil fusulinids, as seen in Figure 2, an optical image of a thin (30 μm) section of the sample taken with transmitted plane-polarized light on a petrographic microscope.

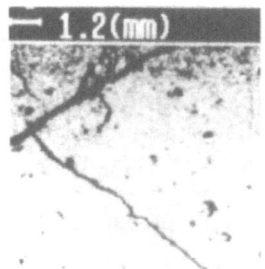

Figure 1. (Right) Successive C-scans (areal scans) of limestone sample. Uppermost image is surface view. Next 3 images are at depths of 1, 2, and 3 mm, respectively. For surface scan, cracks appear dark. In the subsurface scans, brighter regions represent greater amounts of acoustic energy reflected back to the lens, and crack appears bright. Note how the intersection of the image plane with the large crack migrates toward the upper right corner with increasing depth. Trailing, parallel "cracks" are artifacts. A second crack, perpendicular to the major fracture, is visible in the upper left corner.

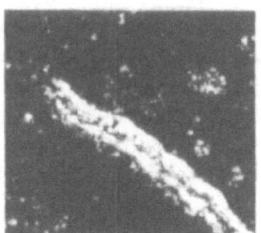

Figure 2. (Below) Optical view of the major crack imaged in Figure 1. Crack runs from upper left to lower right of the field of view. It cuts across the large fossil fusilinid which occupies much of the left side of the image. The fusilinid is approximately 2 mm in longest dimension. The photomicrograph was taken in plane-polarized, transmitted light.

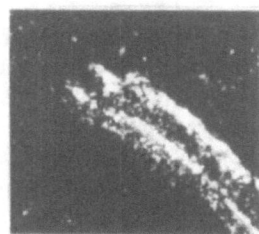

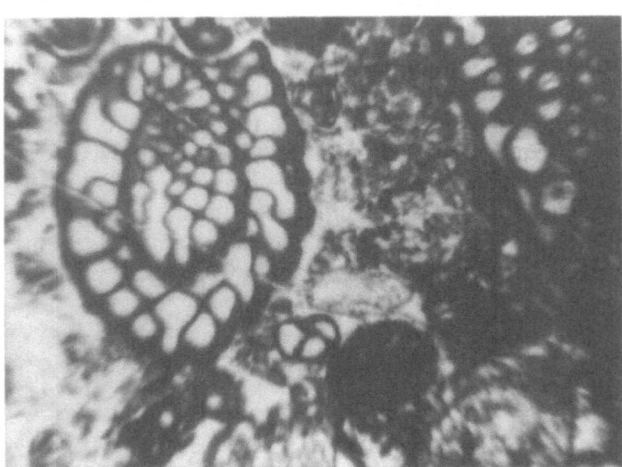

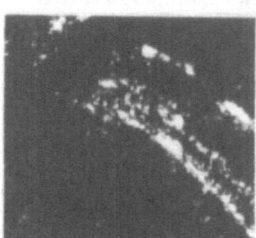

Figure 2 reveals the nature and geometry of the crack imaged acoustically in Figure 1. It is nearly straight in this two-dimensional view. The crack is filled with clear calcite cement, and this fracture zone is narrow, less than 10 μm across in most places. High power optical microscopy and SEM images demonstrate that some open pores are present between the calcite grains, with widths of approximately 1 μm. Cementation and possibly some recrystallization yielded clear calcite, without the darker organic matter which characterizes the fossil grains. Some of this excluded organic matter may have been concentrated at places in the crack to yield the dark line seen in places along the crack zone.

Reflection of energy from this feature undoubtedly is related to its limited porosity, but may also involve grain size and grain packing. The material in the zone provides no inherent acoustic contrast inasmuch as both the bulk rock and the cement are calcite, as verified by optical microscopy, electron probe microanalysis, and x-ray diffraction. The strength and persistence of this reflector was surprising.

Multiple Reflections

Two artifacts were encountered in our images: multiple reflections and angular distortions. Both are relatively simple to detect and correct.

Figure 3 presents synthetic cross sections computed tomographically from 17 C-scans of the sample. Multiples are obvious here. In Figure 1, several multiples can be seen, "trailing" the true intersection of the plunging crack with the gated image plane. False-color images of reflection intensity (not reproduced here) document the decreasing energies of the successive ghost images better than the gray-scale presentation herein.

Figure 3. Synthetic cross-sections of major crack in Figure 1 generated by computed tomography from 17 C-scans taken at evenly spaced depths to 4 mm. Left: Horizontal cross-section. Right: Vertical cross-section. Sections were calculated near the center of Figure 1. Note the two multiple reflections below the primary reflection.

Multiple reflections are generated during subsurface imaging when part of the energy of the returning wave is reflected back into the specimen at the solid-water interface. Subsequent reflection from the feature of interest can produce an apparent feature at a depth/travel time greater than that of the actual feature. With sufficient energy and other appropriate constraints, several such artifacts are possible. Even with a narrow time gate, multiples can appear when the time gate is set later than the two-way travel time to the reflective feature. Multiple reflections are common in geophysical seismic profiles; they are termed "multiples" and can be eliminated during data processing.

Apparent vs. True Slope

It also is necessary to correct the slope of the cracks in Figure 3, in which "depth" is actually the measured two-way travel time. This correction can be calculated from the respective travel times and sonic velocities in the water and the rock. However, many rocks are polymineralic, making it difficult to locate an appropriate sonic velocity in the literature. Furthermore, mineral grains in rock often display preferred crystallographic orientations, rendering average velocity of anisotropic minerals unsuitable. Thus we found it simpler to derive the correction empirically by comparison of the recorded angle with that measured directly on the specimen from a cross-sectional cut or view. This would not pose a problem in the study of a suite of reasonably similar rocks, requiring only an initial measurement to establish the correction procedure. The correction is significant: for example, on one face of the limestone cube the measured angle was 24° whereas the tomographic angle was 14°.

EVALUATION AND CONCLUSIONS

The depth to which an individual crack could be traced and the clarity of the image obtained varied with the type of rock, the size and packing of its grains, the nature of the crack (e.g., open or filled), and the frequency of the transducer. For the specimens illustrated, penetration of approximately 5 mm was achieved with the 25 MHz lens, with perhaps another few millimeters depth of lower quality imaging still possible. The 10 MHz lens produced images of insufficient lateral resolution, and the 50 MHz lens provided insufficient penetration.

Acoustic microscopy presents significant opportunities for research in a broad range of subdisciplines of the geosciences, especially those related to the petroleum industry. We believe that the 3-dimensional images of cracks and fractures generated by the acoustic microscope will have the most impact on determination of the preferred orientation of cracks and fractures in the geo-engineering investigations associated with secondary and tertiary recovery of hydrocarbons (EOR - enhanced oil recovery). Examination of rock with the acoustic microscope is rapid, simple, inexpensive, non-destructive, and requires minimal sample preparation. It therefore is a cost-effective technique for the investigation of cracks and fractures in rocks, especially those which host hydrocarbons.

ACKNOWLEDGEMENT

This material is based upon work supported by the National Science Foundation under Grant No. HRD-9450412.

REFERENCES

Bonner, B.P., 1978, Detecting internal defects in single crystal olivine with the acoustic microscope, *Eos, Trans., Amer. Geophysical Union.* 43:1696 (abs.).

Briggs, A., 1992, "Acoustic Microscopy," Oxford Univ. Press, Oxford.

Briggs, G.A., and Hoppe, M., 1991, Acoustic microscopy, *in:* "Images of Materials," D.B. Williams, A.R. Pelton, and R. Gronsky, eds., Oxford Univ. Press, Oxford.

Carlson, W.D. and Denison, C., 1992, Mechanisms of porphyroblast crystallization: Results from high-resolution computed x-ray tomography: *Science,* 257, p. 1236-1239.

Dansburg, J.S., and Yuhas, D.E., 1978, Acoustic microscope images of rock samples, *Geophysical Res. Letters.* 5:885.

Flannery, B.P., Deckman, H.W., Roberge, W.G., and D'Amico, K.L., 1987, Three-dimensional X-ray Microtomography: *Science,* 237, p. 1439-1444.

Fredrich, J.T., Menendez, B., and Wong, T.-F., 1995, Imaging the pore structure of geomaterials: *Science,* 268, p. 276-279.

Kustov, A.I., Kulakov, M.A., Morozov, A.I., and Erlanger, O.A., 1987, On the potential use of the scanning acoustic microscope in paleontology, *Paleont. Zhur.* (Scripta Technica trans.). 2:117.

Morales, J.G., Rodriguez, R., Durand, J., Ferdj-Allah, H., Hadjoub, Z., Attal, J., and Doghmane, A., 1991, Characterization and identification of berlinite crystals by acoustic microscopy, *J. Materials Res.* 6:2484.

Pingitore, N.E., Gillespie, C.L., Miller, K.C., DuPlessis, L., and Murr, L.E., 1995, Imaging Geological Materials with the Acoustic Microscope: Proceedings, 21st International Symposium on Acoustical Imaging, Plenum, New York (in press).

Rodriguez-Rey, A., Briggs, G.A., Field, and M. Montoto, 1990, Acoustic microscopy of rocks, *Jour. Microscopy.* 160:21.

Scott, A.C., and Hemsley, A.R., 1991, A comparison of new microscopical techniques for the study of fossil spore wall ultrastructure, *Rev. Palaeobotany Palynology.* 67:133.2

PLATE TOMOGRAPHY WITH DRY CONTACT LAMB WAVE TRANSDUCERS

J. Pei, M.I. Yousuf, F.L. Degertekin,
B.V.Honein, and B. T. Khuri-Yakub

Edward L. Ginzton Laboratory
Stanford University
Stanford, CA 94305

ABSTRACT

Ultrasonic Lamb wave techniques are widely used in a number of NDE applications. To excite Lamb waves, mode conversion of bulk waves or photo acoustic excitation are often used. Both of these approaches suffer from the need for liquid couplant or ablation of materials to reach good signal to noise ratio. In this paper, we propose a novel technique that utilizes point source excitation and detection of Lamb waves through dry, elastic contacts to monitor velocity changes. We demonstrate the power of this approach in ultrasonic pipe erosion/corrosion monitoring and its potential application in aircraft skin defect imaging. We present results of measurements of plate thickness, and erosion/corrosion in a section of pipe that was removed from service, as well as imaging of defects in an aluminum thin plate.

MOTIVATION FOR ULTRASONIC LAMB WAVE THICKNESS MEASUREMENT

Ultrasonic measurement of the thickness of plates is most often done by measuring the time of flight of bulk waves in the material[1]. This technique requires a liquid couplant between the transducer and the plate, and is especially difficult when the plate has surface curvature or roughness. Traditional techniques are of limited use in hostile environments because of limitations on the fluid couplant and the piezoelectric transducer material. Plate thickness can also be measured with photo acoustic excitation of Lamb waves[2,3]. However, ablation of the test material is often needed to reach good signal to noise ratio in order to perform accurate measurements[4]. In this paper, we present a new technique that relies on using dry, point contact transducer/buffer pin sets to excite and detect the zeroth order antisymmetric (A_0) Lamb wave mode in plates. Due to the dispersive nature of the Ao mode Lamb wave, its velocity depends on the thickness of the plate. Hence, a measure of the velocity of the Lamb wave yields a measure of the average plate thickness between the transmitter and the receiver. This technique does not require any couplant, and can be applied in hostile environments to monitor erosion/corrosion processes. There is also no particular requirement for the surface condition of the pipe since only point contacts are established for the measurement. The simplicity of the Lamb wave transducer design enables one to construct transducer scan imaging system or even transducer arrays to realize real time defect imaging.

LAMB WAVE PROPAGATION AND A_0 MODE DISPERSION

Lamb waves are elastic perturbations propagating in solid plates with free boundaries. Fig.1 shows the theoretical calculation of the phase velocity of the first few Lamb wave modes in a steel plate as a function of the product of frequency and plate thickness[5]. When the A_0 mode Lamb wave is excited, its phase velocity changes as the thickness of the plate changes. Therefore, this dispersive nature can be used to determine the thickness of a plate given the frequency of operation. In order to excite only A_0 mode, the excitation frequency has to be low enough not to excite any higher propagating modes. In our work which is concerned with steel plates and pipes whose thickness is around 10mm, the Lamb wave transducers are chosen to have their resonant frequency at 70kHz. At this operating point, only the A_0 mode is excited because the method of excitation lacks the symmetry to excite the symmetric S_0 mode, and the operating frequency is well below the cutoff frequency of any higher order mode. Another feature of this operating point is that the phase velocity of the Ao Lamb wave in the plate changes about 100 meters/second (5%) for every millimeter of thickness change. This high sensitivity to thickness is sufficient for our electronics to resolve a change of 0.1mm due to erosion/corrosion.

The experimental setup for pipe erosion/corrosion monitoring is shown in Fig. 2. The Lamb wave is excited using a piezoelectric transducer bonded to a steel buffer pin. Steel is chosen as the buffer pin material to obtain the best acoustic impedance match between the pin and the steel pipe under inspection. The tip of the pin is rounded to have a radius of curvature of 100 μm. The spherical tip gives a dry point contact to the pipe and the transducer/steel rod assembly is spring loaded to insure stable contacts every time the steel pin is pushed against pipe wall surface. An identical transducer/steel pin set is used as a receiver to detect the Lamb wave transmitted through the pipe wall. It is worth noting that different types of pins with different radii of curvature have been used depending on the application and materials involved. Piezoelectric Lead Zirconium Titanate (PZT-5H) is chosen for the transducer material. The PZT-5H is machined to a cylindrical shape with a diameter of 6.35mm and a height of 12.7mm. The resonant frequency of the transducer when bonded to steel is 70 kHz with 50% fractional bandwidth. Fig.3 shows the calculated phase velocity of propagation modes in a steel rod with a diameter of 6.35mm as a function of frequency[6]. The selected 70kHz resonance ensures that only the lowest order extensional mode is generated in the rod, and hence is the only source of Lamb waves in the test structure.

A time of flight(TOF) measurement technique is used to obtain the Lamb wave phase velocity in the pipe wall. The TOF measurement is described in Fig. 4. A short electrical pulse of 500 volts is applied to the transmitter which generates the extensional wave in the steel pin. At the contact interface between pin tip and test structure, part of the extensional mode energy is reflected back to the transmitter and generates an echo electrical pulse, the other part of the energy is coupled into the pipe wall as the A_0 Lamb wave. After propagation through the pipe wall, the Lamb wave is converted back to an extensional mode at the receiver tip and then, to an electrical signal in the receiving transducer. The pin-to-pin time of flight is measured by monitoring the time interval between the transmitted echo and the received signal. The first zero crossing of the echo signal triggers the start of the time delay counter and the first zero crossing of the transmitted signal stops the counter. The time delay measured is the time that takes the Lamb wave to travel through the pipe wall from the transmitter tip to the receiver tip. The effect of the steel pins is eliminated due to the subtraction of the time delay in the pins from both transmitter and receiver.

PIPE WALL THICKNESS MEASUREMENT

Since the time of flight data includes the electronic time delays introduced by the amplifiers and filters, a proper calibration on a sample of the same material is necessary. The calibration is performed on a steel step wedge plate with its thickness varying from 8.25 mm to 9.25 mm. With known distance between the transmitter and receiver pin tips, a TOF data point is taken in a region of the steel plate whose thickness is measured using a

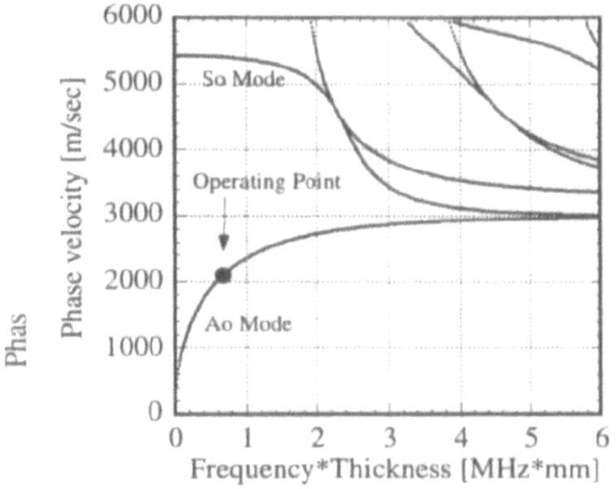

Figure 1. Theoretical calculation of the dispersion relation of steel plates. Ao mode is chosen in our experiment. The Lamb wave transducers are designed to have a frequency well below the cutoff frequency of higher order modes.

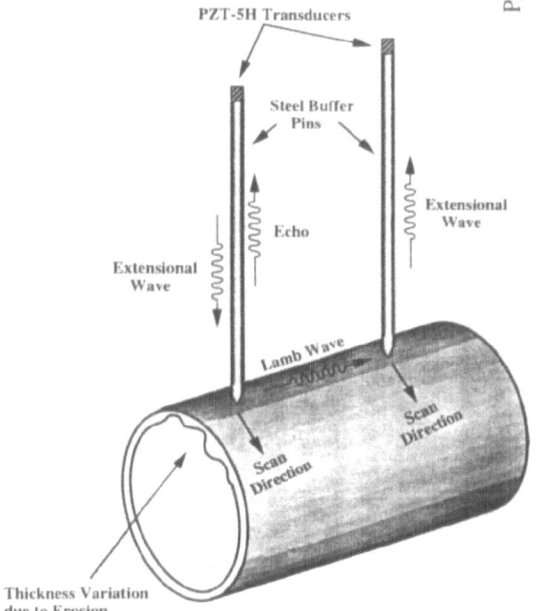

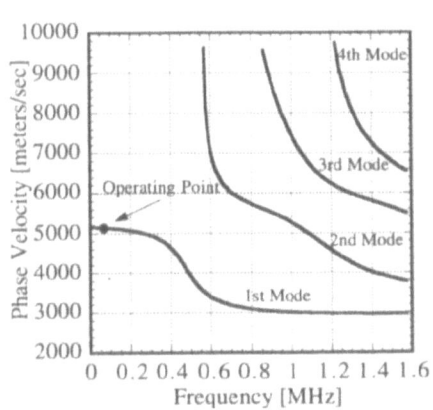

Figure 2. Experimental setup of pipe wall thickness measurement. Two PZT-5H transducer and steel buffer pin sets are used as the Lamb wave transmitter and receiver. The steel pipe under study has a diameter of 35cm and wall thickness around 10mm.

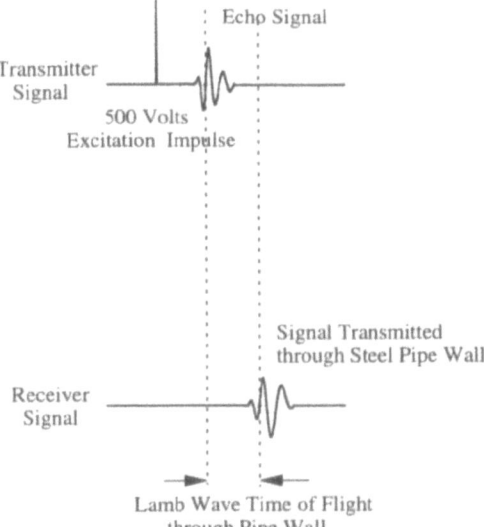

Figure 3. Theoretical dispersion relation of steel rods with diameter of 6.35mm. The frequency PZT transducer is low enough to excite only the first extensional mode.

Figure 4. Timing Diagram for the time of flight measurement. The time delay from the echo signal and transmitted signal is measured. The effect of the two buffer pins is eliminated because the delays of acoustic signals in the transmitter and receiver buffer pins are identical.

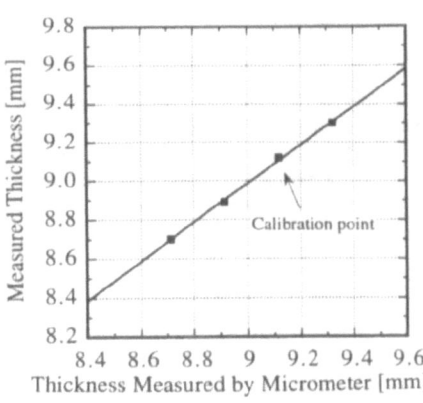

Figure 5. Calibration measurement is performed on a steel step wedge plate with thickness measured using a micrometer. With known distance between the transmitter and receiver pin tips, one point of the time of flight data is corrected for the electronic delay and fitted to the theoretical dispersion curve. Other thickness points fit the calculated curve within 1% error.

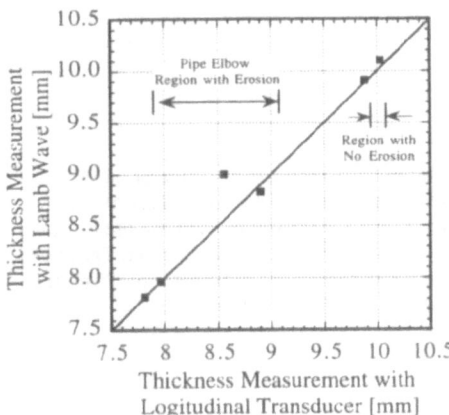

Figure 6. Thickness measurement performed on a section of steel pipe. Thickness variation measured with Lamb wave technique is plotted against the data obtained from a 10mhz longitudinal transducer.

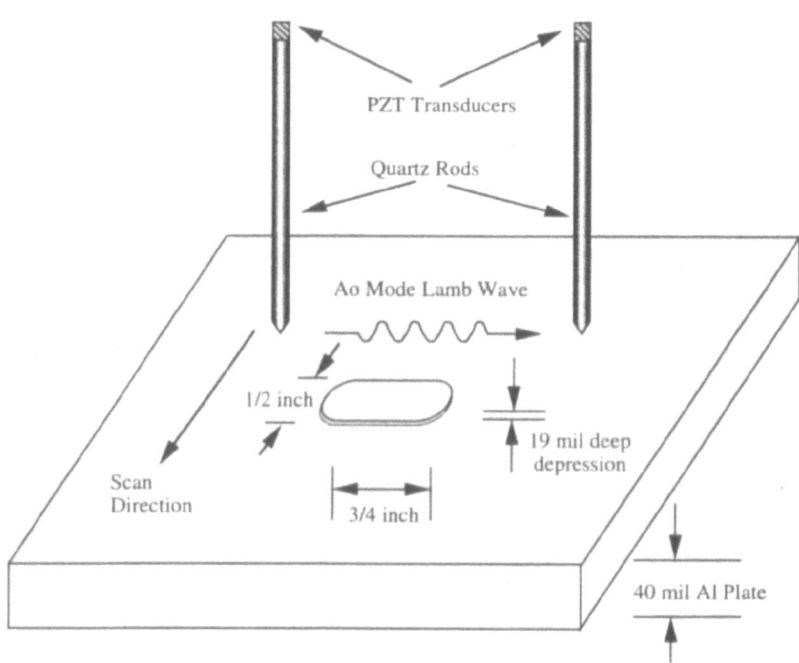

Figure 7. Imaging system for aircraft skin. a 4 mil (0.1mm) depression is milled into the aluminum plate to stimulate a defect. 64 time of flight data point is taken along each scan path. A total of 60 scans are taken at an interval of every 3°.

micrometer caliper. This data point is fitted to the theoretical dispersion curve and fitting parameters corresponding to the electronic delay are obtained. With the same fitting parameters, the TOF data of other regions are converted to plate thickness accordingly. Fig. 5 shows the measurement of regions with different thickness vs. the measurement done with a micrometer. Note that the fit of the data is to better than 1%.

The system was used to measure erosion/corrosion in a steel pipe elbow that was removed from service. The diameter of the pipe is 35 cm, and the wall thickness is of the order of 1 cm. A similar calibration run was made on the steel pipe at a location where the thickness could be independently measured with a micrometer caliper. Wall thickness variations were measured to range from 7.8 mm to 10.1 mm which indicated the presence of extensive corrosion at some locations inside the pipe. This thickness variation was further verified with a traditional pulse echo measurement using a longitudinal wave transducer operating at a frequency of 10 MHz. Fig. 6 shows a comparison of the pipe thickness as measured with both methods. Overall, there is excellent agreement between the two measurements.

SCAN IMAGING SYSTEM

This technique is not limited to the application listed above, indeed, it can also be used for defect detection and imaging in thin plates such as aircraft skins. Fig. 7 shows the experimental setup of a scan imaging system used to detect a depression. The thickness of the aluminum plate under inspection is around 1mm. The transducers, therefore, are chosen according to this thickness to have their operating frequency around 200kHz. Only the A_0 mode Lamb wave can be generated in the plate at this frequency. The PZT-5H piezoelectric material is machined to a cylindrical shape with a diameter of 3.2mm and a height of 5mm to obtain the 200kHz resonance. Quartz rods are used as the buffer pins with their tips sharpened to have a radius of curvature of 100μm. The transducers are spring loaded and pressed on the plate surface to form a dry contact. Again, the sharp tips are not sensitive to surface conditions and do not require any liquid couplant.

A 0.5mm deep rectangular depression is milled into the aluminum plate to simulate a defect. The plate is mounted on an X-Y and rotation state and the transducers scan over the area with defect. Due to the thinning of the defect area, the time of flight increase as the pins scan across the depression. Filtered back projection[7] tomographic inversion method is used to reconstruct the defect image. 64 points are taken along each scan direction and a total of 60 scans are taken at an interval of every 3°. The reconstructed time of flight image is shown in Fig. 8 with a resolution of 64×64 pixels. The depression region is clearly shown in the image.

CONCLUSION

We presented a novel method for measuring the thickness and non destructive evaluation of plate-like structures using the zeroth order antisymmetric Lamb wave. The method employs a Hertzian contact between the buffer pin and the plate. No surface preparation is necessary, and the method can be applied in-situ, in hostile environments such as at high temperature and in the presence of radioactivity, and through insulation. A measurement accuracy of better than 1% was demonstrated experimentally and a scan imaging system is presented. This technique can be applied to a variety of plates for characterizing homogeneous and composite materials. Real time imaging system can be realized with Lamb wave pin transducer arrays.

ACKNOWLEDGMENT

This work was supported by EPRI under contract #W03148-13.

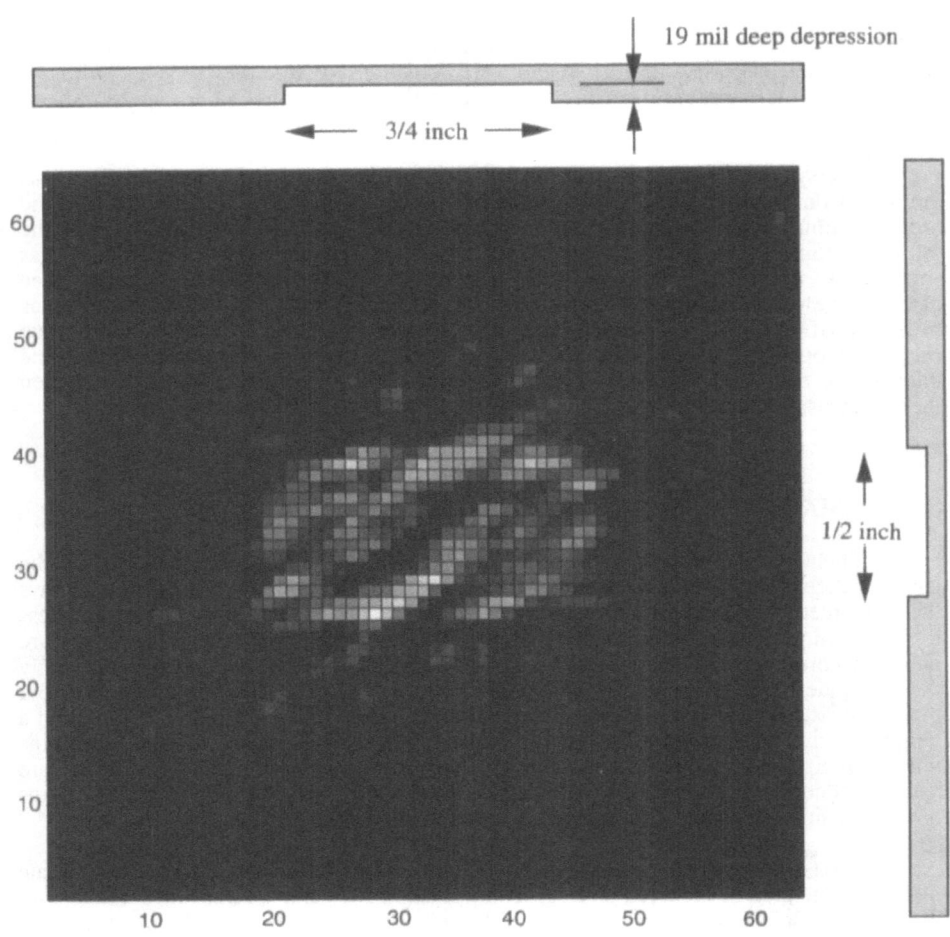

Figure 8. Reconstructed time of flight image of the depression shown in Fig. 7 using the back projection alogrithm. The field of view is 3.2 inches x 3.2 inches (8.1cm x 8.1cm) with 64 x 64 pixels of resolution.

REFERENCES

1. L.C. Lynnworth, *Ultrasonic measurements for Process Control : Theory, Techniques, Applications* (Boston: Academic Press, 1989).
2. D.A. Hutchins, D.P. Jasen and C. Edwards, Proceedings of IEEE 1992 Ultrasonics Symposium. p883-6 vol.2.
3. D.R. Billson, D.A. Hutchins, Nondestructive Testing and Evaluation (1992) vol.10, no.1, p.43-53.
4. D.A. Hutchins, *Physical Acoustics* Vol. 18 (W.P. Mason and R.N. Thurston, eds, Academic Press, New York, 1988).
5. I.A. Victorov, *Rayleigh and Lamb Waves, Physical Theory and Applications* (Plenum, New York , 1967).
6. K.F. Graff, *Wave Motion in Elastic Solids* (Ohio State University Press, Columbus 7 1975).
7. A.C. Kak and M. Slaney, *Principles of Computerized Tomographic Imaging* (IEEE Press, New York, 1988).

MULTI-SAFT: A FLEXIBLE METHOD FOR DEFECT CHARACTERIZATION

Peter Paul van 't Veen and Machteld de Kroon

TNO Institute of Applied Physics
Inspection Technology Department
P.O. Box 155
2600 AD Delft
The Netherlands

INTRODUCTION

Due to improvements in nondestructive testing methods even small defects are detected. Repair of all detected defects is very costly and often unnecessary from a fitness-for-purpose point of view. For this reason discrimination between critical and non-critical defects becomes more important. Ultrasonic techniques are developed to asses the relevant characteristics of defects.

An important development in this respect is the time-of-flight-diffraction technique (TOFD). TOFD[1] can be used to size and localize defects along the weld and normal to the surface in many type of welds including complex geometries. TOFD is however not very sensitive to the type of defect (flat or non-flat).

Another method in development is the synthetic aperture focusing technique (SAFT). SAFT[2] is capable of sizing and localization of defects in three dimensions. Furthermore SAFT can be used for classification of the type of defect.

In this paper results for Multi-SAFT[3], a particular implementation of SAFT, are described. Multi-SAFT is aimed at characterization of defects in welds present in the chemical industry, oil- and gasdistribution industry and conventional power stations. After a description of the data-acquisition and data-processing procedures experimental results are discussed for welds between thin steel plates (6 to 12 mm).

DATA-ACQUISITION

The data-acquisition hardware for Multi-SAFT is designed for characterization of defects detected with other NDT methods. Data-acquisition is performed only at suspect positions along the weld. Data-acquisition for one defect can be performed within 15 minutes. The amount of space required for the scan is limited.

The data-acquisition configuration used on thin steel plates is shown in Figure 1.

Use is made of a mouse scanner and a metal rail for guidance of the mouse. Both source and receiver are 60°, 5 MHz transducers. Because of the limited space available the source and receiver are positioned at an angle to the weld (the split angle). 100 records with a spacing of 1 mm are acquired. The records are digitized by a plug-in board for a personal computer. Each record contains 1024 time samples.

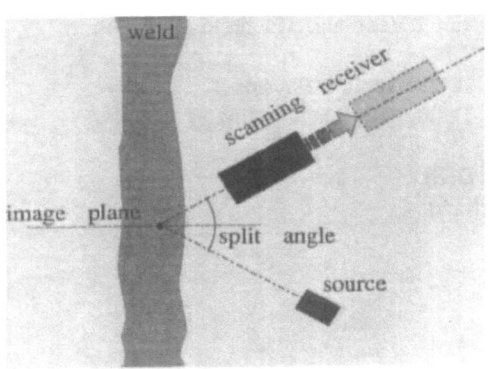

Figure 1. Data-acquisition configuration used on thin steel plates (top view).

DATA-PROCESSING

The Multi-SAFT algorithme is described in detail by Lorenz[3]. Multi-SAFT uses multiple wavepaths to image the defect. The number of surface reflections before and after interaction with the defect strongly depends on the characteristics of the defect. Since these characteristics can only be known after image reconstruction is completed, multiple images assuming different wavepaths are calculated. Moreover different parts of a defect might be imaged by different wavepaths.

More than one indication may apear in the image for one wavepath. Ghost indications origin from different wavepaths with comparable traveltimes. Close excamination of the relation between traveltime, angle of insonification and orientation will lead to the exclusion of the false indications.

For the computation of the traveltimes for the different wavepaths in a plate use can be made of analytical solutions. For more complex geometries use is made of ray-tracing to compute the traveltimes.

CHARACTERIZATION

Classification of flat and voluminous defects is performed by close excamination of the appearance of the unprocessed data. The unprocessed data for a planar defect shows a regular pattern of lines (Figure 2). As the receiver is moved away from the weld the number of reflections after interaction with the defect increases. Every line corresponds to a separate number of reflections. For a volumetric defect the unprocessed data contains an irregular pattern of reflections and diffractions (Figure 3).

To quantify the size and location of the defect the data is processed and the SAFT images are excamined. The relation between traveltime, angle of insonification and orientation is used to exclude false indications. After selection of the correct indications localization and sizing is performed on the contour. The correct contourlevel is determined by callibration on known defects or numerical simulations.

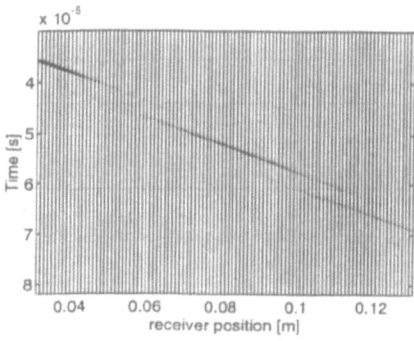

Figure 2. Unprocessed data for a planar defect (lack of fusion).

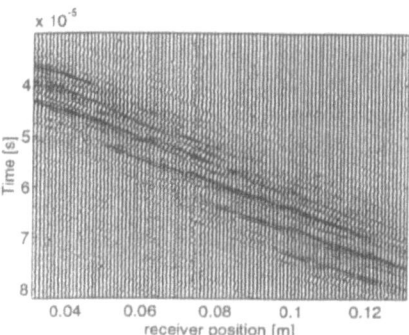

Figure 3. Unprocessed data for a volumetric defect (slag inclusion).

EXPERIMENTAL RESULTS

Multi-SAFT was applied to 20 defects in plates with a thickness in the range from 6 to 12 mm. Defects were deliberatley inserted during welding. Types of defect present were lack of fusion and slag inclusions.

With a mean depth localization error and standard deviation of -0.9 and 2.1 mm respectively and mean heigth sizing error and standard deviation of -0.3 and 1.2 mm respectively, Multi-SAFT was quite comparable to the other ultrasonic techniques.

Unlike most other techniques, Multi-SAFT especially locates (and sizes) perpendicular to the weld. The mean localisation error and standard deviation were 0.2 and 2.3 mm respectively. This proved accurate enough to pinpoint the actual weld flange where the defect was located in all 20 cases.

With respect to defect classification as flat or non-flat, Multi-SAFT proved to be highly reliable. Out of 6 slag inclusions 5 were correctly classified as non-flat, while all 11 lack-of-fusion defects were correctly classified as flat (Table 1).

Table 1. Results of classification for 19 defects.

Type of defect	Number of defects	Flat	Volumetric
Lack of fusion	11	11	0
Slag inclusion	6	1	5

CONCLUSIONS

Multi-SAFT provides a flexible method for characterization of defects in steel welds. Use is made of a simple mouse scanner and standard ultrasonic equipment. Data interpretation is performed on unprocessed data and multiple images.

On thin steel plates sizing and localization of defects is performed with an accuracy comparable to other ultrasonic techniques. Of special interest is the capability to discriminate between flat and volumetric defects.

ACKNOWLEDGEMENTS

The support of this research by the Dutch Electric Power Institute (KEMA) Arnhem, Shell Research Amsterdam, Materiaal Metingen Testgroep Ridderkerk, AEA Sonomatic Oosterhout is gratefully acknowledged. The authors also wish to thank the Dutch Welding Institute (NIL) for placing the specially welded plate sections at their disposal.

REFERENCES

1. J.P. Charlesworth and J.A.G. Temple, "Engineering Applications of Ultrasonic Time-of-Flight Diffraction", Research Studies Press, Somerset (1989).
2. J. Seydel, "Ultrasonic Synthetic Aperture Focusing Techniques in NDT", in: Research Techniques in Nondestructive Testing, R.S. Sharpe (Ed.), Vol. VI, Academic Press, London (1982).
3. M. Lorenz, "Ultrasonic Imaging for the Characterization of Defects in Steel Components", PhD Thesis Delft University of Technology (1993).

QUANTITATIVE NDT BY 3D IMAGE RECONSTRUCTION

V. Schmitz[1], M. Kröning[1], and K.J. Kangenberg[2]

[1]Fraunhofer Institute for Nondestructive Testing
University Building 37
66123 Saarbrücken

[2]University of Kassel
Department of Electrical Engineering
Wilhemshöher Allee 71, 34109 Kassel

INTRODUCTION

One of the classical tasks of ultrasonic NDT is the detection, sizing and characterization of material damages like cracks in welds, lack of fusion, foreign material or delaminations. To increase the reliability and with respect to quality assurance, automatic scanning systems are used together with a μ-processor controlled data acquisition system for the complete high-frequency signal. The applied imaging scheme is based on the backpropagation of the elastic waves back into the component. The application of the implemented method: "Synthetic Aperture Focusing Technique - SAFT" covers a wide spread field of the testing of ferritic and austenitic material used for pipes, turbines, plates or vessels., the investigation on material inhomogeneities within ceramic material, the inspection of glass- or carbon- fibre reinforced composites up to the testing of concrete material. The applied frequencies extend from 100 kHz to 100 MHz. Based upon this experience it has been concluded, that a major task of the evaluation procedure still remains, that is the interpretation of the image. This has lead to a theoretical investigation of „how accurate" real defects can be „reconstructed", to the realization of a 3D-SAFT-imaging system and to a computer aided inspection „CAI" concept to improve further the interpretation capability of acoustic images by applying a combination between theoretical modeling with „Elastodynamic Finite Integration Technique - EFIT" and the SAFT reconstruction capability.

METHODS OF SIZING DEFECTS

The classical method to size defects in components made of steel is the pulse echo method in the frequency range between 1 MHz and 5 MHz. The pulses scattered or reflected by defects are received and the amplitude of the echo is evaluated according to the DGS (distance - gain - equivalent disc size) method. This is a simplified approach and no information will be given concerning depth extension, inclination, shape or roughness of the surface of the defect.

Advanced sizing methods have been developed within research work of nuclear power plants like Synthetic Aperture Focusing Technique (SAFT), ultrasonic tomography using contact technique probes, focusing probes or Phased Array Technology. They need manipulation systems due to the fact that in addition to the amplitude and time-of-flight values which are stored in the memory of a computer, the positions of the probe have to be known. Using one of these methods it is possible to image defects and to evaluate their physical dimensions with respect to the outer or inner surface contour.

The advantage of imaging three-dimensional together with a signal processing method, which allows to eliminate the influence of the diverging sound field provides the opportunity to analyse distributed defects or complex shaped defects. This will be explained schematically in fig. 1 on a simple example - a point like reflector.

Without any signal processing, the C-scan image of the point like reflector with the size Dp will be imaged proportional to the larger beam width Ds, which again is increasing with increasing depth. Scanning in x-direction and applying 2D-SAFT in x-direction shows the ellipsoidal character with the correct width Dp in x-direction and the remaining with Ds of the sound field in y-direction. This holds vice verca for a perpendicular scan in y- direction. Only a 3D-SAFT imaging process yields to a correct disc shaped result of diameter Dp. It is important to mention that this holds only for those cases where the wavelength is small compared to the diameter of the defect which has to be imaged.

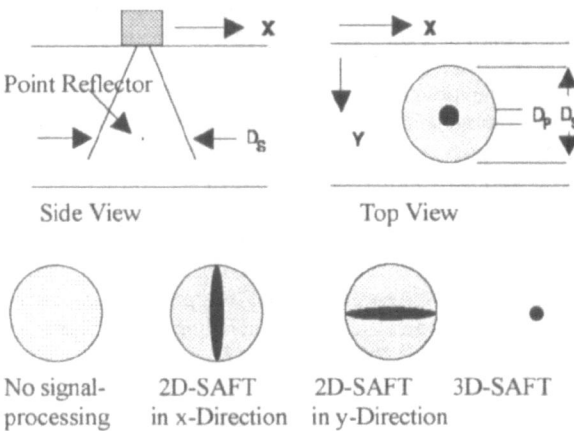

Fig. 1 Image performance using 2D-SAFT and 3D-SAFT demonstrated by a
Top-View presentation of a point like reflector

THEORETICAL BACKGROUND OF ACOUSTIC IMAGING

The Synthetic Aperture Focusing Technique is based upon the heuristic idea concerning the inversion of elastic wave propagation in solid material; it means the „mathematical" observation that in a rectangular coordinate system, the time-of-flight curve of a point like reflector in two dimensions is a hyperbel and for three dimensions a hyperboloid.

In a first step, SAFT relies on the assumption that pressure and shear waves which propagate in solid material can be modelled as scalar waves. Both waves are independent from each other and propagate with different velocities. This is true only for certain idealized situations, but not for each practical situation. In the heuristic prove of the SAFT -algorithm the wave amplitude is described by a scalar, position depending vector $\underline{R}$ and by a time dependent scalar potential $\Phi(\underline{R},t)$. The scalar potential fulfills the wave equation:

(1)
$$\Delta\Phi(\underline{R},t) - \frac{1}{c^2}\frac{\partial^2}{\partial t^2}\Phi(\underline{R},t) = -q(\underline{R},t),$$

with c = velocity of the pressure or of the shear wave
and $q(\underline{R},t)$ as the source of the acoustic wave.

The solution of equation (1) is the three-dimensional integral over the source volume V_q, the so-called retarded potential with R' as the source point coordinate:

(2)
$$\Phi(\underline{R},t) = \iiint_{V_Q} \frac{q\left(\underline{R}',t - \frac{|R-R'|}{c}\right)}{4\pi|\underline{R}-\underline{R}'|}d^3\underline{R}'.$$

It is important to know how many source points $\underline{R}'$ with their source density q contribute at a fixed time point t_q at the fixed time t to the amplitude at a fixed point $\underline{R}$. This holds for

(3)
$$t_q = t - \frac{|\underline{R}-\underline{R}'|}{c}.$$

Rearranging equation (3) leads to:

(4)
$$c(t - t_q) = |\underline{R} - \underline{R}'|$$

which shows clearly that all points $\underline{R}'$ which lie on a sphere around $\underline{R}$ with the fixed time-of-flight radius $c(t - t_q)$ contribute *simultaneously* at the time t to the point $\underline{R}$. The signal amplitude at the spatial point $\underline{R}$. t comes from all source points $q(\underline{R}', t_q)$ on the $\underline{R}'$ - sphere.

With the measuring plane S_M in a distance of $z_0 > 0$ to the x-y-plane of a cartesian coordinate system and the source density $q(\underline{R}', t_q)$ at the time t_q beneath that plane, equation (4) can be written as:

(5)
$$c^2(t - t_q)^2 - (x - x')^2 - (y - y')^2 = (z_0 - z')^2.$$

This is for the time t as a function of the locations x and y a so-called time-of-flight surface and here a rotational hyperboloid. A single point q ($\underline{\mathbf{R}}'$, t_q) in the material defocusses in such a time-of-flight surface. Taking the data from this allows to reconstruct this single point from the measured data, which in fact is the principle of *Synthetic Aperture Focusing Technique*:

- one integrates for each pixel in the reconstruction space along the correspondant time-of-flight surface in the x-y-t- data field
- the resulting value is stored in the pixel and added to values which come from other time-of-flight surfaces.

If the pixel is part of a physical existing scatterer, the summation results in a high value; otherwise there will be a „noise level"-value. In other words, SAFT achieves automatically a time-corrected signal averaging approach.

An example of a typical 2D-SAFT result is presented in fig. 2. The pipe inspection with a wall thickness of 50 mm shows an *image* of the outer and inner surface contour together with a crack starting from the inner surface.

The question if the image of the crack is identically with the reality can be answered through many experiments and comparison with destructive examinations or can be explained by applying the mathematical tools of inverse scattering theory. Here we refer to /1/ and to /2/. A summary of the final conlusion to the application of the heuristic derived SAFT formalism is summarized to be:

- The propagation of the pressure and shear waves is scalar (the problem has been scalarized)
- The defects in the material are either weak scatterer or strong reflecting cracks or pores (the problem has been linearized)
- The probe has an infinite bandwith
- In the case of a plane surface, the scattered field has to be measured, whereas for arbitrary surfaces the scattered field and its normal derivatives have to be measured. The derivative as a function of time can be calculated
- The measuring plane should surround the defect, which means that for plane surfaces both faces have to be used to acquire measuring data.

 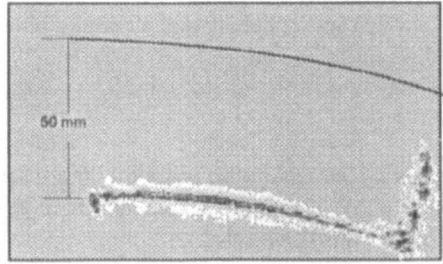

Fig. 2. Defect analysis with 2D-SAFT: micrograph and reconstruction

The nonNDT of components has to deal with a three-dimensional situation. Despite this fact most of the NDE-tasks are solved by acquiring data along a two-dimensional aperture and by presenting two-dimensional images. These are never identical with two-dimensional slices of a three-dimensional image which will be explained in the next section.

EXPERIMENTAL RESULTS

Experimental results with Quasi - 3D-SAFT

In fig. 1 we have explained schematically what happens if the signal processing algorithm is performed in one direction (lateral resolution proportional to the wave length) and no signal signal processing (resolution according to the beam diameter) occured in the direction perpendicular to it. The actual experiment performed on a 145 mm thick testblock with a y-pattern of flat bottom holes (sizes 6.5 mm diameter, distances 1 mm, 2 mm and 3 mm) placed 120 mm beneath the surface confirm that the series of two-dimensional scans are not sufficient to interpret the defect pattern correctly - fig. 3. With a frequency of 5 MHz and normal insonification with longitudinal waves one obtains a lateral resolution of 1 mm which is close to the theoretical limitation of about one wavelength λ.

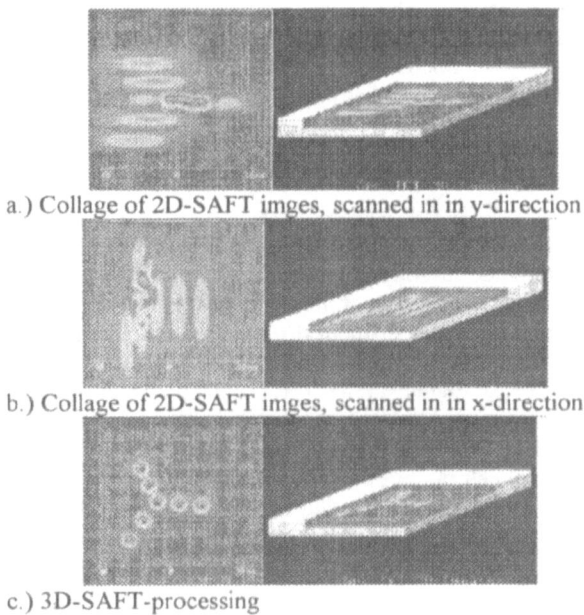

a.) Collage of 2D-SAFT imges, scanned in in y-direction

b.) Collage of 2D-SAFT imges, scanned in in x-direction

c.) 3D-SAFT-processing

Fig. 3 Comparison of series of 2D reconstruction with a true 3D-reconstruction

Experimental result of component inspection in chemical industry

This example demonstrates the efficient application of todays ultrasonic 3D-imaging capabilities together with commercial available application visualization software which allows to combine the ultrasonic image of the inspected volume with the real geometry of the component.

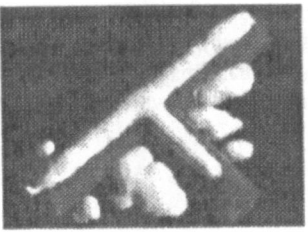

Fig. 4 Isocontour presentation of the three dimensional SAFT-reconstruction including the known T-shaped geometry of the high pressure pipe connection

The high-pressure T-shaped pipe - supplied by the company BASF - has an inner diameter of 7.2 mm, is made of ferritic material and has some cracks starting from the bore holes due to the chemical and high pressure loading. This part has been investigated with longitudinal waves at 2.25 MHz. The probe scanned an area of 44 mm x 44 mm. The data set consisted of 64 x 64 A-scans, each one with a length of 256 time samples. The reconstructed image is a volume of 44 mm x 44 mm x 44 mm with a voxel element sizes of 0.7 mm x 0.7 mm x 0.7 mm. Fig. 4 represents the white appearing reflecting parts from the T-shaped connection together with the cloud like appearance of the cracks starting from the inner surface of the holes into the ferritic material. The known geometry of the T-shaped bore hole has been overlayed to this reconstruction using graphic software packages. The calculation has been performed using a vector computer with 16 MFLOPS and 8 Mbyte main memory. A 3D-SAFT version in the frequency domain has been used by the university of Kassel.

Experimental result of component inspection in steel industry

The detection of the position, size and orientation of nonmetallic inclusions is important for the inspection of turbine rotors. In components with flat surfaces, the lateral resolution is parallel to the surface and axial resolution is in sound propagation direction. This does not hold for rotational symmetric components like turbine shafts. In addition 2D reconstructions of slices can contain unfocussed images of inclusions which are located in neighboured regions and simulate defects at wrong positions. This is shown by mathematical modelling of a situation where one defect in one plane is accompanied by two defects in a neighboured plane. Fig. 5 demonstrates the difference between a 2D- and a 3D-SAFT reconstruction. A 1.5 MHz probe has been used and the turbine has been rotated by 360° during data acquisition. The reconstruction shows a focussed image of the one point like defect lying in this special plane together with two unfocussed images lying in a neighboured plane.

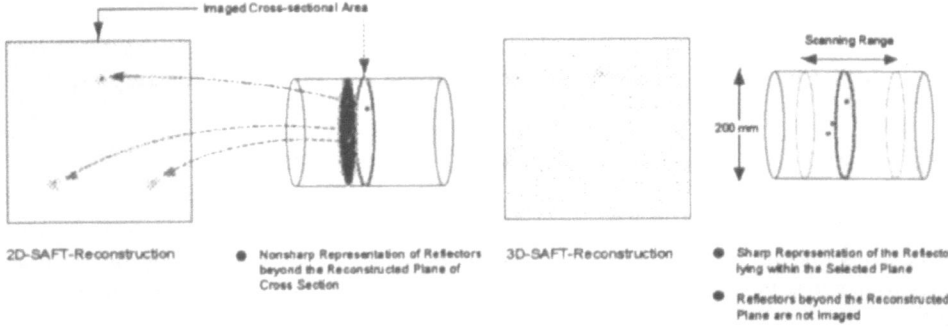

Fig. 5 Comparison of 2D-SAFT with 3D-SAFT results for turbine inspection

This difficulty in interpreting the image can be avoided by using 3D-SAFT. In this special example 128 scans around 360° in a segment of 160 mm has been simulated. The reconstructed volume has a size of 40 mm x 40 mm x 40 mm. The time needed on a Apollo DN 10 000 work station was 13 minutes. Only the defect lying in the plane which has been scanned and reconstructed is correctly displayed.

Experimental result of component inspection in air craft industry

Besides eddy current and thermographic imaging, ultrasonic pulses are launched into carbon-reinforced composite materials used in the air craft industry to detect fibre cracks, delaminations, porosities or impact damages. The following example shows the result of the acoustic 3D imaging process applied to an impact damaged CFC plate of 4 mm thickness. The reconstructed volume of 50 mm x 50 mm x 4 mm is displayed in layers of 0.5 mm thickness. This allows to follow the changing area of the impact damage from the surface through the total thickness - fig. 6. Two additional views show perpendicular slices in the x-z- and in the y-z-plane to allow a continuous depth dependent display of the damage.

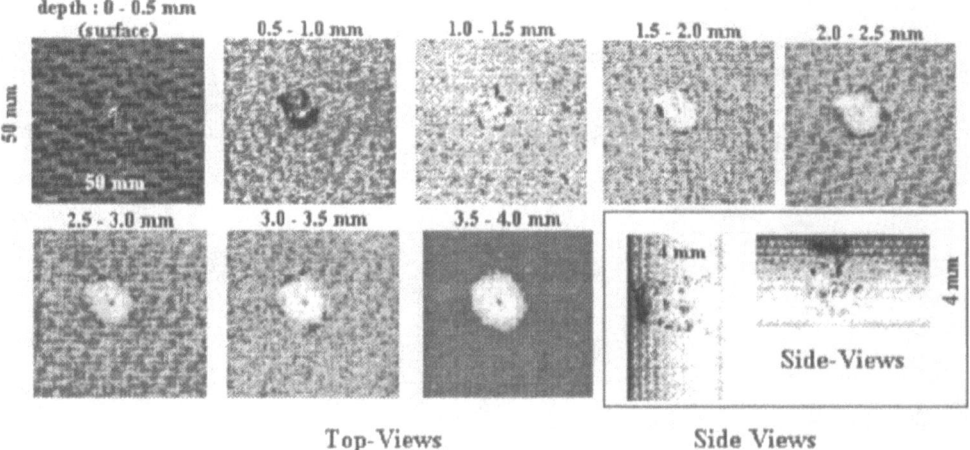

Fig. 6 Analysis of an impact damage on a carbon reinforced composite plate with 3D-SAFT

INSPECTION PROCEDURE ASSESSMENT USING MODELLING CAPABILITIES

Modelling capability to interpret ultrasonic pulses

Despite the advantage of using 3D-imaging software packages versus 2D-data processing techniques, the expert is asked for a physically based interpretation of the presented NDT-result. Scattered waves contain information about location, size, shape and orientation of defects. Predicted signals, transmitted by probes, diffracted or reflected by defects and received again by the same probe can be used as a data set for the 2D- or 3D-SAFT reconstruction procedure. Comparing synthetic images with acutal experimental images is a tool which can be provided using modern computer techniques.

To scope with real life, numerical techniques have to be applied. These have to include arbitrary forces which describe the probes and arbitrary boundary conditions which are needed to include arbitrary geometry of specimens and defects. Direct numerical methods operate directly on the fundamental equation of motion. A technique already applied for a long time in geophysics is to replace all derivatives by finite differences, ending up with a finite difference time domain scheme Another popular method borrows the idea of finite elements from solid mechanics and matches it to elastic wave propagation. According to a procedure which is convenient for Maxwell`s equation, the differential formulation and the isotrpic version of Hooke`s law leads to appropriate integral formulation. These integral formulations are then applied to each cell of a cubic grid superimposing the test specimen under inspection. This procedure yields a numerical scheme which has been called EFIT /3/.

Modelling capabilities are able to assess the inspection procedure and to explain the fine structure of received ultrasonic pulses or the structure of acoustic images. A first example shows the efficency of mathematical modelling techniques in explaining pulses which are received from a surface breaking 14 mm deep crack in a plate of 28 mm. Fig. 7 compares experimental and modelling results for a single A-scan. The coincidence is nearly complete,

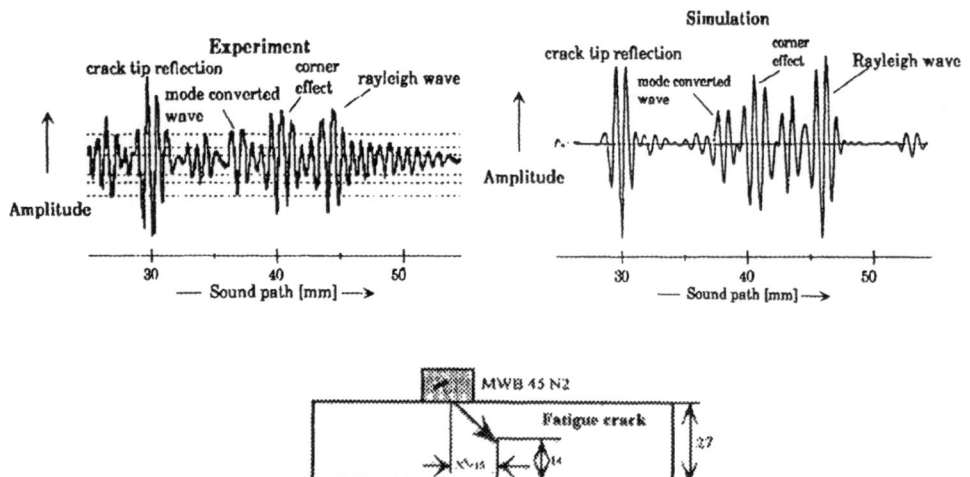

Fig. 7 Experiment and 2D-EFIT simulation of a single A-scan

except for the late pulses associated with Rayleigh waves travelling along the crack faces and being converted into shear waves. The example demonstrated here, is very complex due to mode conversions which occure at the surface of the crack. Identified pulses could be assigned to: shear-shear reflection at the crack tip, shear-pressure-shear mode conversion, shear-shear-shear reflection at the corner, shear-Rayleigh-shear mode conversion and shear Rayleigh-Rayleigh-shear mode conversion.

Still existing differences might be caused by the crack surface; fatique cracks have surface roughnesses the modelled crack was assumed to be perfectly plane and stress free.

Modelling capability to interpret acoustic images

EFIT allows to predict ultrasonic high-frequency data on real life components. Repeating this step for different locations of probes, allows to calculate all the individual A-scans for all transducer positions along the scanning line. All these calculated rf-data are now used as input data for a 2D-SAFT or 3D-SAFT reconstruction procedure. The two images - fig. 8 - correlate not completely due to the fact, that the surface roughness of the crack has not yet been implemented into the model.

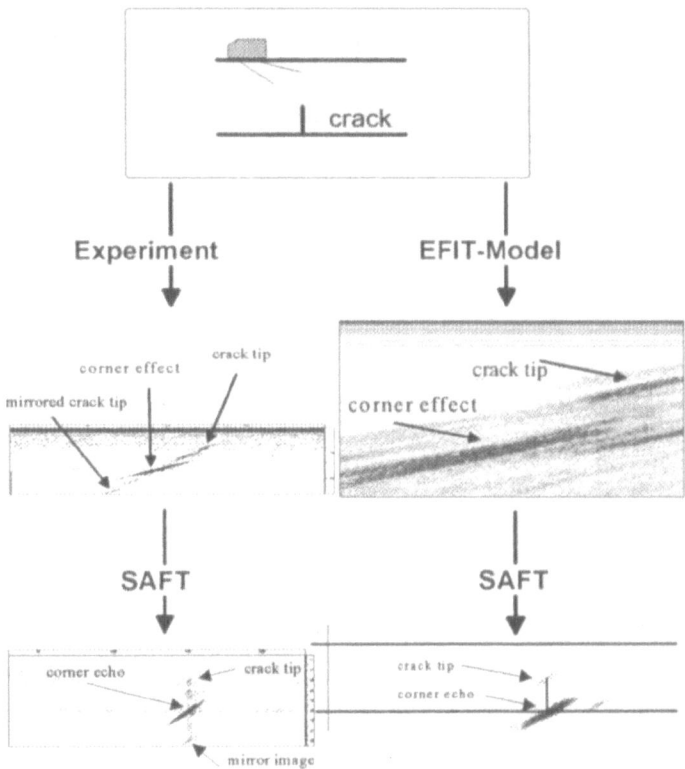

Fig. 8 SAFT images of a backwall breaking crack
obtained from simulated and from experimental data

IMPROVED DEFECT CHARACTERIZATION

Numerical modelling of ultrasonic inspections provides an important element of quantitative NDT by supporting both the phase of inspection and the evaluation of the inspection data. The experimental and the synthetic A-scan data serve as a common data base. These data are used as input for 2D- or 3D reconstruction schemes. After displaying the reconstructions, the expert extracts a flaw model, which again is used to generate new A-scans and with these A-scans new 2D- or 3D-reconstructions. Modelled images and experimental images are iteratively compared until a stop criteria leads to the final evaluation. Now with commercial available image processing algorithms 2D planes can be selected out from 3D-planes, a contour estimation of the reconstructed imag can be performed and line- or area features can be extracted in different planes. These lead to a decision e.g. if a planar defect or a non planar defect or a embedded or a surface connected defect may be the cause for a possible failure of the component.

CONCLUSION

One of the classical tasks of ultrasonic NDT, the detection, sizing and charcterization of material damages like cracks in welds, lack of fusion, foreign material or impact damages can be improved by scanning the surface of the component under inspection in two dimensionas and by applying to the data set a three dimensional imaging algorithm. The imaging algorithm (Synthetic Aperture Focussing Technique - SAFT" is based on the inverse scattering theory and has been implemented in three-dimensional version in the frequency domain on a Apollo work station and in the time domain on a Pentium personal computer. SAFT is based upon some assumptions like infinite bandwidth of the probe, scalar wave propagation or scattering on weak or strong reflecting defects. To find out quantitatively if a violation against these assumptions lead to strong misinterpretation of the flaw evaluation it has been proposed to combine 3D-imaging algorithm with numerical modelling, based upon a elastodynamic finite integration theory „EFIT". This combination allows by comparison of theoretically generated images with the experimental obtained image to extract the most likely flaw model. Based upon this flaw model commercial available image processing programs can be used to extract line- or area features and to make a decision e.g., if the defect is of a planar or non-planar type.

REFERENCES

/1/ G.T. Hermann, H.K.Tuy, K.J.Langenberg, P.Sabatier: „Basic Methods of Tomography and Inverse Problems", Adam Hilger, Bristol 1987

/2/ V.Schmitz, W. Müller, K.J.Langenberg: „HOLOSAFT II - Abschlußbericht"; Technisch-Wissenschaftlicher Bericht Nr. 920206, Jan. 92, Fraunhofer-Institut, Universität, Bau 37, 66123 Saarbrücken

/3/ P. Fellinger, K.J. Langenberg: „Numerical techniques for elastic wave propagation and scattering", in S.K.. Datta, J.D. Achenbach and Y.S. Rajapakse (eds), Elastic Waves and Ultrasonic Non-Destructive Evaluation, North-Holland, Amsterdam, 1990

OBLIQUE PITCH-CATCH IMAGING OF DEFECTS
IN MULTI-LAYERED STRUCTURES

J. D. Achenbach, I. N. Komsky and M. Zhang

Center for Quality Engineering and Failure Prevention
Northwestern University
Evanston, IL 60208, USA

INTRODUCTION

Thin-walled multi-layered structures, with sealants in-between the layers, may develop interface defects on deeper interfaces. For nondestructive inspection it is desirable to generate a visual indication of such defects by an ultrasonic imaging technique. For aircraft structures the imaging must generally be done with one-sided access only and by the use of contact transducers. A conventional normal incidence reflection technique is not practical because the interference of multiple reflections at the interfaces makes it difficult to extract relevant signals for image generation. In this paper these difficulties have been overcome by the use of an oblique pitch-catch transducer configuration. The parameters in such a configuration are wedge angles for the contact transducers, center frequencies, and the spacing of the transducers in the pitch-catch configuration. These parameters must be selected correctly for optimal penetration into the multi-layered structure and optimal reception of ultrasonic signals. The theoretical background of the technique is discussed and images are presented of second layer corrosion in wing structures.

This work was motivated by the interest in a more economical technique to inspect the wingbox of the DC-9 for corrosion, without the need for fuel tank entry. The objectives of the project were formulated as follows: (1) Develop a system for detection of corrosion inside the DC-9 wingbox by ultrasonic inspection, accomplished from outside the wing skin instead of the current open and visually inspect procedure. The system should provide for quantitative characterization of material corrosion loss and detection of exfoliation corrosion and stress corrosion cracks. (2) Develop the transducer fixture and the scanner for ultrasonic inspection by the use of contact transducers. (3) Integrate the ultrasonic technique with a commercially available computerized system for data acquisition and imaging. A full description of the project has been given by Komsky *et al* (1995).

Figure 1 shows that the DC-9 wing structure near the lower T-cap at station Xcw=58.500 is a two-layered metal structure with a layer of sealant in between the wing

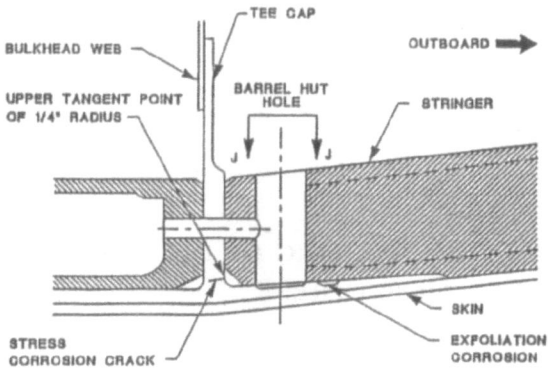

Figure 1. DC-9 wing structure near the lower T-cap.

skin and the tee cap. As a spin-off from earlier work at Northwestern University on a "self-compensating" technique (Achenbach *et al*, 1992) a suitable ultrasonic method was developed by Komsky and Achenbach. In this technique two wedge mounted transducers operating in a pitch-catch configuration as shown in Figure 2, scan a multi-layered structure with inclined-angle beams of transversely polarized ultrasound. With carefully selected wedge angles, transducer spacing and operating frequency, maximum penetration in the second layer can be achieved. Transducer #1 generates ultrasonic signals in both layers. One signal propagates in the bottom layer, and after reflection by the sealant is received by transducer #2. Another signal penetrates through the sealant, and after being reflected either by the top of the second layer or by corrosion spots, is received by transducer #2. Times of flight of both waves can be used for material loss characterization. Even though signals reflected from corrosion spots in deeper layers are of a complicated nature, they can be used to generate images. Optimal selection of time gating for date acquisition is however essential for best image quality.

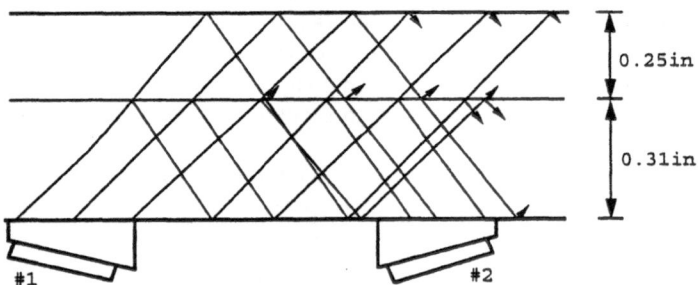

Figure 2. Depiction of sound paths between the probes.

THEORETICAL CONSIDERATIONS

The theoretical model is based on elastodynamic ray theory. Thus we will first consider the reflection and transmission phenomena for a single ray, or equivalently for a plane wave. the analysis will be carried out directly in the time domain. Figure 3 shows some relevant configurations. The transducer which is placed on a perspex

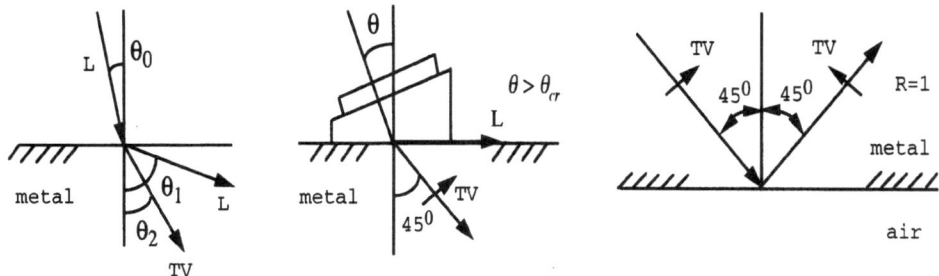

Figure 3. Applications of Snell's law: $(c_L)_w sin\theta_1 = (c_L)_m sin\theta_0$ and $(c_L)_w sin\theta_2 = (c_T)_m sin\theta_0$.

wedge produces a signal of longitudinal wave motion in the wedge, propagating under an angle θ_0 with the normal. The longitudinal pulse in the wedge will produce pulses of longitudinal (L) and vertically polarized transverse (TV) wave motion in the metal. These pulses propagate under angles θ_1 and θ_2 which can be obtained by Snell's law. However the angle θ_0 is chosen to exceed the critical angle θ_{cr}. As a consequence the longitudinal pulse will propagate along the surface of the metal, with an amplitude which decays with depth. Thus, only a pulse of transverse wave motion (TV) will propagate into the metal. In addition θ_0 is chosen such that $\theta_2 = 45^0$. The angle of 45^0 has the advantage that a TV pulse is reflected at a free surface as a TV pulse only with a reflection coefficient, R, equal to unity. This is also shown in Figure 3. The remaining question then is what will happen to a TV pulse when it is incident on a sealant layer under an angle of 45^0. This question will be analyzed in some detail.

Consider a system of incident, reflected and transmitted TV waves relative to the coordinate system shown in Figure 4. At this point we assume that the interface conditions do not give rise to the generation of L waves. It will be shown in the sequel that this is an acceptable assumption provided that the inertia of the sealant layer at the interface may be neglected. In the time domain the *incident* displacement wave is of the form

$$\mathbf{u}^0 = \mathbf{d}^0 f(t - \mathbf{x} \bullet \mathbf{p}^0 / c_T) \tag{1}$$

Here the unit vector $\mathbf{d}^0$ defines the direction of the displacement while the unit vector $\mathbf{p}^0$ defines the propagation direction. The function $f(t)$ defines the propagating pulse and c_T is the transverse wave velocity, $c_T = (\mu/\rho)^{1/2}$, where μ is the shear modulus. The displacement components corresponding to Equation (1) are

$$u_i^{(0)} = d_i^{(0)} f(t - x_i p_i^{(0)} / c_T), \tag{2}$$

and the stress components follow from Hooke's law as

$$\tau_{lm} = \delta_{lm} \lambda u_{j,j} + \mu(u_{l,m} + u_{m,l}) \tag{3}$$

where λ and μ are the Lamé elastic constants.

For an incident transverse wave with both $\mathbf{d}$ and $\mathbf{p}$ in the $x_1 x_2$-plane we have

$$p_1^{(0)} = sin\theta_0, \quad p_2^{(0)} = cos\theta_0, \quad d_1^{(0)} = -cos\theta_0, \quad d_2^{(0)} = sin\theta_0 \tag{4}$$

We then find at $x_2 = 0$

$$u_1^{(0)} = -cos\theta_0 f(t - x_1 sin\theta_0 / c_T), \quad u_2^{(0)} = sin\theta_0 f(t - x_1 sin\theta_0 / c_T) \tag{5a, b}$$

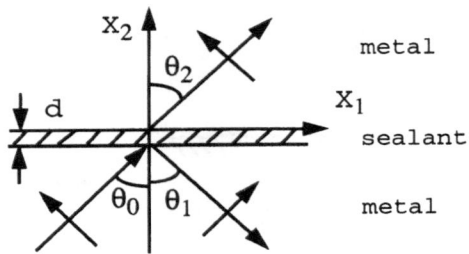

Figure 4. Reflection of TV waves at sealant.

$$\tau_{22}^{(0)} = -\frac{\mu}{c_T}(2\sin\theta_0\cos\theta_0)f'(t - x_1\sin\theta_0/c_T) \qquad (6)$$

$$\tau_{21}^{(0)} = -\frac{\mu}{c_T}(\sin^2\theta_0 - \cos^2\theta_0)f'(t - x_1\sin\theta_0/c_T), \qquad (7)$$

where the prime denotes the derivative with respect to the argument. The reflected transverse wave is defined by

$$u_i^{(1)} = d_i^{(1)}g(t - x_i p_i^{(1)}/c_T) \qquad (8)$$

At $x_2=0$, we have

$$u_1^{(1)} = \cos\theta_1 g(t - x_1\sin\theta_1/c_T), \quad u_2^{(1)} = \sin\theta_1 g(t - x_1\sin\theta_1/c_T) \qquad (9a, b)$$

$$\tau_{22}^{(1)} = \frac{\mu}{c_T}(2\sin\theta_1\cos\theta_1)g'(t - x_1\sin\theta_1/c_T) \qquad (10)$$

$$\tau_{21}^{(1)} = -\frac{\mu}{c_T}(\sin^2\theta_1 - \cos^2\theta_1)g'(t - x_1\sin\theta_1/c_T) \qquad (11)$$

Finally the transmitted transverse wave is

$$u_i^{(2)} = d_i^{(2)}g(t - x_i p_i^{(2)}/c_T) \qquad (12)$$

The corresponding stresses and displacement at $x_2=0$ are

$$u_1^{(2)} = -\cos\theta_2 h(t - x_1\sin\theta_2/c_T), \quad u_2^{(2)} = \sin\theta_2 h(t - x_1\sin\theta_2/c_T) \qquad (13a, b)$$

$$\tau_{22}^{(2)} = -\frac{\mu}{c_T}(2\sin\theta_2\cos\theta_2)h'(t - x_1\sin\theta_2/c_T) \qquad (14)$$

$$\tau_{21}^{(2)} = -\frac{\mu}{c_T}(\sin^2\theta_2 - \cos^2\theta_2)h'(t - x_2\sin\theta_2/c_T) \qquad (15)$$

The sealant is considered as a *very thin* elastic layer of thickness h and elastic constants λ_s, μ_s, whose inertia may be neglected. The absence of inertia implies that the stresses on the sealant due to the incident and reflected waves must be balanced by the stresses corresponding to the transmitted wave:

$$\tau_{22}^{(0)} + \tau_{22}^{(1)} = \tau_{22}^{(2)}, \quad \tau_{21}^{(0)} + \tau_{21}^{(1)} = \tau_{21}^{(2)} \qquad (16a, b)$$

Now let us assume the special case $\theta_0 = \theta_1 = \theta_2 = 45^0$. Equation (16a) then implies

$$-f(t) + g(t) = -h(t) \qquad (17)$$

While (16b) yields

$$\tau_{21}^{(0)} + \tau_{21}^{(1)} = \tau_{21}^{(2)} \equiv 0 \qquad (18)$$

The elastostatic analysis of the thin sealant layer can not be included here due to length limitations. It can, however, be shown that to a first order approximation the relation between τ_{22}, ϵ_{22} and ϵ_{11} yields the following equation

$$h' + \alpha h = \alpha f \tag{19}$$

where

$$\alpha = \frac{\lambda_s + 2\mu_s}{\lambda_s + \mu} \frac{c_T}{d} \frac{1}{sin45^0} \tag{20}$$

Here d is the thickness of the sealant. The solution to Equation (20) is

$$h(t) = \alpha e^{-\alpha t} \int_0^t e^{\alpha s} f(s) ds \tag{21}$$

and g(t) follows from Equation (17).

In summary, a system of TV waves, $f(t), g(t)$ and $h(t)$ propagating under 45^0 in the two metal layers is consistent with the presence of a thin (massless) sealant layer.

The results for g(t), plotted in Figure 5, have been obtained by using 50 transmitted rays with a Gaussian distribution of amplitudes over the beam width. At the receiving transducers the amplitudes of the received rays were again weighted by a Gaussian distribution, and the total signal was normalized by the maximum amplitude of the total signal at the transmitting transducer.

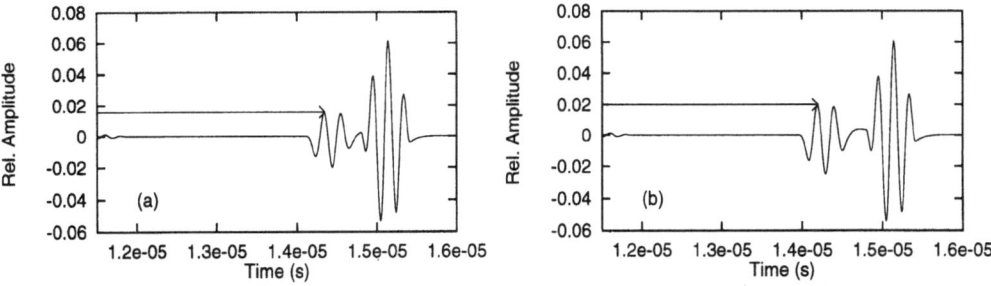

Figure 5. Time trace for received signals for two applications. Transducer spacing: 1.34in. Thicknesses: bottom layer 0.31in; top layer 0.25in for (a) and 0.237in for (b).

RESULTS FROM EXPERIMENTS

Two samples were used for the experiments. The first sample was designed and manufactured by Northwest Airlines. The sample is made from machined plate stock. The two plates are bonded with sealant and clamped together using fasteners. This specimen has flat bottom holes and EDM notches which were intended to simulate corrosion loss and stress corrosion cracks, respectively. Flat bottom holes were machined to represent 20, 10 and 5% material loss. Subsurface EDM notches were cut in the T-cap emanating from the flat bottom holes. The EDM notches were 0.020in wide and ranged from 0.100in to 0.188in in depth and from 0.25in to 0.375in in length. A full set of laboratory tests was carried out by linking the pitch-catch transducer system to an Ultra-Image IV system. The imaging results are shown in Komsky *et al* (1995). All the flat bottom holes were successfully detected including the smallest one of 5% material loss. The EDM notches were also detected. In addition the ultrasonic technique was successfully tested on an actual DC-9 sample shown in Figure 6. The results are shown in Figure 7. Cracks, corrosion, and spot faces were identified.

Further details of the results and comments on the economic benefits from using the technique can be found in the paper by Komsky *et al*(1995).

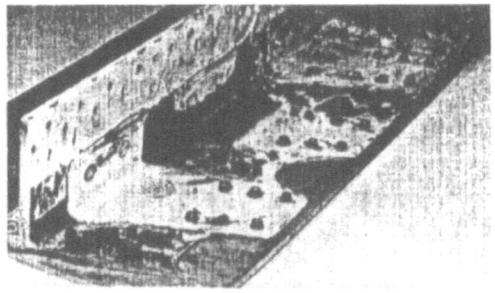

Figure 6. Sample from DC-9 Wingbox.

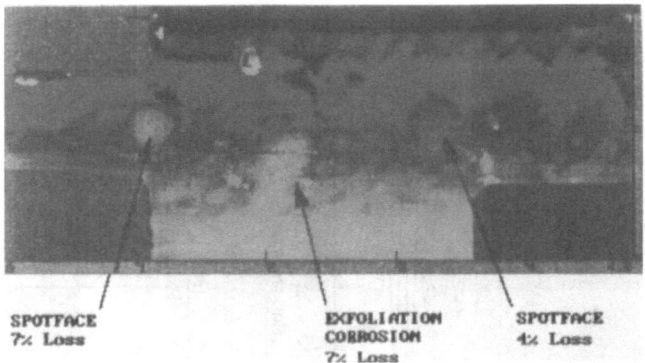

SPOTFACE 7% Loss   EXFOLIATION CORROSION 7% Loss   SPOTFACE 4% Loss

Figure 7. Imaging results showing second-layer thinning from machined spot faces and exfoliation corrosion.

ACKNOWLEDGMENT

This paper is based on work for the Air Force Office of Scientific Research under Contract No. F49620-93-1-0257 and for the FAA Center for Aviation Systems Reliability supported by the Federal Aviation Administration Technical Center, Atlantic City, New Jersey, under Grant number 94G011.

REFERENCES

Jan D. Achenbach, Igor N. Komsky, Y.-C.Lee and Yves C. Angel, "Self-Calibrating Ultrasonic Technique for Crack-Depth Measurement," Journal of Nondestructive Evaluation, Vol. 11, 1992, pp.103-108.

Igor N. Komsky, Jan D. Achenbach, Glenn Andrew, Bob Grills, Jeff Register, Greg Linkert, Gerardo M. Hueto, Al Steinberg, Mike Asbaugh, David G. Moore and Hans Weber, "An Ultrasonic Technique to Detect Corrosion in DC-9 Wing Box–from Concept to Field Application," Materials Evaluation, Vol 53, 1995, pp848-852.

THREE POSSIBLE ULTRASONIC METHODS FOR THE IMAGING OF BONDING FLAWS IN PLASMA-SPRAYED COATINGS

Dominique Lescribaa, Fereydoun Lakestani, Walter J. Vortrefflich, and Jean-François Coste

European Commission, Joint Research Centre
Institute for Advanced Materials, Ultrasonics Development Laboratory, TP 750
21020 Ispra (Italy)

INTRODUCTION

Metallic parts in hot sections of gas turbines require a surface protection in order to prevent their early deterioration by high temperature corrosion. The use of MCrAlY alloys (M=Ni or Co or both) deposited by plasma spraying is the most developed way of providing an efficient protection.

Plasma spraying is a complex elaboration process and gives coatings with a very poor homogeneous microstructure: lamellae (each lamella is the result of a sprayed molten grain), pores, stratifications (due to the torch pass) and cracks within the coating and inside the lamellae (due to the various thermal transitions). Moreover, although the substrate is blasted before spraying in order to create a rough surface allowing an enhancement of the coating adherence, bonding flaws can occur at the substrate-coating interface. Thus, there is a considerable need for non-destructive techniques which can test the integrity of these coatings. Special methods based on the propagation of ultrasonic waves are among the most suitable[1,2] for the quality control of coated components.

However, the interaction of the ultrasonic wave with a plasma coating can provide very heterogeneous responses on successive points, because of the particular microstructure and the roughness of the coating surface. So, the acoustic images built with a parameter supported by the waveform can be very noisy and this noise must be taken into account when discussing the performance of an ultrasonic method.

The aim of the present work is to show three possible ultrasonic methods allowing the bonding flaw detection. These techniques differ on the angle of the incident beam (cylindrical symmetry) and are applied on NiCoCrAlY coatings (200 μm thick) sprayed on a metallic substrate (AISI 316). Artificial bonding flaws have been produced retaining a smooth surface (10x10 mm) on the substrate before spraying. The three ultrasonic methods were tested on these bonding flaws and their results are compared.

EVALUATION OF THE PERFORMANCES

The performances of the various ultrasonic methods used in this work are compared concerning both the detection of the flaw (presence of a contrast) and the resolution of the technique. Because of the unfavorable heterogeneity of the particular material used, the spatial variations of the chosen ultrasonic parameter must be considered, as has already been explained in the introduction.

The performances of each method are evaluated on a 25 mm x line crossing the square defect. The amplitude of the selected parameter is acquired along this line and the noise is evaluated for the sound zone and the defected one. The mean value A of the amplitude and the standard deviation σ are computed for each zone on 60 points (0.1 mm step), away from the flaw border in order to avoid edge effects.

The contrast is due to the amplitude change of the selected parameter, ΔA, between the defected area and the other part of the sample. However, information is clearly given on the image, if the noise level is lower than the previous change. Thus, the contrast is associated to $|\Delta A|/\sigma$.

As small flaws are not available, an indication of the resolution can be given at the flaw border considering the distance d covered by the probe on the x line to obtain the parameter amplitude change from the sound zone to the defected one. A 0.05 mm step has been used on the x line around the flaw border, for the study of the resolution. The standard deviation of the defect width, using the -6 dB drop method, is given by the length $\sqrt{2}\sigma/m$, where m is the slope of the amplitude locus curve when crossing the defect border and is smaller than $\Delta A/d$. Consequently, the length Δx, equal to the ratio of d and the contrast $\Delta A/\sigma$, can be used as a single performance parameter in order to classify the various ultrasonic methods, because it takes into account the contrast and the resolution.

EXPERIMENTAL CONDITIONS

The three ultrasonic methods are applied in echo mode in an immersion tank and broadband focusing probes (highly damped) with large angular apertures are used. These methods are based on the propagation of Rayleigh and Lamb waves for the first technique, on the resonance of compression waves in normal incidence for the second one, and on the resonance of shear waves in angular incidence for the last one. Various masks were superimposed on the usual probes and provide the suitable shape and angle of the beam which are required by each technique. The three configurations are given in Fig.1 with the main nominal characteristics of each probe.

For the first method (see Fig.1a), the probe is made of a piezoelectric composite providing a very large angular aperture (θ=48° in water). It can generate surface waves in most materials even if they offer a relatively low Rayleigh wave velocity, which can be the case for plasma-sprayed coatings[2]. For this study, the Rayleigh angle, θr, is frequency-dependent but stays around 36°. This angle was calculated considering the Rayleigh wave velocity measured elsewhere by a special technique[3] based on the comparison of the experimental and theoretical $\phi(z)$ curves, where ϕ is the phase of the Fourier transform of the echo signal and z the distance between the transducer and the specimen surface. The use of broadband pulses allows separation of the leaky Rayleigh wave signal from the unwanted direct reflection, by defocusing the probe from the specimen surface sufficiently. Dispite this, a 9 mm diameter mask made of a high ultrasonic absorption material is placed at the centre of the probe in order to avoid noises due to the contribution of the end of the central direct reflection. Indeed, this mask suppresses the central beam up to about 34°. However, although the central part of the beam was

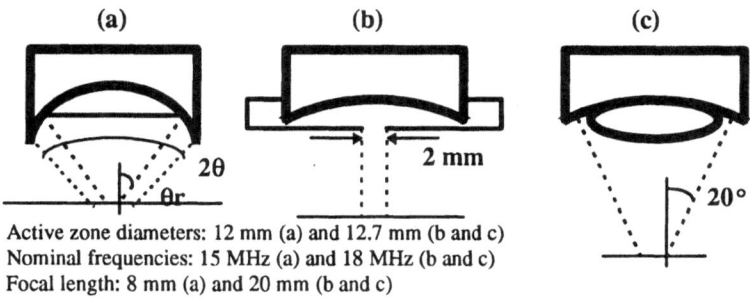

Active zone diameters: 12 mm (a) and 12.7 mm (b and c)
Nominal frequencies: 15 MHz (a) and 18 MHz (b and c)
Focal length: 8 mm (a) and 20 mm (b and c)

Figure 1. Schematic diagram of the three probes with their mask

suppressed, there is specular reflection due to the beam divergence, especially when the transducer/sample distance is near the focal length.

For the second and the third methods (see Fig.1b and Fig.1c), the probe used is made of a PVDF transducer. The normal incidence is obtained by keeping only the centre of the beam thanks to a 2 mm diameter drilled mask. For the third one, the resonance of shear wave in the non-bonded coating is obtained by converting in angular incidence the compression mode in shear mode at the water/coating interface. The central part of the beam was suppressed by an air bubble injected in the cavity of the probe. So, only the external annular part of the beam is kept and provides a 20° incidence angle. This angle was determined on a glass plate (150 µm thick). First, the shear wave velocity was measured normally to the surface with a contact transducer. Then, the incidence angle of the third configuration (Fig.1c) was calculated from the resonance frequency obtained with this configuration under angular incidence and deduced from the velocity in normal incidence. This method can be applied with a glass plate because it is an isotropic material and thus, the velocity is the same for all directions of propagation. Moreover, from the knowledge of the compression and shear wave velocities measured by transmission in a non-bonded (self-standing) NiCoCrAlY coating, this 20° incidence angle is above the first critical angle and corresponds to about a 40° shear wave refraction angle for the studied coating. So, the resonance phenomenon appearing in the non-bonded coating is only due to the propagation of shear waves.

The ultrasonic investigations have been carried out with an immersion bench. The mechanical part enables x,y,z displacements of the probe, driven by stepping motors (10 µm minimum step). The ultrasonic part includes a pulser-receiver and a signal digitizer with a averaging function minimizing the electronic noise effect. A computer drives both the probe displacement and the acquisition of the ultrasonic response. In addition to the amplitude in a given gate, the entire waveform can be acquired either for a point position of the probe (A-scan), or for each position during the scanning of the specimen (B-scan). The acquired information can be transferred via the network to a workstation, for later processing by the PV-WAVE programming language.

RESULTS AND DISCUSSION

All along this study the external noise is minimized in order to keep only the noise due to the material. For example, as already mentioned, a mask avoids the end of the central reflection in method 1, and signals are always averaged in order to minimize the electronic noise contribution. For each method, the ultrasonic parameter acquired on the waveform are explained and performances defined at the beginning of this paper are given in Table 1 at the end of this part. Moreover, for methods 1 and 3, the distance between the transducer and the surface specimen must be set previously. For method 1, the aim is to find the defocus range where the surface wave is the main component of the ultrasonic response, i.e. where there is no interference with the specular reflection[4]. For method 3, the z parameter must be considered because it induces the shape of the waveform directly. On the other hand for method 2, the z parameter is less important, because of the very long focal spot of the almost straight beam. In what follows z=0 is defined when the focal point is on the sample surface and z<0 when it goes under the surface (defocus).

Method 1

The A parameter defined at the beginning is the amplitude of the signal, V. Figure 2 shows this amplitude as a function of z for the sound and defected zones, V_s and V_d, respectively. The $V_s(z)$ and $V_d(z)$ curves result from the interference of the specular reflection with the leaky Rayleigh and Lamb waves, respectively. Indeed, a Lamb mode has been identified on the non-bonded area and is generated for a 40° incidence angle[4].

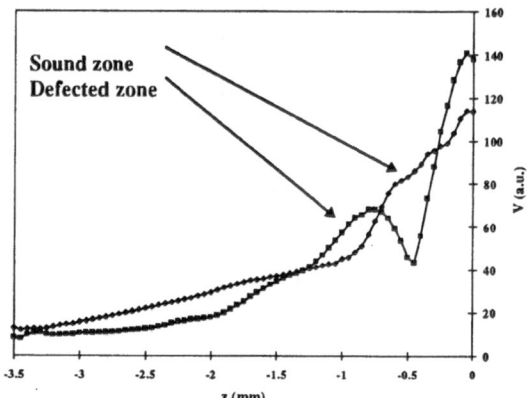

Figure 2. V(z) curves in sound and defected zones

However, by defocusing the probe from the specimen surface sufficiently, the specular reflection no longer affects the ultrasonic response. Because of the almost linear shape of the V(z) curves between z=-2 and z=-3 mm, only the leaky Rayleigh and Lamb waves remain in this range. So, the contrast is due to the difference of amplitude between these two waves. Moreover, the study of the contrast in the all z range (from 0 to -3.5 mm) has shown that the best value is obtained for z=-2.4 mm, in the range of pure Rayleigh and Lamb wave. The study of the resolution in the z range between -2 and -3 mm has shown that the d parameter is about the same for all this range. Performances are given in Table 1 for z=-2.4 mm.

Because of the crossing in several points of the $V_s(z)$ and $V_d(z)$ curves (-0.30, -0.70 and -1.25 mm), three kinds of image are obtained, depending on the V_s-V_d sign. Figure 3 shows C-scans for three interesting z values, -0.90, -1.25 and -2.40 mm, corresponding to $V_s-V_d>0$, $V_s-V_d=0$ and $V_s-V_d<0$, respectively. There is a grey level inversion between the first image and the third one which is associated to a sign change of V_s-V_d. For the second image, there is no contrast as defined in this paper, but the flaw contour clearly appears. In this case, the waveforms obtained on the defect and on the sound zone have the same amplitude, but a destructive interference might explain the amplitude fall for the probe passage above the flaw border.

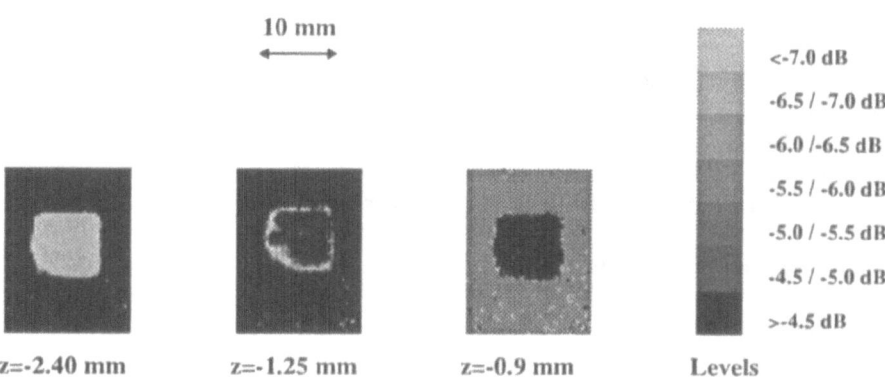

z=-2.40 mm   z=-1.25 mm   z=-0.9 mm   Levels

Figure 3 C-Scan images obtained with method 1 for various z

Method 2

For this method, the contrast results from the compression wave resonance of the non-bonded coating. An example of signals obtained in the sound and defected zones is shown in Fig.4. The z distance is set in order to obtain the highest amplitude (pseudo focus). The presence of the resonance should be detected by acquiring the amplitude voltage in a gate triggered on the main signal component, which corresponds to the specular reflection at the water-coating interface. But the blanking time between the detection of the main component and the beginning of the gate is difficult to set, which can be a source of noise. Indeed, as the resonance amplitude decreases on the signal, the maximum can be placed at the first oscillation and the solution of acquiring the peak to peak in a time gate cannot be applied. Also, another parameter must be taken on the waveform. The frequency spectra of the whole signals given in Fig.4 were computed. Their comparison (see Fig.5) shows an amplitude fall around 13 MHz on the bonding flaw, corresponding to the frequency resonance. This fall is due to an energy loss induced by both the heterogeneities in the coating and the roughness. The fall was detected by acquiring in a frequency gate (10-17 MHz) the difference between the maximum peak and the minimum one. An image was built in these conditions.

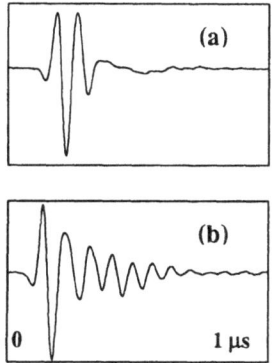
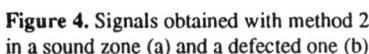

Figure 4. Signals obtained with method 2 in a sound zone (a) and a defected one (b)

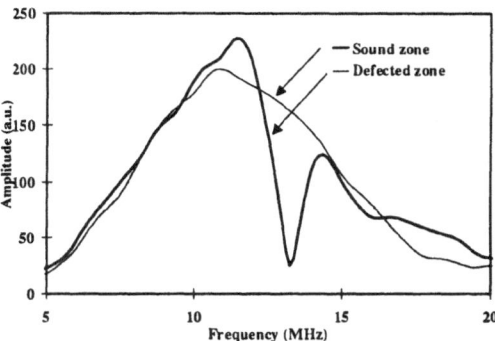

Figure 5. Frequency spectrum amplitude of signals given in Fig. 4

Method 3

The shape of the waveform obtained for this method is very similar to that of method 2. However, the z range where the resonance appears is limited, and z is set at -5 mm. The fall of amplitude in the spectrum is acquired, as in method 2, in the 9-15 MHz gate. The shear frequency resonance (12.7 MHz) is close to that obtained in normal compression mode and can be explained by taking into account the 10 % anisotropy found in these coatings[7].

Result Synthesis and Discussion

Performances of each method are given in Table 1 with the wavelengths in the water and in the coating. Concerning the noise, the surface roughness effect is sensitive to the wavelength in water and the part of the noise due to the microstructure heterogeneity is linked to the wavelength in the coating. Thus, the lowest noise is obtained for method 1, because of the higher wavelengths. On the other hand, method 3 has the shortest wavelengths inducing the worst noise. However, a theoretical study must be performed in order to explain more quantitatively the effect of the interaction of each wave with the roughness and the heterogeneities of the coating. Contrast values are directly dependent on the noise, because the relative amplitude change $\Delta A/A$ is about the same for each technique (between 50 and 60 %). The resolutions are dependent on the beam diameters, and the best results are obtained for methods 1 and 2. However, the Δx parameter, proportional to the precision of the flaw size, is 3 times better for method 1.

Table 1. Synthesis of the results

	σ/A	$|\Delta A|/\sigma$	d (mm)	Δx (mm)	λ (mm) in water	λ (mm) in coating
Method 1	0.042	15.1	0.45	0.03	0.10	0.41
Method 2	0.171	6.3	0.55	0.09	0.08	0.39
Method 3	0.269	4.2	0.70	0.17	0.08	0.27

CONCLUSION

Three ultrasonic methods were used for the Non-Destructive Testing of metallic plasma-sprayed coatings. Artificial bonding flaws allowed the setting of these methods and the determination of their performances, which were defined by taking the noise into account. Results showed that the method involving the Rayleigh waves gives three kinds of image depending on the defocus. Moreover, the best performance is obtained with this method. Bonding flaws smaller than 1 mm would be detected.

ACKNOWLEDGMENT: Much of this work has been performed in the framework of a contract with ENEL Spa.

REFERENCES

1. J.M.Miller, A.Dodgson and R.Granville, "Applications of high frequency ultrasonic imaging" Proc. 4th European Conference on Non-Destructive Testing (1987)
2. D. LESCRIBAA and A. VINCENT "Ultrasonic characterization of plasma sprayed coatings" To be published in Surface and Coatings Technology
3. J.F.Coste and F.Lakestani "Description of a method for the measurement of the velocity of Rayleigh waves: application to the thickness measurement of metallic coatings" 1994 IEEE Ultrasonics Symposium To be published
4. D.LESCRIBAA, W.J.VORTREFFLICHTand F. LAKESTANI "Ultrasonic testing of plasma sprayed coatings with artificial bonding flaws using focused Rayleigh waves" Proceedings of the First Joint Belgian-Hellenic Conference on Non Destructive Testing, Patras, Greece (1995)

NONDESTRUCTIVE TESTING WITH ULTRASOUND:
NUMERICAL MODELING AND IMAGING

R. Marklein, K. J. Langenberg, R. Bärmann, K. Mayer, and S. Klaholz

Department of Electrical Engineering
University of Kassel
D-34109 Kassel, Germany

INTRODUCTION

Our aim of this work is to present the capabilities of combining numerical modeling and imaging in nondestructive testing (NDT) with ultrasound. In order to model NDT situations in the time domain we need a powerful numerical tool. Well-known numerical methods are the Elastodynamic Finite Difference Method (EFDM) (Alterman, 1968, Bond et al., 1988), the Velocity-Stress Finite Difference Method (VS-FDM) (Madariaga, 1976, Virieux, 1986), the Elastodynamic Finite Element Method (EFEM) (Ludwig & Lord, 1986), and the Elastodynamic Finite Integration Technique (EFIT) (Fellinger, 1991). All these methods have been applied to various problems in geophysics and nondestructive testing. In this paper we are concerned about the application of numerical modeling and imaging to the NDT of concrete which is presently a major research subject in civil engineering (NDT-CE) (Marklein et al., 1996). Currently, the aims of the application of numerical time domain tools are threefold: (1) to get a better physical understanding of the wave propagation and scattering, (2) to interprete or analyse measurements, and (3) to generate synthetic data as a testbed for imaging algorithms. The ansatz point of imaging algorithms to solve the inverse problem are twofold: (1) a full nonlinear ansatz, and (2) with the introduction of approximations such as linearizations like Born or Kirchhoff approximation in order to simplify the problem. Here, we are interested in the latter in an inversion scheme with Kirchhoff approximation for perfect stress-free scatterers. Especially, we developed an extended version of the scalar FT-SAFT (Mayer et al., 1990) the vector imaging scheme called EL-FT-SAFT for a planar measurement surface (Bärmann & Langenberg, 1995) which takes advantage of the vector character of elastodynamic waves.

NUMERICAL MODELING OF ULTRASONIC WAVES WITH EFIT

In linear elastodynamics ultrasonic waves (elastodynamic waves) are governed by Cauchy's equation of motion and the equation of deformation rate (Auld, 1973, van der Hijden, 1987). These equations read in differential form

$$\frac{\partial}{\partial t}\underline{p}(\underline{R}, t) = \nabla \cdot \underline{\underline{T}}(\underline{R}, t) + \underline{f}(\underline{R}, t),$$ (1)

$$\frac{\partial}{\partial t}\underline{\underline{\mathbf{S}}}(\mathbf{R},t) = \text{sym}\{\boldsymbol{\nabla}\,\underline{\mathbf{v}}(\mathbf{R},t)\} + \underline{\underline{\mathbf{h}}}(\mathbf{R},t)\,, \tag{2}$$

and in integral form for a finite volume V with the surface S

$$\iiint_V \frac{\partial}{\partial t}\underline{\mathbf{p}}(\mathbf{R},t)\,\mathrm{d}V = \oiint_S \underline{\mathbf{n}}\cdot\underline{\underline{\mathbf{T}}}(\mathbf{R},t)\,\mathrm{d}S + \iiint_V \underline{\mathbf{f}}(\mathbf{R},t)\,\mathrm{d}V\,, \tag{3}$$

$$\iiint_V \frac{\partial}{\partial t}\underline{\underline{\mathbf{S}}}(\mathbf{R},t)\,\mathrm{d}V = \oiint_S \text{sym}\{\underline{\mathbf{n}}\,\underline{\mathbf{v}}(\mathbf{R},t)\}\,\mathrm{d}S + \iiint_V \underline{\underline{\mathbf{h}}}(\mathbf{R},t)\,\mathrm{d}V\,; \tag{4}$$

$\underline{\mathbf{p}}$ is the momentum density vector, $\underline{\underline{\mathbf{T}}}$ is the stress second rank tensor, $\underline{\underline{\mathbf{S}}}$ is the strain second rank tensor, $\underline{\mathbf{v}}$ is the particle velocity vector which is the first time derivative of the displacement vector $\underline{\mathbf{u}}$, $\underline{\mathbf{f}}$ is the source of force density, $\underline{\underline{\mathbf{h}}}$ is the source of deformation rate second rank tensor, $\underline{\mathbf{n}}$ is the outward normal unit vector of S, and sym$\{\underline{\mathbf{n}}\,\underline{\mathbf{v}}\}$ denotes the symmetric part of the dyad $\underline{\mathbf{n}}\,\underline{\mathbf{v}}$. The material properties of a non-dissipative solid are given by the following constitutive equations

$$\underline{\mathbf{p}}(\mathbf{R},t) = \varrho_0(\underline{\mathbf{R}})\underline{\mathbf{v}}(\mathbf{R},t)\,, \qquad \underline{\underline{\mathbf{S}}}(\mathbf{R},t) = \underline{\underline{\mathbf{s}}}(\underline{\mathbf{R}}) : \underline{\underline{\mathbf{T}}}(\mathbf{R},t)\,; \tag{5}$$

ϱ_0 is the volume density of mass at rest and $\underline{\underline{\mathbf{s}}}$ is the compliance tensor of fourth rank. For the isotropic case the compliance tensor reads

$$\underline{\underline{\mathbf{s}}}^{\text{iso}}(\underline{\mathbf{R}}) = \Lambda(\underline{\mathbf{R}})\underline{\underline{\mathbf{I}}\underline{\mathbf{I}}} + M(\underline{\mathbf{R}})\left(\underline{\underline{\mathbf{I}}\underline{\mathbf{I}}}^{1324} + \underline{\underline{\mathbf{I}}\underline{\mathbf{I}}}^{1342}\right)\,, \tag{6}$$

$$\text{with} \quad \Lambda(\underline{\mathbf{R}}) = \frac{\lambda(\underline{\mathbf{R}})}{2\mu(\underline{\mathbf{R}})\left[3\lambda(\underline{\mathbf{R}}) + 2\mu(\underline{\mathbf{R}})\right]}\,, \qquad M(\underline{\mathbf{R}}) = \frac{1}{4\mu(\underline{\mathbf{R}})}\,; \tag{7}$$

λ, μ are Lamé's constants and $\underline{\underline{\mathbf{I}}}$ is the unit dyadic or the idemfactor. Then, the inverse compliance tensor defines the stiffness tensor

$$\underline{\underline{\mathbf{c}}}^{\text{iso}}(\underline{\mathbf{R}}) = \lambda(\underline{\mathbf{R}})\underline{\underline{\mathbf{I}}\underline{\mathbf{I}}} + \mu(\underline{\mathbf{R}})\left(\underline{\underline{\mathbf{I}}\underline{\mathbf{I}}}^{1324} + \underline{\underline{\mathbf{I}}\underline{\mathbf{I}}}^{1342}\right)\,. \tag{8}$$

For homogeneous materials we rewrite equ. (2) with the stiffness tensor in the form

$$\frac{\partial}{\partial t}\underline{\underline{\mathbf{T}}}(\mathbf{R},t) = \lambda\underline{\underline{\mathbf{I}}}\,\boldsymbol{\nabla}\cdot\underline{\mathbf{v}}(\mathbf{R},t) + 2\mu\,\text{sym}\{\boldsymbol{\nabla}\underline{\mathbf{v}}(\mathbf{R},t)\} + \underline{\underline{\mathbf{g}}}(\mathbf{R},t)\,. \tag{9}$$

Substitution of equ. (9) into equ. (1) yields with $\underline{\mathbf{f}}(\mathbf{R},t) = \underline{\mathbf{0}}$ and $\underline{\underline{\mathbf{g}}}(\mathbf{R},t) = \underline{\underline{\mathbf{0}}}$ the Navier equation for the particle displacement vector

$$(\lambda + \mu)\,\boldsymbol{\nabla}\,\boldsymbol{\nabla}\cdot\underline{\mathbf{u}}(\mathbf{R},t) + \mu\Delta\underline{\mathbf{u}}(\mathbf{R},t) - \varrho_0\frac{\partial^2}{\partial t^2}\underline{\mathbf{u}}(\mathbf{R},t) = 0\,. \tag{10}$$

For isotropic materials the elastodynamic wavefield consists of pressure and shear waves denoted by P and S. Plane P- and S-waves are solutions of the wave equations

$$\Delta\Phi(\underline{\mathbf{R}},t) - \frac{1}{c_{\text{P}}^2}\frac{\partial^2}{\partial t^2}\Phi(\mathbf{R},t) = 0\,, \qquad \Delta\underline{\boldsymbol{\Psi}}(\underline{\mathbf{R}},t) - \frac{1}{c_{\text{S}}^2}\frac{\partial^2}{\partial t^2}\underline{\boldsymbol{\Psi}}(\mathbf{R},t) = \underline{\mathbf{0}}\,, \tag{11}$$

in which c_P and c_S are the phase velocities of the P-wave and S-wave, respectively

$$c_\text{P} = \sqrt{\frac{\lambda + 2\mu}{\varrho_0}}\,, \qquad c_\text{S} = \sqrt{\frac{\mu}{\varrho_0}}\,. \tag{12}$$

This representation is due to the introduction of a scalar potential $\Phi(\mathbf{R},t)$ and a vector potential $\underline{\boldsymbol{\Psi}}(\mathbf{R},t)$ which yield the Helmholtz decomposition of $\underline{\mathbf{u}}(\mathbf{R},t)$ according to

$$\underline{\mathbf{u}}(\mathbf{R},t) = \boldsymbol{\nabla}\Phi(\mathbf{R},t) + \boldsymbol{\nabla}\times\underline{\boldsymbol{\Psi}}(\mathbf{R},t)\,, \qquad \text{with} \qquad \boldsymbol{\nabla}\cdot\underline{\boldsymbol{\Psi}}(\mathbf{R},t)\,. \tag{13}$$

EFIT: Elastodynamic Finite Integration Technique

In order to derive EFIT we apply the Finite Integration Technique to equ.'s (3) and (4). We evaluate the volume and surface integrals approximately for each cubic discretization volume $V = \Delta x^3$ assuming constant $\underline{v}$ and $\underline{\underline{T}}$ within V and on each of the six quadratic surfaces $S = \Delta x^2$ of V. Obviously, this requires staggered grids, which are centered around the cartesian components v_1, v_2, v_3 of $\underline{v}$, and around the main and off-diagonal elements $T_{11}, T_{22}, T_{33}, T_{12}, T_{13}, T_{23}$ of $\underline{\underline{T}}$.

In the case of inhomogeneous materials a discretization of the material parameter distribution has to be introduced ensuring continuity of $\underline{n} \cdot \underline{\underline{T}}$ and $\underline{v}$ on its grid surfaces. A more detailed description can be found in a recent publication by Fellinger et al. (1995). For isotropic inhomogeneous materials the following discrete EFIT equations for v_1, T_{11}, and T_{12} components are obtained, where M_i, $i = 1, 2, 3$ denote the node number distance in x_i-direction of the node adjacent to node n:

$$\dot{v}_1^{(n)}(t) = \frac{1}{\Delta x} \frac{2}{\varrho_0^{(n)} + \varrho_0^{(n+M_1)}}$$

$$\times \left[T_{11}^{(n+M_1)}(t) - T_{11}^{(n)}(t) + T_{12}^{(n)}(t) - T_{12}^{(n-M_2)}(t) + T_{13}^{(n)}(t) - T_{13}^{(n-M_3)}(t) \right], \tag{14}$$

$$\dot{T}_{11}^{(n)}(t) = \frac{1}{\Delta x} \left\{ (\lambda^{(n)} + 2\mu^{(n)}) \left[v_1^{(n)}(t) - v_1^{(n-M_1)}(t) \right] + \right.$$

$$\left. + \lambda^{(n)} \left[v_2^{(n)}(t) - v_2^{(n-M_2)}(t) + v_3^{(n)}(t) - v_3^{(n-M_3)}(t) \right] \right\}, \tag{15}$$

$$\dot{T}_{12}^{(n)}(t) = \frac{1}{\Delta x} \frac{4}{\dfrac{1}{\mu^{(n)}} + \dfrac{1}{\mu^{(n+M_1)}} + \dfrac{1}{\mu^{(n+M_2)}} + \dfrac{1}{\mu^{(n+M_1+M_2)}}}$$

$$\times \left[v_1^{(n+M_2)}(t) - v_1^{(n)}(t) + v_2^{(n+M_1)}(t) - v_2^{(n)}(t) \right]. \tag{16}$$

The complete set of discrete equations is given by Fellinger et al. (1995).

Of course, if the medium is inhomogeneously anisotropic, the same ideas as above apply (Marklein et al., 1995).

The time derivative in (14)-(16) is approximated by central differences according to

$$\{v\}_i^{(z)} = \{v\}_i^{(z-1)} + \{\dot{v}\}_i^{(z-1/2)} \Delta t \tag{17}$$

$$\{T\}_{ij}^{(z+1/2)} = \{T\}_{ij}^{(z-1/2)} + \{\dot{T}\}_{ij}^{(z)} \Delta t, \quad i, j = 1, 2, 3 \tag{18}$$

yielding a leap-frog integration scheme. In (17) and (18), $\{v\}_i^{(z)}$ denotes the vector composed of particle velocity components v_i for all nodes n for a certain time $t = z\Delta t$ with integer z; $\{T\}_{ij}^{(z)}$ is similarly defined.

EFIT Modeling of Ultrasound in Concrete

In this section we present the numerical modeling of ultrasound in concrete utilizing the Elastodynamic Finite Integration Technique (EFIT). First applications of such numerical modeling of ultrasonic waves in concrete and an imaging scheme EL-FT-SAFT has been presented by Marklein et al. (1995). Due to the additives and air inclusions which may occur the material concrete is very inhomogeneous. Normally concrete consists of cement as the base material and several additives like basalt, plaster, and biatitgranite. The max. aggregate size, grading curve, and water/air concentration can vary according to circumstances. We modeled two different situations: concrete without air and concrete with 6% air, that is, we used the concrete models given in

Fig. 1 with the base material cement and the additives 25% basalt, 25% plaster, and 25% biatitgranite with a max. aggregate size of 8mm and grading curve B8. In the case with air inclusions we have a model of ca. 175,000 ellipses which are varying randomly in size, orientation and given additive (see Fig. 1 at the right). Every material cell of the EFIT grid can cover different material parameters. In praxis, the total number of the ellipses is only limited by the number of material cells. In the two examples, we have a mesh size of $2,000 \times 2,000 = 4 \times 10^6$ material cells with a cell width of $\Delta x = 250 \mu m$. The right side of Fig. 1 clearly shows the statistical distribution of the ellipses. We used

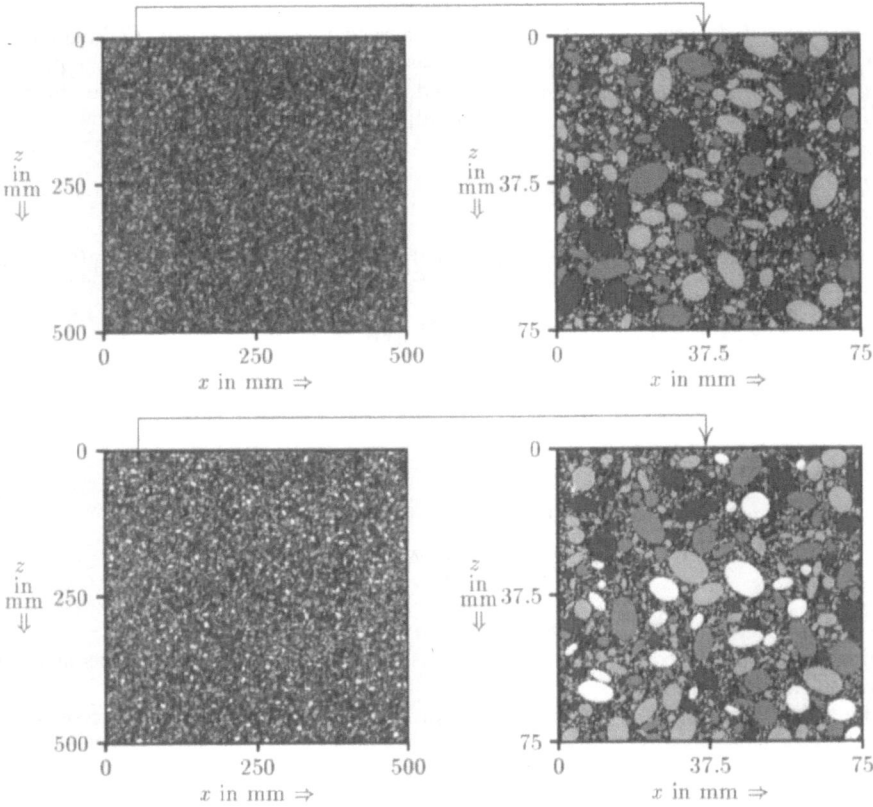

Figure 1. Left: concrete models of the size 50cm×50cm for EFIT modeling; right: detail drawings of the size 7.5cm×7.5cm; top: concrete without air; bottom: concrete with 6% air inclusions (white ellipses).

a 80 kHz normal pressure probe which has a diameter of $D=5$cm applied in pulse-echo technique. The time history of the probe is modeled by a raised cosine with two cycles called RC2. For EFIT modeling we presume a stress-free boundary condition on the top and bottom surface and an open boundary condition (Higdon) at the left and right boundary in order to model an infinite concrete sample. EFIT-$|\underline{v}|$-snapshots of the elastic wavefield for the two examples are displayed in Fig. 2. We recognize in Fig. 2.1 and Fig. 2.5 at t_1 the near-field of the excited wavefield. Fig. 2.2 at t_2 shows a prominent pressure wave pursued by a shear wave, and also we identify head waves, and of course Rayleigh waves traveling at the top surface. In the case with air inclusions (Fig. 2.6-2.8) the wavefield is already totally distorted, this is due to the effects of multiple reflection and mode conversion on the air inclusions. With an applied open boundary condition reflections and mode conversions are suppressed at the vertical boundaries. The backwall echo of the pressure wave shows up very clearly only in Fig. 2.3 at t_3, which

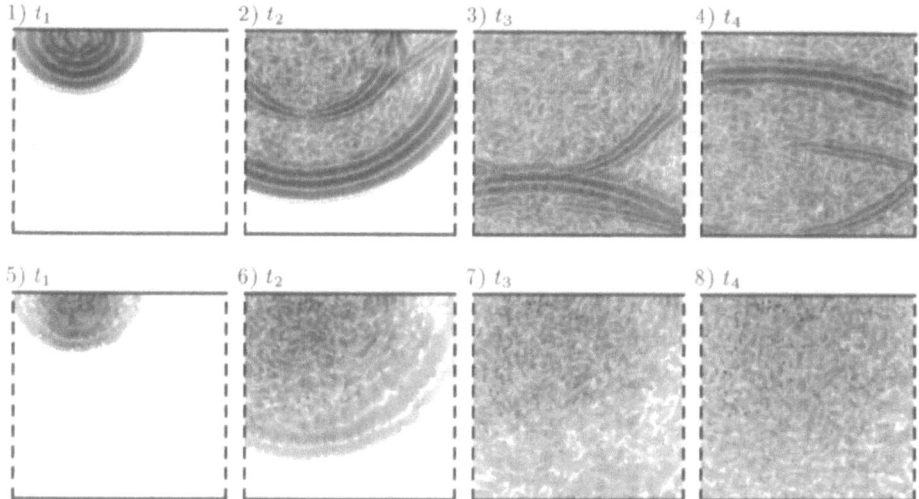

Figure 2. EFIT-$|\underline{v}|$-snapshots of the elastic wavefield in concrete; cement with biatitgranite, basalt, and plaster (max. aggregate size 8mm, grading curve B8); top (2.1-2.4): without air; bottom (2.5-2.8): with 6% air inclusions (black dots).

then propagates back to the top surface, and it is finally recorded by the probe (see Fig. 2.4). Only in concrete without air inclusions the traveling time of the backwall echo can then be further analysed, e.g. for thickness determination.

ELASTODYNAMIC VECTOR IMAGING: EL-FT-SAFT

For inverse scattering in elastodynamics we formulated an elastic vector imaging algorithm which we call ELastic Fourier Transform Synthetic Aperture Focusing Technique (EL-FT-SAFT). It is based on the linear elastic far-field inversion including a near-field far-field transformation. The formulation of the diffraction imaging algorithm is based on Huygens' principle. This reads in linear elastodynamics for a volume V with the closed surface S

$$
\oiint_S \underline{n}' \cdot \left[\underline{\underline{T}}(\underline{R}',\omega) \cdot \underline{\underline{G}}(\underline{R} - \underline{R}',\omega) - \underline{u}(\underline{R}',\omega) \cdot \underline{\underline{\Sigma}}'(\underline{R} - \underline{R}',\omega) \right] d\,S'
$$

$$
+ \iiint_V \underline{f}(\underline{R}',\omega) \cdot \underline{\underline{G}}(\underline{R} - \underline{R}',\omega) \, d\,V' =
\begin{cases}
\underline{u}(\underline{R},\omega) & \underline{R} \in V \setminus S \\
\frac{1}{2}\underline{u}(\underline{R},\omega) & \underline{R} \in S \\
\underline{0} & \underline{R} \in \mathbb{R}^3 \setminus V
\end{cases}
\quad ; \qquad (19)
$$

$\underline{n}'$ is the outward normal unit vector of S, $\underline{\underline{\Sigma}}$ is the triadic and $\underline{\underline{G}}$ the dyadic displacement Green's function of free space (Fellinger, 1991). Application of the equivalence principle leads to a representation of the scattered field on the surface S. Therefore, the scattered field in terms of the particle velocity vector is given by

$$
\underline{u}^{\mathrm{sca}}(\underline{R},\omega) =
$$

$$
\oiint_S \left\{ \underline{u}(\underline{R}',\omega) \cdot \left[\underline{n}' \cdot \underline{\underline{\Sigma}}'(\underline{R} - \underline{R}',\omega) \right] - \underline{n}' \cdot \underline{\underline{T}}(\underline{R}',\omega) \cdot \underline{\underline{G}}(\underline{R} - \underline{R}',\omega) \right\} d\,S' . \quad (20)
$$

Introducing the far-field approximation in (20) we get the definition of the linearized far-field scattering amplitude $\underline{\underline{C}}^{\mathrm{sca}}_{\alpha\beta}(\widehat{\underline{R}},\omega)$ for perfect stress-free ($\underline{n} \cdot \underline{\underline{T}} = \underline{0}$) scatterers. Then, in linear physical elastodynamics (PE) the illuminated part of the singular function of a perfect stress-free scatterer for the bistatic case is given by the elastodynamic far-field algorithm in the K-space (Bärmann & Langenberg, 1995).

$$\widetilde{\gamma}_u^{\mathrm{E}}(\underline{\mathbf{K}}) = \frac{1}{|\underline{\mathbf{V}}_{\alpha\beta}^{\mathrm{PE,bi}}(\widehat{\mathbf{R}}, \widehat{\mathbf{k}}_i, \omega)|^2} \underline{\mathbf{V}}_{\alpha\beta}^{\mathrm{PE,bi}}(\widehat{\mathbf{R}}, \widehat{\mathbf{k}}_i, \omega) \cdot \underline{\mathbf{C}}_{\alpha\beta}^{\mathrm{sca}}(\widehat{\mathbf{R}}, \omega), \qquad \alpha, \beta = \mathrm{P, S}, \qquad (21)$$

with

$$\underline{\mathbf{V}}_{\alpha\beta}^{\mathrm{PE,bi}}(\widehat{\mathbf{R}}, \widehat{\mathbf{k}}_i, \omega) = \frac{1}{2\pi} R_f \, \underline{\mathbf{u}}_{0\alpha}(\widehat{\mathbf{k}}_i, \omega) \, \underline{\mathbf{n}}_{\alpha\beta}(\widehat{\mathbf{R}}, \widehat{\mathbf{k}}_i) : \underline{\underline{\mathbf{T}}}_\beta^{\mathrm{far}}(\widehat{\mathbf{R}}, \omega); \qquad (22)$$

α and β denote the excited and calculated mode, R_f is the reflection coefficient, $\underline{\mathbf{u}}_{0\alpha}$ is the vector frequency spectrum of the transmitted signal, $\underline{\mathbf{n}}_{\alpha\beta}$ is the stationary phase normal, and

$$\underline{\underline{\mathbf{T}}}_{\mathrm{P}}^{\mathrm{far}}(\widehat{\mathbf{R}}, \omega) = -\frac{j\,k_P^3}{\varrho_0\omega^2}\left(\lambda\underline{\mathbf{I}}\,\widehat{\mathbf{R}} + 2\mu\widehat{\mathbf{R}}\,\widehat{\mathbf{R}}\,\widehat{\mathbf{R}}\right), \qquad (23)$$

$$\underline{\underline{\mathbf{T}}}_{\mathrm{S}}^{\mathrm{far}}(\widehat{\mathbf{R}}, \omega) = -\frac{j\,k_S^3}{\varrho_0\omega^2}\mu\left(\widehat{\mathbf{R}}\,\underline{\mathbf{I}} + \widehat{\mathbf{R}}\,\underline{\mathbf{I}}^{213} - 2\widehat{\mathbf{R}}\,\widehat{\mathbf{R}}\,\widehat{\mathbf{R}}\right) \qquad (24)$$

are the triadic prefactors of the triadic far-field Green functions

$$\underline{\underline{\boldsymbol{\Sigma}}}_{\mathrm{P}}^{\mathrm{far}}(\widehat{\mathbf{R}}, \omega) = \underline{\underline{\mathbf{T}}}_{\mathrm{P}}^{\mathrm{far}}(\widehat{\mathbf{R}}, \omega)G_{\mathrm{P}}^{\mathrm{far}}(\widehat{\mathbf{R}}, \omega), \qquad \underline{\underline{\boldsymbol{\Sigma}}}_{\mathrm{S}}^{\mathrm{far}}(\widehat{\mathbf{R}}, \omega) = \underline{\underline{\mathbf{T}}}_{\mathrm{S}}^{\mathrm{far}}(\widehat{\mathbf{R}}, \omega)G_{\mathrm{S}}^{\mathrm{far}}(\widehat{\mathbf{R}}, \omega); \qquad (25)$$

$G_{\mathrm{P}}^{\mathrm{far}}(\widehat{\mathbf{R}}, \omega)$, $G_{\mathrm{S}}^{\mathrm{far}}(\widehat{\mathbf{R}}, \omega)$ are the scalar far-field Green functions. $\gamma_u^{\mathrm{E}}(\underline{\mathbf{R}})$ is then calculated by applying a threedimensional inverse Fourier transform with respect to $\underline{\mathbf{K}}$. Equation (21) represents the essence of the elastic vector imaging algorithm EL-FT-SAFT.

Detection of Delaminations in a Metal Duct of Reinforced Concrete

In the present case we consider the following question: "Is it possible to detect delaminations in a metal duct which is filled with cement"? We purposely chose the geometry of Fig. 3, it shows a concrete sample with a metal duct filled with cement and with three delaminations. The EFIT modeled wavefield is given in Fig. 4, on the left

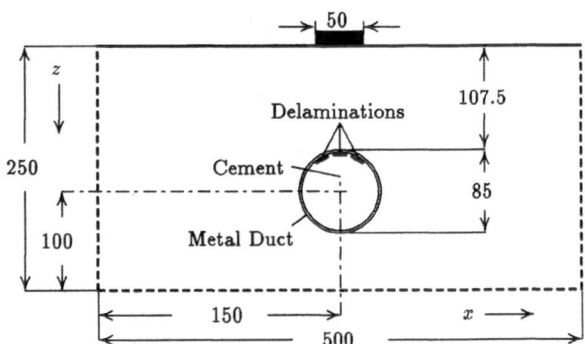

Figure 3. Geometry of reinforced concrete with a metal duct with three delaminations.

side without delaminations and on the right side with delaminations. In the latter case the reflected pressure wave from the metal duct (delaminations) has a higher amplitude. For each situation we „recorded" an rf-data field within a synthetic aperture (indicated by the black bar in Fig. 3). Then we applied the EL-FT-SAFT imaging scheme to the pressure and shear wave part in order to get a P- and S-image separately (see Fig. 5).

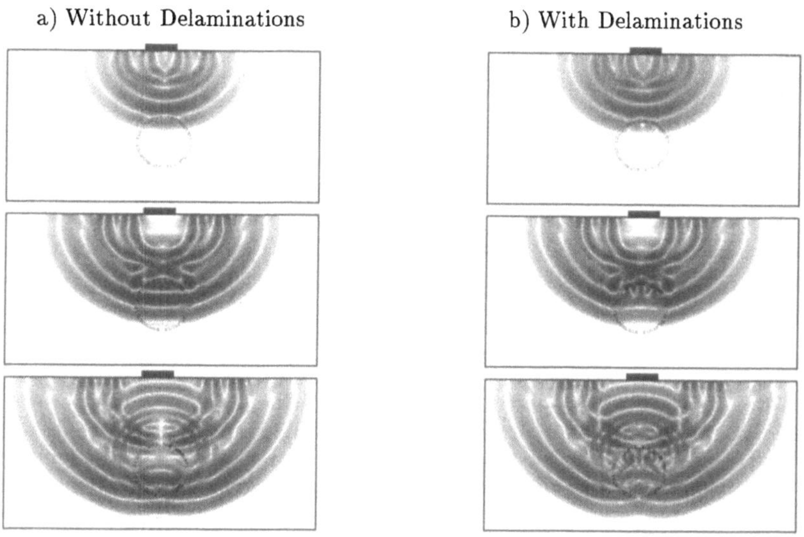

a) Without Delaminations   b) With Delaminations

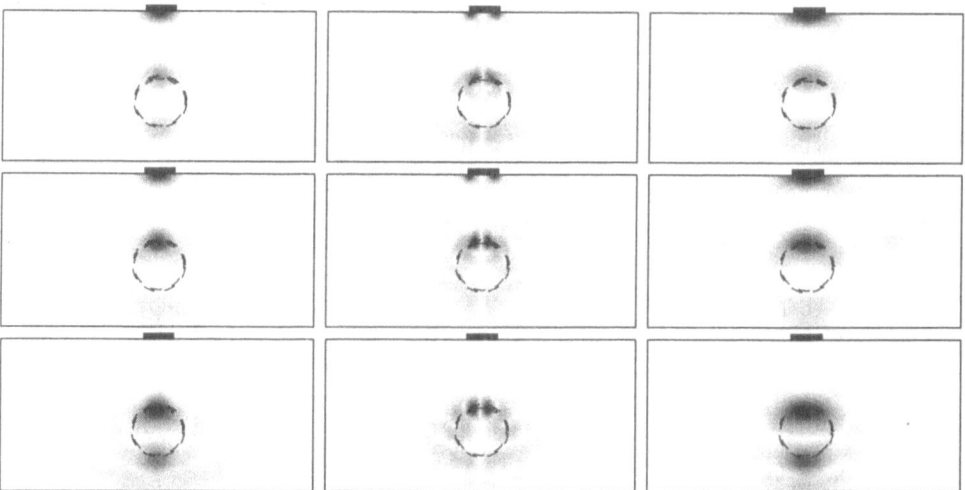

Figure 4. EFIT-$|\underline{v}|$-snapshots of the elastic wavefield in concrete with metal duct; a) without delaminations and b) with delaminations; the probe is indicated by the black bar.

Figure 5. EL-FT-SAFT images: from top to bottom: (1) without delaminations, (2) with delaminations, scattered field (2) minus (1).

If we compare all three columns of images in Fig. 5 we clearly recognize the improvement in the superimposed PS-image. Nevertheless, it is not possible to image the three delaminations itself. But we get an indication of the delaminations in the higher amplitude and a shift of the maximum in z-direction of the reconstructed upper boundary of the metal duct where the delaminations are. The lower boundary of the metal duct is missed if the delaminations are present (see Fig. 6).

CONCLUSIONS

We have presented the numerical modeling of ultrasound in concrete with EFIT. We modeled several NDT situations: (1) the ultrasonic wave propagation and scatter-

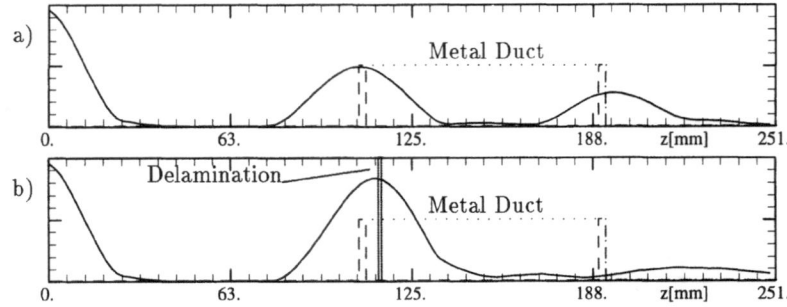

Figure 6. P-image cross sections at $x=0$ with superimposed geometry of the metal duct (dashed line) and delamination (solid line): a) without and b) with delaminations.

ing in concrete with and without air inclusions and (2) the ultrasonic wave scattering by delaminations in a metal duct. In the latter example we recorded modeled A-scans within a synthetic aperture and applied the elastodynamic imaging scheme EL-FT-SAFT taking into account the vector character of elastodynamic waves. We detected delaminations in a metal duct applying ultrasound as an example. These results proves the capabilities of combining numerical modeling and vector imaging in elastodynamics.

REFERENCES

Alterman, Z. S., 1968, "Finite Difference Solutions to Geophysical Problems," *J. Phys. Earth* 16.

Auld, B. A., 1990, "Acoustic Fields and Waves in Solids," Krieger, Malabar.

Bärmann, R., Langenberg, K. J., *in*: "Advances in Acoustic Microscopy 2," G. A. D. Briggs, W. Arnold, eds., Oxford University Press, Oxford.

Bond, L. J., Punjani, M., Saffari, N., 1988, Ultrasonic Wave Propagation and Scattering Using Explicit Finite Difference Methods, *in*: M. Blakemore, G. A. Georgiou, eds., "Mathematical Modelling in Non-Destructive Testing," Clarendon Press, Oxford 81.

Fellinger, P., 1991 Ein Verfahren zur numerischen Lösung elastischer Wellenausbreitungsprobleme im Zeitbereich durch direkte Diskretisierung der elastodynamischen Grundgleichungen, Ph. D. Thesis, University of Kassel, Kassel.

Fellinger, P., Marklein, R., Langenberg, K. J., Klaholz, S., 1995, "Numerical Modeling of Elastic Wave Propagation and Scattering with EFIT — Elastodynamic Finite Integration Technique," *Wave Motion* 21:47.

van der Hijden, J. H. M. T., 1987, "Propagation of Transient Elastic Waves in Stratified Anisotropic Media," North-Holland, Amsterdam.

Langenberg, K. J., Fellinger, P., Marklein, R., Zanger, P., Mayer, K., Kreutter, T., 1993, "Inverse Methods and Imaging," *in*: Evaluation of Materials and Structures by Quantitative Ultrasonics, J. D. Achenbach, ed., Springer-Verlag, Vienna 317

Ludwig, R., Lord, W., 1988, "A Finite-Element Formulation for the Study of Ultrasonic NDT Systems," *IEEE Trans. Ultrasonics, Ferroelectrics, and Frequency Control* 35:809.

Madariaga, R., 1976, "Dynamic of an Expanding Circular Fault," *Bull. Seis. Soc. of Am.* 66:639.

Marklein, R., Bärmann, R., Langenberg, K. J., 1995, The Ultrasonic Modeling Code EFIT as Applied to Inhomogeneous Dissipative Isotropic and Anisotropic Media, *in*: "Review of Progress in Quantitative Nondestructive Evaluation," D. O. Thompson, D. E. Chimenti, eds., Plenum Press, New York.

Marklein, R., Langenberg, K. J., Bärmann, R., 1996, "Ultrasonic and Electromagnetic Wave Propagation and Inverse Scattering Applied to Concrete," *in*: "Review of Progress in Quantitative Nondestructive Evaluation," D. O. Thompson, D. E. Chimenti, eds., Plenum Press, New York.

Mayer, K., Marklein, R., Langenberg, K. J., Kreutter, T., 1990, "Three-dimensional Imaging System based on Fourier Transform Synthetic Aperture Focusing Technique," *Ultrasonics* 28:241.

Virieux, J., 1986, "P-SV Wave Propagation in Heterogeneous Media: Velocity-Stress Finite-Difference Method," *Geophysics* 51:889.

CHARACTERIZATION OF ULTRASONIC SIGNALS USING SYNTHETIC DATA AND NEURAL NETWORKS

B.J.I. Eriksson and T. Stepinski

Uppsala University, Teknikum, Circuits & Systems Group
Box 534
S-751 21 Uppsala
Sweden

INTRODUCTION

Classifying detected flaws is becoming more and more important when applying NDT to modern complex structures and manufactured parts. Once a reflector within some mechanical component is detected, it is very important to know whether it is a sharp crack likely to lead to catastrophic failure. Classification in itself is nontrivial and it becomes a very difficult issue when large data volumes are acquired and large variations in measurement conditions are encountered. Thus it is highly desirable to have at least some decision support. Since there is human expertise available, it may seem natural to use Knowledge Based Systems. However, the problem of formulating the experience and knowledge of a human expert in general as explicit knowledge, is far from trivial in many cases (Waterman 1986), and experts in classification of ultrasonic signals is no exception (McNab 1995). It seems that neural networks, creating their own decision rules from examples, may be better suited for solving ultrasonic flaw classification problems. The problem of acquiring training examples of sufficient quality in sufficient quantities should be solvable by using numerical methods to predict the ultrasonic response from certain types of measurement situations in combination with measurements using real and induced flaws. For this report, a numerical model developed by Boström et al at Chalmers (Boström 1995) is the main source of data. The data volumes required to describe the response from one single flaw can be significant, e.g., using one full A-scan as input to a classifier may lead to a decision space with 1024 degrees of freedom or dimensions. The conventional approach to manage this is to simplify the classifier task by feature extraction. A literature study conducted previously has lead to the use of certain features (Eriksson 1994).

SIGNALS USED

Although most of the results and tools described below should be generally useful, the one basic ultrasonic technique that is tacitly assumed for most of the following is pulse-echo testing. So far all signals used has been generated using the package UTDefect written by A Boström and co-workers (Boström 1995). UTDefect is a program to model ultrasonic non-destructive testing for some simple defects using various types of integral equation methods that yield in principle exact solutions. A basic limitation of the package is that the tested component is assumed to be made of an isotropic and homogenous material. There are other methods that allow both inhomogenous and anistotropic materials as well as more complex flaw geometries, but there are important advantages with the approach taken that more than well compensates for this:

- The input to the program is basically specifications for a measurement situation; there is no need for a complex grid or mesh specification.
- The execution times and demands for computer memory are reasonable, even for large specimens.
- The method is in principle exact and useful for interesting frequency ranges.
- The limitations on geometries that can be simulated are reasonable.

For the calculations temporal variations in the signals are treated in the frequency (Fourier-transform) domain, and the results can be given either as Fourier coefficients or as time-domain signals. It should also be noted that the transducer is modelled as an equivalent traction vector on the component surface (piston model) with a phase factor to model angled probes. The complexities of a piezoelectric crystal or composite, wedge, backing and couplant are not modelled in detail.

An example of the results of a simulation, using a small l-wave transducer and a spherical cavity of diameter 8 mm at depth 40 mm, can be seen in fig. 1 below. There is an l-wave echo around x=40 mm, i.e., 45° beam angle, and two t-wave echoes around x=20 mm, i.e., approximately 25°. Since the transducer is rather small, there are t-waves of significant strength radiated from the edges of the transducer despite the fact that it is an l-wave transducer. The figures are produced using MATLAB and the toolbox described below.

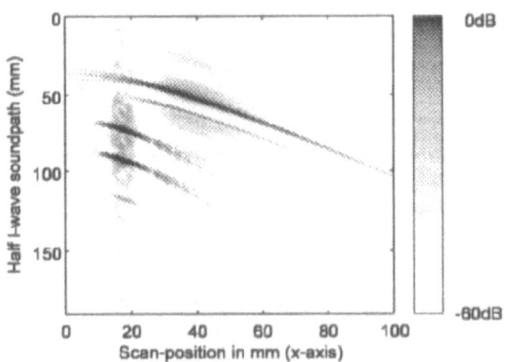

Figure 1. A B-scan for a spherical cavity of diameter 8 mm at depth 40 mm. The l-wave echo is around x=40 mm, i.e., 45°beam angle, while the two t-wave echoes to the left are around x=20 mm, i.e., approximately 25°.

Taking the user's point of view, UTDefect can be described as a software package that accepts a specification of a measurement setup as input, returning the corresponding ultrasonic signals. The input contains information such as the type of material, type of flaw and type of transducer as well as their relative positions and orientations. There are four basic flaw-shapes available: penny-shaped cracks (i.e., infinitesimally thin disc-shaped discontinuities inside the material); strip-like cracks (i.e., half-planes); side-drilled holes (i.e., cylindrical cavities) and spherical inclusions and cavities. The transducer model includes both l- and t-waves, but not the conversion between electrical and ultrasonic energy, i.e., the only transfer functions available are an ideal pass band or a sine-shaped one covering a specified frequency range (Boström 1995).

As mentioned in the introduction, the plan is to use both signals from numerical simulations and measured signals. A few test blocks with side-drilled and flat-bottomed holes as well as machined notches have been fabricated, and new computerized ultrasonic measurement equipment is being taken into use as this is being written (Eriksson 1995).

THE TOOLBOX

In order to go from a measurements specification to a flaw classification we have several distinct subsystems involved. There is UTDefect, written in FORTRAN and currently running on DEC Alpha machines; there is the commercial Astus measurement system (from NDT Systems S.A. in Poitiers, France) including software running under Windows on Intel machines; and there is the classifier/feature extractor currently being developed here using MATLAB (from The Mathworks Inc., Natick, MA, USA). Matlab is available on several different platforms. Most code developed here has been written using the MATLAB m-file language, but there is some software written using C-code and UNIX shell scripts as well (e.g., to reformat data from UTDefect into MATLAB readable format while keeping track of the specification). A graphical illustration of the system can be found in fig. 2 below. The upper path in fig. 2a shows how a measurement specification currently can be used to predict the ultrasonic response, while the lower part shows the corresponding thing for doing measurements on actual specimens. The output from UTDefect is taken as Fourier coefficients to allow full flexibility and control of the ultrasonic measurement system model frequency response (essentially the square of the transfer function of the transducer). Thus predicted signals are stored in the form of Fourier coefficients in MATLAB's binary format. The output to the right in the figure consists of time-signals; typically the spectra are windowed and transformed into the time domain before the actual feature extraction takes place. There are provisions for using actual measured transfer functions, and once the measurement system is fully operational this approach will be evaluated.

In fig. 2b there are two main blocks showing the feature extraction and classification as separate modules. Both of them are modular as illustrated. The graphical ROI (Region Of Interest) selection routines asks for an initial rough estimate of the location of the flaw and then utilises a priori information about the scanning geometry and sound speeds to suggest more precise regions for both l- and t-waves. After selection of the top echo (the l-wave response), the system suggests the region between the hyperbolic lines in fig. 3a as

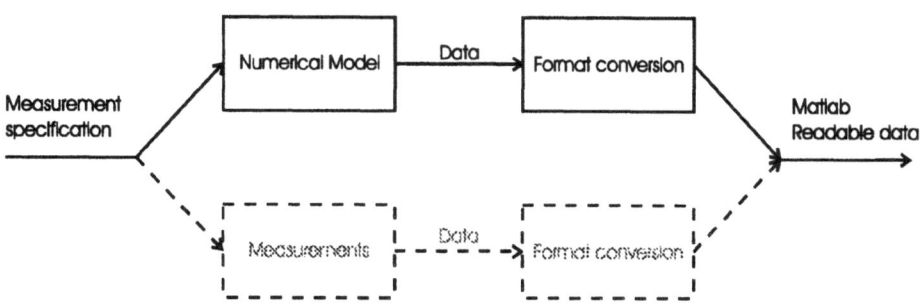

Figure 2a. Current and planned path from measurement specification to MATLAB readable data.

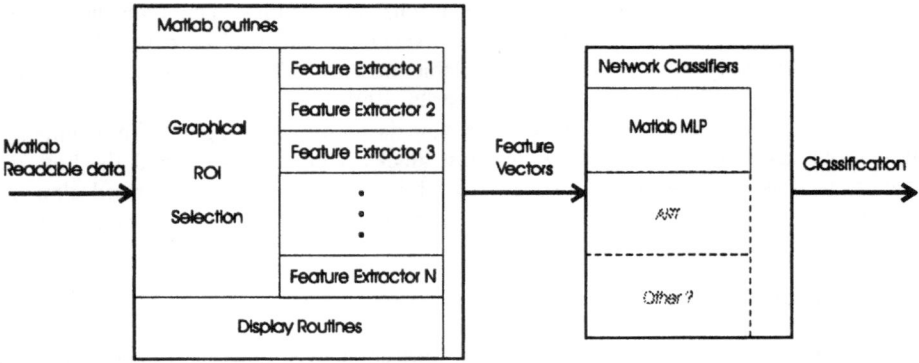

Figure 2b. The structure of feature extraction and classification modules.

the region to do initial feature extraction, and once this is accepted some of the feature extraction results are displayed together with a suggestion (rectangle) on what region to actually use for the feature vector as in fig. 3b. The features are the 35% and 90% points of the envelope of the main pulse in the interesting interval of each A-scan. These features are then used to find rise time, fall time and pulse duration of each A-scan (Rose 1984). As concluded in the literature study (Eriksson 1994), one important aspect of the feature vector is the variation of different features with aspect angle. Since there is a nonlinear relation between scanning position and lookangle, the equidistant scanning normally used leads to non-linear scale in angle. It should also be noted that with flaws on varying depths, the non-linear relation changes. Since it is desirable to have a well defined correspondence between each input and an angle (or angle interval), provisions have been made to compensate for these effects to some degree by non-integer interpolation/decimation and, if necessary, truncation and/or zero padding. The decimation and interpolation is done in the Fourier domain by zero-padding or truncating the Fourier followed by inverse transform.

The system has been designed in order to allow convenient evaluation of many different features, even though only a limited number of rather simple features has been implemented and evaluated so far. One important reason is that in order to do a proper evaluation, it is necessary to have a more challenging classification problem than the noise free synthetic signals used so far.

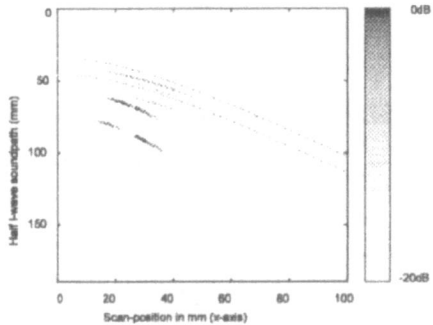

Figure 3a. The hyperbolic region suggested by the ROI-selection routine. The region is displayed in color on the computer screen

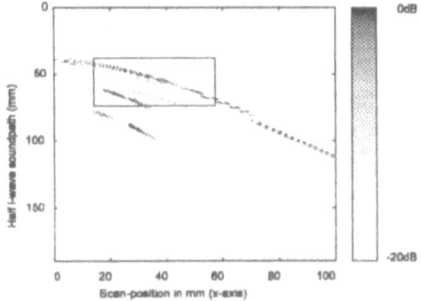

Figure 3b. The interval of A-scans suggested by the ROI-selection routine for the feature vector, including the 35%, 90% 90% and 35% points on the main pulse of each A-scan in the region (in red on computer screen).

The classifier type available so far is a conventional MultiLayer Perceptron (MLP) with one hidden layer with adjustable number of nodes. The number of inputs are determined from the feature vector given. Results so far are encouraging, with 100% correct classification on previously unseen data. It should be noted, however, that this is on synthetic data without noise added. It may be also worth pointing out that the only case with problems in achieving 100% correct classifications was when trying to distinguish penny-shaped cracks from strip-like cracks, two flaw types that both have the sharp edges that are likely to lead to catastrophic failure.

There are m-files both for training and testing, with graphical display of the training and testing error as illustrated in fig. 4a and b. The training error is the RMS value of the difference between the actual and desired classification for each training epoch, i.e., training set presentation (with the weights adjusted accordingly), and the testing error is the RMS error for the testing set. Both figures show the same training sequence, but the left one has fewer epochs to show a little bit more detail. As can be seen from figures this is a schoolbook example of overtraining: around epoch 3-4000 the training error continues to decrease while the testing error starts to increase.The example in fig. 4 is based on a set of 31 training and testing flaws. The training sub-set is is composed of signals from eight

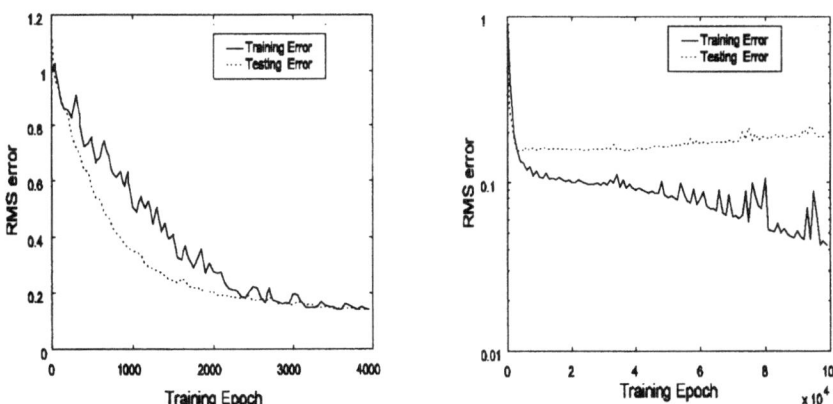

Figure 4. The display of training and testing error for a specific training session. **a (left)** From epoch 0 to 4000. **b (right)** From epoch 0 to 100 000.

spheres (four located far from the backwall and four located near the backwall) and from five surface breaking strip-like cracks, all at depth 40 mm. The testing sub-set consists of signals from eight spheres far from the backwall and ten surface breaking striplike cracks, at 39 and 41 mm depth. The latest and largest set of examples currently consists of the above 31 examples plus 30 examples, with all of the additional 30 penny-shaped cracks at different depths and tilt-angles. At the time of writing, the latest revision of code to run training is set to run 'one-out cross validation', i.e., training for 8000 epochs on each combination of 60 of these 61 examples and then testing on the 61'st. There are some convergence problems - different instances of the random initial weights in the network combined with the exact set of examples used for training sometimes leads to convergence and sometimes to divergence. Therefore this code will have to be run for different sets of initial weights, and possibly with random order of presentation for the examples, different number of nodes in the hidden layer and so on.

CONCLUSIONS AND FUTURE WORK

The toolbox developed should be a good basis both for evaluation of the concept and of the different modules used together. Clearly, the problem of classifying real signals with more random variations and noise is a more challenging problem than what has been done with the system so far, but it should be noted that focus for the system has been on getting a first version of it to go all the way from measurement specification to classification, in order to have a test bed for evaluating different combinations of the various modules available, including the signal sources. Thus, there are many things that seem important and/or promising that have not yet been tried: extending the set of signals (measured and synthetic), adapting methods from image processing to improve the ROI-selection module, using other feature extractors and other classifiers and so on. Combining the last two items in the list includes the possibility to use self-organizing networks to find good features, possibly using the same network for feature extraction as for the final classification - one could well imagine using an Adaptive Resonace Theory (ART) network, e.g., ART2, for this (Carpenter 1987). The area where most effort will be spent in the near future, however, is in acquisition of measured data.

REFERENCES

Boström, A., 1995, UTDefect – A computer program modelling ultrasonic NDT of cracks and other defects. Research Report, Swedish Nuclear Power Inspectorate (SKI)

Burch, S. F. and Bealing, N.K., 1986, A physical approach to the automated ultrasonic characterization of buried weld defects in ferritic steel, *NDT International*, 19: 3.

Eriksson, B.J.I. and Stepinski, T., 1994, Ultrasonic Characterization of Defects Part I. Literature Review, *Research Report, Swedish Nuclear Power Inspectorate (SKI)*

Eriksson, B.J.I. and Stepinski, T., 1995, Ultrasonic Characterization of Defects Part II. Theoretical studies, *Research Report, Swedish Nuclear Power Inspectorate (SKI)*

Carpenter, G.A., and Grossberg, S., 1987, ART2: Self-organization of stable category recognition codes from analog input patterns, Applied Optics, 26, pp 4919-4930, *(Reprinted in:* "Pattern Recognition by Self-Organizing Neural Networks", Carpenter G.A, and Grossberg, S., ed's., MIT Press, London, England, 1991)

McNab, A. and Dunlop, I., 1995, A review of artificial intelligence applied to ultrasonic defect evaluation, *Insight*, 37:1.

Rose, J L, 1984, Elements of a feature based ultrasonic inspection system, *Materials Evaluation* 42 February .

Waterman, D. A., 1986, "A Guide to Expert Systems," Addison-Wesley, Reading, Massachusetts.

INSPECTION DEVICE FOR SPOT WELDED NUGGET

Hajime Yuasa and Kuniyuki Masazumi

Akishima Laboratories (Mitsui Engineering & Shipbuilding) Inc.
1-50, Tsutsujigaoka 1-chome Akishima, Tokyo 196, Japan

INTRODUCTION

The number of spot welds performed to assemble sheet metal components totals several thousand points in an automobile production line. As the number amount to some 3,000 for a cab shell of a truck, devices are required for facilitating inspection of the quality of welds in an assembly line. In a routine line, spot welded nuggets are stripped off by hammer and inspection of the detached portion made, but for periodical inspections, every weld of a whole cab shell is stripped off and inspected, thereby causing to consume much labor and time.

As an alternative, since 1987 FY, research and development have been carried out of the "Mechanical Scanning Type Inspection Device for Spot Welded Nuggets" which can measure spot welded nuggets, as a nondestructive inspection method utilizing the ultrasonic technology. However, as a sole ultrasonic thickness gauge touches the object to the nondestructive test and moved mechanically, problems such as time consuming, operator's skill required, and the wear and friction of the thickness gauge due to the rough surface of nuggets are encountered. Therefore, a new method was searched for conducting the accurate measurement of nugget's diameter with one contact. The result of investigation revealed that any thickness gauge to satisfy such requirements is not available on the market, and so, new development would be required.

The research and development started in 1990 FY. Under the guidance of Professor Koyama of the Material Engineering Faculty of Yamagata University's Department of Engineering, a series of techniques for making copolymer membrane through poling was acquired, resulting in the successful development of an innovative linear array sensor after repeated trial and error.

Now, a brief outline of the "Mechanical Scanning Type Inspection Device for Spot Welded Nuggets", which has been developed since the outset, is given, and at the same time, the history and achievement of research and development concerning ultrasonic linear array with use of copolymer membrane, which is the key technology of this equipment, and its applied "Copolymer Linear Array Type Inspection Device for Spot Welded Nuggets" are explained.

MECHANICAL SCANNING TYPE

Principle

Because cab shells incorporate thin steel plates, a pencil type ultrasonic thickness gauge with a frequency of 20MHz was used. This type of ultrasonic thickness gauge, available on the market, can focus ultrasonic beams to be finer by fixing piezoelectric elements in a film state to the cone-shaped base of the gauge and also the configuration of such cone with copolymer moderators. By having the tip of the cone-shaped copolymer touch the object to be inspected as the ultrasonic thickness gauge, echo from inside can be received.

Coupling liquid (oil or water) is applied to the surfaceenable scanning roughly along the center line of the spot welded nugget, and while the ultrasonic thickness gauge follows the surface, the distance it moves is measured by the encoder. Data concerning the thickness of the plate acquired by the ultrasonic element and the distance measured by the encoder are scanned, inputted in a computer and the diameter of the nugget is calculated by identifying both of its ends. Furthermore, by indicating the distribution of the plate's thickness on the display, it can be confirmed whether any inside defect exists or not. By comparing the pre-inputted allowable nugget diameter and the calculated diameter, the inspection result is also displayed.

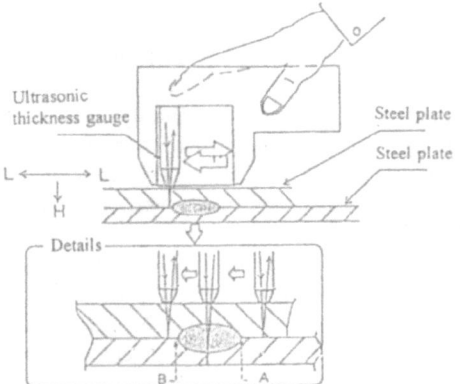 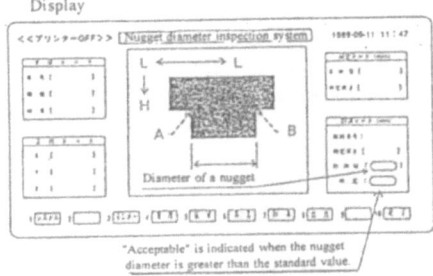

Fig. 1. Illustrated Principle of Mechanical Scanning Type Inspection Device for Spot Welded Nuggets

Fig. 2. Display Example of the Mechanical Scanning Type Inspection Device for Spot Welded Nuggets

Because the nugget's centerline is scanned mechanically, this method is referred to in this paper as the "Mechanical Scanning Type." For this method, experimental models of both manual and motor driven types were manufactured. Illustrated principle of the motor driven type and an example of the display are shown in Fig. 1 and Fig. 2 respectively.

Functions of the Inspection Device

As can be seen from the display in Fig. 2, the thickness distribution of the spot welded nuggets on steel plates is indicated in the display center and the spot weld is expressed as protrusion underneath the two steel plates. The width of the protrusion corresponds to the diameter of the nugget. To the left, the number, part, and control data of side numbers, and the data of X, Y and Z coordinates, and to the right, the reference value of the nugget's diameter, and the inspected thickness of the spot weld, which were inputted as allowable limit values, are indicated. Furthermore, toward the lower right corner, the test number, the measured thickness of the steel plates (inspected thickness), and the diameter of the measured nugget (measured value) are shown as measurement data, and based on comparison with the inspection data, either "Acceptable" or "Rejected" is displayed.

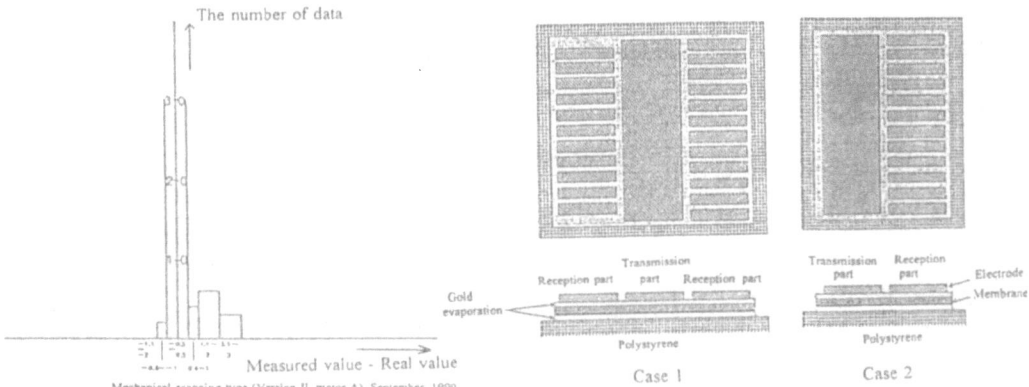

Fig.3. Verification Test Results of Mechanical
Scanning Type Spot Welding

Fig. 4 Configuration of Copolymer Thin
Membrane and Transmission
/Reception electrodes

Specifications, Verification Test and Problematic points

The following result was obtained concerning the specifications of the mechanical scanning type:

Measurable plate thickness	:	0.12-5 mm
Measurable nugget diameter	:	0.2-15 mm
Resolution of measurement	:	0.2 mm
Measurement time	:	Approx. 2sec.
Plate surface condition	:	Smooth
Power source	:	AC100V or battery

To confirm the reliability of the subject device, the diameters of spot welded nuggets were actually measured with a vernier caliper after being stripped off by a hammer. The result is shown in Fig. 3. The axis of abscissa indicates a value as a result of reducing the real value from the measured value, and the axis of ordinate shows the count of such data. In this case, difference between the measured value and the real value is within a range of 2mm to 3mm.

As to whether such accuracy can practically be acceptable or not, its judgment is up to users. The problematic points of this method are such that an error will occur to the nugget's diameter in case the encoder slips on the measuring surface; the tip of the ultrasonic thickness gauge will be susceptible to wear and friction due to burrs, etc. when measuring while plates are detached; and scanning time required is approx. 2 seconds.

LINEAR ARRAY TYPE

Development of the linear array type was tackled to overcome problems inherent in the mechanical scanning type. A search was conducted among sensors available on the market, but the size of the element on the ultrasonic thickness gauge of a 10MHz electronic focus type linear array used in ultrasonic medical examination devices was found to have a width of 0.8mm, a long wavelength, and consequently, judged that it could only identify welding defects in 1mm in the maximum. Ultrasonic sensors which use copolymer membrane, just like other single crystal elements and ceramic elements, are able to transmit and receive with the same element. However, for meeting the required specifications, there were much difficulty and problems in manufacturing an integrated type with functions of both transmitting and receiving, as well as from cost-wise consideration. Consequently, to overcome these problems, the following new concepts were introduced for the early realization.

Principle

As far as the thin copolymer membrane is concerned, the reception part and the transmission part were not separated but integrated, and an independent electrode was installed for each function. The beam receiving surface of the reception part of the membrane was so made as to form an overall earth electrode, while on the opposite surface, a number of comb like electrodes were installed in parallel. To improve the resolution, the comb like receiving electrodes were made as thin as possible. For the transmitting part on the same membrane, by installing only a single transmission element, a certain space was secured so that maximum transmitting energy could be radiated (See Fig. 4).

Plan Specifications

The specifications of the copolymer thin membrane used in the original plan's ultrasonic sensor of the linear array type are as follows:

Material : P(VDF-TrFE)
Transmission frequency : 20 MHZ
Area of thin membrane : Length 20mm, width 6mm
Reception electrode : Length 1mm, width 0.2mm, interval 0.4mm, 56 channels
(High resolution was achieved by placing approximately one half, in cross-stitch style, on both sides of a long and large transmission electrode.)
Transmission electrode : Length 17mm, width 1mm, 1 channel.

In the following chapters, explanations are made regarding the development of propagation materials and piezoelectric copolymer thin membrane which are important components of ultrasonic sensors.

Development of Ultrasonic Propagation Materials

To focus ultrasonic beam and make them come into contact with the objects to be inspected, ultrasonic sensors were constructed by affixing piezoelectric membrane to propagation retarding materials. The propagation materials require functions such as the low damping of ultrasonic propagation, the capability of focusing beams, etc. At the time of making a prototype, materials such as polyimide and polystyrene were used, and for configuration, the three shapes of fan, box and semicylinder were tried. The 4mm wide transmission membrane as illustrated in Fig. 6 and Fig. 7 was first affixed to the semicylindrical propagation materials as illustrated in Fig. 5. Then, difference between both positions of the curved top surface at its center and ends was measured with our own developed pinpoint hydrophone. The pinpoint hydrophone mechanically scanned the bottom of the semicylindrical propagation materials and measured the condition of focus of the ultrasonic beam formation. In either case, it was confirmed that the hydrophone receiving intensity was concentrated at the center of the semicylinder. Thus, unnecessary parts of the propagation materials were deleted and the fan shape type was adopted.

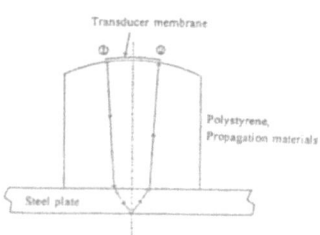

Fig. 5. Experimental Propagation
Materials and Transducer Membrane

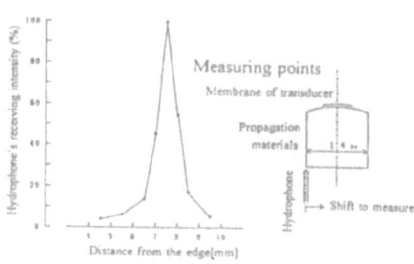

Fig. 6. Affixing Position (Center) of Copolymer
Thin Membrane and Beam Formation

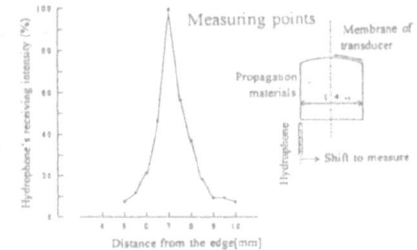

Fig. 7. Affixing Position (Ends) of Copolymer
Thin Membrane and Beam Formation

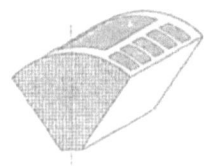

Fig. 8. Shape of Propagation Materials
and Ultrasonic Sensor

Development of Piezoelectric Copolymer Thin Membrane [1]

Using a method which does not require any drawing operation, the formation of membrane was achieved on one hand. On the other, P(VDF-TrFE), which has high piezoelectric property and electromechanical coupling coefficient despite being one of copolymer materials, was used. The prescribed quantity of P(VDF-TrFE)was dissolved in dimethylhormamide (DMF) and after removing fine air bubbles, it was poured into a receptacle, and thus translucent thin membrane was produced through vacuum drying. It was then placed in a vacuum dryer, and through a heating process in nitrogen gas, became opaque membrane. It was stripped from the receptacle, checked and finally confirmed as being of prescribed thickness.

After cutting a sample from part of the thin membrane, gold was coated on both outer and inner sides of it through vapor deposition to make electrodes, and then poling (polarization) was performed. The sample's resonance frequency, impedance, and phase retardation were measured, and a comprehensive judgment of the membrane's properties was made. Thereupon, remaking of the membrane or additional poling as required was performed. Observation of wave shape was performed through the tests of the transducer using samples, and after confirming a satisfactory result, judgment was made that the original membrane also had identical properties. As such, a full-scale manufacture of transmission/reception elements was started.

Specifications

Under the above-mentioned circumstances, the final specifications for the sensor were determined as follows (Fig. 9):

Materials   : P(VDF-TrFE)
Transmission frequency   : 20MHZ
Area of thin membrane   : Length 24mm, width 8mm
Reception electrode   : Length 4mm, width 0.2mm, interval 0.2mm, 56 channels
(A long and large transmission electrode was installed on one side of the fan-shaped arc and the reception electrode on the opposite side.)
Transmission electrode   : Length 23mm, width 4mm, 1 channel

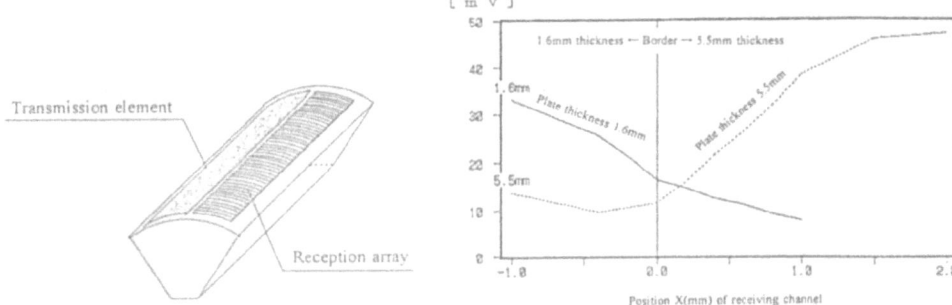

Fig. 9. Configuration of Piezoelectric Copolymer Ultrasonic Sensor

Fig. 10. Change in Plate Thickness and Measurement Accuracy

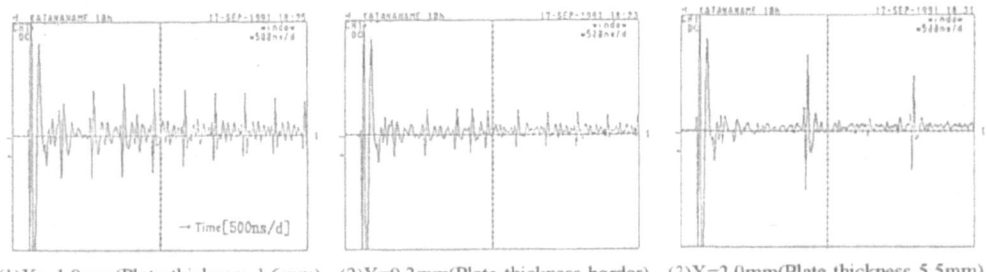

(1)X=-1.0mm(Plate thickness 1.6mm) (2)X=0.2mm(Plate thickness border) (3)X=2.0mm(Plate thickness 5.5mm)

Fig. 11. Change in Plate Thickness and Measurement Wave Shape
(X=Position of receiving channel)

Fundamental Test

To perform a fundamental investigation as to whether detection of a spot welded nugget is possible or not, an ultrasonic sensor was affixed to the top of a polystyrene block, the top and bottom sides of which are parallel. Only one channel of the receiving element was active, but the others were grounded and slid to a position on the steel plate without uniformity in thickness. The lengthwise direction of the transmission element was consistent with the direction of the slide, and measurement was made in such a manner that the reception element could move from the thick plate to the thin plate (5.5mm → 1.6mm).

Fig. 10, with the border of the thick plate "0," shows the receiving voltage based on the position of the receiving channel (mm) on the axis of abscissa of measurement results in the limits of 1.6mm of the thin plate in the negative direction and 5.5mm of the thick plate in the positive direction. Fig. 11. shows the measured wave shapes at three typical receiving positions, -1.0, 0.2 and 2.0mm.

From these figures, it can be seen that reflected waves within the thin plate still exist there even in a case in which a reception channel exceeds the border of plate thickness and enters the thick plate. Such sphere of influence (corresponding to crosstalk), as a result of data analysis, extends from the vertical direction to plus/minus 6 degrees and would pose no problem in any practical way.

System Configuration

As a result of reviewing the foregoing, the following was clarified. When pulse signals transmitted by a single transmission element are detected by a reception array with 56 channel elements after being reflected at the bottom of the steel plate, only the thickness of the sheet directly under the respective reception element can be detected without electronic scanning. This means that signals from the reception elements of the 56 channels would adequately be inputted in personal computers or micro computers, after being converted to plate thickness through the 56 channel reception device. However, because the transmitting pulse is 20MHZ and also time required for transmitting/receiving is extremely short, even with use of a transmitting/receiving device and with switching over and processing signals from receiving elements by multiplexer, it would appear to be real time.

In order to make the device as compact as possible, handy terminals (H.T.) available on the market were used instead of personal computers. The H.T. successively processes the loop shown in Fig. 12. for the 56 channels at several ten KHZ intervals in the transmitting circuit which forms transmitting pulses, and displays, in real time, the cross section, the result of findings, etc., of the object being tested.

Furthermore, because the electrical impedance of the piezoelectric copolymer sensor is low, careful attention should be paid to matching between the cable and the electrical circuit. With this device, the installation of a pre-amplifier adjacent to the receiving element solved the problem (See Fig. 13 and Fig. 14).

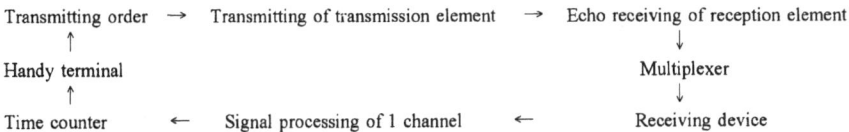

Fig. 12. System Configuration

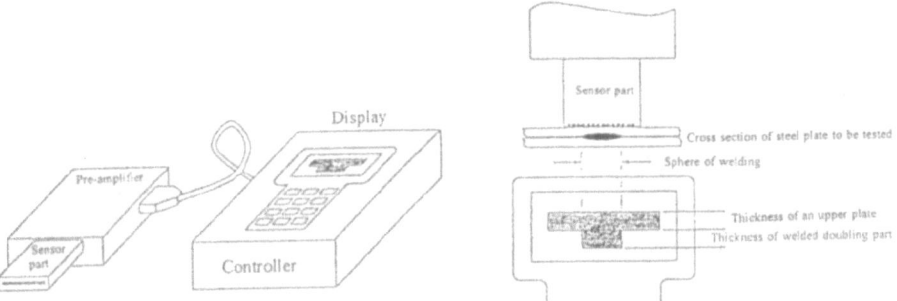

Fig. 13. Total Configuration   Fig. 14. Outline of Measurement

Verification Test

Verification tests were performed of small pieces of uneven surface aluminum and spot welded nuggets using the subject device. Fig. 15. and Fig. 16. show pictures as such results. The uneven surface or the border of plate thickness was clearly identified, thus proving that the performance of the device is very excellent.

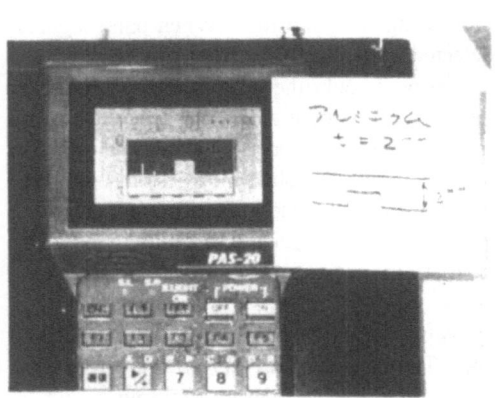

Fig. 15. Cross Section Image of
Small Piece of Aluminum

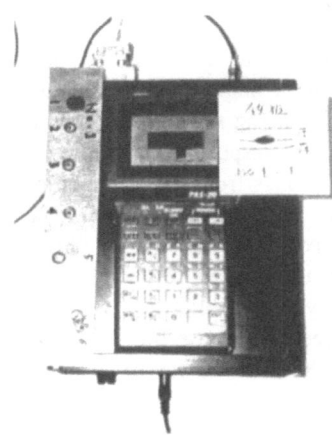

Fig. 16. Cross Section Image of
a Steel Plate Nugget

CONCLUSION

Car bodies are assembled by jointing a number of steel plates and spot welding plays a major role in such jointing. The only method available now for judging whether spot welds are acceptable or not is visual observation by stripping off nuggets with a hammer which involves a great deal of labor and processes. The research and development of this device were performed to improve efficiency of such operation. Prior to commencing work on this device, the mechanical scanning type inspection device for spot welded nuggets with an encoder attached to an ultrasonic thickness gauge, was developed. However, since the encoder used for measuring the distance of travel would slip, depending on the surface condition of the object undergoing tests, erroneous indication of nugget diameters occurred sometimes. Under the circumstances, the development of a device which would not require mechanical scanning has been keenly sought. The development of the linear array type device needed two years, but due to its high resolution and simple configuration, it has satisfied our specifications as originally targetted.

Nevertheless, the existence of burrs and an uneven state of nugget surface do affect the result of measurement even with this new device. Consequently, this "Copolymer Linear Array Type Inspection Device for Spot Welded Nuggets" still has problems and cannot be an all-rounder. It is important, therefore, to employ this device with a thorough understanding of its characteristics and behaviors as mentioned above.

REFERENCES

1. K. Koyama and H. Itoh. "Copolymer ultrasonic transducer," *Ultrasonic Techno.* Dec.(1990)

DEVELOPMENT OF HIGH RESOLUTION ULTRASONIC INSPECTION METHODS FOR WELDING MICRODEFECTOSCOPY

Roman Gr. Maev,[1] Daniel F. Watt,[2] Rong Pan,[2]
Vadim M. Levin,[1] and Konstantin I. Maslov[1]

[1]Acoustic Microscopy Center, Russian Academy of Sciences,
Kosygin st.4, Moscow, Russia 117334
[2]Department of Mechanical and Materials Engineering,
University of Windsor, Windsor, Ontario, Canada N9B 3P4

INTRODUCTION

The hottest NDT problem today is high resolution inspection of welding joints. The problem of inspection for flat many-layered (2 or more layers) joints formed by spot-welding or diffusion welding is under investigation here. Using acoustic microscopes with very short probe pulses provides a way to visualize small-scale failures of contact and other defects at different depths. For spot welding acoustic microscopy makes it possible to inspect fine details of internal areas of joints in spite of the curved outer surfaces of welding spots.

To show the potential of acoustic microscopy we used a wide-field short-pulse acoustic microscope with operation frequencies of 25, 50 and 100 MHz. The time length of the probe pulse was very short (1 - 3 oscillation periods). Short pulses make it possible to resolve fine details of the microstructure in depth (with resolution 50 - 150 μm). The microscope was able to scan areas with sizes of 80x100 mm^2, and with a scanning step of >10 μm along one axis and 50 μm along the other.

THE INSTRUMENTATION

The ultrasonic equipment, the precision scanner, the IBM PC plus the control software for scanning movement, ultrasonic unit adjustment and display processing are all components of the device, and are matched to each other. An electronic microscope system was designed as a standard IBM PC expansion card with a single remote preamplifier unit and with no manually adjustable parts. It is installed in the computer and works independently, although it is fully programmable from the computer. The current state of all the adjustable parameters of the microscope are displayed on the computer monitor simultaneously with an A-scan image of the reflected signal and a help message. The short time length of the probe pulse (1-3 oscillation periods) makes it possible to resolve fine details of the microstructure, in-depth, with a resolution of 20 to 100 μm.

SCANNER

We developed a precise scanner having a drive system step width of 0.05 mm across the fast scanning axis and 10 um across the slow axis, with a scanning area of 80 mm x 120 mm. It is a desktop device with maximum dimensions of 200 (H) x 510 (W) x 430 (D) mm. Rather than rely on the immersion of the investigated sample in a water-filled tank, we use a continuous stream of liquid placed alongside the ultrasonic probe. The light-weight probe holder enables precise manual vertical adjustment of the probe. When scanning in the XY plane there is a minimum of mass forces at the point in time that guarantees a precise and oscillation free movement at a high scanning speed of up to 300 mm/s. If the sample to be investigated is allowed to be in mechanical contact with the probe it is possible to adjust the distance according to the sample surface, and thus, examined samples which are not perfectly flat. If it is not possible, the support on which the test object is positioned can be adjusted in all planes by screw adjusters so that the part is easily and precisely aligned.

FOCUSING ULTRASONIC PROBES

To achieve the maximum possible resolution, we developed a set of various focusing ultrasonic probes carefully designed for different materials and different penetration depths. The Lithium Niobate transducer is used, tightly attached to the quartz buffer rod (25, 50, 100 MHz and higher frequencies). The low quality factor of the transducer allows us to use a very short ultrasonic pulse and thus easily separate different reflected signals from each other. As a result, better sensitivity and resolution are achieved. Probes have a different focal length varying from 2 to 15 mm and different aperture angles from 20° to 60°, decided according to the ultrasonic velocity in tested samples. With all probes the focal diameter of the order of a single wavelength can be achieved. The same probes have been specially developed for transversal wave-only investigation for maximum spatial resolution and maximum penetration depth.

OPERATING THE SCANNING ACOUSTIC MICROSCOPE

Data evolution of the SAM is made on-line by the computer with every scan line. Operating the scanner is especially easy due to the presentation of a C-scan on the display, with a superimposed cursor that shows the actual position of the ultrasonic probe. The corresponding A-scan allows the operator to carefully adjust the receiving system of microscope. The C-scan is presented on a high resolution display in 256 colors. An amplitude, integral, weighted integral, or transit time presentation can be produced. Some special measurement procedures such as 0.005 mm step scanning and B-scan imaging are available. Using an optional 100 MHz oscilloscope it is possible to measure a transit time with precision of less then 0.1 ns, which is enough for precise absolute measurement of surface acoustic wave velocity. The user can also write his/her own data acquisition procedure using the simple built-in command language or, for faster measurements, using any high level computer language.

The original data of C-, B-, and A-scan patterns at a chosen point can be stored along with the collection of all adjustable ultrasonic system values and labels, and can be recalled at any time later. It is possible to change the color allocation for amplitude values, choose the palette, blow up any part of an image, store an image in one of standard formats used by widespread image painting programs or store any image cross section as an ASCII code file for future analysis. All images can be printed by a large variety of printers as half-tone or as quasi 3-D pictures, using our own software or through use of any commercially available program for the printing of black & white or color images.

APPLICATIONS OF THE SHORT-PULSE ACOUSTIC MICROSCOPE FOR WELDING MICRODEFECTOSCOPY

In the research area of applications of high-resolution acoustical non-destructive methods for investigation and optimization of the welding joints technology, control of any inclusions and of the various types of discontinuity are a quality control concern. There may be technological or manufacturing defects, such as cavities (conical, stratification, die etc.) in the welding area as well as defects caused by the failure of the final product by mechanical damage, the corrosion (rusting) process, environmental conditions, etc.

Another very important and popular application area for acoustical high frequency defectoscopy and acoustical microscopy, is in the non-destructive quality control of numerous types of joints in the final production in real time in any of the following industries: automotive, aircraft, aerospace, navy, electronics. The necessity of considering such control systems arises from the fact that often such joints, have reduced mechanical durability, strength, ductility, and at the same time are areas of increased concentration of the mechanical strain.

At the moment, ultrasonic methods are used actively for various types of joints in different industrial applications. First and foremost, the methods are used in the industries where the essential features required of the final product are extremely high (as in the aerospace, aircraft or electronic industries), or in areas where potential damage as a result of possible accidents can be huge (such as in a nuclear power plant).

EXPERIMENTAL

The method was applied to evaluation of spot-welding joints of two steel sheets 1 - 2 mm thick of each. The scanning field was about 40 mm^2. Acoustic lenses with different frequencies (25, 50, 100 MHz) and number apertures (20°, 40°, 60°) were used for visualization of welding area. In spite of the fact that the top surface of specimen was disturbed by welding, high quality acoustic images of the interface between two welded sheets were obtained. The C-scan and B-scan images contain well-shaped defects of joints and the confirm power of the method for non-destructive evaluation of spot-welding joints.

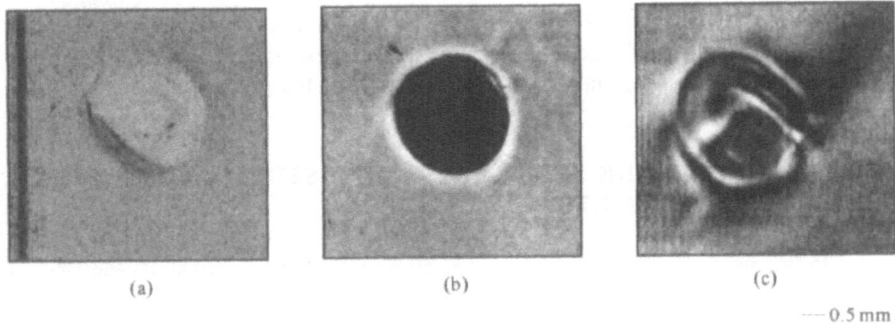

(a)   (b)   (c)

— 0.5 mm

Figure 1. A set of C-scan images of the spot welding sample. (a) top surface; (b) interface, depth penetration,-1mm; (c) back surface, depth penetration,-2mm; Frequency- 50 MHz, Aperture 30°

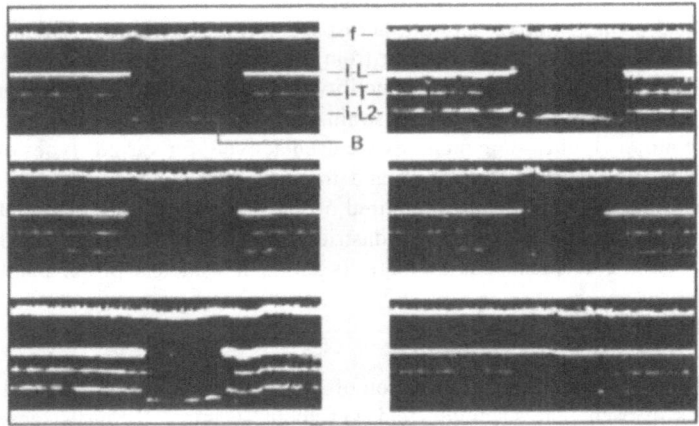

Figure 2. Sequence of B-scans in different cross-sections of the same spot-welding sample as in Figure 1. In B-scans obtained for a set of successive cross-sections of welding area there is complicated system of received signals. It involves signals formed by longitudinal waves reflected from the front surface of a specimen (**f**), from the interface (**i-L**), then a signal produced by transverse waves reflected from the interface (**i-T**) and a signal due to longitudinal waves twice re-reflected from the interface (**i-L2**). And only inside the welding area there exists a signal (**B**) reflected from the bottom (back side of second steel sheet).

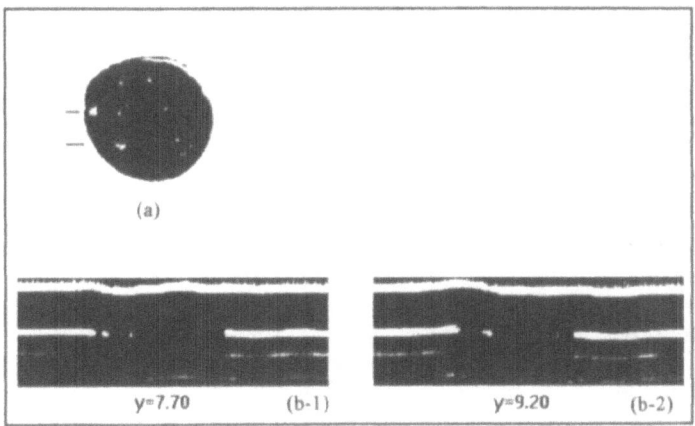

Figure 3. (a) C-scan image of the same interface (depth penetration 1 mm) of the spot welding sample as Figure 1, but obtained in intensified pulse method (Frequency -50 MHz). The white spots presented in this C-scan image are the defects in the spot welding sample, detected by the intensive pulse. The lines on the left side of C-scan image indicate where the B-scan images were positioned. (b-1) and (b-2) are B-scan images. In the B-scans, it is clearly seen that the white spot defects in the cross-section of the welding area, correspond to the defects which appeared in C-scan image. y represents the position in C-scan (a) where the B-scans were made.

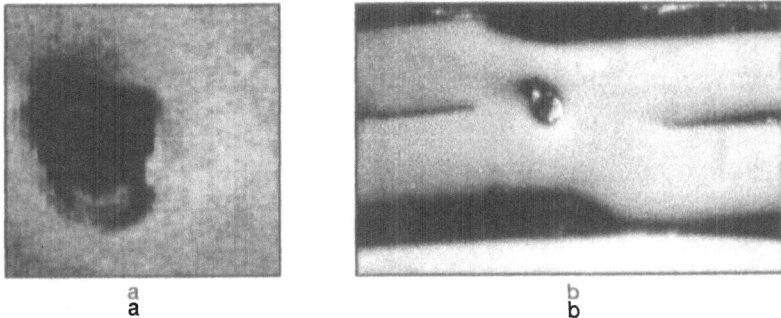

Figure 4. (a) C-scan image of the interface from the spot welding sample. A small white gray area located inside of black spot welding zone represents an unsound welded defect. (b) An optical metallographic picture of the cross-section cut through the welded sample shows the defect which is the same as that detected in the C-scan image (a) rotated 90°.

CONCLUSIONS

The results of this preliminary study show that short-pulse acoustic microscopy can be applied in spot welding as an effective inspection method to detect any kind of defects, and can be used for welding quality control. The combination of the C-scan and the B-scan images in one picture can give a real 3-dimensional picture of the defect distribution inside the welding zone, which is very important from a technological point of view.

ACKNOWLEDGEMENT

The authors are very grateful to Chrysler Canada Ltd. and Centreline Ltd. for providing us the spot welding samples. This work was support by the Center of Acoustic Microscopy, Russian Academy of Sciences and the University of Windsor, Ontario, Canada.

REFERENCES

1. R.G.Maev, Einsatz der Akustomicroscopie in den Materialwissenschaften. Review, *in*: Proc. of BRD-UdSSR Bilaterial Seminar " Microscopy in Material Sciences", Moscow, p.p.35-51, (1988).
2. K.I.Maslov, Short-pulse scanning acoustic microscope, *in*: Acoustic Imaging, v.19, pp. 645-650, Bochum, Plenum Press, N.Y., (1992).
3. K.I.Maslov, R.G.Maev, V.M.Levin, New methods and technical principles of low frequency scanning acoustic microscopy, *in*: Acoustic Imaging, v. 20, Nanjin, China, Plenum Press, N.Y., (1993).
4. K.I.Maslov, R.G.Maev, V.M.Levin, Wide-Band Low Frequency Acoustic Microscope, Russian Patent, SU 5049658/28 from 26.06.92.
5. R.G.Maev, Principles of short-pulse acoustic microscopy for advanced materials bulk study, *in*: Res. of 41st Int. Confer. on Analytical Science and Spectroscopy, pp.53-54, Windsor, Canada, (1995).

WOOD ANATOMIC ELEMENTS WITH

ACOUSTIC MICROSCOPY

Jacques Attal*, Amina Saied*, and Voichita Bucur**

*Université de Montpellier II Place E. Bataillon, F-34095
Montpellier Cedex 5, France

**Université "Henri Poincaré" de Nancy I, B.P. 239, F-54506
Vandoeuvre les Nancy Cedex, France

INTRODUCTION

The main interest in acoustic microscopy arises from the direct interaction between the wave and the elastic properties of the material through which it propagates. Acoustic waves permit the exploration of the region of specimen beneath the surface.

The aim of this research is to report recent investigations of wood using scanning acoustic microscope in reflection mode, at 200 MHz. Acoustic microscopy is proposed for the identification of wood anatomic elements on oak, spruce and pine.

METHODS AND SPECIMENS

The detailed construction of the scanning acoustic microscope is described in different recent articles and books (Attal 1979; Briggs 1985; Johnes 1987).

The experiments described here were performed at 200 MHz with a lens having 1.7mm focal length, focussed approximately at 10 μm below the surface. The resolution was 6 μm.

The samples were small blocks of 20 x20 x(2...10)mm and were oriented following the natural symmetry directions of wood. The surface of the specimens were very carefully prepared with a microtome, currently used for thin sections of wood for studies in optical wood microscopy.

The effective scanned surface of the specimens was 5x5 mm^2, in a scan ranging between 50 μm and 5mm.

In this experiment the anatomic features of wood were observed at the scale 1000 m, 500 µm and 250 µm.

The moisture content of wood specimens was less than 8%.

Three characteristic wood species growing in France were observed, namely oak as a broadleaf species, with very big vessels, rays and fibers and on the other hand, spruce and pine, as softwoods, with annual rings having distinct zones of earlywood and latewood in each annual ring.

The quality of acoustic images is dependent on the polishing quality of the surface of the specimen, the wetting properties of the coupling liquid and on the aperture angle of the acoustic lens. In this experiment we used mercury.

RESULTS

The acoustic micrographs were obtained in the reflection mode, representing imagings of oak, spruce and pine explored in one of the three main anisotropic planes.

The intrinsic contrast of the acoustic micrographs is seen to be good, as can be seen in previous publication (Bucur 1995).

In oak details of the porous growth pattern were observed clearttly. We note the large earlywood pores concentrated in a dark band at the begining of the annual ring growth.

Latewood pores very much smaller, were quite difficult to be observed in the latewood zone of the annual ring. (The term "latewood" refers to the zone of the annual increment, or annual ring, produced by the tree late in the growing season). The zone of fibers and of rays appears as bright zone.

The contrast observed on wood acoustic micrographs is due to the difference in acoustic impedance of tissues.

It is well known from X-ray microdensitometric analysis (Polge et al. 1973) that oak rays have a higher density than pores (vessels) or parenchyma zone. This fact is clearly seen in the acoustic micrographs where the rays appear as a brilliant zone comparing to the dark zone of the vessels.

The anatomic details on softwoods were very well observed.

Generally it is possible to recognize the delineation between annuals rings on the transverse section of wood because of the change in pattern of cell distribution and size as the year's growth proceeds. It is generally accepted that one of the most important anatomic characteristics of these species is the sharp transition between latewood and earlywood in the annual ring.

The abrupt transition between the two zones in transversal anisotropic plane was shown in a very good contrast. The dark zone corresponds to the earlywood zons and the brilliant zone to the latewood, very much dense than earlywood.

Anatomic details observed in the longitudinal - radial anisotropic plane of wood were: the walls of the tracheids, their pits and the rays.

a) Optic image of oak wood

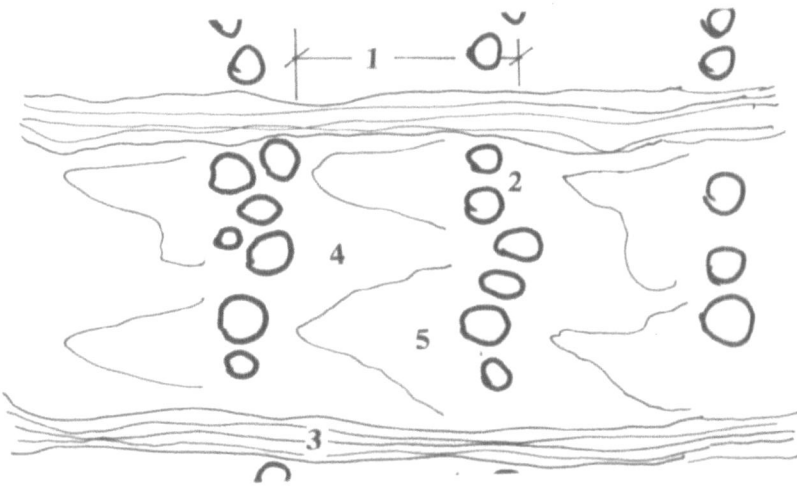

b) Acoustic image of oak wood (Bucur 1995)

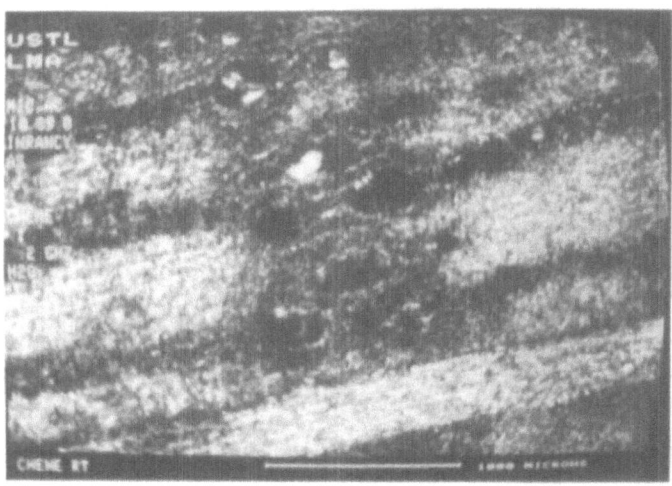

Figure 1 . Microscopic description of oak wood in optic (a) and acoustic microscoppy (b). Anatomic features are observed in transverse anisotropic plane .

Legend: 1 annual ring width, 2 vessel in earlywood, 3 medulary ray, 4 fibers, 5 parenchyma in latewood

The rays appears as discontinous ribbons of tissues of approximately constant width and different lenght.

The pits are the most common sculpturing features of the cellular walls. They appear as depression in the cell wall. Acoustic micrographs furnish views of the pit structure, as for example the torus situated in the center of the pit.

CONCLUDING REMARKS

The challenge in using acoustic microscopy and particularly reflectometry is to find a means of visualizing the wood anatyomic features and their acoustic properties with the highest possible resolution.

The very fact that the acoustic microscope is looking directly at the acoustic properties of anatomic elements may be a chief attribute in the development of a new nondestructive procedure for very fine qualitative anatomic studies of wood.

A second interesting aspect of the technique lies in its subsurface imaging capability.

The acoustic micrograph suggests the utilization of acoustic reflectometry together with microechography as complementary techniques to X-ray microdensitometry for the definition of the local mechanical behavior of wood, i.e. the elastic constants of anatomic elements in every annual ring of wood.

REFERENCES

Attal, J. 1979 : The acoustic microscope, a tool for non-
    destructive testing. *in* "Nondestructive evaluation of semiconductor materials and devices" J. N. Zemel, ed, Plenum, New York 631-676

Briggs,A. 1985: An introduction to scanning acoustic micro-
    scopy. Oxford University Press and Royal Microscopic Society, U.K.

Bucur, V. 1995: Acoustics of Wood. C.R.C. Publ.Boca Raton

Johnes, H.W. 1987 : Acoustic Imaging. Plenum, New York

Polge,H.; Keller,R.; Thiercelin,F.1973: Influence de
    l'élagage de branches vivantes sur la structure des accroissements annuels et sur quelques caractéristiques du bois de Douglas. *Ann. Sci. Forest.* 30,127-140

NON DESTRUCTIVE CHARACTERIZATION OF PHOTOCHEMICALLY GENERATED CROSSLINKING GRADIENTS IN POLYMERS

L. Simonin[1,2], J.J. Hunsinger[1], J.P. Gonnet[1], D.J. Lougnot[2], and B. Cros[1]

[1]Institut Polytechnique de Sévenans
90010 Belfort Cedex, France

[2]Laboratoire de Photochimie Générale
URA CNRS 431, ENSCMu
68093 Mulhouse Cedex, France

INTRODUCTION

Photopolymers are extensively used in the field of coating technology, stereolithography, but also to fabricate optical elements. A good knowledge of the crosslinking process at the local scale is essential to characterize these materials and to control their quality, especially in regard to the optical or the mechanical properties, because the viscoelastic constants of the polymer as also its refractive index depend on the crosslinking density. Microechography provides a means to resolve structure heterogeneities in small volumes down to the wavelength of longitudinal sound waves in the material. In a first approach, microechography was applied to the study of small polymer disks that were photocured either through a two-level amplitude mask or a progressive neutral density filter.

A large number of analytical techniques such as Differential Scanning Calorimetry (DSC), IR or UV spectroscopy, dynamic mechanical thermal analysis or interferometry can only provide information on the macroscopic properties of polymer materials. Most recently, the possibility of producing photopolymerizable materials exhibiting attractive optical properties - i.e. the control of the shape of the irradiated surface or the refractive index of the photoconverted material - through a spatial control of the irradiance has been demonstrated[1-3]. Thus, gradients of conversion or crosslinking density have been found to go along with gradients of refractive index[4]. In addition, many industrial processes as stereolithographic process based on the photocuring of polymerizable layers might suffer some drawbacks resulting from crosslinking gradients caused by a spatially inhomogeneous illumination. The existence of such gradients of crosslinking density however, cannot be revealed by the above-mentioned techniques which only provide information on the macroscopic structure of polymer materials.

The present communication is concerned with the prospects displayed by microacoustic techniques when used to visualize the local variation of Young's modulus[5-9] This technique consists in measuring the maximum amplitude of a reflected acoustical wave in the interface of polymer and coupling fluide (water). Indeed, microechography that evaluates the longitudinal velocity of sound in an elementary volume, the size of which depends on the probing frequency that is used (25 Mhz), provides a means for resolving structure heterogeneities down to 100 μm. In addition, this non-destructive technique is able to provide quantitative information in real time so that it could allow the phototransformation which is used to generate specific properties in polymer materials to be monitored. In a first approach, microechography was applied to the study of polymer disks that were photocured either through a two-level amplitude mask or a progressive neutral density.

Although qualitative, this exploratory study is aimed at identifying the parameters that could reveal the existence of crosslinking gradients and/or measure their intensity. Moreover, futher experiments that are aimed at the quantitative determination of the correspondence between the acoustical reflection coefficient, Youngs modulus, chemical parameters (kinetics of free-radical polymerisation, monomer conversion rate), refractive index and adhesion between a polymer layer and a substrate.

MATERIALS

Compounds

UV curing is primarily the use of ultraviolet radiation to convert a monomeric or oligomeric unsaturated product into a polymer[10,11]. UV formulations usually consist of an oligomer, a polyfunctional monomer and a photoinitiator. A photoinitiated radical polymerization can be described by the following reactions where M is the monomer and I the photoinitiator :

$$I \xrightarrow{h\upsilon} R_1^{\bullet} + R_2^{\bullet} \qquad (1)$$

$$R^{\bullet} + M \rightarrow RM^{\bullet}$$

$$RM^{\bullet} + M \rightarrow RM_2^{\bullet}$$

$$RM_{n-1}^{\bullet} + M \rightarrow RM_n^{\bullet} \qquad (2)$$

The first step is the generation of free radicals through the homolytic splitting of an excited state of the photoinitiator (1). The radicals, then, interact with the monomer or the oligomer and start the reaction chain (2). Generally, the monomer is mixed in various ratios to reduce the viscosity of the reactive system where the oligomer imparts the mechanical, physical and chemical properties of the polymer.

In the experiments reported in this paper, the formulation was based on a mixture of ca 80 w/w % of pentaerythritol triacrylate, 20 w/w % of tetraethylene glycol diacrylate and 0.3 w/w % of hydroxyalkylphenylketone photoinitiator which is sensitive in the near U.V. range. The U.V. source was a high pressure mercury arc, and the available fluences of the actinic beam (λ = 312, 334, 365 nm) were respectively 2.75 and 0.77 mW.cm^{-2} without and with an attenuating neutral density filter.

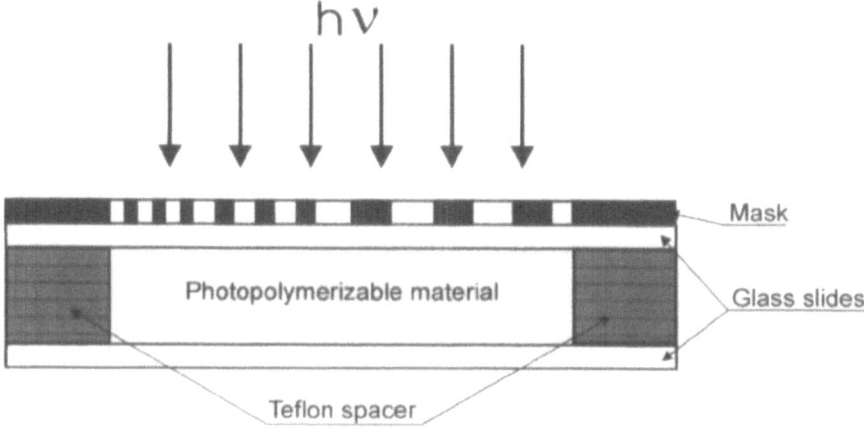

Figure 1. Schematic of the recording setup

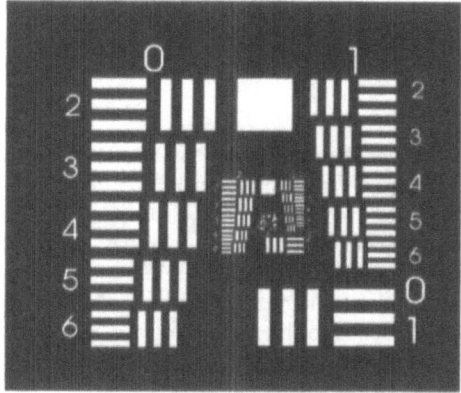

Figure 2. USAF test target (negative)

Recording process for imaging material

The photopolymerizable system was sandwiched between two transparent glass slides. The thickness of the sensitive layer was determined by a 3 mm thick Teflon spacer. A photomask was added on top of the cover glass slide (figure 1). Two kinds of images were constructed into the polymer. The first system involved the negative USAF 1951 test target. As known, in this target the number of line pairs per millimeter doubles with every seventh target element. An element consists of 2 patterns of 3 lines each, at right angles to each other. Six consecutive elements build a group (figure 2). The recording involved a two step process. The former consisted of an imagewise illumination through the mask (irradiation energy : 3.3 J.cm^{-2}), the mask was then removed, and the sample was postpolymerized with an irradiation energy of 0.46 J.cm^{-2}. The second mask is a chromium apodizing filter. The optical density of the filter increases from the center to the periphery (figure 3). To ensure full conversion from monomer to polymer throughout the sample, the irradiation energy was fixed at 1.65 J.cm^{-2}. In that case, no postpolymerization was required.

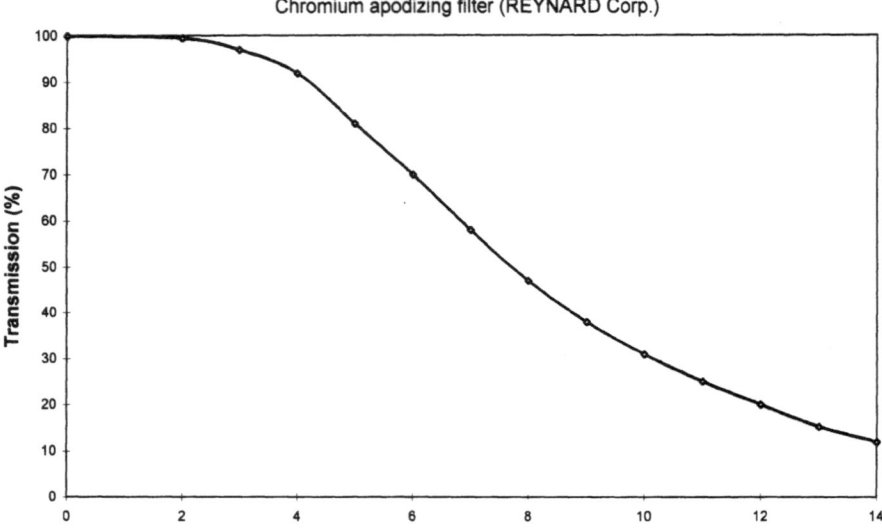

Figure 3. Transmission characteristics of the apodizing filter

MEASUREMENTS

Sound velocities and absorption measurements provide informations for evaluation of structural factors such as the glass transition temperature, crosslinking density, morphology, and the elastic constants of the material. The acoustical microechography was chosen to investigate the mechanical properties of the surface of photopolymer samples to characterize the crosslinking gradient. This technique consists in measuring the maximum amplitude of the acoustic pulsed longitudinal waves reflected by the sample surface. Assuming a perpendicular incidence to the sample surface, the reflection factor is expressed as follows :

$$R = \frac{Z_{mat} - Z_{liq}}{Z_{mat} + Z_{liq}}$$

Z is the acoustical impedance of each propagation medium, and can be expressed by :
$$Z = \rho V_L$$
ρ : Volumetric mass of propagation medium
V_L : Propagation velocity of longitudinal waves

The acoustical impedance Z_{liq} of the coupling fluid is constant. Only the material impedance changes. The propagation velocity V_L can be expressed as follows :

$$V_L = \sqrt{\frac{E}{\rho} \cdot \frac{1 - \upsilon}{(1 + \upsilon)(1 - 2\upsilon)}}$$

The acoustical impedance of a material is a function of E, ρ and Poisson's ration υ, and these terms crucially depend on the crosslinking ratio.

APPARATUS

The apparatus operated in the reflection mode with a simple focusing transducer (PANAMETRICS V324 SU, 6 mm aperture, 13 mm focal length in water and a 25 MHz working frequency). The transducer and the sample were immersed in water that acts as the coupling fluid, and we used a XY moving table to establish a reflectance cartography by scanning a given area (figure 4).

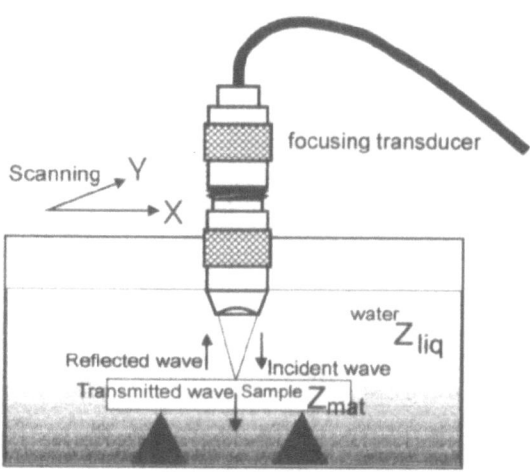

Figure 4. Focusing transducer operating in reflection mode.

The transducer was connected to a pulse emitter-receiver able to send to it very short high voltage pulses (typically 10 ns at 120V). The receiver consisted of a sample-hold amplifier and a 6-bits analog to digital converter which allows an accuracy of ca 2% of the maximum readable value. The digitalized value were sent, for storage and display in false colours, to a desktop computer which also drives the moving table.

The probed area at the focus point has a diameter depending on the frequency, the velocity of the sound in water, the aperture and the focal length of the transducer. This diameter Φ can be evaluated as :

$$\Phi = \frac{V_L}{f} \cdot \frac{F}{D}$$

V_L : Velocity of longitudinal waves

f : Working frequency of the transducer

F : Focal length of the transducer

D : Aperture of the transducer

In our working conditions, we can expect a spot diameter down to 130 µm @ -6 dB.

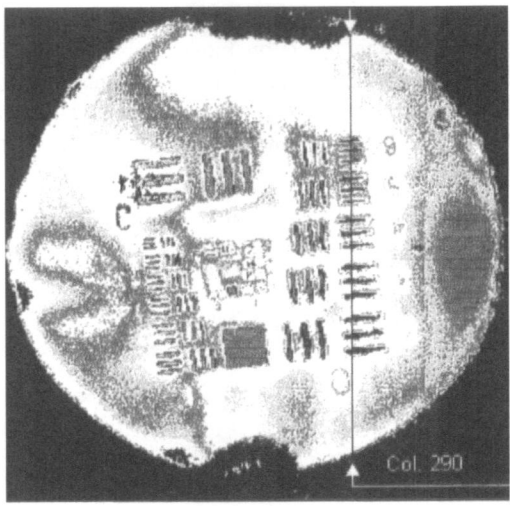

Figure 5. Acoustic image of a polymer sample insulated through the test target. The sample has a diameter of 27 mm.

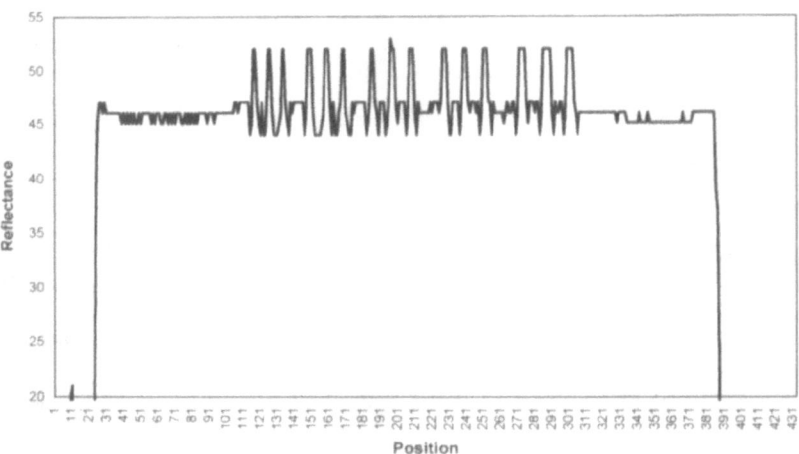

Figure 6. Vertical sampled line extracted from data file of fig. 5 (col. 290). The Y axis values are 6 bits numerical datas. Value 63 corresponds to 16mV peak of the amplitude at the output of the transducer. Step value : 625 µm.

RESULTS

In order to check the capability of microechography, the samples are polymerized through the USAF 1951 test target and undergo a light polishing to avoid any effect due to geometric surface flaw. Figure 5 shows the amplitude image, taken perpendicularly to the beam. The image corresponding to periodically spaced bars (figure 6) highlights the contrasts of crosslinking density. Each region appears to be homogenous, but the contour of bars is not as sharp as in the mask.

Images of the cross-section of a sample (parallel to the insulation beam) show the highly contrasted crosslinking density (figure 7). In order to minimize the interfering reflections on the cover glass, the incidence of the luminous radiations was slighty slanted with regard to the vertical axis defined by the sample surface. Starting from the surface, a light shrinkage of the exposed zones can first be observed, in agreement with the optical effect of the mask apertures. In deeper regions, the widening of the crosslinked domain is related to the changes of the photolinking phenomenon with decreasing beam intensity and to the loss of resolution induced by local gradients of conversion.

Figure 8 shows the image, taken perpendicularly to the beam, of a sample obtained by illumination through the apodizing filter. The concentric zones appearing on the figure indicate a radial gradient of the acoustic impedance Z. Figure 9 reports the value of the reflectivity measured along one diameter. Since the reflectivity is related to Young's modulus and the photopolymerization is carried out up to maximum conversion, it can be concluded that the observed image contrast represents a crosslinkage gradient.

Figure 7. Image of a cross section of the sample of figure 5. The cut has been done near the line drawn on figure 5.

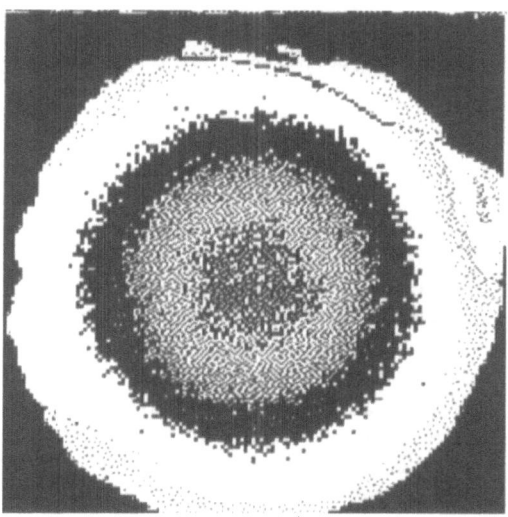

Figure 8. Grey scaled reflectivity map of a photopolymer sample insulated with a radial gradient. The diameter of the sample is 27 mm.

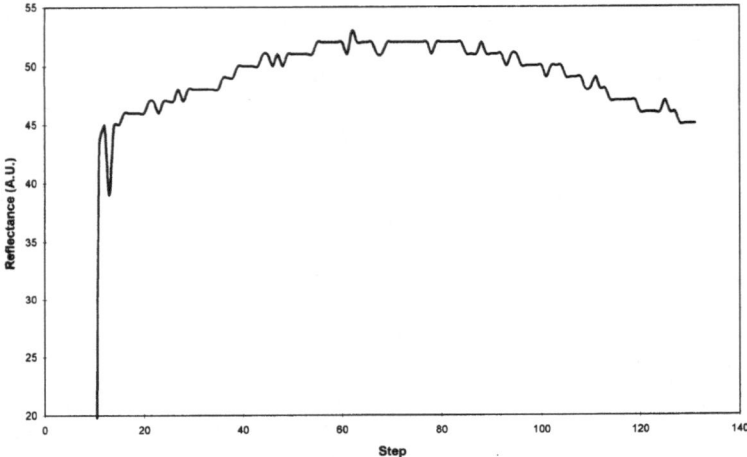

Figure 9. Relative reflectance along a diameter of a photopolymer sample insulated through an apodizing filter.

Figure 10 shows a cross-section parallel to the insulation beam. It reveals very clearly the consequence of combining the radial gradient of incident irradiance and the gradient of absorbed actinic light that takes place along the depth of the sample due to attenuation by the initiator molecules.

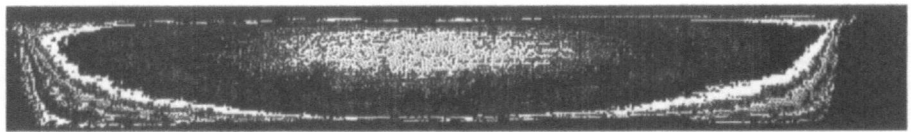

Figure 10. Acoustic image of the cross section of a polymer sample insulated with a radial gradient

In order to explore the possibility of using acoustic reflectance measurements to derive quantitative information on the conversion ratio of polymer materials, three series of samples, the photopolymerization of which was conducted under various illumination density (0.28, 0.73 and 2.33 mW.cm^{-2} respectively), were studied in parallel. Figure 11 represents the acoustic reflectance and conversion ratio of these samples at different stages of the photopolymerization. This latter value was obtained by FTIR analysis of the residual acrylate fonction present in the samples. Although not corresponding each other through a linear transformation, the set of curves shows more than a qualitative correlation which should be auspiciously turned to account for studying polymerization kinetics.

The lineary deviation in the correspondance of these curves must be due to the fact that microechography probes only the surface layer of the polymer sample whereas the information available from FTIR measurements refers to the bulk material. Incidentally, this deviation could well be turned to account to learn about peal effects which are related to oxygen quenching processes that take place when radical polymerizations are carried out in open reactors.

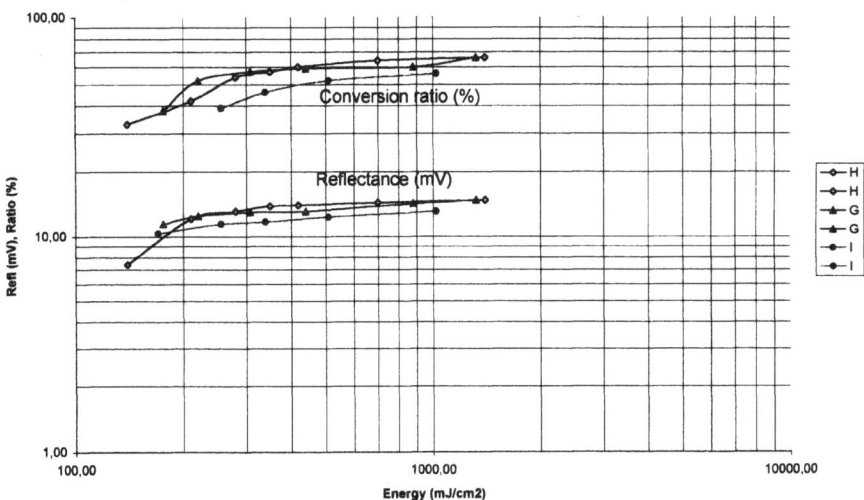

Figure 11. Relation between conversion ratio and Reflectance. Three sets of samples, respectively illuminated under three different illumination density have been studied.

DISCUSSION-CONCLUSION

Acoustic microechography is particularly suitable for the observation of surfaces and defects as well as structure inside polymer materials (penetration 0.1mm). In relation to other available microscopic techniques, microechography is easier to implement and to use. It is a non destructive method which is able to analyse samples without surface preparation, because a surface profile with a planeity lower than 100 µm and a roughness better than 5 µm, have no effect on the amplitude of the echo. Moreover, an adjustable time-delay gate is used to isolate the selected echo, because the signal travelling backwards contains generally a succession of echoes. So the reflected signal corresponds to the surface and amplitude variations are only due to changes in mechanical properties. The pulse method operating at frequencies between 10 MHz and 150 MHz is composed of a single acoustic wave that allows soft materials such as polymeric network to be studied. In fact, such materials which are characterized by a strong acoustical absorption would not lend themselves to acoustic microscopy investigations. In return for that, the use of a low frequency acoustic wave limits the lateral resolution of the instrument (20 - 150 µm).

Besides the possibility of visualizing heterogeneities in polymers, the results obtained suggest that microechography could be a well suited tool for the quantitative study of crosslinking gradients. Such gradients generally escape the sensitivity of classical analytic technique and are transparent to DSC or FTIR spectroscopy. They are detected through the correspondant variations of the Young's modulus at the local scale, a mechanical characteristic which greatly depends on the cohesion of the sample. This technique therefore, exhibits a high sensitivity which is of the greatest interest for the study of polymer materials with almost complete conversion.

The local variations of Young's modulus that can be detected by microechography may also be related to heterogeneities of structure, i.e. difference in the spatial arrangement of the crosslinked polymer network, hence a possibility of visualizing and quantify the associated local changes of refractive index.

This technique appears thus as a very promising means for the quantitative and real time study of polymerization dynamics as well as for the evaluation of polymer optical elements that can be fabricated by using space resolved photopolymerization.

BIBLIOGRAPHY

[1]   D.J. Lougnot, "Photopolymers and Holography", in Radiation Curing in Polymer Science and Technology, Vol 3, J.P. Fouassier, J. Rabek (Eds), Chapman and Hall, Andover (1993)

[2]   Y.B. Boiko, J.R. Moya-Cessa, F. Mendoza-Santoyo, D.J. Lougnot, "Diffractive optics for CO_2 lasers on dry photopolymer : fabrication and testing" Proc. SPIE, San Diego, 2000, 256-269 (1994)

[3]   Y.B. Boiko, V.S. Solovjev, S. Calixto, D.J. Lougnot, "Dry photopolymer films for computer generated infrared radiation focusing elements" Appl. Optics, 33,(5), 787-793 (1994)

[4]   D. Bonvallot, Thèse de l'université de Haute-Alsace, "Développement de photopolymères à gradient d'indice de réfraction en vue d'applications en optique ophtalmique", Mulhouse (1994)

[5]   R.A. Lemons and C.F. Quate, Appl. Phys. Lett., 24, 163 (1974)

[6]   C.F. Quate, A. Atalar, H.K. Wickramasing, Proc. IEEE, 67 (1979) pp. 1092-1194

[7]   A. Briggs, Acoustic Microscopy, Ed. Clarendon Press, Oxford (1992), 103

[8]   M. Issoukis "Scanning acoustic microscopy - Principles and applications" Metallic Materials, Vol.5, N°2, Feb. 1989, pp 63-67

[9]   F. Lisy, A. Hiltner, E. Baer, J.L. Katz, A. Meunier "Application of scanning acoustic microscopy to polymer materials" Journal of Applied Polymer Science Vol. 52, 1994, pp 329-352

[10]   S.P. Pappas, UV Curing : Science and Technology, Technology Marketing Corp., Stamford 1985; Willey, New York 1992

[11]   J.P. Fouassier, J. Rabek (Eds.), New aspects of Radiation Curing in Polymer Science and Technology, Chapman and Hall, London 1993

[12]   R.J.M. da Fonseca, Y.M.B. de Almeida, B. Cros, J.M. Saurel and M.J.M. Abadie, Thin Solid Films, 291 (1994) pp 110,122

[13]   J.J. Hunsinger, L. Simonin, J.P. Gonnet, B. Cros, D.J. Lougnot, Pure Appl. Opt. 4 (1995) pp 529-542

DEVELOPMENT OF HIGH-SPEED ULTRASONIC
TESTING DEVICE USING ELECTRONIC SCANNING

Y.Takishita, H.Kino, S.Yamaguchi, A.Iwasaki, and N.Yamamoto

Technical Research Laboratory, Hitachi Construction
Machinery Co., Ltd., 650 Kandatsu-machi, Tsuchiura,
Ibaraki, 300 Japan

INTRODUCTION

Utilization of high-frequency, ultrasonic Scanning Acoustic Tomograph (SAT) has been expanding. It has been used on various materials such as semi-conductors, automobile parts and target materials etc.. Recently, it has been increasingly used as an in-line flaw detection tool besides as a tool for Research & Development. Since high speed is a prerequisite for in-line flaw detection in production, in some cases, the conventional mechanical scanning method cannot satisfy this demand for high speed. For this reason, we have developed an ultrasonic testing device which uses high-frequency (10 to 25MHz) electronic scanning technology to satisfy the demand for high speed.

We report herein on the outline of our newly developed device as well as on various application examples.

SCANNING METHOD

In Fig.1, the scanning method of our newly developed device is compared to that of a conventional device[1] Using the conventional device (a), two dimensional data is collected by moving a point focused probe in the X-axis and Y-axis directions, performing mechanical scanning. A "C scope image" is formed based on these data. With the conventional device, the overall measurement time is constrained due to this mechanical probe-traveling time.

On the other hand, our newly developed device is equipped with an array probe, performing electronic scanning in the X-axis direction. Since the velocity of the electronic scanning is fast, ranging from 4 to 8 m/s, practically a C scope image can be formed quickly in the time that the array probe mechanically moves a specified distance in the Y-axis direction. In addition, an approximately 0.2 mm diameter spot beam is formed using electronic focusing, which uses the array transducer requiring 16 to 24 simultaneously activated elements to form a focused beam in the X-axis direction, and a line focused acoustic lens focusing in the Y-axis direction.

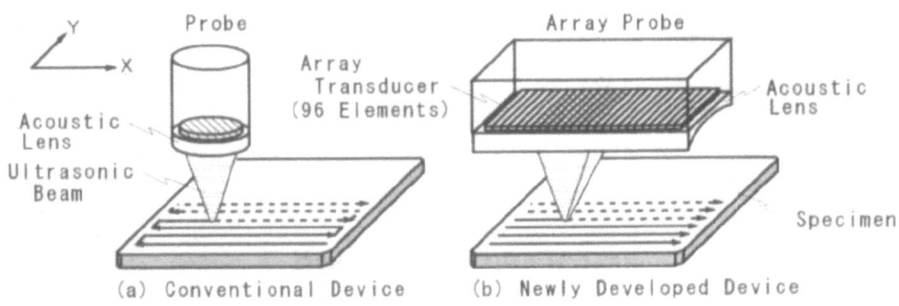

Fig. 1 Comparison of Scanning Methods

(a) Conventional Device (b) Newly Developed Device

DEVELOPMENT OF THE ULTRASONIC TRANSCEIVER

In the ultrasonic transceiver, 96 channels of pulser·receiver circuits are provided, each independently corresponding to one of 96 transducer elements. For this reason, a high speed ultrasonic beam switching of up to 50 μs (the pulse repetition rate=20 kHz) has been realized. In addition, a delay time setting circuit, which is necessary for electronic focusing, is provided in the ultrasonic transceiver, setting 0 to 200 ns of delay times for each of array transducer elements, both in ultrasonic transmission and receiving.

The main specifications of our newly developed array probe are shown in Table 1. As shown in Fig.2, we have succeeded in miniaturizing the array probe by applying ideas such as eliminating wiring connectors.

Table 1 Main Specifications of Array Probes

Center Frequency	10MHz	25MHz
Number of Elements	96	←
Transducer Pitch	0.4 mm	0.2 mm
Transducer Width	8.0 mm	6.0 mm
Underwater Focal Length	15 mm	←
Electronic Scanning Range	32.4 mm	14.6 mm
External Dimensions	50 (W) ×18 (D) ×24 (H) mm	

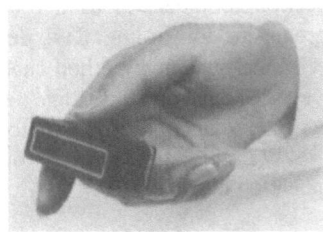

Fig. 2 Array Probe

SCANNING METHOD TO COVER LARGE SCANNING RANGE

In order to expand the electronic scanning range, widening the transducer element pitch or increasing the number of the transducer elements is required. Since the transducer element pitch is equal to the sampling pitch, widening the transducer element pitch deteriorates image resolution. Increasing the number of the transducer elements is technically possible. In fact, we have succeeded in fabricating an array probe[2] having 192 transducer elements. However, it is not the best way to increase the scanning coverage since an array probe has to be made larger.

For the reasons above, we have examined an inspection method to go beyond the fixed electronic scanning range of our 96 element array probe using the 96 element array probe. The inspection method is shown in Fig.3. At the end of each mechanical Y-axis direction scanning stroke, the array probe is moved to the X-axis direction for a distance equal to the electronic scanning range (L), and each area of the Y-axis direction scanning data is joined together, expanding the X-axis direction scanning range.

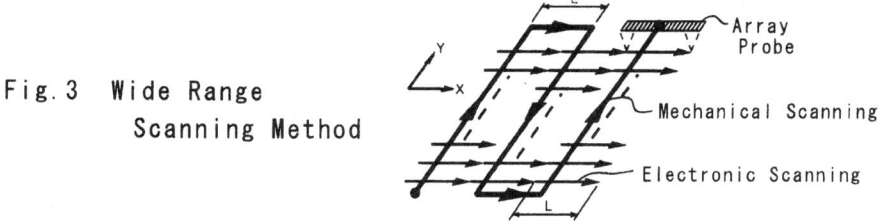

Fig. 3 Wide Range
         Scanning Method

CONSTRUCTION OF THE DEVELOPED DEVICE

Fig.4 shows the overall block diagram of the developed device. After the scanner starts mechanical movement of the array probe, one line of an electronic scanning is activated each time the probe is moved for the specified distance (the sampling pitch). The received ultrasonic signals are input to the microcomputer as digital data via a gated peak detector and an A/D converter, and are displayed in real time on the color monitor after address translation.

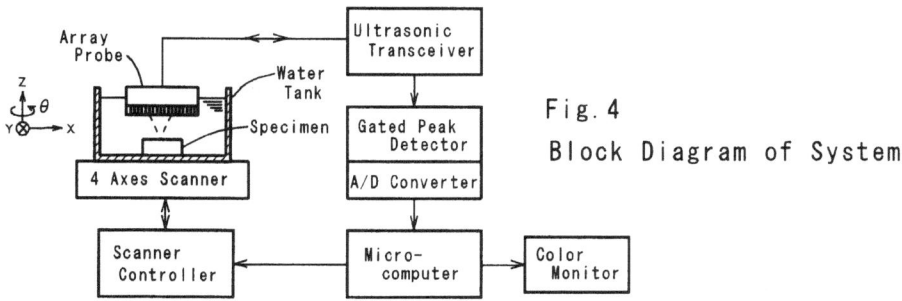

Fig. 4

Block Diagram of System

Up to 512 × 512 data can be displayed with 256 scales of pseudo-colors. In addition, the device is equipped with a digital processing logic function, which is activated in automatic scannings immediately after completing C scope image display.The scanner controller controls up to four axes (X, Y, Z and θ). After selecting the electronic scanning direction from among the X-axis, Y-axis and Z-axis direction, the mechanical scanning direction shall be selected from among the remaining directions including the θ direction.

EXAMPLES OF DEVICE APPLICATION

Fig.5 shows the welded boundary surface of a valve plate consisting of a steel part and a copper part, which is imaged using the array probe at a frequency of 25MHz . The delaminated (failed) area on the boundary surface is shown in bright tone. As the preset overall electronic scanning range was 102.2 mm (the pitch: 0.2 mm), the electronic scanning (14.6 mm in range) was repeated seven times and joined together to obtain the whole welded boundary image of the valve plate. The picture shown in Fig.5 is an image in the middle of the sixth electronic scanning. The overall inspection time was approximately 23 seconds, a speed approximately 10 times faster than that of a conventional device.

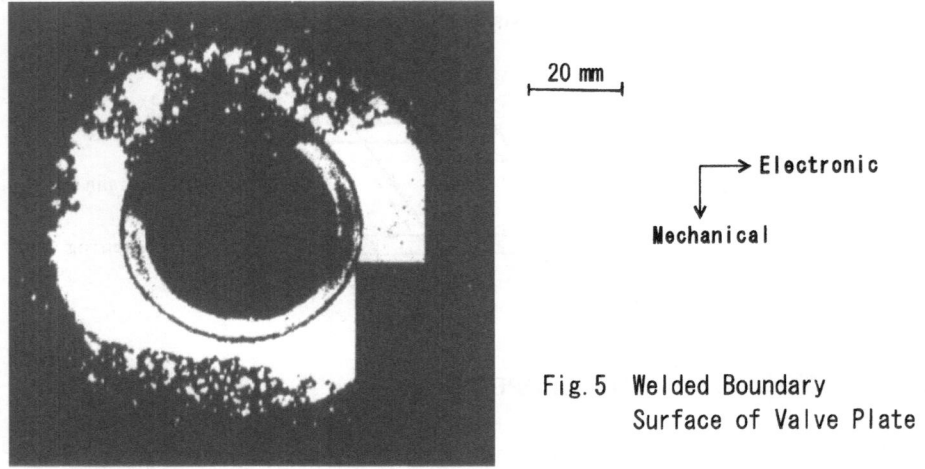

20 mm

→ Electronic

↓
Mechanical

Fig. 5 Welded Boundary
Surface of Valve Plate

Fig.6 is an image of the brazed boundary surface of an aluminum-made honeycomb structure (between the honeycomb core and the top cover), obtained using the array probe at a frequency of 25 MHz . On the image, the brazed lines are shown in black lines and the segment which was not brazed is clearly recognized as a missing line segment. In this instance, the preset overall electronic scanning range was 87.6 mm (the pitch: 0.2 mm) and the electronic scanning was repeated six times to form the image of the whole honeycomb structure brazed boundary. The overall inspection time was approximately 17.5 seconds.

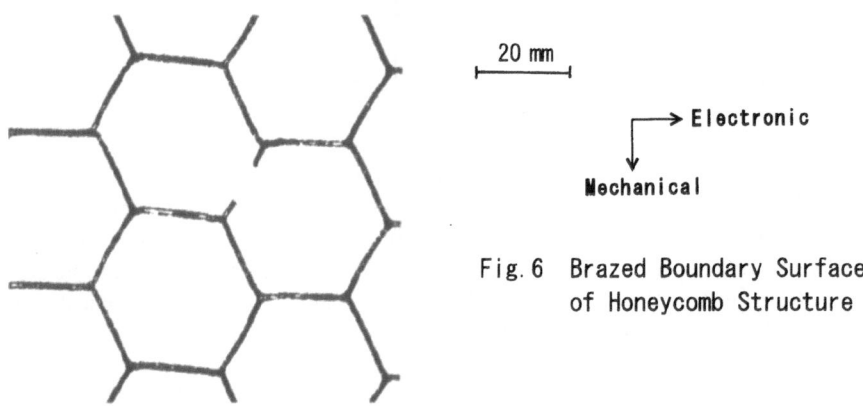

20 mm

→ Electronic

↓
Mechanical

Fig. 6 Brazed Boundary Surface
of Honeycomb Structure

Fig.7 is a picture image of two-fifths of a cylinder-shaped-forged-part inner circumference (I.D.: 53 mm). In the forged part, artificial flaws were planted by electrical discharge machining as shown in Fig.8. The frequency used was 10 MHz and the electronic scanning range was 32.4 mm (the pitch: 0.4 mm). The electronic scanning was performed only one stroke in the specified direction. The required time to complete the shown image was approximately 1.5 seconds. In this instance, artificial flaws ranging from 1 mm to 10 mm are shown clearly because the array probe was positioned at an angle against the inner surface, as shown in Fig.9, so as to extract only flaws as echoes.

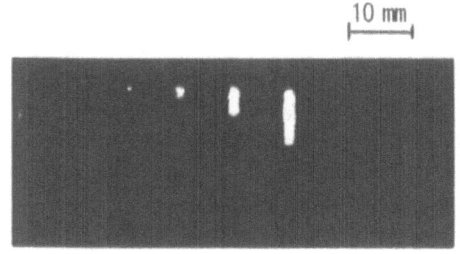

Fig. 7 Image of Artificial Flaws
Detected by Scanning along Inner
Surface of Cylindrical Object

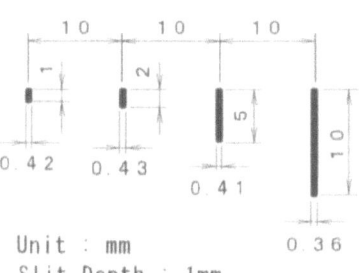

Unit : mm
Slit Depth : 1mm

Fig. 8 Artificial Flaws

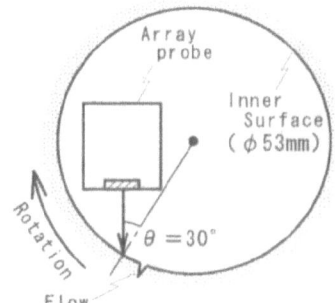

Fig. 9 Flow Detection Method

EXAMINATION OF IMAGE RESOLUTION

We fabricated artificial test pieces to examine fine-flaw detectabilty of the device as follows (Fig. 10) : First, we etched 15 fine flaws, 0.3 μm in depth, varying in width from 50 μm to 1000 μm on a 1 millimeter-thick, 3 inch-diameter silicon wafer. Next, another silicon wafer of the same size was laminated on the etched surface side by pressing and heating. Because the oxide (SiO_2) spread and formed on the boundary surface, the two wafers were bonded sufficiently so that no echoes existed from the boundary surface when ultrasonic waves were applied to the top of the bonded wafer. As a result, the fabricated test piece could have been considered as a 2 mm thick wafer having artificial flaws at the depth of one millimeter from the surface.

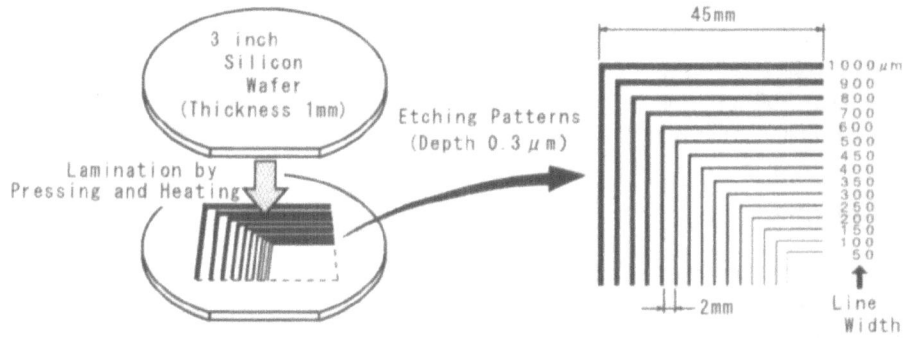

Fig. 10 Test Piece Fabrication Method

An image obtained using the above mentioned test piece is shown in Fig.11. The frequency of the array probe used was 25 MHz and other scanning conditions were the same as the ones in the case of the honeycomb structure (Fig.6). The white ring shown in the circumference of the wafer is the unbonded area of the wafer due to fine warping of the wafers, detected as bottom echoes of the top wafer. Among 15 etched line patterns, lines with the width down to 100 μm were clearly detected. Moreover, the 50 μm width line was slightly imaged, but did not come out clearly as the echo level was low.

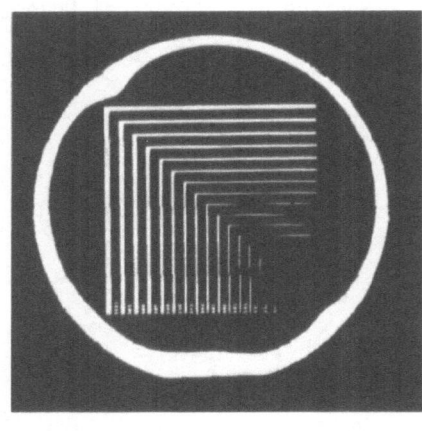

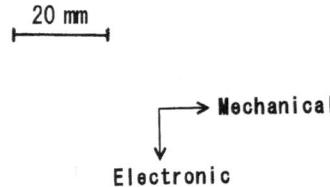

Fig. 11 Image Obtained using
Test Piece Fabricated to
Evaluate Image Resolution

From the above results, we have confirmed that, at a frequency of 25 MHz, flaws down to approximately 100 μm can be recognized on the original data without any image processing. Moreover, inspecting the same test piece with an array probe at 10 MHz, we confirmed that lines with a width down to 150 μm were clearly detected.

SUMMARY

(1) We have developed a high-speed ultrasonic inspection device which produces C scope images using a combined scanning techniques of a high-frequency (10 to 25MHz) electronic scanning and mechanical scanning.
(2) We have succeeded in widening the electronic scanning range by combining the mechanical scanning with the electronic scanning and by introducing a method to join each strip of electronic scanning ranges together.
(3) Through various sample examinations, we have proved that the device has a performance speed 10 times faster than that of conventional devices.
(4) With the artificial-test-piece-used examination, we have confirmed that flaws down to approximately 100 μm can be recognized at a frequency of 25 MHz on the original data without any image processing.

REFERENCES

1. I.G.Park,S.Sasaki and H.Yamada,"Ultrasonic C-scan Technique for Nondestructive Evaluation of Spot Weld Quality",The 1994 Far East Conference on Nondestructive testing and ROCSNT 9th Annual Conference,103-110 (1994).
2. S.Sasaki,Y.Takishita,A.Iwasaki,J.Kubota,Y.Musha and H.Okada,"High-speed C-scan Imaging System with Electronic Scanning of 25MHz Ultrasonic Beam", Acoustical Imaging 19, 251-256 (1992).

WAVE PROPAGATION IN CONCRETE: THEORETICAL AND NUMERICAL RESULTS FOR MEANFIELD ATTENUATION

Eberhard Burr, Norbert Gold, and Ulrich Werner

Geophysical Institute
University of Karlsruhe
Hertzstr. 16
76187 Karlsruhe
GERMANY

INTRODUCTION

The subject of the SFB 381 is characterization of damage-propagation in fiber-reinforced-composites by Non-Destructive-Testing. Our group has the intention to simulate ultrasonic experiments via finite-difference-methods as close to reality as possible. In our presentation we only deal with concrete, but our media simulating program is very variable, and we can also lengthen and straighten the shown ellipses, introduce cracks into the synthetic material and deal with laminates.

We also try to handle the wave-propagation analytically. Describing the elastic medium parameters by correlation functions that fit their true distribution. This is what our poster is dealing with. In our example of concrete we take a linear combination of two exponential correlation functions which fits our true correlation function of concrete very well.

We split the wave u propagating through this medium into a scattered part u_f and a coherent part $< u >$, the so-called Meanfield. The Meanfield is carrying information about the elastic parameters of the medium and the scattered part is telling us more about irregularities in the structure of the medium. Our ability to separate these two effects is giving us the chance to grab for this information.

In our example we only deal with attenuation because of scattering. We compare our twodimensional analytical attenuation coefficient with the one we get from a plane wave transmitted through a 2D concrete model numerically. This plane wave is scattered at several stuffing materials with varying elastic parameters and is recognized at different receiver lines throughout the medium.

Acoustical Imaging, Vol. 22, Edited by P. Tortoli
and L. Masotti, Plenum Press, New York, 1996

surface(m)

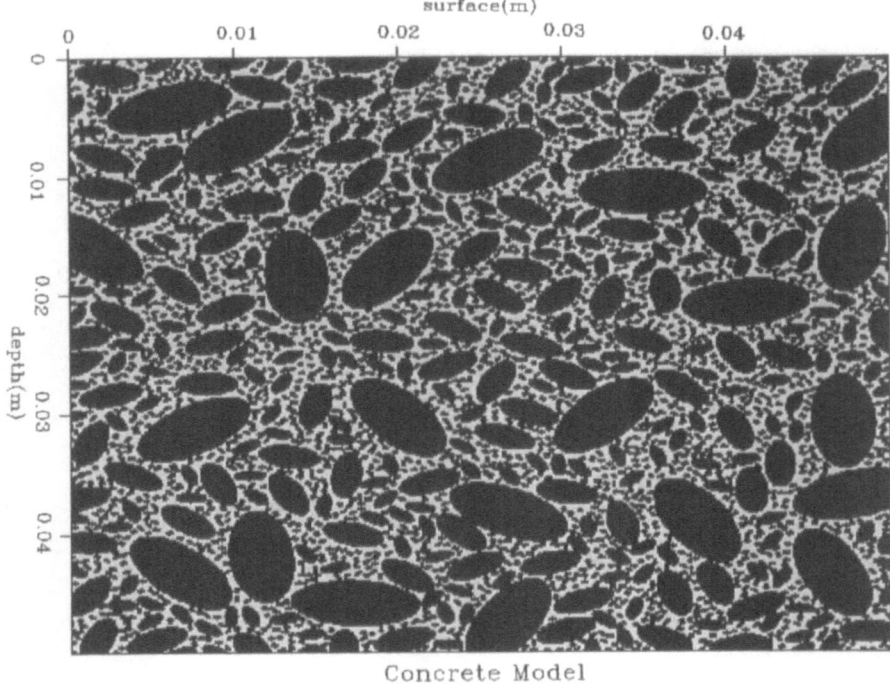

Concrete Model

Figure 1. Synthetic model for concrete with realistic parameters

NUMERICAL EXPERIMENT

The plane-wave-propagation through the 2D concrete model is simulated via very exact finite-difference-operators. We needed 500*500 gridpoints for scanning our 5cm*5cm concrete medium, so that even the very little sand ingredients (0.4 - 0.6mm) could be considered, and it took 2500 timesteps to propagate the wave through the model.

As source we took a Ricker wavelet with the very high frequency of 7.5 MHz because there should be large vertical scattering at the stuffing material inside the synthetic concrete.

There are ten receiver lines in our model spaced every 0.5cm, each consisting of 50 receivers at a distance of 1mm. The receiverlines at the top of the model show an almost unscattered wave. This is the weak fluctuation region where $\frac{|u_f|}{|<u>|} < 1$. During the propagation the wave gets more and more scattered, and after some time we reach the strong fluctuation region, where $\frac{|u_f|}{|<u>|} > 1$.

After this simulation, the received data are processed: we compute the Meanfield by stacking the receiverlines, normalizing them and taking the logarithm of each A-Scan. Then we search for the sum spectrum of these receiverlines and pick the frequencies we need.

Now we can extract the numerical scattering coefficient for several frequencies by plotting the normalized Meanfield $\ln(|\text{Ampl}(< u >)|)$ of each frequency versus the distance. The result is a straight line, because scattering is independent of the distance travelled. The gradient of the straight line is the numerically computed scattering coefficient of the true correlation function for concrete parameters.

Now, our analytical scattering coefficient is computed for a corresponding linear combination of exponential correlation functions, which describe the fluctuation of the elastic parameters of a random medium .

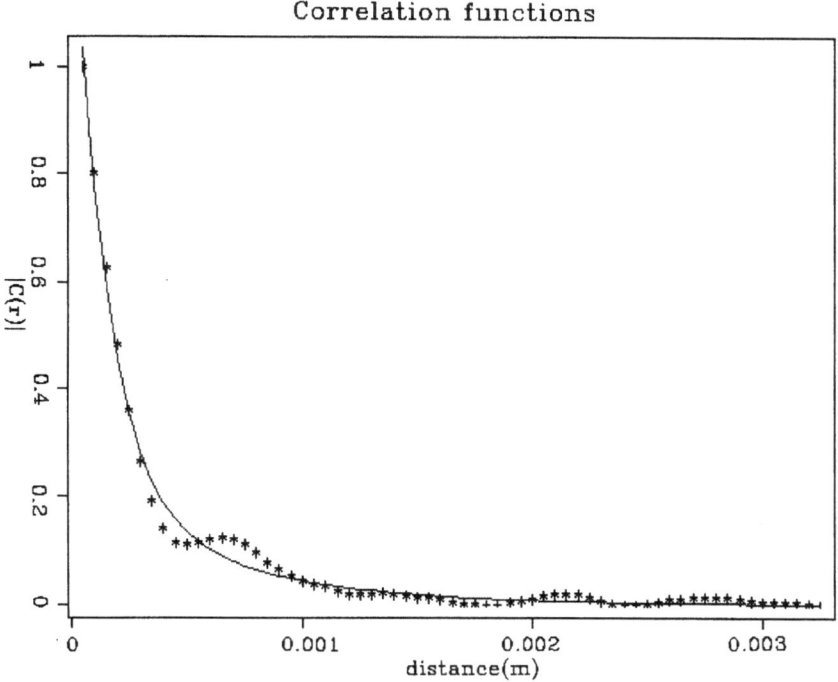

Figure 2. The true correlation function for concrete parameters (dotted line), and a corresponding linear combination of exponential correlation functions (solid line) for the random medium parameters.

For comparing our numerical scattering coeficient succesively with the analytical one, we vary the fluctuating parameters: In the first example take only λ-fluctuations, in the second example we let λ and μ fluctuate and in the third example the true fluctuations of all elastic parameters λ, μ and ρ are used.

At last we can decide if it's a good possibility to describe concrete as a random medium with an exponential distribution. (Look at the table with the results at the end of this article.)

THEORETICAL DESCRIPTION

Throughout this section we use standard tensor index convention. Indices that occur twice imply summation. The subscript $,i$ denotes differentiation with respect to the coordinate i.

To find a theoretical description of the attenuation of the coherent field, it is necessary to get a formula for the scattered part of the incident energy. The power flux across a surface A can be described by (Gubernatis et al., 1977)

$$P = \int_A I_i n_i dA , \tag{1}$$

where n_i is the unit normal to the direction of propagation and I_i is the intensity vector, defined as follows :

$$I_i = \sigma_{ij} \dot{u}_j . \tag{2}$$

In this equation σ_{ij} denotes the stress field, u stands for the wavefield (displacement field). We have for the incident and scattered power flux :

$$j_{E,in} \sim |I_{oi}| A_s = |u_o|^2 V_s / L_s \tag{3}$$

$$j_{E,sc} \sim \int_{A'} I_i(\theta, \phi) n_i(\theta, \phi) . \tag{4}$$

Here L_s is the thickness of the scattering object and A_s the area perpendicular to the direction of wave propagation. We got the scattered power flux by integrating over a sphere A' with radius r around the scatterer.

Now we assume that r is large compared with the extension of the scattering object. This causes no error and has the positive effect that in equation (4) I_i and n_i become parallel. If we take into account that I_i is proportional to $|u|^2$ and use spherical coordinates, we obtain an exponential decrease of the transmitted (unscattered) wavefield with travel distance L_s.

The attenuation coefficient reads :

$$\alpha_u = \pi \int_0^\pi sin(\theta) \left[\frac{r^2}{V_s} \frac{|u_s|^2}{|u_o|^2} \right] d\theta . \tag{5}$$

To determine α_u, it is now sufficient to know the scattered field at points far from the scatterer.

We use the basic equation from Gubernatis et al :

$$
\begin{aligned}
u_i^s(r) = {} & \omega^2 \int_V dV' \delta\rho(r') g_{im}(r - r') u_m(r') \\
& + \int_V dV' \delta C_{jklm} g_{ij,k}(r - r') u_{l,m'}(r') ,
\end{aligned}
\tag{6}
$$

where g_{im} stands for the elastodynamic Green's function and C_{jklm} is the stiffness tensor. Now we make the following assumptions, which are valid for concrete :

1.) The medium is assumed to be isotropic. So C_{jklm} depends only on the lame parameters λ and μ.

2.) We make the Born approximation. This means that in the right side of equation (6) the wavefield u is replaced by the incident (undistorted) field. One can show that the effect on α_u is negligible if the correlation length of the medium fluctuations is small compared with the inverse of α_u.

3.) We assume an isotropic spatial correlation function of the medium fluctuations.

Taking the square of (6), we get after some calculations the attenuation coefficient for the coherent field of a P-wave :

$$
\begin{aligned}
\alpha_u^{PP}(3D) \;=\; & \frac{\alpha^2}{8} \int\limits_{-1}^{1} d\tilde{\theta} \left(\frac{\lambda_o^2}{(\lambda_o + 2\mu_o)^2} F'_{\lambda\lambda} + \tilde{\theta}^2 F'_{\rho\rho} \right. \\
& + \frac{4\mu_o^2}{(\lambda_o + 2\mu_o)^2} \tilde{\theta}^4 F'_{\mu\mu} + \frac{4\lambda_o\mu_o}{(\lambda_o + 2\mu_o)^2} \tilde{\theta}^2 F'_{\mu\lambda} \\
& \left. - \frac{4\mu_o}{\lambda_o + 2\mu_o} \tilde{\theta}^3 F'_{\rho\mu} - \frac{2\lambda_o}{\lambda_o + 2\mu_o} \tilde{\theta} F'_{\rho\lambda} \right)
\end{aligned}
\tag{7}
$$

$$
\begin{aligned}
\alpha_u^{PS}(3D) \;=\; & \frac{\beta^2}{8} \int\limits_{-1}^{1} d\tilde{\theta} \left((1 - \tilde{\theta}^2) F'_{\rho\rho} \right. \\
& \left. - \frac{4\alpha}{\beta} (\tilde{\theta} - \tilde{\theta}^3) F'_{\rho\mu} + \frac{4\alpha^2}{\beta^2} (\tilde{\theta}^2 - \tilde{\theta}^4) F'_{\mu\mu} \right)
\end{aligned}
\tag{8}
$$

In equations (7) and (8) we used :

α $\quad$ = wavenumber of P-wave
β $\quad$ = wavenumber of S-wave
$\tilde{\theta}$ $\quad$ = cos (scattering angle θ)
$F_{ab}(k)$ = 1D Fourier transform of (cross)correlation B_{ab} at
k $\quad$ = $2\alpha sin(\theta/2)$
F' $\quad$ = $\partial/\partial\tilde{\theta} F$.

In the case of a 2D medium, we get the following result :

$$
\begin{aligned}
\alpha_u^{PP}(2D) \;=\; & \frac{\alpha^3}{4} \int\limits_{-\pi}^{\pi} d\theta \left(\frac{\lambda_o^2}{(\lambda_o + 2\mu_o)^2} G_{\lambda\lambda} + \tilde{\theta}^2 G_{\rho\rho} \right. \\
& + \frac{4\mu_o^2}{(\lambda_o + 2\mu_o)^2} \tilde{\theta}^4 G_{\mu\mu} - \frac{4\mu_o}{\lambda_o + 2\mu_o} \tilde{\theta}^3 G_{\rho\mu} \\
& \left. + \frac{4\lambda_o\mu_o}{(\lambda_o + 2\mu_o)^2} \tilde{\theta}^2 G_{\mu\lambda} - \frac{2\lambda_o}{\lambda_o + 2\mu_o} \tilde{\theta} G_{\rho\lambda} \right)
\end{aligned}
\tag{9}
$$

$$
\begin{aligned}
\alpha_u^{PS}(2D) \;=\; & \frac{\alpha\beta^2}{4} \int\limits_{-\pi}^{\pi} d\tilde{\theta} \left(sin^2(\theta) G_{\rho\rho} - \frac{4\alpha}{\beta} cos(\theta) sin^2(\theta) G_{\rho\mu} \right. \\
& \left. + \frac{4\alpha^2}{\beta^2} cos^2(\theta) sin^2(\theta) G_{\mu\mu} \right)
\end{aligned}
\tag{10}
$$

In these equations G_{ab} is defined as follows :

$$
G_{ab} = \int\limits_{0}^{\infty} r J_o(kr) B_{ab}(r) ,
\tag{11}
$$

where $B_{ab}(r)$ is the crosscorrelaton function of the medium parameters a and b.

In the case of media with exponential correlation functions (as often occuring in Geophysics), we have found an analytical solution of equations (7) - (10). We compared our results with the results for acoustical random media (Shapiro, Kneib, 1993) and found an exact agreement in the case of $\mu = 0$ and no fluctuations of ρ, where we can consider our medium as acoustical.

RESULTS

The following table shows the different scattering coefficients, we get in our analytical and numerical computations:

Table 1

fluctuating parameters	f(MHz)	analytical α_E ($\frac{1}{m}$)	numerical α_E ($\frac{1}{m}$)
λ	3,5	11,47	9,62
	7	50,80	46,50
λ and μ	4	0,73	0,46
	7,5	1,54	1,05
λ, μ and ρ	3,5	15,68	12,37
	7	34,91	29,10

CONCLUSIONS

We can see that theoretical and numerical results are of the same order. If we sum up over different realizations of concrete, the numerical scattering coefficient converges against the analytical one. (We did this for λ - fluctuations, that's why these coefficients are closer to each other.)

This is a very good result if we look at the correlation function of concrete and its approximation, if we compare the two different models and if we think about the mistakes in the numerical computation (dispersion, truncation errors,etc.).

So we investigated a simple description (random medium) for a complicated medium like concrete and had the chance to describe the wavepropagation in this medium analytically.

REFERENCES

Gubernatis, J., E., Domany, E., Krumhansl, J., A., 1977: Formal aspects of the theory of the scattering of ultrasound by flaws in elastic materials, Journal of Applied Physics, Vol.48, No.7 (July 1977); P.2804-2811.

Kneib, G., Kerner, C. , 1993: Accurate and efficient seismic modelling in random media, Geophysics, Vol. 58, No. 4 (April 1993); P.576-588.

Shapiro, S.A., Kneib, G., 1993: Seismic attenuation by scattering: theory and numerical results, Geophys. J. Int.(1993), 114, 373-391.

ULTRASONIC VELOCITY MEASUREMENT FOR SUPERCONDUCTING MATERIAL VERSUS CRYOGENIC TEMPERATURES

E. Biagi

Dipartimento Ingegneria Elettronica
Università di Firenze
Via S. Marta 3, 50139, Firenze, Italy

INTRODUCTION

Ultrasonic pulse propagation in polycrystalline $YBa_2Cu_3O_{6+x}$ samples was examined during thermocycling from room temperature to well below the critical temperature Tc. An accurate measurement of ultrasonic velocity is strictly correlated to the knowledge of the propagation modes of the ultrasonic waves which are induced in the material. If the ultrasonic propagation direction is not perfectly aligned with the sample symmetry axis, non-pure propagation modes can be excited. In this case quasi-longitudinal and quasi-shear modes are obtained with different velocity values and consequently different echo pulses are generated which can appear overlapped. In the time domain this overlap appears as a perturbation of the signal shape which is difficult to quantify. In this paper the signal spectral analysis performed on the ultrasonic echoes, at various temperatures, is demonstrated to be a powerful tool to clarify and to detect this occurrence, which would have been impossible to explain by a simple analysis in the time domain. For some samples, spectral analysis has highlighted the presence of two partially overlapped echo pulses. This overlap begins at about 170 K and is observed up to Tc. The presence of these two pulses leads to the interesting hypothesis that non-pure modes can be excited in particular conditions, giving rise to peculiar effects. Phase transitions and changes in the lattice anharmonicity can be ascribed to possible causes which affect the propagation mode as temperature varies. If the sample structural organization changes versus temperature an out-of-axis propagation can occur inducing non-pure modes. Following this approach, derived from the acousto-elasticity theory, the velocity results obtained in previous works are also revised[1,2]. The velocity curves versus temperature confirm the presence of two hysteresis loops during thermocycling for all the investigated samples. This result is interesting and new with respect to data generally reported by other groups. Moreover different hysteresis loops are obtained not only for samples with different structural characteristics, but also for nominally equal samples. This discrepancy could be attributed to different orientation between the crystallographic axes and the geometrical sample axes induced by the manufacturing process.

EXPERIMENTAL SETUP

The employed experimental measurement system was reported in detail in previous work[2]. The thermocycling was performed in the temperature range: 50 to 290 K. Temperature was varied at rates between 60 and 12 K/h both for cooling and heating. Two silicon diode thermometers, whose uncertainty was about 0.5 K, detect possible thermal gradient and decoupling between the specimen and its holder. The ultrasonic investigation was carried out by using a dry contact pulse echo technique. Two broad band transducers for longitudinal and shear waves were used. The longitudinal transducer had a nominal central frequency equal to 20 MHz with a 2.8 mm active diameter surface and a 4 µs delay line (fig.1). The shear wave transducer, with a 5 MHz nominal central frequency, had a 2.8 mm diameter and a 7 µs delay line. Honey was used for coupling the transducers to the sample. The echo signals obtained by the ultrasonic investigation were acquired at a 100 MHz sampling frequency and stored for subsequent processing. Each acquired signal was the time domain average of 64 sight lines. In the present work we make reference to the

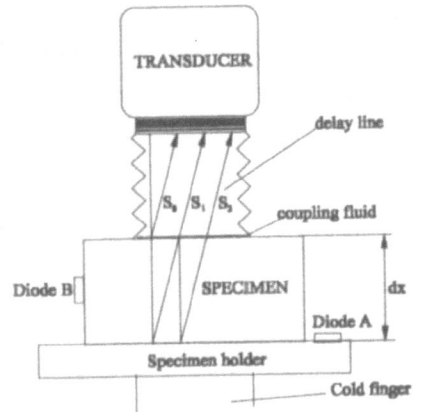

Figure 1. Transducer-specimen configuration.

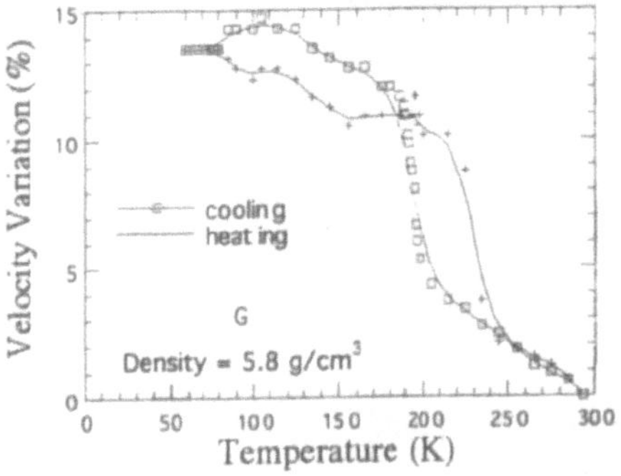

Figure 2. Sample G.Velocity-temperature for longitudinal wave. $\Delta V\% = [(V_T - V_{300})/V_{300}]\%$

investigation on ten polycrystalline $YBa_2Cu_3O_{6+x}$ (x≈0.9) samples with high density (5.8 g/cm³). The Tc values for all the investigated samples were between 89 and 91 K. The specimens, obtained by the method of co-precipitation of oxalate, were bar-shaped with dimensions 11x10x4 mm. The grain average dimension along the specimen maximum direction was about 20 - 30 µm. All samples exhibited a tendency to axial texture with the crystallite c-axes partially aligned with the symmetry axis. The texture degree decreases as porosity increases. The data reported in this work correspond to propagation of ultrasonic signals along the minimum sample dimension which could coincides with the internal symmetry axis, except for manufacturing tolerances.

ULTRASONIC VELOCITY

Evaluation of the velocity was performed by Time-of-flight (TOF) measurements obtained with a 0.1% uncertainty applying a cross correlation algorithm to the echoes coming from multiple paths inside the specimen. The measurement accuracy was computed taking into account the cross correlation resolution and the sample thickness measurement errors. The cross correlation TOF is the time shift τ for which $C(\tau)$, defined in the following expression, reaches its maximum value:

$$C(\tau) = \left| \lim_{T \to \infty} \frac{1}{T} \int_{-T}^{T} S_1(t) S_2(t+\tau) dt \right|$$

The symbols used are indicated in figure 1: $S_1(t)$ and $S_2(t)$ are the first and second back-wall echoes. The curves hereafter reported are evaluated according to the following equation:

$$\Delta V\% = [(V_T - V_{300})/V_{300}]\%$$

where V_T is the velocity at the current temperature and V_{300} is the velocity at room temperature, $\Delta V\%$ is the velocity variation in percent. For all the investigated samples, the velocity-temperature behavior shows, both for longitudinal and shear waves, an interesting peculiarity, generally not observed by other researches. For all the investigated samples two distinct hysteresis loops are detected. For sake of brevity only the results obtained on two investigated samples, labelled G and E, are reported. A first loop is observed between 190 and 70 K, a second one opens and closes itself around 230 and 200 K. The velocity curves, obtained on specimen G, are shown in fig 2 a and b, for longitudinal and shear wave, respectively. It can be noted that the lower temperature (70 to 190) loop results narrower for shear (fig.3) than for longitudinal velocity (fig.2).

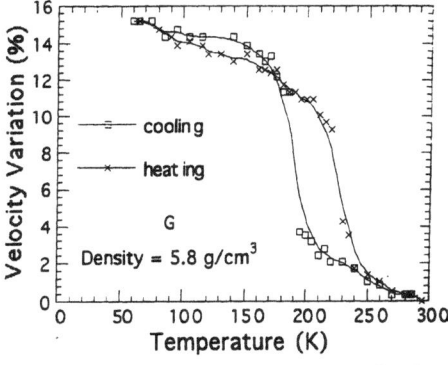

Figure 3. Sample G.Velocity-temperature for shear wave. $\Delta V\% = [(V_T - V_{300})/V_{300}]\%$

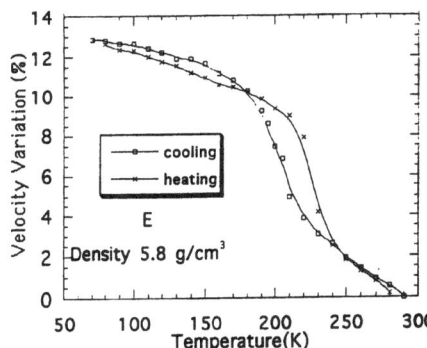

Figure 4. Sample E. Velocity-temperature for longitudinal wave. $\Delta V\% = [(V_T - V_{300})/V_{300}]\%$

Data obtained for longitudinal propagation in sample E, are reported in fig.4. A comparison with the curves of fig. 2 and 3, relative to sample G gives interesting indications. The total velocity variation on thermocycling is practically the same for the two samples (about 15% for G and 14 % for E). The 200 to 230 K loop presents the same shape and temperature interval for both the samples. However, the lower temperature loop obtained on sample E for longitudinal propagation is narrower than that measured for specimen G and resembles the loop here obtained for shear waves (see fig. 3). This behavior, in sample E, could be ascribed to the contemporary generation of quasi-longitudinal and quasi-shear modes, due to a non-perfect parallelism between the ultrasonic propagation direction and the sample symmetry axis. As mentioned above the manufacturing process does not provide sample repeatability in terms of alignment between crystallographic and geometric axes. Samples E and G have the same density but it is probable that sample E exhibits a strong skew of the geometric axes in relation to crystallographic ones, which results in an out-of-axis ultrasonic propa- gation. This hypothesis will be extensively examined and supported by experimental results in the following sections.

SPECTRAL ANALYSIS

If the non-pure mode echo signals are partially overlapped in the time domain a perturbation of the signal shape is produced which is difficult to quantify. The spectral analysis was carried out on the ultrasonic pattern echoes acquired at various temperatures for all the samples. Only sample E exhibits this echo overlapping, as reported in figure 5, which begins to appear at about 170 K, becomes more evident as temperature is further decreased up to Tc, and disappears on heating the sample above 200 K. The signal overlapping between quasi-longitudinal and quasi-shear echoes appears evident on the second echo signal of the quasi-longitudinal propagation S_2, that is, a double path for quasi-longitudinal wave has about the same TOF of a single path S_1 for the quasi-shear one, as it was demonstrate hereafter at room temperature. If two pulses, whose times of flight differ for an interval τ, are considered, it can be easily demonstrated that the total power spectral density, $|X(\omega)|^2$, is related to the single pulse power spectral density, $|S(\omega)|^2$, by the following simplified expression:

$$|X(\omega)|^2 = |S(\omega)|^2 \left(r_1^2 + r_2^2 + r_1 r_2 \cos(\omega \tau)\right)$$

where r_1 and r_2 are coefficients related to the respective intensities of the two overlapped pulses. In the frequency domain, an oscillation is superimposed on the single pulse power spectrum, creating the scalloping effect. The power spectral distribution for the second echo S_2 at 100 K for sample E is shown in fig.6. The scalloping effect is rather weak since the component of the quasi-shear mode, detected by the longitudinal transducer, is much less intense than the component corresponding to the quasi-longitudinal mode. A scalloping with a periodicity of about 2 MHz is observed in figure 6, thus the corresponding delay time τ is equal to 0.5 μs as also appears in figure 5.

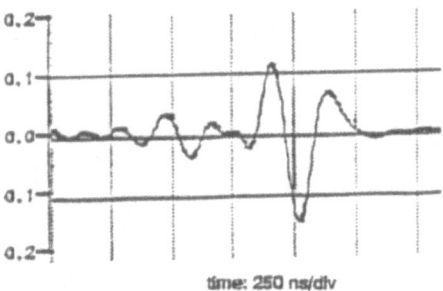

Figure 5. Sample E. Radiofrequency signal 100 K.

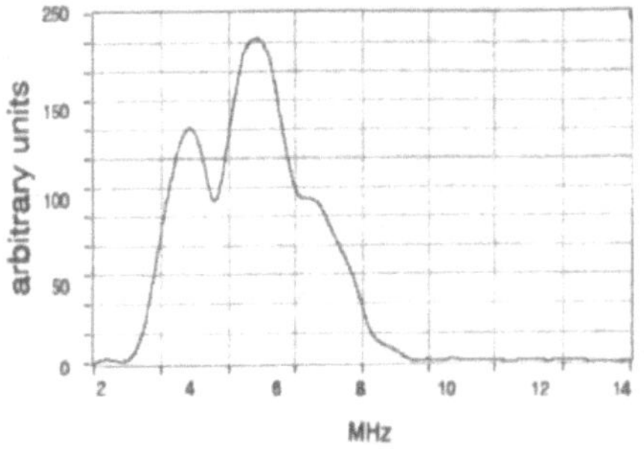

Figure 6. Sample E. Power spectral density $|X(\omega)|^2$ of second echo S_2 at 100 K.

To give further support to the hypothesis that non-pure modes are generated in sample E, room temperature velocity measurements was carried out on the same sample. The following values are obtained: the velocity of the quasi-longitudinal mode is equal to 4.37 km/s and that of the quasi-shear mode to 2.46 km/s. Since the sample thickness is 4.77 mm, the time of flight difference,is equal to 0.53 μs, by considering a double path for longitudinal wave, S_2 and a single path for shear wave S_1. This result is in agreement with the estimation obtained by the scalloping examination of the delay time. The fact that the scalloping effect begins to be evident during cooling at about 170 K, is probably due to two concomitant causes. Firstly, to the reduction of material absorption which makes the quasi-shear mode detectable. The second cause is related to the structural transitions, occurring in this temperature zone. The new material microstructure could better support the propagation of one mode rather than the other.

DISCUSSION AND CONCLUSIONS

Ultrasonic velocity measurement results were performed on YBCO sample during thermocycling up to the critical temperature Tc. Some interesting hypotheses are presented to evaluate a few discrepancies on the experimental data. The existence of non-pure propagation modes, depending on a non-perfect alignment between the propagation direction and the sample symmetry, is pointed out. Spectral analysis proved to be an indispensable tool for the identification of non-pure propagation modes through the observation of the scalloping effect on the spectral distribution. The appearance of non-pure modes has been highlighted in order to correctly evaluated ultrasonic parameters. Since this phenomenon is closely related to the crystallographic order and reticular material configuration, it would be a powerful indicator for investigation of material evolution which leads to superconducting behavior.

REFERENCES

1. E. Biagi, E. Borchi, R. Garré, S. De Gennaro, L. Masi and L.Masotti "Evidence For Two Hysteresis loops In $YBa_2Cu_3O_{6+x}$ Ultrasonic Velocity Measurements " Phys. Stat. Sol. (a) 138, pp. 249-256 July 1993.
2. E.Biagi, E. Borchi, L. Masi and L. Masotti "Ultrasonic Testing For Oxide High Temperatures Superconductors" Acous. Imag.20, 33, ed. Yu Wei and Benli Gu, Plenum Press, New York (1993).

PARTICIPANTS

Pedro Acevedro Contla
Universidad Nacional Autonoma de Mexico -
DEA-IIMAS
Admon no.20
Del.Alvaro Obregon, 20-726 - MEXICO
Fax: +52 525-550-0047
E-mail: pedro@uxdea1.iimas.unam.mx

Jan D. Achenbach
Northwestern University - CTR QEFP
Rm. 324 Catalysis Bldg. - 2137 Sheridan Rd.
Evanston, IL, 60208-3020 - USA
Fax: +1 708-4915227
E-mail: achenbach@nwu.edu

Iwaki Akiyama
Shonan Institute of Technology - Dept. of Electrical
Engineering
1-1-25, Tsujido-nishikaigan
Fujisawa, 251 - JAPAN
Fax: +81 466-358897
E-mail: akiyama@akiyama.elec.shonan-it.ac.jp

Pierre M. Alais
Universite Paris 6 - Laboratoire de Mecanique
Physique
2 Place de la gare de Ceinture
Saint-Cyr L'Ecole, 78210 - FRANCE
Fax: +33 1-34606959

F.W. Albert
Klinikumum Kaiserslautern - Med.KlinikIII
Postfach 3049
Kaiserslautern, 67653 - GERMANY

Mahmoud E. Allam
Middle East Technical and Commercial Office
28 Adly Street, Apt 33
Cairo 11111, EGYPT
E-mail: metaco@idsc.gov.eg

Michael P. Andre
University of California and Veterans Affairs
Medical Center - Department of Radiology
3350 La Jolla Village Drive
San Diego, CA, 92161 - USA
Fax: +1 619-5527452
E-mail: mandre@ucsd.edu

Fabio Andreuccetti
ESAOTE S.P.A. - Dipartimento Ultrasuoni
Via di Caciolle, 15
Firenze, 50127 - ITALY
Fax: +39 55-4223305

Yoshinao Aoki
Hokkaido Unversity - Dept. of Information
Engineering
N13, W8, Sapporo, 060 - JAPAN
Fax: +81 11-7573525
E-mail: aoki@acs5.huie.hokudai.ac.jp

Marcel Arditi
Bracco Research SA
7, route de Drize
Carouge-Genève, 1227 - SWITZERLAND
Fax: +41 22-3040405

Walter Arnold
Fraunhofer-Institute - GR
Bldg. 37, University
Saarbrucken, 66123 - GERMANY
Fax: +49 681-30693
E-mail: arnold@izfp.fhg.de

Carlo Atzeni
University of Florence - Electronic Eng.Dept.
Via S. Marta, 3
Firenze, 50139 - ITALY

Massimo Azzali
C.N.R.-I.R.Pe.M.
Largo della Pesca
Ancona, 60100 - ITALY
Fax: +39 71-55313

Therese Baldeweck
URA CNRS 1458 - Lab. Imagerie Parametrique
15, Rue de l'Ecole de Medicine
Paris, 75006 - FRANCE
Fax: +33 1-46335673

Jeffrey C. Bamber
Royal Maasden Hospital - Institute of Cancer
Research, Physics Dept.
Downs Road
Sutton, Surrey, SM2 5PT - UK
E-mail: jeff@icr.ac.uk

Moreno Bardelli
University of Trieste - Istitute of Medicina Clinica
c/o Ospedale di Cattimara
Trieste, 34125 - ITALY
Fax: +39 40-912881

Zalman M. Benenson
Scientific Council on Cibernetics - Russia Academy
of Science
2-nd Baltijsky per., 3-A.
Moscow, 125315 - RUSSIA
Fax: +7 95-2926511
E-mail: nskmin@glas.apc.org

Jurgen Bereiter-Hahn
J.W. Goethe-Universitat - Cinematic Cell Research
Group - Biozentrum
Marie Curiestr. 9
Frankfurt/Main, 60439 - GERMANY
Fax: +49 69-58009607
E-mail: bereiter@zoology.uni-frankfurt.d400.de

Genevieve Berger
Laboratoire d'Imagerie Parametrique - URA
C.N.R.S. 1458 - Universite PARIS 6
15, rue de l'ecole de medecine
75006 Paris - FRANCE
Fax: +33 1-46335673
E-mail: berger@idf.ext.jussieu.fr

Arthur P. Berkhoff
University Hospital Nijmegen - Biophysics Lab.
P.O. Box 9101
Nijmegen, 6500 - THE NETHERLANDS
Fax: +31 80-540522
E-mail: A.Berkhoff@ohk.azn.nl

Michel Bertrand
University of Montreal - Institute of Biomedical
Engineering
BOX 6079, Station Centre-Ville
Montreal, PQ, H3C 3A7 - CANADA

Elena Biagi
University of Florence - Electronic Eng.Dept.
Via S. Marta, 3
Firenze, 50139 - ITALY
Fax: +39 55-494569
E-mail: uscnd@ingfi1.ing.unifi.it

Gerrit Blacquière
TNO Institute of Applied Physics - Inspection
Technology Department
P.O. Box 155
Delft, 2600 AD - THE NETHERLANDS
Fax: +31 15 692111
E-mail: blacquie@tpd.tno.nl

Leslie J. Bowen
Materials Systems Inc.
521, Great Road
Littleton, MA, 01460 - USA
Fax: +1 508-4860706
E-mail: 76035.1644@compuserve.com

Peter J. Brands
University of Linburg - Department of Biophysics
P.O. Box 616
Maastrict, 6200 MD - THE NETHERLANDS
Fax: +31 43-672287
E-mail: p.brands@bfcrulinburg.nl

Andrew Briggs
University of Oxford - Dept. of Materials
Parks Road
Oxford, OX1 3PH - UK
E-mail: briggs@vax.ox.ac.uk

Valentin A. Burov
Moscow State University - Faculy of Physics -
Department of Acoustics
Moscow, 119899 - RUSSIA
Fax: +7 95-9328820
E-mail: asl@nabla.phys.msu.su

E. Burr
University of Karlsruhe - Geophysical Institute
Hertzstr. 16
Karlsruhe, 76187 - GERMANY
E-mail: ngold@gpirs2.physic.uni-karlsruhe.de

Lorenzo Capineri
University of Florence - Electronic Eng.Dept.
Via S. Marta, 3
Firenze, 50139 - ITALY
Fax: +39 55-494569
E-mail: capineri@ingfi1.ing.unifi.it

Dominique Cathignol
INSERM U-281
151 Cours Albert Thomas
Lyon Cedex 03, 69424 - FRANCE
Fax: +33 72-350509
E-mail: cathignol@lyon151.inserm.fr

Dan Censor
Ben Gurion University of the Negev - Dept. of
Electrical and Computer Engineering
Beer Sheva, - ISRAEL
Fax: +97 27-276338
E-mail: censor@bguee.bgu.ac.il

Roland Jozef Collaris
University of Limburg - Department of Biophysics
P.O. Box 616
Maastricht, 6200 - THE NETHERLANDS
Fax: +31 43-672287

Pier Paolo Delsanto
Politecnico di Torino - Dipartimento di Fisica
Corso Duca degli Abruzzi, 20
Torino, 10129 - ITALY
Fax: +39 11-5647399

Yves Doisy
Thomson-Sintra ASM - BP.57
Sophia-Antipolis Cedex, 06903 - FRANCE
Fax: +33 92-963977
E-mail: plaisant@asm.thomson.fr

Christian Dorme
Universit, Denis Diderot, URA CNRS 1503 -
Laboratoire Ondes et Acoustique E.S.P.C.I.
10, rue Vauquelin
Paris Cedex 05, 75231 - FRANCE
Fax: +33 1-40794425
E-mail: chris@loa.espci.fr

Enrico D'Ottavi
CNR - Instituto di Acustica O.M. Corbino
Via Cassia 1216
Roma, 00189 - ITALY
Fax: +39 6-30365341

Domenico Dotti
Universita' di Pavia - Dip. Informatica e
Sistemistica
Via Abbiategrasso 209
Pavia, 27100 - ITALY
Fax: +39 382-505373
E-mail: microlab@ipv85.unipv.it

Bernard Duchene
Laboratoire des Signaux et Systemes (CNRS-ESE)
- Division Ondes ESE
Plateau de Moulon
Gif sur Yvette Cedex, 91192 - FRANCE

Peter-Christian Eccardt
Siemens AG - Corporate Research and
Development
Otto-Hahn-Ring 6
Munchen, 81730 - GERMANY
Fax: +49 89-63646881
E-mail: eccardt@curry.zfe.siemens.de

Helmut Ermert
Ruhr Universitat Bochum - Institut fur
Hochfrequenztechnik
BLDG. IC6/132
Bochum, 44780 - GERMANY
Fax: +49 234-7094167
E-mail: he@hf.ruhr-uni-bochum.de

Leonard A Ferrari
Virginia Polytechnic and State University -
Electrical Engineering Dept.
340, Whittemore Hall
Blacksburg, VA 24061-0106 - USA
Fax: +1 540 2313362
E-mail: ferrari@vt.edu

Leszek Filipczynski
Polish Academy of Science - Institute of
Fundamental Techn. Research - Ultrasound
Department
ul. Swietokrzyska 21
Warsaw, 00-049 - POLAND
Fax: +48 22-269815
E-mail: tkujaw@hpzu.ippt.gov.pl

Flemming Forsberg
Thomas Jefferson University Hospital - Division of
Diagnostic Ultrasound - Dept. of Radiology
132 S 10th Street, Suite 763 J Main Building
Philadelphia, PA, 19107 - USA
Fax: +1 215-9235793
E-mail: forsberg@esther.rad.tju.edu

Ada Fort
University of Florence - Electronic Eng.Dept.
Via S. Marta, 3
Firenze, 50139 - ITALY
Fax: +39 55-494569

Luciano Fumagalli
Bracco S.P.A. - Responsabile dei Servizi Centro
Ricerche Milano della Bracco
via Egidio Folli 50
Milano, 20134 - ITALY
Fax: +39 2-26410678

Tilio Gaertner
Martin-Luther-University Halle-Wittenberg -
Institute for Medical Physics and Biophysics -
Medical Faculty
Str. der DSF 81, P.O. Box 302
Halle (Saale), 06097 - GERMANY

Woon Siong Gan
Acoustical Services Pte, Ltd.
29 Telok Ayer Street
Singapore, 0104 - REPUBLIC OF SINGAPORE
Fax: +65 7-913665
E-mail: MA2625792@ntuvax.ntu.ac.sg

Hans-Georg von Garssen
Siemens AG - Dep. ZFE T KM 1
Otto-Hahn-Ring 6
Munchen, 81739 - GERMANY
Fax: +49 89-63646881
E-mail: garssen@curry.zfe.siemens.de

Bernhard Gassmann
Inst. f. Medizinische Physik - Klinikum Berlin-
Buch
Hobrechtsfelder Chaussee 100
Berlin, 13122 - GERMANY
Fax: +49 30-9482329

Jean Francois Gelly
Thomson Microsonics - Direction Scientifique et
Technique
399 route des Cretes, BP 232
Sophia-Antipolis Cedex, 06904 - FRANCE
Fax: +33 92-963000

Norbert Gold
University of Karlsruhe - Geophysical Institute
Hertzstr. 16
Karlsruhe, 76187 - GERMANY
Fax: +49 721-71173
E-mail: ngold@gpirs2.physic.uni-karlsruhe.de

Jan Pierre Gonnet
Institut Polytechnique de S,venans - Departement
de Genie Mecanique
Belfort Cedex, 90010 - FRANCE
Fax: +33 84-583030

James F. Greenleaf
Mayo Clinic and Foundation - Dept. of Physiology
and Biophysics -
Biodynamics Research Unit
200 First Street, SW
Rochester, MN, 55905 - USA
Fax: +1 507-2841632
E-mail: jfg@mayo.edu

Elisa Greiff
ATL
P.O.Box 3003 mail stop 264
Bothell, WA, 98041-3003 - USA
Fax: +1 206-4845220
E-mail: egreif@atl.com

Alessandro Gubbini
ESAOTE Biomedica
Via di Caciolle 15
Firenze, 50127 - ITALY

Francesco Guidi
University of Florence - Electronic Eng.Dept.
Via S. Marta, 3
Firenze, 50139 - ITALY
Fax: +39 55-494569
E-mail: f_guidi@ingfi1.ing.unifi.it

Gabriele Guidi
University of Florence - Electronic Eng.Dept.
Via S. Marta, 3
Firenze, 50139 - ITALY
Fax: +39 55-494569
E-mail: g.guidi@ieee.org

Rolf Kahrs Hansen
OmniTech as.
Nedre Aastveit 12, Ovre Ervik
Bergen, 5083 - NORWAY
Fax: +47 55-193205
E-mail: rkh@omnitech.no

Isabelle Hardouin
Deutsches Herzzentrum Berlin - German Heart
Institute Berlin - H3
Augustenburger Platz 1
Berlin, 13353 - GERMANY
Fax: +49 30-45068200
E-mail: hardouin@dhzb.de

Jerry M Harris
Stanford University - Dept of Geophysics
Mitchell Bldg.
Palo Alto, CA, 94305-2215 - USA

Henkjan J. Huisman
University of Nijmegen - Biophysics Laboratory
P.O. Box 9101
Nijmegen, 6500 HB - THE NETHERLANDS
Fax: +31 80-540522
E-mail: H.Huisman@ohk.azn.nl

J.Jacques Hunsinger
Institut Polytechnique de Sevenans - Departement
de Genie Mecanique
Belfort Cedex, 90010 - FRANCE
Fax: +33 84-583030

Kohji Iida
Hokkaido University - Dept. of Fisheries Science
Minato-cho 3-1-1
Hakodate, 041 - JAPAN
Fax: +81 138-435015
E-mail: iidacs@fish.hokudai.ac.jp

Mitsuteru Inoue
Toyohashi University of Technology - Dept. of
Electrical and Electronic Eng.
1-1 Hibari-Ga-Oka, Tempaku, Toyohashi, Aichi,
441 - JAPAN
Fax: +81 532-473275
E-mail: inouem@eee.tut.ac.jp

Michael F. Insana
University of Kansas Medical Center - Dept. of
Diagnostic Radiology
3901 Rainbow Blvd.
Kansas City, KS, 66160-7234 - USA
Fax: +1 913-5887876
E-mail: insana@speedy.mc.ukans.edu

Antonio Iula
Universita' degli Studi di Salerno - Dip. di
Ingegneria dell'informazione e Ingegneria Elettrica
Via Ponte Don Melillo
Fisciano (SA), 84084 - ITALY

Helmar S. Janee
University of California and Veterans Affairs
Medical Center - Dept. of Radiology (114)
3350 La Jolla Village Drive
San Diego, CA, 92121 - USA

Jorgen Arendt Jensen
Technical University of Denmark - Electronics
Institute, Building 349
Lyngby, 2800 - DENMARK
Fax: +45 45-880117
E-mail: drjaj@ei.dtu.dk

David Johnson
Nycomed R & D
Collegeville, PA, 19426 - USA

Joie P. Jones
University of California at Irvine - Dept. of
Radiological Sciences
Irvine, CA, 92717 - USA
Fax: +1 714-8566532
E-mail: jpjones@uci.edu

Faouzi Kallel
University of Montreal - Institute of Biomed. Eng.
BOX 6079, Station Centre-Ville
Montreal, PQ, H3C 3A7 - CANADA
Fax: +1 514-3404611
E-mail: kallel@igb.polymtl.ca

Elena Eugen'evna Kasatkina
Moscow State University - Faculy of Physics -
Department of Acoustics
Moscow, 119899 - RUSSIA
Fax: +7 95-9328820
E-mail: asl@nabla.phys.msu.su

Lawrence W. Kessler
Sonoscan Inc.
530 E Green St.
Bensenville, IL, 60106 - USA
Fax: +1 708-7667088

Pierre Khuri-Yakub
Stanford University - Edward L. Ginzton
Laboratory
Stanford, CA, 94305-4085 - USA
E-mail: pierre@macro.Stanford.edu

Oleg V. Kolosov
University of Oxford - Dept. of Materials
Parks Road
Oxford, OX1 3PH - UK
E-mail: kolosov@vax.ox.ac.uk

Klaus Kosbi
University of Bremen - Institute for Materials
Science and Structure Research - Physics Dept.
P.O. Box 330440
Bremen, 28334 - GERMANY
Fax: +49 421-2187381
E-mail: kosbi@schall.physik.uni-bremen.de

Martin Krueger
Ruhr Universitat Bochum - Institut f.
Hochfrequenztechnik
BLDG 1C/6/132
Bochum, 44780 - GERMANY
Fax: +49 234-7094167
E-mail: mk@hf.ruhr-uni-bochum.de

Elfgard Kuhnicke
Dresden University of Technology - Institute of
Technical Acoustics
Mommsenstr. 13
Dresden, 01062 - GERMANY
Fax: +49 351-4637091
E-mail: kom@eakss1.et.tu-dresden.de

Ewald Lagler
Kretztechnik AG
Tiefenbach 15
Zipf, 4871 - AUSTRIA
Fax: +76 82-226147

Nicola Lamberti
University of Salerno
Via Ponte Don Melillo
Fisciano (SA), 84084 - ITALY
Fax: +39 89-964218
E-mail: lamberti@vaxsa.csied.unisa.it

Charles T. Lancee
Erasmus University Rotterdam EE 2302
P.O. Box 1738
Rotterdam - THE NETHERLANDS

Hua Lee
University of California - Department of Electrical
& Computer Engineering
Santa Barbara, CA, 93106-9560 - USA
Fax: +1 805-8933262
E-mail: hualee@ece.ucsb.edu

Sidney Leeman
King's College Hospital (Dulwich) - Dept. Med.
Eng. & Phys.
East Dulwich Grove
London, SE 228 PT - UK
Fax: +44 81-9092126

Sidney Lees
Forsyth Dental Center - Dept. of Bioengineering
140 Fenway
Boston, MA, 02115 - USA
Fax: +1 617-2624021
E-mail: SLees@forsyth.org

Dominique Lescribaa
European Commission, Joint Research Centre,
Ultrasonics Development Lab.
TP 750
Ispra (VA), 21020 - ITALY
Fax: +39 332-785779
E-mail: dominique.lescriba@jrc.it

Marc Lethiecq
Gip Ultrasons
2 bis Boul. Tonnellè - BP3
37032 Tours Cedex - FRANCE

Wayne R. Lewis
University of California - Department of Electrical
& Computer Engineering
Box 139, Engineering I
Santa Barbara, CA, 93106 - USA
Fax: +1 805-8933262
E-mail: wayne@jumers.ece.ucsb.edu

Jerzy Litniewski
Polish Academy of Science - Institute of
Fundamental Techn. Research - Ultrasound
Department
ul. Swietokrzyska 21
Warsaw, 00-049 - POLAND
Fax: +48 22-269815

Thanasis Loupas
Ultrasonics Laboratory
126 Greville St.
Chatswood, NSW, 2067 - AUSTRALIA
E-mail: tloupas rp.csiro.au

Jian-yu Lu
Mayo Clinic and Foundation - Department of
Physiology and Biophysics - Biodyn. Research Unit
200 First Street, SW
Rochester, MN, 55905 - USA
Fax: +1 507-2841632
E-mail: jian@Mayo.edu

Vincenza Anna Maria Luprano
Pastis-CNRSM
S.S. 7 "Appia" km 714
Brindisi, 72100 - ITALY
Fax: +39 831-507379
E-mail: luprano@cnrsm.it

Roman G. Maev
Acoustic Microscopy Center
Kosygin Str. 4
Moscow, 117977 - RUSSIA
E-mail: maev@server.uwindsor.ca

Maria G. Markaki
Foundation for Research and Tecnology-Hellas -
Institute of Applied & Computational Mathematics
P.O. Box 1527
Heraklion Crete, 71110 - GREECE

Renè Marklein
University of Kassel - Dept. of Electrical
Engineering (FB16)
Wilhemshher Allee 71
Kassel, 34109 - GERMANY
Fax: +49 561-8046489
E-mail: marklein@tet.e-technik.uni-kassel.de

Leonardo Masotti
University of Florence - Electronic Eng.Dept.
Via S. Marta, 3
Firenze, 50139 - ITALY

Celine Maurel
URA CNRS D1881 - Laboratoire d'analyse des
interfaces et de nanophysique
Case Courrier 82 - Place Eugene Bataillon
Montpellier Cedex, 34095 - FRANCE
Fax: +33 67-521584

S.G. Mesohoryanakis
Sofia University "St. Kliment Ochridsky" - Faculty
of Physics - Department of Solid State Physics and
Microelectronics
James Boujer Blvd. 5
Sofia, 1126 - BULGARIA
Fax: +359 2-689085
E-mail: burov@phys.uni-sofia.bg

Veronique Miette
Universite Paris 7 - Laboratoire Ondes et
Acoustique ESPCI
10, Rue Vauquelin
Paris Cedex 05, 75005 - FRANCE

Bruno Migeon
Laboratoire Vision et Robotique
63, avenue de Lattre de Tassigny
Bourges Cedex, 18020 - FRANCE
Fax: +33 48-659961

Toyokatsu Miyashita
Ryukoku University - Dept. Electronics and
Informatics
Seta Ohecho Yokotani 1-5
Ohtsu, 520-21 - JAPAN
Fax: +81 775-437428
E-mail: miya@rins.ryukoku.ac.jp

Francisco R. Montero de Espinosa
Instituto de Acustica, C.S.I.C.
Calle Serrano 144
Madrid, 28006 - SPAIN

Andrzej Nowicki
Institute of Fundamental Technological Research -
Dept. of Ultrasound
ul. Swietokrzyska 21
Warsaw, 00-049 - POLAND
Fax: +48 22-269815
E-mail: anowicki@hpzu.ippt.gov.pl

William D. O'Brien,jr.
University of Illinois - Dept. of Electrical &
Computer Engineering
1406 West Green Street
Urbana, IL, 61801 - USA
Fax: +1 217-2440105
E-mail: wdo@uiuc.edu

Francis Ollivier
Universite Paris 6 - Laboratoire de Mechanique
Physique
2 Place de la gare de ceinture
Saint-Cyr L'Ecole, 78210 - FRANCE

Jonathan Ophir
The University of Texas Medical School -
Department of Radiology, Ultrasonics Laboratory
6431 Fannin Street, MSMB 2.130
Houston, TX, 77030 - USA
Fax: +1 713-7925645
E-mail: jophir@msrad3.med.uth.tmc.edu

Frederic Ossant
GIP Ultrasound
2bis, bd Tonnelle'
Tours Cedex, 37044 - FRANCE
Fax: +33 47-366120

Madhukar Pandit
Universitat Kaiserslautern
Postfach 3049
Kaiserslautern, 67653 - GERMANY
Fax: 49-631-2054205
E-mail: Pandit@rhrk.uni-kl.de

Panagiotis J. Papadakis
Institute of Applied & Computational Mathematics
P.O. Box 1527
Iraklion Crete, 71110 - GREECE
Fax: +30 81-391801
E-mail: panos@iacm.forth.gr

Song Bai Park
Korea Advanced Institute of Science & Technology
- Dept. of Electrical Engineering
P.O.Box 150, Chongyangni
Seoul, 130-650 - KOREA
Fax: +82 822-9602103

Christian Passmann
Ruhr Universitat Bochum - Institut fur HF
Universitats strasse 150
Bochum, 44780 - GERMANY
Fax: +49 234-7094167
E-mail: cp@hf.ruhr-uni-bochum.de

Peter A. Payne
UMIST - DIAS
P.O.Box 88, Blackville Street
Manchester, M60 1QD - UK
Fax: +44 161-2004879
E-mail: p.a.payne@umist.ac.uk

Wagner C.A. Pereira
Federal University of Rio de Janeiro - Biomedical
Engineering Program
P.O. Box 68510
Rio de Janeiro, 21945-970 - BRAZIL
Fax: +55 21-2906626
E-mail: wagner@serv.peb.ufrj.br

Paolo Pignoli
Merate Hospital - Dept. of Surgery
Merate (CO), 22055 - ITALY

Nicholas E. Pingitore
University of Texas at El Paso - Dept. of
Geological Sciences
El Paso, TX, 79968-0555 - USA
Fax: +1 915-7475073
E-mail: nick@geo.utep.edu

Riccardo Pini
University of Firenze - Institute of Gerontology and
Geriatrics
via delle Oblate 4
Firenze, 50141 - ITALY

Mieczyslaw Pluta
Technical University of Wroclaw - Institute of
Physics
Wybrzeze Wyspianskiego 27
Wroclaw, 50-370 - POLAND
Fax: +48 71-222229

Luca Ponziani
University of Florence - Electronic Eng.Dept.
Via S. Marta, 3
Firenze, 50139 - ITALY

John P. Powers
Naval Postgraduate School - Department of
Electrical and Computer
Engineering (Code EC/PO)
833 Dyer Rd., Rm 437
Monterey, CA, 93943-5121 - USA
Fax: +1 408-6563222
E-mail: jpowers@nps.navy.mil

Ute Rabe
Fraunhofer-Institute
Bldg. 37, University
Saarbrucken, 66123 - GERMANY

Santina Rocchi
University of Siena - Engineering Faculty
Via Roma, 56
Siena, 53100 - ITALY

Angelo Maria Sabatini
Sc. Superiore di Studi Universitari S. Anna
Via Carducci, 40
Pisa, 56127 - ITALY
Fax: +39 50-883215
E-mail: sabatini@sssup1.sssup.it

Yoshifumi Saijo
Tohoku University - Dept. of Medical Engineering
and Cardiology - Institute of Development, Aging
and Cancer
4-1 Seiryo-machi, Aoba-ku
Sendai, 980 - JAPAN
Fax: +81 22-2732823
E-mail: saijo@yambe.idac.tohoku.ac.jp

Hidehiko Sasaki
Tohoku University - Dept. of Medical Engineering
and Cardiology
4-1 Seiryomachi, Aoba-ku
Sendai, 980 - JAPAN
Fax: +81 22-2757324
E-mail: saijo@yambe.idac.tohoku.ac.jp

Jaroslav D. Satrapa
TCC
Lenaustrasse, 10
Timelkam, 4850 - AUSTRIA
Fax: +43 7672-27512

Wolfram Schmidt
University of Rostock - Institute of Biomedical
Engineering
E. Heydemann-Str. 6
Rostock, 18055 - GERMANY
Fax: +49 381-4947602
E-mail: wolfram.schmidt@medizin.uni-rostock.de

Rainer Schmitt
Universitat Saarbrucken
Bldg. 37, University
Saarbrucken, 66123 - GERMANY

Natalia N. Shibanova
Moscow Institute of Physics and Technology
9 Dolgoprudny
Moscow region, 141700 - RUSSIA
Fax: +7 95-2038414

Parvinder Singh
P.E.S. (I) - K.No. 116, Phase-4, Sector-59
S.A.S. Nagar
Chandigarh, 160059 - INDIA

Wallace Arden Smith
Office of Naval Research - Materials Division
800, N Quincy St., RM 704
Arlington, VA, 22217 - USA
E-mail: smithw@onrpo2.onr.navy.mil

Andreas Steinmetz
Siemens AG - Medizintechnik
Henkestrasse 127
Erlangen, 91052 - GERMANY
Fax: +49 91-31844771
E-mail: as@erlh.Siemens.de

Tadeusz Stepinski
Uppsala University - Teknikum, Circuits & Systems
Box 534
Uppsala, 751 21 - SWEDEN
Fax: +46 18-555096
E-mail: tas@mercur.teknikum.uu.se

Andrzej Stepnowski
Technical University of Gdansk - Acoustic
Department
ul.Narutowicza 11/12
Gdansk, 80-952 - POLAND
Fax: +48 58-472839
E-mail: astep@sunrise.pg.gda.pl

Yoshihiko Takishita
Hitachi Construction Machinery Co. - Technical
Research Laboratory
650 Kandatsu-machi, Tsuchiura
Ibaraki, 300 - JAPAN
Fax: +81 298-310201
E-mail: fukuchi@HCM-TRL.po.iijnet.or.jp

Lewis W. Thompson
University of California - Department of Electrical
and Computer Engineering
Santa Barbara, CA, 93106 - USA
Fax: +1 805-8933262
E-mail: lewis@zorbas.ece.ucsb.edu

Rosemary S. Thompson
University of Sydney - School of Mathematics and
Statistics
Sydney, NSW, 2006 - AUSTRALIA
Fax: +61 612-3514534
E-mail: thompson_r@maths.su.oz.au

Piero Tortoli
University of Florence - Electronic Eng.Dept.
Via S. Marta, 3
Firenze, 50139 - ITALY
Fax: +39 55-494569
E-mail: ultra@ingfil.ing.unifi.it

Andrea Trucco
University of Genoa - Dept. of Biophysical and
Electronic Engineering (DIBE)
Via Opera Pia 11A
Genova, 16145 - ITALY
Fax: +39 10-3532134
E-mail: dafne@dibe.unige.it

Volkmar Uhlendorf
Schering AG - PH Ultraschall - Research
Laboratories
Mllerstr. 170-178
Berlin, 13342 - GERMANY
Fax: +49 30-46916773

Pascal Vairac
C.N.R.S. - Laboratoire de Physique et de
Metrologie des Oscilateurs
32, Avenue de l'Observatoire
Besancon Cedex, 25044 - FRANCE
Fax: +33 81-666998

Frank M.J. Valckx
University Hospital Nijmegen - Biophysics
Laboratory - Dept. of ophthalmology
P.O. Box 9101
Nijmegen, 6500 - THE NETHERLANDS
Fax: +31 80-540522
E-mail: F.Valckx@ohk.azn.nl

Peter-Paul van 't Veen
TNO Institute of Applied Physics - Inspection
Technology Department
P.O. Box 155
Delft, 2600 AD - THE NETHERLANDS
Fax: +31 15-692111
E-mail: vveen@tpd.tno.nl

Valerio Vignoli
University of Siena - Engineering Faculty
Via Roma, 56
Siena, 53100 - ITALY

Martin Vossiek
Siemens AG - Dept. ZFE T KM 1
Otto-Hahn-Ring 6
Munchen, 81739 - GERMANY
Fax: +49 89-63646881
E-mail: vossiek@curry.zfe.siemens.de

Glen Wade
University of California - Department of Electrical
& Computer Engineering
Santa Barbara, CA, 93106-9560 - USA
Fax: +1 805-8933262
E-mail: DocGWade@aol.com

Wieland Weise
University of Bremen - Institute for Material
Science and Structure
Reserch-Phisics Dept.
Fachbereich 1, Postfach 330440
Bremen, 28334 - GERMANY
Fax: +49 421-2187381
E-mail: weise@diana.physik.uni-bremen.de

Peter N.T. Wells
Bristol General Hospital - Dept. of Medical Physics
Guinea Street
Bristol, BS1 6SY - UK
Fax: +44 117-9286371
E-mail: Peter.Wells@bristol.ac.uk

Ulrich Werner
University of Karlsruhe - Geophysical Institute
Hertzstr. 16
Karlsruhe, 76187 - GERMANY
Fax: +49 721-71173
E-mail: uwerner@gpiwap6.physic.uni-karlsruhe.de

Ralf Wichard
Fraunhofer Institute Biomedical Engineering -
Dept. Ultrasound
Ensheimer Strasse, 48, St. Ingbert, 66386 -
GERMANY
Fax: +49 689-4980400
E-mail: wichard@ibmt.fhg.de

Armand Wirgin
Centre National de la Recherche Scientifique -
Laboratoire de
Mecanique et d'Acoustique
31 Chemin Joseph-Alguier, Marseille Cedex 20,
13402 - FRANCE
Fax: +33 91-220875
E-mail: wirgin@lma.cnrs-mrs.fr

Yoshiki Yamakoshi
Gunma University - Faculty of Engineering
1-5-1, Tenjin-cho, Kiryu-shi, Gunma, 376 - JAPAN
Fax: +81 277-301707
E-mail: yamakosi@el.gunma-u.ac.jp

Samir Yastas
J.W. Goethe-Universitat - Cinematic Cell Research
Group - Biozentrum
Marie Curiestr. 9, Frankfurt/Main, 60439 -
GERMANY
Fax: +49 69-79829607
E-mail: Yastas@zoology.uni-frankfurt.de

Juha Ylitalo
University of Trondheim - Institute for Biomedical
Technology
Olav Kyrres gate 3, Trondheim, 7005 - NORWAY
Fax: +47 73-598613

Masahide Yoneyama
Toyo University - Dept. of Information &
Computer Sciences
2100 Kujirai, Kawagna, Saitama, 350 - JAPAN
Fax: +81 492-339788

Hajime Yuasa
Mitsui Engineering & Shipbuilding - Akishima
Laboratories (MITSUI ZOSEN) Inc.
I-50 Tsutsujigaoka 1- Chome, Akishima Tokyo,
196 - JAPAN
Fax: +81 425-463570
E-mail: nbd02747@niftyserve.or.jp

AUTHORS' INDEX

SUBJECT INDEX